VOLUME ONE HUNDRED AND SIXTEEN

CURRENT TOPICS IN DEVELOPMENTAL BIOLOGY

Essays on Developmental Biology, Part A

CURRENT TOPICS IN DEVELOPMENTAL BIOLOGY

"A meeting-ground for critical review and discussion of developmental processes"
A.A. Moscona and Alberto Monroy (Volume 1, 1966)

VOLUME ONE HUNDRED AND SIXTEEN

Current Topics in DEVELOPMENTAL BIOLOGY

Essays on Developmental Biology, Part A

Edited by

PAUL M. WASSARMAN

Department of Developmental and Regenerative Biology
Icahn School of Medicine at Mount Sinai
New York, NY, USA

ELSEVIER

AMSTERDAM • BOSTON • HEIDELBERG • LONDON
NEW YORK • OXFORD • PARIS • SAN DIEGO
SAN FRANCISCO • SINGAPORE • SYDNEY • TOKYO
Academic Press is an imprint of Elsevier

Academic Press is an imprint of Elsevier
50 Hampshire Street, 5th Floor, Cambridge, MA 02139, USA
525 B Street, Suite 1800, San Diego, CA 92101-4495, USA
The Boulevard, Langford Lane, Kidlington, Oxford OX5 1GB, UK
125 London Wall, London, EC2Y 5AS, UK

First edition 2016

ISBN: 978-0-12-802956-5
ISSN: 0070-2153

For information on all Academic Press publications visit our website at http://store.elsevier.com/

CONTENTS

CONTRIBUTORS

Markus Affolter
Biozentrum der Universität Basel, Basel, Switzerland

Spyros Artavanis-Tsakonas
Department of Cell Biology, Harvard Medical School, Boston; Biogen Idec, Cambridge, Massachusetts, USA

Philip N. Benfey
Biology Department and HHMI, Duke University, Durham, North Carolina, USA

David L.H. Bennett
The Nuffield Department of Clinical Neurosciences, University of Oxford, John Radcliffe Hospital, Oxford, United Kingdom

Carmen Birchmeier
Developmental Biology, Max-Delbrück-Center for Molecular Medicine, Berlin, Germany

Bruce Bowerman
Institute of Molecular Biology, University of Oregon, Eugene, Oregon, USA

Ali H. Brivanlou
Laboratory of Stem Cell Biology and Molecular Embryology, The Rockefeller University, New York, USA

Marianne E. Bronner
Division of Biology and Biological Engineering, California Institute of Technology, Pasadena, California, USA

Margaret Buckingham
Department of Developmental and Stem Cell Biology, Institut Pasteur, Paris, France

Corey Bunce
Department of Cell Biology, Duke University Medical Center, Durham, North Carolina, USA

Blanche Capel
Department of Cell Biology, Duke University Medical Center, Durham, North Carolina, USA

Martin Chalfie
Department of Biological Sciences, Columbia University, New York, USA

Stephen M. Cohen
Department of Cellular and Molecular Medicine, University of Copenhagen, Copenhagen, Denmark

Fabrice Darbellay
School of Life Sciences, Federal Institute of Technology, Lausanne, Switzerland

Edward M. De Robertis
Howard Hughes Medical Institute and Department of Biological Chemistry, University of California, Los Angeles, California, USA

Alessia Deglincerti
Laboratory of Stem Cell Biology and Molecular Embryology, The Rockefeller University, New York, USA

Melvin L. DePamphilis
National Institute of Child Health and Human Development, National Institutes of Health, Bethesda, Maryland, USA

Claude Desplan
Center for Genomics & Systems Biology, New York University, Abu Dhabi, UAE; Department of Biology, New York University, New York, USA

Wolfgang Driever
Developmental Biology Unit, Institute Biology I, Faculty of Biology, and Center for Biological Signaling Studies (BIOSS), Albert-Ludwigs-University, Freiburg, Germany

Denis Duboule
School of Life Sciences, Federal Institute of Technology, Lausanne; Department of Genetics and Evolution, University of Geneva, Geneva, Switzerland

Elisabeth Dupin
INSERM U968; Sorbonne Universités, UPMC Univ Paris 06, UMR S 968, Institut de la Vision; CNRS, UMR 7210, Paris, France

Teresa Eichenlaub
Department of Cellular and Molecular Medicine, University of Copenhagen, Copenhagen, Denmark

Scott W. Emmons
Department of Genetics; Dominick P. Purpura Department of Neuroscience, Albert Einstein College of Medicine, Bronx, New York, USA

Fred Etoc
Laboratory of Stem Cell Biology and Molecular Embryology, The Rockefeller University, New York, USA

Manfred Frasch
Department of Biology, Division of Developmental Biology, Friedrich-Alexander University of Erlangen-Nürnberg, Erlangen, Germany

Elaine Fuchs
Laboratory of Mammalian Development and Cell Biology, Howard Hughes Medical Institute, The Rockefeller University, New York, USA

Margaret T. Fuller
Departments of Developmental Biology and Genetics, Stanford University School of Medicine, Stanford, California, USA

Brigitte Galliot
Department of Genetics and Evolution, Institute of Genetics and Genomics in Geneva (IGe3), Faculty of Sciences, University of Geneva, Geneva, Switzerland

Scott F. Gilbert
Swarthmore College, Swarthmore, Pennsylvania, USA

Thomas Gridley
Center for Molecular Medicine, Maine Medical Center Research Institute, Scarborough, Maine, USA

John B. Gurdon
Wellcome Trust/Cancer Research UK Gurdon Institute; Department of Zoology, University of Cambridge, Cambridge, United Kingdom

Héctor Herranz
Department of Cellular and Molecular Medicine, University of Copenhagen, Copenhagen, Denmark

Diana M. Ho
Department of Cell Biology, Harvard Medical School, Boston, Massachusetts, USA

Oliver Hobert
Department of Biological Sciences, Howard Hughes Medical Institute, Columbia University, New York, USA

Mingi Hong
Department of Developmental and Regenerative Biology, Icahn School of Medicine at Mount Sinai, New York, NY, USA

Philip W. Ingham
Lee Kong Chian School of Medicine, Nanyang Technological University, Singapore; Department of Medicine, Imperial College, London, United Kingdom

William R. Jeffery
Department of Biology, University of Maryland, College Park, Maryland, USA

Alexandra L. Joyner
Developmental Biology Program, Sloan Kettering Institute, New York, USA

David Kimelman
Department of Biochemistry, University of Washington, Seattle, Washington, USA

Peter Koopman
Institute for Molecular Bioscience, The University of Queensland, Brisbane, Queensland, Australia

Raphael Kopan
William K. Schubert Professor of Pediatrics; Division of Developmental Biology, Cincinnati Children's Hospital Medical Center, University of Cincinnati College of Medicine, Cincinnati, OH, USA

Thomas B. Kornberg
Cardiovascular Research Institute, University of California, San Francisco, California, USA

Robert S. Krauss
Department of Developmental and Regenerative Biology, Icahn School of Medicine at Mount Sinai, New York, NY, USA

Robb Krumlauf
Stowers Institute for Medical Research, Missouri; Department of Anatomy & Cell Biology, Kansas University Medical School, Kansas City, Kansas, USA

Michel Labouesse
Laboratoire de Biologie du Développement, CNRS-UMR7622, Institut de Biologie Paris-Seine, Université Pierre et Marie Curie, Sorbonne Universités, Paris, France

Peter A. Lawrence
Department of Zoology, University of Cambridge, Cambridge, United Kingdom

Nicole M. Le Douarin
Collège de France, 3 rue d'Ulm; INSERM U968; Sorbonne Universités, UPMC Univ Paris 06, UMR S 968, Institut de la Vision; CNRS, UMR 7210, Paris, France

Thomas Lecuit
Aix Marseille Université, CNRS, IBDM UMR7288, Marseille, France

Ruth Lehmann
Howard Hughes Medical Institute (HHMI), Department of Cell Biology, Kimmel Center for Biology and Medicine of the Skirball Institute, New York University School of Medicine, New York, USA

Ihor R. Lemischka
Department of Regenerative and Developmental Biology, Icahn School of Medicine at Mount Sinai; Black Family Stem Cell Institute; Department of Pharmacology and System Therapeutics, Mount Sinai School of Medicine, Systems Biology Center New York, New York, USA

Qiyan Mao
Aix Marseille Université, CNRS, IBDM UMR7288, Marseille, France

Yuki Moriyama
Howard Hughes Medical Institute and Department of Biological Chemistry, University of California, Los Angeles, California, USA

Nathalie Nériec
Center for Genomics & Systems Biology, New York University, Abu Dhabi, UAE; Department of Biology, New York University, New York, USA

Daria Onichtchouk
Developmental Biology Unit, Institute Biology I, Faculty of Biology, and Center for Biological Signaling Studies (BIOSS), Albert-Ludwigs-University, Freiburg, Germany

M. Zeeshan Ozair
Laboratory of Stem Cell Biology and Molecular Embryology, The Rockefeller University, New York, USA

Dmitri Papatsenko
Department of Regenerative and Developmental Biology, Icahn School of Medicine at Mount Sinai; Black Family Stem Cell Institute, Mount Sinai School of Medicine, New York, USA

Aaron F. Severson
Department of Biological, Geological, and Environmental Sciences, Center for Gene Regulation in Health and Disease, Cleveland State University, Cleveland, Ohio, USA

Marcos Simões-Costa
Division of Biology and Biological Engineering, California Institute of Technology, Pasadena, California, USA

Matthias C. Vogg
Department of Genetics and Evolution, Institute of Genetics and Genomics in Geneva (IGe3), Faculty of Sciences, University of Geneva, Geneva, Switzerland

George von Dassow
Oregon Institute of Marine Biology, University of Oregon, Charleston, Oregon, USA

Thanh T.K. Vuong-Brender
Laboratoire de Biologie du Développement, CNRS-UMR7622, Institut de Biologie Paris-Seine, Université Pierre et Marie Curie, Sorbonne Universités, Paris, France

Yvan Wenger
Department of Genetics and Evolution, Institute of Genetics and Genomics in Geneva (IGe3), Faculty of Sciences, University of Geneva, Geneva, Switzerland

Xinyi Yang
Laboratoire de Biologie du Développement, CNRS-UMR7622, Institut de Biologie Paris-Seine, Université Pierre et Marie Curie, Sorbonne Universités, Paris, France

Chaogu Zheng
Department of Biological Sciences, Columbia University, New York, USA

PREFACE

ESSAYS ON DEVELOPMENTAL BIOLOGY-2016

These volumes commemorate the 50th or "golden" anniversary of the founding of *Current Topics in Developmental Biology* (CTDB) in the summer of 1966 by Aron Moscona (1921–2009) and Alberto Monroy (1913–1986) with Academic Press as publisher. In 1966 Moscona and Monroy anticipated the changing tide of developmental biology, from a classical to a more molecular science, and intended that CTDB be a venue for the change. From the beginning CTDB was to serve as "a meeting ground for critical review and discussion of developmental processes". More than 115 volumes later, still published by Academic Press (Elsevier), CTDB continues to strive to serve this purpose for developmental biologists.

The idea of publishing these essays came about in late 2013 when I realized that 2016 would be the 50th anniversary of the CTDB series. The object was to assemble essays contributed by a large number of outstanding developmental biologists around the world working on different aspects of development with a variety of model organisms. To achieve this goal more than 120 research scientists were contacted individually by e-mail in January 2014 and invited to contribute to this project. The response was surprisingly good with about 90 invitees agreeing to submit essays by September 15, 2015. Many flattering comments accompanied the acceptances, including "I think it is very appropriate to have an anniversary edition of what has been a very influential series over the years" and "CTDB is a review series that I appreciate. It has been, and is, important for my field of developmental biology". When the late Walter Gehring received my invitation and list of contributors he wrote back immediately saying, "After looking at your impressive list of contributors, I have to join your gang."

In the end more than 70 essays were submitted for publication by some of the most accomplished and well-known developmental biologists in the world. The essays appear here in alphabetical order according to the invited author's last name, not according to topic or on some other basis. Just a glance at the Table of Contents reveals the wide breadth of research topics and variety of model organisms discussed. Many essays relate the history of a particular area of research and/or personal experiences in research, whereas others are quite philosophical about a specific area of research and/or research in general. The authors were encouraged to speculate and in many instances have done so. The essays are not necessarily intended to be comprehensive reviews, however, they provide a window onto the very rich landscape of contemporary research in developmental biology.

I'm grateful to all the authors who devoted a great deal of time and effort preparing their essays. From a personal point of view, it's been a great pleasure to interact via e-mail with so many outstanding developmental biologists during the past 18 months or so. They tolerated my frequent reminders ("nagging") about the essays and, in a few instances, with my pleas to contribute to these volumes of CTDB. Despite a few sleepless nights worrying about who might not deliver an essay on time, overall this turned out to be a gratifying experience. My goal was to assemble 50th anniversary volumes of CTDB that would reflect the current state of developmental biology and be useful to students and investigators for years to come. It's up to you the reader to decide whether this goal was achieved. I can't help but wonder what Moscona and Monroy would think about the essays and the state of developmental biology presented in these 50th anniversary volumes of CTDB.

Finally, I want to thank Zoe Kruze and Sarah Lay at Elsevier, Oxford, and my wife, Eveline Litscher, for their forbearance, advice, and assistance during the course of this lengthy project.

We fondly remember four outstanding developmental biologists, Professors Eric H. Davidson (1937–2015; California Institute of Technology, Pasadena, California), Walter J. Gehring (1939–2014; Biozentrum, University of Basel, Switzerland), Julian H. Lewis (1946–2014; Cancer Research, London Research Institute, England), and Yoshiki Sasai (1962–2014; RIKEN Center for Developmental Biology, Japan) who agreed to be contributors to Essays on Developmental Biology-2016, but died unexpectedly in 2014 or 2015. Our condolences to their families and friends.

Paul M. Wassarman
Series Editor
Current Topics in Developmental Biology

CHAPTER ONE

Seeing is Believing, or How GFP Changed My Approach to Science

Markus Affolter[1]
Biozentrum der Universität Basel, Basel, Switzerland
[1]Corresponding author: e-mail address: markus.affolter@unibas.ch

Contents

Abstract

The field of "Developmental Biology" has dramatically changed over the past three decades. While genetic analysis had been center stage in the 1980s and continues to be a corner stone for investigations, the introduction of green fluorescent protein (GFP) in the 1990s has allowed us to look into living, developing embryos, and see how cells form tissues and how organ morphogenesis proceeds in real time. The introduction of protein binders into developmental studies some years ago has raised the precision yet another step, since it will allow the manipulation and study of how proteins function in real time. This chapter is a personal account on how GFP has, and how protein binders may, change the design of studies in the field of developmental biology.

1. FROM FLY MUTANTS TO DEVELOPMENTAL CONTROL GENES

In the early 1990s, when I was a PhD student studying cytochrome P450 and histone genes at Laval University in Quebec, Canada, I followed the emergence of gene cloning in the field of developmental biology with great interest and fascination from a distance. Upon the cloning of the first homeotic genes in *Drosophila* (McGinnis, Levine, Hafen, Kuroiwa, & Gehring, 1984; Scott & Weiner, 1984), the use of short stretches of

Current Topics in Developmental Biology, Volume 116
ISSN 0070-2153
http://dx.doi.org/10.1016/bs.ctdb.2015.12.001

conserved DNA (such as the homeobox) to screen for additional regulatory genes (Carrasco, McGinnis, Gehring, & De Robertis, 1984; McGinnis, Garber, Wirz, Kuroiwa, & Gehring, 1984) as well as the cloning of developmental regulatory genes isolated from DNA of polytene chromosomes of the salivary gland (Preiss, Rosenberg, Kienlin, Seifert, & Jackle, 1985) resulted in tens of developmental control genes being identified and characterized in a few years only, both in invertebrates and in vertebrates. Developmental biology turned truly molecular as I was struggling to finish my PhD!

Following this research topic, I became more and more interested in how animal morphologies emerge out of the building blocks of living creatures, the constituent cells, and I undoubtedly wanted to go to a laboratory which was involved in these exciting new developments. After a first refusal, I finally got the opportunity to join the laboratory of Walter J. Gehring in Basel as a postdoctoral fellow. Because it had been shown in Walter's lab that ectopic expression of *Drosophila* homeotic genes could redesign the body plan, resulting in flies with altered adult morphology (e.g., ectopic expression of *Antennapedia* resulted in the formation of legs instead of antennae on the head; Schneuwly, Klemenz, & Gehring, 1987), I thought that genes inducing such transformations could be taken as entry points to study the molecular mechanisms underlying tissue morphogenesis, ultimately leading to a comprehensive understanding of cell behavior during organ formation. However, although understanding how an antenna transforms into a leg will provide important insights into developmental control mechanisms and their transcriptional regulation by homeotic genes, understanding how either the leg or the antenna reaches their specific three-dimensional architecture during development requires much more detailed knowledge about the control of cell behavior in time and space. Clearly, the focus of such studies has to be at the cellular and the tissue level, and even today, we are lacking a comprehensive view of the complex processes involved in leg or antennal development.

Taking a more simple tissue to ask questions regarding morphogenesis and focus on individual cells seemed to be an alternative way to go, so upon my postdoctoral stay in Walter Gehring's lab, I started to study tracheal development. This decision was in part triggered by a chance observation; while trying to identify cofactors for homeotic proteins in Walter's lab, we characterized the *Drosophila* homologue of a protein called serum response factor (SRF). It turned out that SRF is expressed in the peripheral cells of the tracheal system in the fly embryo (Affolter et al., 1994). Looking at the beautiful pictures obtained when staining both for the tracheal system and SRF positive cells, which showed that the latter were distributed like grapes at the periphery of the tracheal tree (see Fig. 1), I started to wonder how such a ramified structure was built during development. A new research subject was born in the lab, and over the next 20 years, we and several other labs (headed by Mark Krasnow, Benny Shilo, Reinhard Schuh, Jordi Casanova, Shigeo Hayashi, Debbie Andrew, just to mention a few of the early members of the trachea research field) invested much effort to start to unravel the molecular bases underlying tracheal branching.

Genetic approaches, including the analysis of mutant embryos via antibody staining in fixed tissues, represented the major way toward a better understanding of the mechanisms underlying tracheal branching morphogenesis at that time. However, these early studies lacked a strong focus on

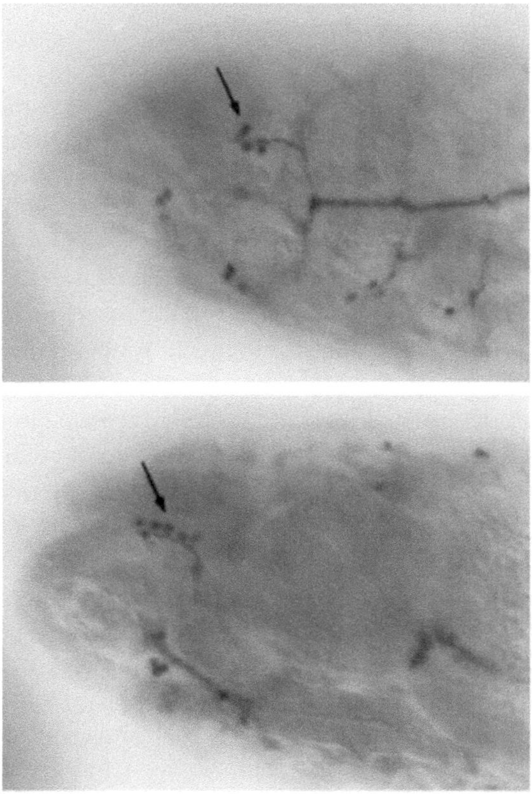

Figure 1 Head region of a *Drosophila* embryo showing the tracheal lumen (brown) and the nuclei of tracheal cells expressing dSRF (blue). *Taken from Affolter et al. (1994).* (A color version of this figure is located at the end of this chapter.)

cells and their dynamic behavior during morphogenesis, which was due to the lack of appropriate methods to investigate single-cell behavior *in vivo*. It turned out that these early studies provided a better understanding of tissue patterning by cell–cell signaling and the resulting transcriptional regulation, rather than directly involved in cell and tissue morphogenesis (Affolter & Shilo, 2000; Metzger & Krasnow, 1999).

2. FROM MUTANTS TO CELL BEHAVIOR

While we followed these rather well-established lines of genetic research during the 1990s, we decided at some point that determining the transcriptome of wild-type and mutant tracheal cells would be another way to go in order to get a deeper look into tracheal cells. Ulrich Certa and colleagues from Hoffmann-La Roche produced the first complete *Drosophila* microarray for the Basel fly community, which, to the great despair of Walter Gehring, turned out to lack sequences complementary to *Antennapedia*! Carlos Ribeiro, a PhD student in the lab, used a transgenic fly line expressing cytoplasmic green fluorescent protein (GFP) in the tracheal system with the aim to isolate these cells via FACS from embryos for RNA isolation (this line was kindly given to us by Shigeo Hayashi; Shiga, Tanaka-Matakatsu, & Hayashi, 1996). Before treating the embryos with trypsin to individualize the cells, he went to a confocal microscope to verify whether tracheal cells were indeed fluorescently labeled. That day, Carlos showed me a beautiful picture a tracheal tip cell with many long and fine filopodial extensions, structures that we had not seen previously in our images of fixed embryos! Making time lapse movies of embryos expressing GFP revealed the extremely dynamic behavior of these tip cell filopodia during branch formation and made the tracheal system become alive (Ribeiro, Ebner, & Affolter, 2002; see Fig. 2). For the first time, we could visualize how tracheal cells behave during the morphogenesis process and how they rearrange to build a highly organized tubular network from a simple, sac-like invagination! We were incredibly excited about what we could see in such *in vivo*-derived movies and decided not to push too hard to isolate these cells via cell sorting, but rather to strengthen our efforts to image the dynamic behavior of tracheal cells. These studies allowed us to analyze our mutant collection using *single-cell behavior* over time from live embryos as a read-out, and thereby get to the essence of what we ultimately wanted to understand, i.e., how cells behave during a complex morphogenesis process *in vivo*

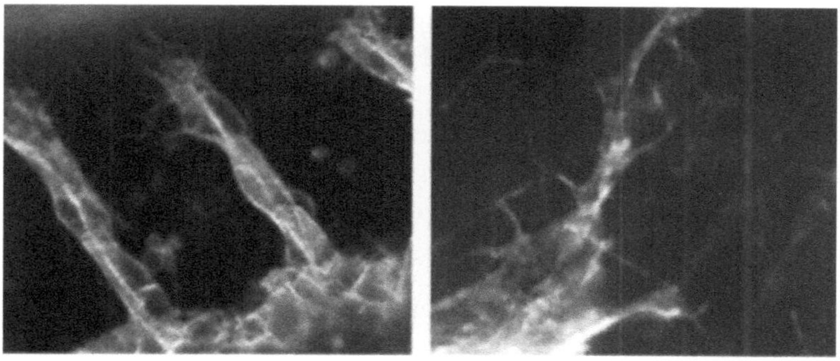

Figure 2 Three-dimensional reconstruction of living *Drosophila* embryos expressing Src-GFP (left panel) or GFP-actin (right panel) in the tracheal system. Left panel: low resolution showing cellular extensions from tip cells of growing dorsal branches. Right panel: high resolution of tip cell extensions. *Taken from Ribeiro et al. (2002).*

and how this cell behavior is controlled at the molecular level! Using this approach, morphogenetic processes did not have to be pieced together by aligning pictures taken from different fixed embryos at different time points, but rather could be visualized by following a single, living embryo as it develops!

We were overwhelmed by the wealth of novel information we could gather using this approach. For example, we had previously described mutants in Dpp/Bmp2 receptor-encoding genes, which lacked certain tracheal branches, while others remained intact (Vincent et al., 1997). However, for many years, we had no explanation as to why certain branches were missing while others developed normally, and we could not ascribe a defined alteration in cell behavior to the observed phenotype. Using live imaging of tracheal development in wild-type embryos and in embryos lacking Dpp/Bmp2 receptors, we found that it is the lack of cell intercalation in the fine tracheal branches, which leads to branch regression and ultimately to the absence of these tracheal branches (Ribeiro et al., 2002; Ribeiro, Neumann, & Affolter, 2004).

From now on, live imaging using different fluorescent proteins as markers became our prime and most important tool to observe cell behavior *in vivo*, and it helped us tremendously to make cell-based models of morphogenesis. These studies, together with the studies done by several other groups interested in tracheal morphogenesis, lead to a cellular description of the process, starting from the invagination of tracheal cells and the formation of a sac-like structure, which, upon complex yet highly regulated cell migrations, cell rearrangements and cell shape changes, resulted in the highly organized branching pattern (reviewed in Caviglia & Luschnig, 2014; Maruyama & Andrew, 2012; Ochoa-Espinosa & Affolter, 2012). The introduction and visualization of GFP and other fluorescent proteins into living animals (Chalfie, Tu, Euskirchen, Ward, & Prasher, 1994; Tsien, 1998) has definitely made a huge impact and has allowed for a much deeper look into living embryos as they undergo morphogenesis!

One of the key aspects of morphogenesis (the making and/or the transforming of shape) must be the generation and interpretation of forces; without unbalanced forces, cells would not reorganize or change shape in time and space. It is therefore a major question as to where and how forces are generated and how the underlying cellular network interprets them. Actin and the actomyosin network play a key role in generating cellular forces, and one of our major interests was to identify the force generation mechanism as well as the resulting responses in the tracheal system. We wanted to have a better idea as to why the sac-like structure was transformed into a branched network! Benny Shilo's group had previously shown that the Fgf receptor Breathless is critically required for branching (Klambt, Glazer, & Shilo, 1992) and Mark Krasnow's group had identified an Fgf ligand they called Branchless, which was expressed in nontracheal cells in those regions toward which branches were extending (Sutherland, Samakovlis, & Krasnow, 1996). These results suggested that an Fgf signaling pathway specifies the branching pattern (and ultimately the morphology) of the tracheal system by guiding tracheal cell migration. Indeed, and as described above, we later showed that tracheal tip cells did migrate actively and did so via cells at the tip of the extending branches, while stalk cells rearranged extensively during branch extension.

In a first step toward identifying the origin of forces controlling tracheal morphogenesis, we resorted to laser ablation in combination with live imaging to visualize inherent forces. Using a high energy, pulsed laser, the death of specific cells in the tracheal system can be induced, leading to a discontinuity in cell connections within extending branches. The resulting alterations in cell migration and/or cell rearrangement could then provide insight into the forces that normally act in the system. Using a series of laser ablation experiments in combination with live imaging, we found that the major force generators during the branching process are the tip cells; their active migration results in a tensile force that triggers cell intercalation in the stalk cells, as long as they are connected to the remainder of the tracheal system (Caussinus, Colombelli, & Affolter, 2008). In contrast to other tissues in the fly, in which forces underlying tissue morphogenesis are produced in each individual cell (as for example, during germ band extension in *Drosophila* embryos; Guillot & Lecuit, 2013), this task is taken up by a few peripheral cells, the tip cells, in the tracheal system.

This result suggests that, in contrast to these other systems, tracheal cells might not use the actomyosin network to generate forces to produce the ordered outgrowth of branches. However, it is very difficult to test whether the actomyosin network is involved in branch outgrowth and elongation, since actomyosin is used beforehand for cell division, and because maternal contribution persists to a large extent throughout embryogenesis, making straightforward genetic analysis of reiteratively used proteins and protein complexes difficult or impossible. For morphogenesis studies to become more tractable, new methods might be needed, and I will come back at the end of the assay on how we try to address this problem.

3. FROM FLIES TO FISH

Becoming increasingly fascinated by the many possibilities to use live imaging in studying morphogenesis processes at the cellular level, we also started to ask ourselves more and more often whether the underlying principles we observed (directed cell migration, stereotyped cell rearrangement, involvement of apical extracellular matrix (Jazwinska, Ribeiro, & Affolter, 2003), etc.) were at work in other tissues undergoing similar processes during development, and whether they were conserved in organ development in other animal species. For example, it has been argued that tracheal development in flies might be similar to lung development in mammals; both organs consist of ramified tubular structures and transport oxygen. In addition, it became apparent that the branching patterns in the trachea and in the lung are both controlled by Fgf molecules (Bellusci, Grindley, Emoto, Itoh, & Hogan, 1997; Sutherland et al., 1996), which extended the similarity from the tissue level to the molecular level. However, due to the problem of imaging mice embryos as they develop *in utero*, it has not been possible yet to obtain much insight into the cellular processes underlying branching morphogenesis in the lung, and the question of similarities in cell behavior in tracheal and lung branching is thus still out there, waiting for an answer.

However, while we were in the midst of our tracheal studies in fly embryos, a novel model system, the zebrafish, had become more and more widely used in the field of developmental genetics and developmental biology. The zebrafish has been studied for a number of years by George Streisinger and Charles Kimmel, and they and their colleagues have shown that the zebrafish can be used to identify mutants and to analyze embryos of such mutants in quite some detail, due to their external development and their transparency (Kimmel, 1989 and references therein). In the early 1990s, Christiane Nüsslein-Volhard reasoned that a large-scale screen similar to the one she had done some years ago with Eric Wieschaus in *Drosophila* (Nusslein-Volhard & Wieschaus, 1980) would be an important contribution to vertebrate genetics, and her lab set up the necessary methods and infrastructures to undertake such large-scale screens (similar screens were also done in Boston by Wolfgang Driever). These screens identified a large number of mutations affecting the development of many tissues (Nusslein-Volhard, 2012) and were the basis for a flood of papers in the following years. However, in these early days of zebrafish developmental genetic studies, it was still rather difficult to make transgenic zebrafish lines (for example, lines expressing GFP in a tissue-restricted manner), and it took a while until the zebrafish was also established as an excellent model system in the field of live imaging.

Inspired by the elegant work done using zebrafish, we also started to use zebrafish embryos as a novel, attractive model system for early vertebrate development in the practical courses we organize every year for our last year bachelor students at the Biozentrum. Looking at these translucent, rapidly developing vertebrate embryos was so fascinating that we started to think seriously about adopting this novel model system for our own scientific research projects in the lab. Understanding branching morphogenesis in the developing tracheal system in flies was still on going, but now we were eager to start to study another branched organ, the vasculature! Zebrafish was the model system of choice for us, since, in principle, it could be used for high-resolution live imaging, the approach we had chosen in the fly to investigate cell behavior. Work in the lab of Brant Weinstein had pioneered the generation and use of transgenic fish lines marking specifically endothelial cells and his group was the first to describe the developing vascular network of the zebrafish in great detail (Lawson & Weinstein, 2002). Meanwhile, Holger Gerhardt in the lab of Christer Betsholtz showed that sprouting vessels in the mouse retina are guided by so-called tip cells, very similar to what is seen in the trachea (Gerhardt et al., 2003). Inspired by these observations and eager to start to work on vertebrate embryos, but without solid knowledge of the zebrafish system, I first needed to recruit researchers with experience in this field, and I was joined by Heinz-Georg Belting and later by Elin Ellertsdottir, both from the lab of Wolfgang Driever in Freiburg, Germany. Together with Yannick Blum, Lukas Herwig, and

Alice Krudewig, we started to generate our own transgenic zebrafish lines expressing fluorescent proteins in the vasculature and used them to either track all the endothelial cells or use single-cell labeling to gain more insight into the behavior of individual cells during angiogenesis. In our *Drosophila* studies, we had experienced that some of the most useful insights for our purpose were obtained by looking at cell junctions, which allowed us to reveal the shape of individual cells. This turned out to be particularly useful in the vasculature, since endothelial cells are rather flat and the Adherens Junctions that link them to neighboring cells outline the cell borders, thus providing rather detailed information about cell shape. Therefore, we used the UAS-Gal4 system, which was adapted from the fly system to the zebrafish system in the meantime, and generated transgenic fish lines expressing the junctional protein ZO-1 and deletion versions of VE-Cadherin in endothelial cells. Using these and many other lines, we started to systematically describe cell behavior in angiogenesis, including the processes of vascular sprouting, anastomosis, and pruning (see Fig. 3). Since such *in vivo* analyses had not been done yet in the mouse, the zebrafish work was the first to provide insight into cell behavior during vascular development in a living animal at high, single-cell resolution.

We found that endothelial cell behavior is extremely dynamic in angiogenesis. During sprouting and anastomosis, the vessel lumen can be formed by different processes, either by hollowing out individual cells or by cell

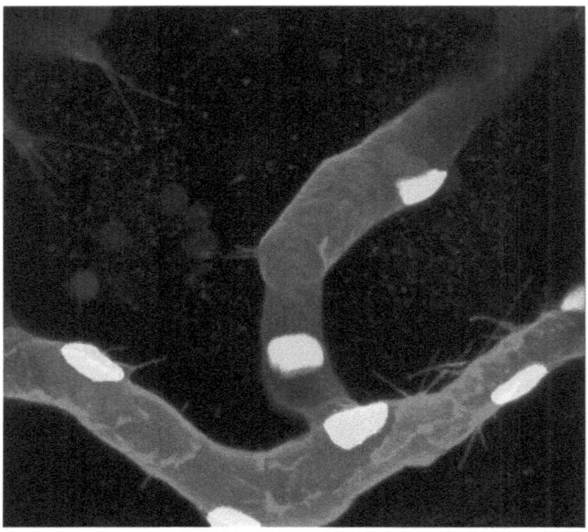

Figure 3 Blood vessels in the head of a zebrafish embryo expressing membrane-tethered RFP and nuclear GFP. *Picture taken by Anna Lenard.* (A color version of this figure is located at the end of this chapter.)

rearrangements (Herwig et al., 2011). During anastomosis, endothelial cells that surround the lumen often break up (undergo a fission event) in order to allow neighboring cells to make new contacts (Lenard et al., 2013). In contrast, endothelia cells often self-fuse during vascular pruning, to seal a regressing branch and prohibit fluid leakage (Lenard et al., 2015). These studies and studies by several other laboratories start to provide a detailed description of cell behavior in angiogenesis (Schuermann, Helker, & Herzog, 2014) and they need to be complemented with functional studies at the molecular level in the years to come.

4. FROM GFP TO GFP BINDERS

One problem we had already faced in our genetic studies of the tracheal system in flies was that for many genes, embryos carrying null mutations were either not developing to the stage at which we wanted to analyze tracheal development, or maternal contribution of the gene product allowed the embryo to complete embryogenesis, making genetic analyses difficult or impossible (conditional mutants in developing fly embryos were not easy to make with the tools available at the time). A method, which would allow acute, tissue-specific protein degradation, would be incredibly useful in order to study morphogenesis processes, since the latter most often rely on proteins also used during cell division (actin, myosin, cytoskeleton, etc.), making a straightforward analysis using mitotic clones almost impossible. Although several such methods exist, none has been used extensively, neither in flies nor in vertebrates.

Emmanuel Caussinus, a postdoc in the lab, designed a novel strategy to degrade proteins using a rather clever and innovative approach (Caussinus, Kanca, & Affolter, 2012). He argued that if a protein could be brought to the proteasome in an inducible or tissue-specific fashion, this protein would be degraded and the loss-of-function phenotype could be scored over time in a given tissue. In order to recognize or target a given protein, he used an anti-GFP nanobody, a small protein of 120 amino acids derived from the heavy chain of a single chain antibody of a camel, which recognizes GFP with high affinity (Rothbauer et al., 2008). For a protein to be recognized by this nanobody (which is also called an intrabody since it folds properly inside cells), the protein simply needed to be tagged by GFP such as to produce a functional, in-frame fusion. In order to bring the GFP fusion protein to the proteasome, Emmanuel attached the F-box of the *Drosophila* Slmb protein to the anti-GFP nanobody. Using this novel fusion protein expressed from a transgene, he showed that nuclear histone proteins fused to GFP could be degraded, both in tissue culture and in developing fly embryos (Caussinus et al., 2012; see Fig. 4). Many other GFP fusion proteins were also degraded in the presence of the F-box containing anti-GFP nanobody, opening the possibility to study loss-of-function phenotypes by protein depletion rather than by gene deletion or inactivation.

In principle, we now have a tool in hand, with which we can target specific proteins for degradation, and document the resulting phenotypes in time and space! Since its initial development, the deGradFP system has been successfully used to elucidate the tissue-specific function of actinomyosin in dorsal closure (D. Brunner and L. Pasakarnis, personal communication) and in tracheal branching (A. Ochoa and M. Affolter, ready for publication), to study the role of maternal contribution of certain centromere proteins (Raychaudhuri et al., 2012), to study the stability and maintenance of protein complexes *in vivo* (Batz, Forster, & Luschnig, 2014), and to investigate the role of proteins in memory formation and maintenance in adult flies (Nagarkar-Jaiswal, Lee, et al., 2015). Large-scale efforts are underway to tag an increasing number of *Drosophila* proteins by inserting GFP at the endogenous locus (Lowe et al., 2014; Nagarkar-Jaiswal, DeLuca, et al., 2015), and this will tremendously enlarge the number of potential target proteins for deGradFP in particular, and for the anti-GFP nanobody (see below) in general.

Being able to recognize GFP-tagged proteins with an intracellular nanobody also allows, in principle, for many other manipulations, such as protein relocalization or trapping (via the fusion of the anti-GFP nanobody to a localization sequence), or the posttranscriptional modification of specific GFP-tagged proteins (by fusing the nanobody to protein domains carrying enzymatic activities, such as kinases, etc.). We are in the process of generating and validating several such novel tools, both in transgenic flies and in zebrafish, as well as in cultured cells.

Since there are different nanobodies against GFP, which bind to non-overlapping epitopes, several studies have reported the generation of cell-specific gene or protein manipulation by using tissue specifically expressed GFP as a scaffolding protein (Tang et al., 2015, 2013). Since nanobodies (Helma, Cardoso, Muyldermans, & Leonhardt, 2015; Kaiser, Maier, Traenkle, Emele, & Rothbauer, 1844) or other protein binders (Pluckthun, 2015) can be isolated against virtually any protein of interest, and functionalized in many different ways (see Fig. 5), such binding proteins might be used much more often in the coming years in developmental and/or cell biological studies.

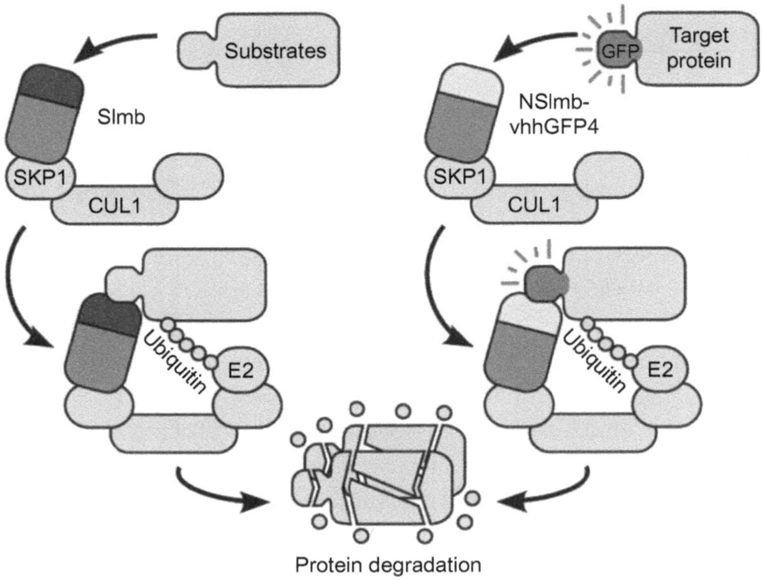

Figure 4 Protein degradation using an anti-GFP nanobody (vhhGFP4) to target GFP fusion proteins. *Taken from Caussinus et al. (2012).* (A color version of this figure is located at the end of this chapter.)

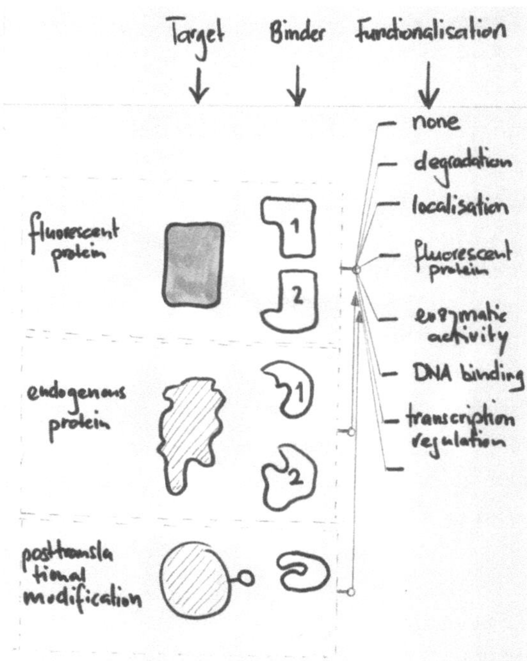

Figure 5 Protein binders (nanobodies, DARPINs, and others) can be isolated against fluorescent proteins, endogenous proteins, or against protein posttranslational modifications. Such protein binders can be functionalized by fusing them with different protein domains. For more information, see Helma et al. (2015), Kaiser et al. (1844), and Pluckthun (2015). (A color version of this figure is located at the end of this chapter.)

5. THE CRISPR REVOLUTION OR BRINGING IT ALL TOGETHER

Much of the exciting work done in zebrafish over the past 20 years has relied on the isolation and characterization of mutants in large-scale genetic screens. However, it is clear that the role of many proteins affecting cell behavior will be difficult to dissect using classical genetic approaches (thinking about proteins involved in cellular trafficking, membrane biogenesis, actin dynamics, etc.). Being able to manipulate genes at their endogenous locus using the Crispr technology (Doudna & Charpentier, 2014) has opened up the way to introduce any desired sequence alterations or sequence tags into the genome. At the same time, the expression of inducible and functionalized nanobodies or other protein binders, either recognizing such tags or recognizing endogenous, unmodified proteins, should in principle allow for a number of interesting scenarios that could help to uncover and better describe the role of certain proteins in cell behavior. One of these scenarios is to follow proteins in their natural environment through the use of so-called chromobodies (proteins consisting of a nanobody recognizing a specific protein fused to a fluorescent protein; Panza, Maier, Schmees, Rothbauer, & Sollner, 2015). Another possibility is to degrade proteins upon adaptation of the deGradFP to zebrafish and other model systems. Additionally, proteins could be relocalized to specific compartments in order to better characterize their functions. We are trying to implement the deGradFP system and validate some of the other tools (Fig. 5) in zebrafish.

In conclusion, the availability of GFP for developmental biology studies has changed our approaches twice in a dramatic manner. First, it allowed us to look at cell dynamics in normal and altered development *in vivo*; the living embryo rather than a fixed tissue section gives us the answer. Second, it allowed us to manipulate proteins in different ways due to the availability of a nanobody directed against GFP. I am very excited about the future developments that are made possible by the combination of Crispr technologies and the introduction of protein binders and protein tags into the zebrafish genome (or any other genome). I hope that our aim to better understand cell behavior in the complex environment of a living multicellular organism will greatly profit from such a novel tool box. Maybe, an account on the progress of protein binders in developmental biology will be published in 50 years in a special issue of *Current Topics in Developmental Biology* celebrating its 100th anniversary!

ACKNOWLEDGMENTS

I would like to thank Heinz-Georg Belting, Helen Preiss, and Loic Sauteur for help with the manuscript. I would also like to thank all present and past members of the lab for making scientific research such a pleasant and stimulating endeavor!

REFERENCES

Affolter, M., Montagne, J., Walldorf, U., Groppe, J., Kloter, U., LaRosa, M., et al. (1994). The Drosophila SRF homolog is expressed in a subset of tracheal cells and maps within a genomic region required for tracheal development. *Development*, *120*, 743–753.

Affolter, M., & Shilo, B. Z. (2000). Genetic control of branching morphogenesis during Drosophila tracheal development. *Current Opinion in Cell Biology*, *12*, 731–735.

Batz, T., Forster, D., & Luschnig, S. (2014). The transmembrane protein Macroglobulin complement-related is essential for septate junction formation and epithelial barrier function in Drosophila. *Development*, *141*, 899–908.

Bellusci, S., Grindley, J., Emoto, H., Itoh, N., & Hogan, B. L. (1997). Fibroblast growth factor 10 (FGF10) and branching morphogenesis in the embryonic mouse lung. *Development*, *124*, 4867–4878.

Carrasco, A. E., McGinnis, W., Gehring, W. J., & De Robertis, E. M. (1984). Cloning of an X. laevis gene expressed during early embryogenesis coding for a peptide region homologous to Drosophila homeotic genes. *Cell*, *37*, 409–414.

Caussinus, E., Colombelli, J., & Affolter, M. (2008). Tip-cell migration controls stalk-cell intercalation during Drosophila tracheal tube elongation. *Current Biology: CB*, *18*, 1727–1734.

Caussinus, E., Kanca, O., & Affolter, M. (2012). Fluorescent fusion protein knockout mediated by anti-GFP nanobody. *Nature Structural & Molecular Biology*, *19*, 117–121.

Caviglia, S., & Luschnig, S. (2014). Tube fusion: Making connections in branched tubular networks. *Seminars in Cell & Developmental Biology*, *31*, 82–90.

Chalfie, M., Tu, Y., Euskirchen, G., Ward, W. W., & Prasher, D. C. (1994). Green fluorescent protein as a marker for gene expression. *Science*, *263*, 802–805.

Doudna, J. A., & Charpentier, E. (2014). Genome editing. The new frontier of genome engineering with CRISPR-Cas9. *Science*, *346*, 1258096.

Gerhardt, H., Golding, M., Fruttiger, M., Ruhrberg, C., Lundkvist, A., Abramsson, A., et al. (2003). VEGF guides angiogenic sprouting utilizing endothelial tip cell filopodia. *The Journal of Cell Biology*, *161*, 1163–1177.

Guillot, C., & Lecuit, T. (2013). Mechanics of epithelial tissue homeostasis and morphogenesis. *Science*, *340*, 1185–1189.

Helma, J., Cardoso, M. C., Muyldermans, S., & Leonhardt, H. (2015). Nanobodies and recombinant binders in cell biology. *The Journal of Cell Biology*, *209*, 633–644.

Herwig, L., Blum, Y., Krudewig, A., Ellertsdottir, E., Lenard, A., Belting, H. G., et al. (2011). Distinct cellular mechanisms of blood vessel fusion in the zebrafish embryo. *Current Biology: CB*, *21*, 1942–1948.

Jazwinska, A., Ribeiro, C., & Affolter, M. (2003). Epithelial tube morphogenesis during Drosophila tracheal development requires Piopio, a luminal ZP protein. *Nature Cell Biology*, *5*, 895–901.

Kaiser, P. D., Maier, J., Traenkle, B., Emele, F., & Rothbauer, U. (1844). Recent progress in generating intracellular functional antibody fragments to target and trace cellular components in living cells. *Biochimica et Biophysica Acta*, *2014*, 1933–1942.

Kimmel, C. B. (1989). Genetics and early development of zebrafish. *Trends in Genetics: TIG*, *5*, 283–288.

Klambt, C., Glazer, L., & Shilo, B. Z. (1992). breathless, a Drosophila FGF receptor homolog, is essential for migration of tracheal and specific midline glial cells. *Genes & Development, 6*, 1668–1678.
Lawson, N. D., & Weinstein, B. M. (2002). In vivo imaging of embryonic vascular development using transgenic zebrafish. *Developmental Biology, 248*, 307–318.
Lenard, A., Daetwyler, S., Betz, C., Ellertsdottir, E., Belting, H. G., Huisken, J., et al. (2015). Endothelial cell self-fusion during vascular pruning. *PLoS Biology, 13*, e1002126.
Lenard, A., Ellertsdottir, E., Herwig, L., Krudewig, A., Sauteur, L., Belting, H. G., et al. (2013). In vivo analysis reveals a highly stereotypic morphogenetic pathway of vascular anastomosis. *Developmental Cell, 25*, 492–506.
Lowe, N., Rees, J. S., Roote, J., Ryder, E., Armean, I. M., Johnson, G., et al. (2014). Analysis of the expression patterns, subcellular localisations and interaction partners of Drosophila proteins using a pigP protein trap library. *Development, 141*, 3994–4005.
Maruyama, R., & Andrew, D. J. (2012). Drosophila as a model for epithelial tube formation. *Developmental Dynamics: An Official Publication of the American Association of Anatomists, 241*, 119–135.
McGinnis, W., Garber, R. L., Wirz, J., Kuroiwa, A., & Gehring, W. J. (1984). A homologous protein-coding sequence in Drosophila homeotic genes and its conservation in other metazoans. *Cell, 37*, 403–408.
McGinnis, W., Levine, M. S., Hafen, E., Kuroiwa, A., & Gehring, W. J. (1984). A conserved DNA sequence in homoeotic genes of the Drosophila Antennapedia and bithorax complexes. *Nature, 308*, 428–433.
Metzger, R. J., & Krasnow, M. A. (1999). Genetic control of branching morphogenesis. *Science, 284*, 1635–1639.
Nagarkar-Jaiswal, S., DeLuca, S. Z., Lee, P. T., Lin, W. W., Pan, H., Zuo, Z., et al. (2015). A genetic toolkit for tagging intronic MiMIC containing genes. *eLife, 4*.
Nagarkar-Jaiswal, S., Lee, P. T., Campbell, M. E., Chen, K., Anguiano-Zarate, S., Gutierrez, M. C., et al. (2015). A library of MiMICs allows tagging of genes and reversible, spatial and temporal knockdown of proteins in Drosophila. *eLife, 4*.
Nusslein-Volhard, C. (2012). The zebrafish issue of Development. *Development, 139*, 4099–4103.
Nusslein-Volhard, C., & Wieschaus, E. (1980). Mutations affecting segment number and polarity in Drosophila. *Nature, 287*, 795–801.
Ochoa-Espinosa, A., & Affolter, M. (2012). Branching morphogenesis: From cells to organs and back. *Cold Spring Harbor Perspectives in Biology, 4*.
Panza, P., Maier, J., Schmees, C., Rothbauer, U., & Sollner, C. (2015). Live imaging of endogenous protein dynamics in zebrafish using chromobodies. *Development, 142*, 1879–1884.
Pluckthun, A. (2015). Designed ankyrin repeat proteins (DARPins): Binding proteins for research, diagnostics, and therapy. *Annual Review of Pharmacology and Toxicology, 55*, 489–511.
Preiss, A., Rosenberg, U. B., Kienlin, A., Seifert, E., & Jackle, H. (1985). Molecular genetics of Kruppel, a gene required for segmentation of the Drosophila embryo. *Nature, 313*, 27–32.
Raychaudhuri, N., Dubruille, R., Orsi, G. A., Bagheri, H. C., Loppin, B., & Lehner, C. F. (2012). Transgenerational propagation and quantitative maintenance of paternal centromeres depends on Cid/Cenp-A presence in Drosophila sperm. *PLoS Biology, 10*, e1001434.
Ribeiro, C., Ebner, A., & Affolter, M. (2002). In vivo imaging reveals different cellular functions for FGF and Dpp signaling in tracheal branching morphogenesis. *Developmental Cell, 2*, 677–683.

Ribeiro, C., Neumann, M., & Affolter, M. (2004). Genetic control of cell intercalation during tracheal morphogenesis in Drosophila. *Current Biology: CB*, *14*, 2197–2207.
Rothbauer, U., Zolghadr, K., Muyldermans, S., Schepers, A., Cardoso, M. C., & Leonhardt, H. (2008). A versatile nanotrap for biochemical and functional studies with fluorescent fusion proteins. *Molecular & Cellular Proteomics: MCP*, 7, 282–289.
Schneuwly, S., Klemenz, R., & Gehring, W. J. (1987). Redesigning the body plan of Drosophila by ectopic expression of the homoeotic gene Antennapedia. *Nature*, *325*, 816–818.
Schuermann, A., Helker, C. S., & Herzog, W. (2014). Angiogenesis in zebrafish. *Seminars in Cell & Developmental Biology*, *31*, 106–114.
Scott, M. P., & Weiner, A. J. (1984). Structural relationships among genes that control development: Sequence homology between the Antennapedia, Ultrabithorax, and fushi tarazu loci of Drosophila. *Proceedings of the National Academy of Sciences of the United States of America*, *81*, 4115–4119.
Shiga, Y., Tanaka-Matakatsu, M., & Hayashi, S. (1996). A nuclear GFP/β-galactosidase fusion protein as a marker for morphogenesis in living Drosophila. *Development, Growth & Differentiation*, *38*, 99–106.
Sutherland, D., Samakovlis, C., & Krasnow, M. A. (1996). branchless encodes a Drosophila FGF homolog that controls tracheal cell migration and the pattern of branching. *Cell*, *87*, 1091–1101.
Tang, J. C., Rudolph, S., Dhande, O. S., Abraira, V. E., Choi, S., Lapan, S. W., et al. (2015). Cell type-specific manipulation with GFP-dependent Cre recombinase. *Nature Neuroscience*, *18*, 1334–1341.
Tang, J. C., Szikra, T., Kozorovitskiy, Y., Teixiera, M., Sabatini, B. L., Roska, B., et al. (2013). A nanobody-based system using fluorescent proteins as scaffolds for cell-specific gene manipulation. *Cell*, *154*, 928–939.
Tsien, R. Y. (1998). The green fluorescent protein. *Annual Review of Biochemistry*, *67*, 509–544.
Vincent, S., Ruberte, E., Grieder, N. C., Chen, C. K., Haerry, T., Schuh, R., et al. (1997). DPP controls tracheal cell migration along the dorsoventral body axis of the Drosophila embryo. *Development*, *124*, 2741–2750.

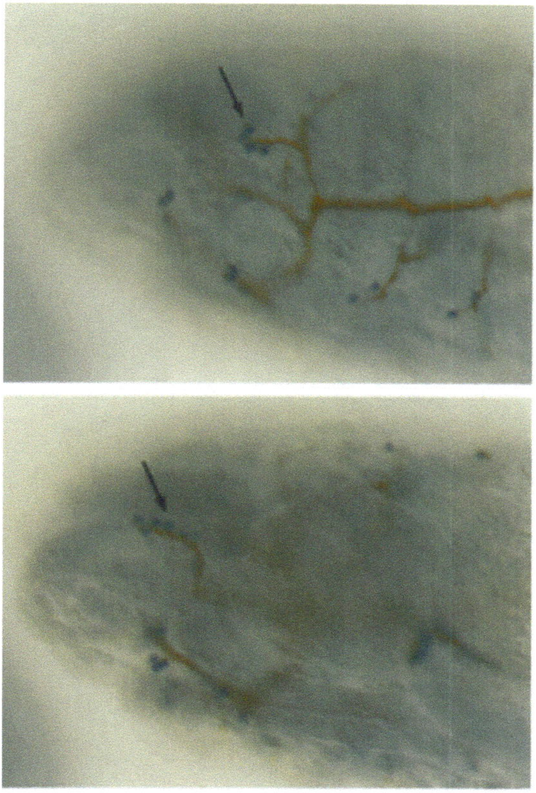

Markus Affolter, Figure 1 Head region of a *Drosophila* embryo showing the tracheal lumen (brown) and the nuclei of tracheal cells expressing dSRF (blue). *Taken from Affolter et al. (1994).*

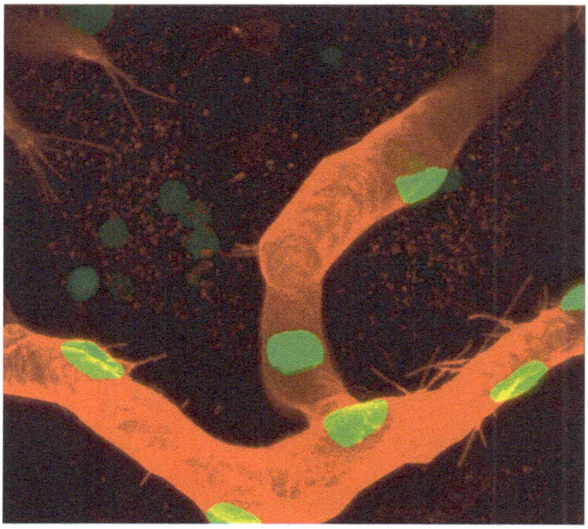

Markus Affolter, Figure 3 Blood vessels in the head of a zebrafish embryo expressing membrane-tethered RFP and nuclear GFP. *Picture taken by Anna Lenard.*

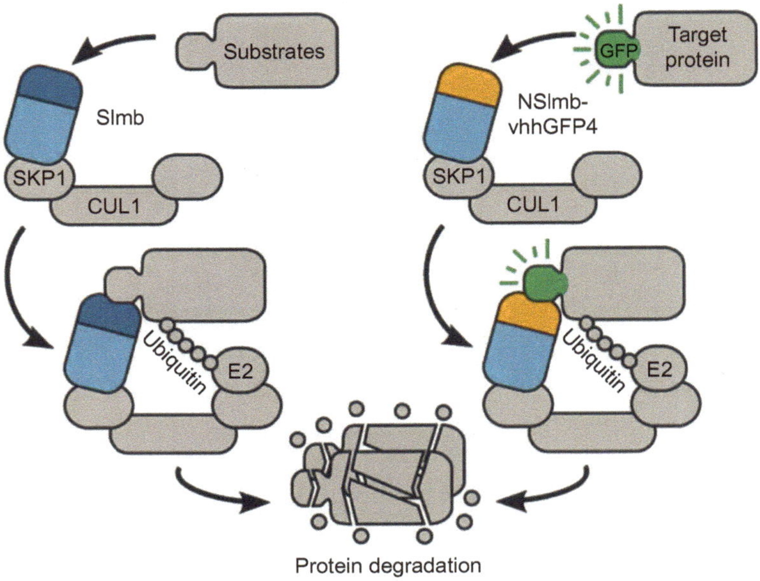

Markus Affolter, Figure 4 Protein degradation using an anti-GFP nanobody (vhhGFP4) to target GFP fusion proteins. *Taken from Caussinus et al. (2012).*

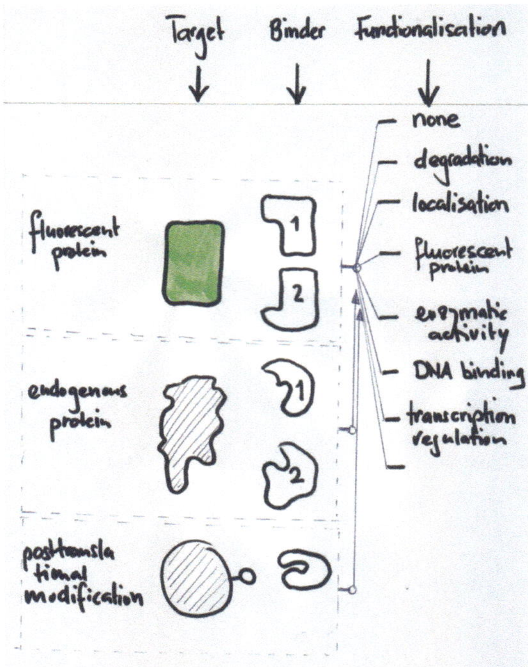

Markus Affolter, Figure 5 Protein binders (nanobodies, DARPINs, and others) can be isolated against fluorescent proteins, endogenous proteins, or against protein post-translational modifications. Such protein binders can be functionalized by fusing them with different protein domains. For more information, see Helma et al. (2015), Kaiser et al. (1844), and Pluckthun (2015).

CHAPTER TWO

The Notch-Mediated Proliferation Circuitry

Diana M. Ho*, Spyros Artavanis-Tsakonas*,†,1
*Department of Cell Biology, Harvard Medical School, Boston, Massachusetts, USA
†Biogen Idec, Cambridge, Massachusetts, USA
[1]Corresponding author: e-mail address: artavanis@hms.harvard.edu

Contents

Abstract

The essential and highly conserved Notch signaling pathway controls a wide range of cell fate decisions during development, including cellular proliferation. Notch mediates both pro- and anti-proliferative effects in development, stem cells, and cancer depending on cellular context. Furthermore, it can induce proliferation in both cell-autonomous and non-cell-autonomous manners. Interacting genes and crosstalking signaling pathways play essential roles in regulating the proliferative response to Notch signals. A large number of genes that participate in the Notch network to influence proliferation have been identified, including several that activate the JNK signaling pathway, which interacts with Notch to induce both hyperplastic and invasive cellular behaviors. It is clear that dissecting the genetic circuitry surrounding Notch is essential to understanding the proliferative response to Notch in both development and cancer.

Notch is one of only a handful of essential, conserved, and ancient signaling pathways that control metazoan development by directing cell fates toward proliferation, differentiation, self-renewal, or cell death. Tightly regulated cellular proliferation is essential for the development of complex multicellular organisms from a single-celled embryo and for the continued

Current Topics in Developmental Biology, Volume 116
ISSN 0070-2153
http://dx.doi.org/10.1016/bs.ctdb.2015.12.016

maintenance of tissues throughout adulthood; deregulation of cell fate control can lead to developmental defects as well as hyperplasia and other pro-oncogenic conditions that may eventually progress to cancer.

The uniqueness of the canonical Notch signaling pathway lies in the fact that it mediates interactions between two neighboring cells, one of which is the signaling cell and the other is the receiving cell, via the physical interaction of the ligand with the Notch receptor at the cell surface. Ligand binding to the Notch receptor results in a series of proteolytic cleavages within the receiving cell that release the active Notch intracellular domain to translocate into the nucleus. There, it forms a transcriptional complex that regulates a cohort of downstream target genes (reviewed in Andersson, Sandberg, & Lendahl, 2011; Artavanis-Tsakonas & Muskavitch, 2010; Artavanis-Tsakonas, Rand, & Lake, 1999; Bray, 2006; Bray & Bernard, 2010; Hori, Sen, & Artavanis-Tsakonas, 2013). The fruit fly *Drosophila melanogaster* harbors a single Notch receptor and two ligands, Delta and Serrate; in contrast, the mammalian genome codes for four Notch paralogs (Notches 1, 2, 3, and 4) and five ligands (Andersson et al., 2011; Artavanis-Tsakonas & Muskavitch, 2010).

Misregulation of the Notch pathway has been associated with a number of developmental disorders and cancer (Louvi & Artavanis-Tsakonas, 2012). Notably, activating mutations in Notch1 have been found to be causative for T-cell acute lymphoblastic leukemia (T-ALL) (Ellisen et al., 1991; Pear et al., 1996; Weng et al., 2004, 2003). Although Notch mutations themselves are uncommon in solid tumors, there is plenty of evidence that the pathway is aberrantly activated or misregulated in many cases (Koch & Radtke, 2010; Ranganathan, Weaver, & Capobianco, 2011). Our own analyses in different tumor models demonstrated that although Notch activation may not always be oncogenic *per se*, it does cause hyperplasia, which increases dramatically the chance of progression to a malignant state through synergies with other genes (Fre et al., 2009; Ho, Pallavi, & Artavanis-Tsakonas, 2015; Kiaris et al., 2004; Pallavi, Ho, Hicks, Miele, & Artavanis-Tsakonas, 2012). The oncogenic behavior of Notch, not unlike other cell fate events driven by Notch signaling, is highly context dependent. For example, Notch can both stimulate and suppress proliferation and can act as both an oncogene and a tumor suppressor (Artavanis-Tsakonas & Muskavitch, 2010; Koch & Radtke, 2010). To add a further level of complexity, there is clear evidence that Notch activation can induce proliferation in cells distant from and indeed not in direct contact with the ligand- and receptor-bearing cells (de Celis, Garcia-Bellido, & Bray, 1996; Giraldez & Cohen, 2003; Go, Eastman, & Artavanis-Tsakonas, 1998). Since the core Notch pathway requires two adjacent cells in direct contact with each other, this observation indicates that additional non-cell-autonomous relay mechanisms must also interact with the Notch network. It is thus clear that the cellular context, most notably the network of genes with which Notch interacts, is a crucial determinant (Bray & Bernard, 2010; Hurlbut, Kankel, Lake, & Artavanis-Tsakonas, 2007).

1. NOTCH CAN HAVE OPPOSING EFFECTS ON PROLIFERATION

In the *Drosophila* wing imaginal disc, a sharply delineated stripe of cell cycle-arrested cells known as the zone of non-proliferating cells (ZNC) appears at the dorsal/ventral (D/V) boundary during the third instar stage (Johnston & Edgar, 1998; O'Brochta & Bryant, 1985). Elimination of Notch signaling results in a loss of the ZNC, indicating that Notch is required for the proliferation arrest (Herranz, Pérez, Martín, & Milán, 2008). However, ectopic activation of the Notch pathway in the wing disc causes severe hyperplasia, demonstrating that Notch can also exert pro-proliferative effects (Go et al., 1998). Spatial analyses revealed that Notch induces hyper-proliferation at the periphery of the wing pouch, whereas it inhibits proliferation in the center of the pouch (Baonza & Garcia-Bellido, 2000; Giraldez & Cohen, 2003; Go et al., 1998), emphasizing the context-dependent developmental interpretation of the Notch signal (Fig. 1A and B).

How are these distinct spatiotemporal zones of Notch responsiveness determined? Establishment of the non-proliferative ZNC requires the concerted action of the Wnt ligand wingless (wg) and Notch, as well as the downstream effector genes *myc* and *bantam* (Herranz et al., 2008). *Scalloped* (*sd*) and *vestigial* (*vg*) are also required in this region, where they interact with Notch (Djiane et al., 2013). Interestingly, wg can also mediate the pro-proliferative activity of Notch in both the wing disc (Giraldez & Cohen, 2003) and the eye (Fre et al., 2009). *Wg* is itself a target of Notch, suggesting the presence of a feed-forward loop that integrates differentially with Notch depending on positional cues (Djiane et al., 2013). The Notch/wg synergy, in both its pro- and anti-proliferative incarnations, seems to act at least partially through modulation of cell cycle progression (Herranz et al., 2008). Cross talk between Notch and Wnt is also essential in mammals, where the two pathways cooperate to stimulate proliferation and oncogenesis in the mouse intestine (Fre et al., 2009).

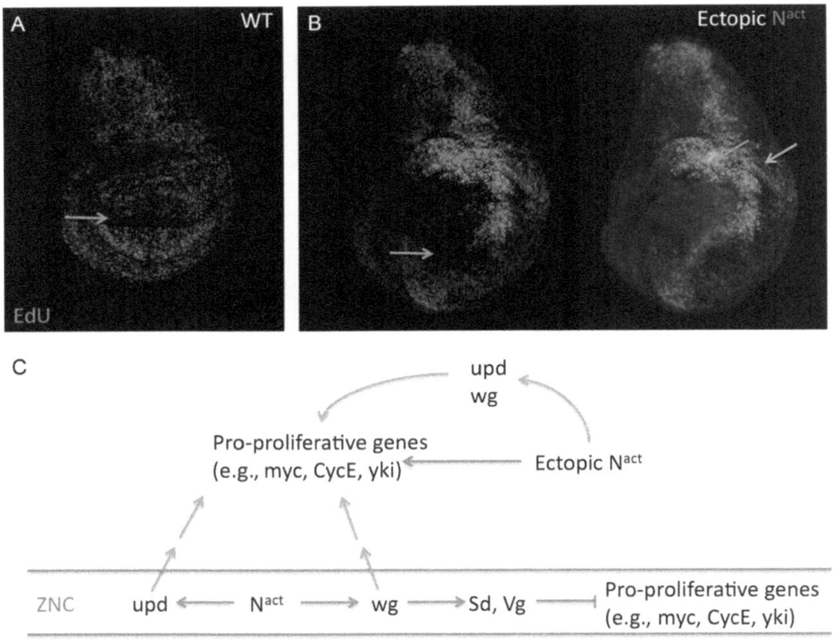

Figure 1 (A and B) Proliferation patterns in late third instar *Drosophila* wing imaginal discs stained with EdU. In wild-type discs (A), the ZNC (gray arrow) can be clearly distinguished by lack of EdU incorporation. When activated Notch (N^{act}) is expressed along the anterior–posterior boundary (labeled with GFP), the ZNC (gray arrow) expands and hyperproliferation occurs in the distal wing pouch and hinge regions. Both cell-autonomous (blue arrow) and non-cell-autonomous (orange arrow) components can be seen. (C) Simplified schematic of Notch gene interactions driving proliferation in the wing disc. Blue arrows represent cell-autonomous events, whereas orange arrows represent non-cell-autonomous events. See text for details. (A color version of this figure is located at the end of this chapter.)

Recent work has shown that a number of both positive and negative regulators, and cell-autonomous and non-cell-autonomous factors, are transcriptional targets of Notch itself and are expressed in combinatorial and complementary patterns to precisely pattern the wing pouch into proliferative and non-proliferative zones (Djiane et al., 2013). For example, Notch is necessary and sufficient to inhibit the activity of the pro-proliferative factor Yorkie (yki) but only in the central part of the wing disc, where it cooperates with vg and sd (Djiane, Zaessinger, Babaoğlan, & Bray, 2014). In contrast, Notch has also been reported to induce yki activity in a non-cell-autonomous manner at the periphery of the disc (Graves, Woodfield, Yang, Halder, & Bergmann, 2012). Taken together, the *Drosophila* data indicate that a complex and interconnected web of genes together with Notch itself determine the nature of the developmental response to the Notch signal at defined positional values.

A similar observation on the dual roles of Notch as both a pro- and anti-proliferative factor has been made in mammalian cells. Mazzone et al. found that in 3D cultures of MCF-10A mammary epithelial cells, constitutively active Notch induced both hyperproliferative and growth-arrested structures in a single culture; this heterogeneous response was found to be dependent upon the dosage of Notch in individual cells. Relatively low doses of Notch induced hyperproliferation that was dependent upon EGF and JAK/STAT signals, in contrast with higher doses, which inhibited proliferation (Mazzone et al., 2010). An additional complicating factor in mammals is that the different Notch paralogs may have opposing effects on proliferation. For example, it has been shown that Notch3 can suppress proliferation in the mammary gland (Lafkas et al., 2013), whereas active Notch1 can promote it in the same tissue (Kiaris et al., 2004; Rodilla et al., 2015), albeit in different subpopulations of cells. Although our understanding of the functional differences between Notch ligands is still fairly unclear (Andersson et al., 2011), the combinatorial possibilities for differential regulatory mechanisms based on specific receptor–ligand pairs should not be discounted either.

2. NOTCH AFFECTS PROLIFERATION IN BOTH CELL-AUTONOMOUS AND NON-CELL-AUTONOMOUS MANNERS

As already alluded to, Notch can induce both cell-autonomous and non-cell-autonomous proliferation in the wing disc (de Celis et al., 1996; Giraldez & Cohen, 2003; Go et al., 1998). The same is true in the eye imaginal disc, whereby Notch expression in only a small subset of eye disc cells is sufficient to direct proliferation patterns throughout the entire disc, again in a spatiotemporally regulated manner (Baonza & Freeman, 2005; Reynolds-Kenneally & Mlodzik, 2005). Observations in mammalian tumor cells suggest that the cell-autonomous component may be mediated through direct regulatory effects on cell cycle modulators, notably D-type cyclins that regulate the G1/S transition (Joshi et al., 2009).

Since the core Notch pathway itself is restricted to pairs of adjacent cells in direct contact with one another, the induction of proliferation in non-neighboring cells must of necessity involve secreted relay factors. At least two pathways, Wnt and JAK/STAT, have been reported to fill this role in *Drosophila* studies. Wg is known to be secreted in response to Notch and can drive target gene expression in the wing disc up to 30 cell diameters away (Zecca & Struhl, 2007); furthermore, wg and Notch synergize to drive proliferation non-cell autonomously (Go et al., 1998). However, *wg* mutant cells expressing activated Notch still induce proliferation in surrounding wild-type cells, indicating the presence of at least one other secreted pro-proliferative factor in the wing disc (Giraldez & Cohen, 2003). There is ample evidence implicating the *unpaired* (*upd*) family of JAK/STAT ligands in this process in both the wing (Moberg, Schelble, Burdick, & Hariharan, 2005; Vaccari & Bilder, 2005) and the eye (Reynolds-Kenneally & Mlodzik, 2005). Interestingly, in the eye at least, JAK/STAT signaling appears to be dispensable for the cell-autonomous component (Reynolds-Kenneally & Mlodzik, 2005). Suppression of the Salvador/warts/hippo growth-inhibitory pathway via yki activation in target cells may be one mechanism by which these factors exert their influence on proliferation; interestingly, this process appeared to be JAK/STAT independent (Graves et al., 2012). Cell cycle regulators such as *Cyclin E* and *myc* are also induced by Notch in non-cell-autonomous patterns, suggesting that they may also be downstream effectors in target cells (Djiane et al., 2013) (Fig. 1).

The existence of an indirect, non-cell-autonomous mechanism by which Notch can induce proliferation not only extends the range of Notch activity but also adds additional levels of regulatory complexity to the Notch proliferation network. It also has clear implications for oncogenesis. The fact that Notch can exert a proliferative effect at a distance and without need for direct cell contact suggests that Notch expressed in the surrounding stroma could influence tumor cells that do not themselves show aberrations in the Notch pathway. It may also be relevant in heterogeneous tumors in which only a few cells with aberrant Notch signaling might be sufficient to promote proliferation throughout an entire tumor via a propagated relay signal such as Wnt or JAK/STAT.

3. NOTCH AND PROLIFERATION IN ONCOGENESIS

Over 50% of T-ALLs display activating mutations in Notch1, which have been shown to be causative for oncogenesis (Ellisen et al., 1991; Pear et al., 1996; Weng et al., 2004, 2003). Notch not only initiates oncogenesis in T-ALL cells but also stimulates their continued proliferation via cyclin D3-mediated G1/S cell cycle transition (Joshi et al., 2009). Aberrant Notch signaling has also been linked to a number of other hematologic malignancies (Liu, Sato, Cerletti, & Wagers, 2010; Lobry, Oh, Mansour, Look, & Aifantis, 2014).

In contrast to hematopoietic cancers, the oncogenic properties of Notch in solid tumors are far less definitive and mutations in Notch itself are less common. Nevertheless, the expression/activity of Notch or Notch pathway members is altered to tumorigenic effect in a wide range of tumors, including those of the breast, colon, lung, pancreas, brain, and liver (reviewed in Koch & Radtke, 2010; Ranganathan et al., 2011). Rare functional rearrangements of the Notch family genes have been identified in breast cancers, suggesting that such mutations, while uncommon, do occur (Robinson et al., 2011).

In several mouse models, activated Notch alone merely causes a hyperplastic but not tumorigenic state, and it is likely that a second or even a third mutation in cooperating genes must occur for oncogenesis to begin. For example, constitutively active Notch1 ($N1^{IC}$) overexpression in the mammary epithelium causes mice to develop Myc-dependent pregnancy/lactation-dependent neoplasms that regress upon gland involution, indicating that they still respond to normal apoptotic stimuli (Kiaris et al., 2004). After multiple pregnancy/lactation cycles, some of these tumors progress to non-regressing, invasive carcinomas, suggesting that they have accumulated additional, cooperating mutations (Kiaris et al., 2004; Klinakis et al., 2006). An analogous situation occurs in the mouse intestine, where $N1^{IC}$ causes increased proliferation but does not induce tumors; however, when $N1^{IC}$ is combined with an $APC^{-/-}$ mutation that results in aberrant activation of the Wnt pathway, numerous adenomas form, resulting in increased lethality (Fre et al., 2005, 2009). Notably, these tumors are more numerous and localized in distinct parts of the intestine from $APC^{-/-}$ tumors without Notch, indicating a synergistic effect of the two pathways (Fre et al., 2009).

It is possible that the causative role for Notch in T-ALL and other hematopoietic tumors may be due to the fact that Notch is a master regulator of normal T-cell development (Koch & Radtke, 2011) and is important for many steps of hematopoiesis (Koch, Lehal, & Radtke, 2013; Liu, Sato, et al., 2010); hence, the precursor cells quite likely already express a proper, permissive network of supporting genes. In other tumor precursors that may not be as primed to respond to Notch signals, additional mutations seem to be required to initiate tumor formation.

Notwithstanding its oncogenic role, Notch can also act as a tumor suppressor, highlighting its context dependence. The best studied case of Notch's tumor suppressive properties is that of the skin, where Notch is thought to promote differentiation at the expense of proliferative cell fates (reviewed in Koch & Radtke, 2010). Loss of *Notch1* in mouse skin results in hyperproliferation and eventual onset of tumor formation; the long latency period between these two processes and the heterogeneity of the tumors suggests that additional cooperating mutations must occur for tumorigenesis to commence (Demehri, Turkoz, & Kopan, 2009; Nicolas et al., 2003). Indeed, the *Notch1* loss-of-function mutation can cooperate with oncogenic *Ras* mutations to promote carcinogenesis (Demehri et al., 2009; Nicolas et al., 2003), highlighting the fact that both positive and negative Notch synergistic interactions can have profound consequences. Interestingly, mice bearing a mesenchymal-specific deletion of the Notch effector *CSL/RBP-Jk* develop highly proliferative multifocal epithelial tumors, suggesting a non-cell-autonomous effect (Hu et al., 2012).

4. PROLIFERATIVE EFFECTS OF NOTCH IN STEM CELLS

Notch has been associated with virtually every stem/precursor cell population studied to date, both embryonic and adult, where it can regulate multiple aspects of stem cell biology, including establishment of a multipotent progenitor population, self-renewal, and differentiation (Koch et al., 2013; Liu, Sato, et al., 2010). Not surprisingly, Notch can be either pro- (e.g., in muscle satellite cells and mouse intestinal stem cells (ISCs)) or anti-proliferative (e.g., in skin stem cells and *Drosophila* ISCs) in different stem cell types (Koch et al., 2013; Liu, Sato, et al., 2010). Cancer stem cells, which display self-renewal capacity and multipotency similar to *bona fide* stem cells, tend to be resistant to therapy, and can repopulate an entire tumor from only a few cells, have been shown in some cases (glioblastoma, colorectal cancer, breast cancer) to be dependent upon Notch for self-renewal (Beck & Blanpain, 2013; Colak & Medema, 2014; Rizzo et al., 2008). Furthermore, several novel potential stem cell populations in the mouse mammary gland marked by specific Notch paralogs have been identified (Lafkas et al., 2013; Rodilla et al., 2015; Sale, Lafkas, & Artavanis-Tsakonas, 2013); given that different Notch paralogs have been associated with different breast tumor subtypes and prognoses (Koch & Radtke, 2010; Rizzo et al., 2008), further understanding of these novel progenitor pools and the circuitry surrounding them may provide insights into the sources of various complex and diverse breast tumors.

Examination of the role of Notch in ISCs in the fly and the mouse provides an interesting case study. In the adult *Drosophila* intestine, Notch is found in the enteroblasts (EBs, which are multipotent transient cycling progenitors derived from ISCs) (Ohlstein & Spradling, 2006, 2007). Notch signals suppress proliferation of ISCs via suppression of JAK/STAT (Liu, Singh, & Hou, 2010; Micchelli & Perrimon, 2006). Interestingly, Notch-defective *Drosophila* ISCs can, in concert with EGF and stress signals, generate stem cell tumors by coopting normal stem cell renewal signals (Patel, Dutta, & Edgar, 2015). In contrast, in mouse intestinal crypts, Notch1 is found in the Lgr5+, mitotically active ISC population, where it promotes proliferation via cell cycle regulation (Fre et al., 2011, 2005; Riccio et al., 2008; van Es et al., 2005). As discussed above, activated Notch1 can cooperate with Wnt signals in this case to promote tumorigenesis (Fre et al., 2009). The differences between the fly and the mouse systems remain unclear at present; one likely possibility is that since the EBs in flies and the Lgr5+ cells in mammals are not functionally equivalent, they also most likely have significantly different cellular contexts (Fre, Bardin, Robine, & Louvard, 2011; Koch et al., 2013). These examples serve once again to emphasize the fundamental property of Notch signals to be interpreted, both in development and disease, in a highly context-dependent manner. Without a very precise molecular understanding of the biochemical relationships between Notch signals and other cellular elements in a given context, one can never assume the developmental outcome of Notch activation or inactivation, nor can one extrapolate an outcome across tissues or species.

5. NOTCH SYNERGIES IN PROLIFERATION

Screens for modifiers of the core Notch pathway have been conducted (Kankel et al., 2007; Mourikis, Lake, Firnhaber, & DeDecker, 2010; Saj et al., 2010); although these screens did not directly score for proliferation effects, they nevertheless would most likely have detected some components of the proliferation circuitry. Notably, genes that control processes such as differentiation or apoptosis that are diametrically opposed to proliferation could also impinge upon proliferation regulation. Genome-wide transcriptional profiling analyses of Notch targets in hyperplastic discs have also identified a diverse range of downstream target genes both alone (Djiane et al., 2013) and in synergy with Ras (Hurlbut, Kankel, & Artavanis-Tsakonas, 2009). Although these studies have been informative, they almost certainly have not identified all of the genes that can synergize with Notch in proliferation.

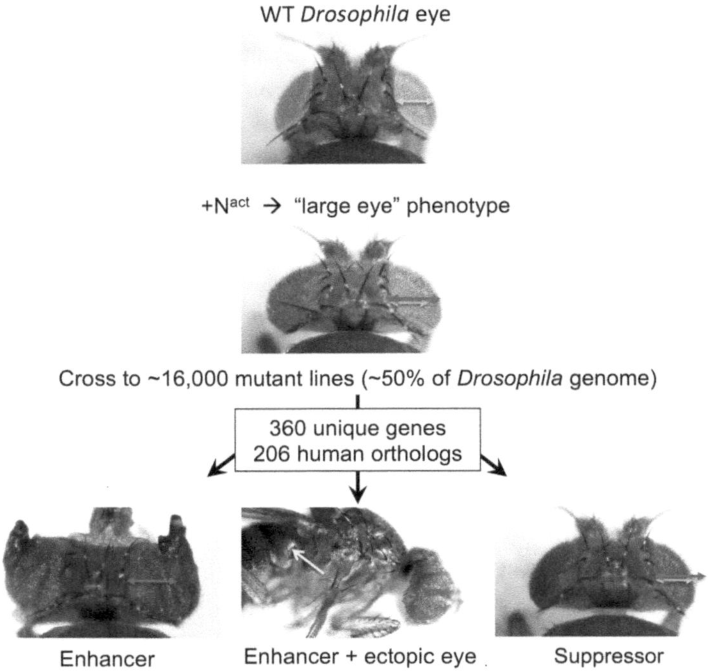

Figure 2 Schematic of and representative phenotypes from our genetic screen for modifiers of Notch-induced proliferation in the *Drosophila* eye. Flies ectopically expressing activated Notch (N^{act}) in the developing eye, which results in a "large eye" phenotype, were crossed to ~16,000 mutant lines representing ~50% of the *Drosophila* genome. Three hundred and sixty unique modifier genes, which caused either enhancement or suppression of the N^{act} phenotype, sometimes in conjunction with ectopic eyes, were identified, of which 206 had clear human orthologs. (A color version of this figure is located at the end of this chapter.)

Overexpression of activated Notch in the developing *Drosophila* eye causes a "large eye" phenotype that is dramatically enhanced or suppressed by the activation or mutation of synergistic genes. In order to identify novel synergistic partners of Notch in proliferation, our laboratory recently took advantage of this dramatic phenotype to conduct an unbiased genetic screen for modifiers of Notch-induced proliferation, and identified 360 potential genetic interactors, 206 of which have clear human orthologs (Ho et al., 2015) (Fig. 2). As we only screened approximately half of the genome, we estimate the true number of interactors to be at least double. A subset of these Notch synergies also caused an additional *de novo* ectopic eye phenotype that has been reported to be an indicator of metastatic behavior, suggesting a link between hyperplasia and metastasis (Ferres-Marco et al.,

2006; Pallavi et al., 2012). Notably, the genes identified in our genetic screen are likely to encompass Notch synergistic interactions at multiple levels of the circuitry and may affect both cell-autonomous and non-cell-autonomous components.

Perhaps surprisingly, this list of modifiers is not only numerous but also extremely diverse and is enriched only for morphogenetic/developmental factors (which can at least partially be explained by the developmental model system we used), transcription factors, and cell cycle genes. An enrichment for cell cycle genes was also reported in a ChIP-SEQ analysis of Notch target genes in hyperplastic wing discs (Djiane et al., 2013). A role for cell cycle regulators as effectors of Notch has been clearly demonstrated in flies (see above), and they are also known to be conserved Notch targets in T-ALL and mammary cancer models (Joshi et al., 2009). The diversity of the genes identified in this and other screens points to the likelihood that many and varied pathways can interface with Notch in order to influence proliferation.

6. A NOTCH–JNK SIGNALING AXIS COUPLES PROLIFERATION AND INVASIVENESS

Of particular interest were two of the synergistic modifiers identified in our screen, the transcription factor *Mef2* and the proto-oncogene *Src42A* (as well as its close relative *Src64B*). Both of these synergies induced immense hyperplasia caused at least in part by deregulation of the cell cycle; we also observed an upregulation of markers of invasiveness (Ho et al., 2015; Pallavi et al., 2012). These and other downstream effects are a consequence of strong upregulation of the JNK signaling pathway (Ho et al., 2015; Pallavi et al., 2012), which has been implicated in control of apoptosis, proliferation, and invasiveness (Ríos-Barrera & Riesgo-Escovar, 2013). A role for JNK in the Notch network had previously been reported in *Drosophila* tissues containing activated Notch along with a mutation for the epithelial cell polarity regulator *scribble*; these tissues display neoplastic overgrowth (Brumby & Richardson, 2003). Taken together, these data indicate that multiple, independent Notch synergies can converge upon a Notch/JNK signaling axis. Notwithstanding the similarities between the Notch/Mef2 and Notch/Src synergies, there are also significant differences, most notably a suppression of JNK-induced apoptosis in Notch/Mef2 but not Notch/Src (Ho et al., 2015). These seemingly subtle changes in the surrounding circuitry may cause profound changes in downstream processes and oncogenic potential.

In addition to cementing the role of the Notch–JNK signaling axis in proliferation, these findings suggest that the balance between diametrically opposed Notch-dependent processes can be shifted and even overthrown completely by certain synergistic interactions. Like Notch, JNK has been reported to act as both a tumor suppressor and an oncogene, at least in part by its ability to induce both apoptosis and proliferation (Enomoto, Kizawa, Ohsawa, & Igaki, 2015; Ríos-Barrera & Riesgo-Escovar, 2013; Shaulian & Karin, 2002). Clones of cells activating JNK alone tend to be eliminated due to excessive cell death, which can be at least partially mitigated by the pro-proliferative synergistic activity of Notch or Ras (Brumby & Richardson, 2003; Igaki, Pastor-Pareja, Aonuma, Miura, & Xu, 2009). In the case of Notch/Mef2, the inhibition of JNK-dependent apoptosis by an as-yet undetermined mechanism in cells that also have an unchecked cell cycle may therefore confer an even greater growth advantage.

It has also been reported that, in MCF-10A cells, lower levels of Notch cause hyperproliferation whereas higher levels cause growth suppression coupled with a loss of basement membrane adhesion (Mazzone et al., 2010). This may represent a mutually protective mechanism in which invasive cells can be outcompeted by wild-type or lower Notch neighbors, while hyperproliferative cells cannot metastasize. In Notch/Src and Notch/Mef2 cells, we observe an increase in *both* proliferation and markers of invasiveness/deadhesion (Ho et al., 2015; Pallavi et al., 2012). One possibility is that, as in the MCF-10A cells, different clonal populations within the wing disc express different levels of the transgenes; however, this is unlikely to be the case, as clonal populations of Notch/Mef2 cells can both overgrow and migrate (Pallavi et al., 2012). A more interesting possibility is that the Notch/Mef2 and Notch/Src synergies have altered the circuitry such that the two processes can occur concurrently. A clue might be found in the observation that Notch/Src causes the suppression of certain Notch targets (e.g., *e(spl) m8*), whereas it enhances the transcription of others (e.g., *e(spl)mγ*) (Ho et al., 2015). A single-cell population that has both a growth advantage and is capable of migrating to distant sites has clear implications for oncogenesis.

7. NOTCH SYNERGIES IN ONCOGENESIS: A HYPOTHESIS

The genetic circuitry surrounding Notch and its role in proliferation is complex, highly interconnected, and exquisitely context dependent. It is also becoming increasingly clear that this circuitry extends not only within the Notch-expressing cells themselves but also into neighboring and distant recipients of downstream signals. This has strong implications both for development and for oncogenesis.

Given that Notch is active in various adult tissues, an alteration in the expression of a gene that might on its own be relatively innocuous could trigger a Notch synergistic response and a subsequent wave of hyperplastic growth. Notably, this may even occur in cells where a normal, physiological level of Notch signaling is already present, eliminating the need for a mutation in Notch itself and perhaps explaining the rarity of Notch mutations in solid tumors. At a minimum, such synergies could dramatically increase the number of cells available to accumulate additional oncogenic mutations while simultaneously providing a continued growth advantage. In more extreme cases (e.g., Notch/Mef2 or Notch/Src), the synergy may also itself confer additional pro-tumorigenic properties such as invasiveness or resistance to cell death signals. Furthermore, the fact that Notch can induce non-cell-autonomous proliferation suggests that such synergies could take place not only in potentially tumor-forming cells themselves but also in the surrounding stroma. It also suggests that only a few cells with the proper synergistic combination (possibly a cancer stem cell population) might be sufficient to drive proliferation throughout an entire large, heterogeneous tumor. This scenario becomes potentially even more "alarming" when one considers that normal active Notch signals are often associated with stem cells and other undifferentiated, self-renewing cell populations that already lack some of the anti-proliferative safeguards found in differentiated cells.

However, factors that interact with Notch are also potential therapeutic targets. The Notch pathway is in theory a good target for therapies because it often seems to control an extremely early step in oncogenesis, but, since Notch is so pleiotropic, the core pathway is a challenging target. Synergistic interactors that are more specifically expressed in cancer cells or niches may be far more amenable as drug targets. It is conceivable, with adequate knowledge of cellular context, that we may in the future be able to specifically target pro-oncogenic synergies in Notch-related cancers.

REFERENCES

Andersson, E. R., Sandberg, R., & Lendahl, U. (2011). Notch signaling: Simplicity in design, versatility in function. *Development, 138*(17), 3593–3612.

Artavanis-Tsakonas, S., & Muskavitch, M. A. T. (2010). Chapter one—Notch: The past, the present, and the future. *Current Topics in Developmental Biology, 92*, 1–29.

Artavanis-Tsakonas, S., Rand, M. D., & Lake, R. J. (1999). Notch signaling: Cell fate control and signal integration in development. *Science*, *284*(5415), 770–776.
Baonza, A., & Freeman, M. (2005). Control of cell proliferation in the Drosophila eye by Notch signaling. *Developmental Cell*, *8*(4), 529–539.
Baonza, A., & Garcia-Bellido, A. (2000). Notch signaling directly controls cell proliferation in the Drosophila wing disc. *Proceedings of the National Academy of Sciences of the United States of America*, *97*(6), 2609–2614.
Beck, B., & Blanpain, C. (2013). Unravelling cancer stem cell potential. *Nature Reviews Cancer*, *13*(10), 727–738.
Bray, S. J. (2006). Notch signalling: A simple pathway becomes complex. *Nature Reviews Molecular Cell Biology*, *7*(9), 678–689.
Bray, S., & Bernard, F. (2010). Notch targets and their regulation. *Current Topics in Developmental Biology*, *92*, 253–275.
Brumby, A. M., & Richardson, H. E. (2003). Scribble mutants cooperate with oncogenic Ras or Notch to cause neoplastic overgrowth in Drosophila. *The EMBO Journal*, *22*(21), 5769–5779.
Colak, S., & Medema, J. P. (2014). Cancer stem cells—Important players in tumor therapy resistance. *The FEBS Journal*, *281*(21), 4779–4791.
de Celis, J. F., Garcia-Bellido, A., & Bray, S. J. (1996). Activation and function of Notch at the dorsal-ventral boundary of the wing imaginal disc. *Development*, *122*(1), 359–369.
Demehri, S., Turkoz, A., & Kopan, R. (2009). Epidermal Notch1 loss promotes skin tumorigenesis by impacting the stromal microenvironment. *Cancer Cell*, *16*(1), 55–66.
Djiane, A., Krejci, A., Bernard, F., Fexova, S., Millen, K., & Bray, S. J. (2013). Dissecting the mechanisms of Notch induced hyperplasia. *The EMBO Journal*, *32*(1), 60–71.
Djiane, A., Zaessinger, S., Babaoğlan, A. B., & Bray, S. J. (2014). Notch inhibits Yorkie activity in Drosophila wing discs. *PLoS ONE*, *9*(8), e106211.
Ellisen, L. W., Bird, J., West, D. C., Soreng, A. L., Reynolds, T. C., Smith, S. D., et al. (1991). TAN-1, the human homolog of the Drosophila notch gene, is broken by chromosomal translocations in T lymphoblastic neoplasms. *Cell*, *66*(4), 649–661.
Enomoto, M., Kizawa, D., Ohsawa, S., & Igaki, T. (2015). JNK signaling is converted from anti- to pro-tumor pathway by Ras-mediated switch of Warts activity. *Developmental Biology*, *403*(2), 162–171.
Ferres-Marco, D., Gutierrez-Garcia, I., Vallejo, D. M., Bolivar, J., Gutierrez-Aviño, F. J., & Dominguez, M. (2006). Epigenetic silencers and Notch collaborate to promote malignant tumours by Rb silencing. *Nature*, *439*(7075), 430–436.
Fre, S., Bardin, A., Robine, S., & Louvard, D. (2011). Notch signaling in intestinal homeostasis across species: The cases of Drosophila, Zebrafish and the mouse. *Experimental Cell Research*, *317*(19), 2740–2747.
Fre, S., Hannezo, E., Sale, S., Huyghe, M., Lafkas, D., Kissel, H., et al. (2011). Notch lineages and activity in intestinal stem cells determined by a new set of knock-in mice. *PLoS One*, *6*(10), e25785.
Fre, S., Huyghe, M., Mourikis, P., Robine, S., Louvard, D., & Artavanis-Tsakonas, S. (2005). Notch signals control the fate of immature progenitor cells in the intestine. *Nature*, *435*(7044), 964–968.
Fre, S., Pallavi, S. K., Huyghe, M., Laé, M., Janssen, K.-P., Robine, S., et al. (2009). Notch and Wnt signals cooperatively control cell proliferation and tumorigenesis in the intestine. *Proceedings of the National Academy of Sciences of the United States of America*, *106*(15), 6309–6314.
Giraldez, A. J., & Cohen, S. M. (2003). Wingless and Notch signaling provide cell survival cues and control cell proliferation during wing development. *Development*, *130*(26), 6533–6543.

Go, M. J., Eastman, D. S., & Artavanis-Tsakonas, S. (1998). Cell proliferation control by Notch signaling in Drosophila development. *Development, 125*(11), 2031–2040.

Graves, H. K., Woodfield, S. E., Yang, C.-C., Halder, G., & Bergmann, A. (2012). Notch signaling activates Yorkie non-cell autonomously in Drosophila. *PLoS ONE, 7*(6), e37615.

Herranz, H., Pérez, L., Martín, F. A., & Milán, M. (2008). A wingless and Notch double-repression mechanism regulates G1–S transition in the Drosophila wing. *The EMBO Journal, 27*(11), 1633–1645.

Ho, D. M., Pallavi, S. K., & Artavanis-Tsakonas, S. (2015). The Notch-mediated hyperplasia circuitry in Drosophila reveals a Src-JNK signaling axis. *eLife, 4*, e05996.

Hori, K., Sen, A., & Artavanis-Tsakonas, S. (2013). Notch signaling at a glance. *Journal of Cell Science, 126*(Pt. 10), 2135–2140.

Hu, B., Castillo, E., Harewood, L., Ostano, P., Reymond, A., Dummer, R., et al. (2012). Multifocal epithelial tumors and field cancerization from loss of mesenchymal CSL signaling. *Cell, 149*(6), 1207–1220.

Hurlbut, G. D., Kankel, M. W., & Artavanis-Tsakonas, S. (2009). Nodal points and complexity of Notch-Ras signal integration. *Proceedings of the National Academy of Sciences of the United States of America, 106*(7), 2218–2223.

Hurlbut, G. D., Kankel, M. W., Lake, R. J., & Artavanis-Tsakonas, S. (2007). Crossing paths with Notch in the hyper-network. *Current Opinion in Cell Biology, 19*(2), 166–175.

Igaki, T., Pastor-Pareja, J. C., Aonuma, H., Miura, M., & Xu, T. (2009). Intrinsic tumor suppression and epithelial maintenance by endocytic activation of Eiger/TNF signaling in Drosophila. *Developmental Cell, 16*(3), 458–465.

Johnston, L. A., & Edgar, B. A. (1998). Wingless and Notch regulate cell-cycle arrest in the developing Drosophila wing. *Nature, 394*(6688), 82–84.

Joshi, I., Minter, L. M., Telfer, J., Demarest, R. M., Capobianco, A. J., Aster, J. C., et al. (2009). Notch signaling mediates G1/S cell-cycle progression in T cells via cyclin D3 and its dependent kinases. *Blood, 113*(8), 1689–1698.

Kankel, M. W., Hurlbut, G. D., Upadhyay, G., Yajnik, V., Yedvobnick, B., & Artavanis-Tsakonas, S. (2007). Investigating the genetic circuitry of mastermind in Drosophila, a notch signal effector. *Genetics, 177*(4), 2493–2505.

Kiaris, H., Politi, K., Grimm, L. M., Szabolcs, M., Fisher, P., Efstratiadis, A., et al. (2004). Modulation of notch signaling elicits signature tumors and inhibits Hras1-induced oncogenesis in the mouse mammary epithelium. *The American Journal of Pathology, 165*(2), 695–705.

Klinakis, A., Szabolcs, M., Politi, K., Kiaris, H., Artavanis-Tsakonas, S., & Efstratiadis, A. (2006). Myc is a Notch1 transcriptional target and a requisite for Notch1-induced mammary tumorigenesis in mice. *Proceedings of the National Academy of Sciences of the United States of America, 103*(24), 9262–9267.

Koch, U., Lehal, R., & Radtke, F. (2013). Stem cells living with a Notch. *Development, 140*(4), 689–704.

Koch, U., & Radtke, F. (2010). Notch signaling in solid tumors. *Current Topics in Developmental Biology, 92*, 411–455.

Koch, U., & Radtke, F. (2011). Mechanisms of T cell development and transformation. *Annual Review of Cell and Developmental Biology, 27*, 539–562.

Lafkas, D., Rodilla, V., Huyghe, M., Mourao, L., Kiaris, H., & Fre, S. (2013). Notch3 marks clonogenic mammary luminal progenitor cells in vivo. *The Journal of Cell Biology, 203*(1), 47–56.

Liu, J., Sato, C., Cerletti, M., & Wagers, A. (2010). Notch signaling in the regulation of stem cell self-renewal and differentiation. *Current Topics in Developmental Biology, 92*, 367–409.

Liu, W., Singh, S. R., & Hou, S. X. (2010). JAK-STAT is restrained by Notch to control cell proliferation of the Drosophila intestinal stem cells. *Journal of Cellular Biochemistry*, *109*(5), 992–999.
Lobry, C., Oh, P., Mansour, M. R., Look, A. T., & Aifantis, I. (2014). Notch signaling: Switching an oncogene to a tumor suppressor. *Blood*, *123*(16), 2451–2459.
Louvi, A., & Artavanis-Tsakonas, S. (2012). Notch and disease: A growing field. *Seminars in Cell & Developmental Biology*, *23*(4), 473–480.
Mazzone, M., Selfors, L. M., Albeck, J., Overholtzer, M., Sale, S., Carroll, D. L., et al. (2010). Dose-dependent induction of distinct phenotypic responses to Notch pathway activation in mammary epithelial cells. *Proceedings of the National Academy of Sciences of the United States of America*, *107*(11), 5012–5017.
Micchelli, C. A., & Perrimon, N. (2006). Evidence that stem cells reside in the adult Drosophila midgut epithelium. *Nature*, *439*(7075), 475–479.
Moberg, K. H., Schelble, S., Burdick, S. K., & Hariharan, I. K. (2005). Mutations in erupted, the Drosophila ortholog of mammalian tumor susceptibility gene 101, elicit non-cell-autonomous overgrowth. *Developmental Cell*, *9*(5), 699–710.
Mourikis, P., Lake, R. J., Firnhaber, C. B., & DeDecker, B. S. (2010). Modifiers of notch transcriptional activity identified by genome-wide RNAi. *BMC Developmental Biology*, *10*, 107.
Nicolas, M., Wolfer, A., Raj, K., Kummer, J. A., Mill, P., Van Noort, M., et al. (2003). Notch1 functions as a tumor suppressor in mouse skin. *Nature Genetics*, *33*(3), 416–421.
O'Brochta, D. A., & Bryant, P. J. (1985). A zone of non-proliferating cells at a lineage restriction boundary in Drosophila. *Nature*, *313*(5998), 138–141.
Ohlstein, B., & Spradling, A. (2006). The adult Drosophila posterior midgut is maintained by pluripotent stem cells. *Nature*, *439*(7075), 470–474.
Ohlstein, B., & Spradling, A. (2007). Multipotent Drosophila intestinal stem cells specify daughter cell fates by differential notch signaling. *Science*, *315*(5814), 988–992.
Pallavi, S. K., Ho, D. M., Hicks, C., Miele, L., & Artavanis-Tsakonas, S. (2012). Notch and Mef2 synergize to promote proliferation and metastasis through JNK signal activation in Drosophila. *The EMBO Journal*, *31*(13), 2895–2907.
Patel, P. H., Dutta, D., & Edgar, B. A. (2015). Niche appropriation by Drosophila intestinal stem cell tumours. *Nature Cell Biology*, *17*(9), 1182–1192.
Pear, W. S., Aster, J. C., Scott, M. L., Hasserjian, R. P., Soffer, B., Sklar, J., et al. (1996). Exclusive development of T cell neoplasms in mice transplanted with bone marrow expressing activated Notch alleles. *The Journal of Experimental Medicine*, *183*(5), 2283–2291.
Ranganathan, P., Weaver, K. L., & Capobianco, A. J. (2011). Notch signalling in solid tumours: A little bit of everything but not all the time. *Nature Reviews Cancer*, *11*(5), 338–351.
Reynolds-Kenneally, J., & Mlodzik, M. (2005). Notch signaling controls proliferation through cell-autonomous and non-autonomous mechanisms in the Drosophila eye. *Developmental Biology*, *285*(1), 38–48.
Riccio, O., van Gijn, M. E., Bezdek, A. C., Pellegrinet, L., van Es, J. H., Zimber-Strobl, U., et al. (2008). Loss of intestinal crypt progenitor cells owing to inactivation of both Notch1 and Notch2 is accompanied by derepression of CDK inhibitors $p27^{Kip1}$ and $p57^{Kip2}$. *EMBO Reports*, *9*(4), 377–383.
Ríos-Barrera, L. D., & Riesgo-Escovar, J. R. (2013). Regulating cell morphogenesis: The Drosophila Jun N-terminal kinase pathway. *Genesis*, *51*(3), 147–162.
Rizzo, P., Osipo, C., Foreman, K., Golde, T., Osborne, B., & Miele, L. (2008). Rational targeting of Notch signaling in cancer. *Oncogene*, *27*(38), 5124–5131.

Robinson, D. R., Kalyana-Sundaram, S., Wu, Y. M., Shankar, S., Cao, X., Ateeq, B., et al. (2011). Functionally recurrent rearrangements of the MAST kinase and Notch gene families in breast cancer. *Nature Medicine*, *17*(12), 1646–1651.

Rodilla, V., Dasti, A., Huyghe, M., Lafkas, D., Laurent, C., Reyal, F., et al. (2015). Luminal progenitors restrict their lineage potential during mammary gland development. *PLoS Biology*, *13*(2), e1002069.

Saj, A., Arziman, Z., Stempfle, D., van Belle, W., Sauder, U., Horn, T., et al. (2010). A combined ex vivo and in vivo RNAi screen for notch regulators in Drosophila reveals an extensive notch interaction network. *Developmental Cell*, *18*(5), 862–876.

Sale, S., Lafkas, D., & Artavanis-Tsakonas, S. (2013). Notch2 genetic fate mapping reveals two previously unrecognized mammary epithelial lineages. *Nature Cell Biology*, *15*(5), 451–460.

Shaulian, E., & Karin, M. (2002). AP-1 as a regulator of cell life and death. *Nature Cell Biology*, *4*(5), E131–E136.

Vaccari, T., & Bilder, D. (2005). The Drosophila tumor suppressor vps25 prevents nonautonomous overproliferation by regulating notch trafficking. *Developmental Cell*, *9*(5), 687–698.

van Es, J. H., van Gijn, M. E., Riccio, O., van den Born, M., Vooijs, M., Begthel, H., et al. (2005). Notch/gamma-secretase inhibition turns proliferative cells in intestinal crypts and adenomas into goblet cells. *Nature*, *435*(7044), 959–963.

Weng, A. P., Ferrando, A. A., Lee, W., Morris, J. P., Silverman, L. B., Sanchez-Irizarry, C., et al. (2004). Activating mutations of NOTCH1 in human T cell acute lymphoblastic leukemia. *Science*, *306*(5694), 269–271.

Weng, A. P., Nam, Y., Wolfe, M. S., Pear, W. S., Griffin, J. D., Blacklow, S. C., et al. (2003). Growth suppression of pre-T acute lymphoblastic leukemia cells by inhibition of notch signaling. *Molecular and Cellular Biology*, *23*(2), 655–664.

Zecca, M., & Struhl, G. (2007). Recruitment of cells into the Drosophila wing primordium by a feed-forward circuit of vestigial autoregulation. *Development*, *134*(16), 3001–3010.

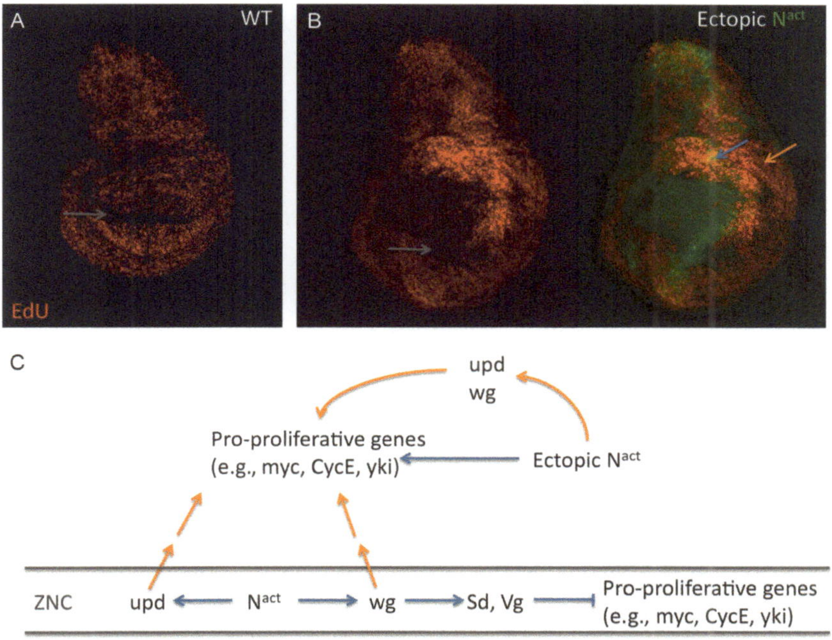

Diana M. Ho and Spyros Artavanis-Tsakonas, Figure 1 (A and B) Proliferation patterns in late third instar *Drosophila* wing imaginal discs stained with EdU. In wild-type discs (A), the ZNC (gray arrow) can be clearly distinguished by lack of EdU incorporation. When activated Notch (N^{act}) is expressed along the anterior–posterior boundary (labeled with GFP), the ZNC (gray arrow) expands and hyperproliferation occurs in the distal wing pouch and hinge regions. Both cell-autonomous (blue arrow) and non-cell-autonomous (orange arrow) components can be seen. (C) Simplified schematic of Notch gene interactions driving proliferation in the wing disc. Blue arrows represent cell-autonomous events, whereas orange arrows represent non-cell-autonomous events. See text for details.

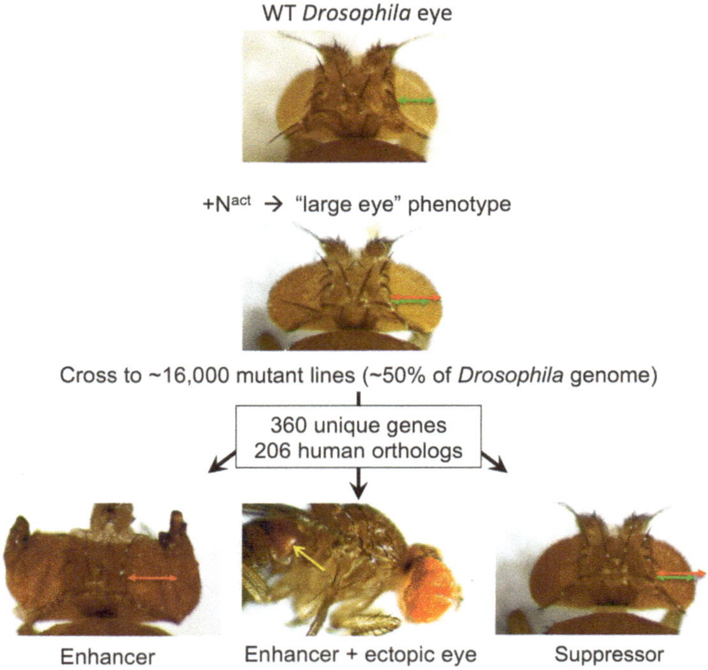

Diana M. Ho and Spyros Artavanis-Tsakonas, Figure 2 Schematic of and representative phenotypes from our genetic screen for modifiers of Notch-induced proliferation in the *Drosophila* eye. Flies ectopically expressing activated Notch (N^{act}) in the developing eye, which results in a "large eye" phenotype, were crossed to ∼16,000 mutant lines representing ∼50% of the *Drosophila* genome. Three hundred and sixty unique modifier genes, which caused either enhancement or suppression of the N^{act} phenotype, sometimes in conjunction with ectopic eyes, were identified, of which 206 had clear human orthologs.

CHAPTER THREE

Defining the Path from Stem Cells to Differentiated Tissue

Philip N. Benfey[1]
Biology Department and HHMI, Duke University, Durham, North Carolina, USA
[1]Corresponding author: e-mail address: philip.benfey@duke.edu

Abstract

The molecular events that control the progression from stem cells to differentiated tissue are largely unknown. The root of *Arabidopsis* is an excellent model to study this process, as cells at every stage of development are present at all times. A network that involves the mobile transcription factor, SHORT-ROOT, controls the asymmetric division of the stem cell that gives rise to endodermis and cortex. After initially interacting with another transcription factor, SCARECROW (SCR), together they activate a D-type cyclin, which specifically regulates this asymmetric cell division. Another direct target of SCR is a MYB transcription factor that regulates formation of the Casparian strip, the signature differentiated feature of the endodermis. Thus, the outline of a regulatory network from stem cell to differentiated tissue has been traced, with much remaining to be filled in.

A common definition of development is progressing from a single cell to a multicellular organism. Fundamental to this process is the passage from a stem cell to differentiated tissue. And yet, we are remarkably ignorant of the molecular steps required for this passage. Textbooks speak of specification and determination as intermediate steps and they tend to describe it as a series of switches, which, once thrown, are irreversible (Gilbert, 2010). The ability to completely reverse this process in mammalian cells with the introduction of four transcription factors (Okita, Ichisaka, & Yamanaka, 2007) shows that there is no such thing as "terminal" differentiation. In plants, the capacity for reverting to a stem cell-like state had been known and manipulated for years. In fact, 25 years ago in one of the major molecular biology textbooks, the only mention of plants was that they had this capacity to regenerate from a single cell (Darnell, Lodish, & Baltimore, 1990).

Current Topics in Developmental Biology, Volume 116
ISSN 0070-2153
http://dx.doi.org/10.1016/bs.ctdb.2015.12.002

If differentiation can be reversed, then maybe the switch analogy is not very informative. An alternative would be to think of this process as a probabilistic movement through a series of cellular states. Presumably, gene regulatory networks would define each state and perturbations of these networks would be required to move from one state to the next. It is also likely that epigenetic factors would play a role, perhaps a major role in mediating the degree of reversibility of specific steps.

An ideal system to study the path from stem cell to differentiated tissue would be one in which all of the stages are present and accessible. This is the case for the root of *Arabidopsis*. A stem cell niche consisting of a core of rarely dividing cells, the quiescent center (QC), is surrounded by four stem cell populations located at the root tip (Fig. 1; Benfey & Scheres, 2000). Divisions of the stem cells generate all of the cell types in the root. In the body of the root, the different cell types are organized as concentric cylinders around the central vascular tissue. At the very tip of the root, the root cap protects the stem cells as the root pushes through soil (Fig. 1).

A well-studied example of stem cell division in the root is the set of divisions that generate the cortex and endodermis, collectively referred to as the ground tissue. The stem cell responsible for ground tissue formation is called

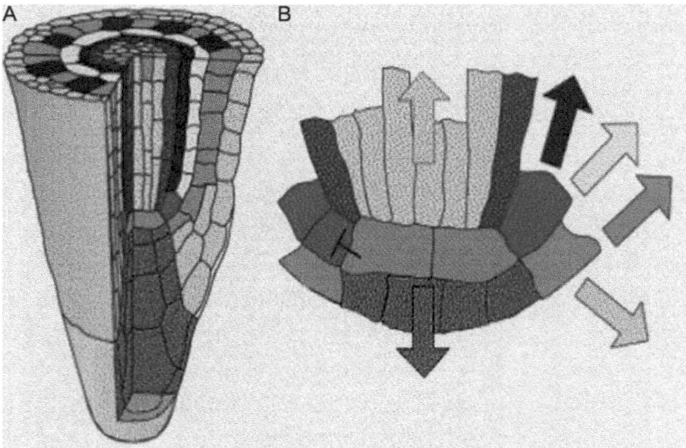

Figure 1 Stem cell niche in the root tip. (A and B) The four stem cell populations are adjacent to the quiescent center (QC) in gray. The cells immediately above the QC in beige give rise to the vascular tissue, the green cells give rise to the ground tissue (endodermis and cortex), the purple cells give rise to the epidermis and lateral root cap, and the pink cells give rise to the columella root cap. *From Benfey and Scheres (2000).* (A color version of this figure is located at the end of this chapter.)

the cortex/endodermal initial. It undergoes a first asymmetric division in the transverse (anticlinal) orientation to regenerate itself and a daughter, which then divides in the longitudinal (periclinal) orientation to produce the first two cells of the endodermal and cortex lineages (Fig. 2; Benfey & Scheres, 2000). These cells then enter a transit-amplifying zone in which they undergo a series of symmetric divisions prior to arresting division and entering the elongation zone where their cell volume dramatically increases. Once they are fully expanded, cells acquire specialized functions. Because plant cells do not move in relation to each other, the progression from stem cell to differentiated tissue can be followed along a single file of cells.

A genetic screen for *Arabidopsis* seedlings with short roots identified two mutants that not only had shorter roots but were also missing a layer of ground tissue (Benfey et al., 1993). They differed in the identity of the remaining cell layer. In *short-root*, the mutant cell layer had characteristics of cortex, while in *scarecrow*, the remaining layer expressed markers normally found in both endodermis and cortex (Benfey et al., 1993). Clearly, both SHORT-ROOT (SHR) and SCARECROW (SCR) are needed for the second asymmetric division of the stem cell, which normally would produce the cortex and endodermal lineages. Moreover, SHR is also needed to specify endodermal identity, since in the mutant, only markers for cortex were detected. By contrast, SCR appeared to be required primarily for the asymmetric division, as both cell fates were specified, they just occurred within the same cell lineage.

Cloning of the affected genes revealed that they belonged to the same family of plant-specific transcription factors. SCR was the first to be cloned and its RNA and protein were localized to the endodermal lineage as well as the QC (Di Laurenzio et al., 1996). When SHR was cloned, its RNA was shown to be expressed in the vascular tissue (Helariutta et al., 2000), while the SHR protein was localized to the vascular tissue as well as the

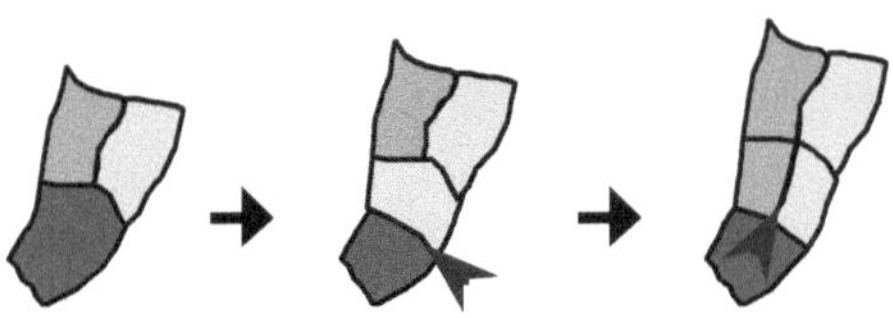

Figure 2 Division of the cortex/endodermal initial. The stem cell that gives rise to the cortex and endodermis divides first in the transverse (anticlinal) orientation to regenerate itself and a daughter, which divides in the longitudinal (periclinal) orientation to give the first cells of the endodermis and cortex lineages. *From Nakajima, Sena, Nawy, and Benfey (2001).* (A color version of this figure is located at the end of this chapter.)

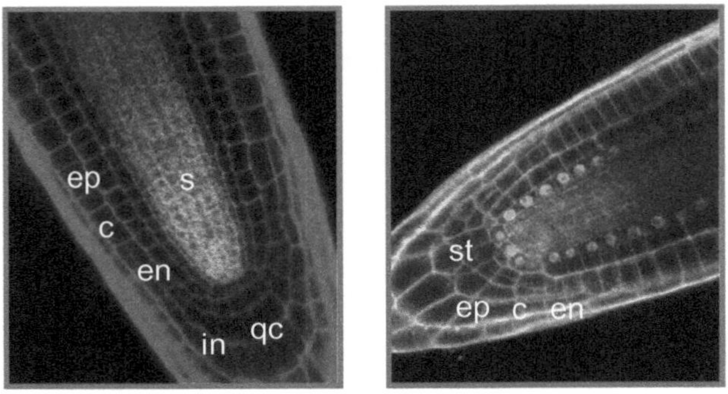

Figure 3 Expression of SHORT-ROOT (SHR) mRNA and protein. Left: the SHR promoter driving GFP. Right: the SHR promoter driving the SHR cDNA fused to GFP. SHR protein moves from the vascular tissue to the adjacent QC and endodermis. *From Nakajima et al. (2001).* (A color version of this figure is located at the end of this chapter.)

endodermal lineage and QC (Fig. 3; Nakajima et al., 2001). Further work demonstrated that the SHR protein is synthesized in the vascular tissue, then is actively transported to the adjacent cell layer, through pores called plasmodesmata (Gallagher & Benfey, 2009; Nakajima et al., 2001). Once in the endodermal lineage, SHR forms a complex with SCR (Cui et al., 2007). This helps to sequester SHR in the nucleus preventing further movement. Direct expression of SHR in the endodermis results in the formation of ectopic cell layers, pointing up the importance of regulating its movement (Helariutta et al., 2000).

To identify downstream effectors of SHR, its activity was induced in the *shr* mutant background (Sozzani et al., 2010). Approximately 6 h after induction, periclinal divisions began to occur in the mutant layer. These became frequent within 7–8 h postinduction. A time course was run, taking samples every 3–6 h after induction. To analyze only the RNA expressed in the mutant layer, fluorescence-activated cell sorting (FACS) was employed. In the *shr* mutant background, there was a green fluorescent protein (GFP) reporter that expressed in the mutant tissue both before and after division. At each time point, roots were harvested, cell walls enzymatically digested, then the resulting protoplasts sorted for GFP fluorescence. RNA was extracted from the sorted cells and used to make probes for hybridization to microarrays. Analysis of the microarrays revealed dramatic increases in expression of several genes around the time of the first asymmetric divisions. These included *SCR* and members of a family of C2H2 zinc-finger transcription factors collectively referred to as the Bird factors. A surprising finding was that a D-type cyclin, CYCD6-1, was upregulated in concert with these other genes (Sozzani et al., 2010).

It turned out that this D-type cyclin is a direct target of both SHR and SCR. In fact, it is expressed specifically in the stem cells just prior to the asymmetric division that generates the first cells of the cortex and endodermal lineages (Fig. 4; Sozzani et al., 2010). Unlike many convoluted developmental pathways, here was a short, direct route between key developmental regulators and a component of the cell cycle machinery. Another protein that plays a role in the cell cycle, RETINOBLASTOMA RELATED (RBR), was then shown to bind SCR, reducing its activity. It was known that CYCD6-1 along with its associated kinase could phosphorylate RBR, which would reduce RBR's affinity for binding to other proteins (Cruz-Ramirez et al., 2012). This posed a conundrum: if RBR binding to SCR reduced its activity, how would enough SCR be made to produce CYCD6-1, which could phosphorylate RBR to reduce its binding to SCR?

Mathematical modeling of this network presented a potential solution. Simulations from a set of ordinary differential equations revealed the possibility of two states (Cruz-Ramirez et al., 2012). At low levels of SHR in the stem cells, SCR would also be low. As SHR levels increased there would be a sudden shift to a higher level of SCR, which would be maintained even if SHR levels were to decrease again. This emergent property of the network

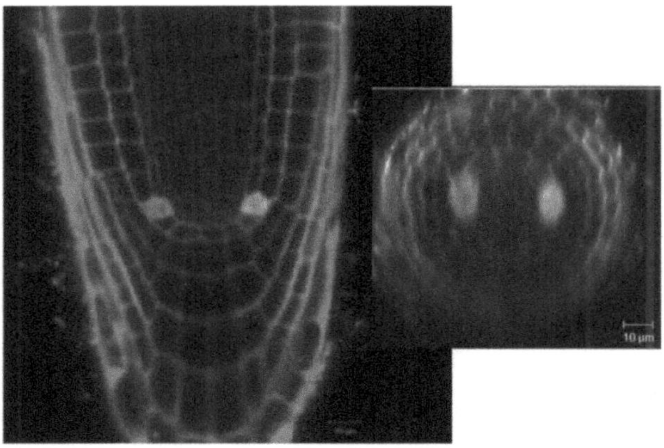

Figure 4 Expression of CYCD6-1. The CYCD6-1 promoter driving GFP exhibits expression in the cortex/endodermal initials prior to their periclinal division. *From Sozzani et al. (2010).* (A color version of this figure is located at the end of this chapter.)

would cause it to act as a bistable switch. With some additional components, this would provide the basis for a network controlling the beginning of the process from stem cell to differentiated tissue.

How are these early steps linked to differentiation? The salient feature of the endodermis is a waterproof barrier known as the Casparian strip. Differentiated endodermal cells localize lignin between them so that water and solutes must traverse the cells as opposed to passing between them. This allows for selective uptake of solutes. It also reduces leakage from the water-transporting xylem and may be involved in protection against pathogens. A set of proteins that are specifically expressed in the endodermis were shown to be required for Casparian strip formation. Expression of these CASP proteins commences shortly before the first detectable sign of lignin deposition. Double mutants of CASP genes result in compromised barrier function. Independently, two groups identified MYB36 as a key regulator of the CASP genes (Fig. 5; Kamiya et al., 2015; Liberman, Sparks, Moreno-Risueno, Petricka, & Benfey, 2015). It was also shown to regulate several other genes known to be involved in Casparian strip formation. Although MYB36 appeared to be able to bind the promoters of some of these genes, use of an inducible version with and without the translational inhibitor, cycloheximide, indicated that it probably needed a partner protein, which was induced by it (Liberman et al., 2015). This would constitute a feed-forward loop. Expression analysis indicated that MYB36 could be a direct target of SCR, which was confirmed through induction assays. This suggested that regulators of stem cell divisions could also be involved in regulating end-stage differentiation.

Thus, we are beginning to piece together the molecular networks that allow passage from stem cell to differentiated tissue in the ground tissue. A set of transcription factors and cell cycle components regulate the asymmetric divisions that generate the first cells of the endodermal and cortex lineages. At least one of these transcription factors, SCR, activates expression of MYB36, which, in turn, regulates expression of a set of proteins critical for formation of the Casparian strip, the signature differentiated feature of the endodermis. However, there are many gaps in our knowledge. SCR cannot be the only regulator of MYB36, as the latter does not express at high levels until the late stages of the elongation zone, while SCR is present throughout the endodermis. MYB36 and the CASPs cannot be the only regulators of Casparian strip formation, as, in their absence, a barrier can still form, only higher up in the root. And, of course, there are many more network components that need to be added to allow us to understand the intermediate steps between stem cell and differentiated tissue.

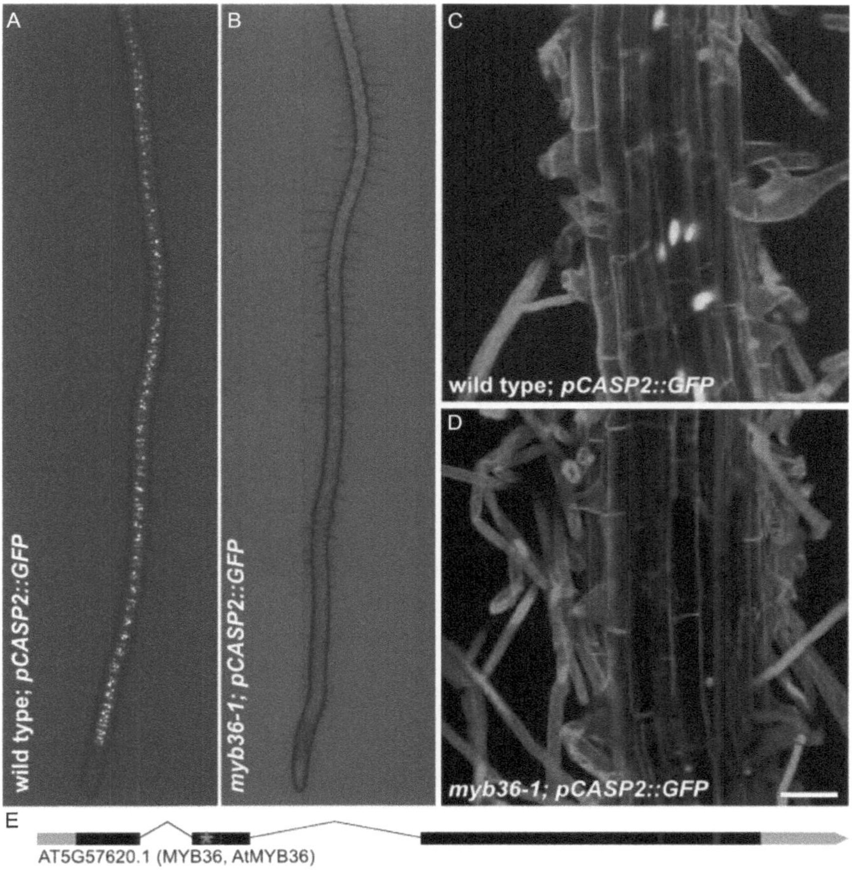

Figure 5 Characterization of a mutation in the MYB36 gene. (A and C) Expression of the CASP 2 gene is specific to the endodermis. (B and D) Expression is greatly reduced in the *myb36* mutant. (E) Structure of the MYB36 locus. *From Liberman et al. (2015).* (A color version of this figure is located at the end of this chapter.)

In addition to filling the gaps in the network, another major challenge is testing the current models. Mathematical modeling of biological processes usually relies on parameters that are either estimated or derived from other contexts. For example, a diffusion parameter can be based on measured diffusion of a protein in bacteria. Within the framework of the model, one can determine how robust it is to change in the different parameters, but most parameters can vary at least 10-fold and no model will be robust across the whole range of parameters. Thus, an urgent need is to experimentally determine critical parameters in the biological context being used for the model. This can be a daunting task, but new image analysis methods such as fluorescence correlated spectroscopy can provide accurate measures of diffusion, stoichiometry, and cell-to-cell protein movement (Levi, Ruan, Kis-Petikova, & Gratton, 2003).

We also urgently need to measure and analyze the dynamics of the molecular processes underlying development. We have long been mesmerized by movies of dividing cells and forming organs, and yet we know little about the kinetics of the regulatory networks that underlie these processes. We need ways to watch, in real time, the production and destruction of transcription factors and downstream effectors. New microscopy modalities such as selected plane illumination microscopy, also called light sheet microscopy, allow for long-term imaging of biological specimens without the phototoxicity that is associated with confocal microscopy.

A central tenet of engineering is that a process is understood only when it can be reproduced. To test our knowledge of the design principles of the networks controlling development, we need to reproduce them with synthetic components. This has been accomplished in single-cell organisms, but rarely in the context of multicellularity. Part of the problem is the lack of synthetic components that come from or have been tested in a multicellular organism. The discovery of DNA-binding and -modifying components made by bacteria, but active in multicellular organisms including TALEs and the CRISPR system, has started to change this. With these components, network motifs, such as feed-forward loops, can be generated. By modifying the expression levels and binding properties of these synthetic components, we should be able to determine the state space in which the networks can function.

In summary, the root of *Arabidopsis* has yielded important insights into the pathways that lead from stem cell to differentiated tissue. The major challenges are now to fill in the missing players and understand the underlying dynamics. Synthetic subnetworks should provide insights into the probabilistic nature of the transitions that cells undergo as they progress from stem cells to differentiated tissue.

REFERENCES

Benfey, P. N., Linstead, P. J., Roberts, K., Schiefelbein, J. W., Hauser, M. T., & Aeschbacher, R. A. (1993). Root development in Arabidopsis: Four mutants with dramatically altered root morphogenesis. *Development*, *119*, 57–70.

Benfey, P. N., & Scheres, B. (2000). Root development. *Current Biology*, *10*, R813–R815.

Cruz-Ramirez, A., Diaz-Trivino, S., Blilou, I., Grieneisen, V. A., Sozzani, R., Zamioudis, C., et al. (2012). A bistable circuit involving SCARECROW-

RETINOBLASTOMA integrates cues to inform asymmetric stem cell division. *Cell, 150*, 1002–1015.

Cui, H., Levesque, M. P., Vernoux, T., Jung, J. W., Paquette, A. J., Gallagher, K. L., et al. (2007). An evolutionarily conserved mechanism delimiting SHR movement defines a single layer of endodermis in plants. *Science, 316*, 421–425.

Darnell, J. E., Lodish, H. F., & Baltimore, D. (1990). *Molecular cell biology*. New York: Scientific American Books: Distributed by W.H. Freeman. pp. xl; 1105 p.

Di Laurenzio, L., Wysocka-Diller, J., Malamy, J. E., Pysh, L., Helariutta, Y., Freshour, G., et al. (1996). The SCARECROW gene regulates an asymmetric cell division that is essential for generating the radial organization of the Arabidopsis root. *Cell, 86*, 423–433.

Gallagher, K. L., & Benfey, P. N. (2009). Both the conserved GRAS domain and nuclear localization are required for SHORT-ROOT movement. *The Plant Journal, 57*, 785–797.

Gilbert, S. F. (2010). *Developmental biology*. Sunderland, MA: Sinauer Associates. pp. xxi, 711, 80 p.

Helariutta, Y., Fukaki, H., Wysocka-Diller, J., Nakajima, K., Jung, J., Sena, G., et al. (2000). The SHORT-ROOT gene controls radial patterning of the Arabidopsis root through radial signaling. *Cell, 101*, 555–567.

Kamiya, T., Borghi, M., Wang, P., Danku, J. M., Kalmbach, L., Hosmani, P. S., et al. (2015). The MYB36 transcription factor orchestrates Casparian strip formation. *Proceedings of the National Academy of Sciences of the United States of America, 112*, 10533–10538.

Levi, V., Ruan, Q., Kis-Petikova, K., & Gratton, E. (2003). Scanning FCS, a novel method for three-dimensional particle tracking. *Biochemical Society Transactions, 31*, 997–1000.

Liberman, L. M., Sparks, E. E., Moreno-Risueno, M. A., Petricka, J. J., & Benfey, P. N. (2015). MYB36 regulates the transition from proliferation to differentiation in the Arabidopsis root. *Proceedings of the National Academy of Sciences of the United States of America, 2015*(112), 12099–12104.

Nakajima, K., Sena, G., Nawy, T., & Benfey, P. N. (2001). Intercellular movement of the putative transcription factor SHR in root patterning. *Nature, 413*, 307–311.

Okita, K., Ichisaka, T., & Yamanaka, S. (2007). Generation of germline-competent induced pluripotent stem cells. *Nature, 448*, 313–317.

Sozzani, R., Cui, H., Moreno-Risueno, M. A., Busch, W., Van Norman, J. M., Vernoux, T., et al. (2010). Spatiotemporal regulation of cell-cycle genes by SHORT-ROOT links patterning and growth. *Nature, 466*, 128–132.

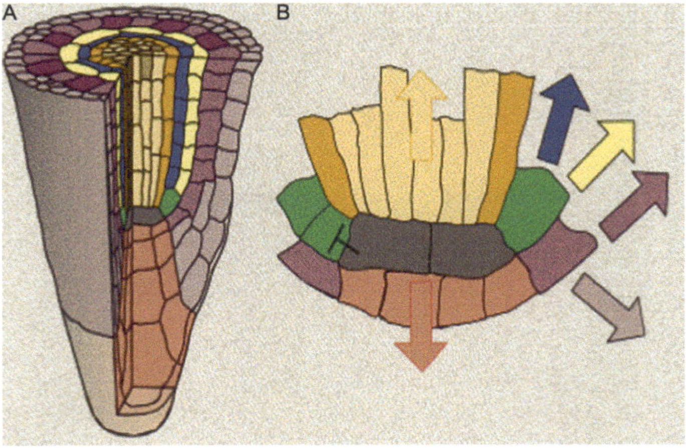

Philip N. Benfey, Figure 1 Stem cell niche in the root tip. (A and B) The four stem cell populations are adjacent to the quiescent center (QC) in gray. The cells immediately above the QC in beige give rise to the vascular tissue, the green cells give rise to the ground tissue (endodermis and cortex), the purple cells give rise to the epidermis and lateral root cap, and the pink cells give rise to the columella root cap. *From Benfey and Scheres (2000).*

Philip N. Benfey, Figure 2 Division of the cortex/endodermal initial. The stem cell that gives rise to the cortex and endodermis divides first in the transverse (anticlinal) orientation to regenerate itself and a daughter, which divides in the longitudinal (periclinal) orientation to give the first cells of the endodermis and cortex lineages. *From Nakajima, Sena, Nawy, and Benfey (2001).*

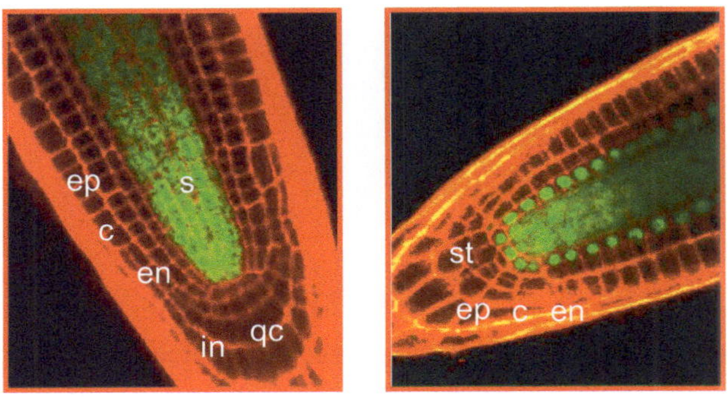

Philip N. Benfey, Figure 3 Expression of SHORT-ROOT (SHR) mRNA and protein. Left: the SHR promoter driving GFP. Right: the SHR promoter driving the SHR cDNA fused to GFP. SHR protein moves from the vascular tissue to the adjacent QC and endodermis. *From Nakajima et al. (2001).*

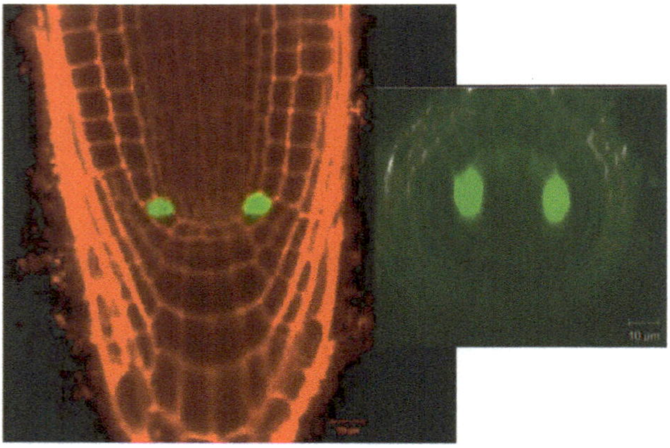

Philip N. Benfey, Figure 4 Expression of CYCD6-1. The CYCD6-1 promoter driving GFP exhibits expression in the cortex/endodermal initials prior to their periclinal division. *From Sozzani et al. (2010).*

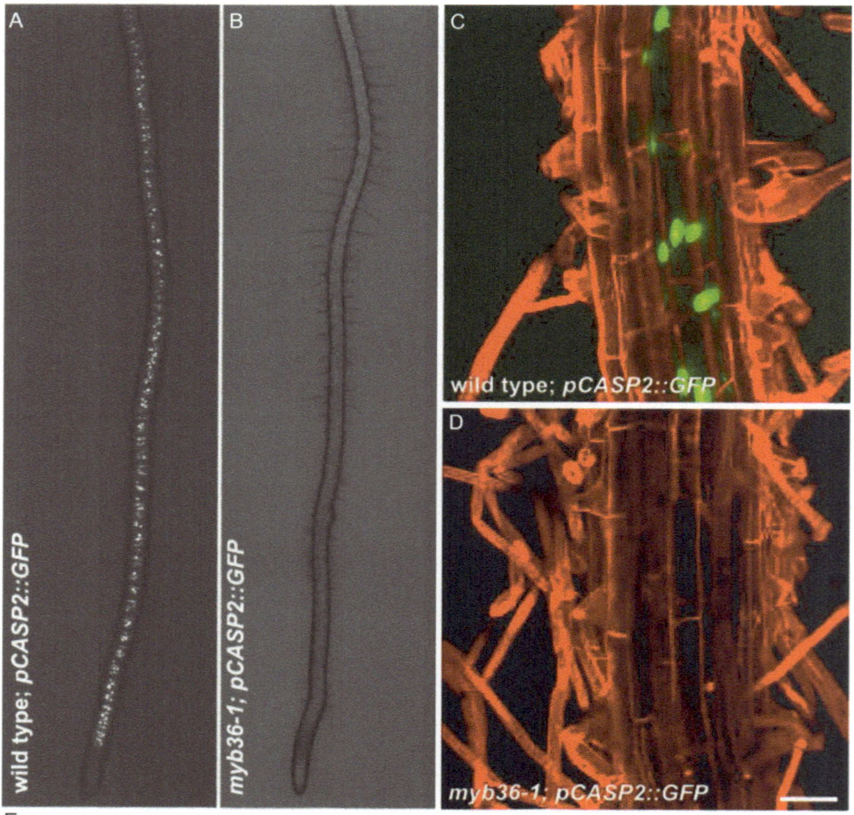

Philip N. Benfey, Figure 5 Characterization of a mutation in the MYB36 gene. (A and C) Expression of the CASP 2 gene is specific to the endodermis. (B and D) Expression is greatly reduced in the *myb36* mutant. (E) Structure of the MYB36 locus. *From Liberman et al. (2015).*

CHAPTER FOUR

Neuregulin/ErbB Signaling in Developmental Myelin Formation and Nerve Repair

Carmen Birchmeier*[,1], David L.H. Bennett†[,1]
*Developmental Biology, Max-Delbrück-Center for Molecular Medicine, Berlin, Germany
†The Nuffield Department of Clinical Neurosciences, University of Oxford, John Radcliffe Hospital, Oxford, United Kingdom
[1]Corresponding authors: e-mail address: cbirch@mdc-berlin.de; david.bennett@ndcn.ox.ac.uk

Contents

Abstract

Myelin is essential for rapid and accurate conduction of electrical impulses by axons in the central and peripheral nervous system (PNS). Myelin is formed in the early postnatal period, and developmental myelination in the PNS depends on axonal signals provided by Nrg1/ErbB receptors. In addition, Nrg1 is required for effective nerve repair and remyelination in adulthood. We discuss here similarities and differences in Nrg1/ErbB functions in developmental myelination and remyelination after nerve injury.

1. INTRODUCTION

Communication between cells is fundamental for formation and regeneration of organs. We know today that only a handful of important signaling systems are used for cell communication during development, and receptor tyrosine kinases are prominent among them. Many receptor tyrosine kinases were identified because of their oncogenic potential when

Current Topics in Developmental Biology, Volume 116
ISSN 0070-2153
http://dx.doi.org/10.1016/bs.ctdb.2015.11.009

mutated, and the nonmutant variants (the proto-oncogenes) function in development (Schlessinger, 2000). Recent evidence demonstrates that these receptors are frequently reused in the adult where they orchestrate regeneration.

Neuregulins (Nrgs, in humans called heregulins HRGs) are a family of growth factors that display sequence similarities with epidermal growth factor (EGF; Falls, 2003). The factors signal via tyrosine kinase receptors of the ErbB (in humans called HER) family (Yarden & Sliwkowski, 2001). Nrgs/ErbBs were identified and intensely studied because of their functions in cancer. Antibodies against the receptors or small molecular weight inhibitors that interfere with their activity are used in cancer therapy and provide excellent tools for functional studies (Hynes & Lane, 2005). Among the different members of the family, the most important and the best-studied is Neuregulin-1 (Nrg1) that is encoded by one of the largest genes in the human genome and spans almost two megabases. From this gene, many different isoforms are produced by alternative splicing and the usage of different promoters (Falls, 2003). All Nrg1 isoforms contain an EGF-like domain embedded in otherwise divergent sequences, and the EGF-like domain is sufficient to bind and activate ErbB receptors. Major classes of Nrg1 isoforms use different promoters and are expressed in a characteristic and distinct pattern, for instance type III mainly by neuronal cells, and type I Nrg1 by a few restricted neuronal cell types and by mesenchymal cells of many organs (Meyer & Birchmeier, 1995). Interestingly, different isoforms take over distinct functions *in vivo*. For instance type III and type I Nrg1 are produced by all or just a subset of sensory neurons, respectively, and have distinct functions in myelination and muscle spindle induction (Cheret et al., 2013; Hippenmeyer et al., 2002; Meyer & Birchmeier, 1995; Perlin, Lush, Stephens, Piotrowski, & Talbot, 2011). Nrg1 isoforms are produced as soluble membrane-bound proteins. Membrane-bound variants can be released by proteases, or they remain bound to the cell surface after proteolysis (Falls, 2003). Because Nrg1 can be shed or remain membrane-bound, the receptors detect signals provided by distant or directly neighboring cells (paracrine signaling). In addition, occasionally the same cell produces and receives the signal (autocrine signaling). Thus, the diversity of the Nrg1 isoforms contributes to the versatility of the signaling system.

Important Nrg1 functions were identified by genetics, mainly in mice but also in zebrafish. Responses to Nrg1 are astonishingly diverse, and encompass migration, cell fate decisions, morphogenesis, proliferation, and the control of cell size. Moreover, glial cells, neurons, muscle, and epithelial cells respond to Nrg1. Prominent among the Nrg1 functions is its role in peripheral nervous system (PNS) development and regeneration. In particular, the myelinating glia of the PNS, Schwann cells, depend on Nrg1, and we concentrate on discussing this function here. In Schwann cells, the Nrg1 signal is mediated by two receptors, ErbB2 and ErbB3 (Woldeyesus et al., 1999), and these receptors are unusual in the respect that ErbB3 lacks tyrosine kinase activity, whereas ErbB2 is a ligand-less receptor (Citri, Skaria, & Yarden, 2003). They function as heterodimers where ErbB2 and ErbB3 provide the tyrosine kinase and the ligand-binding activity, respectively. Nrg1 binding to ErbB2/3 results in receptor tyrosine phosphorylation, which triggers recruitment of adaptors and enzymes and culminates in the activation of signaling cascades (Schlessinger, 2000; Yarden & Sliwkowski, 2001).

2. SCHWANN CELL PRECURSORS

Schwann cells depend on Nrg1 signaling for their development. Schwann cells derive from neural crest and Schwann cell precursors are defined by their anatomical position, i.e., they associate with axons of peripheral nerves. Schwann cells depend on Nrg1/ErbB signaling for migration along the axon tracts and for proliferation, and are basically absent in mice or zebrafish that lack Nrg1, ErbB2, or ErbB3 (Lyons et al., 2005; Meyer & Birchmeier, 1995; Perlin et al., 2011; Riethmacher et al., 1997; Woldeyesus et al., 1999). The Nrg1 signal that drives Schwann cell development is provided by neurons, and presented in the axolemma. Thus, axonal membranes are mitogenic for Schwann cell precursors, and the mitogenic agent is Nrg1 (Dong et al., 1995; Morrissey, Levi, Nuijens, Sliwkowski, & Bunge, 1995; Schroering & Carey, 1998). Genetic evidence demonstrates that type III Nrg1 drives Schwann cell development (Meyer et al., 1997); this isoform can remain membrane-bound after proteolytic cleavage (Cabedo, Luna, Fernandez, Gallar, & Ferrer-Montiel, 2002).

Schwann cell precursors retain characteristics of neural crest cells and remain pluripotent (Morrison, White, Zock, & Anderson, 1999). They generate immature Schwann cells, endoneurial fibroblasts of fetal nerves as well as melanocytes, parasympathetic neurons, and odontoblasts (Adameyko et al., 2009; Dyachuk et al., 2014; Espinosa-Medina et al., 2014; Joseph et al., 2004; Kaukua et al., 2014). Schwann cell progenitors give rise to immature Schwann cells around E16 (Jessen, Mirsky, & Lloyd, 2015). This transition is associated with axon wrapping, appearance of a basal lamina around Schwann cells, and formation of the perineural sheath. Subsequently, Schwann cells sort and wrap axons, and they generate the multilayered myelin sheaths.

3. ENTRY INTO THE MYELINATION PROGRAM

Myelination is essential for neuronal function, and deficits in myelination cause devastating disease (Nave & Trapp, 2008; Quarles, Macklin, & Morell, 2006; Suter & Scherer, 2003). Myelin electrically insulates axons and allows fast propagation of nerve impulses by saltatory conduction. The overall conduction velocity of the myelinated axon is determined by myelin and axonal thickness, internodal length, and myelin integrity (Court et al., 2004; Waxman, 1980). The overall organization of the myelin sheath is similar in the PNS and CNS, but substantial differences exist in development, assembly, and molecular composition of the myelin, or in the extrinsic signals and transcription factors that drive myelination (Brinkmann et al., 2008; Emery, 2013; Hornig et al., 2013; Quarles et al., 2006). We discuss here the molecular processes active during peripheral myelination.

Myelination is tightly controlled by axons. Only medium and large diameter axons are myelinated, and thin axons (smaller than 1 μm) are ensheathed by nonmyelinating Schwann cells and are organized in bundles (called Remak bundles; Hillarp & Olivecrona, 1946). Thin and thick myelinated axons have thin and thick myelin sheaths, respectively. This leads to the constant G-ratio (defined as ratio of axonal diameter to diameter of the myelinated fiber) of 0.67 (Donaldson & Hoke, 1905). Thus, the axon provides cues that determine its myelination fate and the thickness of its myelin sheath. Conversely, Schwann cell-derived signals allow neuronal survival by providing trophic signals and/or metabolic support (Riethmacher et al., 1997; Viader et al., 2013). Interactions between Schwann cells and axons also guide cell adhesion molecules and ion channels into distinct axonal domains, the nodes of Ranvier and internodes, a prerequisite for efficient saltatory conduction (Eshed-Eisenbach & Peles, 2013). Finally, intact myelin provides signals for radial axonal growth and enhances axonal transport, which is impaired in demyelinating diseases (de Waegh, Lee, & Brady, 1992; Kiryu-Seo, Ohno, Kidd, Komuro, & Trapp, 2010; Saxton & Hollenbeck, 2012; Watson, Nachtman, Kuncl, & Griffin, 1994).

The surface membrane of Schwann cells was estimated to increase during myelination up to several thousand-fold. Therefore, myelination is accompanied by the production of huge amounts of myelin proteins and lipids. The entry into the myelination program in Schwann cells is controlled on the transcriptional level by factors like Egr2 (Krox-20), Pou3f1 (Oct-6/Scip/Tst-1), and Pou3f2 (Brn2) that interact with Sox10 to drive expression of myelin genes (Bermingham et al., 1996; Finzsch et al., 2010; Jaegle et al., 2003; Topilko et al., 1994; see also Svaren & Meijer, 2008, and the references herein). In addition, lipid biosynthesis genes are coordinately regulated, and the Srebp/Scap transcription factors and the control of their expression play important roles in lipid production (Verheijen et al., 2009; Norrmen et al., 2014). Extrinsic signals that trigger entry into myelination are provided by Nrg1/ErbB2/3, GPR126, integrins, extracellular matrix components, and ADAM22/Lgi4 (Bermingham et al., 2006; Nave & Salzer, 2006; Ozkaynak et al., 2010; Raphael & Talbot, 2011; Taveggia, Feltri, & Wrabetz, 2010). Thus, the Nrg1/ErbB2/3 system that regulates expansion of the progenitor pool and migration in early Schwann cell development is unexpectedly reused for myelination. It was therefore discussed whether Nrg1 provides instructive or permissive signals during myelination (Lemke, 2006; Nave & Salzer, 2006).

4. Nrg1 AND OTHER EXTRINSIC SIGNALS THAT CONTROL MYELINATION

Nrg1 activates Ras/MAPK/Erk1/2, PI3K/Akt, PLC$_{\gamma}$, focal adhesion kinase (FAK), Rho-GTPases, and JNK, and these pathways have been implicated in myelination using mouse genetics or cell culture experiments (Newbern & Birchmeier, 2010). Cell culture models are frequently used to study entry into myelination and are combined with pharmacological inhibitors, overexpression through viral transduction and interfering RNAs. Results obtained with such methods can vary according to the exact culture conditions. Such studies mainly implicated PI3K/Akt as key signals in myelination, and suggested that MAPK/Erk play a limited role or even drive Schwann cells into dedifferentiation (Harrisingh et al., 2004; Maurel & Salzer, 2000; Ogata, Yamamoto, Nakamura, & Tanaka, 2006). Recent genetic experiments have now shown that MAPK/Erk signaling is a major positive regulator of myelination *in vivo*, indicating that this notion has to be revisited (Grossmann et al., 2009; Ishii, Furusho, & Bansal, 2013; Newbern et al., 2011; Sheean et al., 2014).

Erbin is an intracellular molecule which through binding of ErbB2 promotes Nrg1 signaling and is essential for myelination (Tao et al., 2009).

Shp2 mediates Nrg1 signals downstream of ErbB2/3 and is required for sustained activation of MAPK/Erk1/2 as well as for Nrg1-evoked cellular responses like proliferation, migration, or myelination, which are impaired when Shp2 is ablated *in vivo* (Grossmann et al., 2009). Furthermore, ablation of Erk1/2 in Schwann cells causes deficits that are remarkably similar to those observed in Shp2 or Nrg1/ErbB2/ErbB3 mutants (Newbern et al., 2011). Interestingly, mild activation of MAPK signaling releases Schwann cells from their dependence of Nrg1 signaling during all stages of Schwann cell development and myelination (Sheean et al., 2014). In particular, MAPK/Erk positively regulate entry into myelination but a strong and sustained activation of Raf (and thus the Ras/MAPK/Erk cascade) results in demyelination, indicating that signaling strength determines the cellular response (Napoli et al., 2012).

Pharmacological inhibition of PI3K/Akt blocked myelin gene expression and myelination in neuron/Schwann cell cocultures (Maurel & Salzer, 2000; Ogata et al., 2006). PTEN is a negative regulator of the PI3K/Akt pathway and, accordingly, loss of PTEN results in Akt activation. This leads to mild hypermyelination of small peripheral axons, and to locally restricted hypermyelination close to the nodes of Ranvier (Goebbels et al., 2010, 2012).

The activator of adenylate cyclase, forskolin, has long been known to stimulate myelination in culture (Lemke & Chao, 1988; Morgan, Jessen, & Mirsky, 1991). *In vivo*, cAMP-mediated signals provided by the G protein-coupled receptor GPR126 are essential for myelination, and mutations of GPR126 preclude peripheral myelination in zebrafish and mice (Monk et al., 2009; Monk, Oshima, Jors, Heller, & Talbot, 2011). Myelination in GPR126 mutant fish is rescued by forskolin in the swimming water (Monk et al., 2009). GPR126 is activated by binding to laminin and collagen, providing evidence that the extracellular matrix formed by immature Schwann cells provides signals that control entry into myelination (Paavola, Sidik, Zuchero, Eckart, & Talbot, 2014; Petersen et al., 2015). Integrins are important adhesion and signaling receptors that also recognize extracellular matrix components like laminin, and upon binding to matrix they recruit intracellular molecules like FAK. Laminin, integrin, and FAK provide essential cues for axonal sorting and myelination *in vivo* (Berti et al., 2011; Chen & Strickland, 2003; Feltri et al., 2002; Grove et al., 2007; Pellegatta et al., 2013). Thus, GPR126 and integrins provide links to the extracellular matrix and signal in a matrix-dependent manner to control axonal sorting and myelination.

5. PROTEOLYTIC PROCESSING IS RATE LIMITING FOR Nrg1 ACTIVITY AND FUNCTION

Many Nrg1 isoforms are synthesized as membrane-bound proteins and require proteolytic processing for function. Type I and type III Nrg1 are processed by Bace1, an aspartic-acid protease, and by members of the disintegrin and metalloproteinase (Adam) family (Fleck et al., 2013; Horiuchi, Zhou, Kelly, Manova, & Blobel, 2005; Hu et al., 2006; La Marca et al., 2011; Willem et al., 2006; Yokozeki et al., 2007). A single Bace1-cleavage of type I Nrg1 releases a soluble EGF-domain containing fragment, whereas cleavage of type III Nrg1 at a similar site generates an EGF-domain containing fragment that remains associated with the membrane (Fleck et al., 2013; Hu et al., 2006; Willem et al., 2006). Systematic *in vitro* mapping demonstrates that protease cleavage sites exist that surround the EGF-like domain in type III Nrg1 (Fleck et al., 2013; Hu et al., 2006; Willem et al., 2006). Thus, small soluble EGF-domain containing peptides can be produced that are biologically active *in vitro* and in the zebrafish *in vivo* (Fleck et al., 2013). In addition, ADAM17 might cleave type III Nrg1 in the EGF-domain and thus destroy its activity but this cleavage site was not found in all reported mapping experiments (Fleck et al., 2013; La Marca et al., 2011).

Genetic models demonstrated that Bace1-dependent cleavage of Nrg1 is physiologically important for its function. Thus, Bace1 mutant mice and zebrafish display deficits in developmental myelination. In addition, remyelination of peripheral nerves and formation and maintenance of muscle spindles are impaired in mice, and these phenotypes resemble the well-characterized ones observed when Nrg1/ErbB signaling is ablated (Cheret et al., 2013; Fleck et al., 2013; Hu et al., 2006; Willem et al., 2006). It should be noted that phenotypes in Bace1 mutants are not as strong as the ones observed in Nrg1 signaling mutants, indicating that other proteases participate in shedding. However, overexpression of type I Nrg1 in transgenic mice induces supernumerary muscle spindles in a strictly Bace1-dependent manner, indicating that this isoform requires Bace1 processing for its function (Cheret et al., 2013). The finding that Bace1 cleaves Nrg1 might be of interest for new therapies that are being developed for Alzheimer's disease. Bace1 is essential for APP processing and amyloid peptide formation, and Bace1 inhibition is a strategy for treatment of Alzheimer's disease. Adverse effects due to inhibition of other Bace1 substrates including Nrg1 must be considered (Fleck, Garratt, Haass, & Willem, 2012).

6. RATES OF PROTEIN SYNTHESIS CONTROL THE END OF MYELINATION AND MYELIN THICKNESS

During normal development, myelin is formed rapidly during the postnatal period. In particular, radial myelin growth is fast during the first two postnatal weeks in mice, slows in the maturing animal and ceases in adulthood. Recent data indicate that translational control mechanisms are important determinants that end myelination and thereby control myelin thickness (Pereira et al., 2010; Sheean et al., 2014).

The control of protein synthesis *in vivo* is difficult to analyze due to technical limitations, and synthesis of myelin components have been little investigated. However, new methods are now available that are based on a pulse-labeling of proteins with stable (nonradioactive) isotope-labeled amino acids and on quantification of labeled and unlabeled proteins by mass spectroscopy (Doherty, Hammond, Clague, Gaskell, & Beynon, 2009; Schwanhausser et al., 2011; Selbach et al., 2008; Sheean et al., 2014). With these techniques it was possible to assess translational control mechanisms during myelination in the PNS and their dependence on MAPK/Erk (Sheean et al., 2014). Similar mechanisms appear to be operative in myelination of the CNS (Michel, Zhao, Karl, Lewis, & Fyffe-Maricich, 2015).

Protein synthesis rates are highly regulated. A rate-limiting step is the initiation of translation that is strongly enhanced by growth factor signaling mediated by mTor and MAPK. Sustained activation of MAPK overcomes the signals that end myelination and results in continuous myelin growth achieved by enhanced protein synthesis rates. MAPK exerts this effect in part by mTor-dependent signaling, possibly by regulating the negative regulator of mTor, TSC2, but also by mTor-independent mechanisms (Sheean et al., 2014). Further supporting evidence for the importance of this pathway derives from the analysis of mTor mutant mice that arrest myelin growth prematurely and that form thin myelin (Sherman et al., 2012). Dlg1-PTEN interactions were also reported to control the end of myelination and thus myelin thickness, and might impinge on the control of protein synthesis (Cotter et al., 2010).

7. Nrg1 AND REMYELINATION AFTER INJURY

The potent role of Nrg1 as an axonal signal that drives virtually all stages of Schwann cell development and myelination provides a rationale for studying its function in peripheral nerve injury. Despite its importance in developmental myelination, Nrg1/ErbB signaling is not required for the maintenance of the myelin sheath in adulthood (Atanasoski et al., 2006; Fricker et al., 2011, 2013). Injury radically alters the communication between axons, glia and immune cells, and effective signaling between these cells is crucial for successful nerve repair (Fawcett & Keynes, 1990; Scheib & Hoke, 2013). There is growing evidence that Nrg1 is important for repair and remyelination after injury, but signaling requirements in repair do not merely recapitulate requirements in development. Instead, differences exist in usage of Nrg1 isoforms and cooperating signaling systems.

Peripheral nerve injury has many causes ranging from traumatic damage to metabolic disturbances in diabetes mellitus. The primary pathology can affect axons (nerve transection) or Schwann cells, and their interdependence means that injury to one will ultimately affect the other. Fortunately, the PNS has a significant capacity for repair, and axons are able to regenerate over long distances and to remyelinate. This repair process leads to appreciable recovery of function after damage, but significant challenges remain: the clinical outcome after nerve injury often remains poor and there is no licensed therapeutic (Scheib & Hoke, 2013). The outcome depends on the injury type: complete nerve transection has a worse outcome than nerve crush due to misdirected axon growth, and proximal injuries also have a poorer outcome than distal ones due to the slow (1 mm/day) rate of axon regrowth (Fu & Gordon, 1995). In addition, aged Schwann cells are less supportive of axon outgrowth than young Schwann cells and might produce less efficiently beneficial growth and metabolic factors (Painter et al., 2014).

Traumatic nerve injury is followed by stereotypic events called Wallerian degeneration, i.e., axonal degeneration, Schwann cell dedifferentiation, and proliferation. Subsequently, axons begin to regenerate, Schwann cells redifferentiate and remyelinate and targets are reinnervated. Although similarities between development and repair of myelin are apparent, key differences exist, and for instance remyelinated axons have shorter internodes and thinner myelin sheaths (Fancy, Chan, Baranzini, Franklin, & Rowitch, 2011). Nevertheless, the reformed myelin enables rapid saltatory conduction.

Traumatic nerve injury provides a model for the analyses of biological pathways important for nerve degeneration and repair, and the stereotypic events allow the quantitation of degeneration/regeneration (Conforti, Gilley, & Coleman, 2014). Genetic mutants in invertebrates and vertebrates demonstrated that injury induces an active axon destruction program (Gilley, Orsomando, Nascimento-Ferreira, & Coleman, 2015; Mack et al., 2001; McDonald, Cheng, Chen, & Zochodne, 2006; Osterloh et al., 2012). Schwann cells in the distal nerve stump lose their axonal attachment, proliferate, and resorb myelin. Autophagy of myelin by Schwann cells provides the rate-limiting mechanism for myelin clearance and depends on JNK/c-Jun (Gomez-Sanchez et al., 2015). The initial signal for the Schwann cell response appears to be provided by axons, as Wallerian degeneration slow (Wlds) mouse mutants whose axon destruction is delayed also show gross delays in myelin removal and nerve regeneration. The remaining myelin was proposed to inhibit axon outgrowth (Brown, Perry, Hunt, & Lapper, 1994).

An interesting concept that emerged over the last decade is that Schwann cells in the distal nerve stump display unique characteristics (they were also called "repair cells") and do not simply revert to an earlier developmental stage (Arthur-Farraj et al., 2012; Jessen et al., 2015). These "repair cells" are essential for nerve repair that requires axoglial signaling: axon-derived signals promote differentiation of repair cells and growth factors released by them stimulate axon growth (Fawcett & Keynes, 1990; Jessen et al., 2015).

Traumatic nerve injury results in the dysregulated expression of many components of the Nrg1 signaling pathway like increased expression of ErbB2/ErbB3 receptors and enhanced receptor activation in the distal nerve (Carroll, Miller, Frohnert, Kim, & Corbett, 1997; Cohen, Yachnis, Arai, Davis, & Scherer, 1992; Kwon et al., 1997). Neuronal type III Nrg1 expression declines after axon transection, but is restored as axons reinnervate their targets and may therefore depend on target-derived growth factors (Bermingham-McDonogh, Xu, Marchionni, & Scherer, 1997). After axon destruction, type I Nrg1 expression in Schwann cells is transiently upregulated (Stassart et al., 2013).

Remyelination depends on juxtacrine Nrg1 derived from the axolemma. Mice in which Nrg1 is genetically ablated in a subpopulation of axons (and not from Schwann cells) show absent or grossly impaired remyelination, and the deficit is restricted to those axons that lack Nrg1 (Fricker et al., 2011). Thus, axon-derived type III Nrg1 plays a key role in remyelination. Schwann cell-derived type I Nrg1 modulates remyelination, and selective ablation of Nrg1 in Schwann cells does not impair developmental myelination but impinges on remyelination and functional recovery after injury (Stassart et al., 2013). Thus, developmental myelination exclusively depends on axonal type III Nrg1, whereas Schwann cell-derived type I Nrg1 contributes to remyelination. The protease Bace1 is required for full Nrg1 activity and remyelination is impaired in mice lacking Bace1 (Hu et al., 2008). Erbin expression increases in distal nerve after injury, and Erbin mutant mice display impaired remyelination and functional recovery (Liang et al., 2012). The early phases of the injury response are not Nrg1 dependent: myelin clearance is not delayed when Nrg1 is lacking, and Schwann cell proliferation during regeneration was not impaired in the absence of Nrg1 signaling (Atanasoski et al., 2006; Fricker et al., 2013). Finally, the inflammatory response is not Nrg1-dependent (Fricker et al., 2013).

A further important difference between Nrg1 functions in development and repair exists: remyelination and functional recovery are greatly delayed when Nrg1 is absent, but ultimately large or medium diameter axons are remyelinated, whereas small diameter axons are wrapped but not myelinated (Fricker et al., 2013). Therefore, Nrg1 determines the rate of remyelination after injury, but it neither decides the myelination fate (i.e., selects the axons that are remyelinated) nor the final outcome of remyelination. This is distinct from its developmental function where even a reduction in Nrg1 signaling results in reduced myelin thickness that is not compensated with time (Garratt, Voiculescu, Topilko, Charnay, & Birchmeier, 2000; Michailov et al., 2004). This result implies that signals cooperate with Nrg1 during repair, which are not operative during development (Fricker et al., 2011). Nrg1 might not only modulate Schwann cell function during nerve repair, since the rate of axon regeneration was also reduced when Nrg1 was lacking. "Back signaling" (i.e., the release of the intracellular domain of type III Nrg1 by gamma secretase; Hancock, Nowakowski, Role, Talmage, & Flanagan, 2011) or alternatively, indirect effects due to changed release of glial factors might contribute to this.

Can exogenous Nrg1 promote nerve repair and functional recovery after injury and thus be of therapeutic use? A successful outcome of a Nrg1 therapy may require just the right level of signaling, since excessive stimulation of Schwann cells is known to cause proliferation, demyelination, and formation of myelin that is too thick and possibly unstable (Sheean et al., 2014; Zanazzi et al., 2001). Moreover, Nrg1-dependent activation of ErbB receptors on Schwann cells (and other cell types exposed to Nrg1) could cause oncogenic transformation (Huijbregts, Roth, Schmidt, & Carroll, 2003). The final outcome of such a therapy is therefore likely to depend on the exact mode of Nrg1 delivery including timing of the treatment and dose. Despite these caveats, encouraging findings have emerged. For the treatment of cardiovascular disease, Nrg1 is already being tested in clinical trials and the treatment has a good safety profile (Jabbour et al., 2011). Transgenic overexpression of type I or type III Nrg1 in neurons promotes remyelination after traumatic nerve injury (Stassart et al., 2013). Furthermore, exogenous administration of type I and type II Nrg1 protein or delivery of type I Nrg1 by viral vectors promotes axon outgrowth and nerve repair in experimental settings (Cai, Peng, Nelson, Eberhart, & Smith, 2004; Mahanthappa, Anton, & Matthew, 1996). Thus, the effects were not restricted to improved remyelination but also enhanced axonal outgrowth, which represents probably the critical factor in functional repair.

Charcot-Marie-Tooth disease (CMT) is an inherited neuropathy that affects approximately 1 in 2500 people (Rossor, Polke, Houlden, & Reilly, 2013). CMT1A, the most common variant of CMT, is caused by the duplication of a large region of chromosome 17 that includes the PMP22 gene, i.e., an excess of the myelin protein PMP-22. Thus, CMT1A affects primarily the myelin sheath and is associated with aberrant Schwann cell differentiation, myelination, and demyelination. Administration of exogenous Nrg1 during early postnatal development did overcome impaired peripheral myelination, prevented axon loss, and ameliorated the impairment of nerve function (Fledrich et al., 2014).

Nrg1 also promotes the survival of terminal Schwann cells that envelop the neuromuscular junction, and these cells contribute to remodeling of the neuromuscular synapse after injury and to reinnervation of denervated synapses (Kang, Tian, Mikesh, Lichtman, & Thompson, 2014). The survival of terminal Schwann cells after neonatal denervation can be improved through Nrg1 administration, and they remain responsive to Nrg1 in adulthood (Trachtenberg & Thompson, 1996).

In summary, Nrg1 is required for effective nerve repair in adulthood and acts during remyelination. Both type I and type III Nrg1 isoforms contribute to myelin repair, in contrast to the developmental myelination, which appears to be strictly type III dependent (Fig. 1). In contrast to developmental myelination, during repair the loss of the Nrg1 signal is partly compensated by other mechanisms. Exogenous Nrg1 helps to promote nerve repair in a number of different animal models ranging from traumatic nerve injury to hereditary neuropathy, and promotes axon outgrowth as well as myelin repair. It is thus an encouraging therapeutic candidate. Complex decisions will need to be made prior to its use in clinical trials in regards to optimal isoform/formulation, dose, and timing of administration.

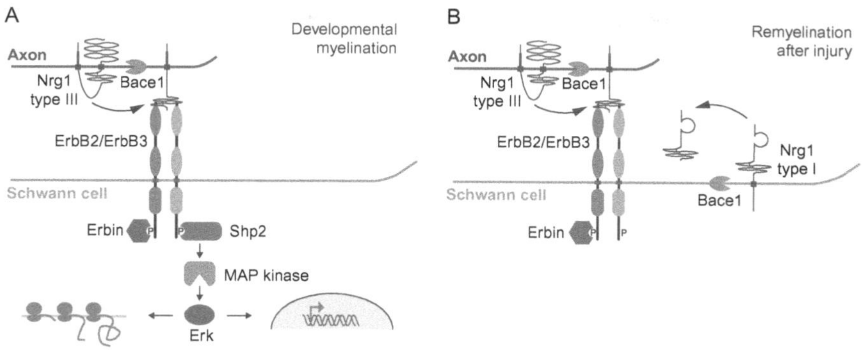

Figure 1 Nrg1 signaling during developmental myelination and remyelination. Developmental myelination (A) depends on type III Nrg1 that is produced by peripheral sensory and motoneurons and presented in the axonal membrane. Nrg1 needs to be processed by Bace1 for full signaling activity. ErbB2/ErbB3 receptors on Schwann cells recognize this signal and transmit it. Erbin, an intracellular ErbB2 interacting protein, is important for full activation of ErbB receptors. Many signaling components act downstream of Nrg1/ErbB, and we display here Shp2 and Erk1/2, because mutations of Nrg1/ErbB, Shp2, and Erk1/2 all produce very similar or even identical changes in Schwann cell development and myelination. Activation of MAPK, i.e., expression of an activated variant, can substitute for Nrg1/ErbB3/Shp2 during Schwann cell development and myelination. MAPK represents thus an important readout of Nrg1/ErbB signaling. Transcriptional and translational responses to the Nrg1 signal are observed in developmental myelination. Remyelination after injury (B) reuses the Nrg1 signaling network, and remyelination and nerve repair are severely delayed in the absence of Nrg1. Two Nrg1 isoforms participate in remyelination, axonal type III and Schwann cell-derived type I. Full ErbB activation and efficient regeneration depend again on Bace1 and Erbin. (A color version of this figure is located at the end of this chapter.)

ACKNOWLEDGMENTS

We thank Michael Strehle for preparing the figure and Dominique Bröhl for critically reading the manuscript. C.B. was supported by a grant (SFB665) of the German Research foundation (DFG). D.L.H.B. is a senior Wellcome clinical scientist fellow (ref. no. 095698z/11/z).

REFERENCES

Adameyko, I., Lallemend, F., Aquino, J. B., Pereira, J. A., Topilko, P., Muller, T., et al. (2009). Schwann cell precursors from nerve innervation are a cellular origin of melanocytes in skin. *Cell*, *139*(2), 366–379.

Arthur-Farraj, P. J., Latouche, M., Wilton, D. K., Quintes, S., Chabrol, E., Banerjee, A., et al. (2012). c-Jun reprograms Schwann cells of injured nerves to generate a repair cell essential for regeneration. *Neuron*, *75*(4), 633–647.

Atanasoski, S., Scherer, S. S., Sirkowski, E., Leone, D., Garratt, A. N., Birchmeier, C., et al. (2006). ErbB2 signaling in Schwann cells is mostly dispensable for maintenance

of myelinated peripheral nerves and proliferation of adult Schwann cells after injury. *The Journal of Neuroscience*, *26*(7), 2124–2131.
Bermingham, J. R., Jr., Scherer, S. S., O'Connell, S., Arroyo, E., Kalla, K. A., Powell, F. L., et al. (1996). Tst-1/Oct-6/SCIP regulates a unique step in peripheral myelination and is required for normal respiration. *Genes & Development*, *10*(14), 1751–1762.
Bermingham, J. R., Jr., Shearin, H., Pennington, J., O'Moore, J., Jaegle, M., Driegen, S., et al. (2006). The claw paw mutation reveals a role for Lgi4 in peripheral nerve development. *Nature Neuroscience*, *9*(1), 76–84.
Bermingham-McDonogh, O., Xu, Y. T., Marchionni, M. A., & Scherer, S. S. (1997). Neuregulin expression in PNS neurons: Isoforms and regulation by target interactions. *Molecular and Cellular Neuroscience*, *10*(3-4), 184–195.
Berti, C., Bartesaghi, L., Ghidinelli, M., Zambroni, D., Figlia, G., Chen, Z. L., et al. (2011). Non-redundant function of dystroglycan and beta1 integrins in radial sorting of axons. *Development*, *138*(18), 4025–4037.
Brinkmann, B. G., Agarwal, A., Sereda, M. W., Garratt, A. N., Muller, T., Wende, H., et al. (2008). Neuregulin-1/ErbB signaling serves distinct functions in myelination of the peripheral and central nervous system. *Neuron*, *59*(4), 581–595.
Brown, M. C., Perry, V. H., Hunt, S. P., & Lapper, S. R. (1994). Further studies on motor and sensory nerve regeneration in mice with delayed Wallerian degeneration. *The European Journal of Neuroscience*, *6*(3), 420–428.
Cabedo, H., Luna, C., Fernandez, A. M., Gallar, J., & Ferrer-Montiel, A. (2002). Molecular determinants of the sensory and motor neuron-derived factor insertion into plasma membrane. *The Journal of Biological Chemistry*, *277*(22), 19905–19912.
Cai, J., Peng, X., Nelson, K. D., Eberhart, R., & Smith, G. M. (2004). Synergistic improvements in cell and axonal migration across sciatic nerve lesion gaps using bioresorbable filaments and heregulin-beta1. *Journal of Biomedical Materials Research Part A*, *69*(2), 247–258.
Carroll, S. L., Miller, M. L., Frohnert, P. W., Kim, S. S., & Corbett, J. A. (1997). Expression of neuregulins and their putative receptors, ErbB2 and ErbB3, is induced during Wallerian degeneration. *The Journal of Neuroscience*, *17*(5), 1642–1659.
Chen, Z. L., & Strickland, S. (2003). Laminin gamma1 is critical for Schwann cell differentiation, axon myelination, and regeneration in the peripheral nerve. *The Journal of Cell Biology*, *163*(4), 889–899.
Cheret, C., Willem, M., Fricker, F. R., Wende, H., Wulf-Goldenberg, A., Tahirovic, S., et al. (2013). Bace1 and Neuregulin-1 cooperate to control formation and maintenance of muscle spindles. *The EMBO Journal*, *32*(14), 2015–2028.
Citri, A., Skaria, K. B., & Yarden, Y. (2003). The deaf and the dumb: The biology of ErbB-2 and ErbB-3. *Experimental Cell Research*, *284*(1), 54–65.
Cohen, J. A., Yachnis, A. T., Arai, M., Davis, J. G., & Scherer, S. S. (1992). Expression of the neu proto-oncogene by Schwann cells during peripheral nerve development and Wallerian degeneration. *Journal of Neuroscience Research*, *31*(4), 622–634.
Conforti, L., Gilley, J., & Coleman, M. P. (2014). Wallerian degeneration: An emerging axon death pathway linking injury and disease. *Nature Reviews Neuroscience*, *15*(6), 394–409.
Cotter, L., Ozcelik, M., Jacob, C., Pereira, J. A., Locher, V., Baumann, R., et al. (2010). Dlg1-PTEN interaction regulates myelin thickness to prevent damaging peripheral nerve overmyelination. *Science*, *328*(5984), 1415–1418.
Court, F. A., Sherman, D. L., Pratt, T., Garry, E. M., Ribchester, R. R., Cottrell, D. F., et al. (2004). Restricted growth of Schwann cells lacking Cajal bands slows conduction in myelinated nerves. *Nature*, *431*(7005), 191–195.
de Waegh, S. M., Lee, V. M., & Brady, S. T. (1992). Local modulation of neurofilament phosphorylation, axonal caliber, and slow axonal transport by myelinating Schwann cells. *Cell*, *68*(3), 451–463.

Doherty, M. K., Hammond, D. E., Clague, M. J., Gaskell, S. J., & Beynon, R. J. (2009). Turnover of the human proteome: Determination of protein intracellular stability by dynamic SILAC. *Journal of Proteome Research*, *8*(1), 104–112.
Donaldson, H. H., & Hoke, G. W. (1905). On the areas of the axis cylinder and medullary sheath as seen in cross sections of the spinal nerves of vertebrates. *Journal of Comparative Neurology and Psychology*, *15*, 1–16.
Dong, Z., Brennan, A., Liu, N., Yarden, Y., Lefkowitz, G., Mirsky, R., et al. (1995). Neu differentiation factor is a neuron-glia signal and regulates survival, proliferation, and maturation of rat Schwann cell precursors. *Neuron*, *15*(3), 585–596.
Dyachuk, V., Furlan, A., Shahidi, M. K., Giovenco, M., Kaukua, N., Konstantinidou, C., et al. (2014). Neurodevelopment. Parasympathetic neurons originate from nerve-associated peripheral glial progenitors. *Science*, *345*(6192), 82–87.
Emery, B. (2013). Playing the field: Sox10 recruits different partners to drive central and peripheral myelination. *PLoS Genetics*, *9*(10), e1003918.
Eshed-Eisenbach, Y., & Peles, E. (2013). The making of a node: A co-production of neurons and glia. *Current Opinion in Neurobiology*, *23*(6), 1049–1056.
Espinosa-Medina, I., Outin, E., Picard, C. A., Chettouh, Z., Dymecki, S., Consalez, G. G., et al. (2014). Neurodevelopment. Parasympathetic ganglia derive from Schwann cell precursors. *Science*, *345*(6192), 87–90.
Falls, D. L. (2003). Neuregulins: Functions, forms, and signaling strategies. *Experimental Cell Research*, *284*(1), 14–30.
Fancy, S. P., Chan, J. R., Baranzini, S. E., Franklin, R. J., & Rowitch, D. H. (2011). Myelin regeneration: A recapitulation of development? *Annual Review of Neuroscience*, *34*, 21–43.
Fawcett, J. W., & Keynes, R. J. (1990). Peripheral nerve regeneration. *Annual Review of Neuroscience*, *13*, 43–60.
Feltri, M. L., Graus Porta, D., Previtali, S. C., Nodari, A., Migliavacca, B., Cassetti, A., et al. (2002). Conditional disruption of beta 1 integrin in Schwann cells impedes interactions with axons. *The Journal of Cell Biology*, *156*(1), 199–209.
Finzsch, M., Schreiner, S., Kichko, T., Reeh, P., Tamm, E. R., Bosl, M. R., et al. (2010). Sox10 is required for Schwann cell identity and progression beyond the immature Schwann cell stage. *The Journal of Cell Biology*, *189*(4), 701–712.
Fleck, D., Garratt, A. N., Haass, C., & Willem, M. (2012). BACE1 dependent neuregulin processing: Review. *Current Alzheimer Research*, *9*(2), 178–183.
Fleck, D., van Bebber, F., Colombo, A., Galante, C., Schwenk, B. M., Rabe, L., et al. (2013). Dual cleavage of neuregulin 1 type III by BACE1 and ADAM17 liberates its EGF-like domain and allows paracrine signaling. *The Journal of Neuroscience*, *33*(18), 7856–7869.
Fledrich, R., Stassart, R. M., Klink, A., Rasch, L. M., Prukop, T., Haag, L., et al. (2014). Soluble neuregulin-1 modulates disease pathogenesis in rodent models of Charcot-Marie-Tooth disease 1A. *Nature Medicine*, *20*(9), 1055–1061.
Fricker, F. R., Antunes-Martins, A., Galino, J., Paramsothy, R., La Russa, F., Perkins, J., et al. (2013). Axonal neuregulin 1 is a rate limiting but not essential factor for nerve remyelination. *Brain*, *136*(Pt 7), 2279–2297.
Fricker, F. R., Lago, N., Balarajah, S., Tsantoulas, C., Tanna, S., Zhu, N., et al. (2011). Axonally derived neuregulin-1 is required for remyelination and regeneration after nerve injury in adulthood. *The Journal of Neuroscience*, *31*(9), 3225–3233.
Fu, S. Y., & Gordon, T. (1995). Contributing factors to poor functional recovery after delayed nerve repair: Prolonged denervation. *The Journal of Neuroscience*, *15*(5 Pt 2), 3886–3895.
Garratt, A. N., Voiculescu, O., Topilko, P., Charnay, P., & Birchmeier, C. (2000). A dual role of erbB2 in myelination and in expansion of the schwann cell precursor pool. *The Journal of Cell Biology*, *148*(5), 1035–1046.

Gilley, J., Orsomando, G., Nascimento-Ferreira, I., & Coleman, M. P. (2015). Absence of SARM1 rescues development and survival of NMNAT2-deficient axons. *Cell Reports*, *10*(12), 1974–1981.
Goebbels, S., Oltrogge, J. H., Kemper, R., Heilmann, I., Bormuth, I., Wolfer, S., et al. (2010). Elevated phosphatidylinositol 3,4,5-trisphosphate in glia triggers cell-autonomous membrane wrapping and myelination. *The Journal of Neuroscience*, *30*(26), 8953–8964.
Goebbels, S., Oltrogge, J. H., Wolfer, S., Wieser, G. L., Nientiedt, T., Pieper, A., et al. (2012). Genetic disruption of Pten in a novel mouse model of tomaculous neuropathy. *EMBO Molecular Medicine*, *4*(6), 486–499.
Gomez-Sanchez, J. A., Carty, L., Iruarrizaga-Lejarreta, M., Palomo-Irigoyen, M., Varela-Rey, M., Griffith, M., et al. (2015). Schwann cell autophagy, myelinophagy, initiates myelin clearance from injured nerves. *The Journal of Cell Biology*, *210*(1), 153–168.
Grossmann, K. S., Wende, H., Paul, F. E., Cheret, C., Garratt, A. N., Zurborg, S., et al. (2009). The tyrosine phosphatase Shp2 (PTPN11) directs Neuregulin-1/ErbB signaling throughout Schwann cell development. *Proceedings of the National Academy of Sciences of the United States of America*, *106*(39), 16704–16709.
Grove, M., Komiyama, N. H., Nave, K. A., Grant, S. G., Sherman, D. L., & Brophy, P. J. (2007). FAK is required for axonal sorting by Schwann cells. *The Journal of Cell Biology*, *176*(3), 277–282.
Hancock, M. L., Nowakowski, D. W., Role, L. W., Talmage, D. A., & Flanagan, J. G. (2011). Type III neuregulin 1 regulates pathfinding of sensory axons in the developing spinal cord and periphery. *Development*, *138*(22), 4887–4898.
Harrisingh, M. C., Perez-Nadales, E., Parkinson, D. B., Malcolm, D. S., Mudge, A. W., & Lloyd, A. C. (2004). The Ras/Raf/ERK signalling pathway drives Schwann cell dedifferentiation. *The EMBO Journal*, *23*(15), 3061–3071.
Hillarp, N. A., & Olivecrona, H. (1946). The role played by the axon and the Schwann cells in the degree of myelination of the peripheral nerve fibre. *Acta Anatomica (Basel)*, *2*(1), 17–32.
Hippenmeyer, S., Shneider, N. A., Birchmeier, C., Burden, S. J., Jessell, T. M., & Arber, S. (2002). A role for neuregulin1 signaling in muscle spindle differentiation. *Neuron*, *36*(6), 1035–1049.
Horiuchi, K., Zhou, H. M., Kelly, K., Manova, K., & Blobel, C. P. (2005). Evaluation of the contributions of ADAMs 9, 12, 15, 17, and 19 to heart development and ectodomain shedding of neuregulins beta1 and beta2. *Developmental Biology*, *283*(2), 459–471.
Hornig, J., Frob, F., Vogl, M. R., Hermans-Borgmeyer, I., Tamm, E. R., & Wegner, M. (2013). The transcription factors Sox10 and Myrf define an essential regulatory network module in differentiating oligodendrocytes. *PLoS Genetics*, *9*(10), e1003907.
Hu, X., He, W., Diaconu, C., Tang, X., Kidd, G. J., Macklin, W. B., et al. (2008). Genetic deletion of BACE1 in mice affects remyelination of sciatic nerves. *The FASEB Journal*, *22*(8), 2970–2980.
Hu, X., Hicks, C. W., He, W., Wong, P., Macklin, W. B., Trapp, B. D., et al. (2006). Bace1 modulates myelination in the central and peripheral nervous system. *Nature Neuroscience*, *9*(12), 1520–1525.
Huijbregts, R. P., Roth, K. A., Schmidt, R. E., & Carroll, S. L. (2003). Hypertrophic neuropathies and malignant peripheral nerve sheath tumors in transgenic mice overexpressing glial growth factor beta3 in myelinating Schwann cells. *The Journal of Neuroscience*, *23*(19), 7269–7280.
Hynes, N. E., & Lane, H. A. (2005). ERBB receptors and cancer: The complexity of targeted inhibitors. *Nature Reviews Cancer*, *5*(5), 341–354.

Ishii, A., Furusho, M., & Bansal, R. (2013). Sustained activation of ERK1/2 MAPK in oligodendrocytes and schwann cells enhances myelin growth and stimulates oligodendrocyte progenitor expansion. *The Journal of Neuroscience*, *33*(1), 175–186.

Jabbour, A., Hayward, C. S., Keogh, A. M., Kotlyar, E., McCrohon, J. A., England, J. F., et al. (2011). Parenteral administration of recombinant human neuregulin-1 to patients with stable chronic heart failure produces favourable acute and chronic haemodynamic responses. *European Journal of Heart Failure*, *13*(1), 83–92.

Jaegle, M., Ghazvini, M., Mandemakers, W., Piirsoo, M., Driegen, S., Levavasseur, F., et al. (2003). The POU proteins Brn-2 and Oct-6 share important functions in Schwann cell development. *Genes & Development*, *17*(11), 1380–1391.

Jessen, K. R., Mirsky, R., & Lloyd, A. C. (2015). Schwann cells: Development and role in nerve repair. *Cold Spring Harbor Perspectives in Biology*, 7(7), a020487.

Joseph, N. M., Mukouyama, Y. S., Mosher, J. T., Jaegle, M., Crone, S. A., Dormand, E. L., et al. (2004). Neural crest stem cells undergo multilineage differentiation in developing peripheral nerves to generate endoneurial fibroblasts in addition to Schwann cells. *Development*, *131*(22), 5599–5612.

Kang, H., Tian, L., Mikesh, M., Lichtman, J. W., & Thompson, W. J. (2014). Terminal Schwann cells participate in neuromuscular synapse remodeling during reinnervation following nerve injury. *The Journal of Neuroscience*, *34*(18), 6323–6333.

Kaukua, N., Shahidi, M. K., Konstantinidou, C., Dyachuk, V., Kaucka, M., Furlan, A., et al. (2014). Glial origin of mesenchymal stem cells in a tooth model system. *Nature*, *513*(7519), 551–554.

Kiryu-Seo, S., Ohno, N., Kidd, G. J., Komuro, H., & Trapp, B. D. (2010). Demyelination increases axonal stationary mitochondrial size and the speed of axonal mitochondrial transport. *The Journal of Neuroscience*, *30*(19), 6658–6666.

Kwon, Y. K., Bhattacharyya, A., Alberta, J. A., Giannobile, W. V., Cheon, K., Stiles, C. D., et al. (1997). Activation of ErbB2 during wallerian degeneration of sciatic nerve. *The Journal of Neuroscience*, *17*(21), 8293–8299.

La Marca, R., Cerri, F., Horiuchi, K., Bachi, A., Feltri, M. L., Wrabetz, L., et al. (2011). TACE (ADAM17) inhibits Schwann cell myelination. *Nature Neuroscience*, *14*(7), 857–865.

Lemke, G. (2006). Neuregulin-1 and myelination. *Science's STKE*, *2006*(325), pe11.

Lemke, G., & Chao, M. (1988). Axons regulate Schwann cell expression of the major myelin and NGF receptor genes. *Development*, *102*(3), 499–504.

Liang, C., Tao, Y., Shen, C., Tan, Z., Xiong, W. C., & Mei, L. (2012). Erbin is required for myelination in regenerated axons after injury. *The Journal of Neuroscience*, *32*(43), 15169–15180.

Lyons, D. A., Pogoda, H. M., Voas, M. G., Woods, I. G., Diamond, B., Nix, R., et al. (2005). erbb3 and erbb2 are essential for schwann cell migration and myelination in zebrafish. *Current Biology*, *15*(6), 513–524.

Mack, T. G., Reiner, M., Beirowski, B., Mi, W., Emanuelli, M., Wagner, D., et al. (2001). Wallerian degeneration of injured axons and synapses is delayed by a Ube4b/Nmnat chimeric gene. *Nature Neuroscience*, *4*(12), 1199–1206.

Mahanthappa, N. K., Anton, E. S., & Matthew, W. D. (1996). Glial growth factor 2, a soluble neuregulin, directly increases Schwann cell motility and indirectly promotes neurite outgrowth. *The Journal of Neuroscience*, *16*(15), 4673–4683.

Maurel, P., & Salzer, J. L. (2000). Axonal regulation of Schwann cell proliferation and survival and the initial events of myelination requires PI 3-kinase activity. *The Journal of Neuroscience*, *20*(12), 4635–4645.

McDonald, D., Cheng, C., Chen, Y., & Zochodne, D. (2006). Early events of peripheral nerve regeneration. *Neuron Glia Biology*, *2*(2), 139–147.

Meyer, D., & Birchmeier, C. (1995). Multiple essential functions of neuregulin in development. *Nature, 378*(6555), 386–390.
Meyer, D., Yamaai, T., Garratt, A., Riethmacher-Sonnenberg, E., Kane, D., Theill, L. E., et al. (1997). Isoform-specific expression and function of neuregulin. *Development, 124*(18), 3575–3586.
Michailov, G. V., Sereda, M. W., Brinkmann, B. G., Fischer, T. M., Haug, B., Birchmeier, C., et al. (2004). Axonal neuregulin-1 regulates myelin sheath thickness. *Science, 304*(5671), 700–703.
Michel, K., Zhao, T., Karl, M., Lewis, K., & Fyffe-Maricich, S. L. (2015). Translational control of myelin basic protein expression by ERK2 MAP kinase regulates timely remyelination in the adult brain. *The Journal of Neuroscience, 35*(20), 7850–7865.
Monk, K. R., Naylor, S. G., Glenn, T. D., Mercurio, S., Perlin, J. R., Dominguez, C., et al. (2009). A G protein-coupled receptor is essential for Schwann cells to initiate myelination. *Science, 325*(5946), 1402–1405.
Monk, K. R., Oshima, K., Jors, S., Heller, S., & Talbot, W. S. (2011). Gpr126 is essential for peripheral nerve development and myelination in mammals. *Development, 138*(13), 2673–2680.
Morgan, L., Jessen, K. R., & Mirsky, R. (1991). The effects of cAMP on differentiation of cultured Schwann cells: Progression from an early phenotype (04+) to a myelin phenotype (P0+, GFAP-, N-CAM-, NGF-receptor-) depends on growth inhibition. *The Journal of Cell Biology, 112*(3), 457–467.
Morrison, S. J., White, P. M., Zock, C., & Anderson, D. J. (1999). Prospective identification, isolation by flow cytometry, and in vivo self-renewal of multipotent mammalian neural crest stem cells. *Cell, 96*(5), 737–749.
Morrissey, T. K., Levi, A. D., Nuijens, A., Sliwkowski, M. X., & Bunge, R. P. (1995). Axon-induced mitogenesis of human Schwann cells involves heregulin and p185erbB2. *Proceedings of the National Academy of Sciences of the United States of America, 92*(5), 1431–1435.
Napoli, I., Noon, L. A., Ribeiro, S., Kerai, A. P., Parrinello, S., Rosenberg, L. H., et al. (2012). A central role for the ERK-signaling pathway in controlling Schwann cell plasticity and peripheral nerve regeneration in vivo. *Neuron, 73*(4), 729–742.
Nave, K. A., & Salzer, J. L. (2006). Axonal regulation of myelination by neuregulin 1. *Current Opinion in Neurobiology, 16*(5), 492–500.
Nave, K. A., & Trapp, B. D. (2008). Axon-glial signaling and the glial support of axon function. *Annual Review of Neuroscience, 31*, 535–561.
Newbern, J., & Birchmeier, C. (2010). Nrg1/ErbB signaling networks in Schwann cell development and myelination. *Seminars in Cell and Developmental Biology, 21*(9), 922–928.
Newbern, J. M., Li, X., Shoemaker, S. E., Zhou, J., Zhong, J., Wu, Y., et al. (2011). Specific functions for ERK/MAPK signaling during PNS development. *Neuron, 69*(1), 91–105.
Norrmen, C., Figlia, G., Lebrun-Julien, F., Pereira, J. A., Trotzmuller, M., Kofeler, H. C., et al. (2014). mTORC1 controls PNS myelination along the mTORC1-RXRgamma-SREBP-lipid biosynthesis axis in Schwann cells. *Cell Reports, 9*(2), 646–660.
Ogata, T., Yamamoto, S., Nakamura, K., & Tanaka, S. (2006). Signaling axis in schwann cell proliferation and differentiation. *Molecular Neurobiology, 33*(1), 51–62.
Osterloh, J. M., Yang, J., Rooney, T. M., Fox, A. N., Adalbert, R., Powell, E. H., et al. (2012). dSarm/Sarm1 is required for activation of an injury-induced axon death pathway. *Science, 337*(6093), 481–484.
Ozkaynak, E., Abello, G., Jaegle, M., van Berge, L., Hamer, D., Kegel, L., et al. (2010). Adam22 is a major neuronal receptor for Lgi4-mediated Schwann cell signaling. *The Journal of Neuroscience, 30*(10), 3857–3864.

Paavola, K. J., Sidik, H., Zuchero, J. B., Eckart, M., & Talbot, W. S. (2014). Type IV collagen is an activating ligand for the adhesion G protein-coupled receptor GPR126. *Science*, 7(338), ra76.

Painter, M. W., Brosius Lutz, A., Cheng, Y. C., Latremoliere, A., Duong, K., Miller, C. M., et al. (2014). Diminished Schwann cell repair responses underlie age-associated impaired axonal regeneration. *Neuron*, *83*(2), 331–343.

Pellegatta, M., De Arcangelis, A., D'Urso, A., Nodari, A., Zambroni, D., Ghidinelli, M., et al. (2013). alpha6beta1 and alpha7beta1 integrins are required in Schwann cells to sort axons. *The Journal of Neuroscience*, *33*(46), 17995–18007.

Pereira, J. A., Baumann, R., Norrmen, C., Somandin, C., Miehe, M., Jacob, C., et al. (2010). Dicer in Schwann cells is required for myelination and axonal integrity. *The Journal of Neuroscience*, *30*(19), 6763–6775.

Perlin, J. R., Lush, M. E., Stephens, W. Z., Piotrowski, T., & Talbot, W. S. (2011). Neuronal Neuregulin 1 type III directs Schwann cell migration. *Development*, *138*(21), 4639–4648.

Petersen, S. C., Luo, R., Liebscher, I., Giera, S., Jeong, S. J., Mogha, A., et al. (2015). The adhesion GPCR GPR126 has distinct, domain-dependent functions in Schwann cell development mediated by interaction with laminin-211. *Neuron*, *85*(4), 755–769.

Quarles, R., Macklin, W., & Morell, P. (2006). Myelin formation, structure and biochemistry. In G. Siegel, R. Albers, S. Brady, & D. Price (Eds.), *Basic neurochemistry: Molecular, cellular and medical aspects*. CA, USA: San Diego.

Raphael, A. R., & Talbot, W. S. (2011). New insights into signaling during myelination in zebrafish. *Current Topics in Developmental Biology*, *97*, 1–19.

Riethmacher, D., Sonnenberg-Riethmacher, E., Brinkmann, V., Yamaai, T., Lewin, G. R., & Birchmeier, C. (1997). Severe neuropathies in mice with targeted mutations in the ErbB3 receptor. *Nature*, *389*(6652), 725–730.

Rossor, A. M., Polke, J. M., Houlden, H., & Reilly, M. M. (2013). Clinical implications of genetic advances in Charcot-Marie-Tooth disease. *Nature Reviews Neurology*, *9*(10), 562–571.

Saxton, W. M., & Hollenbeck, P. J. (2012). The axonal transport of mitochondria. *Journal of Cell Science*, *125*(Pt 9), 2095–2104.

Scheib, J., & Hoke, A. (2013). Advances in peripheral nerve regeneration. *Nature Reviews Neurology*, *9*(12), 668–676.

Schlessinger, J. (2000). Cell signaling by receptor tyrosine kinases. *Cell*, *103*(2), 211–225.

Schroering, A., & Carey, D. J. (1998). Sensory and motor neuron-derived factor is a transmembrane heregulin that is expressed on the plasma membrane with the active domain exposed to the extracellular environment. *The Journal of Biological Chemistry*, *273*(46), 30643–30650.

Schwanhausser, B., Busse, D., Li, N., Dittmar, G., Schuchhardt, J., Wolf, J., et al. (2011). Global quantification of mammalian gene expression control. *Nature*, *473*(7347), 337–342.

Selbach, M., Schwanhausser, B., Thierfelder, N., Fang, Z., Khanin, R., & Rajewsky, N. (2008). Widespread changes in protein synthesis induced by microRNAs. *Nature*, *455*(7209), 58–63.

Sheean, M. E., McShane, E., Cheret, C., Walcher, J., Muller, T., Wulf-Goldenberg, A., et al. (2014). Activation of MAPK overrides the termination of myelin growth and replaces Nrg1/ErbB3 signals during Schwann cell development and myelination. *Genes & Development*, *28*(3), 290–303.

Sherman, D. L., Krols, M., Wu, L. M., Grove, M., Nave, K. A., Gangloff, Y. G., et al. (2012). Arrest of myelination and reduced axon growth when Schwann cells lack mTOR. *The Journal of Neuroscience*, *32*(5), 1817–1825.

Stassart, R. M., Fledrich, R., Velanac, V., Brinkmann, B. G., Schwab, M. H., Meijer, D., et al. (2013). A role for Schwann cell-derived neuregulin-1 in remyelination. *Nature Neuroscience*, *16*(1), 48–54.

Suter, U., & Scherer, S. S. (2003). Disease mechanisms in inherited neuropathies. *Nature Reviews Neuroscience*, *4*(9), 714–726.

Svaren, J., & Meijer, D. (2008). The molecular machinery of myelin gene transcription in Schwann cells. *Glia*, *56*(14), 1541–1551.

Tao, Y., Dai, P., Liu, Y., Marchetto, S., Xiong, W. C., Borg, J. P., et al. (2009). Erbin regulates NRG1 signaling and myelination. *Proceedings of the National Academy of Sciences of the United States of America*, *106*(23), 9477–9482.

Taveggia, C., Feltri, M. L., & Wrabetz, L. (2010). Signals to promote myelin formation and repair. *Nature Reviews Neurology*, *6*(5), 276–287.

Topilko, P., Schneider-Maunoury, S., Levi, G., Baron-Van Evercooren, A., Chennoufi, A. B., Seitanidou, T., et al. (1994). Krox-20 controls myelination in the peripheral nervous system. *Nature*, *371*(6500), 796–799.

Trachtenberg, J. T., & Thompson, W. J. (1996). Schwann cell apoptosis at developing neuromuscular junctions is regulated by glial growth factor. *Nature*, *379*(6561), 174–177.

Verheijen, M. H., Camargo, N., Verdier, V., Nadra, K., de Preux Charles, A. S., Medard, J. J., et al. (2009). SCAP is required for timely and proper myelin membrane synthesis. *Proceedings of the National Academy of Sciences of the United States of America*, *106*(50), 21383–21388.

Viader, A., Sasaki, Y., Kim, S., Strickland, A., Workman, C. S., Yang, K., et al. (2013). Aberrant Schwann cell lipid metabolism linked to mitochondrial deficits leads to axon degeneration and neuropathy. *Neuron*, *77*(5), 886–898.

Watson, D. F., Nachtman, F. N., Kuncl, R. W., & Griffin, J. W. (1994). Altered neurofilament phosphorylation and beta tubulin isotypes in Charcot-Marie-Tooth disease type 1. *Neurology*, *44*(12), 2383–2387.

Waxman, S. G. (1980). Determinants of conduction velocity in myelinated nerve fibers. *Muscle and Nerve*, *3*(2), 141–150.

Willem, M., Garratt, A. N., Novak, B., Citron, M., Kaufmann, S., Rittger, A., et al. (2006). Control of peripheral nerve myelination by the beta-secretase BACE1. *Science*, *314*(5799), 664–666.

Woldeyesus, M. T., Britsch, S., Riethmacher, D., Xu, L., Sonnenberg-Riethmacher, E., Abou-Rebyeh, F., et al. (1999). Peripheral nervous system defects in erbB2 mutants following genetic rescue of heart development. *Genes & Development*, *13*(19), 2538–2548.

Yarden, Y., & Sliwkowski, M. X. (2001). Untangling the ErbB signalling network. *Nature Reviews Molecular Cell Biology*, *2*(2), 127–137.

Yokozeki, T., Wakatsuki, S., Hatsuzawa, K., Black, R. A., Wada, I., & Sehara-Fujisawa, A. (2007). Meltrin beta (ADAM19) mediates ectodomain shedding of Neuregulin beta1 in the Golgi apparatus: Fluorescence correlation spectroscopic observation of the dynamics of ectodomain shedding in living cells. *Genes to Cells*, *12*(3), 329–343.

Zanazzi, G., Einheber, S., Westreich, R., Hannocks, M. J., Bedell-Hogan, D., Marchionni, M. A., et al. (2001). Glial growth factor/neuregulin inhibits Schwann cell myelination and induces demyelination. *The Journal of Cell Biology*, *152*(6), 1289–1299.

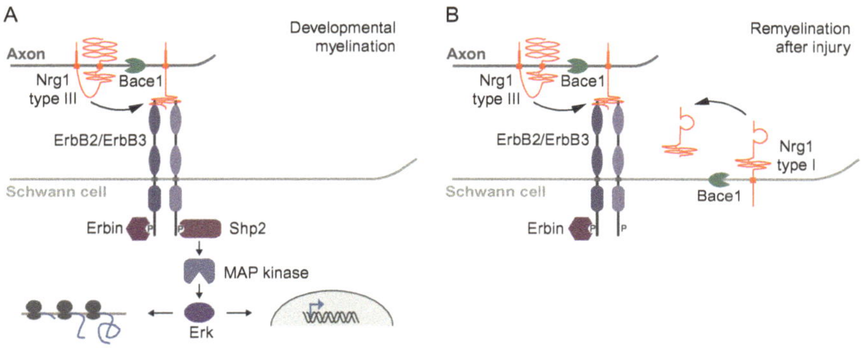

Carmen Birchmeier and David L.H. Bennett, Figure 1 Nrg1 signaling during developmental myelination and remyelination. Developmental myelination (A) depends on type III Nrg1 that is produced by peripheral sensory and motoneurons and presented in the axonal membrane. Nrg1 needs to be processed by Bace1 for full signaling activity. ErbB2/ErbB3 receptors on Schwann cells recognize this signal and transmit it. Erbin, an intracellular ErbB2 interacting protein, is important for full activation of ErbB receptors. Many signaling components act downstream of Nrg1/ErbB, and we display here Shp2 and Erk1/2, because mutations of Nrg1/ErbB, Shp2, and Erk1/2 all produce very similar or even identical changes in Schwann cell development and myelination. Activation of MAPK, i.e., expression of an activated variant, can substitute for Nrg1/ErbB3/Shp2 during Schwann cell development and myelination. MAPK represents thus an important readout of Nrg1/ErbB signaling. Transcriptional and translational responses to the Nrg1 signal are observed in developmental myelination. Remyelination after injury (B) reuses the Nrg1 signaling network, and remyelination and nerve repair are severely delayed in the absence of Nrg1. Two Nrg1 isoforms participate in remyelination, axonal type III and Schwann cell-derived type I. Full ErbB activation and efficient regeneration depend again on Bace1 and Erbin.

CHAPTER FIVE

Oocyte Meiotic Spindle Assembly and Function

Aaron F. Severson*, George von Dassow†, Bruce Bowerman‡,1
*Department of Biological, Geological, and Environmental Sciences, Center for Gene Regulation in Health and Disease, Cleveland State University, Cleveland, Ohio, USA
†Oregon Institute of Marine Biology, University of Oregon, Charleston, Oregon, USA
‡Institute of Molecular Biology, University of Oregon, Eugene, Oregon, USA
[1]Corresponding author: e-mail address: bowerman@uoregon.edu

Contents

Abstract

Gametogenesis in animal oocytes reduces the diploid genome content of germline precursors to a haploid state in gametes by discarding ¾ of the duplicated chromosomes through a sequence of two meiotic cell divisions called meiosis I and II. The assembly of the microtubule-based spindle structure that mediates this reduction in genome content remains poorly understood compared to our knowledge of mitotic spindle assembly and function. In this review, we consider the diversity of oocyte meiotic spindle assembly and structure across animal phylogeny, review recent advances in our understanding of how animal oocytes assemble spindles in the absence of the centriole-based microtubule-organizing centers that dominate mitotic spindle assembly, and

ISSN 0070-2153
http://dx.doi.org/10.1016/bs.ctdb.2015.11.031

discuss different models for how chromosomes are captured and moved to achieve chromosome segregation during oocyte meiotic cell division.

1. INTRODUCTION

A nearly universal feature of animal life is the fusion of two haploid gametes to create a diploid zygote. Most animal cells except gametes have two closely related but genetically distinct copies of each chromosome, called homologous chromosome pairs, one inherited from each parent. The fundamental achievement of gametogenesis is to reduce the diploid genome of germline precursors to a haploid state through two specialized cell divisions, called meiosis I and II (Fig. 1) (Dumont & Desai, 2012; Howe & FitzHarris, 2013; Müller-Reichert et al., 2010; Ohkura, 2015). When two gamete genomes unite after the fertilization of an egg by a sperm, diploidy is restored and life begins anew.

This review focuses on recent advances in our understanding of how an egg reduces its genome content to a haploid state, and more specifically on oocyte meiosis I, the first of the two sequential divisions that produce haploid gametes. While we devote more attention to the nematode *Caenorhabditis elegans*, reference and comparison to other model organisms—including a mammal (*Mus musculus*), an amphibian (*Xenopus laevis*), and an insect (*Drosophila melanogaster*)—will highlight similarities and differences among these widely studied models. After introducing meiosis and how it differs from mitosis (Fig. 1), we will focus on three topics. First, we will consider the diversity of meiotic mechanisms across animal phyla, and our limited understanding of when, how, and why such differences evolved. We will then address structural features of the meiotic cell division machinery—the oocyte meiotic spindle—that distinguish it from the mitotic spindle, and discuss recent progress in understanding the molecular mechanisms that mediate spindle assembly early in oocyte meiosis I. Finally, we will review the spindle dynamics that align and segregate chromosomes later in meiosis I to produce a haploid genome.

2. MAINTAIN PARITY OR PARE DOWN: MITOTIC VERSUS MEIOTIC CELL DIVISION

During both meiotic and mitotic cell division, chromosomes are captured and moved by microtubules, hollow tubes that are 25 nm in diameter

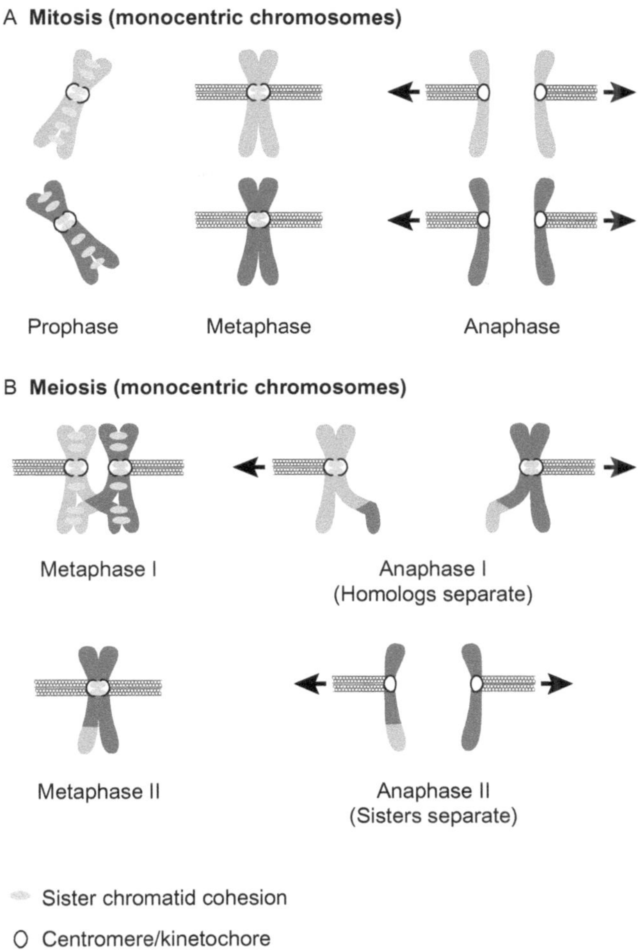

Figure 1 Distinct patterns of chromosome segregation occur during meiosis and mitosis. The progression of a single pair of homologous chromosomes (blue and red) through mitosis (A) and meiosis (B) is shown. Before either division, DNA replication duplicates both homologs. Subsequently, each homolog is composed of two sister chromatids tethered by sister chromatid cohesion (yellow ovals). (A) In many organisms, cohesion along chromosome arms is released during mitotic prophase. At metaphase, sisters remain linked by centromeric cohesion. Bipolar attachment of sister kinetochores to microtubules from opposite spindle poles aligns each homolog on the metaphase plate. In anaphase, centromeric cohesion is released and sister chromatids segregate toward opposite poles. (B) During prophase of meiosis I, homologous chromosomes become linked by reciprocal exchange of DNA during crossover recombination. Both sister kinetochores of one homolog capture microtubules from the same spindle pole; the sister kinetochores of the other homolog attach to microtubules from the opposite spindle pole (for simplicity, microtubule contact with both sister chromatid kinetochores is not depicted). In anaphase of meiosis I, release of cohesion between chromosome arms allows homologs to separate; sisters remain tethered by centromeric cohesion. In anaphase II, centromeric cohesion is released, allowing sisters to separate. (A color version of this figure is located at the end of this chapter.)

and are formed by the polymerization of α- and β-tubulin dimers (Fig. 2A). Microtubules are highly dynamic: they alternate between periods of rapid growth and shrinkage at their plus end, while being more stable at their minus end (Kirschner & Mitchison, 1986; Mitchison & Kirschner, 1984).

Much of our understanding of microtubule and chromosome dynamics during cell division has come from studies of mitosis (Hirano, 2015; Walczak et al., 2010). Before mitotic cell divisions, semiconservative DNA replication duplicates each chromosome in the genome. The resulting copies of each homolog, called sister chromatids, are held together by a ring-shaped protein complex called cohesin, proposed to encircle each pair of sister chromatids. During mitosis, microtubules assemble into a bipolar structure called the spindle and capture each sister chromatid through large protein complexes called kinetochores (Fig. 2B). A kinetochore assembles on each chromatid, and ultimately the sister kinetochores attach to microtubules from opposite spindle poles. Once these so-called "amphitelic attachments" have formed, each pair of sisters aligns on the metaphase plate, a plane midway between the two poles. Next, cell cycle regulated proteolysis breaks open the cohesin rings, freeing the sister chromatids, and allowing their movement toward opposite poles in response to microtubule-based pulling forces (Ohkura, 2015; Petronczki et al., 2003). Each daughter cell receives one copy of each homolog and ploidy is unchanged (Fig. 1A).

The mechanics of meiosis I are fundamentally distinct from those of mitosis. As in mitosis, DNA replication precisely copies each homolog before meiosis I (Fig. 1B). In prophase of meiosis I, the duplicated homologs pair and become joined by an elaborate protein structure called the synaptonemal complex (Ohkura, 2015; Petronczki et al., 2003; Zickler & Kleckner, 2015). The homolog pairs (called bivalents) then become covalently linked through reciprocal exchange of DNA during crossover recombination. Much as in mitosis, microtubules of the oocyte meiotic spindle establish attachments to the bivalents. However, during meiosis I the two sister kinetochores of each duplicated homolog are captured by the same pole, not by opposite poles as occurs during mitosis. Microtubules from the opposite pole capture the sister kinetochores of the other homolog. Once bipolar attachments form and each bivalent aligns on the metaphase plate, a subset of meiotic cohesin complexes are destroyed, allowing the homologs to segregate to opposite poles (Ohkura, 2015; Petronczki et al., 2003). The two sister chromatids of each homolog remain tethered by the surviving cohesin complexes. Meiosis II then proceeds much like mitosis: sister kinetochores are captured by opposite poles and move apart when the remaining cohesin complexes are

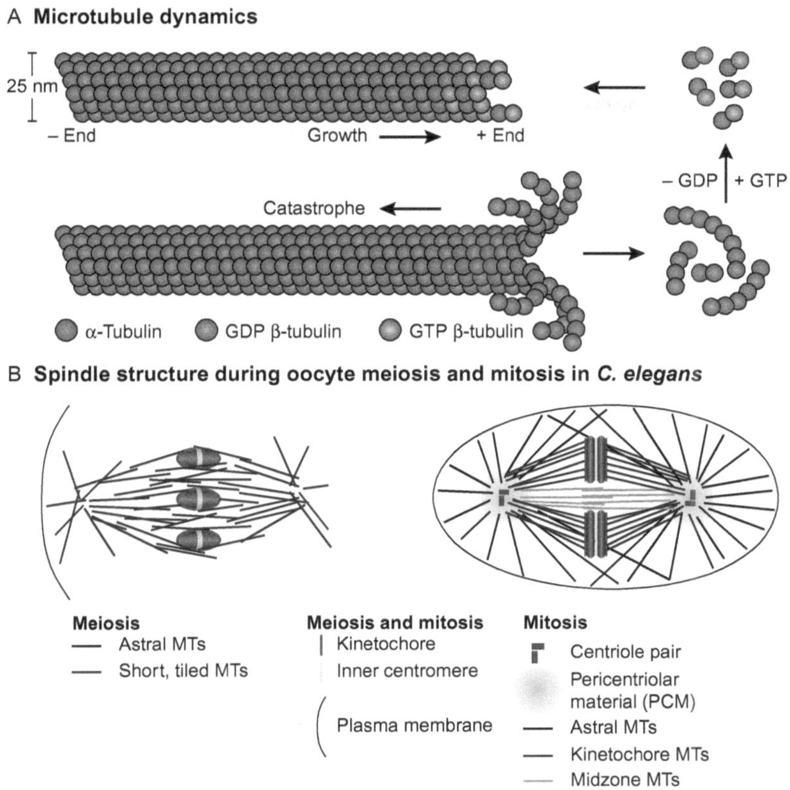

Figure 2 Microtubule dynamics and spindle structures during oocyte meiosis and mitosis. (A) Microtubules are hollow tubes 25 nm in diameter formed by lateral association of 13 protofilaments, linear polymers composed α- and β-tubulin heterodimers. Due to the ordered assembly of α/β-tubulin heterodimers, microtubules have intrinsic polarity. One end (the minus end) has α-tubulin as the terminal subunit. This end is relatively stable. The other end (the plus end) has β-tubulin as the terminal subunit. This highly dynamic end switches between rapid growth and shrinkage depending on whether the exposed β-tubulin subunits are bound to GTP (growth) or GDP (shrinkage). During growth, GTP-tubulin adds to the plus end, forming a GTP cap. GTP is hydrolyzed to GDP, and if GDP-β-tubulin is exposed on the plus end of a microtubule, catastrophic disassembly can occur. (B) Cartoons of spindle structure during meiosis and mitosis in *C. elegans*. During mitosis, long, radial microtubule arrays are nucleated in the pericentriolar material surrounding the centrioles. Some microtubules (blue) from each spindle pole are captured by kinetochores (red). Ultimately, these kinetochore/microtubule attachments align each chromosome on the metaphase plate. Other overlapping, antiparallel microtubules (green) become bundled together in a region called the midzone, a process that stabilizes mitotic spindle structure. In contrast, centrioles are degraded prior to assembly of the oocyte meiotic spindle. Short, often tiled, microtubules form near chromosomes and ultimately coalesce into a bipolar structure. (A color version of this figure is located at the end of this chapter.)

destroyed. While meiotic and mitotic cohesins share several subunits, certain subunits are unique to each type of division and account for the different patterns of chromosome segregation (Herran et al., 2011; Ishiguro et al., 2011, 2014; Klein et al., 1999; Lee & Hirano, 2011; Llano et al., 2012; McNicoll et al., 2013; Severson & Meyer, 2014; Severson et al., 2009).

In summary, meiosis I differs from mitosis in that homologs pair and then segregate to opposite spindle poles, but sister chromatids never separate. Thus, each daughter cell inherits a single homolog composed of two sister chromatids, rather than a single chromatid from both homologs as occurs during mitosis. Meiosis II then simply reduces genome content to a single chromatid from each inherited homolog. While recent advances have improved our understanding of the mechanisms that tether and release sister chromatids to reduce ploidy, this review will focus on the assembly and function of the microtubule-based oocyte meiosis I spindle.

3. CENTROSOMES: THE STARS OF MITOTIC SPINDLE POLES

While the spindles that form in mitotic cells and in oocytes are both bipolar, they can differ substantially in how they form and in microtubule organization (Fig. 2B). In animal cells, centrosome-based cytoplasmic microtubule-organizing centers (MTOCs) dominate mitotic spindle assembly and structure (Nigg & Raff, 2009). At each spindle pole, a centrosome nucleates, anchors, and organizes microtubules, accounting for spindle bipolarity. At the inner core of each centrosome is a pair of short, orthogonally oriented, and very stable microtubule-based structures called centrioles (Gönczy, 2012). One centriole of the pair is an older, mature structure—the "mother"—while the other is newly constructed—the "daughter"—and not yet fully empowered to replicate or organize a new centrosome. To form centrosomes, the centrioles recruit a pericentriolar matrix (PCM) composed of an extensive network of coiled–coil proteins and other associated proteins that expands dramatically early in mitosis, a process called centrosome maturation (Prosser & Pelletier, 2015; Woodruff et al., 2014).

The PCM includes microtubule-related structures called γ-tubulin ring complexes (γ-TURCs), which include a ring of γ-tubulin subunits that nucleate the polarized assembly of α/β-tubulin dimers into microtubules (Oakley et al., 2015). Consequently, microtubules grow with their more stable minus ends anchored in the centrosome and their more dynamic plus ends projecting outward. Because γ-TURCs are distributed throughout the PCM of each centrosome, the microtubules they nucleate project outward in all directions, forming large and radially symmetric asters. Ultimately, some microtubules from each centrosome capture and align chromosomes at the metaphase plate, and then mediate chromosome segregation to the poles during anaphase.

Importantly, centrosome duplication is tightly regulated during the cell cycle (Firat-Karalar & Stearns, 2014). After mitosis, each daughter cell receives one pair of centrioles that separate. Each centriole then produces a new, orthogonally oriented daughter centriole, restoring the seeds of bipolarity for the next mitotic cell division (Gönczy, 2012). Intriguingly, centrioles also serve as the basal bodies from which cilia and flagella grow at the surface of interphase cells. These roles—as centrosome organizers or basal bodies—appear to be mutually exclusive in all animal cells.

4. THE CURIOUS STRUCTURE OF (SOME) OOCYTE MEIOTIC SPINDLES: ACENTRIOLAR POLES

In contrast to the spindles formed during mitosis, oocyte meiotic spindles in many animals—including humans and the laboratory models of mice, frogs, fruit flies, and nematodes—entirely lack centrioles (Dumont & Desai, 2012; Howe & FitzHarris, 2013; Müller-Reichert et al., 2010; Ohkura, 2015). Nevertheless, bipolar spindles still assemble in the absence of these key organizers of the mitotic spindle. Moreover, even oocytes that retain centrioles at meiotic spindle poles typically lose them during or shortly after meiosis, and thus most animal eggs ultimately lack functional centrioles. Over a century ago, Boveri (a codiscoverer of centrioles) noted the absence of centrioles in eggs and that, other than centrioles, sperm provide little that is unique during fertilization. He therefore proposed that eliminating oocyte centrioles might prevent parthenogenetic development, with the sperm-derived centrioles restoring complete cell status to the egg. Consistent with this hypothesis, spindle assembly during the first embryonic mitosis in most animals is mediated by centrioles derived from the sperm basal body. However, both oocytes and sperm are acentriolar in mice and other rodents, and even the mitotic divisions in early mouse embryos lack centrioles (Manandhar et al., 2005; Szollosi et al., 1972). Thus, Boveri's notion of the centriole as an essential mitotic spindle organizer is clearly not absolute. Exceptional cases aside, however, centriole elimination during oogenesis or maturation ensures that cell fusion during fertilization does not increase centrosome count.

But why eliminate centrioles before meiosis? Based on the limited comparative data available, it seems likely that primitive animal oocytes retained centrioles and centrosome-based meiotic spindle pole function, complete with astral microtubules. For example, during meiosis I in sea star and sea urchin (echinoderm) oocytes, each spindle pole harbors a pair of centrioles, as in mitosis, but centriole duplication does not occur after meiosis I. Each pole in meiosis II therefore has only a single centriole (Nakashima & Kato, 2001; Sluder et al., 1989). Experimental analysis of sea star oocytes indicates that the centrioles inherited by polar bodies are fully functional and replicable, whereas the single centriole that remains in the egg after meiosis II is apparently nonreplicable, and presumably eventually degenerates (Sluder et al., 1989; Uetake et al., 2002). The two pairs of centrioles derived from the sperm then mediate the first mitotic division of the zygote. Thus, in sea stars the oocyte's complement of replication-competent centrioles is discarded into polar bodies. Indeed, the failure of polar body emission enables parthenogenetic development due to the retention of functional, reproductive centrosomes (Washitani-Nemoto et al., 1994).

Echinoderms are deuterostomes, like vertebrates, and another echinoderm group, the sea cucumbers, also have centriolar spindles (Holland, 1981). However, chordates, including the tunicates, all appear to assemble acentriolar spindles during oocyte meiosis (Sawada & Schatten, 1988). Which mode is primitive for animals?

Many invertebrate protostomes also retain centrioles through oocyte meiosis. Centrioles have either been detected using transmission electron microscopy (TEM) or have been inferred to exist because of the presence of extensive astral microtubule arrays at meiotic spindle poles. Therefore, centriolar oocyte spindles may be the ancestral condition. In the clam *Spisula solidissima*, the fertilized oocyte contains three centrosomes during prometaphase of meiosis I—one from the sperm and two from the oocyte (Wu & Palazzo, 1999). As in echinoderms, one oocyte centrosome with paired centrioles is extruded into the first polar body at the end of meiosis I. A second oocyte centrosome, with an unpaired centriole, is extruded into the second polar body at the end of meiosis II. Thus, the zygote inherits one maternal centrosome with a single, unduplicated centriole and one paternal centrosome with a duplicated centriole. While the fate of the remaining oocyte centrosome is unknown, subsequent mitotic divisions use the sperm-derived centrioles. Remarkably, the sperm centrosome appears to be repressed during oocyte meiosis: both its γ-tubulin and α-tubulin become undetectable by metaphase of meiosis I. Similar suppression of the sperm centrosome is observed in annelids and echinoderms (Stephano & Gould, 2000). *Spisula* likely typifies the entire Lophotrochozoan supergroup; although TEM has been applied to identify centrioles in only a few species in this group, the presence of large asters around meiotic spindle poles in oocytes of numerous Lophotrochozoan species implies their presence (Crowder et al., 2015b). Meanwhile, although insect and *C. elegans* oocytes feature well-studied acentriolar (and largely anastral) meiotic spindles, this is unlikely to be primitive for the Ecdysozoan supergroup as the oocyte meiotic spindles of some crustaceans have large astral, centriolar poles (Goudeau & Goudeau, 1986; Goudeau & Lachaise, 1980; Lindsay et al., 1992).

Although comparative data are limited, acentriolar spindle assembly pathways likely evolved multiple times in animals, and nearly exclusively for use in oocyte meiosis. This in turn suggests that similar selective pressures may have favored the transition to acentriolar oocyte spindles in these lineages. One adaptive advantage might have been to obviate the book-keeping involved in sorting centrioles from different sources into polar bodies versus eggs. Although this seems like a minor mechanistic burden, any distinctions that enable predictable centriole sorting could in principle be interpreted by other agents, making meiosis more susceptible to cheating (i.e., meiotic drive).

Another often cited rationale for acentriolar meiosis is that small and largely anastral meiotic spindles might minimize the size of polar bodies, the discard products of meiosis. The extreme asymmetry of the two divisions that occur during oocyte meiosis ensures that the zygote inherits almost all of the maternally synthesized gene products, biosynthetic machinery, and fuel that sustain early embryonic development. For example, in *C. elegans* oocytes, very small meiotic spindles assemble in close proximity to the cortex (Fig. 2). While the zygote itself is approximately 50×20 μm, the oocyte meiotic spindles initially are roughly 8 μm in length and by metaphase shorten to only 5 μm (Albertson & Thomson, 1993; McNally & McNally, 2005; McNally et al., 2006; Yang et al., 2003). By contrast, the first mitotic spindle is much larger (Fig. 2), with astral microtubules spanning the entire length and width of the one-cell zygote (Müller-Reichert et al., 2010). Only a small number of short astral microtubules have been detected during oocyte meiosis in *C. elegans*; these microtubules may be important for the microtubule motor-dependent translocation and rotation of the spindle prior to anaphase (Crowder et al., 2015a; Ellefson & McNally, 2009, 2011; Yang et al., 2005). Indeed, spindle positioning in *C. elegans* oocytes is microtubule-dependent but actin-independent (Yang et al., 2003). In contrast, rotation of the acentriolar and likely asterless oocyte spindles in mouse oocytes requires actin and myosin but appears to be independent of any microtubule motors (Maro et al., 1984; Wang et al., 2011). It is not clear whether acentriolar oocyte meiotic spindles in *Drosophila* have any astral microtubules (Skold et al., 2005); however, asters are detectable in *Xenopus* oocytes (Gard, 1992), although it is not known if they are involved in spindle positioning.

While small anastral oocyte meiotic spindles may facilitate the extremely asymmetric division that discards ¾ of the replicated genome into tiny polar bodies, it is clear that this is not the only means to this end. The astral, centriolar meiotic spindles of mollusks and echinoderms accomplish divisions just as asymmetric as the anastral, acentriolar spindles of rodents and nematodes. Moreover, many insects dispense with polar body extrusion altogether, and the discard products of female meiosis are abandoned to their fate in the egg cytoplasm (Foe et al., 1993).

Other rationales for the acentriolar structure of oocyte meiotic spindles warrant consideration. A recent analysis in *C. elegans* indicates that the small spindle size reduces the frequency of aneuploidy due to missegregation of chromosomes (Cortes et al., 2015). Also, centriole loss might relate to the use of centrosomes to establish cell polarity, such as the anterior–posterior body axis in *C. elegans* (Munro & Bowerman, 2009). Perhaps, the premeiotic loss of oocyte centrioles eliminates conflicting signals for polarity establishment. Although nematode sperms are amoeboid and do not use centrioles and flagella for motility, they nevertheless provide the centrioles that polarize the egg and are used to initiate mitotic cell division in the zygote (Nelson et al., 1982). Finally, other key differences between spermatogenesis and oogenesis may explain the loss of centrioles in eggs. For example, the end product of spermatogenesis in most animals is four haploid, flagellated sperm cells. The spermatocyte meiotic spindle is as much a basal body distributor as it is a chromosome sorter: just as each spindle pole captures a haploid genome, each haploid genome captures a spindle pole, the essential seed for the motor that takes it to its fate. In contrast, eggs and polar bodies do not need to swim. Perhaps, Boveri was mistaken to nominate parthenogenesis as the big risk evaded by acentriolar meiosis: instead, what if polar bodies were *too* complete? Animal life cycles depend on the zygote as the one-cell bottleneck both to limit potential conflict among genetically heterogeneous cell lineages within the individual, and to ensure that alleles comprising the genome of each newly formed individual are maximally exposed to selection (Grosberg & Strathmann, 1998). Indeed, if sea star polar bodies inherit enough cytoplasm, they can continue mitosis (Saiki & Hamaguchi, 1997). If such cells were incorporated into the embryo it would potentially violate these imperatives. The chance of this happening is remote in free-spawning invertebrates, but is perhaps higher in encapsulated, brooded, or placental embryos. Depriving polar bodies of centrioles might forestall this possibility. That this is a plausible risk is shown by the rare instances, in humans, of polar body fertilization, giving rise to a chimera or even a twin (Bieber et al., 1981).

Finally, consider the unusual way the freshwater clam *Corbicula leana*, a hermaphroditic triploid, begins a new life. Centrosomes with typical centriole pairs mediate assembly of the oocyte meiosis I spindle, but the spindle remains parallel to the plasma membrane and both poles and all maternal chromosomes are extruded simultaneously into a pair of polar bodies (Komaru et al., 2000). Thus, development in this species is androgenetic: the zygote chromosomes come entirely from the sperm. This is doubtless among the most extreme variants of meiosis, but underscores the extent to which this biological process, so central to all eukaryotic sexuality, evolves in concert with animal life histories.

The intriguing variation of meiotic mechanisms across animal phylogeny seems largely neglected, despite its obvious possible adaptive significance. Nematodes provide one avenue for exploring this remarkable example of evolutionary cell biology. Their oocytes are amenable to live imaging, and many different species have been isolated, often with fully sequenced genomes (Brauchle et al., 2009; Farhadifar et al., 2015; Phillips & Bowerman, 2015). Moreover, there is substantial cryptic genetic variation that impacts embryonic viability among wild *C. elegans* populations (Paaby et al., 2015). Perhaps, some of this genotypic variation influences oocyte spindle assembly and might shed light on the phenotypic variation observed at the family and phyla levels.

5. THE CURIOUS STRUCTURE OF OOCYTE MEIOTIC SPINDLES, PART 2: TILED MICROTUBULES

Another remarkable difference between mitotic and at least some oocyte meiotic spindles is the continuous versus discontinuous nature of the microtubules that span the distance between the poles and the spindle midzone and chromosomes (Fig. 2B). Although mitotic spindles may include some short tiled microtubules (Goshima & Kimura, 2010), long microtubules typically extend from each mitotic centrosome to the midzone and beyond (White & Glotzer, 2012). In some cases, the long microtubules form crosslinked, antiparallel bundles that stabilize the mitotic spindle and ultimately constitute the central spindle and midbody remnant, structures that promote the initiation and completion of cytokinesis, respectively (White & Glotzer, 2012). By contrast, electron tomography studies in *C. elegans* oocytes have shown that individual microtubules do not reach from the poles to the midzone or to the chromosomes (Srayko et al., 2006). Rather a series of short tiled microtubules—presumably bundled together by other proteins—span these distances (Fig. 2B). Similarly, in both *Drosophila* oocytes and *Xenopus* oocyte extracts, meiotic spindles appear to be composed of short tiled microtubules (Burbank et al., 2006; Skold et al., 2005; Yang et al., 2007). Remarkably, the plus ends of microtubules in these arrays grow both from the poles toward the midzone and from the midzone toward the poles (Liang et al., 2009; Yang et al., 2007). In sum, mitotic and oocyte meiotic spindles have remarkably distinct structures and dynamics.

Why oocyte meiotic spindles should be composed of short, discontinuous microtubules is not known, although such an arrangement could provide a traction mechanism for the poleward movement of chromosomes during anaphase as has been proposed for mitotic spindles (Goshima & Kimura, 2010). Nevertheless, it seems ironic that the relatively small oocyte meiotic spindles have such tiling, while the often much larger mitotic spindles include longer, untiled microtubules. One explanation for this curious structure has come from *C. elegans*, where the AAATPase called katanin is critical for assembling oocyte meiotic spindles (Clark-Maguire & Mains, 1994; Connolly et al., 2014; Mains et al., 1990; McNally & McNally, 2011; McNally et al., 2014; Srayko et al., 2000, 2006; Yang et al., 2003). Unlike regulators of microtubule growth that influence polymerization, katanin severs microtubules along their length (Hartman et al., 1998; McNally & Vale, 1993; Srayko et al., 2006). Because acentriolar oocyte meiotic spindles lack the γ-TURCs that nucleate microtubule assembly, microtubule severing might generate substrates for polymerization during oocyte meiosis. Indeed, electron tomography has shown that oocyte spindles in *C. elegans* mutants lacking katanin have fewer but longer microtubules than do wild-type spindles (Srayko et al., 2006). Thus severing apparently promotes microtubule assembly, and perhaps also promotes the use of tiled, discontinuous microtubules in oocyte spindles. Although katanin has been implicated in the scaling of mitotic spindle size relative to cell size in *Xenopus tropicalis* embryos (Loughlin et al., 2011), roles for katanin during oocyte meiotic division have not been found outside of *C. elegans*.

6. ACENTRIOLAR OOCYTE MEIOTIC SPINDLE ASSEMBLY: MICROTUBULE ORIGINS

The role of katanin in generating more microtubules during *C. elegans* meiosis highlights a fundamental issue. Because oocytes in some animals, including humans, lack centrioles and their associated PCM as MTOCs, understanding the origin of oocyte meiotic spindle microtubules is of fundamental importance. A PCM-independent pathway for microtubule nucleation was first discovered when DNA-coated plastic beads, added to *Xenopus* oocyte extracts, were shown to nucleate microtubules that assembled into bipolar spindles in the absence of centrosomes (Heald et al., 1996). This DNA-based microtubule nucleation activity requires the Ran GTPase (Clarke & Zhang, 2008). In the absence of Ran activity, nuclear importins bind to and negatively regulate factors that nucleate microtubule assembly. Chromatin generates a gradient of active Ran that negatively regulates the importins, releasing the bound microtubule nucleators, and thereby promoting microtubule assembly near chromosomes.

While the discovery that chromatin and Ran GTPase can promote microtubule assembly independently of centrosomes has provided insight into spindle assembly during both meiosis and mitosis, oocyte meiotic spindle assembly is at most only partially dependent on this pathway. In *Xenopus* extracts, depletion of Ran only delays oocyte meiotic spindle assembly, and meiosis I spindles assemble in the absence of Ran in mice, *C. elegans* and *Drosophila* (Askjaer et al., 2002; Bamba et al., 2002; Cesario & McKim, 2011; Dumont et al., 2007).

More recently, two additional pathways that contribute to oocyte meiotic spindle assembly have been identified. One requires the augmin complex, which acts through γ-tubulin to nucleate microtubules that branch off the lateral surfaces of existing microtubules. The second pathway requires the chromosomal passenger complex (CPC), which regulates multiple steps in mitotic cell division. The CPC pathway has been proposed to stabilize chromatin-nucleated microtubules (Tulu et al., 2006), and to be required when Ran activity is reduced or absent (Maresca et al., 2009). Although γ-tubulin associates with CPC-dependent MTOCs in mammalian cell culture (Tulu et al., 2006), it remains unclear whether the CPC pathway requires γ-tubulin (Petry & Vale, 2015).

The importance of augmin and the CPC for oocyte spindle assembly appear to differ from organism to organism. Both influence oocyte meiotic spindle assembly in *Xenopus* extracts and *Drosophila* oocytes (Colombie et al., 2008, 2013; Goshima & Kimura, 2010; Meireles et al., 2009; Petry et al., 2011; Radford et al., 2012; Sampath et al., 2004). However, while CPC components are required for oocyte meiotic spindle assembly in mouse and *C. elegans* (Dumont et al., 2010; Sharif et al., 2010), augmin does not appear to be conserved in *C. elegans* (Edzuka et al., 2014), and a role for augmin in mice has not been described.

As with augmin and the CPC, the importance of γ-tubulin for oocyte spindle assembly may vary from organism to organism. In mouse oocytes, γ-tubulin is present at early MTOC foci (Calarco, 2000; Clift & Schuh, 2015; Gueth-Hallonet et al., 1993; Ma et al., 2010; Palacios et al., 1993). While a requirement for γ-tubulin has not been described, mouse oocyte meiotic spindle assembly is severely disrupted in the absence of NEDD1, which is required for γ-tubulin recruitment to early MTOC foci (Ma et al., 2010). Similarly, the fly ortholog of Nedd1, called Dgp71WD, also is required for oocyte meiosis I spindle assembly (Reschen et al., 2012), although the defects in Dgp71WD mutants are more severe than those observed in mutants lacking γ-tubulin (Hughes et al., 2011). While defects in oocyte spindles have not been observed after RNAi-knockdown of *C. elegans* γ-tubulin alone, it is present diffusely throughout the oocyte spindle, and reducing γ-tubulin function in a *mei-1*(−)/*katanin* mutant results in more severe spindle defects with further loss of microtubule density, compared to *mei-1*(−) single mutants (McNally et al., 2006).

The role of γ-tubulin during oocyte spindle assembly clearly warrants further investigation. Although examples of γ-tubulin-independent microtubule assembly have not yet been conclusively identified (Petry & Vale, 2015), ever more examples of γ-tubulin-dependent nucleation of non-centrosomal microtubules are being identified, and minus-end stabilization of microtubules initiated at centrosomes but subsequently released to function elsewhere has also been demonstrated (Feldman & Priess, 2012; Hendershott & Vale, 2014; Keating et al., 1997; Musch, 2004; Oakley et al., 2015; Ori-McKenney et al., 2012; Petry & Vale, 2015; Tanaka et al., 2012; Wang et al., 2015; Yalgin et al., 2015). Whether these other processes contribute to oocyte meiotic spindle assembly remains poorly understood. Nevertheless, it seems likely that multiple pathways contribute to microtubule nucleation and organization during oocyte meiosis, although the relative importance of each pathway may vary from organism to organism.

Given the growing number of proteins known to promote centrosome-independent microtubule assembly during oocyte meiosis, we also need to understand where these factors act. While a gradient of active Ran GTPase can stimulate microtubule assembly around chromosomes, where and how nucleation occurs is not clear. Intriguingly, it appears that some spindle microtubules assemble at sites removed from chromosomes. Cytoplasmic microtubule assembly occurs in mouse, fly, and worm oocytes, and may also occur at the oocyte nuclear envelope in mice and at the oocyte cortex in humans, *Drosophila*, and *C. elegans* (Battaglia et al., 1996; Calarco, 2000; Ellefson & McNally, 2009; Han et al., 2015; Luksza et al., 2013; McNally & McNally, 2005; Schuh & Ellenberg, 2007; Sumiyoshi et al., 2015; Yang et al., 2003; Zou et al., 2008). While γ-tubulin and pericentrin may be important for cytoplasmic microtubule assembly in mouse oocytes, and dynein has been implicated in promoting their movement toward the assembling spindle (Luksza et al., 2013), how microtubules are nucleated at more distant sites and are then transported toward the spindle remain poorly understood and are ripe topics for further investigation.

7. ACENTRIOLAR OOCYTE MEIOTIC SPINDLE ASSEMBLY, PART 2: POLE COMPOSITION

The absence of centrosomes and their impressive microtubule-organizing activity leaves a mechanistic void in our understanding of how oocyte spindles achieve the bipolar structure required to segregate chromosomes in opposite directions. Why are there two poles instead of one, and why only two poles and not more? This intriguing problem is by no means unique to animal oocytes, as higher plants lack centrioles throughout all of their cell divisions (Mineyuki, 2007), but here we limit our discussion to animal oocytes.

While the poles that ultimately form in many animal oocytes lack centrioles, they nevertheless contain PCM components associated with mitotic centrioles. In mouse oocytes, the poles contain pericentrin, γ-tubulin, and Cep192 (Fig. 3), all PCM components in mitosis (Calarco, 2000; Clift & Schuh, 2015; Gueth-Hallonet et al., 1993; Ma et al., 2010; Palacios et al., 1993). In *C. elegans*, the bipolar oocyte spindle poles that ultimately form also contain mitotic PCM proteins, although their dynamics over time in many cases remain poorly understood and some of them also are present elsewhere in the spindle. These include the calponin homology protein ASPM-1, the microtubule severing complex katanin, the kinesin-12 and -13 family members KLP-18 and KLP-7, the minus-end directed microtubule motor dynein, and the NUMA homolog LIN-5 (Connolly et al., 2014,

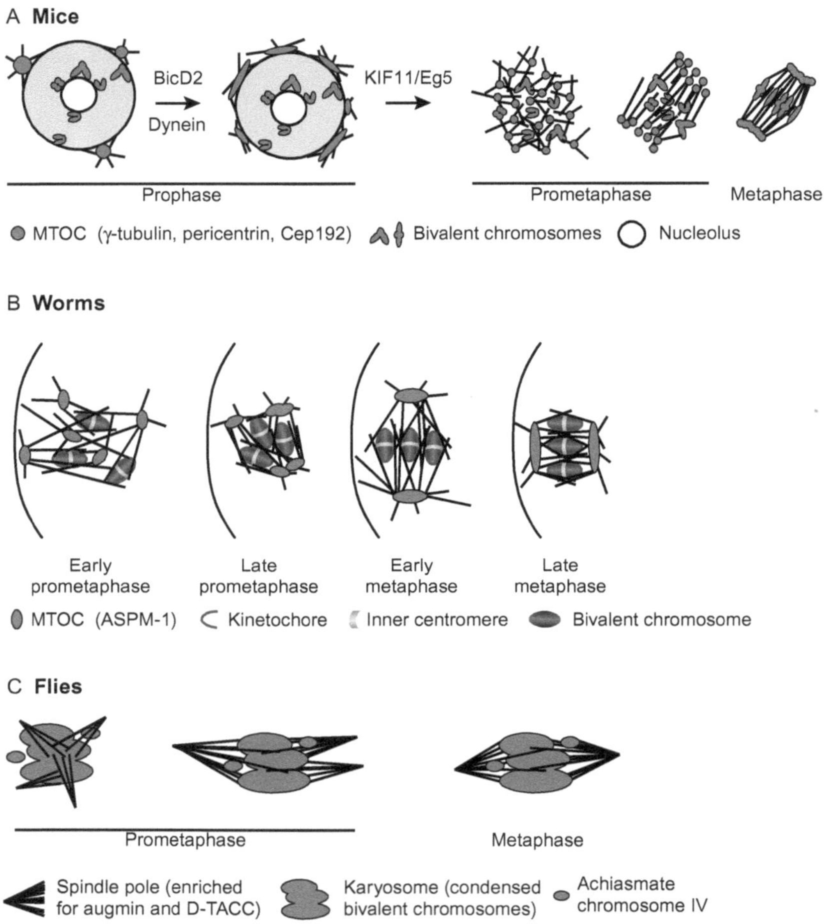

Figure 3 Assembly of meiotic spindles in mouse, worm, and fly oocytes. Live imaging of microtubules and chromosomes during oocyte meiosis in these three model organisms has revealed similarities and differences in spindle assembly. When spindle assembly begins, homolog pairs have undergone crossover recombination to form bivalents, with the exception of *Drosophila* chromosome 4, which does not recombine. (A) In mouse oocytes, small microtubule-organizing centers (MTOCs) form near the nuclear envelope prior to NEB. Subsequently, MTOCs are stretched into thin ribbons, often becoming fragmented. This stretching process requires dynein and BicD2, which anchors dynein at the nuclear envelope. After NEB, the Eg5 homolog KIF11 promotes further MTOC fragmentation, yielding an average of approximately 26 MTOCs per oocyte. These MTOCs then coalesce as bivalents congress, forming a bipolar spindle by metaphase. (B) At NEB in *C. elegans* oocytes, most bivalents are located near the nuclear envelope. Following NEB, a cloud of short microtubules assembles around the bivalents, which subsequently cluster together. Concurrently, several foci of the microtubule-scaffolding protein ASPM-1 appear, presumably present at small MTOCs. The ASPM-1 foci coalesce into two poles as

2015; Ellefson & McNally, 2011; Gomes et al., 2013; McNally et al., 2006, 2014; O'Rourke et al., 2007; van der Voet et al., 2009; Wignall & Villeneuve, 2009). Thus far, γ-tubulin has not been clearly detected at poles but appears to be present diffusely throughout the oocyte spindle (McNally et al., 2006). While less is known about pole composition in *Drosophila*, augmin subunits and the conserved microtubule-associated proteins (MAPs) minispindles and D-TACC have been detected at mature poles (Colombie et al., 2013; Cullen & Ohkura, 2001). *Drosophila* γ-tubulin appears to be somewhat enriched at oocyte spindle poles but also colocalizes with microtubules throughout the prometaphase meiosis I spindle (Endow & Hallen, 2011; Hughes et al., 2011).

8. ACENTRIOLAR OOCYTE MEIOTIC SPINDLE ASSEMBLY, PART 3: POLE ASSEMBLY

A landmark advance in our understanding of oocyte meiotic spindle dynamics came from live cell imaging of pole assembly in mouse oocytes, which showed that multiple small pole foci coalesce to form a bipolar spindle structure (Schuh & Ellenberg, 2007). While earlier studies of fixed oocytes first documented this phenomenon (Carabatsos et al., 2000; Gueth-Hallonet et al., 1993; Palacios et al., 1993), Schuh and Ellenberg used live cell imaging to show that many small MTOCs are initially dispersed throughout the area surrounding the mouse oocyte chromosomes early in meiosis I. Over time these small foci coalesce to form a bipolar structure.

More recently, live imaging of Cep192 fusion to GFP has shown that early pole coalescence in mouse oocytes involves an early dispersal of fewer and larger MTOC foci into more and smaller foci, followed by a still mysterious coalescence (Clift & Schuh, 2015). An early phase of dispersal, prior to nuclear envelope breakdown, requires the polo kinase PLK1 and the minus-end directed microtubule motor dynein, in a process that stretches the early MTOCs out into ribbon-like structures that often fragment. After nuclear envelope breakdown, the late phase further fragments and disperses

bivalents congress onto the metaphase plate. Initially, the bipolar metaphase spindle is oriented parallel to the oocyte cortex, but then rotates to be perpendicular to the cortex. During rotation, the MTOCs widen as the spindle shortens dramatically. (C) In fly oocytes, the recombined bivalents form a compact structure called a karyosome. After NEB, microtubules accumulate around the karyosome and several dynamic, pole-like structures form. These microtubule foci coalesce to form the poles of the bipolar meiotic spindle. Augmin, γ-tubulin, and D-TACC are enriched at the poles. (A color version of this figure is located at the end of this chapter.)

the MTOCs. This latter phase requires the tetrameric kinesin-5 family member KIF11, which presumably promotes fragmentation through its antiparallel microtubule-sliding activity, much as it promotes oocyte meiotic spindle bipolarity in *X. laevis* and mitotic spindle bipolarity in mammalian cell culture (Sawin et al., 1992). When KIF11-dependent dispersal is disrupted, transient monopolar oocyte spindles assemble, although they eventually recover and appear to segregate chromosomes normally (Clift & Schuh, 2015).

Similar processes of pole coalescence have been described during invertebrate oocyte spindle assembly. Time-lapse imaging of a kinesin-GFP fusion that labeled microtubules in live *Drosophila* oocytes suggested that small MTOC foci, initially distributed throughout the oocyte nucleoplasm, coalesce during meiotic spindle assembly (Skold et al., 2005). Microtubule attachment to chromosomes and the crosslinking of such attached microtubules may further promote *Drosophila* oocyte spindle assembly. More recently in *C. elegans*, a GFP fusion to the pole marker ASPM-1 was used for live cell imaging of oocyte meiosis I and II spindle assembly. Multiple small ASPM-1 foci coalesced over time to form a bipolar structure, resembling MTOC dynamics in mouse oocytes (Connolly et al., 2015). While the mechanisms that promote the coalescence of early small pole foci in *C. elegans* also remain unknown, the *C. elegans* kinesin-12 family member KLP-18 promotes spindle bipolarity (Connolly et al., 2014; Segbert et al., 2003; Wignall & Villeneuve, 2009), presumably by promoting the sliding of antiparallel microtubules, much as has been documented for its mammalian orthologs (Tanenbaum et al., 2009; Vanneste et al., 2009). Additionally, the microtubule severing activity of MEI-1/katanin and ASPM-1 itself appear to have roles in pole assembly (Connolly et al., 2014; McNally et al., 2014). Mouse ASPM-1 also is required for oocyte meiotic spindle assembly and is localized to the spindle poles (Xu et al., 2012). Thus, at least some molecular mechanisms of acentriolar pole assembly appear conserved between invertebrates and vertebrates.

Although mice, frogs, nematodes, and fruit flies all appear to assemble oocyte meiotic spindles through the coalescence of multiple early foci, the details of this process vary in each system. In mice, the small MTOC foci initially were shown to coalesce into a ball-like structure, with the MTOCs clustered internally and the chromosomes distributed on the surface (Schuh & Ellenberg, 2007). Subsequently, some MTOCs were ejected peripherally and then coalesced into a bipolar organization, with chromosomes ultimately congressing to form the metaphase plate. However, the more recent study using Cep192 as a marker for MTOC pole foci failed to detect a ball-like structure with MTOCs surrounded by chromosomes; rather the MTOCs and chromosomes were comingled (Clift & Schuh, 2015). In both studies, the subsequent dispersal of MTOCs was shown to require the kinesin-5 family member KIF11 (Fig. 3A) (Clift & Schuh, 2015; Schuh & Ellenberg, 2007). How the multiple, dispersed MTOCs coalesce to form two poles remains unknown.

In *Xenopus* extracts, discrete, small microtubule foci have not been reported as pole intermediates during the assembly of oocyte meiotic spindles. However, acentrosomal bipolar spindle organization does require a kinesin-5 family member (Eg5), as in mouse. Pole assembly also is thought to involve crosslinking of microtubule minus ends mediated by dynein and the coiled–coil protein NUMA (Heald et al., 1996; Merdes et al., 2000; Mitchison et al., 2005).

In *C. elegans*, live-imaging studies indicate that chromosomes are initially dispersed throughout a mass of microtubules, and the small ASPM-1 pole foci formed early in meiosis coalesce to form a bipolar structure (Fig. 3B) (Connolly et al., 2015). Unlike in mice and frogs, kinesin-5 (*C. elegans* BMK-1) is not required for bipolar spindle assembly during meiosis or mitosis (Saunders et al., 2007). Moreover, while the NUMA ortholog LIN-5, together with dynein, is required to position and orient the *C. elegans* oocyte meiotic spindle (Crowder et al., 2015a; Ellefson & McNally, 2009, 2011; van der Voet et al., 2009), roles for either protein in pole assembly have not been documented.

In *Drosophila*, the dynamics of pole coalescence remain less well characterized, but MAPs are clearly involved. Mutational inactivation of the conserved MAP minispindles results in assembly of tripolar meiosis I spindles (Cullen & Ohkura, 2001). Minispindles localization to spindle poles depends on another MAP, called D-TACC. A complex of minispindles and D-TACC, transported to microtubule minus ends by the kinesin Ncd, has been proposed to stabilize microtubule minus ends to promote pole assembly and spindle bipolarity (Cullen & Ohkura, 2001; Matthies et al., 1996). The *C. elegans* D-TACC ortholog TAC-1 acts in a complex with ZYG-9, an XMAP215 ortholog, to promote microtubule stability during mitosis (Bellanger & Gonczy, 2003), and although ZYG-9 is required for meiotic spindle assembly (Yang et al., 2003), how ZYG-9 influences oocyte spindle assembly is not well understood and the role of TAC-1 has not been addressed. Finally, *Drosophila* Subito, a member of the kinesin-6 family that crosslinks antiparallel microtubules, is not required for oocyte spindle assembly. However, *subito* mutants display transient instability of spindle poles, suggesting that Subito-mediated central spindle integrity may be important for promoting spindle bipolarity (Colombie et al., 2013; Jang et al., 2007).

In sum, the molecular pathways that promote pole coalescence and spindle bipolarity appear to partially overlap, but significant differences are also apparent across animal phylogeny. While some of the variation likely reflects multiple, independently evolved appearances of acentriolar spindle assembly, systematic comparisons of the different gene requirements in each model system are needed to better assess the conservation, and divergence of oocyte spindle assembly mechanisms.

9. KINETOCHORE FUNCTION AND POLE COALESCENCE

While an early coalescence of MTOC foci has been implicated in mouse, nematode, and fly oocyte meiotic spindle assembly, the mechanism of MTOC coalescence remains unknown. However, a recent analysis of ASPM-1 dynamics during oocyte meiotic spindle assembly in *C. elegans* suggests that proper microtubule–kinetochore attachments contribute to the coalescence of early pole foci (Fig. 4A) (Connolly et al., 2015). In *C. elegans* mutants that lack the microtubule depolymerizing kinesin KLP-7, called MCAK in vertebrates, ASPM-1 foci fail to coalesce into a bipolar structure and instead often form tripolar or even tetrapolar structures. Both ASPM-1 and MEI-1/katanin mark the poles of *klp*-7(−) mutants, indicating that the supernumerary poles are molecularly similar to those of wild-type bipolar spindles, although these foci have not yet been shown to localize with MTOCs. The extra poles also are functional, as chromosomes often segregate into three discrete masses during anaphase. Thus, KLP-7 is required for the coalescence of early pole foci in *C. elegans* oocytes.

In vertebrates, MCAK regulates the microtubule–kinetochore attachments that mediate mitotic chromosome segregation during anaphase (Ems-McClung & Walczak, 2010; Kline-Smith et al., 2004). In the absence of MCAK in human cell culture lines, improper syntelic (kinetochores of both sister chromatids attached to the same pole) and merotelic (kinetochore of one sister chromatid attached to both poles) microtubule–kinetochore attachments are observed. In wild-type cells, MCAK associates with kinetochores prior to anaphase, and its depolymerase activity may destabilize inappropriate microtubule–kinetochore attachments. In *klp*-7(−)/*MCAK* mutant *C. elegans* oocytes, the persistence of such improper attachments

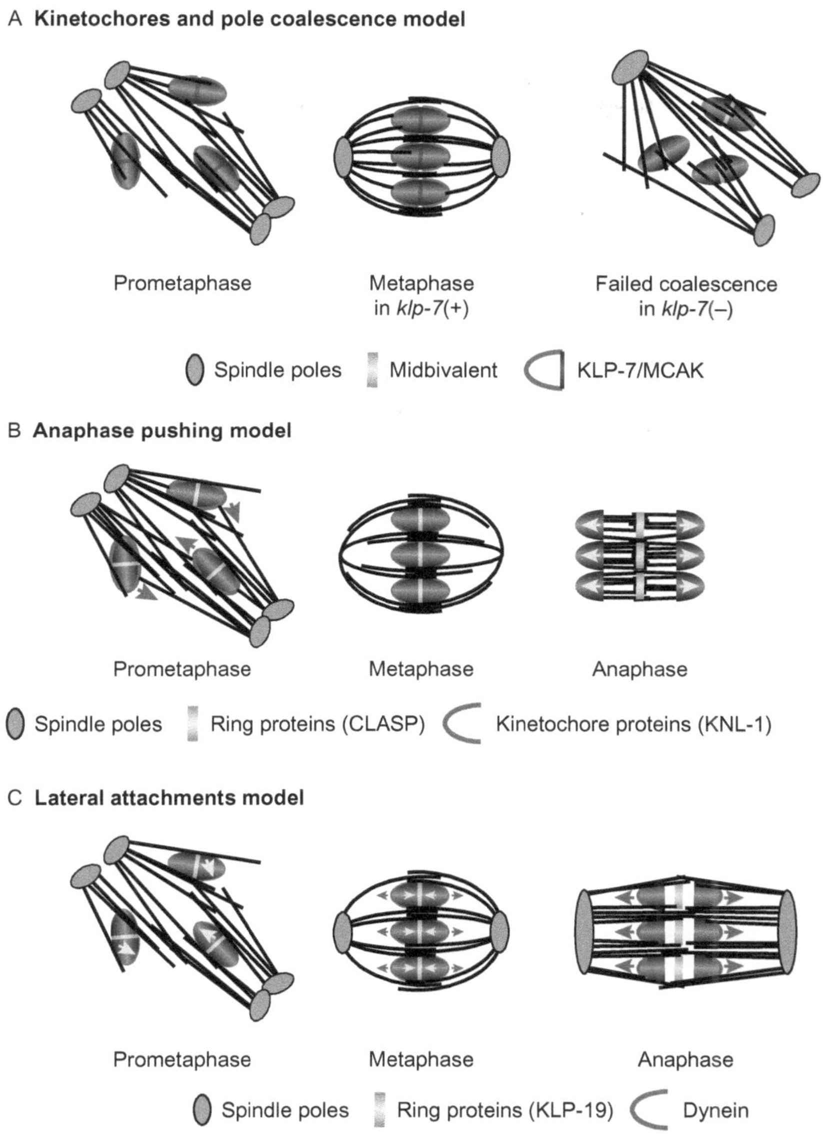

Figure 4 Models for bipolar spindle assembly and chromosome congression and segregation in *C. elegans*. Three models are shown that address pole assembly (A) and chromosome congression and segregation (B and C) in *C. elegans* oocytes. (A) Proper microtubule–kinetochore attachments permit the coalescence of early pole foci to form a bipolar oocyte spindle. In this model, bivalent alignment and spindle bipolarity require that the two homologs of each bivalent attach specifically to microtubules from opposite spindle poles. Incorrect attachments (see text) are eliminated by the microtubule *(Continued)*

during oocyte meiosis might lead to abnormal tension within the assembling spindle, and this imbalance in forces has been proposed to interfere with the coalescence of ASPM-1 foci into a bipolar structure (Connolly et al., 2015). Consistent with such a model, partial knockdown of components of the Ndc80 complex that mediates microtubule–kinetochore attachment rescues spindle bipolarity in *klp-7*(−) mutants.

While the importance of microtubule–kinetochore attachment for oocyte spindle assembly in other species remains largely unknown, roughly 90% of the chromosomes in mouse oocytes transiently form improper syntelic or merotelic attachments (Kitajima et al., 2011). Moreover, knockdown of the Ndc80 complex in both mouse and nematode oocytes results in spindle assembly defects (Dumont et al., 2010; Sun et al., 2010, 2011), suggesting that a role for kinetochores is conserved. Finally, expression of a dominant negative allele of *Drosophila* KLP10A, a kinesin-13/MCAK family member, results in disorganized or extra oocyte spindles poles (Zou et al., 2008), suggesting that a role for kinesin-13/MCAK in limiting pole number also may be conserved. It will be interesting to test whether MCAK-mediated removal of improper microtubule–kinetochore attachments is important for oocyte spindle pole coalescence in other organisms.

Figure 4—Cont'd depolymerase KLP-7/MCAK. Failure to disrupt these incorrect attachments in mutants with reduced KLP-7 function results in an imbalance in spindle tension that interferes with the coalescence of early pole foci and thus results in the assembly of multipolar spindles. (B) Microtubule–kinetochore interactions orient and align bivalents on the metaphase spindle, but at the metaphase-to-anaphase transition, the polar microtubule arrays disassemble, and a new population of microtubules forms between homologs and pushes them apart independently of kinetochores. The assembly of these microtubules is mediated by CLASP and other factors present in ring-shaped structures between each homolog pair at metaphase. When homologs separate, the rings are left behind at the metaphase plate. (C) Lateral interactions between microtubules and microtubule motor proteins mediate both chromosome congression to the metaphase plate and anaphase segregation to the poles. The chromokinesin KLP-19, localized to midbivalent rings, interacts with spindle microtubules to produce a polar ejection force that aligns bivalents at the metaphase plate. Subsequently, during anaphase, polar microtubules interact with dynein, a minus-end directed microtubule motor that accumulates at increasing levels on the poleward/lateral regions of bivalents as meiosis progresses. Dynein-directed motility mediates the poleward movement of chromosomes during anaphase. (A color version of this figure is located at the end of this chapter.)

10. A PUSHY VIEW OF KINETOCHORE FUNCTION AND CHROMOSOME DYNAMICS DURING OOCYTE MEIOSIS

The influence of microtubule–kinetochore attachment on oocyte spindle pole coalescence in *C. elegans* brings us to an intriguing controversy regarding the importance of kinetochores during oocyte meiosis (Fig. 4B and C). An important early observation was that depleting either components of the Ndc80 complex, responsible for the microtubule attachment activity of the kinetochore, or core kinetochore components such as KNL-1, had remarkably minor effects on meiotic spindle assembly and chromosome organization (Dumont et al., 2010). In live-imaging studies of microtubule and chromosome dynamics, using fusions of GFP and mCherry to β-tubulin and a histone, respectively, kinetochore disruption was shown to cause only minor perturbations in spindle morphology, chromosome congression to the metaphase plate, and anaphase chromosome segregation to the poles. During meiosis I in wild-type *C. elegans* oocytes, spindles shorten substantially prior to anaphase, and during anaphase the poles rapidly disassemble and most spindle microtubules are detected between the separating chromosomes (Fig. 4B). Moreover, knockdown of factors required for the assembly or stability of these anaphase microtubules, such as the microtubule-stabilizing protein CLS-2/CLASP, resulted in both a substantial loss of the interchromosomal microtubules and severe defects in anaphase chromosome movements (Dumont et al., 2010). These findings led Dumont and Desai to propose a model in which microtubule polymerization between the segregating chromosomes pushes the chromosomes apart during anaphase of meiosis I and II, with little if any requirement for microtubule–kinetochore interactions and poleward pulling forces during anaphase (Fig. 4B).

Whether the apparent lack of a substantial role for kinetochores during anaphase in *C. elegans* oocytes is relevant to other species is not known. However, TEM analysis of fixed mouse oocytes during meiosis I and II clearly indicate that microtubules attach to kinetochores, and the kinetochore regions lead the anaphase movements of meiotic chromosomes toward the poles (Brunet et al., 1999). Microtubule–kinetochore attachments have also been observed in human oocytes, although the correction of improper syntelic and merotelic attachments appears to be remarkably inefficient (Holubcova et al., 2015). Furthermore, knockdowns of Ndc80 complex components in mouse oocytes cause substantial defects in spindle organization and chromosome segregation (Sun et al., 2010, 2011), and perturbations of mouse meiotic spindle assembly lead to activation of the kinetochore-based spindle assembly checkpoint (Ma et al., 2010; McGuinness et al., 2009). Nevertheless, very few microtubule–kinetochore attachments form prior to metaphase of meiosis I in mouse oocytes and microtubule–kinetochore attachments may not be required for chromosome congression to the metaphase plate (Brunet et al., 1999). Although mutational inactivation of γ-tubulin disrupts microtubule–kinetochore attachments in *Drosophila* oocytes, the role of these attachments in chromosome segregation is unknown (Hughes et al., 2011). Systematic investigations of how microtubule–kinetochore attachments, and kinetochore function more generally, influence spindle assembly and chromosome movement are needed to fully assess and compare the role of these structures during oocyte meiotic cell division in different animal species.

11. AN ALTERNATIVE MODEL FOR CHROMOSOME MOVEMENTS IN *C. ELEGANS* OOCYTES: SIDES MATTER

While kinetochore-independent microtubule polymerization has been proposed to push chromosomes apart during meiotic anaphase in *C. elegans* oocytes (Dumont et al., 2010), two other studies have suggested a very different model for chromosome congression and segregation (Muscat et al., 2015; Wignall & Villeneuve, 2009). In this model (Fig. 4C), kinetochores are proposed to have little if any role during either congression or segregation. Instead, lateral attachments of polar microtubules to chromosomes, and microtubule motor-mediated pushing and pulling forces, were proposed to move chromosomes during both congression to the metaphase plate and anaphase poleward segregation, respectively.

The first of these two studies demonstrated that the chromokinesin KLP-19 localizes to rings that encircle the midregions of each bivalent, that this midbivalent accumulation of KLP-19 requires the CPC, and that KLP-19 is important for chromosome congression to the metaphase plate (Wignall & Villeneuve, 2009). Knockdown of KLP-19 had relatively modest effects on chromosome orientation relative to the spindle poles, but congression of chromosomes to the metaphase plate was often defective. Moreover, when KLP-19 was knocked down in mutants that make monopolar spindles, chromosomes remained clustered around the monopole. In contrast, chromosomes were usually found close to microtubule ends, far from the monopole, in oocytes with wild-type levels of KLP-19. Wignall and Villeneuve therefore proposed that lateral associations of spindle microtubules with meiotic chromosomes, together with KLP-19 motor activity, creates a polar ejection force that moves chromosomes toward the spindle midzone. Upon reaching the midzone, chromosomes encounter overlapping, antiparallel microtubules, and a balance of forces result in no net movement toward either pole, aligning chromosomes at the metaphase plate. Polar ejection forces have also been proposed to move chromosomes toward the metaphase plate in *Drosophila* oocytes, in the context of augmin-mediated microtubule polymerization, although whether these forces involve lateral microtubule attachments or end-on attachment at kinetochores was not addressed (Colombie et al., 2013).

In a more recent study, the KLP-19 rings in *C. elegans* oocytes were shown to detach from bivalents at the metaphase-to-anaphase transition, and this detachment was proposed to promote a dynein-dependent reversal in the direction of chromosome movement during anaphase (Muscat et al., 2015). Because dynein has multiple roles in meiotic spindle assembly, chemical inhibitors and conditional alleles were used to disrupt dynein function during chromosome segregation while minimizing effects on spindle assembly. Although the conditions used likely only partially impair dynein function, lagging chromosomes were observed in most oocytes. Dynein also was shown to be required for the poleward chromosome movement in mutants with monopolar spindles. Because the majority of contacts between chromosomes and microtubules appear to occur at the sides of chromosomes rather than at the poles, Muscat et al. proposed that lateral attachments account not only for congression of chromosomes to the metaphase plate, but also for dynein-mediated poleward movement during anaphase (Fig. 4C). In this model, cell cycle-dependent progression in KLP-19 location and possibly in a transition from plus-end directed motor activity to minus-end directed motor activity, accounts for the reversal in the direction of movement at the metaphase-to-anaphase transition.

The two competing models for how chromosomes move to the poles during anaphase in *C. elegans* oocytes (Fig. 4B and C) both have their strengths and weaknesses. In support of the Wignall and Villeneuve model that lateral attachments mediate chromosome congression and anaphase segregation, most of the oocyte spindle microtubules in *C. elegans* oocytes indeed pass by the chromosomes laterally, with very few microtubules appearing to terminate at the poleward surfaces. KLP-19 depletion disrupted chromosome congression in bipolar spindles and outward movement of chromosomes on monopolar spindles, and reducing dynein function had the opposite effect—lagging chromosomes were observed during anaphase in bipolar spindles and chromosomes failed to move centripetally in monopolar spindles. However, it is important to note that KLP-19 accumulates on both the lateral and poleward faces of meiotic bivalents and on meiotic spindle microtubules (Powers et al., 2004). Similarly, dynein is broadly distributed throughout the meiotic spindle apparatus (Crowder et al., 2015a; Ellefson & McNally, 2009, 2011; van der Voet et al., 2009). In spite of the care taken to disrupt dynein function at the time of chromosome segregation, the methods used likely targeted all dynein, leaving open the possibility that the lagging chromosomes observed resulted from spindle defects rather than from a failure to move chromosomes along lateral microtubules. Finally, *C. elegans* chromosomes are holocentric, with centromeres and kinetochores dispersed throughout the chromosomes (Cheeseman et al., 2004; Dumont et al., 2010; Hagstrom et al., 2002; Howe et al., 2001). Thus even microtubules that pass laterally by microtubules might make contact with kinetochores.

In support of the Dumont and Desai end-on pushing model, all studies of microtubule dynamics during oocyte meiosis in *C. elegans* agree that the polar microtubules observed during prometaphase and early metaphase almost completely disappear during chromosome segregation in anaphase I, while the density of interchromosomal microtubules greatly increases. Moreover, knockdown of CLS-2 and several other factors (HTP-1/2 and BUB-1) clearly disrupted both the assembly of the interchromosomal microtubule network and the poleward movement of homologs (Dumont et al., 2010). However, CLS-2, HTP-1/2, and BUB-1 all associate with lateral and poleward faces of meiotic bivalents in prometaphase and metaphase, as described above for KLP-19, and their depletion results in severe defects in spindle assembly and chromosome congression and/or orientation on the metaphase plate (Dumont et al., 2010). Thus, it remains possible that the defects in anaphase movement are a consequence of the failure to align and orient bivalents within an organized network of microtubules such that they can form productive connections that allow their segregation toward opposite poles, rather than a failure to form a robust interchromosomal microtubule network that pushes chromosomes toward the poles.

It seems likely that a mix of multiple microtubule–microtubule and microtubule–chromosome interactions collaborate and compete with one another to establish the balance of forces that ultimately creates a bipolar spindle, aligns chromosomes at the middle of this spindle, and moves homologs toward opposite spindle poles during anaphase of meiosis I. Elimination of any one mechanism may compromise some but not all movements. Indeed, a recent publication suggests that both lateral and end-on microtubule attachments contribute to chromosome positioning and movement during Drosophila oocyte meiosis I (Radford et al., 2015). More extensive genetic studies that simultaneously reduce the function of different combinations of these alternative pathways may clarify how they are integrated to execute meiotic cell division. But ultimately, a thorough understanding of the pleiotropic functions of these proteins will likely require techniques that selectively inactivate individual proteins in specific subcellular regions at specific times. Given recent advances in optogenetics, such approaches may soon become possible.

12. ADVANCING OUR UNDERSTANDING OF OOCYTE MEIOTIC SPINDLE ASSEMBLY: GENETICS AND LIVE CELL IMAGING

Compared to mitosis, our understanding of oocyte meiotic cell division lags considerably. Classical genetic screens in both *C. elegans* and *Drosophila* have revealed important players in this fundamental process, and a recent RNA interference screen in mouse oocytes promises to provide further insight (Pfender et al., 2015). While genetic screens and live cell imaging thus far indicate that there are common features among organisms, the molecular pathways that mediate meiotic spindle assembly appear to be numerous and diverse. To advance our understanding, more systematic genetic analyses are needed.

But genetic analysis has its limits. For example, many of the genes required for meiotic spindle assembly have earlier essential requirements. Null alleles of genes required for meiosis are often lethal because of mitotic requirements, precluding definitive genetic investigation of gene requirements during oocyte meiosis. While RNA interference, temperature sensitive mutations, and chemical inhibitors can bypass earlier requirements, all are inherently compromised by uncertainty as to how completely or specifically they eliminate gene functions.

The challenges to achieving a conclusive mechanistic understanding of meiotic spindle assembly and function in any one system extend well beyond a need for further genetic studies. Meiotic spindles are often small and in some cases relatively inaccessible to live cell imaging, although recent advances in light microscopy and the application of electron tomography will provide new insight. In addition, the further application of live cell-imaging techniques, with genome editing technology to generate endogenous fluorescent protein fusions, holds much promise. Indeed, a systematic labeling of all meiotic spindle components with fluorescent protein markers, coupled with simultaneous live cell imaging of two or more proteins, is very much needed to provide a foundation of wild-type spindle assembly for comparison to mutant phenotypes. The pace of progress in this field has picked up considerably in the past five years, and recent advances in genetics and microscopy make rapid and significant further progress imminent.

ACKNOWLEDGMENTS

The authors thank Diana Libuda, Kim McKim, Frank McNally, Sarah Radford, Aleesa Schlientz, Asako Sugimoto, and Kenji Sugioka for helpful discussions and insights. GvD was supported by grants from the National Institutes of Health (GM052932) and the National Science Foundation (MCB-0917887). BB was supported by a grant from the National Institutes of Health (GM049869).

REFERENCES

Albertson, D. G., & Thomson, J. N. (1993). Segregation of holocentric chromosomes at meiosis in the nematode, *Caenorhabditis elegans*. *Chromosome Research*, *1*, 15–26.

Askjaer, P., et al. (2002). Ran GTPase cycle and importins alpha and beta are essential for spindle formation and nuclear envelope assembly in living *Caenorhabditis elegans* embryos. *Molecular Biology of the Cell*, *13*, 4355–4370.

Bamba, C., et al. (2002). The GTPase Ran regulates chromosome positioning and nuclear envelope assembly in vivo. *Current Biology*, *12*, 503–507.

Battaglia, D. E., et al. (1996). Changes in centrosomal domains during meiotic maturation in the human oocyte. *Molecular Human Reproduction*, *2*, 845–851.

Bellanger, J. M., & Gonczy, P. (2003). TAC-1 and ZYG-9 form a complex that promotes microtubule assembly in *C. elegans* embryos. *Current Biology*, *13*, 1488–1498.

Bieber, F. R., et al. (1981). Genetic studies of an acardiac monster: Evidence of polar body twinning in man. *Science*, *213*, 775–777.

Brauchle, M., et al. (2009). Evolution of early embryogenesis in rhabditid nematodes. *Developmental Biology*, *335*, 253–262.

Brunet, S., et al. (1999). Kinetochore fibers are not involved in the formation of the first meiotic spindle in mouse oocytes, but control the exit from the first meiotic M phase. *The Journal of Cell Biology*, *146*, 1–12.

Burbank, K. S., et al. (2006). A new method reveals microtubule minus ends throughout the meiotic spindle. *The Journal of Cell Biology*, *175*, 369–375.

Calarco, P. G. (2000). Centrosome precursors in the acentriolar mouse oocyte. *Microscopy Research and Technique*, *49*, 428–434.

Carabatsos, M. J., et al. (2000). Sorting and reorganization of centrosomes during oocyte maturation in the mouse. *Microscopy Research and Technique*, *49*, 435–444.

Cesario, J., & McKim, K. S. (2011). RanGTP is required for meiotic spindle organization and the initiation of embryonic development in *Drosophila*. *Journal of Cell Science*, *124*, 3797–3810.

Cheeseman, I. M., et al. (2004). A conserved protein network controls assembly of the outer kinetochore and its ability to sustain tension. *Genes & Development*, *18*, 2255–2268.

Clarke, P. R., & Zhang, C. (2008). Spatial and temporal coordination of mitosis by Ran GTPase. *Nature Reviews. Molecular Cell Biology, 9*, 464–477.
Clark-Maguire, S., & Mains, P. E. (1994). *mei-1*, a gene required for meiotic spindle formation in *Caenorhabditis elegans*, is a member of a family of ATPases. *Genetics, 136*, 533–546.
Clift, D., & Schuh, M. (2015). A three-step MTOC fragmentation mechanism facilitates bipolar spindle assembly in mouse oocytes. *Nature Communications, 6*, 7217.
Colombie, N., et al. (2008). Dual roles of Incenp crucial to the assembly of the acentrosomal metaphase spindle in female meiosis. *Development, 135*, 3239–3246.
Colombie, N., et al. (2013). Meiosis-specific stable binding of augmin to acentrosomal spindle poles promotes biased microtubule assembly in oocytes. *PLoS Genetics, 9*, e1003562.
Connolly, A. A., et al. (2014). *Caenorhabditis elegans* oocyte meiotic spindle pole assembly requires microtubule severing and the calponin homology domain protein ASPM-1. *Molecular Biology of the Cell, 25*, 1298–1311.
Connolly, A. A., et al. (2015). KLP-7 acts through the Ndc80 complex to limit pole number in *C. elegans* oocyte meiotic spindle assembly. *The Journal of Cell Biology, 210*, 917–932.
Cortes, D. B., et al. (2015). The asymmetry of female meiosis reduces the frequency of inheritance of unpaired chromosomes. *Elife, 4*, e06056.
Crowder, M. E., et al. (2015a). Dynactin-dependent cortical dynein and spherical spindle shape correlate temporally with meiotic spindle rotation in *Caenorhabditis elegans*. *Molecular Biology of the Cell, 26*, 3030–3046.
Crowder, M. E., et al. (2015b). A comparative analysis of spindle morphometrics across metazoans. *Current Biology, 25*, 1542–1550.
Cullen, C. F., & Ohkura, H. (2001). Msps protein is localized to acentrosomal poles to ensure bipolarity of *Drosophila* meiotic spindles. *Nature Cell Biology, 3*, 637–642.
Dumont, J., & Desai, A. (2012). Acentrosomal spindle assembly and chromosome segregation during oocyte meiosis. *Trends in Cell Biology, 22*, 241–249.
Dumont, J., et al. (2007). A centriole- and RanGTP-independent spindle assembly pathway in meiosis I of vertebrate oocytes. *The Journal of Cell Biology, 176*, 295–305.
Dumont, J., et al. (2010). A kinetochore-independent mechanism drives anaphase chromosome separation during acentrosomal meiosis. *Nature Cell Biology, 12*, 894–901.
Edzuka, T., et al. (2014). Identification of the augmin complex in the filamentous fungus *Aspergillus nidulans*. *PLoS One, 9*, e101471.
Ellefson, M. L., & McNally, F. J. (2009). Kinesin-1 and cytoplasmic dynein act sequentially to move the meiotic spindle to the oocyte cortex in *Caenorhabditis elegans*. *Molecular Biology of the Cell, 20*, 2722–2730.
Ellefson, M. L., & McNally, F. J. (2011). CDK-1 inhibits meiotic spindle shortening and dynein-dependent spindle rotation in *C. elegans*. *The Journal of Cell Biology, 193*, 1229–1244.
Ems-McClung, S. C., & Walczak, C. E. (2010). Kinesin-13s in mitosis: Key players in the spatial and temporal organization of spindle microtubules. *Seminars in Cell & Developmental Biology, 21*, 276–282.
Endow, S. A., & Hallen, M. A. (2011). Anastral spindle assembly and gamma-tubulin in *Drosophila* oocytes. *BMC Cell Biology, 12*, 1.
Farhadifar, R., et al. (2015). Scaling, selection, and evolutionary dynamics of the mitotic spindle. *Current Biology, 25*, 732–740.
Feldman, J. L., & Priess, J. R. (2012). A role for the centrosome and PAR-3 in the hand-off of MTOC function during epithelial polarization. *Current Biology, 22*, 575–582.
Firat-Karalar, E. N., & Stearns, T. (2014). The centriole duplication cycle. *Philosophical Transactions of the Royal Society of London. Series B, Biological Sciences, 369*, 20130460.
Foe, V. E., et al. (1993). Mitosis and morphogenesis in the *Drosophila* embryo: Point and counterpoint. *The Development of* Drosophila melanogaster, *1*, 149–300.

Gard, D. L. (1992). Microtubule organization during maturation of *Xenopus* oocytes: Assembly and rotation of the meiotic spindles. *Developmental Biology, 151*, 516–530.

Gomes, J. E., et al. (2013). Microtubule severing by the katanin complex is activated by PPFR-1-dependent MEI-1 dephosphorylation. *The Journal of Cell Biology, 202*, 431–439.

Gönczy, P. (2012). Towards a molecular architecture of centriole assembly. *Nature Reviews. Molecular Cell Biology, 13*, 425–435.

Goshima, G., & Kimura, A. (2010). New look inside the spindle: Microtubule-dependent microtubule generation within the spindle. *Current Opinion in Cell Biology, 22*, 44–49.

Goudeau, M., & Goudeau, H. (1986). The resumption of meiotic maturation of the oocyte of the prawn *Palaemon serratus* is regulated by an increase in extracellular Mg2+ during spawning. *Developmental Biology, 118*, 361–370.

Goudeau, M., & Lachaise, F. (1980). Fine structure and secretion of the capsule enclosing the embryo in a crab (*Carcinus maenas* (L.)). *Tissue and Cell, 12*, 287–308.

Grosberg, R. K., & Strathmann, R. R. (1998). One cell, two cell, red cell, blue cell: The persistence of a unicellular stage in multicellular life histories. *Trends in Ecology & Evolution, 13*, 112–116.

Gueth-Hallonet, C., et al. (1993). gamma-Tubulin is present in acentriolar MTOCs during early mouse development. *Journal of Cell Science, 105*(Pt. 1), 157–166.

Hagstrom, K. A., et al. (2002). *C. elegans* condensin promotes mitotic chromosome architecture, centromere organization, and sister chromatid segregation during mitosis and meiosis. *Genes & Development, 16*, 729–742.

Han, X., et al. (2015). The KLP-7 residue S546 is a putative Aurora kinase site required for microtubule regulation at the centrosome in C. elegans. *PLoS One, 10*, e0132593.

Hartman, J. J., et al. (1998). Katanin, a microtubule-severing protein, is a novel AAA ATPase that targets to the centrosome using a WD40-containing subunit. *Cell, 93*, 277–287.

Heald, R., et al. (1996). Self-organization of microtubules into bipolar spindles around artificial chromosomes in *Xenopus* egg extracts. *Nature, 382*, 420–425.

Hendershott, M. C., & Vale, R. D. (2014). Regulation of microtubule minus-end dynamics by CAMSAPs and patronin. *Proceedings of the National Academy of Sciences of the United States of America, 111*, 5860–5865.

Herran, Y., et al. (2011). The cohesin subunit RAD21L functions in meiotic synapsis and exhibits sexual dimorphism in fertility. *The EMBO Journal, 30*, 3091–3105.

Hirano, T. (2015). Chromosome dynamics during mitosis. *Cold Spring Harbor Perspectives in Biology*, 7, a015792.

Holland, N. D. (1981). Electron microscopic study of development in a sea cucumber, *Stichopus tremulus* (Holothuroidea), from unfertilized egg through hatched blastula. *Acta Zoologica, 62*, 89–111.

Holubcova, Z., et al. (2015). Human oocytes. Error-prone chromosome-mediated spindle assembly favors chromosome segregation defects in human oocytes. *Science, 348*, 1143–1147.

Howe, K., & FitzHarris, G. (2013). Recent insights into spindle function in mammalian oocytes and early embryos. *Biology of Reproduction, 89*, 71.

Howe, M., et al. (2001). HIM-10 is required for kinetochore structure and function on *Caenorhabditis elegans* holocentric chromosomes. *The Journal of Cell Biology, 153*, 1227–1238.

Hughes, S. E., et al. (2011). Gamma-tubulin is required for bipolar spindle assembly and for proper kinetochore microtubule attachments during prometaphase I in *Drosophila* oocytes. *PLoS Genetics*, 7, e1002209.

Ishiguro, K., et al. (2011). A new meiosis-specific cohesin complex implicated in the cohesin code for homologous pairing. *EMBO Reports, 12*, 267–275.

Ishiguro, K., et al. (2014). Meiosis-specific cohesin mediates homolog recognition in mouse spermatocytes. *Genes & Development, 28*, 594–607.

Jang, J. K., et al. (2007). Misregulation of the kinesin-like protein Subito induces meiotic spindle formation in the absence of chromosomes and centrosomes. *Genetics*, *177*, 267–280.
Keating, T. J., et al. (1997). Microtubule release from the centrosome. *Proceedings of the National Academy of Sciences of the United States of America*, *94*, 5078–5083.
Kirschner, M. W., & Mitchison, T. (1986). Microtubule dynamics. *Nature*, *324*, 621.
Kitajima, T. S., et al. (2011). Complete kinetochore tracking reveals error-prone homologous chromosome biorientation in mammalian oocytes. *Cell*, *146*, 568–581.
Klein, F., et al. (1999). A centrole role for cohesins in sister chromatid cohesion, formation of axial elements, and recombination during yeast meiosis. *Cell*, *98*, 91–103.
Kline-Smith, S. L., et al. (2004). Depletion of centromeric MCAK leads to chromosome congression and segregation defects due to improper kinetochore attachments. *Molecular Biology of the Cell*, *15*, 1146–1159.
Komaru, A., et al. (2000). All meiotic chromosomes and both centrosomes at spindle pole in the zygotes discarded as two polar bodies in clam *Corbicula leana*: Unusual polar body formation observed by antitubulin immunofluorescence. *Development Genes and Evolution*, *210*, 263–269.
Lee, J., & Hirano, T. (2011). RAD21L, a novel cohesin subunit implicated in linking homologous chromosomes in mammalian meiosis. *The Journal of Cell Biology*, *192*, 263–276.
Liang, Z. Y., et al. (2009). Mature *Drosophila* meiosis I spindles comprise microtubules of mixed polarity. *Current Biology*, *19*, 163–168.
Lindsay, L. L., et al. (1992). Extracellular Mg2+ induces an intracellular Ca2+ wave during oocyte activation in the marine shrimp *Sicyonia ingentis*. *Developmental Biology*, *152*, 94–102.
Llano, E., et al. (2012). Meiotic cohesin complexes are essential for the formation of the axial element in mice. *The Journal of Cell Biology*, *197*, 877–885.
Loughlin, R., et al. (2011). Katanin contributes to interspecies spindle length scaling in *Xenopus*. *Cell*, *147*, 1397–1407.
Luksza, M., et al. (2013). Rebuilding MTOCs upon centriole loss during mouse oogenesis. *Developmental Biology*, *382*, 48–56.
Ma, W., et al. (2010). NEDD1 is crucial for meiotic spindle stability and accurate chromosome segregation in mammalian oocytes. *Developmental Biology*, *339*, 439–450.
Mains, P. E., et al. (1990). Mutations affecting the meiotic and mitotic divisions of the early *Caenorhabditis elegans* embryo. *Genetics*, *126*, 593–605.
Manandhar, G., et al. (2005). Centrosome reduction during gametogenesis and its significance. *Biology of Reproduction*, *72*, 2–13.
Maresca, T. J., et al. (2009). Spindle assembly in the absence of a RanGTP gradient requires localized CPC activity. *Current Biology*, *19*, 1210–1215.
Maro, B., et al. (1984). Changes in actin distribution during fertilization of the mouse egg. *Journal of Embryology and Experimental Morphology*, *81*, 211–237.
Matthies, H. J., et al. (1996). Anastral meiotic spindle morphogenesis: Role of the non-claret disjunctional kinesin-like protein. *The Journal of Cell Biology*, *134*, 455–464.
McGuinness, B. E., et al. (2009). Regulation of APC/C activity in oocytes by a Bub1-dependent spindle assembly checkpoint. *Current Biology*, *19*, 369–380.
McNally, K. L., & McNally, F. J. (2005). Fertilization initiates the transition from anaphase I to metaphase II during female meiosis in *C. elegans*. *Developmental Biology*, *282*, 218–230.
McNally, K. P., & McNally, F. J. (2011). The spindle assembly function of *Caenorhabditis elegans* katanin does not require microtubule-severing activity. *Molecular Biology of the Cell*, *22*, 1550–1560.
McNally, F. J., & Vale, R. D. (1993). Identification of katanin, an ATPase that severs and disassembles stable microtubules. *Cell*, *75*, 419–429.
McNally, K., et al. (2006). Katanin controls mitotic and meiotic spindle length. *The Journal of Cell Biology*, *175*, 881–891.

McNally, K., et al. (2014). Katanin maintains meiotic metaphase chromosome alignment and spindle structure in vivo and has multiple effects on microtubules in vitro. *Molecular Biology of the Cell, 25*, 1037–1049.
McNicoll, F., et al. (2013). Cohesin in gametogenesis. *Current Topics in Developmental Biology, 102*, 1–34.
Meireles, A. M., et al. (2009). Wac: A new Augmin subunit required for chromosome alignment but not for acentrosomal microtubule assembly in female meiosis. *The Journal of Cell Biology, 184*, 777–784.
Merdes, A., et al. (2000). Formation of spindle poles by dynein/dynactin-dependent transport of NuMA. *The Journal of Cell Biology, 149*, 851–862.
Mineyuki, Y. (2007). Plant microtubule studies: Past and present. *Journal of Plant Research, 120*, 45–51.
Mitchison, T., & Kirschner, M. (1984). Dynamic instability of microtubule growth. *Nature, 312*, 237–242.
Mitchison, T. J., et al. (2005). Roles of polymerization dynamics, opposed motors, and a tensile element in governing the length of *Xenopus* extract meiotic spindles. *Molecular Biology of the Cell, 16*, 3064–3076.
Müller-Reichert, T., et al. (2010). The elegans of spindle assembly. *Cellular and Molecular Life Sciences, 67*, 2195–2213.
Munro, E., & Bowerman, B. (2009). Cellular symmetry breaking during Caenorhabditis elegans development. *Cold Spring Harbor Perspectives in Biology, 1*, a003400.
Muscat, C. C., et al. (2015). Kinetochore-independent chromosome segregation driven by lateral microtubule bundles. *Elife, 4*, e06462.
Musch, A. (2004). Microtubule organization and function in epithelial cells. *Traffic, 5*, 1–9.
Nakashima, S., & Kato, K. H. (2001). Centriole behavior during meiosis in oocytes of the sea urchin *Hemicentrotus pulcherrimus*. *Development, Growth & Differentiation, 43*, 437–445.
Nelson, G. A., et al. (1982). *Caenorhabditis elegans* spermatozoan locomotion: Amoeboid movement with almost no actin. *The Journal of Cell Biology, 92*, 121–131.
Nigg, E. A., & Raff, J. W. (2009). Centrioles, centrosomes, and cilia in health and disease. *Cell, 139*, 663–678.
Oakley, B. R., et al. (2015). gamma-Tubulin complexes in microtubule nucleation and beyond. *Molecular Biology of the Cell, 26*, 2957–2962.
Ohkura, H. (2015). Meiosis: An overview of key differences from mitosis. *Cold Spring Harbor Perspectives in Biology*, 7, a015859.
Ori-McKenney, K. M., et al. (2012). Golgi outposts shape dendrite morphology by functioning as sites of acentrosomal microtubule nucleation in neurons. *Neuron, 76*, 921–930.
O'Rourke, S. M., et al. (2007). Dynein modifiers in C. elegans: Light chains suppress conditional heavy chain mutants. *PLoS Genetics, 3*, e128.
Paaby, A. B., et al. (2015). Wild worm embryogenesis harbors ubiquitous polygenic modifier variation. *Elife, 4*, 309178.
Palacios, M. J., et al. (1993). Gamma-tubulin reorganization during mouse fertilization and early development. *Journal of Cell Science, 104*(Pt. 2), 383–389.
Petronczki, M., et al. (2003). Un menage a quatre: The molecular biology of chromosome segregation in meiosis. *Cell, 112*, 423–440.
Petry, S., & Vale, R. D. (2015). Microtubule nucleation at the centrosome and beyond. *Nature Cell Biology, 17*, 1089–1093.
Petry, S., et al. (2011). Augmin promotes meiotic spindle formation and bipolarity in *Xenopus* egg extracts. *Proceedings of the National Academy of Sciences of the United States of America, 108*, 14473–14478.
Pfender, S., et al. (2015). Live imaging RNAi screen reveals genes essential for meiosis in mammalian oocytes. *Nature, 524*, 239–242.

Phillips, P. C., & Bowerman, B. (2015). Cell biology: Scaling and the emergence of evolutionary cell biology. *Current Biology*, *25*, R223–R225.

Powers, J., et al. (2004). Loss of KLP-19 polar ejection force causes misorientation and missegregation of holocentric chromosomes. *The Journal of Cell Biology*, *166*, 991–1001.

Prosser, S. L., & Pelletier, L. (2015). Centrosome biology: The ins and outs of centrosome assembly. *Current Biology*, *25*, R656–R659.

Radford, S. J., et al. (2012). The chromosomal passenger complex is required for meiotic acentrosomal spindle assembly and chromosome biorientation. *Genetics*, *192*, 417–429.

Radford, S., et al. (2015). Lateral and end-on kinetochore attachments are coordinated to achieve bi-orientation in Drosophila oocytes. *PLOS Genetics*, *11*, e1005605.

Reschen, R. F., et al. (2012). Dgp71WD is required for the assembly of the acentrosomal meiosis I spindle, and is not a general targeting factor for the gamma-TuRC. *Biology Open*, *1*, 422–429.

Saiki, T., & Hamaguchi, Y. (1997). Division of polar bodies induced by their enlargement in the starfish *Asterina pectinifera*. *Experimental Cell Research*, *237*, 142–148.

Sampath, S. C., et al. (2004). The chromosomal passenger complex is required for chromatin-induced microtubule stabilization and spindle assembly. *Cell*, *118*, 187–202.

Saunders, A. M., et al. (2007). Kinesin-5 acts as a brake in anaphase spindle elongation. *Current Biology*, *17*, R453–R454.

Sawada, T., & Schatten, G. (1988). Microtubules in ascidian eggs during meiosis, fertilization, and mitosis. *Cell Motility and the Cytoskeleton*, *9*, 219–230.

Sawin, K. E., et al. (1992). Mitotic spindle organization by a plus-end-directed microtubule motor. *Nature*, *359*, 540–543.

Schuh, M., & Ellenberg, J. (2007). Self-organization of MTOCs replaces centrosome function during acentrosomal spindle assembly in live mouse oocytes. *Cell*, *130*, 484–498.

Segbert, C., et al. (2003). KLP-18, a Klp2 kinesin, is required for assembly of acentrosomal meiotic spindles in *Caenorhabditis elegans*. *Molecular Biology of the Cell*, *14*, 4458–4469.

Severson, A. F., & Meyer, B. J. (2014). Divergent kleisin subunits of cohesin specify mechanisms to tether and release meiotic chromosomes. *Elife*, *3*, e03467.

Severson, A. F., et al. (2009). The axial element protein HTP-3 promotes cohesin loading and meiotic axis assembly in *C. elegans* to implement the meiotic program of chromosome segregation. *Genes & Development*, *23*, 1763–1778.

Sharif, B., et al. (2010). The chromosome passenger complex is required for fidelity of chromosome transmission and cytokinesis in meiosis of mouse oocytes. *Journal of Cell Science*, *123*, 4292–4300.

Skold, H. N., et al. (2005). Assembly pathway of the anastral *Drosophila* oocyte meiosis I spindle. *Journal of Cell Science*, *118*, 1745–1755.

Sluder, G., et al. (1989). Centrosome inheritance in starfish zygotes: Selective loss of the maternal centrosome after fertilization. *Developmental Biology*, *131*, 567–579.

Srayko, M., et al. (2000). MEI-1/MEI-2 katanin-like microtubule severing activity is required for *Caenorhabditis elegans* meiosis. *Genes & Development*, *14*, 1072–1084.

Srayko, M., et al. (2006). Katanin disrupts the microtubule lattice and increases polymer number in *C. elegans* meiosis. *Current Biology*, *16*, 1944–1949.

Stephano, J. L., & Gould, M. C. (2000). MAP kinase, a universal suppressor of sperm centrosomes during meiosis? *Developmental Biology*, *222*, 420–428.

Sumiyoshi, E., et al. (2015). C. elegans Aurora A kinase is required for the formation of spindle microtubules in female meiosis. *Molecular Biology of the Cell*, *26*(23), 4187–4196.

Sun, S. C., et al. (2010). Perturbation of Spc25 expression affects meiotic spindle organization, chromosome alignment and spindle assembly checkpoint in mouse oocytes. *Cell Cycle*, *9*, 4552–4559.

Sun, S. C., et al. (2011). Ndc80 regulates meiotic spindle organization, chromosome alignment, and cell cycle progression in mouse oocytes. *Microscopy and Microanalysis*, *17*, 431–439.

Szollosi, D., et al. (1972). Absence of centrioles in the first and second meiotic spindles of mouse oocytes. *Journal of Cell Science*, *11*, 521–541.
Tanaka, N., et al. (2012). Nezha/CAMSAP3 and CAMSAP2 cooperate in epithelial-specific organization of noncentrosomal microtubules. *Proceedings of the National Academy of Sciences of the United States of America*, *109*, 20029–20034.
Tanenbaum, M. E., et al. (2009). Kif15 cooperates with Eg5 to promote bipolar spindle assembly. *Current Biology*, *19*, 1703–1711.
Tulu, U. S., et al. (2006). Molecular requirements for kinetochore-associated microtubule formation in mammalian cells. *Current Biology*, *16*, 536–541.
Uetake, Y., et al. (2002). Nonequivalence of maternal centrosomes/centrioles in starfish oocytes: Selective casting-off of reproductive centrioles into polar bodies. *Developmental Biology*, *247*, 149–164.
van der Voet, M., et al. (2009). NuMA-related LIN-5, ASPM-1, calmodulin and dynein promote meiotic spindle rotation independently of cortical LIN-5/GPR/Galpha. *Nature Cell Biology*, *11*, 269–277.
Vanneste, D., et al. (2009). The role of Hklp2 in the stabilization and maintenance of spindle bipolarity. *Current Biology*, *19*, 1712–1717.
Walczak, C. E., et al. (2010). Mechanisms of chromosome behaviour during mitosis. *Nature Reviews Molecular Cell Biology*, *11*, 91–102.
Wang, Q., et al. (2011). Mechanism of the chromosome-induced polar body extrusion in mouse eggs. *Cell Division*, *6*, 17.
Wang, S., et al. (2015). The C. elegans ninein-related protein NOCA-1 binds microtubule ends and organizing non-centrosomal microtubule arrays. *Elife*, *4*, e08649.
Washitani-Nemoto, S., et al. (1994). Artificial parthenogenesis in starfish eggs: Behavior of nuclei and chromosomes resulting in tetraploidy of parthenogenotes produced by the suppression of polar body extrusion. *Developmental Biology*, *163*, 293–301.
White, E. A., & Glotzer, M. (2012). Centralspindlin: At the heart of cytokinesis. *Cytoskeleton (Hoboken)*, *69*, 882–892.
Wignall, S. M., & Villeneuve, A. M. (2009). Lateral microtubule bundles promote chromosome alignment during acentrosomal oocyte meiosis. *Nature Cell Biology*, *11*, 839–844.
Woodruff, J. B., et al. (2014). Pericentriolar material structure and dynamics. *Philosophical Transactions of the Royal Society of London. Series B, Biological Sciences*, *369*, 20130459.
Wu, X., & Palazzo, R. E. (1999). Differential regulation of maternal vs. paternal centrosomes. *Proceedings of the National Academy of Sciences of the United States of America*, *96*, 1397–1402.
Xu, X. L., et al. (2012). The microtubule-associated protein ASPM regulates spindle assembly and meiotic progression in mouse oocytes. *PLoS One*, 7, e49303.
Yalgin, C., et al. (2015). Centrosomin represses dendrite branching by orienting microtubule nucleation. *Nature Neuroscience*, *18*, 1437–1445.
Yang, H. Y., et al. (2003). MEI-1/katanin is required for translocation of the meiosis I spindle to the oocyte cortex in *C. elegans*. *Developmental Biology*, *260*, 245–259.
Yang, H. Y., et al. (2005). Kinesin-1 mediates translocation of the meiotic spindle to the oocyte cortex through KCA-1, a novel cargo adapter. *The Journal of Cell Biology*, *169*, 447–457.
Yang, G., et al. (2007). Architectural dynamics of the meiotic spindle revealed by single-fluorophore imaging. *Nature Cell Biology*, *9*, 1233–1242.
Zickler, D., & Kleckner, N. (2015). Recombination, pairing, and synapsis of homologs during meiosis. *Cold Spring Harbor Perspectives in Biology*, 7, a016626.
Zou, J., et al. (2008). A microtubule-destabilizing kinesin motor regulates spindle length and anchoring in oocytes. *The Journal of Cell Biology*, *180*, 459–466.

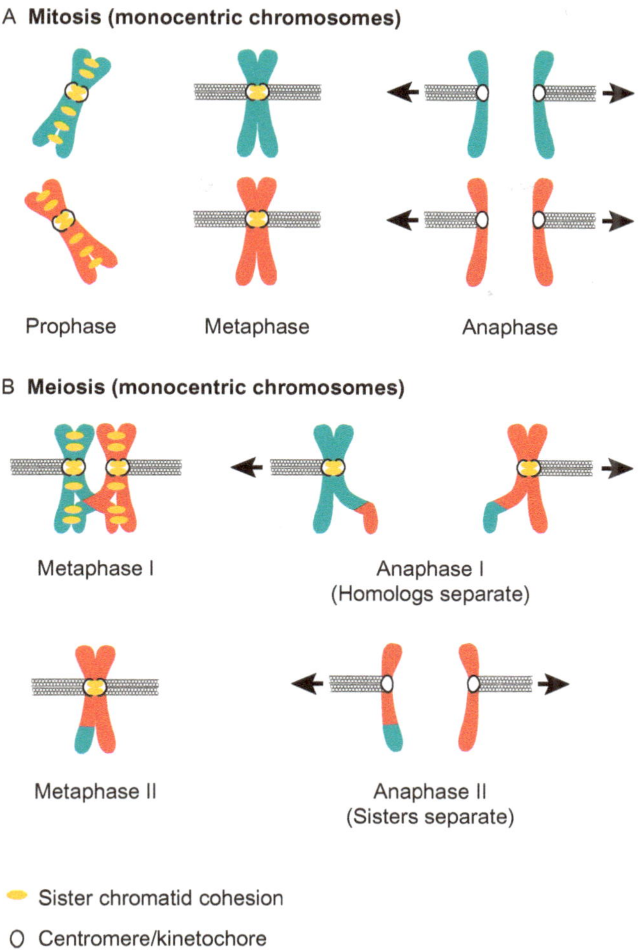

Aaron F. Severson *et al.*, Figure 1 Distinct patterns of chromosome segregation occur during meiosis and mitosis. The progression of a single pair of homologous chromosomes (blue and red) through mitosis (A) and meiosis (B) is shown. Before either division, DNA replication duplicates both homologs. Subsequently, each homolog is composed of two sister chromatids tethered by sister chromatid cohesion (yellow ovals). (A) In many organisms, cohesion along chromosome arms is released during mitotic prophase. At metaphase, sisters remain linked by centromeric cohesion. Bipolar attachment of sister kinetochores to microtubules from opposite spindle poles aligns each homolog on the metaphase plate. In anaphase, centromeric cohesion is released and sister chromatids segregate toward opposite poles. (B) During prophase of meiosis I, homologous chromosomes become linked by reciprocal exchange of DNA during crossover recombination. Both sister kinetochores of one homolog capture microtubules from the same spindle pole; the sister kinetochores of the other homolog attach to microtubules from the opposite spindle pole (for simplicity, microtubule contact with both sister chromatid kinetochores is not depicted). In anaphase of meiosis I, release of cohesion between chromosome arms allows homologs to separate; sisters remain tethered by centromeric cohesion. In anaphase II, centromeric cohesion is released, allowing sisters to separate.

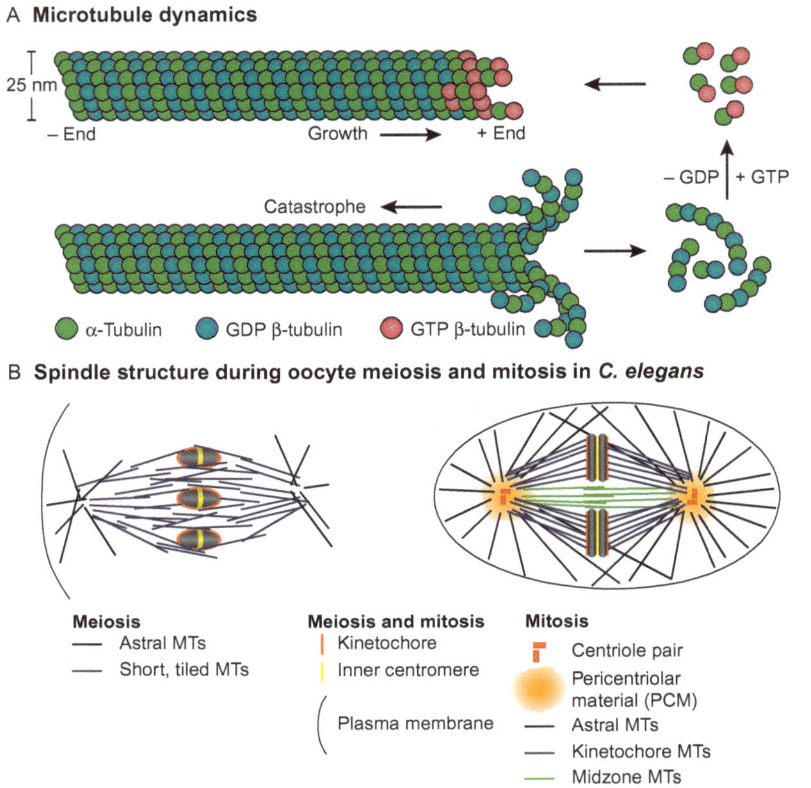

Aaron F. Severson *et al.*, Figure 2 Microtubule dynamics and spindle structures during oocyte meiosis and mitosis. (A) Microtubules are hollow tubes 25 nm in diameter formed by lateral association of 13 protofilaments, linear polymers composed α- and β-tubulin heterodimers. Due to the ordered assembly of α/β-tubulin heterodimers, microtubules have intrinsic polarity. One end (the minus end) has α-tubulin as the terminal subunit. This end is relatively stable. The other end (the plus end) has β-tubulin as the terminal subunit. This highly dynamic end switches between rapid growth and shrinkage depending on whether the exposed β-tubulin subunits are bound to GTP (growth) or GDP (shrinkage). During growth, GTP-tubulin adds to the plus end, forming a GTP cap. GTP is hydrolyzed to GDP, and if GDP-β-tubulin is exposed on the plus end of a microtubule, catastrophic disassembly can occur. (B) Cartoons of spindle structure during meiosis and mitosis in *C. elegans*. During mitosis, long, radial microtubule arrays are nucleated in the pericentriolar material surrounding the centrioles. Some microtubules (blue) from each spindle pole are captured by kinetochores (red). Ultimately, these kinetochore/microtubule attachments align each chromosome on the metaphase plate. Other overlapping, antiparallel microtubules (green) become bundled together in a region called the midzone, a process that stabilizes mitotic spindle structure. In contrast, centrioles are degraded prior to assembly of the oocyte meiotic spindle. Short, often tiled, microtubules form near chromosomes and ultimately coalesce into a bipolar structure.

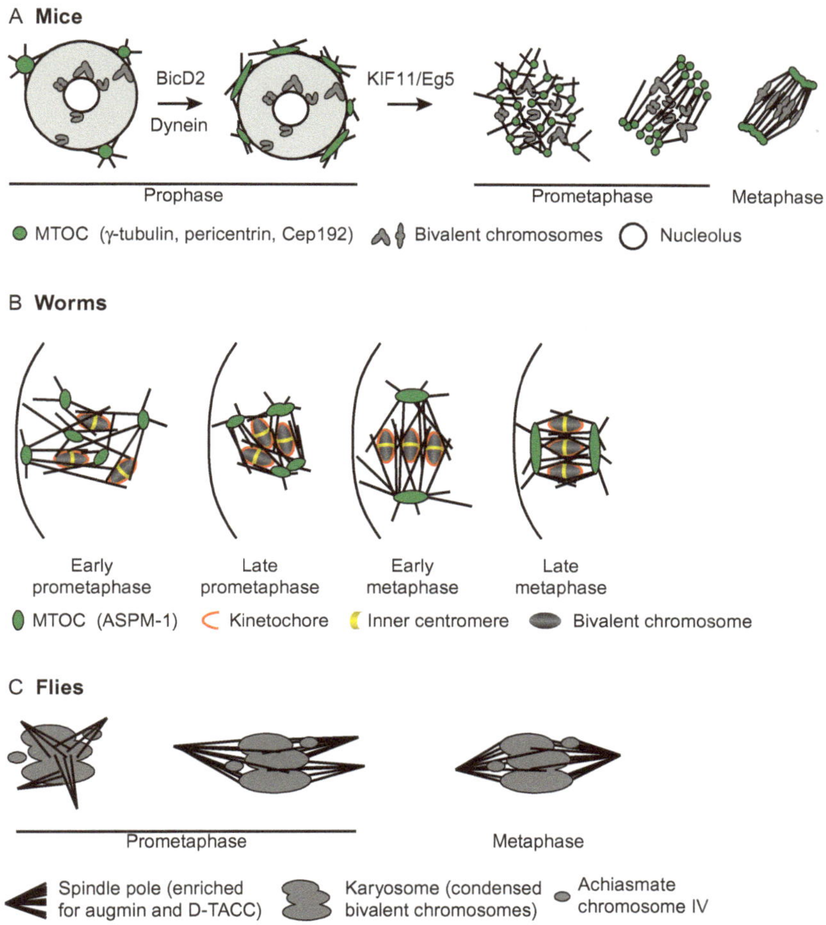

Aaron F. Severson *et al.*, Figure 3 Assembly of meiotic spindles in mouse, worm, and fly oocytes. Live imaging of microtubules and chromosomes during oocyte meiosis in these three model organisms has revealed similarities and differences in spindle assembly. When spindle assembly begins, homolog pairs have undergone crossover recombination to form bivalents, with the exception of *Drosophila* chromosome 4, which does not recombine. (A) In mouse oocytes, small microtubule-organizing centers (MTOCs) form near the nuclear envelope prior to NEB. Subsequently, MTOCs are stretched into thin ribbons, often becoming fragmented. This stretching process requires dynein and BicD2, which anchors dynein at the nuclear envelope. After NEB, the Eg5 homolog KIF11 promotes further MTOC fragmentation, yielding an average of approximately 26 MTOCs per oocyte. These MTOCs then coalesce as bivalents congress, forming a bipolar spindle by metaphase. (B) At NEB in *C. elegans* oocytes, most bivalents are located near the nuclear envelope. Following NEB, a cloud of short microtubules assembles around the bivalents, which subsequently cluster together. Concurrently, several foci of the microtubule-scaffolding protein ASPM-1 appear, presumably present at small MTOCs. The ASPM-1 foci coalesce into two poles as bivalents congress onto the metaphase plate. Initially, the bipolar metaphase spindle is oriented parallel to the oocyte cortex, but then rotates to be perpendicular to the cortex. During rotation, the MTOCs widen as the spindle shortens dramatically. (C) In fly oocytes, the recombined bivalents form a compact structure called a karyosome. After NEB, microtubules accumulate around the karyosome and several dynamic, pole-like structures form. These microtubule foci coalesce to form the poles of the bipolar meiotic spindle. Augmin, γ-tubulin, and D-TACC are enriched at the poles.

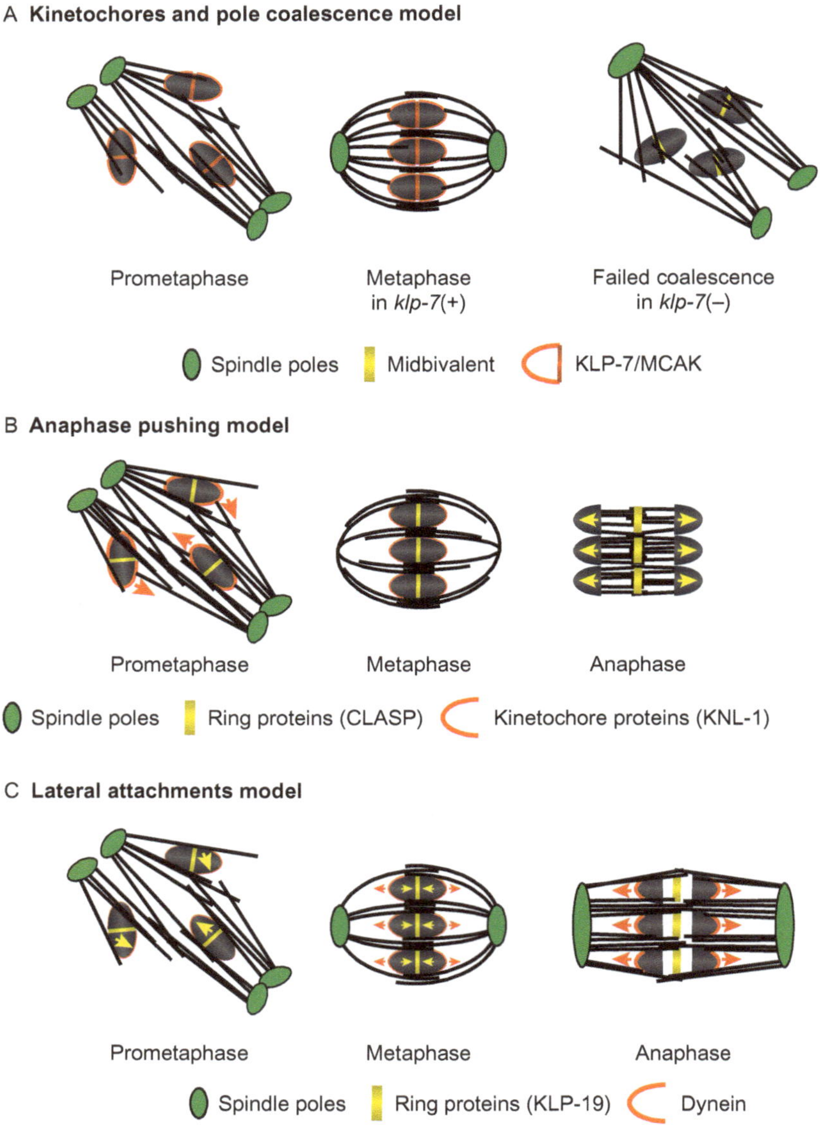

Aaron F. Severson *et al.*, Figure 4 Models for bipolar spindle assembly and chromosome congression and segregation in *C. elegans*. Three models are shown that address pole assembly (A) and chromosome congression and segregation (B and C) in *C. elegans* oocytes. (A) Proper microtubule–kinetochore attachments permit the coalescence of early pole foci to form a bipolar oocyte spindle. In this model, bivalent alignment and spindle bipolarity require that the two homologs of each bivalent attach specifically to microtubules from opposite spindle poles. Incorrect attachments (see text) are

(Continued)

Aaron F. Severson *et al.*, Figure 4—Cont'd eliminated by the microtubule depolymerase KLP-7/MCAK. Failure to disrupt these incorrect attachments in mutants with reduced KLP-7 function results in an imbalance in spindle tension that interferes with the coalescence of early pole foci and thus results in the assembly of multipolar spindles. (B) Microtubule–kinetochore interactions orient and align bivalents on the metaphase spindle, but at the metaphase-to-anaphase transition, the polar microtubule arrays disassemble, and a new population of microtubules forms between homologs and pushes them apart independently of kinetochores. The assembly of these microtubules is mediated by CLASP and other factors present in ring-shaped structures between each homolog pair at metaphase. When homologs separate, the rings are left behind at the metaphase plate. (C) Lateral interactions between microtubules and microtubule motor proteins mediate both chromosome congression to the metaphase plate and anaphase segregation to the poles. The chromokinesin KLP-19, localized to midbivalent rings, interacts with spindle microtubules to produce a polar ejection force that aligns bivalents at the metaphase plate. Subsequently, during anaphase, polar microtubules interact with dynein, a minus-end directed microtubule motor that accumulates at increasing levels on the poleward/lateral regions of bivalents as meiosis progresses. Dynein-directed motility mediates the poleward movement of chromosomes during anaphase.

CHAPTER SIX

Self-Organization of Spatial Patterning in Human Embryonic Stem Cells

Alessia Deglincerti[1], Fred Etoc[1], M. Zeeshan Ozair[1], Ali H. Brivanlou[2]
Laboratory of Stem Cell Biology and Molecular Embryology, The Rockefeller University, New York, USA
[2]Corresponding author: e-mail address: brvnlou@rockefeller.edu

Contents

Abstract

The developing embryo is a remarkable example of self-organization, where functional units are created in a complex spatiotemporal choreography. Recently, human embryonic stem cells (ESCs) have been used to recapitulate *in vitro* the self-organization programs that are executed in the embryo *in vivo*. This represents an unique opportunity to address self-organization in humans that is otherwise not addressable with current technologies. In this chapter, we review the recent literature on self-organization of human ESCs, with a particular focus on two examples: formation of embryonic germ layers and neural rosettes. Intriguingly, both activation and elimination of TGFβ signaling can initiate self-organization, albeit with different molecular underpinnings. We discuss the mechanisms underlying the formation of these structures *in vitro* and explore future challenges in the field.

1. INTRODUCTION

The developing embryo has an extraordinary ability to consistently and reproducibly distribute its cells into the various tissues that give rise to the fully formed organism. This is the result of self-organizing processes where

[1] These authors contributed equally to this work.

Current Topics in Developmental Biology, Volume 116
ISSN 0070-2153
http://dx.doi.org/10.1016/bs.ctdb.2015.11.010

a seemingly homogenous group of cells spontaneously organizes into an overall patterned structure (Fig. 1A). From the earliest stages of development, countless examples of self-organization exist *in vivo*, perhaps best studied in teleosts and amphibians (Movie 1 on http://dx.doi.org/10.1016/bs.ctdb.2015.11.010) but also apparent in mammalian systems (Fig. 1B and C). However, a mechanistic understanding of these processes is lacking, due to the difficulty of integrating multiple events (such as signaling, morphogenetic forces, and movements) over different time and size scales.

The study of self-organization *in vivo* is challenging and limited to a few model organisms that can be easily imaged and manipulated. In mammals, especially humans, *in utero* development represents an experimental barrier that has been difficult to overcome. Recently, the use of embryonic stem cells (ESCs) to recapitulate early developmental processes *in vitro* has started to shed some light on the mechanisms of self-organization. This has highlighted key parameters that control tissue morphogenesis and has provided valuable insights into how cells give rise to higher order architectures.

One of the simplest examples of *in vitro* self-organization occurs when ESCs aggregate into spherical structures known as embryoid bodies (EBs). EBs can differentiate and give rise to the three germ layers (ectoderm,

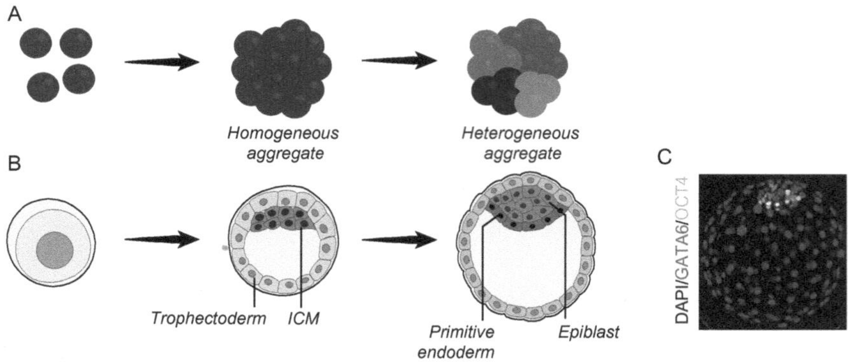

Figure 1 (A) Self-organization is the process by which a population of seemingly homogeneous cells (homogenous aggregate) gives rise to cells with heterogeneous properties (heterogeneous aggregate). (B and C) The developing embryo self-organizes to give rise to cells of different lineages: after a first distinction between trophectoderm (blue) and inner cell mass (ICM) (orange), cells in the ICM further differentiate into hypoblast (the precursor of primitive endoderm, green) and epiblast (which will give rise to the embryo proper, pink). (A color version of this figure is located at the end of this chapter.)

mesoderm, and endoderm), albeit with variability in the localization and relative amounts of each cell type between different EBs. Upon activation of the Wnt pathway, cells in mouse ESC-derived EBs can spontaneously undergo gastrulation-like morphogenetic movements and establish a spatially restricted region that resembles the primitive streak (ten Berge et al., 2008), thus mimicking the anterior–posterior (A–P) organization of the gastrulating embryo. In addition, small clumps of ESCs embedded in matrigel can grow into spheres that polarize and cavitate in the center, mirroring the cavitation in the peri-implantation embryo in both mouse (Bedzhov & Zernicka-Goetz, 2014) and human ESCs (F. Etoc et al., unpublished).

More complex developmental processes have also been reproduced *in vitro*, further confirming the surprising self-organization abilities of ESCs and pluripotent stem cells (PSCs). For example, cell aggregates that have differentiated to retinal epithelium can evaginate to form optic vesicles and subsequently invaginate to give rise to optic cups, thus generating highly ordered structures similar to those observed *in vivo* (Eiraku et al., 2011; Nakano et al., 2012). More recently, the self-organizing properties of ESCs and iPSCs have also been harnessed to generate three-dimensional organ-like structures (so-called "organoids") (Lancaster & Knoblich, 2014).

However, despite the plethora of self-organization examples in biological systems, the molecular control of self-organization remains in its infancy. How does a uniform group of cells become an architecturally complex organized structure? Recent studies, including our own, point to a highly dynamic integration of processes such as cell polarity, signal sensing, and selective competence of cells to integrate and respond to signals. In this review, we use two examples, the formation of germ layers on micropatterns and the generation of neural rosettes, to discuss architectural and molecular aspects of self-organization. We further explore the limitations of current models and touch upon challenges and future directions at the intersection of stem cell biology, bioengineering, and physics.

2. SELF-ORGANIZATION OF HUMAN EMBRYONIC GERM LAYERS

A remarkable example of self-organization within the mammalian embryo occurs at the onset of gastrulation, when each pluripotent epiblast cell is allocated to one of the three germ layers: ectoderm, mesoderm, and endoderm. This process is tightly coordinated in time and space and begins with the formation of a transient structure, the primitive streak, at the posterior end of the embryo (Arnold & Robertson, 2009; Nowotschin & Hadjantonakis, 2010). As cells ingress through the primitive streak, they acquire a mesendodermal fate. Cells that do not go through the streak become ectoderm (Fig. 2A). A correct spatial patterning of the germ layers is critical for the formation of the organism's body plan. Early BMP4 signals deriving from the extra-embryonic tissue setup a cascade of signaling in the epiblast via activation of Wnt, which induces Nodal that in turn maintains BMP4 signals. Integration of signaling activities and cell movement ultimately leads to gastrulation (Arnold & Robertson, 2009).

We have recently discovered a surprising and unexpected self-organization ability of human ESCs cultured in micropatterns of circular geometry and presented with a single BMP4 signal (human gastrulation in a dish, Fig. 2B) (Warmflash et al., 2014). Forty-eight hours after BMP4 application, pluripotent cells differentiate in radially symmetrical patterns of ectoderm, mesoderm, endoderm, and extra-embryonic tissue from the center to the edge, respectively (Fig. 2B). Moreover, mesendodermal cells undergo epithelial–mesenchymal transition, closely paralleling the organization of the primitive streak (Warmflash et al., 2014).

At least, two theoretical concepts can explain the observed spatial patterning. First, a "reaction–diffusion" model based on the signaling suggests that spatial patterns are formed by following a Turing-like mechanism (Turing, 1952). According to this model, patterns arise by extrinsic signals as the result of discrete morphogens that induce, directly or indirectly, their own inhibitors, which diffuse faster than the morphogens themselves. This generates spatial segregation of independent fates (Muller et al., 2012) and has been linked to several developmental contexts including germ-layer formation in zebrafish (Dubrulle et al., 2015). In support of this hypothesis, knockdown of pairs of BMP or Nodal inhibitors has been shown to abolish spatial patterning (Warmflash et al., 2014). We recently proposed a second concept, "edge-sensing." According to this model, cell fate decisions are based on the changes in the intrinsic response of cells as a function of their distance from the edge, relying on forces and architecture rather than diffusion of ligands and inhibitors. This is supported by the observation that reducing the colony size leads only to induction of the outer fate (Fig. 2C; Warmflash et al., 2014).

We found that human ESCs use a combination of both mechanisms (Etoc et al., unpublished). Edge sensing is created within a colony by TGFβ receptor subcellular relocalization. As pluripotent cells proliferate and

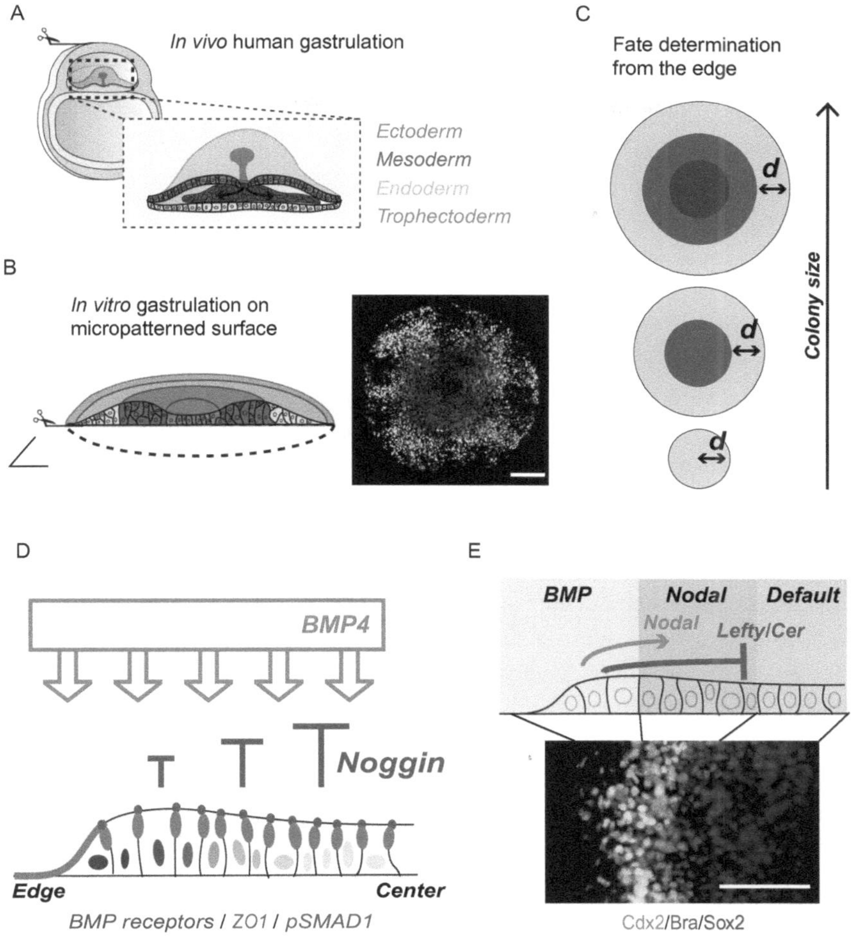

Figure 2 Self-organization of embryonic germ layers. (A) Schematic of a gastrulating human embryo (~ day postfertilization 15) showing the epiblast (light blue), primitive endoderm (yellow), and trophectoderm (green). Inset: Higher magnification representation of the epiblast. Gastrulation starts by formation of the primitive streak (gray). Cells ingress through the streak to form definitive endoderm (yellow) and mesoderm (red). Cells that do not go through the streak become ectoderm (dark blue). (B) *In vitro* gastrulation on micropatterned surfaces. Human ESCs are plated on 1 mm circular substrates. Forty-eight hours after BMP4 presentation, cell fates are radially organized (from center to edge) into ectoderm (blue), mesoderm (red), endoderm (yellow), and trophectoderm (green). A schematic representation of fate organization is shown on the left, while a picture of differentiating cells expressing markers of the three germ *(Continued)*

increase density, cells at the center of the colony gradually relocalize their receptors from the apical to the lateral membrane, immediately below tight junctions, thus losing their capacity to respond to apically presented BMP4 (Fig. 2D) (Nallet-Staub et al., 2015). Reaction–diffusion, instead, is created by BMP4-mediated induction of inhibitors such as Noggin (Etoc et al., unpublished) that progressively reduce the effective BMP4 concentration sensed by the cells (Fig. 2D). As a result of these two dynamic effects—receptor relocalization and secretion of BMP4 inhibitors—BMP4 signaling is abolished at the center and only cells at the outer edge of the colony respond to BMP4 and acquire a trophectodermal fate (Fig. 2E).

At the same time, Nodal and its inhibitors are produced by the peripheral domain at the edge of the colony in response to BMP4. From the edge, Nodal signaling spreads inward, toward the center (Etoc et al., unpublished). Because the domain of Nodal signaling is more extended than that of BMP4 response, some cells only integrate Nodal signaling but no BMP4 (Fig. 2E). These cells become mesendodermal precursors and express Brachyury (mesoderm) or Sox17 (endoderm) between the center and the outer edge. Finally, the cells at the center of the colony that do not receive either BMP4 or Nodal acquire ectodermal fate following the default mechanism (Fig. 2E; Hemmati-Brivanlou & Melton, 1994). While asymmetry in response to TGFβ ligands in micropatterns is controlled by edge-sensing and reaction–diffusion, its relevance to the *in vivo* situation remains to be tested. However, this represents a novel and unexpected mechanism for TGFβ signaling, as well as an exciting model for patterning of the gastrulating embryo.

Figure 2—Cont'd layers is shown on the right. Scale bar 200 μm. (C) Schematic of micropatterned colonies of different sizes: from top to bottom, 1000, 500, and 200 μm diameter. Green: trophectoderm, red: mesendoderm, blue: ectoderm. As the colony size decreases, only the outer fate is induced via an edge-sensing mechanism. (D) Interplay between BMP4 and Noggin in micropatterned colonies. BMP4 is applied homogeneously. BMP receptors (gray) are apically localized in cells at the edge and laterally localized in cells in the center of the colony, below adherens junction (labeled by ZO1, red). Noggin is expressed in response to BMP4 and contributes to inhibition of BMP4 signal at the center of the colony (red arrows). As a result, BMP4 signaling as detected by pSMAD1 intensity, is higher at the edge than at the center (blue nuclei). (E) The domain of maximal BMP4 signaling will give rise to trophectoderm. These cells produce Nodal and its inhibitors that diffuse inward toward the center of the colony. The cells that integrate both Nodal and BMP4 will generate mesendoderm, while the cells that do not respond to either BMP4 or Nodal will give rise to ectoderm. Scale bar 100 μm. (A color version of this figure is located at the end of this chapter.)

3. SELF-ORGANIZATION OF NEURAL ROSETTES

Following gastrulation, the ectoderm develops into, among other tissues, a layer of polarized cells that form the neuroepithelium and will give rise to the central nervous system (CNS). The nascent neuroepithelium is subdivided into regional identities based on its position in the embryonic A–P axis, as well as in the mediolateral plane, which will become the dorso–ventral (D–V) axis. However, all regions share fundamental epithelial characteristics such as apico-basal polarity, pseudostratification, tight and adherens junctions, apical mitosis, and basal lamina. The neural tube arises from the folding of the neuroepithelium along the D–V axis. Following folding, the apical surface of the neuroepithelium lines the inner side of the neural tube, while the basal surface, lined by a basement membrane, faces the outside (Fig. 3A). After neural tube closure, the neuroepithelial cells will elongate to form radial glial (RG) cells (Fig. 3B), the master progenitors of the CNS, while the neural tube will form a fluid-filled cavity, the ventricles. RG cells retain their epithelial characteristics throughout embryonic life; hence, the initial transition from nonpolarized cells to polarized epithelium is a requirement for the generation of every CNS neural cell type in all vertebrates.

It has been shown that ESC-derived neural progenitors—the *in vitro* equivalent of RG cells—can recapitulate the early sequence of events from gastrulation to neural tube closure, either in monolayer or in 3D organoid cultures of both mouse and human ESCs (Elkabetz et al., 2008; Lancaster & Knoblich, 2014; Main et al., 2013; Sasai, 2013). Folding of the neuroepithelium *in vitro* gives rise to structures called neural rosettes (Fig. 3C). Neural rosettes are also found *in vivo* in childhood tumors of the neuroectoderm as well as teratomas, suggesting that they are a fundamental property of primitive neuroepithelium regardless of the development or disease context. Neural rosettes can also be derived *in vitro* from primitive neuroepithelium generated by inhibition of TGFβ signaling from human ESCs (Chambers et al., 2009; Hemmati-Brivanlou & Melton, 1994; Shi et al., 2012; Fig. 3C).

What are the mechanisms that underlie establishment of apical–basal polarity in the neuroepithelium? While a complete mechanistic picture is lacking, there are clues emerging from studies in mouse embryos and neural rosettes. It appears that cilia are essential for establishing the initial direction

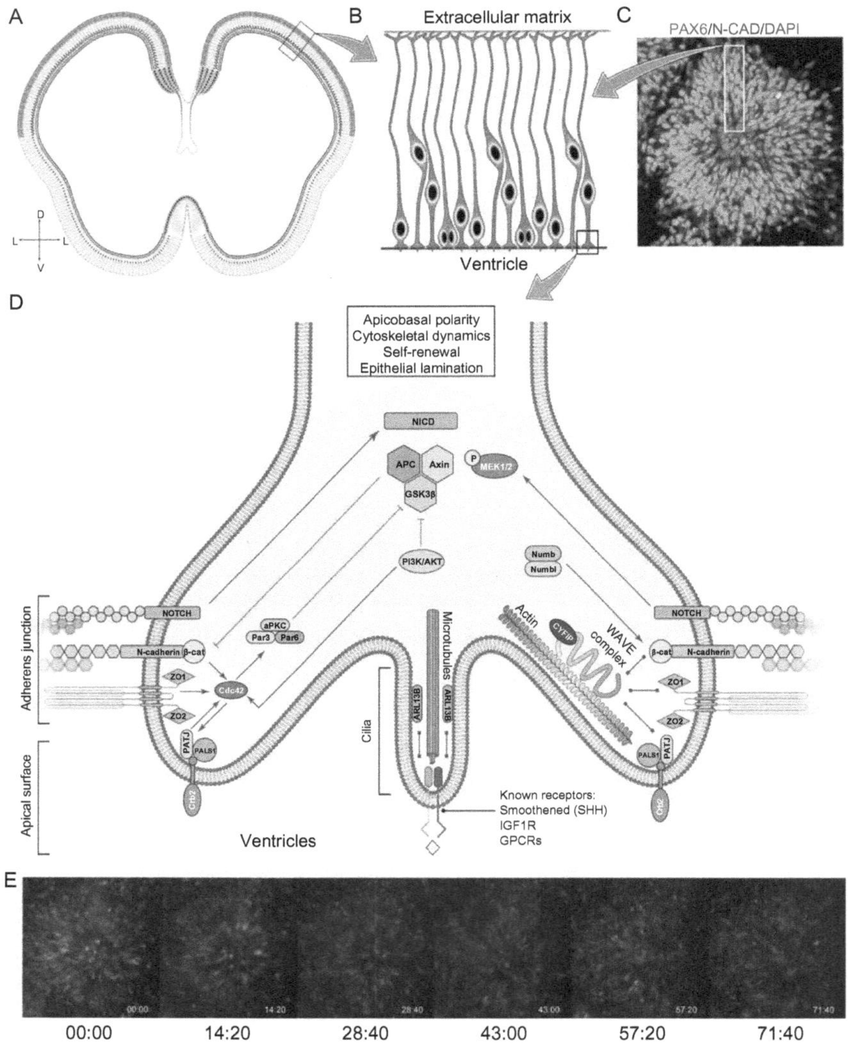

Figure 3 Self-organization of neural rosettes. (A) Schematic of the cortical neuroepithelium in a postconception week 5 human fetus. The neuroepithelium is comprised of radial glial cells. (B) Higher magnification view of the boxed area in (A) representing a schematic of radial glial cells lining the ventricle. The apical surface is the closest to the ventricle while the basal surface is away from it and in contact with the extracellular matrix. (C) Rosette structure at day 22 after neural induction in human ESCs. The neuroepithelium is identified by expression of PAX6 (green), while the central lumen structure (corresponding to the ventricles *in vivo*) is lined by N-cadherin (red), a component of adherens junction. This organization resembles the *in vivo* situation. (D) Schematic of

of neuroepithelial polarity (apical = ventricles, basal = meninges) as knockout of the ciliary-enriched GTPase Arl13b at an early stage results in inversion of the polarity (Higginbotham et al., 2013). In addition, multiple proteins associated with adherens junctions (crumbs, N-cadherin, and β-catenin), polarity proteins (such as aPKC, Cdc42, Par3, and Par6) (Famulski & Solecki, 2013), and the WAVE complex, which connects the cell membrane to the cytoskeleton (Yoon et al., 2014), are required for proper neuroepithelial polarity, as demonstrated by a series of mutation, knockdown, or knockout studies (Table 1). Indeed, adherens junction proteins (such as N-cadherin) as well as tight junction proteins (zona occludens 1, ZO1) are found in the center of rosette structures (Fig. 3C), the correlate of the neural tube-derived ventricles, as are crumbs and polarity complex members (Crb2, aPKC, Par3, Par6) and centriolar markers (pericentrin, ASPM) (Boroviak & Rashbass, 2011; Ohata et al., 2011; Shi et al., 2012). Unlike early embryonic germ-layer specification, described above, the role of geometric confinement in establishment of neuroepithelial properties has not been studied to date. Many of the factors listed above are known to be required for maintenance of polarity as interdependent modules, with all modules being required for maintenance of polarity. While many apical polarity modules have been identified, much less is known about basal polarity modules, and their cross interactions with apical modules. The further study of human ESC-derived neural rosettes will provide a better understanding of polarity and self-organization principles in the developing brain, potentially uncovering species-specific differences between humans and model systems.

---

radial glia end feet and high-magnification view of the boxed region in (B). The various pathways involved in establishment and maintenance of apico-basal polarity are displayed, as reviewed in Table 1. (E) Human ESC-derived neuroepithelium dynamically self-organizes into proliferative and nonproliferative zones. The pictures are still frames from Movie 2 (http://dx.doi.org/10.1016/bs.ctdb.2015.11.010) and depict a human ESC line engineered with FUCCI markers. Green cells in this montage represent actively proliferating radial glia, while the red cells are mostly postmitotic neurons. Initially, green cells are spread across the entire rosette. Over the course of 48 h, radial glia come to lie next to the rosette lumen forming a proliferative zone resembling the "ventricular zone" of the embryonic cortex. Conversely, the postmitotic neurons come to lie on the outside of the rosette; this basal part resembles the "cortical plate" of the embryonic cortex. (A color version of this figure is located at the end of this chapter.)

Table 1 List of Genes Affecting Neuroepithelial Polarity

Gene/ Signaling Pathway	Mechanism of Action	Expression	Cellular/ Developmental Context	Reference(s)
Arl13b	GTPase that maintains ciliary structure and function	Apical surface	Embryonic mouse cortex	Higginbotham et al. (2013)
N-cadherin	Component of adherence junctions, promotes neuroepithelial continuity, and WNT/β-catenin signaling	Apical surface	Embryonic mouse cortex	Kadowaki et al. (2007) and Zhang et al. (2010, 2013)
Crb2	Stabilizes the polarity complex (aPKC/Par6/Cdc42) and promotes notch signaling	Apical surface	Neural rosettes	Boroviak and Rashbass (2011) and Ohata et al. (2011)
β-Catenin	Component of adherence junctions, promotes radial glial self-renewal. Is activated by N-cadherin via AKT. Also part of WNT pathway and promotes self-renewal of radial glia	Apical surface	Embryonic mouse cortex	Kim et al. (2009), Mutch et al. (2010), and Zhang et al. (2010, 2013)
aPKC	Component of polarity complex that is essential for adherens junction integrity and apical polarity	Apical surface	Embryonic mouse cortex	Imai et al. (2006)
Cdc42	A GTPase that is also part of the polarity complex. Required for adherens junction integrity, apical localization of polarity complexes, and radial glial identity	Apical surface	Embryonic mouse cortex	Cappello et al. (2006) and Chen et al. (2006)
Hungtingtin	Scaffolding protein shown to be necessary for neuroepithelial polarization *in vitro*. Interacts with the metalloprotease ADAM10 to regulate N-cadherin cleavage	Apical surface	Neural rosettes	Lo Sardo et al. (2012)

Numb/ Numb-like	Required for adherens junction integrity and polarization by mediating basolateral insertion of cadherins	Apical surface	Embryonic mouse cortex	Rasin et al. (2007)
APC	Required for polarized extension of radial glial endfeet. Regulates β-catenin in the cortex. Also stabilizes microtubules dynamics in radial glia to promote scaffolding function	Cell soma Basal surface	Embryonic mouse cortex	Yokota et al. (2009)
CYFIP1	Component of adherens junctions and part of the WAVE complex, which links actin skeleton to multiple cell surface receptors	Apical surface	Embryonic mouse cortex Human iPSC neural rosettes	Yoon et al. (2014)
GSK3beta		Cytoplasm	Embryonic mouse cortex	Kim et al. (2009) and Morgan-Smith et al. (2014)
Lgl	Regulator of Notch signaling. Involved in maintenance of apical junctional complex and interacts with aPKC. Required for asymmetrical localization of Numb and regulates proliferation	Basal surface	Embryonic mouse cortex	Klezovitch et al. (2004)
FGF	Microlumen	Apical surface	Zebrafish lateral line	Durdu et al. (2014)
Notch		Apical surface	Neural rosettes Embryonic mouse cortex	Main et al. (2013)

4. CHALLENGES AND FUTURE DIRECTIONS

In order to understand the complex local interactions underlying self-organization, several complementary approaches need to be implemented. As self-organization is a dynamic process, live cell-tracking analysis in space and time is required to shed light on the underlying morphogenetic movements. This can be accomplished, for example, by genome editing to mark selective cell type-specific markers (Fig. 2B) or cell cycle reporters such as the FUCCI line (Fig. 3E and Movie 2 on http://dx.doi.org/10.1016/bs.ctdb.2015.11.010) (Sakaue-Sawano et al., 2008). A major question that needs to be addressed is how different territories are created from seemingly homogenous cells. From classical studies of model organisms, at least two mechanisms have been suggested: (1) the sorting of "salt and pepper" cell populations (Steinberg, 1970) and (2) the establishment of spatially precise fate allocation mechanisms (Cohen, Briscoe, & Blassberg, 2013). Moreover, mathematical models that merge dynamic information at the subcellular, cellular, and tissue levels are required for a more complete understanding of these processes. These models will need to integrate large data into comprehensive readouts and provide a quantitative evaluation of their mechanistic basis.

We suggest that the formation of self-organizing structures occurs in three stages: initiation, amplification, and maintenance. First, an initial symmetry-breaking event introduces asymmetry within a seemingly homogeneous structure. In the context of gastrulating micropatterned colonies, this is represented by edge-sensing mechanisms that direct the epithelization of the cells at the colony center. In the case of neuroepithelial rosettes, this may manifest in apical localization of the cilia by a currently unknown mechanism. Second, the early asymmetry is subsequently amplified and propagated to the entire structure. For micropatterned colonies, this is mediated by morphogen/inhibitor combinations that ultimately pattern the tissue from the edge to the center using a reaction–diffusion mechanism. In the case of the neuroepithelium, the initial break in cellular polarity is propagated to the surrounding cells that will line the lumen. Third, the self-organized structure needs to be actively maintained to avoid dissolution of its organization. In micropatterned colonies, hemophilic interactions between differentiated cells of similar type are likely to stabilize differentiated domains. In rosettes, the polarized structures are maintained by adherens junctions, which serve as scaffolds for complexes that mediate proliferative and neurogenic cell divisions (Fig. 3E).

This generic classification of self-organization allows to define specific questions pertaining to its genesis: What triggers the initial symmetry breaking? What is the molecular mechanism underlying this event? Is the exogenous presentation of these factors sufficient for symmetry breaking? What are the feedbacks mechanism(s) that reinforce the initial cue? How are mechanical forces and signaling integrated to generate self-organizing units? What are the mechanisms of dynamic stabilization of self-organized structures that ultimately give rise to functional organs during embryogenesis? And finally, how are self-organizing principles integrated across germ layers to generate the overall embryonic structures? Recent studies using *in vitro* models and *in vivo* genetic approaches are beginning to provide some of the answers to these questions (Eiraku et al., 2011; Muguruma et al., 2015). The dynamic nature of embryogenesis involves ever-changing environments of compressive and geometric forces, which likely contribute to self-organization as well. While modeling these events is currently challenging, the cellular and molecular mechanisms underlying these forces will pave the way for a more developmentally relevant framework of understanding of self-organization.

Finally, despite the progress in the *in vitro* generation of germ layer and neural rosette self-organized structures described above, the establishment of the embryological polarity (i.e., A–P and D–V axis) has yet to be accomplished *in vitro*. The development of novel tools such as microfluidics (Sorre et al., 2014) and optogenetics (Etoc et al., unpublished) will allow the asymmetric presentation of active components with precise spatiotemporal coordinates. Ultimately, this understanding will impact the basic knowledge of embryonic development and will inevitably provide the basis for a rational development of clinical applications.

REFERENCES

Arnold, S. J., & Robertson, E. J. (2009). Making a commitment: Cell lineage allocation and axis patterning in the early mouse embryo. *Nature Reviews. Molecular Cell Biology*, *10*(2), 91–103.

Bedzhov, I., & Zernicka-Goetz, M. (2014). Self-organizing properties of mouse pluripotent cells initiate morphogenesis upon implantation. *Cell*, *156*(5), 1032–1044.

Boroviak, T., & Rashbass, P. (2011). The apical polarity determinant Crumbs 2 is a novel regulator of ESC-derived neural progenitors. *Stem Cells*, *29*(2), 193–205.

Cappello, S., et al. (2006). The Rho-GTPase cdc42 regulates neural progenitor fate at the apical surface. *Nature Neuroscience*, *9*(9), 1099–1107.

Chambers, S. M., et al. (2009). Highly efficient neural conversion of human ES and iPS cells by dual inhibition of SMAD signaling. *Nature Biotechnology*, *27*(3), 275–280.

Chen, L., et al. (2006). Cdc42 deficiency causes Sonic hedgehog-independent holoprosencephaly. *Proceedings of the National Academy of Sciences of the United States of America*, *103*(44), 16520–16525.

Cohen, M., Briscoe, J., & Blassberg, R. (2013). Morphogen interpretation: The transcriptional logic of neural tube patterning. *Current Opinion in Genetics and Development*, *23*(4), 423–428.

Dubrulle, J., et al. (2015). Response to Nodal morphogen gradient is determined by the kinetics of target gene induction. *Elife*, *4*, e05042.

Durdu, S., et al. (2014). Luminal signalling links cell communication to tissue architecture during organogenesis. *Nature*, *515*(7525), 120–124.

Eiraku, M., et al. (2011). Self-organizing optic-cup morphogenesis in three-dimensional culture. *Nature*, *472*(7341), 51–56.

Elkabetz, Y., et al. (2008). Human ES cell-derived neural rosettes reveal a functionally distinct early neural stem cell stage. *Genes and Development*, *22*(2), 152–165.

Famulski, J. K., & Solecki, D. J. (2013). New spin on an old transition: Epithelial parallels in neuronal adhesion control. *Trends in Neurosciences*, *36*(3), 163–173.

Hemmati-Brivanlou, A., & Melton, D. A. (1994). Inhibition of activin receptor signaling promotes neuralization in Xenopus. *Cell*, *77*(2), 273–281.

Higginbotham, H., et al. (2013). Arl13b-regulated cilia activities are essential for polarized radial glial scaffold formation. *Nature Neuroscience*, *16*(8), 1000–1007.

Imai, F., et al. (2006). Inactivation of aPKClambda results in the loss of adherens junctions in neuroepithelial cells without affecting neurogenesis in mouse neocortex. *Development*, *133*(9), 1735–1744.

Kadowaki, M., et al. (2007). N-cadherin mediates cortical organization in the mouse brain. *Developmental Biology*, *304*(1), 22–33.

Kim, W. Y., et al. (2009). GSK-3 is a master regulator of neural progenitor homeostasis. *Nature Neuroscience*, *12*(11), 1390–1397.

Klezovitch, O., et al. (2004). Loss of cell polarity causes severe brain dysplasia in Lgl1 knockout mice. *Genes and Development*, *18*(5), 559–571.

Lancaster, M. A., & Knoblich, J. A. (2014). Organogenesis in a dish: Modeling development and disease using organoid technologies. *Science*, *345*(6194), 1247125.

Lo Sardo, V., et al. (2012). An evolutionary recent neuroepithelial cell adhesion function of huntingtin implicates ADAM10-Ncadherin. *Nature Neuroscience*, *15*(5), 713–721.

Main, H., et al. (2013). Notch signaling maintains neural rosette polarity. *PLoS One*, *8*(5), e62959.

Morgan-Smith, M., et al. (2014). GSK-3 signaling in developing cortical neurons is essential for radial migration and dendritic orientation. *Elife*, *3*, e02663.

Muguruma, K., et al. (2015). Self-organization of polarized cerebellar tissue in 3D culture of human pluripotent stem cells. *Cell Reports*, *10*(4), 537–550.

Muller, P., et al. (2012). Differential diffusivity of Nodal and Lefty underlies a reaction–diffusion patterning system. *Science*, *336*(6082), 721–724.

Mutch, C. A., et al. (2010). Beta-catenin signaling negatively regulates intermediate progenitor population numbers in the developing cortex. *PLoS One*, *5*(8), e12376.

Nakano, T., et al. (2012). Self-formation of optic cups and storable stratified neural retina from human ESCs. *Cell Stem Cell*, *10*(6), 771–785.

Nallet-Staub, F., et al. (2015). Cell density sensing alters TGF-beta signaling in a cell-type-specific manner, independent from Hippo pathway activation. *Developmental Cell*, *32*(5), 640–651.

Nowotschin, S., & Hadjantonakis, A. K. (2010). Cellular dynamics in the early mouse embryo: From axis formation to gastrulation. *Current Opinion in Genetics and Development*, *20*(4), 420–427.

Ohata, S., et al. (2011). Dual roles of Notch in regulation of apically restricted mitosis and apicobasal polarity of neuroepithelial cells. *Neuron*, *69*(2), 215–230.

Rasin, M. R., et al. (2007). Numb and Numbl are required for maintenance of cadherin-based adhesion and polarity of neural progenitors. *Nature Neuroscience*, *10*(7), 819–827.

Sakaue-Sawano, A., et al. (2008). Visualizing spatiotemporal dynamics of multicellular cell-cycle progression. *Cell*, *132*(3), 487–498.
Sasai, Y. (2013). Next-generation regenerative medicine: Organogenesis from stem cells in 3D culture. *Cell Stem Cell*, *12*(5), 520–530.
Shi, Y., et al. (2012). Human cerebral cortex development from pluripotent stem cells to functional excitatory synapses. *Nature Neuroscience*, *15*(3), 477–486. S1.
Sorre, B., et al. (2014). Encoding of temporal signals by the TGF-beta pathway and implications for embryonic patterning. *Developmental Cell*, *30*(3), 334–342.
Steinberg, M. S. (1970). Does differential adhesion govern self-assembly processes in histogenesis? Equilibrium configurations and the emergence of a hierarchy among populations of embryonic cells. *Journal of Experimental Zoology*, *173*(4), 395–433.
ten Berge, D., et al. (2008). Wnt signaling mediates self-organization and axis formation in embryoid bodies. *Cell Stem Cell*, *3*(5), 508–518.
Turing, A. M. (1952). The chemical basis of morphogenesis. *Philosophical Transactions of the Royal Society of London. Series B, Biological Sciences*, *237*(641), 37–72.
Warmflash, A., et al. (2014). A method to recapitulate early embryonic spatial patterning in human embryonic stem cells. *Nature Methods*, *11*(8), 847–854.
Yokota, Y., et al. (2009). The adenomatous polyposis coli protein is an essential regulator of radial glial polarity and construction of the cerebral cortex. *Neuron*, *61*(1), 42–56.
Yoon, K. J., et al. (2014). Modeling a genetic risk for schizophrenia in iPSCs and mice reveals neural stem cell deficits associated with adherens junctions and polarity. *Cell Stem Cell*, *15*(1), 79–91.
Zhang, J., et al. (2010). Cortical neural precursors inhibit their own differentiation via N-cadherin maintenance of beta-catenin signaling. *Developmental Cell*, *18*(3), 472–479.
Zhang, J., et al. (2013). AKT activation by N-cadherin regulates beta-catenin signaling and neuronal differentiation during cortical development. *Neural Development*, *8*, 7.

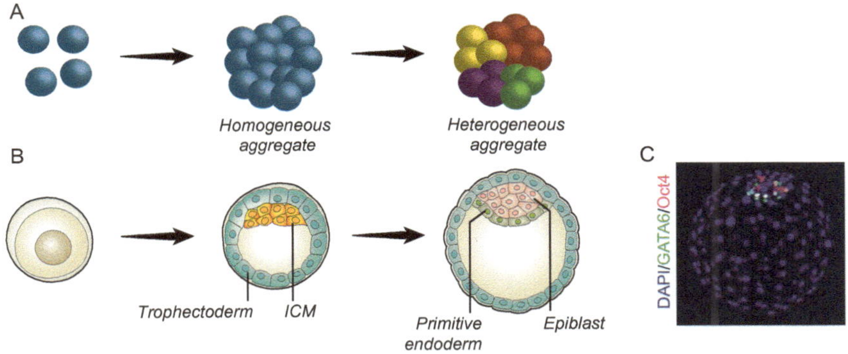

Alessia Deglincerti *et al.*, Figure 1 (A) Self-organization is the process by which a population of seemingly homogeneous cells (homogenous aggregate) gives rise to cells with heterogeneous properties (heterogeneous aggregate). (B and C) The developing embryo self-organizes to give rise to cells of different lineages: after a first distinction between trophectoderm (blue) and inner cell mass (ICM) (orange), cells in the ICM further differentiate into hypoblast (the precursor of primitive endoderm, green) and epiblast (which will give rise to the embryo proper, pink).

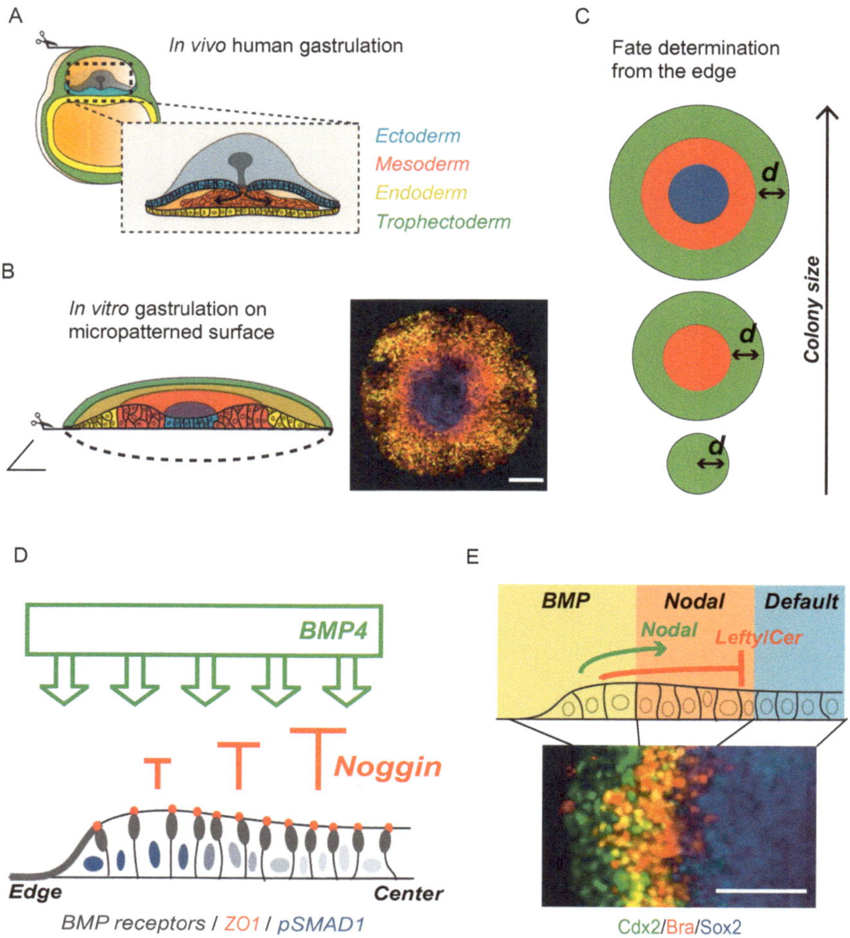

Alessia Deglincerti *et al.*, Figure 2 Self-organization of embryonic germ layers. (A) Schematic of a gastrulating human embryo (~ day postfertilization 15) showing the epiblast (light blue), primitive endoderm (yellow), and trophectoderm (green). Inset: Higher magnification representation of the epiblast. Gastrulation starts by formation of the primitive streak (gray). Cells ingress through the streak to form definitive endoderm (yellow) and mesoderm (red). Cells that do not go through the streak become ectoderm (dark blue). (B) *In vitro* gastrulation on micropatterned surfaces. Human ESCs are plated on 1 mm circular substrates. Forty-eight hours after BMP4 presentation, cell fates are radially organized (from center to edge) into ectoderm (blue), mesoderm (red), endoderm (yellow), and trophectoderm (green). A schematic representation of fate organization is shown on the left, while a picture of differentiating cells expressing markers of

(Continued)

Alessia Deglincerti *et al.*, Figure 2—Cont'd the three germ layers is shown on the right. Scale bar 200 μm. (C) Schematic of micropatterned colonies of different sizes: from top to bottom, 1000, 500, and 200 μm diameter. Green: trophectoderm, red: mesendoderm, blue: ectoderm. As the colony size decreases, only the outer fate is induced via an edge-sensing mechanism. (D) Interplay between BMP4 and Noggin in micropatterned colonies. BMP4 is applied homogeneously. BMP receptors (gray) are apically localized in cells at the edge and laterally localized in cells in the center of the colony, below adherens junction (labeled by ZO1, red). Noggin is expressed in response to BMP4 and contributes to inhibition of BMP4 signal at the center of the colony (red arrows). As a result, BMP4 signaling as detected by pSMAD1 intensity, is higher at the edge than at the center (blue nuclei). (E) The domain of maximal BMP4 signaling will give rise to trophectoderm. These cells produce Nodal and its inhibitors that diffuse inward toward the center of the colony. The cells that integrate both Nodal and BMP4 will generate mesendoderm, while the cells that do not respond to either BMP4 or Nodal will give rise to ectoderm. Scale bar 100 μm.

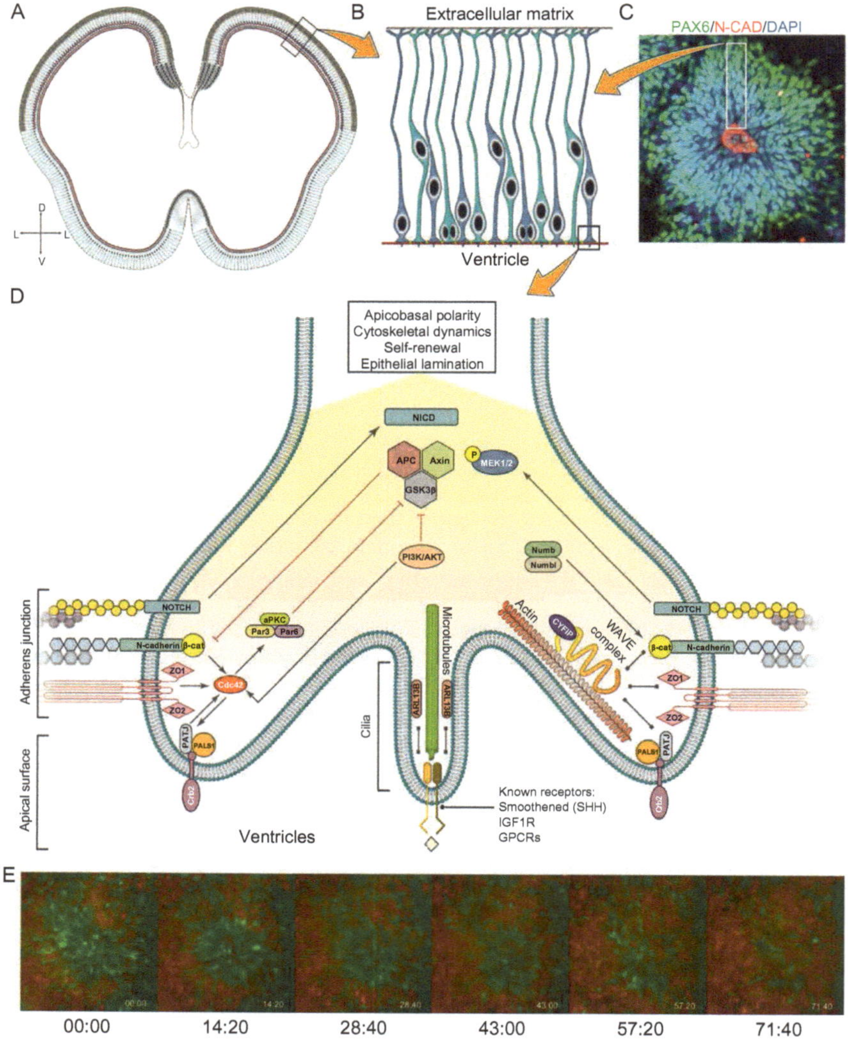

Alessia Deglincerti *et al.*, Figure 3 Self-organization of neural rosettes. (A) Schematic of the cortical neuroepithelium in a postconception week 5 human fetus. The neuroepithelium is comprised of radial glial cells. (B) Higher magnification view of the boxed area in (A) representing a schematic of radial glial cells lining the ventricle. The apical surface is the closest to the ventricle while the basal surface is away from it and in contact with the extracellular matrix. (C) Rosette structure at day 22 after neural induction in human ESCs. The neuroepithelium is identified by expression of PAX6 (green), while the central lumen structure (corresponding to the ventricles *in vivo*) is lined by N-cadherin (red), a component of adherens junction. This organization resembles the *in vivo* situation. (D) Schematic of radial glia end feet and high-magnification view of the boxed

(Continued)

Alessia Deglincerti *et al.*, Figure 3—Cont'd region in (B). The various pathways involved in establishment and maintenance of apico-basal polarity are displayed, as reviewed in Table 1. (E) Human ESC-derived neuroepithelium dynamically self-organizes into proliferative and nonproliferative zones. The pictures are still frames from Movie 2 (http://dx.doi.org/10.1016/bs.ctdb.2015.11.010) and depict a human ESC line engineered with FUCCI markers. Green cells in this montage represent actively proliferating radial glia, while the red cells are mostly postmitotic neurons. Initially, green cells are spread across the entire rosette. Over the course of 48 h, radial glia come to lie next to the rosette lumen forming a proliferative zone resembling the "ventricular zone" of the embryonic cortex. Conversely, the postmitotic neurons come to lie on the outside of the rosette; this basal part resembles the "cortical plate" of the embryonic cortex.

CHAPTER SEVEN

The Neural Crest Migrating into the Twenty-First Century

Marianne E. Bronner[1], Marcos Simões-Costa
Division of Biology and Biological Engineering, California Institute of Technology, Pasadena, California, USA
[1]Corresponding author: e-mail address: mbronner@caltech.edu

Contents

Abstract

From the initial discovery of the neural crest over 150 years ago to the seminal studies of Le Douarin and colleagues in the latter part of the twentieth century, understanding of the neural crest has moved from the descriptive to the experimental. Now, in the twenty-first century, neural crest research has migrated into the genomic age. Here, we reflect upon the major advances in neural crest biology and the open questions that will continue to make research on this incredible vertebrate cell type an important subject in developmental biology for the century to come.

1. INTRODUCTION

The neural crest is a remarkable cell type, unique to vertebrate embryos. Induction of the neural crest occurs at the neural plate border via a signaling interaction between neural and nonneural ectoderm (Selleck & Bronner-Fraser, 1995). After their specification, neural crest precursors reside within the elevating neural folds and dorsal neural tube after its closure. Neural crest cells then undergo an epithelial-to-mesenchymal

Current Topics in Developmental Biology, Volume 116
ISSN 0070-2153
http://dx.doi.org/10.1016/bs.ctdb.2015.12.003

transition (EMT) to emigrate from the neuroepithelium and become a motile cell population (Fig. 1). These cells migrate collectively or individually, navigating along stereotypic pathways to numerous and often distant sites throughout the embryo, where they eventually differentiate into an incredibly diverse array of cell types. Thus, the neural crest is a transient cell type that rapidly transits from multipotent precursors to cell types ranging from neurons and glia of the peripheral nervous system to pigment cells of the skin and cartilage and bone cells of the craniofacial skeleton.

Understanding neural crest development has important medical implications because of the variety of birth defects and cancers that arise from mistakes in neural crest development or mutations that effect neural crest derivatives postnatally. In humans, these "neurocristopathies" are among the most common birth defects in live-born infants, including craniofacial malformations, defects in the cardiac outflow tract, familial dysautonomia,

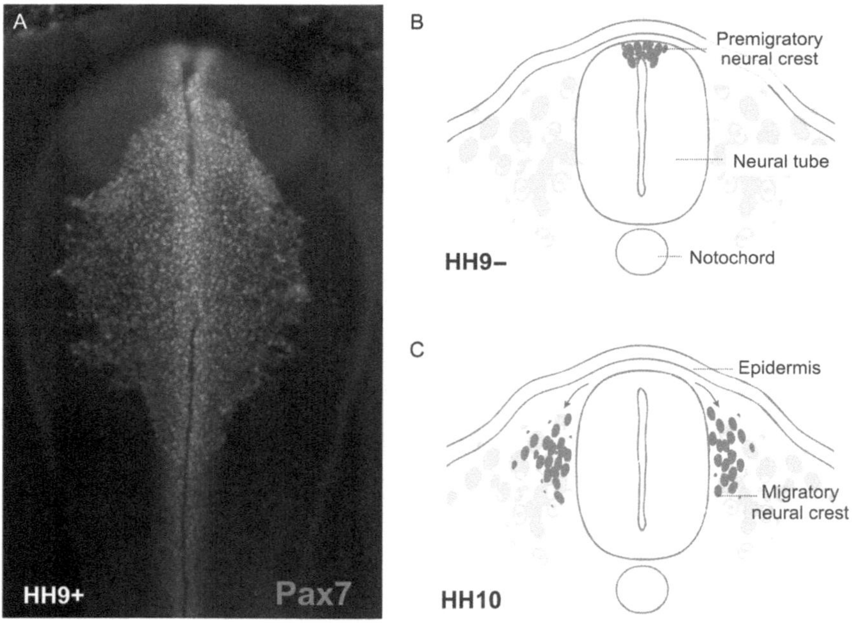

Figure 1 Neural crest cells delaminate from the neural tube and undergo extensive migration. (A) Immunohistochemistry with an antibody for neural crest marker Pax7 highlights neural crest cells delaminating from the neural tube of a stage HH9 chicken embryo. (B) Diagram representing a cross section of a neurulating chicken embryo. At the stage represented, the neural cells occupy the dorsal portion of the neural tube. (C) Neural crest cells eventually undergo epithelial-to-mesenchymal transition to delaminate from the neural tube and undergo extensive migration. (A color version of this figure is located at the end of this chapter.)

and a variety of syndromes including Treacher-Collins, Waardenburg, DiGeorge, and Charge syndromes (see chapter "Developmental Biology: We Are All Walking Mutants" by Trainor). In addition, neural crest-derived cell types seem to be particularly prone to metastasis, giving rise to melanoma, neuroblastoma, and pheochromocytoma, among others.

How these cells arise, migrate, and differentiate has fascinated developmental biologists for over 150 years. In this review, we reflect upon the history of neural crest research, focusing on some major discoveries over the past century and speculating upon the fascinating mysteries that are likely to occupy research on this cell type going forward.

2. A BRIEF HISTORY

In the latter half of the nineteenth century, the invention of the microtome by Wilhelm His together with improved histological techniques enabled visualization of embryonic tissues with previously unprecedented resolution. In 1868, His first described the neural crest as a "Zwischenstrang" or cord between the neural and nonneural ectoderm (Fig. 2A). He also recognized that neural crest cells formed spinal ganglia and for this reason named it "ganglionic crest."

In the early to mid-twentieth century, experimental embryology came to the forefront of research in developmental biology. During this time, "cut and paste" experiments examined the contributions of neural crest cells to various structures, largely by removing the neural crest via surgical ablation not only in amphibians (Hörstadius, 1950) but also in birds (Yntema & Hammond, 1945) as well as by transplantation experiments to examine patterns of pigment cell migration between different species (Twitty, 1936). These experiments revealed many intriguing properties of the neural crest from their extensive migratory ability to their contributions to melanocytes and ganglia of the peripheral nervous system. For a detailed overview of these early experiments, there is a wonderful monograph written by Hörstadius (1950).

The 1960s ushered in a new era of experimentation with the application of novel cell marking techniques to neural crest biology. First, Weston (1963) and Chibon (1967) in avian and amphibian embryos, respectively, radioactively labeled embryos with ^{3}H-thymidine and then transplanted neural tubes from labeled donors to unlabeled host embryos (Fig. 2B). This made it possible to follow the migration of the labeled cells deep into the embryo. In amniotes, Weston found that neural crest cells in the trunk

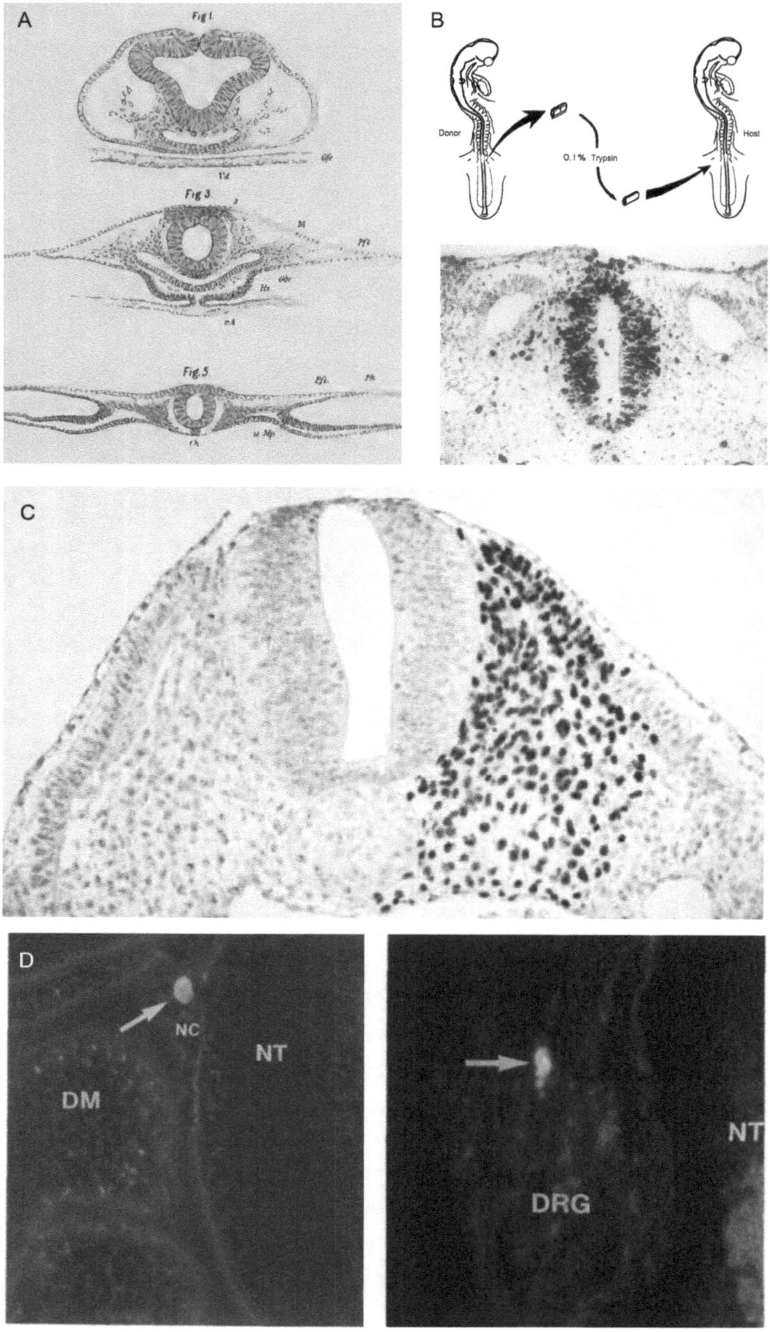

Figure 2 See legend on opposite page.

followed different pathways to contribute to ventral (dorsal root and sympathetic ganglia) or dorsolateral (melanocytes) derivatives. In addition, it was possible to perform these transplantation at different times, revealing that trunk neural crest cells contribute to their derivatives in a ventral to dorsal order, with the first cells to leave the neural tube migrating to the most ventral sites (Weston & Butler, 1966). Later, this was confirmed by experiments using the lipophilic dye, DiI, which made it possible to label the neural tube and neural crest cells without the need for grafting (Serbedzija, Bronner-Fraser, & Fraser, 1989). However, both radioactive and DiI labeling methods suffered from the issue of dilution of the label in the rapidly dividing neural crest population and thus were only amenable to relatively short-term studies.

A major technological advance was made by Le Douarin (1973a, 1973b) with her discovery that quail and chick embryos developed at relatively similar rates, but could be easily and indelibly distinguished since quail cells have a condensed mass of heterochromatin in their nucleolus whereas chick cells are euchromatic. By applying the histological Feulgen stain (Le Douarin & Teillet, 1974) and later using the quail-specific QCPN antibody to these quail–chick chimeras, it was possible to follow the long-term fate of grafted quail neural crest cells (Fig. 2C). Thus, by transplanting neural folds or whole neural tubes from stage-matched donor quails into chick hosts, Le Douarin and Teillet (1973, 1974) could follow quail cells that originated from within the neural tube but subsequently migrated and invaded distant sites in the

Figure 2 Classical experiments in neural crest research utilizing avian embryos. (A) Reproduction of the original drawings of His, including the first depiction of neural crest cells (His, 1868). The neural crest is identified with a "Z" for "Zwischenstrang" or cord between the neural and nonneural ectoderm. (B) Transplant experiments by Weston (1963), where neural tubes from radioactively labeled embryos were transplanted to unlabeled hosts. This allowed for the tracking of neural crest cells after delamination and identification of distinct migratory pathways. (C) The chick–quail chimeras of Nicole Le Douarin. Grafting of quail tissue in chicken embryos permitted extensive fate mapping of progenitor cells (Le Douarin, 2004). This approach was used to map migratory pathways and derivatives of the neural crest at distinct axial levels of the avian embryo (Le Douarin, 1982). (D) Clonal analysis of neural crest cells in the developing embryo. Labeling of single neural crest cells with vital dyes showed that many individual progenitor cells are multipotent and differentiate into diverse cell types *in vivo* (Fraser & Bronner-Fraser, 1991). *Panel (B) Reprinted from Weston (1963). Copyright (1963), with permission from Elsevier. Panel (C) Reprinted from Le Douarin (2004). Copyright (2004), with permission from Elsevier. Panel (D) Reprinted from Fraser and Bronner-Fraserer (1991). Copyright (1991), with permission from Company of Biologists.* (A color version of this figure is located at the end of this chapter.)

periphery. In this way, they were able to map the migration pathways and derivatives formed by neural crest cells at all axial levels of the bird embryo. These seminal experiments (see chapter "The Pluripotency of Neural Crest Cells and Their Role in Brain Development" by Le Douarin and Dupin) lead to the discovery that neural crest cells give rise to a stunning array of derivatives.

Importantly, Le Douarin and colleagues showed that the cell types formed by neural crest cells vary with the axial level from which they originate (summarized in Le Douarin, 1982). At cranial levels, neural crest cells populated cranial ganglia but also contributed to mesenchymal derivatives like connective tissue, cartilage, and bones of the face (Couly & Le Douarin, 1987; Le Lievre & Le Douarin, 1975). In the neck region, the vagal neural crest not only contributed to parts of the cardiac outflow tract but also invaded the foregut and migrated along the entire length of the gut to form enteric ganglia (Burns & Le Douarin, 2001; Le Douarin & Teillet, 1973). A small contribution to the enteric nervous system was also noted for the most caudal (lumbosacral) crest population. At trunk levels, neural crest cells migrated dorsolaterally to form pigment cells and ventrally to form dorsal root, sympathetic, and adrenomedullary cells in agreement with the radioactively labeling experiments of Weston (1963).

The diverse nature of derivatives along the body axis led to the question of whether there were intrinsic differences between neural crest populations at different axial levels or whether the choice of derivative was dictated by the environment. To address this, Le Douarin and colleagues performed "heterotopic" grafts in which the axial level was switched. For example, cranial neural crest cells were grafted to truncal levels and vice versa. The results revealed that there indeed were intrinsic differences between cranial and trunk neural crest cells. While cranial crest grafted to the trunk formed all normal trunk derivatives, they also differentiated into ectopic cartilage nodules. In the reciprocal experiment, trunk neural crest cells grafted to the head formed neuronal and glia derivatives in cranial sensory ganglia, but failed to form cartilage (Le Douarin, 1982). These results showed that, although there is flexibility in the neural crest program, there are also intrinsic, axial level specific differences.

While the grafting experiments were useful for examining the flexibility of neural crest cell fate at a population level, only clonal analysis of the progeny of single cells can inform upon the developmental potential of individual neural crest cells. Clonal analysis *in vitro* was first applied to the neural crest by Cohen and colleagues, who showed that clones derived from individual trunk neural crest cells in culture could indeed form multiple neural crest derivatives like neurons, pigment cells, and support cells (Cohen & Konigsberg, 1975). David Anderson and colleagues went on to show that not only were migratory neural crest cells multipotent but they also had the ability to self-renew in culture (Stemple & Anderson, 1992).

The most extensive clonal analyses of neural crest cells in culture were performed by Dupin, Le Douarin, and colleagues (Baroffio, Dupin, & Le Douarin, 1988; Dupin, Calloni, & Le Douarin, 2010), who showed that some clones of both cranial and trunk neural crest cells can give rise to a very broad array of derivatives, showing a large degree of multipotency, whereas others formed fewer derivatives. Interestingly, they found that both cranial and trunk clones formed some mesenchymal derivatives (i.e., express markers for cartilage and bone) though environmental factors like Shh greatly increased the percentage of trunk neural crest clones with chondrogenic ability (Calloni, Glavieux-Pardanaud, Le Douarin, & Dupin, 2007). These results may suggest that the developmental potential of neural crest cells is broader after exposure to culture conditions *in vitro* than those observed under normal developmental conditions *in vivo*.

An important adjunct to *in vitro* experiments above is clonal analysis in the developing embryo. This was first applied to premigratory (Bronner-Fraser & Fraser, 1988) and migratory (Bronner-Fraser & Fraser, 1989) neural crest cells in avian embryos using intracellular injection of vital dye into single neural crest precursors (Fig. 2D). In agreement with cell culture experiments described above, the results showed that many individual progenitor cells are multipotent, contributing to progeny in distinct anatomical structures and differentiating into diverse cell types *in vivo*. Recently, Sommer and colleagues (Baggiolini et al., 2015) found similar results by driving elegant "confetti" labels in conditional transgenic mouse lines that specifically label either premigratory or early migrating trunk neural crest cells. Given such similar findings in birds and mammals using different but complementary approaches, it is clear that the vast majority of both premigratory and early migrating neural crest cells *in vivo* are multipotent in amniotes and able to contribute to a broad range of neural crest derivatives. This is a lovely illustration of how important questions in neural crest biology (e.g., what is the developmental potential of individual neural crest cells?) remain highly relevant over decades even as the technology for lineage analysis has evolved from using dye labeling to elegant modern confetti technology. The interest of the question remains pertinent and can be examined at ever better resolution as technology advances.

3. NEURAL CREST MIGRATION

Quail–chick chimeras gave a general sense of the pathways of neural crest migration by identifying derivatives and showing that trunk neural crest cells migrated ventrally to form dorsal root and sympathetic ganglia (ventral pathway) or between the ectoderm and somites (dorsolateral pathway) to form melanocytes. However, the trajectories of individual cells could not be followed with the nuclear quail marker. Therefore, there was poor understanding of the mechanisms underlying migration prior to the discovery of neural crest markers that enabled visualization of cell membranes and/or live-cell imaging using vital dyes.

The first insight into the pathways followed by ventrally migrating trunk neural crest cells came from analyses using the HNK-1 antibody, which labels most migratory crest cells (Bronner-Fraser, 1986; Rickmann, Fawcett, & Keynes, 1985). Although it had been assumed previously that trunk neural crest cells migrated around the somites, careful analysis showed that they actually migrated in a segmental pattern through the rostral half of each somitic sclerotome and were restricted from the caudal half. This launched a search for possible rostrally localized attractants and/or caudally localized repellants. Although it was initially thought that ephrins in the caudal sclerotome were primarily responsible for repelling trunk neural crest cells because of their ability to do so *in vitro* (Krull et al., 1997; Wang & Anderson, 1997), the fact that ephrin null mice lacked an obvious neural crest phenotype made this less clear (Wang, Chen, & Anderson, 1998). The discovery that neural crest cells expressed the neuropilin 2 receptor and that Semaphorin3F, its inhibitory ligand, was expressed in the caudal sclerotome solved the puzzle; importantly, mice null for either the receptor or the ligand lost the segmental pattern of neural crest migration in the trunk (Gammill, Gonzalez, Gu, & Bronner-Fraser, 2006). In addition to inhibitory cues, trunk neural crest cells also display chemotactic behavior in response to SDF, which is expressed in a graded manner along the ventral to dorsal axis of the trunk region (Kasemeier-Kulesa, McLennan, Romine, Kulesa, & Lefcort, 2010), suggesting that a combination of attractive and repulsive cues guide their trunk neural crest migration.

In contrast to the relatively simple patterns of neural crest migration in the trunk, cranial neural crest cells migrate along many different pathways, ultimately filling most of the frontonasal process and branchial arches. With the advent of live imaging in chick, frog, and zebrafish embryos, much of what we know about the molecular mechanisms governing neural crest cell migratory behavior has come from analyses of the cranial neural crest population.

The collective migration of the cranial neural crest cells involves a combination of mechanisms, from contact inhibition to chemoattraction and coattraction. As neural crest cells undergo EMT, they switch the nature of their cadherins (Rogers, Saxena, & Bronner, 2013; Scarpa et al., 2015) and transition from nonmotile epithelial cells to migratory mesenchymal cells. This cadherin switch is essential for acquisition of contact inhibition of locomotion and cell repolarization (Scarpa et al., 2015). Contact inhibition of locomotion, originally described for fibroblast cells (Abercrombie & Heaysman, 1954), occurs when two neural crest cells collide; they stop moving at their site of contact and instead move away in the opposite direction. Using Xenopus neural crest cells, Mayor and colleagues have established an important role for contact inhibition of locomotion for cranial neural crest cells both *in vitro* and *in vivo* (Carmona-Fontaine et al., 2011). In addition, they have found that neural crest cells are attracted to each other via a ligand receptor interaction mediated by Complement3a (C3a) and the C3a receptor (Carmona-Fontaine et al., 2011) as well as by chemoattraction to the secreted factor SDF (Theveneau et al., 2010). These three systems of contact inhibition of locomotion, coattraction, and chemotaxis function cooperatively to pattern the cranial neural crest into forward moving streams in the frog embryo.

Other types of chemoattractants for neural crest cells include VEGF, which is important for cranial neural crest migration in the chick (McLennan et al., 2015) and GDNF and endothelin, which are critical for migration of vagal neural crest cells in populating the enteric nervous system (Sasselli, Pachnis, & Burns, 2012). The observation that these chemoattractants sometimes uniformly surround neural crest cells has lead to the idea that there may be a cell-induced gradient in which the "leader" neural crest cells acquire directional information that in turn instructs "trailer" cells to follow (McLennan et al., 2015). This is supported by the fact that leader cells have a different molecular profile that trailers (McLennan et al., 2015).

Studies that combine molecular perturbation and quantitative analysis to examine cell speed, trajectories, and perdurance of migration have provided important insights into neural crest migratory behavior (McLennan et al., 2015; Scarpa et al., 2015). In particular, the blend of imaging, molecular, and mathematical modeling approaches has greatly increased understanding of the molecular mechanisms underlying neural crest motility.

4. GENE REGULATORY NETWORK CONTROLLING NEURAL CREST FORMATION

With the development of perturbation approaches using antisense morpholinos, shRNAs, and dominant negative constructs, many individual genes have been implicated in the process of neural crest induction and migration in several vertebrates, most notably Xenopus, zebrafish, and chick. Taken together, these studies have lead to the assembly of a gene regulatory network (GRN) (Meulemans & Bronner-Fraser, 2004; Simões-Costa & Bronner, 2015) that helps explain the progression of neural crest development (Fig. 3).

In its current formulation, the neural crest GRN posits that inductive signals mediated by Wnts, BMPs, and FGFs establish the neural plate border between neural and nonneural ectoderm. These signals are thought to activate a suite of transcription factors, including Pax3/7, Msx1/2, and Zic1, whose expression overlaps at the neural plate border region. For example, in chick, Pax7 is expressed in the neural plate border at gastrula stages and required for subsequent neural crest formation (Basch, Bronner-Fraser, & Garcia-Castro, 2006). These "neural plate border specifier genes" in turn cooperate with signaling molecules to regulate the next tier of transcription factors, termed "neural crest specifier genes" whose expression initiates in neural crest precursors in the closing neural tube and prior to their emigration. These transcription factors are markers of bona fide neural crest precursors and include genes like *FoxD3*, *Sox9*, *Sox10*, *Myc*, *tfAP2*, *Id2*, and *Ets1* among others. The neural crest specifier genes regulate downstream factors necessary for neural crest EMT and lineage specification. For example, Sox10 is critical for differentiation of neuronal, glial, and melanocytic lineages, whereas Sox9 is involved in cartilage differentiation. The neural crest GRN has been best described at cranial levels and there are some apparent differences between axial levels; for example, *Ets1* is only expressed in cranial but not trunk neural crest cells.

With improved genomic analysis in multiple species and the application of transcriptomics to developmental systems, the neural crest field has migrated into the genomic age. Comparative genome analysis made it possible to identify enhancers that mediate expression of key neural crest genes like *FoxD3* (Simões-Costa, McKeown, Tan-Cabugoa, Sauka-Spengler, & Bronner, 2012) and *Sox10* (Betancur, Bronner-Fraser, & Sauka-Spengler, 2010). Interestingly, these studies have shown that distinct enhancers drive

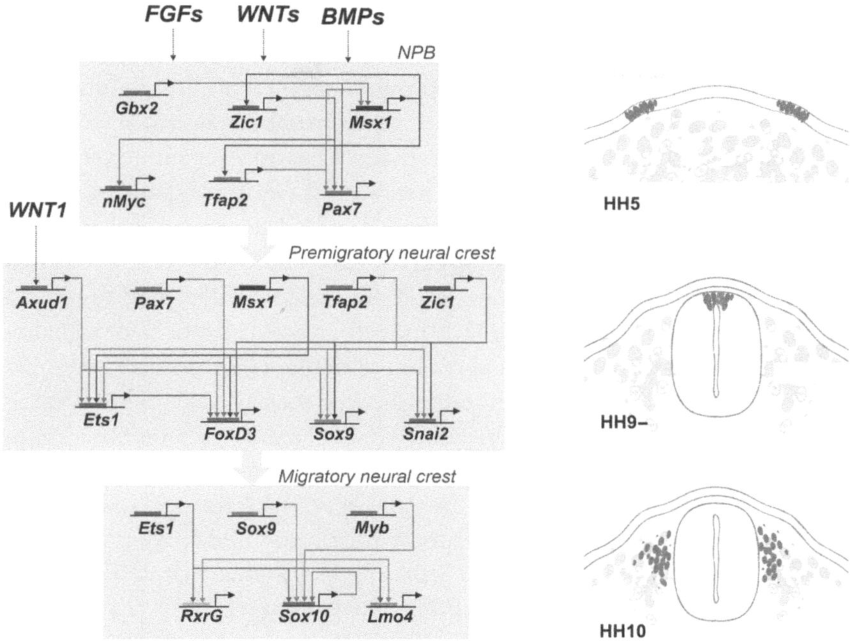

Figure 3 A gene regulatory network (GRN) controls neural crest formation. A complex genetic program comprised of numerous transcription factors underlies neural crest development. These factors are arranged in distinct hierarchical modules corresponding to the different stages of neural crest formation. First, in the neural plate border, the interplay of signaling systems such as Wnts, BMPs, and FGFs activates the neural plate border specifier genes, which include *Gbx2*, *Msx1*, *Zic1*, *Tfap2*, and *Pax7*. These genes drive the transition of the GRN to the neural crest specification module. Signaling systems such as Wnts also feed into the system by activating transcription factor Axud1, which interacts with PAX7 and MSX1 to activate expression of neural crest specifier genes (*FoxD3*, *Ets1*, *Sox9*, etc.) in the dorsal neural folds. The neural crest specifier genes endow the neural crest with its defining features and activate the process of epithelial-to-mesenchymal transition. In the migratory neural crest, there is activation of additional factors that will affect migration and drive differentiation in multiple cell types. (A color version of this figure is located at the end of this chapter.)

expression of these neural crest specifier genes at the cranial and trunk levels. For example, in the case of *FoxD3*, both cranial and trunk enhancers require common inputs from neural plate border genes PAX7 and MSX1/2. Whereas the cranial enhancer also requires ETS1 input, the trunk enhancer instead requires ZIC1 (Simões-Costa et al., 2012). Thus, there are both common and differential inputs mediating *FoxD3* expression in the head versus the trunk.

The availability of neural crest enhancers has facilitated isolation of pure populations of neural crest cells for transcriptome analysis. Importantly, this has greatly expanded the number of known neural crest transcriptional regulators (Simões-Costa, Tan-Cabugao, Antoshechkin, Sauka-Spengler, & Bronner, 2014). As the function of individual factors is examined in detail, the number of known inputs into the neural crest GRN is expanded. As case in point, transcriptome analysis identified *Axud1*, a transcriptional regulator thought to function as a Wnt effector, as highly upregulated in the cranial neural crest (Simões-Costa, Stone, & Bronner, 2015). Detailed functional analysis revealed that the AXUD1 is indeed downstream of Wnt signaling and a direct input required for expression of neural crest specifier genes like *FoxD3* (Fig. 3). Importantly, the results showed that AXUD1 directly interacts with PAX7 and MSX1 on the cranial *FoxD3* enhancer, thus functioning as an integrator of Wnt signaling and transcriptional input from neural plate border genes. This is an important example of how it is possible to go from a broad genomic database, zoom in, and perform detailed analysis of potentially interesting components and gain novel insights that inform upon the structure of the neural crest GRN.

Amazing advances in technology that have rendered sequencing better, faster, and cheaper have enabled analysis of neural crest development at multiple stages, axial levels, and in numerous species by RNA-seq. Moreover, the use of ChIP-seq on human embryonic stem cells differentiated in neural crest cells has identified numerous neural crest enhancers (Rada-Iglesias et al., 2012). Going forward, the combination of transcriptomics and novel methods of enhancer analysis, including ChIP-seq with epigenetic markers and ATAC-seq, will greatly aid in establishing direct connections within the neural crest GRN.

5. DIFFERENTIATION OF NEURAL CREST CELLS INTO MULTIPLE LINEAGES

After completion of migration, neural crest cells settle in discrete sites throughout the embryo where they differentiate into many and diverse cell types that differ depending upon their axial level of origin. The cranial neural crest gives rise to the craniofacial skeleton, including bone, cartilage, and connective tissue as well as most of the dental tissues. In addition to enteric nervous system, the vagal neural crest contributes to the cardiac mesenchyme and smooth muscle of the great vessels. The trunk neural crest also differentiates into a number of secretory cell types such as chromaffin and parafollicular cells of the adrenal and the thyroid glands, respectively. Neural crest cells from all axial levels differentiate into melanocytes and contribute to the peripheral nervous system, giving rise to sensory, sympathetic and parasympathetic neurons, and multiple types of glial cells (Le Douarin, 1982).

The process of cell differentiation requires the deployment of differentiation gene batteries, which are sets of genes that endow each cell with its terminal identity (Erwin & Davidson, 2009). For instance, the differentiation gene battery of a muscle cell includes a large number of contractile proteins, which allow this cell to exert its function. Gene batteries are controlled by differentiation drivers, which are regulators that orchestrate the transition between the progenitor and differentiated states (Peter & Davidson, 2015). In the neural crest, these differentiation drivers are often genes from the neural crest specifier module that are retained during cell migration. In particular, SoxE transcription factors seem to play a central role in activating differentiation gene batteries during neural crest differentiation. Sox10 is involved in the differentiation of the neural crest into neurons, glia, and melanocytes (reviewed by Kelsh, 2006), while Sox9 drives neural crest progenitors toward a chondrocytic fate (Mori-Akiyama, Akiyama, Rowitch, & de Crombrugghe, 2003).

As expected, the activation of gene batteries also requires inputs from signaling systems, which cooperate with differentiation drivers to activate multiple downstream targets. Neural crest differentiation is contingent upon the interactions between the regulatory machinery present in each cell and extrinsic signals that they encounter during migration (Simões-Costa & Bronner, 2015). Since migratory neural crest cells are multipotent, environmental cues play a crucial part in determining the fate of their progeny. These cues are conveyed through signaling systems that are active along the neural crest migratory pathways. For instance, Kit signaling plays a central role in differentiation of the neural crest into melanocytes (Parichy, Rawls, Pratt, Whitfield, & Johnson, 1999), and BMP signaling plays a central role in the differentiation during autonomic nervous system formation (Saito, Takase, Murai, & Takahashi, 2012; Shah, Groves, & Anderson, 1996). To interpret such signals, neural crest cells are endowed with a rich repertoire of receptor molecules (Simões-Costa et al., 2014). Activation of signal transduction pathways impacts gene expression in the migrating cells and influences the fate they will subsequently adopt.

As is often the case in embryonic development, the same signaling systems and transcription factors are used reiteratively during neural crest formation and differentiation. For instance, Sox9 has an early role in neural crest specification (Cheung & Briscoe, 2003) and is later a driver of differentiation to cartilage and bone. Similarly, Wnt signaling plays a critical role first in the determination of the neural plate border territory (reviewed by Prasad, Sauka-Spengler, & LaBonne, 2012), second during neural crest specification (García-Castro, Marcelle, & Bronner-Fraser, 2002; Simões-Costa et al., 2015), and finally in the differentiation of neural crest cells into sensory neurons (Lee et al., 2004). At later stages, Wnts also impact differentiation of neural crest cells into melanocytes, through the activation of the melanocytic driver MITF (Dorsky, Raible, & Moon, 2000). Thus, neural crest cells can interpret similar signals differently in distinct moments of their ontogeny, reflecting the profound shifts in GRN architecture that take place during development.

6. EVOLUTION OF THE NEURAL CREST

All vertebrates including basal jawless vertebrates (agnathans) have neural crest cells that arise from the dorsal tube, undergo an EMT, and migrate extensively. In fact, Gans and Northcutt (1983) proposed that the "New Head" of vertebrates was facilitated by the invention of the neural crest which enabled formation of specific vertebrate features like peripheral ganglia and the craniofacial skeleton. In all vertebrates, there is also a common molecular signature in neural crest cells, such that the premigratory neural crest expresses a suite of transcription factors, most notably *FoxD3* and *SoxE* (Green, Simões-Costa, & Bronner, 2015; Sauka-Spengler, Meulemans, Jones, & Bronner-Fraser, 2007). However, gnathostomes (jawed vertebrates) have some neural crest derivatives that are absent from jawless agnathans like lamprey, including sympathetic ganglia and jaws. Thus, there seems to have been an expansion of cell types formed by the neural crest during the course of vertebrate evolution.

In contrast to vertebrates, basal chordates like amphioxus completely lack neural crest, even though their body plan shares morphological characteristics similar to those of vertebrates, including a dorsal hollow nerve cord, notochord, and segmented mesoderm. From a GRN point of view, amphioxus has neural plate border genes that exhibit similar expression to those of vertebrates. However, neural crest specifier genes, although present in the genome, appear to be absent from the neural folds/dorsal neural tube of amphioxus with the exception AmphiSnail, which is transiently expressed in the lateral portion of the neural tube (Yu, Meulemans, McKeown, & Bronner-Fraser, 2008).

Although cephalochordates like amphioxus resemble vertebrates morphologically, comparative genomic analysis has shown that urochordates, rather than cephalochordates, are the sister group of vertebrates (Holland et al., 2008). Several studies of ascidian embryos have led to the speculation that tunicates may possess cell types with some but not all characteristics present in vertebrate neural crest cells. Most notably, they possess cells that arise at the neural plate border and can differentiate into pigment cells and/or neurons. For example, Abitua, Wagner, Navarrete, and Levine (2012) have found that the a9.49 lineage of *Ciona intestinalis* gives rise to nonmotile pigmented sensory cells. Precursor cells of this lineage express several transcription factors like *Snail*, *ETS*, and *FoxD* genes that are part of the vertebrate neural crest specifier module. Moreover, ectopic expression of Twist in these cells causes them to become migratory. Recently Stolfi, Ryan, Meinertzhagen, and Christiaen (2015) have discovered another lineage in Ciona that gives rise to migratory neuronal precursors. Arising within the posterior neural plate border, these cells subsequently migrate along the mesenchyme and differentiate into BTN sensory neurons. Like neural crest precursors, these cells are migratory and form peripheral neurons. Thus, these cells may represent some type of intermediate that has some but not all of the neural crest GRN. An intriguing possibility is that acquisition of other neural crest specifier genes like SoxE genes to the neural plate border and neural folds may have conferred multipotency and the ability to undergo EMT to these border cells in order to generate a bona fide neural crest.

7. RECENT INSIGHTS INTO NEURAL CREST STEM CELLS IN POSTNATAL TISSUE: "SCHWANN CELL PRECURSORS"

One of the most exciting recent advances in the neural crest field has been the discovery of neural crest stem cells in postnatal tissues. Although it has long been speculated that neural crest cells are "stem" cells, the reality is that the embryonic neural crest is a transient population that quickly transitions from multipotent to restricted progenitors, and their capacity to self-renew is rather limited. This has raised a conundrum in the field: with a cell population like neural crest cells that have a feed-forward developmental program going from multipotency to relatively rapid restriction in cell fate and differentiation, how do vertebrates continue to add differentiated cells to neural crest derivatives with postembryonic growth?

The answer may lie in the fact that a subset of the neural crest population is able to maintain multipotency after embryonic development has concluded. Some evidence suggests that many neural crest derivatives, including skin and peripheral nerves, contain stem cells with the ability to give rise to multiple derivatives (Jinno et al., 2010; Morrison, White, Zock, & Anderson, 1999). Recently, it has been found that "Schwann cell precursors" along nerve processes are able to differentiate into numerous derivatives postnatally. These include pigment cells of the skin (Adameyko et al., 2009), parasympathetic ganglia of the head (Dyachuk et al., 2014; Espinosa-Medina et al., 2014), as well as enteric neurons of the gut (Uesaka, Nagashimada, & Enomoto, 2015). These studies definitively demonstrate the existence of neural crest-derived stem cells with the ability to contribute to melanocytes and neurons at much later times in development than normally associated with neural crest differentiation.

An intriguing possibility is that these may represent the true neural crest stem cells. The degree to which these crest-derived stem cells contribute to derivatives of the adult is not yet known. We speculate that this cell population may contribute to many derivatives classically attributed to the embryonic neural crest. This will be a research area ripe for future discovery.

8. CONCLUSION

This is an incredibly exciting time in biology due to the amazing advances in technology that have emerged in the past decade and continue to develop. We live in an era of high-throughput sequencing, ATAC- and ChIP-seq, CRISPR-Cas9 genome editing, super high-resolution microscopy, and mass-spec. From improvements in genomic analysis, genome editing, imaging, and data acquisition, scientists can examine important question at ever higher resolution. Because of these technologies, the possibility of unraveling mysteries of neural crest formation, migration, differentiation, and evolution is in closer reach than was previously imaginable. And yet the important questions remain unresolved: How do cells at the neural plate border become a neural crest cell rather than part of the CNS, ectoderm, or placodes? Why do some cells undergo EMT and what are the accompanying molecular changes? What determines their pathways of migration at a single cell versus population level? How does a cell decide to differentiate into a neuron versus a glial or pigment cell, etc.? How are neural crest derivatives replaced upon loss in the adult? New tools and innovations hold the promise of answering these long standing questions that have intrigued neural crest investigators for over 100 years.

REFERENCES

Abercrombie, M., & Heaysman, J. E. (1954). Observations on the social behaviour of cells in tissue culture: I. Speed of movement of chick heart fibroblasts in relation to their mutual contacts. *Experimental Cell Research*, *5*, 111–131.

Abitua, P. B., Wagner, E., Navarrete, I. A., & Levine, M. (2012). Identification of a rudimentary neural crest in a non-vertebrate chordate. *Nature*, *492*, 104–107.

Adameyko, I., Lallemend, F., Aquino, J. B., Pereira, J. A., Topilko, P., Müller, T., et al. (2009). Schwann cell precursors from nerve innervation are a cellular origin of melanocytes in skin. *Cell*, *139*, 366–379.

Baggiolini, A., Varum, S., Mateos, J. M., Bettosini, D., Nessy, J., Ziegler, U., et al. (2015). Genetic lineage tracing demonstrates multipotency of premigratory and migratory neural crest cells in vivo. *Cell Stem Cell*, *6*(3), 314–322.

Baroffio, A., Dupin, E., & Le Douarin, N. M. (1988). Clone-forming ability and differentiation potential of migratory neural crest cells. *Proceedings of the National Academy of Sciences of the United States of America*, *85*, 5325–5329.

Basch, M., Bronner-Fraser, M., & Garcia-Castro, M. (2006). Specification of neural crest occurs during gastrulation and requires Pax7. *Nature*, *441*, 218–222.

Betancur, P., Bronner-Fraser, M., & Sauka-Spengler, T. (2010). Genomic code for Sox10, a key regulatory enhancer for cranial neural crest. *Proceedings of the National Academy of Sciences of the United States of America*, *107*, 3570–3575.

Bronner-Fraser, M. (1986). Analysis of the early stages of trunk neural crest migration in avian embryos using the monoclonal antibody HNK-1. *Developmental Biology*, *115*, 44–55.

Bronner-Fraser, M., & Fraser, S. (1988). Cell lineage analysis shows multipotentiality of some avian neural crest cells. *Nature*, *335*(8), 161–164.

Bronner-Fraser, M., & Fraser, S. (1989). Developmental potential of avian trunk neural crest cells in situ. *Neuron*, *3*(6), 755–766.

Burns, A. J., & Le Douarin, N. M. (2001). Enteric nervous system development: Analysis of the selective developmental potentialities of vagal and sacral neural crest cells using quail-chick chimeras. *The Anatomical Record*, *262*(1), 16–28.

Calloni, G. W., Glavieux-Pardanaud, C., Le Douarin, N. M., & Dupin, E. (2007). Sonic Hedgehog promotes the development of multipotent neural crest progenitors endowed with both mesenchymal and neural potentials. *Proceedings of the National Academy of Sciences of the United States of America*, *104*(50), 19879–19884.

Carmona-Fontaine, C., Theveneau, E., Tzekou, A., Tada, M., Woods, M., Page, K. M., et al. (2011). Complement fragment C3a controls mutual cell attraction during collective cell migration. *Developmental Cell*, *21*, 1026–1037.

Cheung, M., & Briscoe, J. (2003). Neural crest development is regulated by the transcription factor Sox9. *Development*, *130*, 5681–5693.

Chibon, P. (1967). Marquage nucléaire par la thymidine tritiée des dérivés de la crête neurale chez l'Amphibien Urodèle *Pleurodeles waltlii* Michah. *Journal of Embryology and Experimental Morphology*, *18*(3), 343–358.

Cohen, A. M., & Konigsberg, I. R. (1975). A clonal approach to the problem of neural crest determination. *Developmental Biology*, *46*(2), 262–280.

Couly, G. F., & Le Douarin, N. M. (1987). Mapping of the early neural primordium in quail-chick chimeras. II. The prosencephalic neural plate and neural folds: Implications for the genesis of cephalic human congenital abnormalities. *Developmental Biology*, *120*(1), 198–214.

Dorsky, R. I., Raible, D. W., & Moon, R. T. (2000). Direct regulation of nacre, a zebrafish MITF homolog required for pigment cell formation, by the Wnt pathway. *Genes & Development*, *14*, 158–162.

Dupin, E., Calloni, G. W., & Le Douarin, N. M. (2010). The cephalic neural crest of amniote vertebrates is composed of a large majority of precursors endowed with neural, melanocytic, chondrogenic and osteogenic potentialities. *Cell Cycle*, *9*(2), 238–249.

Dyachuk, V., Furlan, A., Shahidi, M. K., Giovenco, M., Kaukua, N., Konstantinidou, C., et al. (2014). Neurodevelopment. Parasympathetic neurons originate from nerve-associated peripheral glial progenitors. *Science*, *345*, 82–87.

Erwin, E. H., & Davidson, E. H. (2009). The evolution of hierarchical gene regulatory networks. *Nature Reviews. Genetics*, *10*, 141–148.

Espinosa-Medina, I., Outin, E., Picard, C. A., Chettouh, Z., Dymecki, S., Consalez, G. G., et al. (2014). Neurodevelopment. Parasympathetic ganglia derive from Schwann cell precursors. *Science*, *345*, 87–90.

Fraser, S. E., & Bronner-Fraser, M. (1991). Migrating neural crest cells in the trunk of the avian embryo are multipotent. *Development*, *112*(4), 913–920.

Gammill, L., Gonzalez, C., Gu, C., & Bronner-Fraser, M. (2006). Guidance of trunk neural crest migration requires Neuropilin-2/Semaphorin3F signaling. *Development*, *133*, 99–106.

Gans, C., & Northcutt, R. G. (1983). Neural crest and the origin of vertebrates: A new head. *Science*, *220*, 268–274.

García-Castro, M., Marcelle, C., & Bronner-Fraser, M. (2002). Ectodermal Wnt function as a neural crest inducer. *Science*, *297*, 848–851.

Green, S., Simões-Costa, M., & Bronner, M. E. (2015). Evolution of vertebrates as viewed from the crest. *Nature*, *520*, 474–482.

His, W. (1868). *Untersuchungen über die erste Anlage des Wirbeltierliebes. Die erste Entwicklung des Hünchens im Ei*. Leipzig.

Holland, L., et al. (2008). The amphioxus genome illuminates vertebrate origins and cephalochordate biology. *Genome Research*, *18*, 1100–1111.

Hörstadius, S. (1950). *The neural crest: Its properties and derivatives in the light of experimental research*. London, New York: Oxford University Press.

Jinno, H., Morozova, O., Jones, K. L., Biernaskie, J. A., Paris, M., Hosokawa, R., et al. (2010). Convergent genesis of an adult neural crest-like dermal stem cell from distinct developmental origins. *Stem Cells*, *28*, 2027–2040.

Kasemeier-Kulesa, J. C., McLennan, R., Romine, M. H., Kulesa, P. M., & Lefcort, F. (2010). CXCR4 controls ventral migration of sympathetic precursor cells. *The Journal of Neuroscience*, *30*, 13078–13088.

Kelsh, R. N. (2006). Sorting out Sox10 functions in neural crest development. *BioEssays: News and Reviews in Molecular, Cellular and Developmental Biology*, *28*, 788–798.

Krull, C. E., Lansford, R., Gale, N. W., Marcelle, C., Collazo, A., Yancopoulos, G., et al. (1997). Interactions between Eph-related receptors and ligands confer rostrocaudal polarity to trunk neural crest migration. *Current Biology*, *7*, 571–580.

Le Douarin, N. M. (1973a). A biological cell labeling technique and its use in experimental embryology. *Developmental Biology*, *30*, 217–222.

Le Douarin, N. M. (1973b). A Feulgen-positive nucleolus. *Experimental Cell Research*, *77*(1), 459–468.

Le Douarin, N. (1982). *The neural crest*. Cambridge: Cambridge University Press.

Le Douarin, N. (2004). The avian embryo as a model to study the development of the neural crest: A long and still ongoing story. *Mechanisms of Development*, *121*(9), 1089–1102.

Le Douarin, N. M., & Teillet, M. A. (1973). The migration of neural crest cells to the wall of the digestive tract in avian embryo. *Journal of Embryology and Experimental Morphology*, *1*, 31–48.

Le Douarin, N. M., & Teillet, M. A. (1974). Experimental analysis of the migration and differentiation of neuroblasts of the autonomic nervous system and of neurectodermal mesenchymal derivatives, using a biological cell marking technique. *Developmental Biology*, *41*(1), 162–184.

Le Lievre, C. S., & Le Douarin, N. M. (1975). Mesenchymal derivatives of the neural crest: Analysis of chimaeric quail and chick embryos. *Journal of Embryology and Experimental Morphology*, *34*, 125–154.

Lee, H. Y., Kleber, M., Hari, L., Brault, V., Suter, U., Taketo, M. M., et al. (2004). Instructive role of Wnt/beta-catenin in sensory fate specification in neural crest stem cells. *Science*, *303*, 1020–1023.
McLennan, R., Schumacher, L. J., Morrison, J. A., Teddy, J. M., Ridenour, D. A., Box, A. C., et al. (2015). Neural crest migration is driven by a few trailblazer cells with a unique molecular signature narrowly confined to the invasive front. *Development*, *142*(11), 2014–2025.
Meulemans, D., & Bronner-Fraser, M. (2004). Gene-regulatory interactions in neural crest development and evolution. *Developmental Cell*, *7*, 291–299.
Mori-Akiyama, Y., Akiyama, H., Rowitch, D. H., & de Crombrugghe, B. (2003). Sox9 is required for determination of the chondrogenic cell lineage in the cranial neural crest. *Proceedings of the National Academy of Sciences of the United States of America*, *100*, 9360–9365.
Morrison, S. J., White, P. M., Zock, C., & Anderson, D. J. (1999). Prospective identification, isolation by flow cytometry, and in vivo self-renewal of multipotent mammalian neural crest stem cells. *Cell*, *96*, 737–749.
Parichy, D. M., Rawls, J. F., Pratt, S. J., Whitfield, T. T., & Johnson, S. L. (1999). Zebrafish sparse corresponds to an orthologue of c-kit and is required for the morphogenesis of a subpopulation of melanocytes, but is not essential for hematopoiesis or primordial germ cell development. *Development*, *126*, 3425–3436.
Peter, I., & Davidson, E. H. (2015). *Genomic control process: Development and evolution.* San Diego: Academic Press.
Prasad, M. S., Sauka-Spengler, T., & LaBonne, C. (2012). Induction of the neural crest state: Control of stem cell attributes by gene regulatory, post-transcriptional and epigenetic interactions. *Developmental Biology*, *366*, 10–21.
Rada-Iglesias, A., Bajpai, R., Prescott, S., Brugmann, S. A., Swigut, T., & Wysocka, J. (2012). Epigenomic annotation of enhancers predicts transcriptional regulators of human neural crest. *Cell Stem Cell*, *11*(5), 633–648.
Rickmann, M., Fawcett, J. W., & Keynes, R. J. (1985). The migration of neural crest cells and the growth of motor axons through the rostral half of the chick somite. *Journal of Embryology and Experimental Morphology*, *90*, 437–455.
Rogers, C., Saxena, A., & Bronner, M. E. (2013). Sip1 mediates an E-cadherin-to-N-cadherin switch during cranial neural crest EMT. *The Journal of Cell Biology*, *203*, 835–847.
Saito, D., Takase, Y., Murai, H., & Takahashi, Y. (2012). The dorsal aorta initiates a molecular cascade that instructs sympatho-adrenal specification. *Science*, *336*, 1578–1581.
Sasselli, V., Pachnis, V., & Burns, A. J. (2012). The enteric nervous system. *Developmental Biology*, *366*(1), 64–73.
Sauka-Spengler, T., Meulemans, D., Jones, M., & Bronner-Fraser, M. (2007). Ancient evolutionary origin of the neural crest gene regulatory network. *Developmental Cell*, *13*, 405–420.
Scarpa, E., Szabó, A., Bibonne, A., Theveneau, E., Parsons, M., & Mayor, R. (2015). Cadherin switch during EMT in neural crest cells leads to contact inhibition of locomotion via repolarization of forces. *Developmental Cell*, *34*(4), 421–434.
Selleck, M. A. J., & Bronner-Fraser, M. (1995). Origins of the avian neural crest: The role of neural plate-epidermal interactions. *Development*, *121*, 526–538.
Serbedzija, G., Bronner-Fraser, M., & Fraser, S. E. (1989). Vital dye analysis of the timing and pathways of avian trunk neural crest cell migration. *Development*, *106*, 806–816.
Shah, N. M., Groves, A. K., & Anderson, D. J. (1996). Alternative neural crest cell fates are instructively promoted by TGFbeta superfamily members. *Cell*, *85*, 331–343.
Simões-Costa, M., & Bronner, M. E. (2015). Establishing neural crest identity: A gene regulatory recipe. *Development*, *142*, 242–257.
Simões-Costa, M., McKeown, S., Tan-Cabugoa, J., Sauka-Spengler, T., & Bronner, M. E. (2012). Dynamic and differential regulation of stem cell factor FoxD3 in the neural crest is encrypted in the genome. *PLoS Genetics*, *8*. e1003142.

Simões-Costa, M., Stone, M., & Bronner, M. E. (2015). Axud1 integrates Wnt signaling and transcriptional inputs to drive neural crest formation. *Developmental Cell*, *34*(5), 544–554.

Simões-Costa, M., Tan-Cabugao, J., Antoshechkin, I., Sauka-Spengler, T., & Bronner, M. E. (2014). Transcriptome analysis reveals novel players in the cranial neural crest gene regulatory network. *Genome Research*, *24*, 281–290.

Stemple, D. L., & Anderson, D. J. (1992). Isolation of a stem cell for neurons and glia from the mammalian neural crest. *Cell*, *71*, 973–985.

Stolfi, A., Ryan, K., Meinertzhagen, I. A., & Christiaen, L. (2015). Migratory neuronal progenitors arise from the neural plate borders in tunicates. *Nature*, *527*, 371–374.

Theveneau, E., Marchant, L., Kuriyama, S., Gull, M., Moepps, B., Parsons, M., et al. (2010). Collective chemotaxis requires contact-dependent cell polarity. *Developmental Cell*, *19*(1), 39–53.

Twitty, V. C. (1936). Correlated genetic and embryological experiments in Triturus: I and II. *The Journal of Experimental Zoology*, *74*, 239–302.

Uesaka, T., Nagashimada, M., & Enomoto, H. (2015). Neuronal differentiation in Schwann cell lineage underlies postnatal neurogenesis in the enteric nervous system. *The Journal of Neuroscience*, *35*(27), 9879–9888.

Wang, H. U., & Anderson, D. J. (1997). Eph family transmembrane ligands can mediate repulsive guidance of trunk neural crest migration and motor axon outgrowth. *Neuron*, *18*(3), 383–396.

Wang, H. U., Chen, Z. F., & Anderson, D. J. (1998). Molecular distinction and angiogenic interaction between embryonic arteries and veins revealed by ephrin-B2 and its receptor Eph-B4. *Cell*, *93*(5), 741–753.

Weston, J. A. (1963). A radioautographic analysis of the migration and localization of trunk neural crest cells in the chick. *Developmental Biology*, *6*, 279–310.

Weston, J. A., & Butler, S. L. (1966). Temporal factors affecting localization of neural crest cells in the chicken embryo. *Developmental Biology*, *14*(2), 246–266.

Yntema, C. L., & Hammond, W. S. (1945). Depletions and abnormalities in the cervical sympathetic system of the chick following extirpation of neural crest. *The Journal of Experimental Zoology*, *100*, 237–263.

Yu, J. K., Meulemans, D., McKeown, S., & Bronner-Fraser, M. (2008). Insights from the amphioxus genome on the origin of vertebrate neural crest. *Genome Research*, *18*, 1127–1132.

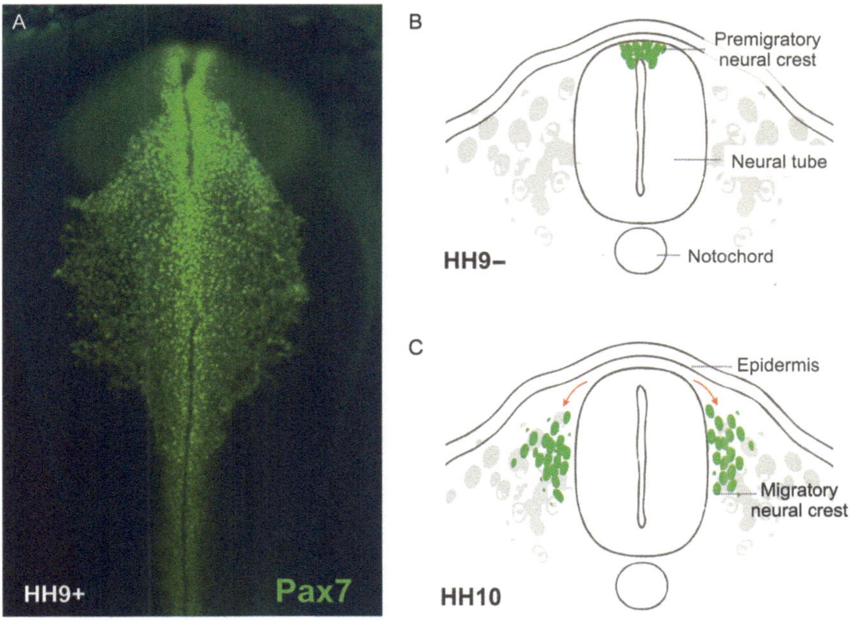

Marianne E. Bronner and Marcos Simões-Costa, Figure 1 Neural crest cells delaminate from the neural tube and undergo extensive migration. (A) Immunohistochemistry with an antibody for neural crest marker Pax7 highlights neural crest cells delaminating from the neural tube of a stage HH9 chicken embryo. (B) Diagram representing a cross section of a neurulating chicken embryo. At the stage represented, the neural cells occupy the dorsal portion of the neural tube. (C) Neural crest cells eventually undergo epithelial-to-mesenchymal transition to delaminate from the neural tube and undergo extensive migration.

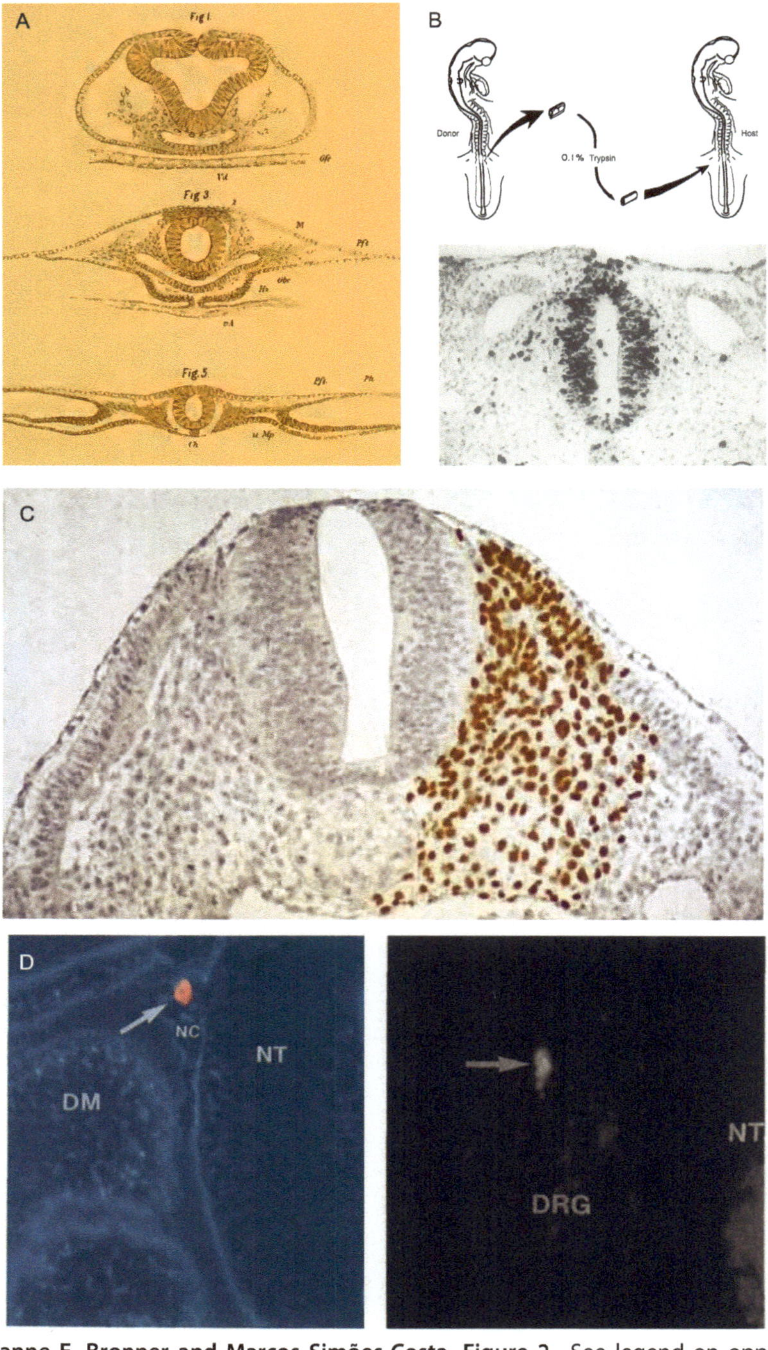

Marianne E. Bronner and Marcos Simões-Costa, Figure 2 See legend on opposite page.

Marianne E. Bronner and Marcos Simões-Costa, Figure 2 Classical experiments in neural crest research utilizing avian embryos. (A) Reproduction of the original drawings of His, including the first depiction of neural crest cells (His, 1868). The neural crest is identified with a "Z" for "Zwischenstrang" or cord between the neural and nonneural ectoderm. (B) Transplant experiments by Weston (1963), where neural tubes from radioactively labeled embryos were transplanted to unlabeled hosts. This allowed for the tracking of neural crest cells after delamination and identification of distinct migratory pathways. (C) The chick–quail chimeras of Nicole Le Douarin. Grafting of quail tissue in chicken embryos permitted extensive fate mapping of progenitor cells (Le Douarin, 2004). This approach was used to map migratory pathways and derivatives of the neural crest at distinct axial levels of the avian embryo (Le Douarin, 1982). (D) Clonal analysis of neural crest cells in the developing embryo. Labeling of single neural crest cells with vital dyes showed that many individual progenitor cells are multipotent and differentiate into diverse cell types *in vivo* (Fraser & Bronner-Fraser, 1991). *Panel (B) Reprinted from Weston (1963). Copyright (1963), with permission from Elsevier. Panel (C) Reprinted from Le Douarin (2004). Copyright (2004), with permission from Elsevier. Panel (D) Reprinted from Fraser and Bronner-Fraserer (1991). Copyright (1991), with permission from Company of Biologists.*

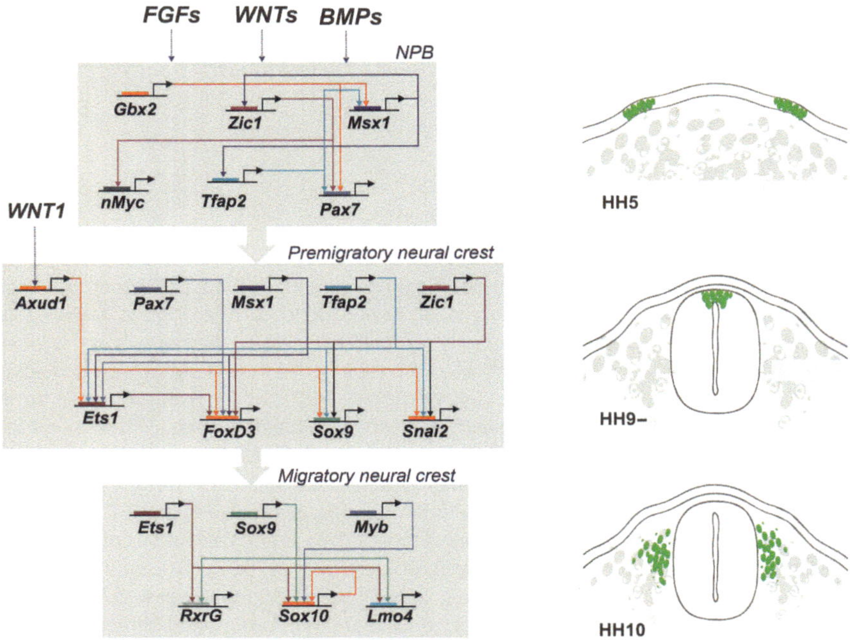

Marianne E. Bronner and Marcos Simões-Costa, Figure 3 A gene regulatory network (GRN) controls neural crest formation. A complex genetic program comprised of numerous transcription factors underlies neural crest development. These factors are arranged in distinct hierarchical modules corresponding to the different stages of neural crest formation. First, in the neural plate border, the interplay of signaling systems such as Wnts, BMPs, and FGFs activates the neural plate border specifier genes, which include *Gbx2*, *Msx1*, *Zic1*, *Tfap2*, and *Pax7*. These genes drive the transition of the GRN to the neural crest specification module. Signaling systems such as Wnts also feed into the system by activating transcription factor Axud1, which interacts with PAX7 and MSX1 to activate expression of neural crest specifier genes (*FoxD3*, *Ets1*, *Sox9*, etc.) in the dorsal neural folds. The neural crest specifier genes endow the neural crest with its defining features and activate the process of epithelial-to-mesenchymal transition. In the migratory neural crest, there is activation of additional factors that will affect migration and drive differentiation in multiple cell types.

CHAPTER EIGHT

Tissue Differentiation: A Personal Account of Research on Myogenesis and Cardiogenesis

Margaret Buckingham[1]
Department of Developmental and Stem Cell Biology, Institut Pasteur, Paris, France
[1]Corresponding author: e-mail address: margaret.buckingham@pasteur.fr

Contents

Abstract

In this essay I trace my own research experience as a developmental biologist, from the study of cell differentiation *in vitro* to tissue formation and regeneration *in vivo*. Beginning with a thesis on histone modifications, I went on to study gene regulation during myogenesis, first in muscle cells in culture and then in the mouse embryo. Later, we also worked on muscle regeneration in the adult. Our work on striated muscle genes also led us into the field of cardiogenesis and characterization of cardiac progenitor cells that form the heart. Comments on the state of the art—changing concepts and the technological advances that underlie scientific progress—accompany this account, with concluding remarks about future directions.

Current Topics in Developmental Biology, Volume 116
ISSN 0070-2153
http://dx.doi.org/10.1016/bs.ctdb.2015.11.025

1. GENE REGULATION DURING CELL DIFFERENTIATION: STARTING WITH A THESIS ON HISTONE MODIFICATIONS

At the end of the 1960s, the discovery of the genetic code and messenger RNA had revolutionized concepts of heredity and how inherited information is used. What applied to prokaryotes was clearly directly relevant to multicellular eukaryotic organisms—to quote Jacques Monod "what is true for E. coli is true for the elephant." Given that all cells of an organism contain the same genetic information, this information must be used differently to generate distinct cell types as development proceeds. Understanding the regulation of such differential gene expression presented a major challenge.

At that time, we knew that DNA was compacted in the nucleus as a nucleoprotein complex, referred to as chromatin, with histones as an important component. These small basic proteins, tightly associated with DNA, were therefore potential players in regulating gene transcription. After obtaining a first degree in biochemistry, I had the possibility of continuing in the Oxford biochemistry department as a D. Phil. student with Dr. Margery Ord. She and Lloyd Stocken were studying histone modifications and my thesis work consisted in following phosphorylation and acetylation of H1 and H3 histones under different cellular conditions. Lymphocyte cultures (the immunologist Rodney Porter was head of the department) provided a model cell system. The techniques were essentially biochemical, with fractionation of different classes of histone and identification of the type of amino acid modified. However, although we could distinguish phosphorylated serines or threonines for example, we did not know where they were located in the protein sequence. We could grossly block a modification, but we did not have the means to manipulate particular amino acids. These experiments showed that histone modifications varied according to different phases of the cell cycle and levels of gene expression (e.g., Cross & Ord, 1970), but did not go beyond the stage of correlation. It would take sophisticated genetic manipulation of histone sequences to obtain functional insights into the role of histone modifications that led to the concept of a histone code (Jenuwein & Allis, 2001).

2. MESSENGER RNA POPULATIONS IN DIFFERENTIATING MUSCLE CELLS

Messenger RNAs, as the immediate readout of gene expression, provide a direct indication of how a cell uses its genetic information as its developmental status changes. Working backward from the mRNA to the gene leads to the underlying question of transcriptional control. After my thesis, I went to Paris to the Pasteur Institute as a postdoc in the lab of François Gros, to work on messenger RNA. At that time he and François Jacob, after the characterization of mRNA in bacteria (Jacob & Monod, 1961), were turning their attention to eukaryotes. Jacob was primarily interested in early embryonic development and was beginning to work with teratocarcinomas. Trying to unravel what might be happening for single genes at the level of the embryo was a herculean task with the tools available, and *in vitro* cell systems appeared more tractable. With F. Gros, we focussed on the progression of a tissue-specific precursor cell to formation of the tissue, so-called terminal differentiation. Muscle tissue provided an attractive model since the precursor cells, myoblasts, could be physically separated from differentiated muscle fibers which were large syncythia formed by myoblast cell fusion. Primary cultures of mononucleated cells were enriched in myoblasts and devoid of differentiated muscle cells. Furthermore, David Yaffe had succeeded in cloning myoblasts and establishing mouse and rat muscle cell lines (Richler & Yaffe, 1970).

Myoblasts spontaneously differentiate when the cultures become confluent. It is thus possible to follow their progression from proliferating precursor cells to the onset of differentiation, followed by the formation of muscle fibers. In our initial studies, we characterized changes in mRNA populations by plotting Rot curves based on hybridization kinetics (Bishop, 1969) and changes in the location of mRNAs in polysomes or ribonucleoprotein particles separated on sucrose gradients. The identification of 3′polyadenylation as a feature of mRNAs had made it possible to separate this class of RNA on polyA/U affinity columns. Radioactive labeling of mRNAs in cultured cells and pulse-chase experiments provided information on mRNA dynamics (e.g., Buckingham, Caput, Cohen, Whalen, & Gros, 1974), showing, for example, that some mRNAs move from the RNP compartment into polysomes at the onset of differentiation when messenger stability also increases. Even this level of general information was conceptually useful in the uncharted domain of cell differentiation. More precise information on the mRNAs came from *in vitro* translation experiments with analysis of the protein products by two-dimensional gel electrophoresis. The major muscle proteins present in fibers were well known and we could now separate different isoforms of contractile proteins such as actins or myosin light chains and follow the switch to those specific for differentiating muscle cells (e.g., Daubas, Caput, Buckingham, & Gros, 1981).

3. MUSCLE GENE EXPRESSION DURING MYOGENESIS *IN VIVO*—THE ADVENT OF CLONED PROBES

Like all experimental research, progress was very dependent on technical advances. The possibility of cloning DNA, which opened up the whole field of genetic engineering at the end of the 1970s, had a huge impact on developmental biology. It became possible to isolate a cDNA sequence for a specific mRNA and to separate genomic DNA of a specific gene. Developments in DNA sequencing technology, initially Maxam and Gilbert, with the fun of reading the nucleotide sequence of bases off the ladder on the gel, and then Sanger, made it possible to characterize the coding sequence. We could thus generate specific probes, often based on noncoding 3′ or 5′ mRNA sequences to avoid cross hybridization between mRNAs coding for similar isoforms. We could describe precisely when different muscle-specific genes began to be expressed as differentiation proceeded in muscle cell cultures. Importantly, specific probes now made it possible to look at myogenesis *in vivo*.

Thus, we identified sequential transitions from the expression of embryonic to fetal to adult myosin genes and also were surprised to find coexpression, during development, of cardiac with skeletal muscle genes, such as those of the actins, before the tissue-specific family member predominated (see Buckingham et al., 1986). The advent of *in situ* hybridization with cDNA probes on muscle sections gave a spatial perspective on gene expression. Classical embryology, coupled with histology, had described the early sites of formation of skeletal muscle and chick-quail chimera grafting experiments had identified the somites as the source of skeletal muscle in the body (see Wachtler & Christ, 1992). We could now map the onset of myogenesis in the mouse embryo in terms of specific marker gene expression.

We could also characterize the structure and organization of the corresponding contractile protein genes. Gene splicing had been documented by then, but the phenomenon of tissue-specific differential splicing was spectacularly demonstrated by genes like Troponin T with up to 64 splice forms (Breitbart et al., 1985). It was, and indeed still is, not clear how significant these all are. The mechanism by which different exons are "chosen" was also a mystery which is only now beginning to be elucidated. Another novel variation on the spicing theme was shown by the fast myosin light chain gene that encodes two functionally distinct isoforms,

MLC1 and MLC3, which are generated by differential spicing with transcription from two promoters (Robert et al., 1984). Genes expressed sequentially during development are clustered at the same chromosomal locus, suggesting shared *cis*-acting regulatory elements, whereas cardiac or skeletal muscle genes that are coexpressed are dispersed in the genome (see Buckingham et al., 1986).

4. MYOGENIC REGULATORY FACTORS OF THE MyoD FAMILY—GENE MANIPULATION IN THE MOUSE

In the 1980s, the next step was to look at the regulation of these genes, to approach the underlying question of how tissue differentiation is initiated. For muscle, as for other tissue models, regulatory sequences were mapped by cell transfection experiments and later by transgenesis when this important *in vivo* approach became accessible. The "holy grail" was a tissue-specific regulatory factor, a tissue determinant, which would orchestrate the activation of tissue-specific genes. In *Drosophila*, genetic approaches had identified *msh* as a muscle-specific homeobox gene. We cloned the mouse homologue *Hox7* (Robert, Sassoon, Jacq, Gehring, & Buckingham, 1989), now known as *Msx1*, which turned out to play a major role in limb development but not in vertebrate myogenesis. The discovery of MyoD, by Hal Weintraub and his collaborators, was a major breakthrough in this respect (see Weintraub et al., 1991). They used an ingenious screen based on myogenic conversion of the multipotent C3H10T1/2 cell line, to isolate and clone cDNA sequences which when overexpressed in these cells led to myogenesis. MyoD was the first tissue determination factor to be isolated. In addition to the 10T1/2 fibroblast cell line, overexpression of MyoD resulted in muscle gene activation and "myogenic conversion" of many cell types. It turned out that there are four members of the muscle-specific MyoD family of basic-helix–loop–helix transcription factors. When we carried out *in situ* hybridization to characterize their expression patterns, we had a bet with Hal Weintraub who was determined that MyoD should be the master regulator. We won since, to his disappointment, other members of the family were expressed in myoblasts well before *MyoD*, at the onset of myogenesis in the mouse embryo (Sassoon et al., 1989; see Tajbakhsh & Buckingham, 2000). Tragically, he died prematurely and it was left to others to follow on with the MyoD story.

The next major technological breakthrough, that revolutionized the study of mouse developmental biology, came from embryonic stem (ES) cell-based techniques to target gene sequences by homologous recombination and then to introduce mutant cells into the blastocyst and to finally obtain a mutant mouse. This opened the way to functional studies *in vivo*. At the beginning of the 1990s, it was shown that in the absence of Myf5 the onset of myogenesis was delayed (Braun, Rudnicki, Arnold, & Jaenisch, 1992) and that in the absence of both Myf5 and MyoD skeletal muscle was totally lacking (Rudnicki et al., 1993). We provided further evidence of the role of Myf5 in myogenic determination by showing that at the onset of myogenesis in $Myf5^{nlacZ/nlacZ}$ embryos, mutant cells, assumed other cell fates (Tajbakhsh, Rocancourt, & Buckingham, 1996). We had been scooped on *Myf5* inactivation, but fortunately had introduced a reporter which allowed us to follow the mutant cells, a facet of gene targeting which has now become common practice. There was a cautionary tale here since mutation of *Myf5* interfered with a second myogenic factor gene, *Mrf4*, present in the same locus and also expressed with *Myf5* at the onset of myogensis (Olson, Arnold, Rigby, & Wold, 1996). Thus, MyoD, Myf5, and Mrf4 were demonstrated to act as determination factors *in vivo*, conferring myogenic potential to myoblasts (Kassar-Duchossoy et al., 2004). The fourth member of the family, Myogenin, is required for myoblast differentiation, a role also played by Mrf4, after its early expression with Myf5 (see Tajbakhsh & Buckingham, 2000). In keeping with this distinction, Myogenin does not have the same chromatin remodeling activity as the three myogenic determination factors (Gerber, Klesert, Bergstrom, & Tapscott, 1997).

For many years, the MyoD family remained a unique example of tissue determinants. This was probably because in most cases tissue determination depends on a combination of factors. For example, it was only in 2008 that a combination of 3 factors was shown to program insulin-producing β-cells (Zhou, Brown, Kanarek, Rajagopal, & Melton, 2008) and, as has often proved to be the case for tissue conversion, only from a closely related pancreatic cell type. One can ask why skeletal muscle should be programed by a single type of transcription factor. Muscle gene activation does of course involve multiple transcription factors but these are either present also in non-myogenic cells or can be directly, or indirectly, activated by the MyoD family. Myogenic conversion depends on the remodeling of repressive chromatin, with recognition of lineage-specific epigenetic marks, and there is now an increasing amount of data on interactions between MyoD and proteins that remodel chromatin structure, although many questions still remain unanswered (see Buckingham & Rigby, 2014).

5. UPSTREAM REGULATORS OF MYOGENESIS: FROM THE *Myf5* GENE TO Pax3

Characterization of the MyoD family of myogenic regulatory factors explained how cells are determined to make skeletal muscle, but upstream of this ultimate step is the regulation of cell fate choices, or lineage commitment, as development proceeds. Experimentally this meant working backward, with regulation of *Myf5*, at the onset of myogenesis in the mouse embryo, as a starting point. It rapidly became apparent that key regulatory sequences were far from the gene itself and we struggled with yeast artificial chromosomes, YACs, for cloning large genomic fragments until BAC technology made this much easier. It is a feature of upstream regulatory genes that they have multiple control elements, often distributed over large distances around the gene, probably reflecting the complex spatiotemporal control required and also the recruitment of regulatory elements during vertebrate radiation. An aberrant example of recent genome shuffling is provided by a sequence present in the mouse *Myf5* locus, not present in other mammalian genomes, that directs transcription to a specific subset of neurons. This inappropriate and potentially dangerous expression of a myogenic determination factor is prevented by posttranscriptional repression of the mRNA by microRNA (Daubas et al., 2009). Presumably, the "neuronal" sequence is destined to mutate and disappear! Where myogenesis is concerned, *Myf5* regulatory sequences, distributed over more than 100 kb 5′ of the gene, direct transcription to different sites of myogenesis, thus providing an exquisite example of fine spatiotemporal control (see Buckingham & Rigby, 2014). As one might expect, regulatory sequences that have been characterized contain response elements to signaling pathways reflecting the microenvironment, as well as regulatory factors of potential interest in directing progression into the myogenic lineage.

Of these factors, Pax3, which is expressed in myogenic progenitor cells prior to activation of the myogenic determination genes, was of particular interest and we concentrated on the function of this transcription factor in the myogenic context. Like other members of the Pax family, Pax3 is not tissue specific—it is also important in the neural crest cell lineage, for example. Pax factors play key upstream roles in the formation of many tissues and organs (see Buckingham & Relaix, 2007). A classic example is provided by Pax6 which is a master control gene for eye formation, a role conserved from *Drosophila* to man (see Gehring & Ikeo, 1999)!

6. GENETIC ANALYSIS OF Pax3 FUNCTION IN THE MOUSE EMBRYO

By the beginning of the twenty-first century when we began making new reporter and mutant alleles—gain as well as loss of function—of *Pax3*, gene targeting in the mouse was a routine procedure. However, it was time consuming and setbacks were potentially disastrous for students or postdocs working in a limited time frame. The recent CRISPR/Cas9 methodology has completely changed the time scale—another key technological revolution for developmental biology! With mutant analysis and transgenic experiments, we could show that Pax3 is required to activate key regulatory elements of the *Myf5* gene. Analysis of *Pax3* "*splotch*" spontaneous mutants had already demonstrated interesting myogenic phenotypes (see Buckingham & Relaix, 2007), notably loss of limb musculature, reflecting the role of Pax3 in activation of the *c-met* gene required for migration of cells from the somite to the limb. Pax3 is also required for the survival of these myogenic progenitors. Other functions of Pax3 in myogenic progenitor cells were not known. Since a classic screen for potential Pax3 targets on a loss-of-function background was complicated by cell death, we carried out a gain-of-function screen, using an allele which encoded a fusion protein with the DNA-binding domain of Pax3 fused to a strong transcriptional activation domain which we had shown would save the mutant phenotype, thus confirming that Pax3 is a transcriptional activator in the myogenic context. For such a screen on embryonic material, which is scarce and even when dissected contains a number of cell types, it was essential to have a tool for separating Pax3-positive myogenic progenitors. This was provided by *GFP* inserted into a *Pax3* allele. From these screens, we isolated and validated a number of Pax3 targets. These act at different stages in the progression toward myogenesis, to regulate progenitor cell behavior (see Buckingham & Relaix, 2015). Prior to activation of *Myf5* and myogenic determination, survival, migration, and self-renewal of progenitor cells are orchestrated by Pax3.

Cell fate choices in the somite depend on tipping the balance of reciprocal negative regulation between *Pax3* and *Foxc2*. Initially, *Pax3* and *Foxc2* are coexpressed throughout the early epithelial somite. Subsequently in the ventral somite, the sclerotome, *Foxc2* (and *Foxc1*) are highly expressed with repression of *Pax3*, leading to the formation of bone and cartilage; whereas in the dorsal somite, the dermomyotome, *Pax3* expression remains high. In this context, upregulation of *Foxc2* promotes nonmyogenic cell fates such as vascular smooth muscle, whereas maintenance of high levels of Pax3 is the hallmark of cells that enter the myogenic lineage (Lagha et al., 2009). Signals, such as Notch acting on the hypaxial dermomyotome, favor nonmyogenic lineages with upregulation of *Foxc2*, as evidenced by the promotion of endothelial cells that contribute to the vasculature of the forelimb, at the expense of myogenic cells (Mayeuf-Louchart et al., 2014). Maintenance of multipotency by reciprocal inhibition between regulatory genes, with cell fate acquisition depending on perturbation of the equilibrium is an important stem cell concept that has also been shown to apply to myeloid lineages in the hematopoietic system (Huang, Guo, May, & Enver, 2007).

Another developmental concept that emerges from work on Pax factors is the multilevel control of lineage progression, discussed here for Pax3. Key upstream regulatory factors are likely to orchestrate the gradual acquisition of tissue identity. The cell survival function of Pax3 is one facet of this and makes sense in that an undirected tissue progenitor cell is a developmental liability. Pax3 is not alone in directing entry into the myogenic program. Other upstream factors, such as Six1/4 or Pitx2 also play a role at certain sites of myogenesis (see Buckingham & Rigby, 2014). As so often happens in vertebrate development, there is also a sister gene, *Pax7*, which is closely related to *Pax3* and has partially overlapping expression in the myogenic context. It is initially expressed with *Pax3* in the central domain of the dermomyotome. This breaks down as development proceeds to release myogenic progenitor cells into all maturing muscle masses, providing the progenitor cell population for subsequent growth. In the *Pax3/Pax7* double mutant, this cell population is lost and there is a major muscle deficit (Relaix, Rocancourt, Mansouri, & Buckingham, 2005), thus demonstrating the key role of these factors in establishing this essential myogenic stem cell population. During fetal and postnatal muscle development *Pax3* is downregulated, whereas Pax7 continues to be expressed in all muscle progenitor cells. This is also the case in the adult.

7. MYOGENIC STEM CELLS—SATELLITE CELLS—AND ADULT MUSCLE REGENERATION

In the last 10 years, adult tissue progenitor cells, often referred to as stem cells, have been a major focus of attention. Stem cell journals have proliferated and impact factors in this area are higher than those of the classic developmental biology journals. This of course partly reflects interest in the potential applications of stem cells to regenerative medicine. For a developmental biologist, the similarities between developmental and regenerative stem cell strategies are very interesting. The stem cells of adult muscle, known as satellite cells because of their location on muscle fibers, are marked by Pax7, and often also by Pax3 (see Buckingham & Relaix, 2015). The demonstration of their role in regeneration depended on their purification, in our case using a $Pax3^{GFP}$ allele and flow cytometry (Montarras et al., 2005). After injury quiescent satellite cells receive activating signals, proliferate and then form new muscle fibers. As in the embryo, this myogenic program depends on MyoD family members. Most satellite cells transcribe the *Myf5* gene, which in the embryo would immediately lead to entry into the myogenic program. A number of mechanisms have now been invoked to explain the maintenance of quiescence. We showed that *Myf5* mRNA is sequestered in ribonucleoprotein granules with repressive microRNA and that these granules break down very rapidly on satellite cell activation, with consequent translation of *Myf5* mRNA and accumulation of the myogenic factor, leading to the regenerative cascade (Crist, Montarras, & Buckingham, 2012). This mechanism thus maintains the tissue identity of the stem cells, present as reserve cells which can very rapidly enter the regenerative program.

Satellite cells have been the focus of much recent research on muscle. Analysis of gene function in the adult often depends on conditional mutations, to bypass earlier mutant lethality, and the development of approaches, such as the tamoxifen-inducible *ERT2-Cre* acting on lox sites inserted around the sequence to be mutated, has made it possible to precisely examine later tissue-specific roles. The role of Pax7 in satellite cells provides a cautionary tale. It is now established that regeneration depends on this regulatory factor (see Buckingham & Relaix, 2015), but initial experiments led to the opposite conclusion because it is difficult to obtain efficient Cre-mediated recombination in all these adult cells and it turned out that even if only a few "escape" they can repopulate the satellite cell pool. It is now clear that, in the absence of Pax7 in the adult, activated satellite cells do not proliferate sufficiently and differentiate prematurely, compromising muscle repair. Pax3 does not save this phenotype although its role awaits detailed analysis of the *Pax3* conditional mutant.

More in depth analysis of Pax3/7 as transcription factors, using modern "omics" technologies on muscle cell lines and satellite cells have provided important new insights into preferred binding sites for Pax3 or Pax7 and cofactors that affect chromatin dynamics. Posttranscriptional regulation at the level of protein turnover and by microRNA interaction with *Pax3/7* mRNAs has also been revealed by work on satellite cells (see Buckingham & Relaix, 2015).

There remain important questions to pursue, such as heterogeneity of satellite cells, molecular characterization of the satellite cell niche on the muscle fiber, signals that trigger activation, the critical regulation of self-renewal versus differentiation, and the emerging field of metabolic control of stem cell states. However many important questions also remain to be addressed concerning myogenesis in the embryo. Insights from the adult stem cell will often be relevant to development also, but there are particular questions such as tissue signaling and spatiotemporal control of tissue formation which are theme tunes of development. Furthermore, genetic manipulations in the embryo will continue to uncover aspects of gene regulatory networks that also apply to the adult.

8. CARDIOGENESIS

In our studies on actin and myosin genes expressed during skeletal muscle differentiation, we also began to look at the heart since there is an overlap between the two striated muscle types. The embryonic myosin light chain typical of developing skeletal muscle is expressed in cardiac atria, for example, and cardiac and skeletal actins are coexpressed before becoming restricted to one tissue. As one might expect, the cardiac actin gene has skeletal as well as cardiac enhancer elements (e.g., Biben, Hadchouel, Tajbakhsh, & Buckingham, 1996), but it is still not clear how the tissue-specific switch is implemented, presumably one enhancer becomes inaccessible because of chromatin modifications as the tissue matures. This rather unfashionable investigation of tissue regulation still merits more research. Cardiac gene transcription does not depend on the MyoD family, although other regulators like the Mef2 family are shared by both tissues, acting with cardiac transcription factors such as Nkx2–5. There is a considerable body of literature on modulation of cardiac gene expression during development (see Olson, 2006) and on the role of microRNAs (see Liu & Olson, 2010).

Our main contribution to cardiogenesis came from an unexpected observation. We were looking at transgene expression with a β-galactosidase reporter under the control of ventricular myosin light chain regulatory elements. To our surprise, in one line the transgene was expressed in cells lying adjacent to the heart instead of in the ventricles. This was the kind of aberrant result which usually is of little interest. However, in this case, it turned out that the transgene had inserted into the *Fgf10* locus and that Fgf10 marks an important population of cardiac progenitor cells which had not been previously characterized in the mouse embryo (Kelly, Brown, & Buckingham, 2001). These cells, that contribute to the arterial pole of the heart, are part of a larger domain that we called the second heart field which feeds into the heart tube as development proceeds. Lineage studies using retrospective clonal analysis based on the rare recombination of a *nlaacZ* genetic marker (Bonnerot & Nicolas, 1993), developed by my colleague J.-F. Nicolas at the Pasteur Institute, demonstrated the myocardial contributions of this second lineage, which forms both the venous and arterial poles of the heart as well as most of the right ventricle and part of the atria (Meilhac, Esner, Kelly, Nicolas, & Buckingham, 2004). It segregates early from the first lineage which is the source of left ventricular myocardium. The so-called first heart field forms the primitive cardiac tube into which second heart field cells feed as development proceeds. Older embryological observations on the chick embryo had pointed to such an addition of adjacent cells, at least posteriorly to the venous pole of the heart. An anterior contribution to the arterial pole of the chick heart was reported at the same time as our initial observations with the mouse transgene. Despite the older avian work, a prevalent model for cardiogenesis in the early 2000s was based on the idea that the different compartments of the heart were prespecified in the early cardiac tube which then grew by proliferation of these cells already *in situ*—the so-called segmentation model (see Buckingham, Meilhac, & Zaffran, 2005). In fact it is only later in development, when second heart field cells have entered the cardiac tube, that the chambers are formed as a result of spatially restricted cell proliferation controlled by the Tbx family and other regulatory factors, as proposed in the conceptually important "ballooning model" (Christoffels et al., 2000).

Identification of the second heart field has important implications for understanding congenital heart disease. Malformations at the arterial pole of the heart, which are frequent in the human population, reflect anomalies in this progenitor cell population which may also result in malformations at the venous pole (Bajolle et al., 2009). Islet1 was first shown to mark the second heart field (Cai et al., 2003), and since then extensive gene regulatory networks have been established (see Vincent & Buckingham, 2010). It is clear that the field is patterned—Fgf10, for example, marks an anterior domain—and clarifying this heterogeneity, how it is established and its functional consequences are on-going challenges (see Meilhac, Lescroart, Blanpain, & Buckingham, 2015). Another challenge is to identify markers of the first lineage and the so-called first heart field which is a useful concept but has not been well characterized. Related to this is the question of the timing of lineage segregation. The *Mesp1* gene, expressed in the primitive streak, marks all cardiac progenitors (Saga, Kitajima, & Miyagawa-Tomita, 2000) and recent genetic tracing experiments, using a conditional system, have given new insight into lineage segregation and also, with the inclusion of a GFP marker, opened the way to fractionation and characterization of the progenitor cells in the first as well as the second lineage (Lescroart et al., 2014).

9. MUSCLES OF THE HEAD AND HEART

An amusing twist to these parallel themes of skeletal and cardiac muscle is their convergence for the upstream regulation of head muscles. Head myogenesis depends on the MyoD family but is controlled by different upstream factors and not by Pax3 which characterizes the paraxial mesoderm that forms somites (see Buckingham & Rigby, 2014). Genetic tracing experiments have shown that regulatory factors expressed in the second heart field, such as Islet1, also mark the progenitor cells of these skeletal muscles. Furthermore, our retrospective clonal analysis established lineage relationships between arterial and venous pole myocardium and different head or neck muscles respectively (see Meilhac et al., 2015). The common origin of muscles of the head and heart has led to the concept of cranio-pharyngeal mesoderm, which has interesting evolutionary implications (Diogo et al., 2015).

10. CONCLUDING COMMENTS

It is a truism to say that technical advances are important drivers of research. In my attempts to address the question of how cells acquire tissue identity, it is clear that major technological breakthroughs like DNA cloning, which made it possible to identify and engineer specific sequences, or transgenesis and gene targeting *in vivo*, which opened up the perspective of functional analyses in the mouse model, had a major impact. CRISPR/Cas9 technology for *in vivo* gene manipulation now constitutes another important technical step forward. In the case of protein detection, new tools are potentially available but have not yet become widely used, for example, aptamer technology (see Tsae & DeRosa, 2015), based on oligonucleotides specific to protein domains, has not replaced antibodies which often pose problems. In my own area, the development of skeletal muscle cell lines gave myogenesis a head start as a model of cell differentiation. The absence of equivalent myocardial cell systems retarded the identification of cardiac regulatory factors. ES cell lines that readily differentiate into cardiac cell types, now provide a model for lineage progression to the cardiomyocyte. Genetic tracing of cell history *in vivo* has become a major tool, with a number of potential pitfalls due to assumptions about expression of the marker gene. Conditional activation in a brief time window increases the power of this approach (see Buckingham & Meilhac, 2011). For the mouse embryo, *in vitro* culture makes it possible to perform manipulations, for example, labeling cells in the second heart field and following their fate over a limited time span (Domínguez, Meilhac, Bland, Buckingham, & Brown, 2012). Live imaging remains problematic, although technical advances should facilitate this. The long-term goal would be to perform single-cell analysis *in vivo* to address questions of cell heterogeneity and cell behavior, even in the mouse embryo. Meanwhile, analysis of isolated single cells has now become feasible. When I began examining gene expression during myogenic differentiation we could only look at populations of mRNAs. Today, we carry out large scale "omics" analyses and construct gene regulatory networks on populations of cells. Eric Davidson, who alas died recently, succeeded in constructing regulatory networks for the sea urchin embryo at the single-cell level (Peter & Davidson, 2011) and this elegant work indicates the potential conceptual pitfalls of averaging out data on gene function. Cell biologists now employ sophisticated fluorescent labeling to look at molecular interactions in real time. In the mouse, with more rapid techniques for gene manipulation, constructing alleles which permit visualization of molecular events within the cell is feasible, but again the *in vivo* imaging at the cellular level is problematic. Some tissues near the surface in cultured embryos may be accessible with current techniques, but otherwise explant cultures are more promising at present for this type of fine analysis. Putting together information on single cells to reconstruct what is happening throughout the tissue will require complex informatics, but, if we are to understand the processes that underlie the spatiotemporal control of progenitor cells and their tissue differentiation, this level of resolution is essential. It will no doubt lead to surprises that will shift today's conceptual framework.

Much of the research first covered by *Current Topics in Developmental Biology* 50 years ago has been superseded, but some of the results and concepts are still applicable and merit attention. It was Andrew Huxley who warned against the exponential rate at which previous publications cease to be read, with loss of precious knowledge and wasteful repetition of experiments.

ACKNOWLEDGMENTS

M.B. thanks the CNRS and the Pasteur Institute for their support.

REFERENCES

Bajolle, F., Zaffran, S., Losay, J., Ou, P., Buckingham, M., & Bonnet, D. (2009). Conotruncal defects associated with anomalous pulmonary venous connections. *Archives of Cardiovascular Diseases, 102*, 105–110.

Biben, C., Hadchouel, J., Tajbakhsh, S., & Buckingham, M. (1996). Developmental and tissue-specific regulation of the murine cardiac actin gene in vivo depends on distinct skeletal and cardiac muscle-specific enhancer elements in addition to the proximal promoter. *Developmental Biology, 173*, 200–212.

Bishop, J. O. (1969). Interpretation of DNA-RNA hybridization data. *Nature, 224*, 600–603.

Bonnerot, C., & Nicolas, J. F. (1993). Clonal analysis in the intact mouse embryo by intragenic homologous recombination. *Comptes Rendus de l'Académie Des Sciences. Série III, Sciences de La Vie, 316*, 1207–1217.

Braun, T., Rudnicki, M. A., Arnold, H. H., & Jaenisch, R. (1992). Targeted inactivation of the muscle regulatory gene Myf-5 results in abnormal rib development and perinatal death. *Cell, 71*, 369–382.

Breitbart, R. E., Nguyen, H. T., Medford, R. M., Destree, A. T., Mahdavi, V., & Nadal-Ginard, B. (1985). Intricate combinatorial patterns of exon splicing generate multiple regulated troponin T isoforms from a single gene. *Cell, 41*, 67–82.

Buckingham, M., Alonso, S., Barton, P., Cohen, A., Daubas, P., Garner, I., et al. (1986). Actin and myosin multigene families: Their expression during the formation and maturation of striated muscle. *American Journal of Medical Genetics, 25*, 623–634.

Buckingham, M. E., Caput, D., Cohen, A., Whalen, R. G., & Gros, F. (1974). The synthesis and stability of cytoplasmic messenger RNA during myoblast differentiation in culture. *Proceedings of the National Academy of Sciences of the United States of America, 71*, 1466–1470.

Buckingham, M. E., & Meilhac, S. M. (2011). Tracing cells for tracking cell lineage and clonal behavior. *Developmental Cell, 21*, 394–409.

Buckingham, M., Meilhac, S., & Zaffran, S. (2005). Building the mammalian heart from two sources of myocardial cells. *Nature Reviews. Genetics, 6*, 826–835.

Buckingham, M., & Relaix, F. (2007). The role of Pax genes in the development of tissues and organs: Pax3 and Pax7 regulate muscle progenitor cell functions. *Annual Review of Cell and Developmental Biology, 23*, 645–673.

Buckingham, M., & Relaix, F. (2015). Pax3 and Pax7 as upstream regulators of myogensis. *Seminars in Cell and Developmental Biology, 4*, 115–125.

Buckingham, M., & Rigby, P. W. J. (2014). Gene regulatory networks and transcriptional mechanisms that control myogenesis. *Developmental Cell, 28*, 225–238.

Cai, C.-L., Liang, X., Shi, Y., Chu, P.-H., Pfaff, S. L., Chen, J., et al. (2003). Isl1 identifies a cardiac progenitor population that proliferates prior to differentiation and contributes a majority of cells to the heart. *Developmental Cell, 5*, 877–889.

Christoffels, V. M., Habets, P. E., Franco, D., Campione, M., de Jong, F., Lamers, W. H., et al. (2000). Chamber formation and morphogenesis in the developing mammalian heart. *Developmental Biology, 223*, 266–278.

Crist, C. G., Montarras, D., & Buckingham, M. (2012). Muscle satellite cells are primed for myogenesis but maintain quiescence with sequestration of Myf5 mRNA targeted by microRNA-31 in mRNP granules. *Cell Stem Cell, 11*, 118–126.

Cross, M. E., & Ord, M. G. (1970). Changes in the phosphorylation and thiol content of histones in phytohaemagglutinin-stimulated lymphocytes. *The Biochemical Journal, 118*, 191–193.

Daubas, P., Caput, D., Buckingham, M., & Gros, F. (1981). A comparison between the synthesis of contractile proteins and the accumulation of their translatable mRNAs during calf myoblast differentiation. *Developmental Biology*, *84*, 133–143.

Daubas, P., Crist, C. G., Bajard, L., Relaix, F., Pecnard, E., Rocancourt, D., et al. (2009). The regulatory mechanisms that underlie inappropriate transcription of the myogenic determination gene Myf5 in the central nervous system. *Developmental Biology*, *327*, 71–82.

Diogo, R., Kelly, R. G., Christiaen, L., Levine, M., Ziermann, J. M., Molnar, J. L., et al. (2015). A new heart for a new head in vertebrate cardiopharyngeal evolution. *Nature*, *520*, 466–473.

Domínguez, J. N., Meilhac, S. M., Bland, Y. S., Buckingham, M. E., & Brown, N. A. (2012). Asymmetric fate of the posterior part of the second heart field results in unexpected left/right contributions to both poles of the heart. *Circulation Research*, *111*, 1323–1335.

Gehring, W. J., & Ikeo, K. (1999). Pax 6: Mastering eye morphogenesis and eye evolution. *Trends in Genetics: TIG*, *15*, 371–377.

Gerber, A. N., Klesert, T. R., Bergstrom, D. A., & Tapscott, S. J. (1997). Two domains of MyoD mediate transcriptional activation of genes in repressive chromatin: A mechanism for lineage determination in myogenesis. *Genes & Development*, *11*, 436–450.

Huang, S., Guo, Y.-P., May, G., & Enver, T. (2007). Bifurcation dynamics in lineage-commitment in bipotent progenitor cells. *Developmental Biology*, *305*, 695–713.

Jacob, F., & Monod, J. (1961). Genetic regulatory mechanisms in the synthesis of proteins. *Journal of Molecular Biology*, *3*, 318–356.

Jenuwein, T., & Allis, C. D. (2001). Translating the histone code. *Science (New York, N.Y.)*, *293*, 1074–1080.

Kassar-Duchossoy, L., Gayraud-Morel, B., Gomès, D., Rocancourt, D., Buckingham, M., Shinin, V., et al. (2004). Mrf4 determines skeletal muscle identity in Myf5:Myod double-mutant mice. *Nature*, *431*, 466–471.

Kelly, R. G., Brown, N. A., & Buckingham, M. E. (2001). The arterial pole of the mouse heart forms from Fgf10-expressing cells in pharyngeal mesoderm. *Developmental Cell*, *1*, 435–440.

Lagha, M., Brunelli, S., Messina, G., Cumano, A., Kume, T., Relaix, F., et al. (2009). Pax3: Foxc2 reciprocal repression in the somite modulates muscular versus vascular cell fate choice in multipotent progenitors. *Developmental Cell*, *17*, 892–899.

Lescroart, F., Chabab, S., Lin, X., Rulands, S., Paulissen, C., Rodolosse, A., et al. (2014). Early lineage restriction in temporally distinct populations of Mesp1 progenitors during mammalian heart development. *Nature Cell Biology*, *16*, 829–840.

Liu, N., & Olson, E. N. (2010). MicroRNA regulatory networks in cardiovascular development. *Developmental Cell*, *18*, 510–525.

Mayeuf-Louchart, A., Lagha, M., Danckaert, A., Rocancourt, D., Relaix, F., Vincent, S. D., et al. (2014). Notch regulation of myogenic versus endothelial fates of cells that migrate from the somite to the limb. *Proceedings of the National Academy of Sciences of the United States of America*, *111*, 8844–8849.

Meilhac, S. M., Esner, M., Kelly, R. G., Nicolas, J.-F., & Buckingham, M. E. (2004). The clonal origin of myocardial cells in different regions of the embryonic mouse heart. *Developmental Cell*, *6*, 685–698.

Meilhac, S. M., Lescroart, F., Blanpain, C., & Buckingham, M. E. (2015). Cardiac cell lineages that form the heart. *Cold Spring Harbor Perspectives in Medicine*, *5*, a026344.

Montarras, D., Morgan, J., Collins, C., Relaix, F., Zaffran, S., Cumano, A., et al. (2005). Direct isolation of satellite cells for skeletal muscle regeneration. *Science (New York, N.Y.)*, *309*, 2064–2067.

Olson, E. N. (2006). Gene regulatory networks in the evolution and development of the heart. *Science (New York, N.Y.)*, *313*, 1922–1927.

Olson, E. N., Arnold, H. H., Rigby, P. W., & Wold, B. J. (1996). Know your neighbors: Three phenotypes in null mutants of the myogenic bHLH gene MRF4. *Cell*, *85*, 1–4.
Peter, I. S., & Davidson, E. H. (2011). A gene regulatory network controlling the embryonic specification of endoderm. *Nature*, *474*, 635–639.
Relaix, F., Rocancourt, D., Mansouri, A., & Buckingham, M. (2005). A Pax3/Pax7-dependent population of skeletal muscle progenitor cells. *Nature*, *435*, 948–953.
Richler, C., & Yaffe, D. (1970). The in vitro cultivation and differentiation capacities of myogenic cell lines. *Developmental Biology*, *23*, 1–22.
Robert, B., Daubas, P., Akimenko, M. A., Cohen, A., Garner, I., Guenet, J. L., et al. (1984). A single locus in the mouse encodes both myosin light chains 1 and 3, a second locus corresponds to a related pseudogene. *Cell*, *39*, 129–140.
Robert, B., Sassoon, D., Jacq, B., Gehring, W., & Buckingham, M. (1989). Hox-7, a mouse homeobox gene with a novel pattern of expression during embryogenesis. *The EMBO Journal*, *8*, 91–100.
Rudnicki, M. A., Schnegelsberg, P. N., Stead, R. H., Braun, T., Arnold, H. H., & Jaenisch, R. (1993). MyoD or Myf-5 is required for the formation of skeletal muscle. *Cell*, *75*, 1351–1359.
Saga, Y., Kitajima, S., & Miyagawa-Tomita, S. (2000). Mesp1 expression is the earliest sign of cardiovascular development. *Trends in Cardiovascular Medicine*, *10*, 345–352.
Sassoon, D., Lyons, G., Wright, W. E., Lin, V., Lassar, A., Weintraub, H., et al. (1989). Expression of two myogenic regulatory factors myogenin and MyoD1 during mouse embryogenesis. *Nature*, *341*, 303–307.
Tajbakhsh, S., & Buckingham, M. (2000). The birth of muscle progenitor cells in the mouse: Spatiotemporal considerations. *Current Topics in Developmental Biology*, *48*, 225–268.
Tajbakhsh, S., Rocancourt, D., & Buckingham, M. (1996). Muscle progenitor cells failing to respond to positional cues adopt non-myogenic fates in myf-5 null mice. *Nature*, *384*, 266–270.
Tsae, P. K., & DeRosa, M. C. (2015). Outlook for aptamers after twenty five years. *Current Topics in Medicinal Chemistry*, *15*, 1153–1159.
Vincent, S. D., & Buckingham, M. E. (2010). How to make a heart: The origin and regulation of cardiac progenitor cells. *Current Topics in Developmental Biology*, *90*, 1–41.
Wachtler, F., & Christ, B. (1992). The basic embryology of skeletal muscle formation in vertebrates: The avian model. *Seminars in Developmental Biology*, *3*, 217–227.
Weintraub, H., Davis, R., Tapscott, S., Thayer, M., Krause, M., Benezra, R., et al. (1991). The myoD gene family: Nodal point during specification of the muscle cell lineage. *Science (New York, N.Y.)*, *251*, 761–766.
Zhou, Q., Brown, J., Kanarek, A., Rajagopal, J., & Melton, D. A. (2008). In vivo reprogramming of adult pancreatic exocrine cells to beta-cells. *Nature*, *455*, 627–632.

CHAPTER NINE

Cycling in the Cell Fate Landscape

Corey Bunce, Blanche Capel[1]
Department of Cell Biology, Duke University Medical Center, Durham, North Carolina, USA
[1]Corresponding author: e-mail address: blanche.capel@duke.edu

Contents

Abstract

The causal relationship between the cell cycle and cell fate has recently been reconsidered in several developmental systems, now including sex determination. Direct links between cell cycle machinery and developmental factors demonstrate how modulating the cell cycle can have a profound influence on cellular decisions, through phenomena largely divided into phase-specific interactions seen as "priming" and "competence," or global changes in transcription or histone modifications during transitions. This integration challenges the common view of cell fate landscapes.

1. INTEGRATION OF SEX DETERMINATION WITH CELL PROLIFERATION

In mammals, sex determination occurs in the bipotential gonad, when somatic cells called "supporting cell precursors" commit to male (Sertoli) or female (granulosa) fate. There is evidence that this fate decision is integrated with proliferation. For example, one of the earliest changes that occurs downstream of the male sex-determining gene, *Sry,* is upregulation of proliferation in the XY gonad (Schmahl, Eicher, Washburn, & Capel, 2000).

Current Topics in Developmental Biology, Volume 116
ISSN 0070-2153
http://dx.doi.org/10.1016/bs.ctdb.2015.10.001

Disruption of several genes known to induce proliferation, such as *Fgf9* (Schmahl, Kim, Colvin, Ornitz, & Capel, 2004), or receptors involved in insulin signaling (Nef et al., 2003; Pitetti et al., 2013) cause male-to-female sex reversal phenotypes. Blocking proliferation during a narrow window of gonad development soon after *Sry* expression begins also leads to a high incidence of sex reversal in XY gonads (Schmahl & Capel, 2003).

This dependence on proliferation to establish the male pathway has been explained by a requirement for a threshold number of Sertoli progenitor cells to induce male development of the entire community of cells in the gonadal field. This idea is derived primarily from the finding that in XX⇔XY chimeric mice, the gonad must be composed of at least 30% XY cells to establish testis development (Burgoyne, Buehr, & McLaren, 1988). However, this may not be the whole story.

Recent findings suggest that the cell cycle may be more directly integrated into the cell fate decision in the bipotential gonad. Surprisingly, the alternative pathway (commitment to granulosa cell fate) is invariably associated with exit from the cell cycle and P27-mediated arrest until after birth (Mork et al., 2012). Loss of female identity of granulosa cells, and transdifferentiation marked by expression of Sertoli markers, rarely occurs before birth, when granulosa cells resume proliferation. In the case of *Wnt4* mutants, which show some expression of male markers just prior to birth, transdifferentiation is preceded by downregulation of P27 and exit from cell cycle arrest (Maatouk, Mork, Chassot, Chaboissier, & Capel, 2013).

These studies raise the possibility that molecular programs guiding the cell cycle and cell fate commitment are integrated during sex determination. In this essay, we consider some of the evidence for a mechanistic intersection of these two processes in other systems.

2. FIRST PRINCIPLES

One of the first molecular windows into the relationship between cell cycle, gene expression, and cell fate came from discovery and investigation of the retinoblastoma gene, *Rb*. In 1971, Knudson proposed that the pediatric tumor, retinoblastoma, arises from mutation of a gene that acts as a tumor suppressor to control the switch between cell proliferation and cell differentiation (Knudson, 1971). Some 40 years of subsequent work have shown that *Rb* controls a major cell cycle checkpoint between G1 and S-phases by directly binding and blocking the activation domain of E2F and other chromatin-modifying factors. Through inhibition of this machinery, Rb blocks entry into S, and promotes withdrawal from the cell cycle and tissue-specific differentiation (Giacinti & Giordano, 2006). Disruption of Rb leads to failures in differentiation of many embryonic tissues (most famously, the retina), and to the formation of many childhood tumors. The idea that the cell cycle machinery is intimately linked with cell fate decisions was further supported by the early finding that MyoD controls cell cycle withdrawal and terminal differentiation of skeletal muscle progenitors by inducing the CDK inhibitor p21, thereby blocking entry into S (Halevy et al., 1995). Since then, in addition to overt regulation of the genes that govern entry into S, many different ideas about how the cell cycle and differentiation are linked have emerged.

3. EARLY VIEWS OF "COMPETENCE"

Early views of the connection between the cell cycle and the potential for differentiation were guided by evidence that cell cycle phases could act as "competence windows" for induction by developmental cues. In cell culture, Mummery and colleagues showed that retinoic acid could induce embryonal carcinoma cells to differentiate only if exposure occurred during G1, but not S phase (Mummery, Van den Brink, & De Laat, 1987). In *Caenorhabditis elegans* embryos, specification of three distinct fates for the six vulva precursor cells (VPCs) was found to depend on the activity of developmental signals, most notably LIN-12, within separate windows of competence before and after VPC S phase (Ambros, 1999). What could be responsible for these "competence" windows?

4. EXAMPLES OF "PRIMING" WINDOWS IN THE CELL CYCLE

Though cell cycle remodeling has long been recognized as a consequence of differentiation, recent reviews have highlighted an increasing number of studies, suggesting that altered cell cycle structure can precede and influence differentiation (Lange & Calegari, 2010; Singh & Dalton, 2009). Cell cycle length can be linked to a propensity for differentiation through interactions between core cell cycle machinery and developmental regulators, typically during G1. Similar to action of Cdk4/cyclin D and Cdk2/cyclin E on Rb during G1, resulting in the release of E2F and expression of genes for the G1/S transition, phosphorylation by cdk/cyclin complexes alters the activity of developmental signaling molecules (reviewed in Lim & Kaldis, 2013).

For example, in the *Xenopus* central nervous system, the proneural transcription factor Neurogenin2 (NGN2) is inhibited by CDK phosphorylation. Increased cell cycle length allows the accumulation of higher levels of CDK inhibitors (CKIs) which suppress cdk activity, reducing inhibitory phosphorylation of NGN2, which can then promote the expression of genes that drive neuronal differentiation (Ali et al., 2011). As CNS development requires a strict balance between proliferation and differentiation, it offers multiple similar examples (Hardwick, Ali, Azzarelli, & Philpott, 2015).

In *Drosophila,* mammalian cell cultures, and *Xenopus* embryos, Wnt signaling is regulated in a cell cycle-dependent manner. Cdk/CyclinY, an oscillating but non-core cell cycle machinery complex, enhances Wnt signaling by phosphorylating Wnt co-receptor Lrp6 during G2/M. The resulting peak in Wnt/β-catenin signaling is enhanced under G2/M arrest (Davidson et al., 2009). In more recent studies of *C. elegans* VPCs, both stabilization and degradation of Notch are regulated by phase-specific cyclins, synchronizing Notch signaling with cell cycle progression to establish G1 windows of Notch priming integral to coordination of the cell fate pattern (Nusser-Stein et al., 2012).

Cellular compartmentalization of developmental transcription factors during G1 also contributes to priming for specific fates in cell cultures. For instance in embryonic stem cells (ESCs), SMAD2/3 localizes to the nucleus during early G1, mediated by cyclins, where it can target endodermal genes, but is cytoplasmic in late G1, potentially limiting an endodermal competence window to G1 (Coronado et al., 2013; Pauklin & Vallier, 2013). TRIM16, a tumor suppressor gene, increases in expression and shifts to the nucleus during G1, where it promotes differentiation and suppresses proliferation by prolonging G1 (Bell, Malyukova, Kavallaris, Marshall, & Cheung, 2013).

Asymmetric allocation of developmental factors during cell division can also be regulated by the cell cycle. For example, in *C. elegans*, the asymmetric division of somatic gonad precursor (SGP) cells initiates sex-specific gonadogenesis in hermaphrodites and males. Mutations in Cyclin D disrupt asymmetry of SGPs and their sex-specific differentiation. While these mutants show a slow G1/S transition, other methods of slowing the cell cycle do not result in the same phenotype. Instead, Cyclin D mediates axis determination by promoting asymmetry of POP-1/TCF in SGP daughter cells, and controlling cell fate specification through E2F repression of *fkh-6*, a male-specific sex determination gene (Tilmann & Kimble, 2005).

5. MIXING MECHANISMS

Fate specification in the *Drosophila* CNS occurs through a combination of cell cycle-dependent and -independent mechanisms. The CNS neuroblasts transition through sequential expression of four genes (Hb, Kr, Pdm, and Cas) important for establishing neuronal cell fates and neural diversity. Kohwi and Doe (2013) call these "temporal identity factors." During each temporal window, in which the cell is thought to be primed for a specific fate, the neuroblast divides asymmetrically, producing a fate-restricted ganglion mother cell (GMC) and a daughter neuroblast that transitions to the next temporal identity. Evidence suggests that the neuroblasts are also transitioning through competence windows, as they undergo genome reorganization and lose the ability to produce earlier primed fates following each temporal window (Kohwi, Lupton, Lai, Miller, & Doe, 2013). Studies arresting the neuroblast cell cycle revealed that the first transition, from Hb to Kr expression, requires cytokinesis, possibly for asymmetric partitioning of transcriptional activators or signaling between the GMC and neuroblast (Isshiki, Pearson, Holbrook, & Doe, 2001). However, the subsequent temporal identity transitions, Kr to Pdm to Cas, show no cell cycle requirement (Grosskortenhaus, Pearson, Marusich, & Doe, 2005), providing an example where a highly reproducible pattern of specification and tissue structure is produced, without the need for a direct link between the cell cycle machinery and developmental factors. This type of "temporal patterning" is typical of neural progenitors and is well studied in mammals as well (reviewed in Kohwi & Doe, 2013).

A shift in the mechanism of specification occurs in the generation of endocrine cells in the pancreas. Multipotent pancreatic progenitors give rise to endocrine, acinar, and ductal cells. The production of endocrine progenitor cells, involving Neurogenin3, occurs through either asymmetric division, which produces a pancreatic progenitor and an endocrine progenitor, or symmetric division, generating two endocrine progenitors. Instead of multiple mechanisms of commitment to the endocrine lineage, a model of stochastic induction has been proposed, whereby the apparent type of division is a result of the time in the cell cycle when the progenitor is induced toward the endocrine fate, with no restriction on the phase in which induction must occur (Kim et al., 2015). In contrast to the production of endocrine progenitors from pancreatic progenitors, the sequential generation of endocrine subtypes, α cells, β cells, and δ and PP cells, from endocrine progenitors, is dependent on a sequence of temporal competence windows (Johansson et al., 2007). The mechanism underlying the competence changes and the level of cell autonomy involved remain unknown.

6. SYNCHRONIZATION AND COMMUNITY EFFECTS

Drosophila eye organogenesis offers an extraordinary example of the importance of coordination of cell cycle regulation and development. In the *Drosophila* third instar larvae, specification of cell fates in the eye imaginal disc occurs in a wave of differentiation that moves anteriorly, marked by the morphogenetic furrow. Dpp and Hh signaling guide the initiation of the process, beginning with adoption of a preproneural fate commitment and coordination of a first mitotic wave, followed by continued synchrony through G1 arrest, furrow depression, a second mitotic wave, and finally differentiation. Several developmental cues have been implicated in regulating cell cycle phase transitions as well as a stepwise specification program, crucial for organization and fate determination of photoreceptors (Escudero & Freeman, 2007; Greenwood & Struhl, 1999). However, mutations in *rux,* a G1 arrest regulator, do not disrupt differentiation per se, but lead to patterning defects, suggesting that the role of cell cycle priming in cell fate commitment can be separated from the importance of synchronizing cell cycles during tissue development (Thomas, Gunning, Cho, & Zipursky, 1994).

The relevance of synchronization of cell cycles for fate determination has also been seen in cell culture. Human ESCs are more prone to spontaneously differentiate during G1 than during G2 or S. This behavior has been linked to reduced Rb phosphorylation levels and transition through cell cycle arrest. Community effects and synchronization are implicated because the propensity to differentiate is increased at lower cell densities, but reduced by the presence of nearby cells in G2 or S phase (Sela, Molotski, Golan, Itskovitz-Eldor, & Soen, 2012).

7. THE MECHANICS OF CYCLE TRANSITIONS: OPPORTUNITIES FOR CHANGE

More detailed investigation of the mechanics of DNA mitosis and synthesis has led to new considerations about how these processes could affect cell fate decisions. For example, there is a global shutdown of transcription at mitosis. Many factors, including Pol II and other components of the transcription machinery, chromatin regulatory complexes, and transcription factors, dissociate from chromatin at mitosis (Martínez-Balbás, Dey, Rabindran, Ozato, & Wu, 1995). This is not merely a result of condensation of chromatin since mitotic transcription is repressed even on chromatin that is nucleosome free and not compacted (Spencer, Kruhlak, Jenkins, Sun, & Bazett-Jones, 2000). Genome-wide eviction of transcriptional complexes at M might facilitate changes in cell fate by providing a window of opportunity for remodeling the transcriptional landscape (Kadauke & Blobel, 2013).

Does the shutdown of transcription during mitosis suggest that the steady-state level of transcripts/proteins oscillates during the cell cycle? Many cell fate decisions are associated with competitive relationships between transcription factors that promote one fate and repress an alternative fate. It remains unclear whether all genes are transcribed at similar levels in G1 and G2, or how this might be integrated with transcript stability or translational control to result in "priming" opportunities. Oscillation of the levels of signals during the cell cycle could provide vulnerability windows, where levels of maintenance factors are low and changes in fate are favored.

Many cells do resume similar patterns of transcription after division, referred to as "bookmarking." At the end of mitosis, transcription is more rapidly activated in cells that previously expressed the gene than in cells where transcription at that locus is initiating *de novo* (Muramoto, Müller, Thomas, Melvin, & Chubb, 2010), suggesting that cells have efficient mechanisms of transcriptional memory. Perhaps more remarkably, transcriptional states are even maintained following somatic cell nuclear transfer and are reprogrammed gradually by the oocyte cytoplasm over the course of several cell divisions (Ng & Gurdon, 2005).

These findings raise questions about how mitotic memory is maintained in cases where cell fate does not change. Methylation of H3K4 is an important bookmarking mechanism. Inheritance of a stable transcriptional program was perturbed in *Dictyostelium discoideum* by disruption of the gene encoding the H3K4 methyltransferase, *Set1* (Muramoto et al., 2010). In addition, BRD4, which rapidly reassociates with acetylated H4K5 residues at previously activated loci, may also be involved in cellular memory of former transcriptional states (Zhao, Nakamura, Fu, Lazar, & Spector, 2011). Although their chromatin association sites are often reduced during mitosis (based on ChIP assays), certain "pioneer" factors seem to escape full eviction from chromatin at M including GATA1 (Kadauke et al., 2012), FOXA1 (Caravaca et al., 2013), and Polycomb (Follmer, Wani, & Francis, 2012). These factors may serve a "bookmarking" function to preserve cell identity by recovering the transcriptional landscape after division (Kadauke & Blobel, 2013). Cohesin binds to "anchor sites" where transcription factors are assembled, which may serve as placeholders for the rapid reassembly of transcription factor complexes (Yan et al., 2013).

Posttranslational modifications of histones vary during the cell cycle. Histones are replaced in a semiconservative manner at S, when parental histones mix with newly synthesized histones behind the replication fork (MacAlpine & Almouzni, 2013; Margueron & Reinberg, 2010). Parental histones carry their di- and trimethyl histone marks to the new strand, but these marks are diluted twofold as a consequence of the incorporation of newly synthesized histones during replication. In general, new histones are modified to mirror the parental histones within one cell cycle. However, K9 and K27 trimethylation marks are acquired continually over several cell generations (Alabert et al., 2015). In HeLa cells, several H3 methylation marks, including trimethylation of H3K27 and H3K9, are not fully restored until after M, which could afford an opportunity for change in G2 (Scharf, Barth, & Imhof, 2009; Xu, Wang, Chen, & Zhu, 2012).

During replication, when histones are acetylated and DNA is accessible, maintenance of a silent gene expression state may be challenging and may provide a vulnerable window for activating polycomb response elements (PREs) in response to developmental or environmental signals (Steffen & Ringrose, 2014). Consistent with this idea, many PcG-repressed genes are transiently upregulated during DNA replication in human keratinocytes (Peña-Diaz et al., 2013).

8. NEW TOOLS

An influx of new tools in the field is likely to lead to many new experiments that address the integration of the cell cycle with cell fate commitment. H2B tagged with fluorescent proteins has been driven by cell-type-specific promoters to reveal dynamic cell divisions in fields of cells using live imaging (Hadjantonakis & Papaioannou, 2004; Kanda, Sullivan, & Wahl, 1998). More recently, a more elaborate reporter of the cell cycle, Fucci2, has been developed that delineates multiple phases of the cell cycle (Abe et al., 2013; Mort et al., 2014). As the cost of transcriptome analysis declines, transcriptomes can be collected from cells synchronized in cell cycle phases, or from single cells in the population. In addition, a method that has been used to label RNA for some time has been adapted to report live transcription of specific genes during development. MS2 stem loops that are incorporated into newly transcribed RNA are rapidly bound by an MS2-GFP coat protein (MSN). This creates a localized accumulation of GFP at the site of transcription that can be visualized above background in living cells (Bertrand et al., 1998). In combination with cell cycle indicators, this approach provides a way to link transcription of specific genes with the phase of the cell cycle.

9. SUMMARY

The tendency to embark on developmental pathways can be dependent on windows of "priming" that result from direct interactions between the cell cycle machinery and developmental factors that affect level, activity, or availability of cell surface receptors, signals, and transcription factors. The resulting oscillations of developmental signal strength can be exaggerated through changing cell cycle phase length, altering a cell's propensity to differentiate. "Competence" windows, periods of increased susceptibility to developmental cues, can result in phase-dependent differentiation. Both phenomena appear as a correlation between cell cycle phase length and probability of differentiation.

Cell fate can be indirectly under cell cycle control through a requirement for global changes in transcription regulation favored at S or M. In general, during replication, acetylated, newly assembled chromatin may favor activation, challenging recovery of the silent state, and favoring recovery of the active state. On the other hand, during mitosis, the eviction of transcriptional complexes and shutdown of transcription may favor the silent state, but challenge reestablishment of the active one. Both S and M phases of the cell cycle may offer opportunities for reorganizing the chromatin landscape and embarking on a new cell fate.

The coordination of growth and differentiation during tissue and organ development clearly involves integrating cell cycle-dependent and -independent pathways. However, a strong argument exists for considering a more direct role of the cell cycle in the adoption of cell fate.

ACKNOWLEDGMENT

We are very grateful to Amanda Bunce for contributing the drawing in Fig. 1.

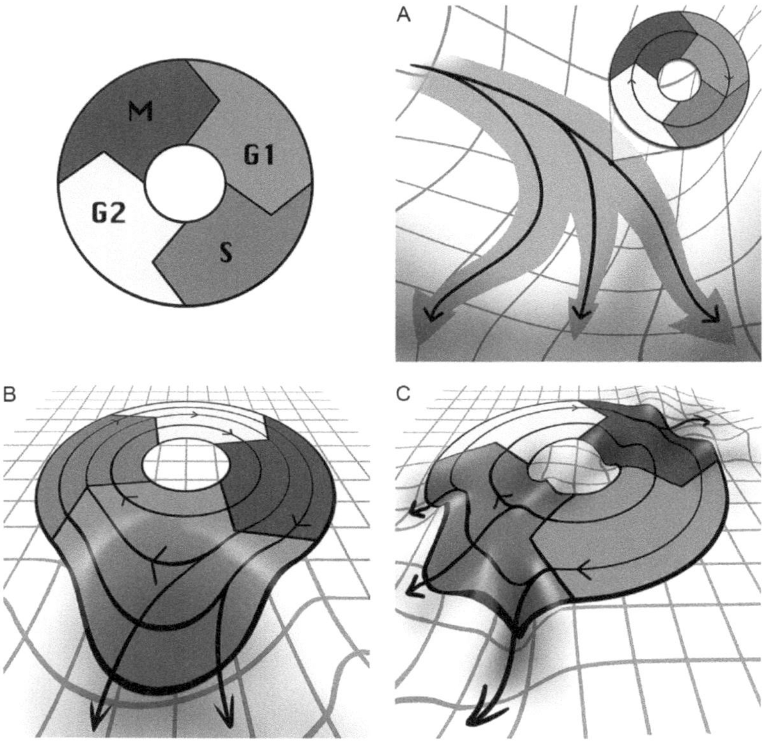

Figure 1 The integration of cell cycle and differentiation pathways depicted as progenitor cell cycles overlaid on epigenetic landscapes. The slope of the landscape represents a cell's propensity to differentiate. (A) The common view, wherein cell movement along a developmental trajectory is independent of its cell cycle dynamics. (B) In phases with "priming" or "competence" windows, changes in path length alter the probability of differentiation. (C) The complexity of nuclear remodeling during S and M phase makes a rugged landscape, where cells are highly susceptible to differentiation cues, suggesting that considerable regulatory programs must be employed to maintain a progenitor cell trajectory. *Figure drawn by Amanda Bunce.* (A color version of this figure is located at the end of this chapter.)

REFERENCES

Abe, T., Sakaue-Sawano, A., Kiyonari, H., Shioi, G., Inoue, K.-i, Horiuchi, T., et al. (2013). Visualization of cell cycle in mouse embryos with Fucci2 reporter directed by Rosa26 promoter. *Development*, *140*(1), 237–246.

Alabert, C., Barth, T. K., Reverón-Gómez, N., Sidoli, S., Schmidt, A., Jensen, O. N., et al. (2015). Two distinct modes for propagation of histone PTMs across the cell cycle. *Genes & Development*, *29*(6), 585–590.

Ali, F., Hindley, C., McDowell, G., Deibler, R., Jones, A., Kirschner, M., et al. (2011). Cell cycle-regulated multi-site phosphorylation of Neurogenin 2 coordinates cell cycling with differentiation during neurogenesis. *Development*, *138*(19), 4267–4277.

Ambros, V. (1999). Cell cycle-dependent sequencing of cell fate decisions in Caenorhabditis elegans vulva precursor cells. *Development*, *126*(9), 1947–1956.

Bell, J. L., Malyukova, A., Kavallaris, M., Marshall, G. M., & Cheung, B. B. (2013). TRIM16 inhibits neuroblastoma cell proliferation through cell cycle regulation and dynamic nuclear localization. *Cell Cycle*, *12*(6), 889–898.

Bertrand, E., Chartrand, P., Schaefer, M., Shenoy, S. M., Singer, R. H., & Long, R. M. (1998). Localization of ASH1 mRNA particles in living yeast. *Molecular Cell*, *2*(4), 437–445.

Burgoyne, P. S., Buehr, M., & McLaren, A. (1988). XY follicle cells in ovaries of XX-XY female mouse chimaeras. *Development*, *104*(4), 683–688.

Caravaca, J. M., Donahue, G., Becker, J. S., He, X., Vinson, C., & Zaret, K. S. (2013). Bookmarking by specific and nonspecific binding of FoxA1 pioneer factor to mitotic chromosomes. *Genes & Development*, *27*(3), 251–260.

Coronado, D., Godet, M., Bourillot, P.-Y., Tapponnier, Y., Bernat, A., Petit, M., et al. (2013). A short G1 phase is an intrinsic determinant of naïve embryonic stem cell pluripotency. *Stem Cell Research*, *10*(1), 118–131.

Davidson, G., Shen, J., Huang, Y.-L., Su, Y., Karaulanov, E., Bartscherer, K., et al. (2009). Cell cycle control of wnt receptor activation. *Developmental Cell*, *17*(6), 788–799.

Escudero, L. M., & Freeman, M. (2007). Mechanism of G1 arrest in the Drosophila eye imaginal disc. *BMC Developmental Biology*, 7(1), 13.

Follmer, N. E., Wani, A. H., & Francis, N. J. (2012). A Polycomb group protein is retained at specific sites on chromatin in mitosis. *PLoS Genetics*, *8*(12), e1003135.

Giacinti, C., & Giordano, A. (2006). RB and cell cycle progression. *Oncogene*, *25*(38), 5220–5227.

Greenwood, S., & Struhl, G. (1999). Progression of the morphogenetic furrow in the Drosophila eye: The roles of Hedgehog, Decapentaplegic and the Raf pathway. *Development*, *126*(24), 5795–5808.

Grosskortenhaus, R., Pearson, B. J., Marusich, A., & Doe, C. Q. (2005). Regulation of temporal identity transitions in Drosophila neuroblasts. *Developmental Cell*, *8*(2), 193–202.

Hadjantonakis, A.-K., & Papaioannou, V. E. (2004). Dynamic in vivo imaging and cell tracking using a histone fluorescent protein fusion in mice. *BMC Biotechnology*, *4*(1), 33.

Halevy, O., Novitch, B. G., Spicer, D. B., Skapek, S. X., Rhee, J., Hannon, G. J., et al. (1995). Correlation of terminal cell cycle arrest of skeletal muscle with induction of p21 by MyoD. *Science*, *267*(5200), 1018–1021.

Hardwick, L. J., Ali, F. R., Azzarelli, R., & Philpott, A. (2015). Cell cycle regulation of proliferation versus differentiation in the central nervous system. *Cell and Tissue Research*, *359*(1), 187–200.

Isshiki, T., Pearson, B., Holbrook, S., & Doe, C. Q. (2001). Drosophila neuroblasts sequentially express transcription factors which specify the temporal identity of their neuronal progeny. *Cell*, *106*(4), 511–521.

Johansson, K. A., Dursun, U., Jordan, N., Gu, G., Beermann, F., Gradwohl, G., et al. (2007). Temporal control of neurogenin3 activity in pancreas progenitors reveals competence windows for the generation of different endocrine cell types. *Developmental Cell*, *12*(3), 457–465.

Kadauke, S., & Blobel, G. A. (2013). Mitotic bookmarking by transcription factors. *Epigenetics & Chromatin*, *6*(6).

Kadauke, S., Udugama, M. I., Pawlicki, J. M., Achtman, J. C., Jain, D. P., Cheng, Y., et al. (2012). Tissue-specific mitotic bookmarking by hematopoietic transcription factor GATA1. *Cell*, *150*(4), 725–737.

Kanda, T., Sullivan, K. F., & Wahl, G. M. (1998). Histone–GFP fusion protein enables sensitive analysis of chromosome dynamics in living mammalian cells. *Current Biology*, *8*(7), 377–385.

Kim, Y. H., Larsen, H. L., Rué, P., Lemaire, L. A., Ferrer, J., & Grapin-Botton, A. (2015). Cell cycle–dependent differentiation dynamics balances growth and endocrine differentiation in the pancreas. *PLoS Biology*, *13*(3).

Knudson, A. G. (1971). Mutation and cancer: Statistical study of retinoblastoma. *Proceedings of the National Academy of Sciences of the United States of America*, *68*(4), 820–823.

Kohwi, M., & Doe, C. Q. (2013). Temporal fate specification and neural progenitor competence during development. *Nature Reviews. Neuroscience*, *14*(12), 823–838.

Kohwi, M., Lupton, J. R., Lai, S.-L., Miller, M. R., & Doe, C. Q. (2013). Developmentally regulated subnuclear genome reorganization restricts neural progenitor competence in Drosophila. *Cell*, *152*(1), 97–108.

Lange, C., & Calegari, F. (2010). Cdks and cyclins link G1 length and differentiation of embryonic, neural and hematopoietic stem cells. *Cell Cycle*, *9*(10), 1893–1900.

Lim, S., & Kaldis, P. (2013). Cdks, cyclins and CKIs: Roles beyond cell cycle regulation. *Development*, *140*(15), 3079–3093.

Maatouk, D. M., Mork, L., Chassot, A.-A., Chaboissier, M.-C., & Capel, B. (2013). Disruption of mitotic arrest precedes precocious differentiation and transdifferentiation of pregranulosa cells in the perinatal Wnt4 mutant ovary. *Developmental Biology*, *383*(2), 295–306.

MacAlpine, D. M., & Almouzni, G. (2013). Chromatin and DNA replication. *Cold Spring Harbor Perspectives in Biology*, *5*(8), a010207.

Margueron, R., & Reinberg, D. (2010). Chromatin structure and the inheritance of epigenetic information. *Nature Reviews. Genetics*, *11*(4), 285–296.

Martínez-Balbás, M. A., Dey, A., Rabindran, S. K., Ozato, K., & Wu, C. (1995). Displacement of sequence-specific transcription factors from mitotic chromatin. *Cell*, *83*(1), 29–38.

Mork, L., Maatouk, D. M., McMahon, J. A., Guo, J. J., Zhang, P., McMahon, A. P., et al. (2012). Temporal differences in granulosa cell specification in the ovary reflect distinct follicle fates in mice. *Biology of Reproduction*, *86*(2), 37.

Mort, R. L., Ford, M. J., Sakaue-Sawano, A., Lindstrom, N. O., Casadio, A., Douglas, A. T., et al. (2014). Fucci2a: A bicistronic cell cycle reporter that allows Cre mediated tissue specific expression in mice. *Cell Cycle*, *13*(17), 2681–2696.

Mummery, C., Van den Brink, C., & De Laat, S. (1987). Commitment to differentiation induced by retinoic acid in P19 embryonal carcinoma cells is cell cycle dependent. *Developmental Biology*, *121*(1), 10–19.

Muramoto, T., Müller, I., Thomas, G., Melvin, A., & Chubb, J. R. (2010). Methylation of H3K4 Is required for inheritance of active transcriptional states. *Current Biology*, *20*(5), 397–406.

Nef, S., Verma-Kurvari, S., Merenmies, J., Vassalli, J.-D., Efstratiadis, A., Accili, D., et al. (2003). Testis determination requires insulin receptor family function in mice. *Nature*, *426*(6964), 291–295.

Ng, R. K., & Gurdon, J. B. (2005). Epigenetic memory of active gene transcription is inherited through somatic cell nuclear transfer. *Proceedings of the National Academy of Sciences of the United States of America*, *102*(6), 1957–1962.

Nusser-Stein, S., Beyer, A., Rimann, I., Adamczyk, M., Piterman, N., Hajnal, A., et al. (2012). Cell-cycle regulation of NOTCH signaling during C. elegans vulval development. *Molecular Systems Biology*, *8*(1), 618.

Pauklin, S., & Vallier, L. (2013). The cell-cycle state of stem cells determines cell fate propensity. *Cell*, *155*(1), 135–147.

Peña-Diaz, J., Hegre, S. A., Anderssen, E., Aas, P. A., Mjelle, R., Gilfillan, G. D., et al. (2013). Transcription profiling during the cell cycle shows that a subset of Polycomb-targeted genes is upregulated during DNA replication. *Nucleic Acids Research*, *41*(5), 2846–2856.

Pitetti, J.-L., Calvel, P., Zimmermann, C., Conne, B., Papaioannou, M. D., Aubry, F., et al. (2013). An essential role for insulin and IGF1 receptors in regulating sertoli cell proliferation, testis size, and FSH action in mice. *Molecular Endocrinology*, *27*(5), 814–827.

Scharf, A. N., Barth, T. K., & Imhof, A. (2009). Establishment of histone modifications after chromatin assembly. *Nucleic Acids Research*, *37*(15), 5032–5040.

Schmahl, J., & Capel, B. (2003). Cell proliferation is necessary for the determination of male fate in the gonad. *Developmental Biology*, *258*(2), 264–276.

Schmahl, J., Eicher, E. M., Washburn, L. L., & Capel, B. (2000). Sry induces cell proliferation in the mouse gonad. *Development*, *127*(1), 65–73.

Schmahl, J., Kim, Y., Colvin, J. S., Ornitz, D. M., & Capel, B. (2004). Fgf9 induces proliferation and nuclear localization of FGFR2 in Sertoli precursors during male sex determination. *Development*, *131*(15), 3627–3636.

Sela, Y., Molotski, N., Golan, S., Itskovitz-Eldor, J., & Soen, Y. (2012). Human embryonic stem cells exhibit increased propensity to differentiate during the G1 phase prior to phosphorylation of retinoblastoma protein. *Stem Cells*, *30*(6), 1097–1108.

Singh, A. M., & Dalton, S. (2009). The cell cycle and Myc intersect with mechanisms that regulate pluripotency and reprogramming. *Cell Stem Cell*, *5*(2), 141–149.

Spencer, C. A., Kruhlak, M. J., Jenkins, H. L., Sun, X., & Bazett-Jones, D. P. (2000). Mitotic transcription repression in vivo in the absence of nucleosomal chromatin condensation. *The Journal of Cell Biology*, *150*(1), 13–26.

Steffen, P. A., & Ringrose, L. (2014). What are memories made of? How Polycomb and Trithorax proteins mediate epigenetic memory. *Nature Reviews. Molecular Cell Biology*, *15*(5), 340–356.

Thomas, B. J., Gunning, D. A., Cho, J., & Zipursky, S. L. (1994). Cell cycle progression in the developing Drosophila eye: Roughex encodes a novel protein required for the establishment of G1. *Cell*, *77*(7), 1003–1014.

Tilmann, C., & Kimble, J. (2005). Cyclin D regulation of a sexually dimorphic asymmetric cell division. *Developmental Cell*, *9*(4), 489–499.

Xu, M., Wang, W., Chen, S., & Zhu, B. (2012). A model for mitotic inheritance of histone lysine methylation. *EMBO Reports*, *13*(1), 60–67.

Yan, J., Enge, M., Whitington, T., Dave, K., Liu, J., Sur, I., et al. (2013). Transcription factor binding in human cells occurs in dense clusters formed around cohesin anchor sites. *Cell*, *154*(4), 801–813.

Zhao, R., Nakamura, T., Fu, Y., Lazar, Z., & Spector, D. L. (2011). Gene bookmarking accelerates the kinetics of post-mitotic transcriptional re-activation. *Nature Cell Biology*, *13*(11), 1295–1304.

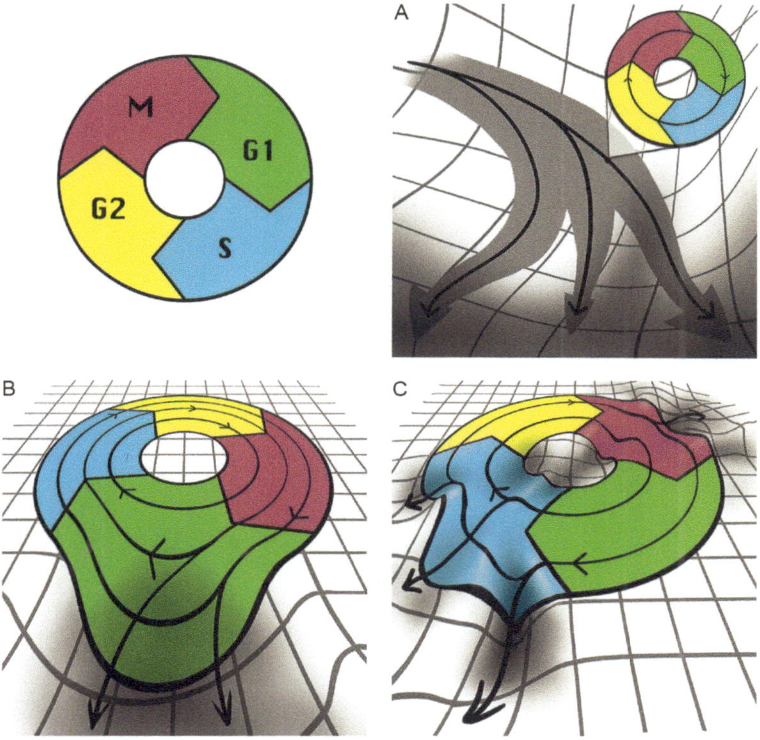

Corey Bunce and Blanche Capel, Figure 1 The integration of cell cycle and differentiation pathways depicted as progenitor cell cycles overlaid on epigenetic landscapes. The slope of the landscape represents a cell's propensity to differentiate. (A) The common view, wherein cell movement along a developmental trajectory is independent of its cell cycle dynamics. (B) In phases with "priming" or "competence" windows, changes in path length alter the probability of differentiation. (C) The complexity of nuclear remodeling during S and M phase makes a rugged landscape, where cells are highly susceptible to differentiation cues, suggesting that considerable regulatory programs must be employed to maintain a progenitor cell trajectory. *Figure drawn by Amanda Bunce.*

CHAPTER TEN

Securing Neuronal Cell Fate in *C. elegans*

Chaogu Zheng, Martin Chalfie[1]
Department of Biological Sciences, Columbia University, New York, USA
[1]Corresponding author: e-mail address: mc21@columbia.edu

Contents

Abstract

Transcription factors control neuronal differentiation by acting as "terminal selectors" that determine the specific cell fates of different types of neurons. The specification of cell fate, however, requires high fidelity, which relies on stable and robust expression of the terminal selectors. Our recent studies in *C. elegans* suggest that a second set of transcription factors function as reinforcing or protecting factors to stabilize the expression and activity of terminal selectors. Some serve as "guarantors" to ensure the activation and continuous expression of the selectors by reducing stochastic fluctuations in gene expression; others safeguard the protein function of selectors by repressing inhibitors that would block their activity. These transcription factors, unlike the terminal selectors, do not induce specification but secure neuronal cell fate and provide reliability in differentiation.

1. INTRODUCTION

The development of any multicellular organism requires the generation of an array of cell types with distinct features and functions. These specific properties define the identity of a cell. Deciphering the mechanisms that control the differentiation of cellular identities is crucial to understand the

Current Topics in Developmental Biology, Volume 116
ISSN 0070-2153
http://dx.doi.org/10.1016/bs.ctdb.2015.11.011

regulation of any developmental process, particularly the development of the nervous system because of the extraordinary variety of neuronal cell types. Each neuronal type assumes a particular neuronal cell fate, which is defined as a differentiation state associated with the expression of a battery of genes that defines the general morphology and function of the neuron.

Building on the work of Garcia-Bellido (1975), Hobert (2008) proposed that the wide variety of cell fates in the nervous system are determined by transcription factors (called "terminal selectors") that directly activate the expression of "terminal differentiation genes," whose products collectively define the specific morphological and functional features associated with the cell fate. This theory suggests that each combination of terminal selectors determines the unique identity of a given neuron type and is essential for the expression of the cell fate. Work in our lab, however, has revealed a second set of transcription factors that promote cell fate acquisition in a way distinct from the action of the conventional selectors. These transcription factors either have a "guarantor" function that reduces expression variability in terminal selector genes or a "protector" function that allows selectors to function by repressing the expression of inhibitors that would block their activities. These factors do not directly induce cell type differentiation but ensure reliable and robust acquisition of cell fates.

2. TOUCH RECEPTOR NEURONS

The nematode *Caenorhabditis elegans* has been very useful for delineating the mechanism of cell fate determination because of the ease of characterizing all of the 302 neurons, which represent 118 morphologically distinct classes, in the adult hermaphrodites (White, Southgate, Thomson, & Brenner, 1986). To date, terminal selectors for more than a dozen of neuron types in *C. elegans* have been identified (Hobert, 2010). Our research focuses on one type of mechanosensory neurons, the touch receptor neurons (Fig. 1A; TRNs). The POU homeodomain transcription factor UNC-86 and LIM homeodomain protein MEC-3 serve as classic examples of terminal selectors for the TRN fate (Fig. 1B; Finney, Ruvkun, & Horvitz, 1988; Way & Chalfie, 1988), since they directly activate the TRN terminal differentiation genes, which include mechanosensory receptors, ion channels, cytoskeletal proteins, and others (Duggan, Ma, & Chalfie, 1998; Zhang et al., 2002). UNC-86 is expressed in the precursors of the TRNs and activates *mec-3* after the generation of the cells (Baumeister, Liu, & Ruvkun, 1996;

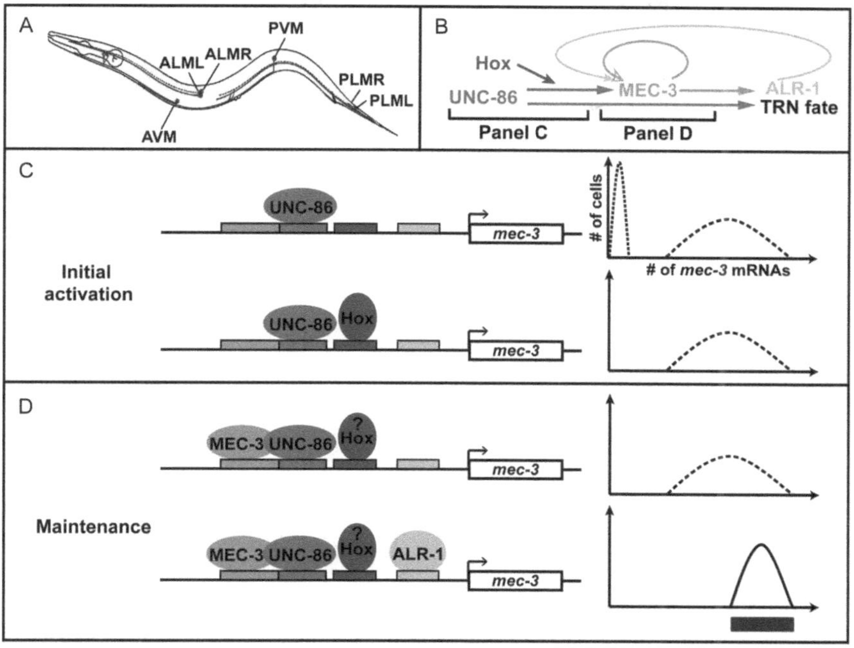

Figure 1 Guarantors ensure the activation and maintained expression of the terminal selector *mec-3* in the TRNs. (A) The six TRNs in *C. elegans*. (B) Proteins involved in TRN specification. UNC-86 initially activates *mec-3*, facilitated by Hox proteins; and the expressed MEC-3 forms a heteromer with UNC-86 that autoregulates *mec-3* and activates the TRN fate. UNC-86/MEC-3 also activates *alr-1*, which potentiates *mec-3* autoregulation. (C) and (D) Modes of *mec-3* transcription in the absence and presence of guarantors. The schematic plots represent quantitative measurements of *mec-3* mRNA molecules in populations of cells using smFISH. (C) During the initiation phase, UNC-86 alone cannot fully activate *mec-3* in all TRNs. In Hox mutants, a binary loss of *mec-3* expression leads to the failure of TRN fate acquisition in some of the cells. (D) *mec-3* autoregulation requires not only UNC-86/MEC-3 but also ALR-1 to reduce the variability and optimize *mec-3* expression. Whether Hox proteins are also needed for the *mec-3* maintenance is unclear. The binding sites of the guarantors are found adjacent to the essential UNC-86/MEC-3 binding site in the proximal promoter of *mec*-3. *Panel (A): Adapted from Bounoutas and Chalfie (2007) with kind permission from Springer Science+Business Media and from Chalfie and Sulston (1981), Copyright (1981), with permission from Elsevier.* (A color version of this figure is located at the end of this chapter.)

Way & Chalfie, 1989). UNC-86 and MEC-3 then form a heteromer that binds to the *cis*-regulatory motifs (UNC-86/MEC-3 sites) in the regulatory regions of TRN genes (Xue, Tu, & Chalfie, 1993). UNC-86/MEC-3 also maintains the expression of *mec-3* through autoregulation (Way & Chalfie, 1989; Xue, Finney, Ruvkun, & Chalfie, 1992).

UNC-86 and MEC-3 are TRN fate determinants because they are absolutely required for the differentiation of the TRNs. In *unc-86* mutants, the expression of *mec-3* and TRN terminal differentiation genes are eliminated (Way & Chalfie, 1989). Mutation of *unc-86* results in lineage defects that prevent the production of four of the six TRNs (the two PLM neurons, AVM, and PVM), but the two cells that would normally develop into the ALM neurons are made but do not differentiate as TRNs (Baumeister et al., 1996; Chalfie, Horvitz, & Sulston, 1981). In *mec-3* mutants, although the cells are generated, none become TRNs and develop the characteristics of mechanosensory neurons, because the activation of the TRN genetic program requires MEC-3 (Chalfie & Au, 1989).

In addition to MEC-3, many LIM homeodomain proteins are important terminal selectors. For instance, Lim3 and Islet form a combinatorial code to control the identity of motor neuron subtypes in *Drosophila* (Thor, Andersson, Tomlinson, & Thomas, 1999); similarly, their homologs in mice, Lhx3 and Isl1, regulate the specification of ventral motor neurons in the spinal cord (Thaler, Lee, Jurata, Gill, & Pfaff, 2002). Therefore, activating and maintaining robust expression of these LIM homeodomain transcription factors is essential for the differentiation of specific neuronal fates. To ensure this robust expression and allow differentiation to occur reliably, cells must overcome inherent stochastic fluctuations in the expression of terminal selectors.

3. HOX PROTEINS ACT AS TRANSCRIPTIONAL GUARANTORS TO ENSURE CELL FATE ACQUISITION

Hox proteins, the conserved homeodomain transcription factors that determine regional identities (McGinnis & Krumlauf, 1992) have often been thought to act as anterior–posterior region-specific cell fate determinants. Such specification does occur (see below), but recent studies from our laboratory suggest that they also act as auxiliary factors to ensure the expression of the terminal selectors. We found that although UNC-86 alone can activate *mec-3*, the full activation of *mec-3* requires the Hox proteins. In the absence of Hox activity, *mec-3* activation does not occur in about 40% of the cells that would normally differentiate into TRNs (Zheng, Jin, & Chalfie, 2015). Quantitative measurements using single molecule *in situ* fluorescent hybridization (smFISH) suggest two distinct modes of *mec-3* transcription in the Hox mutants. 40% of the cells that do not express TRN characteristics show none or very few *mec-3* mRNA molecules;

whereas, the rest 60% of the cells that still express TRN fate markers have wild-type level of *mec-3* mRNA (Zheng, Jin, et al., 2015). This binary loss of *mec-3* expression suggests that Hox proteins facilitate the specification of TRN fate by ensuring robust activation of the terminal selector, which in turn activates the terminal differentiation genes. We refer to proteins that have this ensuring function as guarantors (Fig. 1C; Zheng, Jin, et al., 2015).

One feature of the transcriptional guarantors that distinguishes them from terminal selectors is that they do not activate gene expression on their own. Thus, UNC-86, but not the Hox proteins, activates *mec-3* expression in a heterologous yeast transcription assay. The Hox proteins, however, enhance UNC-86-mediated activation of the *mec-3* in this assay (Zheng, Jin, et al., 2015).

Different TRN subtypes need different Hox proteins as guarantors. The two anterior ALM neurons utilize the anterior Hox protein CEH-13/Lab/Hox1 and the Pbx-class TALE cofactor CEH-20/Exd, whereas the two posterior PLM neurons utilize the posterior Hox protein EGL-5/AbdB/Hox9–13, which appears to act without the aid of Hox cofactors (Zheng, Jin, et al., 2015). CEH-13 and EGL-5 activate *mec-3* through the same *cis*-regulatory element, since the mutation of the Hox/Pbx site reproduces the incomplete loss of *mec-3* expression observed in Hox mutants (except that both ALM and PLM neurons are affected). Strikingly, mutations of this site do not affect *mec-3* expression in the postembryonic TRNs (AVM and PVM), and all of these neurons differentiate into TRNs in any Hox mutants (Zheng, Jin, et al., 2015). Possibly another transcription factor replaces Hox proteins in these cells. Therefore, these results not only confirm that Hox activity is dispensable for the activation of the TRN fate but also suggest that cells belonging to the same neuron type use different mechanisms to ensure the common neuronal fate.

We suspect that the Hox guarantor function may be a general characteristic. Certainly different Hox proteins bind to similar DNA sequences (Gehring et al., 1994), and several different region-specific Hox proteins can facilitate the activation of the same terminal selector in neurons located at different positions along the anterior–posterior axis as they do in *C. elegans*. For instance, the *Drosophila* Hox genes *Ubx* and *Abd-A* both promote the specialization of the neuropeptidergic Va neurons by activating the terminal selector gene *dim* (Suska, Miguel-Aliaga, & Thor, 2011); and multiple genes in the *Hox5–Hox8* paralog groups all contribute to the fate of lateral motor neurons at the brachial level in mouse spinal cord (Lacombe et al., 2013). Moreover, instances where Hox proteins had been thought to specify cell type are better described as instances where they ensure differentiation. For example, although initially Hox5 was needed for the development of phrenic motor neurons that express the fate marker *Scip*, a block of apoptosis revealed that about 80% of the surviving phrenic motor neurons in *Hox5* mutants still expressed *Scip* (Philippidou, Walsh, Aubin, Jeannotte, & Dasen, 2012).

4. HOX PROTEINS ALSO PROMOTE NEURONAL SUBTYPE DIFFERENTIATION

In addition to their guarantor function, which allows disparate cells to efficiently differentiate into the same type of neurons, Hox proteins can also induce the differences among neuronal subtypes that share the same cell fate. Within the TRN subtypes, the posterior Hox protein EGL-5 induces morphological, transcriptional, and functional changes of the PLM neurons that differentiate them from the ALM neurons (Zheng, Diaz-Cuadros, & Chalfie, 2015). ALM neurons maintain a TRN ground state with its unipolar shape, whereas EGL-5 proteins induce the growth of a secondary, posteriorly directed neurite in the PLM neurons. Moreover, EGL-5 modifies the default TRN genetic program and regulates subtype-specific connectivity (Zheng, Diaz-Cuadros, et al., 2015). Similarly, in the *Drosophila* Va neurons, abd-A induces the production of the neuropeptide Capa in the Va neurons of *Drosophila* abdominal segments A2–A4; the Va neurons in segment A1, which does not have abd-A, do not express Capa (Suska et al., 2011). In the mouse lateral motor neurons, individual Hox proteins, such as Hoxc6 and Hoxc8, control unique aspects of subtype identity by regulating the connectivity to specific muscle targets (Dasen, Tice, Brenner-Morton, & Jessell, 2005). Therefore, Hox genes have dual functions in both promoting the convergence onto the common cell fate and inducing subtypes diversification within the same neuronal type.

We do not know how Hox proteins generate subtype features. One model is that they act as subtype terminal selectors. An alternative hypothesis that is an extension of our guarantor model is that Hox proteins act as guarantors that enhance the efficiency of other transcription factors. Our study of Hox effectors in TRN subtype differentiation supports this latter conjecture. In the PLM neurons, EGL-5 promotes neurite outgrowth through activating the expression of recycling endosome-associated protein, RFIP; but EGL-5 alone is not sufficient to activate *rfip-1* promoter in yeast (unpublished results). Although we could not rule out an indirect regulation,

EGL-5 possibly helps another transcription factor activate *rfip-1*, as helping UNC-86 activate *mec-3*.

The guarantor function of Hox proteins might explain why very few of their target genes are known. The loss of a terminal selector is expected to have a qualitative effect on gene expression, with particular downstream genes not being expressed. This complete loss of terminal differentiation gene expression does occur in *mec-3* mutants (Chalfie & Au, 1989; Duggan et al., 1998). In contrast, the loss of a guarantor would be expected to have a quantitative effect on transcription. If so, the appearance of new genes as targets would be rare, but looking for quantitative changes in gene expression may help identify Hox targets.

5. ALR-1, A SECOND GUARANTOR IN TRN DIFFERENTIATION

The differentiation and maintenance of TRN characteristics require not only the activation but also the continuous expression of *mec-3*. MEC-3 itself regulates this maintenance in two ways. First, MEC-3 regulates its own expression (autoregulation) by complexing with UNC-86 and binding to regulatory sequences in the *mec-3* promoter (Way & Chalfie, 1989; Xue et al., 1993). Second, the UNC-86/MEC-3 heteromer activates the expression the homeodomain transcription factor ALR-1, which is homologous to *Drosophila* Aristaless and human Arx (Topalidou, van Oudenaarden, & Chalfie, 2011).

The loss of *alr-1* reveals an increased variability in *mec-3* mRNA expression measured by smFISH, which correlates with variable TRN functions. At the lower end of the range, *alr-1* animals have partially reduced touch sensitivity because of decreased *mec-3* level, and some are completely touch insensitive as *mec-3* level falls below a critical threshold. The variable touch sensation in *alr-1* null mutants suggest that ALR-1 promotes TRN differentiation by reducing stochastic fluctuation in *mec-3* transcription and ensuring consistently strong expression of the terminal selector throughout the development. ALR-1 acts like the Hox proteins in that it does not activate *mec-3* by itself but instead enhances UNC-86/MEC-3-induced *mec-3* transcription presumably through the ALR-1 binding sites found in the *mec-3* promoter (Gerstein et al., 2010; Topalidou et al., 2011). Thus, ALR-1 also appears to be acting as a guarantor, this time of *mec-3* autoregulation (Fig. 1D).

Mutations in *alr-1* also cause defects in gene expression, morphology, and function of the AWA and ASG chemosensory neurons with incomplete penetrance and variable expressivity (Melkman & Sengupta, 2005). Moreover, ALR-1, *Drosophila* Aristaless, and mouse Arx regulate the expression of LIM homeodomain transcription factor LIN-11, Lim-1, and Lhx6/Lhx9, respectively (Kitamura et al., 2002; Melkman & Sengupta, 2005; Tsuji et al., 2000). Therefore, the Aristaless/ARX family proteins may generally serve as stabilizers of gene expression. Similar transcriptional guarantors may be essential for the maintenance of the expression of other terminal selectors.

It is currently unclear how the guarantors facilitate the activity of other transcription factors in gene regulation. One possibility is that the binding of guarantors to the nearby *cis*-regulatory elements stabilizes the binding of the main activators to their own sites and/or potentiates their intrinsic transactivation efficiency. Another possibility is that the guarantors recruit other proteins, e.g., histone-modifying enzymes, which could poise the promoter to be activated. Hox proteins and their cofactors seem to have this second function (Choe, Ladam, & Sagerstrom, 2014; Hassan et al., 2007).

6. SAFEGUARDING MEC-3 ACTIVITY BY ZINC-FINGER HOMEODOMAIN PROTEIN ZAG-1

The robust activation and maintained expression of *mec-3* by the HOX and ALR-1 guarantors is not sufficient for TRN differentiation. After establishing stable *mec-3* expression, the subsequent activation of the downstream TRN differentiation genes requires yet another transcription factor, the zinc-finger homeodomain protein ZAG-1, which is homologous to the transcriptional repressors encoded by *Drosophila zfh1* and the vertebrate *zeb* family genes (Vandewalle, Van Roy, & Berx, 2009). Unlike the Hox genes or *alr-1*, mutations in *zag-1* cause complete loss of the TRN markers without affecting *mec-3* expression (Zheng, Jin, Trippe, & Chalfie, submitted for publication). Although phenotypically similar to MEC-3, ZAG-1 does not act as a terminal selector to directly activate the expression of terminal differentiation genes. Instead, ZAG-1 indirectly promotes the TRN fate by preventing the expression of other proteins that antagonize the activity of UNC-86/MEC-3 (Zheng et al., submitted for publication).

Two such proteins are the TEA domain transcription factor EGL-44 and the zinc-finger protein EGL-46, which form a heterodimer to bind to the same *cis*-regulatory motif recognized by UNC-86 and MEC-3 complex. Having higher affinity for the same DNA sequence, EGL-44/EGL-46 can displace UNC-86/MEC-3 from the promoters of the TRN genes and, thus, silence the TRN program (Fig. 2; Zheng et al., submitted for

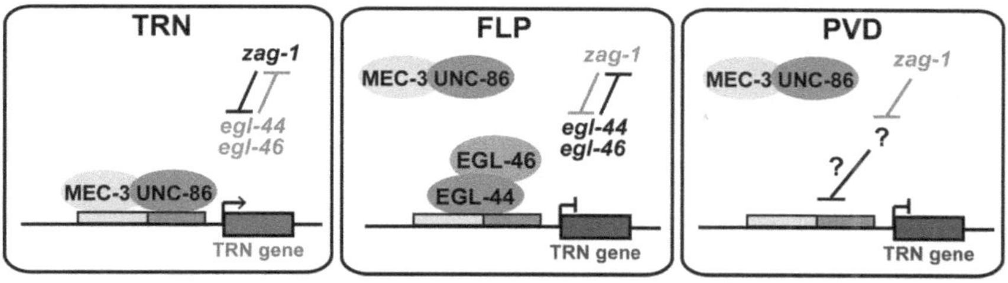

Figure 2 ZAG-1 protects the activity of UNC-86/MEC-3 heteromer. In the TRNs, UNC-86/MEC-3 complex activates the expression of TRN terminal differentiation genes by binding to similar *cis*-regulatory motifs in the promoters. In FLP neurons, EGL-44/EGL-46 complex displaces UNC-86/MEC-3 from its binding sites and thus prevent the expression of TRN fate. Presumably, unknown factor(s) block the activation of TRN genes in PVD neurons. ZAG-1 silences those inhibitors in the TRN neurons but is repressed by them in at least FLPs. This mutual inhibition also enables the switch of cell fate between neurons that express similar terminal selectors. (A color version of this figure is located at the end of this chapter.)

publication). EGL-44/EGL-46 normally prevents the expression of TRN genes in the FLP multimodal sensory neurons, which express both *unc-86* and *mec-3* but do not differentiate as TRNs (Mitani, Du, Hall, Driscoll, & Chalfie, 1993; Wu, Duggan, & Chalfie, 2001). When misexpressed in TRNs, EGL-44/EGL-46 can render UNC-86/MEC-3 inactive and shut down the TRN program. To prevent the expression of such potent inhibitors of the TRN fate, ZAG-1 transcriptionally represses both *egl-44* and *egl-46*, which allows UNC-86/MEC-3 to regulate downstream genes (Zheng et al., submitted for publication). Therefore, ZAG-1 promotes the differentiation of TRN fate by silencing toxic inhibitors and safeguarding the activity of the terminal selectors.

A similar derepression to that caused by ZAG-1 acting on EGL-44/EGL-46 appears to occur in the differentiation of other neurons. For example, Lee and Pfaff (2001) have suggested that neuronal specification in the mouse spinal cord involves the removal of certain repressors. Neocortical projection neurons employs cross-inhibition among transcription factors in a regulatory network that enables mutual exclusion of alternate fates (Greig, Woodworth, Galazo, Padmanabhan, & Macklis, 2013). Moreover, microRNAs enable functional and morphological specification of neurons by targeting chromatin remodeling factors and transcriptional repressors that prevent differentiation (Yoo, Staahl, Chen, & Crabtree, 2009; Zhao, Sun, Li, & Shi, 2009). Similarly, extrinsic signals can also promote neuronal differentiation by altering histone modification and derepressing important developmental genes (Zheng, Karimzadegan, Chiang, & Chalfie, 2013).

Therefore, in addition to the direct activation of cell fate by terminal selectors, derepression mechanisms also contribute to terminal differentiation.

Since EGL-44/EGL-46 also transcriptionally represses *zag-1* in FLP neurons (Zheng et al., submitted for publication), the mutual inhibition between ZAG-1 and EGL-44/EGL-46 establishes a bistable regulatory switch between the TRN and FLP cell fates, both of which requires UNC-86 and MEC-3 as terminal selectors. This mechanism explains why TRN genes are not expressed at high levels in the FLP neurons (a very low level of expression is seen in these cells; Topalidou et al., 2011), but leaves unanswered how UNC-86/MEC-3 can still be needed for FLP differentiation in the presence of EGL-44/EGL-46. One possible explanation is that the nature of the interaction between the two heteromers is target-specific. EGL-44/EGL-46 prevents the activation of TRN genes by UNC-86/MEC-3 but may cooperate with it to activate FLP genes. Supporting this hypothesis, we found that the expression of the gene *sto-5* in FLP neurons requires both EGL-44/EGL-46 and UNC-86/MEC-3 (unpublished results).

Besides TRNs and FLPs, *unc-86* and *mec-3* are also expressed in the PVD neurons, cells that differentiate as multidendritic nociceptors and presumably express other, as yet unknown factors that block the activation of TRN genes (they do not express *egl-44* and *egl-46*). ZAG-1 appears to have the potential to repress such factors, because expression of *zag-1*(+) in the FLP and PVD neurons results in the expression of TRN terminal differentiation genes. Thus, ZAG-1 promotes the TRN fate over the alternative FLP and PVD fates among the three neuronal types that share the common terminal selectors.

7. CONCLUDING REMARKS

Previous studies of neuronal differentiation have mainly focused on identifying terminals selectors that induce the specification of certain neuron types. In the preceding paragraphs, using the differentiation of *C. elegans* mechanosensory TRNs as a model, we provide three examples to illustrate how the activation, maintenance, and function of terminal selectors use previously unrecognized reinforcing or protecting factors. Some, like the Hox proteins and ALR-1, act as guarantor, ensuring the optimal expression of terminal selectors; others, like ZAG-1, allow terminal selectors to act by repressing the expression of inhibitory proteins that block their activity. These different activities point to the need of stabilizing the expression of terminal selectors and protecting their activity. All of these functions are part of the "combinatorial code" that is needed to specify cell fate (Barrera & Ren, 2006; Jessell, 2000; Mitani et al., 1993; Shirasaki & Pfaff, 2002).

One important aspect of cell differentiation is fidelity, the need to have cells develop appropriately despite the inherent variably of gene expression (Munsky, Neuert, & van Oudenaarden, 2012; Ozbudak, Thattai, Kurtser, Grossman, & van Oudenaarden, 2002; Raser & O'Shea, 2004). In a recent review, Lagha, Bothma, and Levine (2012) suggested that three different types of redundancy (redundant enhancers, redundant elements within a single enhancer, and redundant trans-acting transcription factors) can be used to compensate for and overcome this variability, ensure robustness in gene activation, and stabilize differentiation. Our work suggests that guarantors are a fourth mechanism to ensure high fidelity differentiation. The redundancies prevent variability from occurring, but they are not easily modifiable, since the redundant components are always present. In contrast, the guarantors, functioning as single components, can theoretically be modified to act as switches. We envision that they might serve as regulatory control sites shifting differentiation from one cell type to another, and inherited changes in guarantors may give rise to evolutionary diversity.

Terminal selectors that can regulate hundreds of terminal differentiation genes through similar *cis*-regulatory motifs are the central players in neuronal fate determination, but their activities alone are not robust enough to guarantee the induction of specific cell fates in every individual because of stochastic fluctuations. Therefore, a set of auxiliary factors, guarantors, are deployed to increase the efficiency and reliability of cell fate acquisition by enhancing or stabilizing the expression of terminal selectors. Through carefully examining the activity of transcription factors using single-cell techniques and quantitative methods, we expect to identify many more guarantors that contribute to the robustness of cell fate specification.

REFERENCES

Barrera, L. O., & Ren, B. (2006). The transcriptional regulatory code of eukaryotic cells—Insights from genome-wide analysis of chromatin organization and transcription factor binding. *Current Opinion in Cell Biology*, *18*(3), 291–298. http://dx.doi.org/10.1016/j.ceb.2006.04.002.

Baumeister, R., Liu, Y., & Ruvkun, G. (1996). Lineage-specific regulators couple cell lineage asymmetry to the transcription of the Caenorhabditis elegans POU gene unc-86 during neurogenesis. *Genes & Development*, *10*(11), 1395–1410. Retrieved from http://www.ncbi.nlm.nih.gov/pubmed/8647436.

Bounoutas, A., & Chalfie, M. (2007). Touch sensitivity in Caenorhabditis elegans. *Pflügers Archiv*, *454*(5), 691–702. http://dx.doi.org/10.1007/s00424-006-0187-x.

Chalfie, M., & Au, M. (1989). Genetic control of differentiation of the Caenorhabditis elegans touch receptor neurons. *Science, 243*(4894 Pt. 1), 1027–1033. Retrieved from http://www.ncbi.nlm.nih.gov/pubmed/2646709.

Chalfie, M., & Sulston, J. (1981). Developmental genetics of the mechanosensory neurons of Caenorhabditis elegans. *Developmental Biology, 82*(2), 358–370. Retrieved from http://www.ncbi.nlm.nih.gov/pubmed/7227647.

Choe, S. K., Ladam, F., & Sagerstrom, C. G. (2014). TALE factors poise promoters for activation by Hox proteins. *Developmental Cell, 28*(2), 203–211. http://dx.doi.org/10.1016/j.devcel.2013.12.011.

Dasen, J. S., Tice, B. C., Brenner-Morton, S., & Jessell, T. M. (2005). A Hox regulatory network establishes motor neuron pool identity and target-muscle connectivity. *Cell, 123*(3), 477–491. http://dx.doi.org/10.1016/j.cell.2005.09.009.

Duggan, A., Ma, C., & Chalfie, M. (1998). Regulation of touch receptor differentiation by the Caenorhabditis elegans mec-3 and unc-86 genes. *Development, 125*(20), 4107–4119. Retrieved from http://www.ncbi.nlm.nih.gov/pubmed/9735371.

Finney, M., Ruvkun, G., & Horvitz, H. R. (1988). The C. elegans cell lineage and differentiation gene unc-86 encodes a protein with a homeodomain and extended similarity to transcription factors. *Cell, 55*(5), 757–769. Retrieved from http://www.ncbi.nlm.nih.gov/pubmed/2903797.

Garcia-Bellido, A. (1975). Genetic control of wing disc development in Drosophila. *Ciba Foundation Symposium*, (29), 161–182. Retrieved from http://www.ncbi.nlm.nih.gov/pubmed/1039909.

Gehring, W. J., Qian, Y. Q., Billeter, M., Furukubo-Tokunaga, K., Schier, A. F., Resendez-Perez, D., et al. (1994). Homeodomain-DNA recognition. *Cell, 78*(2), 211–223. Retrieved from http://www.ncbi.nlm.nih.gov/pubmed/8044836.

Gerstein, M. B., Lu, Z. J., Van Nostrand, E. L., Cheng, C., Arshinoff, B. I., Liu, T., et al. (2010). Integrative analysis of the Caenorhabditis elegans genome by the modENCODE project. *Science, 330*(6012), 1775–1787. http://dx.doi.org/10.1126/science.1196914.

Greig, L. C., Woodworth, M. B., Galazo, M. J., Padmanabhan, H., & Macklis, J. D. (2013). Molecular logic of neocortical projection neuron specification, development and diversity. *Nature Reviews. Neuroscience, 14*(11), 755–769. http://dx.doi.org/10.1038/nrn3586.

Hassan, M. Q., Tare, R., Lee, S. H., Mandeville, M., Weiner, B., Montecino, M., et al. (2007). HOXA10 controls osteoblastogenesis by directly activating bone regulatory and phenotypic genes. *Molecular and Cellular Biology, 27*(9), 3337–3352. http://dx.doi.org/10.1128/MCB.01544-06.

Hobert, O. (2008). Regulatory logic of neuronal diversity: Terminal selector genes and selector motifs. *Proceedings of the National Academy of Sciences of the United States of America, 105*(51), 20067–20071. http://dx.doi.org/10.1073/pnas.0806070105.

Hobert, O. (2010). Neurogenesis in the nematode Caenorhabditis elegans. *WormBook*, 1–24. http://dx.doi.org/10.1895/wormbook.1.12.2.

Jessell, T. M. (2000). Neuronal specification in the spinal cord: Inductive signals and transcriptional codes. *Nature Reviews. Genetics, 1*(1), 20–29. http://dx.doi.org/10.1038/35049541.

Kitamura, K., Yanazawa, M., Sugiyama, N., Miura, H., Iizuka-Kogo, A., Kusaka, M., et al. (2002). Mutation of ARX causes abnormal development of forebrain and testes in mice and X-linked lissencephaly with abnormal genitalia in humans. *Nature Genetics, 32*(3), 359–369. http://dx.doi.org/10.1038/ng1009.

Lacombe, J., Hanley, O., Jung, H., Philippidou, P., Surmeli, G., Grinstein, J., et al. (2013). Genetic and functional modularity of Hox activities in the specification of limb-innervating motor neurons. *PLoS Genetics, 9*(1), e1003184. http://dx.doi.org/10.1371/journal.pgen.1003184.

Lagha, M., Bothma, J. P., & Levine, M. (2012). Mechanisms of transcriptional precision in animal development. *Trends in Genetics*, *28*(8), 409–416. http://dx.doi.org/10.1016/j.tig.2012.03.006.

Lee, S. K., & Pfaff, S. L. (2001). Transcriptional networks regulating neuronal identity in the developing spinal cord. *Nature Neuroscience*, (4 Suppl.), 1183–1191. http://dx.doi.org/10.1038/nn750.

McGinnis, W., & Krumlauf, R. (1992). Homeobox genes and axial patterning. *Cell*, *68*(2), 283–302. Retrieved from http://www.ncbi.nlm.nih.gov/pubmed/1346368.

Melkman, T., & Sengupta, P. (2005). Regulation of chemosensory and GABAergic motor neuron development by the C. elegans Aristaless/Arx homolog alr-1. *Development*, *132*(8), 1935–1949. http://dx.doi.org/10.1242/dev.01788.

Mitani, S., Du, H., Hall, D. H., Driscoll, M., & Chalfie, M. (1993). Combinatorial control of touch receptor neuron expression in Caenorhabditis elegans. *Development*, *119*(3), 773–783. Retrieved from http://www.ncbi.nlm.nih.gov/pubmed/8187641.

Munsky, B., Neuert, G., & van Oudenaarden, A. (2012). Using gene expression noise to understand gene regulation. *Science*, *336*(6078), 183–187. http://dx.doi.org/10.1126/science.1216379.

Ozbudak, E. M., Thattai, M., Kurtser, I., Grossman, A. D., & van Oudenaarden, A. (2002). Regulation of noise in the expression of a single gene. *Nature Genetics*, *31*(1), 69–73. http://dx.doi.org/10.1038/ng869.

Philippidou, P., Walsh, C. M., Aubin, J., Jeannotte, L., & Dasen, J. S. (2012). Sustained Hox5 gene activity is required for respiratory motor neuron development. *Nature Neuroscience*, *15*(12), 1636–1644. http://dx.doi.org/10.1038/nn.3242.

Raser, J. M., & O'Shea, E. K. (2004). Control of stochasticity in eukaryotic gene expression. *Science*, *304*(5678), 1811–1814. http://dx.doi.org/10.1126/science.1098641.

Shirasaki, R., & Pfaff, S. L. (2002). Transcriptional codes and the control of neuronal identity. *Annual Review of Neuroscience*, *25*, 251–281. http://dx.doi.org/10.1146/annurev.neuro.25.112701.142916.

Suska, A., Miguel-Aliaga, I., & Thor, S. (2011). Segment-specific generation of Drosophila capability neuropeptide neurons by multi-faceted Hox cues. *Developmental Biology*, *353*(1), 72–80. http://dx.doi.org/10.1016/j.ydbio.2011.02.015.

Thaler, J. P., Lee, S. K., Jurata, L. W., Gill, G. N., & Pfaff, S. L. (2002). LIM factor Lhx3 contributes to the specification of motor neuron and interneuron identity through cell-type-specific protein-protein interactions. *Cell*, *110*(2), 237–249. Retrieved from http://www.ncbi.nlm.nih.gov/pubmed/12150931.

Thor, S., Andersson, S. G., Tomlinson, A., & Thomas, J. B. (1999). A LIM-homeodomain combinatorial code for motor-neuron pathway selection. *Nature*, *397*(6714), 76–80. http://dx.doi.org/10.1038/16275.

Topalidou, I., van Oudenaarden, A., & Chalfie, M. (2011). Caenorhabditis elegans aristaless/Arx gene alr-1 restricts variable gene expression. *Proceedings of the National Academy of Sciences of the United States of America*, *108*(10), 4063–4068. http://dx.doi.org/10.1073/pnas.1101329108.

Tsuji, T., Sato, A., Hiratani, I., Taira, M., Saigo, K., & Kojima, T. (2000). Requirements of Lim1, a Drosophila LIM-homeobox gene, for normal leg and antennal development. *Development*, *127*(20), 4315–4323. Retrieved from http://www.ncbi.nlm.nih.gov/pubmed/11003832.

Vandewalle, C., Van Roy, F., & Berx, G. (2009). The role of the ZEB family of transcription factors in development and disease. *Cellular and Molecular Life Sciences*, *66*(5), 773–787. http://dx.doi.org/10.1007/s00018-008-8465-8.

Way, J. C., & Chalfie, M. (1988). mec-3, a homeobox-containing gene that specifies differentiation of the touch receptor neurons in C. elegans. *Cell*, *54*(1), 5–16. Retrieved from http://www.ncbi.nlm.nih.gov/pubmed/2898300.

Way, J. C., & Chalfie, M. (1989). The mec-3 gene of Caenorhabditis elegans requires its own product for maintained expression and is expressed in three neuronal cell types. *Genes & Development*, *3*(12A), 1823–1833. Retrieved from http://www.ncbi.nlm.nih.gov/pubmed/2576011.

White, J. G., Southgate, E., Thomson, J. N., & Brenner, S. (1986). The structure of the nervous system of the nematode Caenorhabditis elegans. *Philosophical Transactions of the Royal Society of London. Series B, Biological Sciences*, *314*(1165), 1–340.

Wu, J., Duggan, A., & Chalfie, M. (2001). Inhibition of touch cell fate by egl-44 and egl-46 in C. elegans. *Genes & Development*, *15*(6), 789–802. http://dx.doi.org/10.1101/gad.857401.

Xue, D., Finney, M., Ruvkun, G., & Chalfie, M. (1992). Regulation of the mec-3 gene by the C. elegans homeoproteins UNC-86 and MEC-3. *EMBO Journal*, *11*(13), 4969–4979. Retrieved from http://www.ncbi.nlm.nih.gov/pubmed/1361171.

Xue, D., Tu, Y., & Chalfie, M. (1993). Cooperative interactions between the Caenorhabditis elegans homeoproteins UNC-86 and MEC-3. *Science*, *261*(5126), 1324–1328. Retrieved from http://www.ncbi.nlm.nih.gov/pubmed/8103239.

Yoo, A. S., Staahl, B. T., Chen, L., & Crabtree, G. R. (2009). MicroRNA-mediated switching of chromatin-remodelling complexes in neural development. *Nature*, *460*(7255), 642–646. http://dx.doi.org/10.1038/nature08139.

Zhang, Y., Ma, C., Delohery, T., Nasipak, B., Foat, B. C., Bounoutas, A., et al. (2002). Identification of genes expressed in C. elegans touch receptor neurons. *Nature*, *418*(6895), 331–335. http://dx.doi.org/10.1038/nature00891.

Zhao, C., Sun, G., Li, S., & Shi, Y. (2009). A feedback regulatory loop involving microRNA-9 and nuclear receptor TLX in neural stem cell fate determination. *Nature Structural & Molecular Biology*, *16*(4), 365–371. http://dx.doi.org/10.1038/nsmb.1576.

Zheng, C., Diaz-Cuadros, M., & Chalfie, M. (2015). Hox genes promote neuronal subtype diversification through posterior induction in Caenorhabditis elegans. *Neuron*, *88*(3), 514–527. http://dx.doi.org/10.1016/j.neuron.2015.09.049.

Zheng, C., Jin, F. Q., & Chalfie, M. (2015). Hox proteins act as transcriptional guarantors to ensure terminal differentiation. *Cell Reports*, *13*(7), 1343–1352. http://dx.doi.org/10.1016/j.celrep.2015.10.044.

Zheng, C., Jin, F. Q., Trippe, B. L., & Chalfie, M. (submitted for publication). Mutual inhibition between Zn-finger homeodomain protein ZAG-1 and TEA domain transcription factors EGL-44 controls a bistable cell fate switch in Caenorhabditis elegans.

Zheng, C., Karimzadegan, S., Chiang, V., & Chalfie, M. (2013). Histone methylation restrains the expression of subtype-specific genes during terminal neuronal differentiation in Caenorhabditis elegans. *PLoS Genetics*, *9*(12). e1004017. http://dx.doi.org/10.1371/journal.pgen.1004017.

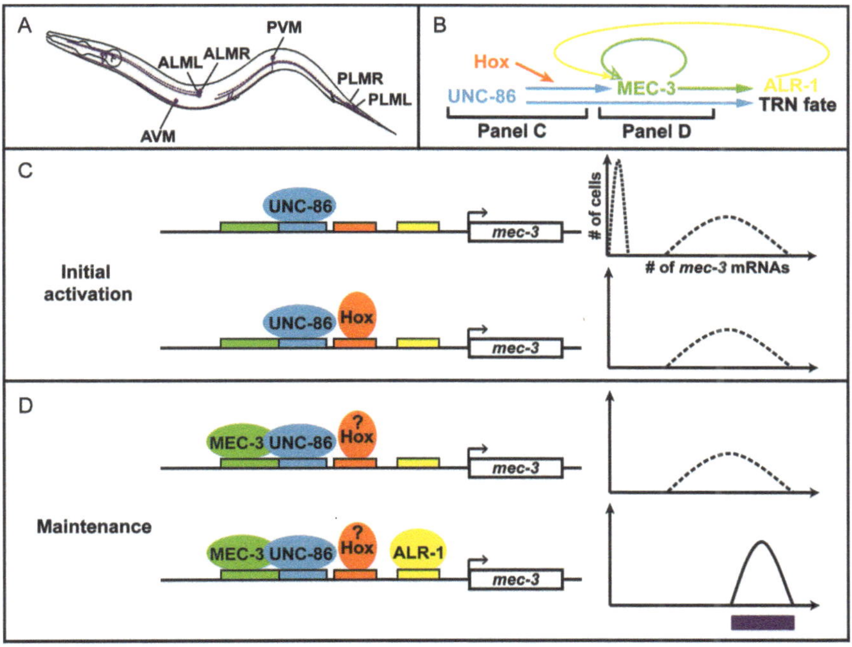

Chaogu Zheng and Martin Chalfie, Figure 1 Guarantors ensure the activation and maintained expression of the terminal selector *mec-3* in the TRNs. (A) The six TRNs in *C. elegans*. (B) Proteins involved in TRN specification. UNC-86 initially activates *mec-3*, facilitated by Hox proteins; and the expressed MEC-3 forms a heteromer with UNC-86 that autoregulates *mec-3* and activates the TRN fate. UNC-86/MEC-3 also activates *alr-1*, which potentiates *mec-3* autoregulation. (C) and (D) Modes of *mec-3* transcription in the absence and presence of guarantors. The schematic plots represent quantitative measurements of *mec-3* mRNA molecules in populations of cells using smFISH. (C) During the initiation phase, UNC-86 alone cannot fully activate *mec-3* in all TRNs. In Hox mutants, a binary loss of *mec-3* expression leads to the failure of TRN fate acquisition in some of the cells. (D) *mec-3* autoregulation requires not only UNC-86/MEC-3 but also ALR-1 to reduce the variability and optimize *mec-3* expression. Whether Hox proteins are also needed for the *mec-3* maintenance is unclear. The binding sites of the guarantors are found adjacent to the essential UNC-86/MEC-3 binding site in the proximal promoter of *mec-3*. *Panel (A): Adapted from Bounoutas and Chalfie (2007) with kind permission from Springer Science+Business Media and from Chalfie and Sulston (1981), Copyright (1981), with permission from Elsevier.*

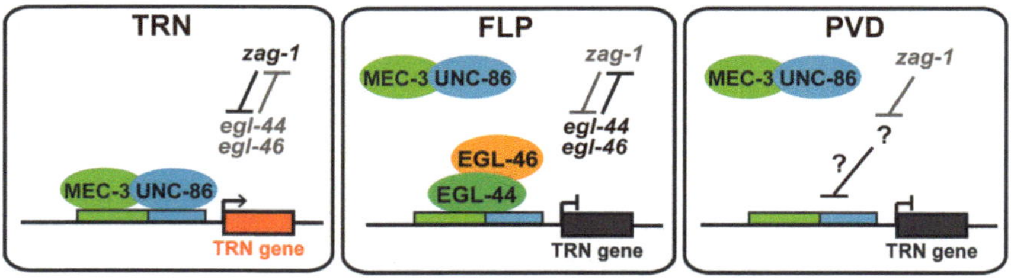

Chaogu Zheng and Martin Chalfie, Figure 2 ZAG-1 protects the activity of UNC-86/MEC-3 heteromer. In the TRNs, UNC-86/MEC-3 complex activates the expression of TRN terminal differentiation genes by binding to similar *cis*-regulatory motifs in the promoters. In FLP neurons, EGL-44/EGL-46 complex displaces UNC-86/MEC-3 from its binding sites and thus prevent the expression of TRN fate. Presumably, unknown factor(s) block the activation of TRN genes in PVD neurons. ZAG-1 silences those inhibitors in the TRN neurons but is repressed by them in at least FLPs. This mutual inhibition also enables the switch of cell fate between neurons that express similar terminal selectors.

CHAPTER ELEVEN

Cancer in *Drosophila*: Imaginal Discs as a Model for Epithelial Tumor Formation

Héctor Herranz, Teresa Eichenlaub, Stephen M. Cohen[1]
Department of Cellular and Molecular Medicine, University of Copenhagen, Copenhagen, Denmark
[1]Corresponding author: e-mail address: scohen@sund.ku.dk

Contents

Abstract

Cancer genomics has greatly increased our understanding of the complexity of the genetic and epigenetic changes found in human tumors. Understanding the functional relationships among these elements calls for the use of flexible genetic models. We discuss the use of *Drosophila* models to study cooperation among genetic factors that contribute to disease progression.

Current Topics in Developmental Biology, Volume 116
ISSN 0070-2153
http://dx.doi.org/10.1016/bs.ctdb.2015.11.037

1. INTRODUCTION

Cancer is a complex disease, driven by the accumulation of mutations and epigenetic changes that alter intrinsic cellular functions and cell–cell interactions in tissues. Mutational signatures are being defined for many cancers (Kandoth et al., 2013), reflecting alteration in a broad range of cellular capabilities that enable disease progression (Hanahan & Weinberg, 2011). The complexity of the genetic landscape reflected in cancer genomes poses a new challenge. Single gene alterations with strong effects have long been identified as cancer drivers (Stratton, 2011). The emerging picture is one in which combinations of genetic modifications may underlie different tumor types and different phases in the development of disease. The complexity of the mutational landscape highlights the need for simple, genetically tractable, models to explore how the numerous (epi)genetic alterations contribute to cancer in the animal.

Conservation of signaling pathways and cellular regulatory systems between humans and flies has made *Drosophila* a useful model for cancer research. Indeed, the first tumor-causing mutation was identified in *Drosophila* (Stark, 1918). Many human oncogenes and tumor-suppressor genes were originally discovered in *Drosophila*, notably, the Hippo tumor-suppressor pathway. A number of interesting *Drosophila* models have been developed to explore different types of cancer (reviewed in Gonzalez, 2013).

The most common type of human tumor, carcinoma, is of epithelial origin. Imaginal disc epithelia are providing good models to study epithelial tumors. Many well-characterized cancer drivers, such as EGFR, Ras, Myc, Notch, YAP or tumor suppressors such as PTEN, and the Hippo pathway genes, result in deregulation of the pathways that control cell growth and proliferation during development (Table 1). Interestingly, activation of these pathways in imaginal disc epithelia leads to benign tissue hyperplasia, but is not on its own sufficient to produce tumors. This has set the stage for use of the fly model to screen for cooperating factors that work together with the identified cancer drivers to promote neoplasia or metastasis. Pioneering work by the Xu and Richardson labs in 2003 used the imaginal discs to identify genes that cooperated with activated forms of Ras and Notch to produce malignant tumors (Brumby & Richardson, 2003; Pagliarini & Xu, 2003). Conceptually similar approaches, using transposon mutagenesis in the mouse, have begun to tackle oncogene cooperation (Copeland & Jenkins, 2010).

Table 1 *Drosophila* Oncogenes and Tumor Suppressors

Pathway	Gene	Class	References
EGFR/ MAPK	*EGFR*	Oncogene	Dominguez, Wasserman, and Freeman (1998) and Wang, Simcox, and Campbell (2000)
	Ras	Oncogene	Karim and Rubin (1998)
Insulin	*PI3K*	Oncogene	Leevers, Weinkove, MacDougall, Hafen, and Waterfield (1996)
	Pten	Tumor suppressor	Gao, Neufeld, and Pan (2000)
	TSC1/ TSC2	Tumor suppressors	Tapon, Ito, Dickson, Treisman, and Hariharan (2001), Potter, Huang, and Xu (2001), Gao and Pan (2001), and Ito and Rubin (1999)
Notch	*Notch*	Oncogene	Go, Eastman, and Artavanis-Tsakonas (1998)
	Delta	Oncogene	Ferres-Marco et al. (2006)
Hedgehog	*hedhehog*	Oncogene	Basler and Struhl (1994)
Wingless/ Wnt	*wingless*	Oncogene	Diaz-Benjumea and Cohen (1995)
Dpp/BMP	*dpp*	Oncogene	Haerry, Khalsa, O'Connor, and Wharton (1998), Lecuit et al. (1996), and Nellen, Burke, Struhl, and Basler (1996)
JAK-STAT	*Upd*	Oncogene	Bach, Vincent, Zeidler, and Perrimon (2003)
Hippo	*fat*	Tumor suppressor	Silva, Tsatskis, Gardano, Tapon, and McNeill (2006)
	Expanded	Tumor suppressor	Hamaratoglu et al. (2006) and Pellock, Buff, White, and Hariharan (2007)
	Merlin	Tumor suppressor	Hamaratoglu et al. (2006) and Pellock et al. (2007)
	Warts	Tumor suppressor	Justice, Zilian, Woods, Noll, and Bryant (1995)
	Salvador	Tumor suppressor	Tapon et al. (2002) and Kango-Singh et al. (2002)

Continued

Table 1 *Drosophila* Oncogenes and Tumor Suppressors—cont'd

Pathway	Gene	Class	References
	Hippo	Tumor suppressor	Harvey, Pfleger, and Hariharan (2003), Udan, Kango-Singh, Nolo, Tao, and Halder (2003), and Wu, Huang, Dong, and Pan (2003)
	yki	Oncogene	Huang, Wu, Barrera, Matthews, and Pan (2005)
Cell polarity	*Scribble*	Tumor suppressor	Bilder, Li, and Perrimon (2000) and Gateff (1978)
	Disc large	Tumor suppressor	
	Lgl	Tumor suppressor	
Vesicle trafficking	*Avalanche*	Tumor suppressor	Lu and Bilder (2005)
	Rab5	Tumor suppressor	Lu and Bilder (2005)
	Tsg101	Tumor suppressor	Moberg, Schelble, Burdick, and Hariharan (2005)
	Vps25	Tumor suppressor	Thompson et al. (2005) and Vaccari and Bilder (2005)
Cellular memory	*Polyhomeotic*	Tumor suppressor	Martinez et al. (2009) and Classen, Bunker, Harvey, Vaccari, and Bilder (2009)
	Psc-Su(z)2	Tumor suppressor	Classen et al. (2009)
	Pc	Tumor suppressor	
	Sce	Tumor suppressor	

Summary of the genes implicated as oncogenes and tumor suppressors in cancer-relevant signaling pathways and cellular processes, using *Drosophila* genetic models.

2. ONCOGENIC COOPERATION

2.1 Ras

The Xu and Richardson labs based their approach on the observation that expression of Ras^{V12} or activated Notch caused tissue hyperplasia that did not progress to form malignant tumors. By producing clones of cells lacking candidate tumor suppressors in these backgrounds, they identified *scribble-disc large-lgl* (*scrib/Dlg/Lgl*) polarity complex as a context-dependent tumor suppressor (Brumby & Richardson, 2003; Pagliarini & Xu, 2003). Mutants for these polarity proteins hyperactivate the Hippo pathway effector Yorkie driving tissue overgrowth (Chen, Schroeder, Kango-Singh, Tao, & Halder, 2012; Doggett, Grusche, Richardson, & Brumby, 2011; Grusche, Degoutin, Richardson, & Harvey, 2011; Grzeschik, Parsons, Allott, Harvey, & Richardson, 2010; Menendez, Perez-Garijo, Calleja, & Morata, 2010) and show defects in epithelial integrity (Bilder, 2004). Human carcinomas lose polarity as they progress toward malignancy. Deregulation of the *scrib/Dlg/Lgl* polarity module is commonly observed in human cancer and correlates with tumor progression (reviewed in Elsum, Yates, Humbert, & Richardson, 2012). Subsequent work by these groups identified RhoGEF and mutants affecting lysosomal proteins as other elements cooperating with oncogenic Ras in metastasis formation (Table 2).

A similar approach was used to identify genes involved in mitochondrial respiratory function as elements cooperating with oncogenic Ras (Ohsawa et al., 2012). Cooperation between Ras^{V12} and mitochondrial dysfunction increased production of reactive oxygen species and Jun N-terminal kinase (JNK) activation. Through upregulation of Interleukin (Upd) and Wnt/wg signaling, these cells were able to induce metastatic tumor formation by nearby cells in which only the Ras pathway was activated. This model proved to be particularly interesting because of the observation that clones of cells with mitochondrial dysfunction and Ras^{V12} were able to induce tumors in nearby cells expressing Ras^{V12} alone. This observation was possible because the experimental design created a specific juxtaposition of cell types. It will be interesting to explore whether other combinations of cooperating cancer genes confer the ability of genetically different cell populations to cooperate in tumor formation.

2.2 Notch

Activation of the Notch pathway has also been observed to drive epithelial hyperplasia, without neoplastic transformation in the eye disc epithelium.

Table 2 Cooperating Oncogenes and Tumor Suppressors

Driver	Cooperating Partners	References
Ras^{V12}	Scribble, Dlg, lgl	Pagliarini and Xu (2003); Brumby and Richardson (2003)
	RhoGEF	Khoo, Allan, Willoughby, Brumby, and Richardson (2013) and Brumby et al. (2011)
	Src/Csk	Vidal, Warner, Read, and Cagan (2007)
	Mitochondrial proteins	Ohsawa et al. (2012)
	Lysosomal proteins	Chi et al. (2010)
EGFR overexpression	bantam/Socs36E	Herranz, Hong, Hung, Voorhoeve, and Cohen (2012)
	mir-375/pipsqueak	Herranz, Weng, and Cohen (2014)
	miR10/pipsqueak	
Notch activation	Scribble	Brumby and Richardson (2003)
	Lola, pipsqueak	Ferres-Marco et al. (2006)
	mir-7/ihog	Da Ros, Gutierrez-Perez, Ferres-Marco, and Dominguez (2013)

This has provided the basis for a genetic screen for cooperating factors that drive neoplasia when coexpressed with the Notch ligand Delta (Ferres-Marco et al., 2006). Upregulation of the Polycomb group epigenetic silencers, Pipsqueak and Lola, cooperated with Delta to promote neoplasia and metastasis. Similarly, the microRNA miR-7 was found to cooperate with Notch in tumor formation. Interestingly, in this model, the effects of Notch activation required downregulation of Hedgehog signaling (Da Ros et al., 2013). Notch activity was found to downregulate the Hh receptor Boi, while miR-7 repressed the Hedgehog receptor ihog. This Hedgehog signaling limited the ability of activated Notch to promote tumor formation in the eye disc. Whether the Hedgehog pathway is also the direct target of the epigenetic silencing effects of Pipsqueak and Lola in this context remains to be determined.

2.3 EGFR

EGFR is overexpressed or mutationally activated in many human cancers. Our lab developed a model to screen for cooperating oncogenes and tumor suppressors based on conditional activation of EGFR in the wing disc (Herranz, Hong, Hung, et al., 2012). EGFR overexpression leads to hyperplasia, which is offset in part by elevated apoptosis. Under normal conditions, this does not progress to neoplasia and epithelial–mesenchymal transition (EMT). We identified several microRNAs that cooperate with EGFR to produce neoplastic transformation and metastasis (Table 2). Targets of those microRNAs behave as conditional tumor suppressors, in the context of EGFR activation (Herranz, Hong, Hung, et al., 2012; Herranz et al., 2014). In each case, the tissue became metastatic. Other qualitative features of the tumors were quite different, suggesting that different pathways were involved.

2.4 Hippo Pathway

The Hippo tumor-suppressor pathway was discovered as a series of mutants that produced tissue hyperplasia (Table 1; reviewed in Pan, 2010). The pathway acts to limit the activity of the transcriptional coactivators Yorkie in *Drosophila*, and mammalian YAP/Taz. There is growing evidence for the importance of the Hippo tumor-suppressor pathway and its effectors YAP/Taz in human cancer (reviewed in Barry & Camargo, 2013; Harvey, Zhang, & Thomas, 2013). Indeed, YAP expression can bypass the requirement for oncogenic Ras in models of human cancer (Hong et al., 2014; Kapoor et al., 2014; Shao et al., 2014).

Although increased Yorkie activity can lead to massive hyperplasia, there have been no reports to date of neoplastic transformation. This contrasts with mouse models, in which prolonged YAP overexpression can lead to neoplasia in liver (Dong et al., 2007). In the fly model, metamorphosis truncates the experiment, precluding analysis of long-term Yorkie overexpression. It is possible that epigenetic changes affecting the expression of cooperating genes leads to the emergence of tumors in mice with long-term YAP overexpression. For example, coexpression of YAP and the apoptosis inhibitor cIAP can lead to liver tumors in the mouse (Zender et al., 2006). We have observed that Yorkie can cooperate with microRNAs to induce neoplasia and metastasis (unpublished observations). The fly model may prove useful to identify genes that cooperate with Yorkie/YAP to promote neoplasia and metastasis.

2.5 Other Pathways

The TSC/tor pathway and Myc have not been examined in depth for oncogenic cooperation using the genetic power of the fly model. These will be interesting areas for future exploration.

3. OTHER HALLMARKS OF CANCER

The examples of oncogenic cooperation discussed in Section 2 mainly focus on signaling pathways that control cell growth and proliferation. Cancer is a complex disease in which many other facets of cell function, cell–cell, and tissue interaction play important roles. Collectively, these features have been called "hallmarks" of cancer (Hanahan & Weinberg, 2011). In this section, we summarize how *Drosophila* models are beginning to shed light on these other important aspects of disease.

3.1 Apoptosis and "Compensatory Cell Proliferation"

Drivers of cell growth and proliferation, such as the EGFR/Ras/MAPK pathway, Src kinases, Yorkie/Yap, and Myc, induce apoptosis in *Drosophila* epithelial models (Bergmann, Tugentman, Shilo, & Steller, 2002; Huang et al., 2005; Karim & Rubin, 1998; Montero, Muller, & Gallant, 2008; Prober & Edgar, 2000; Vidal et al., 2007). Overcoming proliferation-induced apoptosis is a key requirement for human cancers. Studies in *Drosophila* have identified how different signals that can lead to evasion of apoptosis. The Hippo pathway effector Yorkie directly limits apoptosis by controlling the expression of the *Drosophila* inhibitor of apoptosis (DIAP) (Huang et al., 2005) as well as indirectly by activating expression of the *bantam* microRNA (Nolo, Morrison, Tao, Zhang, & Halder, 2006; Thompson & Cohen, 2006). Dpp is a member of the bone morphogenetic protein family and can also inhibit apoptosis by modulating *bantam* levels (Martin, Perez-Garijo, Moreno, & Morata, 2004). *bantam* limits activation of the proapoptotic gene *hid* (Brennecke, Hipfner, Stark, Russell, & Cohen, 2003). The EGFR/Ras/MAPK pathway activity also works both directly and indirectly to limit proliferation-induced apoptosis (Prober & Edgar, 2000; Yang & Baker, 2003). EGFR signals via Ras to target Hid (Bergmann, Agapite, McCall, & Steller, 1998). EGFR acts indirectly on *hid* via bantam (Herranz, Hong, & Cohen, 2012).

Beyond the obvious need for cells to suppress apoptosis in order to be able to continue to proliferate, the imaginal discs have provided important insights into how cell growth and cell death are interconnected. Elimination of unfit, stressed, or damaged cells leads to a process of compensatory proliferation during tissue growth (Martin, Perez-Garijo, & Morata, 2009). When dying cells were prevented from completing the apoptosis program by expression of the baculovirus protein p35, it was discovered that these "undead" cells secreted mitogenic factors that induced proliferation of neighboring cells (Perez-Garijo, Shlevkov, & Morata, 2009). Compensatory proliferation has been observed among cells depleted of the small nucleolar RNP, pseudouridine synthase (Vicidomini et al., 2015). Metabolic imbalance in these cells triggers apoptosis and production of mitogenic cues leading to compensatory proliferation. Interestingly, these cells also activate the JNK pathway leading to induction of matrix metalloproteases and EMT, promoting tumor formation. Whether similar processes apply in human cancer remains to be explored.

3.2 Genome Instability

Genome instability is common in cancer cells. In mammalian models, cytokinesis failure or telomere dysfunction can lead to transient tetraploidy, and to the formation of subtetraploid aneuploid cells with enhanced tumorigenic capacity (Davoli & de Lange, 2012; Davoli, Denchi, & de Lange, 2010; Fujiwara et al., 2005). Tetraploidy has recently been reported to activate the Hippo tumor-suppressor pathway, providing an additional mechanism to limit the proliferative potential of cells that could generate aneuploidy progeny (Ganem et al., 2014). In *Drosophila*, telomere loss can lead to formation of aneuploid cells, which can escape apoptosis and form tumors (Titen & Golic, 2008). Other manipulations involving downregulation of checkpoint genes, exposure to DNA-damaging agents or manipulating centrosome number rarely produced tumors, unless apoptosis was also blocked (Dekanty, Barrio, & Milan, 2015; Dekanty, Barrio, Muzzopappa, Auer, & Milan, 2012; Morais da Silva, Moutinho-Santos, & Sunkel, 2013; Poulton, Cuningham, & Peifer, 2014). Induction of genome instability in epithelial cells activates the apoptotic program and, consequently, these cells are eliminated from the tissue. In mammals, this depends on p53 activity (Li et al., 2010). In *Drosophila*, genome instability leads to the activation of the JNK pathway, which induces apoptosis to remove aneuploid cells (Dekanty et al., 2012). Further study of these models may provide new insights into tumor suppressive mechanisms.

3.3 Tumor–Stroma Interaction

Tumor–stroma interaction provides another dimension of complexity. Tumors are composed of different cell types, intimately associated with ostensibly normal stromal cells, which create a "tumor microenvironment." Although much research has focused on the neoplastic cells, tumors function as tissues in which stromal cells also play critical roles, influencing many facets of disease progression (Bissell & Hines, 2011). Changes in stromal cell gene expression profiles have been reported during tumorigenesis. However, the contribution of these changes to tumorigenesis remains to be understood.

Our lab has developed a *Drosophila* model of tumor–stroma interaction (Herranz et al., 2014). Proliferation of the genetically modified epithelial cells drives proliferation of the unmodified mesenchymal population, and vice versa. These tumors progress from hyperplastic to neoplastic and become invasive and metastatic. The interesting feature of this model is that malignant transformation depends on interaction between genetically modified epithelial cells and the surrounding normal stromal cells. Production of the extracellular matrix protein Perlecan by the tumor cells potentiates bone morphogenetic protein (BMP) signaling to promote stromal cell proliferation. Reciprocally, signals from the stromal compartment were required for epithelial cell transformation. The stromal signal remains to be identified.

3.4 Inflammation

Tumor-related inflammation is an important feature of disease. Transformed epithelial cells secrete chemokines and cytokines which recruit immune cells to the tumor mass. Depending on the context, tumor-related inflammation can either promote or restrict tumor growth.

Tumor formation in the imaginal discs involves activation of JNK, leading to production of unpaired cytokines that activate the JAK/STAT pathway (Pastor-Pareja, Wu, & Xu, 2008; Wu, Pastor-Pareja, & Xu, 2010). Activation of JAK/STAT drives epithelial tumor progression (Wu et al., 2010). These cytokines also recruit circulating immune cells, called hemocytes, which express the *Drosophila* ortholog of tumor necrosis factor-α (TNF-α; Cordero et al., 2010). TNF-α activates JNK in the tumor cells, which in this context induces apoptosis and inhibits tumor growth (Parisi, Stefanatos, Strathdee, Yu, & Vidal, 2014). In fact, when hemocytes are ablated by expression of the proapoptotic gene *hid*, tumors increased in size, indicating that the hemocytes were responsible for limiting tumor growth (Pastor-Pareja et al., 2008). These studies reveal a dual role for the JAK/STAT pathway: promoting tumor growth in one context, while triggering an antitumoral immune response in another.

3.5 Tumor Angiogenesis

Tumors induce the formation of new vasculature to improve nutrient and oxygen supply and clear waste. Angiogenesis plays a critical role in expansion of solid tumors. *Drosophila* has an open circulatory system, but uses a system of tracheal tubes to convey oxygen to internal organs. *Drosophila* epithelial tumors undergo hypoxic stress and secrete tracheogenic factors as fibroblast growth factor that induce the formation of new trachea to promote gas exchange (Grifoni, Sollazzo, Fontana, Froldi, & Pession, 2015). The study of systemic aspects of tumorigenesis is just beginning in the fly model.

3.6 Tumor Metabolism and Wasting Syndrome

Drosophila models have also been used to explore how metabolic dysfunction may be linked to cancer. Using an activated Ras/Src genotype, the Cagan lab found that malignant tumor formation was enhanced in animals on a high-sugar diet (Hirabayashi, Baranski, & Cagan, 2013). Interestingly, Ras/Src tumors were found to retain elevated responsiveness to insulin, despite the acquisition of an insulin resistant diabetic phenotype in the animal.

Most cancer patients suffer from wasting syndrome or cachexia, which involves weight loss and muscle atrophy. Cachexia may be responsible for over 20% of cancer deaths (Fearon, Arends, & Baracos, 2013). Recent studies suggest that the fly may provide a useful model to study cachexia. Malignant tumors were found to induce a cachexia-like phenotype when transplanted into the abdomen of a host fly, due to upregulation ImpL2, a homolog of the insulin growth factor binding protein (IGFBP), an antagonist of insulin signaling. Systemic inhibition of insulin signaling activity resulted in tissue wasting in the host (Figueroa-Clarevega & Bilder, 2015). Similarly, Yorkie-driven intestinal stem cell tumors were found to induce ImpL2, which induced systemic wasting (Kwon et al., 2015). In an interesting parallel to the RAS/SRC model, the tumor cells evade the systemic effects of ImpL2 through activation of the insulin pathway.

Systematic genetic screens for oncogenic cooperation are beginning to identify enzymes involved in other metabolic pathways (unpublished data), suggesting that links between metabolism and neoplasia will be a fruitful avenue for exploration.

4. CELL COMPETITION IN CANCER

During normal tissue growth, cells compete for survival. Genetic studies in the imaginal disc epithelium first showed that faster proliferating cells eliminate slower-dividing but otherwise healthy cells (Simpson, 1979; Simpson & Morata, 1981), suggesting that cell competition is an active process. Subsequent work has shown that subtle differences in expression of Myc are sufficient to induce cell competition: cells with higher Myc levels eliminate the cells with lower Myc levels (De La Cova, Abril, Bellosta, Gallant, & Johnston, 2004; Moreno & Basler, 2004). Cells win in competition in part through inducing apoptosis of their less-fit neighbors. Competition-dependent elimination of cells by induced apoptosis is also observed in mammalian epithelia (Tamori et al., 2010). Evidence is emerging that "loser" cells may then be eliminated by an active process of cell engulfment (Li & Baker, 2007; Sun et al., 2014). These observations have led to the hypothesis that cell competition might contribute to cancer progression (reviewed in Baillon & Basler, 2014; Johnston, 2014; Levayer & Moreno, 2013). Recent unpublished work from our lab described a model in which epithelial tumor cells behave as "super-competitors" inducing apoptosis and engulfment of normal cells to produce tumors (Eichenlaub, Cohen, & Herranz, in press).

There is also a case to be made for cell competition serving as a surveillance system to eliminate tumor cells. For example, cells mutant for the tumor-suppressor genes lgl, scribble, and dlg can be detected and outcompeted by surrounding wild-type cells (Igaki, Pastor-Pareja, Aonuma, Miura, & Xu, 2009; Menendez et al., 2010). A similar behavior has been observed in clones of cells activating the Src oncogene alone (Vidal, Larson, & Cagan, 2006). These studies suggest that these single genetic changes are not enough to confer a neoplastic phenotype. Additional factors, such as suppression of apoptosis, might switch these cells from losers to winners in competition with normal cells to allow tumor formation. These observations highlight the importance of cooperating factors in creating a genetic landscape to drive tumorigenesis.

5. CONCLUDING REMARKS

This short review illustrates some of the ways the fly is being used to explore cancer biology. Systematic genetic investigation using some of the newer models should soon start to suggest interesting new avenues for investigation into human disease.

ACKNOWLEDGMENTS

Work in the authors' laboratory is supported by the NovoNordisk Foundation grant NNF12OC0000552 and the Danish Council for strategic research grant 3048-00002A.

REFERENCES

Bach, E. A., Vincent, S., Zeidler, M. P., & Perrimon, N. (2003). A sensitized genetic screen to identify novel regulators and components of the *Drosophila* janus kinase/signal transducer and activator of transcription pathway. *Genetics*, *165*(3), 1149–1166.

Baillon, L., & Basler, K. (2014). Reflections on cell competition. *Seminars in Cell & Developmental Biology*, *32*, 137–144. http://dx.doi.org/10.1016/j.semcdb.2014.04.034.

Barry, E. R., & Camargo, F. D. (2013). The Hippo superhighway: Signaling crossroads converging on the Hippo/Yap pathway in stem cells and development. *Current Opinion in Cell Biology*, *25*, 247–253. http://dx.doi.org/10.1016/j.ceb.2012.12.006.

Basler, K., & Struhl, G. (1994). Compartment boundaries and the control of *Drosophila* limb pattern by *hedgehog* protein. *Nature*, *368*, 208–214.

Bergmann, A., Agapite, J., McCall, K., & Steller, H. (1998). The *Drosophila* gene hid is a direct molecular target of Ras-dependent survival signaling. *Cell*, *95*(3), 331–341.

Bergmann, A., Tugentman, M., Shilo, B. Z., & Steller, H. (2002). Regulation of cell number by MAPK-dependent control of apoptosis: A mechanism for trophic survival signaling. *Developmental Cell*, *2*(2), 159–170.

Bilder, D. (2004). Epithelial polarity and proliferation control: Links from the *Drosophila* neoplastic tumor suppressors. *Genes & Development*, *18*(16), 1909–1925.

Bilder, D., Li, M., & Perrimon, N. (2000). Cooperative regulation of cell polarity and growth by *Drosophila* tumor suppressors. *Science*, *289*(5476), 113–116.

Bissell, M. J., & Hines, W. C. (2011). Why don't we get more cancer? A proposed role of the microenvironment in restraining cancer progression. *Nature Medicine*, *17*(3), 320–329. http://dx.doi.org/10.1038/nm.2328.

Brennecke, J., Hipfner, D. R., Stark, A., Russell, R. B., & Cohen, S. M. (2003). *bantam* encodes a developmentally regulated microRNA that controls cell proliferation and regulates the pro-apoptotic gene hid in *Drosophila*. *Cell*, *113*, 25–36.

Brumby, A. M., Goulding, K. R., Schlosser, T., Loi, S., Galea, R., Khoo, P., et al. (2011). Identification of novel Ras-cooperating oncogenes in *Drosophila melanogaster*: A RhoGEF/Rho-family/JNK pathway is a central driver of tumorigenesis. *Genetics*, *188*(1), 105–125. http://dx.doi.org/10.1534/genetics.111.127910.

Brumby, A. M., & Richardson, H. E. (2003). *scribble* mutants cooperate with oncogenic Ras or Notch to cause neoplastic overgrowth in *Drosophila*. *The EMBO Journal*, *22*(21), 5769–5779. http://dx.doi.org/10.1093/emboj/cdg548.

Chen, C. L., Schroeder, M. C., Kango-Singh, M., Tao, C., & Halder, G. (2012). Tumor suppression by cell competition through regulation of the Hippo pathway. *Proceedings of the National Academy of Sciences of the United States of America*, *109*(2), 484–489. http://dx.doi.org/10.1073/pnas.1113882109.

Chi, C., Zhu, H., Han, M., Zhuang, Y., Wu, X., & Xu, T. (2010). Disruption of lysosome function promotes tumor growth and metastasis in *Drosophila*. *The Journal of Biological Chemistry*, *285*(28), 21817–21823. http://dx.doi.org/10.1074/jbc.M110.131714.

Classen, A. K., Bunker, B. D., Harvey, K. F., Vaccari, T., & Bilder, D. (2009). A tumor suppressor activity of *Drosophila* Polycomb genes mediated by JAK-STAT signaling. *Nature Genetics*, *41*(10), 1150–1155. http://dx.doi.org/10.1038/ng.445.

Copeland, N. G., & Jenkins, N. A. (2010). Harnessing transposons for cancer gene discovery. *Nature Reviews. Cancer*, *10*(10), 696–706. http://dx.doi.org/10.1038/nrc2916.

Cordero, J. B., Macagno, J. P., Stefanatos, R. K., Strathdee, K. E., Cagan, R. L., & Vidal, M. (2010). Oncogenic Ras diverts a host TNF tumor suppressor activity into tumor promoter. *Developmental Cell*, *18*, 999–1011.

Da Ros, V. G., Gutierrez-Perez, I., Ferres-Marco, D., & Dominguez, M. (2013). Dampening the signals transduced through hedgehog via microRNA miR-7 facilitates notch-induced tumourigenesis. *PLoS Biology*, *11*(5), e1001554. http://dx.doi.org/10.1371/journal.pbio.1001554.

Davoli, T., & de Lange, T. (2012). Telomere-driven tetraploidization occurs in human cells undergoing crisis and promotes transformation of mouse cells. *Cancer Cell*, *21*(6), 765–776. http://dx.doi.org/10.1016/j.ccr.2012.03.044.
Davoli, T., Denchi, E. L., & de Lange, T. (2010). Persistent telomere damage induces bypass of mitosis and tetraploidy. *Cell*, *141*(1), 81–93. http://dx.doi.org/10.1016/j.cell.2010.01.031.
De La Cova, C., Abril, M., Bellosta, P., Gallant, P., & Johnston, L. A. (2004). *Drosophila* Myc regulates organ size by inducing cell competition. *Cell*, *117*(1), 107–116.
Dekanty, A., Barrio, L., & Milan, M. (2015). Contributions of DNA repair, cell cycle checkpoints and cell death to suppressing the DNA damage-induced tumorigenic behavior of *Drosophila* epithelial cells. *Oncogene*, *34*(8), 978–985. http://dx.doi.org/10.1038/onc.2014.42.
Dekanty, A., Barrio, L., Muzzopappa, M., Auer, H., & Milan, M. (2012). Aneuploidy-induced delaminating cells drive tumorigenesis in *Drosophila* epithelia. *Proceedings of the National Academy of Sciences of the United States of America*, *109*(50), 20549–20554. http://dx.doi.org/10.1073/pnas.1206675109.
Diaz-Benjumea, F. J., & Cohen, S. M. (1995). Serrate signals through Notch to establish a Wingless-dependent organizer at the dorsal/ventral compartment boundary of the *Drosophila* wing. *Development*, *121*, 4215–4225.
Doggett, K., Grusche, F. A., Richardson, H. E., & Brumby, A. M. (2011). Loss of the *Drosophila* cell polarity regulator Scribbled promotes epithelial tissue overgrowth and cooperation with oncogenic Ras-Raf through impaired Hippo pathway signaling. *BMC Developmental Biology*, *11*, 57. http://dx.doi.org/10.1186/1471-213X-11-57.
Dominguez, M., Wasserman, J. D., & Freeman, M. (1998). Multiple functions of the EGF receptor in *Drosophila* eye development. *Current Biology*, *8*(19), 1039–1048.
Dong, J., Feldmann, G., Huang, J., Wu, S., Zhang, N., Comerford, S. A., et al. (2007). Elucidation of a universal size-control mechanism in *Drosophila* and mammals. *Cell*, *130*(6), 1120–1133. http://dx.doi.org/10.1016/j.cell.2007.07.019.
Eichenlaub, T., Cohen, S. M., & Herranz, H. (2016). Cell competition drives the formation of metastatic tumors in a *Drosophila* model of epithelial tumor formation. *Current Biology* (in press).
Elsum, I., Yates, L., Humbert, P. O., & Richardson, H. E. (2012). The Scribble-Dlg-Lgl polarity module in development and cancer: From flies to man. *Essays in Biochemistry*, *53*, 141–168. http://dx.doi.org/10.1042/bse0530141.
Fearon, K., Arends, J., & Baracos, V. (2013). Understanding the mechanisms and treatment options in cancer cachexia. *Nature Reviews. Clinical Oncology*, *10*(2), 90–99. http://dx.doi.org/10.1038/nrclinonc.2012.209.
Ferres-Marco, D., Gutierrez-Garcia, I., Vallejo, D. M., Bolivar, J., Gutierrez-Avino, F. J., & Dominguez, M. (2006). Epigenetic silencers and Notch collaborate to promote malignant tumours by Rb silencing. *Nature*, *439*(7075), 430–436. http://dx.doi.org/10.1038/nature04376.
Figueroa-Clarevega, A., & Bilder, D. (2015). Malignant *Drosophila* tumors interrupt insulin signaling to induce cachexia-like wasting. *Developmental Cell*, *33*(1), 47–55. http://dx.doi.org/10.1016/j.devcel.2015.03.001.
Fujiwara, T., Bandi, M., Nitta, M., Ivanova, E. V., Bronson, R. T., & Pellman, D. (2005). Cytokinesis failure generating tetraploids promotes tumorigenesis in p53-null cells. *Nature*, *437*(7061), 1043–1047. http://dx.doi.org/10.1038/nature04217.
Ganem, N. J., Cornils, H., Chiu, S. Y., O'Rourke, K. P., Arnaud, J., Yimlamai, D., et al. (2014). Cytokinesis failure triggers hippo tumor suppressor pathway activation. *Cell*, *158*(4), 833–848. http://dx.doi.org/10.1016/j.cell.2014.06.029.
Gao, X., Neufeld, T. P., & Pan, D. (2000). *Drosophila* PTEN regulates cell growth and proliferation through PI3K-dependent and -independent pathways. *Developmental Biology*, *221*(2), 404–418.

Gao, X., & Pan, D. (2001). TSC1 and TSC2 tumor suppressors antagonize insulin signaling in cell growth. *Genes & Development, 15*(11), 1383–1392.

Gateff, E. (1978). Malignant neoplasms of genetic origin in *Drosophila melanogaster. Science, 200*(4349), 1448–1459.

Go, M. J., Eastman, D. S., & Artavanis-Tsakonas, S. (1998). Cell proliferation control by Notch signaling in *Drosophila* development. *Development, 125*, 2031–2040.

Gonzalez, C. (2013). *Drosophila melanogaster*: A model and a tool to investigate malignancy and identify new therapeutics. *Nature Reviews. Cancer, 13*(3), 172–183. http://dx.doi.org/10.1038/nrc3461.

Grifoni, D., Sollazzo, M., Fontana, E., Froldi, F., & Pession, A. (2015). Multiple strategies of oxygen supply in *Drosophila* malignancies identify tracheogenesis as a novel cancer hallmark. *Scientific Reports, 5*, 9061. http://dx.doi.org/10.1038/srep09061.

Grusche, F. A., Degoutin, J. L., Richardson, H. E., & Harvey, K. F. (2011). The Salvador/Warts/Hippo pathway controls regenerative tissue growth in *Drosophila melanogaster. Developmental Biology, 350*(2), 255–266. http://dx.doi.org/10.1016/j.ydbio.2010.11.020.

Grzeschik, N. A., Parsons, L. M., Allott, M. L., Harvey, K. F., & Richardson, H. E. (2010). Lgl, aPKC, and Crumbs regulate the Salvador/Warts/Hippo pathway through two distinct mechanisms. *Current Biology, 20*(7), 573–581. http://dx.doi.org/10.1016/j.cub.2010.01.055.

Haerry, T. E., Khalsa, O., O'Connor, M. B., & Wharton, K. A. (1998). Synergistic signaling by two BMP ligands through the SAX and TKV receptors controls wing growth and patterning in *Drosophila. Development, 125*, 3977–3987.

Hamaratoglu, F., Willecke, M., Kango-Singh, M., Nolo, R., Hyun, E., Tao, C., et al. (2006). The tumour-suppressor genes NF2/Merlin and Expanded act through Hippo signalling to regulate cell proliferation and apoptosis. *Nature Cell Biology, 8*(1), 27–36.

Hanahan, D., & Weinberg, R. A. (2011). Hallmarks of cancer: The next generation. *Cell, 144*(5), 646–674. http://dx.doi.org/10.1016/j.cell.2011.02.013.

Harvey, K. F., Pfleger, C. M., & Hariharan, I. K. (2003). The *Drosophila* Mst ortholog, hippo, restricts growth and cell proliferation and promotes apoptosis. *Cell, 114*(4), 457–467.

Harvey, K. F., Zhang, X., & Thomas, D. M. (2013). The Hippo pathway and human cancer. *Nature Reviews. Cancer, 13*(4), 246–257. http://dx.doi.org/10.1038/nrc3458.

Herranz, H., Hong, X., & Cohen, S. M. (2012a). Mutual repression by bantam miRNA and Capicua links the EGFR/MAPK and Hippo pathways in growth control. *Current Biology, 22*(8), 651–657. http://dx.doi.org/10.1016/j.cub.2012.02.050.

Herranz, H., Hong, X., Hung, N. T., Voorhoeve, P. M., & Cohen, S. M. (2012b). Oncogenic cooperation between SOCS family proteins and EGFR identified using a *Drosophila* epithelial transformation model. *Genes & Development, 26*(14), 1602–1611. http://dx.doi.org/10.1101/gad.192021.112.

Herranz, H., Weng, R., & Cohen, S. M. (2014). Crosstalk between epithelial and mesenchymal tissues in tumorigenesis and imaginal disc development. *Current Biology, 24*(13), 1476–1484. http://dx.doi.org/10.1016/j.cub.2014.05.043.

Hirabayashi, S., Baranski, T. J., & Cagan, R. L. (2013). Transformed *Drosophila* cells evade diet-mediated insulin resistance through wingless signaling. *Cell, 154*(3), 664–675. http://dx.doi.org/10.1016/j.cell.2013.06.030.

Hong, X., Nguyen, H. T., Chen, Q., Zhang, R., Hagman, Z., Voorhoeve, P. M., et al. (2014). Opposing activities of the Ras and Hippo pathways converge on regulation of YAP protein turnover. *The EMBO Journal, 33*, 2447–2457. http://dx.doi.org/10.15252/embj.201489385.

Huang, J., Wu, S., Barrera, J., Matthews, K., & Pan, D. (2005). The Hippo signaling pathway coordinately regulates cell proliferation and apoptosis by inactivating Yorkie, the *Drosophila* Homolog of YAP. *Cell, 122*(3), 421–434. http://dx.doi.org/10.1016/j.cell.2005.06.007.

Igaki, T., Pastor-Pareja, J. C., Aonuma, H., Miura, M., & Xu, T. (2009). Intrinsic tumor suppression and epithelial maintenance by endocytic activation of Eiger/TNF signaling in *Drosophila*. *Developmental Cell*, *16*(3), 458–465. http://dx.doi.org/10.1016/j.devcel.2009.01.002.

Ito, N., & Rubin, G. M. (1999). gigas, a *Drosophila* homolog of tuberous sclerosis gene product-2, regulates the cell cycle. *Cell*, *96*(4), 529–539.

Johnston, L. A. (2014). Socializing with MYC: Cell competition in development and as a model for premalignant cancer. *Cold Spring Harbor Perspectives in Medicine*, *4*(4), a014274. http://dx.doi.org/10.1101/cshperspect.a014274.

Justice, R. W., Zilian, O., Woods, D. F., Noll, M., & Bryant, P. J. (1995). The *Drosophila* tumor suppressor gene warts encodes a homolog of human myotonic dystrophy kinase and is required for the control of cell shape and proliferation. *Genes & Development*, *9*(5), 534–546.

Kandoth, C., McLellan, M. D., Vandin, F., Ye, K., Niu, B., Lu, C., et al. (2013). Mutational landscape and significance across 12 major cancer types. *Nature*, *502*(7471), 333–339. http://dx.doi.org/10.1038/nature12634.

Kango-Singh, M., Nolo, R., Tao, C., Verstreken, P., Hiesinger, P. R., Bellen, H. J., et al. (2002). Shar-pei mediates cell proliferation arrest during imaginal disc growth in *Drosophila*. *Development*, *129*(24), 5719–5730.

Kapoor, A., Yao, W., Ying, H., Hua, S., Liewen, A., Wang, Q., et al. (2014). Yap1 activation enables bypass of oncogenic Kras addiction in pancreatic cancer. *Cell*, *158*(1), 185–197. http://dx.doi.org/10.1016/j.cell.2014.06.003.

Karim, F. D., & Rubin, G. M. (1998). Ectopic expression of activated Ras1 induces hyperplastic growth and increased cell death in *Drosophila* imaginal tissues. *Development*, *125*(1), 1–9.

Khoo, P., Allan, K., Willoughby, L., Brumby, A. M., & Richardson, H. E. (2013). In *Drosophila*, RhoGEF2 cooperates with activated Ras in tumorigenesis through a pathway involving Rho1-Rok-Myosin-II and JNK signalling. *Disease Models & Mechanisms*, *6*(3), 661–678. http://dx.doi.org/10.1242/dmm.010066.

Kwon, Y., Song, W., Droujinine, I. A., Hu, Y., Asara, J. M., & Perrimon, N. (2015). Systemic organ wasting induced by localized expression of the secreted insulin/IGF antagonist ImpL2. *Developmental Cell*, *33*(1), 36–46. http://dx.doi.org/10.1016/j.devcel.2015.02.012.

Lecuit, T., Brook, W. J., Ng, M., Calleja, M., Sun, H., & Cohen, S. M. (1996). Two distinct mechanisms for long range patterning by dpp in the *Drosophila* wing. *Nature*, *381*, 387–393.

Leevers, S. J., Weinkove, D., MacDougall, L. K., Hafen, E., & Waterfield, M. D. (1996). The *Drosophila* phosphoinositide 3-kinase Dp110 promotes cell growth. *The EMBO Journal*, *15*, 6584–6594.

Levayer, R., & Moreno, E. (2013). Mechanisms of cell competition: Themes and variations. *The Journal of Cell Biology*, *200*(6), 689–698. http://dx.doi.org/10.1083/jcb.201301051.

Li, W., & Baker, N. E. (2007). Engulfment is required for cell competition. *Cell*, *129*(6), 1215–1225. http://dx.doi.org/10.1016/j.cell.2007.03.054.

Li, M., Fang, X., Baker, D. J., Guo, L., Gao, X., Wei, Z., et al. (2010). The ATM-p53 pathway suppresses aneuploidy-induced tumorigenesis. *Proceedings of the National Academy of Sciences of the United States of America*, *107*(32), 14188–14193. http://dx.doi.org/10.1073/pnas.1005960107.

Lu, H., & Bilder, D. (2005). Endocytic control of epithelial polarity and proliferation in *Drosophila*. *Nature Cell Biology*, 7(12), 1232–1239.

Martin, F. A., Perez-Garijo, A., & Morata, G. (2009). Apoptosis in *Drosophila*: Compensatory proliferation and undead cells. *The International Journal of Developmental Biology*, *53*(8–10), 1341–1347. http://dx.doi.org/10.1387/ijdb.072447fm.

Martin, F. A., Perez-Garijo, A., Moreno, E., & Morata, G. (2004). The brinker gradient controls wing growth in *Drosophila*. *Development*, *131*(20), 4921–4930. http://dx.doi.org/10.1242/dev.01385.

Martinez, A. M., Schuettengruber, B., Sakr, S., Janic, A., Gonzalez, C., & Cavalli, G. (2009). Polyhomeotic has a tumor suppressor activity mediated by repression of Notch signaling. *Nature Genetics*, *41*(10), 1076–1082. http://dx.doi.org/10.1038/ng.414.

Menendez, J., Perez-Garijo, A., Calleja, M., & Morata, G. (2010). A tumor-suppressing mechanism in *Drosophila* involving cell competition and the Hippo pathway. *Proceedings of the National Academy of Sciences of the United States of America*, *107*(33), 14651–14656. http://dx.doi.org/10.1073/pnas.1009376107.

Moberg, K. H., Schelble, S., Burdick, S. K., & Hariharan, I. K. (2005). Mutations in erupted, the *Drosophila* ortholog of mammalian tumor susceptibility gene 101, elicit non-cell-autonomous overgrowth. *Developmental Cell*, *9*(5), 699–710.

Montero, L., Muller, N., & Gallant, P. (2008). Induction of apoptosis by *Drosophila* Myc. *Genesis*, *46*(2), 104–111. http://dx.doi.org/10.1002/dvg.20373.

Morais da Silva, S., Moutinho-Santos, T., & Sunkel, C. E. (2013). A tumor suppressor role of the Bub3 spindle checkpoint protein after apoptosis inhibition. *The Journal of Cell Biology*, *201*(3), 385–393. http://dx.doi.org/10.1083/jcb.201210018.

Moreno, E., & Basler, K. (2004). dMyc transforms cells into super-competitors. *Cell*, *117*(1), 117–129.

Nellen, D., Burke, R., Struhl, G., & Basler, K. (1996). Direct and long-range action of a DPP morphogen gradient. *Cell*, *85*(3), 357–368.

Nolo, R., Morrison, C. M., Tao, C., Zhang, X., & Halder, G. (2006). The bantam microRNA is a target of the hippo tumor-suppressor pathway. *Current Biology: CB*, *16*(19), 1895–1904. http://dx.doi.org/10.1016/j.cub.2006.08.057.

Ohsawa, S., Sato, Y., Enomoto, M., Nakamura, M., Betsumiya, A., & Igaki, T. (2012). Mitochondrial defect drives non-autonomous tumour progression through Hippo signalling in *Drosophila*. *Nature*, *490*(7421), 547–551. http://dx.doi.org/10.1038/nature11452.

Pagliarini, R. A., & Xu, T. (2003). A genetic screen in *Drosophila* for metastatic behavior. *Science*, *302*(5648), 1227–1231. http://dx.doi.org/10.1126/science.1088474.

Pan, D. (2010). The hippo signaling pathway in development and cancer. *Developmental Cell*, *19*(4), 491–505. http://dx.doi.org/10.1016/j.devcel.2010.09.011.

Parisi, F., Stefanatos, R. K., Strathdee, K., Yu, Y., & Vidal, M. (2014). Transformed epithelia trigger non-tissue-autonomous tumor suppressor response by adipocytes via activation of Toll and Eiger/TNF signaling. *Cell Reports*, *6*, 855–867.

Pastor-Pareja, J. C., Wu, M., & Xu, T. (2008). An innate immune response of blood cells to tumors and tissue damage in Drosophila. *Disease Models & Mechanisms*, *1*, 144–154.

Pellock, B. J., Buff, E., White, K., & Hariharan, I. K. (2007). The *Drosophila* tumor suppressors Expanded and Merlin differentially regulate cell cycle exit, apoptosis, and Wingless signaling. *Developmental Biology*, *304*(1), 102–115. http://dx.doi.org/10.1016/j.ydbio.2006.12.021.

Perez-Garijo, A., Shlevkov, E., & Morata, G. (2009). The role of Dpp and Wg in compensatory proliferation and in the formation of hyperplastic overgrowths caused by apoptotic cells in the *Drosophila* wing disc. *Development*, *136*(7), 1169–1177. http://dx.doi.org/10.1242/dev.034017.

Potter, C. J., Huang, H., & Xu, T. (2001). *Drosophila* Tsc1 functions with Tsc2 to antagonize insulin signaling in regulating cell growth, cell proliferation, and organ size. *Cell*, *105*(3), 357–368.

Poulton, J. S., Cuningham, J. C., & Peifer, M. (2014). Acentrosomal *Drosophila* epithelial cells exhibit abnormal cell division, leading to cell death and compensatory proliferation. *Developmental Cell*, *30*(6), 731–745. http://dx.doi.org/10.1016/j.devcel.2014.08.007.

Prober, D. A., & Edgar, B. A. (2000). Ras1 promotes cellular growth in the *Drosophila* wing. *Cell, 100*(4), 435–446.

Shao, D. D., Xue, W., Krall, E. B., Bhutkar, A., Piccioni, F., Wang, X., et al. (2014). KRAS and YAP1 converge to regulate EMT and tumor survival. *Cell, 158*(1), 171–184. http://dx.doi.org/10.1016/j.cell.2014.06.004.

Silva, E., Tsatskis, Y., Gardano, L., Tapon, N., & McNeill, H. (2006). The tumor-suppressor gene fat controls tissue growth upstream of expanded in the hippo signaling pathway. *Current Biology: CB, 16*(21), 2081–2089. http://dx.doi.org/10.1016/j.cub.2006.09.004.

Simpson, P. (1979). Parameters of cell competition in the compartments of the wing disc of *Drosophila. Developmental Biology, 69*, 182–193.

Simpson, P., & Morata, G. (1981). Differential mitotic rates and patterns of growth in compartments in the *Drosophila* wing. *Developmental Biology, 85*(2), 299–308.

Stark, M. B. (1918). A hereditary tumor in the fruit fly. *Journal of Cancer Research, 3*, 279–301.

Stratton, M. R. (2011). Exploring the genomes of cancer cells: Progress and promise. *Science, 331*(6024), 1553–1558. http://dx.doi.org/10.1126/science.1204040.

Sun, Q., Luo, T., Ren, Y., Florey, O., Shirasawa, S., Sasazuki, T., et al. (2014). Competition between human cells by entosis. *Cell Research, 24*(11), 1299–1310. http://dx.doi.org/10.1038/cr.2014.138.

Tamori, Y., Bialucha, C. U., Tian, A. G., Kajita, M., Huang, Y. C., Norman, M., et al. (2010). Involvement of Lgl and Mahjong/VprBP in cell competition. *PLoS Biology, 8*(7), e1000422. http://dx.doi.org/10.1371/journal.pbio.1000422.

Tapon, N., Harvey, K., Bell, D., Wahrer, D., Schiripo, T., Haber, D., et al. (2002). salvador promotes both cell cycle exit and apoptosis in *Drosophila* and is mutated in human cancer cell lines. *Cell, 110*(4), 467–478.

Tapon, N., Ito, N., Dickson, B. J., Treisman, J. E., & Hariharan, I. K. (2001). The *Drosophila* tuberous sclerosis complex gene homologs restrict cell growth and cell proliferation. *Cell, 105*(3), 345–355.

Thompson, B. J., & Cohen, S. M. (2006). The Hippo pathway regulates the bantam microRNA to control cell proliferation and apoptosis in *Drosophila. Cell, 126*(4), 767–774. http://dx.doi.org/10.1016/j.cell.2006.07.013.

Thompson, B. J., Mathieu, J., Sung, H. H., Loeser, E., Rorth, P., & Cohen, S. M. (2005). Tumor suppressor properties of the ESCRT-II complex component Vps25 in *Drosophila. Developmental Cell, 9*(5), 711–720.

Titen, S. W., & Golic, K. G. (2008). Telomere loss provokes multiple pathways to apoptosis and produces genomic instability in *Drosophila melanogaster. Genetics, 180*(4), 1821–1832. http://dx.doi.org/10.1534/genetics.108.093625.

Udan, R. S., Kango-Singh, M., Nolo, R., Tao, C., & Halder, G. (2003). Hippo promotes proliferation arrest and apoptosis in the Salvador/Warts pathway. *Nature Cell Biology, 5*(10), 914–920. http://dx.doi.org/10.1038/ncb1050.

Vaccari, T., & Bilder, D. (2005). The *Drosophila* tumor suppressor vps25 prevents non-autonomous overproliferation by regulating notch trafficking. *Developmental Cell, 9*(5), 687–698. http://dx.doi.org/10.1016/j.devcel.2005.09.019.

Vicidomini, R., Di Giovanni, A., Petrizzo, A., Iannucci, L. F., Benvenuto, G., Nagel, A. C., et al. (2015). Loss of *Drosophila* pseudouridine synthase triggers apoptosis-induced proliferation and promotes cell-nonautonomous EMT. *Cell Death & Disease, 6*, e1705. http://dx.doi.org/10.1038/cddis.2015.68.

Vidal, M., Larson, D. E., & Cagan, R. L. (2006). Csk-deficient boundary cells are eliminated from normal *Drosophila* epithelia by exclusion, migration, and apoptosis. *Developmental Cell, 10*(1), 33–44. http://dx.doi.org/10.1016/j.devcel.2005.11.007.

Vidal, M., Warner, S., Read, R., & Cagan, R. L. (2007). Differing Src signaling levels have distinct outcomes in *Drosophila. Cancer Research, 67*(21), 10278–10285. http://dx.doi.org/10.1158/0008-5472.CAN-07-1376.

Wang, S. H., Simcox, A., & Campbell, G. (2000). Dual role for *Drosophila* epidermal growth factor receptor signaling in early wing disc development. *Genes & Development*, *14*(18), 2271–2276.

Wu, S., Huang, J., Dong, J., & Pan, D. (2003). hippo encodes a Ste-20 family protein kinase that restricts cell proliferation and promotes apoptosis in conjunction with salvador and warts. *Cell*, *114*(4), 445–456.

Wu, M., Pastor-Pareja, J. C., & Xu, T. (2010). Interaction between Ras(V12) and scribbled clones induces tumour growth and invasion. *Nature*, *463*(7280), 545–548. http://dx.doi.org/10.1038/nature08702.

Yang, L., & Baker, N. E. (2003). Cell cycle withdrawal, progression, and cell survival regulation by EGFR and its effectors in the differentiating *Drosophila* eye. *Developmental Cell*, *4*(3), 359–369.

Zender, L., Spector, M. S., Xue, W., Flemming, P., Cordon-Cardo, C., Silke, J., et al. (2006). Identification and validation of oncogenes in liver cancer using an integrative oncogenomic approach. *Cell*, *125*(7), 1253–1267. http://dx.doi.org/10.1016/j.cell.2006.05.030.

CHAPTER TWELVE

Genome Duplication: The Heartbeat of Developing Organisms

Melvin L. DePamphilis[1]
National Institute of Child Health and Human Development, National Institutes of Health, Bethesda, Maryland, USA
[1]Corresponding author: e-mail address: depamphm@mail.nih.gov

Contents

Abstract

The mechanism that duplicates the nuclear genome during the trillions of cell divisions required to develop from zygote to adult is the same throughout the eukarya, but the mechanisms that determine where, when and how much nuclear genome duplication occur regulate development and differ among the eukarya. They allow organisms to change the rate of cell proliferation during development, to activate zygotic gene expression independently of DNA replication, and to restrict nuclear DNA replication to once per cell division. They allow specialized cells to exit their mitotic cell cycle and differentiate into polyploid cells, and in some cases, to amplify the number of copies of specific genes. It is genome duplication that drives evolution, by virtue of the errors that inevitably occur when the same process is repeated trillions of times. It is, unfortunately, the same errors that produce age-related genetic disorders such as cancer.

Current Topics in Developmental Biology, Volume 116
ISSN 0070-2153
http://dx.doi.org/10.1016/bs.ctdb.2015.10.002

1. INTRODUCTION

The mechanism by which DNA is replicated is so highly conserved throughout the three domains of life—bacteria, archaea, and eukarya (DePamphilis & Bell, 2010)—that developmental biologists might think that genome duplication is an immutable mechanism that awakes at fertilization and then beats in an unchanging rhythm throughout development. But the real story is more complicated. All living organisms do encode their genomes in DNA, and they do replicate the nuclear/nucleoid portion of their DNA using the "replication fork" mechanism. One DNA strand serves as a template upon which a new complementary strand is synthesized. DNA unwinding is coupled to DNA synthesis that occurs continuously on one template and discontinuously on the other by the repeated synthesis and joining of short RNA-primed nascent DNA chains termed "Okazaki fragments." To be sure, other strategies for genome duplication exist, but they are confined to bacteriophage, bacterial plasmids, animal and plant viruses, and mitochondria. The essential point is that during metazoan development, the mechanism for replicating nuclear DNA is the same from zygote to adult. In fact, the sequence of events at replication forks is the same in all living cells, and the proteins that execute these events in metazoa are orthologs or analogs of the proteins used in archaea and bacteria.

Now consider the problem. A single diploid human cell contains 2.09 m of DNA (6.16 billion base pairs, 6.3 pg) of nuclear DNA that encodes 99.9995% of the genome. The remaining 0.0005% is encoded in mitochondrial DNA, which is maternally inherited, replicates independently of nuclear DNA replication, and varies widely in copy number among cell types. This means that development of a fertilized human egg into an adult of 75 kg requires the production of 29 trillion cells or about 60 trillion meters of DNA. Do the math. That's enough DNA to stretch from the Earth to the Sun and back 200×! Include the approximately 1 trillion hematopoietic cells produced each day in adults, together with the turnover in epithelial cell populations, and the number of cell divisions in a human life span is truly astronomical!

Now consider the solution. Replication fork velocity did not increase to compensate for increasing genome size; in fact, replication forks are about 60 × slower in humans than in bacteria! Instead, the mechanisms that determine where and when DNA replication begins evolved amazing flexibility in order to accommodate the ever-increasing demands of large genomes encoding thousands of different genes that must be turned on and off at different times during cell proliferation and differentiation. In other words, the requirements for DNA replication evolved so that they did not compete with the requirements for DNA transcription. Replication of nuclear DNA became tightly restricted to one duplication for each cell division. This ensured that the trillions of cells in complex metazoa are genetically identical, but mechanisms also evolved to escape this restriction so that specialized cells could produce huge amounts of gene products. Unfortunately, this flexibility came at a price: DNA replication machines make frequent mistakes. The good news is that those mistakes ensure that evolution continues. The bad news is that genetic alterations lead to genomic instability, human disease, and aging.

2. THE FIRST MITOTIC CELL DIVISION IS UNIVERSAL

The mitotic cell cycle is a sequence of events in all eukarya that ensures that nuclear DNA replication (S phase) is always followed by mitosis (M phase) and then cytokinesis, and that the nuclear genome is duplicated once, but only once, during each cell division (Fig. 1). It applies to single-cell organisms as well as metazoa. It applies to zygotes, embryos, and adults. Progress through the mitotic cell cycle in mammals is regulated by at least 34 different genes (Ciemerych & Sicinski, 2005), but only eight protein kinases directly regulate DNA replication, of which only cyclin-dependent kinases Cdk1 and Cdk7, the Dbf4-dependent kinase Cdc7, and the DNA damage checkpoint kinase Chk1/Chek1 are essential for mammalian development (Depamphilis et al., 2012). The other four kinases have roles that can be fulfilled by other proteins. Finally, the mitotic cell cycle contains five major "checkpoints" where progress through the cycle is arrested if cells cannot grow in size, if DNA is damaged, or if mitosis is compromised.

The first mitotic cell division is universal. Proliferating cells prepare for genome duplication by assembling prereplication complexes on chromatin as the cell exits mitosis (Fig. 1). The same event occurs in zygotes (Fig. 2). Oocytes are terminally differentiated cells that are arrested after completing the first stage of meiosis. They support RNA and protein synthesis but not DNA replication. When oocytes complete meiosis, the resulting eggs are not only competent for fertilization; they are competent for DNA replication. Eggs are transcriptionally silent, but translationally active. Cdc6 and Orc6 are the only DNA replication protein missing in the oocytes of flies, frogs, and mice whose translation during meiotic maturation is necessary and

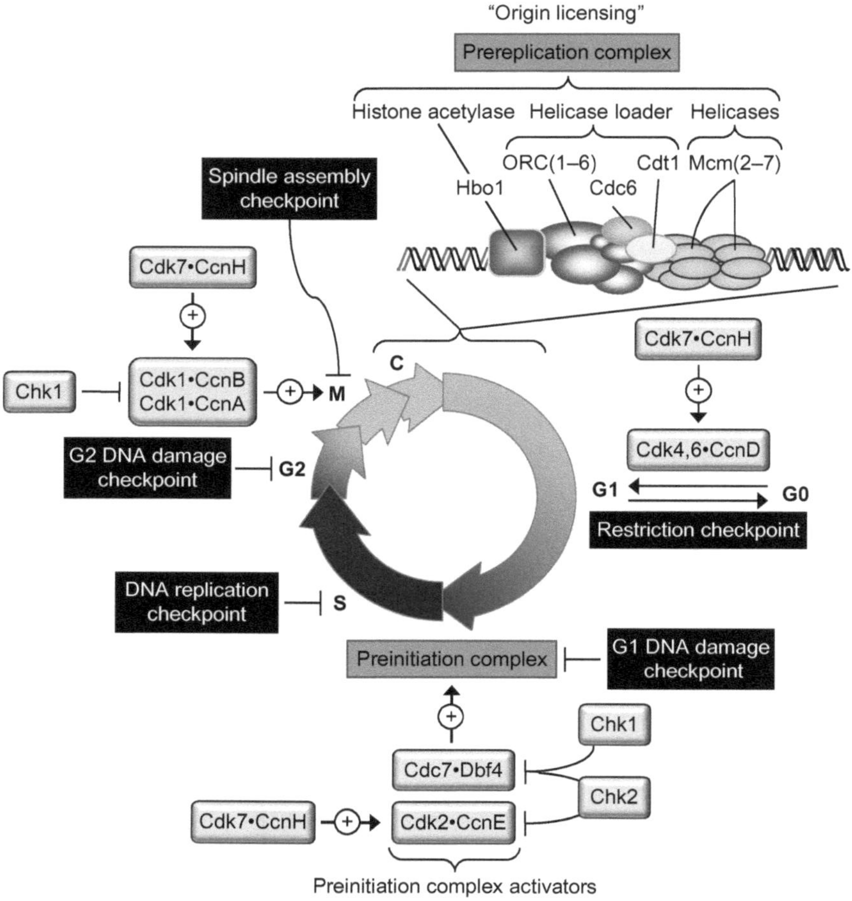

Figure 1 Regulation of DNA replication during mitotic cell cycles. Nuclear DNA replication in the metazoa begins during the anaphase to G1 phase transition of the mitotic cell cycle with the assembly of prereplication complexes onto the DNA in chromatin (Baldinger & Gossen, 2009; Noguchi, Vassilev, Ghosh, Yates, & DePamphilis, 2006). In mammals, this event, termed "origin licensing," begins by assembly of the six subunit "origin recognition complex" [ORC(1–6)] through independent nuclear localization of Orc1, ORC(2–5), and Orc6 (Ghosh, Vassilev, Zhang, Zhao, & DePamphilis, 2011). The BAH domain in human Orc1 facilitates the ability of Orc1 to activate replication origins *in vivo* and triggers assembly of ORC(1–6) (Kara, Hossain, Prasanth, & Stillman, 2015; Noguchi et al., 2006). Mutations in the BAH domain of Orc1 link histone H4 methylation to origin licensing and to Meier–Gorlin syndrome, a genetic defect leading to primordial dwarfism (Kuo et al., 2012). Cdc6 stabilizes the binding of ORC(1–6) to DNA and allows recruitment of Cdt1 to form a "helicase loader." The six subunit "minichromosome maintenance complex" (Mcm2–7) is the eukaryotic replicative DNA helicase. Each replication fork requires one MCM helicase; thus two or more are loaded at each origin of

sufficient to confer DNA replication competence before fertilization (Lemaitre, Bocquet, & Mechali, 2002; Lemaitre et al., 2004; Murai, Stein, Buffone, Yamashita, & Schultz, 2010; Whitmire, Khan, & Coue, 2002). In addition to their role in origin licensing, Cdc6 also behaves as a Cdk1 inhibitor that regulates entry into and progression through mitosis by limiting the level of Cdk1 activity (El Dika et al., 2014). Fertilization triggers formation of two pronuclei, followed by nuclear DNA replication, and

---

bidirectional replication. *In vivo*, the Hbo1/Kat7 histone acetylase activity is also essential origin licensing, because assembly of prereplication complexes is facilitated by histone acetylation (Miotto & Struhl, 2010; Unnikrishnan, Gafken, & Tsukiyama, 2010). However, prereplication complexes cannot unwind DNA; they must first be converted into preinitiation complexes by addition of 10 other protein (Bell & Kaguni, 2013; Tanaka & Araki, 2013). Only then can the MCM helicase unwind the double helix and allow DNA polymerases to synthesize the nascent DNA strand. DNA replication is regulated directly by eight protein kinases (Depamphilis, de Renty, Ullah, & Lee, 2012), five of which are cyclin-dependent kinases (Cdk) that require a cyclin (Ccn) partner. Cdk7·CcnH is essential to activate Cdk1, Cdk2, Cdk4, and Cdk6. Cdk4·CcnD and Cdk6·CcnD drive cells from a quiescent state (G0) into a proliferative state. Cdk1·CcnB1 and Cdk1·CcnA2 initiate and maintain mitosis. Cdk2·CcnE and the Dbf4-dependent kinase Cdc7 are the engines that drive the assembly and activation of preinitiation complexes. Cdk2·CcnA2 prevents DNA rereplication during S phase. Cdk1·CcnA2 prevents premature initiation during G2 phase. The remaining two protein kinases are components of the cell's DNA damage response. Five checkpoints ensure the accuracy of mitotic cell division (DePamphilis & Bell, 2010). In the absence of sufficient nutrients, growth factors or optimal temperature, cells in G1 phase enter a quiescent state (G0). This transition is regulated by the "restriction checkpoint," which consists of two signal detection pathways (Foster, Yellen, Xu, & Saqcena, 2010). One senses growth factors, and the other senses nutrients. The G1 to G0 transition is regulated by action of Cdk4·CcnD and Cdk6·CcnD. When cells are arrested in G0, their DNA replication machines are dismantled, and they eventually enter a senescent state from which they cannot reenter the mitotic cell cycle (Campisi, 2013). Conversely, once a cell passes the restriction checkpoint, it is committed to division; it must divide, or terminally differentiate into a polyploid cell, or die. Nuclear DNA damage is assessed before beginning S phase by the G1 DNA damage checkpoint, during S phase by the DNA replication checkpoint and before beginning M phase by the G2 DNA damage checkpoint, but in reality there are only two DNA damage response systems that basically detect excess single-stranded DNA complexed with replication protein-A that results from different problems. DNA damage and stalled replication forks are detected by the ATR kinase which then activates the Chk1 kinase to phosphorylate and inhibit Cdc7 and Cdc25 (a phosphatase required for activation of Cdk1 and Cdk2). Double-stranded DNA breaks are detected by the ATM kinase, which then activates Cdk2 kinase to phosphorylate Cdc7, Cdc25, and tumor suppressor Tp53. Many other proteins are also targeted. Finally, the spindle assembly checkpoint prevents the onset of anaphase until the kinetochores are attached correctly to the mitotic spindle (Lara-Gonzalez, Westhorpe, & Taylor, 2012). (A color version of this figure is located at the end of this chapter.)

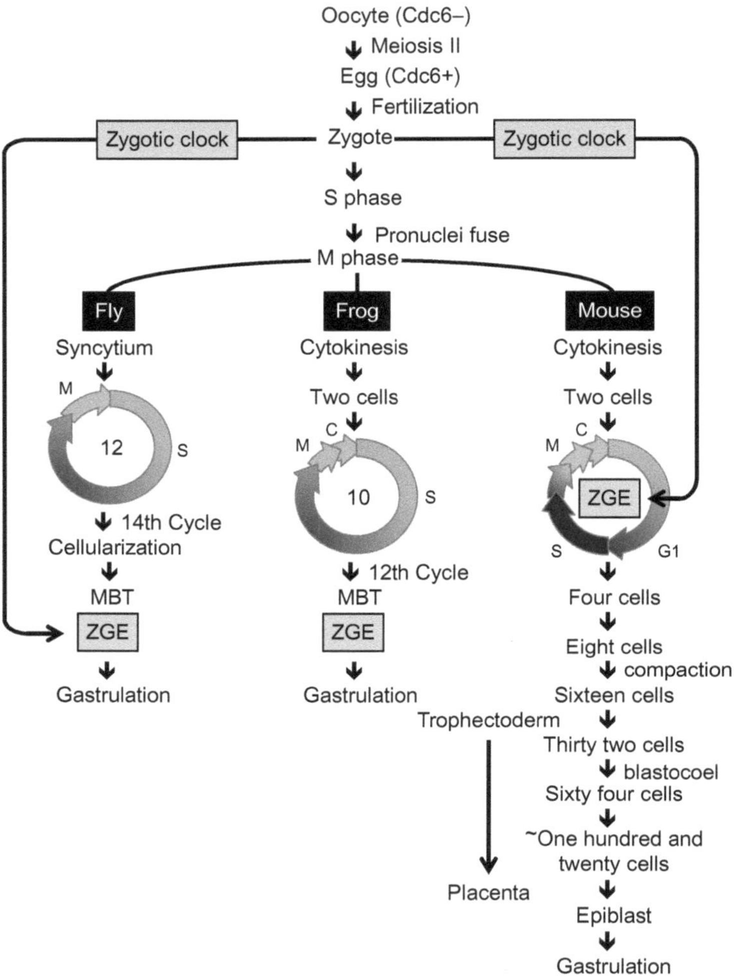

Figure 2 Developmentally programmed changes in the mitotic cell cycle. The first mitotic cell division is universal, as exemplified in *Drosophila melanogaster* (fly), *Xenopus laevis* (frog), and *Mus musculus* (mouse). Oocyte maturation into eggs includes expression of the Cdc6 protein that is essential for initiation of nuclear DNA replication. The second mitotic cell division distinguishes flies from frogs and mammals from nonmammals. Flies and other insects continue to rapidly undergo cycles of DNA replication (S phase) followed by mitosis (M phase) without undergoing cell cleavage. Frogs, fish, and echinoderms continue to rapidly undergo cycles of DNA replication followed by mitosis followed by cell cleavage (cell division in the absence of cell growth). Mice and other mammals continue to undergo slow cycles of DNA replication followed by mitosis followed by cell cleavage. However, mammalian cell cleavage appears to include G phases. The most significant change is that zygotic gene expression (ZGE), the onset of transcription and the degradation of maternally inherited RNA, occurs in two-cell embryos of mice and the 4- to 16-cell embryos of rabbits, cows, pigs, and humans. Nevertheless, activation of the bulk of ZGE in both flies and mice is determined by the time elapsed after fertilization (a "zygotic clock"), rather than by DNA replication or mitosis. The differences in cell division cycles between mammals and nonmammals might well be related to the fact that mammals need to produce both a placenta and an embryo concurrently (O'Farrell, Stumpff, & Su, 2004). (A color version of this figure is located at the end of this chapter.)

then mitosis, all of which happens in the absence of DNA transcription. Here, the first cell division cycles diverge. Whereas the zygotes of frogs and mice undergo cytokinesis to form a two-cell embryo, the zygotes of flies simply continue nuclear DNA replication and mitosis within a syncytium.

3. THE SECOND MITOTIC CELL DIVISION DISTINGUISHES MAMMALS FROM NONMAMMALS

The embryos of flies and frogs, like representatives of all the major phyla that deposit their eggs externally, undergo a series of rapid cleavage cell cycles in the absence of cell growth, so that cells become progressively smaller (Lee, Bonneau, & Giraldez, 2014; O'Farrell et al., 2004). Rapid cell cleavages lack G1 and G2 phases and DNA damage checkpoints. When DNA replication in either *Xenopus* or *Drosophila* embryos is blocked, cells progress to mitosis with catastrophic consequences (Glover et al., 1989; Newport & Dasso, 1989). Mammalian zygotes also undergo a series of cleavage cell cycles, but the first cell cleavage event in mouse zygotes is about 123 × longer than in the zygotes of flies and 37 × longer than those of frogs (Table 1 and Fig. 2). Similarly, fly genomes (represented by *Drosophila melanogaster*) can be replicated in as few as 3.4 min, frog genomes (represented by *Xenopus laevis*) in 20 min, and rodent genomes in 120 min, but the fastest replication rates in flies and frogs occur prior to gastrulation whereas in mammals they occur during gastrulation. As a rule, external fertilization leads to rapid cell cleavage cycles, whereas internal fertilization leads to slow cell cleavage cycles.

Rapid cell cleavage cycles occur in the absence of DNA transcription. Therefore, DNA replication and mitosis depend entirely on maternally inherited proteins and translation of maternally inherited mRNA. Rapid cleavage cycles terminate with the "midblastula transition" (MBT), which is recognized by the onset of transcription and translation of zygotic genes [zygotic gene expression (ZGE)], degradation of maternally inherited RNAs, and an extended cell division cycle (Fig. 1). Frog embryos undergo 12 cleavage events before the MBT. Fly embryos cycle through a series of 14 synchronous S and M phases in the absence of cytokinesis before cellularization occurs, followed quickly by the MBT.

Slow cell cleavages occur in mammals. Fertilization of mouse eggs triggers pronuclei formation, followed by DNA replication, mitosis, and cytokinesis. These events require 17–20 h with subsequent cleavage cell cycles occurring every 10–14 h. These slow cleavage cell cycles exhibit G1, S, G2, and M phases as well as DNA damage checkpoints. When DNA replication

Table 1 Changes in the Time Required for Cell Division During Animal Development

	Time		Velocity	
Organism	**Cell Cycle**	**S Phase**	**Replication (Mb/min)**	**Fork (Kb/min)**
Flies (180 Mbp DNA)				
Drosophila syncytial embryo	8.3→23 m	3.4 m	53	2.6
Kc cells	8 h	6 h		0.8
Frogs (3100 Mbp DNA)				
Xenopus cleavage embryos	30 m	20 m	155	0.3→1.1
Rodents (2600 Mbp DNA)				
Mouse preimplantation embryo				
Cycles 1 and 2	17→20 h	6 h	7.2	
Cycles 3 and 4	10→14 h	6 h		
Mouse gastrula				
Proliferative zone	2.2 h	~1.9 h	~23	
Rat gastrula				
Anterior primitive streak	3→3.5 h	2→2.75 h	16→22	
Mouse fibroblasts				1.8
Human fibroblasts	18→25 h	6→8 h	6.4→8.6	0.6→3.6

Abbreviations: Mega (10^6) bases (Mb), mega base pairs. Kb = "kilo (10^3) bases." Embryo data are from Blumenthal et al. (1974), Bolton et al. (1984), Howlett (1986), Howlett and Bolton (1985), Mac Auley, Werb and Mirkes (1993), O'Farrell et al. (2004), Snow (1977), and references therein. Replication fork data are from Cayrou et al. (2011), Conti, Sacca, et al. (2007), and Conti, Seiler and Pommier (2007). The 3.4 min S phase in flies is for the most rapid early cell cycles. Duration of metaphase/anaphase in mice is ~20 min. Note that DNA replication rates depend strongly on temperature in both animals and cell cultures. Cultured *Xenopus* cells appear similar to cultured mammalian cells, although specific data were not available.

is blocked in mouse preimplantation embryos, cells do not undergo mitosis, cytokinesis, or apoptosis (Bolton, Oades, & Johnson, 1984). ZGE occurs in the two-cell embryos of mice and in the 4- to 16-cell embryos of other mammals. Despite this rapid transition from maternal to zygotic control, cell cleavage continues after activation of ZGE to produce smaller cells with an increased nuclear to cytoplasmic ratio until implantation (Aiken, Swoboda, Skepper, & Johnson, 2004). This change in strategy allows mammals to begin cell differentiation prior to implantation, which insures that placenta and fetus develop concurrently.

4. ACTIVATION OF ZGE DOES NOT DEPEND ON DNA REPLICATION

A small group of zygotic genes are expressed in the embryos of flies and frogs when the nuclear to cytoplasmic ratio reaches a critical stage, presumably because the exponential increase in DNA content during cell cleavages results in the titration and inactivation of a hypothetical cytoplasmic factor that inhibits the MBT (Lee et al., 2014). However, most zygotic transcription, as well as the accompanying maternal RNA degradation, does not rely on the nuclear to cytoplasmic ratio in fly embryos, but instead occurs at a strict time interval following fertilization, indicating the presence of a timing mechanism (Lu, Li, Elemento, Tavazoie, & Wieschaus, 2009). The presence of this "zygotic clock" was demonstrated earlier in mice (Aoki, Worrad, & Schultz, 1997; Bolton et al., 1984; Wiekowski, Miranda, & DePamphilis, 1991). Activation of ZGE occurs in mouse embryos at the same time postfertilization (the time required to form a two-cell embryo) regardless of whether or not DNA replication is inhibited. Zygotes remained morphologically as one-cell embryos, but they activate ZGE as if they had reached the two-cell stage. This clock applies to genes on plasmid expression vectors as well as endogenous genes, to promoters driven by any one of the three RNA polymerases, and to both transcription and translation (Nothias, Miranda, & DePamphilis, 1996).

The zygotic clock delays transcription until both paternal and maternal genomes are replicated and remodeled from a postmeiotic state to one in which transcription is repressed by chromatin structure in a manner that can be relieved by enhancers at appropriate times during development (Nothias, Majumder, Kaneko, & DePamphilis, 1995). Injection of DNA into the nuclei of mouse oocytes and embryos revealed that transcription promoters and origins of DNA replication are repressed in the maternal pronucleus of oocytes and zygotes, and in the nuclei of two-cell embryos, regardless of their parental origin or ploidy. This repression is linked to changes in chromatin, and the ability to relieve this repression with embryo-responsive transcription enhancers requires a unique coactivator that first appears with formation of a two-cell embryo (Majumder, Miranda, & DePamphilis, 1993; Majumder, Zhao, Kaneko, & DePamphilis, 1997). A TATA box was required for enhancer stimulation of promoter activity only after cells began to differentiate (Majumder & DePamphilis, 1994). These enhancers respond to TEAD transcription factors present in preimplantatconventions, genes have toion embryos and embryonic stem cells (Melin, Miranda, Montreau, DePamphilis, & Blangy, 1993). Tead2 is expressed at the beginning of preimplantation development (Kaneko, Cullinan, Latham, & DePamphilis, 1997), and Tead4 is expressed at the eight-cell stage where it is essential for trophectoderm development (Kaneko & DePamphilis, 2013; Yagi et al., 2007). All four TEAD transcription factors bind the same DNA sequence, utilize the Yap1/Yap65 transcriptional coactivator to activate transcription, and are regulated through cytoplasmic localization of Yap1 by 14-3-3 protein (Vassilev, Kaneko, Shu, Zhao, & DePamphilis, 2001). In this manner, Yap1 and its cousin Taz regulate TEAD-dependent transcription in response to mitogenic signals via the HIPPO signal transduction pathway, a regulator of tissue regeneration and stem cell biology (Hansen, Moroishi, & Guan, 2015).

5. REPLICATION ORIGINS ARE DEVELOPMENTALLY REGULATED

In 1974, David Hogness and Harold Callan discovered that the density of replication origins throughout the chromosomes of flies, newts, and frogs decreased significantly as rapidly cleaving nuclei in the fertilized eggs of these organisms developed into adults. At the beginning, replication bubbles were closely spaced, but as cells differentiated, the number of active replication origins in the genome decreased dramatically. With the development of origin mapping techniques, it eventually became clear that selection of initiation sites for DNA replication can change during metazoan development from virtually random initiation throughout the nuclear DNA to site-specific replication origins that can be activated either in clusters or individually. Replication bubbles in *fly* embryos or *frog* eggs or egg extracts are distributed throughout the DNA with periodic spacing of 5–15 kbp that depend on the amount of initiator protein [ORC(1–6)] available (Blow, Gillespie, Francis, & Jackson, 2001; Blumenthal, Kriegstein, & Hogness, 1974; Hyrien & Mechali, 1993). This allows the frog genome to be duplicated at a rate of ~155 Mbp/min (Table 1). As development proceeds, however, some initiation sites disappear while others are enhanced, resulting in a clear pattern of preferred initiation sites at specific genomic loci (Hyrien, Maric, & Mechali, 1995; Sasaki, Sawado, Yamaguchi, & Shinomiya, 1999). Developmental acquisition of origin specification also occurs during mammalian development, although the changes are less dramatic (Borowiec & Schildkraut, 2011).

6. MANY REPLICATION ORIGINS EXIST, BUT FEW ARE ACTIVATED

In 1990, two independent efforts to locate specific DNA loci that acted as replication origins in hamster chromosomes resulted in paradoxical conclusions: one study concluded that initiation events were essentially "randomly" distributed throughout the 55 kbp intergenic region adjacent to the *DHFR* gene(Vaughn, Dijkwel, & Hamlin, 1990), whereas the other study concluded that most initiation events began at a specific "origin of bidirectional replication" located within this intergenic region (Burhans, Vassilev, Caddle, Heintz, & DePamphilis, 1990). How might such disparate results be reconciled? The simplest explanation was that replication complexes assembled at many sites along the chromosomes, but that only a subset of these sites were activated at the onset of S phase (DePamphilis, 1993a,1993b). This hypothesis was termed the "Jesuit Model," because it is the Jesuits who remind us that "many are called, but few are chosen" (Matthew 22:14). Now, 25 years later, the "Jesuit Model" provides an accurate description of how metazoan cells regulate when and where to initiate DNA replication throughout animal development. Cells derived from human and mouse tissues contain 30,000–50,000 fully active replication origins, but deep sequencing of short nascent DNA strands can detect 10 × as many replication sites with a mean interorigin distance of 11 kbp (Besnard et al., 2012; Leonard & Mechali, 2013). Similarly, cells derived from flies contain 62 initiation sites with a mean size of 32 kbp distributed along 22 Mbp of chromatin (MacAlpine, Rodriguez, & Bell, 2004), but the median distance between ORC-binding sites is also 11 kbp (MacAlpine, Gordan, Powell, Hartemink, & MacAlpine, 2010). The spacing of replication origins in cells from mammals and insects is comparable to the spacing during rapid cleavage cycles in frogs. Thus, the bulk of DNA replication is driven by a small subset (~10%) of the total number of available replication origins. Moreover, most of the fully active origins exist in several different cell lines, and more than 80% of the fully active origins used in one mitotic cell cycle are reused in the following cell cycle (Li, Chen, Solessio, & Gilbert, 2003).

The existence of a plethora of minor replication origins provides cells with the ability to respond to genotoxic stress. Replication forks frequently stall, for example, when encountering tightly bound protein–DNA complexes, transcription machinery, repetitive sequences, or DNA lesions. Even changes in deoxyribonucleotide pools can change the frequency of initiation at minor replication origins (Anglana, Apiou, Bensimon, & Debatisse,

2003). Minor replication origins remain dormant until activated in response to replication stress (Ge et al., 2015; McIntosh & Blow, 2012). Dormant replication origins are licensed origins that become fully active only under conditions that inhibit ongoing DNA replication forks. Thus, dormant origins ensure genome integrity by replicating regions between stalled replication forks. Dormant origins confer two important advantages to the development of complex organisms. First, they allow flexibility in determining where DNA replication begins, thereby avoiding conflicts between the need to replicate the entire genome and the need to express specific genes, at specific sites, at a specific time. Second, they allow flexibility in determining how many replication origins are necessary to complete the job.

7. BOTH GENETIC AND EPIGENETIC PARAMETERS DEFINE REPLICATION ORIGINS

DNA replication origins in bacteria, archaea, plasmids, bacteriophage, and viruses are specific sequences essential for initiation of DNA replication, because they contain binding sites for either their cognate helicase loader or cognate DNA helicase (DePamphilis & Bell, 2010; Leonard & Mechali, 2013). In contrast, DNA replication origins in the metazoa are not essential for initiation of replication. Artificial assembly of a transcription complex or binding of an ORC or Cdc6 protein to DNA is sufficient to create a site-specific replication origin in frog eggs and mammalian cells (Crevel & Cotterill, 2012; Danis et al., 2004; Takeda, Shibata, Parvin, & Dutta, 2005). Otherwise, DNA replication in either rapid cleavage embryos or egg extract from either frogs or flies begins at virtually any sequence, foreign or domestic, introduced into these cells (Hyrien & Mechali, 1993; Shinomiya & Ina, 1991). In contrast, DNA injected into the nucleus of mouse zygotes does not replicate unless it contains a known replicator (polyomavirus), and the cognate origin recognition protein (polyomavirus T-antigen) is provided (Martinez-Salas, Cupo, & DePamphilis, 1988). The simplest explanation is that the ratio of replication proteins to DNA in rapid cleavage zygotes is much greater than in mouse zygotes, thereby negating any advantage of specific DNA sequences. The volume of a frog egg (1.2 mm diameter), for example, is about 2800 × greater than a mouse egg (0.085 mm diameter). As cell cleavage reduces the ratio of maternally inherited replication proteins to DNA, some sites will be more effective as replication origins than others.

Metazoan replication origins are simply chromosomal loci that permit assembly of prereplication complexes that can later be converted into preinitiation complexes and activated by Cdk2·CcnE and Cdc7·Dbf4 (Fig. 1). Such sites are determined by both genetic and epigenetic parameters (Aladjem, Falaschi, & Kowalski, 2006; DePamphilis, 1996; Gilbert, 2001; Leonard & Mechali, 2013). Genetic parameters include easily unwound DNA sequences (Aladjem et al., 2006) such as asymmetric A:T-rich sequences (Stanojcic, Lemaitre, Brodolin, Danis, & Mechali, 2008) and consensus G-quadruplex-forming motifs (Besnard et al., 2012; Cayrou et al., 2011; Foulk, Urban, Casella, & Gerbi, 2015; Picard et al., 2014). Epigenetic parameters include nuclear structure (Gilbert, Miyazawa, & DePamphilis, 1995), histone modifications (Dorn & Cook, 2011), chromosome structure, and transcription.

The effect of epigenetic parameters on regulatory DNA sequences begins in mammals during the onset of ZGE when changes in chromatin structure result in the need for enhancers to activate DNA replication replicators as well as transcription promoters. The polyomavirus replicator is active in the paternal pronucleus of a mouse zygote as long as the polyomavirus origin recognition protein (the T-antigen DNA helicase) is provided (Martinez-Salas, Linney, Hassell, & DePamphilis, 1989). However, beginning with the two-cell embryo, the polyomavirus replicator is inactive unless linked to a cell-specific enhancer and provided with T-antigen. Enhancers are required to relieve chromatin-mediated repression of both promoters and replicators (Majumder & DePamphilis, 1995; Majumder et al., 1997; Wiekowski, Miranda, Nothias, & DePamphilis, 1997). Thus, as changes in chromatin structure prevent access to some DNA sites but not to others, the ability to license and then activate replication origins becomes more restricted, as postulated by the "Jesuit Model."

8. RESTRICTING GENOME DUPLICATION TO ONCE PER CELL DIVISION

Proliferating cells must produce one, but only one, copy of their nuclear DNA prior to cell division. Therefore, once S phase begins, origin licensing is blocked until mitosis is completed. If DNA is rereplicated prior to mitosis, the result is DNA damage, particularly in the form of double-stranded breaks that arise from stalled replication forks. If the DNA damage response cannot correct the problem, then apoptosis ensues. To prevent such a catastrophe, at least seven pathways exist that can inactivate the helicase loader during S phase, thereby preventing both the reloading of MCM helicases at activated replication origins, and the licensing of new replication origins. These pathways, which can be grouped into two biological cycles, exist in flies, frogs, nematodes, and mammals (Blow & Dutta, 2005; DePamphilis et al., 2006; Siddiqui, On, & Diffley, 2013; Sonneville, Querenet, Craig, Gartner, & Blow, 2012).

Helicase loaders consist of ORC(1–6), Cdc6, and Cdt1 (Fig. 1). The "ORC Cycle" refers to the inactivation of Orc1 and Cdc6 during S phase by posttranslational CDK-dependent phosphorylation and ubiquitin-dependent degradation, followed by origin licensing during the anaphase to G1 phase transition (Fig. 3). The loss of Orc1 destabilizes the remaining ORC subunits. The "Cdt1 Cycle" refers to the inactivation of Cdt1 by CDK-dependent phosphorylation, ubiquitination of Cdt1-P as well as PCNA–DNA-dependent ubiquitination of Cdt1, and inactivation of Cdt1 by binding to geminin protein. Helicase loader proteins become available again during the anaphase to G1 phase transition followed by origin licensing.

The importance of any one pathway in a particular cell type depends on the relative levels of their respective components. Thus, during early development of frogs and mammals only geminin and the CRL4 ubiquitin ligase are essential to prevent DNA rereplication. Cdt1 is a stable abundant protein in *Xenopus* egg extracts. It is not degraded substantially until cycle 10 (Kisielewska & Blow, 2012). Geminin is only fully degraded by the APC/C after the MBT (cell cycle 12) (Hodgson, Li, Tada, & Blow, 2002; Li & Blow, 2005). Therefore, two concerted pathways prevent DNA rereplication in cleavage stage embryos by regulating both Cdt1 activity and Cdt1 proteolysis. In the beginning, when the nuclear to cytoplasmic ratio is low, Cdt1 activity is blocked by geminin and regulation of geminin activity appears to involve both the APC/C and nuclear import of geminin. As the ratio of nuclei to cytoplasm increases, Cdt1 binds to CRL4·PCNA·DNA complexes that assemble at newly minted replication forks and at sites of DNA repair where Cdt1 is then targeted for ubiquitin-dependent degradation (Arias & Walter, 2005, 2007). Thus, suppressing geminin RNA levels in *Xenopus* egg extracts or two-cell embryos cannot immediately deplete the supply of geminin protein. Consequently, DNA rereplication and apoptosis are not manifested until the MBT and beyond (Kerns, Schultz, Barry, Thorne, & McGarry, 2012; Kerns, Torke, Benjamin, & McGarry, 2007).

Once the developmental program activates ZGE, totipotent and pluripotent cells depend primarily, if not exclusively, on Geminin to prevent DNA rereplication. Geminin is essential for gastrulation in frogs (Kerns et al., 2012), and probably in flies as well (Takada, Kwak,

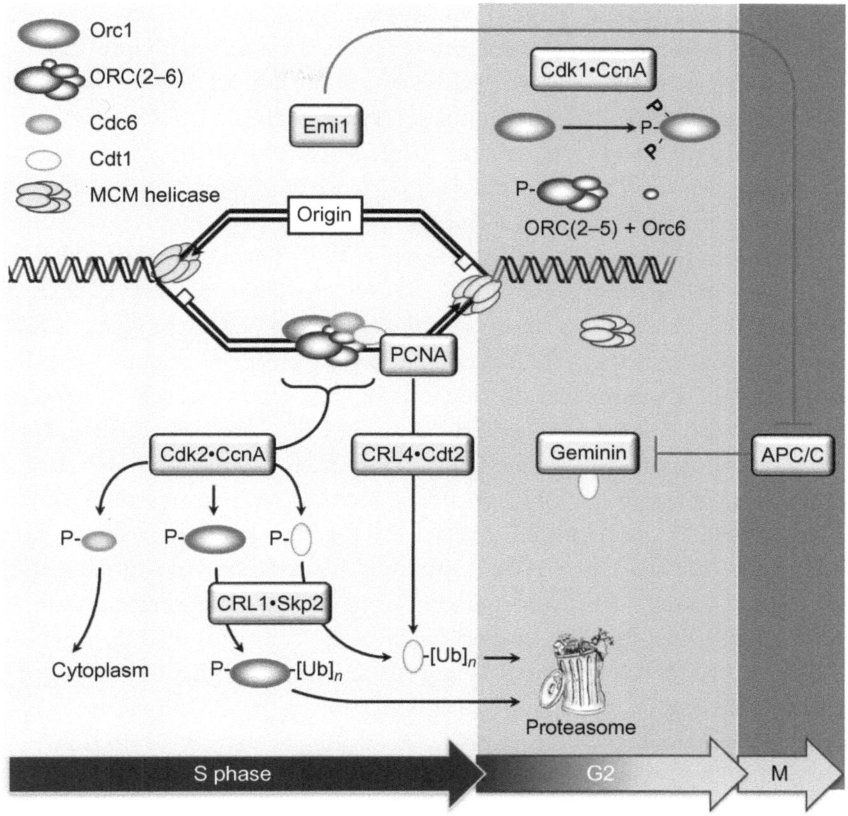

Figure 3 Multiple concerted pathways prevent DNA rereplication. Once metazoan cells have entered S phase, seven concerted pathways can inactivate the helicase loader, thereby preventing rereplication of DNA prior to mitosis. Upon S phase entry, Cdk2·CcnA phosphorylates Orc1, Cdc6, and Cdt1, thereby inhibiting their ability to form helicase loaders. In addition, Cdc6-P is localized to the cytoplasm where it cannot facilitate origin licensing. Orc1-P is ubiquitinated and targeted for destruction, with subsequent dissociation of ORC(2–5) and Orc6 from chromosomes. Cdt1-P is ubiquitinated by the CRL1·Skp2 ubiquitin ligase. Cdt1 that is not phosphorylated is ubiquitinated by the CRL4·Cdt2 ubiquitin ligase bound to PCNA and DNA. PCNA is the eukaryotic sliding clamp protein that facilitates DNA synthesis by DNA polymerases δ and ε. The ubiquitinated proteins are then degraded by the 26S proteasome. Geminin is a protein that binds specifically to Cdt1 and inhibits its activity. These activities are available from S through early M phase. As cells exit mitosis, geminin and cyclin A are ubiquitinated by the anaphase promoting complex [originally termed cyclosome (APC/C)], an activity that is inhibited during S to early M phase by the APC/C-specific inhibitor Emi1. Reassembly of the helicase loader occurs by reassembly of Orc1, ORC(2–5), Orc6, and Cdc6 onto chromatin (Ghosh et al., 2011; Kara et al., 2015; Sonneville et al., 2012), an event that appears to be triggered by the Orc1 subunit (Noguchi et al., 2006). Geminin binding to Cdt1 and CDK-dependent phosphorylation of Cdc6 prevents targeting of Cdt1 and Cdc6 for degradation by the APC/C, thereby allowing them to participate in origin licensing (Ballabeni et al., 2004; Mailand & Diffley, 2005). (A color version of this figure is located at the end of this chapter.)

Koppetsch, & Theurkauf, 2007). Most notably, geminin is essential to prevent DNA rereplication-dependent apoptosis in mouse preimplantation embryos (Hara, Nakayama, & Nakayama, 2006; Huang, Kaneko, Pan, & DePamphilis, 2015), in the epiblast (Huang et al., 2015; Patterson, Waller, & Kroll, 2014), and in embryonic stem cells (Huang et al., 2015), which are derived from the epiblast. However, as pluripotent cells differentiate, geminin is no longer essential for viability, although it continues to contribute to preventing DNA rereplication (Dorn, Chastain, Hall, & Cook, 2009; Huang et al., 2015). In effect, other pathways, such as Cdk2·CcnA (Fig. 3), contribute significantly (Zhu & Depamphilis, 2009). Thus, whether or not a particular cell, such as a cancer cell, requires geminin to prevent DNA rereplication depends on the relative levels of relevant gene products. For example, suppression of CRL4 in HeLa cells induces DNA rereplication (Jin, Arias, Chen, Harper, & Walter, 2006), whereas suppression of geminin does not, unless Cdt1 is overexpressed (Nishitani et al., 2006). Thus, geminin is essential for viability when the cellular levels of Cdt1 are high, as in totipotent blastomeres, pluripotent stem cells, some cancer cells, and perhaps some rapidly proliferating cells in adult animals, but the role of geminin in most cells of the adult organism facilitates proliferation without being essential for viability. Similarly, the primary role of CRL4 in mammals and presumably other animals is to facilitate DNA repair by inhibiting origin licensing (Havens & Walter, 2011).

9. DEVELOPMENTALLY PROGRAMMED POLYPLOIDY

Given the extraordinary problem of preventing excess genome duplication during metazoan development, it is remarkable that some cells are developmentally programmed to exit their mitotic cell cycle in response to environmental signals, injury or stress, and differentiate into nonproliferating, viable, polyploid cells. In fact, developmentally programmed polyploidy is a normal part of animal and plant development that occurs frequently in ferns, flowering plants, mollusks, arthropods, amphibians, and fish, but infrequently in mammals (Lacroix & Maddox, 2012; Ullah, Lee, & Depamphilis, 2009). Some cells become multinucleated by not undergoing cytokinesis, or by fusing together during G1 phase (Gentric & Desdouets, 2014). In contrast, other cells are programmed to undergo multiple S phases in the absence of an intervening mitosis and cytokinesis. These cells become giant cells with one enlarged nucleus containing multiples of 4C DNA, in some cases reaching hundreds of copies

(Orr-Weaver, 2015). DNA replication-dependent polyploidy occurs via endoreplication and endomitosis.

In *Drosophila*, the Notch signal transduction induces endoreplication by inhibiting Cdk1, an activity essential for initiating and maintaining mitosis (Fig. 1). Notch induces expression of Cdh1/Fzr, an activator of the APC/C that triggers ubiquitin-mediated destruction of mitotic cyclins, and Notch suppresses Cdc25/String, a phosphatase that activates the Cdk1·Cyclin complexes essential for mitosis (Lee, Davidson, & Duronio, 2009). Similarly, endoreplication occurs in the trophectoderm of mouse peri-implantation blastocysts in response to the absence of fibroblast growth factor-4 (FGF4) that is produced by the inner cell mass (Roberts & Fisher, 2011). This phenomenon is recapitulated by trophoblast stem cells (TSCs). TSCs undergo self-renewal when cultured in the presence of FGF4, and differentiate into giant cells when cultured in the absence of FGF4 (Fig. 4). Selective inhibition of Cdk1 by RO3306 triggers endoreplication in TSCs undergoing self-renewal. FGF4-deprivation induces expression of the CDK-specific inhibitor p57, which inhibits Cdk1. Since p57 also inhibits Cdk2, p57 levels in giant cells must oscillate during multiple rounds of endoreplication; high during G phase but low during S phase. That permits Cdk2·CcnE to initiate S phase (Fig. 1) and Cdk2·CcnA to prevent DNA rereplication (Fig. 3). FGF4 regulates expression of p57 in TSCs via control of Chk1 levels. Chk1 phosphorylates p57, thereby targeting it for ubiquitin-dependent degradation. FGF4 deprivation in TSCs results in downregulation of geminin, which results in the loss of Chk1 protein with concomitant upregulation of p57 (Fig. 5). Thus, TSCs will also differentiate into giant cells as a result of DNA damage activating Chk1, the DNA damage response kinase that inhibits Cdc25 (required for Cdk1 activity), thereby inhibiting Cdk1 activity (Fig. 4).

Terminally differentiated cells must remain viable, but since they cannot proliferate, they cannot pass on genetic mutations, so repairing DNA damage is not a concern. Induction of DNA rereplication induced DNA damage in *Drosophila* does not induce apoptosis in follicle cells undergoing endoreplication, whereas it does in cells undergoing proliferation (Mehrotra, Maqbool, Kolpakas, Murnen, & Calvi, 2008). Loss of Chk1 protein in TSCs during FGF4-deprivation induces both p57 and p21 expression by the same mechanism (Fig. 5), but p57 localizes to the nucleus where it inhibits Cdk1, and p21 localizes to the cytoplasm where it suppresses DNA damage-dependent apoptosis (de Renty, DePamphilis, et al., 2014).

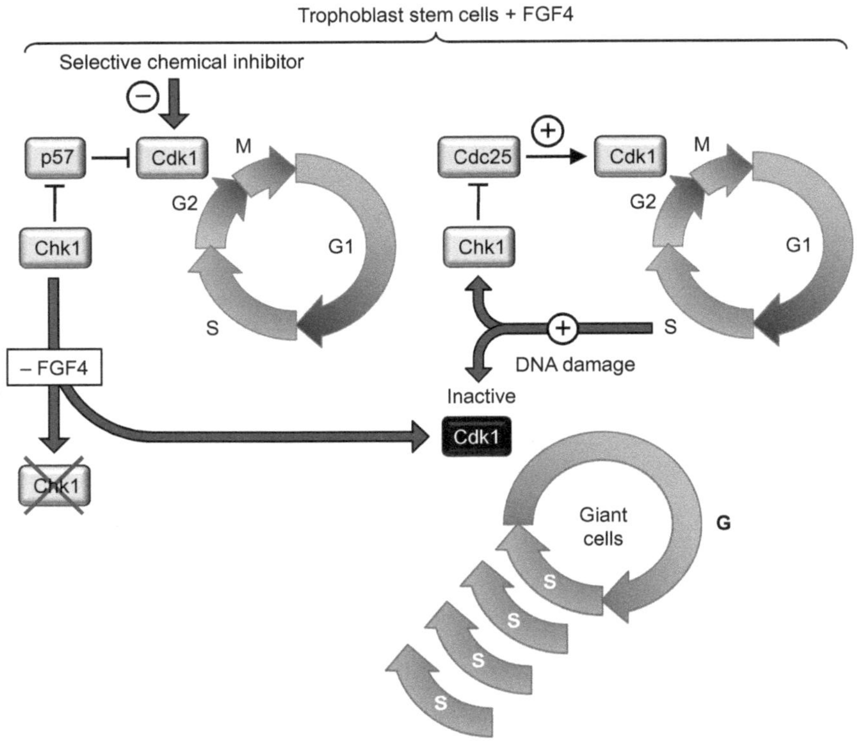

Figure 4 Endoreplication is triggered in trophoblast cells by selective inhibition of Cdk1 activity. Developmentally programmed endoreplication in the metazoa occurs when Cdk1 activity is suppressed so that cells cannot enter mitosis (Edgar, Zielke, & Gutierrez, 2014; Fox & Duronio, 2013; Zielke, Edgar, & DePamphilis, 2013). This results in multiple S phases without an intervening mitosis or cytokinesis to produce giant cells with a single enlarged nucleus containing as many as several hundred copies of each chromosome. In the absence of FGF4, trophoblast cells in mouse blastocysts are triggered to differentiate into giant cells. In the presence of FGF4, the same result can be elicited either by selective inhibition of Cdk1 with RO3306, or by suppressing or inactivating geminin (de Renty, Kaneko, & DePamphilis, 2014; Ullah, Kohn, Yagi, Vassilev, & DePamphilis, 2008). The effect of FGF4 deprivation is to reduce the level of geminin, which results in the loss of checkpoint kinase-1 (Chek1/Chk1). Since Chk1 phosphorylation of the CDK-specific inhibitor Cdkn1c/p57 targets it for ubiquitin-dependent degradation, loss of Chk1 results in p57 inhibition of Cdk1 activity, which leads to endoreplication (Ullah, de Renty, & Depamphilis, 2011). Ironically, DNA damage in trophoblast cells also results in endoreplication, because DNA damage activates Chk1, which then phosphorylates and inactivates Cdc25, a phosphatase that is essential to maintain Cdk1 in its active form. (A color version of this figure is located at the end of this chapter.)

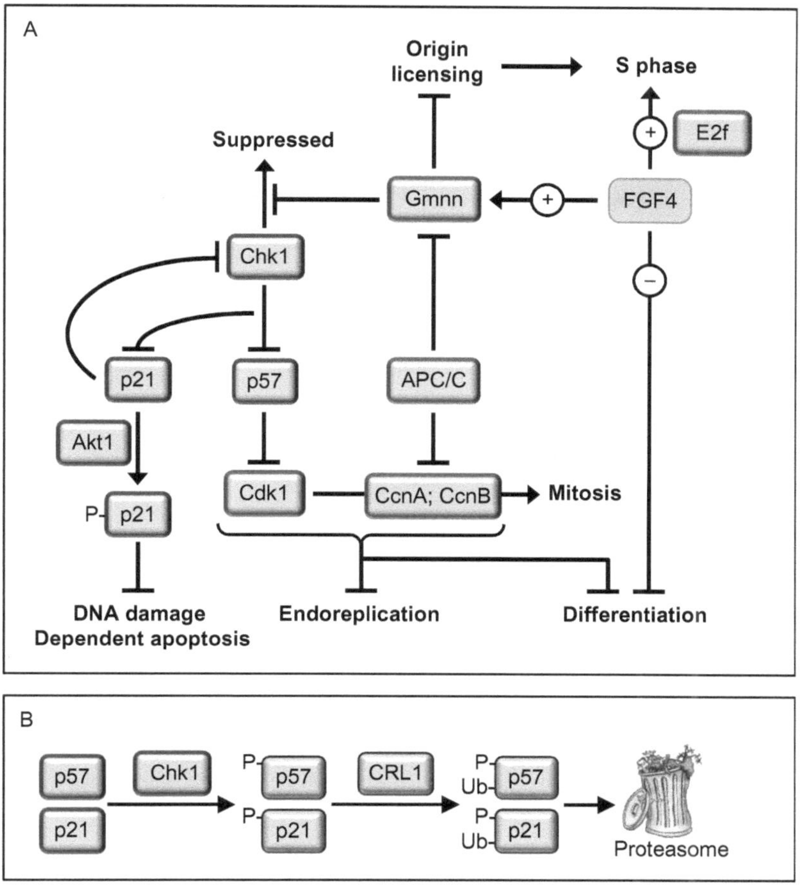

Figure 5 The FGF4 signal transduction pathway governs trophoblast proliferation and differentiation. FGF4 (and probably other mitogenic proteins as well) is essential for trophoblast proliferation. This mitogenic activity is likely mediated by E2F-dependent gene expression (Chen et al., 2010), and possibly directed at regulating the activity of the anaphase-promoting complex (APC) (Yang et al., 2012). FGF4 deprivation results in downregulation of geminin activity to a level that maintains endocycles (de Renty, Kaneko, & DePamphilis, 2014), but that does not prevent downregulation of Chk1 protein. The loss of Chk1 kinase activity results in expression of two CDK-specific inhibitors, p57 and p21 (Ullah et al., 2011). The p57 protein prevents the onset of mitosis by selectively inhibiting Cdk1 activity, thereby triggering the first round of endoreplication (Hattori, Davies, Anson-Cartwright, & Cross, 2000; Ullah et al., 2008). This event activates the G1-phase APC·Cdh1 ubiquitin ligase, which targets geminin, cyclin B, and cyclin A for degradation, thereby allowing licensing of replication origins and the onset of S-phase without passing through mitosis (Ullah, Lee, Lilly, & DePamphilis, 2009). Inhibition of Cdk1 triggers both endoreplication and TSC differentiation. In the absence of p57, FGF4 deprivation produces multinucleated TGCs, revealing the existence of alternative mechanisms to trigger TSC differentiation (Ullah et al., 2008). Endocycles also require p57, which is expressed during G-phase and then suppressed during S-phase to allow sequential assembly and activation of pre-replication complexes (Ullah et al., 2008). Geminin maintains endocycles by preventing DNA rereplication. The p21/Cdkn1a protein localizes to the cytoplasm in TGCs where it prevents DNA damage induced apoptosis (de Renty, DePamphilis, & Ullah, 2014). It might also maintain suppression of Chk1 by reducing *Chk1* RNA levels (Gottifredi, Karni-Schmidt, Shieh, & Prives, 2001), as observed during FGF4 deprivation (Fig. 3). (A color version of this figure is located at the end of this chapter.)

Endomitosis is similar to endoreplication, except that the nuclear envelope breaks down and parts of anaphase occur without nuclear division, thereby producing mononucleated cells with separated chromatids. The mechanism responsible remains speculative, but selective inactivation of Cdk1 readily converts endomitosis into endoreplication (Trakala et al., 2015). Selective inactivation of Cdk1 also induces endoreplication in mouse embryonic fibroblasts and chicken DT40 cells (Diril et al., 2012; Hochegger et al., 2007). Suppression of Cdk1 activity induces endoreplication.

10. DEVELOPMENTALLY PROGRAMMED GENE AMPLIFICATION

Genes encoding products needed in high amounts at specific stages during development are sometimes amplified by selectively replicating those genes multiple times. Developmentally programmed gene amplification is an example of site-specific initiation of DNA replication in terminally differentiated cells (Calvi, 2006; Tower, 2004). Specialized *cis*-acting sequences (amplification origins) allow specific genes to be selectively rereplicated multiple times. In response to ecdysone, the master regulator of insect development, the larval salivary glands of *Sciara coprophila* produce a parallel array of hundreds of chromatids termed "polytene chromosomes." Gene amplification then produces DNA puffs in these chromosomes (not to be confused with RNA puffs, which arise by multiple rounds of transcription). The genes encoding the proteins required for pupation are then amplified many fold. A larval-specific ecdysone receptor binds to a specific DNA sequence adjacent to the origin recognition complex-binding site within a site-specific origin of bidirectional replication (Foulk et al., 2013). Similarly, the follicle cells that envelop each oocyte in *Drosophila* amplify the four genes that produce the proteins for chorion (eggshell) formation. Amplification depends on the interaction of prereplication complexes assembled at two specific DNA sites, ACE3 and Ori-β, within the chorion gene cluster. It appears that amplification origins in flies/gnats use the same proteins required to initiate DNA replication during normal cell proliferation, but in contrast to replication origins activated during S phase of mitotic cell division cycles, these specialized "amplification origins" continue to initiate DNA replication in the absence of mitosis or cytokinesis.

Xenopus oocytes also amplify their rRNA genes, apparently by excising one or more copies from their chromosomes and then replicating them several thousand folds via a "rolling circle" mechanism (Hourcade,

Dressler, & Wolfson, 1973). Although the details of *Xenopus* rRNA amplification remain speculative, the mechanism for rRNA gene amplification in the ciliate protozoan *Tetrahymena thermophila* has been well characterized. This single-cell organism has two nuclei: a transcriptionally silent, diploid, germline micronucleus and a transcriptionally active, polyploid, somatic macronucleus. The chromosomes in the macronucleus are first endoreplicated about 45-fold, and then genome-wide replication ceases, and a 21 kbp linear minichromosome encoding the three rRNAs is generated by excision and rearrangement of the micronuclear rRNA gene locus. The minichromosome is amplified up to 10,000 copies in a single S phase during development, but it is replicated, on average, only once per cell cycle during vegetative growth. DNA replication begins at a specific site within the rRNA gene locus of *Tetrahymena*, as it does within the rRNA gene loci in yeast and mammalian cells (Donti, Datta, Sandoval, & Kapler, 2009).

11. EVOLUTION AND HUMAN DISEASE

The vast majority of mutations are caused by misincorporation of the wrong nucleotide when DNA polymerases replicate DNA and when damaged DNA is repaired. The two human DNA polymerases (δ and ε) responsible for the bulk of nuclear DNA replication have error rates ~1 to 2 per 100,000 nucleotides synthesized, and eight human DNA repair polymerases have error rates of23–72,000 (DePamphilis & Bell, 2010). Thus, the very act of duplicating the genome creates the basis for genomic instability, and the simple fact that natural selection has never completely suppressed the accumulation of errors during genome duplication attests to the theory that biological evolution relies upon these errors. If evolution by natural selection produced error free DNA polymerases, then the evolution of species would cease. It has not, DNA replication continues to drive evolution.

The spontaneous mutation rate in human cells is one mutation in every 10 billion base pairs that are replicated during human development. The reason the rate is not higher is because nature added a "proof reading" mechanism to polymerases δ and ε in the form of an exonuclease that degrades one strand of duplex DNA beginning at its 3′-OH terminus, thereby allowing the DNA polymerase to try again. This reduces the error rate for these enzymes by 4- to 24-fold. In addition, nature developed "mismatch repair" systems to correct single mistakes in base pairing during DNA synthesis. Nevertheless, at an error rate of $1/10^9$ bp, every zygote will receive about

138 new mutations. There are about 400 cell divisions between zygote and spermatocytes and about 30 cell divisions between zygote and oocytes (Crow, 2000).

Unfortunately, the price for continued evolution is disease. Genetic mutations just in human DNA damage response genes, restriction checkpoint genes, DNA helicases, DNA replication proteins, and DNA repair genes are associated with at least 60 different cancers and disease syndromes (DePamphilis, 2006; DePamphilis & Bell, 2010). Mutations are also produced by aberrant events at DNA replication forks that result in synthesis of multiple repeats of the same nucleotide sequence within a gene. This produces a mutated protein containing a series of additional amino acids. At least 19 "expanded repeat diseases" have been identified, including several neurodegenerative disorders.

12. EPILOG

Genome duplication, mitosis, and cytokinesis are all essential for cell proliferation. Changing patterns of gene expression are essential for metazoan development. However, metazoan development could not have evolved without genome duplication becoming flexible enough to accommodate the ever-changing demands of gene expression during cell differentiation and reliable enough to sustain the trillions of cell divisions required from zygote to adult. Although much remains to be discovered, one question has far reaching implications: Are some genes essential for genome duplication in some cell types, but not in others, and if so, can those genes provide therapeutic targets in the treatment of human disease?

REFERENCES

Aiken, C. E., Swoboda, P. P., Skepper, J. N., & Johnson, M. H. (2004). The direct measurement of embryogenic volume and nucleo-cytoplasmic ratio during mouse preimplantation development. *Reproduction*, *128*, 527–535.

Aladjem, M. I., Falaschi, A., & Kowalski, D. (2006). Eukaryotic DNA replication origins. In M. L. DePamphilis (Ed.), *DNA replication and human disease* (pp. 31–62). Cold Spring Harbor, NY: Cold Spring Harbor Laboratory Press.

Anglana, M., Apiou, F., Bensimon, A., & Debatisse, M. (2003). Dynamics of DNA replication in mammalian somatic cells: Nucleotide pool modulates origin choice and interorigin spacing. *Cell*, *114*, 385–394.

Aoki, F., Worrad, D. M., & Schultz, R. M. (1997). Regulation of transcriptional activity during the first and second cell cycles in the preimplantation mouse embryo. *Developmental Biology*, *181*, 296–307.

Arias, E. E., & Walter, J. C. (2005). Replication-dependent destruction of Cdt1 limits DNA replication to a single round per cell cycle in Xenopus egg extracts. *Genes and Development*, *19*, 114–126.

Arias, E. E., & Walter, J. C. (2007). Strength in numbers: Preventing rereplication via multiple mechanisms in eukaryotic cells. *Genes and Development*, *21*, 497–518.

Baldinger, T., & Gossen, M. (2009). Binding of Drosophila ORC proteins to anaphase chromosomes requires cessation of mitotic cyclin-dependent kinase activity. *Molecular and Cellular Biology*, *29*, 140–149.

Ballabeni, A., Melixetian, M., Zamponi, R., Masiero, L., Marinoni, F., & Helin, K. (2004). Human geminin promotes pre-RC formation and DNA replication by stabilizing CDT1 in mitosis. *EMBO Journal*, *23*, 3122–3132.

Bell, S. P., & Kaguni, J. M. (2013). Helicase loading at chromosomal origins of replication. *Cold Spring Harbor Perspectives in Biology*, *5*, 61–80.

Besnard, E., Babled, A., Lapasset, L., Milhavet, O., Parrinello, H., Dantec, C., et al. (2012). Unraveling cell type-specific and reprogrammable human replication origin signatures associated with G-quadruplex consensus motifs. *Nature Structural and Molecular Biology*, *19*, 837–844.

Blow, J. J., & Dutta, A. (2005). Preventing re-replication of chromosomal DNA. *Nature Reviews Molecular Cell Biology*, *6*, 476–486.

Blow, J. J., Gillespie, P. J., Francis, D., & Jackson, D. A. (2001). Replication origins in Xenopus egg extract are 5-15 kilobases apart and are activated in clusters that fire at different times. *Journal of Cell Biology*, *152*, 15–25.

Blumenthal, A. B., Kriegstein, H. J., & Hogness, D. S. (1974). The units of DNA replication in Drosophila melanogaster chromosomes. *Cold Spring Harbor Symposia on Quantitative Biology*, *38*, 205–223.

Bolton, V. N., Oades, P. J., & Johnson, M. H. (1984). The relationship between cleavage, DNA replication, and gene expression in the mouse 2-cell embryo. *Journal of Embryology and Experimental Morphology*, *79*, 139–163.

Borowiec, J. A., & Schildkraut, C. L. (2011). Open sesame: Activating dormant replication origins in the mouse immunoglobulin heavy chain (Igh) locus. *Current Opinion in Cell Biology*, *23*, 284–292.

Burhans, W. C., Vassilev, L. T., Caddle, M. S., Heintz, N. H., & DePamphilis, M. L. (1990). Identification of an origin of bidirectional DNA replication in mammalian chromosomes. *Cell*, *62*, 955–965.

Calvi, B. R. (2006). Developmental gene amplification. In M. L. DePamphilis (Ed.), *DNA replication and human disease* (pp. 233–255). Cold Spring Harbor, NY: Cold Spring Harbor Laboratory Press.

Campisi, J. (2013). Aging, cellular senescence, and cancer. *Annual Review of Physiology*, *75*, 685–705.

Cayrou, C., Coulombe, P., Vigneron, A., Stanojcic, S., Ganier, O., Peiffer, I., et al. (2011). Genome-scale analysis of metazoan replication origins reveals their organization in specific but flexible sites defined by conserved features. *Genome Research*, *21*, 1438–1449.

Chen, L., Chan, S. W., Zhang, X., Walsh, M., Lim, C. J., Hong, W., et al. (2010). Structural basis of YAP recognition by TEAD4 in the hippo pathway. *Genes and Development*, *24*, 290–300.

Ciemerych, M. A., & Sicinski, P. (2005). Cell cycle in mouse development. *Oncogene*, *24*, 2877–2898.

Conti, C., Sacca, B., Herrick, J., Lalou, C., Pommier, Y., & Bensimon, A. (2007). Replication fork velocities at adjacent replication origins are coordinately modified during DNA replication in human cells. *Molecular Biology of the Cell*, *18*, 3059–3067.

Conti, C., Seiler, J. A., & Pommier, Y. (2007). The mammalian DNA replication elongation checkpoint: Implication of Chk1 and relationship with origin firing as determined by single DNA molecule and single cell analyses. *Cell Cycle*, *6*, 2760–2767.

Crevel, G., & Cotterill, S. (2012). Forced binding of the origin of replication complex to chromosomal sites in Drosophila S2 cells creates an origin of replication. *Journal of Cell Science*, *125*, 965–972.

Crow, J. F. (2000). The origins, patterns and implications of human spontaneous mutation. *Nature Reviews Genetics*, *1*, 40–47.

Danis, E., Brodolin, K., Menut, S., Maiorano, D., Girard-Reydet, C., & Mechali, M. (2004). Specification of a DNA replication origin by a transcription complex. *Nature Cell Biology*, *6*, 721–730.

DePamphilis, M. L. (1993a). Eukaryotic DNA replication: Anatomy of an origin. *Annual Review of Biochemistry*, *62*, 29–63.

DePamphilis, M. L. (1993b). Origins of DNA replication in metazoan chromosomes. *Journal of Biological Chemistry*, *268*, 1–4.

DePamphilis, M. L. (1996). Origins of DNA replication. In M. L. DePamphilis (Ed.), *DNA replication in eukaryotic cells* (pp. 45–86). New York: Cold Spring Harbor Laboratory Press.

DePamphilis, M. L. (2006). *DNA replication and human disease* (p. 814). Cold Spring Harbor, NY: Cold Spring Harbor Laboratory Press.

DePamphilis, M. L., & Bell, S. D. (2010). *Genome duplication (concepts, mechanisms, evolution and disease)*. London and New York: Garland Science.

DePamphilis, M. L., Blow, J. J., Ghosh, S., Saha, T., Noguchi, K., & Vassilev, A. (2006). Regulating the licensing of DNA replication origins in metazoa. *Current Opinion in Cell Biology*, *18*, 231–239.

Depamphilis, M. L., de Renty, C. M., Ullah, Z., & Lee, C. Y. (2012). "The octet": Eight protein kinases that control mammalian DNA replication. *Frontiers in Physiology*, *3*, 368.

de Renty, C., DePamphilis, M. L., & Ullah, Z. (2014). Cytoplasmic localization of p21 protects trophoblast giant cells from DNA damage induced apoptosis. *PLoS ONE*, *9*, e97434.

de Renty, C., Kaneko, K. J., & DePamphilis, M. L. (2014). The dual roles of geminin during trophoblast proliferation and differentiation. *Developmental Biology*, *387*, 49–63.

Diril, M. K., Ratnacaram, C. K., Padmakumar, V. C., Du, T., Wasser, M., Coppola, V., et al. (2012). Cyclin-dependent kinase 1 (Cdk1) is essential for cell division and suppression of DNA re-replication but not for liver regeneration. *Proceedings of the National Academy of Sciences of the United States of America*, *109*, 3826–3831.

Donti, T. R., Datta, S., Sandoval, P. Y., & Kapler, G. M. (2009). Differential targeting of Tetrahymena ORC to ribosomal DNA and non-rDNA replication origins. *EMBO Journal*, *28*, 223–233.

Dorn, E. S., Chastain, P. D., 2nd, Hall, J. R., & Cook, J. G. (2009). Analysis of re-replication from deregulated origin licensing by DNA fiber spreading. *Nucleic Acids Research*, *37*, 60–69.

Dorn, E. S., & Cook, J. G. (2011). Nucleosomes in the neighborhood: New roles for chromatin modifications in replication origin control. *Epigenetics*, *6*, 552–559.

Edgar, B. A., Zielke, N., & Gutierrez, C. (2014). Endocycles: A recurrent evolutionary innovation for post-mitotic cell growth. *Nature Reviews Molecular Cell Biology*, *15*, 197–210.

El Dika, M., Laskowska-Kaszub, K., Koryto, M., Dudka, D., Prigent, C., Tassan, J. P., et al. (2014). CDC6 controls dynamics of the first embryonic M-phase entry and progression via CDK1 inhibition. *Developmental Biology*, *396*, 67–80.

Foster, D. A., Yellen, P., Xu, L., & Saqcena, M. (2010). Regulation of G1 cell cycle progression: Distinguishing the restriction point from a nutrient-sensing cell growth checkpoint(s). *Genes & Cancer*, *1*, 1124–1131.

Foulk, M. S., Urban, J. M., Casella, C., & Gerbi, S. A. (2015). Characterizing and controlling intrinsic biases of lambda exonuclease in nascent strand sequencing reveals phasing between nucleosomes and G-quadruplex motifs around a subset of human replication origins. *Genome Research*, *25*, 725–735.

Foulk, M. S., Waggener, J. M., Johnson, J. M., Yamamoto, Y., Liew, G. M., Urnov, F. D., et al. (2013). Isolation and characterization of the ecdysone receptor and its heterodimeric partner ultraspiracle through development in Sciara coprophila. *Chromosoma*, *122*, 103–119.

Fox, D. T., & Duronio, R. J. (2013). Endoreplication and polyploidy: Insights into development and disease. *Development*, *140*, 3–12.

Ge, X. Q., Han, J., Cheng, E. C., Yamaguchi, S., Shima, N., Thomas, J. L., et al. (2015). Embryonic stem cells license a high level of dormant origins to protect the genome against replication stress. *Stem Cell Reports*, *5*(2), 185–194.

Gentric, G., & Desdouets, C. (2014). Polyploidization in liver tissue. *American Journal of Pathology*, *184*, 322–331.

Ghosh, S., Vassilev, A. P., Zhang, J., Zhao, Y., & DePamphilis, M. L. (2011). Assembly of the human origin recognition complex occurs through independent nuclear localization of its components. *Journal of Biological Chemistry*, *286*, 23831–23841.

Gilbert, D. M. (2001). Making sense of eukaryotic DNA replication origins. *Science*, *294*, 96–100.

Gilbert, D. M., Miyazawa, H., & DePamphilis, M. L. (1995). Site-specific initiation of DNA replication in Xenopus egg extract requires nuclear structure. *Molecular and Cellular Biology*, *15*, 2942–2954.

Glover, D. M., Alphey, L., Axton, J. M., Cheshire, A., Dalby, B., Freeman, M., et al. (1989). Mitosis in Drosophila development. *Journal of Cell Science. Supplement*, *12*, 277–291.

Gottifredi, V., Karni-Schmidt, O., Shieh, S. S., & Prives, C. (2001). p53 down-regulates CHK1 through p21 and the retinoblastoma protein. *Molecular and Cellular Biology*, *21*, 1066–1076.

Hansen, C. G., Moroishi, T., & Guan, K. L. (2015). YAP and TAZ: A nexus for Hippo signaling and beyond. *Trends in Cell Biology*, *25*(9), 499–513.

Hara, K., Nakayama, K. I., & Nakayama, K. (2006). Geminin is essential for the development of preimplantation mouse embryos. *Genes to Cells*, *11*, 1281–1293.

Hattori, N., Davies, T. C., Anson-Cartwright, L., & Cross, J. C. (2000). Periodic expression of the cyclin-dependent kinase inhibitor p57(Kip2) in trophoblast giant cells defines a G2-like gap phase of the endocycle. *Molecular Biology of the Cell*, *11*, 1037–1045.

Havens, C. G., & Walter, J. C. (2011). Mechanism of CRL4(Cdt2), a PCNA-dependent E3 ubiquitin ligase. *Genes and Development*, *25*, 1568–1582.

Hochegger, H., Dejsuphong, D., Sonoda, E., Saberi, A., Rajendra, E., Kirk, J., et al. (2007). An essential role for Cdk1 in S phase control is revealed via chemical genetics in vertebrate cells. *Journal of Cell Biology*, *178*, 257–268.

Hodgson, B., Li, A., Tada, S., & Blow, J. J. (2002). Geminin becomes activated as an inhibitor of Cdt1/RLF-B following nuclear import. *Current Biology*, *12*, 678–683.

Hourcade, D., Dressler, D., & Wolfson, J. (1973). The amplification of ribosomal RNA genes involves a rolling circle intermediate. *Proceedings of the National Academy of Sciences of the United States of America*, *70*, 2926–2930.

Howlett, S. K. (1986). A set of proteins showing cell cycle dependent modification in the early mouse embryo. *Cell*, *45*, 387–396.

Howlett, S. K., & Bolton, V. N. (1985). Sequence and regulation of morphological and molecular events during the first cell cycle of mouse embryogenesis. *Journal of Embryology and Experimental Morphology*, *87*, 175–206.

Huang, Y. Y., Kaneko, K. J., Pan, H., & DePamphilis, M. L. (2015). Geminin is essential to prevent DNA re-replication-dependent apoptosis in pluripotent cells, but not in differentiated cells. *Stem Cells*, *33*(11), 3239–3253.

Hyrien, O., Maric, C., & Mechali, M. (1995). Transition in specification of embryonic metazoan DNA replication origins. *Science*, *270*, 994–997.

Hyrien, O., & Mechali, M. (1993). Chromosomal replication initiates and terminates at random sequences but at regular intervals in the ribosomal DNA of Xenopus early embryos. *EMBO Journal*, *12*, 4511–4520.

Jin, J., Arias, E. E., Chen, J., Harper, J. W., & Walter, J. C. (2006). A family of diverse Cul4-Ddb1-interacting proteins includes Cdt2, which is required for S phase destruction of the replication factor Cdt1. *Molecular Cell*, *23*, 709–721.

Kaneko, K. J., Cullinan, E. B., Latham, K. E., & DePamphilis, M. L. (1997). Transcription factor mTEAD-2 is selectively expressed at the beginning of zygotic gene expression in the mouse. *Development*, *124*, 1963–1973.

Kaneko, K. J., & DePamphilis, M. L. (2013). TEAD4 establishes the energy homeostasis essential for blastocoel formation. *Development*, *140*, 3680–3690.

Kara, N., Hossain, M., Prasanth, S. G., & Stillman, B. (2015). Orc1 binding to mitotic chromosomes precedes spatial patterning during G1 phase and assembly of the origin recognition complex in human cells. *Journal of Biological Chemistry*, *290*, 12355–12369.

Kerns, S. L., Schultz, K. M., Barry, K. A., Thorne, T. M., & McGarry, T. J. (2012). Geminin is required for zygotic gene expression at the Xenopus mid-blastula transition. *PLoS ONE*, *7*, e38009.

Kerns, S. L., Torke, S. J., Benjamin, J. M., & McGarry, T. J. (2007). Geminin prevents rereplication during xenopus development. *Journal of Biological Chemistry*, *282*, 5514–5521.

Kisielewska, J., & Blow, J. J. (2012). Dynamic interactions of high Cdt1 and geminin levels regulate S phase in early Xenopus embryos. *Development*, *139*, 63–74.

Kuo, A. J., Song, J., Cheung, P., Ishibe-Murakami, S., Yamazoe, S., Chen, J. K., et al. (2012). The BAH domain of ORC1 links H4K20me2 to DNA replication licensing and Meier-Gorlin syndrome. *Nature*, *484*, 115–119.

Lacroix, B., & Maddox, A. S. (2012). Cytokinesis, ploidy and aneuploidy. *Journal of Pathology*, *226*, 338–351.

Lara-Gonzalez, P., Westhorpe, F. G., & Taylor, S. S. (2012). The spindle assembly checkpoint. *Current Biology*, *22*, R966–R980.

Lee, M. T., Bonneau, A. R., & Giraldez, A. J. (2014). Zygotic genome activation during the maternal-to-zygotic transition. *Annual Review of Cell and Developmental Biology*, *30*, 581–613.

Lee, H. O., Davidson, J. M., & Duronio, R. J. (2009). Endoreplication: Polyploidy with purpose. *Genes and Development*, *23*, 2461–2477.

Lemaitre, J. M., Bocquet, S., & Mechali, M. (2002). Competence to replicate in the unfertilized egg is conferred by Cdc6 during meiotic maturation. *Nature*, *419*, 718–722.

Lemaitre, J. M., Bocquet, S., Terret, M. E., Namdar, M., Ait-Ahmed, O., Kearsey, S., et al. (2004). The regulation of competence to replicate in meiosis by Cdc6 is conserved during evolution. *Molecular Reproduction and Development*, *69*, 94–100.

Leonard, A. C., & Mechali, M. (2013). DNA replication origins. *Cold Spring Harbor Perspectives in Biology*, *5*, a010116.

Li, A., & Blow, J. J. (2005). Cdt1 downregulation by proteolysis and geminin inhibition prevents DNA re-replication in Xenopus. *EMBO Journal*, *24*, 395–404.

Li, F., Chen, J., Solessio, E., & Gilbert, D. M. (2003). Spatial distribution and specification of mammalian replication origins during G1 phase. *Journal of Cell Biology*, *161*, 257–266.

Lu, X., Li, J. M., Elemento, O., Tavazoie, S., & Wieschaus, E. F. (2009). Coupling of zygotic transcription to mitotic control at the Drosophila mid-blastula transition. *Development*, *136*, 2101–2110.

MacAlpine, H. K., Gordan, R., Powell, S. K., Hartemink, A. J., & MacAlpine, D. M. (2010). Drosophila ORC localizes to open chromatin and marks sites of cohesin complex loading. *Genome Research*, *20*, 201–211.
MacAlpine, D. M., Rodriguez, H. K., & Bell, S. P. (2004). Coordination of replication and transcription along a Drosophila chromosome. *Genes and Development*, *18*, 3094–3105.
Mac Auley, A., Werb, Z., & Mirkes, P. E. (1993). Characterization of the unusually rapid cell cycles during rat gastrulation. *Development*, *117*, 873–883.
Mailand, N., & Diffley, J. F. (2005). CDKs promote DNA replication origin licensing in human cells by protecting Cdc6 from APC/C-dependent proteolysis. *Cell*, *122*, 915–926.
Majumder, S., & DePamphilis, M. L. (1994). TATA-dependent enhancer stimulation of promoter activity in mice is developmentally acquired. *Molecular and Cellular Biology*, *14*, 4258–4268.
Majumder, S., & DePamphilis, M. L. (1995). A unique role for enhancers is revealed during early mouse development. *Bioessays*, *17*, 879–889.
Majumder, S., Miranda, M., & DePamphilis, M. L. (1993). Analysis of gene expression in mouse preimplantation embryos demonstrates that the primary role of enhancers is to relieve repression of promoters. *EMBO Journal*, *12*, 1131–1140.
Majumder, S., Zhao, Z., Kaneko, K., & DePamphilis, M. L. (1997). Developmental acquisition of enhancer function requires a unique coactivator activity. *EMBO Journal*, *16*, 1721–1731.
Martinez-Salas, E., Cupo, D. Y., & DePamphilis, M. L. (1988). The need for enhancers is acquired upon formation of a diploid nucleus during early mouse development. *Genes and Development*, *2*, 1115–1126.
Martinez-Salas, E., Linney, E., Hassell, J., & DePamphilis, M. L. (1989). The need for enhancers in gene expression first appears during mouse development with formation of the zygotic nucleus. *Genes and Development*, *3*, 1493–1506.
McIntosh, D., & Blow, J. J. (2012). Dormant origins, the licensing checkpoint, and the response to replicative stresses. *Cold Spring Harbor Perspectives in Biology*, *10*(4), 33–42.
Mehrotra, S., Maqbool, S. B., Kolpakas, A., Murnen, K., & Calvi, B. R. (2008). Endocycling cells do not apoptose in response to DNA rereplication genotoxic stress. *Genes and Development*, *22*, 3158–3171.
Melin, F., Miranda, M., Montreau, N., DePamphilis, M. L., & Blangy, D. (1993). Transcription enhancer factor-1 (TEF-1) DNA binding sites can specifically enhance gene expression at the beginning of mouse development. *EMBO Journal*, *12*, 4657–4666.
Miotto, B., & Struhl, K. (2010). HBO1 histone acetylase activity is essential for DNA replication licensing and inhibited by Geminin. *Molecular Cell*, *37*, 57–66.
Murai, S., Stein, P., Buffone, M. G., Yamashita, S., & Schultz, R. M. (2010). Recruitment of Orc6l, a dormant maternal mRNA in mouse oocytes, is essential for DNA replication in 1-cell embryos. *Developmental Biology*, *341*, 205–212.
Newport, J., & Dasso, M. (1989). On the coupling between DNA replication and mitosis. *Journal of Cell Science. Supplement*, *12*, 149–160.
Nishitani, H., Sugimoto, N., Roukos, V., Nakanishi, Y., Saijo, M., Obuse, C., et al. (2006). Two E3 ubiquitin ligases, SCF-Skp2 and DDB1-Cul4, target human Cdt1 for proteolysis. *EMBO Journal*, *25*, 1126–1136.
Noguchi, K., Vassilev, A., Ghosh, S., Yates, J. L., & DePamphilis, M. L. (2006). The BAH domain facilitates the ability of human Orc1 protein to activate replication origins in vivo. *EMBO Journal*, *25*, 5372–5382.
Nothias, J. Y., Majumder, S., Kaneko, K. J., & DePamphilis, M. L. (1995). Regulation of gene expression at the beginning of mammalian development. *Journal of Biological Chemistry*, *270*, 22077–22080.
Nothias, J. Y., Miranda, M., & DePamphilis, M. L. (1996). Uncoupling of transcription and translation during zygotic gene activation in the mouse. *EMBO Journal*, *15*, 5715–5725.

O'Farrell, P. H., Stumpff, J., & Su, T. T. (2004). Embryonic cleavage cycles: How is a mouse like a fly? *Current Biology*, *14*, R35–R45.
Orr-Weaver, T. L. (2015). When bigger is better: The role of polyploidy in organogenesis. *Trends in Genetics*, *31*, 307–315.
Patterson, E. S., Waller, L. E., & Kroll, K. L. (2014). Geminin loss causes neural tube defects through disrupted progenitor specification and neuronal differentiation. *Developmental Biology*, *393*(1), 44–56.
Picard, F., Cadoret, J. C., Audit, B., Arneodo, A., Alberti, A., Battail, C., et al. (2014). The spatiotemporal program of DNA replication is associated with specific combinations of chromatin marks in human cells. *PLoS Genetics*, *10*, e1004282.
Roberts, R. M., & Fisher, S. J. (2011). Trophoblast stem cells. *Biology of Reproduction*, *84*, 412–421.
Sasaki, T., Sawado, T., Yamaguchi, M., & Shinomiya, T. (1999). Specification of regions of DNA replication initiation during embryogenesis in the 65-kilobase DNApolalpha-dE2F locus of Drosophila melanogaster. *Molecular and Cellular Biology*, *19*, 547–555.
Shinomiya, T., & Ina, S. (1991). Analysis of chromosomal replicons in early embryos of Drosophila melanogaster by two-dimensional gel electrophoresis. *Nucleic Acids Research*, *19*, 3935–3941.
Siddiqui, K., On, K. F., & Diffley, J. F. (2013). Regulating DNA replication in eukarya. *Cold Spring Harbor Perspectives in Biology*, *5*(9), 361–380.
Snow, M. H. L. (1977). Gastrulation in the mouse: Growth and regionalization of the epiblast. *Journal of Embryology and Experimental Morphology*, *42*, 293–303.
Sonneville, R., Querenet, M., Craig, A., Gartner, A., & Blow, J. J. (2012). The dynamics of replication licensing in live Caenorhabditis elegans embryos. *Journal of Cell Biology*, *196*, 233–246.
Stanojcic, S., Lemaitre, J. M., Brodolin, K., Danis, E., & Mechali, M. (2008). In Xenopus egg extracts, DNA replication initiates preferentially at or near asymmetric AT sequences. *Molecular and Cellular Biology*, *28*, 5265–5274.
Takada, S., Kwak, S., Koppetsch, B. S., & Theurkauf, W. E. (2007). grp (chk1) replication-checkpoint mutations and DNA damage trigger a Chk2-dependent block at the Drosophila midblastula transition. *Development*, *134*, 1737–1744.
Takeda, D. Y., Shibata, Y., Parvin, J. D., & Dutta, A. (2005). Recruitment of ORC or CDC6 to DNA is sufficient to create an artificial origin of replication in mammalian cells. *Genes and Development*, *19*, 2827–2836.
Tanaka, S., & Araki, H. (2013). Helicase activation and establishment of replication forks at chromosomal origins of replication. *Cold Spring Harbor Perspectives in Biology*, *5*, a010371.
Tower, J. (2004). Developmental gene amplification and origin regulation. *Annual Review of Genetics*, *38*, 273–304.
Trakala, M., Rodriguez-Acebes, S., Maroto, M., Symonds, C. E., Santamaria, D., Ortega, S., et al. (2015). Functional reprogramming of polyploidization in megakaryocytes. *Developmental Cell*, *32*, 155–167.
Ullah, Z., de Renty, C., & Depamphilis, M. L. (2011). Checkpoint kinase 1 prevents cell cycle exit linked to terminal cell differentiation. *Molecular and Cellular Biology*, *31*, 4129–4143.
Ullah, Z., Kohn, M. J., Yagi, R., Vassilev, L. T., & DePamphilis, M. L. (2008). Differentiation of trophoblast stem cells into giant cells is triggered by p57/Kip2 inhibition of CDK1 activity. *Genes and Development*, *22*, 3024–3036.
Ullah, Z., Lee, C. Y., & Depamphilis, M. L. (2009a). Cip/Kip cyclin-dependent protein kinase inhibitors and the road to polyploidy. *Cell Div*, *4*, 10.
Ullah, Z., Lee, C. Y., Lilly, M. A., & DePamphilis, M. L. (2009b). Developmentally programmed endoreduplication in animals. *Cell Cycle*, *8*, 1501–1509.

Unnikrishnan, A., Gafken, P. R., & Tsukiyama, T. (2010). Dynamic changes in histone acetylation regulate origins of DNA replication. *Nature Structural and Molecular Biology*, *17*, 430–437.

Vassilev, A., Kaneko, K. J., Shu, H., Zhao, Y., & DePamphilis, M. L. (2001). TEAD/TEF transcription factors utilize the activation domain of YAP65, a Src/Yes-associated protein localized in the cytoplasm. *Genes and Development*, *15*, 1229–1241.

Vaughn, J. P., Dijkwel, P. A., & Hamlin, J. L. (1990). Replication initiates in a broad zone in the amplified CHO dihydrofolate reductase domain. *Cell*, *61*, 1075–1087.

Whitmire, E., Khan, B., & Coue, M. (2002). Cdc6 synthesis regulates replication competence in Xenopus oocytes. *Nature*, *419*, 722–725.

Wiekowski, M., Miranda, M., & DePamphilis, M. L. (1991). Regulation of gene expression in preimplantation mouse embryos: Effects of the zygotic clock and the first mitosis on promoter and enhancer activities. *Developmental Biology*, *147*, 403–414.

Wiekowski, M., Miranda, M., Nothias, J. Y., & DePamphilis, M. L. (1997). Changes in histone synthesis and modification at the beginning of mouse development correlate with the establishment of chromatin mediated repression of transcription. *Journal of Cell Science*, *110*(Pt. 10), 1147–1158.

Yagi, R., Kohn, M. J., Karavanova, I., Kaneko, K. J., Vullhorst, D., DePamphilis, M. L., et al. (2007). Transcription factor TEAD4 specifies the trophectoderm lineage at the beginning of mammalian development. *Development*, *134*, 3827–3836.

Yang, V. S., Carter, S. A., Ng, Y., Hyland, S. J., Tachibana-Konwalski, K., Fisher, R. A., et al. (2012). Distinct activities of the anaphase-promoting complex/cyclosome (APC/C) in mouse embryonic cells. *Cell Cycle*, *11*, 846–855.

Zhu, W., & Depamphilis, M. L. (2009). Selective killing of cancer cells by suppression of geminin activity. *Cancer Research*, *69*, 4870–4877.

Zielke, N., Edgar, B. A., & DePamphilis, M. L. (2013). Endoreplication. *Cold Spring Harbor Perspectives in Biology*, *5*, a012948.

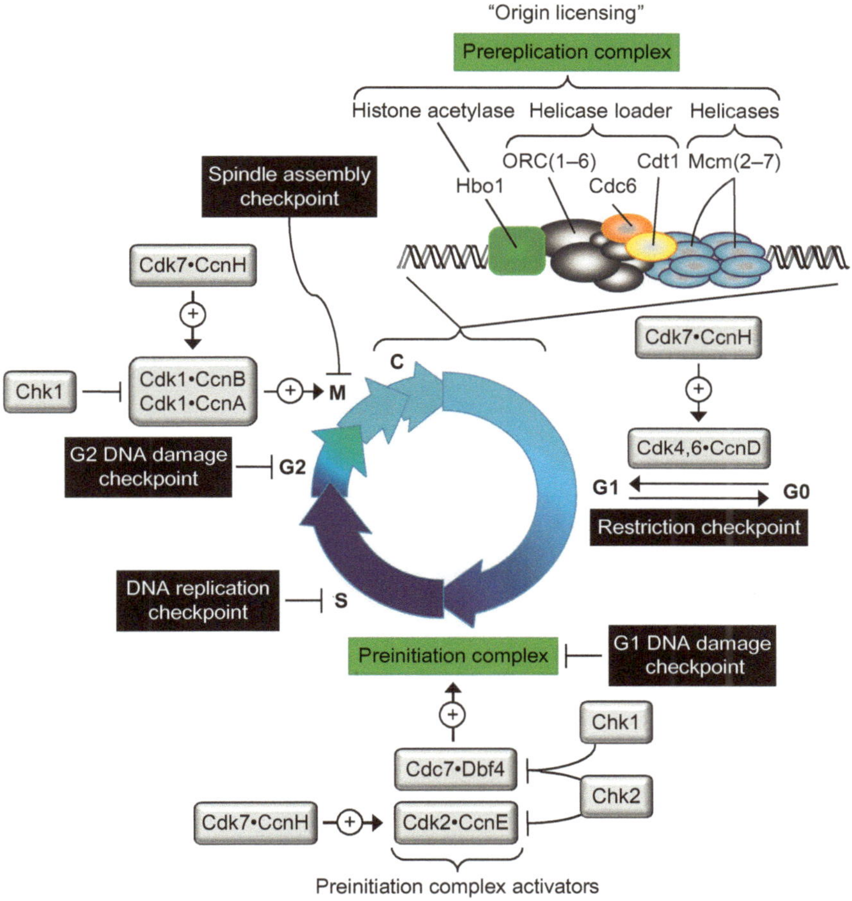

Melvin L. DePamphilis, Figure 1 Regulation of DNA replication during mitotic cell cycles. Nuclear DNA replication in the metazoa begins during the anaphase to G1 phase transition of the mitotic cell cycle with the assembly of prereplication complexes onto the DNA in chromatin (Baldinger & Gossen, 2009; Noguchi, Vassilev, Ghosh, Yates, & DePamphilis, 2006). In mammals, this event, termed "origin licensing," begins by assembly of the six subunit "origin recognition complex" [ORC(1–6)] through independent nuclear localization of Orc1, ORC(2–5), and Orc6 (Ghosh, Vassilev, Zhang, Zhao, & DePamphilis, 2011). The BAH domain in human Orc1 facilitates the ability of Orc1 to activate replication origins *in vivo* and triggers assembly of ORC(1–6) (Kara, Hossain, Prasanth, & Stillman, 2015; Noguchi et al., 2006). Mutations in the BAH domain of Orc1 link histone H4 methylation to origin licensing and to Meier–Gorlin syndrome, a genetic defect leading to primordial dwarfism (Kuo et al., 2012). Cdc6 stabilizes the binding of ORC(1–6) to DNA and allows recruitment of Cdt1 to form a "helicase loader." The six subunit "minichromosome maintenance complex" (Mcm2–7) is the eukaryotic replicative DNA helicase. Each replication fork requires one MCM helicase; thus two or

(Continued)

Melvin L. DePamphilis, Figure 1—Cont'd more are loaded at each origin of bidirectional replication. *In vivo*, the Hbo1/Kat7 histone acetylase activity is also essential origin licensing, because assembly of prereplication complexes is facilitated by histone acetylation (Miotto & Struhl, 2010; Unnikrishnan, Gafken, & Tsukiyama, 2010). However, prereplication complexes cannot unwind DNA; they must first be converted into preinitiation complexes by addition of 10 other protein (Bell & Kaguni, 2013; Tanaka & Araki, 2013). Only then can the MCM helicase unwind the double helix and allow DNA polymerases to synthesize the nascent DNA strand. DNA replication is regulated directly by eight protein kinases (Depamphilis, de Renty, Ullah, & Lee, 2012), five of which are cyclin-dependent kinases (Cdk) that require a cyclin (Ccn) partner. Cdk7·CcnH is essential to activate Cdk1, Cdk2, Cdk4, and Cdk6. Cdk4·CcnD and Cdk6·CcnD drive cells from a quiescent state (G0) into a proliferative state. Cdk1·CcnB1 and Cdk1·CcnA2 initiate and maintain mitosis. Cdk2·CcnE and the Dbf4-dependent kinase Cdc7 are the engines that drive the assembly and activation of preinitiation complexes. Cdk2·CcnA2 prevents DNA rereplication during S phase. Cdk1·CcnA2 prevents premature initiation during G2 phase. The remaining two protein kinases are components of the cell's DNA damage response. Five checkpoints ensure the accuracy of mitotic cell division (DePamphilis & Bell, 2010). In the absence of sufficient nutrients, growth factors or optimal temperature, cells in G1 phase enter a quiescent state (G0). This transition is regulated by the "restriction checkpoint," which consists of two signal detection pathways (Foster, Yellen, Xu, & Saqcena, 2010). One senses growth factors, and the other senses nutrients. The G1 to G0 transition is regulated by action of Cdk4·CcnD and Cdk6·CcnD. When cells are arrested in G0, their DNA replication machines are dismantled, and they eventually enter a senescent state from which they cannot reenter the mitotic cell cycle (Campisi, 2013). Conversely, once a cell passes the restriction checkpoint, it is committed to division; it must divide, or terminally differentiate into a polyploid cell, or die. Nuclear DNA damage is assessed before beginning S phase by the G1 DNA damage checkpoint, during S phase by the DNA replication checkpoint and before beginning M phase by the G2 DNA damage checkpoint, but in reality there are only two DNA damage response systems that basically detect excess single-stranded DNA complexed with replication protein-A that results from different problems. DNA damage and stalled replication forks are detected by the ATR kinase which then activates the Chk1 kinase to phosphorylate and inhibit Cdc7 and Cdc25 (a phosphatase required for activation of Cdk1 and Cdk2). Double-stranded DNA breaks are detected by the ATM kinase, which then activates Cdk2 kinase to phosphorylate Cdc7, Cdc25, and tumor suppressor Tp53. Many other proteins are also targeted. Finally, the spindle assembly checkpoint prevents the onset of anaphase until the kinetochores are attached correctly to the mitotic spindle (Lara-Gonzalez, Westhorpe, & Taylor, 2012).

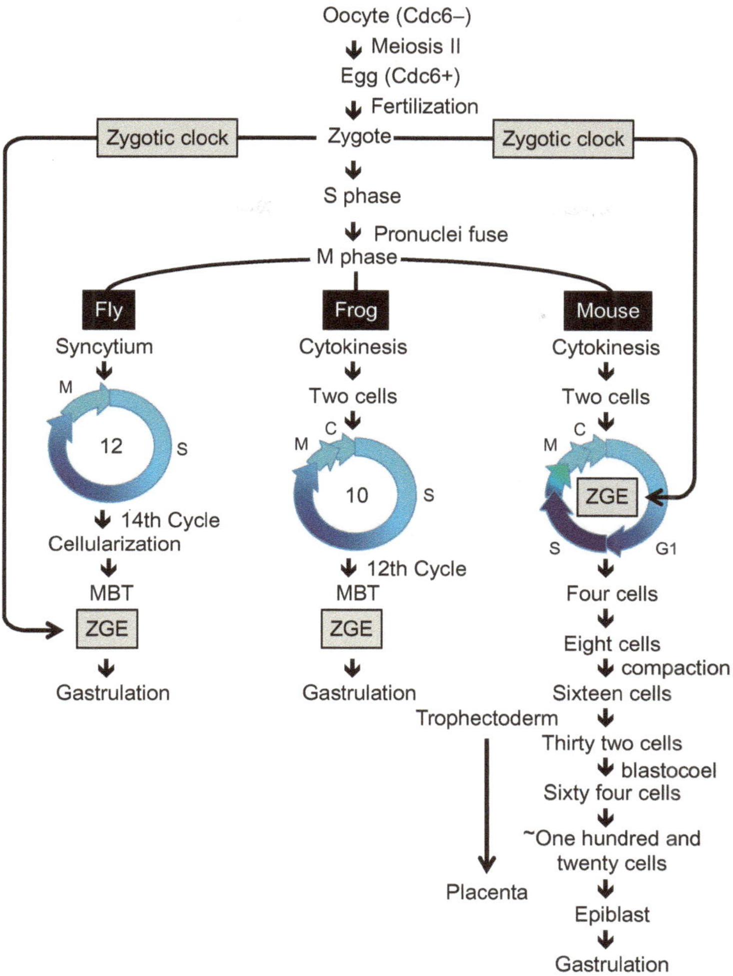

Melvin L. DePamphilis, Figure 2 Developmentally programmed changes in the mitotic cell cycle. The first mitotic cell division is universal, as exemplified in *Drosophila melanogaster* (fly), *Xenopus laevis* (frog), and *Mus musculus* (mouse). Oocyte maturation into eggs includes expression of the Cdc6 protein that is essential for initiation of nuclear DNA replication. The second mitotic cell division distinguishes flies from frogs and mammals from nonmammals. Flies and other insects continue to rapidly undergo cycles of DNA replication (S phase) followed by mitosis (M phase) without undergoing cell cleavage. Frogs, fish, and echinoderms continue to rapidly undergo cycles of DNA replication followed by mitosis followed by cell cleavage (cell division in the absence of cell growth). Mice and other mammals continue to undergo slow cycles of DNA replication followed by mitosis followed by cell cleavage. However, mammalian cell cleavage appears to include G phases. The most significant change is that zygotic gene expression (ZGE), the onset of transcription and the degradation of maternally inherited RNA, occurs in two-cell embryos of mice and the 4- to 16-cell embryos of rabbits, cows, pigs, and humans. Nevertheless, activation of the bulk of ZGE in both flies and mice is determined by the time elapsed after fertilization (a "zygotic clock"), rather than by DNA replication or mitosis. The differences in cell division cycles between mammals and nonmammals might well be related to the fact that mammals need to produce both a placenta and an embryo concurrently (O'Farrell, Stumpff, & Su, 2004).

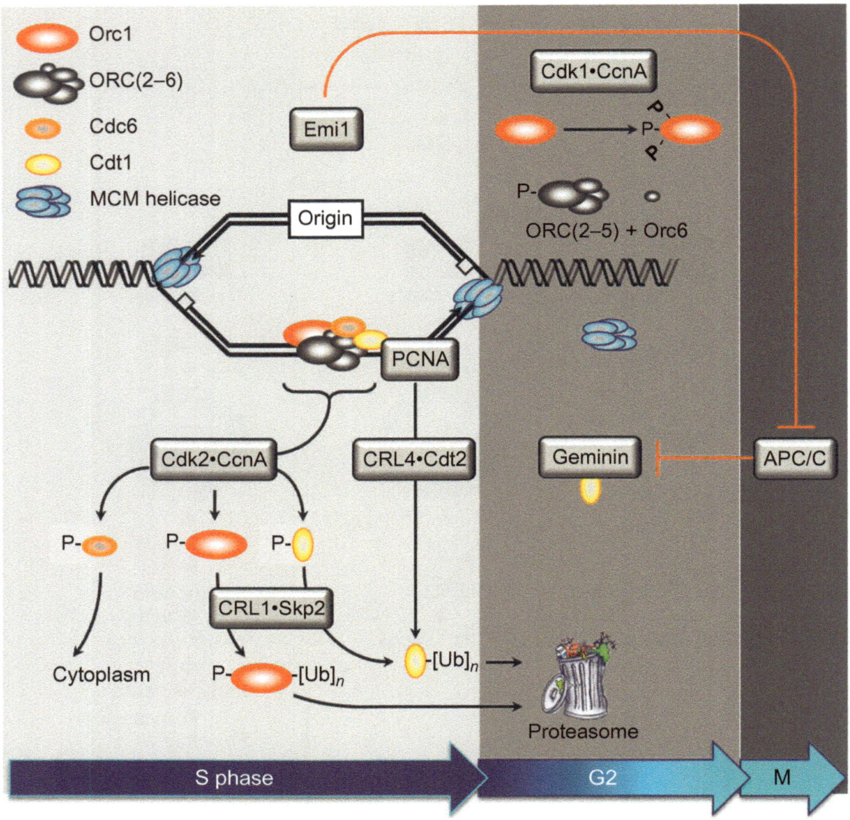

Melvin L. DePamphilis, Figure 3 Multiple concerted pathways prevent DNA rereplication. Once metazoan cells have entered S phase, seven concerted pathways can inactivate the helicase loader, thereby preventing rereplication of DNA prior to mitosis. Upon S phase entry, Cdk2·CcnA phosphorylates Orc1, Cdc6, and Cdt1, thereby inhibiting their ability to form helicase loaders. In addition, Cdc6-P is localized to the cytoplasm where it cannot facilitate origin licensing. Orc1-P is ubiquitinated and targeted for destruction, with subsequent dissociation of ORC(2–5) and Orc6 from chromosomes. Cdt1-P is ubiquitinated by the CRL1·Skp2 ubiquitin ligase. Cdt1 that is not phosphorylated is ubiquitinated by the CRL4·Cdt2 ubiquitin ligase bound to PCNA and DNA. PCNA is the eukaryotic sliding clamp protein that facilitates DNA synthesis by DNA polymerases δ and ε. The ubiquitinated proteins are then degraded by the 26S proteasome. Geminin is a protein that binds specifically to Cdt1 and inhibits its activity. These activities are available from S through early M phase. As cells exit mitosis, geminin and cyclin A are ubiquitinated by the anaphase promoting complex [originally termed cyclosome (APC/C)], an activity that is inhibited during S to early M phase by the APC/C-specific inhibitor Emi1. Reassembly of the helicase loader occurs by reassembly of Orc1, ORC(2–5), Orc6, and Cdc6 onto chromatin (Ghosh et al., 2011; Kara et al., 2015; Sonneville et al., 2012), an event that appears to be triggered by the Orc1 subunit (Noguchi et al., 2006). Geminin binding to Cdt1 and CDK-dependent phosphorylation of Cdc6 prevents targeting of Cdt1 and Cdc6 for degradation by the APC/C, thereby allowing them to participate in origin licensing (Ballabeni et al., 2004; Mailand & Diffley, 2005).

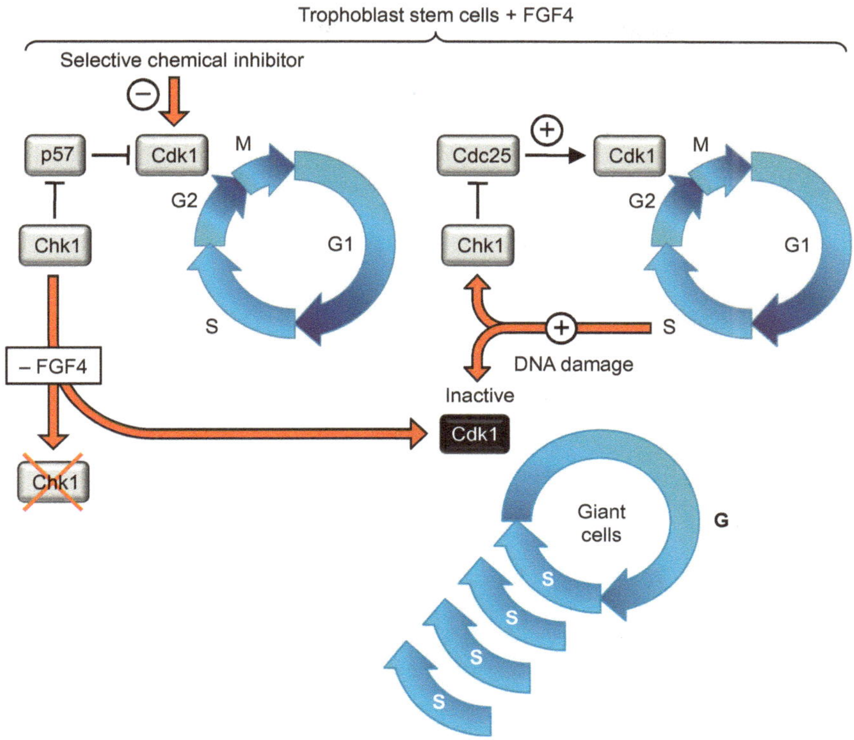

Melvin L. DePamphilis, Figure 4 Endoreplication is triggered in trophoblast cells by selective inhibition of Cdk1 activity. Developmentally programmed endoreplication in the metazoa occurs when Cdk1 activity is suppressed so that cells cannot enter mitosis (Edgar, Zielke, & Gutierrez, 2014; Fox & Duronio, 2013; Zielke, Edgar, & DePamphilis, 2013). This results in multiple S phases without an intervening mitosis or cytokinesis to produce giant cells with a single enlarged nucleus containing as many as several hundred copies of each chromosome. In the absence of FGF4, trophoblast cells in mouse blastocysts are triggered to differentiate into giant cells. In the presence of FGF4, the same result can be elicited either by selective inhibition of Cdk1 with RO3306, or by suppressing or inactivating geminin (de Renty, Kaneko, & DePamphilis, 2014; Ullah, Kohn, Yagi, Vassilev, & DePamphilis, 2008). The effect of FGF4 deprivation is to reduce the level of geminin, which results in the loss of checkpoint kinase-1 (Chek1/Chk1). Since Chk1 phosphorylation of the CDK-specific inhibitor Cdkn1c/p57 targets it for ubiquitin-dependent degradation, loss of Chk1 results in p57 inhibition of Cdk1 activity, which leads to endoreplication (Ullah, de Renty, & Depamphilis, 2011). Ironically, DNA damage in trophoblast cells also results in endoreplication, because DNA damage activates Chk1, which then phosphorylates and inactivates Cdc25, a phosphatase that is essential to maintain Cdk1 in its active form.

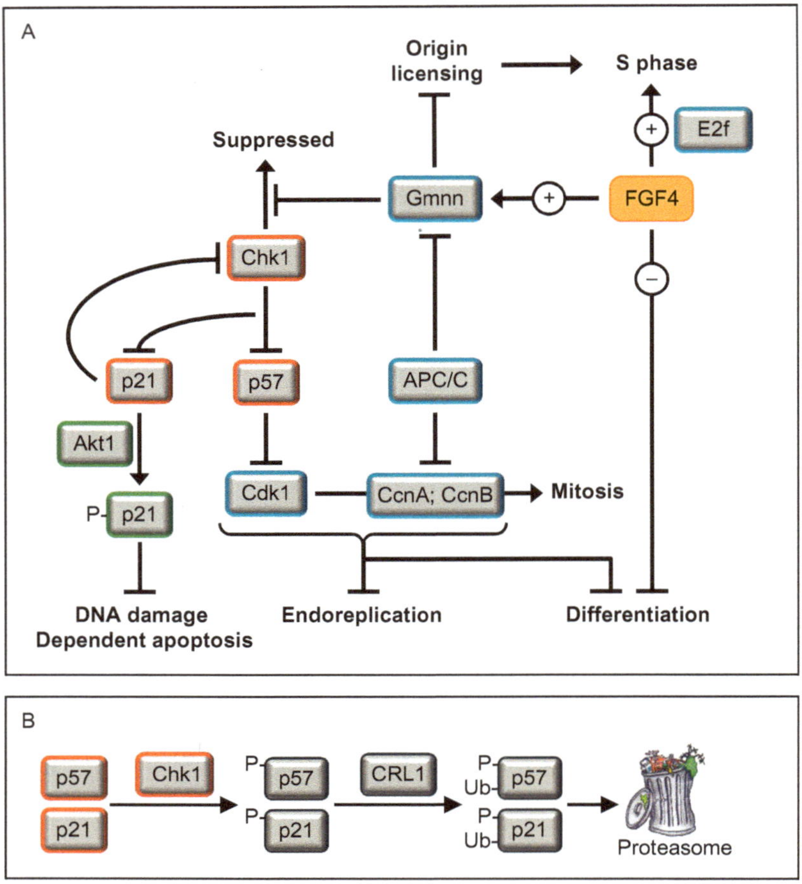

Melvin L. DePamphilis, Figure 5 The FGF4 signal transduction pathway governs trophoblast proliferation and differentiation. FGF4 (and probably other mitogenic proteins as well) is essential for trophoblast proliferation. This mitogenic activity is likely mediated by E2F-dependent gene expression (Chen et al., 2010), and possibly directed at regulating the activity of the anaphase-promoting complex (APC) (Yang et al., 2012). FGF4 deprivation results in downregulation of geminin activity to a level that maintains endocycles (de Renty, Kaneko, & DePamphilis, 2014), but that does not prevent downregulation of Chk1 protein. The loss of Chk1 kinase activity results in expression of two CDK-specific inhibitors, p57 and p21 (Ullah et al., 2011). The p57 protein prevents the onset of mitosis by selectively inhibiting Cdk1 activity, thereby triggering the first round of endoreplication (Hattori, Davies, Anson-Cartwright, & Cross, 2000; Ullah et al., 2008). This event activates the G1-phase APC·Cdh1 ubiquitin ligase, which targets geminin, cyclin B, and cyclin A for degradation, thereby allowing licensing of replication origins and the onset of S-phase without passing through mitosis (Ullah, Lee, Lilly, & DePamphilis, 2009). Inhibition of Cdk1 triggers both endoreplication and TSC differentiation. In the absence of p57, FGF4 deprivation produces multinucleated TGCs, revealing the existence of alternative mechanisms to trigger TSC differentiation (Ullah et al., 2008). Endocycles also require p57, which is expressed during G-phase and then suppressed during S-phase to allow sequential assembly and activation of prereplication complexes (Ullah et al., 2008). Geminin maintains endocycles by preventing DNA rereplication. The p21/Cdkn1a protein localizes to the cytoplasm in TGCs where it prevents DNA damage induced apoptosis (de Renty, DePamphilis, & Ullah, 2014). It might also maintain suppression of Chk1 by reducing *Chk1* RNA levels (Gottifredi, Karni-Schmidt, Shieh, & Prives, 2001), as observed during FGF4 deprivation (Fig. 3).

CHAPTER THIRTEEN

The Chordin Morphogenetic Pathway

Edward M. De Robertis[1], Yuki Moriyama
Howard Hughes Medical Institute and Department of Biological Chemistry, University of California, Los Angeles, California, USA
[1]Corresponding author: e-mail address: ederobertis@mednet.ucla.edu

Contents

Abstract

The ancestral Chordin/bone morphogenetic protein (BMP) signaling pathway that establishes dorsal–ventral (D–V) patterning in animal development is one of the best understood morphogenetic gradients, and is established by multiple proteins that interact with each other in the extracellular space—including several BMPs, Chordin, Tolloid, Ont-1, Crossveinless-2, and Sizzled. The D–V gradient is adjusted redundantly by regulating the synthesis of its components, by direct protein–protein interactions between morphogens, and by long-range diffusion. The entire embryo participates in maintaining the D–V BMP gradient, so that for each action in the dorsal side there is a reaction in the ventral side. A gradient of Chordin is formed in the extracellular matrix that separates ectoderm from endomesoderm, called Brachet's cleft in *Xenopus*. The Chordin/BMP pathway is self-organizing and able to scale pattern in the dorsal half of bisected embryos or in Spemann dorsal lip transplantation experiments.

1. INTRODUCTION

A central question in developmental biology is how a perfectly patterned embryo consisting of many differentiated tissues is reliably generated time after time. A key experiment for understanding embryonic patterning

Current Topics in Developmental Biology, Volume 116
ISSN 0070-2153
http://dx.doi.org/10.1016/bs.ctdb.2015.10.003

was performed by Spemann's graduate student Hilde Mangold, who transplanted a small region of the amphibian gastrula called the dorsal lip of the blastopore into the ventral side of a host embryo (Spemann & Mangold, 1924). The blastopore is the region through which the cells that will form the endoderm and mesoderm layers of the embryo invaginate. The cells that remain on the outside form the third germ layer, the ectoderm. A small graft of this organizer region has a very potent effect on the recipient embryo, dividing the pattern of the whole into two Siamese twins. This experiment established that the dorsal side of the embryo had inductive abilities and that organizer tissue played a crucial role in self-regulation. The mechanism of this phenomenon would have to wait for many decades until gene cloning made molecular embryology a practical possibility. This chapter relates how a biochemical pathway that mediates embryonic induction was elucidated through the testing of a series of successive hypotheses suggested by experiments.

2. A GRADIENT OF GROWTH FACTOR SIGNALING

D–V patterning results from a gradient of activity of a family of secreted growth factors called bone morphogenetic proteins (BMPs). The first correlation between D–V tissue differentiation and BMPs came from the study of *Drosophila* mutations in a gene called decapentaplegic (*Dpp*), which is the homolog of vertebrate BMP2/4 (Ferguson & Anderson, 1992). In mammals, BMPs had a long history of being involved in bone differentiation. It was first noted that bone fragments transplanted subcutaneously or intramuscularly in rabbits could induce bone differentiation even after all cells had been killed with ethanol (Levander, 1938). Marshall Urist, an orthopedic surgeon working at UCLA, found that the proteinaceous bone extracellular matrix (ECM) (obtained by removing calcium from bones by soaking them in hydrochloric acid for several days), had potent ectopic bone morphogenetic activity after transplantation into rabbits or rats (Urist, 1965).The active proteins were purified and cloned by Genetics Institute and found to correspond to growth factors designated BMP2 to BMP7 (Wozney et al., 1988).

In zebrafish and *Xenopus* embryos, a gradient of BMP signaling activity spanning the entire embryo can be visualized indirectly, by staining embryos with an antibody specifically directed against the phosphorylated form of the Smad1 transcription factor. As shown in Fig. 1A and B, BMP signaling

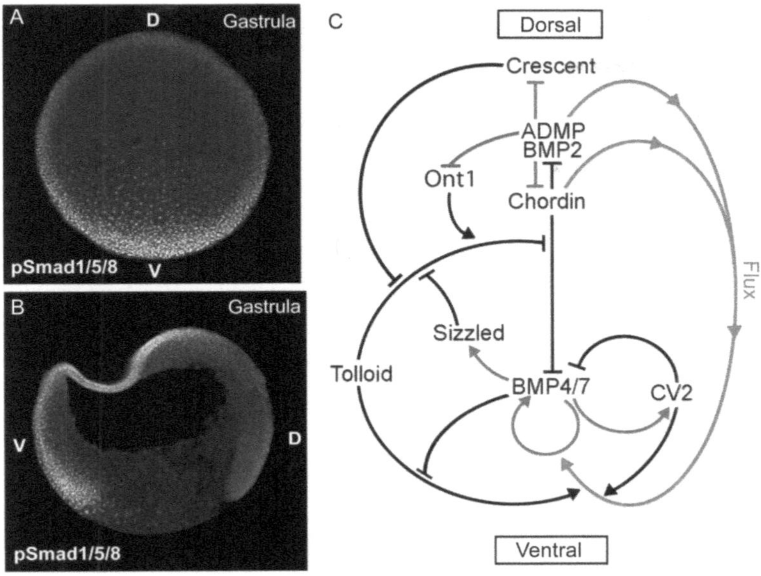

Figure 1 A gradient of nuclear phospho-Smad1 is established by a biochemical network of secreted proteins that interact with each other. (A) Horizontal optical section at gastrula (stage 11) showing a ventral to dorsal gradient of BMP activity using pSmad1/5/8 antibody as the readout. D, dorsal; V, ventral. (B) Gastrula (stage 11) embryo sectioned sagittally, showing higher BMP signaling in ventral nuclei. (C) The D–V morphogenetic signaling pathway results from a series of direct protein–protein interactions between Chordin and other partners (black arrows), transcriptional regulation (blue arrows), and protein flux (red arrows). The entire embryo participates in forming the BMP gradient, which results from the dueling activities of the dorsal and ventral signaling centers. *Images from Plouhinec, Zakin, Moriyama, and De Robertis (2013).* (A color version of this figure is located at the end of this chapter.)

activity is lowest on the dorsal side and gradually increases toward the ventral side, in which cell nuclei accumulate higher amounts of phospho-Smad1.

3. THE CLONING OF CHORDIN

When molecular biology, the great equalizer of modern biology, became practical, the search for the signals produced by the dorsal organizer tissue that mediate embryonic induction was on. In our laboratory, a gene library from manually dissected dorsal blastopore lips from *Xenopus* was prepared. We succeeded in isolating a homeobox gene designated *goosecoid* (Cho, Blumberg, Steinbeisser, & De Robertis, 1991). This gene allowed us, for the first time, to visualize Spemann's organizer using *in situ* hybridizations to *goosecoid* mRNA. Previously, the existence of organizer tissue had to be deduced from its effects after transplantation experiments. Since Goosecoid is a transcription factor, we hypothesized that it might regulate a secreted target gene. In 1994, we isolated *chordin*, a gene activated by *goosecoid*, from our *Xenopus* organizer library (Sasai et al., 1994). Richard Harland had previously isolated *noggin* (Smith & Harland, 1992) and Douglas Melton *follistatin* (Hemmati-Brivanlou, Kelly, & Melton, 1994). We later found that all three gene products acted by antagonizing BMP signaling (Sasai, Lu, Steinbeisser, & De Robertis, 1995). The dorsal side of the gastrula embryo secretes BMP antagonists and the ventral side BMP4 and BMP7. Spemann's organizer proved a fertile fishing ground for novel molecules. Our initial hypothesis had been that dorsal tissue would be the source of novel growth factors but what was found instead was that it secreted a large number of secreted growth factor inhibitors (reviewed in De Robertis & Kuroda, 2004).

Intriguingly, BMP signaling levels are able to regulate cell differentiation in the ectoderm, mesoderm, and endoderm layers simultaneously. Explants of future ectoderm (cells from the animal cap of the embryo) differentiate into CNS at low BMP levels (e.g., in the presence of BMP antagonists such as Chordin, Noggin, or Follistatin) and into epidermis when BMP signaling levels are high. Similarly, in the mesoderm, low BMP gives rise to notochord, at slightly higher levels skeletal muscle (arranged in somites), then kidney (each embryonic segment develops a kidney tubule), then lateral plate (which gives rise to the body wall), and at the highest BMP levels blood islands. These histotypic differentiations represent the invariant body plan shared by all vertebrates. This raises the question of how the D–V development of the different germ layers is integrated. Is there a dedicated BMP gradient for each germ layer? How is each gradient regulated coordinately so that a perfectly harmonious embryo is formed each time? How is the BMP gradient regulated to ensure that each embryo is able to allocate these tissues perfectly?

While we isolated several growth factor antagonists secreted by Spemann's organizer cells, Chordin proved to be the most informative for the regulation of the D–V signaling gradient; it is at the heart of the organizer phenomenon. When Chordin mRNA is microinjected into a ventral cell, it recapitulates the organizer experiment, forming a second neural tube, somites, and even a second gut cavity (Sasai et al., 1994). Chordin can be depleted in *Xenopus* embryos using microinjected antisense morpholino oligos (MOs). Depletion of Chordin produced embryos that developed with smaller heads and dorsal tissues and expanded ventral structures (Oelgeschläger, Kuroda, Reversade, & Robertis, 2003). However, an embryonic axis was still formed. Later, Harland found that the combined depletion of Chordin, Noggin, and Follistatin led to a catastrophic loss of all dorsal tissues (Khokha, Yeh, Grammer, & Harland, 2005). Similarly, in the mouse double knock-out of Chordin and Noggin, the forebrain, midbrain, and notochord are lost (Bachiller et al., 2000). Thus, the BMP antagonists secreted by organizer tissue are able to compensate for each other.

Chordin is essential for the activity of Spemann organizer grafts in *Xenopus*. When the organizer was depleted of Chordin, it remained as a patch of epidermis completely devoid of inducing activity (Oelgeschläger et al., 2003). Transplantation is a very powerful tool in biology. In this case, when dorsal cells were challenged by placing them in a new ventral surrounding their diminished biological capacity due to the loss of the Chordin gene product was strikingly revealed.

We realized very early on that Chordin had to be a very important molecule. A gene of similar sequence had been cloned in the fruit fly *Drosophila* called *short gastrulation* (*sog*) (Francois, Solloway, O'Neill, Emery, & Bier, 1994). We collaborated with F. Michael Hoffman and Edwin "Chip" Ferguson, who had independently cloned *sog*, to show that microinjected *chordin* and *sog* mRNA induced neural tissue both in *Drosophila* and frog embryos (Holley et al., 1995). This led to the realization that we had discovered an ancient D–V patterning molecule that had been conserved during evolution between fruit fly and amphibian embryos (reviewed in Bier & De Robertis, 2015; De Robertis, 2008). Importantly, the phenotypic effects of *sog* mutations were known to be enhanced when the dosage of *dpp*/BMP4 genes was increased in *Drosophila* (Ferguson & Anderson, 1992). This led us to investigate the fruitful working hypothesis that Chordin and BMPs worked on a common signaling pathway. Biochemical experiments showed that Chordin encodes a large protein containing four Cysteine-rich domains that serve as BMP-binding modules (Piccolo, Sasai, Lu, & De Robertis, 1996). Chordin has a cofactor called Twisted gastrulation (Tsg) that helps keep BMPs soluble and binds to both Chordin and BMPs (Little & Mullins, 2004; Oelgeschläger, Larraín, Geissert, & De Robertis, 2000). The ternary complex of Chordin, BMPs, and Tsg prevents binding of BMPs to BMP receptors on the cell surface, explaining why Chordin inhibits BMP signaling.

Chordin mRNA is expressed at high levels in dorsal cells in the exact region that has embryonic inducing activity after transplantation. We measured the amount of Chordin protein secreted by the frog embryo during gastrulation and found that it is produced in prodigious amounts. If distributed uniformly in the extracellular space, Chordin protein would reach concentrations of 33 nM (Lee, Ambrosio, Reversade, & De Robertis, 2006). In the dorsal side, where it is produced, Chordin must reach much higher concentrations. BMP concentrations have not been measured in embryos, but in other tissues, they are active in the picomolar (pM) range. Thus, a vast excess of Chordin is present during early development.

Since Chordin is a BMP antagonist, in principle, its localized expression in the dorsal side could be sufficient to account for the BMP signaling gradient, even if BMP expression were uniform. Using systematic MO knockdown and transplantation experiments, we discovered that Chordin is part of a biochemical network of extracellular proteins that involves the entire embryo (Fig. 1C). It eventually became clear that the organizer effect is not due only to the action of the dorsal side but also to the reaction of ventral cells (Reversade & De Robertis, 2005). The embryo has a dorsal and a ventral center that communicate through secreted proteins that interact with each other. Dorsal genes are expressed when BMP levels are low, while ventral genes are expressed when BMP levels are high (Fig. 1C). The two signaling centers self-regulate after changes in signaling because all the components of the system are under opposing transcriptional control by BMP.

4. FUNCTION OF TOLLOID, SIZZLED, AND CROSSVEINLESS 2

Purified Chordin binds BMPs avidly, with an affinity (dissociation constant, K_D) in the low nM range (Piccolo et al., 1996). However, the inhibitory action of Chordin can be reversed by proteolysis of Chordin by metalloproteinases of the Tolloid family. Tolloid was identified in the classical *Drosophila* genetic screen (Nüsslein-Volhard & Wieschaus, 1980) as a gene that increased Dpp/BMP signaling. We hypothesized that Tolloid might act by inactivating the BMP antagonist Chordin/Sog and were able to show by biochemical experiments that the metalloproteinase Xolloid-related (Xlr) cleaved Chordin at two distinct sites (Piccolo et al., 1997). When this happened, the affinity of the cleaved Chordin for BMP decreased precipitously and the complex of BMP and Tsg was able to bind to BMP receptors, restoring signaling in *Xenopus* explants. Because Tolloid/Xlr is expressed in the ventral side, it serves as a ventral sink that degrades Chordin originating from Spemann's organizer, allowing the flux of BMPs from more dorsal regions to the ventral side in which BMP signaling is maximal (Fig. 1C). Activity of the Tolloid protease is the rate-limiting step in D–V patterning and is subjected to stringent regulation.

Sizzled is a ventral center molecule that is expressed when BMP signaling levels are high (Collavin & Kirschner, 2003). In zebrafish, the *sizzled* mutation (called *ogon/mercedes*) had a phenotype intriguingly similar to that of Chordin mutants (Yabe et al., 2003). We microinjected *sizzled* mRNA into dorsal or ventral half-embryos in *Xenopus* and noted that it had strong dorsalizing effects on dorsal fragments (expansion of the CNS) but none at all in ventral fragments. Thus, *sizzled* function required a dorsal component, giving rise to the working hypothesis that it might inhibit the degradation of Chordin by Tolloid (Lee et al., 2006).

Sizzled encodes a secreted Frizzled-related protein (sFRP). This class of protein is normally involved in inhibiting Wnt signaling. However, in the case of Sizzled, the protein has evolved so that it is able to bind to the Tolloid proteolytic site, but is unable to be cleaved by this enzyme. In this way, Sizzled acts as a competitive inhibitor of Tolloid (Lee et al., 2006). Sizzled depletion by antisense morpholinos results in the same phenotype as Chordin loss-of-function, because in its absence the activity of Tolloid increases and Chordin is degraded. When overexpressed, Sizzled has anti-BMP effects because it inhibits the Tolloid proteinase, leading to the accumulation of Chordin and inhibition of BMP signaling. On the dorsal side of the embryo, another sFRP called Crescent also serves as a Tolloid inhibitor, but under the opposite transcriptional regulation (Ploper, Lee, & De Robertis, 2011; Fig. 1C).

Tolloid activity is also regulated by direct binding of BMP to protein domains located outside of its catalytic region. When BMP levels become high, BMPs bind to domains in Tolloid called CUB domains, inhibiting enzyme activity in a noncompetitive fashion (Lee, Mendes, Plouhinec, & De Robertis, 2009). This new negative feedback loop provided a molecular explanation for an old mystery in the field. Surprisingly, when the first peptide sequences were obtained from extracts with bone-inducing activity the first protein identified (Wozney et al., 1988), designated BMP1, had the sequence of a Tolloid enzyme containing three CUB domains. The reason for the copurification of BMP1 together with the BMP2 to 7 growth factors now had a simple explanation: Tolloids are BMP-binding proteins. This work showed that Tolloid activity is highly regulated and plays a key role in the communication between the dorsal and ventral sides of the embryo.

Ont-1, discovered in the Sasai laboratory, is a secreted protein member of the olfactomedin family that is expressed dorsally where BMP signaling levels are low (Inomata, Haraguchi, & Sasai, 2008). It serves as an adaptor that binds to Tolloid (via its coil–coil domain) and to Chordin (through its olfactomedin domain), facilitating Chordin proteolysis (Fig. 1C).

Another component of the Chordin/BMP pathway is Crossveinless 2 (CV2) which was first described in the *Drosophila* wing (Conley et al., 2000), where it helps reach the maximal BMP signaling required for formation of cross vein structures (Blair, 2007). CV2 has Chordin-like BMP-binding modules. CV2 expression is activated by high BMP signaling in the ventral side of *Xenopus* and is therefore a ventral center protein. When both Chordin and CV2 are depleted, ventral tissues are synergistically expanded (Ambrosio et al., 2008). Although CV2 is a secreted protein, it is unable to diffuse through the extracellular space because it remains anchored by glycosylated proteins (called glypicans) to the surface of the cells that secrete it (Serpe et al., 2008). Using biochemistry with purified proteins and embryo microinjection experiments, we found the mechanism of the CV2 pro-BMP activity. CV2 binds with considerable affinity (dissociation constant 1.4 nM) to Chordin and with even higher affinity to Chordin/BMP complexes (Ambrosio et al., 2008). Thus, in the frog embryo CV2 acts both as a BMP antagonist and as a molecular sink concentrating Chordin/BMP complexes in the ventral side (Fig. 1C). BMP signaling is boosted on the ventral side by the combined action of CV2 and Tolloid, which facilitate the flux of BMPs produced in more dorsal regions of the embryo and transported by Chordin (Ambrosio et al., 2008).

Extensive genetic screens in the zebrafish *Danio rerio* by the group of Christiane Nüsslein-Volhard provided strong support for the biochemical D–V pathway discovered in *Xenopus*. In a striking convergence, zebrafish loss-of-function mutations that specifically affected the allocation of dorsal–ventral tissues all involved genes in the Chordin/BMP pathway. Loss-of-function mutations that increased ventral tissues corresponded to *chordin* (called *chordino* in zebrafish) and *sizzled* (*ogon/mercedes*), while mutations increasing dorsal tissues corresponded to *bmp7* (*snailhouse*), *bmp2b* (*swirl*), the type I BMP receptor *alk8* (*lost-a-fin*), *smad5* (*somitabun*), and *tolloid* (*mini fin*) (Langdon & Mullins, 2011; Little & Mullins, 2006). These genetic experiments demonstrated that a common D–V patterning mechanism was utilized by vertebrate embryos.

Starting with the isolation of Chordin, we were able to identify, one at the time, multiple components that interact with each other in a morphogenetic biochemical pathway that comprises the entire embryo. At the gastrula stage being studied, the embryo is composed of 10,000 cells, raising the question of how the information in the dorsal and ventral centers is transmitted long-range to produce a self-organizing morphogen gradient in different germ layers. We discuss this next.

5. LONG-RANGE DIFFUSION OF CHORDIN IN BRACHET'S CLEFT

We have recently been able to visualize endogenous Chordin protein by immunohistochemistry in the *Xenopus* gastrula (Plouhinec et al., 2013). *Chordin* mRNA is transcribed in an arc of 60° of the dorsal side. However, as shown in Fig. 2A, Chordin protein was present as a gradient extending much further than its mRNA expression domain, spanning from the dorsal side to the ventral-most part of the stage 12 gastrula. The staining was specific because it was eliminated by depletion of Chordin. Unexpectedly, Chordin did not diffuse randomly through the intercellular space but instead was greatly concentrated specifically in the extracellular space that separates the ectoderm from the endomesoderm. In amphibian embryos, this virtual cavity is called Brachet's cleft (in honor of Albert Brachet, the famous

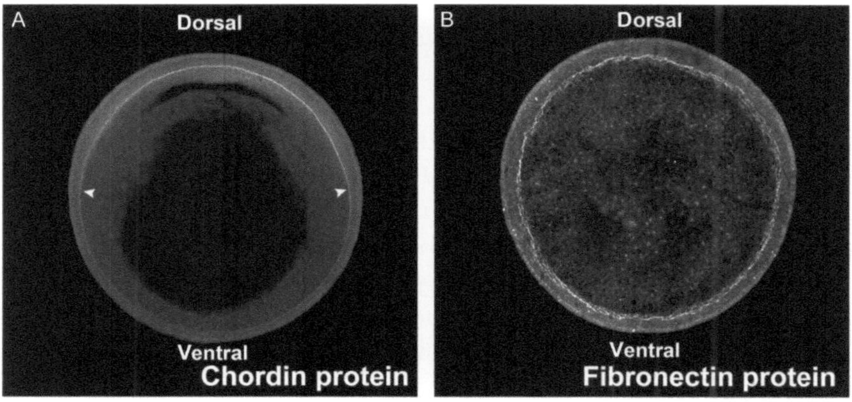

Figure 2 A gradient of endogenous Chordin protein in the extracellular matrix of Brachet's cleft. (A) Visualization of the dorsal to ventral gradient of endogenous Chordin protein in Brachet's cleft (arrowheads). The *Xenopus* embryo has a diameter of 1.3 mm and therefore the half-circumference of Brachet's cleft extends over a range of 2 mm. (B) Fibronectin is abundant in the Brachet's cleft ECM, but is uniformly distributed. *Panel A from Plouhinec et al. (2013). Panel B courtesy of Lise Zakin.*

Belgian embryologist). Brachet's cleft is not an amphibian-specific structure. All vertebrate embryos have an ECM-containing Fibronectin and other proteins separating ectoderm and endomesoderm. Chordin protein forms a D–V gradient in this signaling highway (Fig. 2A, arrowheads), while Fibronectin is uniformly distributed in this ECM (Fig. 2B).

It appears that the embryo may have a single gradient of Chordin that can pattern all three germ layers at the gastrula stage. This provides a simple solution to the question of how many BMP gradients exist: only one. From Brachet's cleft, Chordin/BMP can diffuse into both the ectoderm and the endomesoderm. As mentioned earlier, Chordin reaches high concentrations in the embryo. The fact that it diffuses through a narrow region suggests that Chordin, and perhaps the other components of the D–V patterning pathway, reach very high concentrations in the ECM between ectoderm and endomesoderm. During gastrulation, the germ layers undergo extensive morphogenetic movements; it is possible that cells read the Chordin/BMP gradient contained in this ECM as they move along its surface. Within this narrow cleft, Chordin behaves *in vivo* as predicted by our biochemical pathway in experiments using Tolloid and Sizzled antisense MOs (Plouhinec et al., 2013). We propose that the Chordin gradient is established and maintained by facilitated diffusion in the ECM and driven by the Tolloid protease. This study allowed the visualization, for the first time, of an endogenous gradient of a morphogen protein in a vertebrate embryo.

To demonstrate directly the long-range diffusion of the endogenous Chordin protein in the gastrula embryo, we performed Spemann transplants of lineage-traced dorsal lip tissue (Fig. 3A), which revealed a second gradient of Chordin protein originating from the graft and diffusing for considerable distances along Brachet's cleft (Fig. 3B, arrowheads). In an important precedent, it had been shown that overexpressed Nodal protein diffused along the ECM that separates the lateral plate mesoderm from ectoderm during left–right patterning in tailbud *Xenopus* embryos (Marjoram & Wright, 2011). It is possible that diffusion of morphogens along ECM separating cell layers may constitute a general phenomenon in long-range growth factor signaling.

The most intriguing property of the D–V patterning system is its ability to self-organize. After bisection in the D–V plane, the Chordin gradient in Brachet's cleft regenerated in dorsal halves, but not in the ventral ones (Plouhinec et al., 2013). We are currently investigating the regeneration of morphogenesis in half-embryos after sagittal bisections.

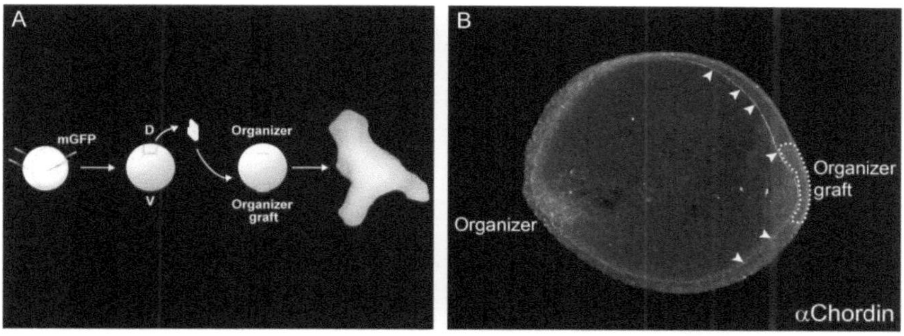

Figure 3 A long-range gradient of Chordin protein emanating from transplanted organizer tissue. (A) Diagram of the experimental procedure; membrane-targeted *mGFP* was injected in the donor embryo (four injections at the four-cell stage) to trace the lineage of grafted Spemann organizer tissue. (B) Embryo transplanted at stage 10 and fixed at stage 12 stained with anti-Chordin antibody shown in optical transverse section. A second gradient of Chordin was observed with Chordin protein staining in Brachet's cleft extending a considerable distance from the graft (arrowheads). The GFP fluorescence of the transplanted organizer cells is outlined by the dotted line. Chordin protein diffuses long-distance from the graft through Brachet's cleft ECM. *Data from Plouhinec et al. (2013).* (A color version of this figure is located at the end of this chapter.)

6. THE ANCESTRAL CHORDIN/BMP MORPHOGENETIC PATHWAY

The Chordin/BMP pathway is ancestral and conserved in bilateral animals as diverse as *Drosophila*, spiders, amphioxus, hemichordates, sea urchins, and vertebrates (Bier & De Robertis, 2015). We proposed that the remarkable conservation of the Hox genes (Carrasco, McGinnis, Gehring, & De Robertis, 1984) and of the BMP/Dpp, and Chordin/Sog systems between *Drosophila* and *Xenopus* suggested that these patterning systems were present in *Urbilateria*, the common ancestor of all bilateral animals (De Robertis & Sasai, 1996). Recently, a remarkable study from the laboratory of Ulrich Technau has revealed that the Chordin morphogenetic network of genes is more ancient than previously thought, as it is present, although in modified form, in cnidarians (Fig. 4).

The sea anemone *Nematostella vectensis* is a diploblast with radial symmetry. During embryogenesis, however, it has traces of bilateral symmetry and a so-called directive axis in which nuclear pSmad1 forms a gradient of BMP signaling. *Chordin* is expressed where BMP signaling is low and coincides with the expression of *Nematostella Dpp* and *BMP5-8* (Fig. 4;

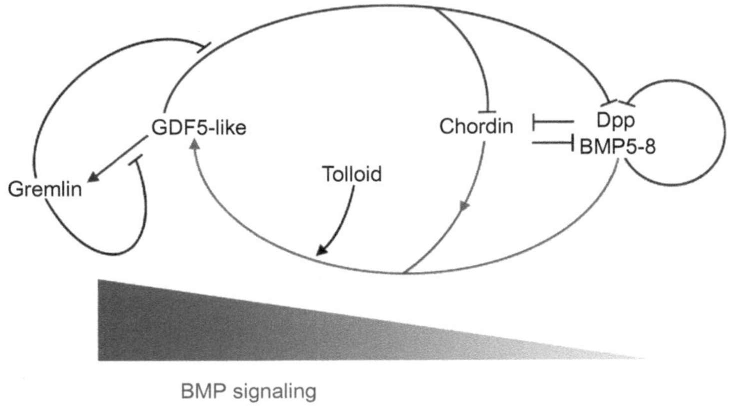

Figure 4 The directive axis of the cnidarian *Nematostella vectensis* is established by a Chordin/BMP gradient. As in *Xenopus*, BMP activity along the body axis results from a series of direct protein–protein interactions between Chordin and other protein partners (black arrows), transcriptional regulation (blue arrows), and protein flux (red arrows) in *Nematostella. Diagram based on Genikhovich et al. (2015).* (A color version of this figure is located at the end of this chapter.)

Genikhovich et al., 2015). Similarly, in *Xenopus*, the dorsal side expresses *Chordin*, *BMP2*, and *ADMP* (ADMP stands for the divergent BMP anti-dorsalizing morphogenetic protein; Moos, Wang, & Krinks, 1995). On the high BMP signaling side, the BMP *GDF5-like* (*growth and differentiation factor* 5) gene is expressed in *Nematostella*, while in *Xenopus* the ventral center expresses BMP4/7. One important difference is that the high BMP side expresses the BMP antagonist Gremlin (Fig. 4). Gremlin is a member of the Cerberus/DAN family of secreted growth factor antagonists first cloned by Harland in *Xenopus*, in which it is not known to play a role during gastrulation (Hsu, Economides, Wang, Eimon, & Harland, 1998). Sizzled, which plays a very prominent role in the *Xenopus* and zebrafish ventral gastrula centers, is not present in *Nematostella* or, to our knowledge, in any invertebrate. Knockdown experiments in sea anemone embryos showed that Chordin expression is the main determinant of where the low BMP side is formed, and mathematical modeling showed that Tolloid and Chordin are the rate-limiting elements in establishing the BMP gradient (Genikhovich et al., 2015). Different animals use variations of the ancestral BMP/Chordin/Tolloid D–V module. In the case of placental mammals, the *ADMP, crescent,* and *sizzled* genes have been lost, probably concomitantly with the loss of yolk in the egg, since these genes play a prominent role in the chick embryo (Ploper et al., 2011).

In conclusion, molecular studies on the Spemann organizer together with biochemical analyses and cell transplantation studies over a period of 20 years led to the formulation of successive working hypotheses that allowed the systematic discovery of a network of extracellular proteins that interact with each other in *Xenopus*. This biochemical pathway forms a morphogenetic field that encompasses the entire embryo and has been utilized since the time of our early animal ancestors to generate long-range gradients of positional information.

ACKNOWLEDGMENTS

We thank the many generations of postdocs and students that elucidated the Chordin pathway, in particular the late Yoshiki Sasai who first isolated Chordin. This work was supported by the Howard Hughes Medical Institute, of which Y.M. is an associate and E. D.R. is an investigator.

REFERENCES

Ambrosio, A. L., Taelman, V. F., Lee, H. X., Metzinger, C. A., Coffinier, C., & De Robertis, E. M. (2008). Crossveinless-2 is a BMP feedback inhibitor that binds Chordin/BMP to regulate *Xenopus* embryonic patterning. *Developmental Cell, 15*, 248–260.

Bachiller, D., Klingensmith, J., Kemp, C., Belo, J. A., Anderson, R. M., May, S. R., et al. (2000). The organizer secreted factors Chordin and Noggin are required for forebrain development in the mouse. *Nature, 403*, 658–661.

Bier, E., & De Robertis, E. M. (2015). BMP gradients: A paradigm for morphogen-mediated developmental patterning. *Science, 348*, aaa5838.

Blair, S. (2007). Wing vein patterning in Drosophila and the analysis of intercellular signaling. *Annual Review of Cell and Developmental Biology, 23*, 293–319.

Carrasco, A. E., McGinnis, W., Gehring, W. J., & De Robertis, E. M. (1984). Cloning of an X. laevis gene expressed during early embryogenesis coding for a peptide region homologous to Drosophila homeotic genes. *Cell, 37*, 409–414.

Cho, K. W. Y., Blumberg, B., Steinbeisser, H., & De Robertis, E. M. (1991). Molecular nature of Spemann's organizer: The role of the *Xenopus* homeobox gene *goosecoid. Cell, 67*, 1111–1120.

Collavin, L., & Kirschner, M. W. (2003). The secreted Frizzled-related protein Sizzled functions as a negative feedback regulator of extreme ventral mesoderm. *Development, 130*, 805–816.

Conley, C. A., Silburn, R., Singer, M. A., Ralston, A., Rohwer-Nutter, D., Olson, D. J., et al. (2000). Crossveinless 2 contains cysteine-rich domains and is required for high levels of BMP-like activity during the formation of the cross veins in Drosophila. *Development, 127*, 3947–3959.

De Robertis, E. M. (2008). Evo-devo: Variations on ancestral themes. *Cell, 132*, 185–195.

De Robertis, E. M., & Kuroda, H. (2004). Dorsal–ventral patterning and neural induction in *Xenopus* embryos. *Annual Review of Cell and Developmental Biology, 20*, 285–308.

De Robertis, E. M., & Sasai, Y. (1996). A common plan for dorsoventral patterning in Bilateria. *Nature, 380*, 37–40.

Ferguson, E. L., & Anderson, K. V. (1992). Localized enhancement and repression of the activity of the TGF-β family member, *decapentaplegic*, is necessary for dorsal–ventral pattern formation in the *Drosophila* embryo. *Development*, *114*, 583–597.

Francois, V., Solloway, M., O'Neill, J. W., Emery, J., & Bier, E. (1994). Dorsal–ventral patterning of the Drosophila embryo depends on a putative negative growth factor encoded by the short gastrulation gene. *Genes and Development*, *8*, 2602–2616.

Genikhovich, G., Fried, P., Prünster, M., Schinko, J., Gilles, A., Fredman, D., et al. (2015). Axis patterning by BMPs: Cnidarian network reveals evolutionary constraints. *Cell Reports*, *10*, 1646–1654.

Hemmati-Brivanlou, A., Kelly, O. G., & Melton, D. A. (1994). Follistatin, an antagonist of activin, is expressed in the Spemann organizer and displays direct neutralizing activity. *Cell*, *77*, 283–295.

Holley, S. A., Jackson, P. D., Sasai, Y., Lu, B., De Robertis, E. M., Hoffman, F. M., et al. (1995). A conserved system for dorsal–ventral patterning in insects and vertebrates involving short gastrulation and chordin. *Nature*, *376*, 249–253.

Hsu, D., Economides, A., Wang, X., Eimon, P., & Harland, R. (1998). The Xenopus dorsalizing factor Gremlin identifies a novel family of secreted proteins that antagonize BMP activities. *Molecular Cell*, *1*, 673–683.

Inomata, H., Haraguchi, T., & Sasai, Y. (2008). Robust stability of the embryonic axial pattern requires a secreted scaffold for chordin degradation. *Cell*, *134*, 854–865.

Khokha, M., Yeh, J., Grammer, T., & Harland, R. (2005). Depletion of three BMP antagonists from Spemann's organizer leads to a catastrophic loss of dorsal structures. *Developmental Cell*, *8*, 401–411.

Langdon, Y., & Mullins, M. (2011). Maternal and zygotic control of zebrafish dorsoventral axial patterning. *Annual Review of Genetics*, *45*, 357–377.

Lee, H. X., Ambrosio, A. L., Reversade, B., & De Robertis, E. M. (2006). Embryonic dorsal–ventral signaling: Secreted Frizzled-related proteins as inhibitors of Tolloid proteinases. *Cell*, *124*, 147–159.

Lee, H. X., Mendes, F. A., Plouhinec, J. L., & De Robertis, E. M. (2009). Enzymatic regulation of pattern: BMP4 binds CUB domains of Tolloids and inhibits proteinase activity. *Genes and Development*, *23*, 2551–2562.

Levander, G. (1938). A study of bone regeneration. *Surgery, Gynecology & Obstetrics*, *67*, 705–714.

Little, S., & Mullins, M. (2004). Twisted gastrulation promotes BMP signaling in zebrafish dorsal–ventral axial patterning. *Development*, *131*, 5825–5835.

Little, S. C., & Mullins, M. C. (2006). Extracellular modulation of BMP activity in patterning the dorsoventral axis. *Birth Defects Research. Part C, Embryo Today: Reviews*, *78*, 224–242.

Marjoram, L., & Wright, C. (2011). Rapid differential transport of Nodal and Lefty on sulfated proteoglycan-rich extracellular matrix regulates left-right asymmetry in Xenopus. *Development*, *138*, 475–485.

Moos, M., Jr., Wang, S., & Krinks, M. (1995). Anti-dorsalizing morphogenetic protein is a novel TGF-beta homolog expressed in the Spemann organizer. *Development*, *121*, 4293–4301.

Nüsslein-Volhard, C., & Wieschaus, E. (1980). Mutations affecting segment number and polarity in Drosophila. *Nature*, *287*, 795–801.

Oelgeschläger, M., Kuroda, H., Reversade, B., & Robertis, E. M. (2003). Chordin is required for the Spemann organizer transplantation phenomenon in Xenopus embryos. *Developmental Cell*, *4*, 219–230.

Oelgeschläger, M., Larraín, J., Geissert, D., & De Robertis, E. M. (2000). The evolutionarily conserved BMP-binding protein Twisted gastrulation promotes BMP signalling. *Nature*, *405*, 757–763.

Piccolo, S., Agius, E., Lu, B., Goodman, S., Dale, L., & De Robertis, E. M. (1997). Cleavage of Chordin by the Xolloid metalloprotease suggests a role for proteolytic processing in the regulation of Spemann organizer activity. *Cell, 91*, 407–416.

Piccolo, S., Sasai, Y., Lu, B., & De Robertis, E. M. (1996). Dorsoventral patterning in Xenopus: Inhibition of ventral signals by direct binding of Chordin to BMP-4. *Cell, 86*, 589–598.

Ploper, D., Lee, H. X., & De Robertis, E. M. (2011). Dorsal–ventral patterning: Crescent is a dorsally secreted Frizzled-related protein that competitively inhibits Tolloid proteases. *Developmental Biology, 352*, 317–328.

Plouhinec, J.-L. L., Zakin, L., Moriyama, Y., & De Robertis, E. M. (2013). Chordin forms a self-organizing morphogen gradient in the extracellular space between ectoderm and mesoderm in the Xenopus embryo. *Proceedings of the National academy of Sciences of the United States of America, 110*, 20372–20379.

Reversade, B., & De Robertis, E. M. (2005). Regulation of ADMP and BMP2/4/7 at opposite embryonic poles generates a self-regulating morphogen field. *Cell, 123*, 1147–1160.

Sasai, Y., Lu, B., Steinbeisser, H., & De Robertis, E. M. (1995). Regulation of neural induction by the Chd and BMP-4 antagonistic patterning signals in *Xenopus*. *Nature, 376*, 333–336.

Sasai, Y., Lu, B., Steinbeisser, H., Geissert, D., Gont, L. K., & De Robertis, E. M. (1994). Xenopus *chordin*: A novel dorsalizing factor activated by organizer-specific homeobox genes. *Cell, 79*, 779–790.

Serpe, M., Umulis, D., Ralston, A., Chen, J., Olson, D. J., Avanesov, A., et al. (2008). The BMP-binding protein Crossveinless 2 is a short-range, concentration-dependent, biphasic modulator of BMP signaling in Drosophila. *Developmental Cell, 14*, 940–953.

Smith, W. C., & Harland, R. M. (1992). Expression cloning of noggin, a new dorsalizing factor localized to the Spemann organizer in Xenopus embryos. *Cell, 70*, 829–840.

Spemann, H., & Mangold, H. (1924). Induction of embryonic primordial by implantation of organizers from a different species. *Roux's Archives fur Mikroskopische Anatomie und Entwicklungsmechanik, 100*, 599–638. (Reprinted and translated in *International Journal of Developmental Biology*. Special Issue, De Robertis, E.M., & Arechaga, J. (Eds.) *45*, 13–31 (2001)).

Urist, M. (1965). Bone: Formation by autoinduction. *Science, 150*, 893–899.

Wozney, J. M., Rosen, V., Celeste, A. J., Mitsock, L. M., Whitters, M. J., Kriz, R. W., et al. (1988). Novel regulators of bone formation: Molecular clones and activities. *Science, 242*, 1528–1534.

Yabe, T., Shimizu, T., Muraoka, O., Bae, Y. K., Hirata, T., Nojima, H., et al. (2003). Ogon/Secreted Frizzled functions as a negative feedback regulator of Bmp signaling. *Development, 130*, 2705–2716.

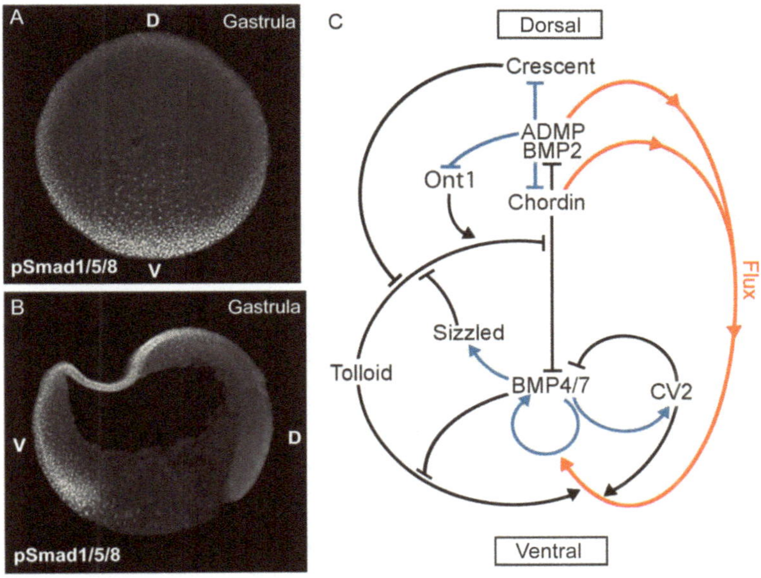

Edward M. De Robertis and Yuki Moriyama, Figure 1 A gradient of nuclear phospho-Smad1 is established by a biochemical network of secreted proteins that interact with each other. (A) Horizontal optical section at gastrula (stage 11) showing a ventral to dorsal gradient of BMP activity using pSmad1/5/8 antibody as the readout. D, dorsal; V, ventral. (B) Gastrula (stage 11) embryo sectioned sagittally, showing higher BMP signaling in ventral nuclei. (C) The D–V morphogenetic signaling pathway results from a series of direct protein–protein interactions between Chordin and other partners (black arrows), transcriptional regulation (blue arrows), and protein flux (red arrows). The entire embryo participates in forming the BMP gradient, which results from the dueling activities of the dorsal and ventral signaling centers. *Images from Plouhinec, Zakin, Moriyama, and De Robertis (2013).*

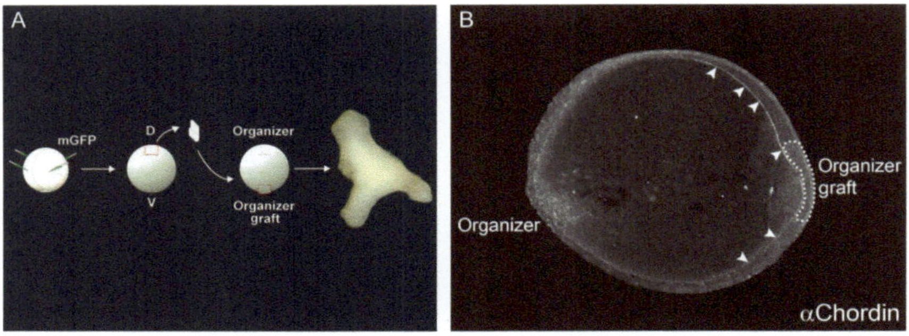

Edward M. De Robertis and Yuki Moriyama, Figure 3 A long-range gradient of Chordin protein emanating from transplanted organizer tissue. (A) Diagram of the experimental procedure; membrane-targeted *mGFP* was injected in the donor embryo (four injections at the four-cell stage) to trace the lineage of grafted Spemann organizer tissue. (B) Embryo transplanted at stage 10 and fixed at stage 12 stained with anti-Chordin antibody shown in optical transverse section. A second gradient of Chordin was observed with Chordin protein staining in Brachet's cleft extending a considerable distance from the graft (arrowheads). The GFP fluorescence of the transplanted organizer cells is outlined by the dotted line. Chordin protein diffuses long-distance from the graft through Brachet's cleft ECM. *Data from Plouhinec et al. (2013).*

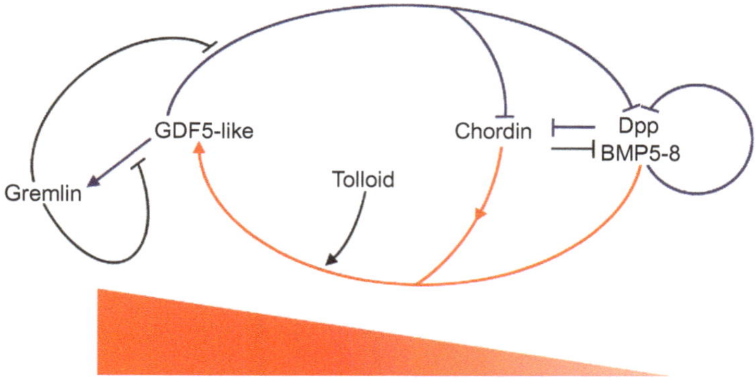

Edward M. De Robertis and Yuki Moriyama, Figure 4 The directive axis of the cnidarian *Nematostella vectensis* is established by a Chordin/BMP gradient. As in *Xenopus*, BMP activity along the body axis results from a series of direct protein–protein interactions between Chordin and other protein partners (black arrows), transcriptional regulation (blue arrows), and protein flux (red arrows) in *Nematostella*. *Diagram based on Genikhovich et al. (2015).*

CHAPTER FOURTEEN

From the Eye to the Brain: Development of the *Drosophila* Visual System

Nathalie Nériec[*,†], **Claude Desplan**[*,†,1]
[*]Center for Genomics & Systems Biology, New York University, Abu Dhabi, UAE
[†]Department of Biology, New York University, New York, USA
[1]Corresponding author: e-mail address: cd38@nyu.edu

Contents

Abstract

How stem cells produce the huge diversity of neurons that form the visual system, and how these cells are assembled in neural circuits are a critical question in developmental neurobiology. Investigations in *Drosophila* have led to the discovery of several basic principles of neural patterning. In this chapter, we provide an overview of the field by describing the development of the *Drosophila* visual system, from the embryo to the adult and from the gross anatomy to the cellular level. We then explore the general molecular mechanisms identified that might apply to other neural structures in flies or in vertebrates. Finally, we discuss the major challenges that remain to be addressed in the field.

Current Topics in Developmental Biology, Volume 116
ISSN 0070-2153
http://dx.doi.org/10.1016/bs.ctdb.2015.11.032

1. INTRODUCTION

The mammalian brain is composed of a very large number of cells belonging to many different cell types, a complexity that poses serious challenges (Kandel, Schwartz, & Jessell, 2000). Our current understanding of the development of neural circuits underlying the computation of different visual stimuli remains highly incomplete across species. Insects provide an attractive model system since their nervous system is relatively simple, yet the animals manifest very sophisticated visual behaviors (Collett, 2008; Zeil, 2012).

How is neuronal diversity achieved during development of the visual system? What are the genes and pathways defining large numbers of different neuronal cell types? How are these cell types connected to form functional circuits in the optic lobes? Thanks to the development of new molecular genetic tools (del Valle Rodriguez, Didiano, & Desplan, 2012), significant progress has been made toward understanding the development of the *Drosophila* visual circuitry.

2. THE *DROSOPHILA* VISUAL SYSTEM

The adult *Drosophila* visual system contains ~150,000 neurons and glia cells (Chiang et al., 2011). Visual information is detected by the retina, while visual processing occurs in the optic lobes that comprise more than 60% of the brain's neurons. The optic lobes are the major centers where neuronal computations extract important features from the visual world, such as shape, motion, color, e-vector orientation of polarized light, which are then transmitted to the central brain (CB) (Fig. 1A–C; Borst & Helmstaedter, 2015; Fischbach & Dittrich, 1989; Homberg, Heinze, Pfeiffer, Kinoshita, & el Jundi, 2011; Meinertzhagen & Hanson, 1993; Meinertzhagen et al., 2009; Otsuna, Shinomiya, & Ito, 2014; Silies, Gohl, & Clandinin, 2014).

2.1 The Compound Eye

The adult *Drosophila* compound eye is made of ~800 independent unit eyes called ommatidia, corresponding to 800 pixels in the animal's visual field. Each ommatidium is composed of eight photoreceptor neurons that project into the optic lobe. The organization of the eye will not be further described here as it has been reviewed extensively in the recent past (Kumar, 2012; Lamb, 2013; Paulk, Millard, & van Swinderen, 2011).

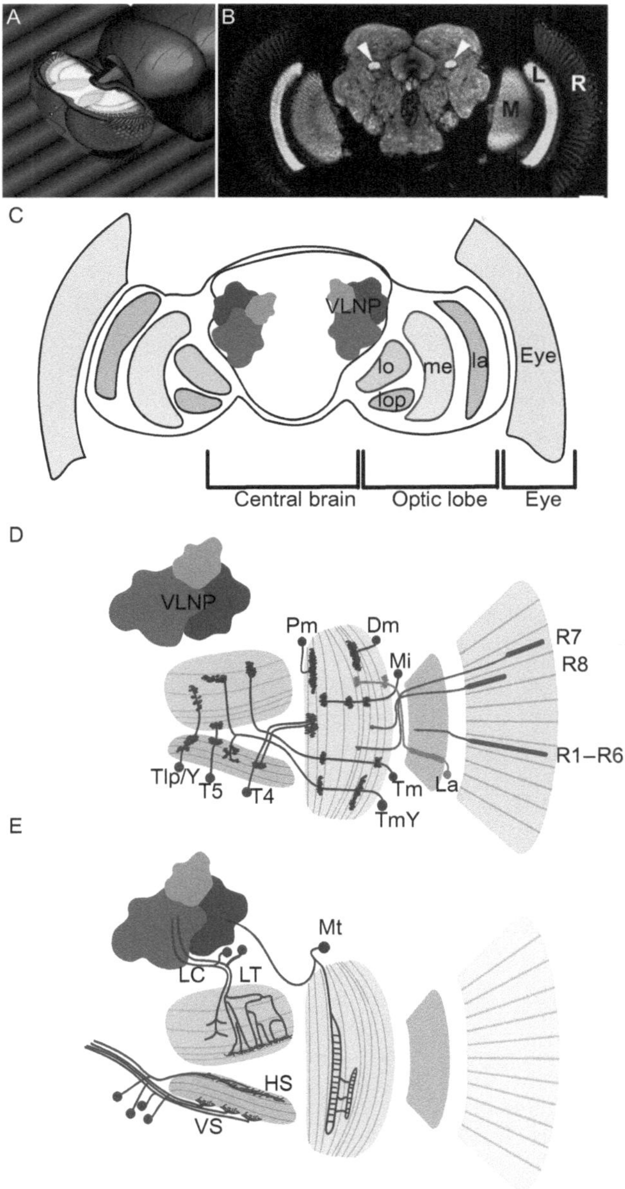

Figure 1 The visual system of *Drosophila*. (A) The fly brain in the head. (B) Neuropils of the optic lobes and central brain. (C) Cross section of the fly brain indicating the different parts of the visual system: eye, lamina (la), medulla (me), lobula (lo), lobula plate (lop), and ventrolateral neuropils (VLNPs). (D) Interneurons of the optic

(Continued)

2.2 Optic Lobe

Fly neurons are organized into approximately 50 areas composed mainly of neuronal processes, called neuropils, with their cell bodies localized at the periphery. Four of these neuropils form the optic lobe: lamina, medulla, and the lobula complex which is further subdivided into the lobula and the lobula plate neuropils (Morante & Desplan, 2004). Two major types of neurons can be identified within the optic lobes: "interneurons" whose cell bodies and projections remain within the optic lobe, and "projection" neurons, which connect the optic lobe to the CB (Fig. 1D and E; Hofbauer & Campos-Ortega, 1990).

2.2.1 Lamina

Photoreceptors from each ommatidium involved in motion vision (outer photoreceptors) first innervate the lamina neuropil, which manifests a columnar organization in which each pixel of the visual field corresponds to one cartridge (Meinertzhagen & Sorra, 2001). The lamina is mostly composed of interneurons, whose projections do not leave the optic lobe with their cell bodies located in the lamina cortex region. Lamina neurons can be divided in two populations: Five types of monopolar neurons that contact a single cartridge and project retinotopically into the medulla, and amacrine cells that contact several cartridges within the lamina (Fischbach & Dittrich, 1989; Hofbauer & Campos-Ortega, 1990; Tuthill, Nern, Holtz, Rubin, & Reiser, 2013).

2.2.2 Medulla

The medulla neuropil receives direct innervation from color (inner) photoreceptors, as well as from lamina monopolar neurons. The medulla neuropil is stratified in 10 layers (M1–M10) with the region between layers M1 and M6 referred to as "distal medulla" that receives these external inputs

Figure 1—Cont'd lobe: outer and inner photoreceptors (pink), lamina neurons (La, green), medulla interneurons (dark blue), distal (Dm) and proximal medulla (Pm); medulla intrinsic (Mi); unicolumnar transmedullary (Tm) or multicolumnar TmY; lobula plate interneurons (T4, T5, Tlp, and Y neurons, dark brown). (E) Projections neurons: medulla tangential (Mt, dark blue); lobula columnar (LC, red) and tangential/tree-like (LT, red); lobula plate tangential cells: HS and VS (dark brown). *Panel (A) adapted from Spalthoff, Gerdes, and Kurtz (2012). Panel (B) from Krzeptowski et al. (2014). Panels (D and E) adapted from Erclik, Hartenstein, McInnes, and Lipshitz (2009).* (A color version of this figure is located at the end of this chapter.)

(Fischbach & Dittrich, 1989; Morante & Desplan, 2008; Takemura, Lu, & Meinertzhagen, 2008). The "proximal medulla" (layers M7–M10) receives information from the distal medulla and further computes visual information. The medulla neuropil is organized in repetitive columnar units, oriented perpendicular to the 10 layers that are reminiscent of lamina cartridges (Fischbach & Dittrich, 1989). Retinotopic connections between medulla columns, photoreceptors, and lamina cartridges ensure an accurate representation of the visual world (Chin, Lin, Fu, Dickson, & Chiang, 2014; Fischbach & Dittrich, 1989; Meinertzhagen & Sorra, 2001; Morante & Desplan, 2008; Zhu, 2013).

The serpentine layer, which separates the distal and proximal medulla, consists of incoming and outgoing axons of projection neurons, which all connect to more than one medulla column. Different types of projection neurons can be identified based on the location of their cell bodies, their dendritic morphology, and their axonal projection patterns (Fischbach & Dittrich, 1989; Strausfeld, 1989), but the functional contribution of most of these neurons remains unclear.

The medulla contains about 40,000 interneurons, representing the largest neuronal subpopulation in the optic lobe (Fischbach & Dittrich, 1989). Their cell bodies are located either within the medulla cortex between the lamina and the medulla neuropils, or within the medulla rim, located below the medulla neuropil near the lobula complex (Bausenwein, Dittrich, & Fischbach, 1992; Fischbach & Dittrich, 1989). Medulla interneurons have been extensively characterized and categorized in over 80 cell types, first by Cajal and Sanchez (1915), followed by Fischbach and Dittrich (1989), and more recently by Morante and Desplan (2008), Raghu, Claussen, and Borst (2013), Raghu, Joesch, Sigrist, Borst, and Reiff (2009), and Varija Raghu, Reiff, and Borst (2011). They can be subdivided into subcategories based on their projections patterns (Fig. 1D): Interneurons that project over a large visual field, across many columns, are called tangential (Dm and Pm in Fig. 1D). Columnar neurons projections are mainly parallel to the medulla columns (Mi, Tm, and TmY in Fig. 1D). They can be unicolumnar and project within a single medulla column, suggesting that they process information from one visual point in space, or multicolumnar with projections spanning several medulla columns. Only columnar neurons project outside the medulla into the lobula and lobula plate, which represent the main output of the medulla neuropil (Fischbach & Dittrich, 1989; Morante & Desplan, 2008).

2.2.3 Lobula Complex: The Lobula

The lobula can be divided into six layers arranged perpendicularly to the columnar structures resulting from the columnar inputs of the medulla neurons (Fischbach & Dittrich, 1989). Almost all lobula cells are projection neurons whose cell bodies are located between the CB and the lobula neuropil. Interestingly, despite the cell bodies' varying distance to the lobula, their projections all merge at the neck of the lobula to form a single fiber tract connecting the lobula to the CB (Fig. 1E; Otsuna & Ito, 2006). Lobula neurons can be divided into two categories: Columnar neurons (LC) receive visual input from 8 to 9 ommatidia (Douglass & Strausfeld, 2003; Mu, Ito, Bacon, & Strausfeld, 2012; Otsuna & Ito, 2006), reminiscent of multicolumnar neurons in the medulla; tangential- and tree-like neurons (LT) receive input from very large visual fields, similar to medulla tangential neurons (Otsuna & Ito, 2006; Fig. 1E).

2.2.4 Lobula Complex: The Lobula Plate

Representing the output center of the neural circuits that process motion, the lobula plate neuropil can be divided in four layers containing dendrites that each manifest maximal sensitivity to motion along one of the four cardinal directions (front-to-back, back-to-front, up, and down) (Fischbach & Dittrich, 1989; Maisak et al., 2013). Two classes of lobula plate interneurons called T4 and T5 cells can each be further subdivided into four subclasses forming dendrites in only one of the four specific layers (hence termed T4a, b, c, d and T5a, b, c, d). Their role in motion detection has been studied and was reviewed in Behnia and Desplan (2015) and Borst and Helmstaedter (2015). Their cell bodies are all located adjacent to the lobula plate neuropil, underneath medulla rim cell bodies (Fig. 1C). Two other classes of interneurons have presynaptic input in the lobula plate and postsynaptic output in the lobula: the translobula plate neurons (Tlp) and Y cells. Both have their cell bodies adjacent to those of T4 and T5 cells. As observed in larger flies, only few intrinsic cells, i.e., which remain within the lobula plate neuropil, can be observed (Fischbach & Dittrich, 1989).

The lobula plate tangential cells (LPTCs) are projection neurons whose characterization has provided great insight into the computation of motion (Hausen, 1984; Maisak et al., 2013). In general, LPTCs are sensitive to visual motion in a direction-selective manner (Hausen, 1984). Historically, the most extensively studied LTPCs belong to the horizontal (HS) and vertical (VS) system (Fig. 1E; Borst & Haag, 2002). They receive their inputs from the T4 and T5 neurons and transmit direction-selective visual information to the ventrolateral neuropils (VLNPs) of the CB (Behnia & Desplan, 2015; Borst, 2014; Borst & Haag, 2002; Silies et al., 2013).

2.3 Visual Centers in the CB

The VLNPs (also called optic glomeruli) are located right underneath the optic lobes and can be considered the next step in visual processing after the optic lobes. All 14 different types of lobula projection neurons project to distinct target regions within the VLNPs (Otsuna & Ito, 2006). They each project into the VLNP as one single bundle in a way reminiscent to olfactory glomeruli (Mu et al., 2012; Otsuna & Ito, 2006). The availability of highly specific Gal4 lines that label most of the optic lobe cell types (Jenett et al., 2012; Li et al., 2014; Pfeiffer et al., 2008) should allow the determination of the visual function performed by each optic glomerulus (Aptekar, Keles, Lu, Zolotova, & Frye, 2015; Mu et al., 2012).

3. DEVELOPMENT OF THE FLY VISUAL SYSTEM

3.1 The Retina

The development of the fly retina is one of the best-understood complex structures. It has been reviewed in many articles (Carthew, 2007; Kumar, 2012; Treisman, 2013) and will not be described further in this review.

3.2 The Ventral Nerve Cord and CB

The ventral nerve cord (VNC) and CB are generated in the embryo from neuroblasts delaminating from the embryonic neuroepithelium. These neuroblasts produce the embryonic nervous system and 10% of the future adult neurons before entering quiescence toward the end of embryonic stages (Fig. 2A). At late L1/early L2 stages, about 100 neuroblasts start dividing again and produce the remaining 90% of adult neurons. The neuroblasts can be divided in two categories: Type I neuroblasts generate all VNC neurons and most of the CB, while eight Type II neuroblasts generate clones of up to 500 cells giving rise to CB neurons. The formation of neurons from the VNC/CB neuroblasts has been recently reviewed in Homem and Knoblich (2012), Kang and Reichert (2015), and Reichert (2011).

3.3 The Optic Lobe

The optic lobe is derived from cells located in the posterior part of the embryonic head (Green, Hartenstein, & Hartenstein, 1993; Nassif et al.,

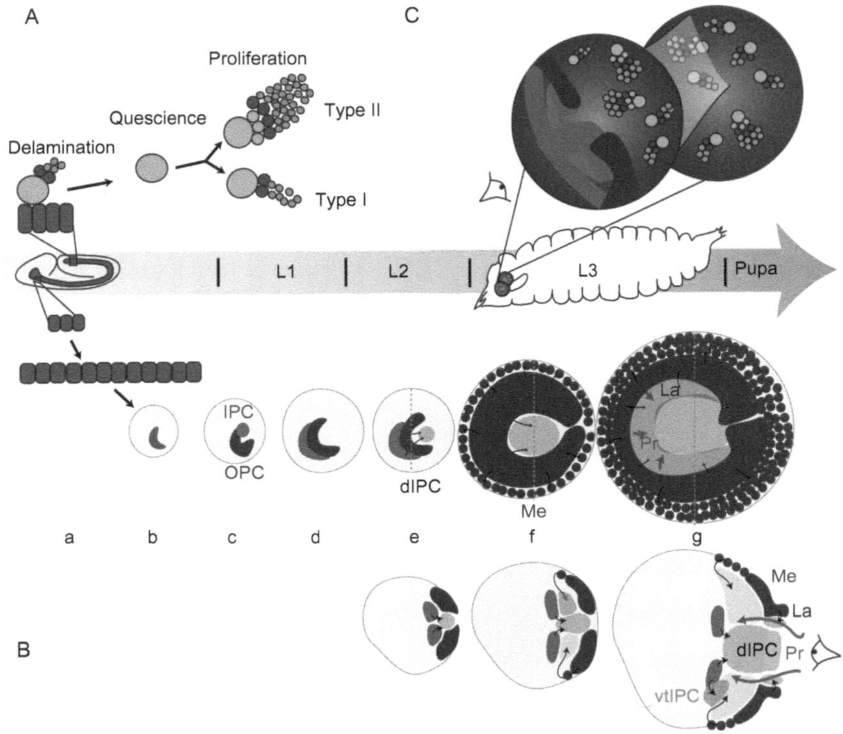

Figure 2 The development of the optic lobe of *Drosophila*. (A) Central brain: neuroblasts (green) delaminate from the neuroepithelium (red) during the embryonic stages and generate GMCs (purple) and larval neurons (blue) before they become quiescent. At L1, Type I and Type II neuroblasts are reactivated and exhibit different modes of proliferation to generate adult neurons. (B) Optic lobe: (a) Cells in the procephalic regions generate the optic placode neuroepithelium (red), which invaginates (b). (c) The optic placode splits into inner proliferation center (IPC, pink) and outer proliferation center (OPC, blue) and adopt a crescent shape (d). (e) Cells migrate out of the IPC to form the dIPC (light brown). (f) The medial edge of the OPC becomes neuroblasts that will produce the medulla (round blue cells). (g) The inner OPC generates lamina progenitors (La, green) that will produce lamina neurons upon induction by photoreceptors (Pr, pink arrows). The ventral tip of the IPC generates lobula complex neurons (vtIPC, dark yellowish green). Lower diagrams are cross sections at the dashed line in upper diagrams. (C) 3D representation of the brain at L3. Central brain neuroblasts and their progeny, OPC (blue) and IPC (pink). The eye symbol shows the view angle of panel (B). *Panels (A and C) adapted from Homem and Knoblich (2012). Panel (B) adapted from Nassif, Noveen, and Hartenstein (2003).* (A color version of this figure is located at the end of this chapter.)

2003). During the same time window when neuroblasts of the VNC and CB delaminate, optic lobe precursor cells undergo four rounds of mitosis to form a plate-like structure that is made of densely packed columnar cells called optic placode (Green et al., 1993; Fig. 2Ba). After the fourth division, the cells of the developing optic lobe remain mitotically quiescent throughout the rest of embryogenesis (Green et al., 1993).

After all neuroblasts of the CB have delaminated from the head ectoderm, the optic lobe placode becomes attached to the ventrolateral surface of the brain and invaginates by apical constriction. During this invagination process, the placode adopts a V-like shape, with an anterior and a posterior tip (Green et al., 1993; Hofbauer & Campos-Ortega, 1990; White & Kankel, 1978; Fig. 2Bc).

Upon hatching of the first instar larva, a small group of cells at the anterior-dorsal tip of the optic placode detaches, splitting the developing optic lobe into two, creating the outer proliferation center (OPC) and the inner proliferation center (IPC) (Meinertzhagen & Hanson, 1993; Nassif et al., 2003; Younossi-Hartenstein, Nassif, Green, & Hartenstein, 1996; Fig. 2Bd).

Toward the end of the second larval instar, the OPC and the IPC both adopt a crescent shape, with the opening of the crescent pointing posteriorly (Fig. 2Bd). Both OPC and IPC anlagens remain in contact with each other until the end of the second instar, when they become separated by newly generated cells that penetrate into the space between the IPC and the OPC (Fig. 2Be). These cells migrate in from the IPC to form a distant proliferative center called the dIPC. The ventral tip of the IPC also proliferates and generates neurons (Apitz & Salecker, 2015; Hofbauer & Campos-Ortega, 1990; Neriec et al., Submitted).

From the end of the second instar onward, cells at the medial edge of the OPC neuroepithelium crescent begin to lose their columnar shape and adherens junctions without delaminating (Fig. 2Bf). They are converted into neuroblasts in a moving proneural wave (reviewed in Apitz & Salecker, 2014). These medulla neuroblasts divide asymmetrically to self-renew and generate their progeny composed of neurons and glia (Egger, Boone, Stevens, Brand, & Doe, 2007). By 20 h after puparium formation (APF), ~40,000 neurons have been generated, most of which will become medulla cortex neurons (Egger, Gold, & Brand, 2010, 2011).

During the third larval instar, a second proliferation zone develops at the opposite edge of the OPC. This second zone is separated from the future medulla neuroblast by the deep lamina furrow (Fig. 2Bg). By mid-third instar, photoreceptor axons start to project into the developing optic lobe via the optic stalk (pink arrow in Fig. 2Bg). Lamina neurons are produced and differentiate in response to Hedgehog (Hh) and EGF produced by photoreceptors axons (Huang & Kunes, 1996, 1998; Huang, Shilo, & Kunes, 1998; Selleck, Gonzalez, Glover, & White, 1992). By 25 h APF, it has generated about 6000 additional cells that become lamina neurons and glia (Apitz & Salecker, 2014). The development of the OPC, which generates the lamina and the medulla, has been studied in much detail, while the development of the IPC, which generates the lobula and lobula plate, has only recently been the topic of investigations (Apitz & Salecker, 2015; Neriec et al., Submitted).

Starting around 25 h APF, the developing medulla starts to rotate, being pulled by the lamina (Langen et al., 2015; White & Kankel, 1978), and inserting itself right under the lamina neuropil. By 40 h APF, the rotation is complete and the optic lobe has adopted its final configuration which the four neuropils: lamina, medulla, lobula, and lobula plate (C. Bertet and K. Fischbach, Personal communications).

3.4 Linking Development to Adult Fate

Due to the extreme morphological changes shaping the optic lobe during pupation, the task of linking cell fates between larval and adult neurons remains incomplete. Within the OPC, the cellular mechanisms that lead to the formation of lamina precursor cells, the induction of their differentiation into laminar monopolar neurons by the photoreceptors and their connections have been well studied (Dearborn & Kunes, 2004; Huang & Kunes, 1996, 1998; Kunes, Wilson, & Steller, 1993). However, how the specification of the five different types of lamina monopolar neurons is achieved remains unknown. The rest of the OPC has been shown to generate medulla cell types as well few lobula and lobula plate neurons (Fig. 3; Bertet et al., 2014). General mechanisms used to specify different types of neurons have been identified by Bertet et al. (2014), Erclik et al. (Under Review), Hasegawa, Kaido, Takayama and Sato (2013), Li, Erclik, et al. (2013), Sato, Suzuki, and Nakai (2013), and Suzuki, Kaido, Takayama, and Sato (2013). A recent study has characterized some of the genetic mechanisms involved in the generation of neurons of the lobula plate and medulla rim from the main part of the IPC (Apitz & Salecker, 2015; Neriec et al., Submitted). Common mechanisms involved in the generation of neuronal diversity have been identified in the optic lobe, the VNC and the CB. They are similar in many regards to those observed in vertebrates and are detailed in Section 4 of this review.

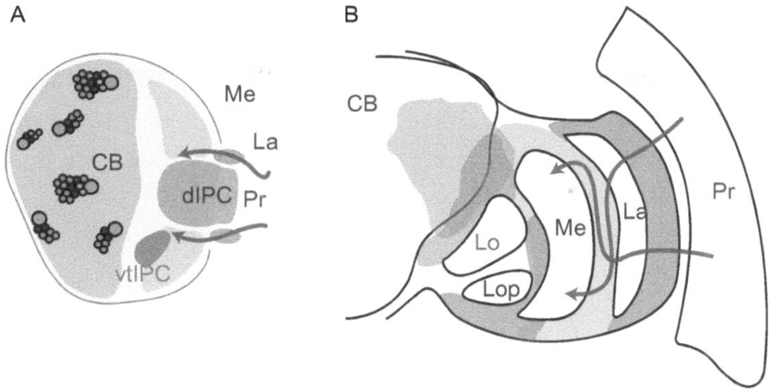

Figure 3 Linking development to adult fate. Fate of the different neuronal populations found at the L3 larval stage (A) in the adult visual system (B). Pr, photoreceptors; La, lamina; Me, medulla; Lo, lobula; Lop, lobula plate; dIPC, distal IPC; vtIPC, ventral tip of the IPC; CB, central brain. (A color version of this figure is located at the end of this chapter.)

The optic lobe projection neurons, including the lobula columnar neurons (LCNs) and the LPTCs, as well as the CB neurons of the ventrolateral neuropils are believed to originate mostly from CB neuroblasts. Recent studies have taken advantage of the fact that CB neurons remain close to their birth place and generate reproducible series of neuronal cell fates that represent the clonal units generated from each of the 100 neuroblasts in the CB (Chiang et al., 2011; Ito, Masuda, Shinomiya, Endo, & Ito, 2013; Lovick et al., 2013; Wong et al., 2013; Yu et al., 2013) including intermediate progenitors of Type II neuroblasts (Wang, Yang, et al., 2014).

Thirteen clonal units of neurons projecting into the optic lobe but also into the CB have been described (Ito et al., 2013). These neurons all project into the VLP where they form distinct nonoverlapping bulbous masses corresponding to functionally relevant optic glomeruli: Neurons originating from the same neuroblast clone share common projection pattern into specific glomeruli, which is indicative of functional similarities (Otsuna & Ito, 2006). This hints toward a direct link between the developmental origin of projection neurons and their adult function. Similarities between the functional connectivity of VLP and olfactory glomeruli have been pointed out (Mu et al., 2012). Extensive work has been undertaken in order to understand the formation of the olfactory glomeruli (for review, see Imai, Sakano, & Vosshall, 2010), and similar future studies on the development of the optic glomeruli will determine if similarities exist between the development of these glomeruli.

4. GENERAL RULES OF NEURONAL DEVELOPMENT

The generation of an adult neuron can be described in two main steps. First, neurons are produced by neuronal stem cells. Then, once produced, the prespecified neurons form axonal and dendritic projections and integrate into developing neural circuits (Fig. 4A). We will review the general

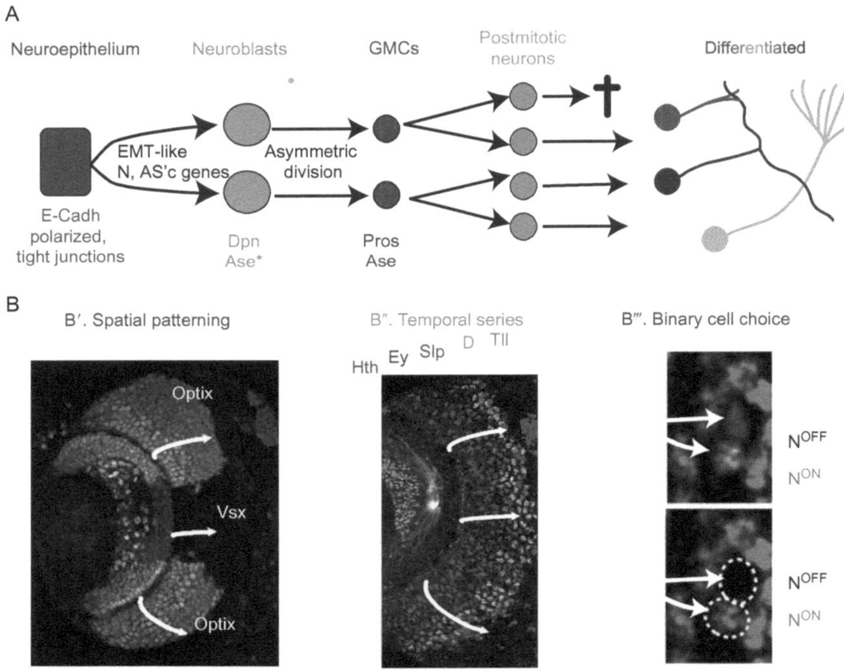

Figure 4 The making of a functional neuronal network. (A) Common mechanisms for neurogenesis: polarized neuroepithelial cells (red) undergo EMT-like transition to become neuroblasts (green) which generate GMCs (purple). GMCs produce postmitotic neurons (blue) that form adult neuronal circuits. * Dpn is found in all neuroblasts while Ase is found in most neuroblasts (see text). (B) Common mechanisms to generate neuronal diversity: (B′) Spatial patterning of the neuroepithelium illustrated by the expression of Optix and Vsx (Gold & Brand, 2014). (B″) Temporal patterning of neuroblasts. In the OPC, neuroblasts sequentially express a series of transcription factors, Hth (red), Ey (blue), Slp (green), D (orange), and Tll (cyan) as they age (arrows) (Li, Chen, & Desplan, 2013; Li, Erclik, et al., 2013). (B‴) Notch-dependent binary cell fate choices during the division of GMCs. Only one of the two daughter cells (red clone, dashed line) receives Notch activity and expresses Apterous (green), while the other does not (Li, Chen, et al., 2013; Li, Erclik, et al., 2013). (A color version of this figure is located at the end of this chapter.)

mechanisms of neurogenesis observed during the production (Section 4.1) and the prespecification (Section 4.2) of optic lobe neurons and compare them with the development of the CB.

4.1 Production of Neurons

4.1.1 Generating Neuroblasts from a Neuroepithelium

Both the *Drosophila* VNC/CB and the optic lobes are generated from cells with epithelial-like characteristics; they are organized as monolayers with cell of rectangular shapes connected by adherens junctions. These cells express high levels of DE-Cadherin (E-Cadh) and manifest apicobasal polarity (Acloque, Adams, Fishwick, Bronner-Fraser, & Nieto, 2009; Doe, 1996).

In the embryonic neuroepithelium, one cell is stochastically singled-out from 6 to 8 neuroepithelial cells via lateral inhibition mediated by Notch. This cell expresses the highest level of AS-C genes and loses its connections to its neighbors as well as with the apical surface just before delaminating and becoming a neuroblast (for review, see Sanes, Reh, & Harris, 2012).

In the OPC, the transition of neuroepithelial cells into neuroblasts occurs in a proneural wave, without the cells undergoing cell movement or delamination: Cells in front of the wave are still neuroepithelial, while cells in the wake of the proneural wave have become neuroblasts. The AS-C protein L(1)sc is expressed in the transition zone between neuroepithelium and neuroblasts and, along with Notch, is necessary for the proneural wave (Apitz & Salecker, 2014; Egger et al., 2007, 2011; Yasugi, Umetsu, Murakami, Sato, & Tabata, 2008). In the IPC, cells undergo EMT and delaminate to later become neuroblasts after migration to the dIPC, most likely also under the control of L(1)sc and Notch (Apitz & Salecker, 2015; Neriec et al., Submitted). Further investigations are needed to determine in more details the mechanisms by which L(1)sc and Notch control the timing of neuroepithelium to neuroblast transition in the optic lobe (Egger et al., 2011; Yasugi, Sugie, Umetsu, & Tabata, 2010).

4.1.2 Generating Neurons from Neuroblasts

New generated neuroblasts undergo asymmetric division which results in one replacement neuroblast and one GMC, which then divides once, generating two postmitotic cells, either neurons or glia. This is the case for 90% of VNC and CB neuroblasts, called Type I neuroblasts, and for most neuroblasts of the OPC. The molecular mechanisms involved in such divisions from the neuroblast to the GMCs have been well characterized and reviewed in Homem and Knoblich (2012), Kang and Reichert (2015), and Sousa-Nunes and Somers (2013). Importantly, neuroblasts express Deadpan (Dpn), GMCs express Prospero (Pros), and both express Asense (Ase) (Brand & Livesey, 2011; Egger, Chell, & Brand, 2008; Egger et al., 2011; Reichert, 2011; Southall & Brand, 2009; Wu, Egger, & Brand, 2008).

In recent years, examples that differ from such classic models have been characterized. During CB neurogenesis, eight specialized Type II neuroblasts undergo asymmetric division to produce intermediate neural progenitors instead of GMCs. These intermediate neural progenitors then divide asymmetrically several times to replace themselves and generate one GMC, leading to lineages that produce a large number of neurons (Bayraktar, Boone, Drummond, & Doe, 2010; Bello, Izergina, Caussinus, & Reichert, 2008; Boone & Doe, 2008; Bowman et al., 2008; Koe et al., 2014; Viktorin, Riebli, & Reichert, 2013; Wallace, Liu, & Vaessin, 2000). Other neuroblasts (Type 0) of the CB and of the optic lobe generate GMCs that do not divide but instead differentiate directly into one neuron (or glia) (Baumgardt et al., 2014; Bertet et al., 2014). Recently, neuroblasts generated by the IPC have been characterized and display atypical features compared to other neuroblasts. First, they migrate between their delamination from the neuroepithelium and their arrival in the dIPC as atypical dividing neuroblasts. Furthermore, these neuroblasts express Ase but not yet Dpn and divide. Finally, they downregulate Ase as they age, thus lacking Ase expression in their GMC progeny (Apitz & Salecker, 2015; Neriec et al., Submitted). Further studies remain necessary to characterize how those differences play a role in the formation of neurons.

4.2 Common Mechanisms for Neuronal Specification

Neuronal cell types can be defined based on cell morphology, cell connectivity, marker expression, or intrinsic properties such as electrophysiological properties. The specification of neuronal identity occurs in three main steps (Lin & Lee, 2012): (a) spatial patterning of the neuroepithelium, (b) temporal patterning of neuroblasts by series of transcription factors, (c) distinct hemilineages from GMCs (Fig. 4B).

4.2.1 Spatial Patterning of the Neuroepithelium

The first mechanism to specify different neuronal identities occurs in the neuroepithelium where spatial patterning cues exist early on. A classic example is the fly embryo, where positional cues are provided by dorsoventral and anteroposterior patterning genes: Gap genes and segment polarity, as well as dorsoventral and Hox genes establish a molecular coordinate system (Technau, 2008; Urbach & Technau, 2003). Such Cartesian grid provides a specific identity to any delaminating neuroblast, and this determines the type of neurons that will be produced (Skeath, 1999; Skeath & Thor, 2003; Technau, Berger, & Urbach, 2006). In the OPC of the optic lobe, the neuroepithelium crescent is also regionalized (Fig. 4B′): The tips of the crescent express Wingless (Wg), which are bordered by a region expressing decapentaplegic (Dpp), and together these two regions express Rx, followed by an Optix region and finally a Vsx-1 expressing region in the center (Erclik et al., Under Review; Gold & Brand, 2014).

Neuroblasts and some of the neurons they produce retain the positional markers of the neuroepithelium in the different regions they are coming from. This can affect their mode of neurogenesis. For instance, young neuroblasts of the Wg region of the OPC do not generate classical Type I like in the rest of the OPC, but instead become Type 0 and generate GMCs that directly differentiate into one neuron (Bertet et al., 2014). More importantly, it ultimately affects the identity of the neurons that are being produced (Bertet et al., 2014; Erclik et al., Under Review). For instance, Pm1 and Pm2 neurons are only generated from the Rx regions of the OPC (Erclik et al., Under Review).

4.2.2 Temporal Patterning of Neuroblasts

The second mechanism by which neuronal diversity is generated temporal patterning of neuroblasts as well as intermediate neuronal progenitors for Type II neuroblasts: These cells sequentially express stereotyped temporal series of transcription factors (reviewed in Li, Chen, et al., 2013; Fig. 4B″). These factors are often transmitted to the postmitotic neurons, affecting the adult identity of the neuronal progeny (Bayraktar et al., 2010; Bertet et al., 2014; Isshiki, Pearson, Holbrook, & Doe, 2001; Li, Erclik, et al., 2013). Despite differences in the transcription factor repertoire used in these temporal series, the general mechanism is very similar: One transcription factor not only induces expression of the next but also represses expression of the previous; changes in transcription factor expression are correlated with changes in adult neuronal progeny (Apitz & Salecker, 2014; Li, Chen, et al., 2013). Studies in vertebrates seem to indicate that similar temporal series also play an important role in mammalian neurogenesis, including the mouse retina (Livesey & Cepko, 2001; Mattar, Ericson, Blackshaw, & Cayouette, 2015; Wang, Sengel, Emerson, & Cepko, 2014).

4.2.3 Hemilineages from the GMCs

Finally, a last source of neuronal diversity arises from molecular asymmetry in the final division of GMCs: Indeed, both in the VNC/CB and during optic lobe neuronal development, each GMC generates two postmitotic cells that receive different levels of the protein Numb, a repressor of Notch signaling. This creates two postmitotic cells with different Notch pathway activity status: $Notch^{ON}$ or $Notch^{OFF}$ (Fig. 4B‴). These two groups of neuroblast progeny are called hemilineages and differ in their adult neuronal identity (Bertet et al., 2014; Lin, Kao, Yu, Huang, & Lee, 2012; Udolph, 2012).

In conclusion, spatial patterning in the NE, temporal series of transcription factors in the neuroblasts, and GMC hemilineages define three main axes through which neuronal diversity is generated (Lin & Lee, 2012). Furthermore, different combinations of transcription factors have also been shown to control cell death or survival (Bertet et al., 2014). For example, in the optic lobe, $Notch^{ON}$ neurons generated from the Wg region of the neuroepithelium undergo apoptosis when they are generated in the time window when the neuroblasts express the transcription factor Eyeless. However, neurons emerging from the same neuroblasts, but during the following Sloppy-paired time window, always survive when they are $Notch^{ON}$ but die when they are $Notch^{OFF}$. These cells then adopt specific adult identities (Bertet et al., 2014). In spite of the contribution of spatial and temporal transcription factors, the general rules that dictate the formation of specific adult neuronal morphologies, interneurons versus projection neurons, for example, remain completely unknown.

5. CONCLUSIONS AND PERSPECTIVES

Since the first description by Cajal over 100 years ago, the fly visual system has been one of the most important models in developmental neurobiology. Progressing from the eye to the brain, a clearer picture emerges regarding not only the connectome and the processing of visual information in the fly brain but also the formation of neurons and establishment of such network during development. Among the remaining studies to be done, three main areas emerge of major interests: The integration of the neuronal specification identities at the molecular level, the mechanisms by which neurons integrate within a neuronal circuits and finally the level of evolutionary conservation between insects and human visual systems.

5.1 Integration at the Molecular Level

Albeit cellular mechanisms remain to be elucidated, such as the detailed correlation of larval cell identities to adult neuronal types, developmental studies of the fly visual system already have provided a blueprint for reaching an understanding of this complex process at a molecular level. What are the molecular mechanisms that control the integration of the three axes of neuronal specification (spatial patterning of the neuroepithelium, window of the temporal series of transcription factors and Notch status) for the expression of specific terminal differentiation genes? Recently, the modules that control the expression of a terminal differentiation gene such as the neurotransmitter V-GLUT have been addressed in *C. elegans* (Serrano-Saiz et al., 2013). What are the modules involved in *Drosophila*? Future genomic and genetic studies will bring great insights in the mechanisms that control the expression of late terminal differentiation genes in large and complex neuronal circuits such as *Drosophila*.

Furthermore, the production of neurons from induced pluripotent stem cells offers one of the most promising therapeutic approaches for the treatment of neurodegenerative disorders (Hallett et al., 2015; Liu & Zhang, 2011). Adult human cells can already be induced into neuronal stem cells and generate neurons *in vitro* that have been successfully transplanted in living animals. However, the similarities and differences between *in vitro*-generated neurons and the *in vivo* neurons they would replace are a major concern. To address this question, *Drosophila* offers a valuable model to understand the molecular mechanisms involved in the formation of a neuronal identity and the role of extrinsic versus intrinsic cues. For example, experiments have shown that Type I neuroblasts from the CB still undergo temporal series when cultivated *in vitro* (Grosskortenhaus, Pearson, Marusich, & Doe, 2005). Similar experiments still remain to be done for other type of neuroblasts, Type II and from the optic lobe.

5.2 Circuits and Retinotopy

While some aspects of neuronal connections have been identified in the formation of the visual system (Lee et al., 2003; Lee, Herman, Clandinin, Lee, & Zipursky, 2001; Sanes & Zipursky, 2010; Timofeev, Joly, Hadjieconomou, & Salecker, 2012), the establishment of connectivity remains largely unknown. A major concept in visual system circuit formation is retinotopy, i.e., the topographic mapping of visual inputs from the retina to optic lobe neurons. Retinotopy starts from the ~800 ommatidia in the eye, each pointing to one point in space and projecting into an equivalent number of cartridges in the lamina, which are therefore representing the same points in space. Similarly, lamina neurons project into ~800 columns of the medulla, in an orderly fashion. Such conservation in the representation of visual information is crucial for any visual system to be able to extract information such as motion and is established very early during development. Due to the direction in which the morphogenetic furrow progresses, lamina neurons acquire their posterior-to-anterior identity based on their time of differentiation. Retinotopy along the dorsoventral axis is also preserved, when rows of photoreceptors generated at a given time point contact developing lamina neurons, thereby forming cartridges. Therefore, time represents an essential component in the generation of retinotopy, with the induction of the lamina by the photoreceptors representing the critical aspect.

However, how retinotopy is established in the medulla and in the higher processing centers during development remains to be further investigated since photoreceptors do not contribute to generating medulla or lobula complex neurons.

5.3 Evolution

The studies described here have generated broad concepts that start to apply to vertebrate neurogenesis. Whether there is a common ancestry of insect and vertebrate nervous systems is a major question in evolutionary biology (Moroz, 2009). Today, the consensus converges toward the concept of a common neuronal origin between flies and vertebrates (Strausfeld, 2009; Strausfeld & Hirth, 2013a, 2013b; Wolff & Strausfeld, 2015), yet little is understood about how much is shared between visual processing systems in terms of development.

To address which features of the fly and vertebrate visual system may have been present in their bilaterian ancestor, comparisons must be made at the anatomical and functional levels (Erclik et al., 2009; Sanes & Zipursky, 2010). However, while similarities between two adult visual systems might very well be due to evolutionary homology, they could also be the result of convergent evolution due to physical constraints on the neuronal circuits processing visual information (Strausfeld & Hirth, 2013a, 2013b).

Programs involved in the formation of neuronal circuits are under specific developmental constraints and adult systems that arise from similar developmental programs are more likely to share a common origin. The observed similarities in adult neuronal systems, likely mirroring similarities in developmental programs, strongly suggest a common origin. As one of the best-understood neuronal system, in flies and in vertebrates, the visual system is well suited to address the question of homology between flies and mammals (Erclik et al., 2009; Sanes & Zipursky, 2010) (_ENREF_92, _ENREF_48).

REFERENCES

Acloque, H., Adams, M. S., Fishwick, K., Bronner-Fraser, M., & Nieto, M. A. (2009). Epithelial-mesenchymal transitions: The importance of changing cell state in development and disease. *The Journal of Clinical Investigation, 119*(6), 1438–1449.

Apitz, H., & Salecker, I. (2014). A challenge of numbers and diversity: Neurogenesis in the Drosophila optic lobe. *Journal of Neurogenetics, 28*(3–4), 233–249.

Apitz, H., & Salecker, I. (2015). A region-specific neurogenesis mode requires migratory progenitors in the Drosophila visual system. *Nature Neuroscience, 18*(1), 46–55.

Aptekar, J. W., Keles, M. F., Lu, P. M., Zolotova, N. M., & Frye, M. A. (2015). Neurons forming optic glomeruli compute figure-ground discriminations in Drosophila. *The Journal of Neuroscience, 35*(19), 7587–7599.

Baumgardt, M., Karlsson, D., Salmani, B. Y., Bivik, C., MacDonald, R. B., Gunnar, E., et al. (2014). Global programmed switch in neural daughter cell proliferation mode triggered by a temporal gene cascade. *Developmental Cell, 30*(2), 192–208.

Bausenwein, B., Dittrich, A. P., & Fischbach, K. F. (1992). The optic lobe of Drosophila melanogaster. II. Sorting of retinotopic pathways in the medulla. *Cell and Tissue Research, 267*(1), 17–28.

Bayraktar, O. A., Boone, J. Q., Drummond, M. L., & Doe, C. Q. (2010). Drosophila type II neuroblast lineages keep Prospero levels low to generate large clones that contribute to the adult brain central complex. *Neural Development, 5*, 26.

Behnia, R., & Desplan, C. (2015). Visual circuits in flies: Beginning to see the whole picture. *Current Opinion in Neurobiology, 34*, 125–132.

Bello, B. C., Izergina, N., Caussinus, E., & Reichert, H. (2008). Amplification of neural stem cell proliferation by intermediate progenitor cells in Drosophila brain development. *Neural Development, 3*, 5.

Bertet, C., Li, X., Erclik, T., Cavey, M., Wells, B., & Desplan, C. (2014). Temporal patterning of neuroblasts controls Notch-mediated cell survival through regulation of Hid or Reaper. *Cell, 158*(5), 1173–1186.

Boone, J. Q., & Doe, C. Q. (2008). Identification of Drosophila type II neuroblast lineages containing transit amplifying ganglion mother cells. *Developmental Neurobiology, 68*(9), 1185–1195.

Borst, A. (2014). Fly visual course control: Behaviour, algorithms and circuits. *Nature Reviews. Neuroscience, 15*(9), 590–599.

Borst, A., & Haag, J. (2002). Neural networks in the cockpit of the fly. *Journal of Comparative Physiology. A, 188*, 419–437.

Borst, A., & Helmstaedter, M. (2015). Common circuit design in fly and mammalian motion vision. *Nature Neuroscience, 18*(8), 1067–1076.

Bowman, S. K., Rolland, V., Betschinger, J., Kinsey, K. A., Emery, G., & Knoblich, J. A. (2008). The tumor suppressors Brat and Numb regulate transit-amplifying neuroblast lineages in Drosophila. *Developmental Cell, 14*(4), 535–546.

Brand, A. H., & Livesey, F. J. (2011). Neural stem cell biology in vertebrates and invertebrates: More alike than different? *Neuron*, *70*, 719–729.
Cajal, S. Ramon Y., & Sanchez, D. (1915). *Contribucion al conociemento de los centros nerviosos de los insectos*. Trabajos del Laboratorio de Investigaciones biológicas de la Universidad de Madrid.
Carthew, R. W. (2007). Pattern formation in the Drosophila eye. *Current Opinion in Genetics & Development*, *17*(4), 309–313.
Chiang, A. S., Lin, C. Y., Chuang, C. C., Chang, H. M., Hsieh, C. H., Yeh, C. W., et al. (2011). Three-dimensional reconstruction of brain-wide wiring networks in Drosophila at single-cell resolution. *Current Biology*, *21*(1), 1–11.
Chin, A. L., Lin, C. Y., Fu, T. F., Dickson, B. J., & Chiang, A. S. (2014). Diversity and wiring variability of visual local neurons in the Drosophila medulla M6 stratum. *The Journal of Comparative Neurology*, *522*(17), 3795–3816.
Collett, T. S. (2008). Insect navigation: Visual panoramas and the sky compass. *Current Biology*, *18*(22), R1058–R1061.
Dearborn, R., Jr., & Kunes, S. (2004). An axon scaffold induced by retinal axons directs glia to destinations in the Drosophila optic lobe. *Development*, *131*(10), 2291–2303.
del Valle Rodriguez, A., Didiano, D., & Desplan, C. (2012). Power tools for gene expression and clonal analysis in Drosophila. *Nature Methods*, *9*(1), 47–55.
Doe, C. Q. (1996). Spindle orientation and asymmetric localization in Drosophila: Both inscuteable? *Cell*, *86*(5), 695–697.
Douglass, J. K., & Strausfeld, N. J. (2003). Anatomical organization of retinotopic motion-sensitive pathways in the optic lobes of flies. *Microscopy Research and Technique*, *62*(2), 132–150.
Egger, B., Boone, J. Q., Stevens, N. R., Brand, A. H., & Doe, C. Q. (2007). Regulation of spindle orientation and neural stem cell fate in the Drosophila optic lobe. *Neural Development*, *2*, 1.
Egger, B., Chell, J. M., & Brand, A. H. (2008). Insights into neural stem cell biology from flies. *Philosophical Transactions of the Royal Society B: Biological Sciences*, *363*(1489), 39–56.
Egger, B., Gold, K. S., & Brand, A. H. (2010). Notch regulates the switch from symmetric to asymmetric neural stem cell division in the Drosophila optic lobe. *Development*, *2987*, 2981–2987. http://dx.doi.org/10.1242/dev.051250.
Egger, B., Gold, K. S., & Brand, A. H. (2011). Regulating the balance between symmetric and asymmetric stem cell division in the developing brain. *Fly (Austin)*, *5*(3), 237–241.
Erclik, T., Hartenstein, V., McInnes, R. R., & Lipshitz, H. D. (2009). Eye evolution at high resolution: The neuron as a unit of homology. *Developmental Biology*, *332*(1), 70–79.
Erclik, T., Li, X., Bertet, C., Baumert, R., Ng, J., Chen, Z., et al., (under review). Selective integration of spatial and temporal inputs generates neural diversity and establishes retinotopy in the Drosophila medulla. Nature.
Fischbach, K. F., & Dittrich, A. P. M. (1989). The optic lobe of Drosophila melanogaster. I. A Golgi analysis of wild-type structure. *Cell and Tissue Research*, *258*(3), 441–475.
Gold, K. S., & Brand, A. H. (2014). Optix defines a neuroepithelial compartment in the optic lobe of the Drosophila brain. *Neural Development*, *9*, 18.
Green, P., Hartenstein, A. Y., & Hartenstein, V. (1993). The embryonic development of the Drosophila visual system. *Cell and Tissue Research*, *273*(3), 583–598.
Grosskortenhaus, R., Pearson, B. J., Marusich, A., & Doe, C. Q. (2005). Regulation of temporal identity transitions in Drosophila neuroblasts. *Developmental Cell*, *8*(2), 193–202.
Hallett, Penelope J., Deleidi, M., Astradsson, A., Smith, Gaynor A., Cooper, O., Osborn, Teresia M., et al. (2015). Successful function of autologous iPSC-Derived dopamine neurons following transplantation in a non-human primate model of Parkinson's disease. *Cell Stem Cell*, *16*, 269–274.

Hasegawa, E., Kaido, M., Takayama, R., & Sato, M. (2013). Brain-specific-homeobox is required for the specification of neuronal types in the Drosophila optic lobe. *Developmental Biology*, *377*(1), 90–99.

Hausen, K. (1984). The lobula-complex of the fly: Structure, function and significance in visual behaviour. In M. A. Ali (Ed.), *Photoreception and vision in invertebrates: Vol. 74* (pp. 523–559). USA: Springer.

Hofbauer, A., & Campos-Ortega, J. A. (1990). Proliferation pattern and early differentiation of the optic lobes in Drosophila melanogaster. *Roux's Archives of Developmental Biology*, *198*, 264–274.

Homberg, U., Heinze, S., Pfeiffer, K., Kinoshita, M., & el Jundi, B. (2011). Central neural coding of sky polarization in insects. *Philosophical Transactions of the Royal Society of London. Series B, Biological Sciences*, *366*(1565), 680–687.

Homem, C. C. F., & Knoblich, J. A. (2012). Drosophila neuroblasts: A model for stem cell biology. *Development*, *139*, 4297–4310.

Huang, Z., & Kunes, S. (1996). Hedgehog, transmitted along retinal axons, triggers neurogenesis in the developing visual centers of the Drosophila brain. *Cell*, *86*(3), 411–422.

Huang, Z., & Kunes, S. (1998). Signals transmitted along retinal axons in Drosophila: Hedgehog signal reception and the cell circuitry of lamina cartridge assembly. *Development*, *125*(19), 3753–3764.

Huang, Z., Shilo, B. Z., & Kunes, S. (1998). A retinal axon fascicle uses spitz, an EGF receptor ligand, to construct a synaptic cartridge in the brain of Drosophila. *Cell*, *95*(5), 693–703.

Imai, T., Sakano, H., & Vosshall, L. B. (2010). Topographic mapping—The olfactory system. *Cold Spring Harbor Perspectives in Biology*, *2*(8), a001776.

Isshiki, T., Pearson, B., Holbrook, S., & Doe, C. Q. (2001). Drosophila neuroblasts sequentially express transcription factors which specify the temporal identity of their neuronal progeny. *Cell*, *106*(4), 511–521.

Ito, M., Masuda, N., Shinomiya, K., Endo, K., & Ito, K. (2013). Systematic analysis of neural projections reveals clonal composition of the Drosophila brain. *Current Biology*, *23*(8), 644–655.

Jenett, A., Rubin, G. M., Ngo, T. T. B., Shepherd, D., Murphy, C., Dionne, H., et al. (2012). A GAL4-driver line resource for Drosophila neurobiology. *Cell Reports*, *2*, 991–1001.

Kandel, E. R., Schwartz, J. H., & Jessell, T. M. (2000). *Principles of neural science* (4th ed.). New York: McGraw-Hill. ISBN: 0-8385-7701-6.

Kang, K. H., & Reichert, H. (2015). Control of neural stem cell self-renewal and differentiation in Drosophila. *Cell and Tissue Research*, *359*(1), 33–45. http://dx.doi.org/10.1007/s00441-014-1914-9. Epub 2014 Jun 6.

Koe, C. T., Li, S., Rossi, F., Wong, J. J. L., Wang, Y., Zhang, Z., et al. (2014). The Brm-HDAC3-Erm repressor complex suppresses dedifferentiation in Drosophila type II neuroblast lineages. *eLife*, *2014*, 1–19.

Krzeptowski, W., Gorska-Andrzejak, J., Kijak, E., Gorlich, A., Guzik, E., Moore, G., et al. (2014). External and circadian inputs modulate synaptic protein expression in the visual system of Drosophila melanogaster. *Frontiers in Physiology*, *5*, 102.

Kumar, J. P. (2012). Building an ommatidium one cell at a time. *Developmental Dynamics*, *241*, 136–149.

Kunes, S., Wilson, C., & Steller, H. (1993). Independent guidance of retinal axons in the developing visual system of Drosophila. *The Journal of Neuroscience*, *13*(2), 752–767.

Lamb, T. D. (2013). Evolution of phototransduction, vertebrate photoreceptors and retina. *Progress in Retinal and Eye Research*, *36*, 52–119.

Langen, M., Agi, E., Altschuler, D. J., Wu, L. F., Altschuler, S. J., & Hiesinger, P. R. (2015). The developmental rules of neural superposition in Drosophila. *Cell, 162*(1), 120–133.
Lee, R. C., Clandinin, T. R., Lee, C.-H., Chen, P.-L., Meinertzhagen, I. A., & Zipursky, S. L. (2003). The protocadherin Flamingo is required for axon target selection in the Drosophila visual system. *Nature Neuroscience, 6*, 557–563.
Lee, C. H., Herman, T., Clandinin, T. R., Lee, R., & Zipursky, S. L. (2001). N-cadherin regulates target specificity in the Drosophila visual system. *Neuron, 30*(2), 437–450.
Li, X., Chen, Z., & Desplan, C. (2013). Temporal patterning of neural progenitors in Drosophila. *Current Topics in Developmental Biology, 105*, 69–96.
Li, X., Erclik, T., Bertet, C., Chen, Z., Voutev, R., Venkatesh, S., et al. (2013). Temporal patterning of Drosophila medulla neuroblasts controls neural fates. *Nature, 498*(7455), 456–462.
Li, H. H., Kroll, Jason R., Lennox, Sara M., Ogundeyi, O., Jeter, J., Depasquale, G., et al. (2014). A GAL4 driver resource for developmental and behavioral studies on the Larval CNS of Drosophila. *Cell Reports, 8*, 897–908.
Lin, S., Kao, C. F., Yu, H. H., Huang, Y., & Lee, T. (2012). Lineage analysis of Drosophila lateral antennal lobe neurons reveals notch-dependent binary temporal fate decisions. *PLoS Biology, 10*(11), e1001425.
Lin, S., & Lee, T. (2012). Generating neuronal diversity in the Drosophila central nervous system. *Developmental Dynamics, 241*(1), 57–68.
Liu, H., & Zhang, S.-C. (2011). Specification of neuronal and glial subtypes from human pluripotent stem cells. *Cellular and Molecular Life Sciences: CMLS, 68*, 3995–4008.
Livesey, F. J., & Cepko, C. L. (2001). Vertebrate neural cell-fate determination: Lessons from the retina. *Nature Reviews. Neuroscience, 2*(2), 109–118.
Lovick, J. K., Ngo, K. T., Omoto, J. J., Wong, D. C., Nguyen, J. D., & Hartenstein, V. (2013). Postembryonic lineages of the Drosophila brain: I. Development of the lineage-associated fiber tracts. *Developmental Biology, 384*(2), 228–257.
Maisak, M. S., Haag, J., Ammer, G., Serbe, E., Meier, M., Leonhardt, A., et al. (2013). A directional tuning map of Drosophila elementary motion detectors. *Nature, 500*(7461), 212–216.
Mattar, P., Ericson, J., Blackshaw, S., & Cayouette, M. (2015). A conserved regulatory logic controls temporal identity in mouse neural progenitors. *Neuron, 85*(3), 497–504.
Meinertzhagen, I. A., & Hanson, T. E. (1993). The development of the optic lobe. In M. Bate & A. Martinez Arias (Eds.), *The development of Drosophila melanogaster* (pp. 1363–1491): New York: Cold Spring Harbor Laboratory Press.
Meinertzhagen, I. A., & Sorra, K. E. (2001). Synaptic organization in the fly's optic lamina: Few cells, many synapses and divergent microcircuits. *Progress in Brain Research, 131*, 53–69. chapter 3.
Meinertzhagen, I. A., Takemura, S. Y., Lu, Z., Huang, S., Gao, S., Ting, C. Y., et al. (2009). From form to function: The ways to know a neuron. *Journal of Neurogenetics, 23*(1–2), 68–77.
Morante, J., & Desplan, C. (2004). Building a projection map for photoreceptor neurons in the Drosophila optic lobes. *Seminars in Cell & Developmental Biology, 15*(1), 137–143.
Morante, J., & Desplan, C. (2008). The color-vision circuit in the medulla of Drosophila. *Current Biology, 18*(8), 553–565.
Moroz, L. L. (2009). On the independent origins of complex brains and neurons. *Brain, Behavior and Evolution, 74*(3), 177–190.
Mu, L., Ito, K., Bacon, J. P., & Strausfeld, N. J. (2012). Optic glomeruli and their inputs in Drosophila share an organizational ground pattern with the antennal lobes. *The Journal of Neuroscience, 32*(18), 6061–6071.

Nassif, C., Noveen, A., & Hartenstein, V. (2003). Early development of the Drosophila brain: III. The pattern of neuropile founder tracts during the larval period. *The Journal of Comparative Neurology*, *455*(4), 417–434.
Neriec, N., Dua, W., Mehmet, O., Pinto Texeira Sousa, F., Li, X. Erclik, T., et al., (Submitted). Cell migration and neuronal diversity during optic lobe neurogenesis in Drosophila. eLife.
Otsuna, H., & Ito, K. (2006). Systematic analysis of the visual projection neurons of Drosophila melanogaster. I. Lobula-specific pathways. *The Journal of Comparative Neurology*, *497*(6), 928–958.
Otsuna, H., Shinomiya, K., & Ito, K. (2014). Parallel neural pathways in higher visual centers of the Drosophila brain that mediate wavelength-specific behavior. *Frontiers in Neural Circuits*, *8*, 8.
Paulk, A., Millard, S. S., & van Swinderen, B. (2011). Vision in Drosophila: Seeing the world through a model's eyes. *Annual Review of Entomology*, *58*, 313–332.
Pfeiffer, B. D., Jenett, A., Hammonds, A. S., Ngo, T. T., Misra, S., Murphy, C., et al. (2008). Tools for neuroanatomy and neurogenetics in Drosophila. *Proceedings of the National Academy of Sciences of the United States of America*, *105*(28), 9715–9720.
Raghu, S. V., Claussen, J., & Borst, A. (2013). Neurons with GABAergic phenotype in the visual system of Drosophila. *The Journal of Comparative Neurology*, *521*(1), 252–265.
Raghu, S. V., Joesch, M., Sigrist, S. J., Borst, A., & Reiff, D. F. (2009). Synaptic organization of lobula plate tangential cells in Drosophila: Dα7 cholinergic receptors. *Journal of Neurogenetics*, *23*(1–2), 200–209.
Reichert, H. (2011). Drosophila neural stem cells: Cell cycle control of self-renewal, differentiation, and termination in brain development. *Results and Problems in Cell Differentiation*, *53*, 529–546.
Sanes, D. H., Reh, T. A., & Harris, W. A. (2012). 1—Neural induction. In D. H. Sanes, T. A. Reh, & W. A. Harris (Eds.), *Development of the nervous system* (3rd ed., pp. 1–22). London: Academic Press.
Sanes, J. R., & Zipursky, S. L. (2010). Design principles of insect and vertebrate visual systems. *Neuron*, *66*(1), 15–36.
Sato, M., Suzuki, T., & Nakai, Y. (2013). Waves of differentiation in the fly visual system. *Developmental Biology*, *380*(1), 1–11.
Selleck, S. B., Gonzalez, C., Glover, D. M., & White, K. (1992). Regulation of the G1-S transition in postembryonic neuronal precursors by axon ingrowth. *Nature*, *355*(6357), 253–255.
Serrano-Saiz, E., Poole, R. J., Felton, T., Zhang, F., De La Cruz, E. D., & Hobert, O. (2013). Modular control of glutamatergic neuronal identity in C. elegans by distinct homeodomain proteins. *Cell*, *155*(3), 659–673.
Silies, M., Gohl, D. M., & Clandinin, T. R. (2014). Motion-detecting circuits in flies: Coming into view. *Annual Review of Neuroscience*, *37*(1), 307–327.
Silies, M., Gohl, D., Fisher, Y., Freifeld, L., Clark, D., & Clandinin, T. R. (2013). Modular use of peripheral input channels tunes motion-detecting circuitry. *Neuron*, *79*, 111–127.
Skeath, J. B. (1999). At the nexus between pattern formation and cell-type specification: The generation of individual neuroblast fates in the drosophila embryonic central nervous system. *BioEssays*, *21*, 922–931.
Skeath, J. B., & Thor, S. (2003). Genetic control of Drosophila nerve cord development. *Current Opinion in Neurobiology*, *13*, 8–15.
Sousa-Nunes, R., & Somers, W. G. G. (2013). Mechanisms of asymmetric progenitor divisions in the Drosophila central nervous system. In G. Hime & H. Abud (Eds.), *Transcriptional and translational regulation of stem cells: Vol. 786* (pp. 79–102). Dordrecht, The Netherlands: Springer.

Southall, T. D., & Brand, A. H. (2009). Neural stem cell transcriptional networks highlight genes essential for nervous system development. *The EMBO Journal*, *28*(24), 3799–3807.
Spalthoff, C., Gerdes, R., & Kurtz, R. (2012). Neuronal representation of visual motion and orientation in the fly medulla. *Frontiers in Neural Circuits*, *6*, 72.
Strausfeld, N. (1989). Insect vision and olfaction: Common design principles of neuronal organization. In N. J. Strausfeld & R. N. Singh (Eds.), *Neurobiology of sensory systems* (pp. 319–353). New York: Plenum.
Strausfeld, N. J. (2009). Brain organization and the origin of insects: An assessment. *Proceedings of the Biological Sciences*, *276*(1664), 1929–1937.
Strausfeld, N. J., & Hirth, F. (2013a). Deep homology of arthropod central complex and vertebrate basal ganglia. *Science*, *340*(6129), 157–161.
Strausfeld, N. J., & Hirth, F. (2013b). Homology versus convergence in resolving transphyletic correspondences of brain organization. *Brain, Behavior and Evolution*, *82*, 215–219.
Suzuki, T., Kaido, M., Takayama, R., & Sato, M. (2013). A temporal mechanism that produces neuronal diversity in the Drosophila visual center. *Developmental Biology*, *380*(1), 12–24.
Takemura, S. Y., Lu, Z., & Meinertzhagen, I. A. (2008). Synaptic circuits of the Drosophila optic lobe: The input terminals to the medulla. *The Journal of Comparative Neurology*, *509*(5), 493–513.
Technau, G. M. (2008). Advances in experimental medicine and biology. Brain development in Drosophila melanogaster. Preface. *Advances in Experimental Medicine and Biology*, *628*, v–vi.
Technau, G. M., Berger, C., & Urbach, R. (2006). Generation of cell diversity and segmental pattern in the embryonic central nervous system of Drosophila. *Developmental Dynamics*, *235*, 861–869.
Timofeev, K., Joly, W., Hadjieconomou, D., & Salecker, I. (2012). Localized netrins act as positional cues to control layer-specific targeting of photoreceptor axons in Drosophila. *Neuron*, *75*(1), 80–93.
Treisman, J. E. (2013). Retinal differentiation in Drosophila. *Wiley Interdisciplinary Reviews. Developmental Biology*, *2*(4), 545–557.
Tuthill, J., Nern, A., Holtz, S., Rubin, G., & Reiser, M. B. (2013). Contributions of the 12 neuron classes in the fly lamina to motion vision. *Neuron*, *79*, 128–140.
Udolph, G. (2012). Notch signaling and the generation of cell diversity in Drosophila neuroblast lineages. *Advances in Experimental Medicine and Biology*, *727*, 47–60.
Urbach, R., & Technau, G. M. (2003). Early steps in building the insect brain: Neuroblast formation and segmental patterning in the developing brain of different insect species. *Arthropod Structure & Development*, *32*, 103–123.
Varija Raghu, S., Reiff, D. F., & Borst, A. (2011). Neurons with cholinergic phenotype in the visual system of Drosophila. *The Journal of Comparative Neurology*, *519*(1), 162–176.
Viktorin, G., Riebli, N., & Reichert, H. (2013). A multipotent transit-amplifying neuroblast lineage in the central brain gives rise to optic lobe glial cells in Drosophila. *Developmental Biology*, *379*(2), 182–194.
Wallace, K., Liu, T. H., & Vaessin, H. (2000). The pan-neural bHLH proteins DEADPAN and ASENSE regulate mitotic activity and cdk inhibitor dacapo expression in the Drosophila larval optic lobes. *Genesis*, *26*(1), 77–85.
Wang, S., Sengel, C., Emerson, M. M., & Cepko, C. L. (2014). A gene regulatory network controls the binary fate decision of rod and bipolar cells in the vertebrate retina. *Developmental Cell*, *30*(5), 513–527.
Wang, Y.-C., Yang, J. S., Johnston, R., Ren, Q., Lee, Y.-J., Luan, H., et al. (2014). Drosophila intermediate neural progenitors produce lineage-dependent related series of diverse neurons. *Development*, *141*, 253–258.

White, K., & Kankel, D. R. (1978). Patterns of cell division and cell movement in the formation of the imaginal nervous system in Drosophila melanogaster. *Developmental Biology*, *65*(2), 296–321.

Wolff, G. H., & Strausfeld, N. J. (2015). Genealogical correspondence of mushroom bodies across invertebrate phyla. *Current Biology*, *25*(1), 38–44.

Wong, D. C., Lovick, J. K., Ngo, K. T., Borisuthirattana, W., Omoto, J. J., & Hartenstein, V. (2013). Postembryonic lineages of the Drosophila brain: II. Identification of lineage projection patterns based on MARCM clones. *Developmental Biology*, *384*(2), 258–289.

Wu, P. S., Egger, B., & Brand, A. H. (2008). Asymmetric stem cell division: Lessons from Drosophila. *Seminars in Cell and Developmental Biology*, *19*, 283–293.

Yasugi, T., Umetsu, D., Murakami, S., Sato, M., & Tabata, T. (2008). Drosophila optic lobe neuroblasts triggered by a wave of proneural gene expression that is negatively regulated by JAK/STAT. *Development*, *135*(8), 1471–1480.

Yasugi, T., Sugie, A., Umetsu, D., & Tabata, T. (2010). Coordinated sequential action of EGFR and Notch signaling pathways regulates proneural wave progression in the Drosophila optic lobe. *Development*, *137*, 3193–3203.

Younossi-Hartenstein, A., Nassif, C., Green, P., & Hartenstein, V. (1996). Early neurogenesis of the Drosophila brain. *The Journal of Comparative Neurology*, *370*(3), 313–329.

Yu, H. H., Awasaki, T., Schroeder, M. D., Long, F., Yang, J. S., He, Y., et al. (2013). Clonal development and organization of the adult Drosophila central brain. *Current Biology*, *23*, 633–643.

Zeil, J. (2012). Visual homing: An insect perspective. *Current Opinion in Neurobiology*, *22*(2), 285–293.

Zhu, Y. (2013). The Drosophila visual system: From neural circuits to behavior. *Cell Adhesion & Migration*, *7*(4), 333–344.

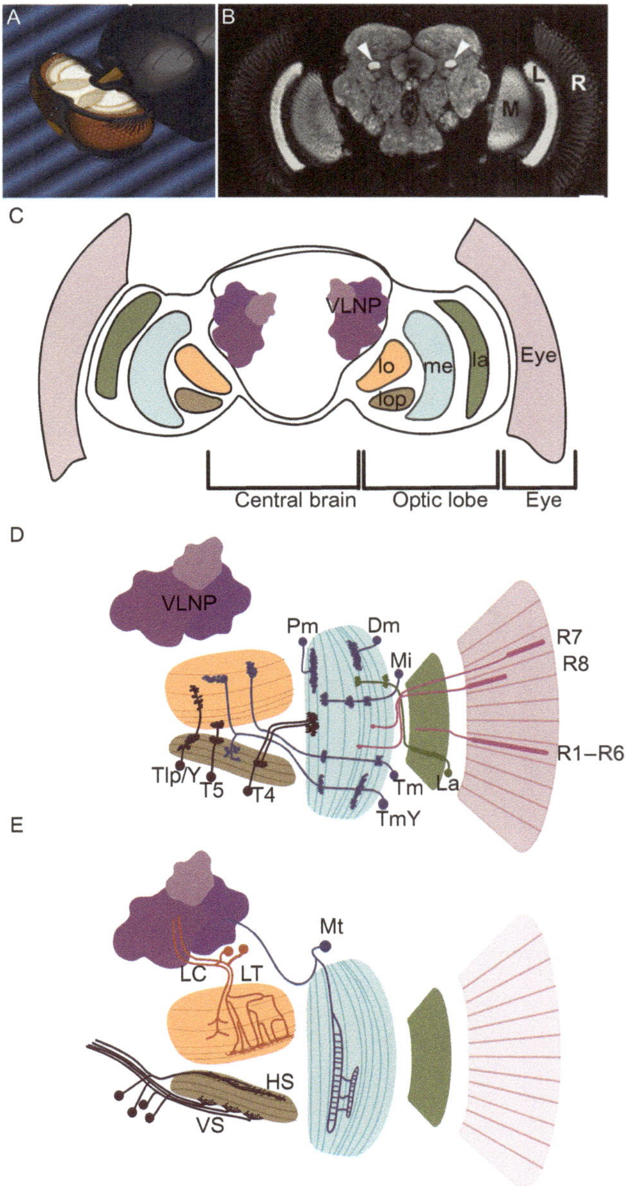

Nathalie Nériec and Claude Desplan, Figure 1 The visual system of *Drosophila*. (A) The fly brain in the head. (B) Neuropils of the optic lobes and central brain. (C) Cross section of the fly brain indicating the different parts of the visual system: eye, lamina (la), medulla (me), lobula (lo), lobula plate (lop), and ventrolateral neuropils (VLNPs). (D) Interneurons of the optic lobe: outer and inner photoreceptors (pink), lamina neurons (La, green), medulla interneurons (dark blue), distal (Dm) and proximal medulla (Pm); medulla intrinsic (Mi); unicolumnar transmedullary (Tm) or multicolumnar TmY; lobula plate interneurons (T4, T5, Tlp, and Y neurons, dark brown). (E) Projections neurons: medulla tangential (Mt, dark blue); lobula columnar (LC, red) and tangential/tree-like (LT, red); lobula plate tangential cells: HS and VS (dark brown). *Panel (A) adapted from Spalthoff, Gerdes, and Kurtz (2012). Panel (B) from Krzeptowski et al. (2014). Panels (D and E) adapted from Erclik, Hartenstein, McInnes, and Lipshitz (2009).*

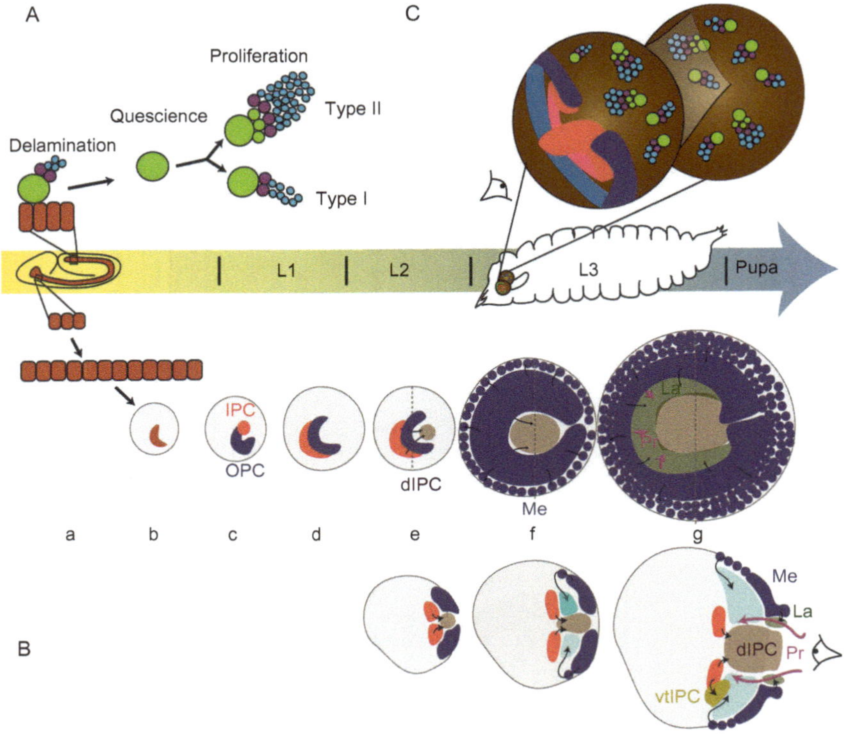

Nathalie Nériec and Claude Desplan, Figure 2 The development of the optic lobe of *Drosophila*. (A) Central brain: neuroblasts (green) delaminate from the neuroepithelium (red) during the embryonic stages and generate GMCs (purple) and larval neurons (blue) before they become quiescent. At L1, Type I and Type II neuroblasts are reactivated and exhibit different modes of proliferation to generate adult neurons. (B) Optic lobe: (a) Cells in the procephalic regions generate the optic placode neuroepithelium (red), which invaginates (b). (c) The optic placode splits into inner proliferation center (IPC, pink) and outer proliferation center (OPC, blue) and adopt a crescent shape (d). (e) Cells migrate out of the IPC to form the dIPC (light brown). (f) The medial edge of the OPC becomes neuroblasts that will produce the medulla (round blue cells). (g) The inner OPC generates lamina progenitors (La, green) that will produce lamina neurons upon induction by photoreceptors (Pr, pink arrows). The ventral tip of the IPC generates lobula complex neurons (vtIPC, dark yellowish green). Lower diagrams are cross sections at the dashed line in upper diagrams. (C) 3D representation of the brain at L3. Central brain neuroblasts and their progeny, OPC (blue) and IPC (pink). The eye symbol shows the view angle of panel (B). *Panels (A and C) adapted from Homem and Knoblich (2012). Panel (B) adapted from Nassif, Noveen, and Hartenstein (2003).*

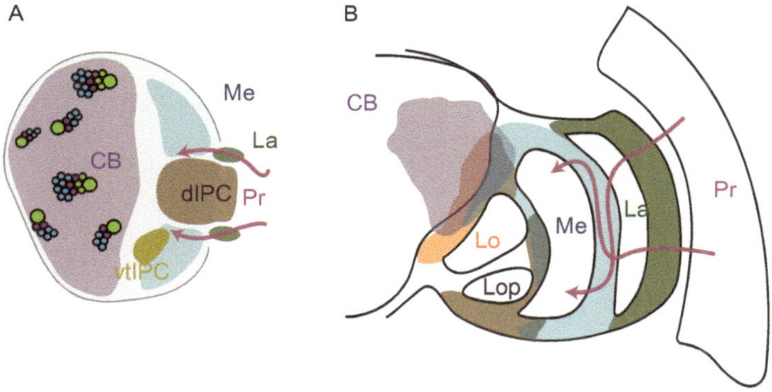

Nathalie Nériec and Claude Desplan, Figure 3 Linking development to adult fate. Fate of the different neuronal populations found at the L3 larval stage (A) in the adult visual system (B). Pr, photoreceptors; La, lamina; Me, medulla; Lo, lobula; Lop, lobula plate; dIPC, distal IPC; vtIPC, ventral tip of the IPC; CB, central brain.

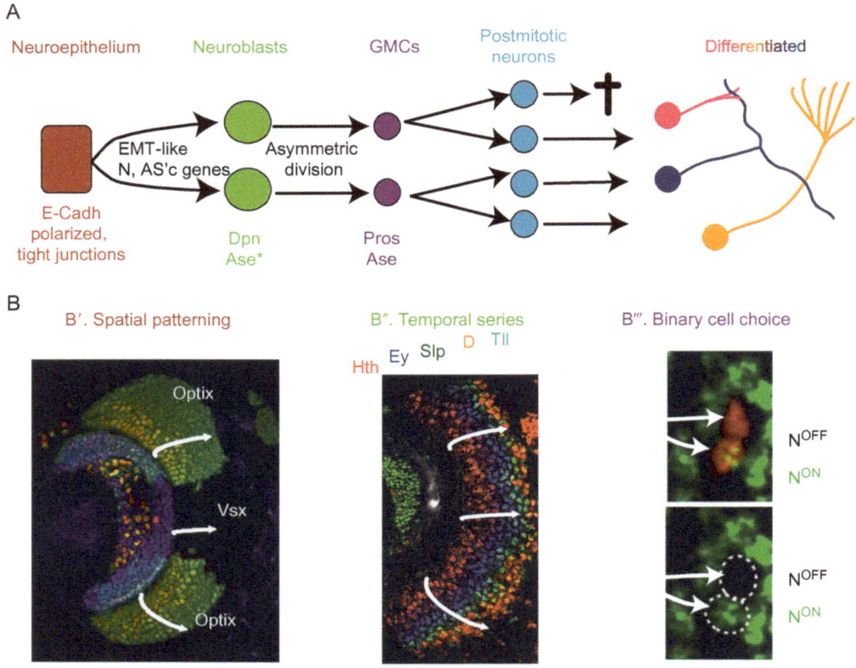

Nathalie Nériec and Claude Desplan, Figure 4 The making of a functional neuronal network. (A) Common mechanisms for neurogenesis: polarized neuroepithelial cells (red) undergo EMT-like transition to become neuroblasts (green) which generate GMCs (purple). GMCs produce postmitotic neurons (blue) that form adult neuronal circuits. * Dpn is found in all neuroblasts while Ase is found in most neuroblasts (see text). (B) Common mechanisms to generate neuronal diversity: (B′) Spatial patterning of the neuroepithelium illustrated by the expression of Optix and Vsx (Gold & Brand, 2014). (B″) Temporal patterning of neuroblasts. In the OPC, neuroblasts sequentially express a series of transcription factors, Hth (red), Ey (blue), Slp (green), D (orange), and Tll (cyan) as they age (arrows) (Li, Chen, & Desplan, 2013; Li, Erclik, et al., 2013). (B‴) Notch-dependent binary cell fate choices during the division of GMCs. Only one of the two daughter cells (red clone, dashed line) receives Notch activity and expresses Apterous (green), while the other does not (Li, Chen, et al., 2013; Li, Erclik, et al., 2013).

CHAPTER FIFTEEN

Zygotic Genome Activators, Developmental Timing, and Pluripotency

Daria Onichtchouk[1], Wolfgang Driever[1]
Developmental Biology Unit, Institute Biology I, Faculty of Biology, and Center for Biological Signaling Studies (BIOSS), Albert-Ludwigs-University, Freiburg, Germany
[1]Corresponding authors: e-mail address: onichtchouk@biologie.uni-freiburg.de; driever@biologie.uni-freiburg.de

Contents

Abstract

The transcription factors Pou5f1, Sox2, and Nanog are central regulators of pluripotency in mammalian ES and iPS cells. In vertebrate embryos, Pou5f1/3, SoxB1, and Nanog control zygotic genome activation and participate in lineage decisions. We review the current knowledge of the roles of these genes in developing vertebrate embryos from fish to mammals and suggest a model for pluripotency gene regulatory network functions in early development.

Current Topics in Developmental Biology, Volume 116
ISSN 0070-2153
http://dx.doi.org/10.1016/bs.ctdb.2015.12.004

1. INTRODUCTION

Zygotic genome activation (ZGA) in developing vertebrates is followed by a pluripotency period, when cells of multicellular organism are developmentally unrestricted and capable of realizing multiple cell fates (Tadros & Lipshitz, 2009). Cell transplantation studies in mice, chick, zebrafish, and frogs have demonstrated that most cells remain pluripotent during morphogenetic gastrulation movements (Domingo & Keller, 2000; Garcia-Martinez & Schoenwolf, 1992; Ho & Kimmel, 1993; Lawson, Meneses, & Pedersen, 1991; Okabayashi & Asashima, 2003). Interest in the genetic control of ZGA in model organisms was recently revived by the finding that close homologues of mammalian embryonic stem (ES)/induced pluripotent (iPS) cell "pluripotency transcription factors" activate zygotic gene expression in zebrafish (Lee et al., 2013; Leichsenring, Maes, Mossner, Driever, & Onichtchouk, 2013) and, presumably, in other vertebrate embryos. ZGA also represents the transition of highly controlled regulatory cell states derived from sperm and oocyte, and often under maternal developmental control, to pluripotent cells. This transition is not a specific feature of vertebrates, but, as studies in *Drosophila* have revealed (Harrison, Li, Kaplan, Botchan, & Eisen, 2011; Liang et al., 2008), may be controlled by similar fundamental mechanisms in metazoans. In this review, we focus on the links between pluripotency, zygotic genome activation, and temporal control in early embryonic development. We summarize the current progress in the field, present a model to link the pluripotency network to early lineage specification in the embryo, and discuss important open issues.

2. PLURIPOTENCY GENE REGULATORY NETWORK: SELF-ACTIVATION OF PLURIPOTENCY GENES, AND LINEAGE PRIMING

It is possible to derive at least two distinct *in vitro* cultures of indefinitely pluripotent cells from mammalian embryos: embryonic stem cells (Evans & Kaufman, 1981; Thomson et al., 1998) and epiblast stem cells (Tesar et al., 2007). These two different pluripotent states exhibit distinct molecular signatures and depend on signaling pathways that often antagonize each other (Boyer et al., 2005; Chenoweth, McKay, & Tesar, 2010; Chenoweth & Tesar, 2010; Gafni et al., 2013; Hanna, Saha, & Jaenisch, 2010; Loh et al., 2006). Both pluripotent cell types express several transcription factors (TFs) playing a central role in the maintenance of pluripotency: Pou5f1/Oct4, Sox2, and Nanog, also called "core pluripotency transcription factors." The pivotal role of TFs in pluripotency was underscored by reprogramming experiments in which the combination of Oct4, Sox2, Klf4, and c-Myc (OKSM) was used to reprogram differentiated cells to pluripotency (Takahashi & Yamanaka, 2006). Subsequently, a number of alternative combinations of reprogramming factors were characterized, which employ various factors that promote the establishment of the core transcriptional circuitry of stem cells: Nanog, Lin28, ESRRB, NR5A2, Prdm14, E-cadherin (Buganim et al., 2014; Chia et al., 2010; Feng et al., 2009; Heng et al., 2010; Ichida et al., 2009; Redmer et al., 2011; Yu et al., 2007). Studies in ES cells led to the predominant view that key pluripotency factors maintain an undifferentiated ES cell phenotype while lineage-specific factors work in opposition to promote cell specialization (Young, 2011). However, this view was recently challenged by reprogramming experiments, where mixtures of lineage-specific factors, not expressed in ES cells, were used successfully to induce pluripotency in mouse and human fibroblasts (Montserrat et al., 2013; Shu et al., 2013). Rather than the previously prevailing model of self-activation of "pluripotency genes," these findings lead to a "seesaw model" of the pluripotent state as a balancing equilibrium that results from continuous mutual competition between alternative lineage specification forces. This view is also supported by recent findings, that apart from maintaining self-renewal in ES cells, Pou5f1/Oct4, Sox2, and Nanog participate in lineage priming in the initial stages of ES cell differentiation (Teo et al., 2011; Thomson et al., 2011) and in mouse embryos (discussed later in Section 5).

Although genetic mechanisms supporting indefinite pluripotency were mostly studied in mammalian systems, they are not a unique property of the mammalian lineage. ES cells were successfully derived from some non-mammalian species, such as birds (Pain et al., 1996) and medaka fish (Hong, He, Schartl, & Hong, 2013; Hong, Winkler, & Schartl, 1996, 1998). Medaka Oct4, the paralog of mammalian Oct4, is indispensable for pluripotency in medaka ES cells (Liu et al., 2015), demonstrating that indefinite pluripotency in lower vertebrate cells is controlled by the same core factor as in mammals. Transfection of mammalian OKSM induces a partially reprogrammed pluripotent stem cell state in birds, fish, and fly, suggesting that some features of the pluripotency gene regulatory network could be dated back to common ancestors which lived 550 million years ago (Rossello et al., 2013). Thus, the experimental advantages of developmental model organisms enable to study regulatory mechanisms used by individual pluripotency factors in the context of embryonic development. Understanding the early gene regulatory networks structure in these organisms may shed a light on mechanisms underlying the reprogramming process and provide a basis for rational optimization of clinically relevant reprogramming efforts.

3. MATERNAL PERIOD IN ANIMAL DEVELOPMENT AND MECHANISMS OF ZYGOTIC GENOME ACTIVATION

The genome of multicellular organism is not immediately used to instruct development of the egg after fertilized. This fact was already recognized in the late nineteenth/early twentieth century, when most biologists working on echinoderms, gastropods, fish, and amphibia believed in maternal origin of pregastrulation development (Boveri, 1918; Conklin, 1908; Godlevski, 1906). Cross-hybridization experiments between different species of echinoderms and amphibia led to the conclusion that, while the fertilized oocyte first develops independently from parental nuclear function, development starting from gastrula period requires the function of both male and female nuclei (Boveri, 1918; Moore, 1941; Rugh & Exener, 1940). This early view was refined by series of X-ray nuclei inactivation experiments performed in teleost fish (loach *Misgurnus fossilis*) by Neyfakh. When zygotic nuclei were inactivated between fertilization and the end of early blastula, the embryo development proceeded till late blastula stage and arrested. Inactivation of nuclei at stages later than midblastula leads to later developmental arrest, indicating that nuclear activity starts from midblastula stage (Neyfakh, 1959, 1964; Neyfakh & Rott, 1968) reviewed in Korzh (2009). The concept of midblastula transition (MBT) was defined by Signoret and Lefresne (1971) reviewed in Gerhart (1980) as a transition from early rapid synchronous to asynchronous cell cycles, occurring at the 10–11th cell division in axolotl. Studies of MBT in *Xenopus* (Newport & Kirschner, 1982a, 1982b), followed by studies in zebrafish (Kane & Kimmel, 1993) and *Fundulus* (Trinkaus, 1992), demonstrated that in parallel to the lengthening of the cell cycle at MBT, the blastomeres become motile and transcriptionally active for the first time. Like in amphibians and fish, lengthening of the cell cycle at *Drosophila* MBT occurs coordinately with both the onset of the bulk transcription from the zygotic genome (Edgar, Kiehle, & Schubiger, 1986) and is marked by cellularization of the syncytial blastoderm. Currently, the term MBT is used mostly for amphibians, fish, and *Drosophila*, where the slowing and progressive desynchronization of the cell cycle occurs at the same time.

The broader definition, maternal-to-zygotic transition (MZT) is applicable to all multicellular animals and relates to the period encompassing zygotic genome activation (ZGA), regulated degradation of maternal transcripts, changes in the cell cycle, acquisition of cell motility, and other events, manifesting transition to zygotic control over development (Giraldez et al., 2006; Langley, Smith, Stemple, & Harvey, 2014; Yartseva & Giraldez, 2015). Although the MZT phenomenon has been studied for more than 100 years in different model organisms, the underlying mechanisms are not yet clear.

Cell cycle duration and ZGA in *Xenopus* and zebrafish depend on reaching a critical ratio of nucleus to cytoplasm, which increases during early cleavages when the cell volume becomes progressively smaller (Kane & Kimmel, 1993; Newport & Kirschner, 1982a; Rott & Sheveleva, 1968). Pre-MBT suppression of transcription in *Xenopus* can be reversed by addition of competing DNA (Newport & Kirschner, 1982b), or by block of DNA synthesis (Kimelman, Kirschner, & Scherson, 1987), suggesting that the pre-MBT egg is fully competent for transcription, and that MBT is triggered by DNA through titration of suppressor components present in the egg. Limiting amounts of four replication factors, titrated by exponentially increasing DNA material, were recently suggested as a reason for slowing the cell cycle in *Xenopus* at MBT (Collart, Allen, Bradshaw, Smith, & Zegerman, 2013), while excess of free histones was suggested to prevent premature transcription before MBT (Almouzni & Wolffe, 1995; Amodeo, Jukam, Straight, & Skotheim, 2015). In *Drosophila*, cell cycle duration is also under control of the nuclear/cytoplasmic ratio (Edgar et al., 1986), but initiation of most zygotically transcribed genes occurs during a limited time interval hours after fertilization, indicating the presence of a separate timing mechanism that operates independently of the nuclear-to-cytoplasmic (N/C) ratio (Lu, Li, Elemento, Tavazoie, & Wieschaus, 2009).

During zygotic genome activation (ZGA), a subset of zygotic genes starts to be transcribed in a rapid and temporally coordinated manner. The estimated number of actively transcribed genes is similar in zebrafish, *Drosophila*, and mice, representing about 12–15% of all genes in the respective genomes (Tadros & Lipshitz, 2009). General permissive mechanisms, allowing bulk transcription, such as lengthening of the cell cycle by increasing N/C ratio, titration of repressory substances, or appearance of new histone modifications, permit ZGA, but cannot fully account for temporal specificity of early transcriptional events, including simultaneous selective and steep activation of several hundreds of genes. Specificity of ZGA is explained by involvement of specific activators of transcription, which were characterized in

Drosophila and zebrafish. In *Drosophila*, the earliest zygotic genes are primed for activation by a single TF, Zelda (Zld). Zld binds to TAGteam motifs, which are specifically present in the enhancers of early genes regulating developmental processes (Kvon, Stampfel, Yanez-Cuna, Dickson, & Stark, 2012; ten Bosch, Benavides, & Cline, 2006; Yanez-Cuna, Dinh, Kvon, Shlyueva, & Stark, 2012). Zebrafish ZGA is controlled by Pou5f3, Nanog, and Sox19b, the maternally expressed member of the functionally redundant SoxB1 group of TFs (Lee et al., 2013; Leichsenring et al., 2013; Okuda, Ogura, Kondoh, & Kamachi, 2010). Pou5f3 and SoxB1 bind preferentially to the regulatory regions of the earliest-expressed zygotic genes in zebrafish, while Nanog occupies regulatory regions of the broader target gene range, including developmental regulators expressed during organogenesis stages (Leichsenring et al., 2013; Xu et al., 2012). A scheme of ZGA in zebrafish is shown in Fig. 1.

Although *Drosophila* and zebrafish zygotic gene activators are not phylogenetically related, they share several common features. High co-occupancy with multiple TFs is a characteristic feature of Zld-binding sites in *Drosophila* (Satija & Bradley, 2012; ten Bosch et al., 2006), binding regions for Pou5f3, SoxB1, and Nanog in zebrafish (Leichsenring et al., 2013) and "core pluripotency factors" in mammalian ES cells (Chen et al., 2008; Kim, Chu, Shen, Wang, & Orkin, 2008). Both Zld and Pou5f3 bind to their target enhancers before MBT and remain bound at least until gastrulation (Harrison et al., 2011; Leichsenring et al., 2013; Nien et al., 2011). During the cell cycles following MBT, Zld sites have been proposed to coordinate expression of early gene regulatory batteries in time and space (Nien et al., 2011). Similarly, Pou5f1 and SoxB1 ensure activation of hundreds of developmental genes at the earliest time point of zygotic transcription, and later control temporal coordination of gene expression programs in zebrafish (Onichtchouk et al., 2010). Thus, gene activation at MBT and further progression of early developmental events are coupled and coordinated by zygotic genome activators in both *Drosophila* and zebrafish.

"Master pluripotency TFs" Oct4, Sox2, and Nanog in ES cells (Mullen et al., 2011; Tallack et al., 2010) and the zygotic gene activator Zld in *Drosophila* (Foo et al., 2014) are thought to facilitate binding of multiple tissue-specific TFs to their binding sites, presumably by "opening" chromatin structure. The nature of the chromatin "opening" mechanisms is not firmly established, and whether the same mechanisms of chromatin modification are used in all systems remains poorly investigated. Similarly to ES cells, early

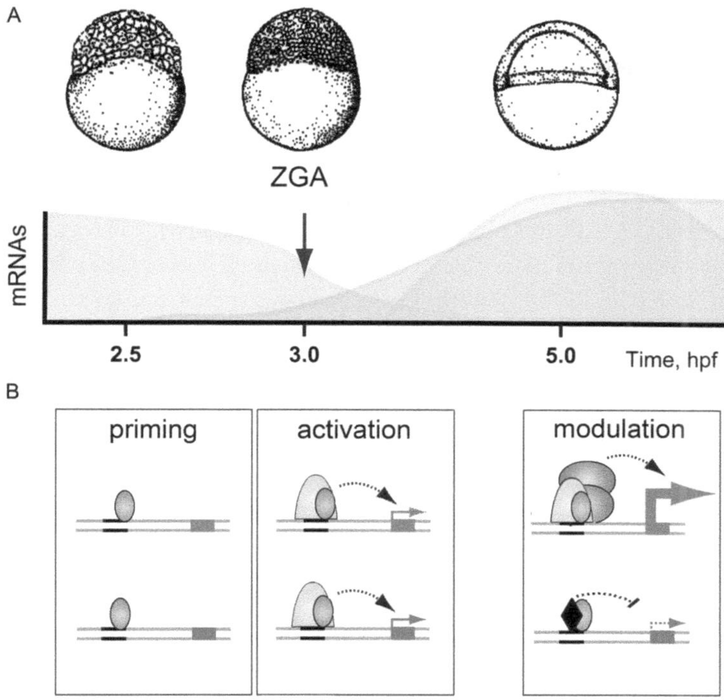

Figure 1 Zygotic genome activation. (A) Schematic drawing of ZGA in zebrafish: starting from 1000 cell stage (middle embryo at the top), early zygotic genes are switched on, and some maternal mRNAs degrade. Hpf, hours postfertilization. (B) Hypothetical scheme of action of zygotic gene activators. Already before ZGA, zygotic genome activators (small oval) bind to defined regions in the genome and prime the neighboring genes for expression (left box, two example genes are shown). Ubiquitous cofactors (RNA polymerase or chromatin modifier enzyme; half oval) are attracted to zygotic genome activator binding site and switch on gene expression (middle box). Later, the binding sites for zygotic genome activators are occupied by additional tissue-specific factors, which either repress (gene at bottom) or enhance (gene at top) activation of transcription (right box). (A color version of this figure is located at the end of this chapter.)

embryos are characterized by "open chromatin" and high transcription levels (Bogdanovic et al., 2012; Mikkelsen et al., 2007; Vastenhouw et al., 2010). Zld (Li, Harrison, Villalta, Kaplan, & Eisen, 2014), Pou5f3, SoxB1 (Leichsenring et al., 2013), and Nanog (Bogdanovic et al., 2012) binding regions are enriched in H3 tail modifications of "open chromatin" (H3K4me3, H3K27ac, H3K4me1), but the role and scope of ZGA TFs in establishing these modifications remain unexplored.

Before zygotic genome activation, the zebrafish genome is already "prepatterned" by modified histones (Lindeman et al., 2011) and DNA methylation at specific loci (Potok, Nix, Parnell, & Cairns, 2013). What is the role of sequence-specific ZGA factors in establishing and changing epigenetic patterns in pre-MBT or MBT stage embryos? One possibility to be tested is a direct recruitment of chromatin remodelers by all or some of the three ZGA TFs to their binding sites, as was shown for regulation of early Wnt-target genes by beta-catenin in *Xenopus* (Blythe, Cha, Tadjuidje, Heasman, & Klein, 2010). Another nonexclusive scenario involves possible "pioneer factor" nucleosome displacement activity at ZGA, mediated by TF binding directly to nucleosomes, analogous to what has been shown for mouse Oct4 and Sox2 during reprogramming process (Soufi, Donahue, & Zaret, 2012; Soufi et al., 2015). Indeed, it was recently demonstrated that Zld is likely acting by displacing nucleosomes at ZGA. In *Drosophila* blastoderm, future Zld-binding enhancers have increased nucleosome density. Zld mediates local depletion of nucleosomes from its binding sites prior to and at ZGA, which facilitates binding of the patterning TF dorsal to the same sites (Sun et al., 2015). It is not clear, if zebrafish zygotic genome activators act in a similar way, as analysis of nucleosome occupancy before and after ZGA on zebrafish gene promoters detected arrays of well-positioned nucleosomes emanating from nucleosome-depleted regions, and forming independently of DNA sequence preference (Zhang et al., 2014). In general, to clarify the mechanism of zebrafish ZGA one needs to resolve causal relationships between three key components: "pluripotency transcription factors," positioning of epigenetic marks, and transcriptional activity of target genes. Creating zebrafish maternal-zygotic mutants for Sox19b and Nanog, if they survive until midgastrula stages of development, will allow to distinguish the unique and shared functions of zygotic genome activators.

4. SoxB1, PouV CLASS, AND NANOG EVOLUTIONARY HISTORY

While SoxB1 class genes appeared already in sponges and are present throughout the metazoan evolutionary tree from sponges to mammals, PouV class genes, and Nanog are vertebrate inventions (Hemmrich & Bosch, 2008), PouV is absent from primitive vertebrates, such as lancelets (Takatori et al., 2008), and is thought to have arisen in common ancestors of jawed vertebrates (gnathostomes). During the second vertebrate-specific genome duplication, which occurred before the gnathostome radiation, the ancestral *pouV* gene was duplicated to give rise to two paralogues (Frankenberg & Renfree, 2013), currently called *pou5f1* and *pou5f3*. Curiously, one of two paralogues was specifically lost in most vertebrate lineages. Only *pou5f3* is retained in zebrafish (*pou2*; Belting et al., 2001; Burgess, Reim, Chen, Hopkins, & Brand, 2002; Takeda, Matsuzaki, Oki, Miyagawa, & Amanuma, 1994), medaka (*Oct4*; Sanchez-Sanchez, Camp, Garcia-Espana, Leal-Tassias, & Mullor, 2010), birds (*cPouV*; Lavial et al., 2007), and *Xenopus* (*XlPou91*, *XlPou25*, and *XlPou60*; Hinkley, Martin, Leibham, & Perry, 1992; Morrison & Brickman, 2006), while only *pou5f1* is retained in lizards and eutherian mammals (Downs, 2008). Both *pou5f1* and *pou5f3* paralogues are present in turtles, coelacanths, marsupials, monotremes, and salamanders (Frankenberg & Renfree, 2013). In all studied vertebrates, *pou5f1* or *pou5f3* gene transcripts are deposited during oogenesis in the egg and show broad ubiquitous expression during gastrulation stages until the end of gastrulation (see references above), coinciding with the pluripotency period. Expression in primordial germ cells is present in mouse, where it is critically required for germ cell survival (Kehler et al., 2004), and also in chick (Lavial et al., 2007) and medaka (Sanchez-Sanchez et al., 2010), but not in zebrafish (Reim & Brand, 2006) or *Xenopus* (Hinkley et al., 1992).

The founding member of the Nanog family arose in the most recent common ancestor of bony fishes. Initially, the Nanog family was considered to consist of only one gene. However, phylogenetic reconstruction and synteny analysis suggest that several independent tandem duplications and lineage-specific losses have taken place in the evolutionary history of the Nanog family (Scerbo et al., 2012).

Existence of two paralogues of PouV and presumably many paralogues of Nanog in the vertebrate lineage poses the question, whether paralogs are functionally equivalent. This question was partially addressed by cross-species rescue assays in frogs, zebrafish, and mice. Mouse Pou5f1 can functionally replace Pou5f3 genes in *Xenopus* and zebrafish (Cao, Siegel, & Knöchel, 2006; Morrison & Brickman, 2006, Onichtchouk et al., 2010), compatible with the view on conserved functions of PouV genes at ZGA and gastrulation stages. PouV family members from different vertebrates rescue pluripotency in mouse ES cells devoid of Oct4/Pou5f1 paralogs with different efficiency. Thus, *Xenopus* Pou5f3 (XlPou91) rescues pluripotency in mouse ES cells devoid of Oct4/Pou5f1 with the same efficiency as mouse Pou5f1, while the rescue efficiency of zebrafish Pou5f3 is very poor

(Morrison & Brickman, 2006; Niwa, Sekita, Tsend-Ayush, & Grutzner, 2008). After discovery of somatic reprogramming of differentiated cell types to induced pluripotent cells (iPS cells; Takahashi & Yamanaka, 2006), vertebrate Pou5f1/3 homologues were tested for their ability to substitute mouse Pou5f1 in the "reprogramming cocktail." Zebrafish Pou5f3 is inefficient in reprogramming mammalian cells to pluripotency, due to the different secondary structure of the nonconserved linker between the two POU and homeodomain DNA-binding domains (Esch et al., 2013). However, this difference is not a feature of all Pou5f3 orthologues: medaka and *Xenopus* Pou5f3 as well as axolotl Pou5f3 could efficiently induce pluripotency in mouse and human fibroblasts (Tapia et al., 2012). As efficiency in cross-species rescue and reprogramming assays of PouV class proteins does not match their assignment as Pou5f1 or Pou5f3 paralogues, there is so far no data to support an evolutionary divergence of their functions.

Mammalian, avian, and teleost homologs of Nanog enabled efficient reprogramming to full pluripotency of mouse Nanog−/− somatic cells, despite sharing as little as 13% sequence identity with mouse Nanog. Surprisingly, the Nanog homeodomain, a short sequence that contains the only recognizable conservation between Nanog homologs, was sufficient to induce naive pluripotency in Nanog-deficient somatic cells (Theunissen et al., 2011). The *Xenopus* genome is lacking Nanog, and it was suggested that Nanog function is partially taken over by related Ventx genes (Scerbo et al., 2012).

Sox2, a SoxB1 class TF, together with Pou5f1 plays a central role in maintenance of pluripotency in mammalian ES cells (Masui et al., 2007). A comparison of the early gene regulatory networks among three mammalian species suggested that bovine POU5F1's interacting partner SOX2 may be replaced by HMGB1, a TF sharing the same DNA-binding domain with SOX2 (Xie et al., 2010). The maternally provided SoxB1-class factor is different in the main vertebrate model systems: Sox2 in mammals, Sox19b in zebrafish (Okuda et al., 2010), and Sox3 in *Xenopus* (Zhang & Klymkowsky, 2007).

5. CONSEQUENCES OF POU5F1/3 LOSS FOR EARLY DEVELOPMENT OF VERTEBRATE EMBRYOS

Pou5f1/3 loss of function was studied in three vertebrate model organisms: zebrafish, *Xenopus*, and mouse. While mouse and zebrafish have only one copy of the *pou5f1* or *pou5f3* gene, respectively, *Xenopus* has three paralogues of Pou5f3 type, arisen from genus-specific duplication event. Although in all three models Pou5f1 homologues are critically required for early embryonic development, phenotypic manifestations of Pou5f1/3 absence substantially differ. The common feature of Pou5f1/3 loss is pleiotropy of the phenotype, indicating that Pou5f1 is involved in many independent processes during early development.

5.1 Pou5f1/Oct4 Role in Mammalian Embryos

Recent experiments exploring the role of Oct4 in blastocyst formation using conditional genetic deletion demonstrated that, in contrast to its role in the maintenance of pluripotency in ES cells, maternal Pou5f1 is not crucial for either the establishment of totipotency in embryos, or the induction of pluripotency in somatic cells using oocytes (Frum et al., 2013; Wu et al., 2013). Induced deletion of Oct4 during the morula-to-blastocyst transition disrupts the ability of inner cell mass cells to adopt lineage-specific identity and acquire the molecular profile characteristic of either epiblast or primitive endoderm (Le Bin et al., 2014). Thus, it is not early pluripotency, but cell fate decisions within the inner cell mass that are critically dependent upon Pou5f1/Oct4 in mammalian embryos.

5.2 Cell Adhesion Defects in Zebrafish Pou5f3 Mutant (MZ*spg*)

Zebrafish maternal and zygotic *pou5f3* null mutants, MZ*spg* (*spiel-ohne-grenzen*), suffer from multiple developmental defects (Lunde, Belting, & Driever, 2004; Reim, Mizoguchi, Stainier, Kikuchi, & Brand, 2004); however, when raised at reduced temperature, the severely abnormal MZ*spg* mutants may survive for several days, MZ*spg* do not develop endoderm (Lunde et al., 2004; Reim et al., 2004), have an excess of dorsal tissues due to reduced bone morphogenetic protein (BMP) signaling (Belting et al., 2011; Reim & Brand, 2006), and have cell motility and cytoskeletal defects leading to abnormal gastrulation (Lachnit, Kur, & Driever, 2008; Reim & Brand, 2006). Maternal and zygotic Pou5f1 functions overlap, and maternal-only mutants (M*spg*) may develop normally (Lachnit et al., 2008; Onichtchouk et al., 2010). Delay in epiboly movements at pregastrulation and gastrulation stages in MZ*spg* is partially due to changes in cell adhesion properties. Reduction of Pou5f3-dependent EGF expression in MZ*spg* leads to abnormal E-cadherin trafficking, such that E-cad accumulates on the plasma membrane instead trafficking through endosomes, causing defects in cell adhesion (Song et al., 2013). Cell adhesion changes in MZ*spg* affect their motility, reducing the migration speed of the cells relative to each other. At the macroscopic level, reduced motility of all embryonic cells results in dramatic epiboly delay in MZ*spg* (Bensch, Song, Ronneberger, & Driever, 2013). Cell adhesion is also severely impaired in Xenopus Pou5f3 morphants (Livigni et al., 2013). It would be interesting to test, if similar adhesion changes occur in mouse blastocyst devoid of Pou5f1. Intriguingly, in mouse ES cells, E-cadherin is a target gene of pluripotency factors, is critical for ES cell pluripotency, and may replace Pou5f1/Oct4 in reprogramming experiments (Redmer et al., 2011). Morpholino knock-down of zebrafish Nanog also results in epiboly delay, but it is currently not known whether the underlying mechanism may be similar to MZ*spg*.

5.3 Pou5f3/Pou5f1 Roles in Endoderm Formation

Pou5f1/3 is critical for endoderm formation at the level of activation of the endoderm specification gene *sox17* in both mice and zebrafish, albeit using apparently different mechanisms. In zebrafish, Pou5f1 acts synergistically with Sox32 to directly bind to the Sox17 promoter (Lunde et al., 2004; Reim et al., 2004). In mouse development, the action of Pou5f1 on Sox17 is not cell-autonomous and relies on FGF4 (Le Bin et al., 2014). In mouse ES cells, genomic redistribution of Pou5f3 by alternative partnering with Sox2 and Sox17 is a fundamental regulatory event of endodermal specification. In contrast to the pluripotent state where Pou5f1 selectively partners with Sox2 at "canonical" binding sites, in ES cells induced to differentiate as primitive endoderm, Pou5f1 partners with Sox17 and binds to a unique "compressed" Sox/Oct motif that earmarks endodermal genes. Evidently, in mice Pou5f1 drives alternative developmental programs by switching Sox partners that affect enhancer selection, leading to either an endodermal or pluripotent cell fate (Aksoy et al., 2013). In contrast to zebrafish and mice, in *Xenopus*, endoderm formation is increased upon knockout of all POUV factors (Morrison & Brickman, 2006). Most likely, this difference between *Xenopus*, zebrafish, and mice may be explained by difference in the structure of the endoderm-specifying networks, rather than species-specific differences in properties of Pou5f1/3 proteins.

5.4 Molecular Interactions of Pou5f3 and the BMP Pathway

The BMP signaling pathway is required for specifying dorsoventral cell fates in invertebrates and vertebrates. Expression of BMP ligands and establishment of a BMP patterning gradient is positively regulated by Pou5f3 homologs in *Xenopus* (Cao et al., 2004) and zebrafish (Khan et al., 2012; Reim & Brand, 2006). The mechanisms involve conserved direct transcriptional activation of BMP pathway genes: in *Xenopus*, XlPou25 induces ventral transcriptional repressor *Xvent2*, and in zebrafish, Pou5f3 directly induces the closely related ventral repressor *vox* (Belting et al., 2011). In addition, slightly later in development *XOct-25* (*XlPou25*) was suggested to play a role in the formation of neural tissue in *Xenopus* by inhibiting competence to BMP-mediated induction of epidermal markers (Takebayashi-Suzuki, Arita, Murasaki, & Suzuki, 2007). In zebrafish, Pou5f3 contributes to dorsoventral patterning by acting synergistically with BMP signaling. Levels of BMP pathway activation and duration of BMP signal are both critical for embryonic patterning: in particular, distinct temporal intervals of BMP signaling are required for correct coupling of dorsoventral and anterioposterior cell fates (Tucker, Mintzer, & Mullins, 2008). In MZ*spg* embryos, due to inhibition by fibroblast growth factor (FGF) signaling shortly after MBT, the strength of BMP signaling is lower and the duration of BMP signaling activity is shorter (Belting et al., 2011). Thus, Pou5f3 is involved in fine-tuning of both the activity levels and signal duration of the BMP network.

6. STRUCTURE OF POU5F3-DEPENDENT Gene Regulatory Network in Zebrafish

Target genes of zebrafish Pou5f3 include signaling molecules and TFs, homologous to known members of the mammalian pluripotency network, such as Klf and Myc family genes (Kotkamp, Kur, et al., 2014; Kotkamp, Mossner, Allen, Onichtchouk, & Driever, 2014). Comparison of genome-wide gene expression in zebrafish wild type and Pou5f3-deficient MZ*spg* mutant embryos during the first 8 h of development revealed time shifts in expression onset for more than 7000 genes in MZ*spg* mutants (Onichtchouk et al., 2010). Strikingly, Pou5f3 deficiency leads to premature expression already during gastrula stages of multiple developmental genes, which are normally only expressed from organogenesis stages onward. Suppression of Pou5f3 direct target mRNAs revealed that Pou5f3 delays the expression of differentiation genes indirectly, presumably by directly activating sets of transcriptional repressors (Onichtchouk et al., 2010). We discuss activation and repression in the Pou5f3-dependent network below.

6.1 Time-Resolved Expression of Direct Pou5f3 Targets: SoxB1 Contribution

SoxB1 and Pou5f3 transcriptional targets in zebrafish (Okuda et al., 2010; Onichtchouk et al., 2010) highly overlap, suggesting transcriptional partnering between Pou5f3 and SoxB1 during activation of a wide range of targets. The levels of Pou5f3 at MBT and throughout gastrulation are relatively constant (Lippok, Song, & Driever, 2014), and not regulated by SoxB1 activity. In contrast, part of SoxB1 activity present in the zebrafish embryo is under control of Pou5f3 and dynamically increases from MBT to midgastrula. In the midgastrula embryo, the majority of both Pou5f3- and SoxB1-regulated transcriptional targets with putative transcriptional repressor activity are expressed in prospective neural (*her3*, *hesx1*) or non-neural ectoderm (*klf2b*, *gata2a*). For most mesendodermal target genes of Pou5f3, activation by SoxB1 was not detected (Onichtchouk et al., 2010). Analyzing the temporal behavior of Pou5f3 transcriptional targets in development, we could identify a correlation between the shape of the temporal profile and dependency on SoxB1. Most SoxB1-independent direct Pou5f3 targets reached half-maximal expression levels soon after the start of zygotic transcription. In contrast, double direct SoxB1- and Pou5f3-dependent genes tend to have biphasic or delayed expression and reach maximum levels after midgastrula. Expression of double Pou5f3- and SoxB1-dependent target genes does not depend on the amount of Pou5f3, as long as Pou5f3 is present above a certain low threshold, but is particularly sensitive to the quantitative level of SoxB1 expression; molecular mechanisms of this behavior are still to be elucidated.

6.2 Downstream of Pou5f3: Repressors of Differentiation or Interterritorial Repressors?

The class of genes prematurely expressed in MZ*spg* embryos, collectively termed "promoters of differentiation" (PODs), includes orphan nuclear receptors (*nr2f1*, *nr2f2*, *nr2f5*), Hox genes (*hoxa2b*, *hoxb8a*), neuron differentiation marker genes (*her2*, *ascl1b*, *elavl3*, *dlb*), and genes encoding patterning TFs such as *tbx2*, *gbx2*, *sox21*, and *pax6a* (Onichtchouk et al., 2010). Presumably, during gastrulation, in normal development PODs are suppressed by transcriptional repressors, collectively termed repressors of differentiation (RODs). RODs are directly activated by Pou5f3, and many of them are evolutionary conserved targets of Pou5f1/Pou5f3 from fish to mammals (Onichtchouk et al., 2010). In zebrafish, the majority of Pou5f3 direct transcriptional targets of this type are tissue restricted: *foxD3* in mesendoderm, *her3* and *hesx1 in* neural ectoderm, *klf2ab* and *klf17* (previously called *klf4b*) in nonneural ectoderm and enveloping layer.

Theoretically, these tissue-specific RODs may act (i) intraterritorial—repressing premature expression of differentiation genes of the same tissue they belong to (e.g., preventing expression of late ectodermal differentiation markers in early ectoderm) or (ii) interterritorial—repressing differentiation of the alternative (adjacent) tissue type (e.g., preventing expression of mesodermal differentiation markers in ectoderm).

In MZ*spg* embryos, *pax6* is expressed in roughly the same anterior neuroectoderm domain, but 2 h earlier than in the wild type, suggesting that some neural-specific RODs act as intraterritorial suppressors of differentiation. However, the identity and specificity of RODs repressing *pax6* and other PODs in the zebrafish embryo remain to be determined.

klf2a and *klf2b* are induced by Pou5f3 and BMP signaling in the prospective ectodermal domain of developing zebrafish embryo and act according to the second scenario as interterritorial repressors. Morpholino knockdown of Klf2b, or double knockdown of Klf2a and Klf2b function led to upregulation of a gene set normally expressed in the mesoderm at later stages of development (Kotkamp, Mossner, et al., 2014). Thus, Klf2 genes contribute to cryptic, interterritorial repression of alternative possible regulatory states within prospective ectodermal domain. This type of regulation is a widespread design feature of developmental gene regulatory networks (Davidson, 2010).

7. MODELS OF TRANSCRIPTIONAL NETWORKS FOR PLURIPOTENCY

Chickarmane, Troein, Nuber, Sauro, and Peterson (2006) suggested an influential "bistable switch" model of transcriptional regulatory circuitry underlying pluripotency in ES cells (Fig. 2A). The model predicted that due to the stability of the self-activating "pluripotency transcription factor" circuit, it should be possible to cultivate ES cells without external growth factors. Indeed, the "naïve" 2i ES cell line able to grow in serum-free and BMP-free conditions were obtained by inhibiting ERK and GSK kinases in 2008 (Ying et al., 2008). In these naïve cells, expression of lineage-specific developmental factors is strongly reduced in comparison to ES cell (Marks et al., 2012).

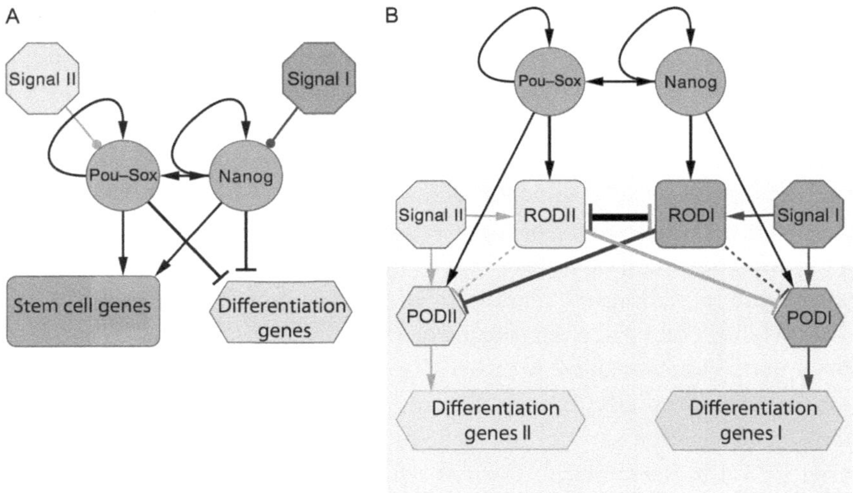

Figure 2 Models of pluripotency. (A) Bistable switch model in ES cell (Chickarmane et al., 2006). Signals I and II can be activatory (arrowheads) or repressory (indicated by rounded arrowheads). The model describes two stable states. In the pluripotent state, in response to environmental signals, Pou5f1, SoxB1, and Nanog genes are switched on, they sustain their own expression by means of positive feedback loops, and the "pluripotency circuit" (Pou–Sox–Nanog) is active. In this state, stem cell genes are on and differentiation genes are switched off. When the positive feedback is interrupted by differentiation signals, the differentiation genes are on. (B) Simplified hypothetical model for transient pluripotency and developmental control in blastula to gastrula stage embryo cells: ROD, inter-(and/or intra-)territorial repressor of differentiation; POD, promoter of differentiation; signals I and II, different tissue specific patterning and differentiation signals. In the late blastula pluripotent state, signals I and II together with Pou–Sox–Nanog control RODII and RODI expression such that their activity is balanced. In this situation, PODs and differentiation genes are off (gray-shaded area). In committed differentiating cell in a specific germ layer or tissue, one RODI is off, POD starts to be expressed, induces differentiation genes, and presumably shuts down the pluripotency circuit. (A color version of this figure is located at the end of this chapter.)

The model of Chickarmane et al. provides a starting point for comparison of ES cells and transiently pluripotent post-ZGA cells in the embryo. In zebrafish, large quantities of Pou5f3, Sox19b, and Nanog are translated at ZGA (Lee et al., 2013), and three factors cobind to their own promoters (Leichsenring et al., 2013; Xu et al., 2012) switching on their own zygotic expression and thus resembling the situation in ES cells at the upper level of hierarchy. In embryos, a significant difference from the steady state of indefinitely proliferating cells is the quick activation of lineage-specific factors. We suggest a hypothetical scheme which takes into account the competition of lineage-specific repressors, or RODs induced by "core stem cell factors" and by patterning signals (Fig. 2B). Among other developmental regulators, RODs are expressed at ZGA. The balance of several RODs within a single cell, receiving differentiation signals to alternative lineages (Signal I and Signal II in Fig. 2B), delays differentiation and keeps the cell pluripotent until one ROD will shut down the other. Subsequently, the differentiation genes of the same lineage (PODs) become derepressed initiate lineage-specific differentiation, and eventually directly or indirectly switch off the "core transcriptional circuit" in this cell. In zebrafish, cells become committed (and thus pluripotency is lost) after 6–8 h of development (Ho & Kimmel, 1993), but in annual killifish, another teleost fish, the pluripotent period of development is significantly expanded. Embryonic blastomeres of annual killifish disperse randomly across the yolk during epiboly, obviously remaining pluripotent, and then reaggregate several days later to start gastrulation. In addition, annual killifish are able to arrest embryonic development at three stages. Recent work demonstrated that *pou5f3*, the *soxB1* group of genes and the developmental genes *chordin* and *noggin* are highly expressed in early killifish embryos (Wagner & Podrabsky, 2015). It would be interesting to find out how the conserved developmental genes and signaling mechanisms shifted in order to promote the annual killifish phenotype of "extended pluripotency," and whether the mechanisms fit into the model suggested in Fig. 2.

8. CONCLUSION

For the mammalian pluripotency control factors, the origin and evolutionary roles during early development of nonmammalian vertebrates have long been a mystery. Recent work has revealed that the ancestral function of the pluripotency factors is zygotic genome activation and control of developmental timing in the blastula and gastrula stage vertebrate embryo. The first process in which the core TFs are involved, ZGA, in a sense is a first major reprogramming event from transcriptionally silent cleavage-stage cells with highly specialized chromatin states to pluripotent post-MBT blastomeres. In this context, Pou5f1 and Sox2 in the embryo do not only establish pluripotency but also contribute to initiation and regulation of all main embryonic regulatory pathways that establish tissue identity. Pou5f1 and Sox2 also contribute to control of developmental timing, i.e., temporal progress of tissue specification and differentiation during gastrulation. Given the significant conservation of components and interactions in the early embryonic and the mammalian pluripotency gene regulatory networks, the zygotic priming-activation-timing mechanism may well have evolved to control the cell pluripotency state in mammalian development by simply temporally expanding the transient regulatory state of inner cell mass "blastula/blastocyst" cells.

The work in vertebrate model organisms opens the unique opportunity to understand the involvement of the pluripotency gene regulatory network during initiation of stable germ layer and tissue-specific cell states. Such knowledge may also be helpful for the currently intense efforts in stem cell-based tissue engineering by providing evolutionary hints on how the embryo accomplishes the task of complex cell state transitions leading to stable tissue precursors and differentiated cells.

ACKNOWLEDGMENTS

D.O. is supported by DFG research grants DFG-ON86/4-1 and EXC294 A7-2, WD by EXC294 BIOSS.

REFERENCES

Aksoy, I., Jauch, R., Chen, J., Dyla, M., Divakar, U., Bogu, G. K., et al. (2013). Oct4 switches partnering from Sox2 to Sox17 to reinterpret the enhancer code and specify endoderm. *The EMBO Journal*, *32*, 938–953.

Almouzni, G., & Wolffe, A. P. (1995). Constraints on transcriptional activator function contribute to transcriptional quiescence during early Xenopus embryogenesis. *The EMBO Journal*, *14*, 1752–1765.

Amodeo, A. A., Jukam, D., Straight, A. F., & Skotheim, J. M. (2015). Histone titration against the genome sets the DNA-to-cytoplasm threshold for the Xenopus midblastula transition. *Proceedings of the National Academy of Sciences of the United States of America*, *112*, E1086–E1095.

Belting, H. G., Hauptmann, G., Meyer, D., Abdelilah-Seyfried, S., Chitnis, A., Eschbach, C., et al. (2001). spiel ohne grenzen/pou2 is required during establishment of the zebrafish midbrain-hindbrain boundary organizer. *Development*, *128*, 4165–4176.

Belting, H. G., Wendik, B., Lunde, K., Leichsenring, M., Mossner, R., Driever, W., et al. (2011). Pou5f1 contributes to dorsoventral patterning by positive regulation of vox and modulation of fgf8a expression. *Developmental Biology*, *356*(2), 323–336.

Bensch, R., Song, S., Ronneberger, O., & Driever, W. (2013). Non-directional radial intercalation dominates deep cell behavior during zebrafish epiboly. *Biology Open*, *2*, 845–854.

Blythe, S. A., Cha, S. W., Tadjuidje, E., Heasman, J., & Klein, P. S. (2010). beta-Catenin primes organizer gene expression by recruiting a histone H3 arginine 8 methyltransferase, Prmt2. *Developmental Cell*, *19*, 220–231.

Bogdanovic, O., Fernandez-Minan, A., Tena, J. J., de la Calle-Mustienes, E., Hidalgo, C., van Kruysbergen, I., et al. (2012). Dynamics of enhancer chromatin signatures mark the transition from pluripotency to cell specification during embryogenesis. *Genome Research*, *22*, 2043–2053.

Boveri, T. (1918). Zwei Fehlerquellen bei Merogonieversuchen und die Entwicklungsfaehigkeit merogonischer und partieller-merogonischer Seeigelbastarde. *Wilhelm Roux' Archiv für Entwicklungsmechanik der Organismen*, *44*, 417–471.

Boyer, L. A., Lee, T. I., Cole, M. F., Johnstone, S. E., Levine, S. S., Zucker, J. P., et al. (2005). Core transcriptional regulatory circuitry in human embryonic stem cells. *Cell*, *122*, 947–956.

Buganim, Y., Markoulaki, S., van Wietmarschen, N., Hoke, H., Wu, T., Ganz, K., et al. (2014). The developmental potential of iPSCs is greatly influenced by reprogramming factor selection. *Cell Stem Cell*, *15*, 295–309.

Burgess, S., Reim, G., Chen, W., Hopkins, N., & Brand, M. (2002). The zebrafish spiel-ohne-grenzen (spg) gene encodes the POU domain protein Pou2 related to mammalian Oct4 and is essential for formation of the midbrain and hindbrain, and for pre-gastrula morphogenesis. *Development*, *129*, 905–916.

Cao, Y., Knochel, S., Donow, C., Miethe, J., Kaufmann, E., & Knochel, W. (2004). The POU factor Oct-25 regulates the Xvent-2B gene and counteracts terminal differentiation in Xenopus embryos. *The Journal of Biological Chemistry*, *279*, 43735–43743. Epub 2004 August 2.

Cao, Y., Siegel, D., & Knöchel, W. (2006). Xenopus POU factors of subclass V inhibit activin/nodal signaling during gastrulation. *Mechanisms of Development*, *123*, 614–625.

Chen, X., Xu, H., Yuan, P., Fang, F., Huss, M., Vega, V. B., et al. (2008). Integration of external signaling pathways with the core transcriptional network in embryonic stem cells. *Cell*, *133*, 1106–1117.

Chenoweth, J. G., McKay, R. D., & Tesar, P. J. (2010). Epiblast stem cells contribute new insight into pluripotency and gastrulation. *Development, Growth & Differentiation*, *52*, 293–301.

Chenoweth, J. G., & Tesar, P. J. (2010). Isolation and maintenance of mouse epiblast stem cells. *Methods in Molecular Biology*, *636*, 25–44.

Chia, N. Y., Chan, Y. S., Feng, B., Lu, X., Orlov, Y. L., Moreau, D., et al. (2010). A genome-wide RNAi screen reveals determinants of human embryonic stem cell identity. *Nature*, *468*, 316–320.

Chickarmane, V., Troein, C., Nuber, U. A., Sauro, H. M., & Peterson, C. (2006). Transcriptional dynamics of the embryonic stem cell switch. *PLoS Computational Biology*, *2*, e123.

Collart, C., Allen, G. E., Bradshaw, C. R., Smith, J. C., & Zegerman, P. (2013). Titration of four replication factors is essential for the Xenopus laevis midblastula transition. *Science*, *341*, 893–896.

Conklin, E. G. (1908). The mechanism of heredity. *Science*, *27*, 89–99.

Davidson, E. H. (2010). Emerging properties of animal gene regulatory networks. *Nature*, *468*, 911–920.

Domingo, C., & Keller, R. (2000). Cells remain competent to respond to mesoderm-inducing signals present during gastrulation in Xenopus laevis. *Developmental Biology*, *225*, 226–240.

Downs, K. M. (2008). Systematic localization of Oct-3/4 to the gastrulating mouse conceptus suggests manifold roles in mammalian development. *Developmental Dynamics*, *237*, 464–475.

Edgar, B. A., Kiehle, C. P., & Schubiger, G. (1986). Cell cycle control by the nucleo-cytoplasmic ratio in early Drosophila development. *Cell*, *44*, 365–372.

Esch, D., Vahokoski, J., Groves, M. R., Pogenberg, V., Cojocaru, V., Vom Bruch, H., et al. (2013). A unique Oct4 interface is crucial for reprogramming to pluripotency. *Nature Cell Biology*, *15*, 295–301.

Evans, M. J., & Kaufman, M. H. (1981). Establishment in culture of pluripotential cells from mouse embryos. *Nature*, *292*, 154–156.

Feng, B., Jiang, J., Kraus, P., Ng, J. H., Heng, J. C., Chan, Y. S., et al. (2009). Reprogramming of fibroblasts into induced pluripotent stem cells with orphan nuclear receptor Esrrb. *Nature Cell Biology*, *11*, 197–203.

Foo, S. M., Sun, Y., Lim, B., Ziukaite, R., O'Brien, K., Nien, C. Y., et al. (2014). Zelda potentiates morphogen activity by increasing chromatin accessibility. *Current Biology*, *24*, 1341–1346. http://dx.doi.org/10.1016/j.cub.2014.04.032. Epub 2014 June 5.

Frankenberg, S., & Renfree, M. B. (2013). On the origin of POU5F1. *BMC Biology*, *11*, 56.

Frum, T., Halbisen, M. A., Wang, C., Amiri, H., Robson, P., & Ralston, A. (2013). Oct4 cell-autonomously promotes primitive endoderm development in the mouse blastocyst. *Developmental Cell*, *25*, 610–622.

Gafni, O., Weinberger, L., Mansour, A. A., Manor, Y. S., Chomsky, E., Ben-Yosef, D., et al. (2013). Derivation of novel human ground state naive pluripotent stem cells. *Nature*, *504*(7479), 282–286.

Garcia-Martinez, V., & Schoenwolf, G. C. (1992). Positional control of mesoderm movement and fate during avian gastrulation and neurulation. *Developmental Dynamics*, *193*, 249–256.

Gerhart, J. (1980). Mechanisms regulating pattern formation in the amphibian egg and early embryo. In R. Goldberg (Ed.), *Biological regulation and development* (pp. 133–315). New York: Plenum Press.

Giraldez, A. J., Mishima, Y., Rihel, J., Grocock, R. J., Van Dongen, S., Inoue, K., et al. (2006). Zebrafish MiR-430 promotes deadenylation and clearance of maternal mRNAs. *Science*, *312*, 75–79.

Godlevski, E. (1906). Untersuchungen ueber die Bastardierung der Echiniden- und Crinoidenfamilie. *Wilhelm Roux' Archiv für Entwicklungsmechanik der Organismen*, *20*, 579–643.

Hanna, J. H., Saha, K., & Jaenisch, R. (2010). Pluripotency and cellular reprogramming: Facts, hypotheses, unresolved issues. *Cell*, *143*, 508–525.

Harrison, M. M., Li, X. Y., Kaplan, T., Botchan, M. R., & Eisen, M. B. (2011). Zelda binding in the early Drosophila melanogaster embryo marks regions subsequently activated at the maternal-to-zygotic transition. *PLoS Genetics*, 7, e1002266.

Hemmrich, G., & Bosch, T. C. (2008). Compagen, a comparative genomics platform for early branching metazoan animals, reveals early origins of genes regulating stem-cell differentiation. *BioEssays: News and Reviews in Molecular, Cellular and Developmental Biology*, *30*, 1010–1018.

Heng, J. C., Feng, B., Han, J., Jiang, J., Kraus, P., Ng, J. H., et al. (2010). The nuclear receptor Nr5a2 can replace Oct4 in the reprogramming of murine somatic cells to pluripotent cells. *Cell Stem Cell*, *6*, 167–174.

Hinkley, C. S., Martin, J. F., Leibham, D., & Perry, M. (1992). Sequential expression of multiple POU proteins during amphibian early development. *Molecular and Cellular Biology*, *12*, 638–649.

Ho, R. K., & Kimmel, C. B. (1993). Commitment of cell fate in the early zebrafish embryo. *Science*, *261*, 109–111.

Hong, N., He, B. P., Schartl, M., & Hong, Y. (2013). Medaka embryonic stem cells are capable of generating entire organs and embryo-like miniatures. *Stem Cells and Development*, *22*, 750–757.

Hong, Y., Winkler, C., & Schartl, M. (1996). Pluripotency and differentiation of embryonic stem cell lines from the medakafish (Oryzias latipes). *Mechanisms of Development*, *60*, 33–44.

Hong, Y., Winkler, C., & Schartl, M. (1998). Production of medakafish chimeras from a stable embryonic stem cell line. *Proceedings of the National Academy of Sciences of the United States of America*, *95*, 3679–3684.

Ichida, J. K., Blanchard, J., Lam, K., Son, E. Y., Chung, J. E., Egli, D., et al. (2009). A small-molecule inhibitor of tgf-Beta signaling replaces sox2 in reprogramming by inducing nanog. *Cell Stem Cell*, *5*, 491–503.

Kane, D. A., & Kimmel, C. B. (1993). The zebrafish midblastula transition. *Development*, *119*, 447–456.

Kehler, J., Tolkunova, E., Koschorz, B., Pesce, M., Gentile, L., Boiani, M., et al. (2004). Oct4 is required for primordial germ cell survival. *EMBO Reports*, *5*, 1078–1083.

Khan, A., Nakamoto, A., Okamoto, S., Tai, M., Nakayama, Y., Kobayashi, K., et al. (2012). Pou2, a class V POU-type transcription factor in zebrafish, regulates dorsoventral patterning and convergent extension movement at different blastula stages. *Mechanisms of Development*, *129*, 219–235.

Kim, J., Chu, J., Shen, X., Wang, J., & Orkin, S. H. (2008). An extended transcriptional network for pluripotency of embryonic stem cells. *Cell*, *132*, 1049–1061.

Kimelman, D., Kirschner, M., & Scherson, T. (1987). The events of the midblastula transition in Xenopus are regulated by changes in the cell cycle. *Cell*, *48*, 399–407.

Korzh, V. (2009). Before maternal-zygotic transition ... There was morphogenetic function of nuclei. *Zebrafish*, *6*, 295–302.

Kotkamp, K., Kur, E., Wendik, B., Polok, B. K., Ben-Dor, S., Onichtchouk, D., et al. (2014). Pou5f1/Oct4 promotes cell survival via direct activation of mych expression during zebrafish gastrulation. *PLoS One*, *9*, e92356.

Kotkamp, K., Mossner, R., Allen, A., Onichtchouk, D., & Driever, W. (2014). A Pou5f1/Oct4 dependent Klf2a, Klf2b, and Klf17 regulatory sub-network contributes to EVL and ectoderm development during zebrafish embryogenesis. *Developmental Biology*, *385*, 433–447.

Kvon, E. Z., Stampfel, G., Yanez-Cuna, J. O., Dickson, B. J., & Stark, A. (2012). HOT regions function as patterned developmental enhancers and have a distinct cis-regulatory signature. *Genes & Development*, *26*, 908–913.

Lachnit, M., Kur, E., & Driever, W. (2008). Alterations of the cytoskeleton in all three embryonic lineages contribute to the epiboly defect of Pou5f1/Oct4 deficient MZspg zebrafish embryos. *Developmental Biology*, *315*, 1–17.

Langley, A. R., Smith, J. C., Stemple, D. L., & Harvey, S. A. (2014). New insights into the maternal to zygotic transition. *Development*, *141*, 3834–3841.

Lavial, F., Acloque, H., Bertocchini, F., Macleod, D. J., Boast, S., Bachelard, E., et al. (2007). The Oct4 homologue PouV and Nanog regulate pluripotency in chicken embryonic stem cells. *Development*, *134*, 3549–3563.

Lawson, K. A., Meneses, J. J., & Pedersen, R. A. (1991). Clonal analysis of epiblast fate during germ layer formation in the mouse embryo. *Development*, *113*, 891–911.

Le Bin, G. C., Munoz-Descalzo, S., Kurowski, A., Leitch, H., Lou, X., Mansfield, W., et al. (2014). Oct4 is required for lineage priming in the developing inner cell mass of the mouse blastocyst. *Development*, *141*, 1001.

Lee, M. T., Bonneau, A. R., Takacs, C. M., Bazzini, A. A., Divito, K. R., Fleming, E. S., et al. (2013). Nanog, Pou5f1 and SoxB1 activate zygotic gene expression during the maternal-to-zygotic transition. *Nature*, *503*, 360–364.

Leichsenring, M., Maes, J., Mossner, R., Driever, W., & Onichtchouk, D. (2013). Pou5f1 transcription factor controls zygotic gene activation in vertebrates. *Science*, *341*, 1005–1009.

Li, X. Y., Harrison, M. M., Villalta, J. E., Kaplan, T., & Eisen, M. B. (2014). Establishment of regions of genomic activity during the Drosophila maternal to zygotic transition. *eLife*, *3*, e03737.

Liang, H. L., Nien, C. Y., Liu, H. Y., Metzstein, M. M., Kirov, N., & Rushlow, C. (2008). The zinc-finger protein Zelda is a key activator of the early zygotic genome in Drosophila. *Nature*, *456*, 400–403.

Lindeman, L. C., Andersen, I. S., Reiner, A. H., Li, N., Aanes, H., Ostrup, O., et al. (2011). Prepatterning of developmental gene expression by modified histones before zygotic genome activation. *Developmental Cell, 21*(6), 993–1004.

Lippok, B., Song, S., & Driever, W. (2014). Pou5f1 protein expression and posttranslational modification during early zebrafish development. *Developmental Dynamics, 243*, 468–477.

Liu, R., Li, M., Li, Z., Hong, N., Xu, H., & Hong, Y. (2015). Medaka Oct4 is essential for pluripotency in blastula formation and ES cell derivation. *Stem Cell Reviews, 11*, 11–23.

Livigni, A., Peradziryi, H., Sharov, A. A., Chia, G., Hammachi, F., Migueles, R. P., et al. (2013). A conserved Oct4/POUV-dependent network links adhesion and migration to progenitor maintenance. *Current biology CB, 23*, 2233–2244.

Loh, Y. H., Wu, Q., Chew, J. L., Vega, V. B., Zhang, W., Chen, X., et al. (2006). The Oct4 and Nanog transcription network regulates pluripotency in mouse embryonic stem cells. *Nature Genetics, 38*, 431–440.

Lu, X., Li, J. M., Elemento, O., Tavazoie, S., & Wieschaus, E. F. (2009). Coupling of zygotic transcription to mitotic control at the Drosophila mid-blastula transition. *Development, 136*, 2101–2110.

Lunde, K., Belting, H. G., & Driever, W. (2004). Zebrafish pou5f1/pou2, homolog of mammalian Oct4, functions in the endoderm specification cascade. *Current Biology: CB, 14*, 48–55.

Marks, H., Kalkan, T., Menafra, R., Denissov, S., Jones, K., Hofemeister, H., et al. (2012). The transcriptional and epigenomic foundations of ground state pluripotency. *Cell, 149*, 590–604.

Masui, S., Nakatake, Y., Toyooka, Y., Shimosato, D., Yagi, R., Takahashi, K., et al. (2007). Pluripotency governed by Sox2 via regulation of Oct3/4 expression in mouse embryonic stem cells. *Nature Cell Biology, 9*, 625–635.

Mikkelsen, T. S., Ku, M., Jaffe, D. B., Issac, B., Lieberman, E., Giannoukos, G., et al. (2007). Genome-wide maps of chromatin state in pluripotent and lineage-committed cells. *Nature, 448*, 553–560. Epub 2007 July 1.

Montserrat, N., Nivet, E., Sancho-Martinez, I., Hishida, T., Kumar, S., Miquel, L., et al. (2013). Reprogramming of human fibroblasts to pluripotency with lineage specifiers. *Cell Stem Cell, 13*, 341–350.

Moore, J. (1941). Developmental rate of hybrid frogs. *Journal of Experimental Zoology, 86*, 405–422.

Morrison, G. M., & Brickman, J. M. (2006). Conserved roles for Oct4 homologues in maintaining multipotency during early vertebrate development. *Development, 133*, 2011–2022.

Mullen, A. C., Orlando, D. A., Newman, J. J., Loven, J., Kumar, R. M., Bilodeau, S., et al. (2011). Master transcription factors determine cell-type-specific responses to TGF-beta signaling. *Cell, 147*, 565–576.

Newport, J., & Kirschner, M. (1982a). A major developmental transition in early Xenopus embryos: I. Characterization and timing of cellular changes at the midblastula stage. *Cell, 30*, 675–686.

Newport, J., & Kirschner, M. (1982b). A major developmental transition in early Xenopus embryos: II. Control of the onset of transcription. *Cell, 30*, 687–696.

Neyfakh, A. A. (1959). X-ray inactivation of nuclei as method for studying their function in the early development of fishes. *Journal of Embryology and Experimental Morphology, 7*, 173–192.

Neyfakh, A. A. (1964). Radiation investigation of nucleo-cytoplasmic interrelations in morphogenesis and biochemical differentiation. *Nature, 201*, 880–884.

Neyfakh, A. A., & Rott, N. N. (1968). A quantitative approach to the detection of nuclear activity after differeetial damage to nucleus and cytoplasm in early development. *Journal of Embryology and Experimental Morphology, 20*, 129–140.

Nien, C. Y., Liang, H. L., Butcher, S., Sun, Y., Fu, S., Gocha, T., et al. (2011). Temporal coordination of gene networks by Zelda in the early Drosophila embryo. *PLoS Genetics*, *7*, e1002339.

Niwa, H., Sekita, Y., Tsend-Ayush, E., & Grutzner, F. (2008). Platypus Pou5f1 reveals the first steps in the evolution of trophectoderm differentiation and pluripotency in mammals. *Evolution & Development*, *10*, 671–682.

Okabayashi, K., & Asashima, M. (2003). Tissue generation from amphibian animal caps. *Current Opinion in Genetics & Development*, *13*, 502–507.

Okuda, Y., Ogura, E., Kondoh, H., & Kamachi, Y. (2010). B1 SOX coordinate cell specification with patterning and morphogenesis in the early zebrafish embryo. *PLoS Genetics*, *6*, e1000936.

Onichtchouk, D., Geier, F., Polok, B., Messerschmidt, D. M., Mossner, R., Wendik, B., et al. (2010). Zebrafish Pou5f1-dependent transcriptional networks in temporal control of early development. *Molecular Systems Biology*, *6*, 354.

Pain, B., Clark, M. E., Shen, M., Nakazawa, H., Sakurai, M., Samarut, J., et al. (1996). Long-term in vitro culture and characterisation of avian embryonic stem cells with multiple morphogenetic potentialities. *Development*, *122*, 2339–2348.

Potok, M. E., Nix, D. A., Parnell, T. J., & Cairns, B. R. (2013). Reprogramming the maternal zebrafish genome after fertilization to match the paternal methylation pattern. *Cell*, *153*, 759–772.

Redmer, T., Diecke, S., Grigoryan, T., Quiroga-Negreira, A., Birchmeier, W., & Besser, D. (2011). E-cadherin is crucial for embryonic stem cell pluripotency and can replace OCT4 during somatic cell reprogramming. *EMBO Reports*, *12*, 720–726.

Reim, G., & Brand, M. (2006). Maternal control of vertebrate dorsoventral axis formation and epiboly by the POU domain protein Spg/Pou2/Oct4. *Development*, *133*, 2757–2770.

Reim, G., Mizoguchi, T., Stainier, D. Y., Kikuchi, Y., & Brand, M. (2004). The POU domain protein spg (pou2/Oct4) is essential for endoderm formation in cooperation with the HMG domain protein casanova. *Developmental Cell*, *6*, 91–101.

Rossello, R. A., Chen, C. C., Dai, R., Howard, J. T., Hochgeschwender, U., & Jarvis, E. D. (2013). Mammalian genes induce partially reprogrammed pluripotent stem cells in non-mammalian vertebrate and invertebrate species. *eLife*, *2*, e00036.

Rott, N. N., & Sheveleva, G. A. (1968). Changes in the rate of cell divisions in the course of early development of diploid and haploid loach embryos. *Journal of Embryology and Experimental Morphology*, *20*, 141–150.

Rugh, R., & Exener, F. (1940). Developmental effects resulting from exposure to x-rays. 11. Development of leopard frog eggs activated by bullfrog sperm. *Proceedings of the American Philosophical Society*, *83*, 607–619.

Sanchez-Sanchez, A. V., Camp, E., Garcia-Espana, A., Leal-Tassias, A., & Mullor, J. L. (2010). Medaka Oct4 is expressed during early embryo development, and in primordial germ cells and adult gonads. *Developmental Dynamics: An Official Publication of the American Association of Anatomists*, *239*, 672–679.

Satija, R., & Bradley, R. K. (2012). The TAGteam motif facilitates binding of 21 sequence-specific transcription factors in the Drosophila embryo. *Genome Research*, *22*, 656–665.

Scerbo, P., Girardot, F., Vivien, C., Markov, G. V., Luxardi, G., Demeneix, B., et al. (2012). Ventx factors function as Nanog-like guardians of developmental potential in Xenopus. *PLoS One*, *7*, e36855.

Shu, J., Wu, C., Wu, Y., Li, Z., Shao, S., Zhao, W., et al. (2013). Induction of pluripotency in mouse somatic cells with lineage specifiers. *Cell*, *153*, 963–975.

Signoret, J., & Lefresne, J. (1971). Contribution a L'etude de la segmentation de l'oeuf d'axolotl: I. Definition de la transition blastuleenne. *Annales d' Embryologie et de Morphogenese*, *4*, 113–123.

Song, S., Eckerle, S., Onichtchouk, D., Marrs, J. A., Nitschke, R., & Driever, W. (2013). Pou5f1-dependent EGF expression controls E-cadherin endocytosis, cell adhesion, and zebrafish epiboly movements. *Developmental Cell*, *24*, 486–501.

Soufi, A., Donahue, G., & Zaret, K. S. (2012). Facilitators and impediments of the pluripotency reprogramming factors' initial engagement with the genome. *Cell*, *151*, 994–1004. http://dx.doi.org/10.1016/j.cell.2012.09.045. Epub 2012 November 15.

Soufi, A., Garcia, M. F., Jaroszewicz, A., Osman, N., Pellegrini, M., & Zaret, K. S. (2015). Pioneer transcription factors target partial DNA motifs on nucleosomes to initiate reprogramming. *Cell*, *161*, 555–568.

Sun, Y., Nien, C. Y., Chen, K., Liu, H. Y., Johnston, J., Zeitlinger, J., et al. (2015). Zelda overcomes the high intrinsic nucleosome barrier at enhancers during Drosophila zygotic genome activation. *Genome Research*, *25*, 1703–1714.

Tadros, W., & Lipshitz, H. D. (2009). The maternal-to-zygotic transition: A play in two acts. *Development*, *136*, 3033–3042.

Takahashi, K., & Yamanaka, S. (2006). Induction of pluripotent stem cells from mouse embryonic and adult fibroblast cultures by defined factors. *Cell*, *126*, 663–676. Epub 2006 August 10.

Takatori, N., Butts, T., Candiani, S., Pestarino, M., Ferrier, D. E., Saiga, H., et al. (2008). Comprehensive survey and classification of homeobox genes in the genome of amphioxus, Branchiostoma floridae. *Development Genes and Evolution*, *218*, 579–590.

Takebayashi-Suzuki, K., Arita, N., Murasaki, E., & Suzuki, A. (2007). The Xenopus POU class V transcription factor XOct-25 inhibits ectodermal competence to respond to bone morphogenetic protein-mediated embryonic induction. *Mechanisms of Development*, *124*, 840–855.

Takeda, H., Matsuzaki, T., Oki, T., Miyagawa, T., & Amanuma, H. (1994). A novel POU domain gene, zebrafish *pou2*: Expression and roles of two alternatively spliced twin products in early development. *Genes & Development*, *8*, 45–59.

Tallack, M. R., Whitington, T., Yuen, W. S., Wainwright, E. N., Keys, J. R., Gardiner, B. B., et al. (2010). A global role for KLF1 in erythropoiesis revealed by ChIP-seq in primary erythroid cells. *Genome Research*, *20*, 1052–1063.

Tapia, N., Reinhardt, P., Duemmler, A., Wu, G., Arauzo-Bravo, M. J., Esch, D., et al. (2012). Reprogramming to pluripotency is an ancient trait of vertebrate Oct4 and Pou2 proteins. *Nature Communications*, *3*, 1279.

ten Bosch, J. R., Benavides, J. A., & Cline, T. W. (2006). The TAGteam DNA motif controls the timing of Drosophila pre-blastoderm transcription. *Development*, *133*, 1967–1977.

Teo, A. K., Arnold, S. J., Trotter, M. W., Brown, S., Ang, L. T., Chng, Z., et al. (2011). Pluripotency factors regulate definitive endoderm specification through eomesodermin. *Genes & Development*, *25*, 238–250.

Tesar, P. J., Chenoweth, J. G., Brook, F. A., Davies, T. J., Evans, E. P., Mack, D. L., et al. (2007). New cell lines from mouse epiblast share defining features with human embryonic stem cells. *Nature*, *448*, 196–199. Epub 2007 June 27.

Theunissen, T. W., Costa, Y., Radzisheuskaya, A., van Oosten, A. L., Lavial, F., Pain, B., et al. (2011). Reprogramming capacity of Nanog is functionally conserved in vertebrates and resides in a unique homeodomain. *Development*, *138*, 4853–4865.

Thomson, J. A., Itskovitz-Eldor, J., Shapiro, S. S., Waknitz, M. A., Swiergiel, J. J., Marshall, V. S., et al. (1998). Embryonic stem cell lines derived from human blastocysts. *Science*, *282*, 1145–1147.

Thomson, M., Liu, S. J., Zou, L. N., Smith, Z., Meissner, A., & Ramanathan, S. (2011). Pluripotency factors in embryonic stem cells regulate differentiation into germ layers. *Cell*, *145*, 875–889.

Trinkaus, J. P. (1992). The midblastula transition, the YSL transition and the onset of gastrulation in *Fundulus*. *Development. Supplement*, *116*, 75–80.

Tucker, J. A., Mintzer, K. A., & Mullins, M. C. (2008). The BMP signaling gradient patterns dorsoventral tissues in a temporally progressive manner along the anteroposterior axis. *Developmental Cell, 14,* 108–119.

Vastenhouw, N. L., Zhang, Y., Woods, I. G., Imam, F., Regev, A., Liu, X. S., et al. (2010). Chromatin signature of embryonic pluripotency is established during genome activation. *Nature, 464,* 922–926.

Wagner, J. T., & Podrabsky, J. E. (2015). Gene expression patterns that support novel developmental stress buffering in embryos of the annual killifish Austrofundulus limnaeus. *EvoDevo, 6,* 1089–1097.

Wu, G., Han, D., Gong, Y., Sebastiano, V., Gentile, L., Singhal, N., et al. (2013). Establishment of totipotency does not depend on Oct4A. *Nature Cell Biology, 15,* 1089–1097.

Xie, D., Chen, C. C., Ptaszek, L. M., Xiao, S., Cao, X., Fang, F., et al. (2010). Rewirable gene regulatory networks in the preimplantation embryonic development of three mammalian species. *Genome Research, 20,* 804–815.

Xu, C., Fan, Z. P., Muller, P., Fogley, R., Dibiase, A., Trompouki, E., et al. (2012). Nanog-like regulates endoderm formation through the Mxtx2-Nodal pathway. *Developmental Cell, 22,* 625–638.

Yanez-Cuna, J. O., Dinh, H. Q., Kvon, E. Z., Shlyueva, D., & Stark, A. (2012). Uncovering cis-regulatory sequence requirements for context-specific transcription factor binding. *Genome Research, 22,* 2018–2030.

Yartseva, V., & Giraldez, A. J. (2015). The maternal-to-zygotic transition during vertebrate development: A model for reprogramming. *Current Topics in Developmental Biology, 113,* 191–232.

Ying, Q. L., Wray, J., Nichols, J., Batlle-Morera, L., Doble, B., Woodgett, J., et al. (2008). The ground state of embryonic stem cell self-renewal. *Nature, 453,* 519–523.

Young, R. A. (2011). Control of the embryonic stem cell state. *Cell, 144,* 940–954.

Yu, J., Vodyanik, M. A., Smuga-Otto, K., Antosiewicz-Bourget, J., Frane, J. L., Tian, S., et al. (2007). Induced pluripotent stem cell lines derived from human somatic cells. *Science, 318,* 1917–1920.

Zhang, C., & Klymkowsky, M. W. (2007). The Sox axis, Nodal signaling, and germ layer specification. *Differentiation; Research in Biological Diversity, 75,* 536–545.

Zhang, Y., Vastenhouw, N. L., Feng, J., Fu, K., Wang, C., Ge, Y., et al. (2014). Canonical nucleosome organization at promoters forms during genome activation. *Genome Research, 24,* 260–266.

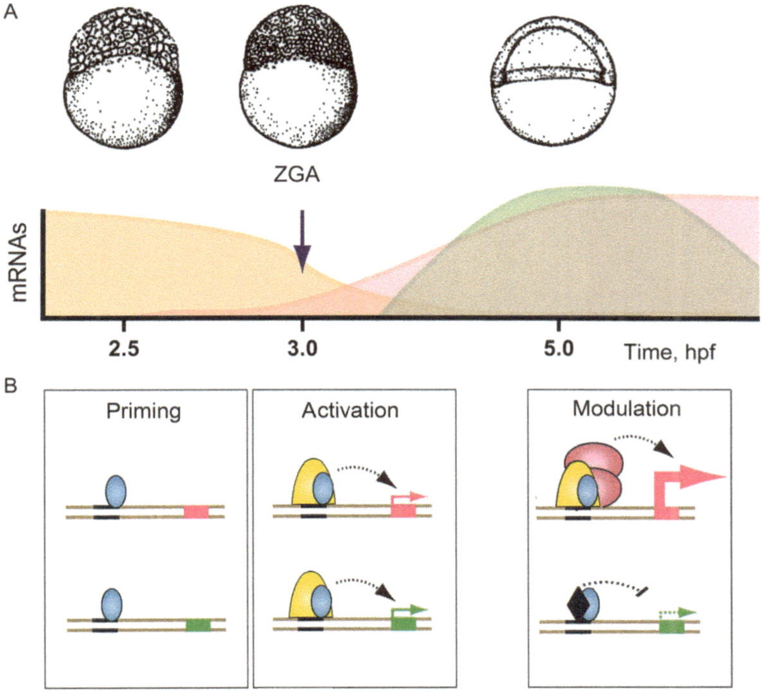

Daria Onichtchouk and Wolfgang Driever, Figure 1 Zygotic genome activation. (A) Schematic drawing of ZGA in zebrafish: starting from 1000 cell stage (middle embryo at the top), early zygotic genes are switched on, and some maternal mRNAs degrade. Hpf, hours postfertilization. (B) Hypothetical scheme of action of zygotic gene activators. Already before ZGA, zygotic genome activators (small oval) bind to defined regions in the genome and prime the neighboring genes for expression (left box, two example genes are shown). Ubiquitous cofactors (RNA polymerase or chromatin modifier enzyme; half oval) are attracted to zygotic genome activator binding site and switch on gene expression (middle box). Later, the binding sites for zygotic genome activators are occupied by additional tissue-specific factors, which either repress (gene at bottom) or enhance (gene at top) activation of transcription (right box).

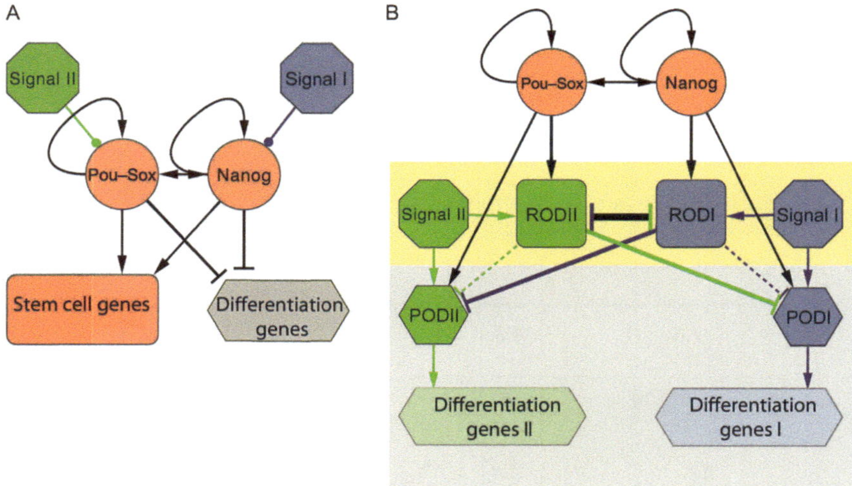

Daria Onichtchouk and Wolfgang Driever, Figure 2 Models of pluripotency. (A) Bistable switch model in ES cell (Chickarmane et al., 2006). Signals I and II can be activatory (arrowheads) or repressory (indicated by rounded arrowheads). The model describes two stable states. In the pluripotent state, in response to environmental signals, Pou5f1, SoxB1, and Nanog genes are switched on, they sustain their own expression by means of positive feedback loops, and the "pluripotency circuit" (Pou–Sox–Nanog) is active. In this state, stem cell genes are on and differentiation genes are switched off. When the positive feedback is interrupted by differentiation signals, the differentiation genes are on. (B) Simplified hypothetical model for transient pluripotency and developmental control in blastula to gastrula stage embryo cells: ROD, inter-(and/or intra-)territorial repressor of differentiation; POD, promoter of differentiation; signals I and II, different tissue specific patterning and differentiation signals. In the late blastula pluripotent state, signals I and II together with Pou–Sox–Nanog control RODII and RODI expression such that their activity is balanced. In this situation, PODs and differentiation genes are off (gray-shaded area). In committed differentiating cell in a specific germ layer or tissue, one RODI is off, POD starts to be expressed, induces differentiation genes, and presumably shuts down the pluripotency circuit.

CHAPTER SIXTEEN

Topological Domains, Metagenes, and the Emergence of Pleiotropic Regulations at *Hox* Loci

Fabrice Darbellay*, Denis Duboule*,†,1
*School of Life Sciences, Federal Institute of Technology, Lausanne, Switzerland
†Department of Genetics and Evolution, University of Geneva, Geneva, Switzerland
1Corresponding author: e-mail addresses: denis.duboule@unige.ch; denis.duboule@epfl.ch

Contents

Abstract

Hox gene clusters of jaw vertebrates display a tight genomic organization, which has no equivalent in any other bilateria genomes sequenced thus far. It was previously argued that such a topological consolidation toward a condensed, metagenic structure was due to the accumulation of long-range regulations flanking Hox loci, a phenomenon made possible by the successive genome duplications that occurred at the root of the vertebrate lineage, similar to gene neofunctionalization but applied to a coordinated multigenic system. Here, we propose that the emergence of such large vertebrate regulatory landscapes containing a range of global enhancers was greatly facilitated by the presence of topologically associating domains (TADs). These chromatin domains, mostly constitutive, may have been used as genomic niches where novel regulations could evolve due to both the preexistence of a structural backbone poised to integrate novel regulatory inputs, and a highly adaptive transcriptional readout. We propose a scenario for the coevolution of such TADs and the emergence of pleiotropy at ancestral vertebrate Hox loci.

Current Topics in Developmental Biology, Volume 116
ISSN 0070-2153
http://dx.doi.org/10.1016/bs.ctdb.2015.11.022

Ever since large *Hox* gene clusters were reported to exist in both protostomes (Bender et al., 1983; Harding, Wedeen, McGinnis, & Levine, 1985; Sanchez-Herrero, Vernos, Marco, & Morata, 1985) and deuterostomes (Boncinelli et al., 1988; Duboule, Baron, Mahl, & Galliot, 1986; Hart, Fainsod, & Ruddle, 1987), thus indicating unexpected commonalities in the developmental strategies of all bilateria (Akam, 1989; Duboule & Dolle, 1989; Graham, Papalopulu, & Krumlauf, 1989), this genetic system, originally described in *Drosophila* (Lewis, 1978) has revealed its remarkable epistemic value. However, the sequencing of genomes belonging to many different groups of animals, such as e.g. cephalopods (Albertin et al., 2015), nematodes (Aboobaker & Blaxter, 2003), platyhelminthes (Pierce et al., 2005), and urochordata (Ikuta, Yoshida, Satoh, & Saiga, 2004; Seo et al., 2004), has shown that *Hox* genes are not always found in a clustered organization, indicating that this particular in-*cis* topology is indeed not necessary for the development of many groups of animals. While some genomes carry compact or disorganized clusters, others have split this series of contiguous genes at one or in multiple positions (see Fig. 1). Nevertheless, at least one

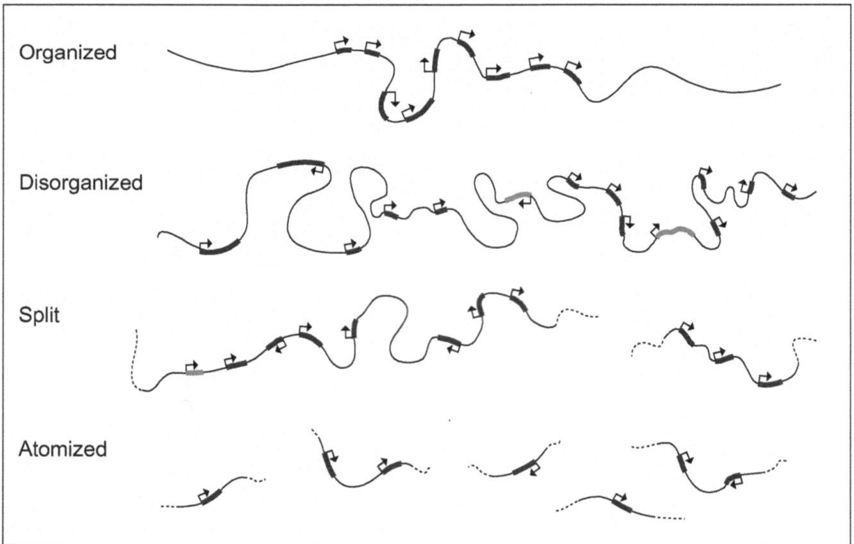

Figure 1 The genomic organization of *Hox* genes varies in different bilaterian animals. *Hox* genes can be found either in tightly organized gene clusters, ca. 100 kb in length, containing few repetitive elements and no foreign genes (organized) or in much larger linkage group, ca. 1 mb in size, including repeats and sometimes foreign genes (disorganized). Such genomic clusters can also be split in various subclusters (split) or even in single unit as in atomized *Hox* (non)clusters (atomized). *Adapted from Duboule (2007).* (A color version of this figure is located at the end of this chapter.)

entire *Hox* gene cluster is present in all metazoans developing their anterior to posterior axis by following a temporal sequence in the production of segments, such as short germ band insects or vertebrates, suggesting that this clustered organization is closely associated with these particular developmental strategies ("the *Hox* clock"; see Duboule, 1994; Kmita & Duboule, 2003).

1. THE RAISE OF *HOX* METAGENES

Of note, among all animal genomes sequenced thus far and containing one or several full *Hox* gene cluster(s), the level of genomic organization and compaction can vary substantially, from rather large gene complexes, close to one megabase in size (e.g., Mehta et al., 2013), to much shorter and compact structures. Surprisingly, jaw vertebrate (gnathostomes) display the highest level of compaction in their *Hox* clusters, with ca. 10 *Hox* genes within 100–120 kilobases (kb) of DNA, very few repetitive elements intermingled, no foreign genes interspersed, and all transcription units encoded by the same DNA strand (see Duboule, 2007). Even in squamata, where *Hox* clusters are larger than those found in fishes, birds, or in mammals (Di-Poï, Montoya-Burgos, & Duboule, 2009; Di-Poi et al., 2010; Vonk et al., 2013), their general aspect is clearly related to that of other jaw vertebrate clusters and their sizes merely 10–20% larger.

As of today, not a single non-jaw vertebrate genome has shown a comparable level of compaction. Conversely, not a single vertebrate genome has revealed a *Hox* cluster resembling the large structures found, for example, in sea urchins or in *Anopheles* (Cameron et al., 2006; Holt et al., 2002) (Fig. 2). Interestingly, the cephalochordate amphioxus *Hox* cluster, while being rather well organized still expands over half a megabase of DNA (see Minguillon et al., 2005). In jaw vertebrates, since neighboring *Hox* genes often share global regulations (see below), can be included into the same transcriptional unit and display redundant functions, this particular clustered organization was referred to as a "metagene" (Duboule, 1994).

2. A GENOMIC SYNAPOMORPHY

In this particular context, the case of vertebrate agnatha (jawless fishes) has remained difficult to evaluate. Jawless vertebrates indeed contain *Hox* complements whose phylogenetic relations with gnathostomes are difficult to establish. In the Japanese lamprey, for instance, the number of *Hox*

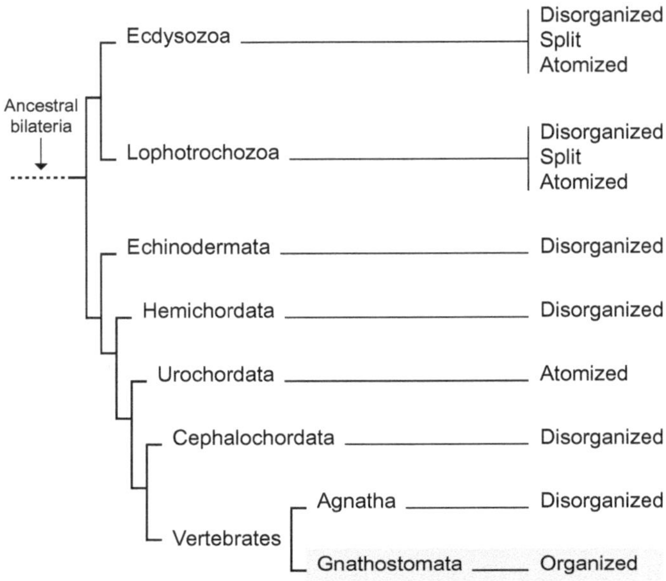

Figure 2 Distribution of various types of *Hox* gene organization among the major groups of bilateria. While invertebrates (Ecdysozoa and Lophotrochozoa) contain taxons with disorganized, split, or atomized *Hox* gene clusters, echinoderms and chordates seems to generally contain disorganized gene clusters, with the exception of urochordates, which do not have distinctive clusters and gnathostomes (jaw vertebrates), which all display tightly organized *Hox* gene clusters. *Adapted from Duboule (2007).*

clusters suggest that at least three events of duplication occurred (Mehta et al., 2013). However, while the R1 and R2 genome duplications (see Putnam et al., 2008) likely predated the separation between agnatha and gnathostomata (Kuraku, Meyer, & Kuratani, 2009), the comparative analysis of potential conserved regulatory sequences raised the possibility that agnatha separated from jaw vertebrates at an earlier stage (discussed in Mehta et al., 2013). In any case, however, the average size of the Japanese lamprey *Hox* clusters, while somewhat variable, is clearly closer to the invertebrate situation than to the gnathostome standard. For example, the *Hoxdelta* cluster (likely related to the tetrapod *HoxD* cluster) is 700 kb large and contains a range of repeats (Mehta et al., 2013). Therefore, by using this particular criterion, lampreys seem to be an interesting exception among vertebrates, which will be further discussed below.

Altogether, compact and well-organized *Hox* clusters thus appear to be a "genomic synapomorphy" of gnathostomes (Fig. 2). As previously argued (Duboule, 2007), there are essentially two exclusive phylogenetic explanations to account for this observation. In the first view, a bilateria ancestor carried a compacted *Hox* gene cluster, which became subsequently disorganized, split, or atomized in every single branch derived from this ancestor, except in jaw vertebrates. Accordingly, gnathostomes would be the only extant direct representatives (*Hox*-wise) of our bilateria ancestor, an alternative that cannot reasonably be considered outside a theological context.

The alternative scenario involves the presence of a large and disorganized *Hox* cluster in the bilateria ancestor, as a result of gene duplications in *cis*. This type of cluster was conserved in many animal groups, due to the existence of in-*cis* constraints in the temporal activation of these genes (Duboule, 1992). As a consequence, all taxa escaping this major developmental constraint via a distinctive ontogenetic strategy such as a cell lineage-dependent developmental process could have split or atomized their gene clusters without damages (discussed in Duboule, 1994, 2007). This scenario, however, implies that *Hox* gene clusters became organized and compacted during the evolution of gnathostomes. At first, this proposal appears slightly counterintuitive, as we tend to associate some entropy to evolutionary mechanisms. In this case indeed, the *Hox* clusters must have evolved into a higher level of genomic organization, while being located within genomes, which are themselves globally less compact than many invertebrate genomes, at least in terms of gene density.

3. REGULATORY PRIMING

Both the proximal cause of such a "consolidation" of *Hox* clusters in gnathostomes and the nature of its underlying mechanism(s) have remained speculative and poorly understood. Based on genetic studies using the mouse *HoxD* cluster as a paradigm, it was proposed that the evolution and progressive accumulation of long-range regulations at the vicinity of the gene cluster may have triggered the cluster to optimize its transcriptional response by becoming better organized. For example, vertebrate *Hox* genes are all encoded by the same DNA strand and large repeats seem to be excluded from these genomic loci, even though some taxa display more than others (see Di-Poï et al., 2009; Duboule, 2007; Simons, Makunin, Pheasant, & Mattick, 2007). It was also proposed that the presence of a strong global enhancer outside the gene cluster itself may "prime" the region for the *de novo* emergence of novel enhancers, as if the former would facilitate or prepare the evolution of the latter, a process which may have been critical for the evolution of jaw vertebrates. This concept of "regulatory priming" (Gonzalez, Duboule, & Spitz, 2007), however, has remained devoid of any mechanistic basis.

Recent advances in the study of chromatin structure and long-range gene regulation now provide us with an emerging conceptual framework that helps reconsidering these issues. In particular, the past 10 years have seen important changes in the way we contemplate developmental gene regulation in vertebrates, whereby the complexity of the transcription unit itself has been progressively replaced by the complexity of its regulatory modalities (e.g., Spitz & Furlong, 2012). The first element to consider is that many transcription factors or signaling molecules of key importance for a given developmental mechanism are generally highly pleiotropic and thus require multiple regulatory inputs. Second—and likely related to this former observation—the various enhancer sequences in charge of these distinct controls in space and time are found over very large distances from the target genes. Such a concentration of long-range transcriptional controls (Lettice et al., 2003), referred to as regulatory landscapes (Spitz, Gonzalez, & Duboule, 2003) have now been described around many such genes (e.g., Montavon & Duboule, 2012; Symmons et al., 2014), usually extending over a genomic interval much larger than the size of the target gene itself (e.g., Amano et al., 2009; Berlivet et al., 2013; Ruf et al., 2011).

4. SO NEAR AND YET SO FAR

Gene regulation via long-range acting enhancers, however, raises a number of questions related both to the mechanism of enhancer–promoter contact (de Laat & Grosveld, 2003; Tolhuis, Palstra, Splinter, Grosveld, & De Laat, 2002) and to the choice and selectivity either of the target gene(s) or of the appropriate enhancer whenever complex regulatory landscapes are involved (reviewed in Spitz & Furlong, 2012). The recent observation that such regulatory landscapes often display constitutive chromatin organizations in 3D, i.e., even in the absence of transcription, allows to address these various issue in a slightly different context. By using chromosome conformation capture (de Laat & Dekker, 2012), a large regulatory landscape containing series of digit enhancers was initially described upstream the *HoxD* gene cluster (Montavon et al., 2011). This particular genetic environment also displayed constitutive contacts, i.e., interactions present in cell types devoid of any *HoxD* transcripts. Because this ca.

1 mb large constitutive chromatin domain is delimited by clear boundaries at both ends, which are shared between different, active, or inactive cell types, the existence of a poised regulatory structure was proposed, whose spatial and functional organization would slightly change in digit cells transcribing the requested target *HoxD* genes.

The presence of related chromatin domains along the entire genome was subsequently observed by using Hi-C and 5C technologies (Dixon et al., 2012; Nora et al., 2012) and referred to as topologically associating domains (TADs; Dixon et al., 2012). TADs tend to be structurally conserved during cell differentiation, even though interactions occurring either inside or in-between TADs may vary (Berlivet et al., 2013; Dixon et al., 2015; Giorgetti et al., 2014). While the potential functions of these chromatin domains, for example, as units of coordinated gene regulation or as a means to prevent regulatory interferences, have been discussed (Nora, Dekker, & Heard, 2013; Symmons et al., 2014), the causality of such processes often remains to be clearly established (see de Laat & Duboule, 2013).

5. WITH A LITTLE HELP FROM MY TAD

A comparison between the ES cells dataset from the Ren laboratory (Dixon et al., 2012) and our *in embryo* study of the *HoxD* loci confirmed that the two large regulatory landscapes flanking the *HoxD* cluster were also scored by Hi-C using embryonic cells and hence they correspond to genuine TADs (Andrey et al., 2013). In addition, functional analysis revealed that these two TADs were active at different times without any temporal overlap and contained various sets of enhancers (Andrey et al., 2013; Delpretti et al., 2013; Lonfat, Montavon, Darbellay, Gitto, & Duboule, 2014). Considering these new elements, how can TADs help us think about potential mechanisms leading to the evolutionary consolidation of *Hox* gene clusters in early vertebrates?

If *Hox* clusters became progressively better organized in response to the accumulation of global enhancers located outside, as proposed above, then the force driving such an accumulation of regulations, i.e., the mechanism underlying this "regulatory priming" ought to be the cause of consolidation. In fact, every additional evolution of a given regulatory sequence within the landscape would help tighten the organization of the gene cluster such as to improve a subset of the global adaptive value of the system. We would like to argue that TADs facilitated this process through their constitutive aspect, i.e., the fact that a poised chromatin structure existed even in the absence of a transcriptional read out. In this view, TADs were used as particular chromatin niches where novel enhancer sequences have evolved (Lonfat & Duboule, 2015) (Fig. 3).

6. TAD HIJACKING

Why should this be the case? At many developmental loci, TADs are often considered as global and independent regulatory units including their target genes (e.g., see Symmons et al., 2014). Accordingly, the *de novo* appearance of a transcription factor specific binding sequence within a given TAD may trigger cells to experience a new differentiation pathway, which may or may not be further elaborated depending on its adaptive value for the organism. In this view, TADs may be seen as regulatory toolkits, as chromatin playgrounds where various factors may have experienced some successful or (mostly) unsuccessful encounters (Lonfat & Duboule, 2015; Lonfat et al., 2014). Understandably, such a trial and error process may function more efficiently should a poised regulatory structure such as TADs already exists, thus merely reducing its cooptation to the progressive evolution of a robust DNA binding sequence inside. Such a structure may favorably "attract" and "sequester" factors and facilitate the latter process, due to a general chromatin environment (a niche) already implementing related types of long-range controls in distinct cellular contexts.

This hypothesis, however, does not explain how this initial protein–DNA recognition can occur. It nevertheless helps thinking about the other necessary factors for a productive interaction to develop, such as the contact with proximal promoter elements and the presence of the required cofactors. Also, the emergence of a novel enhancer within a given TAD likely reenforced the global structure, for example, by evolving novel constitutive contacts in parallel, supporting the new transcription task. This in turn may have triggered new regulatory sequences to emerge, thus leading to a virtuous circle and the observed concentration of enhancers of all kinds within some TADs at developmental loci with high pleiotropic functions (Fig. 3).

7. TAKE IT OR LEAVE IT

Interestingly, *HoxD* cluster is located at the border of two TADs, each chromatin environment regulating a well defined and rather invariant subset of target genes. For instance, the centromeric TAD contains enhancers

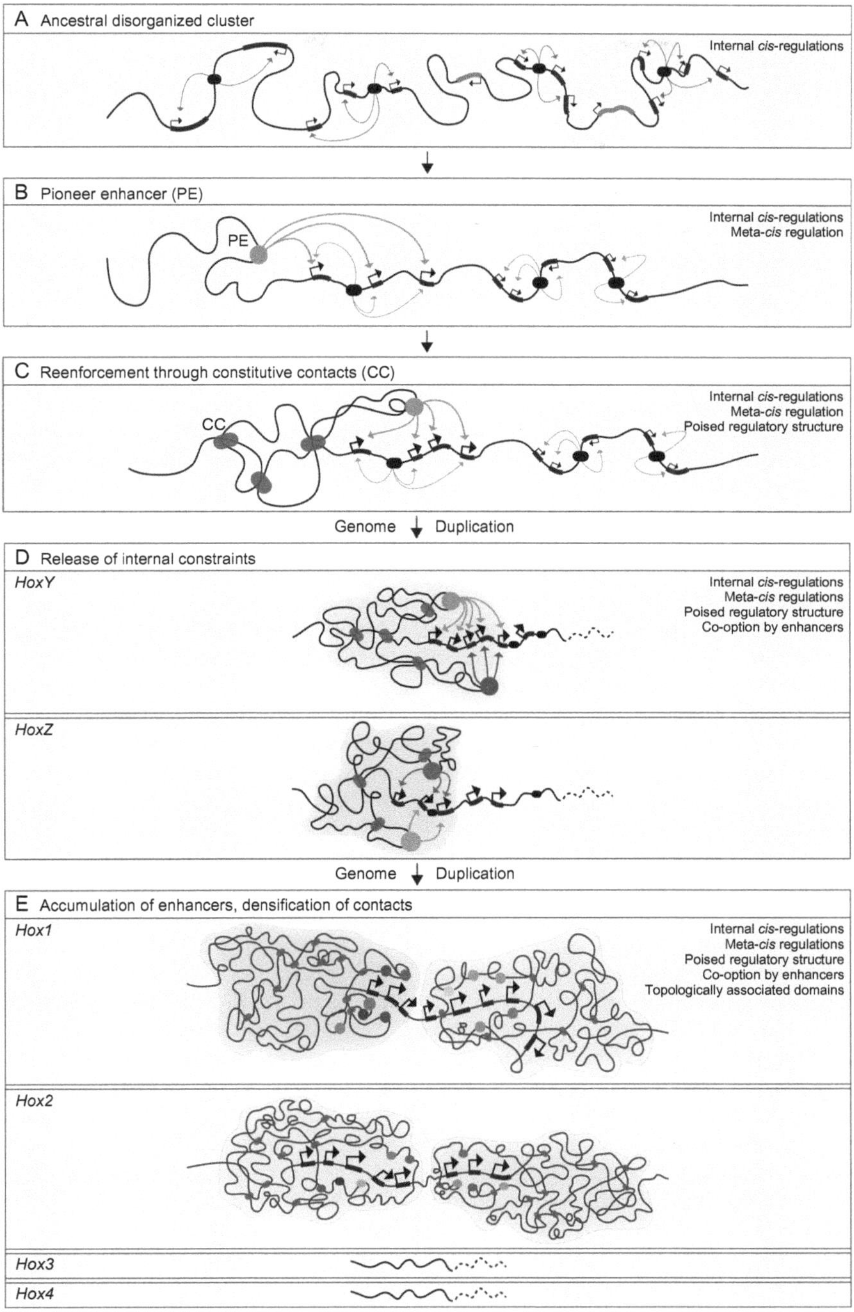

Figure 3 See legend on next page.

active in the developing digits and genitals. In both cases, the *Hoxd13* to *Hoxd9* group of genes is the main target of this regulating landscape (Lonfat et al., 2014; Montavon et al., 2011), even though the function of *Hoxd13* alone would likely be sufficient (Delpretti, Zakany, & Duboule, 2012; Kmita et al., 2005). The same holds true for the telomeric TAD, where the *Hoxd11* to *Hoxd8* subgroup seems to be coordinately regulated, for example, in the developing forearm and intestinal caecum (Andrey et al., 2013; Delpretti et al., 2013). Therefore, it appears that the system itself experiences some internal constraints and that, consequently,

Figure 3 TAD-dependent emergence of multiple global regulations at *Hox* loci and concomitant consolidation of the *Hox* gene clusters in jaw vertebrates. (A) Initially, an ancestral disorganized *Hox* genes (blue) cluster implements its collinear regulation along the rostral to caudal axis through series of internal regulatory sequences (*cis*-regulation; black). The cluster may also contain unrelated genes (orange). In (B), some genes of the cluster become coordinately regulated by the evolution of a strong enhancer located outside the cluster itself, such as not to interfere with the internal regulatory organization. As a consequence, the transcriptional response (the adaptive value) becomes slowly optimized via the progressive genomic rearrangement of the target genes (start of consolidation), in order to optimize their response to this new regulation. This initial meta-*cis* regulation through a "pioneer enhancer" (PE) triggers the formation of a local 3D architecture. (C) Should this coordinated transcriptional control be of some adaptive value, the 3D structure may start to be stabilized or strengthened by the evolution of constitutive interactions (brown ovals), which will help and improve the interactions between the pioneer enhancer and its targets. This global (meta-*cis*) regulation may however interfere with some of the constraints imposed by the internal collinear mechanism. (D) This antagonism between *cis* and meta-*cis* regulations is partly released after genome duplications, which allows some functional compensation to occur between the duplicated clusters and thus restores more flexibility in evolving global regulations. As a consequence of genome duplication, this 3D structure will thus facilitate the evolution of yet other global enhancers on each duplicate (*HoxY* and *HoxZ*), due to the preexistence of a poised transcriptional read out of high adaptive value. This process of "regulatory priming" (Gonzalez et al., 2007) will continue as a virtuous circle and progressively lead to the maximal compaction of the target *Hox* gene cluster, with all genes in the same transcriptional orientation, very few repeats and no longer any foreign genes interspersed. (E) At the same time, large and constitutive regulatory domains will be observed on both sides, reflecting the successful cooption of many global enhancers that accompanied the emergence of most jaw vertebrate innovations (see the text). These domains correspond to topologically associating domains (TADs) (Dixon et al., 2012; Nora et al., 2012). In (B)–(D), only one of the flanking regions of the cluster is shown for sake of simplicity, even though the same mechanism may apply to the other. In (E), only two clusters are shown instead of the four derived from both rounds of genome duplications. (A color version of this figure is located at the end of this chapter.)

not all possibilities exist regarding the evolution of novel regulations. In this view, whichever new tissue specificity may evolve within one of these TADs, the transcription readout will be predetermined according to the internal properties of the system. This fixed and coordinated response of groups of target genes results from the compaction of the gene cluster, which in our hypothesis is itself due to the accumulation of enhancers, illustrating how the evolving system may generate its own set of constraints.

In this oulipian scheme, it is conceptually difficult to imagine the emergence of a subtle, fine-tuned type of regulation, affecting a single gene in a precise manner. Instead, the evolution of global enhancers within either one of these TADs likely resulted in the massive activation of a prefixed battery of neighboring target genes, leading to major modification of the resulting animals. In this context, it is interesting to note that all enhancers located nearby the *HoxD* clusters are involved in structures of major importance for the vertebrate lineage (Spitz, Herkenne, Morris, & Duboule, 2005). Therefore, while TADs may have represented powerful means to evolve pleiotropy at important developmental loci, it may have equally prevented other regulatory input to emerge, due to the constraints imposed by the system itself and the existence of a metagenic structure at its target end.

8. PIONEER ENHANCERS

It is as yet difficult to evaluate whether the constitutive contacts within a given TADs (i.e., contacts not involved in themselves into a transcriptional read out) predated the appearance of specific regulations or instead, if an initial contact between a promoter and an enhancer became strengthened via the subsequent addition of a constitutive architecture. This conundrum may be addressed in the future by comparing TADs either between different species, or between paralogous loci within the same group. Because a constitutive interaction in itself may not carry an adaptive value as strong as a first specific regulatory contact, we favor the latter possibility as a theoretical starting point (see Fig. 3). Accordingly, an initial productive enhancer–promoter interaction had to be established in order to initiate the process by securing a basic constitutive regulatory landscape leading to a tissue-specific enhanced transcription of a group of *Hox* target genes (Fig. 3A and B).

In this view, a pioneer enhancer sequence must have initiated the system and thus facilitated the evolution of subsequent enhancers within the same TAD. This view is somewhat analogous to the concept of pioneer factors

(Zaret & Carroll, 2011), yet it ought to be considered in an evolutionary—rather than developmental—framework; while pioneer transcription factors facilitate the recruitment of additional factors at a given locus by locally reorganizing chromatin during development or reprogramming (Iwafuchi-Doi & Zaret, 2014), pioneer enhancers would have facilitated the emergence of novel regulatory sequences over evolutionary times, by providing an appropriate chromatin structure and transcriptional readout where trial and error processes can occur.

9. REGULATORY PALEONTOLOGY

As a consequence of the previous proposal, TADs containing series of enhancers associated with highly pleiotropic loci should contain traces of their evolutionary histories. A detailed structure–function comparison between orthologous TADs in various species, associated with the presence or absence of the corresponding morphological traits, should allow for a reconstitution of their progressive evolutionary trajectories, concerning both their specific enhancer sequences and their constitutive architectural contacts, up to the original pioneer enhancer sequence, which may not have been necessarily conserved through time, and corresponding binding factors. While such reconstitutions or parts thereof—may be easy whenever regulations will concern characters separated by large evolutionary distances (for example, the emergence of limbs and mammary glands), they may turn out to be more difficult for characters sharing close evolutionary timing and adaptive value, such as digits and genitals (see Lonfat et al., 2014). In this latter case, the question as to which enhancer was first may remain unsolved.

In summary, we propose that initial TAD-like structures, formed by pioneer enhancers, may have helped accumulating global regulations upstream the *Hox* clusters. The concomitant optimization of the transcriptional response triggered a progressive structural consolidation of the gene clusters. Such an accumulation of regulations, however, was possible only after a genome duplication event had occurred such that internal constraints applied to the gene cluster itself (and to its early critical function during trunk development) were partially released as a consequence of functional redundancy (Fig. 3C and D). Some issues, however, remain difficult to integrate into this conceptual framework. For instance, agnatha such as the Japanese lamprey likely experienced several genome duplications and yet the *Hox* clusters have not been consolidated (Mehta et al., 2013) as in other vertebrates. This suggests that the emergence of global regulation may not be a trivial event and that genome duplications were not sufficient to trigger this important step, at least at these loci. Concerning lampreys, it will be interesting in the future to evaluate if at least one pioneer global enhancer exists flanking either one of the *Hox* clusters. In any case, it is striking to see that lampreys do not exhibit any of the morphologies that were associated with global regulations at the *HoxD* cluster, such as for example the proximal and distal limbs, external genitals, intestinal caecum, and the metanephric kidneys.

Equally difficult to understand is how all four *Hox* gene clusters were almost equally consolidated along the gnathostome lineage (as schematized in Fig. 3). A parsimonious answer is that consolidation had started before genome duplications had occurred, under the form of a common pioneer global enhancer, which may have started to control several genes coordinately in one particularly adaptive structure of an ancestral chordate (Fig. 3). The accumulation of regulations triggered by this enhancer, however, may have initiated only after duplications of the locus. In this scenario, the presence of a pioneer enhancer would not only have allowed for pleiotropy and consolidation to develop, but would have made it the rule for jaw vertebrates *Hox* clusters. If true, all four *Hox* clusters should *ad minima* be the target of some global regulations. While this is clearly the case for *HoxA* (Berlivet et al., 2013; Woltering, Noordermeer, Leleu, & Duboule, 2014) and *HoxD* (see, e.g., Andrey et al., 2013; Montavon et al., 2011; Spitz et al., 2003), the situation is less clear either for *HoxB* or for *HoxC*. While the *HoxB* cluster seems to reside at the boundary between two TADs, similar to both *HoxA* and *HoxD* (unpublished observations) suggesting the existence of flanking global regulations, the *HoxC* cluster however does not display this particular organization, at least when an ES cells dataset is used (Dixon et al., 2012). Further studies on the regulations of the *HoxB* and *HoxC* clusters may be informative in this respect.

ACKNOWLEDGMENTS

We thank all the past and present members of the Duboule Laboratories for their numerous contributions to these various issues through their data and discussions and in particular C. Bolt and A. Hintermann for their comments on the chapter. This work is supported by funds from the Swiss SNF, the University of Geneva, The Ecole Polytechnique Federale de Lausanne, and the European Research Council (ERC). F.D. was supported by the EU FP7 project IDEAL.

REFERENCES

Aboobaker, A., & Blaxter, M. (2003). Hox gene evolution in nematodes: Novelty conserved. *Current Opinion in Genetics & Development*, *13*, 593–598.

Akam, M. (1989). Hox and HOM: Homologous gene clusters in insects and vertebrates. *Cell*, *57*, 347–349.

Albertin, C. B., Simakov, O., Mitros, T., Wang, Z. Y., Pungor, J. R., Edsinger-Gonzales, E., et al. (2015). The octopus genome and the evolution of cephalopod neural and morphological novelties. *Nature*, *524*, 220–224.

Amano, T., Sagai, T., Tanabe, H., Mizushina, Y., Nakazawa, H., & Shiroishi, T. (2009). Chromosomal dynamics at the Shh locus: Limb bud-specific differential regulation of competence and active transcription. *Developmental Cell*, *16*, 47–57.

Andrey, G., Montavon, T., Mascrez, B., Gonzalez, F., Noordermeer, D., Leleu, M., et al. (2013). A switch between topological domains underlies HoxD genes collinearity in mouse limbs. *Science*, *340*, 1234167.

Bender, W., Akam, M., Karch, F., Beachy, P. A., Peifer, M., Spierer, P., et al. (1983). Molecular genetics of the bithorax complex in Drosophila melanogaster. *Science*, *221*, 23–29.

Berlivet, S., Paquette, D., Dumouchel, A., Langlais, D., Dostie, J., & Kmita, M. (2013). Clustering of tissue-specific sub-TADs accompanies the regulation of HoxA genes in developing limbs. *PLoS Genetics*, *9*, e1004018.

Boncinelli, E., Somma, R., Acampora, D., Pannese, M., D'Esposito, M., Faiella, A., et al. (1988). Organization of human homeobox genes. *Human Reproduction*, *3*, 880–886.

Cameron, R. A., Rowen, L., Nesbitt, R., Bloom, S., Rast, J. P., Berney, K., et al. (2006). Unusual gene order and organization of the sea urchin hox cluster. *Journal of Experimental Zoology. Part B, Molecular and Developmental Evolution*, *306*, 45–58.

de Laat, W., & Dekker, J. (2012). 3C-based technologies to study the shape of the genome. *Methods*, *58*, 189–191.

de Laat, W., & Duboule, D. (2013). Topology of mammalian developmental enhancers and their regulatory landscapes. *Nature*, *502*, 499–506.

de Laat, W., & Grosveld, F. (2003). Spatial organization of gene expression: The active chromatin hub. *Chromosome Research*, *11*, 447–459.

Delpretti, S., Montavon, T., Leleu, M., Joye, E., Tzika, A., Milinkovitch, M., et al. (2013). Multiple enhancers regulate Hoxd genes and the Hotdog LncRNA during cecum budding. *Cell Reports*, *5*, 137–150.

Delpretti, S., Zakany, J., & Duboule, D. (2012). A function for all posterior Hoxd genes during digit development? *Developmental Dynamics*, *241*, 792–802.

Di-Poï, N., Montoya-Burgos, J., & Duboule, D. (2009). Atypical relaxation of structural constraints in Hox gene clusters of the green anole lizard. *Genome Research*, *19*(4), 602–610.

Di-Poi, N., Montoya-Burgos, J. I., Miller, H., Pourquie, O., Milinkovitch, M. C., & Duboule, D. (2010). Changes in Hox genes' structure and function during the evolution of the squamate body plan. *Nature*, *464*, 99–103.

Dixon, J. R., Jung, I., Selvaraj, S., Shen, Y., Antosiewicz-Bourget, J. E., Lee, A. Y., et al. (2015). Chromatin architecture reorganization during stem cell differentiation. *Nature*, *518*, 331–336.

Dixon, J. R., Selvaraj, S., Yue, F., Kim, A., Li, Y., Shen, Y., et al. (2012). Topological domains in mammalian genomes identified by analysis of chromatin interactions. *Nature*, *485*, 376–380.

Duboule, D. (1992). The vertebrate limb: A model system to study the Hox/HOM gene network during development and evolution. *Bioessays*, *14*, 375–384.

Duboule, D. (1994). Temporal colinearity and the phylotypic progression: A basis for the stability of a vertebrate Bauplan and the evolution of morphologies through heterochrony. *Development* (Suppl.), 135–142.

Duboule, D. (2007). The rise and fall of Hox gene clusters. *Development, 134*, 2549–2560.
Duboule, D., Baron, A., Mahl, P., & Galliot, B. (1986). A new homeo-box is present in overlapping cosmid clones which define the mouse Hox-1 locus. *The EMBO Journal, 5*, 1973–1980.
Duboule, D., & Dolle, P. (1989). The structural and functional organization of the murine HOX gene family resembles that of Drosophila homeotic genes. *The EMBO Journal, 8*, 1497–1505.
Giorgetti, L., Galupa, R., Nora, E. P., Piolot, T., Lam, F., Dekker, J., et al. (2014). Predictive polymer modeling reveals coupled fluctuations in chromosome conformation and transcription. *Cell, 157*, 950–963.
Gonzalez, F., Duboule, D., & Spitz, F. (2007). Transgenic analysis of Hoxd gene regulation during digit development. *Developmental Biology, 306*, 847–859.
Graham, A., Papalopulu, N., & Krumlauf, R. (1989). The murine and Drosophila homeobox gene complexes have common features of organization and expression. *Cell, 57*, 367–378.
Harding, K., Wedeen, C., McGinnis, W., & Levine, M. (1985). Spatially regulated expression of homeotic genes in Drosophila. *Science, 229*, 1236–1242.
Hart, C. P., Fainsod, A., & Ruddle, F. H. (1987). Sequence analysis of the murine Hox-2.2, -2.3, and -2.4 homeo boxes: Evolutionary and structural comparisons. *Genomics, 1*, 182–195.
Holt, R. A., Subramanian, G. M., Halpern, A., Sutton, G. G., Charlab, R., Nusskern, D. R., et al. (2002). The genome sequence of the malaria mosquito Anopheles gambiae. *Science, 298*, 129–149.
Ikuta, T., Yoshida, N., Satoh, N., & Saiga, H. (2004). Ciona intestinalis Hox gene cluster: Its dispersed structure and residual colinear expression in development. *Proceedings of the National Academy of Sciences of the United States of America, 101*, 15118–15123.
Iwafuchi-Doi, M., & Zaret, K. S. (2014). Pioneer transcription factors in cell reprogramming. *Genes & Development, 28*, 2679–2692.
Kmita, M., & Duboule, D. (2003). Organizing axes in time and space; 25 years of colinear tinkering. *Science, 301*, 331–333.
Kmita, M., Tarchini, B., Zakany, J., Logan, M., Tabin, C. J., & Duboule, D. (2005). Early developmental arrest of mammalian limbs lacking HoxA/HoxD gene function. *Nature, 435*, 1113–1116.
Kuraku, S., Meyer, A., & Kuratani, S. (2009). Timing of genome duplications relative to the origin of the vertebrates: Did cyclostomes diverge before or after? *Molecular Biology and Evolution, 26*, 47–59.
Lettice, L. A., Heaney, S. J., Purdie, L. A., Li, L., de Beer, P., Oostra, B. A., et al. (2003). A long-range Shh enhancer regulates expression in the developing limb and fin and is associated with preaxial polydactyly. *Human Molecular Genetics, 12*, 1725–1735.
Lewis, E. B. (1978). A gene complex controlling segmentation in Drosophila. *Nature, 276*, 565–570.
Lonfat, N., & Duboule, D. (2015). Structure, function and evolution of topologically associating domains (TADs) at HOX loci. *FEBS Letters, 589*(20 Pt. A), 2869–2876 (review). http://dx.doi.org/10.1016/j.febslet.2015.04.024.
Lonfat, N., Montavon, T., Darbellay, F., Gitto, S., & Duboule, D. (2014). Convergent evolution of complex regulatory landscapes and pleiotropy at Hox loci. *Science, 346*, 1004–1006.
Mehta, T. K., Ravi, V., Yamasaki, S., Lee, A. P., Lian, M. M., Tay, B. H., et al. (2013). Evidence for at least six Hox clusters in the Japanese lamprey (Lethenteron japonicum). *Proceedings of the National Academy of Sciences of the United States of America, 110*, 16044–16049.

Minguillon, C., Gardenyes, J., Serra, E., Castro, L. F., Hill-Force, A., Holland, P. W., et al. (2005). No more than 14: The end of the amphioxus Hox cluster. *International Journal of Biological Sciences*, *1*, 19–23.

Montavon, T., & Duboule, D. (2012). Landscapes and archipelagos: Spatial organization of gene regulation in vertebrates. *Trends in Cell Biology*, *22*(7), 347–354.

Montavon, T., Soshnikova, N., Mascrez, B., Joye, E., Thevenet, L., Splinter, E., et al. (2011). A regulatory archipelago controls Hox genes transcription in digits. *Cell*, *147*, 1132–1145.

Nora, E. P., Dekker, J., & Heard, E. (2013). Segmental folding of chromosomes: A basis for structural and regulatory chromosomal neighborhoods? *Bioessays*, *35*, 818–828.

Nora, E. P., Lajoie, B. R., Schulz, E. G., Giorgetti, L., Okamoto, I., Servant, N., et al. (2012). Spatial partitioning of the regulatory landscape of the X-inactivation centre. *Nature*, *485*, 381–385.

Pierce, R. J., Wu, W., Hirai, H., Ivens, A., Murphy, L. D., Noel, C., et al. (2005). Evidence for a dispersed Hox gene cluster in the platyhelminth parasite Schistosoma mansoni. *Molecular Biology and Evolution*, *22*, 2491–2503.

Putnam, N. H., Butts, T., Ferrier, D. E., Furlong, R. F., Hellsten, U., Kawashima, T., et al. (2008). The amphioxus genome and the evolution of the chordate karyotype. *Nature*, *453*, 1064–1071.

Ruf, S., Symmons, O., Uslu, V. V., Dolle, D., Hot, C., Ettwiller, L., et al. (2011). Large-scale analysis of the regulatory architecture of the mouse genome with a transposon-associated sensor. *Nature Genetics*, *43*, 379–386.

Sanchez-Herrero, E., Vernos, I., Marco, R., & Morata, G. (1985). Genetic organization of Drosophila bithorax complex. *Nature*, *313*, 108–113.

Seo, H. C., Edvardsen, R. B., Maeland, A. D., Bjordal, M., Jensen, M. F., Hansen, A., et al. (2004). Hox cluster disintegration with persistent anteroposterior order of expression in Oikopleura dioica. *Nature*, *431*, 67–71.

Simons, C., Makunin, I. V., Pheasant, M., & Mattick, J. S. (2007). Maintenance of transposon-free regions throughout vertebrate evolution. *BMC Genomics*, *8*, 470.

Spitz, F., & Furlong, E. E. (2012). Transcription factors: From enhancer binding to developmental control. *Nature Reviews. Genetics*, *13*, 613–626.

Spitz, F., Gonzalez, F., & Duboule, D. (2003). A global control region defines a chromosomal regulatory landscape containing the HoxD cluster. *Cell*, *113*, 405–417.

Spitz, F., Herkenne, C., Morris, M. A., & Duboule, D. (2005). Inversion-induced disruption of the Hoxd cluster leads to the partition of regulatory landscapes. *Nature Genetics*, *37*, 889–893.

Symmons, O., Uslu, V. V., Tsujimura, T., Ruf, S., Nassari, S., Schwarzer, W., et al. (2014). Functional and topological characteristics of mammalian regulatory domains. *Genome Research*, *24*, 390–400.

Tolhuis, B., Palstra, R. J., Splinter, E., Grosveld, F., & De Laat, W. (2002). Looping and interaction between hypersensitive sites in the active beta-globin locus. *Molecular Cell*, *10*, 1453–1465.

Vonk, F. J., Casewell, N. R., Henkel, C. V., Heimberg, A. M., Jansen, H. J., McCleary, R. J., et al. (2013). The king cobra genome reveals dynamic gene evolution and adaptation in the snake venom system. *Proceedings of the National Academy of Sciences of the United States of America*, *110*, 20651–20656.

Woltering, J. M., Noordermeer, D., Leleu, M., & Duboule, D. (2014). Conservation and divergence of regulatory strategies at Hox Loci and the origin of tetrapod digits. *PLoS Biology*, *12*, e1001773.

Zaret, K. S., & Carroll, J. S. (2011). Pioneer transcription factors: Establishing competence for gene expression. *Genes & Development*, *25*, 2227–2241.

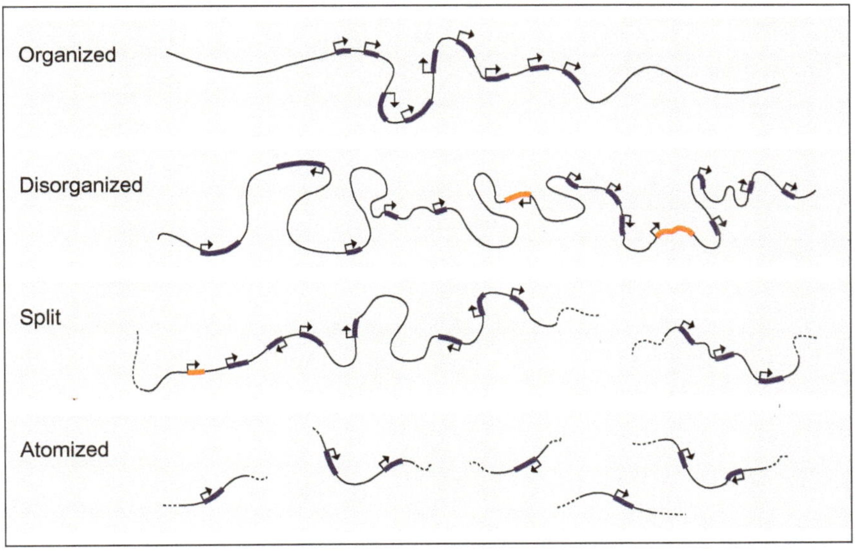

Fabrice Darbellay and Denis Duboule, Figure 1 The genomic organization of *Hox* genes varies in different bilaterian animals. *Hox* genes can be found either in tightly organized gene clusters, ca. 100 kb in length, containing few repetitive elements and no foreign genes (organized) or in much larger linkage group, ca. 1 mb in size, including repeats and sometimes foreign genes (disorganized). Such genomic clusters can also be split in various subclusters (split) or even in single unit as in atomized *Hox* (non)clusters (atomized). *Adapted from Duboule (2007).*

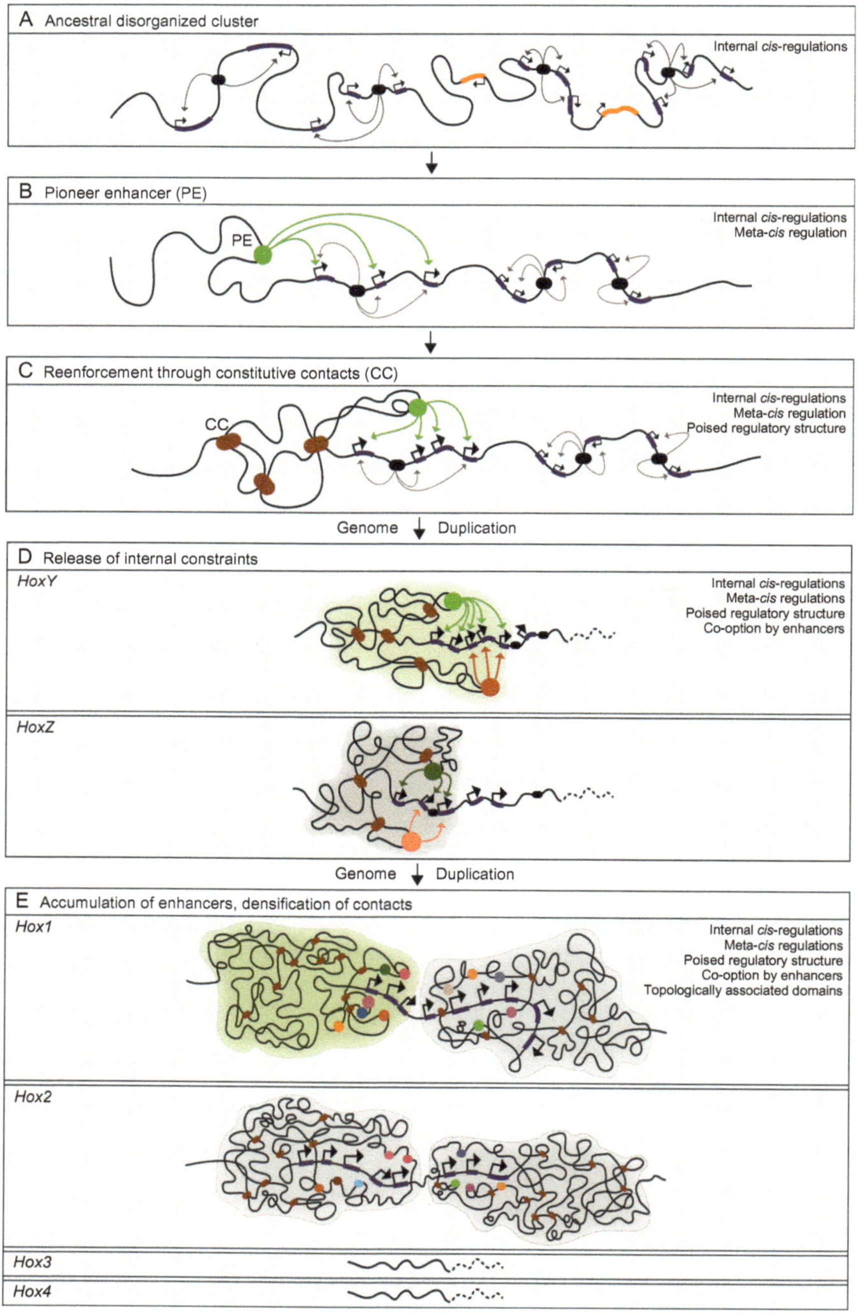

Fabrice Darbellay and Denis Duboule, Figure 3 See legend on opposite page.

Fabrice Darbellay and Denis Duboule, Figure 3 TAD-dependent emergence of multiple global regulations at *Hox* loci and concomitant consolidation of the *Hox* gene clusters in jaw vertebrates. (A) Initially, an ancestral disorganized *Hox* genes (blue) cluster implements its collinear regulation along the rostral to caudal axis through series of internal regulatory sequences (*cis*-regulation; black). The cluster may also contain unrelated genes (orange). In (B), some genes of the cluster become coordinately regulated by the evolution of a strong enhancer located outside the cluster itself, such as not to interfere with the internal regulatory organization. As a consequence, the transcriptional response (the adaptive value) becomes slowly optimized via the progressive genomic rearrangement of the target genes (start of consolidation), in order to optimize their response to this new regulation. This initial meta-*cis* regulation through a "pioneer enhancer" (PE) triggers the formation of a local 3D architecture. (C) Should this coordinated transcriptional control be of some adaptive value, the 3D structure may start to be stabilized or strengthened by the evolution of constitutive interactions (brown ovals), which will help and improve the interactions between the pioneer enhancer and its targets. This global (meta-*cis*) regulation may however interfere with some of the constraints imposed by the internal collinear mechanism. (D) This antagonism between *cis* and meta-*cis* regulations is partly released after genome duplications, which allows some functional compensation to occur between the duplicated clusters and thus restores more flexibility in evolving global regulations. As a consequence of genome duplication, this 3D structure will thus facilitate the evolution of yet other global enhancers on each duplicate (*HoxY* and *HoxZ*), due to the preexistence of a poised transcriptional read out of high adaptive value. This process of "regulatory priming" (Gonzalez et al., 2007) will continue as a virtuous circle and progressively lead to the maximal compaction of the target *Hox* gene cluster, with all genes in the same transcriptional orientation, very few repeats and no longer any foreign genes interspersed. (E) At the same time, large and constitutive regulatory domains will be observed on both sides, reflecting the successful cooption of many global enhancers that accompanied the emergence of most jaw vertebrate innovations (see the text). These domains correspond to topologically associating domains (TADs) (Dixon et al., 2012; Nora et al., 2012). In (B)–(D), only one of the flanking regions of the cluster is shown for sake of simplicity, even though the same mechanism may apply to the other. In (E), only two clusters are shown instead of the four derived from both rounds of genome duplications.

CHAPTER SEVENTEEN

Connectomics, the Final Frontier

Scott W. Emmons[1]
Department of Genetics, Albert Einstein College of Medicine, Bronx, New York, USA
Dominick P. Purpura Department of Neuroscience, Albert Einstein College of Medicine, Bronx, New York, USA
[1]Corresponding author: e-mail address: scott.emmons@einstein.yu.edu

Contents

Abstract

How the synaptic connections in the nervous system are genetically encoded and formed during development remains an unsolved problem. The known connectivity of the nervous system of the nematode *C. elegans* provides an opportunity to search for the genes involved. The circuits for male mating behavior form a complex neural network that would seem to require a large family of molecular cell labels for pre- and postsynaptic cell recognition. It is suggested that a combinatorial code of neural cell adhesion proteins specifying the network of connections may be discovered by comparing the expression patterns of candidate genes with the pattern of connections.

It may seem foolish to use the word "final," like the word "never," to describe anything in science. Nevertheless, from today's perspective, what the connections are in our nervous system, the connectome, seems to be the last major gap in our knowledge of the phenotype, certainly in that aspect of the phenotype which is neuroanatomy. How mind emerges from brain remains one of the great mysteries. Through over a century of neuroscience research, we know that thought, consciousness, and behavior emerge from electrical circuits in the brain. As long as the structure of these circuits remain undescribed, knowledge of the properties of the individual cellular components—the neurons, glia, and muscles—and the nature of

Current Topics in Developmental Biology, Volume 116
ISSN 0070-2153
http://dx.doi.org/10.1016/bs.ctdb.2015.11.001

their pairwise connections taken one-by-one will be insufficient to understand the function of the complex system they create when taken all together. The information required to build this structure lies somewhere in our genes, but we do not know where.

This conceptually elementary aspect of the phenotype remains unknown today because tools are still not available to access it. The essential tool that is required, the electron microscope (EM), is cumbersome and has a tiny viewing window. It is unmatched to the size of the nervous systems of even most small invertebrates. Sydney Brenner, recognizing this, looked for the smallest animal he could find with a nervous system that might be reachable and as is well-known, settled on the free-living nematode worm *Caenorhabditis elegans* (Brenner, 2009; Emmons, 2015). After 15 years of effort, in 1986 the connectivity of each of the 118 classes of neurons in the 302-neuron nervous system of the adult of one sex, the hermaphrodite (the "female" sex of *C. elegans*), was published—the first connectome (White, Southgate, Thomson, & Brenner, 1986). But this was the limit. Even the nervous system of the male worm could not be fully deciphered and no similar work was attempted for over a decade (Sulston, Albertson, & Thomson, 1980). With the male, size was not the issue. The problem was in the accounting. The male nervous system is larger than that of the hermaphrodite and there were too many connections between too many branched neurons in the posterior ganglia that control mating to be managed by the pencil-and-paper methods of the 1970s (D. Albertson, personal communication). While EM imaging and methods to decipher electron micrographs still struggle today to gain traction on larger nervous systems, for the *C. elegans* male the PC was sufficient to overcome the accounting problem and the mating circuits were described in 2012 (Jarrell et al., 2012).

1. HOW DOES IT WORK? HOW IS IT BUILT?

As Brenner pointed out in his first major paper on *C. elegans*, for the nervous system there are two chief and somewhat independent questions: how does it work and how is it built (Brenner, 1974)? It was hoped that knowledge of structure would shed light on both of these questions and this expectation has not met with disappointment. An early experimental study guided by the structure revealed how the worm moves forward and backward and responds to light touch (Chalfie et al., 1985). Since that time a great number of experimental analyses have uncovered the functions of individual neurons and elucidated circuits driving further components of the worm's behavior (see WormBook.org for reviews). In the case of the male, the structure of the mating circuits gave insight into the complex decision-making process by which the male integrates inputs from a large number of sensory neurons to guide its efforts to inseminate the hermaphrodite (Jarrell et al., 2012).

Similarly for how it is built—and, crucially, how it is genetically specified—knowledge of the structure made it possible to contemplate what cellular and molecular mechanisms might underlie its formation. John White considered the issue carefully (White, 1985; White, Southgate, Thomson, & Brenner, 1983; White et al., 1986). His reconstruction revealed that the nervous system of the hermaphrodite has a rather simplified structure compared to those of other animals. The component neurons, constant and stereotyped in number and identity, are largely unbranched and run for long distances in bundles with a constant and small set of neighbors. The neighbors are the same in every animal. This describes the ventral nerve cord and even the more complex nerve ring, which is essentially a large parallel bundle of fibers running around the pharynx (Fig. 1). As a neuron can only establish a synaptic connection to an adjacent neuron, and moreover, since in *C. elegans* it does so with a large fraction of these (average of 45% for four examples examined), the neural cell adhesion proteins that presumptively establish and stabilize these bundles must constitute one set of important genetic determinants of overall connectivity.

In order to adhere to another neuron, a given neuron must first come into contact with it. As pointed out by White, in *C. elegans* opportunities for contact are determined initially by the cell lineage. Only a few cells undertake long-range migrations to relocate from their place of birth. Thus, the neighbors available for association during outgrowth of a neurite are the

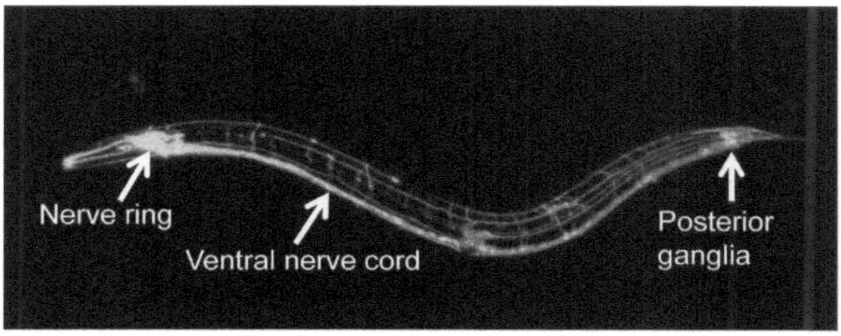

Figure 1 The nerve bundles of the *C. elegans* nervous system, visualized by a reporter gene that expresses the fluorescent protein GFP in all of the neurons. *Image from Hang Ung, Jean-Louis Bessereau laboratory, France.* (A color version of this figure is located at the end of this chapter.)

cells born nearby. In the lineage, which is reproducible from animal to animal, cells of a given type do not in general arise together from specialized branches (the gut, a clone, is the exception that proves this rule). Rather, the distribution of cells across the lineage appears somewhat chaotic with disparate cell types related as siblings or cousins (Sulston, 1983; Sulston, Schierenberg, White, & Thomson, 1983). The underlying principle seems to be that cells are generated near or at the location where they are needed. In the case of neurons, they are born near the neurons they initially fasciculate with. Probably the usually single outgrowing growth cone is drawn from its starting point into the neighborhood that maximizes its set of adhesive interactions. As a result of this mechanism, in addition to the cell recognition and adhesion genes that establish and stabilize process bundles, another set of genes that contribute to specification of nervous system architecture is apparently the set of developmental regulatory genes that underlie determination of cell fate in the lineage.

Cell contact alone is not sufficient to specify the connectome. Although a neuron synapses with a significant fraction of its neighbors, it does not synapse with *all* of them. Since the number of choices is small, White suggested a second signal might simply be neurotransmitter/receptor matching. A simple model would be that for each pair of neurons or neurons and muscles, cell contact triggers a recognition event that results in synapse formation with a certain, genetically specified probability depending on the cells involved. Closer analysis reveals this model is too simple. To address the question, Richard Durbin examined the cell contacts and synapses made in a single animal by presumptively equivalent neurons, the members of homologous left/right pairs (Durbin, 1987). Such pairs exhibit a natural variability in the extent of their contacts with neighbors as well as in the number of synaptic connections they make with them. If synapse formation occurred with a certain fixed probability dependent upon the extent of contact, then the amount of contact with a particular target and the number of synapses made to it should co-vary. But Durbin found that they did not: the member of a pair that made more contact with a particular partner did not necessarily make more synapses with it. A similar result has been found when synapse size as well as synapse number is taken into account (C. Brittin & S.W. Emmons, unpublished observations).

A number of experimental studies in *C. elegans* have revealed something about additional contact-independent mechanisms. First, formation of a presynaptic structure can be regulated—induced or inhibited—by signals from cells not directly involved in the synapse itself. The HSN motor neuron is induced to form a presynaptic structure intended for muscle by a signal from an epithelial cell, and this occurs at a particular location on the HSN neuron whether the muscle target is present or not (Shen & Bargmann, 2003). The AIY interneuron is similarly induced to form presynaptic structures by a signal from a glial cell, while the DA09 motor neuron is prevented from forming presynaptic structures in dendritic regions by signals from epithelial cells (Colón-Ramos, Margeta, & Shen, 2007; Klassen & Shen, 2007; Poon, Klassen, & Shen, 2008; Shao, Watanabe, Christensen, Jorgensen, & Colón-Ramos, 2013).

Second, synapse formation at one location on a neurite can influence the formation of a synapse to a different target elsewhere on the same neurite. In the case of the HSN neuron, if formation of a presynaptic structure at the proper location is not induced, presynapses at inappropriate locations adjacent to inappropriate targets are not pruned away. Pruning requires the protein degradation pathway whose activity is spatially regulated by the synapse-inducing signal (Ding, Chao, Wang, & Shen, 2007). In the case of a class of body wall inhibitory motor neurons, induction of presynaptic sites on the ventral side of the body is necessary to prevent their formation in dendrites on the dorsal side (Howell, White, & Hobert, 2015). It seems there is a balance wherein the multiple synapses formed by a neuron are in competition with one another.

The picture that emerges is one in which neurons are somewhat promiscuous for synapse formation. These synaptogenic tendencies are directed toward the small set of potential partners in contact. Assembly of a presynaptic structure is governed by both internal mechanisms and external signals, likely through regulation of a dynamic balance between biosynthetic and break down pathways or possibly transport pathways within neurites. For neurons that make *en passant* synapses, where to assemble a presynaptic density to communicate with a partner that may be in contact over an extended region is potentially a structural problem that might be solved by an external signal. This is not an issue for *C. elegans* alone. A recent EM reconstruction of a region of mouse neocortex has shown that here too neurites make *en passant* synaptic connections to their targets (Kasthuri et al., 2015).

2. THE *C. ELEGANS* MALE MATING CIRCUITS—A TYPICAL NEUROPIL

All of these features—a small, stereotyped number of unbranched neurons in a highly structured arrangement, constant from animal to animal, forming *en passant* synapses guided by outside signals—might seem to make the *C. elegans* hermaphrodite nervous system a poor representative of most animal nervous systems for the purpose of figuring out how synaptic connections are in general specified. Certainly the structure does not resemble the textbook image in which a neuron sends out an axon behind a growth cone that navigates long distances following pathfinding guidance cues until it reaches its target, which it recognizes and where it forms a synapse.

Such is the typical depiction of a neuron that has guided thinking about the problem of synaptic specificity for many years. In 1963, after he observed in the fish visual system that outgrowing axons bypassed potential target neurons until they arrived at the location of their correct partners, Roger Sperry pointed out that some kind of genetically specified molecular recognition event must occur between pre- and postsynaptic cells (Sperry, 1963). He proposed that this was achieved by expression of complementary chemoaffinity molecules that would serve as molecular cell labels to allow the two cells to recognize each other; he called these "individual identification tags."

Since Sperry, a large number of neural cell adhesion proteins have been identified that are localized to synapses, are required for proper synapse formation or stability, and in some cases have synaptogenic activity. These proteins are good candidates for fulfilling the role of Sperry determinants (Chia, Li, & Shen, 2013; Colon-Ramos, 2009; Hong & Luo, 2014; Langenhan, Aust, & Hamann, 2013; Shapiro, Love, & Colman, 2007; Shen & Scheiffele, 2010; Südhof, 2012; Takahashi & Craig, 2013; Williams, de Wit, & Ghosh, 2010; Zipursky & Sanes, 2010). But it has not been possible to determine in a definitive manner their possible general role as a code for connectivity, as "individual identification tags," because in most systems where they have been studied there is no complete map of connections.

The picture of the *C. elegans* nervous system that emerges when we consider the posterior nervous system of the *C. elegans* male differs significantly from that of the hermaphrodite. In this region, the structures of the neurons and their connections are more typical. Here the complex connectivity would seem to require some type of Sperry-determinant-like functions forming a general molecular code for connectivity.

In the male, the complement of neurons and muscles is augmented by the postembryonic addition of 83 new neurons and 40 new muscles. Altogether 144 neurons and 64 muscles in the tail form a neural network created by over 8000 separate chemical and gap junction synapses resulting in some 3200 connected cell pairs (Jarrell et al., 2012). For comparison, the entire hermaphrodite nervous system has 302 neurons and about 4920 connected cell pairs (not counting muscles) (Varshney, Chen, Paniagua, Hall, & Chklovskii, 2011; White et al., 1986; this laboratory, unpublished). Unlike in the hermaphrodite, where the *en passant* synapses are widely dispersed around the nerve ring, in several ganglia, and along the ventral nerve cord, in the male tail most of the synapses are concentrated within a single ganglion, the preanal ganglion (Fig. 2). Within this ganglion, neurons are not arranged in constant bundles. They branch and intermingle within the 3D volume like strands of spaghetti. In an EM cross-section, as many as 200 neurite profiles may be present and an image looks very much like one of mouse brain.

Three of the key structural features of the hermaphrodite nervous system that seemed to simplify the problem of synaptic specificity do not apply to the male mating circuits. First, the neurons that are generated postembryonically are not necessarily born near where they are needed. While some are born within the preanal ganglion, others are born and reside in ganglia some distance away, from which they will send out their developing processes into the preanal ganglion (Fig. 2). Moreover, some of these neurons lie at positions that vary from animal to animal having been swept into their ganglion by morphogenetic remodeling of the tail region (Sulston et al., 1980). To make their connections, as with neurogenesis in other nervous systems, they send out growth cones that navigate along pathways to reach their targets in the preanal ganglion (Jia & Emmons, 2006; Sulston et al., 1980; Fig. 2). The cell lineage would not seem to play a role here in setting initial cell contacts.

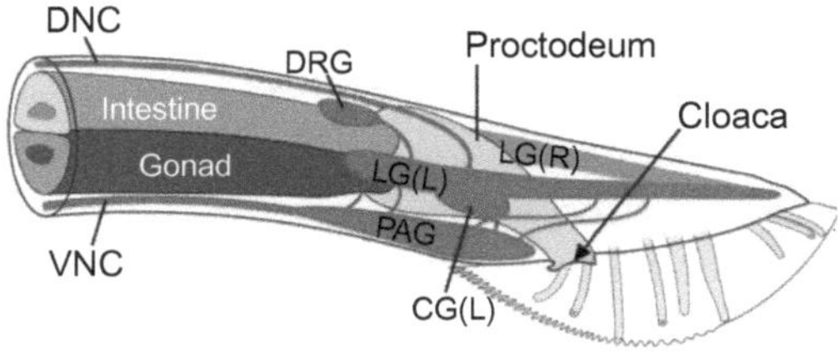

Figure 2 Ganglia in the *C. elegans* male tail. CG(L), left cloacal ganglion; DNC, dorsal nerve cord; DRG, dorsorectal ganglion; LG(L), left lumbar ganglion; LG(R), right lumbar ganglion; PAG, preanal ganglion; VNC, ventral nerve cord; right cloacal ganglion not shown. Cell bodies are distributed in each of the six ganglia. Synapses can be found throughout, but are situated primarily within the PAG. Neurons outside this ganglion send processes through commissures into it. *Image from WormAtlas, http://www.wormatlas.org.* (A color version of this figure is located at the end of this chapter.)

Second, many neurons in the mating circuits are branchy. For those navigating to the target area, upon arriving at the preanal ganglion, which they enter at various points, their processes bifurcate repeatedly and grow among the other processes as they seek their targets. Several embryonic neurons present in both sexes also grow out new branches postembryonically within the ganglion.

Third, neurites are not constrained to a small number of fixed neighborhoods. Neurite outgrowth and branching make it possible in principle for any neuron to come into contact with any other neuron or muscle. In a preliminary study, the frequency was determined with which the neighbors encountered by one large male-specific interneuron in the preanal ganglion, EF3, change along its length. To make a quantitative assessment of the stability of its neighborhoods, cells in contact with EF3 at the ends of 26 segments defined by branch points and endpoints were determined. Fifty-four percent of the 56 neurites in contact at one end of a segment were not present at the other. Thus EF3, rather than running in a constant bundle, experiences a continually changing set of cell contacts. This might reflect active targeting of this cell by the presynaptic neurons. Altogether, 36 different neurons are presynaptic to EF3 over a total of 586 EM sections.

3. A WEALTH OF DETAIL AND COMPLEXITY

PC-based annotation of the electron micrographs of the male tail neuropil made it possible to capture more information about the structure than was previously possible. Most notably, the sizes of the individual synapses were measured. Synapse sizes along with the branchyness of the neurons and the large number of both chemical and gap junction connections each neuron forms with multiple partners present a reproducible structure with a level of complexity that challenges our comprehension. To make some sense of the network, the methods of mathematical graph theory proved to be helpful (Jarrell et al., 2012).

Several features need to be considered. The first is that each neuron in the network interacts with many other neurons and sometimes also muscles. The number of postsynaptic partners of a given neuron ranges continuously from just a few to over 40, with a median of around 10; the number of input and gap junction partners have similar distributions. (These distributions are known as the *degree distributions* of the connectivity graphs in the language of graph theory.) Each type of neuron has a distinct set of pre-, post-, and gap junction synaptic partners. Each of these sets partially overlaps with those of other cell types. How many different such sets are there? Based on the patterns of connectivity as well as other properties (like lineal origin and neurotransmitter expression), the 144 neurons in the network can be assigned to 73 different classes. (This calculation assumes left/right homologs are equivalent cells of a single class and additionally that 73 cells that are not left/right homologs are of 16 types (Jarrell et al., 2012)). Seventy-three different classes mean there are 73 different ways neurons can select targets among the available neurons and muscles.

Once a neuron recognizes an appropriate partner, it must know how to respond. Some pairs of neurons interact exclusively by chemical synapse, some by gap junction synapse, and some by both. Chemical connections are sometimes only in one direction, but a large fraction are reciprocal (albeit usually of different strengths). The prevalence of reciprocal connectivity suggests the possibility that some cell interactions are homophillic.

Inclusion of synapse size in the analysis revealed that the interaction between pre- and postsynaptic cells results in a specified strength of connection (the so-called *edge weight* in the connectivity graph). Jarrell et al. (2012) drew this important conclusion by examining the similarity of connectivity of equivalent neurons. Based on the morphological sizes and numbers of synaptic contacts, each neuron has both strong and weak partners, differing up to 100-fold in connection strength. It was found by comparing the sets of connections made by equivalent neurons that both the strong and weak sets exhibited statistically significant similarity. This observation indicates that not only the set of synaptic partners and the type and direction of each synaptic connection but also the strength of synaptic interaction is genetically specified for each interacting cell pair.

How do growth cones find their targets in the preanal ganglion? Do they follow reproducible pathways adhering to pre-defined neighbors? While a detailed three-dimensional reconstruction has not yet been carried out, comparison of the variable branching architectures of left/right homologs, presumptively equivalent cells, reveals little if any suggestion of such a reproducible architecture. As an example, consider a left/right pair of sensory neurons with dendritic endings in rays, R2BL and R2BR. These neurons are among the ones mentioned above that have cell bodies residing at variable positions in ganglia widely separated from the preanal ganglion. Their growth cones enter the preanal ganglion independently on opposite sides from commissures that, respectively, traverse around the left and right sides of the body. While the neurites invade a similar region of the ganglion, their branching structures bear no similarity and little relationship to synapse formation (Fig. 3). Some branches are formed that make no synapses, while others reach a productive region and make many synapses. Most branches

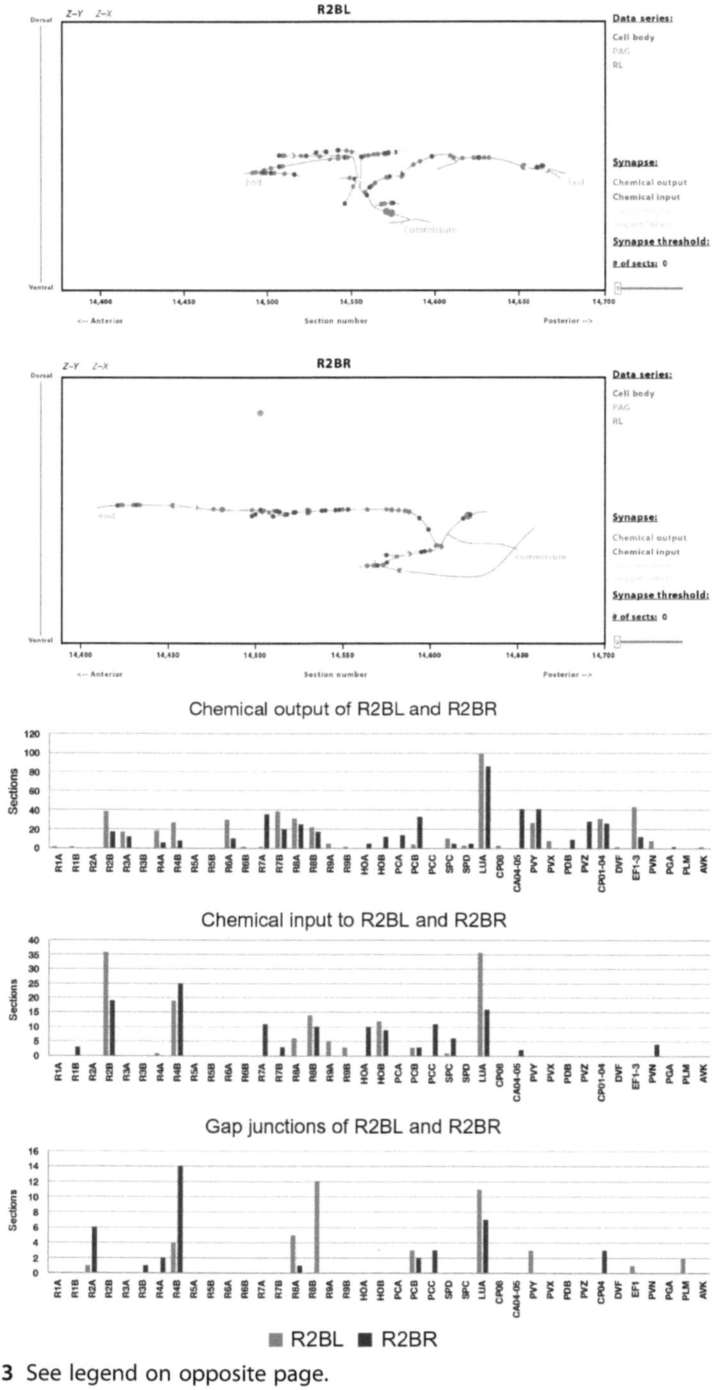

Figure 3 See legend on opposite page.

grow beyond the last synapse. It looks like a random search. Yet in spite of these differences, these neurons make similar (albeit far from identical) sets of connections (Fig. 3).

4. CAN CLUES AS TO THE IDENTITY OF "INDIVIDUAL IDENTIFICATION TAGS" BE FOUND IN THE PATTERNS OF CONNECTIONS?

The complexity strongly implies the existence of a relatively large set of cell labels that collectively carry the instructions for forming the network. As mentioned above, elsewhere in the *C. elegans* nervous system extracellular cell labels that induce or inhibit synapse formation can be expressed in adjacent hypodermal or glial cells. However, most neurites in the male preanal ganglion only contact other neurites, so it is likely that the neurites themselves bear cell labels. It seems there must be many such labels in order to generate the 73 distinct subsets of synaptic connections. White et al. (1986) similarly pointed out that even in the hermaphrodite nervous system the code must be combinatorial since the different sets of postsynaptic cells intersect. Marshaling an argument based on parsimony, can evidence for such overlapping sets of labels be deduced from the patterns of connections?

As an example, consider how the 36 cells that target EF3, which are of 21 different types, recognize it. There could be 21 different labels expressed by EF3, one for each presynaptic cell type. More parsimoniously, EF3 could express a single label (or a set of labels) that is recognized by each of the presynaptic cells. In this case, the members of the set of presynaptic cells should all express the same receptor(s) that recognizes the EF3 label(s). Moreover, the very likely identical EF1 and EF2 neurons, which have similar sets of

Figure 3 Homologous neurons have dissimilar branching structures in the preanal ganglion but similar sets of synaptic partners. Upper panels: the processes and synapses in the PAG of a pair of homologous neurons, the ray sensory neurons R2BL and R2BR. The cell bodies of these neurons are, respectively, in the left and right lumbar ganglia (Fig. 2). Pink circles show sites of chemical synaptic output, magenta circles are chemical input, and blue circles are gap junctions. These maps are from Wormwiring.org. Lower panels: histograms comparing the synaptic relationships made by R2BL and R2BR. The neurons listed along the bottom are the 68 neurons (out of 170) with synaptic interaction of some kind with one or both of these two neurons. Data for left/right homologs of R2BL and R2BR partners are combined; e.g., R1A represents the interaction with R1AL plus R1AR. The vertical axis gives the total morphological strength of interaction in number of EM serial sections. Note the differences in scales. (A color version of this figure is located at the end of this chapter.)

presynaptic partners, should express the same label(s) as EF3. A fruitful experimental approach could be to search for potential extracellular or transmembrane label proteins whose expression patterns match these respective sets of pre- and postsynaptic cells.

As the classes of neurons presynaptic to EF1, EF2, and EF3 are also presynaptic to additional but in each case *different* postsynaptic cells, in addition to sharing recognition for the EF determinant, they must express additional recognition molecules. In a combinatorial code for synapse formation, each neuron might express a number of these determinants. The number that matched those expressed by another cell might determine the probability of synapse formation. It is in this manner that the number of matching isomorphs of DSCAM proteins expressed by neurons in *Drosophila* determines the probability of repulsion between them (Hattori, Millard, Wojtowicz, & Zipursky, 2008). Similarly in mammals, combinations of protocadherin isoforms determine single-cell identities (Thu et al., 2014). Multiple determinants, each contributing to a probability of synapse formation, would provide a flexibility that could allow for gradual changes in synaptic strength and ultimately evolution of connectivity.

In a second example, consider the preanal ganglion interneuron/motorneuron pair PVV and PDB. Male-specific, postembryonic PVV targets ventral bodywall motor circuitry. Sex nonspecific, embryonic PDB innervates dorsal bodywall motorneurons and muscles. Both neurons are extensively branched (the PDB branches are male-specific) and receive significant input, particularly sensory input (Fig. 4). PVV and PDB are clearly not equivalent cells. Yet they receive very similar patterns of synaptic inputs (Fig. 4). Among the 18 ray neuron types, four have no input to either while nine have input to both. Only five have input to just one or the other, and among these, only R5B has relatively strong input to just one, PVV. Once again, a parsimonious explanation is that, with regard to recognition for synapse formation, as far as many neurons are concerned PVV and PDB are indistinguishable. A subset of ray and other neurons have receptors that recognize and interact with the PVV and PDB labels more or less avidly while all other cells do not. A search could be made for genes whose pattern of expression could explain this connectivity.

5. CANDIDATES FOR CELL LABELS

As mentioned above, studies in many systems have identified cell surface recognition and adhesion proteins expressed at synapses. These

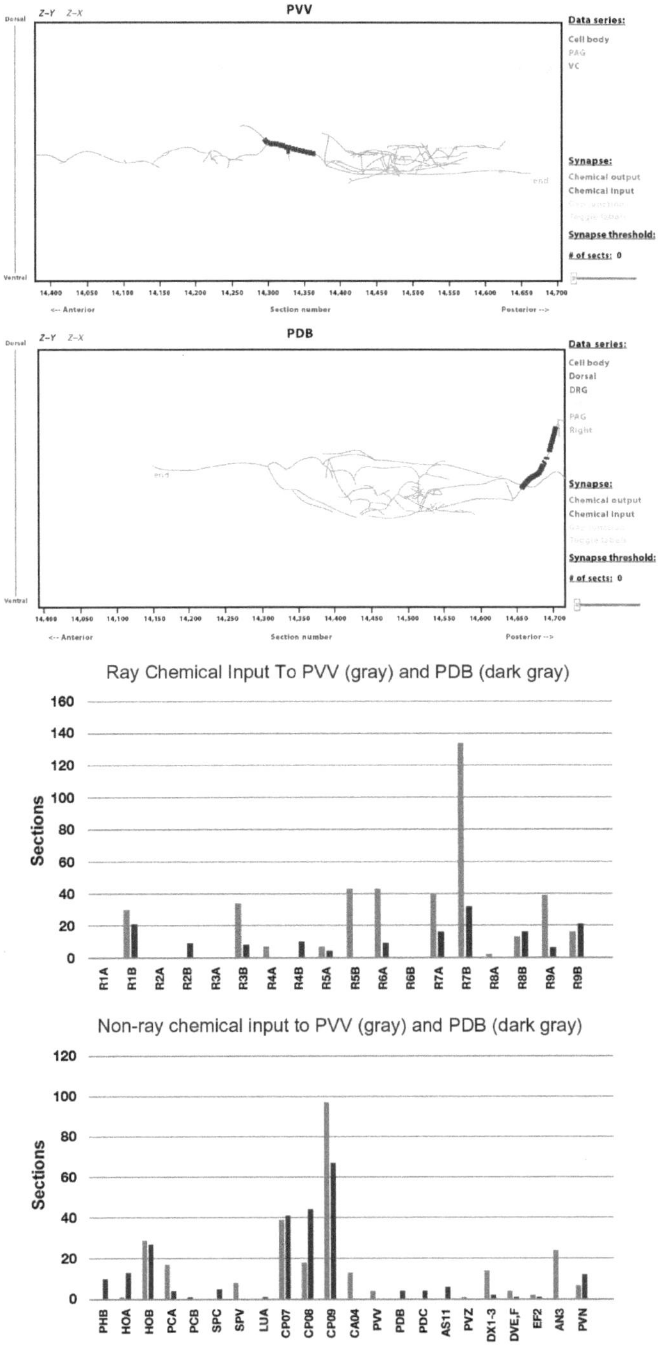

Figure 4 See legend on next page.

variously are necessary for normal synaptogenesis or synapse function or maintenance, and many can induce synapse formation in cell culture experiments. The *C. elegans* genome contains numerous homologs of these genes, as well as genes encoding similar transmembrane proteins with recognizable protein–protein interaction extracellular domains. Hobert has provided a valuable listing (Hobert, 2013). Thus there is no dearth of candidate proteins for subserving the function of Sperry's tags. White and co-workers, nearly 30 years ago, looked forward to the identification of synaptic recognition determinants or critical cell adhesion molecules through analysis of *C. elegans* behavioral mutants (White, 1985; White et al., 1986). It has not happened and one wonders why. The system must be so overdetermined that it is not readily penetrable through forward genetic screens. This should perhaps not be surprising. The wiring is so complex it cannot be constructed exactly the same every time. If that were necessary, developmental error would preclude a functional outcome. We now have direct evidence that some of the genes very likely to be involved (a prominent example is neurexin, which is widely expressed in the *C. elegans* nervous system), have no obvious loss-of-function mutant phenotype. This state of affairs is likely due to redundancy at two levels—each synaptic connection is probably determined by multiple interactions and, second, function of the network is robust to changes in the strengths of individual connections.

6. CONCLUSION

The availability of a complete connectome allows us to view the problem of synaptic specificity both from a global and a detailed perspective. The detailed information makes the wiring problem seem more difficult but the global view provides a way to solve it. By examination of the whole,

Figure 4 PVV and PDB are distinct neurons with similar patterns of synaptic input. Top panels show the overlapping branching structures of PVV and PDB in the preanal ganglion. PVV extends further anteriorly into the ventral cord, while PDB sends a process posteriorly around the end of the tail to the dorsal side. Both have output onto bodywall muscles. The blue boxes are the cell bodies. The horizontal axes show the serial section number of the transverse serial sections (A–P axis); the vertical axis shows position along the D–V axis; the maps are flattened along the L–R axis. The top histogram shows connectivity data for all the ray neurons; R1A(L/R), R2A(L/R), R3A(L/R), and R6B(L/R) make no connection to either PVV or PDB. The bottom histogram shows data for all the remaining neurons that make a chemical connection to either neuron. Data for left/right homologs and other equivalent neurons are combined. (A color version of this figure is located at the end of this chapter.)

including in the analysis both chemical synapses and gap junction connections, as opposed to one synaptic connection at a time, and comparing the entire set of connections to the expression patterns of the neural cell adhesion proteins that are the presumptive cell labels for synapse formation, it should be possible to discern a pattern that reveals the underlying code. The landscape of cell labels is expected to act progressively over a series of steps—to guide growth cones to a general target area, to promote or inhibit contact with other neurons, to adhere partners together, and finally to trigger synapse formation, or possibly to inhibit it—each step making an incremental contribution to specificity. The expression patterns of the cell labels that encode the wiring must map onto the pattern of synaptic connections in a logical way. The wiring diagram is encoded in the protein–protein interaction domains of the cell labels, and in the genomic regulatory sequences that govern their cell-specific expression.

ACKNOWLEDGMENTS

Work in the author's laboratory has been supported by grants from the US National Institutes of Health and the G. Harold and Leila Y. Mathers Charitable Foundation.

REFERENCES

Brenner, S. (1974). The genetics of *Caenorhabditis elegans*. *Genetics*, *77*, 71–94.

Brenner, S. (2009). In the beginning was the worm .... *Genetics*, *182*, 413–415.

Chalfie, M., Sulston, J. E., White, J. G., Southgate, E., Thomson, J. N., & Brenner, S. (1985). The neural circuit for touch sensitivity in *Caenorhabditis elegans*. *The Journal of Neuroscience*, *5*, 956–964.

Chia, P. H., Li, P., & Shen, K. (2013). Cellular and molecular mechanisms underlying presynapse formation. *The Journal of Cell Biology*, *203*, 11–22.

Colon-Ramos, D. A. (2009). Chapter 2: Synapse formation in developing neural circuits. In *Current topics in developmental biology* (pp. 53–79). Waltham, USA: Academic Press.

Colón-Ramos, D. A., Margeta, M. A., & Shen, K. (2007). Glia promote local synaptogenesis through UNC-6 (Netrin) signaling in C. elegans. *Science*, *318*, 103–106.

Ding, M., Chao, D., Wang, G., & Shen, K. (2007). Spatial regulation of an E3 ubiquitin ligase directs selective synapse elimination. *Science*, *317*, 947–951.

Durbin, R. M. (1987). *Studies on the development and organization of the nervous system of* Caenorhabditis elegans. Cambridge, UK: Cambridge University.

Emmons, S. W. (2015). The beginning of connectomics: A commentary on White et al. (1986) '*The structure of the nervous system of the nematode* Caenorhabditis elegans'. *Philosophical Transactions in Royal Society of London B*, *370*(1666). http://dx.doi.org/10.1098/rstb.2014.0309.

Hattori, D., Millard, S. S., Wojtowicz, W. M., & Zipursky, S. L. (2008). Dscam-mediated cell recognition regulates neural circuit formation. *Annual Review of Cell and Developmental Biology*, *24*, 597–620.

Hobert, O. (2013). The neuronal genome of *Caenorhabditis elegans*. In WormBook (Ed.), *The C. elegans research community*. WormBook. http://dx.doi.org/10.1895/wormbook.1.161.1, http://www.wormbook.org.

Hong, W., & Luo, L. (2014). Genetic control of wiring specificity in the fly olfactory system. *Genetics*, *196*, 17–29.

Howell, K., White, J. G., & Hobert, O. (2015). Spatiotemporal control of a novel synaptic organizer molecule. *Nature, 523*, 83–87.

Jarrell, T. A., Wang, Y., Bloniarz, A. E., Brittin, C. A., Xu, M., Thomson, J. N., et al. (2012). The connectome of a decision-making neural network. *Science, 337*, 437–444.

Jia, L., & Emmons, S. W. (2006). Genes that control ray sensory neuron axon development in the *Caenorhabditis elegans* male. *Genetics, 173*, 1241–1258.

Kasthuri, N., Hayworth, Kenneth J., Berger, Daniel R., Schalek, Richard L., Conchello, José A., Knowles-Barley, S., et al. (2015). Saturated reconstruction of a volume of neocortex. *Cell, 162*, 648–661.

Klassen, M. P., & Shen, K. (2007). Wnt signaling positions neuromuscular connectivity by inhibiting synapse formation in C. elegans. *Cell, 130*, 704–716.

Langenhan, T., Aust, G., & Hamann, J. (2013). Sticky signaling—Adhesion class G protein-coupled receptors take the stage. *Science Signal, 6*, re3.

Poon, V. Y., Klassen, M. P., & Shen, K. (2008). UNC-6/netrin and its receptor UNC-5 locally exclude presynaptic components from dendrites. *Nature, 455*, 669–673.

Shao, Z., Watanabe, S., Christensen, R., Jorgensen, Erik M., & Colón-Ramos, Daniel A. (2013). Synapse location during growth depends on glia location. *Cell, 154*, 337–350.

Shapiro, L., Love, J., & Colman, D. R. (2007). Adhesion molecules in the nervous system: Structural insights into function and diversity. *Annual Review of Neuroscience, 30*, 451–474.

Shen, K., & Bargmann, C. I. (2003). The immunoglobulin superfamily protein SYG-1 determines the location of specific synapses in *C. elegans*. *Cell, 112*, 619–630.

Shen, K., & Scheiffele, P. (2010). Genetics and cell biology of building specific synaptic connectivity. *Annual Review of Neuroscience, 33*, 473–507.

Sperry, R. W. (1963). Chemoaffinity in the orderly growth of nerve fiber patterns and connections. *Proceedings of National Academy of Sciences of the United States of America, 50*, 703–710.

Südhof, Thomas C. (2012). The presynaptic active zone. *Neuron, 75*, 11–25.

Sulston, J. E. (1983). Neuronal cell lineages in the nematode Caenorhabditis elegans. *Cold Spring Harbor Symposia on Quantitative Biology, 48*, 443–452.

Sulston, J. E., Albertson, D. G., & Thomson, J. N. (1980). The *Caenorhabditis elegans* male: Postembryonic development of nongonadal structures. *Developmental Biology, 78*, 542–576.

Sulston, J. E., Schierenberg, E., White, J. G., & Thomson, J. N. (1983). The embryonic cell lineage of the nematode *Caenorhabditis elegans*. *Developmental Biology, 100*, 64–119.

Takahashi, H., & Craig, A. M. (2013). Protein tyrosine phosphatases PTPδ, PTPσ, and LAR: Presynaptic hubs for synapse organization. *Trends in Neurosciences, 36*, 522–534.

Thu, Chan A., Chen, Weisheng V., Rubinstein, R., Chevee, M., Wolcott, Holly N., Felsovalyi, Klara O., et al. (2014). Single-cell identity generated by combinatorial homophilic interactions between α, β, and γ protocadherins. *Cell, 158*, 1045–1059.

Varshney, L. R., Chen, B. L., Paniagua, E., Hall, D. H., & Chklovskii, D. B. (2011). Structural properties of the *Caenorhabditis elegans* neuronal network. *PLoS Computational Biology*, 7e1001066.

White, J. G. (1985). Neuronal connectivity in *Caenorhabditis elegans*. *Trends in Neuroscience, 8*, 277–283.

White, J. G., Southgate, E., Thomson, J. N., & Brenner, S. (1983). Factors that determine connectivity in the nervous system of Caenorhabditis elegans. *Cold Spring Harbor Symposia on Quantitative Biology, 48*(Pt. 2), 633–640.

White, J. G., Southgate, E., Thomson, J. N., & Brenner, S. (1986). The structure of the nervous system of the nematode *Caenorhabditis elegans*. *Philosophical Transactions of the Royal Society, Series B Biological Sciences, 314*, 1–340.

Williams, M. E., de Wit, J., & Ghosh, A. (2010). Molecular mechanisms of synaptic specificity in developing neural circuits. *Neuron, 68*, 9–18.

Zipursky, S. L., & Sanes, J. R. (2010). Chemoaffinity revisited: Dscams, protocadherins, and neural circuit assembly. *Cell, 143*, 343–353.

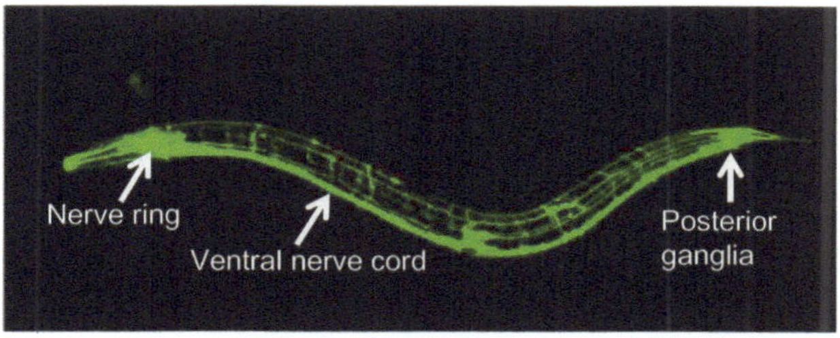

Scott W. Emmons, Figure 1 The nerve bundles of the *C. elegans* nervous system, visualized by a reporter gene that expresses the fluorescent protein GFP in all of the neurons. *Image from Hang Ung, Jean-Louis Bessereau laboratory, France.*

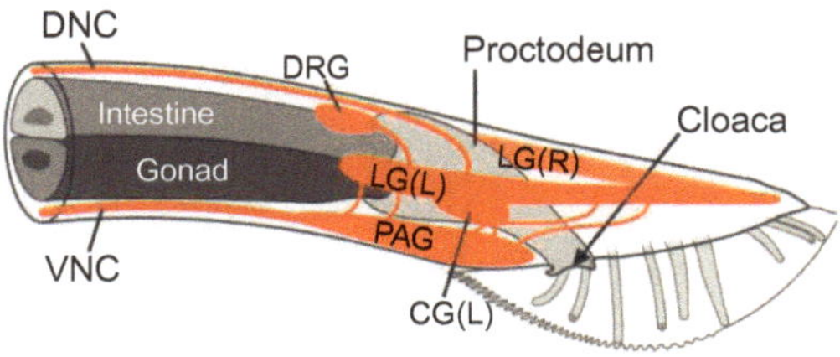

Scott W. Emmons, Figure 2 Ganglia in the *C. elegans* male tail. CG(L), left cloacal ganglion; DNC, dorsal nerve cord; DRG, dorsorectal ganglion; LG(L), left lumbar ganglion; LG(R), right lumbar ganglion; PAG, preanal ganglion; VNC, ventral nerve cord; right cloacal ganglion not shown. Cell bodies are distributed in each of the six ganglia. Synapses can be found throughout, but are situated primarily within the PAG. Neurons outside this ganglion send processes through commissures into it. *Image from WormAtlas, http://www.wormatlas.org.*

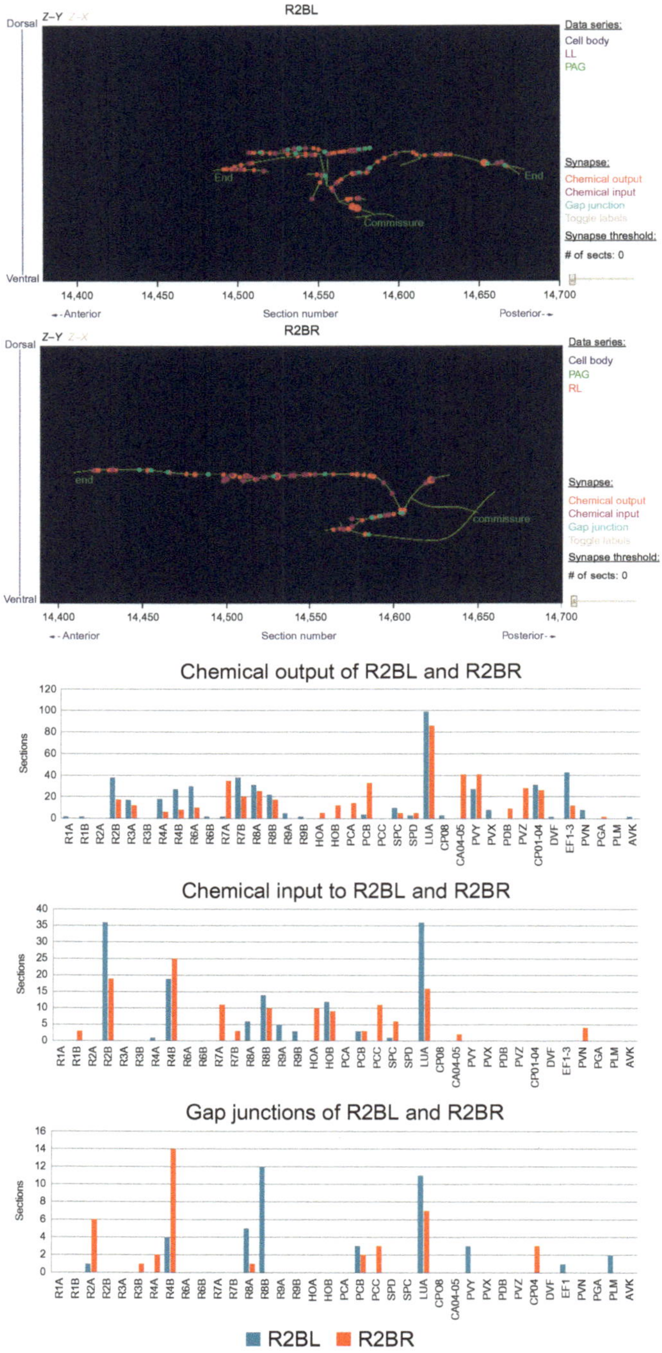

Scott W. Emmons, Figure 3 See legend on opposite page.

Scott W. Emmons, Figure 3 Homologous neurons have dissimilar branching structures in the preanal ganglion but similar sets of synaptic partners. Upper panels: the processes and synapses in the PAG of a pair of homologous neurons, the ray sensory neurons R2BL and R2BR. The cell bodies of these neurons are, respectively, in the left and right lumbar ganglia (Fig. 2). Pink circles show sites of chemical synaptic output, magenta circles are chemical input, and blue circles are gap junctions. These maps are from Wormwiring.org. Lower panels: histograms comparing the synaptic relationships made by R2BL and R2BR. The neurons listed along the bottom are the 68 neurons (out of 170) with synaptic interaction of some kind with one or both of these two neurons. Data for left/right homologs of R2BL and R2BR partners are combined; e.g., R1A represents the interaction with R1AL plus R1AR. The vertical axis gives the total morphological strength of interaction in number of EM serial sections. Note the differences in scales.

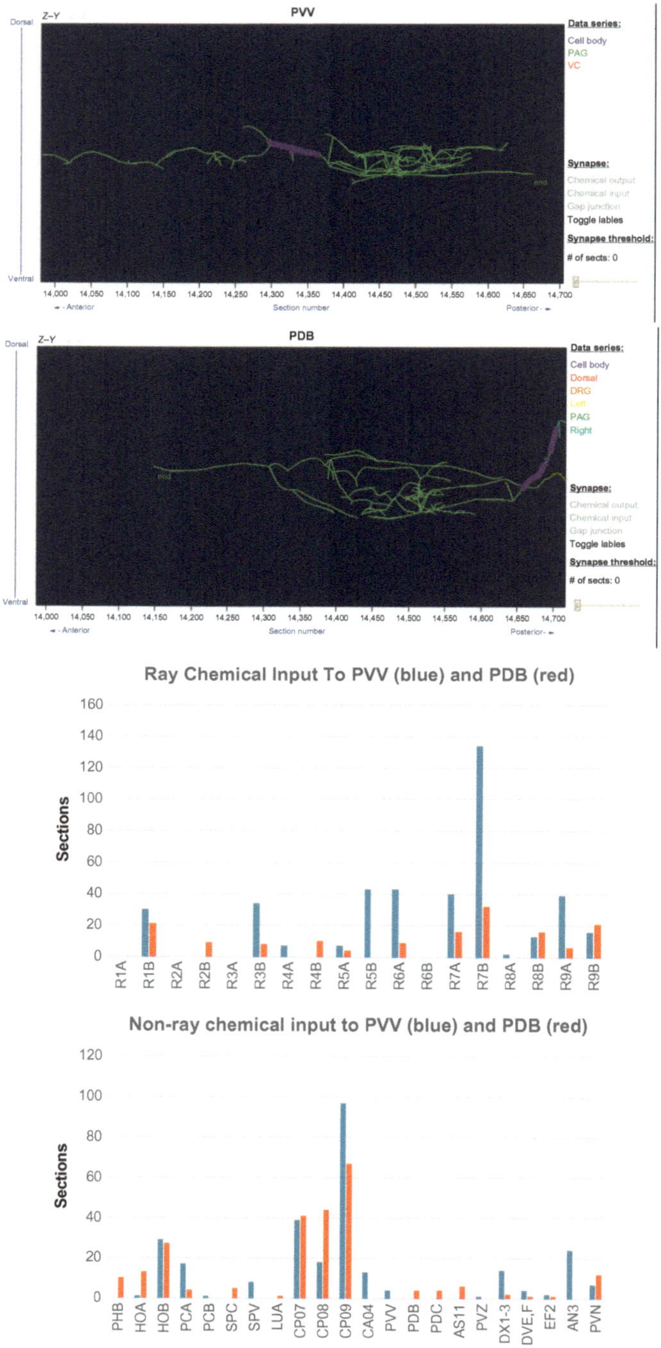

Scott W. Emmons and Dominick P. Purpura, Figure 4 See legend on opposite page.

Scott W. Emmons and Dominick P. Purpura, Figure 4 PVV and PDB are distinct neurons with similar patterns of synaptic input. Top panels show the overlapping branching structures of PVV and PDB in the preanal ganglion. PVV extends further anteriorly into the ventral cord, while PDB sends a process posteriorly around the end of the tail to the dorsal side. Both have output onto bodywall muscles. The blue boxes are the cell bodies. The horizontal axes show the serial section number of the transverse serial sections (A–P axis); the vertical axis shows position along the D–V axis; the maps are flattened along the L–R axis. The top histogram shows connectivity data for all the ray neurons; R1A(L/R), R2A(L/R), R3A(L/R), and R6B(L/R) make no connection to either PVV or PDB. The bottom histogram shows data for all the remaining neurons that make a chemical connection to either neuron. Data for left/right homologs and other equivalent neurons are combined.

CHAPTER EIGHTEEN

Dedifferentiation, Redifferentiation, and Transdifferentiation of Striated Muscles During Regeneration and Development

Manfred Frasch[1]
Department of Biology, Division of Developmental Biology, Friedrich-Alexander University of Erlangen-Nürnberg, Erlangen, Germany
[1]Corresponding author: e-mail address: manfred.frasch@fau.de

Contents

Abstract

In some rare and striking cases, striated muscle fibers of the skeleton or body wall, which consist of terminally differentiated syncytia with complex ultrastructures, were found to be capable of dedifferentiating and fragmenting into mononucleate cells. Examples of such events will be discussed in which the dedifferentiated cells reenter the cell cycle, proliferate, and rebuilt damaged muscle fibers during limb regeneration or transdifferentiate to generate new types of muscles during normal development.

Current Topics in Developmental Biology, Volume 116
ISSN 0070-2153
http://dx.doi.org/10.1016/bs.ctdb.2015.12.005

1. INTRODUCTION

Terminally differentiated striated muscle fibers of the skeleton or body wall consist of very long syncytia that feature a complex ultrastructural organization of the contractile apparatus and sarcoplasmatic reticulum and contain numerous postmitotic nuclei. The differentiation state of striated muscles is by no means static and processes by which muscles undergo hypertrophy-, atrophy-, or fiber-type switches have been studied extensively (Blaauw, Schiaffino, & Reggiani, 2013). In comparison to these phenotypic modulations, it is evident that dedifferentiation of such complex cells, and in particular a reversal of the multinucleate state of the syncytial myofiber to mononucleate cells that may enter the cell cycle, would raise much greater challenges for the cell. The challenges are expected to be even more pronounced for processes of muscle transdifferentiation,[1] in which a terminally differentiated muscle changes its fate into a completely different type of muscle or even a different type of tissue. While there are several examples where muscle dedifferentiation or transdetermination could be achieved via forced expression of pluripotency or tissue determination factors (e.g., Hjiantoniou et al., 2008; Veldman et al., 2013), this review will mainly focus on the rare examples that occur in more natural settings, either during normal development, regeneration, or transplantation and explantation of muscle fibers, in the absence of any molecular manipulations.

2. DEDIFFERENTIATING AND REDIFFERENTIATING MUSCLES DURING REGENERATION

2.1 Muscle Regeneration via Dedifferentiation Versus Stem Cell-Based Repair in Amphibians

The repair and regeneration of skeletal muscles upon injury in principle can utilize two different mechanisms. The first involves the activation of stem cells residing within the basement membrane of the myofibers, called satellite cells, which upon injury start to proliferate and then differentiate into myogenic progenitors. These cells fuse to the damaged myofibers or fuse with one another to form fibers *de novo*, which leads to the repair of the

[1] Transdifferentiation, also called cell conversion or metaplasia (the latter term often being used in pathological situations), is defined as a change of a cell or its progeny from one differentiated phenotype to another without reverting to pluripotency. Lineage reprogramming includes both transdifferentiation and transdetermination (Jopling, Boue, & Izpisúa Belmonte, 2011).

damaged muscles. In mammalians, this process is the predominant one that is used for muscle repair, although specific nonsatellite progenitor cells could also contribute (Yin, Price, & Rudnicki, 2013). By contrast, the second mechanism for muscle regeneration does not involve any stem cells, but rather involves dedifferentiation of injured muscle fibers. In this case, it is the dedifferentiated, mononucleate myoblasts that proliferate and undergo myoblast fusion to repair the damaged skeletal muscles. This second mechanism is used much more rarely in the animal kingdom and, to date, has only been documented in newts.

Amphibians exhibit an astonishing ability to regenerate their limbs and tails upon amputation. This regenerative capacity is particularly effective in urodeles (salamanders and newts). It includes the skeletal muscles, which are regenerated almost fully to their original state (Diogo, Nacu, & Tanaka, 2014). In all cases, the regeneration of limb and tail structures, including the muscles, is accomplished by undifferentiated, proliferative progenitor cells that form a mass of mesenchymal cells covering the amputation site, called the blastema. These observations have raised a number of intriguing questions. The most basic ones include: where do the blastema cells originate from? What is the developmental state and potency of the blastema cells, i.e., are they multipotent such that the descendants of a single cell are able to restore several different tissues, or is each cell already committed to a single fate? And how do these cells (re)differentiate and undergo proper morphogenesis to restore the lost structures? During the past decades, these questions have been addressed with increasingly sophisticated techniques and recently, some important answers have emerged.

With regard to the muscles, early studies performed by using light and electron microscopy with larvae of the spotted salamander, *Ambystoma punctatum* (mole salamander family), showed that massive muscle dedifferentation occurred within the first 6 days after limb amputation, as indicated by the disintegration and loss of myofibrils and, concomitantly, the fragmentation of the muscle fibers into mononucleate cells (Hay, 1959; Thornton, 1938). The fragmentation of the syncytia was thought to involve rows of vesicles seen between nuclei, which were proposed to form the future cell boundaries. These observations led to the assumption that the blastema cells that will regenerate muscles arise via dedifferentiation and fragmentation of muscle fibers at the amputation site. Nuclei in syncytial myofibers within 1–2 mm proximally to the amputation site and, more commonly, in dedifferentiating mononucleate muscle fragments increased in size and incorporated ^{3}H-thymidine, indicating that the muscle nuclei enter the cell cycle prior to their full dedifferentiation (Hay & Fischman, 1961). More recently, high-resolution analyses using 3D reconstructions of regenerating limbs in the red spotted newt *Notophtalmus viridescens* (family of true salamanders and newts) confirmed cell cycle reentry of nuclei within the syncytial myofibers near the amputation site, as evidenced by the incorporation of the thymidine analog EdU (Calve, Odelberg, & Simon, 2010; Calve & Simon, 2011). In addition, these studies showed that shortly after amputation the sarcolemma of injured myofibers self-healed, which presumably ensured cell survival, and subsequently the extracellular matrix proximal to the amputation plane was reorganized around the muscle fibers and contained strongly increased amounts of the ECM component tenascin-C. The presence of this transitional, tenascin-C-rich ECM, spatially correlated with the presence of EdU-containing nuclei within the fibers and with myofiber fragmentation. Because *in vitro* experiments showed that tenascin-C can promote the cell cycle of primary newt myoblasts as well as the fragmentation of syncytial myofibers, it was proposed that the transitional tenascin-C-rich ECM has analogous activities upon amputation *in vivo* and contributes to cell cycle entry and myofiber fragmentation (Calve et al., 2010). *In vivo* imaging of dye-injected muscle fibers in amputated tails of the axolotl, *Ambystoma mexicanum*, detected myofiber dedifferentiation and fragmentation into mononucleate cells, which morphologically resembled blastema cells (Echeverri, Clarke, & Tanaka, 2001). In a different approach for obtaining evidence for muscle dedifferentiation upon limb amputation in newts, it was shown that some of the newly generated mononucleate cells were double positive for distinct muscle-specific and blastema cell-specific antigens, thus indicating that the fragmented myoblasts were on their way to becoming blastema cells (Kintner & Brockes, 1984).

Because the aforementioned *in vivo* data were derived from static images, they did not provide firm proof for the presumed dedifferentiation and fragmentation of muscle fibers into mononucleate myoblasts, nor could they ascertain whether such cells will reconstitute new muscles. However, at least for partially differentiated myotubes that were generated *in vitro* from primary limb cell cultures (Ferretti & Brockes, 1988), dedifferentiation in regenerating limbs of newts could be confirmed through *in vivo* fate-mapping experiments. In *Notophtalmus*, multinucleate, contractile myotubes of this type, which were either labeled with fluorescent dyes or via infection with a retrovirus carrying a marker gene, were implanted after limb amputation. These myotubes formed mononucleate cells that turned off muscle-specific antigens and proliferated in the blastema, as shown by

^{3}H-thymidine or BrdU incorporation within the labeled cells (Kumar, Velloso, Imokawa, & Brockes, 2000; Lo, Allen, & Brockes, 1993). Due to technical limitations, it was however not possible to determine whether these cells formed new muscles upon regeneration. In addition, these experiments could not address the question of whether endogenous, terminally differentiated myofibers behave in the same way as the implanted "artificial" myotubes generated *in vitro*.

Although the observations described thus far made a strong case for muscle dedifferentiation upon limb and tail amputation in newts and salamanders, the possibility that the regenerated muscles are derived partially or even completely from satellite cells could not be ruled out. Indeed, satellite cells were first discovered and proposed to confer muscle regeneration in frogs before they were found in mammals (Mauro, 1961). Subsequently, satellite cells were also described in urodeles, which expressed markers that are typical for mammalian satellite cells, such as the Pax7 transcription factor and, once activated, the myogenic factor Myf5 (Cameron, Hilgers, & Hinterberger, 1986; Morrison, Lööf, He, & Simon, 2006; Popiela, 1976). Importantly, in the case of tail regeneration in *Xenopus* tadpoles, it was shown that it does not involve muscle degeneration, but rather relies on Pax7-dependent satellite cells (Chen, Lin, & Slack, 2006; Gargioli & Slack, 2004). In urodeles, both the newt *Notophtalmus* and the mole salamander *A. mexicanum* (axolotl), proliferating Pax7^{+} cells, presumed to be satellite cells, were observed in the blastema upon amputation of limbs and tails, respectively; however, it was not known whether these cells engage in muscle regeneration (Morrison et al., 2006; Schnapp, Kragl, Rubin, & Tanaka, 2005).

A recent study employing elegant genetic fate-mapping strategies appears to settle the question of whether muscle regeneration in amputated limbs of urodele species involves dedifferentiation of damaged muscles or activation of muscle satellite cells (Sandoval-Guzmán et al., 2014). The experiments used a loxP reporter (nuclear YFP) that was integrated into the genome of limb cells via electroporation. Reporter expression was activated 2 weeks prior to amputation with a muscle-specific Cre expression vector, MCK (muscle creatine kinase)-Cre, which is expressed transiently only in differentiated muscle fibers but not in satellite cells. Therefore, any cells that were positive for reporter gene expression after amputation and regeneration must have descended from differentiated muscle fibers.

When this fate-mapping approach was used in the newt *Notophtalmus*, the YFP reporter was indeed found in the nuclei of myofibers along the entire proximal–distal extent of the regenerated portion of the limb. Nonmuscle tissues of the regenerated portion were completely negative. Thus, the regenerated muscles must have formed from the preexisting differentiated muscle fibers in a tissue-restricted fashion. A closer analysis at early stages of regeneration revealed that $nYFP^+$ skeletal muscle fragments near the amputation plane still contained the muscle differentiation marker MHC (myosin heavy chain), whereas the $nYFP^+$ cells in the distal blastema had lost MHC. In addition, the YFP^+ nuclei of cells in the blastema could be labeled with the proliferation markers PCNA and EdU, thus demonstrating cell cycle reentry. Altogether, these findings corroborated that upon limb amputation, myofiber dedifferentiation generates proliferative blastema cells that contribute to *de novo* muscle reconstitution in newts. The dedifferentiated progenitors are committed to muscle fates and do not contribute to any other cell types of the regenerated limb.

Surprisingly, when analogous experiments were performed in the axolotl, the outcome was very different. In contrast to the newt, after full limb regeneration in the axolotl, YFP^+ nuclei were restricted to muscles proximally to the amputation plane, whereas the newly generated muscles in the distal portions of the limb were completely devoid of this muscle lineage marker. The result was confirmed with an independent technique using a germ line incorporated reporter construct, as well as with animals that were forced to undergo metamorphosis instead of remaining in their typical neotenic (larval-like) state. Correspondingly, although fragmented cells containing the lineage marker were present near the amputation plane, no such cells were detected in the distal blastema, and those near the amputation plane did not stain for proliferation markers. Thus, even though muscle dedifferentiation does seem to take place (at least to some degree, in agreement with the older data), the dedifferentiated cells do not proliferate and regenerate muscles in this species. Instead, it was shown that all cells from myogenic lineages in the distal blastema were $Pax7^+$ and many of these cells proliferated. It appears likely that the same is true for tail muscle regeneration and, thus, the fate or survival of the mononucleate cells produced from dedifferentiated tail muscles as described in Echeverri et al. (2001) is currently not clear. Together, these observations led to the conclusion that in the axolotl, dedifferentiated muscle cells do not contribute to limb muscle regeneration and instead, this process is solely carried out by satellite cells. Conversely, in newts, the limb blastema were devoid of $Pax7^+$ cells, except at very early stages after amputation, which is consistent with the finding that in this species dedifferentiating muscles provide the major source of cells for muscle regeneration (Sandoval-Guzmán et al., 2014). Whether satellite cells make any contribution could not be addressed in these experiments.

In conclusion, these two quite closely-related urodele species use fundamentally different mechanisms for skeletal muscle regeneration (Fig. 1). In the axolotl, the exclusive recruitment of stem cells during this process is similar to the situation observed in frog tadpole tail muscle regeneration and is

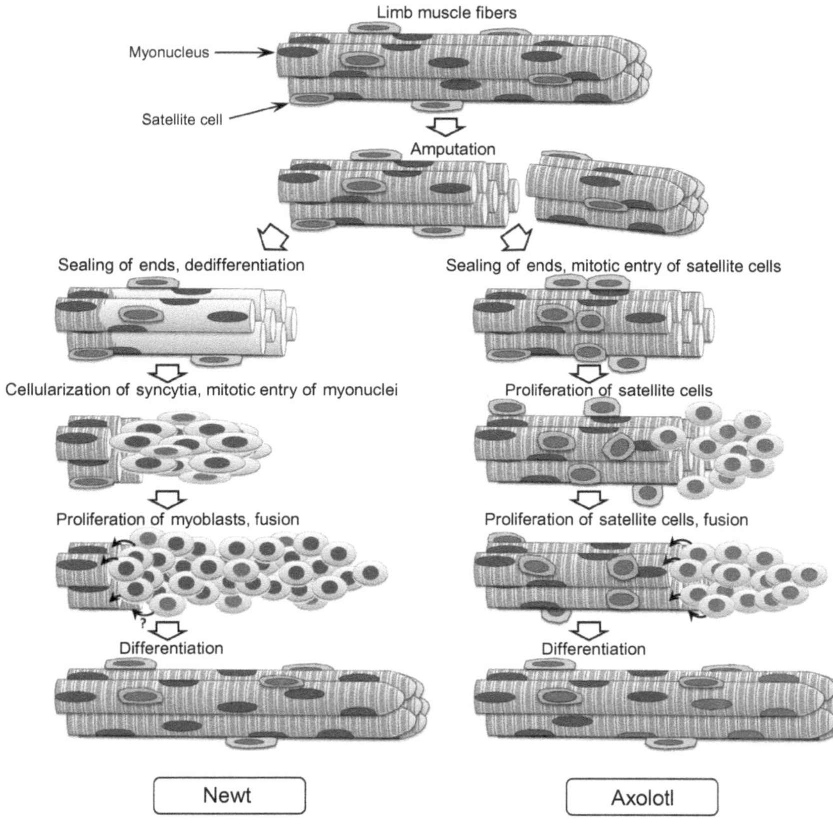

Figure 1 Schematic comparison of limb muscle regeneration via muscle fiber fragmentation versus satellite cell activation. Examples of the distal ends of seven myofibers before, during, and after muscle regeneration are shown from top to bottom. In newt limbs (left), muscle fibers dedifferentiate following amputation and fragment into mononucleate cells, which enter the cell cycle, proliferate, and fuse with the damaged muscle fibers and, potentially, with one another. The contribution of satellite cells to muscle regeneration in this context is not known. In axolotl limbs (and likewise, *Xenopus* tadpole tails; right), amputation activates satellite cells, which proliferate and fuse to regenerate muscles. (A color version of this figure is located at the end of this chapter.)

reminiscent of the mechanisms of muscle repair in mammals. By contrast, in the newt, the muscle progenitors driving muscle regeneration are produced from the preexisting muscle remnants via muscle fragmentation and dedifferentiation (Fig. 1). Consequently, muscle regeneration via muscle dedifferentiation seems to be quite rare within the animal kingdom. It will be interesting to see whether it is also operative in the salamander species that were used in the classical regeneration studies described herein and whether examples for this strategy can be found in other amphibian species or beyond. In this regard, it is interesting to note that a recent study in the mouse using an analogous Cre/Lox strategy as the one described above for marking differentiated muscle fibers in urodeles did detect labeled mononucleate cells upon injury; these cells expressed Pax7, proliferated, and fused with myofibers upon retransplantation (Mu, Peng, Pan, Huard, & Li, 2011). Although in *Xenopus* and zebrafish tail muscle, as in axolotl limb muscle regeneration, there is no lineage continuity between the injured and the newly formed muscles, the injured muscles do partially dedifferentiate in these three species, albeit without fragmenting and entering the cell cycle (Rodrigues, Christen, Martí, & Izpisúa Belmonte, 2012; Wu, Huang, Chen, Chiou, & Lee, 2015). Based on these observations, it cannot be completely ruled out that partial skeletal muscle dedifferentiation contributes to muscle repair upon injury in other vertebrate species, including mammals.

2.2 Molecular Events During Muscle Dedifferentiation and Cell Cycle Reentry

The demonstration that regeneration of amputated newt limb muscles involved muscle dedifferentiation raises important questions about the molecular mechanisms involved in this process and why these are not operative in mammalian species. For example, what are the signals that initiate dedifferentiation and what are the activities that implement it? How is cell cycle reentry controlled and what mechanisms trigger redifferentiation, including myoblast fusion? Until now, only limited information is available for these issues.

One question that has been addressed is whether cell cycle entry is a cause or a consequence of muscle dedifferentiation into mononucleate cells. In the newt model with implanted myotubes, it was demonstrated that the formation of mononucleate cells occurred even in the absence of mitosis, which was blocked either by X-irradiation or by introducing the cell cycle inhibitor p16 into myotubes before their implantation into amputated limbs (Velloso, Kumar, Tanaka, & Brockes, 2000). Hence, myotube fragmentation does not require cell cycle entry and seems to differ fundamentally from cytokinesis.

Some insight into the processes driving myofiber fragmentation into mononucleate cells (also called cellularization) has come from a compound that was identified in an *in vitro* chemical genetics screen for agents affecting the morphology of myotubes. This small molecule, named myoseverin, causes the fission of multinucleate myotubes (generated from murine C2C12 myoblasts) into mononucleate fragments in a reversible fashion (Rosania et al., 2000). Although myoseverin is able to inhibit microtubule polymerization, its role in disrupting the actomyosin filaments of the myofibrils seems to be more pertinent for its effect in myotube cellularization (Ng, Gebski, Grounds, & Bogoyevitch, 2008). Hence, myoseverin may mimic endogenous activities that are needed to degrade the cytoskeleton and/or the myofibrils during muscle dedifferentiation, which in turn may trigger myofiber fragmentation. Because the mononucleate cells that are generated upon myoseverin treatment continue to express muscle differentiation markers, fail to restore centrosomes needed for mitosis, and do not enter S-phase, it was concluded that disruption of the cytoskeleton (and likely of myofibrils) is only one of the several effects that occur in parallel to cause dedifferentiation and fragmentation of postmitotic syncytial muscle fibers into mononucleate and proliferating cells (Duckmanton, Kumar, Chang, & Brockes, 2005; Kim et al., 2012). Furthermore, the inability of myoseverin to cellularize mature myofibers shows that *in vivo*, additional cellular blocks must be overcome in order to achieve myofiber fragmentation (Milner & Cameron, 2013).

A new and important observation, initially made in newts and extended to mouse myotubes, was that myoseverin-triggered myotube fragmentation was preceded by hallmarks of the programmed cell death response, including the activation of caspase 3 (Wang et al., 2015). Because a second proapoptic substance, staurosporine, also caused fragmentation of newt and mouse myotubes whereas apoptosis inhibitors prevented it, entry into the apoptotic pathway appeared to be instrumental for myotube fragmentation. The fragmented, mononucleate cells obtained in this manner from newt or mouse C2C12 myotubes did not divide and ultimately underwent apoptosis, in accordance with previous results. However, when apoptosis was blocked some time after its initiation by myoseverin or staurosporine treatment, the mononucleate cells derived from both species completed their dedifferentiation, entered the cell cycle, and proliferated. These cells produced myotubes and, in a mouse injury model, fused with syncytial muscle fibers

(Wang et al., 2015). The *in vivo* relevance of these findings was demonstrated in newts, where it was shown that limb amputation caused upregulation of caspase 3 and other caspases, as well as downregulation of the caspase inhibitor XIAP. Because DNA fragmentation only occurred in a small and temporally decreasing fraction of the caspase 3-positive cells, muscle dedifferentiation appears to involve only a partial activation of the apoptotic cascade. Importantly, the forced expression of the apoptosis inhibitor XIAP before and during newt limb regeneration caused a strong reduction of mononucleate cells derived from marked muscle fibers (Wang et al., 2015). Altogether, these findings led to the conclusion that the transient induction of components of the apoptotic pathway, particularly caspase activation, upon newt limb amputation leads to myotube fragmentation into mononucleate cells without causing terminal apoptosis, while the subsequent inactivation of these components allows the cells to dedifferentiate fully, proliferate, and ultimately produce new muscles.

The reasons for why the nuclei of injured myofibers are able to enter the cell cycle only in newts but not in mammals have been investigated further. A major gatekeeper of cell cycle entry at the G1/S transition is the retinoblastoma protein (pRb), which can bind and inactivate the E2F transcription factor that would otherwise promote the expression of particular cell cycle regulators. Phosphorylation of pRb by CDK4/cyclin D1 inactivates it, thus creating a necessary condition for cell cycle entry. In accordance with this scenario, newt myotubes from limb cells that were maintained in low serum conditions, which favor the differentiated state, contained pRb in the unphosphorylated form, whereas a switch to high serum conditions correlated with pRb phosphorylation and led to reentry into S-phase (Tanaka, Gann, Gates, & Brockes, 1997). Preventing pRb phosphorylation or introducing a nonphosphorylatable version of pRb into the myotubes blocked S-phase entry upon high serum treatment, thus showing that pRb activity needs to fall below a critical threshold to allow S-phase entry. The situation differs in mouse myotubes, in which myotube differentiation also relies on unphosphorylated pRb, but high serum conditions are unable to relieve this state and fail to cause S-phase entry (reviewed in Lassar, Skapek, & Novitch, 1994). When pRb was inactivated after mouse myotube differentiation (e.g., by conditional *Rb* knockout, including in the absence of the genes for the other two related pocket proteins p107 and p130, or by various other means), these myotubes were still not able to enter S-phase upon serum stimulation (Camarda et al., 2004; Huh, Parker, Scimè, Parks, & Rudnicki, 2004; and references therein). These observations showed that in mouse myotubes generated from primary cells, there must be at least one other mechanism in addition to the one exerted by the pocket proteins of the Rb family that blocks the reversal of the postmitotic state.

A key player in this second mechanism that blocks cell cycle reentry in differentiated mouse myotubes was identified as the tumor suppressor p19ARF, a protein that exists in mammals and chicken but not in lower vertebrates such as amphibians. Indeed, when *p19ARF* was knocked down together with conditionally knocking out *Rb* in differentiated primary mouse myotubes, their nuclei progressed through S-phase into G2, although they did not enter M-phase (Pajcini, Corbel, Sage, Pomerantz, & Blau, 2010). These myotubes showed reduced expression of the myogenic differentiation factor myogenin and terminal muscle differentiation genes, displayed disrupted MHC networks, and collapsed into amorphous syncytia with clustered nuclei. Thus, these myotubes showed significant signs of dedifferentiation in addition to their cell cycle reentry. Cells that were terminally differentiated in the absence of fusion (by seeding them at low density) and then transiently depleted of pRb and p19ARF could be clonally expanded in growth medium. Upon reintroduction of *pRb* and reexpression of *p19ARF*, these cells redifferentiated *in vitro* and even incorporated into myofibers *in vivo* (Pajcini et al., 2010; Wang et al., 2015).

p19ARF is encoded by the *Ink4a* locus jointly with p16INK4a (the latter being upstream of pRb via inhibition of CDK4), but it was shown that the described effects are largely due to p19ARF rather than p16INK4a. The main mode of action of p19ARF is believed to be the inhibition of the E3 ligase MDM2, which causes the stabilization of the MDM2 target p53 (Sherr et al., 2005). Among other roles, p53 is antiproliferative, in part by activating the expression of the CDK inhibitor p21. Thus, in its simplest form, a model emerges in which cell cycle reentry and dedifferentiation of mammalian myotubes require the simultaneous inactivation of pRb (via CDK4/cyclin D-mediated phosphorylation) and of p53 (by relieving the p19ARF-mediated block of its negative regulator MDM2). Together, this would enable E2F to activate regulators of G1/S entry and prevent p53 from activating cell cycle inhibitors. By contrast, amphibians lack p19ARF, which could allow their myotubes to enter the cell cycle more readily because only the pRb branch needs to be inactivated, e.g., via pRb phosphorylation, whereas the relief of the p19ARF/p53 branch is not necessary.[2] During newt limb regeneration, MDM2-mediated downregulation of p53 activity

[2] The situation is similar in myotubes from immortalized C2C12 cells, which exhibit a deletion of the *ink4a* locus and therefore can reenter the cell cycle solely by removing or inactivating pRb (Gu et al., 1993; Pajcini et al., 2010).

can thus occur, which is important for postmitotic cell cycle reentry of myotubes (Yun, Gates, & Brockes, 2013). It has been proposed that in mammals, tumor suppression mediated by the *ink4a* locus has arisen at the expense of regeneration (Pajcini et al., 2010); this could be one of the numerous evolutionary changes responsible for the observed diminution in their regenerative capacities. In addition to the mechanisms that were gleaned from myotubes generated *in vitro* from primary cells or immortalized C2C12 cells, terminally differentiated myofibers *in vivo* undoubtedly receive additional yet unknown regulatory inputs that promote or prevent cell cycle reentry upon injury in newts or mammalians, respectively.

The "holy grail" in the muscle regeneration field consists of the identification of the upstream mechanisms, particularly extracellular signals, that can trigger both muscle dedifferentiation (including caspase-dependent fragmentation) and cell cycle reentry of terminally differentiated muscle fibers. One candidate for such an activity, albeit acting within cells, has been the homeodomain transcription factor Msx1. This factor is expressed specifically in the progress zone of rapidly proliferating, undifferentiated mesenchymal cells in normal as well as in regenerating limb buds, and moreover, it inhibits terminal differentiation of myoblasts (Hill et al., 1989; Robert, Sassoon, Jacq, Gehring, & Buckingham, 1989; Simon et al., 1995; Song, Wang, & Sassoon, 1992). Of note, forced expression of *msx1* in combination with growth medium was able to drive dedifferentiation and (much less efficiently) fragmentation of mouse C2C12 myotubes (Odelberg, Kollhoff, & Keating, 2000). The derived cells proliferated and showed signs of pluripotency, as they could be differentiated into cells that express myogenic, chondrogenic, adipogenic, or osteogenic markers. Forced expression of *msx1* in the presence of growth medium could even drive postmitotic, multinucleated primary mouse myotubes into forming proliferating mononucleate cells. Upon implantation into normal or degenerating *mdx* muscles, these cells fused with existing fibers and efficiently regenerated myofibers, in this case without contributing to any other tissues (Yang et al., 2014).

It has been proposed that the documented role of *msx1* in repressing the *p21* and activating the *cyclin D* genes, now believed to be instrumental in cell cycle reentry, together with its repressive activity on muscle differentiation genes, may partly account for its observed potency for myofiber dedifferentiation (Odelberg, 2002). Further support for the relevance of Msx1 during muscle dedifferentiation and regeneration has been obtained in axolotls. Although it now appears that dedifferentiating muscle fibers do not contribute to regenerating muscles in this species (Sandoval-Guzmán et al., 2014), terminally differentiated myofibers isolated from larval axolotl limb muscles and kept in tissue culture were able to dedifferentiate and fragment into mononucleate, proliferating cells. Morpholino-induced knockdown of Msx1 expression strongly interfered with this process (Kumar, Velloso, Imokawa, & Brockes, 2004). Altogether, these intriguing data suggest that Msx1 is an upstream regulator of muscle dedifferentiation. This can now be tested further, particularly with *in vivo* experiments in the mouse and newts using marked myofibers.

In addition to the proposed cell autonomous role of Msx1, there must exist external signals—chemical and/or mechanical—that trigger muscle dedifferentiation upon amputation in newts. Based upon the ability of serum-containing media to induce cell cycle reentry in primary newt myoblasts, and the occurrence of blood clotting during limb amputation, experiments were initiated to identify the responsible serum factor(s). It was shown that a proteinaceous fraction that was activated by the blood clotting proteases thrombin and plasmin could mimic the activity of serum (Tanaka, Drechsel, & Brockes, 1999). This activity promoted cell cycle reentry of postmitotic newt myotubes, but it did not stimulate proliferation of mononucleate newt myoblasts, both of which distinguished it from known mitogenic serum growth factors, such as PDGF and FGF that were not active in this assay. Although this activity was able to activate specific immediate early gene responses in mouse C2C12 myotubes, it was unable to drive them into S-phase, confirming the presence of a cell-intrinsic block of cell cycle reentry in mouse (Lööf, Straube, Drechsel, Tanaka, & Simon, 2007). In newts, serum stimulation triggered long-term activation of extracellular signal-regulated kinase (ERK), which was required for cell cycle reentry of primary myotubes, in part by downregulating p53. By contrast, in mouse myotubes, serum incubation caused only weak and transient ERK activation, which correlated with their inability to reenter the cell cycle (Yun, Gates, & Brockes, 2014). Of note, an intriguing study that has, however, not been followed up upon had found that protein extracts from regenerating newt limbs (unlike those from intact limbs) were able to cellularize both primary newt and C2C12 mouse myotubes and generate proliferating myoblasts (McGann, Odelberg, & Keating, 2001). The identification of the thrombin-activated serum protein(s) and of putative dedifferentiation factors in amputated limb extracts, as well as the demonstration of their *in vivo* relevance during muscle regeneration, must await additional experiments. The identification of endogenous (not necessarily soluble) factors with such activities, which likely require new and innovative approaches, will undoubtedly herald a major breakthrough in the field.

3. TRANSDIFFERENTIATING MUSCLES UPON MUSCLE FIBER EXPLANTATION AND DURING NORMAL DEVELOPMENT

3.1 Muscle Transdifferentiation in Explanted Muscles from Jellyfish

In comparison to muscle regeneration via de- and redifferentiation, muscle transdifferentiation involves an additional step, namely a fate change of the dedifferentiated cells. It is presumed that cells need to revert to a more primitive state before they can undergo a change in fate. To date, there are only few known examples of transdifferentiation of striated muscle fibers of the skeleton or body wall. An early experimental system for this phenomenon was the jellyfish, *Podocoryne carnea* and related hydrozoans, which contain swimming muscles at the inner surface of the bell. These muscles consist of spindle-shaped mononucleate striated cells, so their dedifferentiation does not require fragmentation of syncytia. When these postmitotic, terminally differentiated muscle cells were isolated and the extracellular matrix was removed enzymatically, the cultured muscle cells began dedifferentiating, as evident from their rounding up and disassembly, followed by progressive disappearance of the myofibrils and formation of flagella. Subsequently, the cells acquired ultrastructural and antigenic features of jellyfish smooth muscles and entered mitosis (Schmid & Alder, 1984; Schmid, Alder, Plickert, & Weber, 1988). These observations pointed to a process of transdifferentiation from postmitotic striated to proliferative smooth muscle fates via dedifferentiation. This process occurred even when cell cycle entry was blocked with DNA replication inhibitors, but it required gene expression. Remarkably, these smooth muscle-like cells continued to exhibit plasticity because during each round of division, each of these cells appeared to generate one daughter resembling a nerve/sensory cell (based on ultrastructural and antigenic features) that stopped dividing and another daughter cell retaining the smooth muscle features of the mother cell that continued to divide asymmetrically. Hence, these cells displayed typical features of stem cells and, *in toto*, there is transdifferentiation from striated muscle to neuronal fates (Alder & Schmid, 1987; Schmid et al., 1988). In a minority of cases, when separate cell layers were formed after two or more divisions, the stem cells were also capable of producing a regenerate that contained fully differentiated digestive cells, gland cells, secretory cells, and nematoblasts (Schmid et al., 1988). These data implied that the striated muscle-derived stem cells can transdifferentiate into another cell type as well as undergo pluripotent reprogramming in the presence of proper cellular interactions. Whether these transdifferentiation or pluripotent reprogramming processes play a role during normal development or regeneration *in vivo* is currently not known. There is also very little information about the molecular regulation, and jellyfish may not be ideally suited for functional studies. Interestingly, it was shown that, unlike its vertebrate orthologs, the cnidarian homeobox gene *Msx* remains expressed in the terminally differentiated striated muscles, which may be connected with the high dedifferentiation potential of these muscles (Galle, Yanze, & Seipel, 2005).

3.2 Potential *In Vivo* Transdifferentiation During Muscle-Related Organ Development in Fish

A number of species of fish feature organs that are derived phylogenetically and/or ontogenetically from multinucleate striated muscles. For example, in billfishes (marlin, etc.), an extraocular muscle is modified to form a thermogenic organ beneath the brain that contains heater cells, which lack organized myofibrils but retain other features of myocytes such as a modified sarcoplasmatic reticulum (Block, O'Brien, & Meissner, 1994). In particular deep sea fishes, luminescent tissue is derived from striated muscle and features cells with fragments of decaying myofibrils together with bioluminescent vesicles and other ultrastructural features that are atypical for muscle fibers (Johnston & Herring, 1985). The residual presence of features that are characteristic of terminally differentiated muscle fibers along with novel features (and presumably molecules), as well as the absence of contractility and the unique physiological functions of these cells, could be indicative of their origin from (at least partially) dedifferentiated and transdifferentiated striated muscle fibers. While this assumption has not been tested in these cases, preliminary evidence for transdifferentiation was associated with the formation of electric organs in fish. Electric organs in different species have evolved independently at least six times and, based upon light and electron microscopy studies, it appears that their electrocytes are myogenically derived in all but one family of fish (Gallant et al., 2014). Although electrocytes are also syncytial, they are very different from muscle fibers in terms of their structure, function, and gene expression profiles. They are much larger and lack the contractile apparatus. Moreover, they are organized asymmetrically such that one side features an innervated surface enriched in cation-specific ion channels and the other side an invaginated plasma membrane enriched in sodium pumps; thus, they resemble units of batteries stacked in series. The global gene expression profiles were compared between striated muscles and electrocytes in five different species and showed strong differences between the tissues, including strongly decreased expression of myogenic regulators and muscle differentiation genes, as well as significantly increased expression of genes for proteins connected with electric excitation and cell growth in electrocytes (Gallant et al., 2014). If electrocytes are indeed being formed from differentiated muscle cells, these data and the histological observations would strongly support the notion that they arise via transdifferentiation.

Developmental studies on electrocyte formation have been performed mostly in a neotropical, weakly electric fish, *Sternopygus macrurus* (knifefish, Gymnotiformes), in the context of regeneration following tail amputation. Histological and immunohistochemical analyses initially provided support for transdifferentiating muscles being the source of regenerating electrocytes. Specifically, it was reported that, during regeneration, fast muscle fibers fused with larger cells (containing residual clusters of myofibrils and presumed to be products of muscle fiber fragmentation), which was followed by the disassembly of their myofibril arrangement, an increase in cell size, and a concomitant loss of muscle differentiation proteins (Patterson & Zakon, 1997; Unguez & Zakon, 1998; Zakon & Unguez, 1999). However, more recent studies, which included lineage tracings of muscle fibers, disproved these interpretations and did not detect any signs of muscle dedifferentiation during regeneration in *S. macrurus*. Instead, regeneration of electric organs appears to rely exclusively on $Pax7^+$ stem cells that are associated with electrocytes and are akin to satellite cells associated with myofibers (Unguez, 2013; Weber, Martindale, Tapscott, & Unguez, 2012). Furthermore, recent investigations of the embryonic origin of electrocytes in a different gymnotiform fish, *Electrophorus electricus*, showed that their progenitors directly arise from a separate portion of the myotome during late embryogenesis and fuse with one another to generate syncytial electrocytes, without any apparent involvement of dedifferentiating or transdifferentiating muscle fibers (Schwassmann, Assunção, & Kirschbaum, 2014). The newer data call into question whether any of the specialized muscle-related organs in fish arise via muscle transdifferentiation, and they reinforce the importance of lineage tracing for the interpretation of presumptive transdifferentiation events.

3.3 Dedifferentiation, Fragmentation, and Transdifferentiation of Muscles During Drosophila Metamorphosis

A radical reconstruction of the multinucleate, striated body wall musculature occurs during insect metamorphosis, which allows the adult muscles to fulfill their highly distinctive physiological roles in the adult and to conform with the size and shape requirements of the adult body parts. In the best-studied insect model organism, the fruit fly *Drosophila melanogaster*, it was found that the vast majority of larval body wall muscles become histolyzed and disappear during metamorphosis, and almost all adult muscles are then newly rebuilt from stem cell-like cells, called adult muscle precursors (AMPs). These AMPs are set aside during embryogenesis at specific locations underneath the epidermis of the body wall and remain undifferentiated and dormant until late larval stages (Bate, Rushton, & Currie, 1991; Figeac, Jagla, Aradhya, Da Ponte, & Jagla, 2011). Shortly before pupariation, they enter the cell cycle and proliferate during early pupal development. After degradation of the larval muscles in the pupa, the amplified AMPs undergo myoblast fusion and form syncytia that give rise to the striated muscles of the abdominal body wall, leg muscles, direct flight muscles, etc. Hence, the reorganization of the musculature during *Drosophila* metamorphosis typically does not involve transdifferentiation but rather the replacement of larval muscles by the progeny of stem cells. With regard to the indirect flight muscles, which power the flight, this "canonical" sequence of events is only used for the development of the dorsoventral set of muscles, which carry out the upward strokes of the wings. The development of the dorsal longitudinal flight muscles (DLMs), which are responsible for the downward stroke, is different in that it relies on three larval muscles located on each side in the thorax that are not degraded and serve as templates for the adult DLMs. DLM development involves the fusion of these templates with AMPs, which is accompanied by significant size increases and the splitting of each muscle along its length followed by terminal differentiation. This doubling results in six DLMs on either side (Dutta & VijayRaghavan, 2004). The entire process leads to new types of muscles with totally different functions, physiological and ultrastructural features, and gene expression profiles (Bryantsev et al., 2012; Roy & VijayRaghavan, 1998; Schönbauer et al., 2011). Thus, in the case of DLMs, the complete reorganization of certain larval muscles after fusion with stem cell-derived myoblasts, which involves dedifferentiation of the larval templates (Reedy & Beall, 1993) and their duplication by splitting, could be considered as a transdifferentiation event.

Recently, a third mode of development of specific adult syncytial muscle fibers was discovered that involves dedifferentiation and fragmentation of larval syncytial muscles, followed by their transdifferentiation into new types of adult muscles. This mode applies to the muscles that are tightly associated with the ventral surface of the adult heart tube and are called ventral longitudinal muscles (VLMs). These are ca. 12–16 syncytial striated muscle fibers, which are aligned in parallel such that many of them span the entire length of the heart tube. Together, the VLMs form a single sheet of musculature attached to the heart tube, which may be required for proper functioning of the heart and forms part of the dorsal diaphragm that also has a role in channeling the hemolymph flow (Lehmacher, Abeln, & Paululat, 2012; Wasserthal, 2007). VLMs do not exist in the embryonic and larval heart and only arise during pupal stages. Strikingly, they develop from distinct larval muscles of a different type, which are called alary muscles (Schaub, März, Reim, & Frasch, 2015). In larvae, there are seven and in adults four bilateral pairs of alary muscles, which are arranged segmentally such that their medial, fan-shaped ends attach to the heart tube while their lateral ends attach to epidermal attachment sites. These striated muscles are also syncytial (typically containing three nuclei each) and are thought to provide tension during heart contraction/expansion by supporting the diastolic phase, as well as to anchor the heart at the body wall.

The derivation of the VLMs from embryonic alary muscles was demonstrated by genetically induced lineage tracing and by ablations of the embryonic alary muscles, which led to a complete failure to form adult VLMs (Schaub et al., 2015). Moreover, examination of early pupal events via live imaging and immunohistochemical analysis revealed that, in a first step of VLM development, the three anterior pairs of alary muscles undergo dedifferentiation, as judged by the disappearance of the striations, and subsequently fragment into mononucleate migrating cells. Without entering the cell cycle, each of these cells then serves as a founder cell of one VLM by fusing with fusion-competent myoblasts.[3] The source of fusion-competent myoblasts corresponds to dorsal AMPs (see above) that have proliferated and migrated toward the heart tube. The growing syncytia then migrate toward the ventral surface of the heart tube, elongate and align in parallel, extend toward the posterior end of the tube, and finally differentiate into functional, striated muscle fibers (VLMs; Fig. 2). This sequence of events also explains why only four pairs of alary muscles remain in the adult and why the number of VLM fibers is roughly equal to the numbers of nuclei within the three anterior pairs of alary muscles.

[3] In *Drosophila* and other insects, each syncytial muscle is formed by the asymmetric fusion of one muscle founder cell with a number of fusion-competent myoblasts. During and after the specification of a founder cell, a genetic program is activated that leads to the expression of a specific set of muscle identity genes in it. These genes encode various transcription factors that ultimately determine the specific identity of the muscle formed by a particular founder cell. By contrast, the fusion-competent cells and their nuclei lack differential identities prior to myoblast fusion (Dobi, Schulman, & Baylies, 2015).

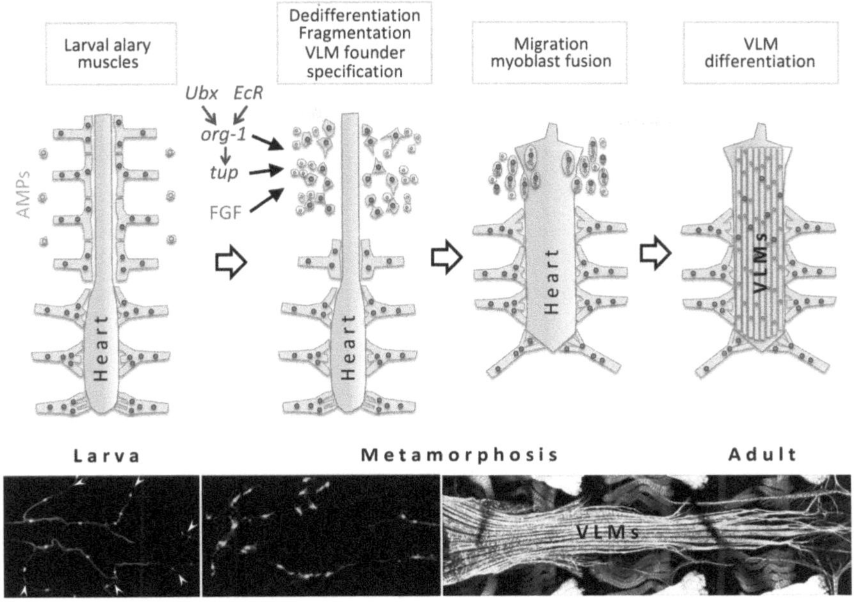

Figure 2 Transdifferentiation of *Drosophila* larval alary muscles into ventral longitudinal muscles of the adult heart. Top: Schematic drawings of the fragmentation of specific larval alary muscles into mononucleate cells, which become specified as founder cells of the ventral longitudinal muscles (VLMs) of the adult heart. Upon myoblast fusion with fusion-competent myoblasts (derived from adult muscle progenitors [AMPs], green), they undergo morphogenesis and differentiate into VLMs. The known genetic inputs are depicted as described in the text. Bottom: Anterior three pairs of larval alary muscles (arrowheads) are shown on the left, fragmented mononucleate cells from these muscles in the center, and fully differentiated VLMs on the right. (All three panels show an *org-1* reporter in green and the right hand panel in addition shows F-actin in white.) (A color version of this figure is located at the end of this chapter.)

Drosophila provides the great advantage that the process of muscle de- and transdifferentiation can be dissected by genetic analysis. Some important components of the genetic program that regulates the development of VLMs from alary muscles have already been identified. Two transcription factors, namely the T-box factor Org-1 (*Drosophila* Tbx1) and the LIM homeodomain factor Tup (*Drosophila* Islet1), cooperate to initiate the process. These two factors have been shown to be coexpressed already in the founder cells and precursors of all alary muscles in the embryo and to have essential role in larval alary muscle specification (Boukhatmi et al., 2014; Schaub, Nagaso, Jin, & Frasch, 2012). Notably, during early pupal stages, these two factors are reexpressed specifically in the three anterior pairs of alary muscles that will transdifferentiate into VLMs but not in the four posterior pairs that will remain as alary muscles. Org-1 acts as a direct upstream activator of the *tup* gene in both contexts (Schaub et al., 2015, 2012). Importantly, functional knockdowns of either *org-1* or *tup* within the anterior early pupal alary muscles prevented dedifferentiation and fragmentation of these muscles and, as a consequence, led to a near absence of VLMs along the heart in the adult (Schaub et al., 2015). These observations indicate that Org-1 and Tup normally activate yet unknown genes within these anterior alary muscles that execute the processes of muscle dedifferentiation and fragmentation into mononucleate cells.

The data have raised the questions of why Org-1 and Tup act differently during early pupal stages as compared to embryonic stages, and what restricts their action to the anterior three pairs of alary muscles during metamorphosis. The questions have been partially answered by the identification of additional transcriptional and signaling activities within these particular muscles that are part of the regulatory network (Schaub et al., 2015). The spatial restriction is due to the action of Hox genes. The expression of the Hox gene Ubx is restricted to the anterior three pairs of larval alary muscles, whereas that of Abd-A is active in the posterior four pairs (Labeau, Trujillo, & Cripps, 2009). Functional downregulation of Ubx by forcing Abd-A expression in the anterior three pairs abrogated reexpression of *org-1* (and hence, presumably *tup*), which prevented their dedifferentiation and fragmentation. Likewise, the temporal regulation is due to stage-specific hormonal inputs, as shown by the observation that knockdown of the receptor of the molting hormone ecdysone (EcR) within the anterior three pairs of alary muscles in early pupae elicited the same blocking effects. Therefore, the initiation of muscle dedifferentiation and fragmentation requires the combinatorial action of at least Org-1, Tup, Ubx, and EcR. In addition, the specification of the mononucleate cells that are derived from alary muscle fragmentation as VLM founder cells, as well as VLM differentiation, is likely to require the continued action of these regulators together with FGF receptor-mediated signals and additional inputs from the heart tube (Schaub et al., 2015; Fig. 2).

4. CONCLUSION

It has long been recognized that dedifferentiation of striated syncytial muscles, their fragmentation into mononucleate cells, or their transdifferentiation into different types of muscles or tissues during normal development and regeneration are very rare events. Given the highly ordered and complex ultrastructural organization and multinucleate state of muscle fibers, this was not unexpected. During recent years, more detailed reexaminations of some developmental and regeneration processes that originally were thought to involve muscle de- and regeneration or transdifferentiation, such as muscle regeneration upon amputation in axolotl and formation of electric organs in electric fish, have found that these processes instead involve the repair of the damaged tissues by activated satellite cells. In vertebrates, currently this leaves only newts, in which muscle regeneration upon amputation was proven to involve muscle fiber dedifferentiation, followed by cellularization, cell cycle reentry, and redifferentiation. Within invertebrates, an example for transdifferentiation of one type of striated syncytial muscle into another via muscle dedifferentiation and cellularization was recently discovered in *Drosophila*. This transdifferentiation process occurs in the context of the rebuilding of the musculature during metamorphosis. The regulatory pathway involved in this process has been partially defined. These systems can now be utilized to define the signals that trigger dedifferentiation and cellularization of muscles, as well as the molecules that exert these processes, to generate knowledge that can also be harnessed for future application in muscle repair in regenerative medicine.

ACKNOWLEDGMENTS

I thank Hanh Nguyen for comments on the manuscript. Special thanks go to the members of my lab, particularly Christoph Schaub, Johannes März, and Ingolf Reim, whose research led to the insights described in parts of this review article. This research was funded by Deutsche Forschungsgemeinschaft (DFG).

REFERENCES

Alder, H., & Schmid, V. (1987). Cell cycles and in vitro transdifferentiation and regeneration of isolated, striated muscle of jellyfish. *Developmental Biology*, *124*, 358–369.

Bate, M., Rushton, E., & Currie, D. A. (1991). Cells with persistent twist expression are the embryonic precursors of adult muscles in Drosophila. *Development*, *113*, 79–89.

Blaauw, B., Schiaffino, S., & Reggiani, C. (2013). Mechanisms modulating skeletal muscle phenotype. *Comprehensive Physiology*, *3*, 1645–1687.

Block, B. A., O'Brien, J., & Meissner, G. (1994). Characterization of the sarcoplasmic reticulum proteins in the thermogenic muscles of fish. *The Journal of Cell Biology*, *127*, 1275–1287.

Boukhatmi, H., Schaub, C., Bataillé, L., Reim, I., Frendo, J.-L., Frasch, M., et al. (2014). An Org-1-Tup transcriptional cascade reveals different types of alary muscles connecting internal organs in Drosophila. *Development*, *141*, 3761–3771.

Bryantsev, A. L., Duong, S., Brunetti, T. M., Chechenova, M. B., Lovato, T. L., Nelson, C., et al. (2012). Extradenticle and homothorax control adult muscle fiber identity in Drosophila. *Developmental Cell*, *23*, 664–673.

Calve, S., Odelberg, S. J., & Simon, H.-G. (2010). A transitional extracellular matrix instructs cell behavior during muscle regeneration. *Developmental Biology, 344*, 259–271.
Calve, S., & Simon, H.-G. (2011). High resolution three-dimensional imaging: Evidence for cell cycle reentry in regenerating skeletal muscle. *Developmental Dynamics, 240*, 1233–1239.
Camarda, G., Siepi, F., Pajalunga, D., Bernardini, C., Rossi, R., Montecucco, A., et al. (2004). A pRb-independent mechanism preserves the postmitotic state in terminally differentiated skeletal muscle cells. *The Journal of Cell Biology, 167*, 417–423.
Cameron, J. A., Hilgers, A. R., & Hinterberger, T. J. (1986). Evidence that reserve cells are a source of regenerated adult newt muscle in vitro. *Nature, 321*, 607–610.
Chen, Y., Lin, G., & Slack, J. M. W. (2006). Control of muscle regeneration in the Xenopus tadpole tail by Pax7. *Development, 133*, 2303–2313.
Diogo, R., Nacu, E., & Tanaka, E. M. (2014). Is salamander limb regeneration really perfect? Anatomical and morphogenetic analysis of forelimb muscle regeneration in GFP-transgenic axolotls as a basis for regenerative, developmental, and evolutionary studies. *Anatomical record (Hoboken), 297*, 1076–1089.
Dobi, K. C., Schulman, V. K., & Baylies, M. K. (2015). Specification of the somatic musculature in Drosophila. *Wiley Interdisciplinary Reviews: Developmental Biology, 4*, 357–375.
Duckmanton, A., Kumar, A., Chang, Y.-T., & Brockes, J. P. (2005). A single-cell analysis of myogenic dedifferentiation induced by small molecules. *Chemistry & Biology, 12*, 1117–1126.
Dutta, D., & VijayRaghavan, K. (2004). Metamorphosis and the formation of the adult musculature. In H. Sink (Ed.), *Muscle development in Drosophila* (pp. 125–142). Georgetown, TX: Landes Bioscience.
Echeverri, K., Clarke, J. D., & Tanaka, E. M. (2001). In vivo imaging indicates muscle fiber dedifferentiation is a major contributor to the regenerating tail blastema. *Developmental Biology, 236*, 151–164.
Ferretti, P., & Brockes, J. P. (1988). Culture of newt cells from different tissues and their expression of a regeneration-associated antigen. *The Journal of Experimental Zoology, 247*, 77–91.
Figeac, N., Jagla, T., Aradhya, R., Da Ponte, J. P., & Jagla, K. (2011). Specification and behavior of AMPs, muscle-committed transient Drosophila stem cells. *Fly (Austin), 5*, 7–9.
Gallant, J. R., Traeger, L. L., Volkening, J. D., Moffett, H., Chen, P.-H., Novina, C. D., et al. (2014). Nonhuman genetics. Genomic basis for the convergent evolution of electric organs. *Science, 344*, 1522–1525.
Galle, S., Yanze, N., & Seipel, K. (2005). The homeobox gene Msx in development and transdifferentiation of jellyfish striated muscle. *The International Journal of Developmental Biology, 49*, 961–967.
Gargioli, C., & Slack, J. M. W. (2004). Cell lineage tracing during Xenopus tail regeneration. *Development, 131*, 2669–2679.
Gu, W., Schneider, J. W., Condorelli, G., Kaushal, S., Mahdavi, V., & Nadal-Ginard, B. (1993). Interaction of myogenic factors and the retinoblastoma protein mediates muscle cell commitment and differentiation. *Cell, 72*, 309–324.
Hay, E. D. (1959). Electron microscopic observations of muscle dedifferentiation in regenerating Amblystoma limbs. *Developmental Biology, 1*, 555–585.
Hay, E. D., & Fischman, D. A. (1961). Origin of the blastema in regenerating limbs of the newt *Triturus viridescens*. An autoradiographic study using tritiated thymidine to follow cell proliferation and migration. *Developmental Biology, 3*, 26–59.
Hill, R. E., Jones, P. F., Rees, A. R., Sime, C. M., Justice, M. J., Copeland, N. G., et al. (1989). A new family of mouse homeo box-containing genes: Molecular structure,

chromosomal location, and developmental expression of Hox-7.1. *Genes & Development*, *3*, 26–37.
Hjiantoniou, E., Anayasa, M., Nicolaou, P., Bantounas, I., Saito, M., Iseki, S., et al. (2008). Twist induces reversal of myotube formation. *Differentiation*, *76*, 182–192.
Huh, M. S., Parker, M. H., Scimè, A., Parks, R., & Rudnicki, M. A. (2004). Rb is required for progression through myogenic differentiation but not maintenance of terminal differentiation. *The Journal of Cell Biology*, *166*, 865–876.
Johnston, I. A., & Herring, P. J. (1985). The transformation of muscle into bioluminescent tissue in the fish Benthalbella infans Zagmayer. *Proceedings of the Royal Society B: Biological Sciences*, *225*, 213–218.
Jopling, C., Boue, S., & Izpisúa Belmonte, J. C. (2011). Dedifferentiation, transdifferentiation and reprogramming: Three routes to regeneration. *Nature Reviews. Molecular Cell Biology*, *12*, 79–89.
Kim, W.-H., Jung, D.-W., Kim, J., Im, S.-H., Hwang, S. Y., & Williams, D. R. (2012). Small molecules that recapitulate the early steps of urodele amphibian limb regeneration and confer multipotency. *ACS Chemical Biology*, *7*, 732–743.
Kintner, C. R., & Brockes, J. P. (1984). Monoclonal antibodies identify blastemal cells derived from dedifferentiating limb regeneration. *Nature*, *308*, 67–69.
Kumar, A., Velloso, C. P., Imokawa, Y., & Brockes, J. P. (2000). Plasticity of retrovirus-labelled myotubes in the newt limb regeneration blastema. *Developmental Biology*, *218*, 125–136.
Kumar, A., Velloso, C. P., Imokawa, Y., & Brockes, J. P. (2004). The regenerative plasticity of isolated urodele myofibers and its dependence on *MSX1*. *PLoS Biology*, *2*, E218.
Labeau, E., Trujillo, D. L., & Cripps, R. M. (2009). Bithorax complex genes control alary muscle patterning along the cardiac tube of *Drosophila*. *Mechanisms of Development*, *126*, 478–486.
Lassar, A. B., Skapek, S. X., & Novitch, B. (1994). Regulatory mechanisms that coordinate skeletal muscle differentiation and cell cycle withdrawal. *Current Opinion in Cell Biology*, *6*, 788–794.
Lehmacher, C., Abeln, B., & Paululat, A. (2012). The ultrastructure of *Drosophila* heart cells. *Arthropod Structure & Development*, *41*, 459–474.
Lo, D. C., Allen, F., & Brockes, J. P. (1993). Reversal of muscle differentiation during urodele limb regeneration. *Proceedings of the National Academy of Sciences of the United States of America*, *90*, 7230–7234.
Lööf, S., Straube, W. L., Drechsel, D., Tanaka, E. M., & Simon, A. (2007). Plasticity of mammalian myotubes upon stimulation with a thrombin-activated serum factor. *Cell Cycle*, *6*, 1096–1101.
Mauro, A. (1961). Satellite cell of skeletal muscle fibers. *The Journal of Biophysical and Biochemical Cytology*, *9*, 493–495.
McGann, C. J., Odelberg, S. J., & Keating, M. T. (2001). Mammalian myotube dedifferentiation induced by newt regeneration extract. *Proceedings of the National Academy of Sciences of the United States of America*, *98*, 13699–13704.
Milner, D. J., & Cameron, J. A. (2013). Muscle repair and regeneration: Stem cells, scaffolds, and the contributions of skeletal muscle to amphibian limb regeneration. *Current Topics in Microbiology and Immunology*, *367*, 133–159.
Morrison, J. I., Lööf, S., He, P., & Simon, A. (2006). Salamander limb regeneration involves the activation of a multipotent skeletal muscle satellite cell population. *The Journal of Cell Biology*, *172*, 433–440.
Mu, X., Peng, H., Pan, H., Huard, J., & Li, Y. (2011). Study of muscle cell dedifferentiation after skeletal muscle injury of mice with a Cre-Lox system. *PloS One*, *6*, e16699.

Ng, D. C. H., Gebski, B. L., Grounds, M. D., & Bogoyevitch, M. A. (2008). Myoseverin disrupts sarcomeric organization in myocytes: An effect independent of microtubule assembly inhibition. *Cell Motility and the Cytoskeleton, 65*, 40–58.

Odelberg, S. J. (2002). Inducing cellular dedifferentiation: A potential method for enhancing endogenous regeneration in mammals. *Seminars in Cell & Developmental Biology, 13*, 335–343.

Odelberg, S. J., Kollhoff, A., & Keating, M. T. (2000). Dedifferentiation of mammalian myotubes induced by *msx1*. *Cell, 103*, 1099–1109.

Pajcini, K. V., Corbel, S. Y., Sage, J., Pomerantz, J. H., & Blau, H. M. (2010). Transient inactivation of Rb and ARF yields regenerative cells from postmitotic mammalian muscle. *Cell Stem Cell, 7*, 198–213.

Patterson, J. M., & Zakon, H. H. (1997). Transdifferentiation of muscle to electric organ: Regulation of muscle-specific proteins is independent of patterned nerve activity. *Developmental Biology, 186*, 115–126.

Popiela, H. (1976). Muscle satellite cells in urodele amphibians: Facilitated identification of satellite cells using ruthenium red staining. *The Journal of Experimental Zoology, 198*, 57–64.

Reedy, M. C., & Beall, C. (1993). Ultrastructure of developing flight muscle in *Drosophila*. I. Assembly of myofibrils. *Developmental Biology, 160*, 443–465.

Robert, B., Sassoon, D., Jacq, B., Gehring, W., & Buckingham, M. (1989). *Hox-7*, a mouse homeobox gene with a novel pattern of expression during embryogenesis. *The EMBO Journal, 8*, 91–100.

Rodrigues, A. M. C., Christen, B., Martí, M., & Izpisúa Belmonte, J. C. (2012). Skeletal muscle regeneration in *Xenopus* tadpoles and zebrafish larvae. *BMC Developmental Biology, 12*, 9.

Rosania, G. R., Chang, Y. T., Perez, O., Sutherlin, D., Dong, H., Lockhart, D. J., et al. (2000). Myoseverin, a microtubule-binding molecule with novel cellular effects. *Nature Biotechnology, 18*, 304–308.

Roy, S., & VijayRaghavan, K. (1998). Patterning muscles using organizers: Larval muscle templates and adult myoblasts actively interact to pattern the dorsal longitudinal flight muscles of *Drosophila*. *The Journal of Cell Biology, 141*, 1135–1145.

Sandoval-Guzmán, T., Wang, H., Khattak, S., Schuez, M., Roensch, K., Nacu, E., et al. (2014). Fundamental differences in dedifferentiation and stem cell recruitment during skeletal muscle regeneration in two salamander species. *Cell Stem Cell, 14*, 174–187.

Schaub, C., März, J., Reim, I., & Frasch, M. (2015). *Org-1*-dependent lineage reprogramming generates the ventral longitudinal musculature of the *Drosophila* heart. *Current Biology, 25*, 488–494.

Schaub, C., Nagaso, H., Jin, H., & Frasch, M. (2012). *Org-1*, the *Drosophila* ortholog of *Tbx1*, is a direct activator of known identity genes during muscle specification. *Development, 139*, 1001–1012.

Schmid, V., & Alder, H. (1984). Isolated, mononucleated, striated muscle can undergo pluripotent transdifferentiation and form a complex regenerate. *Cell, 38*, 801–809.

Schmid, V., Alder, H., Plickert, G., & Weber, C. (1988). Transdifferentiation from striated muscle of medusae in vitro. *Cell Differentiation and Development, 25*(Suppl.), 137–146.

Schnapp, E., Kragl, M., Rubin, L., & Tanaka, E. M. (2005). Hedgehog signaling controls dorsoventral patterning, blastema cell proliferation and cartilage induction during axolotl tail regeneration. *Development, 132*, 3243–3253.

Schönbauer, C., Distler, J., Jährling, N., Radolf, M., Dodt, H.-U., Frasch, M., et al. (2011). Spalt mediates an evolutionarily conserved switch to fibrillar muscle fate in insects. *Nature, 479*, 406–409.

Schwassmann, H. O., Assunção, M. I. S., & Kirschbaum, F. (2014). Ontogeny of the electric organs in the electric eel, *Electrophorus electricus*: Physiological, histological, and fine structural investigations. *Brain, Behavior and Evolution, 84*, 288–302.

Sherr, C. J., Bertwistle, D., den Besten, W., Kuo, M.-L., Sugimoto, M., Tago, K., et al. (2005). p53-Dependent and -independent functions of the Arf tumor suppressor. *Cold Spring Harbor Symposia on Quantitative Biology*, *70*, 129–137.

Simon, H. G., Nelson, C., Goff, D., Laufer, E., Morgan, B. A., & Tabin, C. (1995). Differential expression of myogenic regulatory genes and *Msx-1* during dedifferentiation and redifferentiation of regenerating amphibian limbs. *Developmental Dynamics*, *202*, 1–12.

Song, K., Wang, Y., & Sassoon, D. (1992). Expression of *Hox-7.1* in myoblasts inhibits terminal differentiation and induces cell transformation. *Nature*, *360*, 477–481.

Tanaka, E. M., Drechsel, D. N., & Brockes, J. P. (1999). Thrombin regulates S-phase re-entry by cultured newt myotubes. *Current Biology*, *9*, 792–799.

Tanaka, E. M., Gann, A. A., Gates, P. B., & Brockes, J. P. (1997). Newt myotubes reenter the cell cycle by phosphorylation of the retinoblastoma protein. *The Journal of Cell Biology*, *136*, 155–165.

Thornton, C. R. (1938). The histogenesis of muscle in the regenerating forelimb of larval *Amblystoma punctatum*. *Journal of Morphology*, *62*, 17–47.

Unguez, G. A. (2013). Electric fish: New insights into conserved processes of adult tissue regeneration. *The Journal of Experimental Biology*, *216*, 2478–2486.

Unguez, G. A., & Zakon, H. H. (1998). Phenotypic conversion of distinct muscle fiber populations to electrocytes in a weakly electric fish. *The Journal of Comparative Neurology*, *399*, 20–34.

Veldman, M. B., Zhao, C., Gomez, G. A., Lindgren, A. G., Huang, H., Yang, H., et al. (2013). Transdifferentiation of fast skeletal muscle into functional endothelium *in vivo* by transcription factor Etv2. *PLoS Biology*, *11*, e1001590.

Velloso, C. P., Kumar, A., Tanaka, E. M., & Brockes, J. P. (2000). Generation of mononucleate cells from post-mitotic myotubes proceeds in the absence of cell cycle progression. *Differentiation*, *66*, 239–246.

Wang, H., Lööf, S., Borg, P., Nader, G. A., Blau, H. M., & Simon, A. (2015). Turning terminally differentiated skeletal muscle cells into regenerative progenitors. *Nature Communications*, *6*, 7916.

Wasserthal, L. T. (2007). *Drosophila* flies combine periodic heartbeat reversal with a circulation in the anterior body mediated by a newly discovered anterior pair of ostial valves and "venous" channels. *The Journal of Experimental Biology*, *210*, 3707–3719.

Weber, C. M., Martindale, M. Q., Tapscott, S. J., & Unguez, G. A. (2012). Activation of Pax7-positive cells in a non-contractile tissue contributes to regeneration of myogenic tissues in the electric fish S. macrurus. *PLoS One*, 7, e36819.

Wu, C.-H., Huang, T.-Y., Chen, B.-S., Chiou, L.-L., & Lee, H.-S. (2015). Long-duration muscle dedifferentiation during limb regeneration in axolotls. *PloS One*, *10*, e0116068.

Yang, Z., Liu, Q., Mannix, R. J., Xu, X., Li, H., Ma, Z., et al. (2014). Mononuclear cells from dedifferentiation of mouse myotubes display remarkable regenerative capability. *Stem Cells*, *32*, 2492–2501.

Yin, H., Price, F., & Rudnicki, M. A. (2013). Satellite cells and the muscle stem cell niche. *Physiological Reviews*, *93*, 23–67.

Yun, M. H., Gates, P. B., & Brockes, J. P. (2013). Regulation of p53 is critical for vertebrate limb regeneration. *Proceedings of the National Academy of Sciences of the United States of America*, *110*, 17392–17397.

Yun, M. H., Gates, P. B., & Brockes, J. P. (2014). Sustained ERK activation underlies reprogramming in regeneration-competent salamander cells and distinguishes them from their mammalian counterparts. *Stem Cell Reports*, *3*, 15–23.

Zakon, H. H., & Unguez, G. A. (1999). Development and regeneration of the electric organ. *The Journal of Experimental Biology*, *202*, 1427–1434.

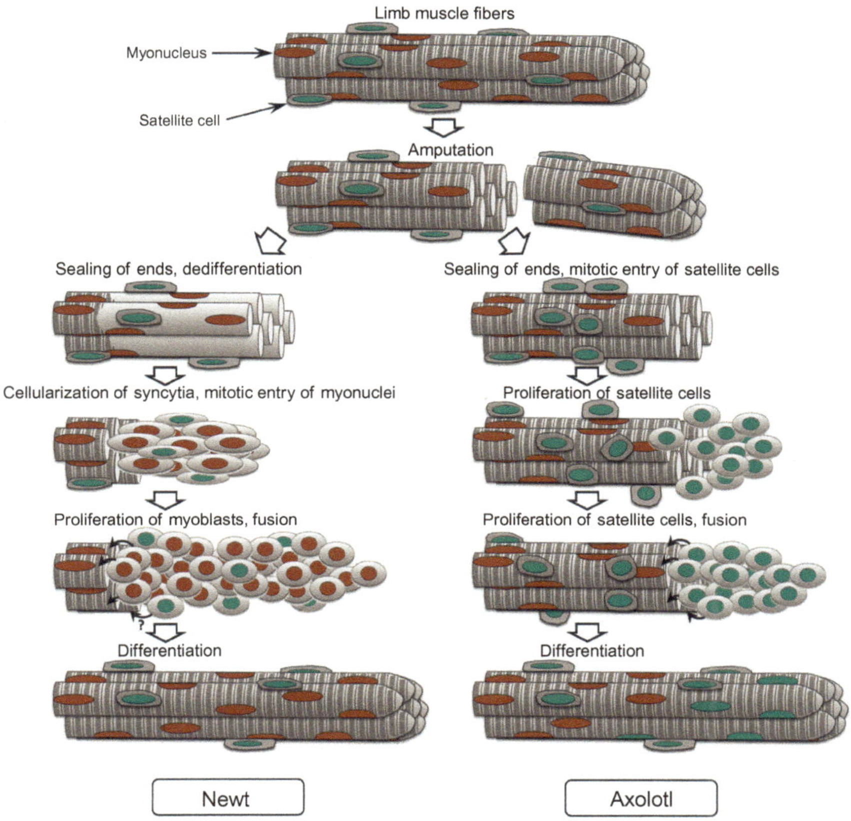

Manfred Frasch, Figure 1 Schematic comparison of limb muscle regeneration via muscle fiber fragmentation versus satellite cell activation. Examples of the distal ends of seven myofibers before, during, and after muscle regeneration are shown from top to bottom. In newt limbs (left), muscle fibers dedifferentiate following amputation and fragment into mononucleate cells, which enter the cell cycle, proliferate, and fuse with the damaged muscle fibers and, potentially, with one another. The contribution of satellite cells to muscle regeneration in this context is not known. In axolotl limbs (and likewise, *Xenopus* tadpole tails; right), amputation activates satellite cells, which proliferate and fuse to regenerate muscles.

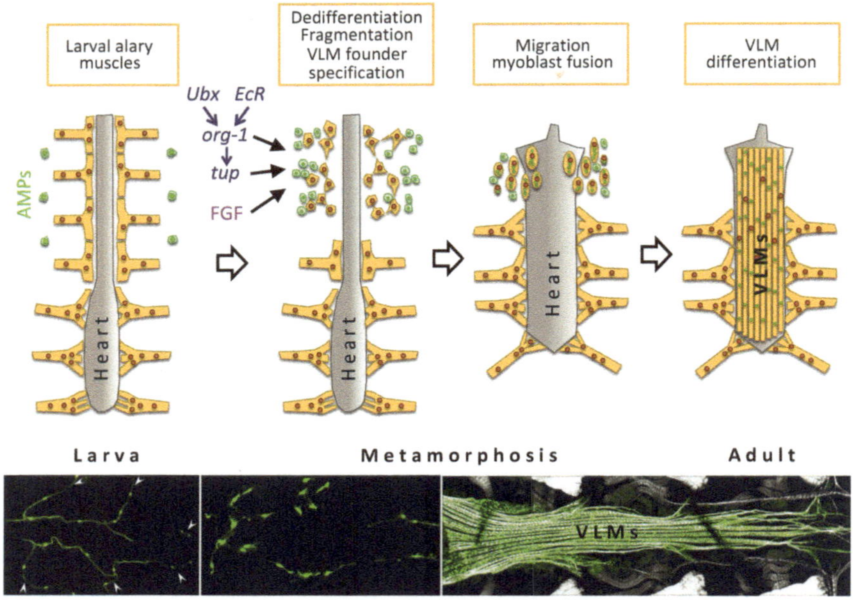

Manfred Frasch, Figure 2 Transdifferentiation of *Drosophila* larval alary muscles into ventral longitudinal muscles of the adult heart. Top: Schematic drawings of the fragmentation of specific larval alary muscles into mononucleate cells, which become specified as founder cells of the ventral longitudinal muscles (VLMs) of the adult heart. Upon myoblast fusion with fusion-competent myoblasts (derived from adult muscle progenitors [AMPs], green), they undergo morphogenesis and differentiate into VLMs. The known genetic inputs are depicted as described in the text. Bottom: Anterior three pairs of larval alary muscles (arrowheads) are shown on the left, fragmented mononucleate cells from these muscles in the center, and fully differentiated VLMs on the right. (All three panels show an *org-1* reporter in green and the right hand panel in addition shows F-actin in white.)

CHAPTER NINETEEN

Epithelial Skin Biology: Three Decades of Developmental Biology, a Hundred Questions Answered and a Thousand New Ones to Address

Elaine Fuchs[1]

Laboratory of Mammalian Development and Cell Biology, Howard Hughes Medical Institute, The Rockefeller University, New York, USA
[1]Corresponding author: e-mail address: fuchslb@rockefeller.edu

Contents

Abstract

The mammalian skin epidermis and its hair and sweat gland appendages provide a protective barrier that retains essential body fluids, guards against invasion by harmful microbes, and regulates body temperature through the ability to sweat. At the interface between the external environment and the body, skin is constantly subjected to physical trauma and must also be primed to repair wounds in response to injury. In adults, the skin maintains epidermal homeostasis, hair regeneration, and wound repair through the use of its stem cells. This essay focuses on when stem cells become established during skin development and where these cells reside in adult epithelial tissues of the skin. I explore how skin stem cells maintain tissue homeostasis and repair wounds and how they regulate the delicate balance between proliferation and differentiation. Finally, I tackle the relation between skin cancer and mutations that perturb the regulation of stem cells.

Current Topics in Developmental Biology, Volume 116
ISSN 0070-2153
http://dx.doi.org/10.1016/bs.ctdb.2015.11.033

The mammalian skin epidermis and its hair and sweat gland appendages provide a protective barrier that retains essential body fluids, guards against invasion by harmful microbes, and regulates body temperature through the ability to sweat. At the interface between the external environment and the body, skin is constantly subjected to physical trauma and must also be primed to repair wounds in response to injury. In adults, the skin maintains epidermal homeostasis, hair regeneration, and wound repair through the use of its stem cells. For over three decades, my laboratory has been interested in the properties of skin stem cells. Where do these cells reside? When do they become established during embryogenesis, and how are they able to build tissues with such remarkably distinct architectures? How do stem cells maintain tissue homeostasis and repair wounds and how do they regulate the delicate balance between proliferation and differentiation? What is the relationship between skin cancer and mutations that perturbs the regulation of stem cells? Here, I will summarize our progress on these fronts, with emphasis on my own laboratory's contributions to the fascinating world of skin stem cells.

1. MAMMALIAN EPIDERMIS, HOMEOSTASIS, AND STEM CELLS

The epidermis is a stratified epithelium whose innermost (basal) layer is attached to an underlying basement membrane rich in extracellular matrix and growth factors. The basal layer is the only layer with proliferative potential. Periodically, these cells will detach from the underlying basement membrane and embark upon a differentiation journey that culminates in the production of dead flattened squames that are then sloughed from the skin surface and continually replaced by inner cells moving outward.

The process of terminal differentiation can be morphologically and molecularly subdivided into four stages. The proliferative basal progenitors express high levels of integrins as well as keratins K14 and K5, which form an extensive filamentous network that provides these cells with mechanical resistance (Fuchs & Green, 1980; Jones, Harper, & Watt, 1995). Spinous suprabasal cells are transcriptionally dynamic, shutting off expression of K14 and K5, and inducing expression of K1 and K10, which form higher ordered cables that add to this mechanical infrastructure (Fuchs & Cleveland, 1998; Fuchs & Green, 1980). These cells also form elaborate desmosomes which reinforce intercellular contacts and provide the spinous-like morphology (Johnson, Najor, & Green, 2014). As spinous cells are pushed upward, they enter the granular layer, forming an elaborate array of filaggrin-rich keratohyalin granules and lamellar granules packed with lipid bilayers. Both spinous and granular cells remain transcriptionally active and make glutamine and lysine-rich proteins, including loricrin and involucrin, which are deposited beneath the plasma membrane. As terminal differentiation nears its endpoint, a calcium influx activates transglutaminases to generate γ-glutamyl-ε-lysine crosslinks that form the cornified envelope. As the organelles including the nucleus are lost, the lamellar granules extrude their lipids to form a lipid-filled sandwich of dead squames. The squames extruded from the surface are indestructible cornified sacs of keratin filaments.

Human epidermis is considerably thicker than mouse epidermis, a feature which most likely explained by the exposed status of the epidermis, and hence the need for an elaborate epidermal skin barrier, when the hair coat is thin. Consistent with this notion is the similarity of human epidermis to early neonatal mouse epidermis, i.e., before the protrusion of the hair from the skin surface. In human epidermis, which typically contains multiple spinous, granular, and stratum corneum layers, it takes ~4 weeks for cells to exit the basal layer and be shed from the skin surface. In adult mouse skin, while turnover still takes place, the thin epidermis typically consists of a barely more than four layers. An exception is the paw skin, which like human palmar and plantar skin, offers the thickest epidermis of the body.

The skin epidermis replenishes itself through a process called tissue homeostasis, in which the number of cell divisions within a tissue compensates for the number of cells lost (Xie & Spradling, 2000). Tissue homeostasis is achieved through stem cells located within the basal layer. The ability of the epidermis to turnover every few weeks places a constant demand on its stem cells. Two distinct models have been proposed to explain the behavior of basal cells. The hierarchical model suggests that the epidermis is composed of discrete proliferative units consisting of a slow-cycling stem cell that gives rise to short-lived, transit-amplifying cells which then depart the basal layer after several divisions to generate upward columnar units of differentiating cells. The stochastic model suggests that basal epidermis is composed of a single type of proliferative progenitor, whose daughter cells choose randomly to differentiate or remain as progenitors.

Lineage tracing involves the genetic marking of one or a group of cells in their normal physiological context in a way that their subsequent progeny retain marker expression. This method has proven to be powerful in evaluating the contribution of stem cells to tissue homeostasis (Xie & Spradling, 2000).

When individually labeled basal cells in tail, ear, or hindpaw epidermis were marked by lineage tracing and their progeny were then monitored long term, the clonal fate data were compatible with the stochastic model (Clayton et al., 2007; Doupe, Klein, Simons, & Jones, 2010; Lim et al., 2013). However, a crucial issue still unresolved is whether the basal cells in these various labeling strategies were marked randomly or selectively. In contrast, a recent study on tailskin employed two inducible Cre-lineage tracers: one was driven by a *Krt14*-promoter, active in all basal cells, and the other by an *Involucrin*-promoter, active only in a discrete subset of K14+ basal cells that precociously express this differentiation-specific gene. In this study, purportedly slow-cycling, basal cells marked by *Krt14-Cre*^{ER} but not *Involucrin-Cre*^{ER} behaved like long-lived stem cells, which gave rise to the *Involucrin-Cre*^{ER} subset that displayed features of more committed basal progenitors (Mascre et al., 2012). This study supports a hierarchical model, and yet shows that progenitors within the basal layer exist and can behave in a stochastic manner. While it is tempting to speculate that the differences in these collective studies arise from the tools used to mark the basal cells, it is still possible that variation in the epidermis at different body sites accounts for the differences. An additional confounding problem for analyzing epidermal homeostasis and lineage tracing in the laboratory mouse is their behavioral traits of scratching and grooming, which can inadvertently affect the outcome. At first glance, it may seem that the elegant live-imaging studies recently performed on mouse ear skin (Pineda et al., 2015; Rompolas, Mesa, & Greco, 2013) will offer a better avenue for tracing clonal behavior. Here, however is the issue of compressing the ear between glass plates in order to capture the image. These mechanical stresses may also affect the outcome of clonal analyses. Single-cell analyses may provide the best current route to a more definitive answer in the future.

While the exact number and characteristics of the epidermal stem cell still remain elusive, their existence and reliance upon underlying extracellular matrix have long been known. Indeed, the first functional demonstration of an epidermal stem cell was made when methods were identified to culture human keratinocytes under conditions where they could be maintained and propagated for hundreds of generations without losing stemness (Rheinwald & Green, 1975, 1977). When grown from an unaffected region of a burn patient, expanded epidermal cultures could be stably engrafted onto the damaged skin (Green, 1991). Some 30 years later, the engrafted epidermis had not developed cancer or other abnormalities, which indicated that, under the right conditions—in this case, coculture with irradiated dermal fibroblasts—*in vitro* stem cell expansion and differentiation can be achieved without deleterious consequence.

The requirement of dermal neighbors for successful culturing of epidermal stem cells highlights the reliance of stem cells upon crosstalk with their niche microenvironment. Indeed, by elucidating key heterologous niche components and/or the crosstalk involved, stem cells from many different epithelia have since been cultured. The key to long-term self-renewal of tissue stem cells *in vitro* is predicated on unraveling the complexities of signaling circuitry governing stem cell behavior. While efforts on the epidermal front are ever improving, we still cannot make an epidermal appendage in a culture dish. Interestingly, the epidermis of the engrafted cultured keratinocytes also never grew hair nor produced sweat. There is still much research to do!

2. MAMMALIAN EPIDERMAL APPENDAGES: HAIR FOLLICLES AND SWEAT GLANDS

Most mammals have a thick hair coat and devote much of their time to taking care of it for warmth and protection. They do so by undergoing cyclical bouts of hair growth (anagen), followed by destruction (catagen) in which the hair stops growing and the lower two thirds of the follicle degenerates, and then rest (telogen) in which not much happens. Such cycles can operate sequentially from a single signaling (dermal papilla), which can activate a single stem cell niche (the bulge) that overlies it (Fig. 1). Each time a hair follicle cycles, a new hair is born and a new stem cell niche is produced at the end of the cycle (Hsu, Pasolli, & Fuchs, 2011). If the old hair is released at the end of each hair cycle, a process known as exogen, only one hair will protrude from each orifice. If, however, the old hair is tightly anchored to its niche, then by the time the hair cycle is launched, the hair coat becomes thicker than it was as a juvenile.

One of the major questions still unanswered is what controls hair anchorage. There are plenty of us in the world who would like to unlock this mystery, not just for scientific purposes! One clue we might be able to glean about this process is hormonal and/or metabolic influence. A musk ox must have a very thick protective coat when it is out on the tundra, but it sheds this coat in the summer. To date, studies of hair shedding en masse have been limited largely to sheep and larger mammals, where genetics and biochemistry are largely impractical. That said, the process of exogen occurs in mice and would be worth exploring in the future.

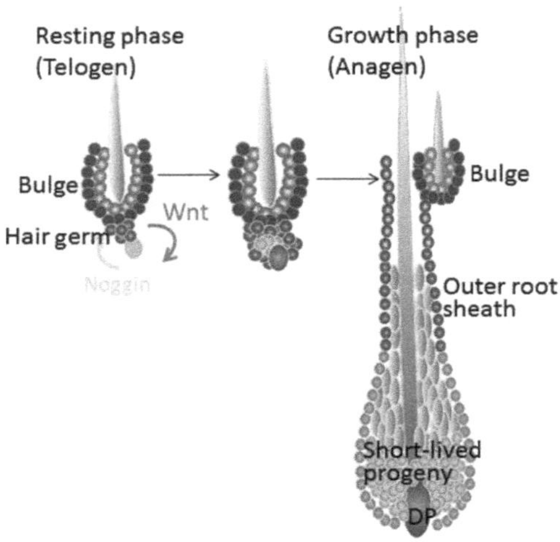

Figure 1 Schematic of the resting and growth phases of the hair follicle. The stem cells (purple) reside in the outer layer of the bulge niche, which anchors the hair from the previous hair cycle. The bulge cells at the base are referred to as the hair germ, which during telogen, is abutted next to the dermal papilla (DP), the mesenchymal stimulus of the hair follicle. Following a buildup of BMP inhibitors (Noggin) in the DP and WNTs in the hair germ, the stem cells at the base are activated and generate the short-lived progeny (orange), which ultimately will differentiate into the hair and its channel. *This figure is patterned after Figure S5 in Hsu, Li, and Fuchs (2014b). Elsevier Press.* (A color version of this figure is located at the end of this chapter.)

When the skin epithelium uses only hair follicles to contribute to thermal regulation, the process is inefficient and most mammals are confined to small climate variations. Horses and camels are among the best examples of animals able to withstand greater temperature extremes. They are able to secrete an oily like fluid through apocrine sweat glands which are an appendage of the hair follicle. This sweating function is critical for their survival and performance, through the ability of apocrine secretions to dissipate heat (McEwan Jenkinson, Elder, & Bovell, 2006; Schmidt-Nielsen, Schmidt-Nielsen, Houpt, & Jarnum, 1957; Schmidt-Nielsen, Schmidt-Nielsen, Jarnum, & Houpt, 1957).

By contrast, higher primates, including humans, have elaborated upon a new type of sweat gland, the eccrine gland, which protrudes directly onto the skin surface rather than from the hair follicle orifice, and secrete a salt–water mixture as opposed to oily sweat. As a result of having a high density of sweat glands over our body, we are able to secrete as much as 3 l of sweat during a marathon, and can regulate body temperature far better than other animals. In humans, as many as 700/cm^2 sweat glands can be found in the skin from the palms and soles.

As the most commonly used laboratory mammal, mouse has eccrine sweat glands exclusively present in the pads of their paws, and its trunk skin lacks sweat glands altogether. Although animals such as this are sensitive to extremes in climate, the sweat glands they do have provide a segue to dissecting the genetics underlying glandular development. In particular, mutant mice lacking either ligand EDA (*Tabby*) or receptor EDAR (*Downless*) exhibit a loss of sweat glands as well as several other curious features, including tooth loss and a loss of the coarsest type of hair, the guard hair, on the mouse's coat (Srivastava et al., 2001). Excessive activation of signaling leads to supernumerary sweat glands and guard hairs (Mustonen et al., 2003). Intriguingly, a natural EDAR variant (V370A) has been found within an East Asian population of humans who have especially coarse hair and an increase in sweat glands (Cluzeau et al., 2012; Fujimoto, Kimura, et al., 2008; Fujimoto, Ohashi, et al., 2008; Kimura et al., 2009; Sabeti et al., 2007). This variant arose more than 30,000 years ago and persists in an ever-expanding population (Kamberov et al., 2013). Thus, the increased sweat gland density confers a selective advantage in the hot and humid climate of this region (Lu & Fuchs, 2014).

Can we simply activate a V370A EDAR variant and humanize the mouse skin? While it would clearly make it easier to unlock the molecular mysteries of sweat glands as well as making a better laboratory model for investigating the biology of skin disorders, the goal is not so simple. Thus, the EDA/EDAR pathway controls the numbers of appendages, but their regional positioning remains largely unaffected by overactivating the pathway. Similarly, many of the distinctive characteristics of different appendages/glands are preserved even under excessive signaling. One exception may be in branching morphology, which occurs in mammary and salivary glands but not in sweat glands or hair follicles. It has been observed that both mammary and salivary glands display defective branching in Tabby mice (Blecher, Debertin, & Murphy, 1983; Voutilainen et al., 2012), and that ductal branching appears to be enhanced on excessive EDA/EDAR signaling (Chang, Jobling, Brennan, & Headon, 2009; Mustonen et al., 2003). It will be interesting in the future to determine whether the "nonbranched" morphologies of sweat glands and hair follicles are achieved by modulating EDA/EDAR signaling during development. Most likely, a complex interplay of signaling pathways, involving EDA/EDAR but not exclusive of them, will be needed before the perfect model in all its elegant regalia will emerge.

3. THE HAIR FOLLICLE AND ITS STEM CELLS

Given the abundance of hair follicles in the laboratory mouse, they have emerged as one of the best systems to explore the key facets of homeostasis and stem cells within an adult tissue. The wonderful aspect about mouse hair follicles is their ability to undergo synchronized cyclical bouts of regeneration and rest. This process is fueled by stem cells located in a region known as the bulge, and in a cluster of cells at the bulge base known as the hair germ. Melanocyte stem cells are intermingled with hair follicle stem cells in the bulge and the hair germ (Chang et al., 2013; Nishimura et al., 2002; Rabbani et al., 2011). When a new hair cycle begins, melanocyte and hair follicle stem cells must synchronize their differentiation programs so that the mature melanocytes that produce melanin will do so at the right time to transfer their melanin pigment to the differentiating hair cells so that they have their distinctive coloring.

The synchronization of the first three hair cycles in the mouse hair coat provided a unique means of studying hair follicle stem cells in their resting, quiescent phase, and comparing them to their anagen phase when they are actively fueling hair follicle regeneration and hair growth. Cells with proliferative potential that spend extended periods in quiescence can thus be marked and monitored by nucleotide analog pulse-chase experiments, which were useful before the advent of lineage tracing with fluorescent markers. Such label-retaining cells reside at the base of the resting telogen follicle, within the outermost layer of the bulge and its associated hair germ (Cotsarelis, Sun, & Lavker, 1990). Label-retaining cells are stem cells, as later demonstrated by using a regulatable fluorescent histone to label the resting bulge cells and then monitor their cell divisions, during normal hair regeneration and following wounding to the skin surface (Tumbar et al., 2004). Similar studies with classical lineage-tracing markers bolstered these findings (Morris et al., 2004).

Both bulge and hair germ share many molecular features of stemness, including expression of *Lgr5* and *Sox9* (Greco et al., 2009; Sennett et al., 2015). However, hair germ cells are always the first to be activated at the start of each new hair cycle, and they undergo more divisions than bulge cells (Greco et al., 2009; Rompolas et al., 2013). Their close proximity to the underlying mesenchymal signaling center, the dermal papillae (DP), impacts the early activation process through activating WNT signals from the germ and inhibitory BMP signals from the DP.

Activated hair germ cells do not maintain stemness *in vitro* (Chen et al., 2012) and *in vivo*, they generate the short-lived progeny that produce the hair and its channel (Greco et al., 2009; Hsu, Li, & Fuchs, 2014a, 2014b). By contrast, once the new hair cycle initiates, some bulge cells leave their niche and form an inverse proliferative gradient along the emerging outer root sheath of the follicle. Early in the hair growth phase, a subset of the short-lived progeny produce a Sonic Hedgehog signal that stimulates the remaining bulge cells to proliferate and replenish the niche (Hsu et al., 2014a, 2014b). SHH also undergoes crosstalk with the DP that fuels the proliferation of the bulb cells (matrix) that will form the seven concentric cones of differentiating cells that will form the central hair shaft, its inner root sheath, and the companion layer that separates it from the outer root sheath.

As the hair follicle grows downward, the SHH-expressing signaling cells and the DP move too far away from the niche and the bulge cells return to quiescence. The outer root sheath cells that are in closest proximity to the bulge soon follow suit, returning to quiescence as well. Soon thereafter, these cells form a new bulge and hair germ for the next cycle (Hsu et al., 2011). Many of these features, first established by staggered pulse-chase and lineage-tracing experiments (Hsu et al., 2011), have been more recently corroborated by elegant live-imaging studies (Rompolas et al., 2013). Moreover, even though bulge normally gives rise to the hair germ, the hair germ can conversely replenish an emptied bulge niche, as shown by laser ablation and live imaging (Rompolas et al., 2013). This not only underscores the close relation between the bulge and hair germ but also the plasticity of the system. Intriguingly, when the DP is ablated the stem cells cannot function. Taken together, these findings suggest a greater importance for the stem cell niche than its residents.

These roots of stem cell plasticity appear to be governed at the level of transcriptional landscaping as my laboratory recently showed (Adam et al., 2015). Several years ago, we showed that hair follicle lineage progression is governed in part by dynamic regulation of Polycomb (PcG)-mediated repression/derepression of chromatin as typified by a trimethylation mark on lysine 27 of histone H3 (H3K27me3) (Ezhkova et al., 2009, 2011). However, hair follicle stem cell identity and function are mainly independent of PcG-regulated genes, indicating that additional epigenetic mechanisms underlie the governance of critical cell identity genes.

Recent studies on cultured embryonic stem cells showed that genes controlling unique cellular identities are driven by so-called super-enhancers, which are marked by acetylation of the H3K27 site over large chromatin regulatory domains bountiful in the pluripotency transcription factor-binding motifs (Whyte et al., 2013). This mechanism enables the factors to bind cooperatively, and since the genes encoding these factors are themselves regulated by super-enhancers, this provides a feed-forward loop that sustains stemness.

The availability of large quantities of hair follicle stem cells allowed us to test this paradigm for the first time in an adult tissue stem cell in its native niche. By conducting super-enhancer analyses, we learned that SOX9, LHX2, TCF3/4, and NFATc1, all bind to small regions (epicenters) within these open chromatin domains of hair follicle bulge stem cells (Adam et al., 2015). All of these factors are genetically essential for maintaining aspects of the bulge stem cells. Some are required to maintain stemness, while others are important in controlling quiescence (Horsley, Aliprantis, Polak, Glimcher, & Fuchs, 2008; Lien & Fuchs, 2014; Lien et al., 2014; Nguyen et al., 2009; Nowak, Polak, Pasolli, & Fuchs, 2008; Rhee, Polak, & Fuchs, 2006; Vidal et al., 2005). These epicenters harbor all of the information necessary for targeting a gene to the stem cell niche, as evidenced by using GFP as a reporter (Fig. 2).

Interestingly, when hair follicle stem cells are removed from their native niche and placed into culture under conditions where engraftment could restore their multilineage capacity (Blanpain, Lowry, Geoghegan, Polak, & Fuchs, 2004), the chromatin landscape changes dramatically, as most stemness transcription factor genes are silenced, and many super-

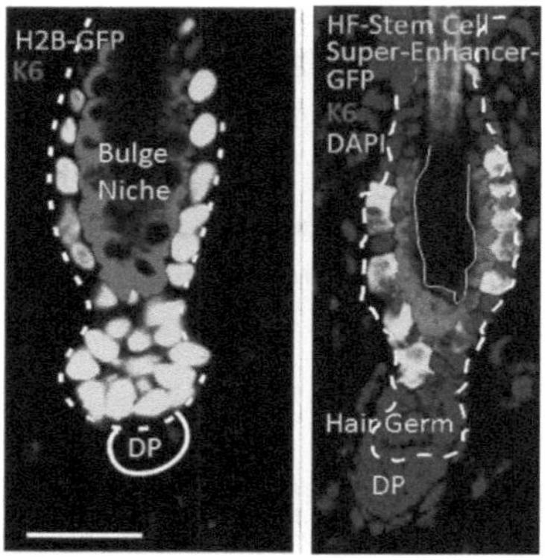

Figure 2 Immunofluorescence images showing a resting phase hair follicle, replete with bulge stem cells and keratin 6+ inner bulge nonstem cells. At left, bulge and hair germ cells are label-retaining and marked by H2B-GFP used in a pulse-chase labeling experiment. At right, bulge is labeled with GFP driven by a HF stem cell-specific super-enhancer epicenter element. *Left panel is Figure 5 Frame C of Hsu et al. (2011). Elsevier Press. Right panel is Figure 2e of Adam et al. (2015). Nature Press.* (A color version of this figure is located at the end of this chapter.)

enhancers are decommissioned (Adam et al., 2015). Intriguingly, new super-enhancers form and regulate a new set of transcription factors that allow these cells to survive in a new environment rich for serum and growth factors. Perhaps not surprisingly, the genes and regulatory factors induced *in vitro* are also similar to those acquired when stem cells exit the bulge and participate in wound repair. Together, these findings further underscore the overpowering importance of the stem cell niche.

4. ADDITIONAL LESSONS LEARNED FROM STUDYING SKIN STEM CELLS: SWEAT GLAND STEM CELLS

Like mammary glands, eccrine sweat glands consist of an outer myoepithelial layer of cells, attached to an underlying basement membrane and an inner layer of luminal cells (Lu & Fuchs, 2014). The sweat gland duct is a single tubular structure that protrudes through the skin surface to secrete sweat and cool the body surface (Fig. 3). A major difference between mammary glands and sweat glands is their activity: whereas mammary glands undergo enormous branching and leafing morphogenesis following lactation, sweat glands remain relatively inert in size and simply crank up sweat output to match the demands on the gland. This has led to speculation as to whether sweat glands have stem cells.

Several years ago, we set out to address this question. Using lineage tracing, we first established the existence of multipotent and unipotent progenitors in the sweat glands (Lu et al., 2012). At the initiation of sweat bud

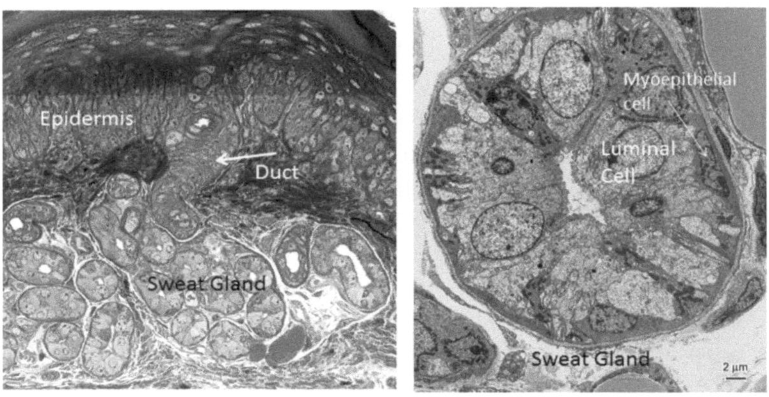

Figure 3 Semi-thin section and ultrastructure of a sweat gland from the mouse paw pad. The gland is a bilayered epithelium consisting of myoepithelial and luminal epithelial cells. The duct opens out onto the surface of the epidermis. *This figure is courtesy of H.A. Pasolli.* (A color version of this figure is located at the end of this chapter.)

downgrowth during development, labeling of K14+ or K5+ basal cells in the overlying epidermis generated first the duct and then the sweat glands. These data confirm that the sweat glands are derived from multipotent epidermal basal progenitors. Labeling of emerging K15+/K18+ luminal cells followed by lineage tracing into mature glands showed that they clonally expand nearly exclusively as luminal cells, indicating that once they form, luminal progenitors are unipotent. Similar lineage tracing of emerging myoepithelial cells showed their unipotency. These findings revealed a switch from multipotency to unipotency in the progenitor properties during sweat gland morphogenesis (Lu et al., 2012).

Interestingly, when we looked for stem cell activity in the adult sweat gland, we did not see signs of proliferation, even though this was readily detected in the basal layer of the overlying epidermis as well as in the basal layer of the ductal epithelium. When we introduced superficial epidermal wounds, the basal progenitors repaired the epidermis, while the ductal epithelial progenitors repaired the orifice of the duct. The gland remained quiescent. Deeper wounds mobilized the gland progenitors: myoepithelial progenitors repaired the myoepithelium while luminal progenitors repaired the damaged lumen (Lu et al., 2012). When myoepithelial progenitors were purified and then engrafted into a mammary gland fat pad, however, now these progenitors could regenerate not only a functional sweat gland but also the duct. These progenitors could also repair the epidermis.

These studies illustrate an important paradigm that we have since seen in many epithelial stem cell populations and which again illuminate our knowledge of plasticity. As long as resident tissue stem cells are present, they function to replenish dying cells and repair local injuries. When the resident tissue stem cells are lost, niches further distant from the wound site are mobilized. With massive damage, a 911 call is placed—any stem cell with the capacity to make and repair the tissue. Alas, for the skin epithelium, the limits appear to be epidermis, sebaceous and sweat glands and hair follicles. However, with a boost of a transcription factor or two, new choices can be coaxed, as illustrated by the recent studies on corneal stem cells (Ouyang et al., 2014).

4.1 Development

During embryogenesis, the single layer of K14+ epidermal basal progenitors undergoes a spindle orientation shift from >90% lateral to ~70% perpendicular cell divisions, which leaves one daughter in the basal layer and one suprabasal differentiating daughter cell (Lechler & Fuchs, 2005) These divisions result in activated Notch signaling in the differentiating daughter cells (Williams, Beronja, Pasolli, & Fuchs, 2011), and involve a mechanism with ancient roots in *Drosophila* neuroblast divisions (Lechler & Fuchs, 2005; Williams, Ratliff, Postiglione, Knoblich, & Fuchs, 2014).

Notably, although all of the epidermal appendages are distinct, they all begin development in a similar way (Mikkola, 2007), and knowledge of this process has been gained by studying the embryonic skin at a stage at which it exists as a single-layered epithelium (Ahtiainen et al., 2014). Before hair follicle morphogenesis, a uniform layer of epithelial cells overlies a disperse population of dermal cells. At about E14.5 of mouse development, mesenchymal–epithelial interactions result in the formation of the first wave of hair placodes that appear as small epidermal invaginations into the dermis (DasGupta & Fuchs, 1999; Hardy, 1992).

Once specified, signals from the epithelium cause dermal cells to aggregate and form a dermal condensate under each placode. This specification process occurs in four overlapping waves, with rare primary guard hairs being specified at E14.5 and the remainder of follicles being specified from E15.5 to P0. Once placodes have formed, they become highly proliferative and grow downward into the dermis, forming hair germs and then hair pegs. Maturation is completed toward the end of the first postnatal week, when follicle downgrowth stops and the upward-moving, terminally differentiated hairs break through the skin surface (Schmidt-Ullrich & Paus, 2005).

The developmental decision to form a hair follicle is the result of mesenchymal–epithelial cross talk that integrates multiple instructive signals necessary to initiate hair follicle morphogenesis. Activation of Wnt/β-catenin signaling in epithelial cells is a key initial step in placode formation. The bipartite transcription factor complex composed of the DNA-binding protein LEF1 and its activation partner (stabilized β-catenin) can be readily detected in the nuclei of developing placode cells, and over a decade ago, the "Wnt reporter" gene *TOPGAL*, containing an enhancer element composed of multimerized LEF1-binding sites, confirmed the specific transcriptional activity of these complexes in the placode (DasGupta & Fuchs, 1999). Notably, when the Wnt inhibitor Dickoff 1 (Dkk1) is expressed ectopically or when β-catenin is conditionally targeted for ablation in epithelial cells, hair follicle morphogenesis is blocked altogether, whereas mice lacking LEF1 or β-catenin display fewer or no follicles (Andl, Reddy, Gaddapara, & Millar, 2002; Huelsken, Vogel, Erdmann, Cotsarelis, & Birchmeier, 2001; van Genderen et al., 1994). Strong evidence for Wnt signaling as a sufficient, instructive cue for placode formation comes from experiments expressing excessive stabilized β-catenin in the epithelium, which result in super-furry mice exhibiting ectopic hair follicles within their interfollicular epidermis (Gat, DasGupta, Degenstein, & Fuchs, 1998; Jamora, DasGupta, Kocieniewski, & Fuchs, 2003). Many studies since then have corroborated these findings in various ways, including chromatin landscaping, lineage-tracing analyses, and genetics (see, for example, Deschene et al., 2014; Genander et al., 2014; Lien & Fuchs, 2014; Lien et al., 2014). That said, surprisingly, despite all of these years of investigation, we still do not fully understand how WNT signaling is involved and how it exerts its instructive powers.

Another major question in hair follicle biology is where and when the adult stem cells become established. Several years ago, my laboratory addressed this question for the bulge niche by exploiting the relatively slow-cycling behavior of adult follicle stem cells to trace their developmental origins (Nowak et al., 2008). Our embryonic pulse-chase studies revealed that, surprisingly, label-retaining cells are specified early in skin development and later become adult bulge stem cells. These early label-retaining cells express SOX9, TCF3/4, and LHX2, which have emerged as key transcriptional regulators of bulge stem cell niche identity. However, our recent studies that show that bulge stem cells can retain stemness outside their niche by mobilizing a new transcriptional network to adapt to a new microenvironment (Adam et al., 2015) now raises an interesting question. Are stem cells formed before their niche or are they formed only once the niche is established? Given the importance of the niche, it is tempting to speculate that the niche must form first. However, it is also formally possible that a microenvironment where stem cells can be born occurs much earlier in skin development. This fascinating question awaits future investigation.

5. SUMMARY

In this synopsis, I have highlighted the findings from my group that I think will be particularly interesting to developmental biologists. I have also touched upon the boundaries of our knowledge, and where pushing these is likely to yield exciting new areas for future exploration. In the three decades of skin epithelial research, we have learned a lot about skin development, homeostasis, and wound repair. However, many of the complex interactions that occur between stem cells and their microenvironment remain unexplored. Our new lentiviral delivery technology that allows us to knockdown expression of any desired gene specifically in the skin epidermis and its appendages (Adam et al., 2015; Beronja, Livshits, Williams, & Fuchs, 2010; Williams et al., 2011) expedites our ability to carry out complex genetics and interrogate the physiological relevance of genes in the skin in a matter of days a great improvement over conventional conditional knockout approaches. We will need these and other new advances in order to address the ever-escalating questions that emerge from exploring this mesmerizing system in all of its dimensions.

I have had the good fortune to mentor a number of outstanding students and postdoctoral fellows, many of whom have continued in this field and have made exceptional contributions to it. I have also had the pleasure to interact with a cohort of wonderful skin biologists and dermatologists, and even though my concentration has been on our own work, their research contributions are enormous and highly insightful. All this aside, however, the frontiers are still vast for the field of skin biology, and the questions will keep us all and those who follow, busy for the decades to come.

REFERENCES

Adam, R. C., Yang, H., Rockowitz, S., Larsen, S. B., Nikolova, M., Oristian, D. S., et al. (2015). Pioneer factors govern super-enhancer dynamics in stem cell plasticity and lineage choice. *Nature, 521*, 366–370.

Ahtiainen, L., Lefebvre, S., Lindfors, P. H., Renvoise, E., Shirokova, V., Vartiainen, M. K., et al. (2014). Directional cell migration, but not proliferation, drives hair placode morphogenesis. *Developmental Cell, 28*, 588–602.

Andl, T., Reddy, S. T., Gaddapara, T., & Millar, S. E. (2002). WNT signals are required for the initiation of hair follicle development. *Developmental Cell, 2*, 643–653.

Beronja, S., Livshits, G., Williams, S., & Fuchs, E. (2010). Rapid functional dissection of genetic networks via tissue-specific transduction and RNAi in mouse embryos. *Nature Medicine, 16*, 821–827.

Blanpain, C., Lowry, W. E., Geoghegan, A., Polak, L., & Fuchs, E. (2004). Self-renewal, multipotency, and the existence of two cell populations within an epithelial stem cell niche. *Cell, 118*, 635–648.

Blecher, S. R., Debertin, M., & Murphy, J. S. (1983). Pleiotropic effect of Tabby gene on epidermal growth factor-containing cells of mouse submandibular gland. *The Anatomical Record, 207*, 25–29.

Chang, S. H., Jobling, S., Brennan, K., & Headon, D. J. (2009). Enhanced Edar signalling has pleiotropic effects on craniofacial and cutaneous glands. *PLoS One, 4*, e7591.

Chang, C. Y., Pasolli, H. A., Giannopoulou, E. G., Guasch, G., Gronostajski, R. M., Elemento, O., et al. (2013). NFIB is a governor of epithelial-melanocyte stem cell behaviour in a shared niche. *Nature, 495*, 98–102.

Chen, T., Heller, E., Beronja, S., Oshimori, N., Stokes, N., & Fuchs, E. (2012). An RNA interference screen uncovers a new molecule in stem cell self-renewal and long-term regeneration. *Nature, 485*, 104–108.

Clayton, E., Doupe, D. P., Klein, A. M., Winton, D. J., Simons, B. D., & Jones, P. H. (2007). A single type of progenitor cell maintains normal epidermis. *Nature, 446*, 185–189.

Cluzeau, C., Hadj-Rabia, S., Bal, E., Clauss, F., Munnich, A., Bodemer, C., et al. (2012). The EDAR370A allele attenuates the severity of hypohidrotic ectodermal dysplasia caused by EDA gene mutation. *The British Journal of Dermatology*, *166*, 678–681.
Cotsarelis, G., Sun, T. T., & Lavker, R. M. (1990). Label-retaining cells reside in the bulge area of pilosebaceous unit: Implications for follicular stem cells, hair cycle, and skin carcinogenesis. *Cell*, *61*, 1329–1337.
DasGupta, R., & Fuchs, E. (1999). Multiple roles for activated LEF/TCF transcription complexes during hair follicle development and differentiation. *Development*, *126*, 4557–4568.
Deschene, E. R., Myung, P., Rompolas, P., Zito, G., Sun, T. Y., Taketo, M. M., et al. (2014). β-Catenin activation regulates tissue growth non-cell autonomously in the hair stem cell niche. *Science*, *343*, 1353–1356.
Doupe, D. P., Klein, A. M., Simons, B. D., & Jones, P. H. (2010). The ordered architecture of murine ear epidermis is maintained by progenitor cells with random fate. *Developmental Cell*, *18*, 317–323.
Ezhkova, E., Lien, W. H., Stokes, N., Pasolli, H. A., Silva, J. M., & Fuchs, E. (2011). EZH1 and EZH2 cogovern histone H3K27 trimethylation and are essential for hair follicle homeostasis and wound repair. *Genes & Development*, *25*, 485–498.
Ezhkova, E., Pasolli, H. A., Parker, J. S., Stokes, N., Su, I. H., Hannon, G., et al. (2009). Ezh2 orchestrates gene expression for the stepwise differentiation of tissue-specific stem cells. *Cell*, *136*, 1122–1135.
Fuchs, E., & Cleveland, D. W. (1998). A structural scaffolding of intermediate filaments in health and disease. *Science*, *279*, 514–519.
Fuchs, E., & Green, H. (1980). Changes in keratin gene expression during terminal differentiation of the keratinocyte. *Cell*, *19*, 1033–1042.
Fujimoto, A., Kimura, R., Ohashi, J., Omi, K., Yuliwulandari, R., Batubara, L., et al. (2008). A scan for genetic determinants of human hair morphology: EDAR is associated with Asian hair thickness. *Human Molecular Genetics*, *17*, 835–843.
Fujimoto, A., Ohashi, J., Nishida, N., Miyagawa, T., Morishita, Y., Tsunoda, T., et al. (2008). A replication study confirmed the EDAR gene to be a major contributor to population differentiation regarding head hair thickness in Asia. *Human Genetics*, *124*, 179–185.
Gat, U., DasGupta, R., Degenstein, L., & Fuchs, E. (1998). De novo hair follicle morphogenesis and hair tumors in mice expressing a truncated beta-catenin in skin. *Cell*, *95*, 605–614.
Genander, M., Cook, P. J., Ramskold, D., Keyes, B. E., Mertz, A. F., Sandberg, R., et al. (2014). BMP signaling and its pSMAD1/5 target genes differentially regulate hair follicle stem cell lineages. *Cell Stem Cell*, *15*, 619–633.
Greco, V., Chen, T., Rendl, M., Schober, M., Pasolli, H. A., Stokes, N., et al. (2009). A two-step mechanism for stem cell activation during hair regeneration. *Cell Stem Cell*, *4*, 155–169.
Green, H. (1991). Cultured cells for the treatment of disease. *Scientific American*, *265*, 96–102.
Hardy, M. H. (1992). The secret life of the hair follicle. *Trends in Genetics*, *8*, 55–61.
Horsley, V., Aliprantis, A. O., Polak, L., Glimcher, L. H., & Fuchs, E. (2008). NFATc1 balances quiescence and proliferation of skin stem cells. *Cell*, *132*, 299–310.
Hsu, Y. C., Li, L., & Fuchs, E. (2014a). Emerging interactions between skin stem cells and their niches. *Nature Medicine*, *20*, 847–856.
Hsu, Y. C., Li, L., & Fuchs, E. (2014b). Transit-amplifying cells orchestrate stem cell activity and tissue regeneration. *Cell*, *157*, 935–949.
Hsu, Y. C., Pasolli, H. A., & Fuchs, E. (2011). Dynamics between stem cells, niche, and progeny in the hair follicle. *Cell*, *144*, 92–105.
Huelsken, J., Vogel, R., Erdmann, B., Cotsarelis, G., & Birchmeier, W. (2001). β-Catenin controls hair follicle morphogenesis and stem cell differentiation in the skin. *Cell*, *105*, 533–545.

Jamora, C., DasGupta, R., Kocieniewski, P., & Fuchs, E. (2003). Links between signal transduction, transcription and adhesion in epithelial bud development. *Nature*, *422*, 317–322.
Johnson, J. L., Najor, N. A., & Green, K. J. (2014). Desmosomes: Regulators of cellular signaling and adhesion in epidermal health and disease. *Cold Spring Harbor Perspectives in Medicine*, *4*, a015297.
Jones, P. H., Harper, S., & Watt, F. M. (1995). Stem cell patterning and fate in human epidermis. *Cell*, *80*, 83–93.
Kamberov, Y. G., Wang, S., Tan, J., Gerbault, P., Wark, A., Tan, L., et al. (2013). Modeling recent human evolution in mice by expression of a selected EDAR variant. *Cell*, *152*, 691–702.
Kimura, R., Yamaguchi, T., Takeda, M., Kondo, O., Toma, T., Haneji, K., et al. (2009). A common variation in EDAR is a genetic determinant of shovel-shaped incisors. *American Journal of Human Genetics*, *85*, 528–535.
Lechler, T., & Fuchs, E. (2005). Asymmetric cell divisions promote stratification and differentiation of mammalian skin. *Nature*, *437*, 275–280.
Lien, W. H., & Fuchs, E. (2014). Wnt some lose some: Transcriptional governance of stem cells by Wnt/beta-catenin signaling. *Genes & Development*, *28*, 1517–1532.
Lien, W. H., Polak, L., Lin, M., Lay, K., Zheng, D., & Fuchs, E. (2014). In vivo transcriptional governance of hair follicle stem cells by canonical Wnt regulators. *Nature Cell Biology*, *16*, 179–190.
Lim, X., Tan, S. H., Koh, W. L., Chau, R. M., Yan, K. S., Kuo, C. J., et al. (2013). Interfollicular epidermal stem cells self-renew via autocrine Wnt signaling. *Science*, *342*, 1226–1230.
Lu, C., & Fuchs, E. (2014). Sweat gland progenitors in development, homeostasis, and wound repair. *Cold Spring Harbor Perspectives in Medicine*, *4*, 1–17.
Lu, C. P., Polak, L., Rocha, A. S., Pasolli, H. A., Chen, S. C., Sharma, N., et al. (2012). Identification of stem cell populations in sweat glands and ducts reveals roles in homeostasis and wound repair. *Cell*, *150*, 136–150.
Mascre, G., Dekoninck, S., Drogat, B., Youssef, K. K., Brohee, S., Sotiropoulou, P. A., et al. (2012). Distinct contribution of stem and progenitor cells to epidermal maintenance. *Nature*, *489*, 257–262.
McEwan Jenkinson, D., Elder, H. Y., & Bovell, D. L. (2006). Equine sweating and anhidrosis Part 1—Equine sweating. *Veterinary Dermatology*, *17*, 361–392.
Mikkola, M. L. (2007). p63 in skin appendage development. *Cell Cycle*, *6*, 285–290.
Morris, R. J., Liu, Y., Marles, L., Yang, Z., Trempus, C., Li, S., et al. (2004). Capturing and profiling adult hair follicle stem cells. *Nature Biotechnology*, *22*, 411–417.
Mustonen, T., Pispa, J., Mikkola, M. L., Pummila, M., Kangas, A. T., Pakkasjarvi, L., et al. (2003). Stimulation of ectodermal organ development by Ectodysplasin-A1. *Developmental Biology*, *259*, 123–136.
Nguyen, H., Merrill, B. J., Polak, L., Nikolova, M., Rendl, M., Shaver, T. M., et al. (2009). Tcf3 and Tcf4 are essential for long-term homeostasis of skin epithelia. *Nature Genetics*, *41*, 1068–1075.
Nishimura, E. K., Jordan, S. A., Oshima, H., Yoshida, H., Osawa, M., Moriyama, M., et al. (2002). Dominant role of the niche in melanocyte stem-cell fate determination. *Nature*, *416*, 854–860.
Nowak, J. A., Polak, L., Pasolli, H. A., & Fuchs, E. (2008). Hair follicle stem cells are specified and function in early skin morphogenesis. *Cell Stem Cell*, *3*, 33–43.
Ouyang, H., Xue, Y., Lin, Y., Zhang, X., Xi, L., Patel, S., et al. (2014). WNT7A and PAX6 define corneal epithelium homeostasis and pathogenesis. *Nature*, *511*, 358–361.
Pineda, C. M., Park, S., Mesa, K. R., Wolfel, M., Gonzalez, D. G., Haberman, A. M., et al. (2015). Intravital imaging of hair follicle regeneration in the mouse. *Nature Protocols*, *10*, 1116–1130.

Rabbani, P., Takeo, M., Chou, W., Myung, P., Bosenberg, M., Chin, L., et al. (2011). Coordinated activation of Wnt in epithelial and melanocyte stem cells initiates pigmented hair regeneration. *Cell*, *145*, 941–955.
Rhee, H., Polak, L., & Fuchs, E. (2006). Lhx2 maintains stem cell character in hair follicles. *Science*, *312*, 1946–1949.
Rheinwald, J. G., & Green, H. (1975). Serial cultivation of strains of human epidermal keratinocytes: The formation of keratinizing colonies from single cells. *Cell*, *6*, 331–343.
Rheinwald, J. G., & Green, H. (1977). Epidermal growth factor and the multiplication of cultured human epidermal keratinocytes. *Nature*, *265*, 421–424.
Rompolas, P., Mesa, K. R., & Greco, V. (2013). Spatial organization within a niche as a determinant of stem-cell fate. *Nature*, *502*, 513–518.
Sabeti, P. C., Varilly, P., Fry, B., Lohmueller, J., Hostetter, E., Cotsapas, C., et al. (2007). Genome-wide detection and characterization of positive selection in human populations. *Nature*, *449*, 913–918.
Schmidt-Nielsen, B., Schmidt-Nielsen, K., Houpt, T. R., & Jarnum, S. A. (1957a). Urea excretion in the camel. *The American Journal of Physiology*, *188*, 477–484.
Schmidt-Nielsen, K., Schmidt-Nielsen, B., Jarnum, S. A., & Houpt, T. R. (1957b). Body temperature of the camel and its relation to water economy. *The American Journal of Physiology*, *188*, 103–112.
Schmidt-Ullrich, R., & Paus, R. (2005). Molecular principles of hair follicle induction and morphogenesis. *Bioessays*, *27*, 247–261.
Sennett, R., Wang, Z., Rezza, A., Grisanti, L., Roitershtein, N., Sicchio, C., et al. (2015). An integrated transcriptome atlas of embryonic hair follicle progenitors, their niche, and the developing skin. *Developmental Cell*, *34*, S1534–S5807.
Srivastava, A. K., Durmowicz, M. C., Hartung, A. J., Hudson, J., Ouzts, L. V., Donovan, D. M., et al. (2001). Ectodysplasin-A1 is sufficient to rescue both hair growth and sweat glands in Tabby mice. *Human Molecular Genetics*, *10*, 2973–2981.
Tumbar, T., Guasch, G., Greco, V., Blanpain, C., Lowry, W. E., Rendl, M., et al. (2004). Defining the epithelial stem cell niche in skin. *Science*, *303*, 359–363.
van Genderen, C., Okamura, R. M., Farinas, I., Quo, R. G., Parslow, T. G., Bruhn, L., et al. (1994). Development of several organs that require inductive epithelial-mesenchymal interactions is impaired in LEF-1-deficient mice. *Genes & Development*, *8*, 2691–2703.
Vidal, V. P., Chaboissier, M. C., Lutzkendorf, S., Cotsarelis, G., Mill, P., Hui, C. C., et al. (2005). Sox9 is essential for outer root sheath differentiation and the formation of the hair stem cell compartment. *Current Biology*, *15*, 1340–1351.
Voutilainen, M., Lindfors, P. H., Lefebvre, S., Ahtiainen, L., Fliniaux, I., Rysti, E., et al. (2012). Ectodysplasin regulates hormone-independent mammary ductal morphogenesis via NF-kappaB. *Proceedings of the National Academy of Sciences of the United States of America*, *109*, 5744–5749.
Whyte, W. A., Orlando, D. A., Hnisz, D., Abraham, B. J., Lin, C. Y., Kagey, M. H., et al. (2013). Master transcription factors and mediator establish super-enhancers at key cell identity genes. *Cell*, *153*, 307–319.
Williams, S. E., Beronja, S., Pasolli, H. A., & Fuchs, E. (2011). Asymmetric cell divisions promote Notch-dependent epidermal differentiation. *Nature*, *470*, 353–358.
Williams, S. E., Ratliff, L. A., Postiglione, M. P., Knoblich, J. A., & Fuchs, E. (2014). Par3-mInsc and Gαi3 cooperate to promote oriented epidermal cell divisions through LGN. *Nature Cell Biology*, *16*, 758–769.
Xie, T., & Spradling, A. C. (2000). A niche maintaining germ line stem cells in the Drosophila ovary. *Science*, *290*, 328–330.

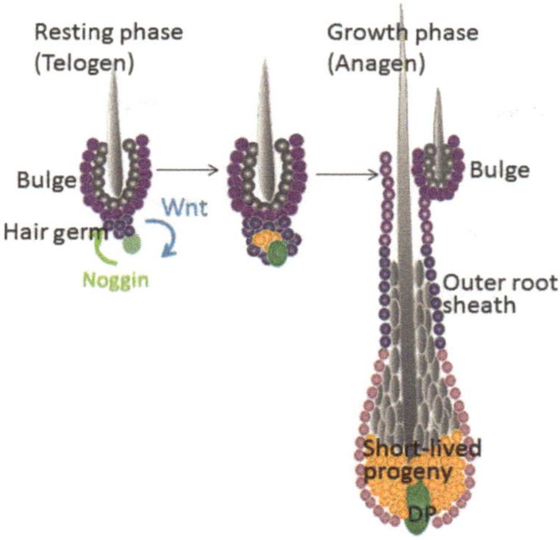

Elaine Fuchs, Figure 1 Schematic of the resting and growth phases of the hair follicle. The stem cells (purple) reside in the outer layer of the bulge niche, which anchors the hair from the previous hair cycle. The bulge cells at the base are referred to as the hair germ, which during telogen, is abutted next to the dermal papilla (DP), the mesenchymal stimulus of the hair follicle. Following a buildup of BMP inhibitors (Noggin) in the DP and WNTs in the hair germ, the stem cells at the base are activated and generate the short-lived progeny (orange), which ultimately will differentiate into the hair and its channel. *This figure is patterned after Figure S5 in Hsu, Li, and Fuchs (2014b). Elsevier Press.*

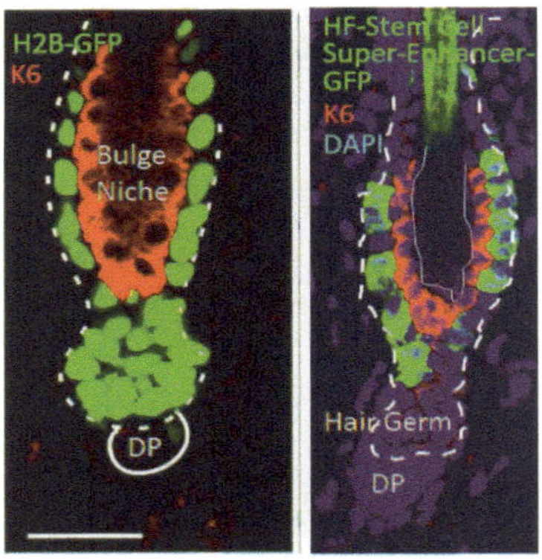

Elaine Fuchs, Figure 2 Immunofluorescence images showing a resting phase hair follicle, replete with bulge stem cells and keratin 6+ inner bulge nonstem cells. At left, bulge and hair germ cells are label-retaining and marked by H2B-GFP used in a pulse-chase labeling experiment. At right, bulge is labeled with GFP driven by a HF stem cell-specific super-enhancer epicenter element. *Left panel is Figure 5 Frame C of Hsu et al. (2011). Elsevier Press. Right panel is Figure 2e of Adam et al. (2015). Nature Press.*

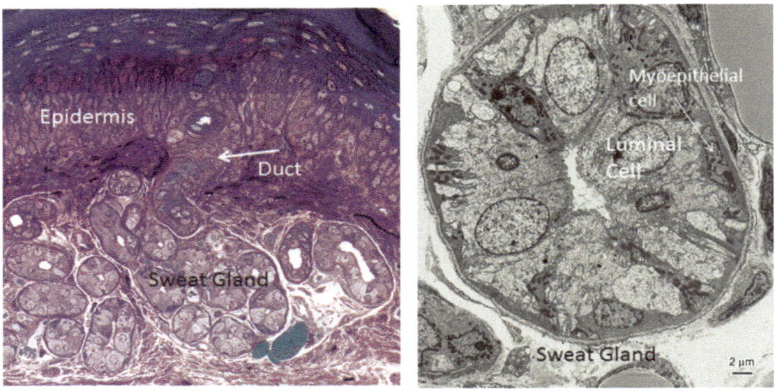

Elaine Fuchs, Figure 3 Semi-thin section and ultrastructure of a sweat gland from the mouse paw pad. The gland is a bilayered epithelium consisting of myoepithelial and luminal epithelial cells. The duct opens out onto the surface of the epidermis. *This figure is courtesy of H.A. Pasolli.*

CHAPTER TWENTY

Differentiation in Stem Cell Lineages and in Life: Explorations in the Male Germ Line Stem Cell Lineage

Margaret T. Fuller[1]
Departments of Developmental Biology and Genetics, Stanford University School of Medicine, Stanford, California, USA
[1]Corresponding author: e-mail address: mtfuller@stanford.edu

Contents

Abstract

I have been privileged to work on cellular differentiation during a great surge of discovery that has revealed the molecular mechanisms and genetic regulatory circuitry that control embryonic development and adult tissue maintenance and repair. Studying the regulation of proliferation and differentiation in the male germ line stem cell lineage has allowed us investigate how the developmental program imposes layers of additional controls on fundamental cellular processes like cell cycle progression and gene expression to give rise to the huge variety of specialized cell types in our bodies. We are beginning to understand how local signals from somatic support cells specify self-renewal versus differentiation in the stem cell niche at the apical tip of the testis. We are discovering the molecular events that block cell proliferation and initiate

Current Topics in Developmental Biology, Volume 116
ISSN 0070-2153
http://dx.doi.org/10.1016/bs.ctdb.2015.11.041

terminal differentiation at the switch from mitosis to meiosis—a signature event of the germ cell program. Our work is beginning to reveal how the developmental program that sets up the dramatic new cell type-specific transcription program that prepares germ cells for meiotic division and spermatid differentiation is turned on when cells become spermatocytes. I have had the privilege of working with incredible students, postdocs, and colleagues who have discovered, brainstormed, challenged, and refined our science and our ideas of how developmental pathways and cellular mechanisms work together to drive differentiation.

1. DIFFERENTIATION IN STEM CELL LINEAGES AND IN LIFE

We scientists have one of the greatest privileges: work that grabs us, obsesses us, fascinates us, and drives us. Every discovery opens more questions. Each step forward opens a new vista revealing the amazing beauty of our natural world. Not views of mountains, stars, and sea, which indeed stir our hearts, but, equally amazing, in the case of developmental biologists, insights into the tiny machines that build the stuff of life and the intricate regulatory circuitry that controls how they work and how they themselves are assembled. How is all this encoded in the genome so that an embryo emerges from a single (albeit very special) cell, the fertilized egg? How does the developmental program ensure a repeat performance every generation by producing the sperm and eggs that will combine to start the process over again?

Starting my independent research career in the 1980s, like many of the authors in this volume, I had the privilege of being a part of the flowering of knowledge that has led to our current understanding of the molecular and genetic mechanisms that underlie developmental biology. I have been able to participate and teach as the field uncovered the sheer beauty and elegance of the regulatory circuitry that guides how living animals develop. Over the past few decades, in the span of my life as a research scientist, the field of Developmental Biology has learned how embryos are built. We have learned how embryonic pattern is set up, in part from spatial information down in the egg and in part generated by local cues and from self-reinforcing positive or negative feedback and feed-forward circuits. We have discovered how cells talk to one another, whispering at close range, and how developmental history controls how the same small set of signals are interpreted differently in different cells and at different times in development.

I think one of the most amazing things about our field is that we never run out of questions. Even though so much has been learned in the last 35 years, there are still many, many compellingly interesting questions. Indeed, the more we know, the more questions arise. But *how* does it work? Each PhD thesis and each postdoctoral research achievement opens up 10 times more new questions. This ability to burrow deeper and deeper into the richness of biological mechanism is constantly enhanced by the new technologies coming on line: genomics, chromatin immunoprecipitation, advanced microscopy, to name a few. One of the challenges of Developmental Biology has always been that the most interesting events—the processes that you really want to understand—usually take place in a very small number of cells in an early precursor, and so are hard to study. Now microfluidics, single-cell analysis, the ability to express and track fluorescent proteins in living cells, and the ability to make populations of cells differentiate along their normal developmental pathway in synchrony are providing the power to investigate these mysteries. Each new tool opens the opportunity to tackle classic but previously unsolvable questions. This is truly an exciting time to be a Developmental Biologist.

I have had the privilege of working with incredible students and postdoctoral research associates and colleagues who have discovered, brainstormed, challenged, and refined our science and our ideas of how developmental pathways and cellular mechanisms work together to drive differentiation. These are the people who made the discoveries. Interacting with them is what makes a career in science so much fun. One of the consequences of our system of training is that we establish deep partnerships, working intensively with talented students and postdocs who become colleagues over 4–6 years, then they move on. This is a wrenching and bittersweet moment—yet we are so proud to foster their careers and follow what they will do next. The flip side is that we scientists develop a community of colleagues, former trainees, and friends all over the world. Often, former students and postdocs become collaborators, sharing reagents and results, thinking through models, writing joint papers, and moving the field forward together even after they have gone on to their independent careers.

1.1 Differentiation of a Developmental Biologist

Many mentors helped me on my path. I credit my father, who used some mason jars, plastic tubing, and the yellow kitchen step stool to teach me how to set up a siphon, with piquing my interest in science. Entering college as a Physics major, I took a job sorting pipettes and making media in a biology lab to earn book and spending money. The first afternoon on the job, I broke a 2-liter flask of hot agar, which immediately gelled all over the lab floor and cabinets. With an introductory biology course under my belt, the summer following my sophomore year I worked doing independent research in a biology lab. During that summer I realized that, in laboratory biology, one could discover something new about the world using just one's mind, some test tubes, and a water bath. I was hooked.

Entering the PhD program in the Biology Department at MIT in 1974, I was immersed in a group of talented fellow graduate students who became not only friends but also coexplorers and colleagues. My PhD advisor Jonathan King taught me how to frame a question and how to give a talk. At Indiana University, my postdoctoral mentors Elizabeth Raff and Thomas Kaufman introduced me to the cell and developmental biology of *Drosophila* and to spermatogenesis as a model cell differentiation system. Another mentor, Tony Mahowald, ran a night school for postdocs to teach us newbies about *Drosophila* development, and Tom Blumenthal gave sage advice to a young postdoc going out on the job market. William Wood gave me my first job (and so many songs) and taught me how to teach, and Dick MacIntosh opened his lab supply closet, equipment, and feast of cell biological questions to me as I first started up own lab. I thank Lucy Shapiro for believing in me, fostering my career as a member of the Searle committee, and giving me the opportunity to join the newly founded Department of Developmental Biology at Stanford in 1990. We pay back our teachers and mentors by teaching, learning from, and mentoring our students.

I love genetics as an approach to research. The awesome power of genetics (APoG) was drilled into me as a graduate student in microbiology at MIT. When thinking about what I wanted to do as a postdoc, I decided that I wanted to apply that APoG to cell biology to learn how proteins build machines to do work inside eukaryotic cells. Moving to Indiana University at Bloomington in 1980 to work on the genetics of the microtubule cytoskeleton as a postdoc, I was plunged into a rich tradition of Developmental Biology and Developmental Genetics, and this set my course for my research career. I love genetics because I like surprises. With a genetic approach, making mutations at random and scoping out what mutants are defective in the process you are interested in, you are not limited by your preconceived notions of how things work. Whether you like it or not, you are dragged kicking and screaming into learning about translational control, or differential splicing, or cell type-specific general transcription machinery, or membrane trafficking, because the mutants you made told you that these processes are important for the biology that you are studying. Phenotype trumps all. What you absolutely need is a good biological system rich with interesting but as yet not understood processes and with an easy and powerful way to score mutant phenotypes. The more deeply you observe and can score disparate events in wild type, the more powerful your genetic screen. This is part of Barbara McClintock's "feeling for the organism." The biggest thrill for me is when new data, say from discovering the molecular identity of a mutant gene that forces you to think about a new mechanism, lead to a model that ties together a series of previously unconnected seeming observations to explain how a process might work.

In 1983, Matthew Scott and I moved to Boulder, Colorado, to set up our own independent labs as assistant professors. I must say that the 10 major things I had to do in the first 2 years as an assistant professor were all things that I had not done before and had never been trained for: setting up a lab, hiring, firing, writing grants, teaching, motivating and guiding graduate student projects, dealing with undergrads crying in my office ... Boulder was great. Generous colleagues mentored us there, and the cell biology community was inspiring. We were joined at Boulder by a fantastic group of postdoctoral fellows and students, and through their efforts our two new labs got on their feet. We also learned to telemark ski and river kayak and gloried in the mountain trails that started five short blocks from our house.

1.2 Teaching Inspires Research: Research Informs Teaching

I was assigned to teach the senior capstone course of the MCDB major, a semester-long course in Developmental Biology, a topic in which I myself had never had a formal course. Luckily, my teaching partner was Bill Wood. Throughout my career, I have found that some of my most rewarding professional interactions have been with the people with whom I teach, as we grapple with identifying what are the most important ideas to impart. I learned so much from teaching, building on the conceptual framework laid out in the original edition of Slack's *Egg to Embryo*. The exceptional series of breakthrough discoveries that led to the current understanding of the genetic control of embryonic pattern formation in *Drosophila* and *Caenorhabditis elegans* was going on at this time, and I remember coming home from scientific meetings and synthesizing the latest unpublished work to explain in class.

Teaching Developmental Biology had profound effects on my own research. Assigned to teach the lectures on Bone and Blood, I learned about the importance of adult stem cells in tissue renewal and the unanswered questions about how adult stem cell self-renewal is regulated. This got me thinking about how the idea of the stem cell niche could be demonstrated experimentally: to test the hypothesis that adult stem cells were maintained by a niche, one had to identify and study stem cells *in vivo*, in the context of their normal anatomy. I realized that my own experimental system, spermatogenesis in *Drosophila*, offered a prime opportunity to investigate the regulatory mechanisms that control stem cell self-renewal and the proliferation and differentiation of stem cell progeny. I also realized that spermatogenesis was an excellent system in which to investigate the question that had originally interested me as an undergraduate in the lab: how do cells change shape? With these goals in mind, I gradually converted my research from probing the assembly and function of the cytoskeleton, where male germ cells undergoing meiotic division were just the cell type in which the studies were being conducted, toward the process of spermatogenesis itself, with the goal of understanding the regulatory circuitry that controls differentiation in adult stem cell lineages. We began screening male-sterile mutants using a rapid yet powerful phase-contrast light microscopy assay, focusing first on "meiotic arrest" mutants, which were likely to affect a key regulatory point as, unlike most classic male-sterile mutants, they blocked progression of spermatogenesis, in this case at the G2/M transition of meiosis I. In addition, with our move to the Department of Developmental Biology at Stanford in 1990, we initiated the "tiny testis" genetic screen to identify genes required for male germ line stem cell (GSC) maintenance or the differentiation of spermatogonia into spermatocytes.

2. DIFFERENTIATION IN THE MALE GERM LINE ADULT STEM CELL LINEAGE

I think that many of the most interesting questions lie at the interface of Cell and Developmental Biology. How does the developmental program impose layer upon layer of additional controls on fundamental cellular processes like assembly and disassembly of the cytoskeleton, cell cycle progression, and gene expression to give rise to the huge variety of specialized cell types in our bodies? The germ line is a case in point. The generation of sperm and eggs, both highly specialized terminally differentiated cell types, is both an end and a beginning, bringing life full circle to prepare for the next generation. The germ line does all the things that other cells do—genome replication and repair, transcription, translation, cell division, cellular morphogenesis, etc.—but always with a special, cell type-specific twist.

My lab studies spermatogenesis to understand how specialized cells differentiate from precursors in adult stem cell lineages. Working at several key points in the process, we are beginning to understand how local signals from somatic support cells specify self-renewal versus differentiation in the stem cell niche at the apical tip of the testis. We are investigating how transit-amplifying (TA) spermatogonia limit the number of mitotic divisions and the molecular events that block cell proliferation and initiate terminal differentiation at the switch from mitosis to meiosis. We have learned much about how the dramatic new cell type-specific transcription program that prepares germ cells for meiotic division and spermatid differentiation is turned on when cells become spermatocytes.

My ambition is to someday be able to draw out the wiring diagram of how male gametes differentiate from stem cells, to the TA spermatogonia, to spermatocytes and the meiotic divisions. I believe that many of the fundamental strategies we discover in *Drosophila* will provide paradigms for thinking about how these processes are regulated in mammals. For example, by studying *Drosophila* male-sterile mutants that fail to undergo the meiotic divisions, we hope to identify and understand genes and molecular pathways that may underlie one of the common forms of human male infertility: meiosis I maturation arrest.

As summarized below under headings highlighting topics of investigation in my lab over the past 25 years, so far this quest has led us into work on cell type-specific transcriptional regulators, translational control, reciprocal cell–cell communication, membrane trafficking, and regulated mitochondrial fusion, to name just some of the molecular mechanisms. The story is not over yet. Many fascinating questions remain and we have much work to do.

2.1 Regulation of Adult Stem Cell Self-Renewal and Differentiation by the Niche

My lab pioneered study of male GSCs starting in 1992. Our discoveries in the *Drosophila* system, along with parallel efforts in the DiNardo and Matunis and other laboratories, have set paradigms for research on mammalian adult stem cells *in vivo*. Based on Bob Hardy's careful ultrastructural studies of the testis apical tip (Hardy, Tokuyasu, Lindsley, & Garavito, 1979), where male GSCs lie in a rosette attached to a plug of normally postmitotic somatic hub cells and flanked by stem cells for the somatic cyst cell lineage (CySC), graduate student Amy Kiger surmised that cell–cell communication might be involved and began to systematically test mutations in known close-range signaling pathways for defects in male GSC self-renewal or differentiation. Shifting males bearing a temperature-sensitive allele of the EGF Receptor (EGFR) to nonpermissive temperature revealed that activation of the EGFR signal transduction pathway in somatic cyst cells was required for male germ cells to properly differentiate into TA spermatogonia, as well as for spermatogonia to become spermatocytes (Kiger, White-Cooper, & Fuller, 2000). Analyzing the *stet* mutant from our "tiny testes" screen, postdoc Cordula Shulz showed that the EGF signal arises from the germ line and that activation of the EGFR signal transduction pathway in cyst cells is required for them to enclose the gonialblast, the daughter of each stem cell division that is displaced away from the apical hub and initiates the TA program as the first step toward differentiation (Sarkar et al., 2007; Schulz, Wood, Jones, Tazuke, & Fuller, 2002). Amy Kiger and postdoc Leanne Jones, also discovered a key role of JAK-STAT signaling, activated in stem cells by a cytokine-like signal from the somatic cells of the apical hub (Kiger, Jones, Schulz, Rogers, & Fuller, 2001). We now know from beautiful work by Leatherman and Dinardo (2008, 2010) that activation of STAT in CySCs is required to maintain stem cell identity in the somatic cyst cell lineage, and that this in turn allows GSCs to maintain the stem cell state. A major direct response of GSCs to the cytokine-like signal from the hub is maintenance of attachment of GSCs to the hub via adherens junctions (Shields, Spence, Yamashita, Davies, & Fuller, 2014; Yamashita, Jones, & Fuller, 2003). Function of the receptor tyrosine phosphatase *lar* is required cell autonomously in male GSCs for maintenance of stem cell identity, likely through maintenance of attachment to the hub, keeping the GSCs in their privileged microenvironment (Srinivasan, Mahowald, & Fuller, 2012). Genetic screens and the tools of cell type-specific RNAi are revealing that, in addition to signals from the local microenvironment, cell intrinsic factors that affect chromatin state play important roles in protecting male GSCs from loss to differentiation (Davies, Lim, Joo, Tam, & Fuller, 2013; Morillo Prado, Srinivasan, & Fuller, 2013).

Male GSCs attached to the hub normally undergo mitosis with an oriented mitotic spindle. As a result, one of the two daughter cells inherits the attachment to the hub and remains in the shelter of the flanking somatic cyst stem cells, while the other daughter is displaced away and initiates differentiation. Postdoc Yukiko Yamashita developed a novel pulse-chase assay to mark newly assembled centrioles and used it to show that in male GSCs, the oldest centrosome remains anchored next to the adherens junctional complex with the niche throughout interphase, setting up the stereotyped spindle orientation (Yamashita, Mahowald, Perlin, & Fuller, 2007) that ensures that the two daughters of each stem cell division are normally born into different environments, specifying the normally asymmetric outcome of each male GSC division. Graduate student Jaclyn Lim showed that without CySCs or the somatic cyst cells they generate, male GSCs continue to proliferate in a stem cell or gonialblast-like state (Lim & Fuller, 2012), suggesting that somatic cyst cells play a key role in permitting or instructing the germ cells they enclose to differentiate.

2.2 Regulation of the Switch from Mitosis to Meiosis by Translational Control

The switch from the TA cell proliferation program to onset of terminal differentiation is a key regulatory point in adult stem cell lineages, setting the number of differentiating progeny produced from one round of stem cell division. Failure to limit TA divisions and make the switch can lead to cancer: indeed the cells of origin for some leukemias and glioblastomas have been shown to be the TA intermediates. Using *Drosophila* spermatogenesis as a model, graduate student Megan Insco showed that the number of rounds of spermatogonial TA divisions is set by accumulation of a key regulator, *bag-of-marbles (bam)* to a critical threshold (Insco, Leon, Tam, McKearin, & Fuller, 2009). Megan, working with fellow graduate students Alexis Bailey and Jongmin Kim and postdoc Gonzalo Olivares, showed that Bam and its binding partner, Bgcn, required for spermatogonia to stop division and enter the spermatocyte program in *Drosophila*, act as translational repressors, directly binding to the 3′ UTR of a target mRNA (Insco et al., 2012). These results, together with Judith Kimble and colleagues' findings that a network of translational control regulates the switch from mitosis to meiosis in *C. elegans* (Kimble, 2011), point toward a deeply conserved mechanism involving translational control underlying this signature event in germ cell differentiation.

2.3 Regulation of the Transcription Program for Germ Cell Differentiation by Spermatocyte-Specific Transcriptional Machinery

My laboratory led discovery of the mechanisms that turn on the transcription program for terminal differentiation once germ cells stop mitotic division and become spermatocytes. Starting from male-sterile mutants that cause meiosis I maturation arrest, graduate student Ting Yi Lin and postdocs Mark Hiller and Xin Chen discovered that five tissue-specific homologs of components of the "general" transcription factor $TF_{II}D$ expressed only in spermatocytes (the tTAFs) are required to upregulate transcription in spermatocytes of hundreds of genes for the spermatid differentiation program (Chen, Lu, Prado, Eun, & Fuller, 2011; Hiller et al., 2004; Hiller, Lin, Wood, & Fuller, 2001; Lin et al., 1996). Postdocs Helen White-Cooper and Lucia Perezgasga showed that other "meiotic arrest" genes encoded testis-specific components of the tMAC complex, a spermatocyte-specific alternate form of the Mip/dREAM wide-spread transcriptional repressor, that are required for tMAC to activate transcription of the differentiation genes in spermatocytes and recruit the tTAFs to promoters (Chen et al., 2011; Perezgasga et al., 2004; White-Cooper, Leroy, MacQueen, & Fuller, 2000). Working in her own lab, Helen has gone on to identify genetically and investigate many additional components of tMAC (Ayyar, Jiang, Collu, White-Cooper, & White, 2003; Doggett, Jiang, Aleti, & White-Cooper, 2011; Jiang, Benson, Bausek, Doggett, & White-Cooper, 2007; Jiang & White-Cooper, 2003; Perezgasga et al., 2004), with the complex being identified biochemically in the laboratory of Michael Botchen at Berkeley (Beall et al., 2007). Together our results reveal a striking picture in which cell type-specific homologs expressed only in spermatocytes of general transcriptional factor and chromatin regulatory machinery complex components act to turn on transcription of over 1000 genes to set up the terminal differentiation program for male gametogenesis, forming the largest cluster of coregulated genes in the *Drosophila* developmental program (Arbeitman et al., 2002). Recent data from postdoc Chenggang Lu indicate that the testis-specific tMAC complex machinery recruits the transcriptional coactivator Mediator, which in turn helps recruit the testis-specific TAFs (Lu & Fuller, 2015).

2.4 Regulation of Cell Cycle Progression During Male Meiosis

Postdoctoral research associate Catherine Baker discovered that some of the earliest transcripts upregulated when spermatogonia become spermatocytes encode male germ cell-specific translational regulators that control the specialized cell cycle of meiosis in the male. Two of these early spermatocyte transcripts encode translational repressors that prevent premature expression of Cyclin B1 during the extended G2 phase of meiotic prophase (Baker,

Gim, & Fuller, 2015). In addition, we discovered that a spermatocyte-specific alternative form of the translation initiation complex core component eIF4G is required for translation of core cell cycle regulators such as Cyclin B1 in maturing spermatocytes (Baker & Fuller, 2007). Together, our findings are beginning to reveal how the germ cell developmental program imposes layers of control on the core cell cycle machinery to set up the specialized cell cycle of meiosis. Analyzing male-sterile mutants that failed to undergo the meiotic divisions, postdoc Chenggang Lu found that loss of function of the major polyubiquitin locus causes meiosis I maturation arrest infertility (Lu, Kim, & Fuller, 2013), a phenotype conserved from yeast sporulation to mouse spermatogenesis.

2.5 Regulation of Cytokinesis During Male Meiosis

With our long-time collaborators in Rome, Maria Grazia Giansanti and Maurizio Gatti, we screened for male-sterile mutants with normal chromosome segregation but defects in cytokinesis during male meiosis. Molecular identification of the genes affected in this treasure trove of mutants has revealed many molecular pathways required for the setup and constriction of the contractile ring. Cytokinesis in male germ cells is incomplete, and postdocs Gary Hime and Julie Brill initiated our work on cell division with a careful analysis of how structural components of the stalled contractile ring are converted into a permanent ring canal connecting mitotic and meiotic sisters as male germ cells undergo the TA spermatogonial divisions, then differentiate into spermatocytes, and then spermatids (Hime, Brill, & Fuller, 1996). Analyzing mutants where meiotic chromosomes segregate but cytokinesis fails in both the first and second meiotic divisions, producing a giant onion stage spermatid with one large mitochondrial derivative and four nuclei rather than the normal four haploid spermatids, Julie Brill discovered the role of the phospholipid kinase *four-wheel drive (fwd)* in cytokinesis during male meiosis (Brill, Hime, Scharer-Schuksz, & Fuller, 2000). Julie went on to found her independent laboratory working on the role of the phosphatidylinositol phosphate (PIP) class of membrane lipids and their regulatory kinases and phosphatases in controlling cellular morphogenesis during development.

Graduate student Rebecca Farkas discovered the important role of membrane trafficking for cytokinesis during male meiosis through her work on the Cog5 homolog *four-way stop (fws)* (Farkas, Giansanti, Gatti, & Fuller, 2003). Further investigation by postdocs Todd Blankenship, Carmen

Robinett, and Maria Grazia Giansanti in my lab then in their own independent groups fleshed out the picture that many of the genes required for cytokinesis during male meiosis encode components of the vesicle trafficking pathway, revealing the key role of membrane addition at the cleavage furrow in spermatocytes undergoing cell division (Belloni et al., 2012; Giansanti et al., 2006, 2004, 2015; Robinett, Giansanti, Gatti, & Fuller, 2009; Sechi et al., 2014; Szafer-Glusman et al., 2008). Postdoc Edith Safer-Glusman, working with Maria Grazia, exploited a missense mutation to uncover the role of the chromosomal passenger protein and tumor suppressor Survivin in triggering when and where the contractile ring assembles (Szafer-Glusman, Fuller, & Giansanti, 2011).

2.6 Discovery of the First Known Protein Mediator of Mitochondrial Fusion

Graduate student Karen Hales carried out a genetic screen for male-sterile *Drosophila* mutants that cause defects in formation of the specialized mitochondrial structure in elongating spermatids. In an illustration of the power of genetic analysis in developmental systems in which cells carry out fundamental processes in an exaggerated way, her work led to discovery of Mitofusins, the first known protein mediators of mitochondrial fusion (Hales & Fuller, 1997). These transmembrane GTPase proteins in the mitochondrial outer membrane have since been shown to drive mitochondrial fusion in many cell types and from yeast to man (Hermann et al., 1998; Santel & Fuller, 2001), with defects in mitochondrial fusion underlying a number of human diseases.

3. VIEW FROM THE PRESENT

As is clear from the brief summary above of our work on self-renewal, proliferation, and differentiation in the male germ line adult stem cell lineage, I have had a phenomenal group of students and postdocs in my lab. The current lab group is just as good, creative, and hard working. People in my lab are constantly helping each other master new techniques, interpret surprising new data, and brainstorm ideas. Right now, I am tremendously excited about the new results emerging from the current work in my lab on how somatic cells in the testis guide differentiation of the germ cells they enclose, how the switch from mitosis to meiosis is regulated by translational control, how the cell type-specific transcription program for germ cell differentiation is turned on in spermatocytes, and how the developmental program regulates core cell cycle machinery components to specify meiotic prophase in males.

Last Words

When I was a senior in college, I remember trying to decide whether to become a painter or a research biologist. I remember thinking: well, I can paint as a hobby, but I cannot do science as a hobby. Actually, I was wrong. You cannot do either as a hobby. Art and science both require that you dive in with your whole heart and mind. Both tap springs of creativity. As an artist, you get people to see everyday things in a whole new way (think Georgia O'Keefe and flowers). As a scientist, you conceive of an explanation that ties together for the first time disparate-seeming bits of information into a model of how something works—a model that can be tested experimentally. Both of these are creative acts; both are great joys. And both the artist and the scientist labor intensively at the leading edge, pushing the envelope, obsessed with that demanding, fascinating, and consuming work, because we love what we do.

ACKNOWLEDGMENTS

I sincerely and deeply thank all the students, postdocs, and research associates who have worked with me over the years. I have not been able to name you all in this brief essay, but you are all in my heart and mind. You are the ones who made these discoveries and my career in science possible and it has been the greatest privilege to work with you. I thank my early funding sources, the Searle Foundation and the American Cancer Society, for believing in me, NIGMS and NICHD for their support of my research program over the years, and the Reed-Hodgson Professorship in Human Biology, which supports my research and teaching endeavors at Stanford.

REFERENCES

Arbeitman, M. N., Furlong, E. E., Imam, F., Johnson, E., Null, B. H., Baker, B. S., et al. (2002). Gene expression during the life cycle of Drosophila melanogaster. *Science*, *297*(5590), 2270–2275.

Ayyar, S., Jiang, J., Collu, A., White-Cooper, H., & White, R. A. (2003). Drosophila TGIF is essential for developmentally regulated transcription in spermatogenesis. *Development*, *130*(13), 2841–2852.

Baker, C. C., & Fuller, M. T. (2007). Translational control of meiotic cell cycle progression and spermatid differentiation in male germ cells by a novel eIF4G homolog. *Development*, *134*(15), 2863–2869.

Baker, C. C., Gim, B.-S., & Fuller, M. T. (2015). Cell-type specific translational repression of cyclin B during meiosis in males. *Development*, *142*, 3394–3402.

Beall, E. L., Lewis, P. W., Bell, M., Rocha, M., Jones, D. L., & Botchan, M. R. (2007). Discovery of tMAC: A Drosophila testis-specific meiotic arrest complex paralogous to Myb-Muv B. *Genes & Development*, *21*(8), 904–919.

Belloni, G., Sechi, S., Riparbelli, M. G., Fuller, M. T., Callaini, G., & Giansanti, M. G. (2012). Mutations in Cog7 affect Golgi structure, meiotic cytokinesis and sperm development during Drosophila spermatogenesis. *Journal of Cell Science, 125*(Pt. 22), 5441–5452.

Brill, J. A., Hime, G. R., Scharer-Schuksz, M., & Fuller, M. T. (2000). A phospholipid kinase regulates actin organization and intercellular bridge formation during germline cytokinesis. *Development, 127*(17), 3855–3864.

Chen, X., Lu, C., Prado, J. R., Eun, S. H., & Fuller, M. T. (2011). Sequential changes at differentiation gene promoters as they become active in a stem cell lineage. *Development, 138*(12), 2441–2450.

Davies, E. L., Lim, J. G., Joo, W. J., Tam, C. H., & Fuller, M. T. (2013). The transcriptional regulator lola is required for stem cell maintenance and germ cell differentiation in the Drosophila testis. *Developmental Biology, 373*(2), 310–321.

Doggett, K., Jiang, J., Aleti, G., & White-Cooper, H. (2011). Wake-up-call, a lin-52 paralogue, and Always early, a lin-9 homologue physically interact, but have opposing functions in regulating testis-specific gene expression. *Developmental Biology, 355*(2), 381–393.

Farkas, R. M., Giansanti, M. G., Gatti, M., & Fuller, M. T. (2003). The Drosophila Cog5 homologue is required for cytokinesis, cell elongation, and assembly of specialized Golgi architecture during spermatogenesis. *Molecular Biology of the Cell, 14*(1), 190–200.

Giansanti, M. G., Bonaccorsi, S., Kurek, R., Farkas, R. M., Dimitri, P., Fuller, M. T., et al. (2006). The class I PITP giotto is required for Drosophila cytokinesis. *Current Biology, 16*(2), 195–201.

Giansanti, M. G., Farkas, R. M., Bonaccorsi, S., Lindsley, D. L., Wakimoto, B. T., Fuller, M. T., et al. (2004). Genetic dissection of meiotic cytokinesis in Drosophila males. *Molecular Biology of the Cell, 15*, 2509–2522.

Giansanti, M. G., Vanderleest, T. E., Jewett, C. E., Sechi, S., Frappaolo, A., Fabian, L., et al. (2015). Exocyst-dependent membrane addition is required for anaphase cell elongation and cytokinesis in Drosophila. *PLoS Genetics, 11*(11), e1005632.

Hales, K. G., & Fuller, M. T. (1997). Developmentally regulated mitochondrial fusion mediated by a conserved, novel, predicted GTPase. *Cell, 90*(1), 121–129.

Hardy, R. W., Tokuyasu, K. T., Lindsley, D. L., & Garavito, M. (1979). The germinal proliferation center in the testis of Drosophila melanogaster. *Journal of Ultrastructure Research, 69*(2), 180–190.

Hermann, G. J., Thatcher, J. W., Mills, J. P., Hales, K. G., Fuller, M. T., Nunnari, J., et al. (1998). Mitochondrial fusion in yeast requires the transmembrane GTPase Fzo1p. *The Journal of Cell Biology, 143*(2), 359–373.

Hiller, M., Chen, X., Pringle, M. J., Suchorolski, M., Sancak, Y., Viswanathan, S., et al. (2004). Testis-specific TAF homologs collaborate to control a tissue-specific transcription program. *Development, 131*(21), 5297–5308.

Hiller, M. A., Lin, T. Y., Wood, C., & Fuller, M. T. (2001). Developmental regulation of transcription by a tissue-specific TAF homolog. *Genes & Development, 15*(8), 1021–1030.

Hime, G. R., Brill, J. A., & Fuller, M. T. (1996). Assembly of ring canals in the male germ line from structural components of the contractile ring. *Journal of Cell Science, 109*(Pt. 12), 2779–2788.

Insco, M. L., Bailey, A. S., Kim, J., Olivares, G. H., Wapinski, O. L., Tam, C. H., et al. (2012). A self-limiting switch based on translational control regulates the transition from proliferation to differentiation in an adult stem cell lineage. *Cell Stem Cell, 11*(5), 689–700.

Insco, M., Leon, A., Tam, C., McKearin, D., & Fuller, M. T. (2009). Accumulation of a differentiation regulator specifies transit amplifying division number in an adult stem cell

lineage. *Proceedings of the National Academy of Sciences of the United States of America*, *106*, 22311–22316.
Jiang, J., Benson, E., Bausek, N., Doggett, K., & White-Cooper, H. (2007). Tombola, a tesmin/TSO1-family protein, regulates transcriptional activation in the Drosophila male germline and physically interacts with Always early. *Development*, *134*(8), 1549–1559.
Jiang, J., & White-Cooper, H. (2003). Transcriptional activation in Drosophila spermatogenesis involves the mutually dependent function of aly and a novel meiotic arrest gene cookie monster. *Development*, *130*(3), 563–573.
Kiger, A. A., Jones, D. L., Schulz, C., Rogers, M. B., & Fuller, M. T. (2001). Stem cell self-renewal specified by JAK-STAT activation in response to a support cell cue. *Science*, *294*(5551), 2542–2545.
Kiger, A. A., White-Cooper, H., & Fuller, M. T. (2000). Somatic support cells restrict germline stem cell self-renewal and promote differentiation. *Nature*, *407*(6805), 750–754.
Kimble, J. (2011). Molecular regulation of the mitosis/meiosis decision in multicellular organisms. *Cold Spring Harbor Perspectives in Biology*, *3*(8), a002683.
Leatherman, J. L., & Dinardo, S. (2008). Zfh-1 controls somatic stem cell self-renewal in the Drosophila testis and nonautonomously influences germline stem cell self-renewal. *Cell Stem Cell*, *3*(1), 44–54.
Leatherman, J. L., & Dinardo, S. (2010). Germline self-renewal requires cyst stem cells and stat regulates niche adhesion in Drosophila testes. *Nature Cell Biology*, *12*(8), 806–811.
Lim, J. G., & Fuller, M. T. (2012). Somatic cell lineage is required for differentiation and not maintenance of germline stem cells in Drosophila testes. *Proceedings of the National Academy of Sciences of the United States of America*, *109*(45), 18477–18481.
Lin, T. Y., Viswanathan, S., Wood, C., Wilson, P. G., Wolf, N., & Fuller, M. T. (1996). Coordinate developmental control of the meiotic cell cycle and spermatid differentiation in Drosophila males. *Development*, *122*(4), 1331–1341.
Lu, C., & Fuller, M. T. (2015). Recruitment of mediator complex by cell type and stage-specific factors required for tissue-specific TAF dependent gene activation in an adult stem cell lineage. *PLoS Genetics*, *11*, e1005701.
Lu, C., Kim, J., & Fuller, M. T. (2013). The polyubiquitin gene Ubi-p63E is essential for male meiotic cell cycle progression and germ cell differentiation in Drosophila. *Development*, *140*(17), 3522–3531.
Morillo Prado, J. R., Srinivasan, S., & Fuller, M. T. (2013). The histone variant His2Av is required for adult stem cell maintenance in the Drosophila testis. *PLoS Genetics*, *9*(11), e1003903.
Perezgasga, L., Jiang, J., Bolival, B., Jr., Hiller, M., Benson, E., Fuller, M. T., et al. (2004). Regulation of transcription of meiotic cell cycle and terminal differentiation genes by the testis-specific Zn-finger protein matotopetli. *Development*, *131*(8), 1691–1702.
Robinett, C. C., Giansanti, M. G., Gatti, M., & Fuller, M. T. (2009). TRAPPII is required for cleavage furrow ingression and localization of Rab11 in dividing male meiotic cells of Drosophila. *Journal of Cell Science*, *122*(Pt. 24), 4526–4534.
Santel, A., & Fuller, M. (2001). Control of mitochondrial morphology by a human mitofusin. *Journal of Cell Science*, *114*(Pt. 5), 867–874.
Sarkar, A., Parikh, N., Hearn, S. A., Fuller, M. T., Tazuke, S. I., & Schulz, C. (2007). Antagonistic roles of Rac and Rho in organizing the germ cell microenvironment. *Current Biology*, *17*(14), 1253–1258.
Schulz, C., Wood, C. G., Jones, D. L., Tazuke, S. I., & Fuller, M. T. (2002). Signaling from germ cells mediated by the rhomboid homolog stet organizes encapsulation by somatic support cells. *Development*, *129*(19), 4523–4534.

Sechi, S., Colotti, G., Belloni, G., Mattei, V., Frappaolo, A., Raffa, G. D., et al. (2014). GOLPH3 is essential for contractile ring formation and Rab11 localization to the cleavage site during cytokinesis in Drosophila melanogaster. *PLoS Genetics*, *10*(5), e1004305.

Shields, A. R., Spence, A. C., Yamashita, Y. M., Davies, E. L., & Fuller, M. T. (2014). The actin-binding protein profilin is required for germline stem cell maintenance and germ cell enclosure by somatic cyst cells. *Development*, *141*(1), 73–82.

Srinivasan, S., Mahowald, A. P., & Fuller, M. T. (2012). The receptor tyrosine phosphatase Lar regulates adhesion between Drosophila male germline stem cells and the niche. *Development*, *139*(8), 1381–1390.

Szafer-Glusman, E., Fuller, M. T., & Giansanti, M. G. (2011). Role of Survivin in cytokinesis revealed by a separation-of-function allele. *Molecular Biology of the Cell*, *22*(20), 3779–3790.

Szafer-Glusman, E., Giansanti, M. G., Nishihama, R., Bolival, B., Pringle, J., Gatti, M., et al. (2008). A role for very-long-chain fatty acids in furrow ingression during cytokinesis in Drosophila spermatocytes. *Current Biology*, *18*(18), 1426–1431.

White-Cooper, H., Leroy, D., MacQueen, A., & Fuller, M. T. (2000). Transcription of meiotic cell cycle and terminal differentiation genes depends on a conserved chromatin associated protein, whose nuclear localisation is regulated. *Development*, *127*(24), 5463–5473.

Yamashita, Y. M., Jones, D. L., & Fuller, M. T. (2003). Orientation of asymmetric stem cell division by the APC tumor suppressor and centrosome. *Science*, *301*(5639), 1547–1550.

Yamashita, Y. M., Mahowald, A. P., Perlin, J. R., & Fuller, M. T. (2007). Asymmetric inheritance of mother versus daughter centrosome in stem cell division. *Science*, *315*(5811), 518–521.

CHAPTER TWENTY-ONE

How Somatic Adult Tissues Develop Organizer Activity

Matthias C. Vogg[1], Yvan Wenger[1], Brigitte Galliot[2]
Department of Genetics and Evolution, Institute of Genetics and Genomics in Geneva (IGe3), Faculty of Sciences, University of Geneva, Geneva, Switzerland
[1]Both authors equally contributed.
[2]Corresponding author: e-mail address: brigitte.galliot@unige.ch

Contents

Abstract

The growth and patterning of anatomical structures from specific cellular fields in developing organisms relies on organizing centers that instruct surrounding cells to modify their behavior, namely migration, proliferation, and differentiation. We discuss here how organizers can form in adult organisms, a process of utmost interest for regenerative medicine. Animals like *Hydra* and planarians, which maintain their shape and fitness thanks to a highly dynamic homeostasis, offer a useful paradigm to study adult organizers in steady-state conditions. Beside the homeostatic context, these model systems also offer the possibility to study how organizers form *de novo* from somatic adult tissues. Both extracellular matrix remodeling and caspase activation play a key role in this transition, acting as promoters of organizer formation in the vicinity of the wound. Their respective roles and the crosstalk between them just start to be deciphered.

Current Topics in Developmental Biology, Volume 116
ISSN 0070-2153
http://dx.doi.org/10.1016/bs.ctdb.2015.11.002

1. INTRODUCTION TO ORGANIZERS

1.1 The Discovery of Induction and the Organizer Phenomenon

In 1924, Hans Spemann and Hilde Mangold demonstrated that the dorsal blastopore lip of an amphibian embryo induces the formation of a secondary body axis when transplanted into the ventral side of another embryo (De Robertis, 2009; Spemann & Mangold, 1924). To trace the fate of transplanted cells, Spemann and Mangold used two newt species, the pigmented *Triton cristatus* as a donor, and, the unpigmented *Triton taeniatus* as a host (Spemann & Mangold, 1924). They observed that within the ectopic body axis, the graft predominantly gave rise to notochord, whereas somites and the central nervous system were mainly of host origin. This surprising result demonstrated that the donor tissue, here the dorsal lip, acts as an inducer, recruiting and organizing cells from the host so that they form a secondary axis (Spemann & Mangold, 1924). Spemann coined the term "organizer" for the dorsal blastopore lip and received the Nobel Prize in medicine or physiology for the discovery of the induction phenomenon in 1932.

In fact, this phenomenon had already been identified 15 years earlier by Ethel Browne who reported that a fragment of the head of a *Hydra* polyp induces the formation of a secondary body axis when transplanted on the body column of a host polyp (Browne, 1909; Fig. 1). Surprisingly, Hans Spemann who had *Hydra* cultures in his own lab and was aware of Ethel Browne's findings, never cited her work (Lenhoff, 1991). Even though Browne never used the words induction and organizer, she convincingly showed that the grafted piece of *Hydra* head recruits and organizes cells of the host into a secondary body axis (Browne, 1909). To do so, she transplanted fragments of polyps from the *Hydra viridissima* species whose endodermal epithelial cells contain a symbiotic green algae, and thus are naturally pigmented, on polyps of the same species artificially depigmented. Using this approach, she demonstrated that the ectopic body axis was predominantly formed of cells from the host and not from the graft. Similarly, she showed that the head-regenerating tip and the presumptive head of growing buds induce the formation of ectopic axis on the host (Fig. 1). Later on, it was shown that the *Hydra* basal region also carries an organizer (Hicklin & Wolpert, 1973; Kadu, Ghaskadbi, & Ghaskadbi, 2012).

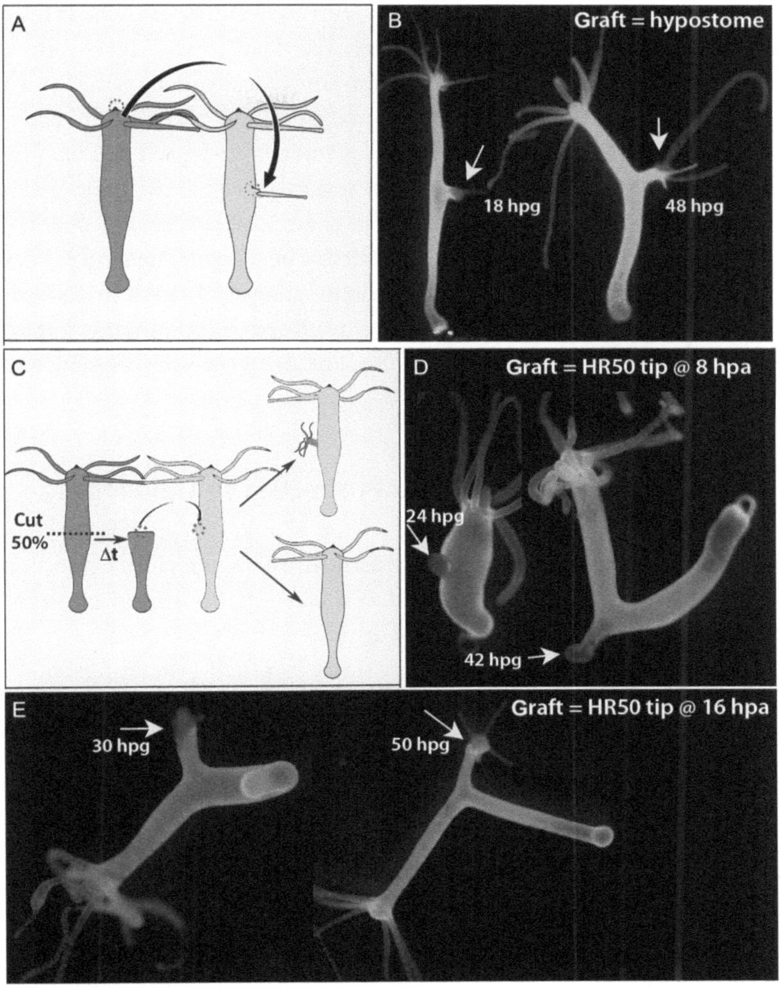

Figure 1 Homeostatic- and regeneration-induced head organizers in *Hydra* as initially characterized by Browne (1909). Reproduction of Ethel Browne's original experiments demonstrating the ability of the *Hydra* head or the *Hydra* head-regenerating tip to induce a complete ectopic axis (Browne, 1909). Here, we used as hosts two *AEP* actin::eGFP transgenic strains that constitutively expresses GFP either in the endodermal (Wittlieb, Khalturin, Lohmann, Anton-Erxleben, & Bosch, 2006) or in the ectodermal (Anton-Erxleben, Thomas, Wittlieb, Fraune, & Bosch, 2009) epithelial cells, named Endo-GFP and Ecto-GFP, respectively. (A) A fragment of head bearing a tentacle was dissected from a nonfluorescent *Hydra AEP* and immediately transplanted to the lateral side of the body column of the Endo-GFP host. As in Browne's experiment, the tentacle was used as a marker, which does not have organizing activity itself. (B) An ectopic axis that contains endodermal fluorescent cells from the host is rapidly induced, observed here at 18 and

(Continued)

An important distinction was made, with the finding that the transplantation of pieces of the body column can also lead to the development of ectopic structures, but this time without any recruitment of the host cells (Broun & Bode, 2002). This phenomenon, which relies on the self-organizing activity of the grafted tissue, is referred to as self-repair. Hence, a precise cell-lineage tracing strategy is necessary to discriminate between the grafted cells and the host cells within the ectopic structure and thus to distinguish self-repair from organizing activity. The elegant method pioneered by Ethel Browne to perform lineage-tracing analysis, thanks to natural differences in pigmentation, is nowadays replaced by transgenic strategies where cells or tissues expressing fluorescent proteins such as GFP enable the tracing of cell fate (Fig. 1).

1.2 Organizers Share Common Properties in Developing Vertebrates

The discovery of the amphibian gastrulation organizer (also named dorsal organizer) by Spemann and Mangold was a major finding in embryology as it proved to be a common feature among vertebrates (Joubin & Stern, 2001), corresponding to the Hensen's node in chick (Waddington, 1932, 1933), the node in mouse (Beddington, 1994), and the shield in zebrafish (Oppenheimer, 1936; Shih & Fraser, 1996). More recently, Yu et al. showed that the blastopore lip in the cephalochordate amphioxus acts as a gastrulation organizer (Yu et al., 2007). Several properties emerged from the comparative analysis of these vertebrate gastrulation organizers: these are transient structures, which pattern neurectodermal and mesodermal derivatives, without showing species specificity as evidenced from successful cross-species transplantations; they prevent the formation of another organizer in their vicinity, and they can regenerate (Joubin & Stern, 2001). Transplantation experiments identified other organizers with similar properties in

Figure 1—Cont'd 48 hours postgrafting (hpg, arrows). (C) A nonfluorescent *Hydra AEP* was bisected at midgastric position (50%), the lower part was allowed to regenerate for 8 (D) or 16 (E) hours, and the tip was transplanted to the lateral side of the body column of a Ecto-GFP host. (D) Ectopic axis induced by a 8 h postamputation (hpa) graft visualized at 24 and 42 hpg (arrows). (E) Ectopic axis induced by a 16 hpa graft visualized at 30 and 50 hpa (arrows). In (D and E), the progressive growth and patterning of the ectopic axis is obvious. However, the development of the structure induced by the 16 hpa graft is more advanced than the one induced by the 8 hpa graft, illustrating the fact that the inductive capability of the tip increases within the first 12 h of head regeneration (MacWilliams, 1983b). (A color version of this figure is located at the end of this chapter.)

vertebrate embryos, namely the limb organizer that induces the formation of ectopic skeletal structures upon transplantation (Saunders, 1948; Tickle, Summerbell, & Wolpert, 1975), and the isthmic organizer at the midbrain–hindbrain boundary (Alvarado-Mallart, Martinez, & Lance-Jones, 1990; Nakamura, 1990). At the cellular level, organizers do not act on single cells but rather on fields of cells, which locally modify their behaviors such as cell proliferation, cell migration, and cell differentiation, and thus undergo morphogenesis. Hans Meinhardt proposed the "boundary model" where the confrontation of boundaries of distinct primary fields is necessary for the formation of organizing regions, which in turn drive the development of secondary fields (Meinhardt, 1983).

1.3 Molecular Activity of Developmental Organizers

How a limited number of cells forming organizers can direct changes in cell behavior remained mysterious for decades. In 1952, Alan Turing proposed an interesting molecular representation of morphogenetic processes, "a system of chemical substances, called morphogens, reacting together and diffusing through a tissue, adequate to account for the main phenomena of morphogenesis" (Turing, 1952). Over the past decades, it was indeed proven that organizers produce morphogens, characterized as a limited number of growth factors whose activity is tightly regulated by specific antagonists, and the production of both is spatially regulated by transcription factors (De Robertis & Kuroda, 2004; Gurdon & Bourillot, 2001). To investigate the molecular regulation of the *Xenopus* gastrulation organizer, the De Robertis lab used its regeneration property, as the *Xenopus* embryo, once bisected, completely regenerates from its dorsal half that contains part of the organizer, while the ventral half only produces ventral tissues (Reversade & De Robertis, 2005).

A sophisticated circuitry between the dorsal and ventral poles was progressively uncovered, maintaining BMP signaling activity high at the ventral side of the embryo, sufficient to trigger ventral differentiation, whereas high levels of BMP inhibitors at the dorsal side (chordin, noggin, and follistatin) maintain BMP signaling low and, together with Wnt inhibitors drive dorsal differentiation, i.e., the differentiation of the neural tube and of adjacent mesenchyme in somites (De Robertis & Kuroda, 2004). However, additional actors were subsequently identified. Recent results point to the important role played by extracellular components asymmetrically produced, such as proteases and protease inhibitors that regulate the stability of the BMP inhibitor chordin (Inomata, Shibata, Haraguchi, & Sasai, 2013; Plouhinec, Zakin, Moriyama, & De Robertis, 2013). Hence, the extracellular matrix (ECM) now appears as an essential partner to regulate the dynamic flux of signals (De Robertis & Colozza, 2013; Fleury, Boardman, & Swartz, 2006).

In the zebrafish embryo, the organizer is actually more diffuse than the dorsal lip of the *Xenopus* blastopore, spreading along the margin of the blastula/gastrula (Thisse & Thisse, 2015). A recent study characterized BMP and Nodal as the minimal molecular machinery required to transform a round shaped blastula/gastrula to an elongated zygote that rapidly differentiates specific structures along its main body axis. BMP and Nodal form two inverted gradients along the dorsoventral axis, and experimental variations of the BMP/Nodal ratio at ectopic positions appear sufficient to direct morphogenesis (Xu, Houssin, Ferri-Lagneau, Thisse, & Thisse, 2014). As in *Xenopus*, BMP is maximal at the ventral side and Nodal at the dorsal side, and a specific ratio of these two morphogens defines at each point along the axis the differentiation of specific structures or tissues. Hence, vertebrate gastrulation organizers might be homologous, building their activity on a similar molecular set up.

1.4 Organizers in Adult Tissues

A major difference between the vertebrate gastrulation organizers and the *Hydra* organizers is that the formers are transient structures that form within developing tissues, whereas the latters are continuously active in an adult organism. Furthermore, adult *Hydra* organizers can get reestablished from adult somatic tissues during regeneration or budding, implying a complete *de novo* formation at any age. An important question is whether *Hydra* is the only animal able to form and maintain organizers in adulthood. In numerous species, blastemas spontaneously forming after amputation of adult appendages or organs could be considered as adult organizers. Indeed, regeneration occurs through blastema formation in planarians regenerating body parts (Newmark & Sanchez Alvarado, 2000; Salo & Baguna, 1984); in polychaete annelids regenerating missing body segments (Paulus & Muller, 2006); in crustaceans regenerating limbs (Konstantinides & Averof, 2014); in echinoderms regenerating arms or gut (Candia Carnevali, Bonasoro, & Biale, 1997; Kondo & Akasaka, 2010; Mashanov, Zueva, & Garcia-Arraras, 2014); and in zebrafish, urodeles, and lizards regenerating appendages (Brockes & Kumar, 2008). Here, we propose to review several aspects of the regulation and the function of animal adult organizers: (i) in the freshwater polyp *Hydra* (Cnidaria) that hosts two active organizers in homeostatic conditions, both rapidly regenerated upon removal, (ii) in the planarian *Schmidtea mediterranea* (Bilateria) whose anterior and posterior poles likely act as organizers for patterning maintenance as well as for body regeneration, and (iii) in vertebrates that form regenerative blastemas, some recent studies investigating how they act as organizers.

2. ORGANIZERS IN HOMEOSTATIC ADULT TISSUES

2.1 Head and Foot Organizers Control Homeostatic Axial Patterning in *Hydra*

Over the past two centuries, *Hydra* was the focus of interest for a number of important questions linked to cell and developmental biology, toxicology, and ecology (Galliot, 2012). In the second half of the twentieth century, *Hydra* became a model system of choice for mathematicians, physicists, and engineers who selected this simple model to propose a dynamic mathematical representation of the mechanisms of morphogenesis (Gierer & Meinhardt, 1972; Turing, 1952; Wolpert, Clarke, & Hornbruch, 1972). Indeed, transplantation experiments provided the basis for mathematical models, based on reaction and diffusion of molecules in developing tissues.

As described above, adult *Hydra* polyps maintain differentiated basal and apical regions thanks to two active organizers: the head organizer located at the apical tip, a dome-shape region named hypostome, and the foot organizer located in the basal region (Yao, 1945). These two organizers are separated by the body column of the animal, corresponding to a distance of approximately 0.5–1 cm. A series of transplantation experiments, either axial or lateral, identified graded activities distributed along the *Hydra* body axis between the basal and apical organizers (reviewed in Shimizu, 2012; Takano & Sugiyama, 1983). These biological gradients were identified as two pairs of parallel but opposed activities, one pair named head activation and head inhibition, maximal at the apex, and another pair named foot activation and foot inhibition maximal at the basis (Fig. 2A).

Concerning the head inhibition capacity, evidences came first from axial grafting procedure, whereby a hypostome grafted on a head-regenerating animal prevents head regeneration (Wilby & Webster, 1970). The graded head inhibition activity was subsequently confirmed by lateral grafting procedure where same apical pieces of tissue were grafted at different heights

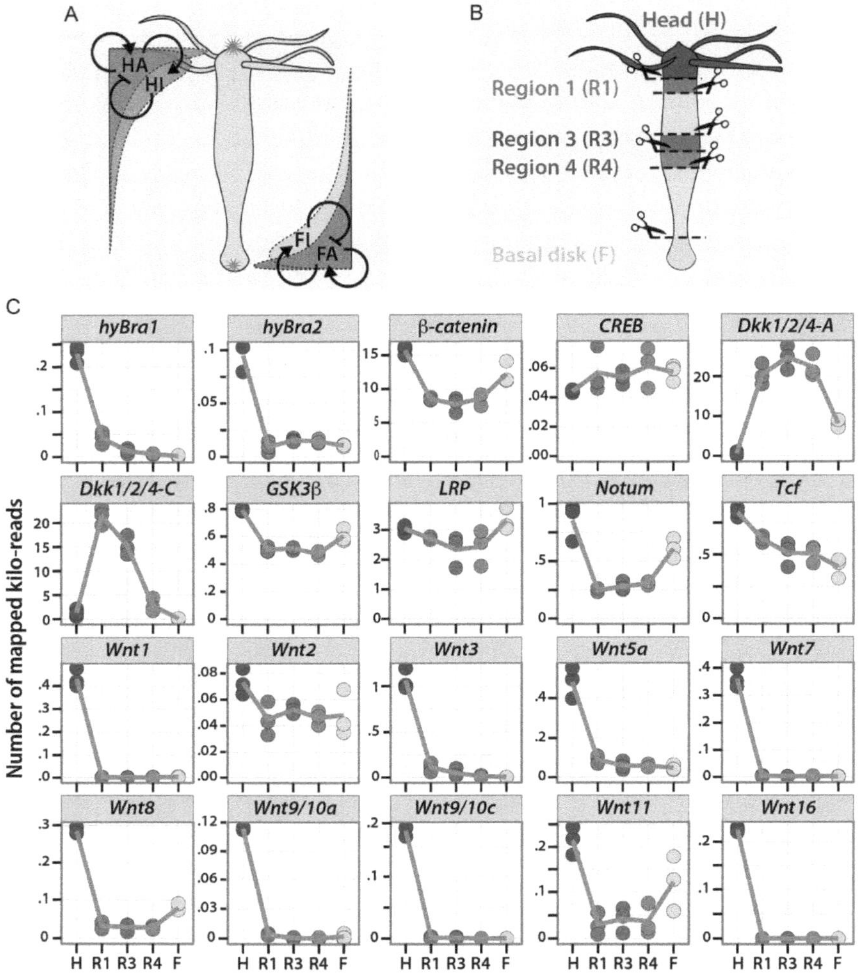

Figure 2 Molecular components of organizer maintenance in homeostatic *Hydra*. (A) Schematic representation of the two pairs of graded biological activities, head activation (HA, green) and head inhibition (HI, red), foot activation (HA, blue) and foot inhibition (FI, golden) as detected by lateral transplantation experiments (see Shimizu, 2012). Note their inverted graded distribution, maximal at the apical pole for HA/HI, at the basal pole for FA/FI. Green and blue stars represent head and foot organizers, respectively. (B) Scheme depicting the procedure used to perform quantitative RNA-seq transcriptomics at five distinct levels along the *Hydra* body axis (for details, see in Wenger, Buzgariu, & Galliot, 2015). (C) RNA-seq profiles of components, putative regulators and target genes of the Wnt/β-catenin pathway, involved in the maintenance and regeneration of the head organizer (Chera et al., 2009; Chera, Ghila, Wenger, & Galliot, 2011; Chera, Kaloulis, & Galliot, 2007; Guder et al., 2006; Hobmayer et al., 2000; Lengfeld et al., 2009; Nakamura, Tsiairis, Ozbek, & Holstein, 2011; Petersen et al., 2015; Technau et al., 2000). The color code for each position corresponds to colors selected in panel (B). Note that wnt signaling in *Hydra* specifies head identity whereas it specifies tail identity in planarians (see Fig. 3). (A color version of this figure is located at the end of this chapter.)

along the body column of the host, revealing that grafting at more basal position in the host produces more frequently an ectopic secondary axis (MacWilliams, 1983a; Shimizu, 2012; Takano & Sugiyama, 1983). Hence, two organizers and two pairs of graded capacities can easily be detected at the poles and along the body column, respectively, each pair fitting the autocatalytic-inhibitory Turing model adapted by Meinhardt and Gierer to explain both the active maintenance of body patterning and the *de novo* morphogenesis in *Hydra* polyps (Gierer, 2012; Gierer & Meinhardt, 1972; Meinhardt, 2012).

A series of strong evidences point to the Wnt pathway as the main component of the head organizer (Broun, Gee, Reinhardt, & Bode, 2005; Hobmayer et al., 2000; Lengfeld et al., 2009; Nakamura et al., 2011). Indeed, most *Wnt* genes including *Wnt3* are strongly or exclusively expressed at the tip of the head (Lengfeld et al., 2009; Fig. 2B and C) where β-catenin is mainly nuclear, which is not the case at other positions along the *Hydra* body column (Nakamura et al., 2011). Furthermore, when the Wnt signaling inhibitor GSK3β is inhibited either by drugs (Broun et al., 2005) or by RNAi (M. Vogg, unpublished), the ectopic formation of head characters, in this case tentacles, are induced in the body column, indicating that Wnt signaling is actively repressed in other regions than the tip of the hypostome in homeostatic *Hydra*. Autocatalytic and repressive *cis*-regulatory elements likely responsible for this tight spatial control have been identified in the regulatory region of *Wnt3* (Nakamura et al., 2011). The Dickkopf gene *Dkk1/2/4-C* (but not *Dkk1/2/4-A*) exhibits a graded expression along the body column; however, the absence of expression in the head region of both of them excludes them as head inhibitor (Augustin et al., 2006; Guder et al., 2006; Fig. 2C). Although *Notum* exhibits a homeostatic pattern consistent with a head inhibition function (Fig. 2C), the head inhibitor component remains uncharacterized, as well as the components of the foot organizer.

2.2 Head and Tail Poles Might Behave as Homeostatic Organizers in Planarians

Even though planarians and *Hydra* are both capable of regenerating missing body parts such as head or foot (tail for planaria), there is currently no definitive evidence for the presence of head or tail organizers in planarians. The softness of the planarian tissue, the mucus around the animals, and the mobility of planarians hamper grafting experiments, which are prerequisites to demonstrate organizer properties. In addition, tools such as differences in pigmentation or transgenic strains are missing to definitely assess the presence of organizers. However, Saito and colleagues succeeded in transplanting the region between the eyes of a donor planarian into the body wall of a host planarian and observed the induction of ectopic head structures (Saito et al., 2003; Fig. 3A). Importantly, the ectopic head was mainly host-derived (K. Agata, personal communication). One possibility would be that anterior pole cells present in the transplanted tissue, induced the ectopic structures by recruiting and organizing cells of the host planarian. Although never reported, one anticipates that transplanting tissue, containing posterior pole cells would similarly result in the induction of an ectopic tail (Fig. 3A).

How do poles mediate tissue organization? As organizers, poles should release signaling molecules that would suffice to (i) provide migratory cues, (ii) drive proliferation of planarian stem cells (named neoblasts), and (iii) induce changes in gene expression of neoblasts that lead to differentiation, so that the surrounding tissue is properly patterned. The Reddien laboratory showed that proliferation of neoblasts, is abrogated after the knockdown of the pole cell gene *follistatin*, which acts as an Activin inhibitor (Gavino et al., 2013). Nonetheless, a bottleneck of the planarian model is the current lack of cell-lineage tracing strategies, meaning that the instructive role of these pole cells on the surrounding tissues cannot be directly traced. Such strategies are definitely required to confirm the organizer role of planarian poles.

3. *DE NOVO* FORMATION OF ORGANIZERS FROM ADULT SOMATIC TISSUES

3.1 Regeneration of the Apical Organizer in *Hydra*

In *Hydra*, any region of the body column can acquire the properties of an organizer after amputation and the region located below the wound shows increasing organizing properties within 12 h postamputation (hpa) after midgastric bisection (Fig. 4A and B; MacWilliams, 1983b). Whether the regeneration of the *Hydra* head organizer is due to a mere continuation of the homeostatic process is still an open question. However, several lines of evidence indicate that this is likely not the case. First, some transcription factors such as the paired-like gene *prdl-a*, are expressed in the apical neuronal lineage in the homeostatic head but in endodermal epithelial cells of the head-regenerating tip at the time of organizer formation, suggesting a distinct regulation during early head regeneration (Gauchat, Kreger, Holstein, & Galliot, 1998). Second,

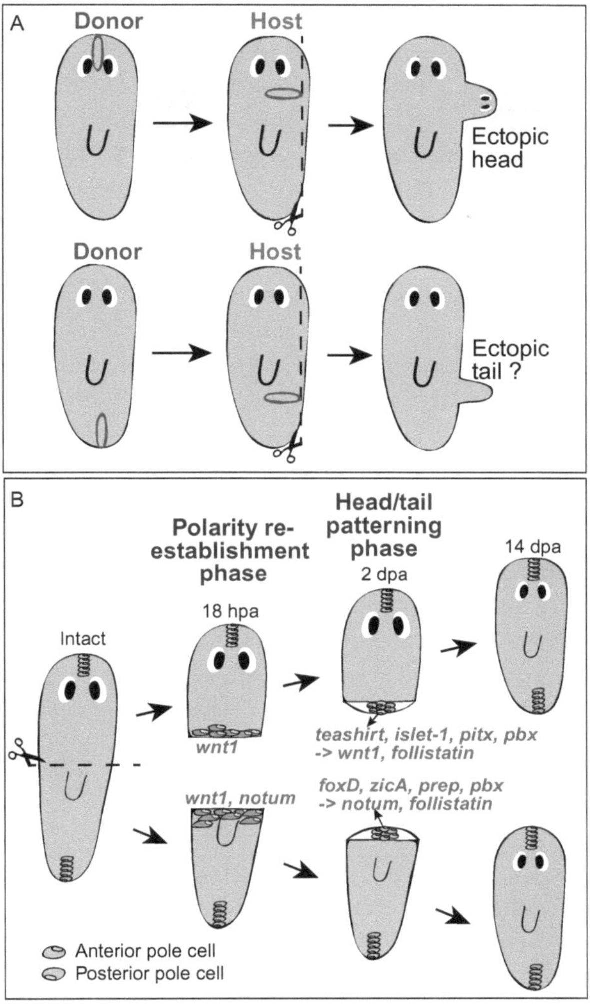

Figure 3 Anterior and posterior poles might represent putative head and tail organizers in homeostatic and regenerating planarians. (A) Transplantation experiment according to Saito, Koinuma, Watanabe, and Agata (2003). Tissue of the region between the eyes was transplanted into a host planarian that was subsequently cut laterally. Head outgrowth was observed in transplanted animals, suggesting that the transplanted tissue has head organizer activity. A similar scenario is expected to be true for the transplantation of midline tail tissue: the outgrowth of a tail on the host would indicate organizer activity of the tail transplant. (B) Amputation at any level along the animal induces *wnt1* expression at anterior- and posterior-facing wound sites at 18 hours postamputation (hpa) (Petersen & Reddien, 2009). At anterior-facing wound sites, the activity of Wnt1

(Continued)

injury-induced signaling and injury-induced cell death are required for the formation of a *de novo* head organizer after midgastric bisection (see in Fig. 4B), but apparently not for the maintenance of the apical organizer.

Briefly, midgastric bisection triggers an asymmetric activation of the MAPK/CREB pathway, which together with ROS signals triggers injury-induced cell death (Chera et al., 2011; Kaloulis et al., 2004). Apoptotic cells, which all belong to the interstitial lineage, transiently release signals including Wnt3, leading to the activation of β-catenin in surrounding S-phase cells and promotes their rapid mitotic division (Chera et al., 2009). In parallel, apoptotic bodies are engulfed by endodermal epithelial cells, which also rapidly upregulate *Wnt3* to launch head organizer activity (Chera et al., 2009). Interestingly, blocking cell death with caspase inhibitors prevents the division of cycling cells as well as the upregulation of Wnt3 in endodermal epithelial cells. These results indicate that caspase activation and signaling from apoptotic cells contribute to head organizer formation after midgastric bisection in *Hydra* (Chera et al., 2009). Interestingly, when a *Hydra* is decapitated, organizer formation occurs independently of apoptosis and *Wnt3* is directly upregulated in endodermal epithelial cells (Hobmayer et al., 2000; Lengfeld et al., 2009). Altogether, Wnt signaling is necessary for organizer maintenance in homeostatic context and for organizer formation in regenerative conditions; however, the formation of the head organizer from the midgastric position relies on additional signaling such as the ones produced by dying cells.

Figure 3—Cont'd is antagonized by the secreted Wnt inhibitor Notum, which triggers anterior polarity reestablishment (Petersen & Reddien, 2011). In contrary, at posterior wounds, active Wnt signaling is necessary for posterior polarity reestablishment. Note that at 18 hpa *notum* and *wnt1* expression is induced in fully differentiated cells, whereas at 48 hpa on the posterior side, the transcription factor genes *islet-1* (Hayashi et al., 2011), *pitx* (Currie & Pearson, 2013; Marz, Seebeck, & Bartscherer, 2013), *pbx* (Blassberg, Felix, Tejada-Romero, & Aboobaker, 2013; Chen, Wang, & Reddien, 2013), and *teashirt* (Owen, Wagner, Chen, Petersen, & Reddien, 2015; Reuter et al., 2015) (blue) are upregulated in neoblasts that subsequently differentiate in pole cells that express *wnt1* (Petersen & Reddien, 2009) and *follistatin* (Gavino, Wenemoser, Wang, & Reddien, 2013). Similarly on the anterior side, the transcription factor genes *foxD, zicA, prep,* and *pbx* (Blassberg et al., 2013; Chen et al., 2013; Felix & Aboobaker, 2010; Scimone, Lapan, & Reddien, 2014; Vasquez-Doorman & Petersen, 2014; Vogg et al., 2014) are upregulated in neoblasts that provide the pole cells expressing *notum* (Petersen & Reddien, 2011) and *follistatin* (Gavino et al., 2013; Roberts-Galbraith & Newmark, 2013). (A color version of this figure is located at the end of this chapter.)

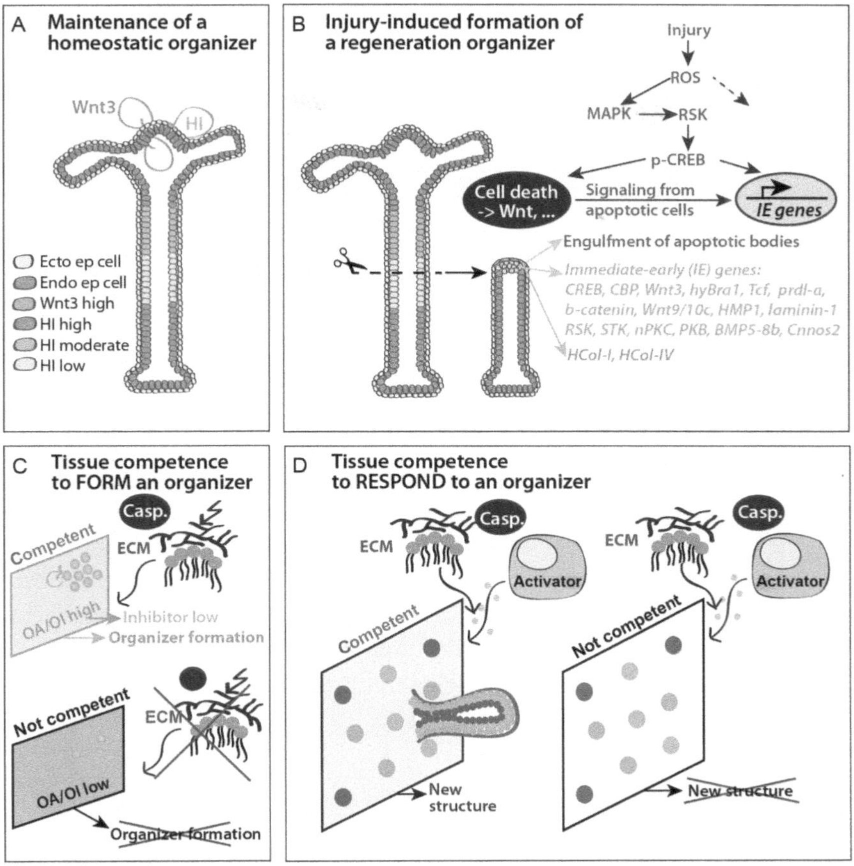

Figure 4 Organizer maintenance, organizer *de novo* formation upon injury, and competence to respond to organizer activity in an adult organism. (A) Organizer activity is maintained in homeostatic *Hydra* thanks to the head activator Wnt3 that maintains active β-catenin signaling in the head through a positive feedback loop (Nakamura et al., 2011). Note that head inhibitor (HI), which is expressed under the control of Wnt3 and antagonizes Wnt3 signaling, is not characterized yet. (B) *De novo* formation of a head organizer in head-regenerating tips after midgastric bisection relies on the asymmetric activation of the MAPK/CREB pathway (Galliot, Welschof, Schuckert, Hoffmeister, & Schaller, 1995; Kaloulis, Chera, Hassel, Gauchat, & Galliot, 2004), which together with injury-induced ROS signaling, triggers local cell death in the interstitial lineage. Apoptotic cells transiently release signaling molecules, including Wnt3, which activates the β-catenin pathway in surrounding cells, inducing a wave of mitotic division (Chera et al., 2009). At the same time, the engulfment of the apoptotic bodies by the endodermal epithelial cells (green) is followed by the upregulation of immediate-early (IE) genes, including *wnt3* that contains CREB response elements (CREs) in its regulatory sequence (Galliot, 2013; Nakamura et al., 2011).

(Continued)

3.2 Regeneration of the Putative Pole Organizers in Planarians

If we assume that pole cells correspond to organizers in planarians, how is pole formation achieved and what are the signaling pathways involved in this process? These questions can be addressed in the context of regeneration, as pole cells can be identified already 2 days after amputation in anterior and posterior blastemas (Fig. 3B). Soon after bisection, *wnt1* is upregulated in differentiated cells at anterior and posterior wound sites (Petersen & Reddien, 2009); however, these tissues rapidly take different fates as at anterior-facing sites, the Wnt inhibitor Notum antagonizes Wnt1 already at 18 hpa, thus enabling the establishment of anterior polarity (Petersen & Reddien, 2011; Fig. 3B). In the absence of *notum*, all regenerating structures become posterior as Wnt signaling triggers posterior polarity reestablishment. One day later, *notum* continues to be detected at the anterior regeneration pole (Petersen & Reddien, 2011), *wnt1* at the posterior regeneration pole (Adell, Salo, Boutros, & Bartscherer, 2009; Petersen & Reddien, 2009), and the Activin inhibitor *follistatin* at both poles (Gavino et al., 2013; Roberts-Galbraith & Newmark, 2013; Fig. 3B). From there, *notum* contributes to head formation (Scimone et al., 2014; Vogg et al., 2014), *wnt1* to tail formation (Petersen & Reddien, 2009), and *follistatin* to head and tail formation (Gavino et al., 2013; Roberts-Galbraith & Newmark, 2013).

Figure 4—Cont'd (C) Tissue competence to form an organizer relies on the respective activities of Activator (OA) and Inhibitor (OI) components, where OA are extracellular signal(s) and OI inhibitors that restrict OA extracellular activity. Only tissues with a high OA/OI ratio can form an organizer. Activators (i.e., morphogens) can be secreted as soluble molecules from the producing cells but also released from the ECM by proteolytic cleavage (Fleury et al., 2006). Once in the extracellular space, several mechanisms can promote their long-range graded activity such as the diffusion and the accumulation of proteases or protease inhibitors that ultimately regulate spatially the stability of proteins involved in morphogenesis (Inomata et al., 2013). Note that in *Hydra*, injury-induced modifications of the extracellular matrix (ECM, red flash) are necessary for regeneration (Sarras, 2012), possibly to increase the OA/OI ratio in the vicinity of the wound, up to the point where tissues become competent to form an organizer. Similarly in adult zebrafish regenerating its fins, signaling from the apoptotic cells and/or from the apoptotic bodies is necessary for the upregulation of IE genes in surrounding cells and subsequent tissue regeneration (Vriz, Reiter, & Galliot, 2014). (D) Tissue competence to respond to an organizer relies on the strength of the intracellular signaling pathway(s) activated by the activator. Tissues with a high signaling activity (blue background) are able to induce ectopic structures in response to OA (left), while tissues with a low signaling activity (white background) are unable to respond (right). Casp., caspase activation. (A color version of this figure is located at the end of this chapter.)

At that stage, the early upregulation of transcription factors such as *pbx*, *prep*, *foxD*, and *zicA* (also named *zic-1*) takes place in neoblasts, necessary for anterior pole formation. The synergy between ZicA and FoxD promotes the upregulation of *notum* and *follistatin* (Blassberg et al., 2013; Chen et al., 2013; Felix & Aboobaker, 2010; Gavino et al., 2013; Roberts-Galbraith & Newmark, 2013; Scimone et al., 2014; Vasquez-Doorman & Petersen, 2014; Vogg et al., 2014). For posterior pole formation, a different set of transcription factors is upregulated: *teashirt* (Owen et al., 2015; Reuter et al., 2015), *islet-1* (Hayashi et al., 2011), *pitx* (Currie & Pearson, 2013; Marz et al., 2013), and *pbx* (Blassberg et al., 2013; Chen et al., 2013; Fig. 3B).

As in *Hydra*, cell death appears crucial to set up organizer activity in planarians. Cell death takes place in two waves after amputation, the earliest one, maximal at 4 hpa, is linked to injury, while the second one, occurring at 72 hpa, is systemic and linked to regeneration (Pellettieri et al., 2010). Cell death is required for the expression of pole cell genes after amputation (Almuedo-Castillo et al., 2014). One possible scenario would be that a subset of neoblasts are more permissive to apoptotic signals, triggering pole formation. What signaling pathways are activated by apoptotic signals? Most likely, β-catenin signaling as *β-catenin-1* is expressed at the anterior pole (Gurley, Rink, & Sanchez Alvarado, 2008; Iglesias, Gomez-Skarmeta, Salo, & Adell, 2008), contributing to the expression of its own inhibitor *notum*, and hence to the formation of a functional anterior pole. But β-catenin signaling is also crucial for posterior differentiation (Gurley et al., 2008; Iglesias et al., 2008; Petersen & Reddien, 2008); indeed, knockdown of *β-catenin-1* leads to the loss of posterior pole cell gene expression (Petersen & Reddien, 2009). Hence, regulators of polarity remain to be uncovered but the proposition that these two regions, and more specifically the pole cells, form *de novo* adult organizers is well supported. One important component of organizer formation is the ECM, which likely plays a key role in planarian homeostasis and regeneration (Balestrini et al., 2014), although its specific role in organizer formation is currently unknown.

3.3 Regenerative Blastemas as Adult Organizers?

Regenerative blastemas are classically defined as a mass of mesenchymal cells that form, thanks to interactions with the wound epidermis and the input of peripheral nerves. Blastemas regulate in a coordinated fashion different cell behaviors such as cell migration, proliferation of undifferentiated cells, and differentiation into new structures (Bely & Nyberg, 2010;

Tanaka & Reddien, 2011). However, some species follow a different scheme, as planarian blastemas that do not proliferate (Salo & Baguna, 1984), or echinoderms blastemas that appear mostly epithelial (J. Garcia-Arraras, personal communication).

Blastema transplantation experiments convincingly showed that the blastema is an autonomous structure that instructs the surrounding tissues once established (Nye, Cameron, Chernoff, & Stocum, 2003), thus acting as an organizer. However, the process of recruitment and instruction of host cells by the blastema was rarely the focus of interest in these studies. A recent cell-lineage tracing study showed that early and late blastemas behave differently once transplanted: late blastemas seem to exhibit only self-repair activity, i.e., no longer able to instruct and recruit cells from the host tissue, whereas early blastemas are able to do so (McCusker & Gardiner, 2013). This suggests that the organizer properties of blastemas vanish with time.

At the molecular level, a crucial question is how a blastema gets established in amputated tissues. Recent transcriptomic analyses showed that, as expected, gene regulations are different from a simple wound response, with some limb-specific and nerve-specific signatures (Monaghan et al., 2012). If actors are roughly the same in developmental and regenerative blastemas, i.e., Wnt, FGF, Hh, and BMP pathways, the scenario is actually different. Indeed, injury signals and cell death play a key role in blastema formation during regeneration: in the amputated zebrafish fin, ROS is rapidly produced, triggering cell death and the rapid proliferation of the wound epidermis, an obligatory step toward blastema formation (Gauron et al., 2013), whereas in the newt regenerating its limb, caspase activation is required for the dedifferentiation of skeletal muscle cells and blastema formation (Wang et al., 2015). Concerning the organizer activity of regenerative blastema, a recent study of zebrafish fin regeneration identified nonproliferative cells with high Wnt/β-catenin activity that act as an organizer, directing the activities of the surrounding epithelial and mesenchymal tissues (Wehner et al., 2014). Hence, a specific limited Wnt/β-catenin area within the regenerative region might actually act as an organizer. Whether this mechanism is evolutionarily conserved remains unknown.

Mammals are a special case as they rarely form blastema-like structures after amputation, with the exception of the MRL mouse (Gourevitch, Clark, Bedelbaeva, Leferovich, & Heber-Katz, 2009; Rajnoch et al., 2003). However, blastemas might not be the exclusive way to induce morphogenetic processes in adult organisms. The only mammalian species known so far to regenerate appendages is the male deer that regrows its antlers each year from a precise spot of the periosteum of its frontal bone (Li, Zhao, Liu, & McMahon, 2014; Price, Allen, Faucheux, Althnaian, & Mount, 2005). This epimorphic regeneration process does not rely on blastema formation but rather on few cells that express stem cell markers and activate the Wnt/β-catenin pathway (Li, Harper, Puddick, Wang, & McMahon, 2012; Mount et al., 2006). Indeed, grafting experiments have clearly shown that these cells act as organizer, inducing the formation of ectopic structures by recruiting cells from the host tissue (Goss & Powel, 1985).

4. CONCLUSIONS AND PERSPECTIVES

Developmental organizers recruit and organize cells so that the surrounding tissue is eventually properly patterned in space and time. This is achieved not only by setting up morphogenetic gradients that result from a precise transcriptional control of the production of growth factors and their inhibitors but also by tightly regulating the activity of these factors and inhibitors thanks to protease and protease inhibitors anchored in the ECM (Fig. 4). Much less is known concerning the organizers that are active in adult organisms. Stable homeostatic organizers appear to be rare, so far only identified in animals with a dynamic patterning in homeostatic conditions such as *Hydra* and possibly planarians. However, the formation of organizers in regenerative contexts is more frequent, as discussed here about zebrafish, urodeles, planarians, or *Hydra*. These organizers share some properties with developmental organizers such as remodeling of the ECM, Wnt/β-catenin signaling regulated by inhibitors, but they also show major differences, which are the role of injury-induced signaling and caspase activation in dying cells or in undead cells.

A series of questions arise from the findings reported here: (1) What is the role of caspase activation on organizer formation from differentiated cells of adult tissues? Caspase activation seems to be linked to regeneration in multiple contexts (Fuchs & Steller, 2015; Miura, 2012; Morata, Shlevkov, & Perez-Garijo, 2011; Vriz et al., 2014), independently of cell death as in the context of fibroblast reprogramming (Li et al., 2010) or myotube dedifferentiation and cell cycle reentry (Wang et al., 2015). (2) What is the impact of the injury-induced modulations of the ECM on the formation and/or the activity of the organizer? The modulations of the metalloproteinases might play multiple roles, releasing anchored proproteins by proteolytic cleavage, scaling by maintaining stable protease activities (Fleury et al., 2006; Inomata et al., 2013). (3) What factors define the competence of the host tissue? This aspect is probably the most unknown, even in developmental organizers. (4) What are the signaling pathways beside the Wnt/β-catenin pathway that supports the regeneration of an organizer? Wnt signaling emerges as a common theme in regenerative contexts, however, its partners and functions might vary, highlighting independent evolutionary events. (5) Are cells active in organizers stem cells or rather nonproliferative cells as in planarians, or as recently identified in zebrafish (Wehner et al., 2014)? (6) Will it be possible to use organoids to study organizer formation and dissect the mechanisms that drive morphogenetic processes? Methods that use human embryonic stem cells to recapitulate embryonic spatial patterning are now available (Warmflash, Sorre, Etoc, Siggia, & Brivanlou, 2014), opening thus the possibility to investigate organizer formation. Answers to these various questions should help understand how regenerative organizers form and work, and to which extent differentiated cells can be manipulated to produce organizers useful in regenerative medicine.

ACKNOWLEDGMENTS

We are grateful to Kiyokazu Agata for providing unpublished information and José Garcia-Arraras for fruitful discussions. This work was supported by the Canton of Geneva, the Swiss National Foundation (SNF 31003A_149630), the NIH (grant number R01AG037962), the Human Frontier Science Program (grant RGP0016/2010), and the Claraz donation.

REFERENCES

Adell, T., Salo, E., Boutros, M., & Bartscherer, K. (2009). Smed-Evi/Wntless is required for beta-catenin-dependent and -independent processes during planarian regeneration. *Development*, *136*, 905–910.

Almuedo-Castillo, M., Crespo, X., Seebeck, F., Bartscherer, K., Salo, E., & Adell, T. (2014). JNK controls the onset of mitosis in planarian stem cells and triggers apoptotic cell death required for regeneration and remodeling. *PLoS Genetics*, *10*, e1004400.

Alvarado-Mallart, R. M., Martinez, S., & Lance-Jones, C. C. (1990). Pluripotentiality of the 2-day-old avian germinative neuroepithelium. *Developmental Biology*, *139*, 75–88.

Anton-Erxleben, F., Thomas, A., Wittlieb, J., Fraune, S., & Bosch, T. C. (2009). Plasticity of epithelial cell shape in response to upstream signals: A whole-organism study using transgenic Hydra. *Zoology (Jena, Germany)*, *112*, 185–194.

Augustin, R., Franke, A., Khalturin, K., Kiko, R., Siebert, S., Hemmrich, G., et al. (2006). Dickkopf related genes are components of the positional value gradient in Hydra. *Developmental Biology*, *296*, 62–70.

Balestrini, L., Isolani, M. E., Pietra, D., Borghini, A., Bianucci, A. M., Deri, P., et al. (2014). Berberine exposure triggers developmental effects on planarian regeneration. *Scientific Reports*, *4*, 4914.

Beddington, R. S. (1994). Induction of a second neural axis by the mouse node. *Development, 120*, 613–620.
Bely, A. E., & Nyberg, K. G. (2010). Evolution of animal regeneration: Re-emergence of a field. *Trends in Ecology & Evolution, 25*, 161–170.
Blassberg, R. A., Felix, D. A., Tejada-Romero, B., & Aboobaker, A. A. (2013). PBX/extradenticle is required to re-establish axial structures and polarity during planarian regeneration. *Development, 140*, 730–739.
Brockes, J. P., & Kumar, A. (2008). Comparative aspects of animal regeneration. *Annual Review of Cell and Developmental Biology, 24*, 525–549.
Broun, M., & Bode, H. R. (2002). Characterization of the head organizer in hydra. *Development, 129*, 875–884.
Broun, M., Gee, L., Reinhardt, B., & Bode, H. R. (2005). Formation of the head organizer in hydra involves the canonical Wnt pathway. *Development, 132*, 2907–2916.
Browne, E. N. (1909). The production of new hydranths in hydra by the insertion of small grafts. *Journal of Experimental Zoology*, 7, 1–37.
Candia Carnevali, M. D., Bonasoro, F., & Biale, A. (1997). Pattern of bromodeoxyuridine incorporation in the advanced stages of arm regeneration in the feather star Antedon mediterranea. *Cell and Tissue Research, 289*, 363–374.
Chen, C. C., Wang, I. E., & Reddien, P. W. (2013). pbx is required for pole and eye regeneration in planarians. *Development, 140*, 719–729.
Chera, S., Ghila, L., Dobretz, K., Wenger, Y., Bauer, C., Buzgariu, W., et al. (2009). Apoptotic cells provide an unexpected source of Wnt3 signaling to drive hydra head regeneration. *Developmental Cell, 17*, 279–289.
Chera, S., Ghila, L., Wenger, Y., & Galliot, B. (2011). Injury-induced activation of the MAPK/CREB pathway triggers apoptosis-induced compensatory proliferation in hydra head regeneration. *Development Growth and Differentiation, 53*, 186–201.
Chera, S., Kaloulis, K., & Galliot, B. (2007). The cAMP response element binding protein (CREB) as an integrative HUB selector in metazoans: Clues from the hydra model system. *Biosystems, 87*, 191–203.
Currie, K. W., & Pearson, B. J. (2013). Transcription factors lhx1/5-1 and pitx are required for the maintenance and regeneration of serotonergic neurons in planarians. *Development, 140*, 3577–3588.
De Robertis, E. M. (2009). Spemann's organizer and the self-regulation of embryonic fields. *Mechanisms of Development, 126*, 925–941.
De Robertis, E. M., & Colozza, G. (2013). Development: Scaling to size by protease inhibition. *Current Biology, 23*, R652–R654.
De Robertis, E. M., & Kuroda, H. (2004). Dorsal-ventral patterning and neural induction in Xenopus embryos. *Annual Review of Cell and Developmental Biology, 20*, 285–308.
Felix, D. A., & Aboobaker, A. A. (2010). The TALE class homeobox gene Smed-prep defines the anterior compartment for head regeneration. *PLoS Genetics, 6*, e1000915.
Fleury, M. E., Boardman, K. C., & Swartz, M. A. (2006). Autologous morphogen gradients by subtle interstitial flow and matrix interactions. *Biophysical Journal, 91*, 113–121.
Fuchs, Y., & Steller, H. (2015). Live to die another way: Modes of programmed cell death and the signals emanating from dying cells. *Nature Reviews. Molecular Cell Biology, 16*, 329–344.
Galliot, B. (2012). Hydra, a fruitful model system for 270 years. *International Journal of Developmental Biology, 56*, 411–423.
Galliot, B. (2013). Injury-induced asymmetric cell death as a driving force for head regeneration in Hydra. *Development Genes & Evolution, 223*, 39–52.
Galliot, B., Welschof, M., Schuckert, O., Hoffmeister, S., & Schaller, H. C. (1995). The cAMP response element binding protein is involved in hydra regeneration. *Development, 121*, 1205–1216.

Gauchat, D., Kreger, S., Holstein, T., & Galliot, B. (1998). prdl-a, a gene marker for hydra apical differentiation related to triploblastic paired-like head-specific genes. *Development, 125*, 1637–1645.
Gauron, C., Rampon, C., Bouzaffour, M., Ipendey, E., Teillon, J., Volovitch, M., et al. (2013). Sustained production of ROS triggers compensatory proliferation and is required for regeneration to proceed. *Scientific Reports, 3*, 2084.
Gavino, M. A., Wenemoser, D., Wang, I. E., & Reddien, P. W. (2013). Tissue absence initiates regeneration through follistatin-mediated inhibition of activin signaling. *eLife, 2*, e00247.
Gierer, A. (2012). The Hydra model—A model for what? *International Journal of Developmental Biology, 56*, 437–445.
Gierer, A., & Meinhardt, H. (1972). A theory of biological pattern formation. *Kybernetik, 12*, 30–39.
Goss, R. J., & Powel, R. S. (1985). Induction of deer antlers by transplanted periosteum. I. Graft size and shape. *Journal of Experimental Zoology, 235*, 359–373.
Gourevitch, D. L., Clark, L., Bedelbaeva, K., Leferovich, J., & Heber-Katz, E. (2009). Dynamic changes after murine digit amputation: The MRL mouse digit shows waves of tissue remodeling, growth, and apoptosis. *Wound Repair Regeneration, 17*, 447–455.
Guder, C., Pinho, S., Nacak, T. G., Schmidt, H. A., Hobmayer, B., Niehrs, C., et al. (2006). An ancient Wnt-dickkopf antagonism in Hydra. *Development, 133*, 901–911.
Gurdon, J. B., & Bourillot, P. Y. (2001). Morphogen gradient interpretation. *Nature, 413*, 797–803.
Gurley, K. A., Rink, J. C., & Sanchez Alvarado, A. (2008). Beta-catenin defines head versus tail identity during planarian regeneration and homeostasis. *Science, 319*, 323–327.
Hayashi, T., Motoishi, M., Yazawa, S., Itomi, K., Tanegashima, C., Nishimura, O., et al. (2011). A LIM-homeobox gene is required for differentiation of Wnt-expressing cells at the posterior end of the planarian body. *Development, 138*, 3679–3688.
Hicklin, J., & Wolpert, L. (1973). Positional information and pattern regulation in hydra: Formation of the foot end. *Journal of Embryology and Experimental Morphology, 30*, 727–740.
Hobmayer, B., Rentzsch, F., Kuhn, K., Happel, C. M., von Laue, C. C., Snyder, P., et al. (2000). WNT signalling molecules act in axis formation in the diploblastic metazoan Hydra. *Nature, 407*, 186–189.
Iglesias, M., Gomez-Skarmeta, J. L., Salo, E., & Adell, T. (2008). Silencing of Smed-betacatenin1 generates radial-like hypercephalized planarians. *Development, 135*, 1215–1221.
Inomata, H., Shibata, T., Haraguchi, T., & Sasai, Y. (2013). Scaling of dorsal-ventral patterning by embryo size-dependent degradation of Spemann's organizer signals. *Cell, 153*, 1296–1311.
Joubin, K., & Stern, C. D. (2001). Formation and maintenance of the organizer among the vertebrates. *International Journal of Developmental Biology, 45*, 165–175.
Kadu, V., Ghaskadbi, S. S., & Ghaskadbi, S. (2012). Induction of secondary axis in hydra revisited: New insights into pattern formation. *International Journal of Molecular and Cellular Medicine, 1*, 11–20.
Kaloulis, K., Chera, S., Hassel, M., Gauchat, D., & Galliot, B. (2004). Reactivation of developmental programs: The cAMP-response element-binding protein pathway is involved in hydra head regeneration. *Proceedings of the National Academy of Sciences of the United States of America, 101*, 2363–2368.
Kondo, M., & Akasaka, K. (2010). Regeneration in crinoids. *Development Growth and Differentiation, 52*, 57–68.
Konstantinides, N., & Averof, M. (2014). A common cellular basis for muscle regeneration in arthropods and vertebrates. *Science, 343*, 788–791.

Lengfeld, T., Watanabe, H., Simakov, O., Lindgens, D., Gee, L., Law, L., et al. (2009). Multiple Wnts are involved in Hydra organizer formation and regeneration. *Developmental Biology, 330*, 186–199.
Lenhoff, H. M. (1991). Ethel Browne, Hans Spemann, and the discovery of the organizer phenomenon. *Biological Bulletin, 181*, 72–80.
Li, C., Harper, A., Puddick, J., Wang, W., & McMahon, C. (2012). Proteomes and signalling pathways of antler stem cells. *PLoS One, 7*, e30026.
Li, F., He, Z., Shen, J., Huang, Q., Li, W., Liu, X., et al. (2010). Apoptotic caspases regulate induction of iPSCs from human fibroblasts. *Cell Stem Cell, 7*, 508–520.
Li, C., Zhao, H., Liu, Z., & McMahon, C. (2014). Deer antler—A novel model for studying organ regeneration in mammals. *International Journal of Biochemistry & Cell Biology, 56*, 111–122.
MacWilliams, H. K. (1983a). Hydra transplantation phenomena and the mechanism of hydra head regeneration. I. Properties of the head inhibition. *Developmental Biology, 96*, 217–238.
MacWilliams, H. K. (1983b). Hydra transplantation phenomena and the mechanism of Hydra head regeneration. II. Properties of the head activation. *Developmental Biology, 96*, 239–257.
Marz, M., Seebeck, F., & Bartscherer, K. (2013). A Pitx transcription factor controls the establishment and maintenance of the serotonergic lineage in planarians. *Development, 140*, 4499–4509.
Mashanov, V. S., Zueva, O., & Garcia-Arraras, J. E. (2014). Postembryonic organogenesis of the digestive tube: Why does it occur in worms and sea cucumbers but fail in humans? *Current Topics in Developmental Biology, 108*, 185–216.
McCusker, C. D., & Gardiner, D. M. (2013). Positional information is reprogrammed in blastema cells of the regenerating limb of the axolotl (Ambystoma mexicanum). *PLoS One, 8*, e77064.
Meinhardt, H. (1983). Cell determination boundaries as organizing regions for secondary embryonic fields. *Developmental Biology, 96*, 375–385.
Meinhardt, H. (2012). Modeling pattern formation in hydra: A route to understanding essential steps in development. *International Journal of Developmental Biology, 56*, 447–462.
Miura, M. (2012). Apoptotic and nonapoptotic caspase functions in animal development. *Cold Spring Harbor Perspective in Biology, 4*, 1–16.
Monaghan, J. R., Athippozhy, A., Seifert, A. W., Putta, S., Stromberg, A. J., Maden, M., et al. (2012). Gene expression patterns specific to the regenerating limb of the Mexican axolotl. *Biology Open, 1*, 937–948.
Morata, G., Shlevkov, E., & Perez-Garijo, A. (2011). Mitogenic signaling from apoptotic cells in Drosophila. *Development Growth and Differentiation, 53*, 168–176.
Mount, J. G., Muzylak, M., Allen, S., Althnaian, T., McGonnell, I. M., & Price, J. S. (2006). Evidence that the canonical Wnt signalling pathway regulates deer antler regeneration. *Developmental Dynamics, 235*, 1390–1399.
Nakamura, H. (1990). Do CNS anlagen have plasticity in differentiation? Analysis in quail-chick chimera. *Brain Research, 511*, 122–128.
Nakamura, Y., Tsiairis, C. D., Ozbek, S., & Holstein, T. W. (2011). Autoregulatory and repressive inputs localize Hydra Wnt3 to the head organizer. *Proceedings of the National Academy of Sciences of the United States of America, 108*, 9137–9142.
Newmark, P. A., & Sanchez Alvarado, A. (2000). Bromodeoxyuridine specifically labels the regenerative stem cells of planarians. *Developmental Biology, 220*, 142–153.
Nye, H. L., Cameron, J. A., Chernoff, E. A., & Stocum, D. L. (2003). Regeneration of the urodele limb: A review. *Developmental Dynamics, 226*, 280–294.
Oppenheimer, J. M. (1936). Structures developed in amphibians by implantation of living fish organizers. *Proceedings of the Society for Experimental Biology and Medicine, 34*, 461–463.

Owen, J. H., Wagner, D. E., Chen, C. C., Petersen, C. P., & Reddien, P. W. (2015). Teashirt is required for head-versus-tail regeneration polarity in planarians. *Development, 142*, 1062–1072.

Paulus, T., & Muller, M. C. (2006). Cell proliferation dynamics and morphological differentiation during regeneration in Dorvillea bermudensis (Polychaeta, Dorvilleidae). *Journal of Morphology, 267*, 393–403.

Pellettieri, J., Fitzgerald, P., Watanabe, S., Mancuso, J., Green, D. R., & Sanchez Alvarado, A. (2010). Cell death and tissue remodeling in planarian regeneration. *Developmental Biology, 338*, 76–85.

Petersen, H. O., Hoger, S. K., Looso, M., Lengfeld, T., Kuhn, A., Warnken, U., et al. (2015). A comprehensive transcriptomic and proteomic analysis of Hydra head regeneration. *Molecular Biology and Evolution, 32*, 1928–1947.

Petersen, C. P., & Reddien, P. W. (2008). Smed-betacatenin-1 is required for anteroposterior blastema polarity in planarian regeneration. *Science, 319*, 327–330.

Petersen, C. P., & Reddien, P. W. (2009). A wound-induced Wnt expression program controls planarian regeneration polarity. *Proceedings of the National Academy of Sciences of the United States of America, 106*, 17061–17066.

Petersen, C. P., & Reddien, P. W. (2011). Polarized notum activation at wounds inhibits Wnt function to promote planarian head regeneration. *Science, 332*, 852–855.

Plouhinec, J. L., Zakin, L., Moriyama, Y., & De Robertis, E. M. (2013). Chordin forms a self-organizing morphogen gradient in the extracellular space between ectoderm and mesoderm in the Xenopus embryo. *Proceedings of the National Academy of Sciences of the United States of America*, 20372–20379.

Price, J. S., Allen, S., Faucheux, C., Althnaian, T., & Mount, J. G. (2005). Deer antlers: A zoological curiosity or the key to understanding organ regeneration in mammals? *Journal of Anatomy, 207*, 603–618.

Rajnoch, C., Ferguson, S., Metcalfe, A. D., Herrick, S. E., Willis, H. S., & Ferguson, M. W. (2003). Regeneration of the ear after wounding in different mouse strains is dependent on the severity of wound trauma. *Developmental Dynamics, 226*, 388–397.

Reuter, H., Marz, M., Vogg, M. C., Eccles, D., Grifol-Boldu, L., Wehner, D., et al. (2015). Beta-catenin-dependent control of positional information along the AP body axis in planarians involves a teashirt family member. *Cell Reports, 10*, 253–265.

Reversade, B., & De Robertis, E. M. (2005). Regulation of ADMP and BMP2/4/7 at opposite embryonic poles generates a self-regulating morphogenetic field. *Cell, 123*, 1147–1160.

Roberts-Galbraith, R. H., & Newmark, P. A. (2013). Follistatin antagonizes activin signaling and acts with notum to direct planarian head regeneration. *Proceedings of the National Academy of Sciences of the United States of America, 110*, 1363–1368.

Saito, Y., Koinuma, S., Watanabe, K., & Agata, K. (2003). Mediolateral intercalation in planarians revealed by grafting experiments. *Developmental Dynamics, 226*, 334–340.

Salo, E., & Baguna, J. (1984). Regeneration and pattern formation in planarians. I. The pattern of mitosis in anterior and posterior regeneration in Dugesia (G) tigrina, and a new proposal for blastema formation. *Journal of Embryology and Experimental Morphology, 83*, 63–80.

Sarras, M. P., Jr. (2012). Components, structure, biogenesis and function of the Hydra extracellular matrix in regeneration, pattern formation and cell differentiation. *International Journal of Developmental Biology, 56*, 567–576.

Saunders, J. W., Jr. (1948). The proximo-distal sequence of origin of the parts of the chick wing and the role of the ectoderm. *Journal of Experimental Zoology, 108*, 363–403.

Scimone, M. L., Lapan, S. W., & Reddien, P. W. (2014). A forkhead transcription factor is wound-induced at the planarian midline and required for anterior pole regeneration. *PLoS Genetics, 10*, e1003999.

Shih, J., & Fraser, S. E. (1996). Characterizing the zebrafish organizer: Microsurgical analysis at the early-shield stage. *Development, 122*, 1313–1322.
Shimizu, H. (2012). Transplantation analysis of developmental mechanisms in Hydra. *International Journal of Developmental Biology, 56*, 463–472.
Spemann, H., & Mangold, H. (1924). Über die induktion von embryonalanlagen durch implantation artfremder organisatoren. *Wilhelm Roux's Archiv für mikroskopische Anatomie und Entwicklungsmechanik, 100*, 599–638.
Takano, J., & Sugiyama, T. (1983). Genetic analysis of developmental mechanisms in hydra. VIII. Head-activation and head-inhibition potentials of a slow-budding strain (L4). *Journal of Embryology and Experimental Morphology, 78*, 141–168.
Tanaka, E. M., & Reddien, P. W. (2011). The cellular basis for animal regeneration. *Developmental Cell, 21*, 172–185.
Technau, U., Cramer von Laue, C., Rentzsch, F., Luft, S., Hobmayer, B., Bode, H. R., et al. (2000). Parameters of self-organization in Hydra aggregates. *Proceedings of the National Academy of Sciences of the United States of America, 97*, 12127–12131.
Thisse, B., & Thisse, C. (2015). Formation of the vertebrate embryo: Moving beyond the Spemann organizer. *Seminars in Cell & Developmental Biology, 42*, 94–102.
Tickle, C., Summerbell, D., & Wolpert, L. (1975). Positional signalling and specification of digits in chick limb morphogenesis. *Nature, 254*, 199–202.
Turing, A. M. (1952). The chemical basis of morphogenesis. *Philosophical Transactions of the Royal Society B, 237*, 37–72.
Vasquez-Doorman, C., & Petersen, C. P. (2014). zic-1 Expression in Planarian neoblasts after injury controls anterior pole regeneration. *PLoS Genetics, 10*, e1004452.
Vogg, M. C., Owlarn, S., Perez Rico, Y. A., Xie, J., Suzuki, Y., Gentile, L., et al. (2014). Stem cell-dependent formation of a functional anterior regeneration pole in planarians requires Zic and Forkhead transcription factors. *Developmental Biology, 390*, 136–148.
Vriz, S., Reiter, S., & Galliot, B. (2014). Cell death: A program to regenerate. *Current Topics in Developmental Biology, 108*, 121–151.
Waddington, C. H. (1932). Experiments on the development of chick and duck embryos, cultivated in vitro. *Philosophical Transactions of the Royal Society B, 221*, 179–230.
Waddington, C. H. (1933). Induction by the primitive streak and its derivatives in the chick. *Journal of Experimental Biology, 10*, 38–46.
Wang, H., Loof, S., Borg, P., Nader, G. A., Blau, H. M., & Simon, A. (2015). Turning terminally differentiated skeletal muscle cells into regenerative progenitors. *Nature Communications, 6*, 7916.
Warmflash, A., Sorre, B., Etoc, F., Siggia, E. D., & Brivanlou, A. H. (2014). A method to recapitulate early embryonic spatial patterning in human embryonic stem cells. *Nature Methods, 11*, 847–854.
Wehner, D., Cizelsky, W., Vasudevaro, M. D., Ozhan, G., Haase, C., Kagermeier-Schenk, B., et al. (2014). Wnt/beta-catenin signaling defines organizing centers that orchestrate growth and differentiation of the regenerating zebrafish caudal fin. *Cell Reports, 6*, 467–481.
Wenger, Y., Buzgariu, W., & Galliot, B. (2015). Loss of neurogenesis in Hydra leads to compensatory regulation of neurogenic and neurotransmission genes in epithelial cells. *Philosophical Transactions of the Royal Society B, 371*, 1–14.
Wilby, O. K., & Webster, G. (1970). Studies on the transmission of hypostome inhibition in hydra. *Journal of Embryology and Experimental Morphology, 24*, 583–593.
Wittlieb, J., Khalturin, K., Lohmann, J. U., Anton-Erxleben, F., & Bosch, T. C. (2006). Transgenic Hydra allow in vivo tracking of individual stem cells during morphogenesis. *Proceedings of the National Academy of Sciences of the United States of America, 103*, 6208–6211.

Wolpert, L., Clarke, M. R., & Hornbruch, A. (1972). Positional signalling along hydra. *Nature: New Biology*, *239*, 101–105.

Xu, P. F., Houssin, N., Ferri-Lagneau, K. F., Thisse, B., & Thisse, C. (2014). Construction of a vertebrate embryo from two opposing morphogen gradients. *Science*, *344*, 87–89.

Yao, T. (1945). Studies on the organizer problem in Pelmatohydra oligactis: II. The effect of som respiratory inhibitors and stimulants and of oxygen deficiency on the induction potency of the hypostome. *Journal of Experimental Biology*, *21*, 151–154.

Yu, J. K., Satou, Y., Holland, N. D., Shin, I. T., Kohara, Y., Satoh, N., et al. (2007). Axial patterning in cephalochordates and the evolution of the organizer. *Nature*, *445*, 613–617.

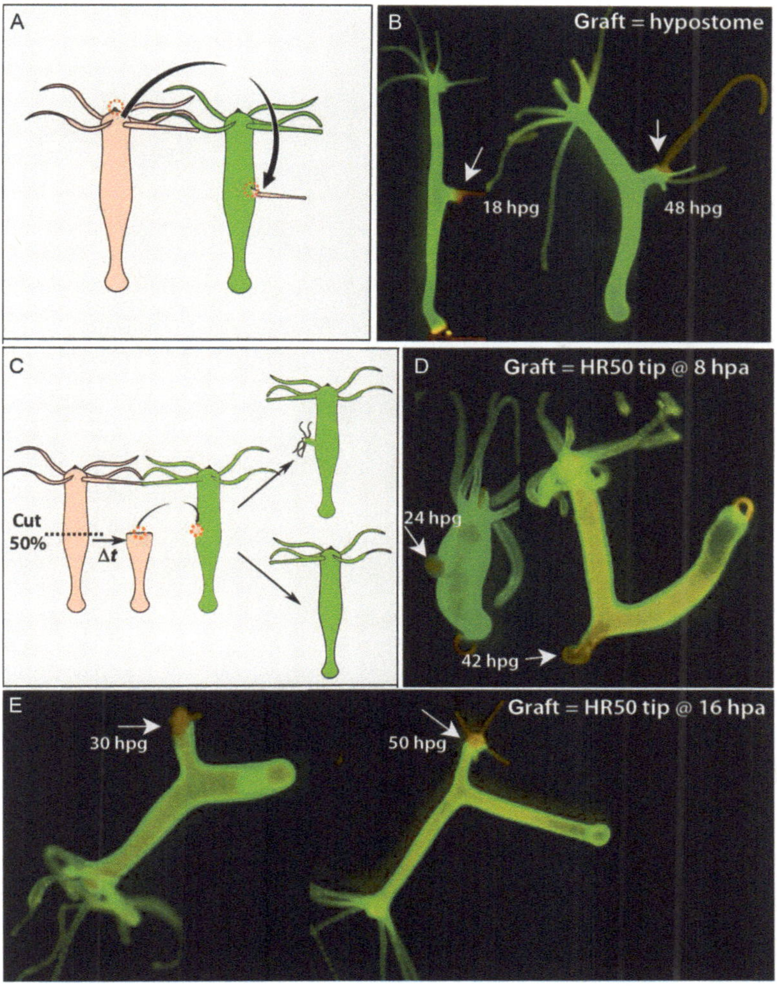

Matthias C. Vogg *et al.*, Figure 1 Homeostatic- and regeneration-induced head organizers in *Hydra* as initially characterized by Browne (1909). Reproduction of Ethel Browne's original experiments demonstrating the ability of the *Hydra* head or the *Hydra* head-regenerating tip to induce a complete ectopic axis (Browne, 1909). Here, we used as hosts two *AEP* actin::eGFP transgenic strains that constitutively expresses GFP either in the endodermal (Wittlieb, Khalturin, Lohmann, Anton-Erxleben, & Bosch, 2006) or in the ectodermal (Anton-Erxleben, Thomas, Wittlieb, Fraune, & Bosch, 2009) epithelial cells, named Endo-GFP and Ecto-GFP, respectively. (A) A fragment of head bearing a tentacle was dissected from a nonfluorescent *Hydra AEP* and immediately transplanted to the lateral side of the body column of the Endo-GFP host. As in Browne's experiment, the tentacle was used as a marker, which does not have organizing activity itself. (B) An ectopic axis that contains endodermal fluorescent cells from the host is rapidly induced, observed here at 18 and 48 hours postgrafting (hpg, arrows). (C) A nonfluorescent *Hydra AEP* was bisected at midgastric position (50%), the lower part was allowed to regenerate for 8 (D) or 16 (E) hours, and the tip was transplanted to the lateral side of the body column of a Ecto-GFP host. (D) Ectopic axis induced by a 8 h postamputation (hpa) graft visualized at 24 and 42 hpg (arrows). (E) Ectopic axis induced by a 16 hpa graft visualized at 30 and 50 hpa (arrows). In (D and E), the progressive growth and patterning of the ectopic axis is obvious. However, the development of the structure induced by the 16 hpa graft is more advanced than the one induced by the 8 hpa graft, illustrating the fact that the inductive capability of the tip increases within the first 12 h of head regeneration (MacWilliams, 1983b).

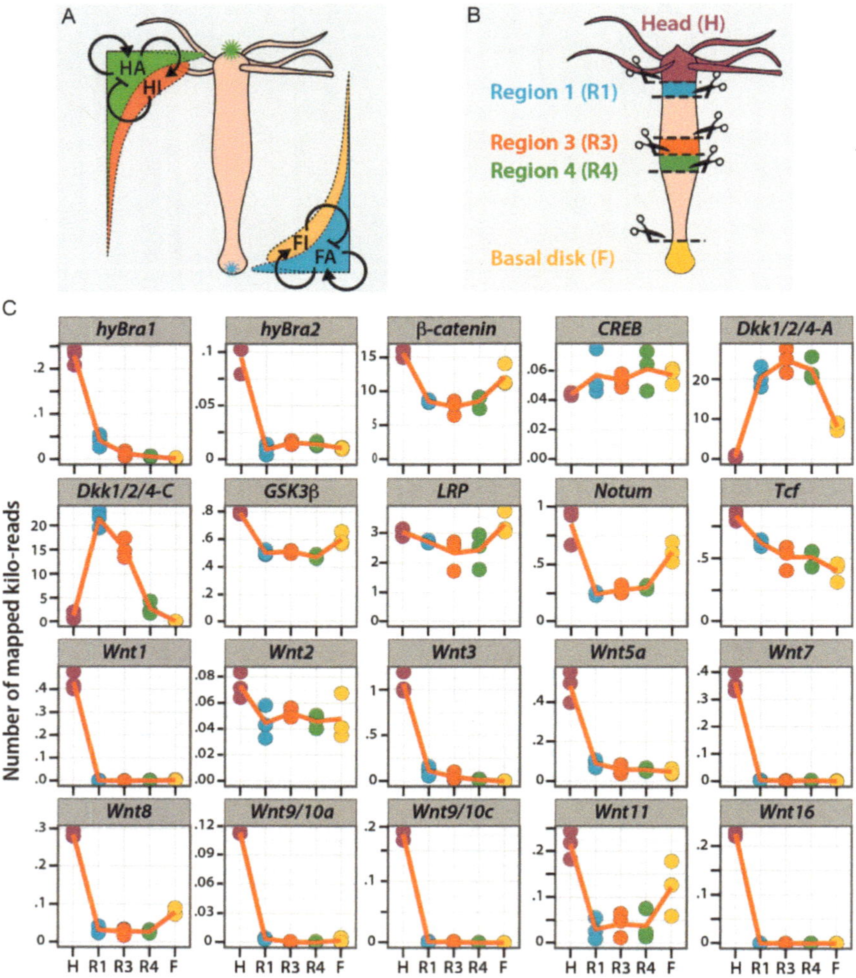

Matthias C. Vogg *et al.*, Figure 2 Molecular components of organizer maintenance in homeostatic *Hydra*. (A) Schematic representation of the two pairs of graded biological activities, head activation (HA, green) and head inhibition (HI, red), foot activation (HA, blue) and foot inhibition (FI, golden) as detected by lateral transplantation experiments (see Shimizu, 2012). Note their inverted graded distribution, maximal at the apical pole for HA/HI, at the basal pole for FA/FI. Green and blue stars represent head and foot organizers, respectively. (B) Scheme depicting the procedure used to perform quantitative RNA-seq transcriptomics at five distinct levels along the *Hydra* body axis (for details, see in Wenger, Buzgariu, & Galliot, 2015). (C) RNA-seq profiles of components, putative regulators and target genes of the Wnt/β-catenin pathway, involved in the maintenance and regeneration of the head organizer (Chera et al., 2009; Chera, Ghila, Wenger, & Galliot, 2011; Chera, Kaloulis, & Galliot, 2007; Guder et al., 2006; Hobmayer et al., 2000; Lengfeld et al., 2009; Nakamura, Tsiairis, Ozbek, & Holstein, 2011; Petersen et al., 2015; Technau et al., 2000). The color code for each position corresponds to colors selected in panel (B). Note that wnt signaling in *Hydra* specifies head identity whereas it specifies tail identity in planarians (see Fig. 3).

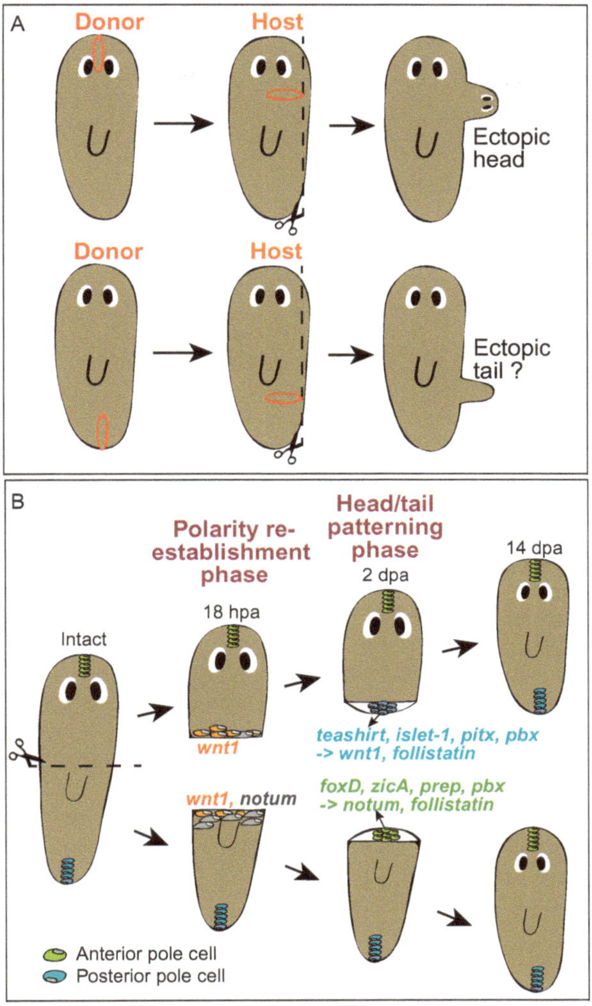

Matthias C. Vogg *et al.*, Figure 3 Anterior and posterior poles might represent putative head and tail organizers in homeostatic and regenerating planarians. (A) Transplantation experiment according to Saito, Koinuma, Watanabe, and Agata (2003). Tissue of the region between the eyes was transplanted into a host planarian that was subsequently cut laterally. Head outgrowth was observed in transplanted animals, suggesting that the transplanted tissue has head organizer activity. A similar scenario is expected to be true for the transplantation of midline tail tissue: the outgrowth of a tail on the host would indicate organizer activity of the tail transplant. (B) Amputation at any level along the animal induces *wnt1* expression at anterior- and posterior-facing wound sites at 18 hours postamputation (hpa) (Petersen & Reddien, 2009). At anterior-facing wound sites, the activity of Wnt1 is antagonized by the secreted Wnt inhibitor Notum, which triggers

(Continued)

Matthias C. Vogg *et al.*, Figure 3—Cont'd anterior polarity reestablishment (Petersen & Reddien, 2011). In contrary, at posterior wounds, active Wnt signaling is necessary for posterior polarity reestablishment. Note that at 18 hpa *notum* and *wnt1* expression is induced in fully differentiated cells, whereas at 48 hpa on the posterior side, the transcription factor genes *islet-1* (Hayashi et al., 2011), *pitx* (Currie & Pearson, 2013; Marz, Seebeck, & Bartscherer, 2013), *pbx* (Blassberg, Felix, Tejada-Romero, & Aboobaker, 2013; Chen, Wang, & Reddien, 2013), and *teashirt* (Owen, Wagner, Chen, Petersen, & Reddien, 2015; Reuter et al., 2015) (blue) are upregulated in neoblasts that subsequently differentiate in pole cells that express *wnt1* (Petersen & Reddien, 2009) and *follistatin* (Gavino, Wenemoser, Wang, & Reddien, 2013). Similarly on the anterior side, the transcription factor genes *foxD, zicA, prep,* and *pbx* (Blassberg et al., 2013; Chen et al., 2013; Felix & Aboobaker, 2010; Scimone, Lapan, & Reddien, 2014; Vasquez-Doorman & Petersen, 2014; Vogg et al., 2014) are upregulated in neoblasts that provide the pole cells expressing *notum* (Petersen & Reddien, 2011) and *follistatin* (Gavino et al., 2013; Roberts-Galbraith & Newmark, 2013).

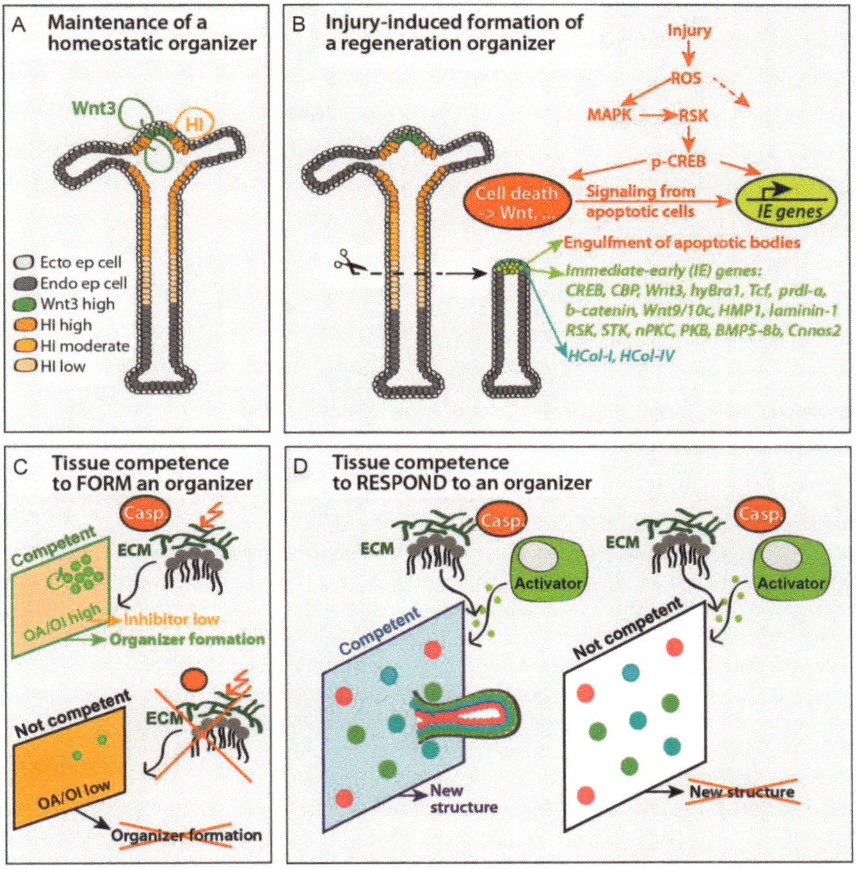

Matthias C. Vogg *et al.*, Figure 4 Organizer maintenance, organizer *de novo* formation upon injury, and competence to respond to organizer activity in an adult organism. (A) Organizer activity is maintained in homeostatic *Hydra* thanks to the head activator Wnt3 that maintains active β-catenin signaling in the head through a positive feedback loop (Nakamura et al., 2011). Note that head inhibitor (HI), which is expressed under the control of Wnt3 and antagonizes Wnt3 signaling, is not characterized yet. (B) *De novo* formation of a head organizer in head-regenerating tips after midgastric bisection relies on the asymmetric activation of the MAPK/CREB pathway (Galliot, Welschof, Schuckert, Hoffmeister, & Schaller, 1995; Kaloulis, Chera, Hassel, Gauchat, & Galliot, 2004), which together with injury-induced ROS signaling, triggers local cell death in the interstitial lineage. Apoptotic cells transiently release signaling molecules, including Wnt3, which activates the β-catenin pathway in surrounding cells, inducing a wave of mitotic division (Chera et al., 2009). At the same time, the engulfment of the apoptotic bodies by the endodermal epithelial cells (green) is followed by the upregulation of immediate-early (IE) genes, including *wnt3* that contains CREB response elements (CREs) in its regulatory sequence (Galliot, 2013; Nakamura et al., 2011). (C) Tissue competence to form an

(Continued)

Matthias C. Vogg *et al.*, Figure 4—Cont'd organizer relies on the respective activities of Activator (OA) and Inhibitor (OI) components, where OA are extracellular signal(s) and OI inhibitors that restrict OA extracellular activity. Only tissues with a high OA/OI ratio can form an organizer. Activators (i.e., morphogens) can be secreted as soluble molecules from the producing cells but also released from the ECM by proteolytic cleavage (Fleury et al., 2006). Once in the extracellular space, several mechanisms can promote their long-range graded activity such as the diffusion and the accumulation of proteases or protease inhibitors that ultimately regulate spatially the stability of proteins involved in morphogenesis (Inomata et al., 2013). Note that in *Hydra*, injury-induced modifications of the extracellular matrix (ECM, red flash) are necessary for regeneration (Sarras, 2012), possibly to increase the OA/OI ratio in the vicinity of the wound, up to the point where tissues become competent to form an organizer. Similarly in adult zebrafish regenerating its fins, signaling from the apoptotic cells and/or from the apoptotic bodies is necessary for the upregulation of IE genes in surrounding cells and subsequent tissue regeneration (Vriz, Reiter, & Galliot, 2014). (D) Tissue competence to respond to an organizer relies on the strength of the intracellular signaling pathway(s) activated by the activator. Tissues with a high signaling activity (blue background) are able to induce ectopic structures in response to OA (left), while tissues with a low signaling activity (white background) are unable to respond (right). Casp., caspase activation.

CHAPTER TWENTY-TWO

Developmental Plasticity and Developmental Symbiosis: The Return of Eco-Devo

Scott F. Gilbert[1]
Swarthmore College, Swarthmore, Pennsylvania, USA
[1]Corresponding author: e-mail address: sgilber1@swarthmore.edu

Contents

Abstract

Ecological developmental biology is the study of the interactions between developing organisms and their environments. Organisms have evolved to use the environment as a source of important cues that can alter the trajectory of their development. First, *developmental plasticity* enables the genome to generate a repertoire of possible phenotypes, and environmental cues are often used to select the phenotype that appears most adaptive at that time. This facilitates evolutionary strategies such as phenotypic accommodation, genetic assimilation, and niche construction. Second, *developmental symbiosis*, wherein the developing animal utilizes cues from other organisms for normal cell differentiation and morphogenesis, has been found to be ubiquitous. The coevolution of symbiotic microbes and animal cells has often led to the dependency of an animal's development on particular microbial signals, making these cues essential and expected components of normal development.

1. ECOLOGICAL DEVELOPMENTAL BIOLOGY

Ecological developmental biology (Eco-Devo; Gilbert, 2001) studies the relationships between developing organisms and their environments. It is where developmental biology leaves the well-controlled

Current Topics in Developmental Biology, Volume 116
ISSN 0070-2153
http://dx.doi.org/10.1016/bs.ctdb.2015.12.006

conditions of the laboratory to study development in the more natural contexts of predators, conspecifics, temperatures, diets, microbes, and stresses. What one finds is that development trajectories in the wild are not necessarily the same as developmental trajectories in the laboratory. Moreover, the ability of the environment to alter such developmental trajectories may be critically important in evolution, conservation biology, and medicine.

Ecological developmental biology is one of the oldest and one of the newest areas of experimental embryology. Indeed, it was probably the first major developmental research tradition outside that of regeneration (Nyhart, 1995). Presently, eco-devo consists of two disciplinary subsets. The first involves environmental disruptions of development. In this category, are the disciplines of teratology, endocrine disruption, and the developmental origins of adult health and diseases (DOAHD). These are largely medical fields and are associated with human developmental pathologies. The second set of disciplines are those studying the roles of environmental factors in facilitating normal development. These fields include those of developmental plasticity and developmental symbiosis. Developmental plasticity is the concept that a single genome can generate several different possible phenotypes, and that the environment plays an active role in eliciting a particular phenotype from this repertoire (Nijhout, 2003; Schlichting & Pigliucci, 1998; West-Eberhard, 2003). Developmental symbiosis is the concept that certain processes of normal development are mediated by signals coming from different species of organisms (McFall-Ngai, 2002). In other words, animal bodies are, in part, constructed by expected signals coming from different, symbiotic, species. This chapter will focus primarily on developmental symbiosis and developmental plasticity in animals. Such symbioses and plasticity have played major roles in the study of plant development; but their importance in animal development has only been recently appreciated (see Gilbert & Epel, 2015).

2. THE CENTURY OF THE GENE: THE DECLINE OF ECO-DEVO

Ecological developmental biology gives the environment instructive agency during development. The genotype can generate several different phenotypes, depending on environmental conditions. This was actually one of the first models proposed to explain development.

The last decade of the 1800s saw a controversy between August Weismann, who proposed that all development was what we would call a read-out from the nuclear genes, and Oscar Hertwig who hypothesized that development involved not only the interactions between embryonic cells but also the interactions between developing organisms and their environments. Hertwig's (1894) volume *The Biological Problem of Today: Preformation or Epigenesis?* cataloged cases of developmental plasticity, including context-dependent sex determination in *Bonellia viridis*, temperature-dependent sex determination in rotifers, nutrition-dependent production of workers and queens in ant colonies, and the temperature-dependent pigmentation patterns of butterfly wings. Hertwig wrote (1894, p. 122), "These seem to me to show how very different final results may grow from identical rudiments, if these, in their early stages of development, be subjected to different external influences."

Weisman's notions were still being fought in 1909, when two publications brought the concept of phenotypic plasticity to the awareness of many biologists. The Danish biologist Wilhelm Johannsen's (1909) *Elemente der Exakten Erblichkeitslehre* made clear the distinction between the genotype and phenotype. Rejecting Weismann's (1893) proposal that all the causes of an embryo's development were compressed into the nucleus of the egg, Johannsen stated specifically that phenotype depends on the interactions of inherited genes with components of the environment. Like Hertwig, Johannsen felt that early development was genetically controlled but that the environment could effect changes in the later developmental stages (Moss, 2003; Roll-Hansen, 2007, 2014).

Also that same year, the German biologist Richard Woltereck reported that genetically identical lines of *Daphnia* (a water flea that reproduces asexually) could produce different phenotypes during different times of the year. Woltereck argued that what actually was inherited was the potential to generate an almost infinite number of small variations in phenotype in response to environmental cues. He called this potential the *Reaktionsnorm* (reaction norm or norm of reaction). Moreover, different genetically pure lines responded differently to seasonal cues. This suggested to Woltereck the *Reaktionsnorm* was what was inherited from generation to generation, and that hereditary change was an alteration of the *Reaktionsnorm*. Woltereck went so far as to identify the *Reaktionsnorm* with the genotype: "*Der 'Genotypus' ... eines Quantitativmerkmals ist die vereberte Reaktionsnorm*" (Woltereck, 1909, p. 136; Sarkar, 1999). Johannsen agreed, saying that Woltereck's *Reaktionsnorm* was "nearly synonymous" with his own conception of genotype (Johannsen, 1911).

This view that the genotype could generate multiple context-dependent phenotypes was an active research program that was soon to be eclipsed. Developmental biology was going indoors. Embryology was to emulate anatomy and physiology, not natural history and ecology. First, the microtome had revolutionized developmental *anatomy*, and shortly afterward, the new science of *Entwicklungsmechanik* would look at the *physiological* mechanisms through which the fertilized egg generated the newborn. As Roux (1894) made clear, these questions of how a fertilized egg developed its cells and organs were important in their own right and did not need to be placed into an evolutionary or ecological framework.

The older embryologists lamented this. Haeckel (1880) wrote to Wilhelm Peters that he feared that their students "will only know cross-sections and stained tissues, but neither the entire animal nor its mode of life." Siebold (1876) was particularly agitated, writing that he was astonished that such a newly trained embryologist could cut an animal embryo into myriads of sections, but "doesn't even know what a Phryganide's casing looks like."[1] Siebold also said that he could understand the urge to go indoors. His students were under such demands to publish and these indoor studies took less time (Nyhart, 1995). The more things change...

As embryology went indoors, two research programs became paramount. One was to determine whether the nucleus or the cytoplasm controlled developmental outcomes. The second concerned the mechanisms through which morphogenesis occurred. The investigators working on the first problem included embryologists such as Thomas Hunt Morgan. Eventually, they settled on the fruit fly as their model system, began calling themselves geneticists, and studied the inheritance of nuclear genes (Gilbert, 1978, 1998). The researchers investigating the second question called themselves experimental embryologists and looked at the questions of "*Gestaltungsgeschichte*," the rules of morphological order (Needham, 1936). Thus, the field was split into two disciplines that competed for explaining development. When Sturtevant (1923) reanalyzed in a genetic manner Boycott's embryological data on snail coiling, he got results that seemed to show genetic methods superior, and in a particularly nasty review paper, Morgan (1932) proclaimed that genetics had superseded embryology and that genetics was, in fact, the truest way to solve embryological problems. The fields went their separate ways, one into fruit fly genes, the other into vertebrate organizers.

[1] Phryganides are the large caddisflies whose larvae make impressive casings by collecting small pebbles and bonding together with silk extruded from its mouth. Their study demanded being near slow moving streams or ponds. I had to look this up, too; so I guess Siebold was right.

Plasticity was largely eliminated from the early research programs of genetics, since such deviations were considered noise. Plasticity was the enemy of those "hereditarians" who sought a genetic determinism. Among them was another embryologist-turned-geneticist, Charles Davenport. Davenport became one of the major proponents of eugenics in the United States, campaigning actively against developmental plasticity, the inheritance of acquired characters, and the importation of "inferior" non-northern European genes into the United States (M. Rich, unpublished). "We are what our chromosomes make us," he declared. Plasticity remained a viable topic in experimental embryology, especially in Austria and the Soviet Union. However, after the World War II, the German and Austrian laboratories (especially the Prater Vivarium of Vienna, which had been the center of plasticity studies) were destroyed. Many of the scientists had committed suicide or killed in Nazi concentration camps. The Soviet research programs, which, under A. Severtsov and I.I. Schmalhausen, had attempted to integrate genetics, embryology, ecology, and evolution (see Adams, 1980), became dominated by the followers of Trofim Lysenko, and provided ideologically suspect and poor science. Lysenko thought genes to be the product of the capitalist imagination and taught that variation was directed completely through environmental factors. In this, he had the full support of Stalin. Thus, genetics in the communist countries crumbled and was replaced by an environmentalist pseudoscience (see Krementsov, 1998; Roll-Hansen, 2005).

Meanwhile, in the West, the reaction against Lysenko caused a further hardening against environmental determination of phenotypes. Nongenetic sources of variation became marginalized and stopped being taught, funded, or discussed (Lewontin & Levins, 1976; Lindegren, 1966; Sapp, 1987). In 1950, the Genetics Society of America hired the public relations firm of Pendrey and Liebert to spread the word that genes provided the basis for organismal characters. The strategy was not to condemn Lysenkoism (which might have disastrous effects on any geneticists still alive in Russia), but to inoculate Americans against the Communist threat. Stressing the agricultural and medical benefits of Mendelian genetics, this was a hugely successful campaign (Wolfe, 2010), and Americans came to have a view of genetics as a friendly, useful, American, science, that explained variation. As genetic explanations became more and more important, genetics became melded with evolution and physiology, and *developmental genetics* was seen as the best way to explain animal development (Gilbert, 1996). Environmental contributions to development were ignored, as molecular biology became increasingly successful at explaining cell differentiation.

Current Topics in Developmental Biology was formed in this milieu, and the term "developmental biology" was to indicate the more molecular aspects of the field, distinguishing it from "embryology." The preface by Monroy and Moscona (1966) emphasizes this change toward a new molecular and biochemical approach to development, and the opening remarks and *raison d'être* for these volumes were written not by a developmental biologist, but by microbial geneticist Joshua Lederberg. In his remarkably prescient essay (Lederberg, 1966), he welcomes embryology into the new world of molecular biology, saying that it is about time. He was dismissive of non-genetic aspects to development, wondering if any really do exist, and he proposed that developmental biologists emulate the phage and microbial geneticists, forming research communities around a few model genetically tractable organisms such as a nematode and a mouse. The developmental genetic approach to development has been incredibly successful, and in 1991, Lewis Wolpert could enthusiastically tell the public the generally accepted view that "DNA provides the programme which controls the development of the embryo." Thus, population genetics explained evolution, and developmental genetic explained the formation of the embryo. As Keller (2000) has claimed, it was the "Century of the Gene."

3. DEVELOPMENTAL PLASTICITY

Toward the end of that century, evolutionary developmental biology had started challenging the notion that evolutionary biology was fully contained within and sufficiently explained by population genetics (Gilbert, Opitz, & Raff, 1996; Raff, 1996; Raff & Kaufman, 1983). Much of this evo-devo sought to explain evolutionary change through changes if the expression of developmental regulatory genes. This eventually become integrated into evolutionary biology (Laland et al., 2014). Another subset of evo-devo, however, was influenced by studies of reaction norms, genetic assimilation, and other phenomena linking development to the environment (Schlichting & Pigliucci, 1998; West-Eberhard, 2003). This program was set to challenge the idea that everything needed for normal development was contained in the fertilized egg. By the beginning of the new millennium, new trends, including environmental consciousness, conservation biology, endocrine disruption, and techniques of PCR and high-throughput RNA analysis, converged to bring forth ecological developmental biology (Gilbert, 2001; Gilbert & Bolker, 2003). The first conference on Ecological Developmental Biology was organized at the American Society of Zoology meeting in Anaheim, January 2002. This symposium brought together many of the disparate research groups in this area. First, the studies by Fred Nijhout and Ehab Abouheif showed how butterfly and ant polyphenisms (alternative phenotypes) could be brought about through the environment's activation of hormones. The neuroendocrine signaling pathway enabled the outside environment to create changes in the internal hormonal milieu, which could then alter gene expression. Work by John Gilbert, David Crews, and John Godwin showed that sex could be determined by environmental factors in vertebrates and invertebrates, and Sonia Sultan and Lynda Delph demonstrated that developmental plasticity enabled plants to expand their ranges by allocating resources differently to germ line and soma. Even physical forces from the environment were involved in phenotype production, as shown by Steve Moore and Gerd Muller. In addition, the mechanisms for possible pathological relationships between the developing organism and its environment were presented by the laboratories of Ana Soto and Anne McNabb. David Epel, Lisa Belden, and Andrew Blaustein presented evidence on how animal embryos prepare defenses against environmental assaults. Carl Schlicting, Trevon Fuller, and Sahotra Sarkar presented articles on the mechanisms for evolving developmental plasticity. The success of the meeting could be measured in its Q&A session having to be continued in a new auditorium, the exchange of business cards among the speakers and listeners, and a review in *Nature* (Duscheck, 2002).

While this meeting highlighted developmental plasticity, two other paradigmatic programs were coming into their own, and each of these highlighted developmental symbiosis. Margaret McFall-Ngai was showing that the light organ of the bobtail squid was not generated without bacteria, and such specific bacteria became an integral part of normal squid development. Her 2002 review, "Unseen forces: the influence of bacteria on animal development" showed that in addition to the traditional notions of developmental plasticity, there was an intimate "other" involved in producing animal phenotypes—the microbial symbiont. That message was being forcibly emphasized in mammals by work by L. Hooper and T.S. Stappenbeck in Jeffrey Gordon's laboratory. Using high-throughput RNA analysis, they showed that intestinal microbes were acting just as normal embryonic cells do: they produced chemicals that induced different patterns of gene expression in the neighboring epithelial cells (Hooper et al., 2001; Stappenbeck, Hooper, & Gordon, 2002). These genes were critical in producing the gut-associated lymphoid tissues and the villi capillary network. In other words, mammalian development was not complete without microbial interaction.

Thus, by 2003, ecological developmental biology was being consolidated as a new part of developmental biology. Plasticity remains a major part of that field. Seasonal polyphenism (the ability of an organism to have two different phenotypes, depending on the temperature) has become well documented, especially in cases such as the wing pattern of the butterfly *Bicyclus anynana*, where the behaviors and wing pattern change to give protection during both seasons (Beldade, Mateus, & Keller, 2011; Monteiro et al., 2013). The basis of temperature-induced sex determination in reptiles is being placed on a molecular level with new discoveries about aromatase and DMRT genes (Bieser & Wibbels, 2014; Matsumoto, Buemio, Chu, Vafaee, & Crews, 2013). The environmental bases of caste formation in bees and in ants have been taken to new levels, as the mechanisms by which hormones can induce different patterns of gene expression in different environmental contexts takes on new urgency (Kamakura, 2011; Rajakumar et al., 2012). Predator-induced changes in morphologies and behaviors have become documented across the animal kingdom and have been linked to critical environmentally induced changes in developmental physiology (Dennis, LeBlanc, & Beckerman, 2014; Middlemis Maher, Werner, & Denver, 2013).

Moreover, as more examples of developmental plasticity were documented molecularly, it appeared that the observable changes in morphology were tightly integrated with changes in the animals' behaviors. This is seen in the environmental determination of male dung beetle horns and mating behaviors (Emlen & Nijhout, 1999; Moczek & Emlen, 2000), in the determination of locust morphological and behavioral phenotypes due to conspecific crowding (Rogers et al., 2004; Simões, Niven, & Ott, 2013), and in the changes in fish that alter sex in response to the population structure (Godwin, Sawby, Warner, Crews, & Grober, 2000; Marsh-Hunkin, Heinz, Hawkins, & Godwin, 2013).

Some of the most interesting environmental changes in behavioral development have shown the importance of maternal care on mice (Cameron et al., 2008). Maternal licking and grooming establish hormonal circuits that eventually demethylate a brain-specific enhancer of the glucocorticoid receptor gene, enabling it to be transcriptionally active in the brain (Weaver et al., 2004). This helps the mouse to relieve anxiety by down-regulating the stress response. Such maternal care also enables to demethylation of certain promoters of the estrogen receptor gene, which actually stimulates maternal licking and grooming (Champagne et al., 2006). So here, the hereditary transmission of an anxious predisposition is not due to DNA alleles. It is due to whether specific regions of DNA are demethylated by hormones initiated or by environmental stimuli.

Developmental plasticity has profound implications for evolution and conservation biology. For example, developmental plasticity is the precondition for phenotypic accommodation, genetic accommodation, and niche construction. If development is more than changes in gene activity and involves environmental input, then evo-devo has to reflect this environmental component as well (Gilbert, Bosch, & Ledón-Rettig, 2015; Hall, 1992; Moczek, 2015; West-Eberhard, 2003). This has spawned a new field, ecological evolutionary developmental biology (eco-evo-devo; Abouheif et al., 2014; EDEN, 2010; Moczek, 2012).

One of the components of eco-evo-devo is phenotypic accommodation, the ability of embryonic or juvenile parts to self-regulate as they form, such that the environment can elicit coordinated and often adaptive changes in morphology. One popular example concerns mammals who are born with deformed forelimbs (West-Eberhard, 2003) and who adopt a bipedal mode of locomotion. Such bipedal dogs and goats develop muscular and skeletal connections appropriate to bipedal species. Moreover, when an extant species possessing ancestral traits has been challenged to elicit a phenotypic response by an environmental change that mimics an evolutionarily relevant environmental transition, these "ancestral" species can sometimes elicit phenotypic responses that mirror novel traits that are standard for derived lineages. For instance, one recent investigation studied the gate of the Nile birchir (*Polypterus birchir*), a fish that possesses both primitive and derived traits with respect to the evolution of terrestrial locomotion. Although this fish has the fins and body form typical of aquatic vertebrates, it can live on land and, when it does, emulate terrestrial locomotion with its pectoral fins. If plasticity facilitated the transition of limbed vertebrates from water to land, a species like *Polypterus* might display phenotypic responses to a terrestrial environment that mimic phenotypes observed in extant, terrestrial vertebrates. Standen, Du, and Larsson (2014) found that the behaviors of land-raised *Polypterus* resembled terrestrial motion relative much more than the land gates of those raised in the water. Most importantly, these behavioral differences in walking instigated alterations in skeletal growth, producing novel bone morphologies that mirrored evolutionary configurations seen in the stem tetrapods of the Devonian period. Thus, plasticity has the potential to generate, in as little as one generation, behavioral and anatomical changes that are adaptive for a major environmental transition.

But can such environmentally triggered changes in phenotype be transmitted genetically? This is the province of genetic assimilation, where there is intense selection placed on the environmentally induced phenotype such that it eventually becomes induced internally as part of the wild-type genome. There are numerous examples of this is the laboratory (Waddington, 1952, 1956; see Gilbert & Epel, 2015; Schmalhausen, 1949).

One of the best studied examples looked at the selection of the bithorax phenotype (wherein the third thoracic segment of a fly becomes a repeat of the second thoracic segment, creating two sets of wings) on flies whose larvae had been exposed to ether. Gibson and Hogness (1996) found that, under selection, ether sensitivity increased as did the prevalence of four specific mutant alleles of the *Ultrabithorax* (*Ubx*) gene in the biothorax complex. More recently, Suzuki and Nijhout (2006) performed selection experiments on plastic phenotypes using a variety of tobacco hornworm (*Manduca sexta*) that produces a gradation of black, pigmented cuticle at warmer temperatures. With selection, they were able to produce lineages that were either liberated from environmental control (monophenic black lines that were not temperature sensitive), or polyphenic lines that were very plastic and responded to temperature in a threshold fashion.

In nature, evidence for such phenotypic accommodation followed by genetic assimilation can be seen in several isolated populations that have been exposed to extreme selection pressure. One such candidate for natural genetic assimilation is the tiger snake (*Notechis scutatus*). Like many vertebrate species, this snake has a jaw structure that can be altered by diet. When faced with large prey, the tiger snake can develop a bigger head. This plasticity in head and jaw anatomy is in evidence when the snakes' diet includes both large and small mice. However, on some islands where the diet contains only large mice, the snakes are born with large heads, and there is no plasticity. Thus, Aubret and Shine (2009) conclude that their data may provide empirical evidence of genetic assimilation, "with the elaboration of an adaptive trait shifting from phenotypically plastic expression through to canalization within a few thousand years."

Developmental plasticity also promotes niche construction, wherein the phenotypes of organisms are plastic in response to the developmental environment that they, themselves, have altered (Moczek, 2015). For example, the larvae of goldenrod gall flies (*Eurosta solidaginis*) secrete factors that induce the stem of the goldenrod plant to form the gall in which it can reside and eat (Williams & Lee, 2005). Niche construction is evolutionarily relevant because it alters the selective environment of the organism eliciting the plastic response and, moreover (since there is genetic variation in niche constructing abilities), some genotypes are more likely than others to experience certain selective environments (Callahan, Fukami, & Fisher, 2014; Odling-Smee, Erwin, Palkovacs, Feldman, & Laland, 2013).

Niche construction can also generate fascinating population dynamics if the different species coconstruct each other's environments. For instance, in certain ponds in northern Japan, the larval salamander, *Hynobius retardatus*, lives in close proximity with the larva of the frog, *Rana pirica*. The salamander larvae will eat the frog larvae, and the frog larvae have a predator-induced polyphenism to become larger if the salamander larvae are present. The salamander larvae, moreover, can respond developmentally to the enlarged (but not the unenlarged) frog larvae by developing an enormous mouth that is capable of ingesting the bigger salamander larvae (Kishida, Costa, Tezuka, & Michimae, 2013; Kishida, Mizuta, & Nishimura, 2006). Niche construction highlights the evolutionary importance of non-genetic inheritance for establishing ecological legacies that can effect entire ecosystems (Laland et al., 2015).

Thus, phenotypic accommodation, genetic assimilation, and niche construction are three ways that developmental plasticity can be important in evolutionary and ecological processes.

This means that developmental plasticity can also play important roles in conservation biology. One reason for this importance is due to the fact that the metabolism of one environmentally induced form may differ substantially from that of the other. For example, the solitary form of the desert locust has small green wings and is not a major plant pest. However, the gregarious form, with its large brown wings, has a different metabolism, is not longer deterred by poisons made by the plants, and can become a major plague (Simões et al., 2013). More subtly, some frog species alter their detoxification metabolisms when challenged by predators. This makes them more susceptible to the toxic consequences of herbicides and pesticides. Thus, a pesticide that does not kill tadpoles in aquaria can wipe out a population that is in a lake and exposed to potential predators. This has major consequences for testing the weed-killers and bug-killers we spray on our fields (Relyea, 2003, 2004).

A second reason for the importance of developmental plasticity in ecology concerns global climate change. As early as 1982, turtle conservation biologists were being warned that incubating all one's eggs at one temperature would produce a monosexual species of turtles that probably would not have the best consequences for prolonging the species (Morreale, Ruiz, Spotila, & Standora, 1982). In many marine turtles, where females are produced at higher temperatures, the sex ratio is already skewed toward producing females, and in some rookeries of the endangered green turtle (*Chelonia mydas*), the sex ratio is 95:5 female:male. Changes in phenology (for instance the timing of insect eclosion relative to the timing of flower formation) and in the alkalinity of sea water may also have important developmental consequences (Hawkes, Broderick, Godfrey, & Godley, 2009; Rafferty et al., 2013; Telemeco, Abbott, & Janzen, 2013).

Developmental plasticity, the ability of the environment to control the specificity of the phenotype, is thus a critically important concept in development, evolution, and conservation biology.

4. DEVELOPMENTAL SYMBIOSIS

One part of the environment that deserves special attention is its microbes. We not only live in a mutualistic symbiosis with our microbes but also *develop* in mutual symbiosis with them (Gilbert, Sapp, & Tauber , 2012; McFall-Ngai et al., 2013). Moreover, bacterial cells are part of the animal developmental program. As mentioned earlier, the pioneering work of the McFall-Ngai/Ruby laboratory showed that a particular organ in the bobtail squid, the light organ, was not formed by uninfected squids. Rather, it was formed by the interactions of the ventral epithelium of the juvenile squid with one particular bacterial species, *Vibrio fisheri* (McFall-Ngai, Heath-Heckman, Gillette, Peyer, & Harvie, 2012). In some invertebrates, such as the nematode *Brugia malayi*, the microbes are responsible for the proper formation of the anterior–posterior axis (Landmann, Foster, Michalski, Slatko, & Sullivan, 2014). In other invertebrates, such as the parasitoid wasp *Asobara*, symbionts are critical for maintaining the ovaries (Pannebakker, Loppin, Elemans, Humblot, & Vavre, 2007). Animals not only coevolve with symbiotic microbes, we codevelop with them. This is sometimes called *symbiopoiesis*.

Animals usually transmit symbionts through the mother (Funkhouser & Bordenstein, 2013). In many cases, the transmission is through the germ line, and *Drosophila* nurse cells can be seen supplying the growing oocyte with *Wolbachia* bacteria just like they transfer ribosomes and the *bicoid*, *nanos*, and *hunchback* messages. Alternatively, as in mammals, the transmission is mediated by infection. Once the amnion breaks, the fetus is open for colonization by the bacteria of the mother's distal reproductive and gastrointestinal tracts. In humans, we have codeveloped and coevolved intimately with these bacteria. During the third trimester, the hormones of the pregnant woman alter the bacterial species of these regions. These bacteria cause some of the characteristic symptoms of late pregnancy including weight gain, insulin resistance, and the promotion of lactation. When placed into germ-free mice, third trimester human gut bacteria (but not that of first-trimester women) cause such conditions in the mice (Koren et al., 2012). Moreover, when the mother feeds the infant, some of the sugars cannot be digested by the baby. Rather, they provide nutrition for a group of bacteria whose presence is beneficial for the continued development of the gut (Chiu & Gilbert, 2015; Sela et al., 2011; Zivkovic, German, Lebrilla, & Mills, 2011).

Once in our guts, these selected bacteria aid their hosts' development. In various vertebrates, the gut microbiome has been seen to be essential for organizing the gut-associated lymphoid tissue, expanding the immune cell repertoire, generating the T_{17} class of helper-T cell, constructing the gut capillary network, and inducing normal division in the intestinal stem cells (Ardeshir et al., 2014; Olszak et al., 2012; Rawls, Samuel, & Gordon, 2004; Rhee, Sethupathi, Driks, Lanning, & Knight, 2004; Stappenbeck et al., 2002; Wesemann et al., 2013). Moreover, they do this by secreting factors that induce gene expression. One remarkable example of such symbiotic development is the induction of angiogenin-4 by *Bacteroides thetaiotaomicron*. This particular gut microbe helps induce the transcription of numerous genes in the Paneth cells of the mouse intestine. One of them is angiogenin-4, which, as its name implies, helps form the nutrient-absorbing capillary network of the intestine. This is important for the normal development of the gut and the normal physiology of the newborn and adult mouse. However, angiogenin-4 has a second function. It selectively destroys *Listeria* bacteria, which in addition to being pernicious for human health, is a major competitor of *Bacteroides* (Cash, Whitham, Behrendt, & Hooper, 2006; Hooper, Stappenbeck, Hong, & Gordon, 2003). In response, the vertebrate also changes the gene expression profile of the bacteria, allowing them to form biofilms. We have coevolved to codevelop. This results in interspecies epigenesis, i.e., the developmental induction between cells of two different kingdoms (Gilbert, 2003). What we also have is a fusion of developmental biology and ecology, reciprocal niche construction within the organism (Chiu & Gilbert, 2015; Laland et al., 2008).

There is also startling evidence that the development of our brain might be, in part, instructed by our microbes. One must remember that about 30% of the metabolites in our blood, including short-chain fatty acids and several hormones such as serotonin, are derived from microbes (McFall-Ngai et al., 2013; Nicholson et al., 2012). So even if our bodies are not exposed directly to bacteria, we mammals are probably exposed to bacterial products. Germ-free mice have altered brain development and express much less *NGF1-A* and *BDNF* accumulation in their striata (Diaz Heijtz et al., 2011). Germ-free mice and mice whose mothers have been given viral stress have autism-like symptoms, and several of these symptoms can be reduced or completely alleviated by exposure to certain microbes (Hsiao et al., 2013). As Diaz Heijtz et al. (2011) note, "during evolution, the colonization of gut microbiota has become integrated into the programming of brain development..."

The environment plays important roles during animal development. We do not inherit a text to be decoded. Rather, we inherit a composition to be interpreted. Each organism is a unique performance, a unique interpretation of that inherited score. Moreover, the score is not written for solo performance: the symbionts have long-established parts to play with us. And plasticity allows us to improvise and to coordinate our performance with those of the other organisms around us. It becomes clear that to develop is to interact with as well as actively construct environmental conditions, often in tight interdependency with other similarly responsive and constructing organisms. To evolve, then, means to change these interdependencies in a heritable manner, a process achievable in many more ways than just changes in allele frequency within a population.

ACKNOWLEDGMENTS

Three critically important professors must be acknowledged here: Michael Somers, N.J. Berrill, and Donna Haraway. Each transmitted the fascination with the organismal aspects of animal development. Funding is from NSF grant IOS 1457177 and a faculty research grant from Swarthmore College.

REFERENCES

Abouheif, E., Favé, M. J., Ibarrarán-Viniegra, A. S., Lesoway, M. P., Rafiqi, A. M., & Rajakumar, R. (2014). Eco-evo-devo: The time has come. *Advances in Experimental Medicine and Biology*, *781*, 107–125.

Adams, M. B. (1980). Severtsov and Schmalhausen: Russian morphology and the evolutionary synthesis. In E. Mayr & W. B. Provine (Eds.), *The evolutionary synthesis: Perspectives on the unification of biology* (pp. 193–225). Cambridge, MA: Harvard University Press.

Ardeshir, A., et al. (2014). Breast-fed and bottle-fed infant rhesus macaques develop distinct gut microbiotas and immune systems. *Science Translational Medicine*, *6*, 252r120.

Aubret, F., & Shine, R. (2009). Genetic assimilation and the postcolonization erosion of phenotypic plasticity in island tiger snakes. *Current Biology*, *19*, 1932–1936.

Beldade, P., Mateus, A. R., & Keller, R. A. (2011). Evolution and molecular mechanisms of adaptive developmental plasticity. *Molecular Ecology*, *20*, 1347–1363.

Bieser, K. L., & Wibbels, T. (2014). Chronology, magnitude, and duration of expression of putative sex-determining/differentiation genes in a turtle with temperature-dependent sex determination. *Sexual Development*, *8*, 364–375.

Callahan, B. J., Fukami, T., & Fisher, D. S. (2014). Rapid evolution of adaptive niche construction in experimental microbial populations. *Evolution*, *68*(11), 3307–3316.

Cameron, N. M., Shahrokh, D., Del Corpo, A., Dhir, S. K., Szyf, M., Champagne, F. A., et al. (2008). Epigenetic programming of phenotypic variations in reproductive strategies in the rat through maternal care. *Journal of Neuroendocrinology*, *20*, 795–801.

Cash, H. L., Whitham, C. V., Behrendt, C. L., & Hooper, L. V. (2006). Symbiotic bacteria direct expression of an intestinal bactericidal lectin. *Science*, *313*(5790), 1126–1130.
Champagne, F. A., Weaver, I. C., Diorio, J., Dymov, S., Szyf, M., & Meaney, M. J. (2006). Maternal care associated with methylation of the estrogen receptor-alpha1β promoter and estrogen receptor-alpha expression in the medial preoptic area of female offspring. *Endocrinology*, *147*, 2909–2915.
Chiu, L., & Gilbert, S. F. (2015). The birth of the holobiont: Multi-species birthing through mutual scaffolding and niche construction. *Biosemiotics*, http://dx.doi.org/10.1007/s12304-015-9232-5.
Dennis, S. R., LeBlanc, G. A., & Beckerman, A. P. (2014). Endocrine regulation of predator-induced phenotypic plasticity. *Oecologia*, *176*, 625–635.
Diaz Heijtz, R. D., et al. (2011). Normal gut microbiota modulates brain development and behavior. *Proceedings of the National Academy of Sciences of the United States of America*, *108*, 3047–3052.
Duscheck, J. (2002). Evolutionary developmental biology: It's the ecology, stupid. *Nature*, *418*, 578–579.
EDEN. (2010). *Evo-Devo-Eco Network*. http://edenrcn.com/.
Emlen, D. J., & Nijhout, H. F. (1999). Hormonal control of male horn length dimorphism in the horned beetle *Onthophagus taurus*. *Journal of Insect Physiology*, *45*, 45–53.
Funkhouser, L. J., & Bordenstein, S. R. (2013). Mom knows best: The universality of microbial transmission. *PLoS Biology*, *11*(8), e1001631.
Gibson, G., & Hogness, D. S. (1996). Effect of polymorphism in the *Drosophila* regulatory gene *Ultrabithorax* on homeotic stability. *Science*, *271*, 200–203.
Gilbert, S. F. (1978). The embryological origins of the gene theory. *Journal of the History of Biology*, *11*, 307–351.
Gilbert, S. F. (1996). Enzyme adaptation and the entrance of molecular biology into embryology. In S. Sarkar (Ed.), *The philosophy and history of molecular biology: New perspectives* (pp. 101–112). Dordrecht: Kluwer Academic Publishers.
Gilbert, S. F. (1998). Bearing crosses: The historiography of genetics and embryology. *American Journal of Medical Genetics*, *76*, 168–182.
Gilbert, S. F. (2001). Ecological developmental biology: Developmental biology meets the real world. *Developmental Biology*, *233*, 1–12.
Gilbert, S. F. (2003). The genome in its ecological context: Philosophical perspectives on interspecies epigenesis. *Annals of the New York Academy of Science*, *981*, 202–218.
Gilbert, S. F., & Bolker, J. A. (2003). Ecological developmental biology: Preface to the symposium. *Evolution & Development*, *5*, 3–8.
Gilbert, S. F., Bosch, T. C. G., & Ledón-Rettig, C. (2015). Eco-devo: Developmental symbiosis and developmental plasticity as evolutionary agents. *Nature Reviews Genetics*, *16*, 611–622.
Gilbert, S. F., & Epel, D. (2015). *Ecological developmental biology* (2nd ed.). Sunderland, MA: Sinauer Associates.
Gilbert, S. F., Opitz, J., & Raff, R. A. (1996). Resynthesizing evolutionary and developmental biology. *Developmental Biology*, *173*, 357–372.
Gilbert, S. F., Sapp, J., & Tauber, A. I. (2012). A symbiotic view of life: We have never been individuals. *The Quarterly Review of Biology*, *87*, 325–341.
Godwin, J., Sawby, R., Warner, R. R., Crews, D., & Grober, M. S. (2000). Hypothalamic arginine vasotocin mRNA abundance variation across sexes and with sex change in a coral reef fish. *Brain Behavior & Evolution*, *55*, 77–84.
Haeckel, E. (1880). Letter to W. Peters, 6 March 1880, in Nyhart, *op cit.* 1995, p. 203.
Hall, B. K. (1992). *Evolutionary developmental biology*. London: Chapman & Hall.
Hawkes, L. A., Broderick, A. C., Godfrey, M. H., & Godley, B. J. (2009). Climate change and marine turtles. *Endangered Species Research*, 7, 137–154.

Hertwig, O. (1894). *The biological problem of today: Preformationism or epigenesis?* (P. C. Mitchell, Trans.). New York: Macmillan.
Hooper, L. V., Wong, M. H., Thelin, A., Hansson, L., Falk, P. G., & Gordon, J. I. (2001). Molecular analysis of commensal host-microbial relationships in the intestine. *Science, 291*, 881–884.
Hooper, L. V., Stappenbeck, T. S., Hong, C. V., & Gordon, J. I. (2003). Angiogenins: A new class of microbiocidal proteins involved in innate immunity. *Nature Immunology, 4*(3), 269–273.
Hsiao, E. Y., McBride, S. W., Hsien, S., Sharon, G., Hyde, E. R., McCue, T., et al. (2013). Microbiota modulate behavioral and physiological abnormalities associated with neurodevelopmental disorders. *Cell, 155*, 1451–1463.
Johannsen, W. (1909). *Elemente der Exakten Erblichkeitslehre*. Jena: Gustav Fischer.
Johannsen, W. (1911). The genotype conception of heredity. *American Naturalist, 45*, 129–159. p. 133.
Kamakura, M. (2011). Royalactin induces queen differentiation in honeybees. *Nature, 473*, 478–483.
Keller, E. F. (2000). *The century of the gene*. Cambridge, MA: Harvard University Press.
Kishida, O., Costa, Z., Tezuka, A., & Michimae, H. (2013). Inducible offences affect predator-prey interactions and life-history plasticity in both predators and prey. *Journal of Animal Ecology, 83*, 899–906.
Kishida, O., Mizuta, Y., & Nishimura, K. (2006). Reciprocal phenotypic plasticity in a predator-prey interaction between larval amphibians. *Ecology, 87*, 1599–1604.
Koren, O., Goodrich, J. K., Cullender, T. C., Spor, A., Laitinen, K., Kling Bäckhed, H., et al. (2012). Host remodeling of the gut microbiome and metabolic changes during pregnancy. *Cell, 150*(3), 470–480.
Krementsov, N. (1998). *Stalinist science*. Princeton, NJ: Princeton University Press.
Laland, K. N., et al. (2008). Evo-devo and niche construction: Building bridges. *Journal of Experimental Zoology, 310*, 549–566.
Laland, K., Uller, T., Feldman, M., Sterelny, K., Müller, G. B., Moczek, A., et al. (2014). Does evolutionary theory need a rethink? *Nature, 514*, 161–164.
Laland, K. N., Uller, T., Feldman, M. W., Sterelny, K., Müller, G. B., Moczek, A., et al. (2015). The extended evolutionary synthesis: Its structure, assumptions and predictions. *Proceedings of the Royal Society B: Biological Sciences, 282*, 20151019.
Landmann, F., Foster, J. M., Michalski, M. L., Slatko, B. E., & Sullivan, W. (2014). Co-evolution between an endosymbiont and its nematode host: *Wolbachia* asymmetric posterior localization and AP polarity establishment. *PLoS Negected Tropical Diseases, 8*(8),e3096.
Lederberg, J. (1966). Remarks. *Current Topics in Developmental Biology, 1*, ix–xiii.
Lewontin, R. C., & Levins, R. (1976). The problem of Lysenkoism. In H. Rose & S. Rose (Eds.), *The radicalisation of science. Ideology of/in the natural sciences* (pp. 32–64). London: Macmillan.
Lindegren, C. C. (1966). *The cold war in biology*. Ann Arbor, MI: Planarian Press.
Marsh-Hunkin, K. E., Heinz, H. M., Hawkins, M. B., & Godwin, J. (2013). Estrogenic control of behavioral sex change in the bluehead wrasse, *Thalassoma bifasciatum*. *Integrative & Comparative Biology, 53*, 951–959.
Matsumoto, Y., Buemio, A., Chu, R., Vafaee, M., & Crews, D. (2013). Epigenetic control of gonadal aromatase (cyp19a1) in temperature-dependent sex determination of red-eared slider turtles. *PLoS One, 8*(6), e63599.
McFall-Ngai, M. J. (2002). Unseen forces: The influence of bacteria on animal development. *Developmental Biology, 242*, 1–14.
McFall-Ngai, M., Hadfield, M. G., Bosch, T. C. G., Carey, H. V., et al. (2013). Animals in a bacterial world, a new imperative for the life sciences. *Proceedings of the National Academy of Science of the United States of America, 110*, 3229–3236.

McFall-Ngai, M., Heath-Heckman, E. A. C., Gillette, A. A., Peyer, S. M., & Harvie, E. A. (2012). The secret languages of coevolved symbioses: Insights from the Euprymna scolopes-Vibrio fischeri symbiosis. *Seminars in Immunology*, *24*, 3–8.

Middlemis Maher, J., Werner, E. E., & Denver, R. J. (2013). Stress hormones mediate predator-induced phenotypic plasticity in amphibian tadpoles. *Proceedings of the Royal Society B. Biological Science*, *280*, 20123075.

Moczek, A. P. (2012). The nature of nurture and the future of evodevo: Toward a theory of developmental evolution. *Integrative and Comparative Biology*, *52*, 108–119.

Moczek, A. P. (2015). Re-evaluating the environment in evolution. *Frontiers in Ecology and Evolution*, *3*(1), 7.

Moczek, A. P., & Emlen, D. J. (2000). Male horn dimorphism in the scarab beetle *Onthophagus taurus*: Do alternative tactics favor alternative phenotypes? *Animal Behavior*, *59*, 459–466.

Monroy, A., & Moscona, A. (1966). Preface. *Current Topics in Developmental Biology*, *1*, xvii–xviii.

Monteiro, A., et al. (2013). Distal-less regulates eyespot patterns and melanization in *Bicyclus* butterflies. *Journal of Experimental. Zoology B. Molecular and Developmental Evolution*, *320*, 321–331.

Morgan, T. H. (1932). The rise of genetics. *Science*, *76*, 261–288.

Morreale, S. J., Ruiz, G. J., Spotila, J. R., & Standora, E. A. (1982). Temperature-dependent sex determination: Current practices threaten conservation of sea turtles. *Science*, *216*, 1245–1247.

Moss, L. (2003). *What genes can't do*. Cambridge, MA: MIT Press.

Needham, J. (1936). *Order and life*. Yale University Press: New Haven. p. 99.

Nicholson, J. K., Holmes, E., Kinross, J., Burcelin, R., Gibson, G., Jia, W., et al. (2012). Host-gut microbiota metabolic interactions. *Science*, *336*, 1262–1267.

Nijhout, H. F. (2003). Development and evolution of adaptive polyphenisms. *Evolution and Development*, *5*, 9–18.

Nyhart, L. (1995). *Biology takes form*. Chicago: University of Chicago Press.

Odling-Smee, J., Erwin, D. H., Palkovacs, E. P., Feldman, M. W., & Laland, K. N. (2013). Niche construction theory: A practical guide for ecologists. *Quarterly Review of Biology*, *88*, 4–28.

Olszak, T., et al. (2012). Microbial exposure during early life has persistent effects on natural killer T cell function. *Science*, *336*, 489–493.

Pannebakker, B. A., Loppin, B., Elemans, C. P. H., Humblot, L., & Vavre, F. (2007). Parasitic inhibition of cell death facilitates symbiosis. *Proceedings of the National Academy of Sciences of the United States of America*, *104*, 213–215.

Raff, R. A. (1996). *The shape of life: Genes, development, and the evolution of animal form*. Chicago: University of Chicago Press.

Raff, R. A., & Kaufman, T. C. (1983). *Embryos, genes, and evolution: The developmental-genetic basis of evolutionary change*. New York: Macmillan.

Rafferty, N. E., et al. (2013). Phenological overlap of interacting species in a changing climate: An assessment of available approaches. *Ecology & Evolution*, *3*, 3183–3193.

Rajakumar, R., San Mauro, D., Dijkstra, M. B., Huang, M. H., Wheeler, D. E., Hiou-Tim,-F., et al. (2012). Ancestral developmental potential facilitates parallel evolution in ants. *Science*, *335*(6064), 79–82.

Rawls, J. F., Samuel, B. S., & Gordon, J. I. (2004). Gnotobiotic zebrafish reveal evolutionarily conserved responses to the gut microbiota. *Proceedings of the National Academy of Sciences of the United States of America*, *101*, 4596–4601.

Relyea, R. A. (2003). Predator cues and pesticides: A double dose of danger for amphibians. *Ecological Applications*, *13*, 1515–1521.

Relyea, R. (2004). Fine-tuned phenotypes: Tadpole plasticity under 16 combinations of predators and competitors. *Ecology*, *85*, 172–179.

Rhee, K. J., Sethupathi, P., Driks, A., Lanning, D. K., & Knight, K. L. (2004). Role of commensal bacteria in development of gut-associated lymphoid tissue and preimmune antibody repertoire. *Journal of Immunology*, *172*, 1118–1124.

Rogers, S. M., Matheson, T., Sasaki, K., Kendsrick, K., Simpson, S. J., & Burrows, M. (2004). Substantial changes in central nervous system neurotransmitters and neuromodulators accompany phase change in the locust. *Journal of Experimental Biology*, *207*, 3603–3617.

Roll-Hansen, N. (2005). *The Lysenko effect: The politics of science*. Amherst, NY: Humanity Books.

Roll-Hansen, N. (2007). *Sources of Johannsen's genotype theory*. Meeting of the International Society for the History, Philosophy, and Social Studies of Biology, Exeter, England.

Roll-Hansen, N. (2014). The holist tradition in twentieth century genetics. Wilhelm Johannsen's genotype concept. *The Journal of Physiology*, *592*(11), 2431–2438.

Roux W., The problems, methods and scope of developmental mechanics, In: W.M. Wheeler (Transl.), Biological lectures of the marine biology laboratory, Woods Hole, 1894, Boston; Ginn, 149–190.

Sapp, J. (1987). *Beyond the gene*. New York: Oxford University Press.

Sarkar, S. (1999). From the *Reaktionsnorm* to the adaptive norm: The norm of reaction, 1906–1960. *Biology & Philosophy*, *14*, 235–252.

Schlichting, C. D., & Pigliucci, M. (1998). *Phenotypic evolution. A reaction norm approach*. Sunderland, MA: Sinauer Associates.

Schmalhausen, I. I. (1949). *Factors of evolution*. Philadelphia: Blakiston Press.

Sela, D. A., et al. (2011). An infant-associated bacterial commensal utilizes breast milk sialyloligosaccharides. *Journal of Biological Chemistry*, *286*, 11909–11918.

Siebold, C. (1876). Letter to Ehlers, 16 November 1876; quoted in Nyhart, *op. cit*. 1995, p. 203.

Simões, P. M., Niven, J. E., & Ott, S. R. (2013). Phenotypic transformation affects associative learning in the desert locust. *Current Biology*, *23*(23), 2407–2412.

Standen, E. M., Du, T. Y., & Larsson, H. C. (2014). Developmental plasticity and the origin of tetrapods. *Nature*, *513*, 54–58.

Stappenbeck, T. S., Hooper, L. V., & Gordon, J. I. (2002). Developmental regulation of intestinal angiogenesis by indigenous microbes via Paneth cells. *Proceedings of the National Academy of Sciences of the United States of America*, *99*, 15451–15455.

Sturtevant, A. (1923). Inheritance of direction of coiling in *Limnaea*. *Science*, *58*, 269–270.

Suzuki, Y., & Nijhout, H. F. (2006). Evolution of a polyphenism by genetic assimilation. *Science*, *311*, 650–652.

Telemeco, R. S., Abbott, K. C., & Janzen, F. J. (2013). Modeling the effects of climate change-induced shifts in reproductive phenology on temperature-dependent traits. *American Naturalist*, *181*, 637–648.

Waddington, C. H. (1952). Selection of the genetic basis for an acquired character. *Nature*, *169*, 278.

Waddington, C. H. (1956). Genetic assimilation of the *bithorax* phenotype. *Evolution*, *10*, 1–13.

Weaver, I. C., et al. (2004). Epigenetic programming by maternal behavior. *Nature Neuroscience*, *7*, 847–854.

Weismann, A. (1893). *The germ plasm* (W. N. Parker & H. Rönnfeldt, Trans.). New York: Charles Scribners Sons.

Wesemann, D. R., et al. (2013). Microbial colonization influences early B-lineage development in the gut lamina propria. *Nature*, *501*, 112–115.

West-Eberhard, M. J. (2003). *Developmental plasticity and evolution*. New York: Oxford University Press.

Williams, J. B., & Lee, R. E., Jr. (2005). Plant senescence cues entry into diapause in the gall fly *Eurosta solidaginis*: Resulting metabolic depression is critical for water conservation. *Journal of Experimental Biology, 208*, 4437–4444.

Wolfe, A. J. (2010). What does it mean to go public? The American response to Lysenko, reconsidered. *Historical Studies in the Natural Sciences, 40*(1), 48–78.

Wolpert, L. (1991). *The triumph of the embryo*. New York: Oxford University Press.

Woltereck, R. (1909). Weitere experimentelle Untersuchungen über Artveränderung, speziell über das Wesen quantitativer Artunderscheide bei Daphniden. *Verhandlungen der deutschen Zoologischen Gesellschaft, 1909*, 110–172.

Zivkovic, A. M., German, J. B., Lebrilla, C. B., & Mills, D. A. (2011). Human milk glycobiome and its impact on the infant gastrointestinal microbiota. *Proceedings of the National Academy of Sciences of the United States of America, 108*(Suppl. 1), 4653–4658.

CHAPTER TWENTY-THREE

Twenty Years in Maine: Integrating Insights from Developmental Biology into Translational Medicine in a Small State

Thomas Gridley[1]

Center for Molecular Medicine, Maine Medical Center Research Institute, Scarborough, Maine, USA

[1]Corresponding author: e-mail address: gridlt@mmc.org

Contents

Abstract

In this chapter, I give my personal reflections on more than 30 years of studying developmental biology in the mouse model, spending 20 of those years doing research in Maine, a small rural state. I also give my thoughts on my recent experience transitioning to a large medical center in Maine, and the issues involved with integrating insights from developmental biology and regenerative medicine into the fabric of translational and clinical patient care in such an environment.

1. INTRODUCTION

How does one bring insights from modern developmental biology and regenerative medicine into translational and clinical research studies, particularly in a relatively small, rural state? I have lived in Maine, and performed biomedical research here, for more than 20 years. Maine has a

Current Topics in Developmental Biology, Volume 116
ISSN 0070-2153
http://dx.doi.org/10.1016/bs.ctdb.2015.10.004

population of approximately 1.3 million people concentrated in the southern and coastal areas of the state. Areas in western and northern regions of the state are very sparsely populated. In addition, Maine's population has the oldest average age in the United States. While my laboratory still does basic research in mouse genetics, with a focus on many roles the Notch pathway plays in both embryonic development and adult physiology, much of my time on these days is spent trying to address this question: how does one bring insights from modern developmental biology and regenerative medicine into translational and clinical research studies in a relatively small state with a large rural population?

2. PERSONAL RESEARCH HISTORY

To provide background for my journey to Maine, I will first describe my research training and history. I did my Ph.D. thesis work at the Massachusetts Institute of Technology, training in molecular immunology with Malcolm Gefter, followed by postdoctoral training with Rudolf Jaenisch at the then newly opened Whitehead Institute for Biomedical Research at MIT. It was in Rudolf's lab that I first started working on mouse development, collaborating with Phil Soriano on retroviral insertional mutagenesis to generate mutants with developmental phenotypes (Gridley et al., 1990; Soriano, Gridley, & Jaenisch, 1987). I also began deriving my own mouse embryonic stem cell lines in the Jaenisch lab.

My first independent faculty position was at the Roche Institute of Molecular Biology. The Roche Institute, located on the site of the recently closed Hoffmann-La Roche facility in Nutley, New Jersey, was an unique and invigorating research environment. The principal investigators at the Institute had the same freedom to choose basic research projects as we would have had in a traditional academic research environment, with Hoffmann-La Roche providing all resources for our research. I was recruited to this position by the Editor of this volume, Paul Wassarman. Paul was Chair of the Department of Cell and Developmental Biology at the Roche Institute. Paul's vision, which was relatively unique at the time (late 1980s), was to create a department that largely concentrated on studying the developmental biology of the mouse. Other members of the department studying various aspects of mouse development were Mel DePamphilis, Andy McMahon, and Colin Stewart. Our department later added two researchers studying Drosophila development, Chris Rushlow and Ken Howard.

I was originally drawn into work on the Notch pathway around 1990, as part of collaboration with Andy McMahon and Ralph Greenspan, who was a Roche Institute colleague in the Department of Neuroscience, and have continued to work on the Notch pathway ever since. In fact, my decision to begin working on the Notch pathway was one of those serendipitous events that, in the final analysis, shape one's career. The Roche Institute was having an off-site retreat, and I needed a ride, so I got a lift with Andy McMahon. It was a 2-hour drive to the retreat site, so we talked a lot of science on the way. Andy told me he had recently done a low stringency genomic library screen with a Xenopus Notch homolog. This probe had been generously provided by Chris Kintner prior to publication to Ralph Greenspan who, early in his career, had done some influential work on Drosophila Notch signaling (Hoppe & Greenspan, 1986). There were a large number of phage hybridizing to the Xenopus probe, but Andy's lab had just cloned six novel mouse Wnt genes that he preferred to spend his time on (Gavin, McMahon, & McMahon, 1990), so he suggested that I characterize the mouse Notch clones. The first paper from my laboratory on a mouse Notch homolog was published in 1992 as a collaboration with Andy and Ralph (Franco Del Amo et al., 1992). The title of the paper also reveals that we named our initial mouse Notch homolog Motch (which turned out to be the *Notch1* gene). This reflected the custom at the time to put a letter or two from the species name at the beginning of the new gene name. For example, Kintner called his Xenopus clone Xotch (Coffman, Harris, & Kintner, 1990), so we named our gene Motch. The lab joke at the time was that we really had to obtain some crayfish genomic DNA. Fortunately, the various organism nomenclature committees soon put a stop to that method of naming species homologs.

My lab concentrated our efforts on using gene targeting in embryonic stem cells to generate and to analyze novel mouse mutants with developmental phenotypes. We generated several interesting mutants, particularly some affecting cranial development (Gendron-Maguire, Mallo, Zhang, & Gridley, 1993; Swiatek & Gridley, 1993), but after we made the first Notch receptor mutant (Swiatek, Lindsell, del Amo, Weinmaster, & Gridley, 1994). I decided to concentrate on our future efforts on genes in the Notch pathway and in the Snail family of transcriptional repressors. Unfortunately, Hoffman-La Roche decided to close the Roche Institute in 1995 after 27 years, although at the time this closure was portrayed in the scientific press as a refocusing effort, with transfer of the Roche Institute to California (Macilwain, 1995; Marshall, 1994). In any event, all members of the Institute had to find other employment. Toward the end of 1995, I moved to the Jackson Laboratory in Bar Harbor, Maine, as it seemed the ideal environment to continue our work on the Notch pathway, as well as the other developmental signaling pathways, in mice. I spent 15 productive and enjoyable years at the Jackson Laboratory (1995–2010), continuing our work on the Notch pathway. Much of our work concentrated on the roles, the Notch pathway plays during embryonic vascular development (Feng, Krebs, & Gridley, 2010; Gridley, 2010; Krebs et al., 2004, 2000; Krebs, Starling, Chervonsky, & Gridley, 2010), although our work addressed many other phenotypic areas as well.

By 2010, I decided that I wanted to move my laboratory to an institution more focused on translational and clinical research. I felt that both personal and professional goals would be better met by staying in Maine, so I moved to Maine Medical Center in Portland, and its Research Institute in nearby Scarborough. Maine Medical Center is the state's largest hospital and tertiary medical center, and it is a growing academic medical center. Maine is one of the few states in the United States lacking an allopathic medical school, so Maine Medical Center's primary academic affiliation is with Tufts University School of Medicine (TUSM) in Boston. TUSM and Maine Medical Center have established a Maine Track MD Education Program, which offers an innovative curriculum that offers clinical training experiences in Maine and exposes medical students to the unique aspects of rural practice as well as training in a major tertiary medical center.

3. PERSONALIZED AND PRECISION MEDICINE (IN MAINE)

In his January 2015 State of the Union address, President Obama announced the creation of a new Precision Medicine Initiative. Precision medicine involves the development and utilization of disease prevention and treatment strategies that take into account individual patient variability. In a commentary on this initiative in the *New England Journal of Medicine*, Francis Collins and Harold Varmus describe the two major components of this initiative: a near term focus on cancer research and a longer term focus on the generation of knowledge applicable to disease prevention and treatment strategies for the entire spectrum of health and disease (Collins & Varmus, 2015).

One obvious concern with the advent of personalized and precision medicine is how does one implement precision medicine in a state without a widespread genomics and biomedical informatics infrastructure? The Jackson Laboratory has considerable strength in genomics, but other parts of the state do not. One very valuable resource for building biomedical and translational research infrastructure in a state such as Maine has been the NIH Institutional Development Award (IDeA) program. This program, established in 1993 and currently supported by the National Institute for General Medical Sciences, supports programs designed to build biomedical research infrastructure and enhance faculty development in states that historically have received lower aggregate levels of NIH funding. Two of the components of the IDeA program are the establishment of Centers of Biomedical Research Excellence (COBRE) and IDeA Networks of Biomedical Research Excellence (INBRE). COBREs are thematic, multidisciplinary research centers at individual institutions. The dual goals of the COBRE program are to develop and mentor junior faculty and to enhance institutional research capabilities and infrastructure.

The INBRE program, on the other hand, consists of state-wide networks (with only one INBRE awarded per state). INBRE applications constitute collaborations between research-intensive universities and other institutions in the state, concentrating primarily on undergraduate institutions and nonprofit research institutes. INBREs enhance research capacity, strengthen the research capabilities of faculty, and provide students with access to biomedical resources.

Maine has had five IDeA awards: the Maine INBRE network in Comparative Functional Genomics and four COBRE grants. The four COBRE awards have had a strong translational medicine focus. Two have been awarded to Maine Medical Center: the COBRE in Vascular Biology, and the COBRE in Stem and Progenitor Cell Biology and Regenerative Medicine. A third COBRE, in the Comparative Biology of Tissue Repair, Regeneration, and Aging, was awarded to the Mount Desert Island Biological Laboratory. The fourth COBRE, the Interdisciplinary Center of Excellence for the Study of Pain and Sensory Function, was awarded to the University of New England. Another key component of the research infrastructure in Maine is the Jackson Laboratory, which has been a National Cancer Institute-designated basic Cancer Center since 1983. These federal awards have been instrumental in enhancing the capacity and infrastructure for basic and translational researches in Maine.

A newer IDeA program initiative, and one that will be key in the ability of Maine to implement personalized and precision medical practice in the state, is the establishment of Clinical and Translational Research (CTR) centers.

The IDeA-CTR program, modeled on the Clinical and Translational Science Award (CTSA) program administered by the NIH National Center for Advancing Translational Sciences (NCATS), provides support for the development of infrastructure and capacity to conduct CTR on diseases that affect medically underserved populations, and diseases that are prevalent in the IDeA states. The IDeA-CTR program further provides for mentoring and career development activities in CTR. Establishment of a CTR center in Maine, administratively located at Maine Medical Center but including collaborating institutions in Maine and in the northern New England region, is a key component of the strategy for implementing personalized and precision medicine here. I am heavily involved in Maine Medical Center's efforts to establish a Northern New England CTR Center. Also, as Associate Chair of the Institutional Review Board and Associate Director of the Center for Clinical and Translational Research at Maine Medical Center, I am very interested in facilitating the growth of investigator-initiated clinical trials at Maine Medical Center, and throughout the state. One important goal for a Maine CTR center would be to facilitate multisite clinical trial networks that would enhance recruitment of rural inhabitants into clinical trials.

4. USE OF NOTCH PATHWAY MODULATORS IN CLINICAL TRIALS

Throughout the world, there has been a greatly enhanced effort in the last decade or so to increase and to accelerate the translation of insights gained from basic biomedical research into improvements in clinical practice and outcomes. This focus is appropriate, as much biomedical research is funded by individuals, either through their taxes or through their donations to private foundations. With respect to Notch signaling and the attempts to move developmental biology research insights into the clinic, the most activity is taking place in the area of utilizing Notch pathway modulators in clinical trials.

Disruptions of Notch signaling have a well-demonstrated role in numerous types of cancer. The first mammalian Notch homolog isolated was cloned from a translocation break point of a T-cell acute lymphoblastic leukemia cell line (Ellisen et al., 1991). Depending on the type of mutation and the tissue involved, Notch signaling mutations either can be directly oncogenic or can disrupt a Notch signaling-mediated tumor suppression function (Lobry, Oh, Mansour, Look, & Aifantis, 2014; South, Cho, & Aster, 2012).

Both agonists and antagonists of Notch signaling are being studied, but the development of Notch pathway inhibitors as therapeutics is more advanced than the Notch agonists. Numerous Notch pathway antagonists have been developed, including both small molecules and biologics. These are being tested in preclinical studies and clinical trials in tumor types where Notch signaling is functioning in an oncogenic role, or to inhibit tumor angiogenesis (Andersson & Lendahl, 2014; Espinoza & Miele, 2013; Golde, Koo, Felsenstein, Osborne, & Miele, 2013; Kangsamaksin, Tattersall, & Kitajewski, 2014; Liu, Fan, Wang, Zheng, & Lu, 2014). Two general strategies have been utilized for therapeutic inhibition of Notch signaling: inhibition of proteolytic processing of Notch family receptors and inhibition of the ligand–receptor interaction. γ-Secretase inhibitors, which inhibit the proteolytic processing of Notch receptors and thereby prevent signal transmission, have been tested in clinical trials directed primarily at either Alzheimer's disease or various types of cancer, including both solid tumors and hematological malignancies. However, the γ-secretase complex is not specific for cleavage of Notch family receptors, and in fact, cleaves more than 100 different transmembrane proteins (Golde et al., 2013). In some trials, patients have experienced severe gastrointestinal toxicities from γ-secretase inhibitors due to the major role Notch signaling plays in regulating intestinal stem cell differentiation (Noah & Shroyer, 2013). Strategies to inhibit the ligand–receptor interaction include neutralizing antibodies to either ligands or receptors, and decoys (soluble forms of the extracellular domains of Notch ligands or receptors). Inhibition of the ligand–receptor interaction may provide improved specificity compared to more broad-spectrum γ-secretase inhibitors.

Notch therapeutics have potential uses in diseases other than cancer. There is a growing evidence that the Notch pathway, so essential during multiple stages of mammalian embryonic development also plays critical roles in adult physiology. For example, an exciting recent study in mice suggests that pharmacological inhibition of Notch signal transduction in adipocytes may have beneficial metabolic effects in humans by promoting adipose browning and regulating energy homeostasis (Bi & Kuang, 2015; Bi et al., 2014; Gridley & Kajimura, 2014). The potential benefits of doing so, however, must be balanced against the many biological roles played by the Notch pathway, and by the fact that, as discussed above, mutations of Notch pathway components can act either directly as oncogenes, or to inactivate tumor suppressor networks. Much work remains to be done before Notch pathway inhibitors could safely be tested as antiobesity therapeutics, but the future is promising.

5. CONCLUDING REMARKS

Having grown up on Long Island, New York, I never thought I would end up in Maine, but I have spent 20 rewarding and productive years here. Our work has evolved from a singular focus on characterizing signaling pathways essential for embryonic development in mice, to studying these pathways in adult physiology and disease. In addition, I spend large portions of my day trying to facilitate and grow the translational and clinical research enterprise in Maine, attempting to better deliver the insights and advances of developmental biology and regenerative medicine research to the residents of Maine and northern New England. I cannot think of a better place to do so.

ACKNOWLEDGMENTS

I would like to thank Paul Wassarman for hiring me for my first job, Andy McMahon for the ride to the Roche Institute retreat (leading to more than 25 years of work on Notch signaling), and all the talented postdocs, graduate students, and researchers who have contributed to the work from my laboratory. In particular, I would like to dedicate this chapter to the memory of Luke Krebs, who made pioneering contributions to our understanding of Notch signaling in vascular biology.

REFERENCES

Andersson, E. R., & Lendahl, U. (2014). Therapeutic modulation of Notch signalling—Are we there yet? *Nature Reviews Drug Discovery*, *13*, 357–378.

Bi, P., & Kuang, S. (2015). Notch signaling as a novel regulator of metabolism. *Trends in Endocrinology and Metabolism*, *26*, 248–255.

Bi, P., Shan, T., Liu, W., Yue, F., Yang, X., Liang, X. R., et al. (2014). Inhibition of Notch signaling promotes browning of white adipose tissue and ameliorates obesity. *Nature Medicine*, *20*, 911–918.

Coffman, C., Harris, W., & Kintner, C. (1990). Xotch, the Xenopus homolog of Drosophila Notch. *Science*, *249*, 1438–1441.

Collins, F. S., & Varmus, H. (2015). A new initiative on precision medicine. *New England Journal of Medicine*, *372*, 793–795.

Ellisen, L. W., Bird, J., West, D. C., Soreng, A. L., Reynolds, T. C., Smith, S. D., et al. (1991). TAN-1, the human homolog of the Drosophila notch gene, is broken by chromosomal translocations in T lymphoblastic neoplasms. *Cell*, *66*, 649–661.

Espinoza, I., & Miele, L. (2013). Notch inhibitors for cancer treatment. *Pharmacology & Therapeutics*, *139*, 95–110.

Feng, X., Krebs, L. T., & Gridley, T. (2010). Patent ductus arteriosus in mice with smooth muscle-specific Jag1 deletion. *Development*, *137*, 4191–4199.

Franco Del Amo, F. F., Smith, D. E., Swiatek, P. J., Gendron-Maguire, M., Greenspan, R. J., McMahon, A. P., et al. (1992). Expression pattern of Motch, a mouse homolog of Drosophila Notch, suggests an important role in early postimplantation mouse development. *Development*, *115*, 737–744.

Gavin, B. J., McMahon, J. A., & McMahon, A. P. (1990). Expression of multiple novel Wnt-1/int-1-related genes during fetal and adult mouse development. *Genes and Development, 4*, 2319–2332.

Gendron-Maguire, M., Mallo, M., Zhang, M., & Gridley, T. (1993). Hoxa-2 mutant mice exhibit homeotic transformation of skeletal elements derived from cranial neural crest. *Cell, 75*, 1317–1331.

Golde, T. E., Koo, E. H., Felsenstein, K. M., Osborne, B. A., & Miele, L. (2013). Gamma-secretase inhibitors and modulators. *Biochimica et Biophysica Acta, 1828*, 2898–2907.

Gridley, T. (2010). Notch signaling in the vasculature. *Current Topics in Developmental Biology, 92*, 277–309.

Gridley, T., Gray, D. A., Orr-Weaver, T., Soriano, P., Barton, D. E., Francke, U., et al. (1990). Molecular analysis of the Mov 34 mutation: Transcript disrupted by proviral integration in mice is conserved in Drosophila. *Development, 109*, 235–242.

Gridley, T., & Kajimura, S. (2014). Lightening up a notch: Notch regulation of energy metabolism. *Nature Medicine, 20*, 811–812.

Hoppe, P. E., & Greenspan, R. J. (1986). Local function of the Notch gene for embryonic ectodermal pathway choice in Drosophila. *Cell, 46*, 773–783.

Kangsamaksin, T., Tattersall, I. W., & Kitajewski, J. (2014). Notch functions in developmental and tumour angiogenesis by diverse mechanisms. *Biochemical Society Transactions, 42*, 1563–1568.

Krebs, L. T., Shutter, J. R., Tanigaki, K., Honjo, T., Stark, K. L., & Gridley, T. (2004). Haploinsufficient lethality and formation of arteriovenous malformations in Notch pathway mutants. *Genes and Development, 18*, 2469–2473.

Krebs, L. T., Starling, C., Chervonsky, A. V., & Gridley, T. (2010). Notch1 activation in mice causes arteriovenous malformations phenocopied by ephrinB2 and EphB4 mutants. *Genesis, 48*, 146–150.

Krebs, L. T., Xue, Y., Norton, C. R., Shutter, J. R., Maguire, M., Sundberg, J. P., et al. (2000). Notch signaling is essential for vascular morphogenesis in mice. *Genes and Development, 14*, 1343–1352.

Liu, Z., Fan, F., Wang, A., Zheng, S., & Lu, Y. (2014). Dll4-Notch signaling in regulation of tumor angiogenesis. *Journal of Cancer Research and Clinical Oncology, 140*, 525–536.

Lobry, C., Oh, P., Mansour, M. R., Look, A. T., & Aifantis, I. (2014). Notch signaling: Switching an oncogene to a tumor suppressor. *Blood, 123*, 2451–2459.

Macilwain, C. (1995). Roche pledges independence to new institute. *Nature, 373*, 94.

Marshall, E. (1994). Roche Institute moves west. *Science, 266*, 723.

Noah, T. K., & Shroyer, N. F. (2013). Notch in the intestine: Regulation of homeostasis and pathogenesis. *Annual Review of Physiology, 75*, 263–288.

Soriano, P., Gridley, T., & Jaenisch, R. (1987). Retroviruses and insertional mutagenesis in mice: Proviral integration at the Mov 34 locus leads to early embryonic death. *Genes and Development, 1*, 366–375.

South, A. P., Cho, R. J., & Aster, J. C. (2012). The double-edged sword of Notch signaling in cancer. *Seminars in Cell and Developmental Biology, 23*, 458–464.

Swiatek, P. J., & Gridley, T. (1993). Perinatal lethality and defects in hindbrain development in mice homozygous for a targeted mutation of the zinc finger gene Krox20. *Genes and Development, 7*, 2071–2084.

Swiatek, P. J., Lindsell, C. E., del Amo, F. F., Weinmaster, G., & Gridley, T. (1994). Notch1 is essential for postimplantation development in mice. *Genes and Development, 8*, 707–719.

CHAPTER TWENTY-FOUR

Cell Fate Determination by Transcription Factors

John B. Gurdon[1]
Wellcome Trust/Cancer Research UK Gurdon Institute, University of Cambridge, Cambridge, United Kingdom
Department of Zoology, University of Cambridge, Cambridge, United Kingdom
[1]Corresponding author: e-mail address: jbg1000@cam.ac.uk

Contents

Abstract

Transcription factors fulfill a key role in the formation and maintenance of different cell-types during development. It is known that transcription factors largely dissociate from chromosomes during mitosis. We found, previously, that mitosis is also a time when somatic nuclei can be far more easily reprogrammed after nuclear transfer than the nuclei of interphase cells. We refer to this as a mitotic advantage. Here, the rate of exchange of a transcription factor on its designated DNA-binding site is discussed. It is proposed that the Xenopus oocyte could serve as an experimental system in which the duration of binding site occupancy could be usefully analyzed. In particular, the Xenopus oocyte has several characteristics which make it possible to determine accurately the concentration and duration of transcription factor binding. It is proposed that the concentration and time are the key variables which govern the action of transcription factors when they activate genes needed for cell lineage determination.

In all animals except mammals, a fertilized one-cell egg is able to develop into a functional moving larva, without any guidance from its environment. Two fundamental mechanisms are largely responsible for initiating this loss of symmetry (Gurdon, 1992): (1) the asymmetric distribution of components of the egg so that differences come to exist between cells of the early

Current Topics in Developmental Biology, Volume 116
ISSN 0070-2153
http://dx.doi.org/10.1016/bs.ctdb.2015.10.005

embryo and (2) signaling from cells in one region of the embryo to cells in another region. To progress from these early events to the formation of functional cell-types in different regions of the embryo, it is necessary for transcription factors to function. These factors bind to the promoter region of cell-type-specific genes and activate their transcription, and hence the synthesis of cell-type-specific proteins. Thus, transcription factors have an essential role to play in development.

Cascades of transcription factors are known for many cell-types (Cui, Siriwon, Li, Davidson, & Peter, 2014; Martin, Hauer, Messmer, Orvain, & Gradwohl, 2007; Pagliuca et al., 2014) in normal development. They are also fundamental in the iPS reprogramming of somatic cells back to an embryonic state from which new cell-types can be formed (Takahashi & Yamanaka, 2006; Weintraub et al., 1989). But how these factors work in iPS is not understood, and it must be remembered that only a very small proportion (<1%) of cells receiving transcription factors subsequently make new cell-types (Takahashi & Yamanaka, 2006).

Another route by which cells can be reprogrammed from a differentiated to an embryonic state is by transplanting the nucleus of a differentiated cell to an unfertilized egg; in this case, the active components of an egg can reset the gene expression program of a somatic nucleus within a few hours (Gurdon, 2006).

1. MITOTIC ADVANTAGE

In the course of our nuclear transfer experiments, we found that at a particular stage in a cell cycle, namely mitosis, cultured cell nuclei become highly susceptible to reprogramming for a brief period (Halley-Stott, Jullien, Pasque, & Gurdon, 2014). This is therefore a window of opportunity for transcription factor access to genes. Mitosis may be an important step in nuclear reprogramming and in cell lineage determination. It is difficult to synchronize embryo cells, but this can work well for cultured cells. For these experiments, we have used *Xenopus* oocytes in first meiosis as recipient cells since they provide a sensitive assay for the efficiency of experimentally induced changes in gene expression and do not induce mitosis in transplanted nuclei (Jullien, Halley-Stott, Miyamoto, Pasque, & Gurdon, 2011).

The mitotic advantage experiments were as follows. Nuclei from the mouse mesodermal cell line C2C12 were effectively synchronized in mitosis. After nuclear transfer to oocytes, mitotic nuclei were induced to transcribe Sox2 about 100 times more rapidly than the same kinds of nuclei in the interphase. By RNA seq analysis, over 500 genes showed this mitotic advantage. It is the rate rather than eventual level of gene transcription that is enhanced. A number of tests excluded the idea that the lack of a nuclear membrane in mitotic cells could provide a more rapid access of reprogramming factors to chromatin than the interphase cells. Using nuclear transfer to oocytes in first meiosis, it is possible to obtain a frequent and recurrent reinitiation of transcription in a suspension of isolated nuclei; this does not happen in any cell-free system. We were able to remove a large proportion of chromosomal proteins from isolated C2C12 nuclei and then monitor on-going transcription for up to 2 days. We found that the mitotic advantage persisted in spite of the lack of many chromosomal proteins which are not therefore responsible for the lack of mitotic advantage during interphase. The major chromosomal components not removed by salt-Triton treatment were core histones. After testing a range of histone modifications, we found that changes in the acetylation, phosphorylation, and methylation associated with mitosis were not responsible for the mitotic advantage. We did, however, find that histone H2A monoubiquitination was at least in part responsible. The most likely interpretation of these results is that mitosis includes histone and other chromosomal protein changes that temporarily alter access to reprogramming factors, including transcription factors. The best explanation for this mitotic advantage is that, during mitosis, many chromosomal proteins dissociate from chromosomes and give access to genes that are not accessible during interphase in many somatic cell-types. These mitotic components, including transcription factors, dissociate from chromatin during mitosis, and then reassociate at telophase.

This phenomenon could be of considerable importance in cell fate determination. It appears to be generally true that, as cells progress from an uncommitted embryonic state through different levels of commitment, they always go through a cell division before reaching a differentiated state with specialized gene expression. Cells never seem to switch from one kind of gene expression to another, or from a progenitor to a mature state, without undergoing mitosis. It has already been shown that chromosomal proteins including most transcription factors are lost from chromosomes during mitosis before being taken up again at the next interphase of the cell cycle (Martinez-Balbas, Dey, Rabindran, Ozato, & Wu, 1995).

2. MODE OF ACTION OF TRANSCRIPTION FACTORS

Transcription factors are usually proteins encoded by single copy genes. These proteins often form dimers (homodimers or heterodimers) and bind specifically to a short DNA sequence of 10–15 nucleotides. Having done so, they make a gene accessible to the transcriptional activity of the cell in which they are present.

It is important, now, to explain what we do NOT understand about transcription factor action. The first uncertainty is how a transcription factor finds its specific sequence in DNA in view of the size of a typical genome (6×10^9 nucleotides for a mouse). The binding sequence for a transcription factor is the promoter region of a downstream gene. It is usually assumed that a downstream gene's promoter needs to be continually occupied by a transcription factor during transcription of that gene. It is possible that the transcription factor can bind and cause a change of structure that persists in the absence of the factor, so that the factor is no longer required at that site once transcription has been initiated. However, it is assumed here that a transcription factor needs to persist at its binding site, though even if it did not, there is still the problem of how it finds its binding site in the first place.

There are two likely ways in which the search for a binding site can be explained. One is for the concentration of the factor in a nucleus to be sufficiently high that even if most such factors are bound to inappropriate sequences in the genome, there will always be enough for some to bind to the designated sequence. This would require transcription factors to be among the most abundant molecules in a nucleus, and this is not the usual assumption. The extent of this problem could be reduced if much of the genome is not accessible; but even if all nucleosomed DNA, for example, is not accessible this would make a difference of only a few fold. Some 50,000 molecules of a factor per nucleus would be needed, a number considerably higher than is usually estimated. One published figure is for Xbra, an especially abundant factor in *Xenopus* embryos, is its concentration in gastrula embryo nuclei which has been measured as 2.9 μM (Gentsch et al., 2013).

Another solution to the transcription factor search problem would be if the dwell time for a factor is very long at its designated site, compared to that for other sites. The affinity of a transcription factor for its site is usually found to be 10^4 times higher than for other sites. The Lac repressor is unusual in having a much higher affinity. However, this difference in affinity is not sufficient to explain specific binding. Dwell time could be very greatly extended if cofactors assist binding, as is often believed to be the case, since most transcription factors are thought to bind in this way.

In either of these scenarios, we need to know the actual concentration of factor in a nucleus where a factor is functional, and/or the actual dwell time. Both of these values are hard to determine in normal tissues, where the cell population is often heterogeneous and the time spent in a nucleus as opposed to cytoplasm is not always constant. Likewise, the dwell time of a factor is best estimated in a competition experiment, but it is difficult, in somatic cells, to introduce a competitor molecule at a desired concentration which is constant enough to test on-going transcription over that time.

3. AN AMPHIBIAN OOCYTE AS AN ASSAY FOR TRANSCRIPTION

To analyze the mechanism of transcription, there is a need for a system in which known cell components can be added or removed, and the effect seen on continuing transcription. One idea would be to use a cell-free preparation where these requirements are met. However, cell-free systems reported so far do not provide on-going transcription in which a gene is transcribed recurrently, that is where new transcripts are continually made and accumulated as a measure of RNA polymerase activity. If whole nuclei are incubated in a cell-free extract, transcripts that are already in progress will readily reach the end of a gene, but transcription will not be reinitiated. This may be because it is not possible to constitute in a cell-free preparation, the enormously high concentration of components that exist in living nuclei where protein concentration is about 100 mg/mL.

From experiments initiated over half a century ago, it became evident that the cell nucleus can be taken out of the somatic cell and transplanted by injection into an egg (Briggs & King, 1952; Gurdon, 1960). Since normal embryos can result from this experiment, the transplanted nucleus and its mitotic products must be transcriptionally active for several days or more. Attempts have been made over time to convert this nuclear transfer system into one in which the role of individual kinds of molecules can be tested for function in transcription. Nearly all of this work has been carried out using eggs (in second, meiotic metaphase) and oocytes (in first, meiotic prophase) of *Xenopus* because of their large size and easy availability. The *Xenopus* oocyte is particularly valuable in this respect and has the following characteristics.

The oocyte is the progenitor of an egg, and hence, embryo and it is related to embryo development via the completion of meiosis, as shown in Fig. 1. An oocyte takes a few months to grow in the ovary from a germ cell to its full size, at which stage it remains for many months in a suspended state until progesterone release in the mother frog causes it to complete first meiosis and progress in the next few hours to the metaphase of second meiosis. It is then released from the ovary; it is laid by the frog and can be fertilized and begin development.

The oocyte has the following properties that make it suitable as a transcription assay. Its own nucleus, known as the germinal vesicle (GV), contains a tetraploid chromosome set as well as a huge reserve of developmentally important components required as maternal contributions to early embryo development. The GV has a volume of 50 nL and is about 100 times larger than a somatic cell nucleus. This nucleus is transcriptionally extremely active, its RNA production being about 1000 times greater than that of a somatic cell nucleus (Davidson, 1986). It remains in this transcriptionally active state for as long as it exists in the parental ovary. Furthermore, such oocytes can be isolated from the ovary and, importantly, will continue to be transcriptionally active for days or even weeks when cultured in a simple salt medium. Its exceptional size and extraordinary tolerance to microinjection means that quite large amounts of any material can be injected into it. For example, purified mRNA can be injected into its cytoplasm, where it will continue to be translated into protein at the rate of up to 200 proteins/mRNA molecule/day (Halley-Stott et al., 2010). Since it is

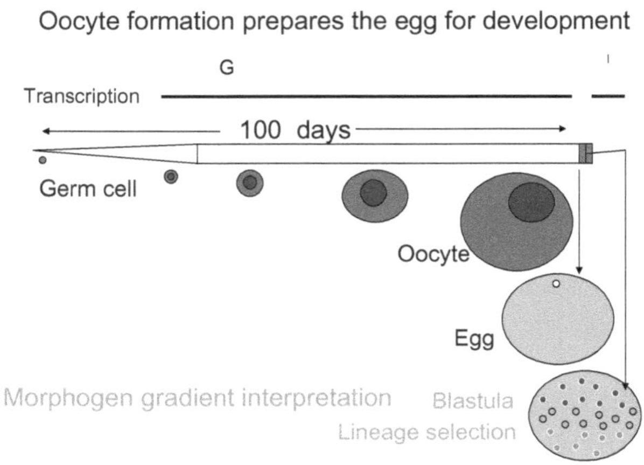

Figure 1 A diagram to illustrate the relationship of an oocyte to the embryo in *Xenopus*. (A color version of this figure is located at the end of this chapter.)

possible to harmlessly inject at least 30 ng of mRNA to an oocyte, this can lead to the synthesis of 10^9 molecules of the encoded protein per day. Surprisingly, it is also possible to microinject up to 15 nL of material into the GV. Even more surprisingly, the GV, if isolated in oil, can remain transcriptionally active for 24 h. It is therefore possible to introduce known amounts of any macromolecule or particle into the living cell, and hence to test the effect of this on transcription. It is important to appreciate that this takes place in single cells all of which are the same, thereby avoiding the complexity of large numbers of cells which can be heterogeneous in cell-type if isolated from normal tissue.

This oocyte assay has been used extensively to analyze the transcription of transplanted nuclei. About 250 somatic nuclei can be transplanted to the GV of an oocyte, and transcripts made over the next few days can be analyzed by increasingly sensitive and accurate molecular methods. Nuclei from mouse embryo fibroblasts undergo substantial changes in transcription in the 48 h after injection into the GV. Of over 400 genes analyzed, 20% were downregulated in an oocyte compared to donor cells, 35% continued transcription, and 45% were newly transcribed in oocytes. Most of the latter were genes that are normally transcribed by the oocyte's own chromosomes (Jullien et al., 2011). The oocyte, therefore, has components that can reprogram gene expression in somatic nuclei toward a set of genes normally expressed in oocytes. A simplifying feature of this nuclear reprogramming system is that it takes place rapidly, within 48 h at 16 °C (equivalent to 12 h at 37 °C), with no DNA synthesis or cell division and in the absence of any necessary protein synthesis. The maternal content of an oocyte is therefore fully sufficient to initiate and to maintain transcription. Notably, there is sufficient RNA polymerase II to provide for the transcription of 10,000 nuclei, no new polymerase being synthesized from fertilization to the blastula stage of embryos. The amount of pol II in transplanted nuclei increased about eightfold during reprogramming compared to donor nuclei before injection (Jullien et al., 2011).

4. DNA INJECTION TO OOCYTES AS AN ASSAY FOR TRANSCRIPTION FACTOR ACTION

The characteristics by which an oocyte can transcribe transplanted nuclei suggest that this procedure could be usefully adapted to analyze the details of transcription action on single genes injected as DNA. This would have the advantage of permitting the analysis of very large numbers of one kind of gene and hence of clarifying what takes place at that gene in the absence of thousands of other genes which could create a confusing background.

Important papers by Wyllie, Gurdon, and Price (1977) and Wyllie, Laskey, Finch, and Gurdon (1978) showed that circular molecules of plasmid DNA are rapidly converted by the oocyte into nucleosomed chromatin, so as to resemble the same gene in a normal chromosome context; up to 12 ng DNA ($=10^9$ molecules of SV40 DNA) can be assembled into nucleosomed chromatin by an oocyte nucleus. Conveniently, this is true only if DNA is injected into the GV. DNA deposited in the oocyte cytoplasm is soon degraded and it is not transcribed. DNA injected to the GV is well transcribed. Furthermore, subsequent work showed that 5S genes and ribosomal genes are also accurately transcribed in the oocyte GV. Some transcripts by polymerase II are correctly initiated and spliced.

In order to provide a valid and useful experimental system for transcription factor analysis, the following would be needed. The gene being analyzed should be one that is normally expressed in development and which has a function determining cell-type lineage. This gene should be in a plasmid for the convenience of production and mutation. The gene should be associated with its normal promoter sequence in DNA to which the normal transcription factor for this gene binds. The binding of a transcription factor should induce a strong transcription of the gene, which should have a downstream reporter for convenience of assay. The reporter would best be a protein to show that the transcription of the gene is meaningful. The reporter response needs to be quantitative and in proportion to the amount of transcription factor provided. The concentration of transcription factor needs to be able to be measured over the duration of the assay. The determination of dwell time is usually tested by adding a competitor to the transcription factor, and this needs to be able to be introduced rapidly and at a desired concentration. Importantly, the assay has to record expression over the time for which the competition is present. The reporter therefore has to have a short half-life, so that the reporter protein synthesized before the provision of the competitor is not scored. With a short half-life reporter protein only that made, since the time when this is supplied would be recorded and hence the dwell time after competitor supply would be monitored. To identify cofactors that could be responsible for a long dwell time, it would be necessary to be able to reisolate the injected plasmid DNA with proteins that were attached during the expression time.

In principle, these conditions can be met by the characteristics of injected oocytes, as described by the account above of nuclear transplant experiments on oocytes. This is a direction of current work, which seems to be likely to provide the details of transcription factor action described at the beginning of this contribution.

ACKNOWLEDGMENTS

J.B.G. acknowledges the core funding provided by the Wellcome Trust (101050) and MRC (MR/K011022/1).

REFERENCES

Briggs, R., & King, T. J. (1952). Transplantation of living nuclei from blastula cells into enucleated frogs' eggs. *Proceedings of the National Academy of Sciences, 38*, 455–463.

Cui, M., Siriwon, N., Li, E., Davidson, E. H., & Peter, I. S. (2014). Specific functions of the Wnt signaling system in gene regulatory networks throughout the early sea urchin embryo. *Proceedings of the National Academy of Sciences of the United States of America, 111*(47), E5029–E5038.

Davidson, E. H. (1986). *Gene activity in early development*. New York: Academic Press.

Gentsch, G. E., Owens, N. D., Martin, S. R., Piccinelli, P., Faial, T., Trotter, M. W., et al. (2013). In vivo T-box transcription factor profiling reveals joint regulation of embryonic neuromesodermal bipotency. *Cell Reports, 4*(6), 1158–1196.

Gurdon, J. B. (1960). The developmental capacity of nuclei taken from differentiating endoderm cells of *Xenopus laevis*. *Journal of Embryology and Experimental Morphology, 8*, 505–526.

Gurdon, J. B. (1992). The generation of diversity and pattern in animal development. *Cell, 68*, 185–199.

Gurdon, J. B. (2006). From nuclear transfer to nuclear reprogramming: The reversal of cell differentiation. *Annual Review of Cell and Developmental Biology, 22*, 1–22.

Halley-Stott, R. P., Jullien, J., Pasque, V., & Gurdon, J. B. (2014). Mitosis gives a brief window of opportunity for a change in gene transcription. *PLoS Biology, 12*(7), e1001914.

Halley-Stott, R. P., Pasque, V., Astrand, C., Miyamoto, K., Simeoni, I., Jullien, J., et al. (2010). Mammalian nuclear transplantation to germinal vesicle stage Xenopus oocytes—A method for quantitative transcriptional reprogramming. *Methods, 51*, 56–65.

Jullien, J., Halley-Stott, R. P., Miyamoto, K., Pasque, V., & Gurdon, J. B. (2011). Mechanisms of nuclear reprogramming by eggs and oocytes: A deterministic process? *Nature Reviews Molecular Cell Biology, 12*, 453–459.

Martin, M., Hauer, V., Messmer, M., Orvain, C., & Gradwohl, G. (2007). Transcription factors in pancreatic development. In R. Scharfmann & J. P. H. Shield (Eds.), *Endocrine development: Vol. 12. Development of the pancreas and neonatal diabetes* (pp. 24–32). Basel: Karger.

Martinez-Balbas, M. A., Dey, A., Rabindran, S. K., Ozato, K., & Wu, C. (1995). Displacement of sequence-specific transcription factors from mitotic chromatin. *Cell, 83*(1), 29–38.

Pagliuca, F. W., Millman, J. R., Gurtler, M., Segel, M., Van Dervort, A., Ryu, J. H., et al. (2014). Generation of functional human pancreatic β cells in vitro. *Cell, 159*(2), 428–439.

Takahashi, K., & Yamanaka, S. (2006). Induction of pluripotent stem cells from mouse embryonic and adult fibroblast cultures by defined factors. *Cell, 126*, 663–676.

Weintraub, H., Tapscott, S. J., Davis, R. L., Thayer, M. J., Adam, M. A., Lassar, A. B., et al. (1989). Activation of muscle-specific genes in pigment, nerve, fat, liver, and fibroblast cell lines by forced expression of MyoD. *Proceedings of the National Academy of Sciences of the United States of America*, *86*, 5434–5438.

Wyllie, A. H., Gurdon, J. B., & Price, J. (1977). Nuclear localization of an oocyte component required for the stability of injected DNA. *Nature*, *268*, 150–152.

Wyllie, A. H., Laskey, R. A., Finch, J., & Gurdon, J. B. (1978). Selective DNA conservation and chromatin assembly after injection of SV40 DNA into *Xenopus* oocytes. *Developmental Biology*, *64*, 178–188.

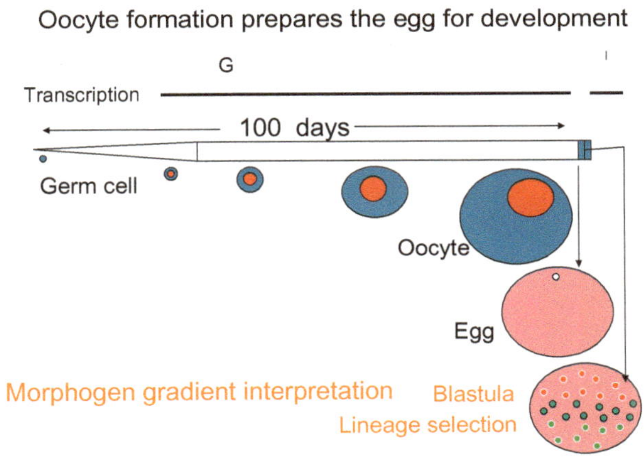

John B. Gurdon, Figure 1 A diagram to illustrate the relationship of an oocyte to the embryo in *Xenopus*.

CHAPTER TWENTY-FIVE

Terminal Selectors of Neuronal Identity

Oliver Hobert[1]
Department of Biological Sciences, Howard Hughes Medical Institute, Columbia University, New York, USA
[1]Corresponding author: e-mail address: or38@columbia.edu

Contents

Abstract

The analysis of the developmental programs that define many different neuron types in the nematode *Caenorhabditis elegans* has revealed common themes in how distinct terminal differentiation programs are controlled. Rather than being controlled in a piece-meal manner, terminal identity features of a mature neuron are often coregulated by so-called terminal selector transcription factors. Here, I summarize the terminal selector concept and emphasize core features of this concept in the *C. elegans* system such as coregulation of terminal effector batteries, combinatorial control mechanisms, and the coupling of initiation and maintenance of neuronal identity.

1. HISTORICAL BACKGROUND

Nervous system development has traditionally been studied with a "top-down" focus, defining inductive events early in embryonic patterning and elucidating signaling events and transcriptional regulatory cascades that

Current Topics in Developmental Biology, Volume 116
ISSN 0070-2153
http://dx.doi.org/10.1016/bs.ctdb.2015.12.007

progressively restrict developmental fates as embryogenesis proceeds (Edlund & Jessell, 1999; Hemmati-Brivanlou & Melton, 1997). This "top-down" approach is well illustrated by Waddington's landscape model in which a ball travels down specific valleys, making several choices along the way (Waddington, 1957). An alternative perspective, the "bottom-up" angle, views the problem of neuronal development from the standpoint of the end product of neuronal development, asking how the terminal features of specific neuron types in a mature nervous system are genetically programmed. The bottom-up perspective has been extensively taken in the *Caenorhabditis elegans* field, not by design but as the result of the initial efforts of Sydney Brenner to identify *C. elegans* mutants that affect specific animal behaviors. Brenner's first behavioral mutants displayed various types of uncoordinated (*unc*) locomotory behavior (Brenner, 1974). The ensuing establishment of various assays that measure the response of animals to specific sensory cues then lead to the identification of viable mutants that selectively affected very specific behaviors such as chemotaxis behavior (Lewis & Hodgkin, 1977), thermotaxis behavior (Hedgecock & Russell, 1975), mechanosensory behavior (Chalfie & Sulston, 1981), or odortaxis behavior (Sengupta, Colbert, & Bargmann, 1994). Strikingly, the molecular characterization of these behavioral mutants showed that many of them affect transcription factors that turned out to act at the terminal step of differentiation of postmitotic neurons (listed in Table 1; Baran et al., 1999; Finney, Ruvkun, & Horvitz, 1988; Hobert et al., 1997; Jin et al., 1994; Prasad et al., 1998; Satterlee et al., 2001; Sengupta et al., 1994; Uchida et al., 2003; Way & Chalfie, 1988).

The phenotypic analysis of these mutants profited from some of the key defining features of the *C. elegans* system: (a) an exceptionally well-described cell lineage history (Sulston, 1983), (b) a most detailed anatomical description of all individual neuron types (White, Southgate, Thomson, & Brenner, 1986), (c) the knowledge of functional features of many individual neuron types based on laser ablation (e.g., Bargmann & Horvitz, 1991; Mori & Ohshima, 1995), and (d) the availability of a vast collection of molecular markers in the form of *gfp* reporter transgenes (Chalfie, Tu, Euskirchen, Ward, & Prasher, 1994). These molecular markers, built by the community over the past 20 years (curated at www.wormbase.org), monitor the neuron-type specific expression of hundreds of genes that define terminal properties of a neuron type (e.g., genes encoding neurotransmitter synthesis, receptors, ion channel, etc.). Together, these anatomical, functional, and molecular features assign precise and unique "phenotypic spaces" to

Table 1 Classic Behavioral Mutants that Originally Defined Terminal Selectors

Gene	Transcription Factor Type	Behavioral Phenotype Leading to Identification	Neuron Class Affected
unc-3	EBF/COE	Uncoordinated locomotion[a]	Cholinergic ventral cord motorneurons[b]
unc-30	Pitx-type homeodomain	Uncoordinated locomotion[a]	GABAergic ventral cord motorneurons[c]
			Cholinergic PVP interneurons[d]
unc-42	Homeodomain	Uncoordinated locomotion[a]	Glutamatergic ASH sensory neurons[e]
			Glutamatergic AIB interneurons[f]
unc-86	Brn3-type POU homeodomain	Mechanosensory defects[g]	Mechanosensory ALM/AVM/PLM/PVM neurons[h]
			Serotonergic HSN motor neurons[i]
			Serotonergic NSM neurons[j]
			Inner labial IL2 sensory-motor neurons[j]
			URA motor neurons[j]
			URB interneurons[j]
			Glutamatergic URX sensory neurons[k]
			Glutamatergic AQR/PQR sensory neurons[k]
			Glutamatergic PVR interneuron[k]
mec-3	Lhx1/5 LIM homeodomain	Mechanosensory defects[g]	Mechanosensory ALM/AVM/PLM/PVM neurons[l]
che-1	C2H2 Zn finger	Chemotaxis defects[m]	Gustatory ASE sensory neurons[n]
ttx-1	Otx homeodomain	Thermotaxis defects[o]	Thermosensory AFD neuron[p]

Continued

Table 1 Classic Behavioral Mutants that Originally Defined Terminal Selectors—cont'd

Gene	Transcription Factor Type	Behavioral Phenotype Leading to Identification	Neuron Class Affected
ttx-3	Lhx2/7 LIM homeodomain	Thermotaxis defects[o]	Cholinergic AIY interneuron[q]
			Cholinergic AIA interneuron[j]
			Serotonergic NSM neuron[j]
			Glutamatergic ASK sensory neuron[k]
odr-7	C4 Zn finger	Odortaxis defects[r]	Olfactory AWA neuron[s]

[a]Brenner (1974).
[b]Kratsios, Stolfi, Levine, and Hobert (2011), Kratsios et al. (2015), and Prasad et al. (1998). *unc-3* locomotory defects match the effect expected from ablating of A- and B-type motor neurons (Chalfie et al., 1985).
[c]Jin, Hoskins, and Horvitz (1994). *unc-30* locomotory defects match the defects observed upon losing signaling by GABAergic D-type motor neurons (McIntire, Jorgensen, & Horvitz, 1993).
[d]Pereira et al. (2015).
[e]Baran, Aronoff, and Garriga (1999) and Serrano-Saiz et al. (2013). *unc-42* mutants phenocopy the sensory defects observed in ASH-ablated animals (Baran et al., 1999).
[f]E. Serrano-Saiz and O. Hobert (unpublished data).
[g]Chalfie and Sulston (1981).
[h]Duggan, Ma, and Chalfie (1998). *unc-86* mutants phenocopy ablation of mechanosensory neurons (Chalfie et al., 1985).
[i]Desai, Garriga, McIntire, and Horvitz (1988). *unc-86* mutants phenocopy loss of HSN neurons.
[j]Zhang et al. (2014).
[k]Serrano-Saiz et al. (2013).
[l]Duggan et al. (1998), Way and Chalfie (1988), and Zhang et al. (2002). *mec-3* mutants phenocopy ablation of mechanosensory neurons (Chalfie et al., 1985).
[m]Lewis and Hodgkin (1977).
[n]Etchberger et al. (2007) and Uchida, Nakano, Koga, and Ohshima (2003). *che-1* mutants phenocopy the ablation of the ASE neurons (Bargmann & Horvitz, 1991).
[o]Hedgecock and Russell (1975).
[p]Satterlee et al. (2001). *ttx-1* mutants phenocopy ablation of the AFD neurons (Mori & Ohshima, 1995).
[q]Altun-Gultekin et al. (2001). *ttx-3* mutants phenocopy ablation of the AFD neurons (Mori & Ohshima, 1995; Tsalik & Hobert, 2003).
[r]Sengupta et al. (1994).
[s]*odr-7* mutants phenocopy ablation of the AWA neurons (Bargmann, Hartwieg, & Horvitz, 1993).
Many additional terminal selectors were identified through forward and reverse genetic approaches and are not listed here. See an upcoming review (O. Hobert, WIREs Dev. Biol., in preparation) for a more comprehensive summary. As illustrated in Fig. 2, one terminal selector can control distinct neuronal differentiation programs through its pairing with distinct cofactors.

individual neuron types which then permitted a detailed assessment of how the regulatory factors defined by the classic behavioral mutants (listed in Table 1) affect neuronal differentiation. Strikingly, all of the transcription factors listed in Table 1 were found to have very broad effects on the adoption of the correct terminal identity of very specific neuron types. They affect the functional properties of the respective neuron types in the sense that the behavioral phenotypes of these mutants match the phenotypic consequences of laser ablating these neuron types and they broadly affect many—if not all—known molecular markers normally expressed specifically in the respective neuron; however, they do not affect the generation of these neurons. Given the neuron identity-defining properties, I proposed to call these transcription factors "terminal selectors" (Hobert, 2008, 2011). This terminology is an extension of the selector gene concept, originally proposed by Garcia-Bellido (1975). "Classic" selector genes are genes that define the identity of specific domains of a developing organism and they act transiently during specific phases of development. In contrast, terminal selectors control the terminally differentiated state of a neuron, i.e., they initiate and maintain the stable identity of nondividing differentiated cells.

I will provide here a concise summary of the properties of terminal selectors as revealed primarily by studies in *C. elegans*, and I will organize these properties around specific themes that are also highlighted in Fig. 1. The description of these properties is based on the few original examples that led to the original terminal selector definition (Hobert, 2008, 2011) but also covers a plethora of recent examples that have allowed further generalization of this concept. The themes that I will discuss are

- Identity-defining terminal effector genes are coregulated.
- Terminal selectors initiate and maintain the terminally differentiated state.
- Terminal selectors act in reiteratively used combinations.
- Terminal selectors control regulatory subroutines.
- Parallel regulatory routines define specific aspects of neuronal identity.
- Diversification of neuronal identity is achieved via selective gene repression.
- Generality of the terminal selector concept.

2. COREGULATION OF TERMINAL EFFECTOR GENES BY TERMINAL SELECTORS

Every fully differentiated neuron type is uniquely defined by the expression of a unique combination of "nuts and bolts," "terminal effector" genes that code for proteins that define the functional features of a mature neuron type, such as neurotransmitter receptors, neurotransmitter synthesizing enzymes, neuropeptides, ion channels, sensory receptors, and synaptic recognition molecules (see Table 2 for an exemplary list of terminal effector genes). As illustrated in Fig. 1, the identification of *cis*-regulatory elements

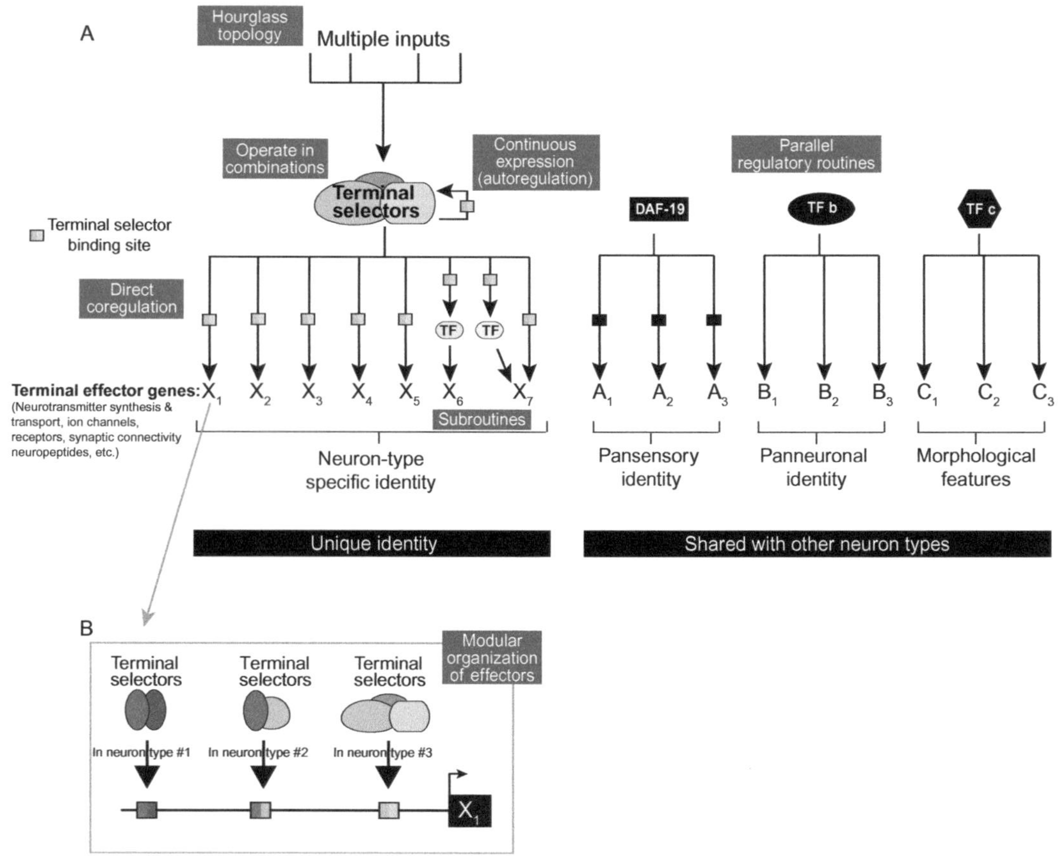

Figure 1 See legend on opposite page.

required for the coexpression of terminal effector genes in specific neuron types, and the analysis of the expression of many terminal effector genes in many different terminal selector mutant backgrounds demonstrated a simple common theme: neuron-type-specific terminal effector batteries are organized into regulons, i.e., they are all coregulated, and this coregulation is mediated by shared *cis*-regulatory motifs and their cognate transcription factors in the form of specific terminal selector transcription factors (or combination thereof; as discussed below) (Cinar et al., 2005; Eastman et al., 1999; Etchberger et al., 2007; Flames & Hobert, 2009; Gordon & Hobert, 2015; Kratsios et al., 2011; Wenick & Hobert, 2004; Zhang et al., 2002).

The broadness of the effect that terminal selectors exert on neuronal identity is illustrated by the way that these effects have been measured in many different neurons types: the repertoire of *gfp* reporter transgenes that the *C. elegans* community has built over the past two decades is large enough that for the majority of the 118 anatomically defined neuron types anything between ~3 and several dozen molecular markers have been described. These reporters most often mark biochemically unlinked molecular features of a neuron and provide a "random" snapshot of the identity of a neuron. In most cases examined, terminal selectors affect either all or most of the markers examined. In several instances many dozen markers were tested (Etchberger et al., 2007; Kratsios et al., 2011; Wenick & Hobert, 2004), in others just a handful of markers were tested (e.g., Serrano-Saiz et al.,

Figure 1 Key features of gene expression programs in mature neuron types. (A) Terminal selector regulons. Red squares highlight key concepts. Sustained expression of terminal selectors is often, but not always ensured by autoregulation. Small rectangles illustrate binding sites for transcription factors and indicate direct regulation of effector genes by terminal selectors. Terminal selectors shown on the left control neuron-type specific identity features that assign a unique identity to a neuron. See Table 2 for an exemplary list of effector genes. Parallel regulatory routines such as those that regulate pansensory features (via DAF-19) are controlled by factors that could also considered to be terminal selectors; but they do not assign unique identities. Morphology regulators control generic aspect of a neuron's morphology, such as placement of axons/dendrites into specific fascicles or axo/dendritic polarity. TF, transcription factor. (B) Modular organization of terminal effector genes. X_1 is a representative of any effector gene expressed in more than one neuron type (e.g., the vesicular transporter of glutamate, which is expressed in all glutamatergic neurons, but not in GABA or cholinergic neurons). This schematic also indicates the key principle of combinatorial "reusage" of the same terminal selector in different neuron types, as schematized in more detail in Fig. 2. (A color version of this figure is located at the end of this chapter.)

Table 2 Examples of Effector Genes of Terminal Selectors

Categories	**Examples**
Neurotransmitter pathway synthesis	GAD, TPH, TH, etc.
Vesicular neurotransmitter transporters	VGLUT, VGAT, VMAT, VAChT
Neurotransmitter receptors	ACh-gated, Glu-gated, GABA-gated ion channels, metabotropic ACh-, Glu-receptors, biogenic amine receptors
Ion channels	DEG/EnaC, TRP, TWK, Ca^{2+}, K^{+} channels
Gap junctions	Innexins
Sensory receptors	Olfactory GPCRs, receptor-type guanylyl cyclases, mechanosensory MEC channels
Neuropeptides	FMRFamides, NLPs
Neuropeptide receptors	FMRFamide-, tachykinin-, somatostatin-like receptors (and others)
Synaptic organizers	*oig-1*, *madd-4*, neuroligin
Signaling	TGFβ (*dbl-1*, *unc-129*), Slit/*slt-1*, G-proteins
Others	Tetraspanins, IgSF, etc.

The table is meant to illustrate the range of biochemical activities that are under terminal selector control. The examples shown were derived from the following studies: Cinar, Keles, and Jin (2005), Eastman, Horvitz, and Jin (1999), Etchberger et al. (2007), Flames and Hobert (2009), Gordon and Hobert (2015), Howell, White, and Hobert (2015), Kratsios et al. (2015, 2011), Serrano-Saiz et al. (2013), Wenick and Hobert (2004), and Zhang et al. (2002).

2013). Ideally, one would want to be able to profile neurons in wild-type and terminal selector-deficient neurons to assess the full extent of the effects. While this has not been done yet, the random sampling of the markers examined, combined with the broad defects observed in terminal selector mutants, make the assessment of effects of terminal selector on neuronal identity only an extrapolation, albeit a reasonable one.

Coregulation of a multitude of identity features is not self-evident. *A priori*, one could easily envision that individual identity features could be independently regulated via distinct transcriptional regulatory factors. Transcriptome profiling studies of individual neuron types (in invertebrates or vertebrates) usually reveal dozens of transcription factors expressed in a mature neuron type (e.g., Etchberger et al., 2007), hence the job of initiating and maintaining the expression of terminal gene batteries could, in theory, be distributed over a number of distinct factors. However, as studies on the transcription factors shown in Table 1 (as well as ensuing analysis of additional terminal selectors) revealed, the expression of many, perhaps most terminal markers for neuronal identity is affected in terminal selector mutants, indicating that coregulation is achieved via a limited number of regulatory factors.

Even in a nervous system as comparatively simple as *C. elegans*, only very few terminal effector genes are exclusively expressed in a single neuron type (e.g., olfactory receptor proteins). The much more prevalent scenario is that genes, like neurotransmitter receptor subunits, neurotransmitter synthesizing enzymes, neuropeptides, or various types of ion channels are expressed in several distinct neuron types. In each specific neuron type, these effector genes are under control of neuron-type specific terminal selectors. This effectively means that such effector genes harbor a modular array of response elements each activated by the different terminal selectors present in each neuron class where it is expressed (Fig. 1B). This quite intuitive concept has been experimentally validated in a number of cases, including the loci encoding the vesicular glutamate transporter (Serrano-Saiz et al., 2013) or the vesicular acetylcholine or choline transporter (Kratsios et al., 2011; Wenick & Hobert, 2004).

3. TERMINAL SELECTORS INITIATE AND MAINTAIN THE TERMINALLY DIFFERENTIATED STATE

A key defining feature of terminal selectors is that they are not only required to initiate terminal differentiation programs but are continuously expressed throughout the life of the neuron (and animal) to *maintain* the differentiated state of a neuron (Doitsidou et al., 2013; Etchberger, Flowers, Poole, Bashllari, & Hobert, 2009; Flames & Hobert, 2009; Kratsios et al., 2011). This has been amply illustrated through temporally controlled, late removal of terminal selectors not just in *C. elegans* but also in mice (Deneris & Hobert, 2014; Holmberg & Perlmann, 2012; Laguna et al., 2015). The continuous expression of terminal selectors is often ensured through autoregulatory feedback mechanisms (Baran et al., 1999; Etchberger et al., 2007; Hobert et al., 1997; Sagasti, Hobert, Troemel, Ruvkun, & Bargmann, 1999; Way & Chalfie, 1989).

This continuous role sets terminal selectors clearly apart from earlier and transiently acting transcription factors. For example, the *C. elegans* Zic gene ortholog REF-2 is required for the terminal differentiation of the AIY interneurons, but it is only transiently expressed in the progenitor of the AIY interneurons, where it acts to initiate the expression of the two cooperatively acting terminal selectors of AIY identity, *ttx-3* and *ceh-10* (Bertrand & Hobert, 2009). If the only role of REF-2 is to turn a terminal selector, REF-2 should not be considered a terminal selector because it is not required to maintain the differentiated state and it cannot be considered to be the "terminal" link of a regulatory cascade to a terminal effector gene (such as a neurotransmitter pathway gene). In principle, one could envision a scenario in which expression of a terminal effector is initiated by a transiently acting factor binding to some *cis*-regulatory element in the effector gene locus and its expression is then maintained via a separate *cis*-regulatory element that is controlled by a terminal selector. However, such a scenario has not yet been documented.

4. TERMINAL SELECTORS ACT IN REITERATIVELY USED COMBINATIONS

Terminal selectors are usually not single proteins that operate in complete isolation. Terminal selectors rather function in the context of combinations of cooperating factors (Fig. 2; Wenick & Hobert, 2004; Xue, Tu, & Chalfie, 1993; Zhang et al., 2014). In one recently published example, four terminal selectors cooperate in a given neuron type (Doitsidou et al., 2013). In an unpublished example, seven terminal selectors cooperate on the level of the *cis*-regulatory elements of the downstream effector genes (N. Flames and O. Hobert, unpublished data). In many cases, no partners for terminal selectors have been identified, but this should not be taken as evidence that terminal selectors can act completely alone.

The biochemical basis for cooperation of terminal selectors appears to be distinct in different cellular contexts. In some cases, terminal selectors exhibit cooperative DNA binding (Wenick & Hobert, 2004; Xue et al., 1993). For example, the UNC-86 and MEC-3 homeodomain proteins bind cooperatively to *cis*-regulatory elements required for the expression of effector genes that define touch neuron identity (Duggan et al., 1998; Xue et al., 1993). In other cases, terminal selectors may bind DNA in a noncooperative manner, but exert cooperative effects possibly through the cooperative recruitment of cofactors (Doitsidou et al., 2013; Gordon & Hobert, 2015). In yet other cases, the type and extent of cooperation is unclear. For example, the serotonergic NSM neurons fail to terminally differentiate in double mutant

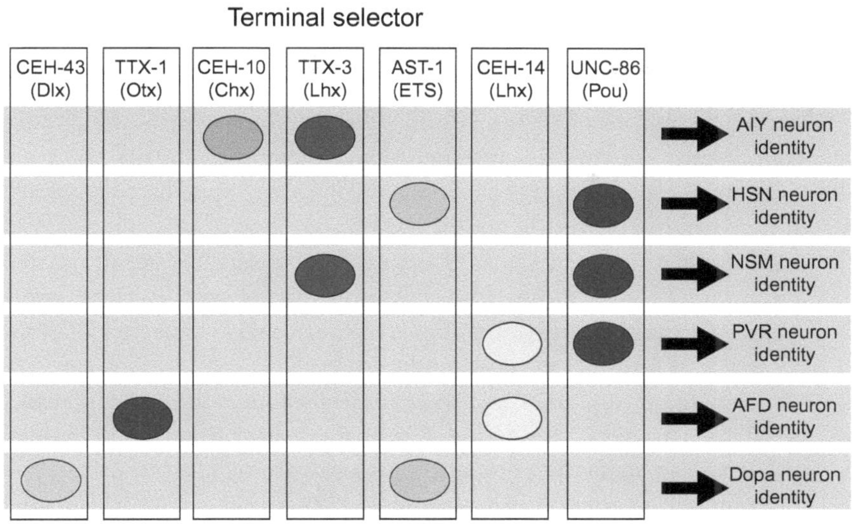

Figure 2 Combinatorial activity of terminal selectors. Selected examples are shown, some not showing the complete set of factors (e.g., dopaminergic neurons are controlled by the combined action of AST-1, CEH-43 plus a Pbx-type homeodomain factor). Note that all except one factor (AST-1) are homeodomain transcription factors. (A color version of this figure is located at the end of this chapter.)

animals lacking the POU homeobox gene *unc-86* and the LIM homeobox gene *ttx-3* (Zhang et al., 2014). However, in *ttx-3* and *unc-86* single mutants, some identity features are either not affected at all or are already affected as strongly as in the double mutant. There may be a spectrum of distinct types of cooperation between terminal selectors, depending on *cis*-regulatory architecture of individual effector genes.

There are a number of key implications of the combinatorial activity of terminal selectors:

(1) It explains how one and the same terminal selector can control very distinct fates. For example, UNC-86 is required to define the terminal identity of vastly different neuron types (Table 1) and in several cases we understand how. UNC-86 works cooperatively with MEC-3 to control touch neuron effector genes via a well-defined UNC-86/MEC-3 binding site (Xue et al., 1993). In contrast, in the BDU neurons, UNC-86 binds to different sites and collaborates with another factor, the Zn finger transcription factor PAG-3 (Gordon & Hobert, 2015).

(2) Conversely, the combinatorial logic also explains why one factor can operate as a terminal selector in one cell type, while not having any discernible effect on the same set of target genes in another cell type.

For example, the *ttx-3* LIM homeobox genes, expressed in the three cholinergic interneuron types AIY, AIA, and AIN, controls the cholinergic properties of the AIY and AIA interneurons (as well as all other known properties of these neurons), but does not control the cholinergic identity of the AIN interneurons (Zhang et al., 2014).

(3) Combinatorial activities provide a solution to the "specificity problem" of transcription factors. Single transcription factors usually interact with only about 4–12 nucleotides. Because such small sites are found too commonly in the genome, the activation of very specific target genes is likely mediated by the much more restrictive co-occurrence of combinations of terminal selector binding sites.

(4) Combinatorial function can also provide a solution to the "sufficiency problem." In many cases tested, terminal selectors found to be *required* to control a specific differentiation program are *sufficient* to induce this differentiation program in only some cellular contexts (e.g., Tursun, Patel, Kratsios, & Hobert, 2011; Wenick & Hobert, 2004). A broader effect may be achieved only upon misexpression of the complete set of combinatorially acting terminal selectors.

The examination of more and more *C. elegans* neuron types has uncovered another striking feature of transcription factor combinations. Even though the *C. elegans* genome has a repertoire of almost 1000 transcription factors, most terminal selectors are homeodomain transcription factors, which constitute only 10% of all *C. elegans* transcription factors. In those cases where non-homeodomain terminal selectors have been identified (e.g., the ETS domain factor AST-1), the non-homeodomain factor cooperates with a homeodomain transcription factor (Doitsidou et al., 2013; Serrano-Saiz et al., 2013). In cases where no homeodomain factor has been identified, it is conceivable that a homeodomain factor still awaits identification. The preponderance of homeodomain transcription factors may be a reflection of their very ancient use as drivers of neuronal identity.

5. TERMINAL SELECTORS CONTROL REGULATORY SUBROUTINES

A number of transcription factors have been identified that control only some, and perhaps in total very few genes in a terminally differentiated neuron. One example is the *ceh-23* homeobox gene which affects expression of a GPCR gene, but not other identity aspects of the cholinergic AIY interneuron (Altun-Gultekin et al., 2001); similarly, the *ceh-14* homeobox gene affects expression of neuropeptides in the BDU neurons, but does not affect other BDU identity features (Gordon & Hobert, 2015). Notably, in all these cases, these transcriptional subroutines are under direct control of a terminal selector (schematized in Fig. 1A). That is, the *ceh-23* gene, as well as its target, the GPCR gene *sra-11*, are under *direct* control of the terminal selectors of AIY identity, *ttx-3* and *ceh-10*, which control scores of other terminal identity features of the AIY interneuron (Altun-Gultekin et al., 2001; Wenick & Hobert, 2004). Similarly, *ceh-14* and its neuropeptide targets are controlled by the terminal selectors of BDU identity, *unc-86* and *pag-3* (Gordon & Hobert, 2015). Hence, in systems biology parlance, terminal selector frequently acts in feedforward loops (Alon, 2006) to control effector genes (Fig. 1A). It is easily conceivable that any terminal selector may regulate scores of regulatory subroutines. For example, the terminal selector *che-1* directly controls not only scores of nuts- and bolts-type terminal identity features (e.g., chemoreceptors, neuropeptides, and channels) via well-defined CHE-1 binding sites but also controls the expression of dozens of transcription factors in the ASE gustatory neurons; each of these transcription factors may control specific regulatory subroutines (Etchberger et al., 2007). Indeed, some of these transcription factors ensure the left/right asymmetric expression of chemoreceptors in the left or right ASE neurons, yet these factors do not control other identity features of the ASE neurons (Chang, Johnston, & Hobert, 2003; Hobert, 2014).

Should those subroutine regulators also be considered as "terminal selectors" of these subroutines? The most striking difference is that in all known cases, loss of these subroutine regulators still leaves the affected neuron with a recognizable identity. For example, while loss of *ceh-14* leads to loss of neuropeptide genes in BDU, the neuron is still recognizable as a BDU neuron in terms of intact expression of many other marker genes and overall morphology. In contrast, loss of *pag-3*, the terminal selector of BDU results in loss of all known molecular markers and alterations in neuron morphology. Even in cases where some terminal identity features are unaffected in a terminal selector mutant, these features are not sufficient to assign the neuron a specific identity. Therefore, it seems reasonable to propose that the terminal selector term should be restricted to regulatory factors whose loss results in a neuron not expressing a recognizable identity (or having switched to an alternative identity; Gordon & Hobert, 2015).

6. PARALLEL REGULATORY ROUTINES

Terminal selectors control the expression of many if not all terminal effector genes that are expressed in a *neuron-type-specific manner* and which, in a combinatorial manner define the unique identity of a neuron type. Such effector genes include biosynthetic enzymes and transporters for specific neurotransmitter systems, neurotransmitter receptors, ion channel, synaptic adhesion molecules, neuropeptide, sensory receptors, and others (Table 2). However, neuron identity-determining terminal selectors do *not* control the expression of *all* genes in a terminally differentiated neuron type. Extensive phenotypic analysis has shown that features that are shared by many or all neuron types are *not* affected by removal of terminal selectors that control neuron-type-specific features. The most striking feature that are commonly unaffected by terminal selector are *panneuronal* genes (e.g., genes expressed in *all* neurons, such as those encoding for synaptic vesicle proteins) or *pansensory* genes (genes involved in building the ciliated structures of sensory neurons) (Altun-Gultekin et al., 2001; Etchberger et al., 2007; Flames & Hobert, 2009; Kratsios et al., 2011). This argues that regulatory routines exist that act in parallel to terminal selectors of neuron-type-specific identity (Fig. 1A). While regulators of panneuronal gene expression are presently ill defined (Stefanakis, Carrera, & Hobert, 2015), the parallel regulatory routine that controls pansensory identity features has been well documented. The RFX-type transcription factor *daf-19* controls the coordinated expression of genes involved in building cilia (Burghoorn et al., 2012; Swoboda, Adler, & Thomas, 2000). *daf-19* can therefore be considered to be a terminal selector of ciliated identity.

While terminal selectors have profound effects on synaptic connectivity of a neuron (Howell et al., 2015; Kratsios et al., 2015; Zhang et al., 2014; Pereira et al., 2015), some overt morphological features remain relatively intact in terminal selector mutants. For example, in *unc-3* or *unc-30* terminal selector mutants, ventral nerve cord motor neurons are miswired, motor axons still project in the ventral and dorsal nerve cord. Similarly, in *che-1* terminal selector mutants, dendrites of the otherwise misspecified ASE chemosensory neurons are still present and its cell body is correctly placed (Uchida et al., 2003). Similarly, in *unc-86* terminal selector mutants, the HSN motor neuron fails to adopt terminal identity features such as its neurotransmitter identity, but still migrates normally (Desai et al., 1988). Hence, there must be factors that act in parallel to terminal selectors to determine some basic morphological features. There are few examples of such regulators. In the case of the HSN neurons, the transiently acting Zn finger transcription factor *egl-43* controls migration but not other features HSN identity (Baum, Guenther, Frank, Pham, & Garriga, 1999). A recent study in flies has shown that transcription factors called "morphology TF" (mTFs) define axonal/dendritic patterning features of motor neurons, but not other terminal features, such as neurotransmitter identity (Enriquez et al., 2015). Transiently acting mTFs are selectors of terminal morphological features of a neuron, but their transient function in regulating effector genes that are likely just transiently expressed (e.g., genes involved in axon pathfinding or cell migration) clearly sets them apart from terminal selectors that are constitutively present in a neuron to initiate and maintain constitutively expressed terminal identity features of a neuron.

Taken together, gene expression programs in terminally differentiated neurons may be controlled by parallel acting "regulons" that define distinct identity aspects; some of these identity aspects define the *unique* identity of a neuron, while others define more generic aspects of a neuron identity, such as pansensory features or some overt morphological features (Fig. 1A).

7. DIVERSIFICATION OF NEURONAL IDENTITY VIA SELECTIVE REPRESSIBILITY

Not every single neuron type appears to have its own, unique complement of terminal selector(s). Rather, distinct but related neuronal subtypes can be diversified by modulating the activity of a terminal selector on specific subsets of targets. This has been illustrated in two different contexts: the left and right ASE neurons are two neuronal subtypes that are very similar to one another but differ in their complement of chemoreceptor expression. Their identity is not controlled by distinct terminal selectors, but is controlled by the same terminal selector, *che-1* (Etchberger et al., 2007). However, the ability of *che-1* to activate some target genes (that contain CHE-1 binding sites) is prevented by the presence of repressor proteins in one of the two ASE subtypes (Etchberger et al., 2009; Hobert, 2014). This selective repression results in the ASEL- and ASER-expressing distinct subsets of genes, while simultaneously also expressing a vast amount of similar genes (under control of *che-1*). Similarly, the terminal selector *unc-3* controls the expression of multiple genes that are expressed differentially in distinct motor neuron subtypes. The ability of UNC-3 to activate some genes in some subtypes is restrained by subtype-specific repressor proteins (Kratsios et al., 2011; Winnier et al., 1999; our unpublished data).

8. GENERALITY OF THE TERMINAL SELECTOR CONCEPT

At the time of writing the original essay on the terminal selector concept, only a handful of examples of such selector genes has been described (Hobert, 2008). By now, transcriptional regulators that control terminal

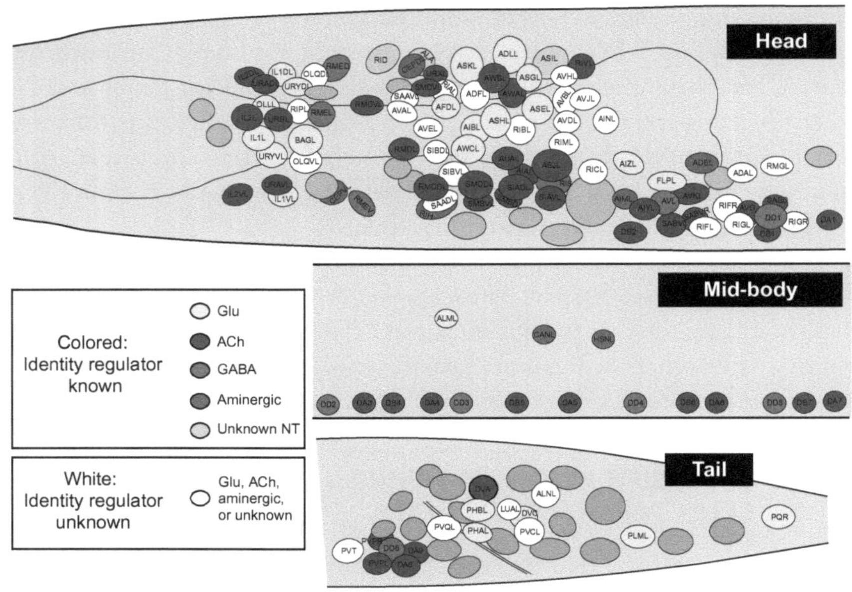

Figure 3 An overview of identity regulator in an L1 stage animal. Colored neurons are those in which terminal selector-type transcription factors have been identified that affect multiple identity features, including (in those cases known) the neurotransmitter identity of the neuron (color-coded). The only currently ambiguous cases included in this figure are the command interneurons, in which one factor (*unc-3*) affects neurotransmitter identity, but not other identity features, while another factor (*unc-42*) controls other identity features, but not neurotransmitter identity (Pereira et al., 2015). Pharyngeal neurons and postembryonically generated neurons are not shown. References: Altun, Chen, Wang, and Hall (2009), Cassata et al. (2000), Doitsidou et al. (2013), Etchberger et al. (2007), Gonzalez-Barrios et al. (2015), Gordon and Hobert (2015), Guillermin, Castelletto, and Hallem (2011), Jin et al. (1994), Hutter (2003), Kim, Kim, and Sengupta (2010), Kratsios et al. (2015, 2011), Lanjuin, VanHoven, Bargmann, Thompson, and Sengupta (2003), Nokes et al. (2009), Pereira et al. (2015), Satterlee et al. (2001), Sengupta et al. (1994), Serrano-Saiz et al. (2013), Sze, Zhang, Li, and Ruvkun (2002), Van Buskirk and Sternberg (2010), Zhang et al. (2014), and Zhang et al. (2002). (A color version of this figure is located at the end of this chapter.)

identity features of more than two-thirds of the 118 neuron classes have been identified (Fig. 3). Obviously, there is no other nervous system that is even close in the extent to which neuronal differentiation programs have been described throughout the nervous system. A concise summary of all these will be provided in an upcoming comprehensive review that is currently in preparation. Not all the regulators have been described in the same depth as others. In contrast to the original, almost "all or nothing" examples in which a mutation in a terminal selector affects the expression of dozens of terminal markers, two "50/50 examples" have now been described in which some terminal markers are affected, while others are not (Pereira et al., 2015; Zhang et al., 2014). It is conceivable that in these apparently exceptional cases, the respective neuron may harbor several terminal selectors that can partially compensate for the loss of each other, as proposed for the case of the NSM neurons (Zhang et al., 2014).

9. REGULATION OF TERMINAL SELECTOR GENES

As exemplified by the case of the terminal selector *ttx-3*, the induction of terminal selector gene expression is complex. Regulatory control regions of *ttx-3* integrate a number of distinct autonomous and nonautonomous inputs (Bertrand & Hobert, 2009; Murgan et al., 2015). Terminal selectors could therefore perhaps be viewed as critical junctions in a regulatory hourglass topology, sampling a diversity of upstream inputs and then branching out again to control a multitude of targets (Fig. 1A). The regulation of more terminal selectors needs to be studied to assess how broadly applicable this concept is.

10. CONCLUSIONS

The purpose of this chapter has been to provide a description of the general features of gene regulatory programs that operate in terminally differentiated neuron types throughout the *C. elegans* nervous system. I have focused here entirely on the *C. elegans* system. I will leave it to researchers of other model systems to assess the extent of similarity between gene expression programs in *C. elegans* in their favorite system and I hope that some of the findings made in *C. elegans* may provide experimental testable hypotheses for researchers in other systems.

ACKNOWLEDGMENTS

I thank former and current lab members whose work has been instrumental in building the terminal selector concept. I thank lab members as well as Nuria Flames, Evan Deneris, Richard Mann, Doug Allan and Stefan Thor for comments on the manuscript. Work in my laboratory is funded by the National Institutes of Health and the Howard Hughes Medical Institute.

REFERENCES

Alon, U. (2006). *An introduction to systems biology: Design principles of biological circuits. 1st ed.* London, UK: Chapman & Hall/CRC.

Altun, Z. F., Chen, B., Wang, Z. W., & Hall, D. H. (2009). High resolution map of Caenorhabditis elegans gap junction proteins. *Developmental Dynamics*, *238*, 1936–1950.

Altun-Gultekin, Z., Andachi, Y., Tsalik, E. L., Pilgrim, D., Kohara, Y., & Hobert, O. (2001). A regulatory cascade of three homeobox genes, ceh-10, ttx-3 and ceh-23, controls cell fate specification of a defined interneuron class in C. elegans. *Development*, *128*, 1951–1969.

Baran, R., Aronoff, R., & Garriga, G. (1999). The C. elegans homeodomain gene unc-42 regulates chemosensory and glutamate receptor expression. *Development*, *126*, 2241–2251.

Bargmann, C. I., Hartwieg, E., & Horvitz, H. R. (1993). Odorant-selective genes and neurons mediate olfaction in C. elegans. *Cell*, *74*, 515–527.

Bargmann, C. I., & Horvitz, H. R. (1991). Chemosensory neurons with overlapping functions direct chemotaxis to multiple chemicals in C. elegans. *Neuron*, *7*, 729–742.

Baum, P. D., Guenther, C., Frank, C. A., Pham, B. V., & Garriga, G. (1999). The Caenorhabditis elegans gene ham-2 links Hox patterning to migration of the HSN motor neuron. *Genes & Development*, *13*, 472–483.

Bertrand, V., & Hobert, O. (2009). Linking asymmetric cell division to the terminal differentiation program of postmitotic neurons in C. elegans. *Developmental Cell*, *16*, 563–575.

Brenner, S. (1974). The genetics of Caenorhabditis elegans. *Genetics*, *77*, 71–94.

Burghoorn, J., Piasecki, B. P., Crona, F., Phirke, P., Jeppsson, K. E., & Swoboda, P. (2012). The in vivo dissection of direct RFX-target gene promoters in C. elegans reveals a novel cis-regulatory element, the C-box. *Developmental Biology*, *368*, 415–426.

Cassata, G., Kagoshima, H., Andachi, Y., Kohara, Y., Durrenberger, M. B., Hall, D. H., et al. (2000). The LIM homeobox gene ceh-14 confers thermosensory function to the AFD neurons in Caenorhabditis elegans. *Neuron*, *25*, 587–597.

Chalfie, M., & Sulston, J. (1981). Developmental genetics of the mechanosensory neurons of Caenorhabditis elegans. *Developmental Biology*, *82*, 358–370.

Chalfie, M., Sulston, J. E., White, J. G., Southgate, E., Thomson, J. N., & Brenner, S. (1985). The neural circuit for touch sensitivity in Caenorhabditis elegans. *The Journal of Neuroscience*, *5*, 956–964.

Chalfie, M., Tu, Y., Euskirchen, G., Ward, W. W., & Prasher, D. C. (1994). Green fluorescent protein as a marker for gene expression. *Science*, *263*, 802–805.

Chang, S., Johnston, R. J., Jr., & Hobert, O. (2003). A transcriptional regulatory cascade that controls left/right asymmetry in chemosensory neurons of C. elegans. *Genes & Development*, *17*, 2123–2137.

Cinar, H., Keles, S., & Jin, Y. (2005). Expression profiling of GABAergic motor neurons in Caenorhabditis elegans. *Current Biology*, *15*, 340–346.

Deneris, E. S., & Hobert, O. (2014). Maintenance of postmitotic neuronal cell identity. *Nature Neuroscience*, *17*, 899–907.

Desai, C., Garriga, G., McIntire, S. L., & Horvitz, H. R. (1988). A genetic pathway for the development of the Caenorhabditis elegans HSN motor neurons. *Nature*, *336*, 638–646.

Doitsidou, M., Flames, N., Topalidou, I., Abe, N., Felton, T., Remesal, L., et al. (2013). A combinatorial regulatory signature controls terminal differentiation of the dopaminergic nervous system in C. elegans. *Genes & Development*, *27*, 1391–1405.

Duggan, A., Ma, C., & Chalfie, M. (1998). Regulation of touch receptor differentiation by the Caenorhabditis elegans mec-3 and unc-86 genes. *Development*, *125*, 4107–4119.

Eastman, C., Horvitz, H. R., & Jin, Y. (1999). Coordinated transcriptional regulation of the unc-25 glutamic acid decarboxylase and the unc-47 GABA vesicular transporter by the

Caenorhabditis elegans UNC-30 homeodomain protein. *The Journal of Neuroscience*, *19*, 6225–6234.
Edlund, T., & Jessell, T. M. (1999). Progression from extrinsic to intrinsic signaling in cell fate specification: A view from the nervous system. *Cell*, *96*, 211–224.
Enriquez, J., Venkatasubramanian, L., Baek, M., Peterson, M., Aghayeva, U., & Mann, R. S. (2015). Specification of individual adult motor neuron morphologies by combinatorial transcription factor codes. *Neuron*, *86*, 955–970.
Etchberger, J. F., Flowers, E. B., Poole, R. J., Bashllari, E., & Hobert, O. (2009). Cis-regulatory mechanisms of left/right asymmetric neuron-subtype specification in C. elegans. *Development*, *136*, 147–160.
Etchberger, J. F., Lorch, A., Sleumer, M. C., Zapf, R., Jones, S. J., Marra, M. A., et al. (2007). The molecular signature and cis-regulatory architecture of a C. elegans gustatory neuron. *Genes & Development*, *21*, 1653–1674.
Finney, M., Ruvkun, G., & Horvitz, H. R. (1988). The C. elegans cell lineage and differentiation gene unc-86 encodes a protein with a homeodomain and extended similarity to transcription factors. *Cell*, *55*, 757–769.
Flames, N., & Hobert, O. (2009). Gene regulatory logic of dopamine neuron differentiation. *Nature*, *458*, 885–889.
Garcia-Bellido, A. (1975). Genetic control of wing disc development in Drosophila. *Ciba Foundation Symposium*, *29*, 161–182.
Gonzalez-Barrios, M., Fierro-Gonzalez, J. C., Krpelanova, E., Mora-Lorca, J. A., Pedrajas, J. R., Penate, X., et al. (2015). Cis- and trans-regulatory mechanisms of gene expression in the ASJ sensory neuron of Caenorhabditis elegans. *Genetics*, *200*, 123–134.
Gordon, P. M., & Hobert, O. (2015). A competition mechanism for a homeotic neuron identity transformation in C. elegans. *Developmental Cell*, *34*, 206–219.
Guillermin, M. L., Castelletto, M. L., & Hallem, E. A. (2011). Differentiation of carbon dioxide-sensing neurons in Caenorhabditis elegans requires the ETS-5 transcription factor. *Genetics*, *189*, 1327–1339.
Hedgecock, E. M., & Russell, R. L. (1975). Normal and mutant thermotaxis in the nematode Caenorhabditis elegans. *Proceedings of the National Academy of Sciences of the United States of America*, *72*, 4061–4065.
Hemmati-Brivanlou, A., & Melton, D. (1997). Vertebrate neural induction. *Annual Review of Neuroscience*, *20*, 43–60.
Hobert, O. (2008). Regulatory logic of neuronal diversity: Terminal selector genes and selector motifs. *Proceedings of the National Academy of Sciences of the United States of America*, *105*, 20067–20071.
Hobert, O. (2011). Regulation of terminal differentiation programs in the nervous system. *Annual Review of Cell and Developmental Biology*, *27*, 681–696.
Hobert, O. (2014). Development of left/right asymmetry in the Caenorhabditis elegans nervous system: From zygote to postmitotic neuron. *Genesis*, *52*, 528–543.
Hobert, O., Mori, I., Yamashita, Y., Honda, H., Ohshima, Y., Liu, Y., et al. (1997). Regulation of interneuron function in the C. elegans thermoregulatory pathway by the ttx-3 LIM homeobox gene. *Neuron*, *19*, 345–357.
Holmberg, J., & Perlmann, T. (2012). Maintaining differentiated cellular identity. *Nature Reviews. Genetics*, *13*, 429–439.
Howell, K., White, J. G., & Hobert, O. (2015). Spatiotemporal control of a novel synaptic organizer molecule. *Nature*, *523*, 83–87.
Hutter, H. (2003). Extracellular cues and pioneers act together to guide axons in the ventral cord of C. elegans. *Development*, *130*, 5307–5318.
Jin, Y., Hoskins, R., & Horvitz, H. R. (1994). Control of type-D GABAergic neuron differentiation by C. elegans UNC-30 homeodomain protein. *Nature*, *372*, 780–783.

Kim, K., Kim, R., & Sengupta, P. (2010). The HMX/NKX homeodomain protein MLS-2 specifies the identity of the AWC sensory neuron type via regulation of the ceh-36 Otx gene in C. elegans. *Development*, *137*, 963–974.

Kratsios, P., Pinan-Lucarre, B., Kerk, S. Y., Weinreb, A., Bessereau, J. L., & Hobert, O. (2015). Transcriptional coordination of synaptogenesis and neurotransmitter signaling. *Current Biology*, *25*, 1282–1295.

Kratsios, P., Stolfi, A., Levine, M., & Hobert, O. (2011). Coordinated regulation of cholinergic motor neuron traits through a conserved terminal selector gene. *Nature Neuroscience*, *15*, 205–214.

Laguna, A., Schintu, N., Nobre, A., Alvarsson, A., Volakakis, N., Jacobsen, J. K., et al. (2015). Dopaminergic control of autophagic-lysosomal function implicates Lmx1b in Parkinson's disease. *Nature Neuroscience*, *18*, 826–835.

Lanjuin, A., VanHoven, M. K., Bargmann, C. I., Thompson, J. K., & Sengupta, P. (2003). Otx/otd homeobox genes specify distinct sensory neuron identities in C. elegans. *Developmental Cell*, *5*, 621–633.

Lewis, J. A., & Hodgkin, J. A. (1977). Specific neuroanatomical changes in chemosensory mutants of the nematode Caenorhabditis elegans. *The Journal of Comparative Neurology*, *172*, 489–510.

McIntire, S. L., Jorgensen, E., & Horvitz, H. R. (1993). Genes required for GABA function in Caenorhabditis elegans. *Nature*, *364*, 334–337.

Mori, I., & Ohshima, Y. (1995). Neural regulation of thermotaxis in Caenorhabditis elegans. *Nature*, *376*, 344–348.

Murgan, S., Kari, W., Rothbacher, U., Iche-Torres, M., Melenec, P., Hobert, O., et al. (2015). Atypical transcriptional activation by TCF via a Zic transcription factor in C. elegans neuronal precursors. *Developmental Cell*, *33*, 737–745.

Nokes, E. B., Van Der Linden, A. M., Winslow, C., Mukhopadhyay, S., Ma, K., & Sengupta, P. (2009). Cis-regulatory mechanisms of gene expression in an olfactory neuron type in Caenorhabditis elegans. *Developmental Dynamics*, *238*, 3080–3092.

Pereira, L., Kratsios, P., Serrano-Saiz, E., Sheftel, H., Mayo, A., Hall, D. H., et al. (2015). A cellular and regulatory map of the cholinergic nervous system of *C. elegans*. *Elife*, *4*. pii: e12432. doi: 10.7554/eLife.12432.

Prasad, B. C., Ye, B., Zackhary, R., Schrader, K., Seydoux, G., & Reed, R. R. (1998). unc-3, a gene required for axonal guidance in Caenorhabditis elegans, encodes a member of the O/E family of transcription factors. *Development*, *125*, 1561–1568.

Sagasti, A., Hobert, O., Troemel, E. R., Ruvkun, G., & Bargmann, C. I. (1999). Alternative olfactory neuron fates are specified by the LIM homeobox gene lim-4. *Genes & Development*, *13*, 1794–1806.

Satterlee, J. S., Sasakura, H., Kuhara, A., Berkeley, M., Mori, I., & Sengupta, P. (2001). Specification of thermosensory neuron fate in C. elegans requires ttx-1, a homolog of otd/Otx. *Neuron*, *31*, 943–956.

Sengupta, P., Colbert, H. A., & Bargmann, C. I. (1994). The C. elegans gene odr-7 encodes an olfactory-specific member of the nuclear receptor superfamily. *Cell*, *79*, 971–980.

Serrano-Saiz, E., Poole, R. J., Felton, T., Zhang, F., de la Cruz, E. D., & Hobert, O. (2013). Modular control of glutamatergic neuronal identity in C. elegans by distinct homeodomain proteins. *Cell*, *155*, 659–673.

Stefanakis, N., Carrera, I., & Hobert, O. (2015). Regulatory logic of pan-neuronal gene expression in C. elegans. *Neuron*, *87*, 733–750.

Sulston, J. E. (1983). Neuronal cell lineages in the nematode Caenorhabditis elegans. *Cold Spring Harbor Symposia on Quantitative Biology*, *48*, 443–452.

Swoboda, P., Adler, H. T., & Thomas, J. H. (2000). The RFX-type transcription factor DAF-19 regulates sensory neuron cilium formation in C. elegans. *Molecular Cell*, *5*, 411–421.

Sze, J. Y., Zhang, S., Li, J., & Ruvkun, G. (2002). The C. elegans POU-domain transcription factor UNC-86 regulates the tph-1 tryptophan hydroxylase gene and neurite outgrowth in specific serotonergic neurons. *Development*, *129*, 3901–3911.

Tsalik, E. L., & Hobert, O. (2003). Functional mapping of neurons that control locomotory behavior in Caenorhabditis elegans. *Journal of Neurobiology*, *56*, 178–197.

Tursun, B., Patel, T., Kratsios, P., & Hobert, O. (2011). Direct conversion of C. elegans germ cells into specific neuron types. *Science*, *331*, 304–308.

Uchida, O., Nakano, H., Koga, M., & Ohshima, Y. (2003). The C. elegans che-1 gene encodes a zinc finger transcription factor required for specification of the ASE chemosensory neurons. *Development*, *130*, 1215–1224.

Van Buskirk, C., & Sternberg, P. W. (2010). Paired and LIM class homeodomain proteins coordinate differentiation of the C. elegans ALA neuron. *Development*, *137*, 2065–2074.

Waddington, C. H. (1957). *The strategy of the genes*. London, UK: Allen and Unwin.

Way, J. C., & Chalfie, M. (1988). mec-3, a homeobox-containing gene that specifies differentiation of the touch receptor neurons in C. elegans. *Cell*, *54*, 5–16.

Way, J. C., & Chalfie, M. (1989). The mec-3 gene of Caenorhabditis elegans requires its own product for maintained expression and is expressed in three neuronal cell types. *Genes & Development*, *3*, 1823–1833.

Wenick, A. S., & Hobert, O. (2004). Genomic cis-regulatory architecture and trans-acting regulators of a single interneuron-specific gene battery in C. elegans. *Developmental Cell*, *6*, 757–770.

White, J. G., Southgate, E., Thomson, J. N., & Brenner, S. (1986). The structure of the nervous system of the nematode *Caenorhabditis elegans*. *Philosophical Transactions of the Royal Society of London. Series B, Biological Sciences*, *314*, 1–340.

Winnier, A. R., Meir, J. Y., Ross, J. M., Tavernarakis, N., Driscoll, M., Ishihara, T., et al. (1999). UNC-4/UNC-37-dependent repression of motor neuron-specific genes controls synaptic choice in Caenorhabditis elegans. *Genes & Development*, *13*, 2774–2786.

Xue, D., Tu, Y., & Chalfie, M. (1993). Cooperative interactions between the Caenorhabditis elegans homeoproteins UNC-86 and MEC-3. *Science*, *261*, 1324–1328.

Zhang, F., Bhattacharya, A., Nelson, J. C., Abe, N., Gordon, P., Lloret-Fernandez, C., et al. (2014). The LIM and POU homeobox genes ttx-3 and unc-86 act as terminal selectors in distinct cholinergic and serotonergic neuron types. *Development*, *141*, 422–435.

Zhang, Y., Ma, C., Delohery, T., Nasipak, B., Foat, B. C., Bounoutas, A., et al. (2002). Identification of genes expressed in C. elegans touch receptor neurons. *Nature*, *418*, 331–335.

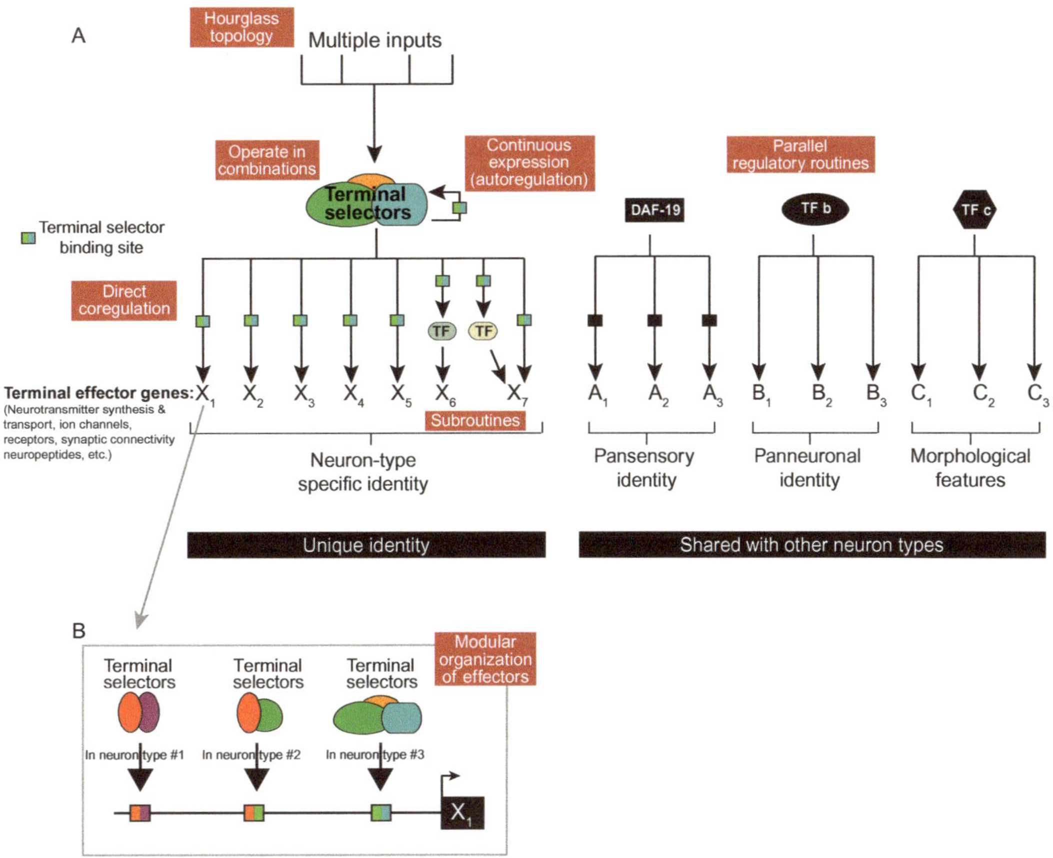

Oliver Hobert, Figure 1 See legend on next page.

Oliver Hobert, Figure 1 Key features of gene expression programs in mature neuron types. (A) Terminal selector regulons. Red squares highlight key concepts. Sustained expression of terminal selectors is often, but not always ensured by autoregulation. Small rectangles illustrate binding sites for transcription factors and indicate direct regulation of effector genes by terminal selectors. Terminal selectors shown on the left control neuron-type specific identity features that assign a unique identity to a neuron. See Table 2 for an exemplary list of effector genes. Parallel regulatory routines such as those that regulate pansensory features (via DAF-19) are controlled by factors that could also considered to be terminal selectors; but they do not assign unique identities. Morphology regulators control generic aspect of a neuron's morphology, such as placement of axons/dendrites into specific fascicles or axo/dendritic polarity. TF, transcription factor. (B) Modular organization of terminal effector genes. X_1 is a representative of any effector gene expressed in more than one neuron type (e.g., the vesicular transporter of glutamate, which is expressed in all glutamatergic neurons, but not in GABA or cholinergic neurons). This schematic also indicates the key principle of combinatorial "reusage" of the same terminal selector in different neuron types, as schematized in more detail in Fig. 2.

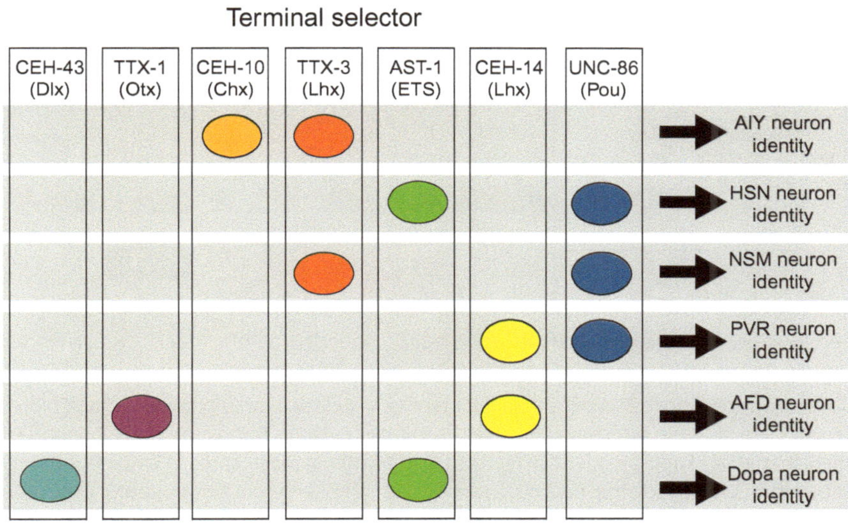

Oliver Hobert, Figure 2 Combinatorial activity of terminal selectors. Selected examples are shown, some not showing the complete set of factors (e.g., dopaminergic neurons are controlled by the combined action of AST-1, CEH-43 plus a Pbx-type homeodomain factor). Note that all except one factor (AST-1) are homeodomain transcription factors.

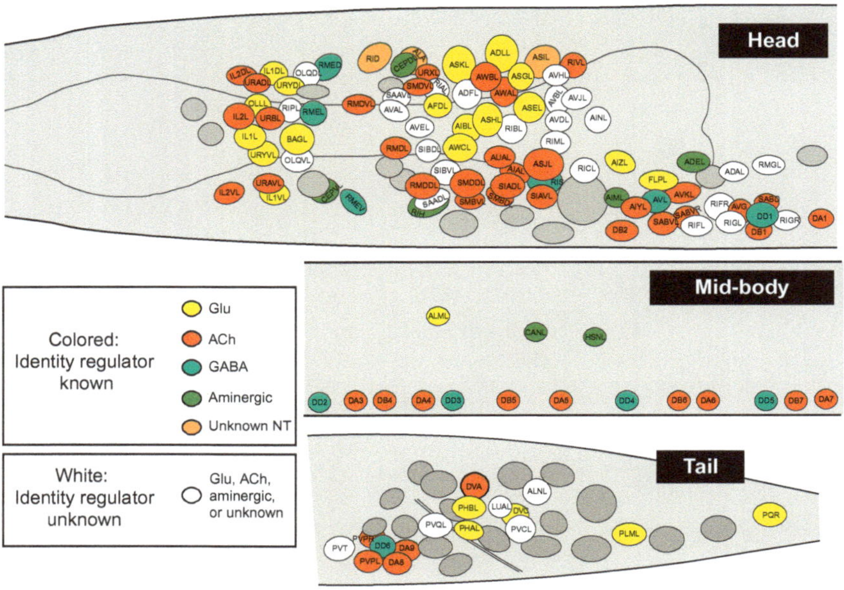

Oliver Hobert, Figure 3 An overview of identity regulator in an L1 stage animal. Colored neurons are those in which terminal selector-type transcription factors have been identified that affect multiple identity features, including (in those cases known) the neurotransmitter identity of the neuron (color-coded). The only currently ambiguous cases included in this figure are the command interneurons, in which one factor (*unc-3*) affects neurotransmitter identity, but not other identity features, while another factor (*unc-42*) controls other identity features, but not neurotransmitter identity (Pereira et al., 2015). Pharyngeal neurons and postembryonically generated neurons are not shown. References: Altun, Chen, Wang, and Hall (2009), Cassata et al. (2000), Doitsidou et al. (2013), Etchberger et al. (2007), Gonzalez-Barrios et al. (2015), Gordon and Hobert (2015), Guillermin, Castelletto, and Hallem (2011), Jin et al. (1994), Hutter (2003), Kim, Kim, and Sengupta (2010), Kratsios et al. (2015, 2011), Lanjuin, VanHoven, Bargmann, Thompson, and Sengupta (2003), Nokes et al. (2009), Pereira et al. (2015), Satterlee et al. (2001), Sengupta et al. (1994), Serrano-Saiz et al. (2013), Sze, Zhang, Li, and Ruvkun (2002), Van Buskirk and Sternberg (2010), Zhang et al. (2014), and Zhang et al. (2002).

CHAPTER TWENTY-SIX

Drosophila Segment Polarity Mutants and the Rediscovery of the Hedgehog Pathway Genes

Philip W. Ingham[1]
Lee Kong Chian School of Medicine, Nanyang Technological University, Singapore
Department of Medicine, Imperial College, London, United Kingdom
[1]Corresponding author: e-mail address: pingham@ntu.edu.sg

Abstract

The Nüsslein-Volhard and Wieschaus screen for mutations disrupting the segmentation of the *Drosophila* embryo revolutionized developmental genetics, leading the way to the identification of many of the transcription factors and signaling pathways that orchestrate development, not just in the fruit fly but across the animal kingdom. The Hedgehog signaling pathway is a case in point: yet remarkably, all but one of the genes encoding the Hedgehog pathway components—including Hedgehog itself—had previously been discovered, in some cases decades earlier. Here I review the original identification of these genes and consider why their significance remained obscure until the Nobel Prize winning screen.

Embryonic development is orchestrated by a small set of secreted signaling molecules that together mediate the inductive interactions between cell populations that sculpt the form of an animal. Among these signals are the members of the Hedgehog (Hh) protein family. Hh proteins regulate a plethora of processes, ranging from gonadal stem cell proliferation in *Drosophila* (Forbes et al., 1996; Michel et al., 2012; Zhang & Kalderon, 2001) to the specification of digit number and identity in tetrapods (Chiang et al., 2001; Harfe et al., 2004; Riddle et al., 1993). Most of the roles of Hh proteins have been uncovered in the past two decades following the isolation of vertebrate homologues of the *Drosophila hh* gene in 1993 (Echelard et al., 1993; Krauss, Concordet, & Ingham, 1993; Riddle et al., 1993). Prior to this, analysis of Hh signaling was something of an esoteric pursuit, focused

Current Topics in Developmental Biology, Volume 116
ISSN 0070-2153
http://dx.doi.org/10.1016/bs.ctdb.2016.01.007

primarily on its role in patterning the body segments of the *Drosophila* embryo (Forbes et al., 1993; Hidalgo, 1991; Mohler, 1988). This stemmed from the identification of *hh* alleles among the segment polarity class of mutants in the Nobel Prize winning embryonic lethal screens performed by Nüsslein-Volhard and Wieschaus in the late 1970s and described in papers that they published in 1980 and 1984 (Nüsslein Volhard, Wieschaus, & Kluding, 1984; Nüsslein-Volhard & Wieschaus, 1980). Indeed, these screens are commonly credited with the discovery of the genes encoding the components of Hh signaling pathway; yet in point of fact only one such gene, *smoothened* (*smo*), was identified for the first time by these screens (Nüsslein Volhard et al., 1984). The genius of Nüsslein-Volhard and Wieschaus was to design a screen that allowed the functional relationship between these genes to be revealed by demonstrating that their null mutations belong to a common phenotypic class.

The first—and viable—mutant allele of *hh*, hh^1, was isolated on the basis of its bar-shaped eyes by Ives (1950), some 30 years prior to publication of the famous 1980 *Nature* paper. This phenotype is similar to that caused by mutations of the eponymous X-linked *Bar* locus, the product of which is required for photoreceptor differentiation and expression of which has subsequently been shown to be regulated by Hh (Lim & Choi, 2004). This phenotypic similarity to Bar, together with its mapping to the 3rd chromosome, led to the hh^1 allele originally being designated *bar-3*. We now know that the *bar-3* mutation is associated with the deletion of an eye-specific enhancer within the first intron of the *hh* gene (Pauli et al., 2005; Rogers et al., 2005); the consequent eye-specific loss of *hh* expression results in the failure of morphogenetic furrow progression and hence the reduction of ommatidia that characterizes the mutant animals (Heberlein, Wolff, & Rubin, 1993; Lee et al., 1992; Ma et al., 1996; Mohler, 1988).

Hh proteins elicit their responses by binding to Patched (Ptc), a multipass transmembrane protein (Hooper & Scott, 1989; Nakano et al., 1989). Ptc is an unusual receptor for a secreted polypeptide: it belongs to the RND family of transporter proteins (Taipale et al., 2002) and constitutively suppresses the Hh signaling pathway by inhibiting the activity of Smo, a member of the G-protein-coupled receptor (GPCR) family (Quirk et al., 1997). Binding of Hh ligand to Ptc abrogates its activity, thereby releasing Smo from inhibition. In *Drosophila*, the activation of Smo is accompanied by its dimerization and recruitment of the kinesin family member, Costal-2 (Cos-2), and the consequent activation of the serine-threonine kinase, Fused (Fu) (Shi et al., 2011). The latter plays a key role in activating the transcription

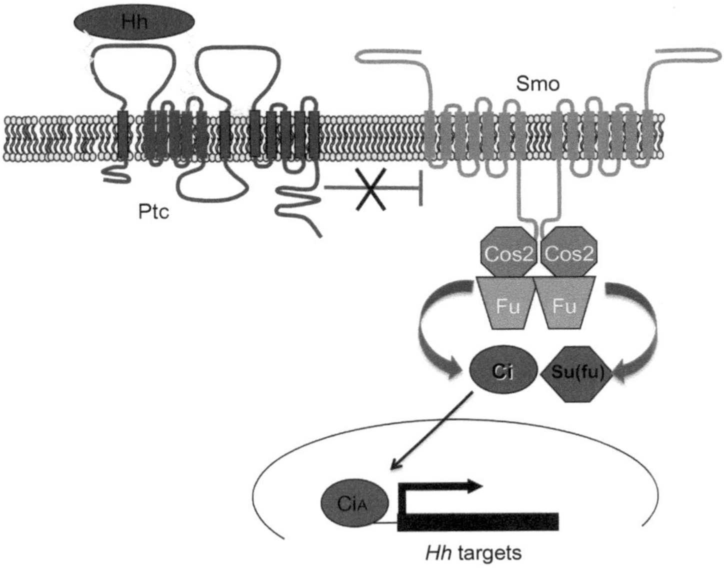

Figure 1 Simplified schema of the *Drosophila* Hh signal transduction pathway showing the core components. Negative regulators and repressive interactions are shown in red, positive regulators and activating interactions are shown in green. (A color version of this figure is located at the end of this chapter.)

factor Cubitus interruptus (Ci), in part by promoting its dissociation from the inhibitory Suppressor of Fused (Su(Fu)) protein, facilitating its entry into the nucleus where it activates transcription of Hh target genes (see Fig. 1).

The isolation of mutations in genes encoding three of these core pathway components: viz., *ptc*, *ci*, and *fu*, predated the characterization of the hh^{1}/*bar-3* allele—but there was little in their phenotypes to suggest a connection between them. The first analysis of *fu* was published by Clara Lynch, a student of TH Morgan, almost a century ago (Lynch, 1919). Although originally identified on the basis of the fusion of the third and fourth wing veins, an effect now understood to reflect the requirement for Hh signaling in specifying this region of the wing (Vervoort et al., 1999), Lynch's analysis focused on the striking effects of the mutation on female fertility. Her studies showed that homozygous females are effectively sterile when crossed to mutant males, but that this sterility can be rescued by crossing the females to wild-type males. We now understand that this reflects the contribution of both maternal and zygotic gene expression to the provision of Fu kinase activity that mediates the response of cells to Hh signals during embryogenesis. Nearly 40 years elapsed before, the requirement for *fu* in embryonic development was analyzed by Counce (1956); her illustration and description of the cuticle phenotype of maternal-zygotic *fu* mutant embryos can readily be recognized as typical of the segment polarity class of phenotypes (see Fig. 2).

Like *fused*, cubitus interruptus was originally discovered through the effects of a viable mutant allele on wing venation. This dominant allele, designated ci^D, was described by two Russian geneticists Tiniakow and Ternetieva in 1932 (Tiniakow & Terentieva, 1933). The name derives from

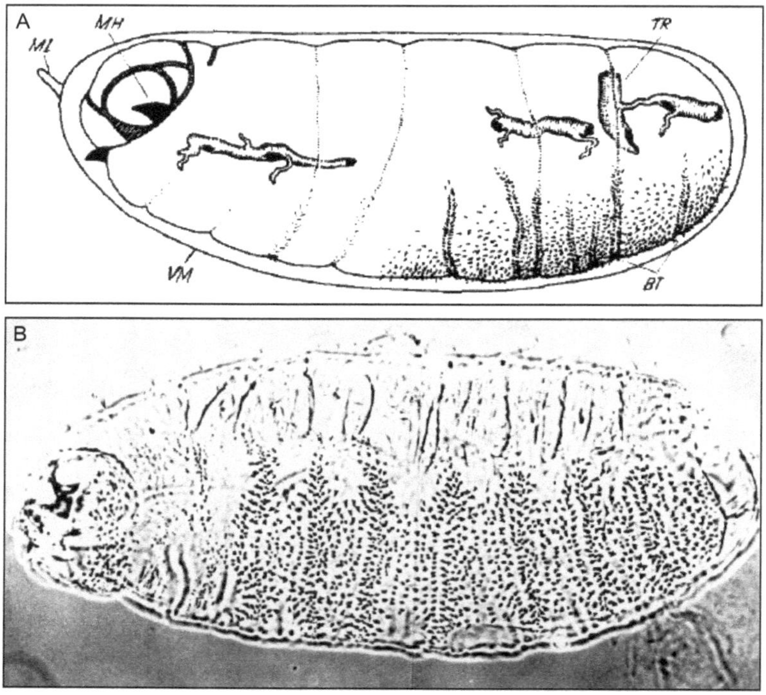

Figure 2 (A) Drawing of a *fused* mutant embryo by Counce (1956). Her description is as follows: "Individual segments are often incompletely formed, the number apparent ventrally exceeding that dorsally. They may also vary in width and number laterally. Rings of chitinous setae, such as are found at the anterior edge of each segment in normal embryos, are only partially formed. Instead, heavy bristles are distributed more or less at random over the ventral posterior regions, but tend to reflect the abnormal segmentation." (B) Phase contrast image of a cuticle preparation of a *fused* mutant embryo. *Panel (A): With kind permission from Counce (1956), figure 1. Panel (B): Adapted from Martinez-Arias, A. The development of fused-embryos of Drosophila melanogaster.* Development, 87, *99–114, Copyright (1985), with permission from The Company of Biologists.*

the fact that mutant animals display interruptions in the L4—also known as the cubitus—and L5 wing veins. The first mutant allele of *ptc* to be isolated was described by Sturtevant in 1948 (Sturetvant, 1948); originally known as *tufted* (*tuf*) due to the small ectopic tufts of hairs that form between the eyes and antennae, it was also noted to cause small duplications in the anterior half of the wing (Lindsley & Grell, 1968). Screens for new alleles of *tuf* performed by Robert Whittle in the 1970s, resulted in the isolation of tuf^{S2}, which in *trans* with the original tuf^{1} allele caused much more dramatic duplications of the anterior wing than those typical of tuf^{1} homozygotes (see Fig. 3). Unlike tuf^{1}, tuf^{S2} was found to be homozygous lethal with an embryonic phenotype identical to that of the *ptc* alleles described by Nüsslein-Volhard and Wieschaus (1980). Complementation analysis established the allelism of *ptc* and *tuf*, leading to the redesignation of the tuf^{1} and tuf^{S2} mutations as *ptc* alleles (Nakano et al., 1989).

Wing duplications similar to those caused by the *tuf* mutations (Fig. 3) also led Whittle to isolate the first mutant alleles of the *costal-2* (*cos-2*) gene in 1976 (Whittle, 1976). Notably, this phenotype was initially only uncovered in animals simultaneously heterozygous for mutations of both *cos-2* and

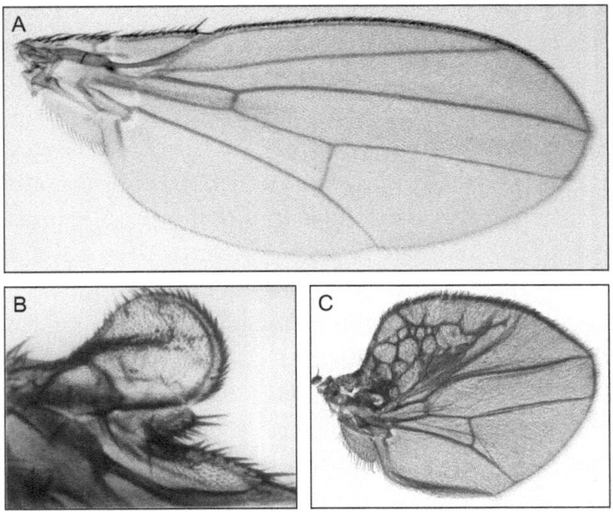

Figure 3 (A) Wing from a wild-type *Drosophila*. (B) Wing from a *cos1/+; cos2/+* double heterozygote showing the outgrowth from the costal region, from which the mutations derive their name. (C) Wing from a tuf^{1}/tuf^{S2} (ptc^{1}/ptc^{S2}) transheterozygote in which the anterior region of the wing is replaced by a mirror image duplication of more medial wing tissue. *Panel (B): Reprinted from Whittle (1976), Copyright (1976), with permission from Elsevier.*

a distinct locus, named *cos-1*, that was subsequently shown to correspond to the gene encoding the catalytic subunit of PKA (Collier et al., 2004). It would be another 20 years after the discovery of *cos-2* before the key role played by PKA in Hh signaling would be discovered through the isolation of new alleles of *cpka* (Jiang & Struhl, 1995; Lepage et al., 1995; Li et al., 1995; Pan & Rubin, 1995; Strutt, Wiersdorff, & Mlodzik, 1995). The *cos-2* and *ptc* wing phenotypes bear striking analogy to the mirror image digit duplications caused by transplantation of the so-called Zone of Polarizing Activity or ZPA, first described by Saunders in 1968 (Saunders & Gasseling, 1968) and the subject of intense investigation by Lewis Wolpert and his colleagues in London, contemporaneously with Whittle's *Drosophila* studies in the 1970s (Tickle, Summerbell, & Wolpert, 1975; Tickle et al., 1976). Of course, it is now well understood that both phenomena reflect the ectopic activity of the Hh pathway: loss of *cos-2* or *ptc* results in the ectopic expression of *decapentaplegic* (*dpp*), the *Drosophila* orthologue of the genes encoding vertebrate bone morphogenetic proteins 2 and 4, which controls both the proliferation and patterning of imaginal disc cells (Chen & Struhl, 1996; Ho et al., 2005); the ZPA is the source of Sonic Hedgehog (Shh) protein, one of the vertebrate homologues of *Drosophila* Hh, and transplantation of the ZPA to the anterior margin of the limb bud establishes a second source of Shh that signals inappropriately to the surrounding cells (Riddle et al., 1993). The discovery of *cos-2* thus took Whittle tantalizingly close to uncovering the similarities between the patterning of insect appendages and tetrapod limbs. At the time, few would have suspected that misexpression of the same signaling pathway could cause both phenomena, although the notion that common principles underly the patterning of insect and vertebrate limbs had recently been posited by Vernon French and Peter and Susan Bryant, whose "Polar Coordinate Model" set out universal rules describing the regenerative behavior of Axolotl limbs, cockroach legs, and *Drosophila* imaginal discs (French, Bryant, & Bryant, 1976). In the absence of any knowledge of the spatially restricted gene expression that presages and underpins the emergence of pattern, or the tools that allow the exquisite manipulation of the expression of these genes, unraveling the basis of such common principles of pattern formation would remain out of reach for another 15 years.

Instead, it was the prevailing preoccupation with lineage restrictions and developmental compartments that dominated thinking about pattern formation in *Drosophila* imaginal discs in the 1970s. As it turned out, however, this focus on compartments would provide an important tool in the unraveling of the Hh pathway, namely the DNA sequence encoding the Engrailed protein. The *engrailed* gene had become a central focus of *Drosophila* developmental genetics by the late 1970s, thanks to its identification as the selector gene that distinguishes between the anterior and posterior compartments into which each imaginal disc—and indeed each body segment—is subdivided (Morata & Lawrence, 1975). As such, it became a prime target for the fledgling efforts to apply the power of molecular biology to the analysis of *Drosophila* development. Cloning of *en* was achieved in the early 1980s both through the deliberate effort to isolate the gene by positional cloning (Kuner et al., 1985) and somewhat serendipitously, through its possession of the recently identified homeobox sequence (Fjose, McGinnis, & Gehring, 1985). This achievement, together with the development of the *in situ* hybridization technique, which for the first time allowed the spatial distribution of gene transcripts to be visualized in the intact organism (Hafen, Kuroiwa, & Gehring, 1984), shifted the focus of attention on *en* from its role in imaginal discs to its function in embryogenesis. As for *hh*, null alleles of *en* had been isolated in the Nüsslein-Volhard and Wieschaus screens, but in contrast to *hh*, their phenotype had led to *en* being classified as one of the so-called "pair-rule" genes. But whereas the other pair-rule genes, such as *fushi-tarazu*, *even-skipped* and *hairy*, were found to be transiently expressed in alternating segment-wide stripes at the blastoderm stage of development (Hafen et al., 1984; Ingham, Howard, & Ish-Horowicz, 1985; Macdonald, Ingham, & Struhl, 1986), expression of *en* was shown to initiate in a series of narrow stripes that define the boundaries of every segment and are maintained throughout embryogenesis in each posterior compartment (Fjose et al., 1985; Ingham et al., 1985; Kornberg et al., 1985). The *en* transcript (and protein) thus provided the first molecular readout for positional identity within each embryonic segment, stimulating the first analyses of the molecular basis of the segment polarity mutant phenotypes (DiNardo et al., 1988; Martizez Arias, Baker, & Ingham, 1988). These studies revealed how segment polarity genes control segmental patterning in part by maintaining the expression of *en*; the molecular cloning of one of the segment polarity genes, *wingless* (*wg*) (Baker, 1987), provided an additional probe for positional identify within each segment. Simultaneous analysis of both *wg* and *en* expression led to the discovery that the Hh and Wg proteins mediate reciprocal interactions, which maintain their expression in adjacent stripes of cells defining the boundaries between the anterior and posterior compartments of each segment (Hidalgo & Ingham, 1990; Martizez Arias et al., 1988).

The establishment of these simple gene expression assays provided the basis for ordering the segment polarity genes into pathways by epistasis analysis. For example, embryos mutant for *hh*, *fu*, and *ci* were each found to lack *wg* expression, implicating these genes in *wg* maintenance, whereas embryos mutant for *ptc* exhibited ectopic *wg* expression, implicating *ptc* in the repression of *wg* transcription (Forbes et al., 1993; Ingham, 1993; Ingham, Taylor, & Nakano, 1991). Combining the *hh* and *ptc* mutations in the same embryo revealed that *wg* transcription is independent of *hh* function in the absence of *ptc*, implying that Hh acts upstream of Ptc. By contrast, *hh* transcription is lacking in embryos double mutant for *ptc* and *fu or ptc* and *ci*, implying that both Fu and Ci act downstream of Ptc to maintain *wg* transcription (Forbes et al., 1993).

Thus, it was through the identification of the segment polarity genes as a phenotypic class, and in particular the isolation of null alleles of *hh*, that the Nüsslein-Volhard and Wieschaus screen made possible the elucidation of the Hh—and indeed the Wg—signaling pathways. Once established in the embryo, understanding of the Hh pathway provided a framework for interpreting the adult mutant phenotypes through which most of the genes had originally been identified. At the same time, the wing imaginal disc provided an assay system for the analysis of the mechanisms of Hh pathway activity at the cellular and molecular level as well as for the identification of additional components of the pathway—such as PKA and the SCF ubiquitin ligase protein Slimb—through genetic mosaic screens (Jiang & Struhl, 1995, 1998; Lepage et al., 1995; Li et al., 1995; Pan & Rubin, 1995; Strutt et al., 1995).

Two decades after the rediscovery of the Hh pathway genes in *Drosophila*, another forward genetic screen, this time in mice, made a further major contribution to the understanding of Hh signaling, through the isolation of mutant alleles of genes involved in the assembly and function of the Primary Cilium. Quite unexpectedly, the phenotypes of these mutants revealed them to disrupt Hh signaling, implicating the Primary Cilium as a key hub for the regulation of the Hh signal transduction pathway (Huangfu et al., 2003). It is interesting to reflect that this remarkable discovery, like that of the *Drosophila* Hh pathway before it, was also preempted by the isolation of a spontaneously arising mutation, *talpid*, in chicken some 60 years earlier (Cole, 1942) and first characterized in detail by Donald Ede in the 1960s (Ede & Kelly, 1964) (see Fig. 4). Subsequent analysis of this mutation using molecular markers by Cheryl Tickle and colleagues, revealed that it disrupts Hh signaling (Lewis et al., 1999) and its eventual molecular cloning

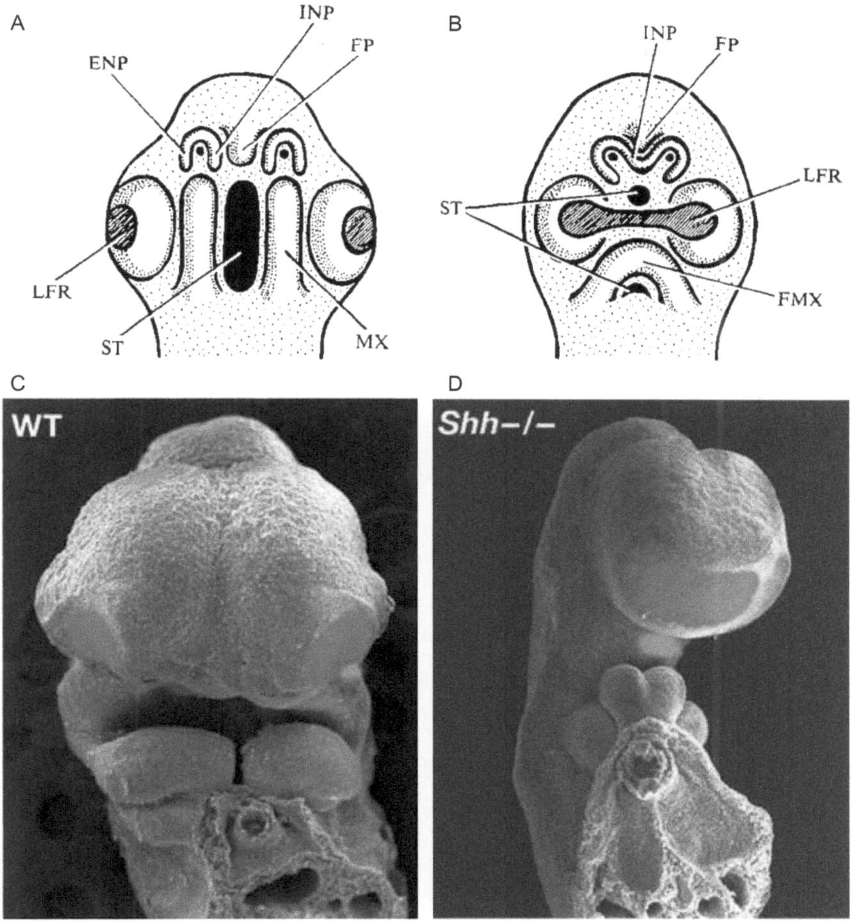

Figure 4 (A and B) Schematic representations of the facial structures in a wild-type chick embryo (A) and a *talpid3* mutant chick embryo (B). ENP, external nasal process; FMX, fused maxillary processes; FP, frontal process; INP, internal nasal process; LFR, lens-forming region; MX, maxillary process; ST, stomodaeum. (C and D) Scanning electron micrographs of a wild-type (C) and Shh (D) mutant mouse embryo. Note the similarities between the *talpid3* and *Shh* mutant phenotypes. *Panels (A) and (B): Reprinted from Ede and Kelly (1964), Copyright (1964), with permission from The Company of Biologists. Panels (C) and (D): Reprinted from Aota, K., et al. (2009). Mouse Shh is required for prechordal plate maintenance during brain and craniofacial morphogenesis.* Developmental Biology, 327, *106–120. doi:10.1016/j.ydbio.2008.11.022. Copyright (2009), with permission from Elsevier.* (A color version of this figure is located at the end of this chapter.)

showed that it encodes a centriolar protein essential for Primary Ciliogenesis (Yin et al., 2009). Intriguingly, Ede had previously worked together with Sheila Counce on the analysis of the *Drosophila fused* mutation at CH Waddington's Institute of Animal Genetics in Edinburgh; thus, Ede had been the first person to analyze—albeit unwittingly—the effects of disrupting the Hh signaling pathway in both *Drosophila* and vertebrates.

REFERENCES

Baker, N. E. (1987). Molecular cloning of sequences from wingless, a segment polarity gene in *Drosophila*: The spatial distribution of a transcript in embryos. *The EMBO Journal*, *6*(6), 1765–1773.

Chen, Y., & Struhl, G. (1996). Dual roles for patched in sequestering and transducing Hedgehog. *Cell*, *87*(3), 553–563.

Chiang, C., et al. (2001). Manifestation of the limb prepattern: Limb development in the absence of sonic hedgehog function. *Developmental Biology*, *236*(2), 421–435.

Cole, R. K. (1942). The "talpid lethal" in the domestic fowl. *Journal of Heredity*, *33*, 82–86.

Collier, L. S., et al. (2004). Drosophila Costal1 mutations are alleles of protein kinase A that modulate hedgehog signaling. *Genetics*, *167*(2), 783–796.

Counce, S. J. (1956). Studies on female-sterility genes in *Drosophila melanogaster*. II. The effects of the gene fused on embryonic development. *Zeitschrift für Induktive Abstammungs- und Vererbungslehre*, *87*(3), 462–481.

DiNardo, S., et al. (1988). Two-tiered regulation of spatially patterned engrailed gene expression during *Drosophila* embryogenesis. *Nature*, *332*(6165), 604–609.

Echelard, Y., et al. (1993). Sonic hedgehog, a member of a family of putative signaling molecules, is implicated in the regulation of CNS polarity. *Cell*, *75*(7), 1417–1430.

Ede, D. A., & Kelly, W. A. (1964). Developmental abnormalities in the head region of the talpid mutant of the fowl. *Journal of Embryology and Experimental Morphology*, *12*, 161–182.

Fjose, A., McGinnis, W. J., & Gehring, W. J. (1985). Isolation of a homoeo box-containing gene from the engrailed region of *Drosophila* and the spatial distribution of its transcripts. *Nature*, *313*(6000), 284–289.

Forbes, A. J., et al. (1993). Genetic analysis of hedgehog signalling in the *Drosophila* embryo. *Development Supplement*, 115–124.

Forbes, A. J., et al. (1996). Hedgehog is required for the proliferation and specification of ovarian somatic cells prior to egg chamber formation in *Drosophila*. *Development*, *122*(4), 1125–1135.

French, V., Bryant, P. J., & Bryant, S. V. (1976). Pattern regulation in epimorphic fields. *Science*, *193*(4257), 969–981.

Hafen, E., Kuroiwa, A., & Gehring, W. J. (1984). Spatial distribution of transcripts from the segmentation gene fushi tarazu during *Drosophila* embryonic development. *Cell*, *37*(3), 833–841.

Harfe, B. D., et al. (2004). Evidence for an expansion-based temporal Shh gradient in specifying vertebrate digit identities. *Cell*, *118*(4), 517–528.

Heberlein, U., Wolff, T., & Rubin, G. M. (1993). The TGF beta homolog dpp and the segment polarity gene hedgehog are required for propagation of a morphogenetic wave in the *Drosophila* retina. *Cell*, *75*(5), 913–926.

Hidalgo, A. (1991). Interactions between segment polarity genes and the generation of the segmental pattern in *Drosophila*. *Mechanisms of Development*, *35*(2), 77–87.

Hidalgo, A., & Ingham, P. (1990). Cell patterning in the *Drosophila* segment: Spatial regulation of the segment polarity gene patched. *Development, 110*(1), 291–301.
Ho, K. S., et al. (2005). Differential regulation of Hedgehog target gene transcription by Costal2 and Suppressor of Fused. *Development, 132*(6), 1401–1412.
Hooper, J. E., & Scott, M. P. (1989). The *Drosophila* patched gene encodes a putative membrane protein required for segmental patterning. *Cell, 59*(4), 751–765.
Huangfu, D., et al. (2003). Hedgehog signalling in the mouse requires intraflagellar transport proteins. *Nature, 426*(6962), 83–87.
Ingham, P. W. (1993). Localized hedgehog activity controls spatial limits of wingless transcription in the *Drosophila* embryo. *Nature, 366*(6455), 560–562.
Ingham, P. W., Howard, K. R., & Ish-Horowicz, D. (1985). Transcription pattern of the *Drosophila* segmentation gene hairy. *Nature, 318*(5), 439–445.
Ingham, P. W., Taylor, A. M., & Nakano, Y. (1991). Role of the *Drosophila* patched gene in positional signalling. *Nature, 353*(6340), 184–187.
Ingham, P., et al. (1985). Expression of engrailed in the parasegment of *Drosophila*. *Nature, 318*, 437–440.
Ives, P. (1950). New mutants report: Bar-3. *Drosophila Information Service, 24*, 58.
Jiang, J., & Struhl, G. (1995). Protein kinase A and hedgehog signaling in *Drosophila* limb development. *Cell, 80*(4), 563–572.
Jiang, J., & Struhl, G. (1998). Regulation of the Hedgehog and Wingless signalling pathways by the F-box/WD40-repeat protein Slimb. *Nature, 391*(6666), 493–496.
Kornberg, T., et al. (1985). The engrailed locus of *Drosophila*: In situ localization of transcripts reveals compartment-specific expression. *Cell, 40*(1), 45–53.
Krauss, S., Concordet, J. P., & Ingham, P. W. (1993). A functionally conserved homolog of the *Drosophila* segment polarity gene hh is expressed in tissues with polarizing activity in zebrafish embryos. *Cell, 75*(7), 1431–1444.
Kuner, J. M., et al. (1985). Molecular cloning of engrailed: A gene involved in the development of pattern in *Drosophila melanogaster*. *Cell, 42*(1), 309–316.
Lee, J. J., et al. (1992). Secretion and localized transcription suggest a role in positional signaling for products of the segmentation gene hedgehog. *Cell, 71*(1), 33–50.
Lepage, T., et al. (1995). Signal transduction by cAMP-dependent protein kinase A in *Drosophila* limb patterning. *Nature, 373*(6516), 711–715.
Lewis, K. E., et al. (1999). Expression of ptc and gli genes in talpid3 suggests bifurcation in Shh pathway. *Development, 126*(11), 2397–2407.
Li, W., et al. (1995). Function of protein kinase A in hedgehog signal transduction and *Drosophila* imaginal disc development. *Cell, 80*(4), 553–562.
Lim, J., & Choi, K. W. (2004). Induction and autoregulation of the anti-proneural gene Bar during retinal neurogenesis in *Drosophila*. *Development, 131*(22), 5573–5580.
Lindsley, D. L., & Grell, E. H. (1968). *Genetic variations of* Drosophila melanogaster. Washington, DC: Carnegie Institute Pub. 627.
Lynch, C. J. (1919). An analysis of certain cases of intra-specific sterility. *Genetics, 4*(6), 501–533.
Ma, C., et al. (1996). Identification and characterization of autosomal genes that interact with glass in the developing *Drosophila* eye. *Genetics, 142*(4), 1199–1213.
Macdonald, P. M., Ingham, P., & Struhl, G. (1986). Isolation, structure, and expression of even-skipped: A second pair-rule gene of *Drosophila* containing a homeo box. *Cell, 47*(5), 721–734.
Martizez Arias, A., Baker, N. E., & Ingham, P. W. (1988). Role of segment polarity genes in the definition and maintenance of cell states in the *Drosophila* embryo. *Development, 103*(1), 157–170.
Michel, M., et al. (2012). Hh signalling is essential for somatic stem cell maintenance in the *Drosophila* testis niche. *Development, 139*(15), 2663–2669.

Mohler, J. (1988). Requirements for hedgehog, a segmental polarity gene, in patterning larval and adult cuticle of *Drosophila. Genetics, 120*(4), 1061–1072.
Morata, G., & Lawrence, P. A. (1975). Control of compartment development by the engrailed gene in *Drosophila. Nature, 255*(5510), 614–617.
Nakano, Y., et al. (1989). A protein with several possible membrane-spanning domains encoded by the *Drosophila* segment polarity gene patched. *Nature, 341*(6242), 508–513.
Nüsslein Volhard, C., Wieschaus, E., & Kluding, H. (1984). Mutations affecting the pattern of the larval cuticle in *Drosophila* melanogaster. *Wilhelm Roux's Archives of Developmental Biology, 193*, 267–282.
Nüsslein-Volhard, C., & Wieschaus, E. (1980). Mutations affecting segment number and polarity in *Drosophila. Nature, 287*(5785), 795–801.
Pan, D., & Rubin, G. M. (1995). cAMP-dependent protein kinase and hedgehog act antagonistically in regulating decapentaplegic transcription in *Drosophila* imaginal discs. *Cell, 80*(4), 543–552.
Pauli, T., et al. (2005). Identification of functional sine oculis motifs in the autoregulatory element of its own gene, in the eyeless enhancer and in the signalling gene hedgehog. *Development, 132*(12), 2771–2782.
Quirk, J., et al. (1997). The smoothened gene and hedgehog signal transduction in *Drosophila* and vertebrate development. *Cold Spring Harbor Symposia on Quantitative Biology, 62*, 217–226.
Riddle, R. D., et al. (1993). Sonic hedgehog mediates the polarizing activity of the ZPA. *Cell, 75*(7), 1401–1416.
Rogers, E. M., et al. (2005). Pointed regulates an eye-specific transcriptional enhancer in the *Drosophila* hedgehog gene, which is required for the movement of the morphogenetic furrow. *Development, 132*(21), 4833–4843.
Saunders, J. W., & Gasseling, M. T. (1968). Ectodermal-mesenchymal interactions in the origins of limb symmetry. In R. Fleischmajer & R. E. Billingham (Eds.), *Epithelial-mesenchymal interactions* (pp. 78–97). Baltimore: Williams & Wilkins.
Shi, Q., et al. (2011). The Hedgehog-induced smoothened conformational switch assembles a signaling complex that activates fused by promoting its dimerization and phosphorylation. *Development, 138*(19), 4219–4231.
Strutt, D. I., Wiersdorff, V., & Mlodzik, M. (1995). Regulation of furrow progression in the *Drosophila* eye by cAMP-dependent protein kinase A. *Nature, 373*(6516), 705–709.
Sturetvant, A. H. (1948). Sequence of loci near the centromere of chromosome 2. *Drosophila Information Service, 22*, 56.
Taipale, J., et al. (2002). Patched acts catalytically to suppress the activity of smoothened. *Nature, 418*(6900), 892–897.
Tickle, C., Summerbell, D., & Wolpert, L. (1975). Positional signalling and specification of digits in chick limb morphogenesis. *Nature, 254*(5497), 199–202.
Tickle, C., et al. (1976). Positional signalling by mouse limb polarising region in the chick wing bud. *Nature, 259*(5542), 396–397.
Tiniakow, G. G., & Terentieva, E. L. (1933). Cubitus interruptus, a new genovariation of the fourth chromosome of *Drosophila melanogaster. Genetics, 18*(2), 117–120.
Vervoort, M., et al. (1999). The COE transcription factor Collier is a mediator of short-range Hedgehog-induced patterning of the *Drosophila* wing. *Current Biology, 9*(12), 632–639.
Whittle, J. R. (1976). Clonal analysis of a genetically caused duplication of the anterior wing in *Drosophila melanogaster. Developmental Biology, 51*(2), 257–268.
Yin, Y., et al. (2009). The Talpid3 gene (KIAA0586) encodes a centrosomal protein that is essential for primary cilia formation. *Development, 136*(4), 655–664.
Zhang, Y., & Kalderon, D. (2001). Hedgehog acts as a somatic stem cell factor in the *Drosophila* ovary. *Nature, 410*(6828), 599–604.

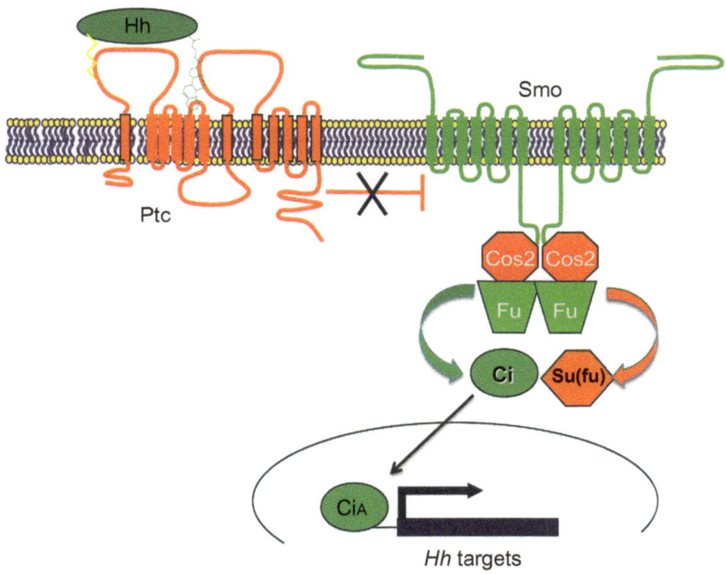

Philip W. Ingham, Figure 1 Simplified schema of the *Drosophila* Hh signal transduction pathway showing the core components. Negative regulators and repressive interactions are shown in red, positive regulators and activating interactions are shown in green.

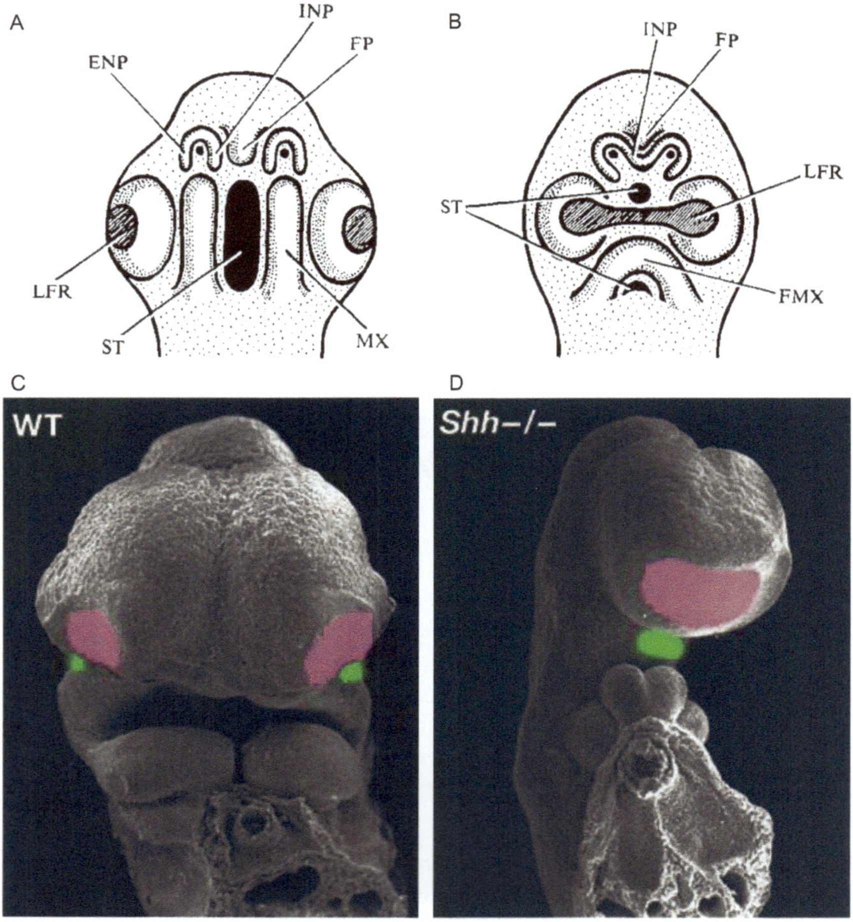

Philip W. Ingham, Figure 4 (A and B) Schematic representations of the facial structures in a wild-type chick embryo (A) and a *talpid3* mutant chick embryo (B). ENP, external nasal process; FMX, fused maxillary processes; FP, frontal process; INP, internal nasal process; LFR, lens-forming region; MX, maxillary process; ST, stomodaeum. (C and D) Scanning electron micrographs of a wild-type (C) and Shh (D) mutant mouse embryo. Note the similarities between the *talpid3* and *Shh* mutant phenotypes. *Panels (A) and (B): Reprinted from Ede and Kelly (1964), Copyright (1964), with permission from The Company of Biologists. Panels (C) and (D): Reprinted from Aota, K., et al. (2009). Mouse Shh is required for prechordal plate maintenance during brain and craniofacial morphogenesis.* Developmental Biology, 327, *106–120. doi:10.1016/j.ydbio.2008.11.022. Copyright (2009), with permission from Elsevier.*

CHAPTER TWENTY-SEVEN

The Comparative Organismal Approach in Evolutionary Developmental Biology: Insights from Ascidians and Cavefish

William R. Jeffery[1]
Department of Biology, University of Maryland, College Park, Maryland, USA
[1]Corresponding author: e-mail address: jeffery@umd.edu

Contents

Abstract

Important contributions to evolutionary developmental biology have been made using the comparative organismal approach. As examples, I describe insights obtained from studies of *Molgula* ascidians and *Astyanax* cavefish.

1. INTRODUCTION

During the past several decades, considerable progress has been made in understanding the processes and genes underlying the evolution of development using the comparative organismal approach. In this approach, organisms are compared to explore the molecular, cellular, and morphological basis for similarities and differences that have evolved between them. One version of the comparative organismal approach employs distantly related species to understand the extent of conservation in developmental processes. This version revealed a common set of conserved genes and signaling pathways controlling animal development (Scott, 1994). In a few cases, genes from one species have been shown to control the development

Current Topics in Developmental Biology, Volume 116
ISSN 0070-2153
http://dx.doi.org/10.1016/bs.ctdb.2015.10.006

of similar phenotypes after their expression in a distantly related species. The most famous case is the formation of ectopic eyes on *Drosophila* legs after misexpression of the squid, ascidian, frog, or mouse *pax6* genes (Gehring & Ikeo, 1999). The discovery of conserved genes and mechanisms unified developmental biology and provided a background for most of the contemporary research in evolutionary developmental biology.

A second version of the comparative organismal approach employs closely related species with different modes of development or the same species exhibiting different morphs to probe the developmental basis of evolutionary divergence. The second version targets the heart of evolutionary developmental biology and is especially compelling when the species or morphs can be hybridized, permitting the use of genetic techniques to study the evolution of development. This is the case for the model systems highlighted in this essay: (1) closely related species of molgulid ascidians with different modes of development (Fig. 1A–E) and (2) conspecific surface- and cave-dwelling morphs of the teleost *Astyanax mexicanus* (Fig. 1F–K). Using these models as examples, I discuss how they have revealed some of the key processes underlying the evolution of developmental diversity.

2. MOLGULID ASCIDIANS: EVOLVING DIFFERENT MODES OF DEVELOPMENT

Ascidians are tunicates: invertebrate chordates inferred to be the closest living relatives of vertebrates (Bourlat et al., 2006; Delsuc, Brinkmann, Chourrout, & Phillipe, 2006). They have two stages in their life cycle, a nonfeeding larval stage, the product of embryonic development, and an adult stage, produced by modification and reorganization of the larva during metamorphosis. With the exception of branchial gill slits, adult ascidians have no resemblances to vertebrates; their sac-like body is specialized for filter feeding and producing gametes. However, the motile larval stage, a tadpole-like organism consisting of a trunk (or head) and tail, has definitive chordate features: a dorsal central nervous system, a notochord of exactly 40 cells, and a postanal tail equipped with striated muscle cells (reviewed by Satoh, 1994).

The majority of ascidian species form tadpole-like swimming larvae prior to metamorphosis, which is called urodele (tailed) development (Fig. 1A). In contrast, a small subset of ascidian species has lost the tailed larva (Fig. 1C; reviewed by Jeffery & Swalla, 1990). The tailless (or anural) ascidian species, which are mostly found in the Family Molgulidae, develop more or less

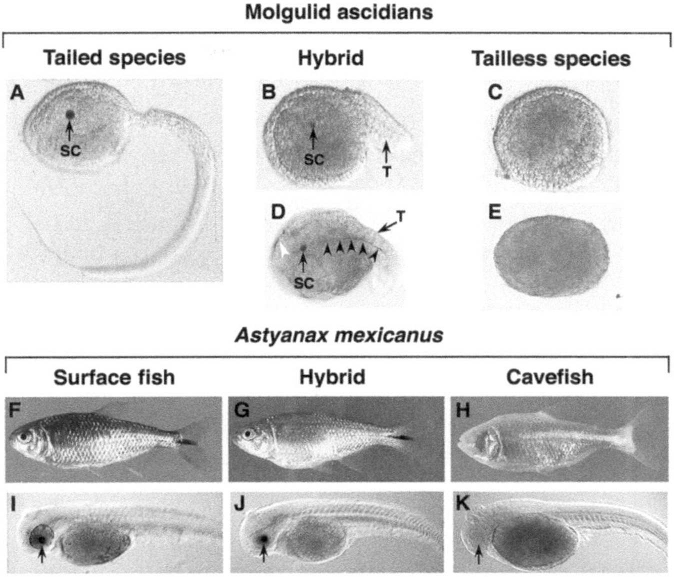

Figure 1 The Molgulid ascidian and *Astyanax mexicanus* models. *Top*: molgulid ascidians. (A) Tailed urodele larva of *Molgula oculata* (the tailed species). (B) Short-tailed urodele larva of a cross between *Molgula occulta* (the tailless species) eggs and *M. oculata* (the tailed species) sperm. (C) Tailless anural larva of the *M. occulta* (the tailless species). (D and E) Alpha-acetylated tubulin antibody staining showing the restoration of anterior and tail nerves, which is absent in the tailless species larva (E), in a tailless × tailed species hybrid (D). SC, brain sensory cell. White arrowhead: part of a restored anterior nerve. Black arrowheads: restored tail nerve. *Bottom*: *A. mexicanus*. (F) Adult surface fish with eyes and pigment. (G) Adult surface fish × cavefish hybrid with eyes and pigment. (H) Adult cavefish lacking eyes and pigment. (I) Larval surface fish showing *alphaA-crystallin* gene expression in the lens. (J) Larval surface fish × cavefish hybrid showing *alphaA-crystallin* gene expression in the lens. (K) Larval cavefish lacking *alphaA-crystallin* gene expression in the lens. Arrows: lens. *Panels (A) and (B) from Swalla and Jeffery (1996), (C) and (D) from Y. Yamamoto and W.R. Jeffery (unpublished), and (I)–(K) from Ma, Parkhurst, and Jeffery (2014).* (A color version of this figure is located at the end of this chapter.)

directly into the adult. The anural developers lack a melanin pigmented sensory cell in their brain, an extended notochord, differentiated tail muscle cells, and nerves leading from the brain into the head and tail (Fig. 1C and E), which are present in the tadpole larvae of all urodele developers. The anural mode of development has evolved multiple times from ancestors with conventional urodele development (Hadfield, Swalla, & Jeffery, 1995). The anural ascidian *Molgula occulta* is closely related to the urodele ascidian *Molgula oculata*. To avoid confusion due to the similarity in names, we will refer to them as the tailed (urodele *M. oculata*) and the tailless (anural *M. occulta*) species. The tailed and tailless species are codistributed on subtidal sand flats along the northwest coast of Europe, where they may have evolved recently from a common urodele ancestor. They are indeed so closely related that unlike most other ascidians can produce hybrids when crossed (Fig. 1B and D; Swalla & Jeffery, 1991). Because of radically different modes of development and capacity for interspecific hybridization, the *Molgula* species are an excellent comparative model in evolutionary developmental biology.

Ascidian embryonic development is controlled by maternal factors localized in the egg and segregated into stereotypical cell lineages during cleavage, as well as zygotic processes, such as induction (reviewed by Satoh, 1994). Do maternal processes, zygotic processes, or both control the evolution of different modes of ascidian development? Hybridization between the two *Molgula* species has addressed this question (Swalla & Jeffery, 1991). Remarkably, when eggs of the tailless species are fertilized with sperm of the tailed species, some of the lost urodele features are restored in the hybrid embryos (tailless × tailed species hybrids; Fig. 1B and D). The restored urodele features include a melanin-containing brain sensory cell, a short tail containing an extended notochord, and nerves emanating from the brain into the head and tail (Fig. 1B–E). The rescue of urodele features in interspecific hybrids shows that anural development is controlled in large part by changes in zygotic processes. However, maternal processes may also be involved, as evidenced by the inability of the tailed species genome to restore the entire program of terminal cell divisions in the notochord and muscle lineages (Swalla & Jeffery, 1991) and by differences in cytoplasmic localization in eggs of the tailless species (Swalla, Badgett, & Jeffery, 1991).

The *Molgula* model also illustrates the importance of heterochrony (relative changes in the timing of developmental events) and cell death (apoptosis) in the evolution of development. In urodele developers, the notochord, tail muscle cells, and parts of the brain that function in larval development undergo apoptosis during metamorphosis. In the tailless species, however, the precursors of larval brain and tail tissues undergo apoptosis prematurely during the embryonic period (Jeffery, 2002, 2006). In tailless × tailed hybrids, the temporal relationship between apoptosis and metamorphosis typical of urodele developers is restored, suggesting that zygotic processes also regulate heterochrony between the two *Molgula* species (Jeffery, 2002). Some of the genes involved in anural development through effects on cell survival have been identified. The *Manx* (Swalla & Jeffery, 1996) and *foxA5* (Olsen, Natzle, & Jeffery, 1999) genes are expressed in the brain, notochord, and muscle cells of the tailed species but are downregulated or show abbreviated expression, respectively, in the tailless species, and their urodele expression patterns are restored in tailless × tailed hybrids. Functional studies have shown that the*se genes* protect the urodele-related brain and tail tissues from premature apoptosis (Jeffery, 2002). Many other zygotic genes are also likely to be involved in the urodele to anural developmental transition, which are currently being investigated by transcriptome and genomic analysis.

Does anural development involve changes at the beginning or the end of gene regulatory cascades? Ascidian tail muscle cell development is controlled by a chain of events beginning with Macho-1, a transcription factor encoded by localized maternal mRNA (reviewed by Nishida, 2012), and ending with the synthesis of muscle structural proteins in the tail muscle cell lineages. The ability of a tailed species' muscle actin gene construct to be expressed in the undifferentiated muscle cells of the tailless species suggests that upstream regulatory factors in the muscle pathway are functional in the tailless species (Kusakabe, Swalla, Satoh, & Jeffery, 1996). Although there is no published information about *macho-1* mRNA in the tailless species, maternal *macho-1* transcripts are present and segregated to the muscle lineage in another anural ascidian species, *Molgula tectiformis* (Gyoja, 2006). Furthermore, *Tbx6*, a Macho-1 target gene, is expressed in the muscle lineage of the tailless species (Takada, Satoh, & Swalla, 2002). Accordingly, instead of changing upstream regulatory genes, their downstream targets in the muscle differentiation pathway are affected; the muscle actin genes of the tailless species contain stop codons rendering them nonfunctional (Kusakabe et al., 1996). The conversion of larval muscle actin genes into pseudogenes has also occurred in two other anural species *Molgula bleizi* and *M. tectiformis* (Gyoja et al., 2007; Jeffery, Swalla, Ewing, & Kusakabe, 1999), suggesting that muscle structural gene decay may be a common theme in anural development. Thus, the evolutionary change(s) responsible for the absence of muscle cell differentiation in anural developers is controlled by changes at the end of the gene cascade. A similar situation may occur for the notochord gene cascade, as the transcription factor Brachyury, the initiator of the ascidian notochord developmental pathway (Yasuo & Satoh, 1994), is also expressed normally in the notochord cell lineage of the tailless species (Takada, York, et al., 2002).

Anural development is relatively common in the Molgulidae, but it is encountered only rarely in other ascidian families (reviewed by Jeffery & Swalla, 1990). The repeated alteration of muscle development in anural

Molgula species via the decay of muscle structural genes may be one of the reasons why molgluid ascidians are so readily subject to this evolutionary change. Kusakabe (2001) has proposed an intriguing hypothesis to explain this paradox based on low muscle actin gene copy number and pseudogene formation. Most urodele ascidians have many copies of larval muscle actin genes (Kusakabe, Araki, Satoh, & Jeffery, 1997), which are probably needed to produce large amounts of structural protein during rapid larval myogenesis. However, molgluid ascidians are proposed to have a lower number of redundant muscle actin genes, making muscle cell differentiation more susceptible to evolutionary change via pseudogene formation. Because there are specific sets of larval and adult muscle structural genes in ascidians, the decay of the larval genes in anural developers would not have adverse effects in adults, possibly "preadapting" molgluid ascidians for anural development.

3. CAVEFISH: EVOLUTION IN THE DARK

The teleost *A. mexicanus* is a single species with an eyed and pigmented surface dwelling morph (surface fish) and many different eyeless and depigmented cave-dwelling (cavefish) morphs (Fig. 1F and H; see reviews in Keene, Yoshizawa, & McGaugh, 2015). The cavefish morphs spend their entire life cycle in perpetual darkness. As a consequence of relaxed selection over many generations, traits essential for survival in lighted habitats, such as eyes and pigmentation, have been reduced or lost (Fig. 1F and H). Constructive differences have also evolved between the surface fish and cavefish morphs. For instance, cavefish show substantially modified craniofacial morphology, have larger jaws, increased numbers of taste buds, and specialized behaviors, such as an attraction to water vibrations (Yoshizawa, Goricki, Soares, & Jeffery, 2010). Different cavefish populations have evolved multiple times from surface fish in an extensive karst region in northeastern Mexico (Gross, 2012). The time since the separation of cavefish from surface fish ancestors is estimated to be 5 million to less than a million years (Mitchell, Russell, & Elliott, 1977; Porter, Dittmar, & Pérez-Losada, 2007).

The *Astyanax* conspecific morphs can be hybridized to produce viable F1 progeny (Fig. 1F–H). As in the tailed and tailless ascidian species, hybridization has shown that zygotic processes are changed in cavefish. For example, surface fish × cavefish F1 hybrids develop eyes and pigmentation (Wilkens, 1988) and express genes that are downregulated during cavefish development, such as lens *alphaA-crystallin* (Fig. 1I–K; Ma et al., 2014). Intercrossing between the progeny of surface fish and cavefish has allowed the cavefish phenotypes to be mapped by genetic analysis (Kowalko et al., 2013; O'Quin, Yoshizawa, Doshi, & Jeffery, 2013; Protas et al., 2008). The genetic mapping studies revealed multiple genes controlling most of the cavefish phenotypes. An exception is albinism, the loss of melanin pigmentation, which evolved by mutations in the *oca2* gene (Protas et al., 2006). Deletions in the *oca2* coding regions are responsible for albinism in independently evolved cavefish morphs, implying that this gene is a frequent target of evolutionary change. The sequenced cavefish genome (McGaugh et al., 2014) will allow further progress in identifying and characterizing the genes underlying gained and lost traits during cavefish evolution. Below, I provide examples using the eye and pigment phenotypes to illustrate how the *Astyanax* model has contributed to evolutionary developmental biology.

Although cavefish adults lack external eyes, eye primordia with an embryonic lens and optic cup are formed during embryogenesis. After larval hatching the cavefish lens undergoes massive apoptosis, and further lens development is arrested (Jeffery & Martasian, 1998; Soares, Yamamoto, Strickler, & Jeffery, 2004). Lens dysfunction has widespread consequences on optic development, as demonstrated by lens transplantation experiments (Yamamoto & Jeffery, 2000). Transplantation of a surface fish embryonic lens into a cavefish optic cup rescues many aspects of eye development and growth, leading to the recovery of an external eye in adult cavefish. Reciprocally, transplantation of a cavefish embryonic lens into the surface fish optic cup triggers eye degeneration, resulting in the disappearance of an external eye in adult surface fish. Two specific consequences of the absence of a functional cavefish lens are worthy of further attention in the context of this essay. First, the cornea and iris, which are induced by the lens during normal eye development, fail to form in cavefish (Strickler, Yamamoto, & Jeffery, 2001). Second, the stem cell progeny that contribute differentiating cells to the growing retina also undergo apoptosis in the absence of a functional lens, retarding the growth of this optic tissue (Strickler, Yamamoto, & Jeffery, 2007). Both cornea induction and retinal growth can be rescued by transplanting a functional surface fish lens into the cavefish optic cup (Yamamoto & Jeffery, 2000), showing effects on inductive processes and cell survival during development are important in the evolution of cavefish eye degeneration.

During cavefish larval development, the small degenerating eye does not keep pace with growth of the body and is buried in the adult orbit as a tiny nonfunctional vestige. Why would the eye be formed during embryogenesis only to subsequently degenerate in the larva? The probable cause is a developmental constraint: the early steps of eye development are also required to control the development of other crucial parts of the embryo, such as the brain, and are thus indispensible. Another example of a developmental constraint is that retinal development is not caused by the loss of stem cells or by blocking their proliferation division, which would be the most energetically efficient mode of suppressing growth, but instead by inducing apoptosis after new cells are already produced by the retinal stem cells. The retina is actually a part of the brain. Thus, apoptosis of stem cell progeny may be more permissible than losing their function, because they are likely important in regulating other parts of the growing brain. Accordingly, only developmental pathways or their parts that are not strongly intertwined with other essential pathways may be subject to change during evolution.

The cavefish model also illustrates the principle of developmental amplification: small changes in early development can be increased incrementally as development proceeds to result in major changes in adult morphology. Surface fish and cavefish have significant differences in head shape and craniofacial morphology. Some of these differences, namely the shape and number of circumorbital bones, are linked to eye loss (Yamamoto, Espinasa, Stock, & Jeffery, 2003). Accordingly, transplantation of a surface fish lens into a cavefish optic cup can promote the formation of a surface fish-like craniofacial skeleton. The sequence of developmental amplification is the functional surface fish lens promotes retinal growth, producing a large eye on the external surface of the head, and the eye controls the surrounding craniofacial morphology to develop the surface fish phenotype. In cavefish, loss of lens function (probably along with other factors) promotes the arrest of retinal growth, and subsequent eye degeneration leads to development of the cavefish-type craniofacial morphology.

At the molecular level, the *Astyanax* model has highlighted the importance of antagonistic pleiotropy (Williams, 1957) in the evolution of development. The Sonic Hedgehog (Shh) midline-signaling pathway has critical roles in cavefish development and evolution. The cavefish *shh* genes are overexpressed along the anterior embryonic midline relative to surface fish counterparts (Yamamoto, Stock, & Jeffery, 2004), and this expansion of Shh signaling impacts the expression of downstream developmental targets. For example, negative regulation of *pax6* through expanded Shh signaling causes size reduction in the cavefish retina (Menuet, Alunni, Joly, Jeffery, & Rétaux, 2007) and *shh* overexpression triggers lens apoptosis and eye degeneration in surface fish embryos, producing a phenocopy of the cavefish eye (Yamamoto et al., 2004). Cavefish larvae differ from those of surface fish in having a larger mouth with a wider jaw span and more taste buds. Shh signaling is also involved in the enhancement of these oral features. In a direct demonstration of *shh* gene pleiotropy, experimental *shh* overexpression causes enhancement of jaws and taste buds at the expense of eyes in the same surface fish larvae (Yamamoto, Byerly, Jackman, & Jeffery, 2009). The trade-off between eye and oral development may have adaptive significance due to benefits associated with increased feeding efficiency in the dark cave environment.

A metabolic trade-off involving the melanin and catecholamine (CAT) synthesis pathways could be important in the evolution of cavefish albinism. In the melanin synthesis pathway, L-tyrosine is converted to L-DOPA, and then in several steps to melanin pigment. Melanin is deposited in the eye and body pigment cells of surface fish, but is absent or decreased in cavefish. In the CAT pathway, L-tyrosine is converted to L-DOPA, which is then metabolized to dopamine, epinephrine, and norepinephrine. Dopamine and norepinephrine are increased in albino cavefish compared to pigmented surface fish (Bilandžija, Ma, Parkhurst, & Jeffery, 2013), possibly to support CAT-related neurological and behavioral changes. The cavefish melanin synthesis pathway is blocked at the conversion of L-tyrosine to L-DOPA (McCauley, Hixon, & Jeffery, 2004), making excess L-tyrosine available for utilization in the CAT synthesis pathway. The close relationship between these two pathways has been confirmed by *oca2* knockdown in surface fish, which abolishes melanin synthesis while simultaneously increasing L-tyrosine and dopamine (Bilandžija et al., 2013). Thus, a potential benefit of cavefish albinism may be the use of an L-tyrosine stockpile for purposes that are more suitable in the dark cave environment.

4. CONCLUSIONS

Insights from the comparative organismal approach indicate that many aspects of development are flexible and subject to potential modification during evolution. I have described some examples in ascidians and cavefish, including changes in cytoplasmic localization, zygotic control processes, the complete development of cell lineages, embryonic inductive events, the balance between cell survival and death, and the heterochrony of developmental events. Moreover, the cavefish model demonstrates that developmental processes are subject to trade-offs through the function of pleiotropic genes, which may be important factors in the evolution of development. As appreciated early in the history of genetic research, most genes have pleiotropic functions (Stearns, 2010). Thus, further studies of multifunctional genes may be helpful in understanding the rapid evolution of morphological traits. The ascidian and cavefish models also highlight developmental processes that are not readily subject to change, defining inflexible developmental constraints that limit and channel the possibilities for evolution.

The comparative organismal approach in ascidian, cavefish, and other model systems (see, for example, Goldstein, 2001) has provided a foundation for continued progress to understand how development is changed during evolution. However, little is known about the environmental cues and the evolutionary forces leading to these changes. Therefore, one of the future challenges is to determine not only "how" but also "why" development is changed during evolution. The comparative organismal approach will be an important tool in solving the next generation of questions in evolutionary developmental biology.

REFERENCES

Bilandžija, H., Ma, L., Parkhurst, A., & Jeffery, W. R. (2013). A potential benefit of albinism in *Astyanax* cavefish: Downregulation of the *oca2* gene increases L-tyrosine and catecholamine levels as an alternative to melanin synthesis. *PloS One*, *8*, e80823.

Bourlat, S. J., Juliusdottir, T., Lowe, C. J., Freeman, R., Aronowicz, J., Kirschner, M., et al. (2006). Deuterostome phylogeny reveals monophyletic chordates and the new phylum Xenoturbellida. *Nature*, *444*, 85–88.

Delsuc, F., Brinkmann, H., Chourrout, D., & Phillipe, H. (2006). Tunicates and not cephalochordates are the closest living relatives of vertebrates. *Nature*, *439*, 965–968.

Gehring, W. J., & Ikeo, K. (1999). Pax 6: Mastering eye morphogenesis and eye evolution. *Trends in Genetics*, *15*, 371–377.

Goldstein, B. (2001). On the evolution of early development in the Nematoda. *Philosophical Transactions of the Royal Society of London. Series B, Biological Sciences*, *356*, 1521–1531.

Gross, J. B. (2012). The complex origin of *Astyanax* cavefish. *BMC Evolutionary Biology*, *12*, 105. http://dx.doi.org/10.1186/1471-2148-12-105.

Gyoja, F. (2006). Expression of a muscle determination gene, *macho 1*, in the anural ascidian *Molgula tectiformis*. *Development Genes and Evolution*, *216*, 285–289.

Gyoja, F., Satou, Y., Shin-I, T., Kohara, Y., Swalla, B. J., & Satoh, N. (2007). Analysis of large scale expression sequenced tags (ESTs) from the ascidian, *Molgula tectiformis*. *Developmental Biology*, *307*, 460–482.

Hadfield, K. A., Swalla, B. J., & Jeffery, W. R. (1995). Multiple origins of anural development in ascidians inferred from rDNA sequences. *Journal of Molecular Evolution*, *40*, 413–427.

Jeffery, W. R. (2002). Programmed cell death in the ascidian embryo: Modulation by Fox2A and Manx and roles in the evolution of larval development. *Mechanisms of Development*, *118*, 118–124.

Jeffery, W. R. (2006). Evolution and development of brain sensory organs in molgulid ascidians. *Evolution & Development*, *6*, 170–179.

Jeffery, W. R., & Martasian, D. P. (1998). Evolution of eye regression in the cavefish *Astyanax*: Apoptosis and the Pax-6 gene. *American Zoologist, 38*, 685–696.

Jeffery, W. R., & Swalla, B. J. (1990). Anural development in ascidians: Evolutionary modification and elimination of the tadpole larva. *Seminars in Developmental Biology, 96*, 125–143.

Jeffery, W. R., Swalla, B. J., Ewing, N., & Kusakabe, T. (1999). Evolution of the ascidian anural larva: Evidence from embryos and molecules. *Molecular and Biological Evolution, 16*, 646–654.

Keene, A., Yoshizawa, M., & McGaugh, S. (2015). *Biology and evolution of the Mexican cavefish*. New York: Elsevier, pp. 1–401.

Kowalko, J. E., Rohner, N., Rompani, S. B., Peterson, B. K., Linden, T., Yoshizawa, M., et al. (2013). Genetic analysis of the loss of schooling behavior in cavefish reveals both sight-dependent and independent mechanisms. *Current Biology, 23*, 1874–1883.

Kusakabe, T. (2001). Evolution of anural development in ascidians: Role of muscle specific differentiation genes. In H. Sawada, H. Yokosawa, & C. C. Lambert (Eds.), *Biology of Ascidians* (pp. 225–229). Tokyo: Springer.

Kusakabe, T., Araki, I., Satoh, N., & Jeffery, W. R. (1997). Evolution of chordate actin genes: Evidence from genomic organization and amino acid sequences. *Journal of Molecular Evolution, 44*, 289–298.

Kusakabe, T., Swalla, B. J., Satoh, N., & Jeffery, W. R. (1996). Mechanism of an evolutionary change in muscle differentiation in ascidians with different modes of development. *Developmental Biology, 174*, 379–392.

Ma, L., Parkhurst, A., & Jeffery, W. R. (2014). The role of a lens survival pathway including *sox2* and *αA-crystallin* in the evolution of cavefish eye degeneration. *EvoDevo, 5*, 28. http://dx.doi.org/10.1186/2041-9139-5-28.

McCauley, D. W., Hixon, E., & Jeffery, W. R. (2004). Evolution of pigment cell regression in the cavefish *Astyanax*: A late step in melanogenesis. *Evolution & Development, 6*, 209–218.

McGaugh, S. E., Gross, J. B., Aken, B., Blin, M., Borowsky, R., Chalopin, D., et al. (2014). The cavefish genome reveals candidate genes for eye loss. *Nature Communications, 5*, 5307. http://dx.doi.org/10.1038/natcommun630.

Menuet, A., Alunni, A., Joly, J.-S., Jeffery, W. R., & Rétaux, S. (2007). Shh overexpression in *Astyanax* cavefish: Multiple consequences on forebrain development and evolution. *Development, 134*, 845–855.

Mitchell, R. W., Russell, W. H., & Elliott, W. R. (1977). Mexican eyeless Characin fishes, genus *Astyanax*: Environment, distribution, and evolution. In *Special Publication of the Museum. Texas Tech University* (pp. 1–89).Lubbock, Texas: Texas Tech University. Special Publication No. 12.

Nishida, H. (2012). The maternal muscle determinant in the ascidian egg. *Wiley Interdisciplinary Reviews. Developmental Biology, 3*, 425–433.

O'Quin, K. E., Yoshizawa, M., Doshi, P., & Jeffery, W. R. (2013). Quantitative genetic analysis of retinal degeneration in the blind cavefish. *PloS One, 8*(2), e57281. http://dx.doi.org/10.1371/journal.pone.0057281.

Olsen, C. L., Natzle, J. E., & Jeffery, W. R. (1999). The forkhead gene FH1 is involved in evolutionary modification of the ascidian tadpole larva. *Mechanisms of Development, 85*, 49–58.

Porter, M. L., Dittmar, K., & Pérez-Losada, M. (2007). How long does evolution of the troglomorphic form take? Estimating divergence times in *Astyanax mexicanus*. *Acta Carsologica, 36*, 173–182.

Protas, M. E., Hersey, C., Kochanek, D., Zhou, Y., Wilkens, H., Jeffery, W. R., et al. (2006). Genetic analysis of cavefish reveals molecular convergence in the evolution of albinism. *Nature Genetics, 38*, 107–111.

Protas, M. E., Tabansky, I., Conrad, M., Gross, J. B., Vidal, O., Tabin, C. J., et al. (2008). Multi-trait evolution in a cave fish, *Astyanax mexicanus*. *Evolution & Development, 10*, 196–209.

Satoh, N. (1994). *Developmental biology of Ascidians*. Cambridge, UK: Cambridge University Press.

Scott, M. P. (1994). Intimations of a creature. *Cell, 79*, 1121–1124.

Soares, D., Yamamoto, Y., Strickler, A. G., & Jeffery, W. R. (2004). The lens has a specific influence on optic nerve and tectum development in the blind cavefish *Astyanax*. *Developmental Neuroscience, 26*, 308–319.

Stearns, F. W. (2010). One hundred years of pleiotropy: A retrospective. *Genetics, 186*, 767–773.

Strickler, A. G., Yamamoto, Y., & Jeffery, W. R. (2001). Early and late changes in *Pax 6* expression accompany eye degeneration during cavefish development. *Development Genes and Evolution, 211*, 138–144.

Strickler, A. G., Yamamoto, Y., & Jeffery, W. R. (2007). The lens controls cell survival in the retina: Evidence from the blind cavefish *Astyanax*. *Developmental Biology, 311*, 512–523.

Swalla, B. J., Badgett, M. R., & Jeffery, W. R. (1991). Identification of a cytoskeletal protein localized in the myoplasm of ascidian eggs: Localization is modified during anural development. *Development, 111*, 425–436.

Swalla, B. J., & Jeffery, W. R. (1991). Interspecific hybridization between an anural and urodele ascidian: Differential expression of urodele features suggests multiple mechanisms control anural development. *Developmental Biology, 143*, 319–334.

Swalla, B. J., & Jeffery, W. R. (1996). Requirement of the *Manx* gene for expression of chordate features in a tailless ascidian larva. *Science, 274*, 1205–1208.

Takada, N., Satoh, N., & Swalla, B. J. (2002). Expression of Tbx6, a muscle lineage T-box gene, in the anural ascidian *Molgula tectiformis*. *Development Genes and Evolution, 212*, 354–356.

Takada, N., York, J., Davis, J. M., Schumpert, B., Yasuo, H., Satoh, N., et al. (2002). Brachury expression in tailless Molgulid ascidian embryos. *Evolution & Development, 4*, 205–211.

Wilkens, H. (1988). Evolution and genetics of epigean and cave *Astyanax fasciatus* (Characidae, Pisces)—Support for the neutral mutation hypothesis. *Evolutionary Biology, 23*, 271–367.

Williams, G. C. (1957). Pleiotropy, natural selection, and the evolution of senescence. *Evolution, 11*, 398–411.

Yamamoto, Y., Byerly, M. S., Jackman, W. R., & Jeffery, W. R. (2009). Pleiotropic functions of embryonic sonic hedgehog expression link jaw and taste bud amplification with eye loss during cavefish evolution. *Developmental Biology, 330*, 200–211.

Yamamoto, Y., Espinasa, L., Stock, D. W., & Jeffery, W. R. (2003). Development and evolution of craniofacial patterning is mediated by eye-dependent and -independent processes in the cavefish *Astyanax*. *Evolution & Development, 5*, 435–446.

Yamamoto, Y., & Jeffery, W. R. (2000). Central role for the lens in cave fish eye degeneration. *Science, 289*, 631–633.

Yamamoto, Y., Stock, D. W., & Jeffery, W. R. (2004). Hedgehog signalling controls eye degeneration in blind cavefish. *Nature, 431*, 844–847.

Yasuo, H., & Satoh, N. (1994). An ascidian homolog of the mouse Brachyury (T) gene is expressed exclusively in notochord cells at the fate restricted stage. *Development, Growth & Differentiation, 36*, 9–18.

Yoshizawa, M., Gorički, Š., Soares, D., & Jeffery, W. R. (2010). Evolution of a behavioral shift mediated by superficial neuromasts helps cavefish find food in darkness. *Current Biology, 20*, 1631–1636.

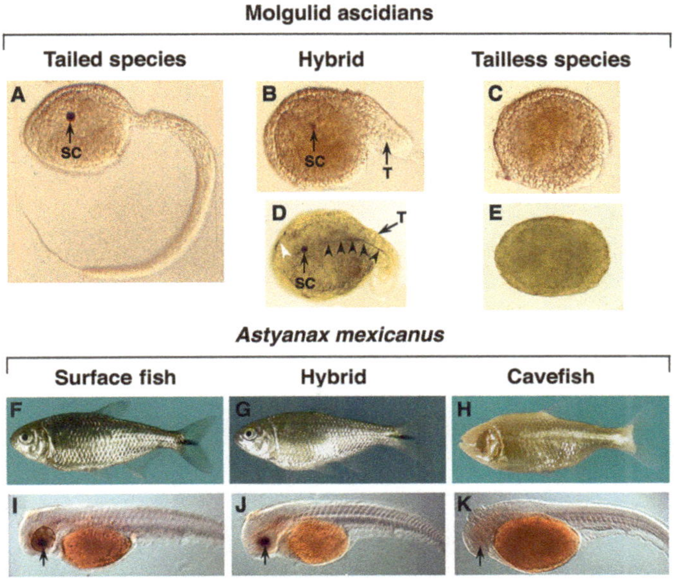

William R. Jeffery, Figure 1 The Molgulid ascidian and *Astyanax mexicanus* models. *Top*: molgulid ascidians. (A) Tailed urodele larva of *Molgula oculata* (the tailed species). (B) Short-tailed urodele larva of a cross between *Molgula occulta* (the tailless species) eggs and *M. oculata* (the tailed species) sperm. (C) Tailless anural larva of the *M. occulta* (the tailless species). (D and E) Alpha-acetylated tubulin antibody staining showing the restoration of anterior and tail nerves, which is absent in the tailless species larva (E), in a tailless × tailed species hybrid (D). SC, brain sensory cell. White arrowhead: part of a restored anterior nerve. Black arrowheads: restored tail nerve. *Bottom*: *A. mexicanus*. (F) Adult surface fish with eyes and pigment. (G) Adult surface fish × cavefish hybrid with eyes and pigment. (H) Adult cavefish lacking eyes and pigment. (I) Larval surface fish showing *alphaA-crystallin* gene expression in the lens. (J) Larval surface fish × cavefish hybrid showing *alphaA-crystallin* gene expression in the lens. (K) Larval cavefish lacking *alphaA-crystallin* gene expression in the lens. Arrows: lens. *Panels (A) and (B) from Swalla and Jeffery (1996), (C) and (D) from Y. Yamamoto and W.R. Jeffery (unpublished), and (I)–(K) from Ma, Parkhurst, and Jeffery (2014).*

CHAPTER TWENTY-EIGHT

From Cloning Neural Development Genes to Functional Studies in Mice, 30 Years of Advancements

Alexandra L. Joyner[1]
Developmental Biology Program, Sloan Kettering Institute, New York, USA
[1]Corresponding author: e-mail address: joynera@mskcc.org

Contents

Abstract

The invention of new mouse molecular genetics techniques, initiated in the 1980s, has repeatedly expanded our ability to tackle exciting developmental biology problems. The brain is the most complex organ, and as such the more sophisticated the molecular genetics technique, the more impact they have on uncovering new insights into how our brain functions. I provide a general time line for the introduction of new techniques over the past 30 years and give examples of new discoveries in the neural development field that emanated from them. I include a look to what the future holds and argue that we are at the dawn of a very exciting age for young scientists interested in studying how the nervous system is constructed and functions with such precision.

1. ESSAY

Having joined the developmental biology research community in the early 1980s, I am very fortunate to have been able to witness many revolutions in both the way we consider developmental concepts and the research tools available to us for probing key questions. These two aspects of research, of course, go hand in hand. New technologies allow questions, which previously could not be approached, to be answered, and at the same time the novel insights gained allow us to revise basic developmental concepts and

Current Topics in Developmental Biology, Volume 116
ISSN 0070-2153
http://dx.doi.org/10.1016/bs.ctdb.2015.11.035

generate new questions. In my mind, the most exciting studies in the field of developmental biology in the early 1980s were the fruit fly developmental genetics studies whereby mutant genetic screens led to identification and cloning of the first genes regulating developmental processes (e.g., Garber, Kuroiwa, & Gehring, 1983; Nusslein & Wieschaus, 1980). Progress was fast paced, and while I did not want to become a fly geneticist, my dream was to apply equally powerful genetics to the study of *in vivo* developmental processes in mice. With respect to a key discovery made in the mouse genetics field around the same time, in 1981 the first papers describing the establishment of mouse embryonic stem cell (ESC) lines from blastocysts were published, independently by my postdoctoral fellow mentor, Martin (1981) and her postdoctoral mentor, Evans (Evans & Kaufman, 1981). I doubt whether anyone at the time realized the enormous impact ESCs would have on the biomedical research community as a whole, let alone to mouse developmental genetics, nor that ESCs would 20 years later be the foundation of a new and very trendy research area, in part due to the establishment of human ESC lines and their relatives. Some think of, or have advocated for "Stem Cell" research being a new field of research; however, it is actually an area within the greater field of Developmental Biology, and the highest quality research being published is based on developmental biology concepts discovered by research in animal models. It seems possible, however, that new developmental concepts will be gained soon from using ESCs to generate specific cell types and most excitingly 3D organ cultures to study how multiple cell types are assembled to form functioning tissues. This might be particularly true for human ESC studies, if there are indeed developmental processes or concepts that are specific to humans, and that are not only new twists on existing theories. This approach has the added benefit that it is amenable to live imaging to identify and track the cell behaviors responsible for morphogenesis.

By the mid 1980s, a growing number of mouse developmental genes were being cloned, many based on the novel finding that a high degree of homology exists between fly and mammalian developmental genes (McGinnis, Garber, Wirz, Kuroiwa, & Gehring, 1984; McGinnis, Hart, Gehring, & Ruddle, 1984). This is an example of a fact that is presently taken for granted, but in the 1980s was a hotly debated idea. The discovery that gene sequences, often restricted to critical functional domains, were conserved between invertebrates and mammals provided a powerful approach for cloning mouse and human genes (Gehring, 1985). In a study I was involved in, it was only in retrospect that we realized it was very fortuitous that Tom Kornberg's lab happened to give me the part of the fly engrailed gene that they later found contained a divergent homeobox sequence, to use as a probe to clone two related genes from mouse (Joyner, Kornberg, Coleman, Cox, & Martin, 1985). Now such a gene could be identified in a few minutes using a sequence search on a computer, and then a cDNA purchased or generated in a few days! Perhaps the most famous example of cloning by homology is the *Hox* family of homeobox-containing genes. The ensuing discovery of colinearity of gene position on the chromosome and anterior–posterior position of gene expression within an organism was a very exciting fact discovered by the end of the 1980s (reviewed in Alexander, Nolte, & Krumlauf, 2009; Maconochie, Nonchev, Morrison, & Krumlauf, 1996).

A turning point for the explosion of applications of mouse ESCs for genetics studies was the finding that when injected into a mouse blastocyst, the descendants of ESCs could populate the germline of chimeras and produce ESC-derived mice (Bradley, Evans, Kaufman, & Robertson, 1984), much like inner cell mass cells in the blastocyst (Gardner & Rossant, 1979). It was a relatively easy next step to alter the genome in ESCs using established transfection protocols (Gossler, Doetschman, Korn, Serfling, & Kemler, 1986). However, the development of gene targeting by Oliver Smithies and Mario Capecchi took manipulating the mouse genome to a new level, *en par* with yeast genetics at the time (Doetschman et al., 1987; Joyner, Skarnes, & Rossant, 1989; Koller et al., 1989, Koller, Marrach, Kappler, & Smithies, 1990; Smithies, Gregg, Boggs, Koralewski, & Kucherlapati, 1985; Thomas & Capecchi, 1987, 1990). It is only due to the most recent revolution, spurred by the application of CRISPR/Cas9 and related technologies for editing the genomes (Peng et al., 2014), that the way has been paved for sophisticated genome editing to be engineered in any model organism. Thus, the great advantage mouse has held for nearly three decades, as the "queen" of reverse genetics for multicellular model organisms has only recently begun to wane. Applications of CRISPR/Cas9 technologies are likely to expand beyond what is currently imagined, as with ESCs in the 1980s, and it seems will have an equally profound impact on diverse research fields, in part by allowing a greater depth of genetic comparisons between experimental organisms. Until a year ago, I never imagined a mouse developmental biologist's genetic toolbox would someday not include mouse ESCs. It now seems a likelihood, since all genetic manipulations might soon be done through injection of zygotes (see Skarnes, 2015).

In the 1990s, the ease with which the mouse ESC genome could be manipulated, and turned into mutant or transgenic mice, lead to the development of critical tools for developmental biologists. Null mutations were replaced with conditional mutations to allow gene functions to be examined in any organ and at any time during development. Reporter lines were generated (Hadjantonakis, Gertsenstein, Ikawa, Okabe, & Nagy, 1998; Mansour, Thomas, Deng, & Capecchi, 1990) to visualize gene expression at single cell resolution and allow isolation of cell populations or lineages by FACS (fluorescence-activated cell sorting), and gene knock-in approaches to compare protein functions (Hanks, Wurst, Anson-Cartwright, Auerbach, & Joyner, 1995; Wang, Schnegelsberg, Dausman, & Jaenisch, 1996). Site-specific recombinases, especially using Cre/loxP, were used to generate conditional and reporter alleles with precision, and enabled genome editing down to a single nucleotide (plus the insertion of a loxP site). A critical application of site-specific recombinases for developmental biologists has been for fate-mapping studies, especially those incorporating genetic inducible fate mapping (GIFM) using regulatable forms of Cre and Flpo (Joyner & Zervas, 2006). The ability to label a group of spatially defined cells, or a particular lineage at any developmental stage, and characterize the initial population and then follow the movement and/or fate of the cells through normal or altered developmental processes has revolutionized our understanding of lineage allocation during development of all organ systems. An example from my field of cerebellum development is the finding that the two progenitor zones of the cerebellum, the ventricular zone (VZ) lining the ventricle and the rhombic lip (RL), at the posterior edge of the cerebellar anlage in rhombomere 1, give rise to two distinct classes of neurons, the inhibitory or gamma aminobutyric acid neurons and excitatory (glutamatergic neurons) neurons, respectively (Hoshino et al., 2005; Machold & Fishell, 2005). A previous fate-mapping study in chick embryos involving physical labeling of the rhombic lip had defined the cell types derived from the rhombic lip (Wingate & Hatten, 1999); however, the ability to genetically label the cells in a reproducible manner and/or mutate genes in the cell population in mice has been extremely powerful. For example, using inducible labeling with CreER, the temporal sequence of generation of embryonic cell types was defined for both the cells derived from the rhombic lip (*Atoh1-CreER*; Machold & Fishell, 2005) and ventricular zone (*Ascl1*^{CreER}; Sudarov et al., 2011; Fig. 1A and B). In elegant recent studies using *Tnc*$^{YFP\text{-}CreER}$ and *Glast*^{CreER} knock-in alleles (Fleming et al., 2013; Parmigiani et al., 2015), the lineage of progenitors in the white matter (WM), that are derived from

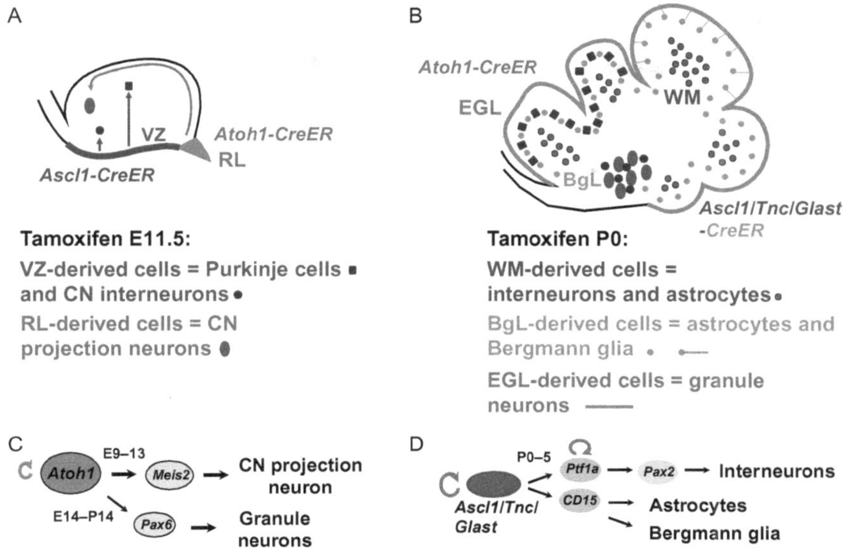

Figure 1 Genetic inducible fate-mapping studies define distinct progenitor pools in the cerebellum separated spatially and/or in time. (A and B) Schematic drawings of midsagittal sections of cerebella at E11.5 and P0 are shown. The cells that can be labeled using a reporter allele such as *R26*^{Yfp} and an *Atoh1-CreER* transgene or *Ascl1*^{CreER} knock-in allele (*Ascl1-CreER*) are indicated by orange and purple, respectively, and the cells they give rise to when tamoxifen is given at E11.5 (A) or (P0) are indicated below. In (B), Purkinje cells are only shown in the anterior lobules (folds) and glial fibers of Bergmann glial progenitors only in one posterior lobule for simplicity. At E12.5–17.5, the rhombic lip gives rise to the external granule cell layer (orange) shown in (B) made up of *Atoh1*-expressing progenitors that can divide until P16. Three knock-in alleles have been used to mark progenitors in the WM and BgL (*Ascl1*^{CreER}, *Tnc*^{CreER}, *Glast*^{CreER}). Round arrows indicate proliferating progenitors. VZ, ventricular zone; RL, rhombic lip; CN, cerebellar nuclei; EGL, external granule cell layer; BgL, Bergmann glial layer; WM, white matter. (C and D) The *Atoh1* and *Ascl1/Tnc/Glast* lineages are show. *Data are based on Fleming et al. (2013), Machold and Fishell (2005), Parmigiani et al. (2015), and Sudarov et al. (2011).* (A color version of this figure is located at the end of this chapter.)

the ventricular zone, was dissected and revealed the existence of bipotent and lineage-restricted progenitors during the first week after birth (Fig. 1B–D). In one study, tamoxifen was applied to the surface of the cerebellum to induce *Glast*^{CreER} cell labeling only in the Bergmann glial layer (BgL) progenitor pool and not those in the WM, based on the fact that Bergmann glia cells project a glial fiber to the surface of the cerebellum and thus only these cells are accessible to tamoxifen. Using the spatial separation of two progenitor pools, the Rossi group (Parmigiani et al., 2015) was able to demonstrate that interneurons are only generated from WM progenitors, whereas both Bergmann glial and WM progenitors produce astrocytes (including Bergmann glia).

A burgeoning area of stem cell research is examination of the potency or plasticity of quiescent cells (progenitor/stem cells) in adult organs. GIFM has been particularly valuable as a tool to identify and characterize endogenous stem cells in adult organs based on the ability of the initial population of marked cells to expand over time. In the adult forebrain, we used a CreER-knock-in allele of the sonic hedgehog target gene, *Gli1*, to mark quiescent neural stem cells in the subventricular zone and dentate gyrus of the adult forebrain, and found that whereas only a small number of cells are initially labeled, within a few weeks the population expands massively, because the *Gli1*-expressing quiescent stem cells give rise to a transient amplifying population of cells (Ahn & Joyner, 2005). Moreover, the migration of the labeled cells derived from the quiescent subventricular zone stem cells could be traced into the olfactory bulbs, where they differentiate into several classes of interneurons. In addition, the marked cells were shown to produce new neurons for over a year, indicating that the quiescent neural stem cells are either long-lived or only a subset of stem cells expand at any given time and then are exhausted. Follow-up studies have provided evidence for both models (Bonaguidi et al., 2011; Encinas et al., 2011). Since our first study in the brain, the same approach has been used to identify adult stem cells in many organs. One novel finding in several adult epithelial organs was that basal and luminal cells can be generated from lineage-restricted progenitors (e.g., Choi, Zhang, Zhang, Ittmann, & Xin, 2012; de Visser et al., 2012; Van Keymeulen et al., 2011; Wang et al., 2013), although during development a bipotent stem cell gives rise to both cell types (e.g., de Visser et al., 2012; Ousset et al., 2012; van Amerongen, Bowman, & Nusse, 2012) and rare bipotent cells likely also exist in the adult (Rios, Fu, Lindeman, & Visvader, 2014; Wang et al., 2009). There is similar evidence for subtype-restricted progenitors in the adult mesenchyme, at least during regeneration of the prostate (Peng, Levine, Zahid, Wilson, & Joyner, 2013). The addition of multicolor lineage reporters (Bronner, 2015; Rinkevich, Lindau, Ueno, Longaker, & Weissman, 2011; Rios et al., 2014; Snippert et al., 2010), use of the MADM (mosaic analysis with double markers) technique (Espinosa & Luo, 2008; Zong, Espinosa, Su, Muzumdar, & Luo, 2005), or reducing the dose of tamoxifen to a clonal level (Brownell, Guevara, Bai, Loomis, & Joyner, 2011; Legue & Nicolas, 2005; Legue, Riedel, & Joyner, 2015) has provided a means to test the potential of individual cells (clonal analysis), rather than whole populations. The powerful combination of genetic fate mapping and clonal approaches with live imaging *in vivo* (Ritsma et al., 2014) or in organ cultures systems (Packard et al., 2013) and organoids derived from ESCs is an exciting next step for these technologies for addressing basic questions of stem cell potency, lineage allocation, and the most challenging questions of determining the sequence of cell behaviors that underlie 3D morphogenesis of organs.

Recent clonal analyses in the developing brain have revealed interesting rules as to how circuits are assembled. Using intraventricular injection of GFP-expressing retroviruses to label only a few well-separated radial glial progenitor cells in the neocortex, it was found that clonally related excitatory neurons have a radial spatial organization, and preferentially form synapses with each other compared to nonclonally related neighboring neurons (Yu, Bultje, Wang, & Shi, 2009). Transient electrical coupling during embryonic and neonatal development was found to lead to the formation and/or stabilization of chemical synapses between clonally related excitatory neurons (Gao, Sultan, Zhang, & Shi, 2013; He, Li, Ge, Yu, & Shi, 2015; Yu et al., 2012). Moreover, in the visual cortex clonally related excitatory neurons are preferentially stimulated by a similar orientation of a bar moving across the visual field (Li et al., 2012). A recent study used the MADM approach to follow the progeny of sister radial progenitors, as they can be marked with different fluorescent labels, and also applied mathematical modeling to conclude that lineage progression of proliferation behaviors of radial glial progenitors in the developing neocortex is largely deterministic and neurons are generated before glia (Fig. 1A; Gao et al., 2014). In contrast, clonal analysis of hippocampal excitatory neuron development using retroviral infection and MADM revealed horizontal clusters of neurons that are not electrically coupled. Interestingly, instead the activity of clonally related hippocampal neurons is preferentially regulated by common synaptic input from nearby inhibitory interneurons (Xu et al., 2014). Finally, several studies have analyzed development of clones of neocortical interneurons, which derive from a distinct location in the forebrain compared to excitatory neurons. Using the RCAS (Replication-Competent Avian SLV Long Terminal Repeat with a Splice acceptor) system to specifically infect dividing radial glial progenitors in the interneuron progenitor region (i.e., the medial ganglionic eminence—MGE) based on their conditional expression of the avian tumor virus receptor A, clonally related interneurons were observed to not be randomly distributed but to tend to be near each other in the cortex (Brown et al., 2011). Grouping of cells were seen in a second study using a Cre-dependent retroviral reporter, but with some differences in layer distribution (Ciceri et al., 2013). In more recent studies using a bar-coded library of retroviruses to randomly label progenitors in the MGE, two studies concluded that there is wider dispersion of clonally related neocortical interneurons, and because cells in a clone can be identified based on their unique bar code, these studies also revealed that some clones cross into other forebrain regions (Harwell et al., 2015; Mayer et al., 2015). The reasons for the different conclusions with respect to the degree of clustering in the neocortex may in part be due to technical limitations of each approach, including distinct tropism for subsets of radial glia progenitors and reporter gene silencing (Gil-Sanz & Muller, 2015). Mouse *in vivo* genetics-based approaches need to be developed to circumvent these problems, and larger numbers of clones need to be analyzed to fully reveal the diversity of interneuron clonal behaviors, and to determine whether their integration into neural circuits is primarily driven by environmental cues in the neocortex rather than clonal lineage, and whether subtypes of interneurons with preferentially distinct behaviors derive from molecularly distinct radial glial progenitors.

Unlike the neocortex where progenitors undergo largely asymmetric cell divisions, several clonal studies in the cerebellum have revealed the progenitors of excitatory granule cells divide almost exclusively symmetrically, and remarkably, clonally related progenitors (with clones sizes of >200 cells) differentiate *en masse* over a few days (Fig. 2B; Espinosa & Luo, 2008; Legue et al., 2015; Zong et al., 2005). Given the relationships discovered in the forebrain between clonal lineage and wiring, and thus how circuits function, it will be interesting to determine whether clonally related granule cells participate in related circuits that receive similar inputs or have analogous properties of stimulation of target Purkinje cells.

Mosaic mutant analysis is a very powerful approach that has been utilized extensively in flies for studying mutant cell behaviors, but only recently exploited in mouse in a few studies. The basic idea is to mutate only scattered cells in a tissue or lineage and to mark the cells (often fluorescently) so that the mutant behavior of the cells can be determined. The most widely used approach for mosaic mutant analysis is to place a MADM allele on the same chromosome as a null allele and to induce interchromosomal recombination using a *Cre* allele to generate clones carrying two null alleles (Espinosa, Tea, & Luo, 2014; Zong, 2014). An alternative that can easily be applied at any floxed allele is the MASTR (mosaic analysis with spatial and temporal regulation) approach whereby FlpoER induces expression of a GFP–CRE fusion protein that then deletes homozygous floxed alleles (Lao, Raju, Bai, &

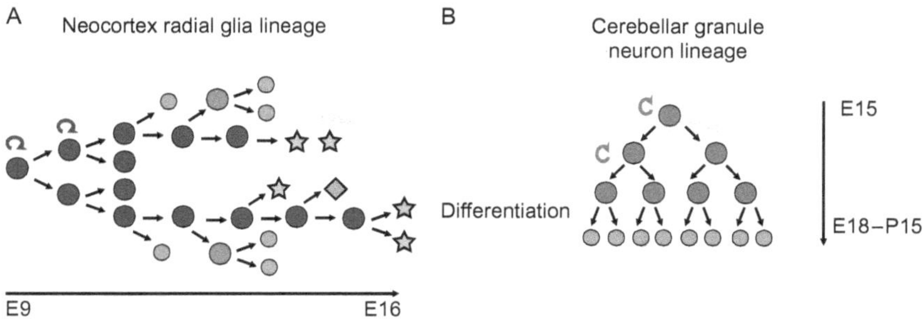

Figure 2 Clonal analyses of glutamatergic progenitors in the forebrain (neocortex) and cerebellum reveal different modes of generating neurons and glia. (A) Radial glial progenitors (blue circles) in the ventricular zone of the neocortex initially divide symmetrically to expand the progenitors and then divide mainly asymmetrically to produce a progenitor and a differentiated projection neuron (gray circle), astrocyte (star), or oligodendrocyte (diamond), or a progenitor with more limited proliferation capacity (light blue) that gives rise often to two neurons. Some symmetric divisions produce two neurons or astrocytes (differentiation divisions). Round arrows indicate symmetrical divisions producing two progenitors. Gliogenesis (production of astrocytes and oligodendrocytes) occurs after neurogenesis. (B) In the cerebellum, granule cell progenitors in the external granule cell layer (EGL in Fig. 1B) divide only symmetrically to expand the number of progenitors and then to produce postmitotic granule neurons. *Data are based on Espinosa and Luo (2008), Gao et al. (2014), and Legue et al. (2015).* (A color version of this figure is located at the end of this chapter.)

Joyner, 2012). Such studies are distinct from ones in which a gene is mutated in all cells in a whole tissue or lineage, where compensatory processes (such as upregulation of a related gene or pathway) sometimes mask cellular phenotypes. A recent study beautifully illustrates this point (Joo, Hippenmeyer, & Luo, 2014). Ablating the neurotrophin receptor gene, tropomyosin-related kinase C (*TrkC*), mosaically in cerebellar Purkinje cells was found to lead to a defect in dendrite growth, whereas deletion of *TrkC* in all Purkinje cells produced no phenotype. Ablation of the ligand NT3 rescued the mosaic mutant phenotype, suggesting competition between wild-type and TRKC-signaling mutant Purkinje cells results in TRKC-dependent stimulation of dendrite morphogenesis in wild-type cells. For the MADM approach to be widely used, the double marker alleles must be placed on every chromosome. A limitation of the MASTR approach is that few *FlpoER*-expressing transgenes have been generated. Development of second-generation MASTR alleles is a high priority.

In order to understand how individual and interconnecting neural circuits form during development and ultimately function in a highly specific and effective manner to regulate animal behaviors, it will be necessary to define the projections of every neuron, or at the very least those of functionally related neurons. This is not a trivial undertaking, as in the cerebellum of mice alone, there are 70 million cells, which accounts for nearly 60% of the cells in the entire brain (Herculano-Houzel, Mota, & Lent, 2006). Furthermore, the projection neurons of the cerebellar nuclei project throughout the brain, when both direct and indirect (through intermediate target neurons) connections are considered. In addition, poorly defined closed-loop circuits have been described that project from the cerebellum to distinct regions of the neocortex via the thalamus and then return to the cerebellum via the pontine nucleus (Strick, Dum, & Fiez, 2009). Like the human genome project initiated two decades ago, an ambitious goal for the neuroscience community is to perform large-scale fine mapping of long-range neural circuits (between brain regions) in the adult. Ultimately, this must be applied to developing circuits that undergo refinement, including pruning back of axons that do not form robust synapses. Such data will provide the building blocks, or critical foundation, for most future neuroscience studies, just as the sequences of whole genomes of many species have transformed all of biological research. However, insufficient technologies for neural circuit tracing have been a major limitation. The NIH BRAIN (Brain Research through Advancing Innovative Neurotechnologies) initiative (http://braininitiative.nih.gov/) provides an important step toward developing strategies to map circuits and determine the molecular properties of most neurons. Extending the mapping to include trans-synaptic mapping (Yook, Druckmann, & Kim, 2013), for example, for closed-loop circuits, is an even more challenging but critical goal. In recent years, great progress has been made in developing viruses expressing fluorescent proteins for circuit mapping, genetically encoded indicators of neural activity (voltage, glutamate, or calcium indicators), and powerful optogenetic effectors that allow silencing or stimulation of neurons for functional studies (e.g., Arenkiel et al., 2007; Chen et al., 2013). The Allen Institute for Brain Science has expended substantial effort into developing mouse lines that conditionally express a variety of fluorescent protein indicators of neural activity and optogenetic effectors (Madisen et al., 2015). The advantage of genetically controlled systems is that theoretically, using intersectional approaches (Dymecki, Ray, & Kim, 2010), the expression domain of the proteins can be limited to not only subtype-specific groups of neurons but cells that are spatially restricted and labeled at a particular developmental stage. However, a major challenge for such studies will be identifying enhancer elements that convey spatial resolution in addition to cell-type specificity in their expression. A combination of spatially targeted viral injections to induce combinatorial expression from alleles engineered into mice might be the most practical solution in the short run. The ability to rapidly engineer new alleles using CRISPR/Cas9-mediated integration of neural circuit reporter alleles should allow rapid testing of large numbers of enhancers and reporter alleles. Given the importance that new technologies have in advancing scientific breakthroughs, and the state of public research funding in the United States, it is encouraging that the NIH BRAIN initiative underscores the critical impact that technology development has on basic science, and their combined importance for advancing translational medicine. Beyond my years, these approaches will help solve the fascinating question of how new neural circuits have evolved in mammals, or ancient ones have become more sophisticated during evolution. It certainly seems that it is again (after 30 years) an exciting time for young scientists to enter the field of neural development.

ACKNOWLEDGMENTS

I would like to thank Drs. Dan Turnbull (NYUSoM) and Songhai Shi (MSKCC) for their helpful comments on the essay. Our work is partially funded by grants from the NIH (R37MH085726, R01NS092096, R21HD083860, and NIH/NCI Cancer Center Support Grant P30 CA008748). I apologize to the many people who contributed to the development of modern mouse genetics techniques and who's work was not mentioned. This essay is meant to convey the general time line for how the field has changed, with some specific examples in the area of brain development.

REFERENCES

Ahn, S., & Joyner, A. L. (2005). In vivo analysis of quiescent adult neural stem cells responding to Sonic hedgehog. *Nature*, *437*, 894–897.

Alexander, T., Nolte, C., & Krumlauf, R. (2009). Hox genes and segmentation of the hindbrain and axial skeleton. *Annual Review of Cell and Developmental Biology*, *25*, 431–456.

Arenkiel, B. R., Peca, J., Davison, I. G., Feliciano, C., Deisseroth, K., Augustine, G. J., et al. (2007). In vivo light-induced activation of neural circuitry in transgenic mice expressing channelrhodopsin-2. *Neuron*, *54*, 205–218.

Bonaguidi, M. A., Wheeler, M. A., Shapiro, J. S., Stadel, R. P., Sun, G. J., Ming, G. L., et al. (2011). In vivo clonal analysis reveals self-renewing and multipotent adult neural stem cell characteristics. *Cell*, *145*, 1142–1155.

Bradley, A., Evans, M., Kaufman, M. H., & Robertson, E. (1984). Formation of germ-line chimaeras from embryo-derived teratocarcinoma cell lines. *Nature*, *309*, 255–256.

Bronner, M. (2015). Confetti clarifies controversy: Neural crest stem cells are multipotent. *Cell Stem Cell*, *16*, 217–218.

Brown, K. N., Chen, S., Han, Z., Lu, C. H., Tan, X., Zhang, X. J., et al. (2011). Clonal production and organization of inhibitory interneurons in the neocortex. *Science*, *334*, 480–486.

Brownell, I., Guevara, E., Bai, C. B., Loomis, C. A., & Joyner, A. L. (2011). Nerve-derived sonic hedgehog defines a niche for hair follicle stem cells capable of becoming epidermal stem cells. *Cell Stem Cell*, *8*, 552–565.

Chen, T. W., Wardill, T. J., Sun, Y., Pulver, S. R., Renninger, S. L., Baohan, A., et al. (2013). Ultrasensitive fluorescent proteins for imaging neuronal activity. *Nature*, *499*, 295–300.

Choi, N., Zhang, B., Zhang, L., Ittmann, M., & Xin, L. (2012). Adult murine prostate basal and luminal cells are self-sustained lineages that can both serve as targets for prostate cancer initiation. *Cancer Cell*, *21*, 253–265.

Ciceri, G., Dehorter, N., Sols, I., Huang, Z. J., Maravall, M., & Marin, O. (2013). Lineage-specific laminar organization of cortical GABAergic interneurons. *Nature Neuroscience*, *16*, 1199–1210.

de Visser, K. E., Ciampricotti, M., Michalak, E. M., Tan, D. W., Speksnijder, E. N., Hau, C. S., et al. (2012). Developmental stage-specific contribution of LGR5(+) cells to basal and luminal epithelial lineages in the postnatal mammary gland. *The Journal of Pathology*, *228*, 300–309.

Doetschman, T., Gregg, R. G., Maeda, N., Hooper, M. L., Melton, D. W., Thompson, S., et al. (1987). Targetted correction of a mutant HPRT gene in mouse embryonic stem cells. *Nature*, *330*, 576–578.

Dymecki, S. M., Ray, R. S., & Kim, J. C. (2010). Mapping cell fate and function using recombinase-based intersectional strategies. *Methods in Enzymology*, *477*, 183–213.

Encinas, J. M., Michurina, T. V., Peunova, N., Park, J. H., Tordo, J., Peterson, D. A., et al. (2011). Division-coupled astrocytic differentiation and age-related depletion of neural stem cells in the adult hippocampus. *Cell Stem Cell*, *8*, 566–579.

Espinosa, J. S., & Luo, L. (2008). Timing neurogenesis and differentiation: Insights from quantitative clonal analyses of cerebellar granule cells. *The Journal of Neuroscience*, *28*, 2301–2312.

Espinosa, J. S., Tea, J. S., & Luo, L. (2014). Mosaic analysis with double markers (MADM) in mice. *Cold Spring Harbor Protocols*, *2014*, 182–189.

Evans, M. J., & Kaufman, M. H. (1981). Establishment in culture of pluripotential cells from mouse embryos. *Nature*, *292*, 154–156.

Fleming, J. T., He, W., Hao, C., Ketova, T., Pan, F. C., Wright, C. C., et al. (2013). The Purkinje neuron acts as a central regulator of spatially and functionally distinct cerebellar precursors. *Developmental Cell*, *27*, 278–292.

Gao, P., Postiglione, M. P., Krieger, T. G., Hernandez, L., Wang, C., Han, Z., et al. (2014). Deterministic progenitor behavior and unitary production of neurons in the neocortex. *Cell*, *159*, 775–788.

Gao, P., Sultan, K. T., Zhang, X. J., & Shi, S. H. (2013). Lineage-dependent circuit assembly in the neocortex. *Development*, *140*, 2645–2655.

Garber, R. L., Kuroiwa, A., & Gehring, W. J. (1983). Genomic and cDNA clones of the homeotic locus Antennapedia in Drosophila. *The EMBO Journal*, *2*, 2027–2036.

Gardner, R. L., & Rossant, J. (1979). Investigation of the fate of 4–5 day post-coitum mouse inner cell mass cells by blastocyst injection. *Journal of Embryology and Experimental Morphology*, *52*, 141–152.

Gehring, W. J. (1985). The Homeo Box: A key to the understanding of development. *Cell*, *40*, 3–5.

Gil-Sanz, C., & Muller, U. (2015). A new chapter in the life of Cajal's short-axon neurons: Separation of interneuron siblings after birth. *Neuron*, *87*, 909–911.

Gossler, A., Doetschman, T., Korn, A., Serfling, E., & Kemler, R. (1986). Transgenesis by means of blastocyst-derived embryonic stem cell lines. *Proceedings of the National Academy of Sciences of the United States of America*, *83*, 9065–9069.

Hadjantonakis, A. K., Gertsenstein, M., Ikawa, M., Okabe, M., & Nagy, A. (1998). Generating green fluorescent mice by germline transmission of green fluorescent ES cells. *Mechanisms of Development*, *76*, 79–90.

Hanks, M., Wurst, W., Anson-Cartwright, L., Auerbach, A. B., & Joyner, A. L. (1995). Rescue of the En-1 mutant phenotype by replacement of En-1 with En-2. *Science*, *269*, 679–682.

Harwell, C. C., Fuentealba, L. C., Gonzalez-Cerrillo, A., Parker, P. R., Gertz, C. C., Mazzola, E., et al. (2015). Wide dispersion and diversity of clonally related inhibitory interneurons. *Neuron*, *87*, 999–1007.

He, S., Li, Z., Ge, S., Yu, Y. C., & Shi, S. H. (2015). Inside-out radial migration facilitates lineage-dependent neocortical microcircuit assembly. *Neuron*, *86*, 1159–1166.

Herculano-Houzel, S., Mota, B., & Lent, R. (2006). Cellular scaling rules for rodent brains. *Proceedings of the National Academy of Sciences of the United States of America*, *103*, 12138–12143.

Hoshino, M., Nakamura, S., Mori, K., Kawauchi, T., Terao, M., Nishimura, Y. V., et al. (2005). Ptf1a, a bHLH transcriptional gene, defines GABAergic neuronal fates in cerebellum. *Neuron*, *47*, 201–213.

Joo, W., Hippenmeyer, S., & Luo, L. (2014). Neurodevelopment. Dendrite morphogenesis depends on relative levels of NT-3/TrkC signaling. *Science*, *346*, 626–629.

Joyner, A. L., Kornberg, T., Coleman, K. G., Cox, D. R., & Martin, G. R. (1985). Expression during embryogenesis of a mouse gene with sequence homology to the Drosophila engrailed gene. *Cell*, *43*, 29–37.

Joyner, A. L., Skarnes, W. C., & Rossant, J. (1989). Production of a mutation in mouse En-2 gene by homologous recombination in embryonic stem cells. *Nature*, *338*, 153–156.

Joyner, A. L., & Zervas, M. (2006). Genetic inducible fate mapping in mouse: Establishing genetic lineages and defining genetic neuroanatomy in the nervous system. *Developmental Dynamics*, *235*, 2376–2385.

Koller, B. H., Hagemann, L. J., Doetschman, T., Hagaman, J. R., Huang, S., Williams, P. J., et al. (1989). Germ-line transmission of a planned alteration made in a hypoxanthine phosphoribosyltransferase gene by homologous recombination in embryonic stem cells. *Proceedings of the National Academy of Sciences of the United States of America*, *86*, 8927–8931.

Koller, B. H., Marrach, P., Kappler, J. W., & Smithies, O. (1990). Normal development of mice deficient in beta 2M, MHC class I proteins, and CD8+ T cells. *Science*, *248*, 1227–1230.

Lao, Z., Raju, G. P., Bai, C. B., & Joyner, A. L. (2012). MASTR: A technique for mosaic mutant analysis with spatial and temporal control of recombination using conditional floxed alleles in mice. *Cell Reports*, *2*, 386–396.

Legue, E., & Nicolas, J. F. (2005). Hair follicle renewal: Organization of stem cells in the matrix and the role of stereotyped lineages and behaviors. *Development*, *132*, 4143–4154.

Legue, E., Riedel, E., & Joyner, A. L. (2015). Clonal analysis reveals granule cell behaviors and compartmentalization that determine the folded morphology of the cerebellum. *Development*, *142*, 1661–1671.

Li, Y., Lu, H., Cheng, P. L., Ge, S., Xu, H., Shi, S. H., et al. (2012). Clonally related visual cortical neurons show similar stimulus feature selectivity. *Nature*, *486*, 118–121.

Machold, R., & Fishell, G. (2005). Math1 is expressed in temporally discrete pools of cerebellar rhombic-lip neural progenitors. *Neuron*, *48*, 17–24.

Maconochie, M., Nonchev, S., Morrison, A., & Krumlauf, R. (1996). Paralogous Hox genes: Function and regulation. *Annual Review of Genetics*, *30*, 529–556.

Madisen, L., Garner, A. R., Shimaoka, D., Chuong, A. S., Klapoetke, N. C., Li, L., et al. (2015). Transgenic mice for intersectional targeting of neural sensors and effectors with high specificity and performance. *Neuron*, *85*, 942–958.

Mansour, S. L., Thomas, K. R., Deng, C. X., & Capecchi, M. R. (1990). Introduction of a lacZ reporter gene into the mouse int-2 locus by homologous recombination. *Proceedings of the National Academy of Sciences of the United States of America*, *87*, 7688–7692.

Martin, G. R. (1981). Isolation of a pluripotent cell line from early mouse embryos cultured in medium conditioned by teratocarcinoma stem cells. *Proceedings of the National Academy of Sciences of the United States of America*, *78*, 7634–7638.

Mayer, C., Jaglin, X. H., Cobbs, L. V., Bandler, R. C., Streicher, C., Cepko, C. L., et al. (2015). Clonally related forebrain interneurons disperse broadly across both functional areas and structural boundaries. *Neuron*, *87*, 989–998.

McGinnis, W., Garber, R. L., Wirz, J., Kuroiwa, A., & Gehring, W. J. (1984). A homologous protein-coding sequence in Drosophila homeotic genes and its conservation in other metazoans. *Cell*, *37*, 403–408.

McGinnis, W., Hart, C. P., Gehring, W. J., & Ruddle, F. H. (1984). Molecular cloning and chromosome mapping of a mouse DNA sequence homologous to homeotic genes of Drosophila. *Cell*, *38*, 675–680.

Nusslein, V. C., & Wieschaus, E. (1980). Mutations affecting segment number and polarity in Drosophila. *Nature*, *287*, 795–801.

Ousset, M., Van Keymeulen, A., Bouvencourt, G., Sharma, N., Achouri, Y., Simons, B. D., et al. (2012). Multipotent and unipotent progenitors contribute to prostate postnatal development. *Nature Cell Biology*, *14*, 1131–1138.

Packard, A., Georgas, K., Michos, O., Riccio, P., Cebrian, C., Combes, A. N., et al. (2013). Luminal mitosis drives epithelial cell dispersal within the branching ureteric bud. *Developmental Cell*, *27*, 319–330.

Parmigiani, E., Leto, K., Rolando, C., Figueres-Onate, M., Lopez-Mascaraque, L., Buffo, A., et al. (2015). Heterogeneity and bipotency of astroglial-like cerebellar progenitors along the interneuron and glial lineages. *The Journal of Neuroscience*, *35*, 7388–7402.

Peng, Y., Clark, K. J., Campbell, J. M., Panetta, M. R., Guo, Y., & Ekker, S. C. (2014). Making designer mutants in model organisms. *Development*, *141*, 4042–4054.

Peng, Y. C., Levine, C. M., Zahid, S., Wilson, E. L., & Joyner, A. L. (2013). Sonic hedgehog signals to multiple prostate stromal stem cells that replenish distinct stromal subtypes during regeneration. *Proceedings of the National Academy of Sciences of the United States of America*, *110*, 20611–20616.

Rinkevich, Y., Lindau, P., Ueno, H., Longaker, M. T., & Weissman, I. L. (2011). Germ-layer and lineage-restricted stem/progenitors regenerate the mouse digit tip. *Nature*, *476*, 409–413.

Rios, A. C., Fu, N. Y., Lindeman, G. J., & Visvader, J. E. (2014). In situ identification of bipotent stem cells in the mammary gland. *Nature*, *506*, 322–327.

Ritsma, L., Ellenbroek, S. I., Zomer, A., Snippert, H. J., de Sauvage, F. J., Simons, B. D., et al. (2014). Intestinal crypt homeostasis revealed at single-stem-cell level by in vivo live imaging. *Nature*, *507*, 362–365.

Skarnes, W. C. (2015). Is mouse embryonic stem cell technology obsolete? *Genome Biology*, *16*, 109.

Smithies, O., Gregg, R. G., Boggs, S. S., Koralewski, M. A., & Kucherlapati, R. S. (1985). Insertion of DNA sequences into the human chromosomal beta-globin locus by homologous recombination. *Nature*, *317*, 230–234.

Snippert, H. J., van der Flier, L. G., Sato, T., van Es, J. H., van den Born, M., Kroon-Veenboer, C., et al. (2010). Intestinal crypt homeostasis results from neutral competition between symmetrically dividing Lgr5 stem cells. *Cell*, *143*, 134–144.

Strick, P. L., Dum, R. P., & Fiez, J. A. (2009). Cerebellum and nonmotor function. *Annual Review of Neuroscience*, *32*, 413–434.

Sudarov, A., Turnbull, R. K., Kim, E. J., Lebel-Potter, M., Guillemot, F., & Joyner, A. L. (2011). Ascl1 genetics reveals insights into cerebellum local circuit assembly. *The Journal of Neuroscience*, *31*, 11055–11069.

Thomas, K. R., & Capecchi, M. R. (1987). Site-directed mutagenesis by gene targeting in mouse embryo-derived stem cells. *Cell*, *51*, 503–512.

Thomas, K. R., & Capecchi, M. R. (1990). Targeted disruption of the murine *int-1* proto-oncogene resulting in severe abnormalities in midbrain and cerebellar development. *Nature*, *346*, 847–850.

van Amerongen, R., Bowman, A. N., & Nusse, R. (2012). Developmental stage and time dictate the fate of Wnt/β-catenin-responsive stem cells in the mammary gland. *Cell Stem Cell*, *11*, 387–400.

Van Keymeulen, A., Rocha, A. S., Ousset, M., Beck, B., Bouvencourt, G., Rock, J., et al. (2011). Distinct stem cells contribute to mammary gland development and maintenance. *Nature*, *479*, 189–193.

Wang, X., Kruithof-de Julio, M., Economides, K. D., Walker, D., Yu, H., Halili, M. V., et al. (2009). A luminal epithelial stem cell that is a cell of origin for prostate cancer. *Nature*, *461*, 495–500.

Wang, Z. A., Mitrofanova, A., Bergren, S. K., Abate-Shen, C., Cardiff, R. D., Califano, A., et al. (2013). Lineage analysis of basal epithelial cells reveals their unexpected plasticity and supports a cell-of-origin model for prostate cancer heterogeneity. *Nature Cell Biology*, *15*, 274–283.

Wang, Y., Schnegelsberg, P., Dausman, J., & Jaenisch, R. (1996). Functional redundancy of the muscle-specific transcription factors Myf5 and myogenin. *Nature*, *379*, 823–825.

Wingate, R. J., & Hatten, M. E. (1999). The role of the rhombic lip in avian cerebellum development. *Development*, *126*, 4395–4404.

Xu, H. T., Han, Z., Gao, P., He, S., Li, Z., Shi, W., et al. (2014). Distinct lineage-dependent structural and functional organization of the hippocampus. *Cell*, *157*, 1552–1564.

Yook, C., Druckmann, S., & Kim, J. (2013). Mapping mammalian synaptic connectivity. *Cellular and Molecular Life Sciences*, *70*, 4747–4757.

Yu, Y. C., Bultje, R. S., Wang, X., & Shi, S. H. (2009). Specific synapses develop preferentially among sister excitatory neurons in the neocortex. *Nature*, *458*, 501–504.

Yu, Y. C., He, S., Chen, S., Fu, Y., Brown, K. N., Yao, X. H., et al. (2012). Preferential electrical coupling regulates neocortical lineage-dependent microcircuit assembly. *Nature*, *486*, 113–117.

Zong, H. (2014). Generation and applications of MADM-based mouse genetic mosaic system. *Methods in Molecular Biology*, *1194*, 187–201.

Zong, H., Espinosa, J. S., Su, H. H., Muzumdar, M. D., & Luo, L. (2005). Mosaic analysis with double markers in mice. *Cell*, *121*, 479–492.

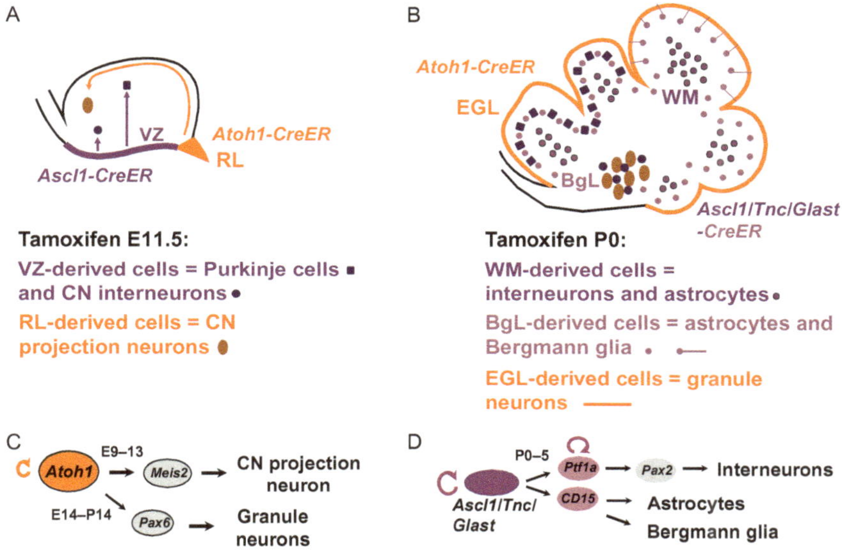

Alexandra L. Joyner, Figure 1 Genetic inducible fate-mapping studies define distinct progenitor pools in the cerebellum separated spatially and/or in time. (A and B) Schematic drawings of midsagittal sections of cerebella at E11.5 and P0 are shown. The cells that can be labeled using a reporter allele such as *R26*^{Yfp} and an *Atoh1-CreER* transgene or *Ascl1*^{CreER} knock-in allele (*Ascl1-CreER*) are indicated by orange and purple, respectively, and the cells they give rise to when tamoxifen is given at E11.5 (A) or (P0) are indicated below. In (B), Purkinje cells are only shown in the anterior lobules (folds) and glial fibers of Bergmann glial progenitors only in one posterior lobule for simplicity. At E12.5–17.5, the rhombic lip gives rise to the external granule cell layer (orange) shown in (B) made up of *Atoh1*-expressing progenitors that can divide until P16. Three knock-in alleles have been used to mark progenitors in the WM and BgL (*Ascl1*^{CreER}, *Tnc*^{CreER}, *Glast*^{CreER}). Round arrows indicate proliferating progenitors. VZ, ventricular zone; RL, rhombic lip; CN, cerebellar nuclei; EGL, external granule cell layer; BgL, Bergmann glial layer; WM, white matter. (C and D) The *Atoh1* and *Ascl1/Tnc/Glast* lineages are show. *Data are based on Fleming et al. (2013), Machold and Fishell (2005), Parmigiani et al. (2015), and Sudarov et al. (2011).*

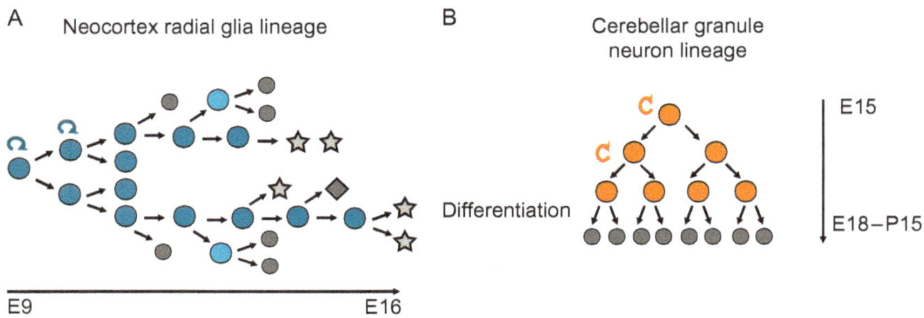

Alexandra L. Joyner, Figure 2 Clonal analyses of glutamatergic progenitors in the forebrain (neocortex) and cerebellum reveal different modes of generating neurons and glia. (A) Radial glial progenitors (blue circles) in the ventricular zone of the neocortex initially divide symmetrically to expand the progenitors and then divide mainly asymmetrically to produce a progenitor and a differentiated projection neuron (gray circle), astrocyte (star), or oligodendrocyte (diamond), or a progenitor with more limited proliferation capacity (light blue) that gives rise often to two neurons. Some symmetric divisions produce two neurons or astrocytes (differentiation divisions). Round arrows indicate symmetrical divisions producing two progenitors. Gliogenesis (production of astrocytes and oligodendrocytes) occurs after neurogenesis. (B) In the cerebellum, granule cell progenitors in the external granule cell layer (EGL in Fig. 1B) divide only symmetrically to expand the number of progenitors and then to produce postmitotic granule neurons. *Data are based on Espinosa and Luo (2008), Gao et al. (2014), and Legue et al. (2015).*

CHAPTER TWENTY-NINE

Tales of Tails (and Trunks): Forming the Posterior Body in Vertebrate Embryos

David Kimelman[1]
Department of Biochemistry, University of Washington, Seattle, Washington, USA
[1]Corresponding author: e-mail address: kimelman@uw.edu

Contents

Abstract

A major question in developmental biology is how the early embryonic axes are established. Recent studies using different model organisms and mammalian *in vitro* systems have revealed the surprising result that most of the early posterior embryonic body forms from a Wnt-regulated bipotential neuromesodermal progenitor population that escapes early germ layer patterning. Part of the regulatory network that drives the maintenance and differentiation of these progenitors has recently been determined, but much remains to be discovered. This review discusses some of the common features present in all vertebrates, as well as unique aspects that different species utilize to establish their anterior–posterior (A–P) axis.

Current Topics in Developmental Biology, Volume 116
ISSN 0070-2153
http://dx.doi.org/10.1016/bs.ctdb.2015.12.008

1. FORMATION OF THE POSTERIOR BODY

In all vertebrates, the posterior body comprises everything except the head and is subdivided into two major domains: from the head to the anus or cloaca is the trunk and from the anus or cloaca to the most posterior end is the tail (Fig. 1B). The tissues present in both regions of the embryonic posterior body include the somites, which produce musculature, dermis, bone, and cartilage; the neural tube; the notochord; and the vasculature (Fig. 1A), whereas the trunk region includes all of the aforementioned tissues as well as the gut, which terminates at the anus, and the pronephros, which ends at the cloaca. During early development, the posterior body grows progressively from the anterior to posterior, which is seen most obviously with the somites that form at regular intervals starting with the most anterior somite, just posterior to the head, and ending with the most posterior somite (Benazeraf & Pourquie, 2013; Holley, 2007). Although this period when the axis of the embryo elongates and the anterior–posterior (A–P) axis of the vertebrate embryo is completely established has different names in different model systems, here it will be referred to as the somitogenesis stages. This progressive mode of growth from anterior to posterior is observed in a broad range of species, and while not used in long germ-band insects such as the widely

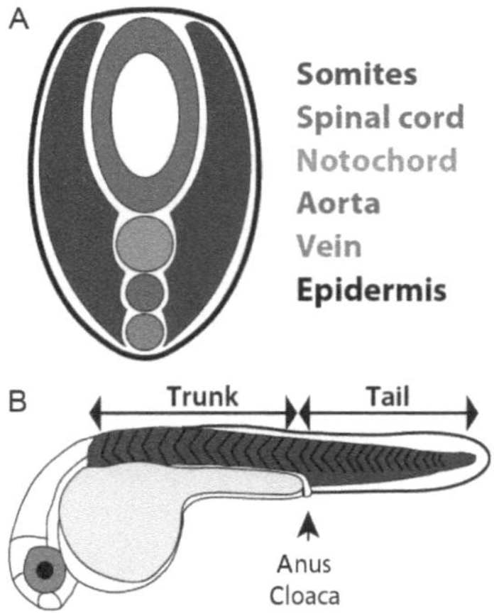

Figure 1 Morphological features of the vertebrate posterior body. (A) A generic cross section through the postanal region of a vertebrate embryo. In the preanal region, the pronephric tubules and gut tube would also be present. (B) The posterior of the embryo is divided into preanal trunk and postanal tail. Shown is a zebrafish embryo but the same division applies to all vertebrates. (A color version of this figure is located at the end of this chapter.)

studied *Drosophila*, is found in the less well studied but more ubiquitous short germ-band insects and may represent the ancestral mode of posterior body development (Martin & Kimelman, 2009).

The formation of the posterior body has fascinated scientists for well over a century, and consequently, there is an enormous body of work investigating this issue, yet as recent studies have shown, there is much more to learn and many new tools from embryonic stem (ES) cells to improved transgenic technology are now available to advance our understanding. However, because of the different morphologies of various vertebrate embryos, as well as the types of studies performed in each of the common model organisms, it is often difficult to piece together shared features from the literature. In this review, I will focus on the mechanisms that regulate differentiation of the early embryonic progenitors during A–P axis formation, emphasizing features that unite the vertebrates from fish to humans while also pointing out interesting species-specific differences that reflect evolutionary changes.

2. DIFFERING MORPHOLOGICAL FEATURES VERSUS A COMMON MECHANISM

Until relatively recently, one of the major ways to study embryos was to focus on morphological features visible under white light, but these structures often represent physical adaptations used by different species of embryos and are less revealing of the underlying mechanism than is often ascribed to them. For example, much of the literature refers to the tailbud, a group of cells located at the most posterior end of the growing embryonic body during the postgastrula stages. However, despite the name, the tailbud does not necessarily produce the true tail. For example, the zebrafish tailbud forms just as gastrulation ends and contributes cells to both the trunk and the tail (Kanki & Ho, 1997), whereas the chick tailbud forms later and does represent the cells forming the tail (Criley, 1969). Other visibly defined features in posterior growth include a postgastrula region of amniote embryos (e.g., chick and mouse) comprising the caudal end of the node and the rostral end of the primitive streak called the node-streak border (Wilson, Olivera-Martinez, & Storey, 2009), but this feature is not found in nonamniotes (e.g., fish and frog). Similarly, cells of the chordoneural hinge play an essential role in embryonic development in many species (Wilson et al., 2009), but this structure does not exist in fish which have the equivalent cells located in a somewhat different region (Martin & Kimelman, 2012). Thus, while these different and species-specific features provide useful landmarks for investigators, recent evidence discussed below reveals that all vertebrate embryos use a common mechanism to form the posterior body.

3. THE NEUROMESODERMAL PROGENITOR

While it is clear that cells at the very posterior end of the extending body contribute to a variety of cell types (Fig. 1A), a long-standing and major question has been whether these derive from a pluripotent population or from a set of progenitors with very restricted fates. As with many issues in biology, the answer lies somewhere in the middle. The general view has been that germ layer patterning occurs very early in development, and one of the major accomplishments in developmental biology over the past 30 years has been defining the many signals and pathways that establish these fates (Kimelman, 2006; Ozair, Kintner, & Brivanlou, 2013; Zorn & Wells, 2009). However, recent work has clearly shown that progenitors of the posterior body remain in a multipotent state during the somitogenesis stages. A cell lineage study in frogs intriguingly suggested that the cells at the most posterior end of the embryo could contribute to multiple tissues such as muscle, notochord, neural, and epidermis, but this study was limited by an inability of the investigators to label single cells (Davis & Kirschner, 2000). However, a lineage study using transgenic mouse embryos definitively showed that many cells of the embryonic posterior body remain in a bipotential neuromesodermal state late into development that produces cells contributing to the somites and spinal cord, whereas endoderm, epidermis, and cranial and cardiac mesoderm segregate early as independent lineages (Fig. 2; Tzouanacou et al., 2009). The existence of this neuromesodermal population has also been found in fish, demonstrating that this is a conserved feature of vertebrate embryonic development (Martin & Kimelman, 2012). Very recent studies have shown that it is possible to produce the neuromesodermal progenitors (NMPs) *in vitro* using mouse and human stem cells (Gouti et al., 2014; Tsakiridis et al., 2014; Turner et al., 2014), opening up the possibility of detailed genomic analysis of the neuromesodermal state.

Of the other tissues in the posterior body, the data are still incomplete. Some of the tail vasculature come from the posterior progenitor population, at least in fish (Martin & Kimelman, 2012), but most of the vasculature originates separately in the embryo (Saha, Cox, & Sipe, 2004). The origin of the notochord is also uncertain (Tzouanacou et al., 2009), although a recent study in mouse indicates that the NMPs and notochord are separate

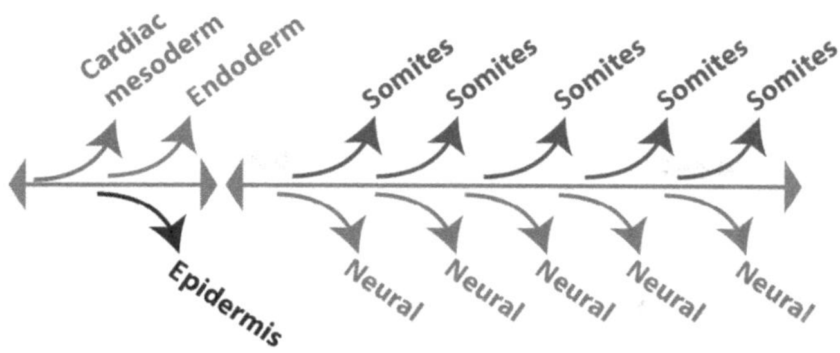

Figure 2 Maintenance of the neuromesodermal progenitors. Lineage labeling studies in the mouse show that while many germ layer decisions are made early, a neuromesodermal progenitor population continues to produce both neural and mesodermal cells during the somitogenesis stages (Tzouanacou, Wegener, Wymeersch, Wilson, & Nicolas, 2009). The times at which the gastrula stage fate decisions are made are approximate. (A color version of this figure is located at the end of this chapter.)

populations (Garriock et al., 2015). In the nonamniotes, the notochord originates from the dorsal marginal zone, separate from the lateral and ventral marginal zones where most of the posterior body originates (Dale & Slack, 1987; Kimelman & Martin, 2012; Warga & Nusslein-Volhard, 1999), and these different cells are only brought near each other by the mechanics of gastrulation, consistent with an ancestral separation of notochord and NMP fates. With the recent availability of *in vitro* models (Gouti, Metzis, & Briscoe, 2015), the exact nature of multipotency in the mammalian NMPs should now be testable.

4. STEM OR PROGENITOR CELLS?

While it is now clear that there are multipotent progenitor cells that form the posterior body, an important question is whether or not these progenitor cells are also stem cells, with self-renewing capacity. In amniotes, the evidence is increasingly in favor of a stem cell capacity (Nicolas, Mathis, Bonnerot, & Saurin, 1996; Selleck & Stern, 1992; Wilson et al., 2009). For example, the chordoneural hinge regions of mouse and chick embryos could be transplanted into the primitive streak region of earlier host embryos through multiple passages and they continued to contribute cells to the somites, neural tube, and notochord (Cambray & Wilson, 2002; McGrew et al., 2008; Tam & Tan, 1992). In nonamniotes, the issue is not resolved since these types of transplantation experiments have thus far not been technically feasible. Since it is not yet clear whether lower vertebrates have self-renewing capacity, at this point, it is safest to say that all vertebrates have a population of posterior progenitors that at least in amniotes has stem-like self-renewing capabilities.

In amniotes, the progenitor cells that contribute to the medial and lateral regions of the somites have separate origins. Cells destined for the medial region of somites (closer to the midline) have stem cell properties based on lineage studies, whereas the lateral somite cells have a much more restricted developmental potential (Freitas, Rodrigues, Charrier, Teillet, & Palmeirim, 2001; Iimura, Yang, Weijer, & Pourquie, 2007; Selleck & Stern, 1991). What regulates these two different modes of development is unclear, but potentially tied to signaling events occurring during gastrulation. It has been suggested that this division of the somitic mesodermal precursors is preserved in the lower vertebrates (Iimura et al., 2007) since zebrafish have a discrete population of medial muscle cells that form next to the notochord (the adaxial cells; Stickney, Barresi, & Devoto, 2000). However, there is no evidence that the zebrafish adaxial cells, which form a small amount of slow muscle in the embryo unlike the bulk of the somites that form fast muscle, have stem cell properties. Moreover, the adaxial progenitors have little proliferative capacity compared to the fast muscle progenitors (Bouldin, Snelson, Farr, & Kimelman, 2014; Hirsinger, Stellabotte, Devoto, & Westerfield, 2004), and there is as of yet no evidence that they have both neural and mesodermal potential. Thus, it may be that the common vertebrate ancestor had NMPs with a limited proliferative potential that formed all of the somites and that amniotes developed both NMPs with stem-like self-renewing capacity and non-NMP-derived cells to form the lateral parts of the somites.

5. BRACHYURY AND THE NMP NICHE

The *brachyury* mouse mutant was discovered almost 90 years ago as recessive gene causing a short-tail phenotype in heterozygotes (*brachyury* is Greek for short tail; Dobrovolskaia-Zavadskaia, 1927), and a lack of notochord and an almost complete absence of somitic mesoderm in homozygotes (Chesley, 1935; Gluecksohn-Schoenheimer, 1944). Since this gene has different names in different species (e.g., *T* in the mouse), I will use the name *brachyury* to refer to it in all vertebrates. In zebrafish, a *brachyury* mutant was discovered that produced homozygous embryos with a normal head and trunk but lacking a tail, suggesting that it has a less important role in lower vertebrates (Halpern, Ho, Walker, & Kimmel, 1993; Schulte-Merker, van Eeden, Halpern, Kimmel, & Nusslein-Volhard, 1994). Later studies revealed that fish have a second *brachyury* gene and that embryos unable to express both genes recapitulate the mouse mutant phenotype, demonstrating a conserved and crucial ancestral role for this gene among the vertebrates (Martin & Kimelman, 2008).

Brachyury is the founding member of the T-box transcription factor family (Herrmann, Labeit, Poustka, King, & Lehrach, 1990) and is coexpressed with *sox2* in the NMPs (Fig. 3; Garriock et al., 2015; Martin & Kimelman, 2012; Olivera-Martinez, Harada, Halley, & Storey, 2012; Tsakiridis et al., 2014), allowing the NMPs to adopt either a mesodermal fate (*brachyury+/sox2−*) or a neural fate (*brachyury−/sox2+*) when they begin to differentiate. Despite *brachyury* being essential for formation of the notochord and almost all somite mesoderm, a cell transplantation study revealed the surprising result that individual cells lacking *brachyury* function are able to

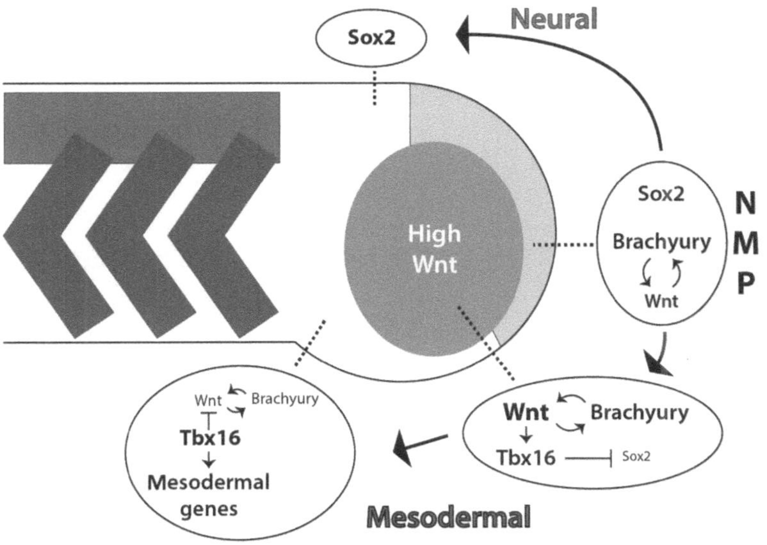

Figure 3 Model for regulation of mesoderm formation in zebrafish. Sox2-expressing NMP cells are in a zone of moderate Wnt signaling (light orange) that maintains Brachyury. As cells leave this region, if they are not exposed to continue Wnt signaling, they express Sox2 but not Brachyury and become neural. Instead, if they are exposed to high levels of Wnt (dark orange), they express Tbx16, which represses Sox2. Tbx16 also represses Wnts so as cells continue to move anteriorly (left side), they shut off Brachyury and turn on mesodermal differentiation genes, thus locking in the mesodermal choice. (A color version of this figure is located at the end of this chapter.)

normally differentiate into muscle cells (Martin & Kimelman, 2008), although they are not able to differentiate into notochord (Halpern et al., 1997). These results revealed that the major function of *brachyury* is noncell autonomous, and led to the idea that *brachyury* is essential in the whole embryo for forming the somites because it creates the niche that allows the NMPs to remain as progenitors (Martin & Kimelman, 2010). That key targets of Brachyury are specific Wnt ligands (Martin & Kimelman, 2008), together with previous results showing that *brachyury* is regulated by Wnt (Arnold et al., 2000; Vonica & Gumbiner, 2002; Yamaguchi, Takada, Yoshikawa, Wu, & McMahon, 1999), demonstrates that Brachyury and Wnt participate in an autoregulatory loop required to maintain the NMPs (Fig. 3).

A second important function of Brachyury is to induce expression of the retinoic acid (RA)-degrading enzyme Cyp26a in order to prevent RA, which is produced within the somites, from spreading posteriorly and thus disrupting the Brachyury–Wnt autoregulatory loop within the NMPs since RA inhibits *brachyury* expression (Iulianella, Beckett, Petkovich, & Lohnes, 1999; Martin & Kimelman, 2010; Sakai et al., 2001). Although Cyp26a would not be typically thought of as a noncell autonomous factor since it is not secreted, the NMP cells must work as a community to degrade RA, preventing it from altering the fate of the NMPs during the stages of axis elongation.

6. WNT SIGNALING AND THE NEUROMESODERMAL FATE CHOICE

How do the NMP cells decide which fate to choose between neural and mesodermal? Evidence from fish to ES cells now strongly supports the view that Wnt is a critical regulator of this decision. In fish embryos, single NMP cells with Wnt signaling blocked become neural, whereas NMP cells expressing a constitutively active form of the downstream Wnt effector β-catenin (ca-β-catenin) are directed to the mesodermal fate (Martin & Kimelman, 2012). Similar results have been observed in mammalian ES cells (Gouti et al., 2015, 2014). Moreover, mice lacking either the key posterior Wnt, Wnt3a, β-catenin, or the Wnt-responsive transcription factors Tcf1 and Lef1, show an expansion of neural tissue and a severe deficit in mesoderm (Cunningham, Kumar, Yamaguchi, & Duester, 2015; Dunty et al., 2008; Galceran, Farinas, Depew, Clevers, & Grosschedl, 1999; Garriock et al., 2015; Nowotschin, Ferrer-Vaquer,

Concepcion, Papaioannou, & Hadjantonakis, 2012; Takada et al., 1994; Yoshikawa, Fujimori, McMahon, & Takada, 1997), all of which support a key role for Wnt signaling in suppressing neural fate and activating the mesodermal fate. Recent gain-of-function studies in mouse are, however, somewhat contradictory. One study showed that overexpression of Wnt3a inhibits neural tissue formation but does not promote mesoderm (Jurberg, Aires, Novoa, Rowland, & Mallo, 2014), while another showed that overexpression of ca-β-catenin promotes mesoderm but does not fully block neural fates (Garriock et al., 2015). These discrepancies may be due to the domain-specific expression levels of the different promoters used in the two studies (Jurberg et al., 2014), as well as the nature of the Wnt pathway activators since ca-β-catenin is not subject to the types of feedback control that the Wnt ligand is (MacDonald, Tamai, & He, 2009). While it remains possible that there are some differences among the vertebrates in precisely how signaling controls differentiation of the NMPs, thus far the preponderance of the evidence indicates that as the posterior body extends, NMPs are making neural or mesodermal choices depending on their exposure to Wnt signaling. We have proposed that this system allows the embryo to allocate cells to either the mesodermal or the neural lineages as the body extends by dynamically regulating Wnt levels in the most posterior region of the embryo, ensuring that at every axial position, the embryo produces the correct ratio of spinal cord to somite-derived mesoderm (Martin & Kimelman, 2010).

7. WNT SIGNALING AND ESTABLISHMENT OF THE NMPs

Wnt signaling has an additional important role in the establishment of the NMPs, being essential for their formation *in vivo* (Garriock et al., 2015) and for their formation *in vitro* together with Fgf (Gouti et al., 2014; Tsakiridis et al., 2014; Turner et al., 2014). How is Wnt able to both establish the NMPs and direct NMPs toward a mesodermal fate? In zebrafish, we have found that Wnt signaling is present in the NMP-containing progenitor zone, and then increased in the cells that have just begun the commitment to mesoderm (Bouldin et al., 2015). Based on this and other data, we propose that NMP cells are maintained by an intermediate level of Wnt, and then if they see elevated levels of Wnt they commit to mesoderm, whereas if they see decreased Wnt they adopt the neural fate (Fig. 3). Intriguingly, mathematical modeling and experiments in *Xenopus* have shown that cells respond to fold changes in Wnt signaling not absolute levels (Goentoro & Kirschner,

2009), which would allow cells to change fate as they move between regions with different levels of Wnt signaling even if the absolute levels of Wnt signaling changes as the axis elongates. The new *in vitro* models of mammalian NMPs (Gouti et al., 2015) will provide an excellent system to determine if activation of the mesodermal program in mammalian embryos is similarly regulated by fold changes in the level of Wnt pathway activation.

8. TBX6/16 AND THE ACQUISITION OF MESODERMAL FATE

If cells leave the progenitor zone and are no longer exposed to Wnt signaling, they maintain *sox2* expression and commit to the neural fate, which has been well covered in an excellent recent review and is therefore not discussed here (Gouti et al., 2015). In contrast, if cells leaving the progenitor zone are exposed to high Wnt signaling, they adopt the mesodermal fate by activating expression of particular members of the *tbx6/16* family, which all derive from a common ancestral gene (Ahn, You, & Kim, 2012), but which have varying roles depending on the species. For example, mouse embryos lacking *tbx6* or zebrafish embryos lacking *tbx16* have a severe deficit in somite-derived mesoderm, and an enlarged tailbud that contains the cells that normally would contribute to the somites (Chapman, Cooper-Morgan, Harrelson, & Papaioannou, 2003; Chapman & Papaioannou, 1998; Griffin, Amacher, Kimmel, & Kimelman, 1998; Kimmel, Kane, Walker, Warga, & Rothman, 1989). Remarkably, mouse *tbx6* mutants form ectopic neural tubes (Chapman & Papaioannou, 1998), and both mouse *tbx6* mutants and zebrafish *tbx16* mutants upregulate the neural marker *sox2*, demonstrating that cells leaving the NMPs adopt a neural fate at the expense of mesoderm when *tbx6/16* gene function is absent (Fig. 3; Bouldin et al., 2015; Takemoto et al., 2011).

In addition to being necessary for the activation of genes involved in somite mesoderm formation (Bouldin et al., 2015; Chapman & Papaioannou, 1998; Griffin & Kimelman, 2002; Takemoto et al., 2011), Tbx6/16 directly activates expression of the *delta* and *mesp* genes that regulate formation of the boundaries between somites (Garnett et al., 2009; Jahangiri, Nelson, & Wardle, 2012; White & Chapman, 2005; Yasuhiko et al., 2008). In fish, Tbx16 acts as a dual function activator/repressor, and through its repressive function is able to inhibit *sox2*, thus blocking the neural fate (Fig. 3; Bouldin et al., 2015). In mouse, detailed analysis has identified a small fragment of the *sox2* enhancer (enhancer N1) that mediates Sox2 repression within the newly forming mesoderm downstream of Tbx6 (Takemoto, Uchikawa, Kamachi, & Kondoh, 2006; Takemoto et al., 2011). *In vitro* analysis indicates that Tbx6 does not bind *sox2* enhancer N1, and thus in mouse, the repression of *sox2* may be indirect (Kondoh & Takemoto, 2012; Takemoto et al., 2011), although the function of Tbx6 *in vivo* and on the whole *sox2* promoter has thus far not yet been studied. However, zebrafish does not obviously have the N1 enhancer element (Okuda et al., 2006), and so it is possible that the mechanism of *sox2* regulation by Tbx6/16 has diverged over time.

Tbx6/16 turn off the progenitor fate by also inhibiting *brachyury* expression (Chapman & Papaioannou, 1998; Griffin & Kimelman, 2002), although in this case the repression is indirect, mediated by inhibition of *wnt* gene expression (Bouldin et al., 2015; Kondoh & Takemoto, 2012; Takemoto et al., 2011). Thus, as cells adopt the mesodermal fate, Tbx6/16 locks them into the mesodermal fate by turning off the neural fate and progenitor state (Fig. 3). Why is this important? Wnt signaling is confined to the most posterior end of the embryo, which is clearly essential because forced expansion of Wnt signaling anteriorly causes serious defects in mesodermal differentiation, somite formation, and A/P patterning of the somites (Aulehla et al., 2008; Dunty et al., 2008; Garriock et al., 2015; Jurberg et al., 2014). Thus, while Wnt signaling directly turns on *tbx16* in the most posterior end of the fish embryo (Bouldin et al., 2015), and potentially also *tbx6* in the mouse (Dunty et al., 2008), cells rapidly leave the Wnt zone as they move anteriorly. These cells therefore could potentially revert to a progenitor state or switch to becoming neural if they did not continue to experience high Wnt levels. However, using the regulatory logic shown in Fig. 3, once the cells commit to mesoderm, they will not adopt another fate.

9. MESOGENIN1 AND MESODERMAL FATE

In addition to *tbx6/16*, another important mesodermal regulator activated by Wnt in the newly forming mesoderm is *mesogenin1* (*msgn1*; Nowotschin et al., 2012; Wittler et al., 2007; Yoon & Wold, 2000). In the mouse, the *msgn1* mutant phenotype is similar to *tbx6* mutants with an enlarged tailbud and a loss of all but the anterior somites (Yoon & Wold, 2000). Interestingly, Msgn1 is at least partially necessary for *tbx6* expression, and Tbx6 is at least partially necessary for *msgn1* expression, demonstrating that these factors act in an autoregulatory loop and that both are necessary for mesoderm formation (Fig. 4; Chalamalasetty et al., 2014;

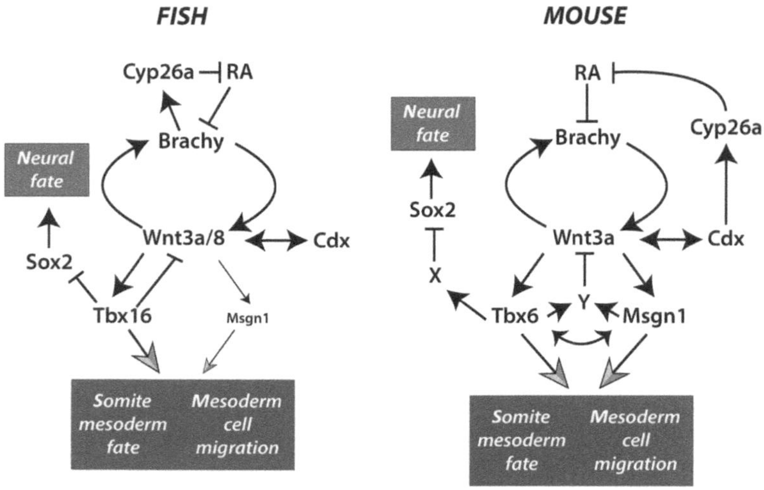

Figure 4 Comparison of the regulatory logic of fish and mouse posterior cell fate choices. A schematic showing the regulatory relationships among signaling and transcription factors discussed in this review. Tbx6 is shown here to cause repressive effects indirectly, but as with Tbx16, it may turn out to be able to directly repress target genes. Fgf is omitted for simplicity. (A color version of this figure is located at the end of this chapter.)

Nowotschin et al., 2012; Wittler et al., 2007). Msgn1 overexpression and loss-of-function studies show that it is unable to suppress neural fates even as it promotes mesoderm formation from the NMPs (Chalamalasetty et al., 2014; Nowotschin et al., 2012), although it does act to inhibit *wnt3a* expression, presumably indirectly (Nowotschin et al., 2012). Thus, it has both similarities and differences to Tbx6 (Fig. 4).

In zebrafish, the regulatory logic is somewhat different since loss of *msgn1* has a quite mild phenotype causing an average 15% reduction in the size of the somites (Fior et al., 2012; Yabe & Takada, 2012). While *tbx16* mutants have a complete lack of trunk somites, they do produce small tail somites, whereas a loss of *tbx16* and *msgn1* causes a complete absence of all somites (Fior et al., 2012; Yabe & Takada, 2012). This demonstrates that *msgn1* in fish, as in mouse, can control mesoderm differentiation. The reason for the trunk/tail difference is that Tbx16 is necessary for *msgn1* expression during the trunk forming stages but not during the tail stages of somitogenesis (Griffin & Kimelman, 2002). Thus, unlike in mouse, Msgn1 has a subsidiary role to the Tbx6/16 factors, which is most clearly revealed when Tbx16 is not present (Fig. 4). The role of *msgn1* in other vertebrates remains to be determined.

10. CDX CONTRIBUTES TO POSTERIOR DEVELOPMENT

The *Caudal*-related genes (*Cdx* in vertebrates) also play essential roles in axial elongation. In mouse, three *Cdx* genes are likely to be involved directly in this process through the direct regulation of *Wnt3a*, *Brachyury*, and *Cyp26a1* (Savory et al., 2009; Young et al., 2009), each of which promotes the NMP state (Fig. 4). At least two of the *Cdx* genes are also direct targets of Wnt signaling (Lickert et al., 2000; Pilon et al., 2006), and so the Cdx factors operate in an autoregulatory loop that parallels the one between Wnt and Brachyury. Intriguingly, whereas $Cdx2^{+/-};Cdx4^{-/-}$ embryos have strong axis extension defects, individual mutant cells with this genotype behave normally in a wild-type background, indicating that a major role for the Cdx factors is to noncell autonomously regulate axis extension (Bialecka, Wilson, & Deschamps, 2010), which is very reminiscent of the essential noncell autonomous role for Brachyury, discussed above.

In zebrafish, knockdown of the two *cdx* genes expressed during the gastrula and somitogenesis stages causes strong posterior defects, similar to those observed when Wnt signaling is inhibited (Davidson & Zon, 2006; Shimizu, Bae, Muraoka, & Hibi, 2005; Thorpe, Weidinger, & Moon, 2005). Similar results are observed in *Xenopus*, where experiments indicate that the *Cdx* genes, as in mouse, act in a positive feedback loop with Wnt signaling (Faas & Isaacs, 2009). Thus, the evidence indicates that in all vertebrates the *Cdx* genes play a key role in preserving the signaling environment necessary for maintenance of the NMPs.

11. TERMINATION OF BODY ELONGATION

The posterior body will continue to form somites and spinal cord as long as there are NMPs to produce these tissues. However, at some point, the progenitors are exhausted and the extension of the body terminates, completing formation of the embryonic A–P axis. In zebrafish, the NMPs are proliferative during the gastrula stage but then become quiescent during the somitogenesis stages (Bouldin et al., 2014), so termination of the axis may simply occur when the last of the NMPs begins differentiation. Evidence in chick indicates that as the embryo reaches the full extent of axis elongation, *brachyury* expression disappears from the NMPs, leading them to switch to a neural fate (Olivera-Martinez et al., 2012). The cause of *brachyury* downregulation was attributed to an increase of RA at the posterior end of the embryo during the late stages of axis elongation, which in turn causes inhibition of Fgf signaling in the posterior end of the embryo, and consequent *brachyury* downregulation. However, in fish and mouse, blocking RA synthesis does not affect termination of axis formation (Begemann, Schilling, Rauch, Geisler, & Ingham, 2001; Cunningham, Zhao, & Duester, 2011) so this mechanism may be specific to only some species.

In addition, a recent study in chick has shown that a subset of posterior *Hox* genes can inhibit Wnt (and Fgf) signaling, with the most posterior *Hox* genes having the greatest strength of repression (Denans, Iimura, & Pourquie, 2015). Since Wnt and Fgf promote *brachyury* expression and since Brachyury promotes expression of the RA-degrading enzyme Cyp26a (Martin & Kimelman, 2010), the most posterior *Hox* genes in chick could cause a rise in RA in the NMPs, and thus the two models for axis termination in chick could be connected (Fig. 5; Denans et al., 2015; Olivera-Martinez et al., 2012). Intriguingly, *Hoxb13* mutant mice have supernumerary vertebrae and an overgrowth of the posterior spinal cord, although this was attributed to increased proliferation and decreased apoptosis (Economides, Zeltser, & Capecchi, 2003). It will be interesting to examine the signaling and transcription changes during the stages of axis termination in these mice in light of the recent studies in chick.

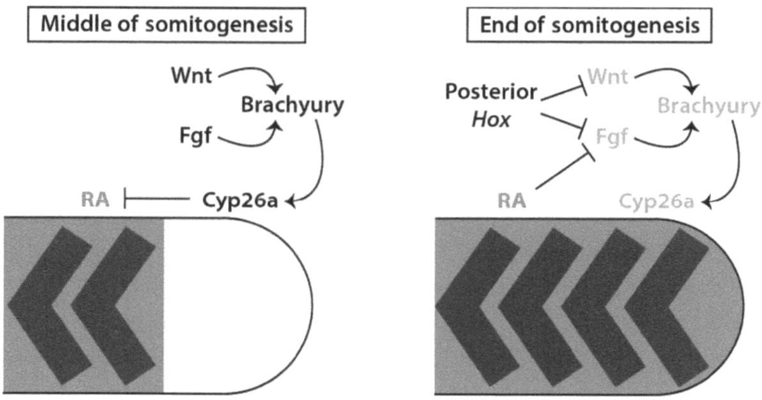

Figure 5 Model for termination of axis extension in chick. During somitogenesis, the progenitor domain excludes RA. Near the end of somitogenesis, the most posterior *Hox* genes, possibly together with other molecular changes, causes RA to expand posteriorly, thus inhibiting the progenitor cell fate. (A color version of this figure is located at the end of this chapter.)

12. CONCLUSIONS AND FUTURE DIRECTIONS

The discovery of the bipotential NMPs has opened up a major new way of understanding how the vertebrate posterior axis elongates with the discovery that these cells escape the early mechanisms of germ layer patterning. The use of several model systems has been instrumental in elucidating common features of posterior growth in the vertebrates, demonstrating similar mechanisms operating from fish to humans. With the recent development of mammalian *in vitro* systems as well as constant improvements in *in vivo* approaches, it will now be possible to make rapid progress in understanding both common features and species-specific differences in the regulation of these progenitors as they make their different cell fate decisions. But these decisions must also be coupled with precise cell movements to actually form the posterior body, and this is an exciting area that is being aided with improving optics and methodology (Benazeraf & Pourquie, 2013; Lawton et al., 2013; Uriu, Morelli, & Oates, 2014). Combining an understanding of intercellular signaling, genomic regulation, and morphogenesis is certain to bring exciting new discoveries to the long-standing question of how the vertebrate embryonic body forms.

ACKNOWLEDGMENTS

I thank Cort Bouldin and Alyssa Manning for critical comments on the manuscript, and Ben Martin and Val Wilson for providing valuable insight. D.K. is supported by a NIH grant (GM079203).

REFERENCES

Ahn, D., You, K. H., & Kim, C. H. (2012). Evolution of the *tbx6/16* subfamily genes in vertebrates: Insights from zebrafish. *Molecular Biology and Evolution, 29*, 3959–3983.

Arnold, S. J., Stappert, J., Bauer, A., Kispert, A., Herrmann, B. G., & Kemler, R. (2000). *Brachyury* is a target gene of the Wnt/beta-catenin signaling pathway. *Mechanisms of Development, 91*, 249–258.

Aulehla, A., Wiegraebe, W., Baubet, V., Wahl, M. B., Deng, C., Taketo, M., et al. (2008). A beta-catenin gradient links the clock and wavefront systems in mouse embryo segmentation. *Nature Cell Biology, 10*, 186–193.

Begemann, G., Schilling, T. F., Rauch, G. J., Geisler, R., & Ingham, P. W. (2001). The zebrafish *neckless* mutation reveals a requirement for *raldh2* in mesodermal signals that pattern the hindbrain. *Development, 128*, 3081–3094.

Benazeraf, B., & Pourquie, O. (2013). Formation and segmentation of the vertebrate body axis. *Annual Review of Cell and Developmental Biology, 29*, 1–26.

Bialecka, M., Wilson, V., & Deschamps, J. (2010). *Cdx* mutant axial progenitor cells are rescued by grafting to a wild type environment. *Developmental Biology, 347*, 228–234.

Bouldin, C. M., Manning, A. J., Peng, Y.-H., Farr, G. H., Hung, K. L., Dong, A., et al. (2015). Wnt signaling and *tbx16* form a bistable switch to commit bipotential progenitors to mesoderm. *Development*, *142*, 2499–2507.

Bouldin, C. M., Snelson, C. D., Farr, G. H., 3rd., & Kimelman, D. (2014). Restricted expression of *cdc25a* in the tailbud is essential for formation of the zebrafish posterior body. *Genes & Development*, *28*, 384–395.

Cambray, N., & Wilson, V. (2002). Axial progenitors with extensive potency are localised to the mouse chordoneural hinge. *Development*, *129*, 4855–4866.

Chalamalasetty, R. B., Garriock, R. J., Dunty, W. C., Jr., Kennedy, M. W., Jailwala, P., Si, H., et al. (2014). Mesogenin 1 is a master regulator of paraxial presomitic mesoderm differentiation. *Development*, *141*, 4285–4297.

Chapman, D. L., Cooper-Morgan, A., Harrelson, Z., & Papaioannou, V. E. (2003). Critical role for *Tbx6* in mesoderm specification in the mouse embryo. *Mechanisms of Development*, *120*, 837–847.

Chapman, D. L., & Papaioannou, V. E. (1998). Three neural tubes in mouse embryos with mutations in the T-box gene *Tbx6*. *Nature*, *391*, 695–697.

Chesley, P. (1935). Development of the short-tailed mutant in the house mouse. *The Journal of Experimental Zoology*, *70*, 429–435.

Criley, B. B. (1969). Analysis of embryonic sources and mechanisms of development of posterior levels of chick neural tubes. *Journal of Morphology*, *128*, 465–501.

Cunningham, T. J., Kumar, S., Yamaguchi, T. P., & Duester, G. (2015). Wnt8a and Wnt3a cooperate in the axial stem cell niche to promote mammalian body axis extension. *Developmental Dynamics*, *244*, 797–807.

Cunningham, T. J., Zhao, X., & Duester, G. (2011). Uncoupling of retinoic acid signaling from tailbud development before termination of body axis extension. *Genesis*, *49*, 776–783.

Dale, L., & Slack, J. M. (1987). Fate map for the 32-cell stage of *Xenopus laevis*. *Development*, *99*, 527–551.

Davidson, A. J., & Zon, L. I. (2006). The *caudal*-related homeobox genes *cdx1a* and *cdx4* act redundantly to regulate *hox* gene expression and the formation of putative hematopoietic stem cells during zebrafish embryogenesis. *Developmental Biology*, *292*, 506–518.

Davis, R. L., & Kirschner, M. W. (2000). The fate of cells in the tailbud of *Xenopus laevis*. *Development*, *127*, 255–267.

Denans, N., Iimura, T., & Pourquie, O. (2015). *Hox* genes control vertebrate body elongation by collinear Wnt repression. *eLife*, *4*, e04379.

Dobrovolskaia-Zavadskaia, N. (1927). Sur la mortification spontanee de la queue chez la souris nouveau-nee et sur l'existence d'un caractere (facteur) hereditaire non-viable. *Critical Reviews of Social Biology*, *97*, 114–116.

Dunty, W. C., Jr., Biris, K. K., Chalamalasetty, R. B., Taketo, M. M., Lewandoski, M., & Yamaguchi, T. P. (2008). Wnt3a/beta-catenin signaling controls posterior body development by coordinating mesoderm formation and segmentation. *Development*, *135*, 85–94.

Economides, K. D., Zeltser, L., & Capecchi, M. R. (2003). *Hoxb13* mutations cause overgrowth of caudal spinal cord and tail vertebrae. *Developmental Biology*, *256*, 317–330.

Faas, L., & Isaacs, H. V. (2009). Overlapping functions of *Cdx1*, *Cdx2*, and *Cdx4* in the development of the amphibian *Xenopus tropicalis*. *Developmental Dynamics*, *238*, 835–852.

Fior, R., Maxwell, A. A., Ma, T. P., Vezzaro, A., Moens, C. B., Amacher, S. L., et al. (2012). The differentiation and movement of presomitic mesoderm progenitor cells are controlled by Mesogenin 1. *Development*, *139*, 4656–4665.

Freitas, C., Rodrigues, S., Charrier, J. B., Teillet, M. A., & Palmeirim, I. (2001). Evidence for medial/lateral specification and positional information within the presomitic mesoderm. *Development*, *128*, 5139–5147.

Galceran, J., Farinas, I., Depew, M. J., Clevers, H., & Grosschedl, R. (1999). *Wnt3a−/−-like phenotype and limb deficiency in Lef1(−/−)Tcf1(−/−) mice*. *Genes & Development, 13*, 709–717.

Garnett, A. T., Han, T. M., Gilchrist, M. J., Smith, J. C., Eisen, M. B., Wardle, F. C., et al. (2009). Identification of direct T-box target genes in the developing zebrafish mesoderm. *Development, 136*, 749–760.

Garriock, R. J., Chalamalasetty, R. B., Kennedy, M. W., Canizales, L. C., Lewandoski, M., & Yamaguchi, T. P. (2015). Lineage tracing of neuromesodermal progenitors reveals novel Wnt-dependent roles in trunk progenitor cell maintenance and differentiation. *Development, 142*, 1628–1638.

Gluecksohn-Schoenheimer, S. (1944). The development of normal and homozygous *Brachy* (*T/T*) mouse embryos in the extraembryonic coelom of the chick. *Proceedings of the National Academy of Sciences of the United States of America, 30*, 134–140.

Goentoro, L., & Kirschner, M. W. (2009). Evidence that fold-change, and not absolute level, of beta-catenin dictates Wnt signaling. *Molecular Cell, 36*, 872–884.

Gouti, M., Metzis, V., & Briscoe, J. (2015). The route to spinal cord cell types: A tale of signals and switches. *Trends in Genetics, 31*, 282–289.

Gouti, M., Tsakiridis, A., Wymeersch, F. J., Huang, Y., Kleinjung, J., Wilson, V., et al. (2014). In vitro generation of neuromesodermal progenitors reveals distinct roles for Wnt signalling in the specification of spinal cord and paraxial mesoderm identity. *PLoS Biology, 12*, e1001937.

Griffin, K. J., Amacher, S. L., Kimmel, C. B., & Kimelman, D. (1998). Molecular identification of *spadetail*: Regulation of zebrafish trunk and tail mesoderm formation by T-box genes. *Development, 125*, 3379–3388.

Griffin, K. J., & Kimelman, D. (2002). One-eyed pinhead and spadetail are essential for heart and somite formation. *Nature Cell Biology, 4*, 821–825.

Halpern, M. E., Hatta, K., Amacher, S. L., Talbot, W. S., Yan, Y. L., Thisse, B., et al. (1997). Genetic interactions in zebrafish midline development. *Developmental Biology, 187*, 154–170.

Halpern, M. E., Ho, R. K., Walker, C., & Kimmel, C. B. (1993). Induction of muscle pioneers and floor plate is distinguished by the zebrafish *no tail* mutation. *Cell, 75*, 99–111.

Herrmann, B. G., Labeit, S., Poustka, A., King, T. R., & Lehrach, H. (1990). Cloning of the *T* gene required in mesoderm formation in the mouse. *Nature, 343*, 617–622.

Hirsinger, E., Stellabotte, F., Devoto, S. H., & Westerfield, M. (2004). Hedgehog signaling is required for commitment but not initial induction of slow muscle precursors. *Developmental Biology, 275*, 143–157.

Holley, S. A. (2007). The genetics and embryology of zebrafish metamerism. *Developmental Dynamics, 236*, 1422–1449.

Iimura, T., Yang, X., Weijer, C. J., & Pourquie, O. (2007). Dual mode of paraxial mesoderm formation during chick gastrulation. *Proceedings of the National Academy of Sciences of the United States of America, 104*, 2744–2749.

Iulianella, A., Beckett, B., Petkovich, M., & Lohnes, D. (1999). A molecular basis for retinoic acid-induced axial truncation. *Developmental Biology, 205*, 33–48.

Jahangiri, L., Nelson, A. C., & Wardle, F. C. (2012). A cis-regulatory module upstream of *deltaC* regulated by Ntla and Tbx16 drives expression in the tailbud, presomitic mesoderm and somites. *Developmental Biology, 371*, 110–120.

Jurberg, A. D., Aires, R., Novoa, A., Rowland, J. E., & Mallo, M. (2014). Compartment-dependent activities of Wnt3a/beta-catenin signaling during vertebrate axial extension. *Developmental Biology, 394*, 253–263.

Kanki, J. P., & Ho, R. K. (1997). The development of the posterior body in zebrafish. *Development, 124*, 881–893.

Kimelman, D. (2006). Mesoderm induction: From caps to chips. *Nature Reviews. Genetics*, *7*, 360–372.
Kimelman, D., & Martin, B. L. (2012). Anterior–posterior patterning in early development: Three strategies. *WIREs Developmental Biology*, *1*, 253–266.
Kimmel, C. B., Kane, D. A., Walker, C., Warga, R. M., & Rothman, M. B. (1989). A mutation that changes cell movement and cell fate in the zebrafish embryo. *Nature*, *337*, 358–362.
Kondoh, H., & Takemoto, T. (2012). Axial stem cells deriving both posterior neural and mesodermal tissues during gastrulation. *Current Opinion in Genetics & Development*, *22*, 374–380.
Lawton, A. K., Nandi, A., Stulberg, M. J., Dray, N., Sneddon, M. W., Pontius, W., et al. (2013). Regulated tissue fluidity steers zebrafish body elongation. *Development*, *140*, 573–582.
Lickert, H., Domon, C., Huls, G., Wehrle, C., Duluc, I., Clevers, H., et al. (2000). Wnt/(beta)-catenin signaling regulates the expression of the homeobox gene *Cdx1* in embryonic intestine. *Development*, *127*, 3805–3813.
MacDonald, B. T., Tamai, K., & He, X. (2009). Wnt/beta-catenin signaling: Components, mechanisms, and diseases. *Developmental Cell*, *17*, 9–26.
Martin, B. L., & Kimelman, D. (2008). Regulation of canonical Wnt signaling by Brachyury is essential for posterior mesoderm formation. *Developmental Cell*, *15*, 121–133.
Martin, B. L., & Kimelman, D. (2009). Wnt signaling and the evolution of embryonic posterior development. *Current Biology*, *19*, R215–R219.
Martin, B. L., & Kimelman, D. (2010). Brachyury establishes the embryonic mesodermal progenitor niche. *Genes & Development*, *24*, 2778–2783.
Martin, B. L., & Kimelman, D. (2012). Canonical Wnt signaling dynamically controls multiple stem cell fate decisions during vertebrate body formation. *Developmental Cell*, *22*, 223–232.
McGrew, M. J., Sherman, A., Lillico, S. G., Ellard, F. M., Radcliffe, P. A., Gilhooley, H. J., et al. (2008). Localised axial progenitor cell populations in the avian tail bud are not committed to a posterior Hox identity. *Development*, *135*, 2289–2299.
Nicolas, J. F., Mathis, L., Bonnerot, C., & Saurin, W. (1996). Evidence in the mouse for self-renewing stem cells in the formation of a segmented longitudinal structure, the myotome. *Development*, *122*, 2933–2946.
Nowotschin, S., Ferrer-Vaquer, A., Concepcion, D., Papaioannou, V. E., & Hadjantonakis, A. K. (2012). Interaction of Wnt3a, Msgn1 and Tbx6 in neural versus paraxial mesoderm lineage commitment and paraxial mesoderm differentiation in the mouse embryo. *Developmental Biology*, *367*, 1–14.
Okuda, Y., Yoda, H., Uchikawa, M., Furutani-Seiki, M., Takeda, H., Kondoh, H., et al. (2006). Comparative genomic and expression analysis of group B1 *sox* genes in zebrafish indicates their diversification during vertebrate evolution. *Developmental Dynamics*, *235*, 811–825.
Olivera-Martinez, I., Harada, H., Halley, P. A., & Storey, K. G. (2012). Loss of FGF-dependent mesoderm identity and rise of endogenous retinoid signalling determine cessation of body axis elongation. *PLoS Biology*, *10*, e1001415.
Ozair, M. Z., Kintner, C., & Brivanlou, A. H. (2013). Neural induction and early patterning in vertebrates. *Wiley Interdisciplinary Reviews. Developmental Biology*, *2*, 479–498.
Pilon, N., Oh, K., Sylvestre, J. R., Bouchard, N., Savory, J., & Lohnes, D. (2006). *Cdx4* is a direct target of the canonical Wnt pathway. *Developmental Biology*, *289*, 55–63.
Saha, M. S., Cox, E. A., & Sipe, C. W. (2004). Mechanisms regulating the origins of the vertebrate vascular system. *Journal of Cellular Biochemistry*, *93*, 46–56.
Sakai, Y., Meno, C., Fujii, H., Nishino, J., Shiratori, H., Saijoh, Y., et al. (2001). The retinoic acid-inactivating enzyme CYP26 is essential for establishing an uneven distribution

of retinoic acid along the anterio–posterior axis within the mouse embryo. *Genes & Development, 15*, 213–225.

Savory, J. G., Bouchard, N., Pierre, V., Rijli, F. M., De Repentigny, Y., Kothary, R., et al. (2009). Cdx2 regulation of posterior development through non-Hox targets. *Development, 136*, 4099–4110.

Schulte-Merker, S., van Eeden, F. J., Halpern, M. E., Kimmel, C. B., & Nusslein-Volhard, C. (1994). *no tail* (*ntl*) is the zebrafish homologue of the mouse *T* (*Brachyury*) gene. *Development, 120*, 1009–1015.

Selleck, M. A., & Stern, C. D. (1991). Fate mapping and cell lineage analysis of Hensen's node in the chick embryo. *Development, 112*, 615–626.

Selleck, M. A., & Stern, C. D. (1992). Evidence for stem cells in the mesoderm of Hensen's node and their role in embryonic pattern formation. In R. Bellairs, E. J. Sanders, & J. W. Lash (Eds.), *Formation and differentiation of early embryonic mesoderm* (pp. 23–31). New York: Plenum.

Shimizu, T., Bae, Y. K., Muraoka, O., & Hibi, M. (2005). Interaction of Wnt and *caudal-related* genes in zebrafish posterior body formation. *Developmental Biology, 279*, 125–141.

Stickney, H. L., Barresi, M. J., & Devoto, S. H. (2000). Somite development in zebrafish. *Developmental Dynamics, 219*, 287–303.

Takada, S., Stark, K. L., Shea, M. J., Vassileva, G., McMahon, J. A., & McMahon, A. P. (1994). Wnt-3a regulates somite and tailbud formation in the mouse embryo. *Genes & Development, 8*, 174–189.

Takemoto, T., Uchikawa, M., Kamachi, Y., & Kondoh, H. (2006). Convergence of Wnt and FGF signals in the genesis of posterior neural plate through activation of the *Sox2* enhancer N-1. *Development, 133*, 297–306.

Takemoto, T., Uchikawa, M., Yoshida, M., Bell, D. M., Lovell-Badge, R., Papaioannou, V. E., et al. (2011). Tbx6-dependent *Sox2* regulation determines neural or mesodermal fate in axial stem cells. *Nature, 470*, 394–398.

Tam, P. P., & Tan, S. S. (1992). The somitogenetic potential of cells in the primitive streak and the tail bud of the organogenesis-stage mouse embryo. *Development, 115*, 703–715.

Thorpe, C. J., Weidinger, G., & Moon, R. T. (2005). Wnt/beta-catenin regulation of the Sp1-related transcription factor *sp5l* promotes tail development in zebrafish. *Development, 132*, 1763–1772.

Tsakiridis, A., Huang, Y., Blin, G., Skylaki, S., Wymeersch, F., Osorno, R., et al. (2014). Distinct Wnt-driven primitive streak-like populations reflect in vivo lineage precursors. *Development, 141*, 1209–1221.

Turner, D. A., Hayward, P. C., Baillie-Johnson, P., Rue, P., Broome, R., Faunes, F., et al. (2014). Wnt/beta-catenin and FGF signalling direct the specification and maintenance of a neuromesodermal axial progenitor in ensembles of mouse embryonic stem cells. *Development, 141*, 4243–4253.

Tzouanacou, E., Wegener, A., Wymeersch, F. J., Wilson, V., & Nicolas, J. F. (2009). Redefining the progression of lineage segregations during mammalian embryogenesis by clonal analysis. *Developmental Cell, 17*, 365–376.

Uriu, K., Morelli, L. G., & Oates, A. C. (2014). Interplay between intercellular signaling and cell movement in development. *Seminars in Cell & Developmental Biology, 35*, 66–72.

Vonica, A., & Gumbiner, B. M. (2002). Zygotic Wnt activity is required for *Brachyury* expression in the early *Xenopus laevis* embryo. *Developmental Biology, 250*, 112–127.

Warga, R. M., & Nusslein-Volhard, C. (1999). Origin and development of the zebrafish endoderm. *Development, 126*, 827–838.

White, P. H., & Chapman, D. L. (2005). *Dll1* is a downstream target of Tbx6 in the paraxial mesoderm. *Genesis, 42*, 193–202.

Wilson, V., Olivera-Martinez, I., & Storey, K. G. (2009). Stem cells, signals and vertebrate body axis extension. *Development, 136*, 1591–1604.

Wittler, L., Shin, E. H., Grote, P., Kispert, A., Beckers, A., Gossler, A., et al. (2007). Expression of *Msgn1* in the presomitic mesoderm is controlled by synergism of WNT signalling and Tbx6. *EMBO Reports*, *8*, 784–789.

Yabe, T., & Takada, S. (2012). Mesogenin causes embryonic mesoderm progenitors to differentiate during development of zebrafish tail somites. *Developmental Biology*, *370*, 213–222.

Yamaguchi, T. P., Takada, S., Yoshikawa, Y., Wu, N., & McMahon, A. P. (1999). *T* (*Brachyury*) is a direct target of Wnt3a during paraxial mesoderm specification. *Genes & Development*, *13*, 3185–3190.

Yasuhiko, Y., Kitajima, S., Takahashi, Y., Oginuma, M., Kagiwada, H., Kanno, J., et al. (2008). Functional importance of evolutionally conserved Tbx6 binding sites in the presomitic mesoderm-specific enhancer of *Mesp2*. *Development*, *135*, 3511–3519.

Yoon, J. K., & Wold, B. (2000). The bHLH regulator *pMesogenin1* is required for maturation and segmentation of paraxial mesoderm. *Genes & Development*, *14*, 3204–3214.

Yoshikawa, Y., Fujimori, T., McMahon, A. P., & Takada, S. (1997). Evidence that absence of Wnt-3a signaling promotes neuralization instead of paraxial mesoderm development in the mouse. *Developmental Biology*, *183*, 234–242.

Young, T., Rowland, J. E., van de Ven, C., Bialecka, M., Novoa, A., Carapuco, M., et al. (2009). *Cdx* and *Hox* genes differentially regulate posterior axial growth in mammalian embryos. *Developmental Cell*, *17*, 516–526.

Zorn, A. M., & Wells, J. M. (2009). Vertebrate endoderm development and organ formation. *Annual Review of Cell and Developmental Biology*, *25*, 221–251.

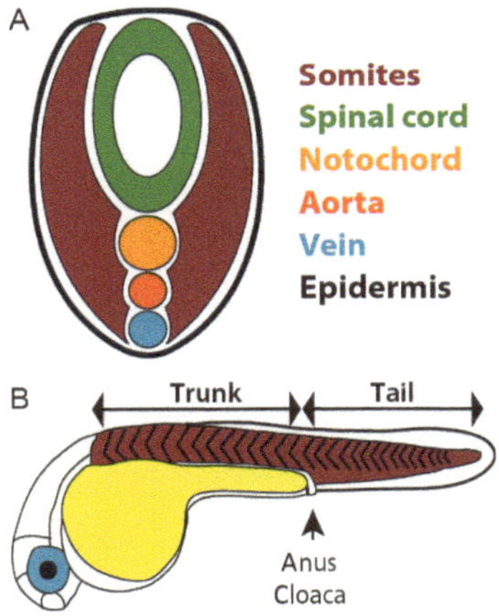

David Kimelman, Figure 1 Morphological features of the vertebrate posterior body. (A) A generic cross section through the postanal region of a vertebrate embryo. In the preanal region, the pronephric tubules and gut tube would also be present. (B) The posterior of the embryo is divided into preanal trunk and postanal tail. Shown is a zebrafish embryo but the same division applies to all vertebrates.

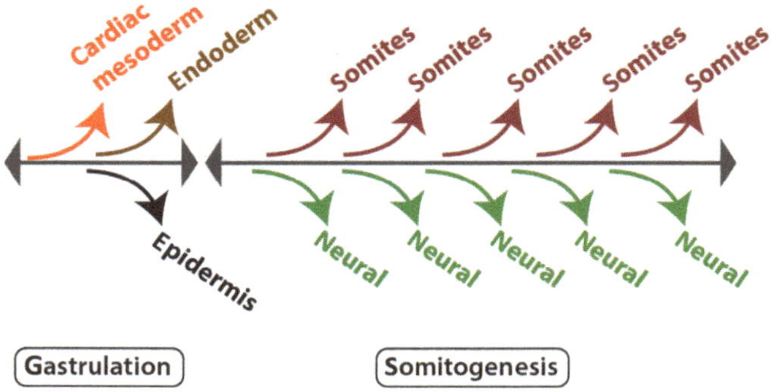

David Kimelman, Figure 2 Maintenance of the neuromesodermal progenitors. Lineage labeling studies in the mouse show that while many germ layer decisions are made early, a neuromesodermal progenitor population continues to produce both neural and mesodermal cells during the somitogenesis stages (Tzouanacou, Wegener, Wymeersch, Wilson, & Nicolas, 2009). The times at which the gastrula stage fate decisions are made are approximate.

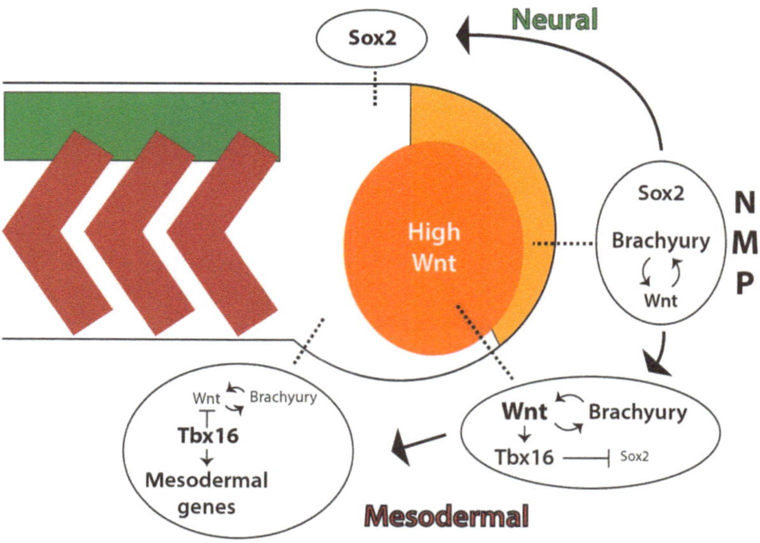

David Kimelman, Figure 3 Model for regulation of mesoderm formation in zebrafish. Sox2-expressing NMP cells are in a zone of moderate Wnt signaling (light orange) that maintains Brachyury. As cells leave this region, if they are not exposed to continue Wnt signaling, they express Sox2 but not Brachyury and become neural. Instead, if they are exposed to high levels of Wnt (dark orange), they express Tbx16, which represses Sox2. Tbx16 also represses Wnts so as cells continue to move anteriorly (left side), they shut off Brachyury and turn on mesodermal differentiation genes, thus locking in the mesodermal choice.

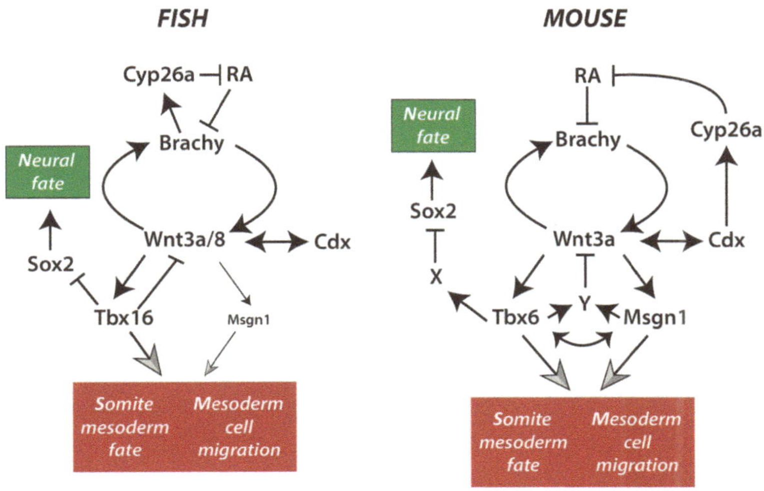

David Kimelman, Figure 4 Comparison of the regulatory logic of fish and mouse posterior cell fate choices. A schematic showing the regulatory relationships among signaling and transcription factors discussed in this review. Tbx6 is shown here to cause repressive effects indirectly, but as with Tbx16, it may turn out to be able to directly repress target genes. Fgf is omitted for simplicity.

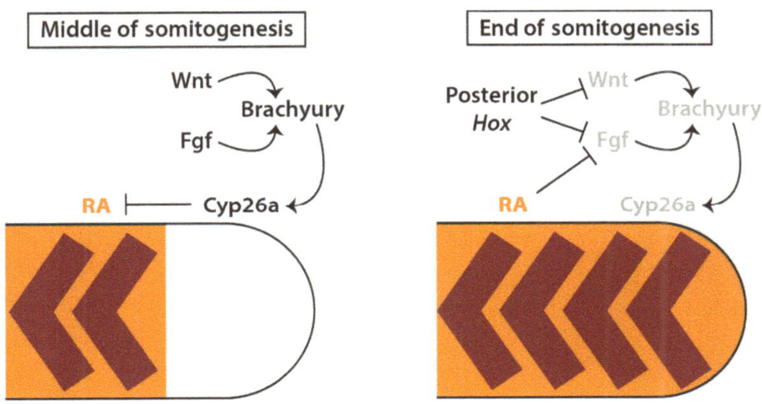

David Kimelman, Figure 5 Model for termination of axis extension in chick. During somitogenesis, the progenitor domain excludes RA. Near the end of somitogenesis, the most posterior *Hox* genes, possibly together with other molecular changes, causes RA to expand posteriorly, thus inhibiting the progenitor cell fate.

CHAPTER THIRTY

The Curious World of Gonadal Development in Mammals

Peter Koopman[1]
Institute for Molecular Bioscience, The University of Queensland, Brisbane, Queensland, Australia
[1]Corresponding author: e-mail address: p.koopman@imb.uq.edu.au

Contents

Abstract

The distinct sex phenotypes of male and female hinge on the development of the fetal gonads as testes or ovaries, in turn, regulated by the molecular genetic machinery of sex determination. Here, I discuss five aspects of mammalian gonadal development that distinguish it from other examples of organogenesis, and continue to surprise and fascinate.

Let's face it: males and females are very different animals—so much so, that for any species there are really two developmental biologies, not one. Humans have been intrigued by the differences between men and women since the beginning of recorded history, and presumably long before. As a developmental biologist, it is especially fascinating to ask how the differences between the sexes arise. Finding the answers involves a stimulating mix of molecular genetics, cell biology, and developmental anatomy. Since our sex phenotype depends critically on the formation of testes or ovaries in the embryo, research efforts focus largely on the genetic control of sex determination and the organogenesis of the gonads. After half a lifetime, I am still busy delving into these issues. In this chapter, I attempt to rationalize this enduring fascination by describing five aspects of sex development that continue to captivate.

Current Topics in Developmental Biology, Volume 116
ISSN 0070-2153
http://dx.doi.org/10.1016/bs.ctdb.2015.12.009

1. DURING SEX DEVELOPMENT, THE GONADAL ANLAGEN CAN FORM EITHER OF TWO COMPLETELY DIFFERENT ORGANS

Our journey as males or females begins with the differentiation of the paired gonadal anlagen as either testes or ovaries, respectively. This situation presents an entirely different logistical proposition compared with the development of any other organ system, where the ultimate fate of the organ rudiment is cast from the beginning. Although testes and ovaries serve analogous functions in males and females, the mechanics of producing and delivering two quite distinct, highly specialized types of gametes, in spectacularly different quantities, calls for two quite different organs in terms of size, cellular composition, and structural arrangement.

What this means is that the gonadal anlagen must contain a number of undifferentiated cell types that are poised to differentiate into one of two counterparts, depending on the molecular signals received. It is currently thought that one precursor cell type can differentiate into either Sertoli cells in the testis or granulosa cells in the ovary, another can differentiate into either hormone-producing testicular Leydig cells or ovarian theca cells, and that primordial germ cells can either enter meiosis during fetal life and embark on oogenesis, or mitotically arrest, heralding spermatogenesis (Svingen & Koopman, 2013).

Curiously, the logistics of how these two different gonadal outcomes are achieved differs from the strategy used to generate other structures that distinguish males and females. The reproductive tract, for example, does not arise from a single bipotential primordium: instead, two sets of ducts are laid down, the Wolffian (mesonephric) and Müllerian (paramesonephric) ducts, which have the potential to develop into the female and male internal genitalia, respectively. In males, Müllerian ducts degenerate under the influence of anti-Müllerian hormone (AMH, or Müllerian-inhibiting substance, MIS), while Wolffian ducts mature to form the epidymides, vasa deferentia, and seminal vesicles. In females, the opposite happens, resulting in formation of oviducts, uterus, cervix, and upper vagina. In other words, the embryo plans for its two possible outcomes by making both primordia, one being retained and the other discarded in each sex. For the external genitalia, similarly constituted but differently shaped structures are made in each sex by shaping and plumbing a common set of primordia in different ways. Thus, the genital tubercle, urogenital folds, urogenital sinus, and labioscrotal swelling morph into penis and scrotum in males, or clitoris and vulva in females. Finally, primordial breast tissue is present in both males and females, but is encouraged to thrive during puberty in females, but remain vestigial in males. Here, it is a hormone-driven stop-go decision that differentiates males from females.

Clearly, as with any complex developing organ system, there is much to do in a short space of time in order to generate functional testes or ovaries and, later, the internal ducts and external genitalia. This hurry is especially evident in the mouse, where the gonads do not begin to differentiate until over halfway through gestation, leaving only 10 days for prepubertal sex differentiation to be all but completed. One reason for the late start is that differentiation of the gonads needs to be delayed until they are populated by primordial germ cells—there is no point making testis cords until they have the germ cells inside. Primordial germ cells themselves arise surprisingly late in the piece, and need to migrate from their site of genesis at the posterior of the gastrula-stage embryo, along the hindgut, toward a target that continues to move further away by the hour as the embryo grows. This arrangement begs the question of why such a complex and potentially error-prone mechanism is used to get germ cells to where they need to be during embryo development, but that is a question for another time.

2. THE DIFFERENTIATION PROGRAM OF THE TESTIS IS COORDINATED BY ONE CELL TYPE

Because almost all mammals have a sex-determining mechanism based on the presence or absence of the Y chromosome, attention has focused on how the Y-linked testis-determining gene acts to initiate the formation of testes. Even before the cloning of *Sry*, Burgoyne, Buehr, Koopman, Rossant, and McLaren (1988) wondered whether the testis-determining locus (*Tdy*) induces the differentiation of all of the cell lineages in the genital ridge cell-autonomously. To address this question, they generated chimeric embryos composed of both XX and XY cells, and used molecular markers to study the sex chromosome genotype of each cell type present in the testes. They found that while other cell types could be a mixture of XX and XY cells, Sertoli cells were exclusively XY. They concluded that the *Tdy* must act cell-autonomously to induce differentiation only of the Sertoli cell lineage, which then secretes molecular signals to induce other cells, either XX or XY, to differentiate into testicular cell types. (Later analysis using more sophisticated techniques showed that in fact, about 10% of the Sertoli cell population was XX in these chimeras, forcing the reinterpretation that "one of the steps in the *Tdy*-initiated process of Sertoli cell determination is capable of locally recruiting XX cells" (Palmer & Burgoyne, 1991). This recruitment mechanism is discussed below.)

Again, it is pertinent to ask why such a seemingly complicated cascading signaling system is used. The answer may lie partly in the vulnerability of *Sry*, such that its expression levels appear to be dangerously close to the threshold of inactivity (as discussed below). If each cell type had to independently differentiate under the influence of *Sry*, the possibility exists that in any given gonad, some cells might succeed and others might fail to differentiate appropriately, resulting in a chaotic mixture of testicular and ovarian cell types. By having the system depends on the differentiation of one cell type, a more robust mechanism is established whereby, provided Sertoli cells differentiate appropriately, all other cell types will follow suit. In other words, a mechanism is provided to canalize differentiation of the gonad into either a testis or an ovary, not a mixture of both.

3. THE KEY GENE THAT DIRECTS SEX DETERMINATION IS VULNERABLE AND FEEBLE

There is no doubt that *Sry* has a uniquely important role—it underpins the entire system of sexual reproduction in nearly all mammals, including of course humans. Despite this, *Sry* has an Achilles heel. Its location on the Y chromosome means that, like all Y-borne genes, it is unable to repair mutations using conventional mechanisms available to other genes during meiosis. As a result, *Sry* is accumulating mutations at a high rate, as evidenced by the fact that some 20% of 46, XY gonadal dysgenesis in humans is caused by mutations in the *SRY* coding region. Presumably, the *cis*-regulatory sequences important for appropriate timing, levels, and location of *Sry* transcription are also being pelted with mutations. What may once have been a lusty gush of expression now seems to have been reduced, at least in mice, to a brief and relatively weak dribble. As a result, *Sry* probably has been incapacitated to the point where it is barely able to function (Kashimada & Koopman, 2010; Polanco & Koopman, 2007).

How does the system cope with this situation? It seems that a number of mechanisms have evolved to amplify the signal generated by Sry. First, a number of lines of evidence indicate that the cell differentiation relies on robust and continued expression of *Sox9*, not of *Sry*. *Sox9* is the immediate downstream target of Sry (Sekido & Lovell-Badge, 2008), and XX transgenic mice overexpressing Sox9 in the genital ridges develop testes (Vidal, Chaboissier, de Rooij, & Schedl, 2001), suggesting that the only important function of Sry in sex development is to activate *Sox9*, which then does the heavy lifting in sex determination, activating a variety of targets required for Sertoli cell phenotype (Li, Zheng, & Lau, 2014). Once *Sox9* is activated, its gene product is able to maintain *Sox9* expression in a positive autoregulatory loop that does not require continued expression of *Sry* (Sekido & Lovell-Badge, 2008), the situation is analogous to a short pulse from a relatively flimsy starter motor which awakens a powerful engine that keeps itself running.

Second, it is likely that some Sertoli cell precursors in XY genital ridges express enough *Sry* to kick start *Sox9* expression, but others do not (Wilhelm et al., 2005). Cell mixing experiments, combined with molecular studies, have indicated that prostaglandin D2, produced as a result of *Sox9* expression in Sertoli cells is able to induce *Sox9* expression in XX cells of the same lineage (Wilhelm et al., 2005). This mechanism almost certainly explains why some XX cells were found to contribute to the Sertoli cell population in the experiments of Burgoyne and colleagues described above (Palmer & Burgoyne, 1991). By extension, it may be supposed that during normal testis development, Sertoli cell precursors that express sufficient *Sry* to switch on *Sox9* cell-autonomously are able to recruit cells expressing lower levels of *Sry* into the cell population and prevent them from differentiating as ovarian granulosa cells.

Interestingly, there seems to be a limit to how far this recruitment mechanism can be pushed. In XX–XY chimeras where less than about 10% of gonadal cells are XY, ovaries develop, whereas between about 10% and 35% XY cells result in ovotestes (Burgoyne et al., 1988). Only higher proportions of XY cells were able to induce complete testis development in XX–XY chimeras.

4. THE MOLECULAR PATHWAYS DRIVING TESTIS AND OVARY DEVELOPMENT ARE MUTUALLY ANTAGONISTIC

After *Sry* was initially isolated, the key experiment was to add it as a transgene to XX mouse embryos. *Sry* alone was able to divert the course of development of the resulting mice from female to male, thus proving its role as the testis-determining gene (Koopman, Gubbay, Vivian, Goodfellow, & Lovell-Badge, 1991). This result further highlighted two important points:

first, females have all of the genes required for making testes, it is just that the absence of *Sry* from the XX genome usually prevents them from being activated. Second, males have all of the genes required for making ovaries, including whatever it is that activates the ovarian-determining pathway in the first place. Logically, therefore, genes of the testis-determining pathway must actively suppress the ovarian-determining pathway.

While this is known to be true, exact details of how it is accomplished remain unclear. FGF9, a product of Sertoli cells and a key component of the testis-determining pathway, is able to antagonize the action of the ovarian-determining factor Wnt4; in mouse gonads lacking *Fgf9*, *Wnt4* is ectopically upregulated (Kim et al., 2006). Presumably, key testicular transcription factors play a part in negatively regulating transcription of key ovarian genes. Because Sox9 is a transcription factor and orchestrates a wide diversity of activities during testis differentiation, it is tempting to speculate that it directly represses key ovarian-determining genes. Although Sox9 is conventionally viewed as an activating factor, it is also known to repress transcription of some genes, including *Spp1*, *alpha-sarcoglycan*, *Cdx2*, and *Muc2*, in other contexts (Blache et al., 2004; Hernandez-Hernandez et al., 2009; Peacock, Huk, Ediriweera, & Lincoln, 2011). In sum, the exact mechanism(s) by which the testis-determining pathway suppresses the ovarian-determining pathway are not clear at present.

Perhaps more surprisingly, it seems that the ovarian-determining pathway also actively suppresses the testis-determining pathway. It has been demonstrated that in *Wnt4*-knockout mice, FGF9 transcription is ectopically upregulated (Kim et al., 2006). Further, loss of function of several ovary genes including *Rspo1* (Tomizuka et al., 2008), *Foxl2* (Garcia-Ortiz et al., 2009), and ER-α (Couse et al., 1999) has been found to activate the male pathway, implying that their normal functions include keeping male sex-determining genes turned off. Accordingly, FOXL2 and ER-α both bind to a testis-specific enhancer of *Sox9* and synergistically repress its activity *in vivo*. Conditional deletion of either leads to *Sox9* upregulation. β-Catenin mediates *Wnt4* and *Rspo1* activity; stabilization or activation of β-catenin also silences *Sox9* expression and causes XY sex reversal (Maatouk et al., 2008). Because Sox9 is expressed at low levels in gonads of both sexes at an early stage, presumably this antagonism helps to ensure that the *Sox9* signal is not amplified by autoregulation to the point where the testis-determining pathway becomes activated.

A further surprise is that these mechanisms of mutual antagonism between the testicular and ovarian developmental pathways continue to be important for the life of the animal. It had been assumed once the decision has been made by the gonad to develop as a testis or an ovary that decision was locked in. It was therefore surprising to find that experimental inactivation of the testis gene *Dmrt1* (Minkina et al., 2014), or of the ovary gene *Foxl2* (Uhlenhaut et al., 2009), postnatally, causes gonadal sex reversal. These experiments revealed an unusual level of plasticity of gonadal fate in mammals, perhaps related in some evolutionary way to the ability of some species of fish to change sex as adults in response to social cues.

5. TESTIS CORD FORMATION: A UNIQUE MODE OF TUBULOGENESIS

One of the most striking morphological features of the developing testes, when viewed under the microscope with transmitted light, is their prominent transverse stripes, which correspond to the developing testis cords. When examined more closely, for example, after *in situ* hybridization for a Sertoli or germ-cell marker, it becomes clear that the cords in their early stages resemble a stack of toroids or doughnuts, joined together along one edge (Combes, Lesieur, et al., 2009). The line along which they are joined is known as the rete, and it is positioned adjacent to the mesonephros. Eventually, the mesonephric tubules connect with the rete, thus providing an outlet for sperm from the testis tubules into the vas deferens.

How do these cords form, what determines the shape, orientation, and number? Since testis cords are principally composed of Sertoli cells surrounding germ cells, the most obvious mechanism for cord formation might involve germ cells acting as nucleation points around which Sertoli cells assemble. However, mutant testes that lack germ cells are able to form cords. Therefore, the mechanism of cord formation must involve Sertoli cells recognizing each other irrespective of the presence or absence of germ cells. That being so, the question arises as to how the Sertoli cells form tubular structures rather than clumps, and also how the germ cells end up inside these tubular structures.

Careful observation has shown that testis cord formation represents a mode of tubulogenesis that is quite unlike any other used in the embryo (Combes, Lesieur, et al., 2009). At a certain point, the genital ridge contains clusters of pre-Sertoli and germ cells. Shortly thereafter, irregularly spaced "proto-cords" can be seen, separated by migrating endothelial cells making their way from the mesonephros to the opposite face of the gonad. These shafts of migrating endothelial cells are evidently critical for the formation and remodeling of the cords into regularly spaced units, because experimentally interfering with endothelial cell migration in cultured fetal testes disrupts cord formation (Combes, Wilhelm, et al., 2009; Cool, DeFalco, & Capel, 2011). Viewed in two dimensions, it makes perfect sense that the migrating endothelial cells "carve up" the gonadal field into regularly spaced segments, each of which becomes occupied by a single testis cord, and that the cords and the nascent vasculature form parallel stripes across the developing testis. Viewed in three dimensions, however, it remains a mystery how thin columns of endothelial cells can orchestrate the formation of toroidal cords parallel to them.

We recently studied cord formation in a mouse mutant that has unusually long genital ridges (Wainwright, Svingen, Ng, Wicking, & Koopman, 2014). In this mutant, we found an increased number of testis cords, demonstrating that the number of testis cords depends on the available space, rather than being stereotypically predetermined.

5.1 Closing Comments

As a younger person, I was captivated by the concept of microcosms: electrons whizzing around nuclei somehow miraculously echoing the graceful dance of moons around planets, and planets around the sun. So it is with embryo development: each organ is a miniature version of the embryo, and each cell a tiny organ. The fascination of observing new life unfold in the embryo is the essence of developmental biology. Take a close look. There is a whole world in there.

REFERENCES

Blache, P., van de Wetering, M., Duluc, I., Domon, C., Berta, P., Freund, J. N., et al. (2004). SOX9 is an intestine crypt transcription factor, is regulated by the Wnt pathway, and represses the CDX2 and MUC2 genes. *The Journal of Cell Biology*, *166*, 37–47.

Burgoyne, P. S., Buehr, M., Koopman, P., Rossant, J., & McLaren, A. (1988). Cell-autonomous action of the testis-determining gene: Sertoli cells are exclusively XY in XX-XY chimaeric mouse testes. *Development*, *102*, 443–450.

Combes, A. N., Lesieur, E., Harley, V. R., Sinclair, A. H., Little, M. H., Wilhelm, D., et al. (2009). Three-dimensional visualization of testis cord morphogenesis, a novel tubulogenic mechanism in development. *Developmental Dynamics*, *238*, 1033–1041.

Combes, A. N., Wilhelm, D., Davidson, T., Dejana, E., Harley, V., Sinclair, A., et al. (2009). Endothelial cell migration directs testis cord formation. *Developmental Biology*, *326*, 112–120.

Cool, J., DeFalco, T. J., & Capel, B. (2011). Vascular-mesenchymal cross-talk through Vegf and Pdgf drives organ patterning. *Proceedings of the National Academy of Sciences of the United States of America*, *108*, 167–172.

Couse, J. F., Hewitt, S. C., Bunch, D. O., Sar, M., Walker, V. R., Davis, B. J., et al. (1999). Postnatal sex reversal of the ovaries in mice lacking estrogen receptors alpha and beta. *Science (New York)*, *286*, 2328–2331.

Garcia-Ortiz, J. E., Pelosi, E., Omari, S., Nedorezov, T., Piao, Y., Karmazin, J., et al. (2009). Foxl2 functions in sex determination and histogenesis throughout mouse ovary development. *BMC Developmental Biology*, *9*, 36.

Hernandez-Hernandez, J. M., Delgado-Olguin, P., Aguillon-Huerta, V., Furlan-Magaril, M., Recillas-Targa, F., & Coral-Vazquez, R. M. (2009). Sox9 represses alpha-sarcoglycan gene expression in early myogenic differentiation. *Journal of Molecular Biology*, *394*, 1–14.

Kashimada, K., & Koopman, P. (2010). Sry: The master switch in mammalian sex determination. *Development*, *137*, 3921–3930.

Kim, Y., Kobayashi, A., Sekido, R., DiNapoli, L., Brennan, J., Chaboissier, M. C., et al. (2006). Fgf9 and Wnt4 act as antagonistic signals to regulate mammalian sex determination. *PLoS Biology*, *4*, e187.

Koopman, P., Gubbay, J., Vivian, N., Goodfellow, P., & Lovell-Badge, R. (1991). Male development of chromosomally female mice transgenic for *Sry*. *Nature*, *351*, 117–121.

Li, Y., Zheng, M., & Lau, Y. F. (2014). The sex-determining factors SRY and SOX9 regulate similar target genes and promote testis cord formation during testicular differentiation. *Cell Reports*, *8*, 723–733.

Maatouk, D. M., DiNapoli, L., Alvers, A., Parker, K. L., Taketo, M. M., & Capel, B. (2008). Stabilization of beta-catenin in XY gonads causes male-to-female sex-reversal. *Human Molecular Genetics*, *17*, 2949–2955.

Minkina, A., Matson, C. K., Lindeman, R. E., Ghyselinck, N. B., Bardwell, V. J., & Zarkower, D. (2014). DMRT1 protects male gonadal cells from retinoid-dependent sexual transdifferentiation. *Developmental Cell*, *29*, 511–520.

Palmer, S. J., & Burgoyne, P. S. (1991). *In situ* analysis of fetal, prepuberal and adult XX-XY chimaeric mouse testes: Sertoli cells are predominantly, but not exclusively, XY. *Development*, *112*, 265–268.

Peacock, J. D., Huk, D. J., Ediriweera, H. N., & Lincoln, J. (2011). Sox9 transcriptionally represses Spp1 to prevent matrix mineralization in maturing heart valves and chondrocytes. *PloS One*, *6*, e26769.

Polanco, J. C., & Koopman, P. (2007). Sry and the hesitant beginnings of male development. *Developmental Biology*, *302*, 13–24.

Sekido, R., & Lovell-Badge, R. (2008). Sex determination involves synergistic action of SRY and SF1 on a specific Sox9 enhancer. *Nature*, *453*, 930–934.

Svingen, T., & Koopman, P. (2013). Building the mammalian testis: Origins, differentiation, and assembly of the component cell populations. *Genes & Development*, *27*, 2409–2426.

Tomizuka, K., Horikoshi, K., Kitada, R., Sugawara, Y., Iba, Y., Kojima, A., et al. (2008). R-spondin1 plays an essential role in ovarian development through positively regulating Wnt-4 signaling. *Human Molecular Genetics*, *17*, 1278–1291.

Uhlenhaut, N. H., Jakob, S., Anlag, K., Eisenberger, T., Sekido, R., Kress, J., et al. (2009). Somatic sex reprogramming of adult ovaries to testes by FOXL2 ablation. *Cell*, *139*, 1130–1142.

Vidal, V. P. I., Chaboissier, M. C., de Rooij, D. G., & Schedl, A. (2001). *Sox9* induces testis development in XX transgenic mice. *Nature Genetics*, *28*, 216–217.

Wainwright, E. N., Svingen, T., Ng, E. T., Wicking, C., & Koopman, P. (2014). Primary cilia function regulates the length of the embryonic trunk axis and urogenital field in mice. *Developmental Biology*, *395*, 342–354.

Wilhelm, D., Martinson, F., Bradford, S., Wilson, M. J., Combes, A. N., Beverdam, A., et al. (2005). Sertoli cell differentiation is induced both cell-autonomously and through prostaglandin signaling during mammalian sex determination. *Developmental Biology*, *287*, 111–124.

CHAPTER THIRTY-ONE

The Unaimed Arrow Never Misses

Raphael Kopan[1]
William K. Schubert Professor of Pediatrics, Cincinnati Children's Hospital Medical Center, University of Cincinnati College of Medicine, Cincinnati, OH, USA
Division of Developmental Biology, Cincinnati Children's Hospital Medical Center, University of Cincinnati College of Medicine, Cincinnati, OH, USA
[1]Corresponding author: e-mail address: raphael.kopan@cchmc.org

Abstract

In this assay, Raphael Kopan argues that focused emphasis on disease and translation stifles innovation, and outline the reasons why, in my opinion, developmental biologists are more likely to produce new and important discoveries than their more "focused" colleagues.

We are celebrating the 50th anniversary of *Current Topics in Developmental Biology*, started by Drs. Aaron A. Moscona and Alberto Monroy. I had the unique privilege to have been mentored by the late Aaron Moscona, bringing with it the opportunity to spend valuable time with him and his wife Malka, both of whom were exceptional developmental biologists, teachers, and colleagues. They are responsible for transforming my budding fascination in Developmental Biology into a life-long pursuit. It is highly appropriate therefore to contribute this essay in part to commemorate one of the founding editors of this series.

In the last 50 years, we have seen our field ascend to a peak with the elucidation of the mechanisms underlying the major signaling pathways, a fundamental understanding of the genetic pathways regulating pattern formation, which was recognized by the 1995 Nobel Prize in Physiology or Medicine to Edward (Ed) Lewis, Christian (Janni) Nusslein-Volhrd, and Eric Wieschaus, and the seminal work of John Gurdon, who laid the foundation to one of the most transformative discoveries of the twenty-first century, induced pluripotency (shared 2012 Nobel Prize in Physiology or Medicine with Dr. Shinya Yamanaka). Sadly, by that time, many

Current Topics in Developmental Biology, Volume 116
ISSN 0070-2153
http://dx.doi.org/10.1016/bs.ctdb.2015.10.007

developmental biologists interested in stem cells defected to create their own specialized discipline, publications, and meetings. This latest blow together with the overall decline in funding driven by the blunted, short-sighted vision of Congress and thus, the NIH, are the most important factors in the present state of our beloved discipline driving students and postdocs away. But, why should we care? Isn't the "wisdom" of group science and disease focus self-evident?

Developmental biologists are generalists. To paraphrase Sydney Brenner, I see developmental biologists as retaining a working multidimensional zoom, allowing them to travel in scale and in time and through any appropriate model organism, to continually convert original observations into discoveries of unexpected new mechanisms with relevance ranging from basic biological mechanisms to human disease. In this analogy, cell biologists rarely zoom up to the organ and the organism levels, while the stem-cell biologist are not always interested to zoom forward in time to look at the differentiated cells or organ. A stem cell biologist may look at stem cell aging only as an obstacle limiting the function of a perfect regenerative engine. In contrast, a developmental biologist would consider stem cell aging as a component in a complex developmental program that matches lifespan with the mode of reproduction (sexual vs. asexual), reproductive strategy (K vs. R), and niche availability. Only developmental biologists with a working zoom lens and a self-issued license to ponder any and all of these subjects, would consider all of them fair game.

Perhaps I convinced you that this analogy has some merit, or perhaps not. Yet you wonder—so what? What is the advantage of a working multidimensional zoom? This seems so unfocused, slow, and counterproductive. We live in an era with heroes driven by focused, fast paced progress. Fixing on stem cells or neurons or T-cells only accelerates our ability to translate, utilize, commoditize, and monetize. True, but focus comes at a price. It reduces our ability to innovate. Innovation occurs in our peripheral vision, and only if we are willing (or feel entitled to do so because we are developmental biologists) to avert our gaze from the target and explore that observational flicker. Sure enough, it slows progress towards the target. But, it increases our chance to experience serendipity. Horace Walpole coined this term in 1754 to describe those "always making discoveries, by accidents and sagacity, of things which they were not in quest of." Things you are not in quest of? So unfocused.

Serendipity gets us involved in areas where we lack expertise, amplifying its benefits. Again paraphrasing Sydney Brenner, this allows one to re-embrace one's ignorance, to use it as a sharp scalpel, to think outside the box and ask "stupid" questions. Ignorance was instrumental in my own work on Notch signaling. I simply did not know anything about how signals were supposed to be propagated, and I worked with a mentor (the late Harold Weintraub) that felt no need to educate me on that point, allowing instead for the experimental path I took to unfold as it may. I learned first hand that when we apply a rigorous scientific method, and keep an open mind, we can make important discoveries we were not in the quest of: my target in those early days was to find genes involved in hair follicle development (Kopan & Weintraub, 1993). Residual Notch expression in the presomitic mesoderm (unrelated to hair biology) lead to an assay, which lead to the observation that active Notch constructs that could repress myogenesis had to reside in the nucleus, that lead to the hypothesis that Notch somehow had to get there in order to function, and so on.

Of the many examples I can choose, lets consider how RNAi was discovered. The use of antisense DNA or RNA was commonplace at the time Kenneth Kemphues was investigating the PAR genes. However, Su Guo and Kemphues were the first to note the absence of strand specificity and the effectiveness of dsRNA in gene silencing (Guo & Kemphues, 1995). That was not what they were in the quest of, but they reported it anyway. The significance of this sidebar was not lost on a pair of developmental biologists, Drs. Craig Mello and Andrew Fire, already distracted from their respective quests by peripheral observations. As Dr. Mello recalled[1]: "The power of this gene-silencing approach accelerated our studies and we began to make rapid progress in understanding the developmental mechanisms that specify cell fate in the early embryo. However, we also became interested in the silencing phenomenon itself. The first observation that truly galvanized my interest occurred when, having injected RNA targeting apx-1, a gene essential for embryogenesis, I observed by chance that some embryos hatched and matured to adulthood only to produce 100% apx-1 dead embryos. The silencing phenomenon had skipped a generation and had been passed on via the germline to the next generation! This was truly amazing and prompted further studies that demonstrated the transmission of silencing for multiple generations via both the sperm and the egg... discovered, in part by accident while [a graduate student was] learning to inject...These findings, along with the inheritance properties, and the lack of strand specificity (first noted by Guo and Kemphues), prompted us to

[1] http://www.nobelprize.org/nobel_prizes/medicine/laureates/2006/mello-bio.html.

recognize the silencing phenomenon as an active response in the organism to the RNA." This paragraph twice highlights the gravitational pull of a peripheral event, which combined with a prepared mind propelled these un-focused colleagues to a transformative discovery recognized by the award of the Nobel prize in Medicine, 2006, to Drs. Mello and Fire. The odds that a scientist with complete focus on human health would have discovered RNAi are minuscule. The larger the umbrella, the greater the opportunity we may engage in exploring things we were not in the quest of.

I have gotten to the age when students decades younger than me lament that in my student years, discoveries were easier to come by (I confess to having the same concern learning about the birth of molecular biology (Weiner, 1999)). Now, they say, all the "good stuff" has been discovered. To that I reply: nonsense! We have only scratched the surface of what is yet to be discovered. Your problem is that you are too narrowly focused. Become a developmental biologist; claim all of life as your domain, and follow your peripheral vision.

ACKNOWLEDGMENTS

The author wishes to thank Drs. Brian Gebelein, Kenny Campbell, Aaron Zorn, and Jeff Whitest for commenting on the chapter, and Tal Kopan, for expert advice and encouragement. R.K is supported by NIH RO1 GM55479.

REFERENCES

Guo, S., & Kemphues, K. J. (1995). par-1, a gene required for establishing polarity in C. elegans embryos, encodes a putative Ser/Thr kinase that is asymmetrically distributed. *Cell*, *81*, 611–620.

Kopan, R., & Weintraub, H. (1993). Mouse notch: Expression in hair follicles correlates with cell fate determination. *Journal of Cell Biology*, *121*, 631–641.

Weiner, J. (1999). *Time, love, memory: A great biologist and his quest for the origins of behavior.* New York: Knopf.

CHAPTER THIRTY-TWO

A Path to Pattern

Thomas B. Kornberg[1]
Cardiovascular Research Institute, University of California, San Francisco, California, USA
[1]Corresponding author: e-mail address: tkornberg@ucsf.edu

Contents

Abstract

The field of developmental biology is not the same one that I entered in 1975. At that time, it seemed that most of its practitioners used various kinds of microscopes to watch animals as they matured, described morphological details with impressive temporal and spatial resolution, and recorded responses to physical and genetic insults. The number of genes whose mutant phenotypes offered insights into developmental mechanisms was small, the expression and functionalities of these genes were unknown, and because the extent of evolutionary conservation between different animals or even different organs in the same animal was also unknown, the vocabularies that were used to describe development were unique to each system. The distance between the descriptors and inferred molecular mechanisms was vast; it was a descriptive discipline. Today genome sequences are available for the animals that developmental biologists study, saturation genetic screens are possible, transgenesis offers powerful ways to modify genomes, and the proteins that direct and implement developmental processes can be imaged in real time. These advances have transformed the field into one that merges with cell biology, physiology, neurobiology, and immunology, and they have transformed our understanding of development. In this essay, I offer my perspectives and my sense of some principles that have emerged.

Current Topics in Developmental Biology, Volume 116
ISSN 0070-2153
http://dx.doi.org/10.1016/bs.ctdb.2015.11.026

1. THEMES AND VARIATIONS

The universality of the genetic code was understood and admired by the mid 1970s, but no one assumed or expected evolutionary conservation to extend extensively to the processes of development. It does, and more so than anyone imagined. Many organ systems provide similar functionality in distantly related animals, but their glaring differences suggest that they are constructed from different plans and arise through distinct processes that reflect their separate and independent origins. Their similar functionality is therefore judged to be a consequence of convergent evolution. One early surprise was the genetic basis that generates metameres in mammalian and insect embryos. The absence of a common ancestor in the fossil record was taken as strong evidence that the metameric subdivisions that become prominent in vertebrate and invertebrate gastrulas originated independently in different ancestral species. I recall a colleague's firm declaration: "Only a fool would suggest that the somites of vertebrates and the segments of insects are evolutionarily related." Today, the many examples of evolutionary kinship suggest that the reverse is more apt—only a fool assumes convergent evolution for yet untested systems!

The common origin of mammalian somitogenesis and invertebrate segmentation was revealed when the *Drosophila* Hox genes were discovered and their sequences were used as probes for nucleic acid hybridization to clone their mammalian orthologs. The vertebrate and invertebrate genes have extensive regions of sequence conservation, both sets of genes are expressed in spatial domains that correlate with the metameric subdivisions and both code for functions that establish the developmental identity of cells in their respective domains. This strong molecular and genetic evidence removed all doubt that the most recent common ancestor was segmented. It also illustrates the power of factual molecular and genetic evidence, in contrast to conclusions based on anatomical comparisons that are in fact just opinions, no matter the expertise they might represent.

Similar scenarios played out when the *Drosophila* genes responsible for eye, muscle, heart, appendage, posterior gut, nervous system, immune system, brain and head development were identified, and the sequences of these genes were used to isolate vertebrate counterparts. Evidently, although evolution has changed the organ systems in ways that largely obscure their common origins, cells in the organ systems assume their destiny under the guidance of conserved genetic networks. The evolutionary conservation suggests that the most recent common ancestor had all the major organ systems that make up the body plans of the descendants we know, and the failure to find evidence for the common ancestor in the fossil record is just that. The history of false attributions to convergent evolution should serve as a reminder to be cautious when basing conclusions on negative results. Today, the many examples of common origins suggest that until proven otherwise, the informed position should assume that all organ systems with similar functionalities in distantly related animals derive from common precursors and are most likely the products of evolutionary drive to diversification. Granting agencies and reviewers, please take note.

2. ELEGANT SOLUTIONS

Developmental biology tries to decipher how the vast morphological diversity of animal tissues is encoded in genomes and is generated during development. If we consider the many variations of legs or wings, for instance, we might accept the premise that the genetic networks that assign leg or wing identity are conserved and are shared across the animal kingdom, but we might not assume that the processes that lead to the diversity of legs and wings have much in common—perhaps every structure represents the product of a novel routine that evolved specifically to produce each variation. In the mid 1970s, I recall Sydney Brenner pondering over this possibility and wondering whether "the genetic code was Nature's last elegant solution." He later addressed it with his uniquely humorous and imaginative way in a column he wrote for *Current Biology* (Brenner, 1996).

> *I shared an office with Francis Crick for twenty years in Cambridge. At one time he was interested in embryology and spent a lot of time thinking about imaginal discs in Drosophila. One day, he threw the book he was reading down onto his desk with an exasperated cry. "God knows how these imaginal discs work." In a flash I saw the whole story of Francis arriving in heaven and Peter welcoming him with "Oh Dr Crick, you must be tired after your long journey. Do sit down, have a drink and relax." "No", says Francis, "I must see this fellow, God; I have to ask him a question." After some persuasion, the angel agrees to take Francis to God. They cross the middle part of heaven, and finally right at the back, across the railway tracks, they come to a shed, with a corrugated iron roof, surrounded by junk. And in the back part, there is a little man in overalls with a large spanner in his back pocket. "God", says the angel, "This is Dr Crick; Dr Crick, this is God." "I am so pleased to meet you", says Francis. "I must ask you this question. How do imaginal discs work?" "Well", comes the reply, "We took a little bit of this stuff and we added some things to it and ... actually, we don't know, but I can tell you that we've been building flies up here for 200 million years and we have had no complaints.*

One of the cornerstone principles of development is that growth and patterning are directed by specially endowed groups of cells called signaling centers (developmental organizers), and that every organ has one or more that regulate adjacent regions. Such signaling centers were discovered more than 100 years ago in studies of hydra regeneration (Browne, 1909), and the ideas were subsequently crafted into a general theory in studies of amphibian embryogenesis (Spemann & Mangold, 1924). These studies investigated how cells respond when they are transplanted to various ectopic locations in host animals and they established that not only can some cells remember where they come from to make the same structures that they would have made if not for the insult of surgery and relocation, but they can persuade neighboring host cells to change identity and join them in making ectopic structures (Fig. 1). These observations led to the idea that some cells function as signaling centers by producing "inducers," and that cells in the surrounding regions detect these inducers and behave according to the information the inducers impart.

The images in Figs. 2 and 3 are presented as examples of the degree to which biology can control time and space during development. The panels in Fig. 2 are frames from a movie taken of *Xenopus* embryos in which the synchronous and ordered cleavage divisions are visible. Although the timing of cell divisions, movements and other structural changes become more variable at subsequent stages, the precision of these early cleavage stages shows that the sequence and timing of complex developmental processes can be controlled with great precision. The panels in Fig. 3 illustrate *Drosophila* cuticle structures whose reproducible features also suggest precise developmental control. The micrographs are of the same portion of the wing margin of eight *Drosophila* wings. Note the oriented wing hairs and bristles, wing hairs and the sensillae of the peripheral nervous system (arrows), and the relative

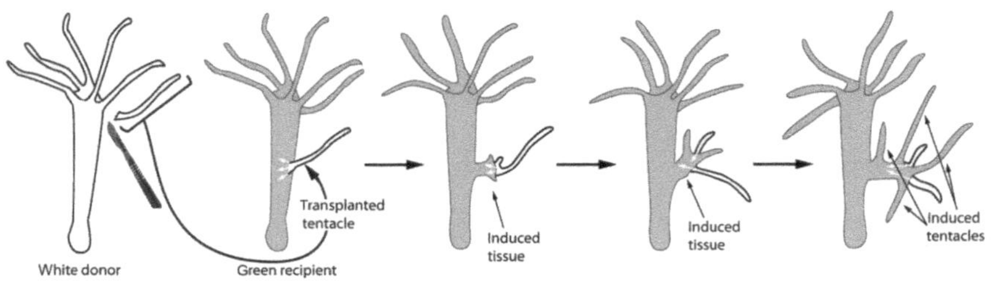

Figure 1 Grafting experiments reveal developmental induction. A tentacle transplanted from a white hydra to the stalk of a green recipient induces the host to produce ectopic tentacles. (A color version of this figure is located at the end of this chapter.)

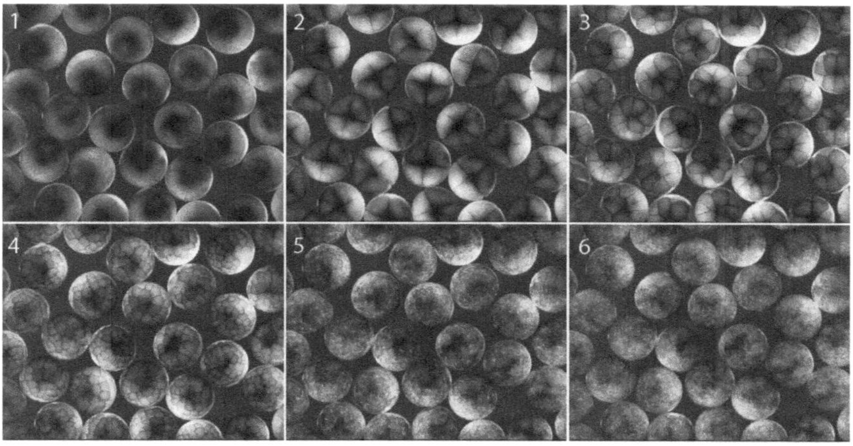

Figure 2 *Xenopus* embryos undergoing cleavage stage divisions. Fertilization of this field of embryos was initiated simultaneously, and the synchrony and regularity of the early stages of development are apparent in these six frames of a movie taken by J. Smith. (A color version of this figure is located at the end of this chapter.)

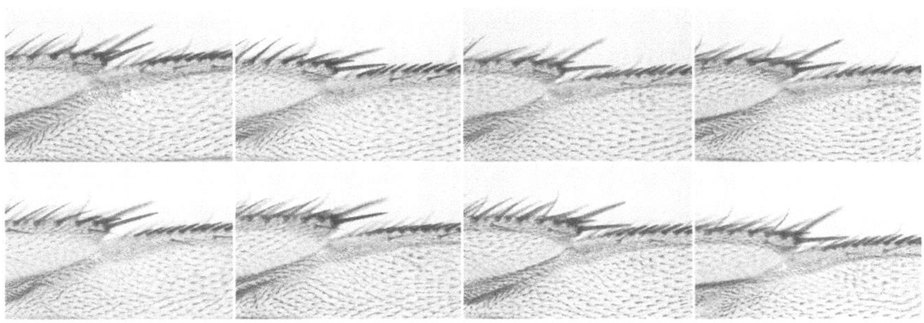

Figure 3 Images of the *Drosophila* wing. Micrographs showing a portion of the wing margin of wings dissected from eight *Drosophila* adults. The orientation, relative size, and color of all the cuticular elements are reproducibly patterned in these specimens. Arrows point to two sensillae, sense organs of the peripheral nervous system. (A color version of this figure is located at the end of this chapter.)

colors and sizes of wing hairs and bristles. In each wing, these features are patterned reproducibly with great fidelity, suggesting that they are encoded in a plan that development accurately executes in each fly.

These are the features that the inducer molecules specify, and the quest to identify them occupied (and frustrated) investigators for decades. Although it was assumed from the outset that they are small organic molecules that diffuse in and out of cells, today we know that they are proteins such as the proteins of the bone morphogenic protein (BMP), Wnts, Hedgehog

(Hh), fibroblast growth factor (FGF), and epidermal growth factor (EGF) families. This group of proteins may represent most of the repertoire of inducers, its size small not apparently because it is incomplete, but because the roles of these proteins are conserved and universal. In *Drosophila*, various combinations of these proteins—Decapentaplegic (Dpp, a BMP homolog), Wingless (Wg, a Wnt homolog), Branchless (a FGF homolog), and Spitz and Gürken (EGF homologs)—have been implicated at every developmental stage in the development of every organ system that has been studied. Homologs with similar roles have been implicated in other animals. If these signaling proteins can be considered to make up the language of development, then this language has a vocabulary that is remarkably concise. Although our ability to interpret this language is not yet very advanced, the tools that we have to decipher it are improving, and as described below, some of its general features are now understood. We can also marvel at the elegance of the mechanisms that elaborate such vast diversity from so limited a vocabulary, and recognize the power of evolution that found ways to exploit it. Sydney, there is elegance beyond the code.

3. REGULATION, REGULATION, REGULATION

When I was enrolled as a student, my lecturers and textbooks described cellular processes such as mRNA export from the nucleus, mRNA degradation, protein folding, and protein degradation as unregulated, random events. Earlier generations understood that the amino acids of proteins and the nucleotides of nucleic acids were randomly ordered and that proteins were undefined coils that lacked defined structures. Today, we understand that order of amino acids in proteins and of nucleotides in nucleic acids are defined and spell out specific sequences, that proteins adopt specific structures, that cells have complex protein machines that orchestrate mRNA maturation, that cells transport different mRNAs to designated intracellular destinations for local translation, sometimes in response to external stimuli, that mRNA engagement with ribosomes and translational efficiency are controlled, sometimes by interactions with other RNAs (siRNA, miRNA, and piRNA), and mRNAs are degraded by regulated processes. Molecular chaperones facilitate protein folding, unfolding and assembly, and proteasome complexes scavenge proteins that might be damaged or that have been targeted for removal. History thus informs us that quantitative, spatial, and temporal regulation in biology is greater than we appreciate at any given juncture, and therefore that we

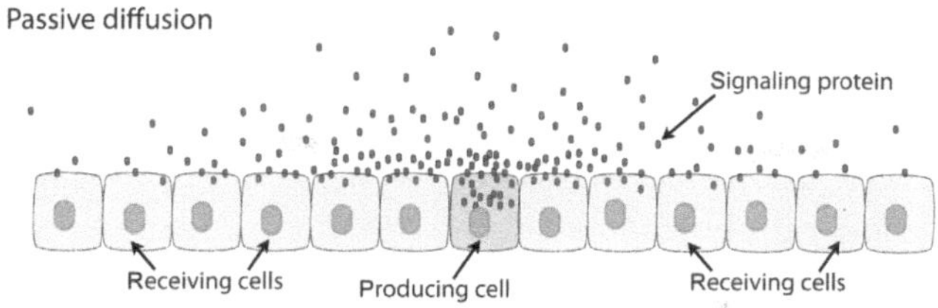

Figure 4 Passive diffusion of secreted signaling protein. This drawing of a hypothetical epithelial sheet depicts a single source cell that secretes signaling protein into extracellular space, generating a distribution of signaling protein that declines with increasing distance from the source. (A color version of this figure is located at the end of this chapter.)

should assume that processes that we study today are more structured, localized, regimented, and regulated than we know.

The identification of BMPs, Hhs, Wnts, FGFs, and EGFs as the "inducers" that regulate growth and morphogenesis did not reveal how these proteins exit the cells that make them or how they move across tissues to engage and instruct target cells. When their biological activities were first recognized and it was understood that they act at a distance, it was assumed that their expression is regulated but that they are released constitutively, and that they passively diffuse in order to engage target cells at various distances from their source (Fig. 4). From the perspective that biology maximizes regulation, this makes little sense. Quantitative, spatial, and temporal parameters of signaling proteins becomes challenging (if not impossible) to control after release into extracellular fluid, yet regulation must be exquisitely precise for the signals to convey the information that so exactly and reproducibly defines what the cells should do. A mechanism that endows signaling protein dispersal, targeting, and activation with such strict regulation is therefore essential.

4. CELL TALK AND CROSS-TALK

Two modes of cell–cell signaling by secreted extracellular signaling molecules have been thoroughly investigated and characterized—endocrine and synaptic. Both involve synthesis and release of signaling molecules by producing cells, binding to receptors on the surface of target cells and activation of signal transduction pathways, but they differ in important ways. Endocrine signals distribute systemically, circulating in the blood of animals that have a vascular system, or in the case of insects, in the hemolymph, the body fluid that bathes the internal organs. Dissemination of endocrine signaling molecules is in the range of seconds to minutes, and generates essentially uniform distributions throughout the animal so that cells near or far from the producing cells are exposed to similar concentrations. In mammals, endocrine hormones such as insulin are made in specific cells (pancreatic β-cells) as a proprotein that is inactive, and insulin remains stored and inactive in granules in the β-cells even after proteolysis converts it to its mature form. Insulin activity is regulated primarily by limiting the availability and exposure of receptors on target cells to circulating hormone—by controlling release and clearing. The insulin is biologically active only after circulating glucose stimulates β-cells to release it into the bloodstream, and insulin activity declines as insulin is removed from the bloodstream by endocytosis of the insulin–receptor complex or by clearing the hormone from the bloodstream in the liver and kidney. Insulin administered by injection has equivalent biological activity and can substitute for endogenously produced hormone because it can also distribute uniformly in the bloodstream and because the temporal and quantitative parameters that determine how cells respond are broadly defined and do not depend on the source.

Synaptic signaling is a form of juxtacrine signaling—signaling that is dependent on cell–cell contact (Fig. 5). The synapses that precisely appose the plasma membranes of contacting cells make it possible to precisely regulate exchanges of chemical and electrical signals. Key features of synaptic exchange are spatial specificity, temporal precision, quantitative control

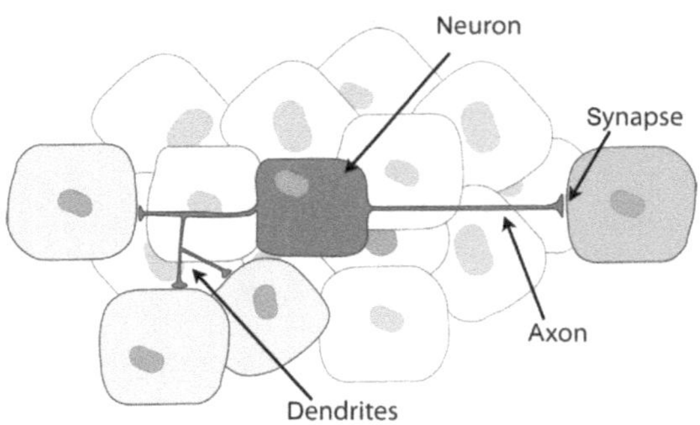

Figure 5 Cytoneme-mediated dispersion of signaling protein. This drawing of a hypothetical epithelial sheet depicts a single source cell that released signaling protein only at sites of direct contact between cytonemes that extend from cells that receive the signaling protein. Signaling protein moves across the epithelial sheet in a cytoneme-associated form and is extracellular only at the point of transfer where cytonemes make direct contact. (A color version of this figure is located at the end of this chapter.)

and rapidity, all of which contrast with the endocrine mechanism. Synaptic signaling is the principal way that neurons pass signals to other cells and it is also the way lymphocytes of the vertebrate immune system interface with target cells. Neuronal synapses are the best characterized, and although neurons make many different types, most of their synapses have complex assemblages of proteins on both the presynaptic and postsynaptic sides that regulate the release and uptake of active signals. At a chemical synapse, specialized synaptic vesicles loaded with neurotransmitters concentrate near the synaptic terminal, and when they fuse with the presynaptic membrane, they release active neurotransmitters into the synaptic cleft. This membrane fusion reaction has been called the fastest and most tightly regulated in biology. Neurotransmitter that is released moves freely in the synaptic 30–50 nm cleft, and elicits a postsynaptic response in as little as 60 μs. The amount of free neurotransmitter depends on the number of synaptic vesicles that release their contents after fusing with the presynaptic membrane and the actions of reuptake transporters and various types of inhibitors that clear free neurotransmitter from synaptic cleft. The postsynaptic response is sensitive to the number of activated neurotransmitter receptors at the postsynaptic membrane.

Neurons can signal to specific targets without affecting other cells because free neurotransmitter is confined to the synaptic clef where it is deposited. In contrast to endocrine hormones that disseminate in a vasculature that distributes the signals uniformly throughout the body, neurotransmitter signaling is limited to the synapse where the neurotransmitter is released, every signaling target predetermined by its synaptic connections.

Axon wiring involves distinct stages of pathfinding and synapse formation. Pathfinding refers to the process that growing axons use to navigate in a directed manner toward a target cell. For several reasons, pathfinding remains a fascinating mystery despite the identification of a group of proteins that function as attractant or repellent "guidance cues." These proteins include the netrins, Slits, semaphorins, and ephrins as well as morphogens such as the Wnts, BMPs, and Hhs. The conceptual problem is that we do not yet understand how the spatial information that dictates the stereospecific wiring patterns is encoded: whereas the patterns of connectivity are immensely complex, the number of guidance cues is small and seemingly insufficient to encode the vast repertoire of instructions needed to generate the complex spatial patterns. The mechanistic problem is that the guidance cues are assumed to be secreted, extracellular signals that disseminate to form chemotactic concentration gradients, but there is no experimental evidence that they are secreted into extracellular space, that they generate such distributions or for their mechanism of action. There is much to learn, but it is also important to note that although the processes that generate the stereospecific and complex neuronal connectivity networks are not yet adequately explained, our understanding of neurotransmitter release and action is strong because the mechanisms involved in neuron physiology and development are separable. We can understand how neurons communicate even if we do not understand how they identify their targets.

The second stage in the connectivity process is the construction of synapses, complex structures made up of more than 1000 different proteins in the presynaptic and postsynaptic terminals. These proteins include transsynaptic adhesion proteins, ion channels and signaling receptors that participate in a dialog between the presynaptic and postsynaptic cells. Although we know only some molecular details of the choreography that ensures that the correct partners connect and that the types of synapses are appropriate, there is strong genetic and cell biological evidence that both the presynaptic and postsynaptic cells contribute to a multistep process that leads to functional synapses, and that release of signals depends on the "synaptic handshake" by the producing and receiving cells.

5. IS EVERY CELL A "NEURON"?

The features that are said to define a neuron are its unique shape, its synapses and its electrical excitability. The anatomy of a neuron is certainly exceptional and distinctive. Some neurons have dendrites that extend for hundreds of microns and branch many times to form complex dendritic trees. Some axons are meters long, and some axons also branch hundreds of times. No other cell has extensions that rival these, but other cells do have extensions. Filopodia are one type that has been identified in many kinds of cells and in many developmental contexts. Antigen-presenting dendritic cells of the mammalian immune system, for instance, surround themselves with a latticework made of hundreds of filopodia, an arbor that recalls the dendritic trees of neurons. In recent years, studies of filopodia have benefited from improvements in imaging, especially from the new types of microscopes and systems for robust expression of fluorescent proteins, and it is now apparent that filopodia are dynamic, more numerous, more varied in structure and function, and more prevalent that had been appreciated. Indeed, filopodial extensions may be a universal feature of cells and the

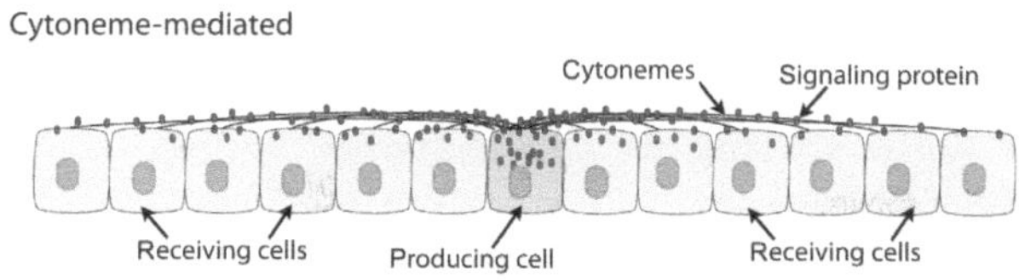

Figure 6 Synaptic signaling by a neuron. This drawing of a hypothetical neuron with an axon and dendritic extensions, making multiple synaptic contacts with target cells. (A color version of this figure is located at the end of this chapter.)

textbook description of a cell with static and smooth plasma membrane borders may have little relevance to development.

Studies of the *Drosophila* wing imaginal disc have identified filopodia that extend from three types of cells—disc-associated tracheal cells, disc-associated myoblasts, and the highly columnar epithelial cells of the disc proper (Huang & Kornberg, 2015; Roy, Hsiung, & Kornberg, 2011; Roy, Huang, Liu, & Kornberg, 2014). These filopodia are specialized for signaling (Fig. 6) and to distinguish them from filopodia that are involved in force generation or migration, they are termed cytonemes (cytoplasmic threads) (Ramírez-Weber & Kornberg, 1999). They mediate and are essential for signaling by Dpp, Hh, Wg, FGF, and Delta/Notch, and two types of cytonemes have been described that emanate from each of the three cell types. Tracheal cells extend cytonemes that are either populated by motile vesicles that contain the Dpp receptor (Thickveins, Tkv) or the FGF receptor (FGFR), but no vesicles containing both receptors have been found; nor have cytonemes containing both types of vesicles been observed. Properties that implicate cytonemes in signaling include the ability of Tkv-containing cytonemes to synapse with and take up Dpp from Dpp-producing disc cells. The lack of Dpp uptake or of Dpp signal transduction in tracheal cells that have defective cytonemes and are unable to form synapses testifies to the importance of these filopodia and to the role of synaptic signaling in a non-neuronal context. The presence of cytonemes that are specific to the Dpp and FGF signaling indicates that these signaling filopodia are specialized for particular a signaling system and reveals that filopodia that have a similar appearance may have distinct components and roles. If specificity for a signaling system is a general property of cytonemes, then every signaling system has a dedicated and unique set of cytonemes that mediates transport and exchange of its particular signaling protein.

Disc-associated myoblasts also extend two types of cytonemes—cytonemes that are populated by Delta-containing vesicles or by Wg receptor (Frizzled, Fz)-containing vesicles (Huang & Kornberg, 2015). These cytonemes also make synapses—with either Notch-expressing tracheal cells or Wg-expressing disc cells, respectively (Fig. 7). These synapses are also essential for signaling in the postsynaptic cells.

The third example may be the one with the most obvious similarity to neurons (Fig. 8). Disc columnar epithelial cells that express Hh and activate

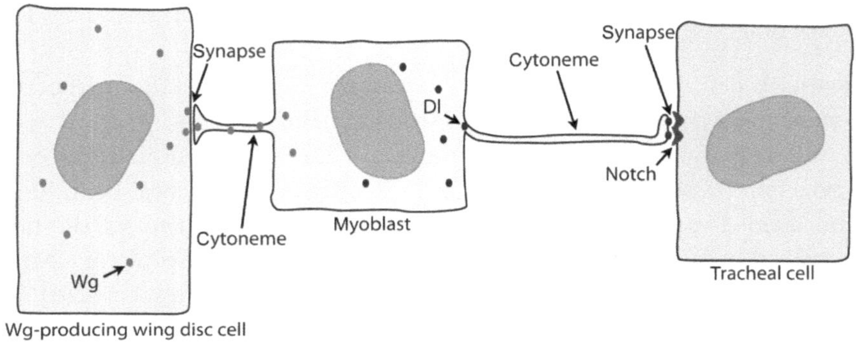

Figure 7 Synaptic signaling by columnar, tracheal, and myoblast cells of the *Drosophila* wing imaginal disc. The three cell types depicted are a wing disc columnar epithelial cell (light red) that expresses Wg, a myoblast (light blue) that picks up Wg from the disc cell and expresses Dl, and a tracheal cell (light purple) that expresses Notch and picks up Dl from the myoblast. Exchange of Wg and signaling by Dl takes place at the synapses at the tips of the two types of myoblast cytonemes. (A color version of this figure is located at the end of this chapter.)

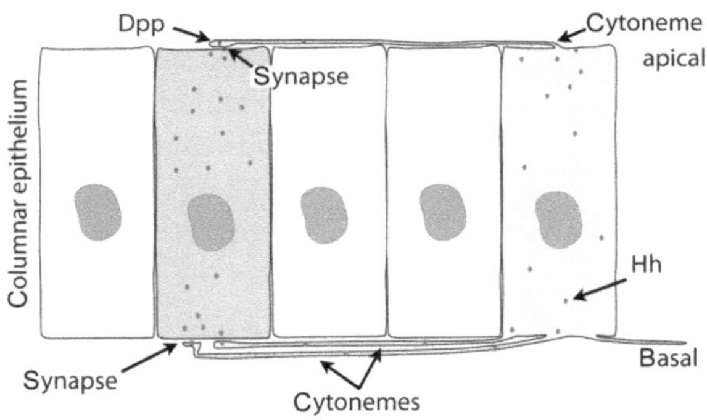

Figure 8 Synaptic signaling by columnar cells of the *Drosophila* wing imaginal disc. This drawing of a portion of the wing disc columnar epithelium depicts a Hh-expressing cell (light green) that extends Hh-carrying cytonemes from its basal surface and apical cytonemes that pick up and transport Dpp from a Dpp-expressing cell (light red). Exchange of Dpp and Hh takes place at the synapses at the tips of the cytonemes. (A color version of this figure is located at the end of this chapter.)

Dpp signal transduction extend cytonemes from their basal membrane that deliver Hh to Hh-receiving cells elsewhere in the disc, and the same cells extend cytonemes from their apical membrane that take up Dpp from Dpp-expressing disc cells. By comparison to a neuron that has an axon that receives signals and dendrites that deliver signals, these epithelial cells can also be described as having asymmetric extensions that either receive or deliver signals. There are differences. For instance, microtubules constitute the structural core of dendrites and axons, but filopodia contain actin filaments and lack detectable tubulin. However, the different structural features of these extensions may not signify fundamental functional distinctions: dendrites grow by sprouting filopodia, some of which mature into dendrite branches that make synapses, and the growth cones of axons are guided by pioneering growth cone filopodia. Moreover, several proteins have been identified (Capricious, Diaphanous, and Neuroglian) that localize to the synapses of both axons and cytonemes and that contribute essential functions to both types of synapses. Thus, the differences between cytonemes, dendrites, and axons may mask fundamental homologies of form and function. Although it remains for further work to determine whether electrical excitability is a feature that distinguishes neurons from other cells, it is already clear that there are important signaling functions that neuronal and cytoneme synapses share.

6. ONE ELEGANT SOLUTION

The term positional information represents the idea that cells are informed of their relative position by genetically encoded coordinate systems, and that by interpreting their relative positional value, neurons make their immensely complex patterns of connections and all other cells contribute appropriately to the specific and complex multicellular forms in tissues (Wolpert, 1969). The model posits that information in these coordinate systems is encoded in the identity and quantity of signaling molecules that distribute in space to form concentration gradients, and that cells adopt fates at a particular time and place based on their position in the gradients. This positional information model is supported by the distributions and activities of morphogen signaling proteins that have been characterized in a number of different systems. However, the existence of these concentration gradients does not reveal how the gradients form or how a continuous distribution might reproducibly specify the many detailed and complex outcomes.

The "inducers" of Spemann (Spemann & Mangold, 1924) and positional information substances of Wolpert (Wolpert, 1969) were assumed to move across tissues by passive diffusion and to engage their targets where they find them. This would be an example of paracrine signaling, a form of cell–cell signaling that contrasts with synaptic and endocrine signaling. Whereas synaptic signaling is spatially specific, temporally precise, rapid and quantitative, paracrine signaling is modeled as involving signals that are secreted into extracellular space and spread out to the neighborhood by diffusion; it is spatially limited, slow, and temporally and quantitatively imprecise. Endocrine signaling also lacks rapidity and either temporal or quantitative precision, but generates uniform distributions in contrast to paracrine signaling. A positional information system that precisely specifies forms and patterns would seem to demand great spatial, temporal, and quantitative control of the distributions of its active substances, and therefore would seem to be incompatible with imprecisely defined concentration gradients that develop by passive diffusion of secreted morphogens. The same argument applies to secreted guidance cues, which specify the branching patterns of neuronal networks.

The observed distributions of morphogen signaling proteins show that these proteins are endocytosed by and induce responses in target cells, some of which may be separated from producing cells by long distances—several hundred microns (dozens of cells) in the chick limb bud, for example. They act at a distance. However, these distributions do not establish where the proteins are released from producing cells or what state they are in when they move between producing and target cells. The paracrine model assumes that they are secreted from the plasma membrane of a producing cell and that they disperse in extracellular space after release, but there has been no direct experimental evidence for a site of release or for extracellular protein that supports the paracrine mechanism, despite the many studies that have characterized the properties of signaling proteins. For example, there have been many studies of signaling proteins produced by cultured cells. These signaling proteins can be found in the medium in which the cells are grown, and have been shown to have biological activity, but cells in culture are not in a normal context and it is not clear whether the *ex vivo* behavior of cultured cells might inform us about normal development where constraint is critical. Constitutive release of the proteins in culture does not show that producing cells release them similarly *in vivo*. Another example is the analysis of beads that are impregnated with signaling proteins and are implanted in embryos. Although signaling proteins diffuse from such beads and spread to surrounding areas and activate signal transduction in the neighboring cells, the relevance to the way that cells *in vivo* present and release signaling proteins is unknown. The cell culture and implanted bead experiments show that cells can respond when presented with an activating ligand, but they do not show that the signaling proteins normally find their receptors by passive diffusion. Distributions that are observed in fixed specimens are also uninformative if the structures that move signaling proteins do not survive the procedures of fixation and preparation, or if the analysis lacks sufficient resolution.

Cytoneme-mediated transport and cytoneme synaptic signaling have the level of control that we expect of a positional information system. The roles of cytonemes as structures that move signaling proteins between cells and present signaling proteins for direct exchange are also supported by experiment. Cytonemes extend between signal producing cells and their targets, making direct synaptic contacts that link signaling cells. In studies of Dpp signaling, receptor bound Dpp has been observed in vesicles that move along cytonemes that link signaling cells at synaptic contacts. These motile vesicles transport Dpp from producing to receiving cells. This is the mechanism of Dpp dissemination. Dpp transport along cytonemes follows the secretion and release of Dpp at the synapse and uptake of Dpp by the receiving cell. Transit across the synaptic cleft is presumably by passive diffusion, but the essential feature is that Dpp is delivered directly to the target cell and exchange is focused to the controlled environment of the synaptic gap. This mechanism is likely endowed with features that regulate the number of cytonemes, the number and constitution of the Dpp and receptor-containing vesicles, and the rate at which the vesicles move along the cytonemes and release their contents at the synapse. These parameters presumably determine the contours of Dpp distribution across the field of target cells.

This cytoneme mechanism of dispersion and delivery applies not only to Dpp, but has also been shown for Hh, Wg, FGF, and Delta/Notch in the wing imaginal disc, EGF in the eye imaginal disc (Callejo et al., 2011; Huang & Kornberg, 2015; Ramírez-Weber & Kornberg, 1999; Roy et al., 2011, 2014), Dpp in the stem cell niche of the *Drosophila* testis (Inaba, Buszczak, & Yamashita, 2015), Wnt (Stanganello et al., 2015), and Notch (Hamada et al., 2014) in the zebrafish embryo and Hh in the chick embryo (Sanders, Llagostera, & Barna, 2013). There are many other tissues in which these proteins are known to signal in the *Drosophila*, zebrafish, and chick embryos and these would seem to be likely candidates for cytoneme-mediated transport as well. There are, in addition, other contexts in which the presence and distribution of filopodia correlates with cell–cell signaling and with cell-to-cell transport—the spread of viruses from infected to uninfected cells (Sherer et al., 2007) and signaling between tumor and stromal cells (Calleja & Kornberg, 2015) to cite two. Cytonemes may be conduits that are relevant generally to nonendocrine cell–cell communication.

7. RECALCULATING THE HORIZON

The model of positional information as a coordinate system that is elaborated in the distributions of signaling molecules has been validated by the discoveries of signaling molecules that spread across tissues in concentration gradients. Although these signaling proteins induce various concentration-dependent responses, their properties and the contexts in which the gradients form make it implausible that passive diffusion can specify information with the precision and robustness that biological systems require (Kerszberg & Wolpert, 2007). In contrast, many features of the cytoneme mechanism make it an attractive alternative.

Cytonemes move signaling proteins across tissues using membrane-encased extensions and transfer the signaling proteins between cells at synapses, thereby constraining signal dispersion to selected targets with a level of precision that has previously only been associated with the nervous system. Moreover, in transit signaling proteins are present in vesicles that move along extensions that have backbones composed of cytoskeletal filaments, so the formation, maturation, movement, and composition of these vesicles offers many steps that might be regulated. And because experience teaches us that most biological processes are controlled at multiple levels and that few opportunities for regulation are missed, we assume that cytoneme formation, synapse formation and stability, efficiency of synaptic transfer, and flux of signaling protein-containing vesicles are regulated in order to control the contours and characteristics of the concentration gradients. These features of the cytoneme mechanism of signal dispersion change the paradigm of long distance cell–cell signaling.

Although the evidence that cytonemes are a means for transport and transfer is strong, it remains for future work to determine where and when signaling protein activates signal transduction in the responding cell and how cytonemes navigate to their targets to make the correct connections. Clearly there is much to learn and many fundamental questions left to answer, but the questions can now be posed from the perspective of a known mechanism.

REFERENCES

Brenner, S. (1996). Francisco Crick in Paradiso. *Current Biology*, *6*, 1202.

Browne, N. E. (1909). The production of new hydranths in Hydra by the insertion of small grafts. *Journal of Experimental Zoology*, *7*, 1–23.

Calleja, M. & Kornberg, T. B. (2015). Unpublished observations.

Callejo, A., Bilioni, A., Mollica, E., Gorfinkiel, N., Andres, G., Ibanez, C., et al. (2011). Dispatched mediates Hedgehog basolateral release to form the long-range morphogenetic gradient in the Drosophila wing disk epithelium. *Proceedings of the National Academy of Sciences of the United States of America*, *108*, 12591–12598.

Hamada, H., Watanabe, M., Lau, H. E., Nishida, T., Hasegawa, T., Parichy, D. M., et al. (2014). Involvement of delta/notch signaling in zebrafish adult pigment stripe patterning. *Development*, *141*, 318–324.

Huang, H., & Kornberg, T. B. (2015). Myoblast cytonemes mediate Wg signaling from the wing imaginal disc and delta-notch signaling to the air sac primordium. *eLife*, *4*, e06114.

Inaba, M., Buszczak, M., & Yamashita, Y. M. (2015). Nanotubes mediate niche-stem cell signaling in the Drosophila testis. *Nature*, *523*(7560), 329–332.

Kerszberg, M., & Wolpert, L. (2007). Specifying positional information in the embryo: Looking beyond morphogens. *Cell*, *130*, 205–209.

Ramírez-Weber, F. A., & Kornberg, T. B. (1999). Cytonemes: Cellular processes that project to the principal signaling center in Drosophila imaginal discs. *Cell*, *97*, 599–607.

Roy, S., Hsiung, F., & Kornberg, T. B. (2011). Specificity of Drosophila cytonemes for distinct signaling pathways. *Science*, *332*, 354–358.

Roy, S., Huang, H., Liu, S., & Kornberg, T. B. (2014). Cytoneme-mediated contact-dependent transport of the Drosophila decapentaplegic signaling protein. *Science*, *343*, 1244624.

Sanders, T. A., Llagostera, E., & Barna, M. (2013). Specialized filopodia direct long-range transport of SHH during vertebrate tissue patterning. *Nature*, *497*, 628–632.

Sherer, N. M., Lehmann, M. J., Jimenez-Soto, L. F., Horensavitz, C., Pypaert, M., & Mothes, W. (2007). Retroviruses can establish filopodial bridges for efficient cell-to-cell transmission. *Nature Cell Biology*, *9*, 310–315.

Spemann, H., & Mangold, H. (1924). Induction of embryonic primordia by implantation of organizers from a different species. *Roux's Arch*, *109*, 599–638.

Stanganello, E., Hagemann, A. I., Mattes, B., Sinner, C., Meyen, D., Weber, S., et al. (2015). Filopodia-based Wnt transport during vertebrate tissue patterning. *Nature Communications*, *6*, 5846.

Wolpert, L. (1969). Positional information and the spatial pattern of cellular differentiation. *Journal of Theoretical Biology*, *25*, 1–47.

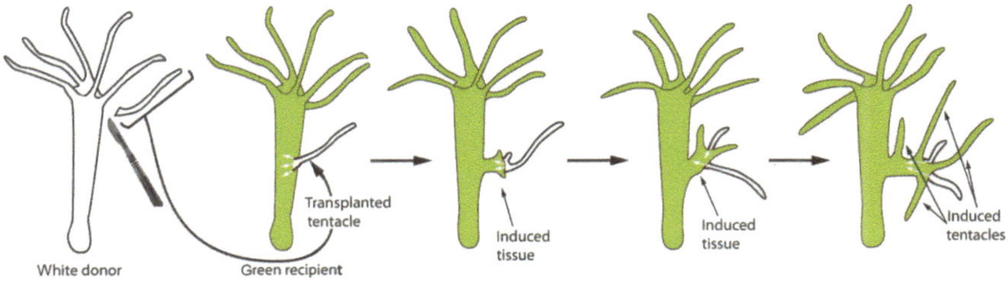

Thomas B. Kornberg, Figure 1 Grafting experiments reveal developmental induction. A tentacle transplanted from a white hydra to the stalk of a green recipient induces the host to produce ectopic tentacles.

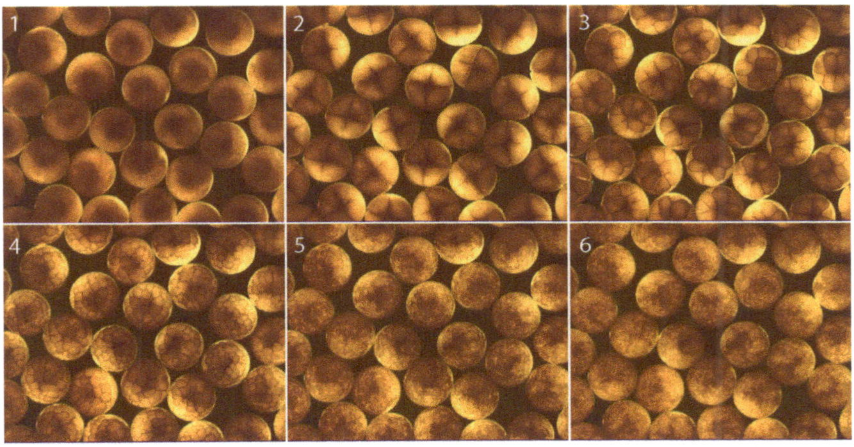

Thomas B. Kornberg, Figure 2 *Xenopus* embryos undergoing cleavage stage divisions. Fertilization of this field of embryos was initiated simultaneously, and the synchrony and regularity of the early stages of development are apparent in these six frames of a movie taken by J. Smith.

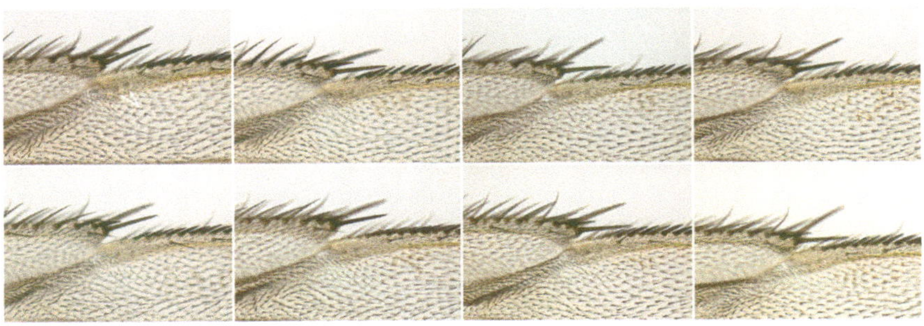

Thomas B. Kornberg, Figure 3 Images of the *Drosophila* wing. Micrographs showing a portion of the wing margin of wings dissected from eight *Drosophila* adults. The orientation, relative size, and color of all the cuticular elements are reproducibly patterned in these specimens. Arrows point to two sensillae, sense organs of the peripheral nervous system.

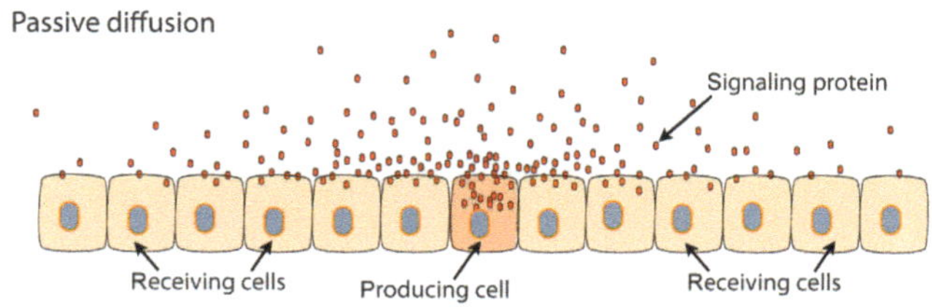

Thomas B. Kornberg, Figure 4 Passive diffusion of secreted signaling protein. This drawing of a hypothetical epithelial sheet depicts a single source cell that secretes signaling protein into extracellular space, generating a distribution of signaling protein that declines with increasing distance from the source.

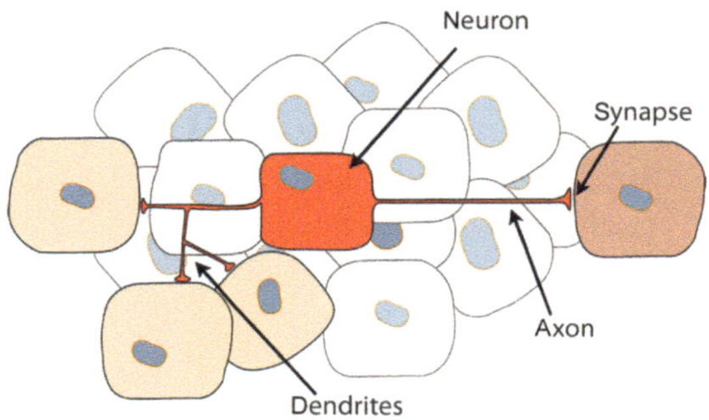

Thomas B. Kornberg, Figure 5 Cytoneme-mediated dispersion of signaling protein. This drawing of a hypothetical epithelial sheet depicts a single source cell that released signaling protein only at sites of direct contact between cytonemes that extend from cells that receive the signaling protein. Signaling protein moves across the epithelial sheet in a cytoneme-associated form and is extracellular only at the point of transfer where cytonemes make direct contact.

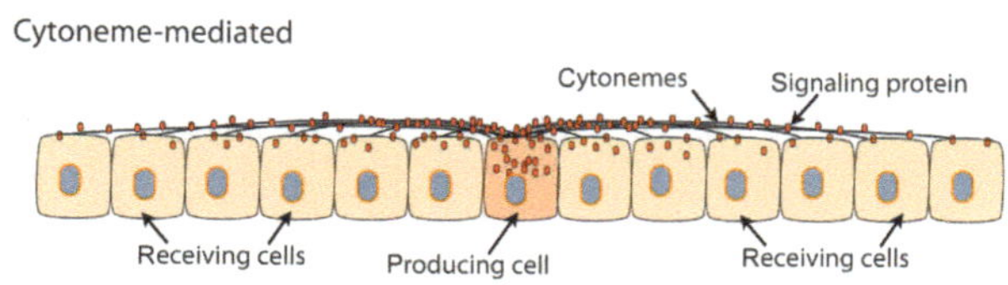

Thomas B. Kornberg, Figure 6 Synaptic signaling by a neuron. This drawing of a hypothetical neuron with an axon and dendritic extensions, making multiple synaptic contacts with target cells.

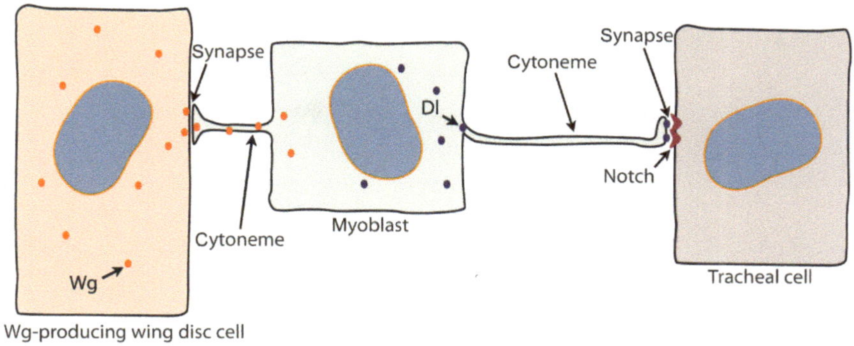

Thomas B. Kornberg, Figure 7 Synaptic signaling by columnar, tracheal, and myoblast cells of the *Drosophila* wing imaginal disc. The three cell types depicted are a wing disc columnar epithelial cell (light red) that expresses Wg, a myoblast (light blue) that picks up Wg from the disc cell and expresses Dl, and a tracheal cell (light purple) that expresses Notch and picks up Dl from the myoblast. Exchange of Wg and signaling by Dl takes place at the synapses at the tips of the two types of myoblast cytonemes.

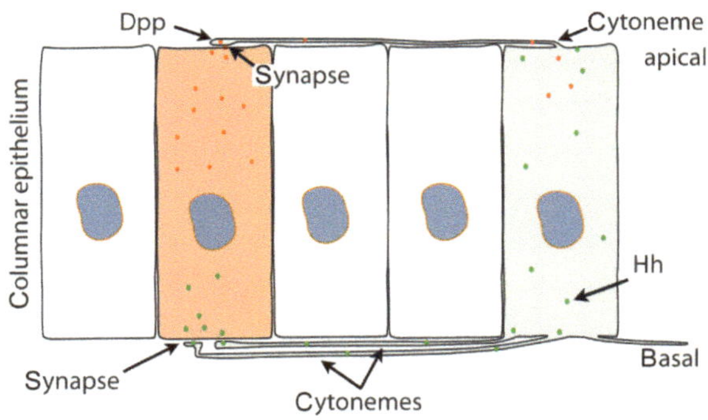

Thomas B. Kornberg, Figure 8 Synaptic signaling by columnar cells of the *Drosophila* wing imaginal disc. This drawing of a portion of the wing disc columnar epithelium depicts a Hh-expressing cell (light green) that extends Hh-carrying cytonemes from its basal surface and apical cytonemes that pick up and transport Dpp from a Dpp-expressing cell (light red). Exchange of Dpp and Hh takes place at the synapses at the tips of the cytonemes.

CHAPTER THIRTY-THREE

Gene–Environment Interactions and the Etiology of Birth Defects

Robert S. Krauss[1], Mingi Hong
Department of Developmental and Regenerative Biology, Icahn School of Medicine at Mount Sinai, New York, NY, USA
[1]Corresponding author: e-mail address: robert.krauss@mssm.edu

Contents

Abstract

It is thought that most structural birth defects are caused by a complex combination of genetic and environmental factors that interact to interfere with morphogenetic processes. It is important not only to identify individual genetic and environmental risk factors for particular defects but also to identify which environmental factors interact specifically with which genetic variants that predispose to the same defect. Genomic and epidemiological studies are critical to this end. Development and analysis of model systems will also be essential for this goal, as well as for understanding the mechanisms that underlie specific gene–environment interactions.

1. INTRODUCTION

Neural tube defects are caused by a little bit of this and a little bit of that.

F. Clarke Fraser, International Neural Tube Defect Conference, September 12, 2009

Serious structural birth defects affect approximately 8 million newborns each year, worldwide (Christianson, Howson, & Modell, 2006). The Centers for Disease Control and Prevention estimates that about 1 in 33 newborns in the United States are so affected (Centers for Disease Control, 2008). Among the most common such defects are congenital heart defects (CHD), cleft lip and/or palate (CL/P), neural tube defects (NTD), and

Current Topics in Developmental Biology, Volume 116
ISSN 0070-2153
http://dx.doi.org/10.1016/bs.ctdb.2015.12.010

holoprosencephaly (HPE). Each of these defects may occur as part of a known syndrome, but also in isolated, nonsyndromic form; it is the latter with which this essay is most concerned. Many infants with CHD will die in infancy, and the great majority of fetuses with HPE die *in utero* (Gilboa, Salemi, Nembhard, Fixler, & Correa, 2010; Shiota, Yamada, Komada, & Ishibashi, 2007). Children with CHD, NTD, and CL/P may face multiple surgeries. Even in the case of those with CL/P, where the frequency of infant mortality is low, lifelong physiological and psychological problems may occur (Dixon, Marazita, Beaty, & Murray, 2011). Consequently, birth defects cause considerable suffering to patients and their families, and they are a sizable public health concern.

It is widely believed that most such defects arise from a complex and ill-defined combination of genetic and environmental factors. Therefore, it is by no means flippant to state that the etiology of such defects involves "a little bit of this and a little bit of that." A major challenge to clinicians and scientists is to identify the genetic and nongenetic factors that contribute to birth defects, sort out which such factors interact in a specific manner, and understand the mechanistic bases of these interactions. It is already clear that preventive measures can be successful in reducing the frequency of some birth defects. Accordingly, it is to be expected that in-depth knowledge of gene–environment interactions that underlie birth defects should produce more of these successes. In this essay, we discuss some of the challenges in deciphering gene–environment interactions and also recent work with model systems that is beginning to provide insights into this problem. Readers are referred to the following articles for detailed reviews on the birth defects named above (Dixon et al., 2011; Gelb & Chung, 2014; Juriloff & Harris, 2008; Petryk, Graf, & Marcucio, 2015; Roessler & Muenke, 2010; van der Linde et al., 2011; Wilde, Petersen, & Niswander, 2014).

2. ETIOLOGY OF BIRTH DEFECTS

Observations on familial recurrence, and studies with twins, of CHD, NTD, CL/P, HPE, and other structural birth defects indicated a genetic component to these anomalies (Webber et al., 2015). Subsequently, analyses of candidate genes, genome-wide association studies with single-nucleotide polymorphism (SNP) arrays, and whole exome/genome sequencing have revealed variants in numerous genes in each of these categories of birth defects (Webber et al., 2015). Such studies have reinforced the notion that gene mutations that fall within specific pathways are often associated with specific birth defects. Examples include the planar cell polarity pathway in NTDs, and the Sonic hedgehog (SHH) and Nodal pathways in HPE (Petryk et al., 2015; Wilde et al., 2014). However, the fraction of birth defects that have a known genetic defect ascribed to them remains small. Additionally, it is likely that rare variants are important, meaning that analysis of large numbers of cases will be required for a comprehensive understanding. Nevertheless, with the ever-increasing sophistication and ever-decreasing costs of genomic analyses, one can visualize a path toward a thorough survey of the genetic component in causation of these birth defects.

There is compelling evidence that a genetic component is insufficient to fully explain the etiology of many birth defects. For example, non-Mendelian inheritance patterns are observed in CL/P and HPE (Dixon et al., 2011; Roessler, Vélez, Zhou, & Muenke, 2012). This issue has been well studied for HPE. In individuals with mutations in the HPE-associated genes, *SHH* and *SIX3*, the mutant allele was inherited more than 80% of the time, but HPE penetrance was incomplete and clinical presentation highly variable (Lacbawan et al., 2009; Solomon et al., 2012). Importantly, HPE-associated mutations are always heterozygous. Statistical modeling of this scenario has suggested that such mutations are necessary but not sufficient for severe defects and that multiple modifiers, genetic and/or environmental, likely dictate the phenotypic response to haploinsufficiency (Roessler et al., 2012).

Environmental influences are also important variables in the etiology of structural birth defects (Gilbert-Barness, 2010). Feldkamp et al. have recently suggested a distinction between true teratogens and "risk factors" (Feldkamp, Botto, & Carey, 2015). This view states that teratogens are sufficient to induce structural defects, whereas risk factors are defined as demographic characteristics and exposures that play a predisposing role in multifactorial causation. This is a useful distinction, and some birth defects may be caused directly and solely by teratogen exposure, while other cases occur via a confluence of factors. Nevertheless, the boundaries of such categories may not be fully clean, as a subthreshold dose of a true teratogen could conceivably cooperate with or modify other genetic or nongenetic insults in a multifactorial etiology. Many human teratogens/nongenetic risk factors have been identified, including infectious agents, environmental toxins, prescription drugs, and maternal health (e.g., prepregnancy diabetes and/or obesity) (Gilbert-Barness, 2010). Further work in this area is of high importance, as nongenetic risk factors may offer opportunities for prevention in a way that genetic predispositions generally will not. The classic example here is maternal dietary supplementation of folic acid (FA) in reducing the risk for NTD (Wallingford, Niswander, Shaw, & Finnell, 2013; Zhu, Kartiko, & Finnell, 2009). Despite these successes, identification of nongenetic risk factors for birth defects remains challenging. Whereas whole genome sequencing will presumably eventually allow an unbiased survey of genetic contributions, epidemiological investigation of nongenetic risk factors requires assessment of preselected conditions. Such potential risk factors as ubiquitous environmental pollutants are hard to study in this way.

If most birth defects are caused by the interaction of multiple genetic and environmental factors, it will be important not only to identify individual genetic and environmental risk factors but also to identify which environmental factors may interact specifically with which genetic variants that predispose to the same defect. It seems highly likely that such specific interactions occur. For example, 30–50% of NTD are not preventable by FA supplementation (Blom, Shaw, den Heijer, & Finnell, 2006; Wallingford et al., 2013). The mechanisms that underlie this observation are unknown. More than 300 genes, encompassing diverse functions and pathways, are required for neural tube closure (Wilde et al., 2014), so FA-responsive and FA-nonresponsive NTD seem likely to arise through distinct pathophysiological mechanisms. These mechanisms may in turn have etiologies comprising distinct genetic and environmental risk factors. Studies on birth defects that simultaneously assess both types of risk factors have been performed and suggested some relatively specific connections. For example, interactions between specific gene variants and maternal cigarette smoking may potentially occur in CL/P and CHD (Tang et al., 2015; Zeiger, Beaty, & Liang, 2005). In general, however, it is difficult to acquire information on both from the same patients (and, ideally, their parents). Furthermore, very large sample sizes will be required to reliably detect specific gene–environment interactions. The estimable efforts of the National Birth Defects Prevention Study are expected to make progress toward this critical goal. Model systems are an additional way to probe gene–environment interactions. Although not a substitute for human population-based studies, they are of high value in addressing specific questions and offer unparalleled ability to explore mechanisms.

3. A ROLE FOR MODEL SYSTEMS IN UNDERSTANDING THE COMPLEX ETIOLOGY OF BIRTH DEFECTS

Animal models have been a rich source of information about structural birth defects. In many cases, targeted mutations in mice provided initial leads to the identification of genetic variants associated with human birth defects. Rodents have also long been used in teratology research. More recently, genetically tractable organisms like mice and zebrafish have allowed testing of hypotheses about specific gene–environment interactions under controlled conditions, something that is rarely achievable for the human population. Importantly, screens for such interactions are also feasible. An exhaustive review of this topic is beyond the scope of this essay, and illustrative examples of recent progress are discussed here.

Alcohol (ethanol) is recognized as a human teratogen (Gilbert-Barness, 2010; Mead & Sarkar, 2014). Maternal consumption is associated with fetal alcohol spectrum disorders (FASD), which include anatomical, behavioral, and cognitive abnormalities; additionally, it has been associated with various craniofacial defects and HPE. Studies with mice are consistent with alcohol functioning as a teratogen and also as a contributing component to multifactorial causation of birth defects. Fetal alcohol induces craniofacial anomalies similar to those seen in FASD and also causes HPE in mice, but in a strain-dependent manner (Downing, Balderrama-Durbin, Broncucia, Gilliam, & Johnson, 2009; Hong & Krauss, 2012; Lipinski, Godin, O'leary-Moore, Parnell, & Sulik, 2010; O'Connor, Steeds, Wiseman, & Pavlath, 2008). C57BL/6 mice are sensitive, whereas 129S6 mice are resistant to these effects of alcohol (Downing et al., 2009; Hong & Krauss, 2012; Sulik, Johnston, & Webb, 1981). Defects in the SHH pathway are the main known cause of HPE, including mutations in *SHH*, and its receptors *PTCH1*, *CDON*, and *GAS1* (Bae et al., 2011; Pineda-Alvarez et al., 2012; Roessler & Muenke, 2010). 129S6 mice with a targeted mutation in *Cdon* have a subthreshold defect in SHH signaling due to expression of additional receptors that provide a redundant function (Allen et al., 2011; Allen, Tenzen, & McMahon, 2007; Zhang, Hong, Bae, Kang, & Krauss, 2011; Zhang, Kang, Cole, Yi, & Krauss, 2006). Consequently, *Cdon*-null mice on a 129S6 background have only mild HPE phenotypes at low penetrance. However, these animals are sensitized to fetal alcohol-induced HPE at high penetrance, including the most severe forms of this defect (Hong & Krauss, 2012). Furthermore, loss of Cdon and fetal alcohol exposure combined to inhibit SHH signaling in the developing rostroventral midline of these embryos. This model demonstrated a strongly synergistic interaction between mutation of a bona fide human HPE gene and a suspected environmental risk factor for HPE. Heterozygosity for *Shh*, the most frequent known mutation in human HPE, also interacted with fetal alcohol to exacerbate HPE, even in C57BL/6 mice (Kietzman, Everson, Sulik, & Lipinski, 2014).

The ease of use and low cost of zebrafish make them a model system amenable to screens for gene–environment interactions in production of structural birth defects. Eberhart and colleagues screened a panel of mutant zebrafish lines for sensitivity to alcohol-induced craniofacial abnormalities (McCarthy & Eberhart, 2014; McCarthy et al., 2013; Swartz et al., 2014). Several loci that showed such interactions were identified, notably *Pdgfra* (encoding platelet-derived growth factor receptor-α). A genome-wide association study of FASD individuals provided preliminary evidence that this interaction may be relevant in humans (McCarthy et al., 2013). From these studies, one can envision that larger scale zebrafish screens may be a practical way to identify gene–environment interactions, either screening a single important mutant with many potentially relevant environmental agents or a single environmental agent against a large panel of mutants.

Hypoxia is a suspected environmental condition that may contribute to certain structural developmental defects (Hutter, Kingdom, & Jaeggi, 2010). Among these are vertebral malformations such as congenital scoliosis, a lateral curvature of the spine (Giampietro et al., 2013). Recessive mutations in *HES7* and *MESP2* have been identified in the more severe vertebral malformation, spondylothoracic dysostosis (Sparrow, Chapman, & Dunwoodie, 2011). Mice heterozygous for mutations in *Hes7* or *Mesp2* have mild or no obvious spinal deformation, respectively, and exposure of pregnant dams to mild short-term hypoxia also produced only mild defects in wild-type offspring. In contrast, the same hypoxic exposure strongly exacerbated formation of vertebral defects in *Hes7*$^{+/-}$ and *Mesp2*$^{+/-}$ offspring (Sparrow et al., 2012). *DLL3* is also mutated in spondylothoracic dysostosis, but in contrast to *Hes7*$^{+/-}$ and *Mesp2*$^{+/-}$ mice, *Dll3*$^{+/-}$ mice were not sensitized to hypoxia, indicating the specificity of the response in the former two lines (Sparrow et al., 2011, 2012). Vertebrae are derived from the ventral portion of the somite, the sclerotome. Further experimentation led to the hypothesis that hypoxia inhibited FGF signaling during somite formation, delaying this process (Sparrow et al., 2012). By this notion, haploinsufficiency for a gene important for the timing of somitogenesis (i.e., *Hes7* or *Mesp2*) acted in combination with a temporary, hypoxia-induced halt to somite formation to produce an irreversible negative outcome.

Molecular and developmental analysis of mouse mutants may also provide insight into unanticipated, potential gene–environment interactions. Mice lacking both the paired-box transcription factors, Pax3 and Pax7, have facial clefts, whereas *Pax3* and *Pax7* single mutants do not display clefts (Zalc, Rattenbach, Auradé, Cadot, & Relaix, 2015). Genome-wide association studies have identified SNPs in the *PAX3* locus that are associated with distance between the eyes and positioning of the nasion (Liu et al., 2012). Additionally, *PAX3* is mutated in one form of Waardenburg syndrome, which features craniofacial abnormalities (Wang et al., 2008). Transcriptome analysis of the facial prominences of embryos lacking Pax3/Pax7 function revealed strongly elevated expression of the aryl hydrocarbon receptor (*AhR*) and its target genes (Zalc et al., 2015). 2,3,7,8-Tetrachlorodibenzo-*p*-dioxin (TCDD) is a potent ligand for AhR and causes cleft palate in sensitive strains of mice (Bock & Köhle, 2006; Moriguchi et al., 2003). Loss of *Pax3* in mice of a TCDD-resistant genetic background sensitized these mice to TCDD-induced facial clefting (Zalc et al., 2015). Humans are less sensitive to TCDD than mice, but it is conceivable that individuals with weak *PAX3* variant alleles may be sensitized to craniofacial defects by exposure to TCDD or other AhR ligands. Interestingly, facial clefting in embryos lacking Pax3/Pax7 function was substantially rescued by treatment with an AhR antagonist (Zalc et al., 2015).

4. POTENTIAL MECHANISMS OF GENE–ENVIRONMENT INTERACTIONS

The use of animal models to study gene–environment interactions offers opportunities to understand the mechanistic bases of such interactions. For example, it is possible to determine the window of sensitivity to a given environmental agent, which in turn may provide important leads as to mechanism of synergy with a specific mutation. Many critical genes function reiteratively in development, even within morphogenesis of a single organ or structure. Identification of a window of sensitivity to an environmental exposure defines a temporal limitation as to which functions of the gene in question could be the key targets in production of a defect. Additionally, the ability to perform *in situ* hybridization and other analyses of gene activity in mouse and zebrafish embryos allows for specific readouts of localized effects on gene expression and pathway activity in affected embryos.

A priori, several potential types of interaction may be considered. As mentioned above, defects in specific pathways are often associated with specific structural defects. If a threshold of activity of a signaling pathway is required for successful development of the structure in question, it is logical to envision that a heterozygous mutation in a given pathway would interact with an environmental agent that targeted the same pathway. The environmental agent could conceivably act directly on the protein produced by the remaining wild-type allele (i.e., the genetic and nongenetic insults have an identical target), or it could act on another factor within the pathway, rendering total signal strength below the level needed to avoid defects in the developmental process. An additional possibility is that submaximal inhibition of two pathways that both promote a similar developmental process could synergize to result in a defective outcome. Although this type of interaction is less intuitive than the previously described types, experimental evidence in a chick model of HPE suggests that it could occur. Heterozygous mutations in the NODAL and SHH signaling pathways are each associated with HPE, but thought to be insufficient on their own to cause severe defects (Roessler et al., 2012). Treatment of chick embryos at appropriate developmental stages with individual chemical inhibitors of these two pathways produced HPE, consistent with the phenotypes of recessive mutations in these pathways in both mice and zebrafish (Mercier et al., 2013). Simultaneous treatment of chick embryos with suboptimal doses of the two inhibitors led to a synergistic interaction that produced more severe forms of HPE than could be accounted for by a simply additive effect (Mercier et al., 2013).

More complicated scenarios are also possible and are indeed likely. Many, if not most, morphogenetic signaling pathways alter gene expression programs as a direct outcome of signaling. These gene expression programs are, in turn, critical for promoting the biological responses to signaling, usually in a cell type-specific manner. The genes comprising these programs work together in a network to dictate cell fate and morphogenesis. The Soriano lab performed a screen that identified downstream transcriptional targets of PDGF signaling and then permitted the rapid genetic analysis of these target genes in mice. It was demonstrated that mutation of some PDGF pathway target genes yielded no or mild phenotypes, but further genetic removal of one copy of *Pdgfra* produced strong phenotypes, including cleft palate (Chen, Delrow, Corrin, Frazier, & Soriano, 2004; Schmahl, Raymond, & Soriano, 2007). Translating these results to a gene–environment interaction scenario, it is conceivable that haploinsufficiency for *PDGFRA* could synergize with nongenetic risk factors that inhibited the function of specific downstream gene products of this receptor. Transcription factors are themselves often targets of mutation in structural birth defects, and they directly control expression of large gene expression networks. A similar, if even more complex, set of interactions might therefore apply in situations where the genetic component of a multifactorial etiology is a heterozygous mutation in a transcription factor and the nongenetic component hits a target gene of this transcription factor. Moreover, it has been suggested that epigenetic modifications of DNA or core histones may be targets of environmental insults, resulting in abnormal development (Friedman, 2011). While these modifications are ubiquitous (if distinct in different cell lineages and stages within those lineages), a combination of the appropriate timing of environmental exposure and cell type-specific effects of genetic insults could provide the necessary degree of specificity, so this notion fits in well with the ideas put forward here.

In summary, interaction between genetic and nongenetic risk factors is important in the etiology of structural birth defects. Genomic and epidemiological studies of large numbers of human subjects will be required to understand this problem. Studies with animal model systems are expected to provide critical leads and penetrating insights into the mechanisms that underlie specific gene–environment interactions. Together, these approaches should ultimately provide bases for rational prevention efforts.

ACKNOWLEDGMENTS

We thank Rick Finnell for sharing the quotation from F. Clarke Fraser that opens this essay. Work in the authors' laboratory on this subject has been supported by NIDCR, NIAAA, and the March of Dimes.

REFERENCES

Allen, B. L., Song, J. Y., Izzi, L., Althaus, I. W., Kang, J.-S., Charron, F., et al. (2011). Overlapping roles and collective requirement for the co-receptors Gas1, Cdo and Boc in Shh pathway function. *Developmental Cell, 20*, 775–787.

Allen, B. L., Tenzen, T., & McMahon, A. P. (2007). The Hedgehog-binding proteins Gas1 and Cdo cooperate to positively regulate Shh signaling during mouse development. *Genes & Development, 21*, 1244–1257.

Bae, G. U., Domené, S., Roessler, E., Schachter, K., Kang, J. S., Muenke, M., et al. (2011). Mutations in CDON, encoding a hedgehog receptor, result in holoprosencephaly and defective interactions with other hedgehog receptors. *American Journal of Human Genetics, 89*, 231–240.

Blom, H. J., Shaw, G. M., den Heijer, M., & Finnell, R. H. (2006). Neural tube defects and folate: Case far from closed. *Nature Reviews. Neuroscience*, 7, 724–731.

Bock, K. W., & Köhle, C. (2006). Ah receptor: Dioxin-mediated toxic responses as hints to deregulated physiologic functions. *Biochemical Pharmacology, 72*, 393–404.

Centers for Disease Control and Prevention. (2008). Update on overall prevalence of major birth defects. *MMWR. Morbidity and Mortality Weekly Report, 57*, 1–5.

Chen, W. V., Delrow, J., Corrin, P. D., Frazier, J. P., & Soriano, P. (2004). Identification and validation of PDGF transcriptional targets by microarray-coupled gene-trap mutagenesis. *Nature Genetics, 36*, 304–312.

Christianson, A., Howson, C. P., & Modell, B. (2006). *Global report on birth defects. In March of Dimes Global Report on Birth Defects: The Hidden Toll of Dying and Disabled Children*

(pp. 1–85). http://wwwmarchofdimesorg/materials/global-report-on-birth-defects-the-hidden-toll-of-dying-and-disabled-children-full-reportpdf.
Dixon, M. J., Marazita, M. L., Beaty, T. H., & Murray, J. C. (2011). Cleft lip and palate: Understanding genetic and environmental influences. *Nature Reviews. Genetics*, *12*, 167–178.
Downing, C., Balderrama-Durbin, C., Broncucia, H., Gilliam, D., & Johnson, T. E. (2009). Ethanol teratogenesis in five inbred strains of mice. *Alcoholism: Clinical and Experimental Research*, *33*, 1238–1245.
Feldkamp, M. L., Botto, L. D., & Carey, J. C. (2015). Reflections on the etiology of structural birth defects: Established teratogens and risk factors. *Birth Defects Research, Part A: Clinical and Molecular Teratology*, *103*, 652–655.
Friedman, J. M. (2011). Using genomics for birth defects epidemiology: Can epigenetics cut the GxE gordian knot? *Birth Defects Research, Part A: Clinical and Molecular Teratology*, *91*, 986–989.
Gelb, B. D., & Chung, W. K. (2014). Complex genetics and the etiology of human congenital heart disease. *Cold Spring Harbor Perspectives in Medicine*, *4*, a013953.
Giampietro, P. F., Raggio, C. L., Blank, R. D., McCarty, C., Broeckel, U., & Pickart, M. A. (2013). Clinical, genetic and environmental factors associated with congenital vertebral malformations. *Molecular Syndromology*, *4*, 94–105.
Gilbert-Barness, E. (2010). Teratogenic causes of malformations. *Annals of Clinical and Laboratory Science*, *40*, 99–114.
Gilboa, S. M., Salemi, J. L., Nembhard, W. N., Fixler, D. E., & Correa, A. (2010). Mortality resulting from congenital heart disease among children and adults in the United States, 1999 to 2006. *Circulation*, *122*, 2254–2263.
Hong, M., & Krauss, R. S. (2012). Cdon mutation and fetal ethanol exposure synergize to produce midline signaling defects and holoprosencephaly spectrum disorders in mice. *PLoS Genetics*, *8*, e1002999.
Hutter, D., Kingdom, J., & Jaeggi, E. (2010). Causes and mechanisms of intrauterine hypoxia and its impact on the fetal cardiovascular system: A review. *International Journal of Pediatrics*, *2010*, 9.
Juriloff, D. M., & Harris, M. J. (2008). Mouse genetic models of cleft lip with or without cleft palate. *Birth Defects Research, Part A: Clinical and Molecular Teratology*, *82*, 63–77.
Kietzman, H. W., Everson, J. L., Sulik, K. K., & Lipinski, R. J. (2014). The teratogenic effects of prenatal ethanol exposure are exacerbated by Sonic Hedgehog or GLI2 haploinsufficiency in the mouse. *PLoS One*, *9*(2), e89448.
Lacbawan, F., Solomon, B. D., Roessler, E., El-Jaick, K., Domené, S., Vélez, J. I., et al. (2009). Clinical spectrum of SIX3-associated mutations in holoprosencephaly: Correlation between genotype, phenotype and function. *Journal of Medical Genetics*, *46*, 389–398.
Lipinski, R. J., Godin, E. A., O'leary-Moore, S. K., Parnell, S. E., & Sulik, K. K. (2010). Genesis of teratogen-induced holoprosencephaly in mice. *American Journal of Medical Genetics Part C: Seminars in Medical Genetics*, *154C*, 29–42.
Liu, F., van der Lijn, F., Schurmann, C., Zhu, G., Chakravarty, M., Hysi, P., et al. (2012). A genome-wide association study identifies five loci influencing facial morphology in Europeans. *PLoS Genetics*, *8*, e1002932.
McCarthy, N., & Eberhart, J. (2014). Gene–ethanol interactions underlying fetal alcohol spectrum disorders. *Cellular and Molecular Life Sciences*, *71*, 2699–2706.
McCarthy, N., Wetherill, L., Lovely, C. B., Swartz, M. E., Foroud, T. M., & Eberhart, J. K. (2013). Pdgfra protects against ethanol-induced craniofacial defects in a zebrafish model of FASD. *Development*, *140*, 3254–3265.
Mead, E. A., & Sarkar, D. (2014). Fetal alcohol spectrum disorders and their transmission through genetic and epigenetic mechanisms. *Frontiers in Genetics*, *5*, 154.

Mercier, S., David, V., Ratié, L., Gicquel, I., Odent, S., & Dupé, V. (2013). NODAL and SHH dose-dependent double inhibition promotes an HPE-like phenotype in chick embryos. *Disease Models & Mechanisms, 6*, 537–543.
Moriguchi, T., Motohashi, H., Hosoya, T., Nakajima, O., Takahashi, S., Ohsako, S., et al. (2003). Distinct response to dioxin in an arylhydrocarbon receptor (AHR)-humanized mouse. *Proceedings of the National Academy of Sciences of the United States of America, 100*, 5652–5657.
O'Connor, R. S., Steeds, C. M., Wiseman, R. W., & Pavlath, G. K. (2008). Phosphocreatine as an energy source for actin cytoskeletal rearrangements during myoblast fusion. *The Journal of Physiology, 586*, 2841–2853.
Petryk, A., Graf, D., & Marcucio, R. (2015). Holoprosencephaly: Signaling interactions between the brain and the face, the environment and the genes, and the phenotypic variability in animal models and humans. *Wiley Interdisciplinary Reviews. Developmental Biology, 4*, 17–32.
Pineda-Alvarez, D. E., Roessler, E., Hu, P., Srivastava, K., Solomon, B. D., Siple, C. E., et al. (2012). Missense substitutions in the GAS1 protein present in holoprosencephaly patients reduce the affinity for its ligand, SHH. *Human Genetics, 131*, 301–310.
Roessler, E., & Muenke, M. (2010). The molecular genetics of holoprosencephaly. *American Journal of Medical Genetics Part C: Seminars in Medical Genetics, 154C*, 52–61.
Roessler, E., Vélez, J. I., Zhou, N., & Muenke, M. (2012). Utilizing prospective sequence analysis of SHH, ZIC2, SIX3 and TGIF in holoprosencephaly probands to describe the parameters limiting the observed frequency of mutant gene × gene interactions. *Molecular Genetics and Metabolism, 105*, 658–664.
Schmahl, J., Raymond, C. S., & Soriano, P. (2007). PDGF signaling specificity is mediated through multiple immediate early genes. *Nature Genetics, 39*, 52–60.
Shiota, K., Yamada, S., Komada, M., & Ishibashi, M. (2007). Embryogenesis of holoprosencephaly. *American Journal of Medical Genetics Part A, 143A*, 3079–3087.
Solomon, B. D., Bear, K. A., Wyllie, A., Keaton, A. A., Dubourg, C., David, V., et al. (2012). Genotypic and phenotypic analysis of 396 individuals with mutations in Sonic Hedgehog. *Journal of Medical Genetics, 49*, 473–479.
Sparrow, D., Chapman, G., & Dunwoodie, S. (2011). The mouse notches up another success: Understanding the causes of human vertebral malformation. *Mammalian Genome, 22*, 362–376.
Sparrow, D. B., Chapman, G., Smith, A. J., Mattar, M. Z., Major, J. A., O'Reilly, V. C., et al. (2012). A mechanism for gene-environment interaction in the etiology of congenital scoliosis. *Cell, 149*, 295–306.
Sulik, K. K., Johnston, M. C., & Webb, M. A. (1981). Fetal alcohol syndrome: Embryogenesis in a mouse model. *Science, 214*, 936–938.
Swartz, M. E., Wells, M. B., Griffin, M., McCarthy, N., Lovely, C. B., McGurk, P., et al. (2014). A screen of zebrafish mutants identifies ethanol-sensitive genetic loci. *Alcoholism: Clinical and Experimental Research, 38*, 694–703.
Tang, X., Hobbs, C. A., Cleves, M. A., Erickson, S. W., MacLeod, S. L., Malik, S., et al. (2015). Genetic variation affects congenital heart defect susceptibility in offspring exposed to maternal tobacco use. *Birth Defects Research, Part A: Clinical and Molecular Teratology, 103*, 834–842.
van der Linde, D., Konings, E. E. M., Slager, M. A., Witsenburg, M., Helbing, W. A., Takkenberg, J. J. M., et al. (2011). Birth prevalence of congenital heart disease worldwide: A systematic review and meta-analysis. *Journal of the American College of Cardiology, 58*, 2241–2247.
Wallingford, J. B., Niswander, L. A., Shaw, G. M., & Finnell, R. H. (2013). The continuing challenge of understanding, preventing, and treating neural tube defects. *Science, 339*, 1222002.

Wang, Q., Fang, W.-H., Krupinski, J., Kumar, S., Slevin, M., & Kumar, P. (2008). Pax genes in embryogenesis and oncogenesis. *Journal of Cellular and Molecular Medicine, 12*, 2281–2294.

Webber, D. M., MacLeod, S. L., Bamshad, M. J., Shaw, G. M., Finnell, R. H., Shete, S. S., et al. (2015). Developments in our understanding of the genetic basis of birth defects. *Birth Defects Research, Part A: Clinical and Molecular Teratology, 4*, 680–691.

Wilde, J. J., Petersen, J. R., & Niswander, L. (2014). Genetic, epigenetic, and environmental contributions to neural tube closure. *Annual Review of Genetics, 48*, 583–611.

Zalc, A., Rattenbach, R., Auradé, F., Cadot, B., & Relaix, F. (2015). Pax3 and Pax7 play essential safeguard functions against environmental stress-induced birth defects. *Developmental Cell, 33*, 56–66.

Zeiger, J. S., Beaty, T. H., & Liang, K. Y. (2005). Oral clefts, maternal smoking, and TGFA: A meta-analysis of gene-environment interaction. *The Cleft Palate-Craniofacial Journal, 42*, 58–63.

Zhang, W., Hong, M., Bae, G.-U., Kang, J.-S., & Krauss, R. S. (2011). *Boc* modifies the holoprosencephaly spectrum of *Cdo* mutant mice. *Disease Models & Mechanisms, 4*, 368–380.

Zhang, W., Kang, J.-S., Cole, F., Yi, M. J., & Krauss, R. S. (2006). Cdo functions at multiple points in the Sonic Hedgehog pathway, and Cdo-deficient mice accurately model human holoprosencephaly. *Developmental Cell, 10*, 657–665.

Zhu, H., Kartiko, S., & Finnell, R. H. (2009). Importance of gene–environment interactions in the etiology of selected birth defects. *Clinical Genetics, 75*, 409–423.

CHAPTER THIRTY-FOUR

Hox Genes and the Hindbrain: A Study in Segments

Robb Krumlauf[1]
Stowers Institute for Medical Research, Kansas City, Missouri, USA
Department of Anatomy & Cell Biology, Kansas University Medical School, Kansas City, Kansas, USA
[1]Corresponding author: e-mail address: rek@stowers.org

Abstract

The hindbrain develops through a process of segmentation which is coupled with the ordered expression of *Hox* genes to generate regional diversity of key neural and craniofacial derivatives during head development. This is a fundamental feature governed by a gene regulatory network conserved to the base of vertebrate evolution.

The vertebrate hindbrain and its relationship to head development are a good model system for understanding fundamental mechanisms of patterning and morphogenesis during development. The vertebrate central nervous system (CNS) has long been postulated to be organized into segments, referred to as neuromeres, based on morphological features. For over a century and a half wide, variation in observations between animals and stages of development led to considerable debate on the validity of this hypothesis. However, a little over 25 years ago, several studies on properties of the hindbrain in chick and mouse embryos reawakened interest in the idea of metameric segments as basic units of organization for some regions of the CNS. The hindbrain is a highly conserved complex coordination center in the CNS. Neurofilament immunostaining and DiI tracing of cranial nerve roots in the chick embryo revealed a segmental organization of the branchiomotor nerves and that nerve roots develop in a two-segment periodicity (Lumsden & Keynes, 1989). Gene expression studies in the mouse revealed that the zinc-finger transcription factor Krox20 is expressed in two alternate segments and that members of the *Hox* homeobox gene clusters are

Current Topics in Developmental Biology, Volume 116
ISSN 0070-2153
http://dx.doi.org/10.1016/bs.ctdb.2015.12.011

coordinately expressed in nested domains of the hindbrain, in a manner that correlates with specific segments (Hunt et al., 1991; Murphy, Davidson, & Hill, 1989; Wilkinson, Bhatt, Chavrier, Bravo, & Charnay, 1989; Wilkinson, Bhatt, Cook, Boncinelli, & Krumlauf, 1989). This opened the flood gates for a large number of cellular and molecular studies in different vertebrate model systems that have convincingly demonstrated that the formation of regional diversity in the hindbrain is achieved through a process of segmentation, which ultimately gives rise to well-defined regions of the adult brain. This segmental organization is critical for patterning of the cranial neural crest which generates most of the bone and connective tissues of head and facial structures. The Hox family of transcription factors is coupled to this process and provides a molecular framework for specifying the unique identities of hindbrain segments (rhombomeres) and facial structures. This body of research has collectively provided mechanistic insight into how the hindbrain arises through a process of segmentation to generate and pattern a series of distinct neural and craniofacial structures (as reviewed in Alexander, Nolte, & Krumlauf, 2009; Lumsden, 2004; Lumsden & Krumlauf, 1996; Moens & Prince, 2002; Santagati & Rijli, 2003; Trainor & Krumlauf, 2000; Tumpel, Wiedemann, & Krumlauf, 2009).

The dynamic process of hindbrain segmentation divides the developing neural plate in this region into eight lineage-restricted cellular compartments, termed rhombomeres (r). Cells freely mix along the anterioposterior (AP) axis of the neural epithelium during early stages (Fraser, Keynes, & Lumsden, 1990), but changes in cell adhesive properties (Guthrie, Prince, & Lumsden, 1993), Eph/ephrin signaling (Cooke & Moens, 2002; Mellitzer, Xu, & Wilkinson, 1999; Sela-Donenfeld & Wilkinson, 2005; Xu, Mellitzer, Robinson, & Wilkinson, 1999), and the formation of distinct boundary cell populations with signaling properties (Cheng et al., 2004; Gonzalez-Quevedo, Lee, Poss, & Wilkinson, 2010) progressively limit intermingling between rhombomeres. During this process, *Hox* gene expression becomes tightly coupled to specific segments and provides a combinatorial code which modulates differentiation programs and leads to the generation of unique identities in each rhombomere (Hunt et al., 1991; Wilkinson, Bhatt, Cook, et al., 1989). This segmentation has a major impact on patterning craniofacial development through its effects on formation, migration, and patterning of cranial neural crest cells, whose derivatives form most of the bone and connective tissue of the vertebrate head (Birgbauer, Sechrist, Bronner-Fraser, & Fraser, 1995; Nieto, Sechrist, Wilkinson, & Bronner-Fraser, 1995; Santagati & Rijli, 2003;

Simoes-Costa & Bronner, 2015; Trainor & Krumlauf, 2000; Trainor, Melton, & Manzanares, 2003). Studies employing gain and loss of function analyses in mouse, zebra fish, and other vertebrate models have validated the diverse roles of *Hox* genes in regulating multiple aspects of segmentation, segmental identity, and neural crest patterning (reviewed in Alexander et al., 2009; Maconochie, Nonchev, Morrison, & Krumlauf, 1996; Moens & Prince, 2002; Santagati & Rijli, 2003).

A major focus of my group's research over many years has been to develop a deep level of understanding of the *cis*-regulatory components of *Hox* clusters, their cognate upstream factors, and the signaling inputs which govern the segment-restricted expression and function of *Hox* genes in hindbrain development. Because the segmental processes of head development are highly conserved among jawed vertebrates, comparative studies between different species have greatly facilitated the characterization of gene regulatory cascades that control early hindbrain development. Investigation of the mechanistic basis for regulation of expression of the nine *Hox* genes implicated in hindbrain segmentation, using mouse, chicken, and zebra fish experimental models, has enabled the identification and functional characterization of the majority of *cis*-regulatory modules which direct restricted expression of these *Hox* genes in individual rhombomeres (reviewed in Alexander et al., 2009; Parrish, Nolte, & Krumlauf, 2009; Tümpel, Maconochie, Wiedemann, & Krumlauf, 2002; Tumpel, Wiedemann, & Krumlauf, 2009).

The organization and properties of this *cis*-circuitry have helped to build a picture of the *Hox* gene regulatory network (GRN) for the hindbrain and several general principles emerged from these analyses (Tümpel et al., 2009). The overall pattern of segmental expression for any given gene is the result of the combined activity of multiple independent *cis*-modules which control specific subsets of the expression domain. Each module is subject to its own distinct regulatory inputs and timing, resulting in unique regulatory circuits in each segment that confer the differential regulation of *Hox* genes. Retinoid and Fgf signaling are involved in transiently inducing early nested domains of *Hox* expression in the CNS (Bel-Vialar, Itasaki, & Krumlauf, 2002; Deschamps & van Nes, 2005; Hernandez, Rikhof, Bachmann, & Moens, 2004), which later become segmentally restricted through the combined action of transcription factors (e.g., Kreisler, Krox20, vhnf1, and Hox) in multiple modules. Direct auto- and cross-regulatory interactions between the *Hox* genes themselves play a significant role in maintaining the rhombomere-restricted domains of expression

(Gould, Morrison, Sproat, White, & Krumlauf, 1997; Manzanares et al., 2001; Tümpel et al., 2007). This modular organization of regulatory elements helps to explain variations in the levels of expression within specific rhombomeres and the complex temporal dynamics associated with establishing *Hox* expression. It also provides insight into how *Hox* genes integrate cues from the key signaling pathways involved in control of axial elongation and patterning.

The dynamic process through which *Hoxb1* expression becomes segmentally restricted to r4 in vertebrates (Frohman, Boyle, & Martin, 1990; Murphy et al., 1989; Sundin & Eichele, 1990; Wilkinson, Bhatt, Cook, et al., 1989) is illustrated in Fig. 1A and serves as a good example to illustrate some of the general regulatory principles of the *Hox* GRN. Figure 1B represents a schematic of the *Hoxb1* locus marking some of the enhancers, binding sites, and upstream factors involved in generating the r4-restricted domain of expression. These coordinate the temporal series of steps leading to the r4-restricted domain, as illustrated at the bottom of Fig. 1B. The process is triggered, by the newly forming somites adjacent to the caudal hindbrain. They express the enzyme Raldh2, which synthesizes retinoids that spread into the hindbrain to activate transcription (Begemann, Schilling, Rauch, Geisler, & Ingham, 2001; Molotkova, Molotkov, Sirbu, & Duester, 2005; Niederreither et al., 2003). A neural enhancer 3′ of *Hoxb1* contains a retinoic acid response element (RARE) (Marshall et al., 1994) which is directly stimulated by the somite-derived retinoids to activate expression of *Hoxb1* which extends anteriorly toward the midbrain (Mb). The *Hoxa1* gene also contains an analogous 3′ RARE and is activated in the same manner as *Hoxb1* but slightly earlier (Dupé et al., 1997). During this stage, a signaling center in the Mb involving the Fgf pathway induces the expression of *Cyp26a*, which encodes an enzyme that degrades retinoids (Duester, 2008; Hernandez, Putzke, Myers, Margaretha, & Moens, 2007; Sirbu, Gresh, Barra, & Duester, 2005). These opposing influences limit the initial anterior expansion of *Hoxb1* and *Hoxa1* expression to the r2/r3 boundary (Sirbu et al., 2005). In the next phase, *Krox20* is induced specifically in future r3 and r5 (Giudicelli, Taillebourg, Charnay, & Gilardi-Hebenstreit, 2001; Schneider-Maunoury et al., 1993; Wilkinson, Bhatt, Chavrier, et al., 1989) and binds to an r3/r5 repressor element upstream of *Hoxb1* (Fig. 1B). This repressor also contains an RARE, and together with Krox20, it inhibits expression specifically in r3 and r5 (Studer, Popperl, Marshall, Kuroiwa, & Krumlauf, 1994). Concurrently, levels of *Hoxb1* expression in r4 are also stimulated by Hoxa1 via a Hox response

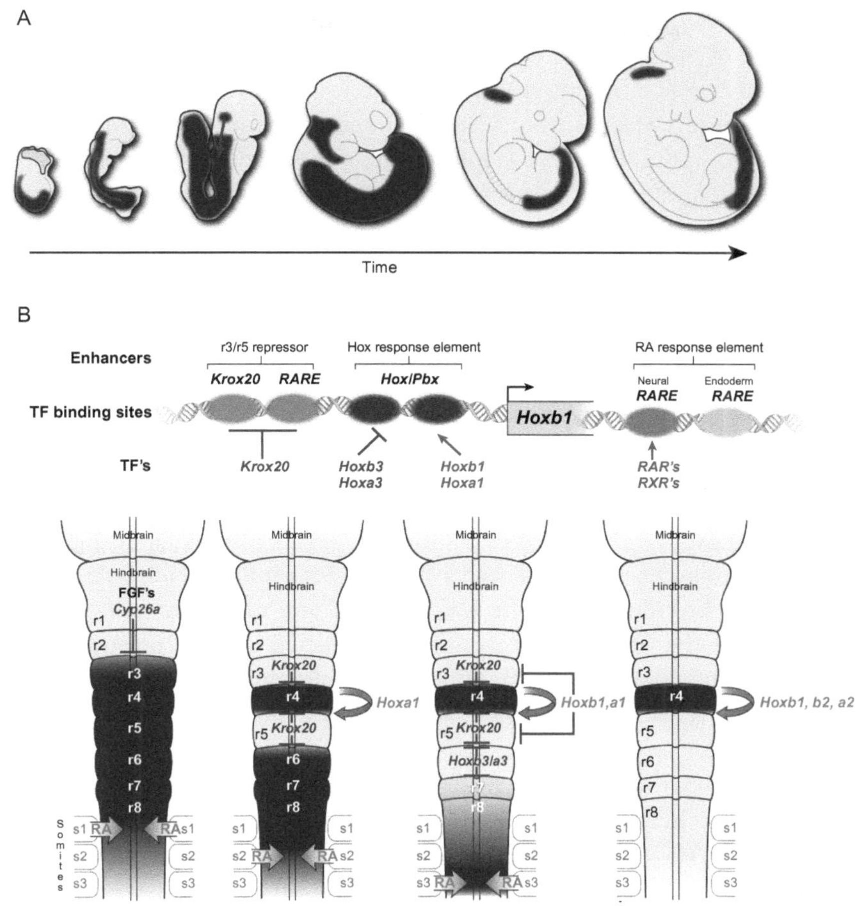

Figure 1 Temporal dynamics in the segmentally restricted expression and regulation of *Hoxb1* in r4. (A) Sketches of *in situ* expression profiles of *Hoxb1* in mouse embryos showing changes over time. (B) At the top is a diagram of the *Hoxb1* gene with enhancer regions, binding sites, and regulatory inputs from transcription factors. At the bottom is a drawing of how the initial broad domain of *Hoxb1* becomes progressively restricted to r4 through the input of repressive and activation inputs. (A color version of this figure is located at the end of this chapter.)

element upstream of the gene (Pöpperl et al., 1995; Studer et al., 1998b; Tvrdik & Capecchi, 2006).

In the next step, Hoxb1 autoregulates its own expression in conjunction with the cofactors Pbx and Meis through its highly conserved Hox response element (Gavalas et al., 1998; Moens & Selleri, 2006; Pöpperl et al., 1995; Studer et al., 1998b; Vitobello et al., 2011; Waskiewicz, Rikhof,

Hernandez, & Moens, 2001). *Hoxb1* and *Hoxa1* repress the expression of *Krox20* in r4 (Helmbacher et al., 1998). In the r6 region, expression of *Hoxb3* and *Hoxa3* is activated and they bind to the same Hox response element and repress *Hoxb1* in r6 (Gaufo, Thomas, & Capecchi, 2003; Wong et al., 2011). In the final step, as somites differentiate they cease production of Raldh2; hence, there is no longer broad stimulation of *Hoxb1* by retinoids in the hindbrain region. The *Hoxa1* gene does not possess an autoregulatory element and its expression rapidly declines (Dupé et al., 1997; Studer et al., 1998b; Tvrdik & Capecchi, 2006). Through the repressive and auto- and cross-regulatory feedback loops, *Hoxb1* maintains its own expression in r4, and also begins to receive positive inputs from *Hoxb2* and *Hoxa2*, which it activates in r4 (Davenne et al., 1999; Gavalas, Ruhrberg, Livet, Henderson, & Krumlauf, 2003; Maconochie et al., 1997; Pattyn et al., 2003; Tümpel et al., 2007). This illustrates how rhombomere-restricted expression is achieved through the integration of activation in a broad domains followed by successive sculpting by repression and activation.

These studies have underscored a key role for retinoic acid (RA) signaling in transient induction of the early ordered and nested domains of *Hox* expression in the CNS. RA signaling is implicated in early positioning of the anterior boundaries of 3′ *HoxB* genes (paralog groups 1–5) (Bel-Vialar et al., 2002; Marshall et al., 1994; Sirbu et al., 2005; Studer et al., 1998a) and later in the rostral expansion of the expression domains of 5′ genes in the cluster (Ahn, Mullan, & Krumlauf, 2014). Even though no RAREs have been identified in the 5′ region of the *HoxB* cluster, several RAREs are present around *Hoxb4* and *Hoxb5* (Gould, Itasaki, & Krumlauf, 1998; Oosterveen et al., 2003; Sharpe, Nonchev, Gould, Whiting, & Krumlauf, 1998). Recent experiments have demonstrated that RAREs embedded within and adjacent to the HoxB cluster act on several genes through enhancer sharing and long-range regulation (Ahn et al., 2014; Nolte, Jinks, Wang, Martinez Pastor, & Krumlauf, 2013). These studies identify the positions of highly conserved *cis*-elements that mediate the responses of *Hox* clusters to retinoids and suggest that they may be a contributing factor in maintaining the clustered organization and coupling of *Hox* genes to axial patterning in the CNS.

The knowledge gained from this GRN framework (Alexander et al., 2009; Tümpel et al., 2009) enables exploration of the degree to which it is conserved during evolution of vertebrates (Nolte, Ahn, & Krumlauf, 2012). By integrating the information on *Hox* enhancers, regulatory networks, and inputs from signaling pathways (RA, Fgfs, and Wnts) in vertebrates, it is possible to investigate their conservation or divergence during chordate evolution. For example, amphioxus, a cephalochordate, contains a single *Hox* cluster, and a screen for regulatory elements in the amphioxus *Hox* genes that function in transgenic mice or chicken embryos revealed no *cis*-elements capable of mediating segmental expression in the vertebrate hindbrain (Manzanares et al., 2000). However, several neural enhancers containing RAREs were found in the amphiHox cluster and these displayed a dependence upon retinoid signaling (Manzanares et al., 2000; Wada, Escriva, Zhang, & Laudet, 2006). It is intriguing that they are located in the same relative positions that RARE-dependent neural enhancers are found in several vertebrate *Hox* clusters (Ahn et al., 2014). This suggests that these RARE-based control modules might represent conserved regulatory features of an ancient cluster present before the emergence of vertebrates. In accord with this idea, studies by Chris Lowe and colleagues on the acorn worm *Saccoglossus kowalevskii*, a hemichordate, have shown that the A/P domain map of transcription factors (including Hox) and signaling ligands can surprisingly be used to align the bodies of hemichordates and chordates (Gerhart, Lowe, & Kirschner, 2005; Lowe et al., 2003; Pani et al., 2012). These findings imply that the axial signaling centers must have evolved long ago in a common chordate ancestor. Hence, the conserved positions of RAREs in amphiHox and vertebrate clusters might reflect an ancient *cis*-signature of retinoid signaling involved in generating ordered domains of *Hox* neural expression. It will be interesting to determine whether such RAREs exist in the hemichordate *Hox* cluster (Freeman et al., 2012).

While analyses in amphioxus and *Ciona* have found that none of the *cis*-elements responsible for the segmental expression of *Hox* genes in the vertebrate hindbrain appear to be present in these chordates (Manzanares et al., 2000; Natale et al., 2011), comparative studies in lamprey, zebra fish, and mice have begun to provide insight on the question of when ordered domains of *Hox* expression became coupled to hindbrain segmentation in the origin of chordates. As detailed above, all jawed vertebrates pass through an embryonic stage in which the hindbrain is divided into rhombomeric compartments and the hindbrain "*Hox code*" and GRN shows remarkable conservation from zebra fish to mammals. Due to its position at the base of the vertebrate family tree, the sea lamprey, a jawless fish (cyclostome), can give unique insights into the ancestry of vertebrates (Shimeld & Donoghue, 2012). Lampreys have neural crest cells with a conserved neural crest GRN (Sauka-Spengler, Meulemans, Jones, & Bronner-Fraser, 2007).

However, it has been unclear whether lamprey embryos develop a segmented hindbrain associated with segmental *Hox* expression (Cohn, 2002; Murakami et al., 2004; Murakami, Uchida, Rijli, & Kuratani, 2005; Pascual-Anaya et al., 2012; Takio et al., 2004, 2007). However, a recent study showing that enhancers from the *Hox* clusters of jawed vertebrates mediated rhombomere-restricted domains of reporter expression in the lamprey hindbrain indicates that the lamprey must have the upstream regulatory factors of the hindbrain GRN capable of directing segmental expression (Parker, Bronner, & Krumlauf, 2014). This work also uncovered ordered segmental expression of endogenous lamprey *Hox* genes and identified an enhancer that mediates the rhombomeric expression of a group 2 *Hox* gene in r2 and r4 (Parker et al., 2014). This enhancer is located in the same relative position as the mouse *Hoxa2* r2/r4 segmental enhancer (Tümpel et al., 2007; Tümpel, Cambronero, Sims, Krumlauf, & Wiedemann, 2008; Tümpel, Cambronero, Wiedemann, & Krumlauf, 2006). This insight from regulatory analysis suggests that key aspects of the hindbrain GRN, including *Hox* genes, were already present in the common ancestor of lamprey and jawed vertebrates. Therefore, during the evolution of the vertebrate head, segmental *Hox* expression in the hindbrain appears to be a fundamental innovation wired into the GRNs of the developmental program at the base of vertebrate tree. Consistent with this idea, analysis of conserved noncoding regions in the lamprey genome has shown that many of these are associated with *cis*-elements that direct *Hox*-dependent gene regulatory activities that may pattern head development (Parker, Piccinelli, Sauka-Spengler, Bronner, & Elgar, 2011).

A major challenge for the future is that though the function of Hox proteins is critical for specification of regional properties of hindbrain segments and other tissues, very little is known about the downstream target loci where they are recruited as protein complexes to exert their functional activities. The 39 mammalian Hox proteins have very similar structures and *in vitro* binding specificities (Slattery et al., 2011), so their individual specificity is likely to be modulated by subtle differences in cofactors, interacting proteins, target sites, or other unknown processes (Mann, Lelli, & Joshi, 2009; Merabet, Hudry, Saadaoui, & Graba, 2009; Saadaoui et al., 2011). It seems likely that differences in the downstream targets of *Hox* genes between species may be a central aspect of how they regulate common processes in a distinct manner. It will be essential to identify *in vivo* relevant Hox response elements to gain insight into the sites, binding partners, and mechanisms of transcriptional regulation essential for their function in hindbrain and other tissues and to link these loci to the pathways and cellular processes they modulate.

While we understand a great deal about the ground plan of segmentation and how distinct segmental identities are established, there is a major gap in our knowledge on how this modulates the programs that control the formation and identity of distinct neuronal populations within the segments. Most of the focus on expression of *Hox* genes in the CNS has been on their AP domains and boundaries. However, each of the four mouse Hox clusters displays distinct dorsoventral (DV) patterns of expression that correlate with the birth of different classes of neurons (Graham, Maden, & Krumlauf, 1991). This is illustrated in Fig. 2A. All of the *HoxB* genes are initially expressed uniformly throughout the DV axis of the neural tube, but over time they become progressively restricted with the highest levels in dorsal domains. Figure 2C illustrates how the r4-restricted domain of *Hoxb1* is further limited in later stages along the DV axis to specific populations of differentiating neurons in r4. These stripes of neuronal expression are significant, as in *Hox* mutants these cell populations fail to develop properly (Gavalas et al., 2003; Pattyn et al., 2003). Different patterns of DV expression are observed for the other *Hox* clusters (Fig. 2B). Elegant work from the groups of Jessell and Dasen have detailed the roles of *Hox* genes in regulating diversity and identities of motor neuron pools in the spinal cord (Dasen, De Camilli, Wang, Tucker, & Jessell, 2008; Dasen & Jessell, 2009; Dasen, Liu, & Jessell, 2003; Dasen, Tice, Brenner-Morton, & Jessell, 2005; Jung et al., 2010; Philippidou, Walsh, Aubin, Jeannotte, & Dasen, 2012). This clearly demonstrates that the DV restrictions to expression in later stages are functionally meaningful in the spinal cord; however, little is known about whether and how these dynamic DV patterns of expression translate to regulation of neuronal differentiation programs in the hindbrain. Dissecting these cellular programs during the establishment and differentiation of neuronal populations will be critical for understanding how the hindbrain generates the components and circuits associated with its role as a complex coordination center of the CNS.

Finally, another important direction for the future is to understand how hindbrain segmentation and *Hox* genes play roles in later stages of elaboration of the CNS to generate adult structures and circuits. Insight is arising from conditional mouse mutants that reveals roles in somatosensory circuits, respiratory circuits, and generation of neural progenitor pools (Briscoe & Wilkinson, 2004; Chatonnet et al., 2003; Davenne et al., 1999; del Toro et al., 2001; Di Bonito, Glover, & Studer, 2013; Di Bonito, Narita,

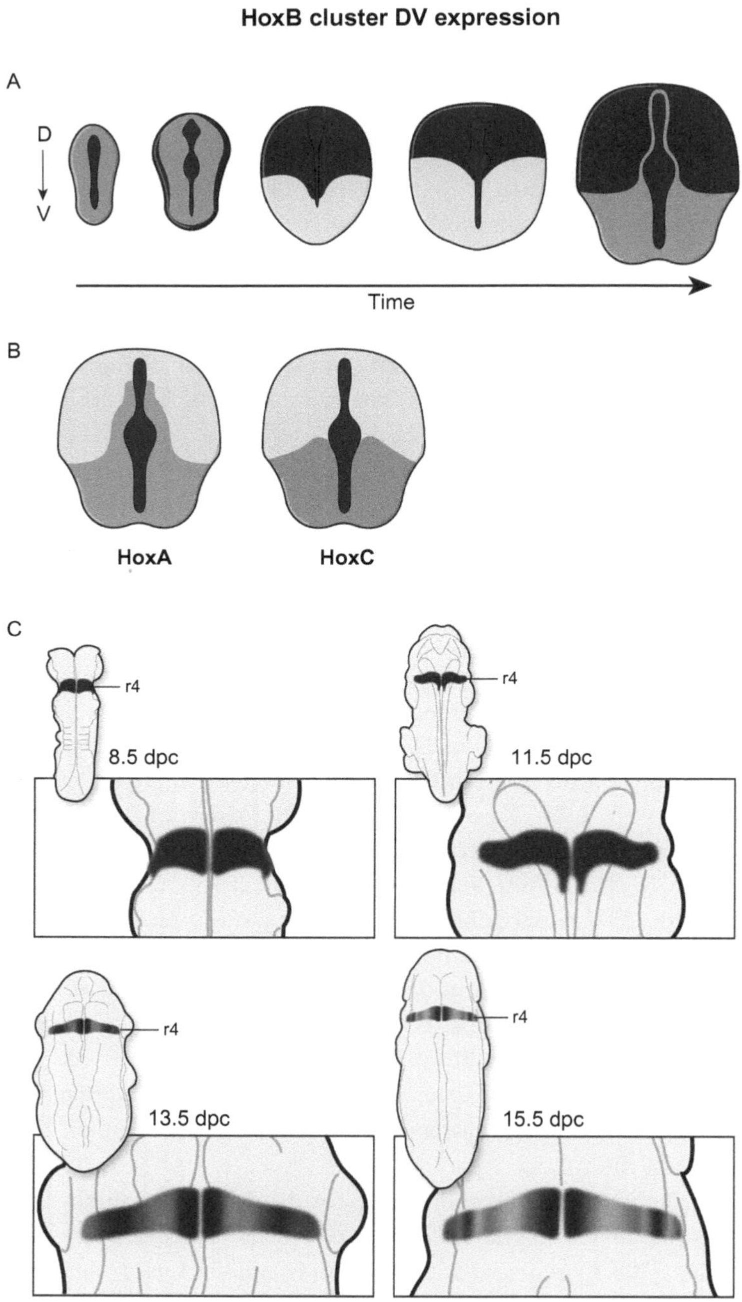

Figure 2 Dorsoventral changes in *Hox* expression in the neural tube. (A) Cross-sectional diagrams of the developing neural tube showing how *HoxB* genes become enriched in dorsal regions. (B) Comparison of the different dorsoventral distributions of genes from separate Hox clusters. (C) Sketch showing that the r4 expression of *Hoxb1* becomes restricted to small longitudinal stripes of neurons in r4. (A color version of this figure is located at the end of this chapter.)

et al., 2013; Oury et al., 2006; Pasqualetti, Diaz, Renaud, Rijli, & Glover, 2007; Pattyn et al., 2003). With the development of new tools and technologies, there is hope that efforts will be successful in understanding the full implications of how hindbrain segmentation eventually leads to the full elaboration of this region of the CNS and assess how variations in the process impact development, disease, and evolution.

ACKNOWLEDGMENTS

I would like to thank Mark Miller for the illustrations and the many friends, colleagues, and lab members who over the years have participated in the research and exchange of ideas that have made this such a wonderful area to study.

REFERENCES

Ahn, Y., Mullan, H. E., & Krumlauf, R. (2014). Long-range regulation by shared retinoic acid response elements modulates dynamic expression of posterior Hoxb genes in CNS development. *Developmental Biology*, *388*, 134–144.

Alexander, T., Nolte, C., & Krumlauf, R. (2009). Hox genes and segmentation of the hindbrain and axial skeleton. *Annual Review of Cell and Developmental Biology*, *25*, 431–456.

Begemann, G., Schilling, T. F., Rauch, G. J., Geisler, R., & Ingham, P. W. (2001). The zebrafish neckless mutation reveals a requirement for raldh2 in mesodermal signals that pattern the hindbrain. *Development*, *128*, 3081–3094.

Bel-Vialar, S., Itasaki, N., & Krumlauf, R. (2002). Initiating Hox gene expression: In the early chick neural tube differential sensitivity to FGF and RA signaling subdivides the HoxB genes in two distinct groups. *Development*, *129*, 5103–5115.

Birgbauer, E., Sechrist, J., Bronner-Fraser, M., & Fraser, S. (1995). Rhombomeric origin and rostrocaudal reassortment of neural crest cells revealed by intravital microscopy. *Development*, *121*, 935–945.

Briscoe, J., & Wilkinson, D. G. (2004). Establishing neuronal circuitry: Hox genes make the connection. *Genes and Development*, *18*, 1643–1648.

Chatonnet, F., Dominguez del Toro, E., Thoby-Brisson, M., Champagnat, J., Fortin, G., Rijli, F. M., et al. (2003). From hindbrain segmentation to breathing after birth: Developmental patterning in rhombomeres 3 and 4. *Molecular Neurobiology*, *28*, 277–294.

Cheng, Y. C., Amoyel, M., Qiu, X., Jiang, Y. J., Xu, Q., & Wilkinson, D. G. (2004). Notch activation regulates the segregation and differentiation of rhombomere boundary cells in the zebrafish hindbrain. *Developmental Cell*, *6*, 539–550.

Cohn, M. J. (2002). Evolutionary biology: Lamprey Hox genes and the origin of jaws. *Nature*, *416*, 386–387.

Cooke, J. E., & Moens, C. B. (2002). Boundary formation in the hindbrain: Eph only it were simple. *Trends in Neurosciences*, *25*, 260–267.

Dasen, J. S., De Camilli, A., Wang, B., Tucker, P. W., & Jessell, T. M. (2008). Hox repertoires for motor neuron diversity and connectivity gated by a single accessory factor, FoxP1. *Cell*, *134*, 304–316.

Dasen, J. S., & Jessell, T. M. (2009). Hox networks and the origins of motor neuron diversity. *Current Topics in Developmental Biology*, *88*, 169–200.

Dasen, J. S., Liu, J. P., & Jessell, T. M. (2003). Motor neuron columnar fate imposed by sequential phases of Hox-c activity. *Nature*, *425*, 926–933.

Dasen, J. S., Tice, B. C., Brenner-Morton, S., & Jessell, T. M. (2005). A Hox regulatory network establishes motor neuron pool identity and target-muscle connectivity. *Cell*, *123*, 477–491.
Davenne, M., Maconochie, M. K., Neun, R., Pattyn, A., Chambon, P., Krumlauf, R., et al. (1999). *Hoxa2* and *Hoxb2* control dorsoventral patterns of neuronal development in the rostral hindbrain. *Neuron*, *22*, 677–691.
del Toro, E. D., Borday, V., Davenne, M., Neun, R., Rijli, F. M., & Champagnat, J. (2001). Generation of a novel functional neuronal circuit in Hoxa1 mutant mice. *Journal of Neuroscience*, *21*, 5637–5642.
Deschamps, J., & van Nes, J. (2005). Developmental regulation of the Hox genes during axial morphogenesis in the mouse. *Development*, *132*, 2931–2942.
Di Bonito, M., Glover, J. C., & Studer, M. (2013). Hox genes and region-specific sensorimotor circuit formation in the hindbrain and spinal cord. *Developmental Dynamics*, *242*, 1348–1368.
Di Bonito, M., Narita, Y., Avallone, B., Sequino, L., Mancuso, M., Andolfi, G., et al. (2013). Assembly of the auditory circuitry by a Hox genetic network in the mouse brainstem. *PLoS Genetics*, *9*, e1003249.
Duester, G. (2008). Retinoic acid synthesis and signaling during early organogenesis. *Cell*, *134*, 921–931.
Dupé, V., Davenne, M., Brocard, J., Dollé, P., Mark, M., Dierich, A., et al. (1997). *In vivo* functional analysis of the *Hoxa1* 3′ retinoid response element (3′ RARE). *Development*, *124*, 399–410.
Fraser, S., Keynes, R., & Lumsden, A. (1990). Segmentation in the chick embryo hindbrain is defined by cell lineage restrictions. *Nature*, *344*, 431–435.
Freeman, R., Ikuta, T., Wu, M., Koyanagi, R., Kawashima, T., Tagawa, K., et al. (2012). Identical genomic organization of two hemichordate hox clusters. *Current Biology*, *22*, 2053–2058.
Frohman, M. A., Boyle, M., & Martin, G. R. (1990). Isolation of the mouse *Hox-2.9* gene; analysis of embryonic expression suggests that positional information along the anterior-posterior axis is specified by mesoderm. *Development*, *110*, 589–607.
Gaufo, G. O., Thomas, K. R., & Capecchi, M. R. (2003). Hox3 genes coordinate mechanisms of genetic suppression and activation in the generation of branchial and somatic motoneurons. *Development*, *130*, 5191–5201.
Gavalas, A., Ruhrberg, C., Livet, J., Henderson, C. E., & Krumlauf, R. (2003). Neuronal defects in the hindbrain of Hoxa1, Hoxb1 and Hoxb2 mutants reflect regulatory interactions among these Hox genes. *Development*, *130*, 5663–5679.
Gavalas, A., Studer, M., Lumsden, A., Rijli, F. M., Krumlauf, R., & Chambon, P. (1998). *Hoxa1* and *Hoxb1* synergize in patterning the hindbrain, cranial nerves and second pharyngeal arch. *Development*, *125*, 1123–1136.
Gerhart, J., Lowe, C., & Kirschner, M. (2005). Hemichordates and the origin of chordates. *Current Opinion in Genetics and Development*, *15*, 461–467.
Giudicelli, F., Taillebourg, E., Charnay, P., & Gilardi-Hebenstreit, P. (2001). Krox-20 patterns the hindbrain through both cell-autonomous and non cell-autonomous mechanisms. *Genes and Development*, *15*, 567–580.
Gonzalez-Quevedo, R., Lee, Y., Poss, K. D., & Wilkinson, D. G. (2010). Neuronal regulation of the spatial patterning of neurogenesis. *Developmental Cell*, *18*, 136–147.
Gould, A., Itasaki, N., & Krumlauf, R. (1998). Initiation of rhombomeric *Hoxb4* expression requires induction by somites and a retinoid pathway. *Neuron*, *21*, 39–51.
Gould, A., Morrison, A., Sproat, G., White, R. A., & Krumlauf, R. (1997). Positive cross-regulation and enhancer sharing: Two mechanisms for specifying overlapping *Hox* expression patterns. *Genes & Development*, *11*, 900–913.

Graham, A., Maden, M., & Krumlauf, R. (1991). The murine Hox-2 genes display dynamic dorsoventral patterns of expression during central nervous system development. *Development*, *112*, 255–264.
Guthrie, S., Prince, V., & Lumsden, A. (1993). Selective dispersal of avian rhombomere cells in orthotopic and heterotopic grafts. *Development*, *118*, 527–538.
Helmbacher, F., Pujades, C., Desmarquet, C., Frain, M., Rijli, F. M., Chambon, P., et al. (1998). *Hoxa1* and *Krox20* synergize to control the development of rhombomere 3. *Development*, *125*, 4739–4748.
Hernandez, R. E., Putzke, A. P., Myers, J. P., Margaretha, L., & Moens, C. B. (2007). Cyp26 enzymes generate the retinoic acid response pattern necessary for hindbrain development. *Development*, *134*, 177–187.
Hernandez, R. E., Rikhof, H. A., Bachmann, R., & Moens, C. B. (2004). vhnf1 integrates global RA patterning and local FGF signals to direct posterior hindbrain development in zebrafish. *Development*, *131*, 4511–4520.
Hunt, P., Gulisano, M., Cook, M., Sham, M. H., Faiella, A., Wilkinson, D., et al. (1991). A distinct Hox code for the branchial region of the vertebrate head. *Nature*, *353*, 861–864.
Jung, H., Lacombe, J., Mazzoni, E. O., Liem, K. F., Jr., Grinstein, J., Mahony, S., et al. (2010). Global control of motor neuron topography mediated by the repressive actions of a single hox gene. *Neuron*, *67*, 781–796.
Lowe, C. J., Wu, M., Salic, A., Evans, L., Lander, E., Stange-Thomann, N., et al. (2003). Anteroposterior patterning in hemichordates and the origins of the chordate nervous system. *Cell*, *113*, 853–865.
Lumsden, A. (2004). Segmentation and compartition in the early avian hindbrain. *Mechanisms of Development*, *121*, 1081–1088.
Lumsden, A., & Keynes, R. (1989). Segmental patterns of neuronal development in the chick hindbrain. *Nature*, *337*, 424–428.
Lumsden, A., & Krumlauf, R. (1996). Patterning the vertebrate neuraxis. *Science*, *274*, 1109–1115.
Maconochie, M., Nonchev, S., Morrison, A., & Krumlauf, R. (1996). Paralogous *Hox* genes: Function and regulation. *Annual Review of Genetics*, *30*, 529–556.
Maconochie, M. K., Nonchev, S., Studer, M., Chan, S. K., Popperl, H., Sham, M. H., et al. (1997). Cross-regulation in the mouse *HoxB* complex: The expression of *Hoxb2* in rhombomere 4 is regulated by *Hoxb1*. *Genes and Development*, *11*, 1885–1896.
Mann, R. S., Lelli, K. M., & Joshi, R. (2009). Hox specificity unique roles for cofactors and collaborators. *Current Topics in Developmental Biology*, *88*, 63–101.
Manzanares, M., Bel-Vialer, S., Ariza-McNaughton, L., Ferretti, E., Marshall, H., Maconochie, M. K., et al. (2001). Independent regulation of initiation and maintenance phases of *Hoxa3* expression in the vertebrate hindbrain involves auto and cross-regulatory mechanisms. *Development*, *128*, 3595–3607.
Manzanares, M., Wada, H., Itasaki, N., Trainor, P. A., Krumlauf, R., & Holland, P. W. (2000). Conservation and elaboration of Hox gene regulation during evolution of the vertebrate head. *Nature*, *408*, 854–857.
Marshall, H., Studer, M., Pöpperl, H., Aparicio, S., Kuroiwa, A., Brenner, S., et al. (1994). A conserved retinoic acid response element required for early expression of the homeobox gene *Hoxb-1*. *Nature*, *370*, 567–571.
Mellitzer, G., Xu, Q., & Wilkinson, D. (1999). Eph receptors and ephrins restrict cell intermingling and communication. *Nature*, *400*, 77–81.
Merabet, S., Hudry, B., Saadaoui, M., & Graba, Y. (2009). Classification of sequence signatures: A guide to Hox protein function. *Bioessays*, *31*, 500–511.
Moens, C. B., & Prince, V. E. (2002). Constructing the hindbrain: Insights from the zebrafish. *Developmental Dynamics*, *224*, 1–17.

Moens, C. B., & Selleri, L. (2006). Hox cofactors in vertebrate development. *Developmental Biology, 291*, 193–206.
Molotkova, N., Molotkov, A., Sirbu, I. O., & Duester, G. (2005). Requirement of mesodermal retinoic acid generated by Raldh2 for posterior neural transformation. *Mechanisms of Development, 122*, 145–155.
Murakami, Y., Pasqualetti, M., Takio, Y., Hirano, S., Rijli, F. M., & Kuratani, S. (2004). Segmental development of reticulospinal and branchiomotor neurons in lamprey: Insights into the evolution of the vertebrate hindbrain. *Development, 131*, 983–995.
Murakami, Y., Uchida, K., Rijli, F. M., & Kuratani, S. (2005). Evolution of the brain developmental plan: Insights from agnathans. *Developmental Biology, 280*, 249–259.
Murphy, P., Davidson, D. R., & Hill, R. E. (1989). Segment-specific expression of a homeobox-containing gene in the mouse hindbrain. *Nature, 341*, 156–159.
Natale, A., Sims, C., Chiusano, M. L., Amoroso, A., D'Aniello, E., Fucci, L., et al. (2011). Evolution of anterior Hox regulatory elements among chordates. *BMC Evolutionary Biology, 11*, 330.
Niederreither, K., Vermot, J., Le Roux, I., Schuhbaur, B., Chambon, P., & Dolle, P. (2003). The regional pattern of retinoic acid synthesis by RALDH2 is essential for the development of posterior pharyngeal arches and the enteric nervous system. *Development, 130*, 2525–2534.
Nieto, M. A., Sechrist, J., Wilkinson, D. G., & Bronner-Fraser, M. (1995). Relationship between spatially restricted Krox-20 gene expression in branchial neural crest and segmentation in the chick embryo hindbrain. *The EMBO Journal, 14*, 1697–1710.
Nolte, C., Ahn, Y., & Krumlauf, R. (2012). Evolutionary developmental biology: *Hox* gene evolution. *eLS citable reviews in life sciences*. Chichester: John Wiley and Sons Ltd.
Nolte, C., Jinks, T., Wang, X., Martinez Pastor, M. T., & Krumlauf, R. (2013). Shadow enhancers flanking the HoxB cluster direct dynamic Hox expression in early heart and endoderm development. *Developmental Biology, 383*, 158–173.
Oosterveen, T., Niederreither, K., Dolle, P., Chambon, P., Meijlink, F., & Deschamps, J. (2003). Retinoids regulate the anterior expression boundaries of 5′ Hoxb genes in posterior hindbrain. *The EMBO Journal, 22*, 262–269.
Oury, F., Murakami, Y., Renaud, J. S., Pasqualetti, M., Charnay, P., Ren, S. Y., et al. (2006). Hoxa2- and rhombomere-dependent development of the mouse facial somatosensory map. *Science, 313*, 1408–1413.
Pani, A. M., Mullarkey, E. E., Aronowicz, J., Assimacopoulos, S., Grove, E. A., & Lowe, C. J. (2012). Ancient deuterostome origins of vertebrate brain signalling centres. *Nature, 483*, 289–294.
Parker, H. J., Bronner, M. E., & Krumlauf, R. (2014). A Hox regulatory network of hindbrain segmentation is conserved to the base of vertebrates. *Nature, 514*, 490–493.
Parker, H. J., Piccinelli, P., Sauka-Spengler, T., Bronner, M., & Elgar, G. (2011). Ancient Pbx-Hox signatures define hundreds of vertebrate developmental enhancers. *BMC Genomics, 12*, 637.
Parrish, M., Nolte, C., & Krumlauf, R. (2009). Hox genes expression. In L. R. Squire (Ed.), *Encyclopedia of neuroscience* (pp. 1221–1231). Oxford: Academic Press.
Pascual-Anaya, J., Adachi, N., Alvarez, S., Kuratani, S., D'Aniello, S., & Garcia-Fernandez, J. (2012). Broken colinearity of the amphioxus Hox cluster. *EvoDevo, 3*, 28.
Pasqualetti, M., Diaz, C., Renaud, J. S., Rijli, F. M., & Glover, J. C. (2007). Fate-mapping the mammalian hindbrain: Segmental origins of vestibular projection neurons assessed using rhombomere-specific Hoxa2 enhancer elements in the mouse embryo. *Journal of Neuroscience, 27*, 9670–9681.
Pattyn, A., Vallstedt, A., Dias, J. M., Samad, O. A., Krumlauf, R., Rijli, F. M., et al. (2003). Coordinated temporal and spatial control of motor neuron and serotonergic neuron generation from a common pool of CNS progenitors. *Genes & Development, 17*, 729–737.

Philippidou, P., Walsh, C. M., Aubin, J., Jeannotte, L., & Dasen, J. S. (2012). Sustained Hox5 gene activity is required for respiratory motor neuron development. *Nature Neuroscience, 15*, 1636–1644.

Pöpperl, H., Bienz, M., Studer, M., Chan, S., Aparicio, S., Brenner, S., et al. (1995). Segmental expression of *Hoxb1* is controlled by a highly conserved autoregulatory loop dependent upon *exd/Pbx*. *Cell, 81*, 1031–1042.

Saadaoui, M., Merabet, S., Litim-Mecheri, I., Arbeille, E., Sambrani, N., Damen, W., et al. (2011). Selection of distinct Hox-Extradenticle interaction modes fine-tunes Hox protein activity. *Proceedings of the National Academy of Sciences of the United States of America, 108*, 2276–2281.

Santagati, F., & Rijli, F. M. (2003). Cranial neural crest and the building of the vertebrate head. *Nature Review Neuroscience, 4*, 806–818.

Sauka-Spengler, T., Meulemans, D., Jones, M., & Bronner-Fraser, M. (2007). Ancient evolutionary origin of the neural crest gene regulatory network. *Developmental Cell, 13*, 405–420.

Schneider-Maunoury, S., Topilko, P., Seitandou, T., Levi, G., Cohen-Tannoudji, M., Pournin, S., et al. (1993). Disruption of *Krox-20* results in alteration of rhombomeres 3 and 5 in the developing hindbrain. *Cell, 75*, 1199–1214.

Sela-Donenfeld, D., & Wilkinson, D. G. (2005). Eph receptors: Two ways to sharpen boundaries. *Current Biology, 15*, R210–R212.

Sharpe, J., Nonchev, S., Gould, A., Whiting, J., & Krumlauf, R. (1998). Selectivity, sharing and competitive interactions in the regulation of *Hoxb* genes. *EMBO Journal, 17*, 1788–1798.

Shimeld, S. M., & Donoghue, P. C. (2012). Evolutionary crossroads in developmental biology: Cyclostomes (lamprey and hagfish). *Development, 139*, 2091–2099.

Simoes-Costa, M., & Bronner, M. E. (2015). Establishing neural crest identity: A gene regulatory recipe. *Development, 142*, 242–257.

Sirbu, I. O., Gresh, L., Barra, J., & Duester, G. (2005). Shifting boundaries of retinoic acid activity control hindbrain segmental gene expression. *Development, 132*, 2611–2622.

Slattery, M., Riley, T., Liu, P., Abe, N., Gomez-Alcala, P., Dror, I., et al. (2011). Cofactor binding evokes latent differences in DNA binding specificity between Hox proteins. *Cell, 147*, 1270–1282.

Studer, M., Gavalas, A., Marshall, H., Ariza-McNaughton, L., Rijli, F., Chambon, P., et al. (1998a). Genetic interaction between *Hoxa1* and *Hoxb1* reveal new roles in regulation of early hindbrain patterning. *Development, 125*, 1025–1036.

Studer, M., Gavalas, A., Marshall, H., Ariza-McNaughton, L., Rijli, F. M., Chambon, P., et al. (1998b). Genetic interactions between Hoxa1 and Hoxb1 reveal new roles in regulation of early hindbrain patterning. *Development, 125*, 1025–1036.

Studer, M., Popperl, H., Marshall, H., Kuroiwa, A., & Krumlauf, R. (1994). Role of a conserved retinoic acid response element in rhombomere restriction of *Hoxb-1*. *Science, 265*, 1728–1732.

Sundin, O., & Eichele, G. (1990). A homeo domain protein reveals the metameric nature of the developing chick hindbrain. *Genes and Development, 4*, 1267–1276.

Takio, Y., Kuraku, S., Murakami, Y., Pasqualetti, M., Rijli, F. M., Narita, Y., et al. (2007). Hox gene expression patterns in *Lethenteron japonicum* embryos-Insights into the evolution of the vertebrate Hox code. *Developmental Biology, 308*, 606–620.

Takio, Y., Pasqualetti, M., Kuraku, S., Hirano, S., Rijli, F. M., & Kuratani, S. (2004). Evolutionary biology: Lamprey Hox genes and the evolution of jaws. *Nature, 429*. 1 p following 262.

Trainor, P. A., & Krumlauf, R. (2000). Patterning the cranial neural crest: Hindbrain segmentation and *Hox* gene plasticity. *Nature Reviews Neuroscience, 1*, 116–124.

Trainor, P. A., Melton, K. R., & Manzanares, M. (2003). Origins and plasticity of neural crest cells and their roles in jaw and craniofacial evolution. *International Journal of Developmental Biology*, *47*, 541–553.

Tümpel, S., Cambronero, F., Ferretti, E., Blasi, F., Wiedemann, L. M., & Krumlauf, R. (2007). Expression of *Hoxa2* in rhombomere 4 is regulated by a conserved cross-regulatory mechanism dependent upon *Hoxb1*. *Developmental Biology*, *302*, 646–660.

Tümpel, S., Cambronero, F., Sims, C., Krumlauf, R., & Wiedemann, L. M. (2008). A regulatory module embedded in the coding region of Hoxa2 controls expression in rhombomere 2. *Proceedings of the National Academy of Sciences of the United States of America*, *105*, 20077–20082.

Tümpel, S., Cambronero, F., Wiedemann, L. M., & Krumlauf, R. (2006). Evolution of *cis* elements in the differential expression of two *Hoxa2* coparalogous genes in pufferfish (*Takifugu rubripes*). *Proceedings of the National Academy of Sciences of the United States of America*, *103*, 5419–5424.

Tümpel, S., Maconochie, M., Wiedemann, L. M., & Krumlauf, R. (2002). Conservation and diversity in the cis-regulatory networks that integrate information controlling expression of Hoxa2 in hindbrain and cranial neural crest cells in vertebrates. *Developmental Biology*, *246*, 45–56.

Tumpel, S., Wiedemann, L. M., & Krumlauf, R. (2009). Hox genes and segmentation of the vertebrate hindbrain. *Current Topics in Developmental Biology*, *88*, 103–137.

Tvrdik, P., & Capecchi, M. R. (2006). Reversal of hox1 gene subfunctionalization in the mouse. *Developmental Cell*, *11*, 239–250.

Vitobello, A., Ferretti, E., Lampe, X., Vilain, N., Ducret, S., Ori, M., et al. (2011). Hox and Pbx factors control retinoic acid synthesis during hindbrain segmentation. *Developmental Cell*, *20*, 469–482.

Wada, H., Escriva, H., Zhang, S., & Laudet, V. (2006). Conserved RARE localization in amphioxus Hox clusters and implications for Hox code evolution in the vertebrate neural crest. *Developmental Dynamics*, *235*, 1522–1531.

Waskiewicz, A. J., Rikhof, H. A., Hernandez, R. E., & Moens, C. B. (2001). Zebrafish Meis functions to stabilize Pbx proteins and regulate hindbrain patterning. *Development*, *128*, 4139–4151.

Wilkinson, D. G., Bhatt, S., Chavrier, P., Bravo, R., & Charnay, P. (1989). Segment-specific expression of a zinc-finger gene in the developing nervous system of the mouse. *Nature*, *337*, 461–464.

Wilkinson, D. G., Bhatt, S., Cook, M., Boncinelli, E., & Krumlauf, R. (1989). Segmental expression of Hox-2 homeobox-containing genes in the developing mouse hindbrain. *Nature*, *341*, 405–409.

Wong, E. Y., Wang, X. A., Mak, S. S., Sae-Pang, J. J., Ling, K. W., Fritzsch, B., et al. (2011). Hoxb3 negatively regulates Hoxb1 expression in mouse hindbrain patterning. *Developmental Biology*, *352*, 382–392.

Xu, Q., Mellitzer, G., Robinson, V., & Wilkinson, D. G. (1999). *In vivo* cell sorting in complementary segmental domains mediated by Eph receptors and ephrins. *Nature*, *399*, 267–271.

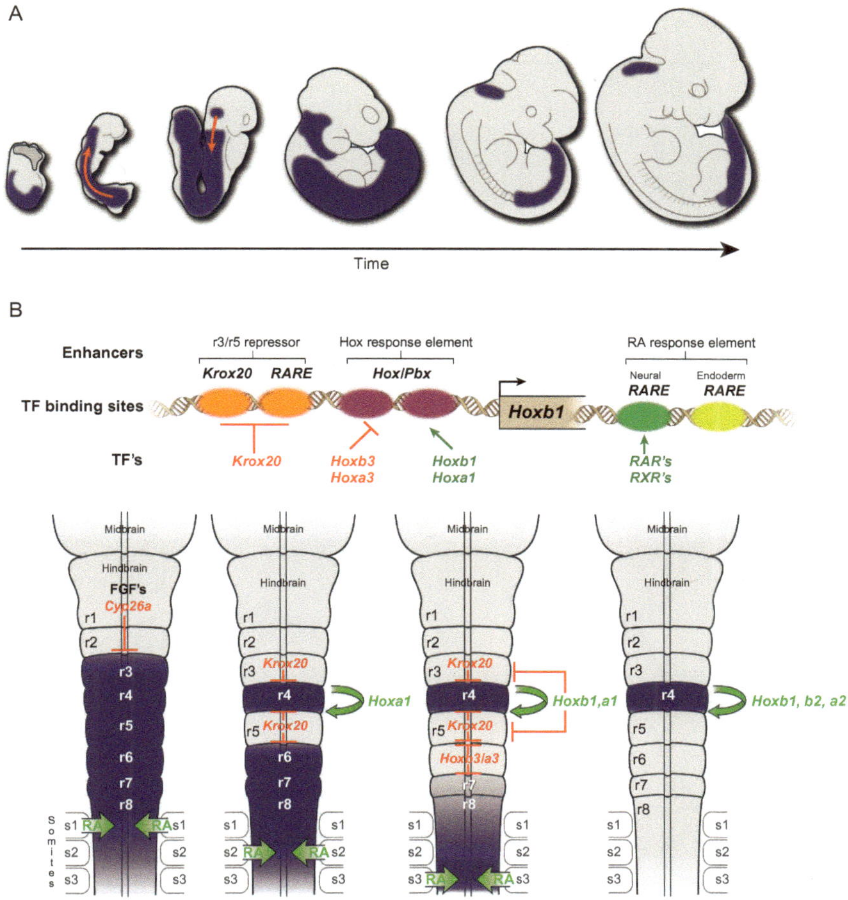

Robb Krumlauf, Figure 1 Temporal dynamics in the segmentally restricted expression and regulation of *Hoxb1* in r4. (A) Sketches of *in situ* expression profiles of *Hoxb1* in mouse embryos showing changes over time. (B) At the top is a diagram of the *Hoxb1* gene with enhancer regions, binding sites, and regulatory inputs from transcription factors. At the bottom is a drawing of how the initial broad domain of *Hoxb1* becomes progressively restricted to r4 through the input of repressive and activation inputs.

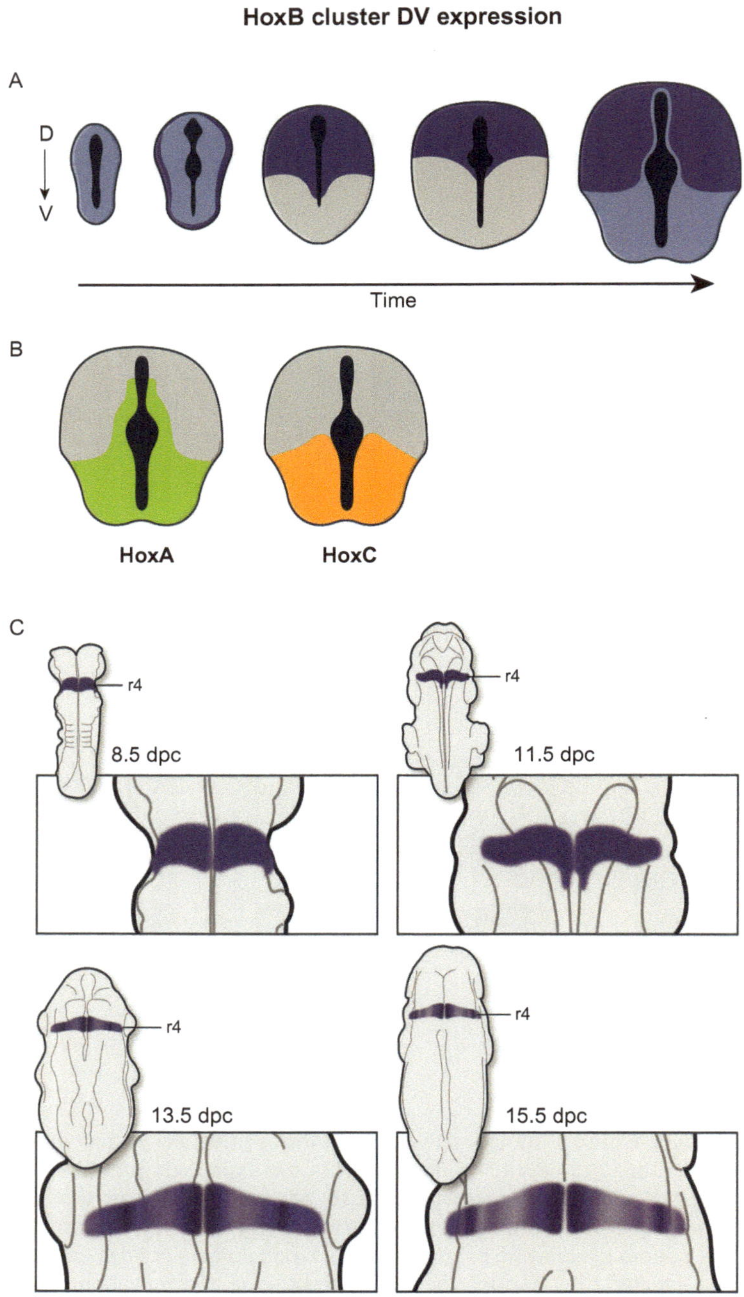

Robb Krumlauf, Figure 2 Dorsoventral changes in *Hox* expression in the neural tube. (A) Cross-sectional diagrams of the developing neural tube showing how *HoxB* genes become enriched in dorsal regions. (B) Comparison of the different dorsoventral distributions of genes from separate Hox clusters. (C) Sketch showing that the r4 expression of *Hoxb1* becomes restricted to small longitudinal stripes of neurons in r4.

CHAPTER THIRTY-FIVE

C. elegans Embryonic Morphogenesis

Thanh T.K. Vuong-Brender[1], Xinyi Yang, Michel Labouesse[1]
Laboratoire de Biologie du Développement, CNRS-UMR7622, Institut de Biologie Paris-Seine, Université Pierre et Marie Curie, Sorbonne Universités, Paris, France
[1]Corresponding authors: e-mail address: thanh.vuong@upmc.fr; michel.labouesse@upmc.fr

Contents

Abstract

Morphogenesis is a four-dimensional process which involves the crucial interplay between signaling, mechanical forces, and spatial changes. *Caenorhabditis elegans* presents a simple yet versatile model to study morphogenesis. Here, we review recent progress on cellular and molecular drivers of morphological changes during *C. elegans* epiboly and embryonic elongation: actin dynamics and actomyosin contractility, migration guidance cues and junction remodeling. In addition, we discuss how mechanical forces contribute to the process.

The free-living nematode *Caenorhabditis elegans* is well known for its simplicity as a model organism (Ankeny, 2001; Antoshechkin & Sternberg, 2007). In particular, its tube-like shape represents one of the most basic forms in the animal kingdom. How the animal acquires its shape results from the morphogenesis of the epidermal sheet, which spreads out to enclose the embryo and undergoes striking shape changes to become cylindrical over a period of less than 3 h. Analysis of these processes (called ventral enclosure and embryonic elongation, respectively) has shown the importance of precise regulation of signaling and mechanical forces at the cellular scale. Here, we review these processes and discuss what recent findings bring to our understanding of cell migration through actin-based protrusions in response to guidance cues, cell shape changes based on actomyosin contractility, and junction remodeling.

Current Topics in Developmental Biology, Volume 116
ISSN 0070-2153
http://dx.doi.org/10.1016/bs.ctdb.2015.11.012

1. VENTRAL ENCLOSURE

Epidermal cells are born at the end of gastrulation as a group of cells on the posterior and dorsal surface of the embryo. They form six rows of cells: two dorsal, two lateral, and two ventral rows (Heid, Raich, et al., 2001; Sulston, Schierenberg, et al., 1983). The first morphogenetic event to occur in the epidermis corresponds to the intercalation of dorsal epidermal cells which we will not discuss (for review, see Bone, Tapley, et al., 2014; Chisholm & Hardin, 2005; Fridolfsson, Ly, et al., 2010; Meyerzon, Fridolfsson, et al., 2009; Walck-Shannon, Reiner, et al., 2015). Dorsal intercalation leads to a lengthening of the epithelial epidermal cell sheet and to the establishment of correct circumferentially oriented actin bundles, which are essential for the elongation phase (Chisholm & Hardin, 2005; Heid et al., 2001). Very soon after completion of dorsal intercalation, the ventral epidermal cells on each side of the embryo migrate toward the ventral midline and seal up to enclose the embryo.

Ventral enclosure takes place in two steps (Raich, Agbunag, et al., 1999; Williams-Masson, Malik, et al., 1997; Fig. 1). First, the two most anterior epidermal cells on each side, called the "leading cells," initiate the migration by extending long protrusions toward the ventral midline (Fig. 1A). They eventually meet and establish contacts. In the second step, the posterior cells form a continuous actin-rich border around the "ventral pocket" (Fig. 1B). The pocket cells become wedge-shaped and initiate a coordinated

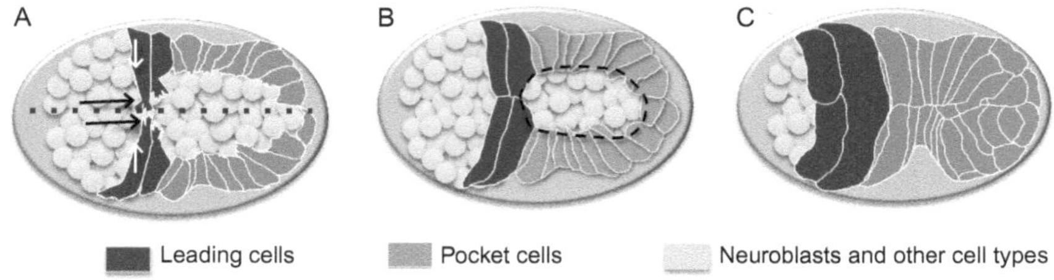

Figure 1 The two steps of epidermal ventral enclosure. (A) Schematic ventral view of an embryo (anterior to the left). First, the leading cells (dark red) extend long protrusions (yellow arrows), migrate (white arrows show the direction of migration) toward the ventral midline (dotted line), meet and establish adherens junctions. (B) The pocket cells (pink) exert coordinated contraction to close the open pouch (dashed line). (C) End of ventral closure; the head is not totally enclosed (some head epidermal cells anterior to the leading cells are not illustrated). Neuroblasts and other cell types are depicted in blue. (A color version of this figure is located at the end of this chapter.)

contraction to fully enclose the posterior part through a purse-string mechanism reminiscent of *D. melanogaster* dorsal closure (Fig. 1B; Gorfinkiel, Schamberg, et al., 2011). The anterior part of the embryo is finally enclosed by head epidermal cells (Fig. 1C; Chisholm & Hardin, 2005). In the next paragraphs, we discuss how imaging and drug assays helped uncover the cytoskeletal proteins and the cells required for enclosure. We also review the results of genetic screens that identified mutants that either fail to migrate, or that do not seal properly.

Laser inactivation experiments, in which epidermal cells are not completely killed but lose their capacity to migrate, have confirmed the role of leading cells in migration initiation and the cooperative action of more posterior cells in pocket closure (Williams-Masson et al., 1997). In particular, inactivation of leading cells on both sides of the ventral midline induces a retraction of the whole epidermis dorsally. Likewise, inactivation of the ventral pocket cells even on one side of the ventral midline leads to pocket cell retraction and their inability to reinitiate enclosure (Williams-Masson et al., 1997).

Mechanistically, pharmacological experiments have established the primary role of actin dynamics during ventral enclosure, while genetic studies helped identify conserved actin remodeling factors. Inhibition of actin polymerization with cytochalasin-D immediately blocks migration and causes the leading cells to regress dorsally (Williams-Masson et al., 1997). Affecting actin dynamics by mutations in the WAVE/SCAR and ARP2/3 complex, which act downstream of CED-10/Rac1 (Fig. 2A), strongly reduces actin at the leading edge, resulting in fewer and smaller protrusions in epidermal cells, and in severe ventral enclosure phenotypes (Bernadskaya, Wallace, et al., 2012; Patel, Bernadskaya, et al., 2008; Sawa, Suetsugu, et al., 2003; Soto, Qadota, et al., 2002). The WAVE/SCAR complex is known to activate the Arp2/3 complex, which in turn promotes actin filament branching to enable cell migration (Blanchoin, Boujemaa-Paterski, et al., 2014). In *C. elegans*, the dynamics of actin assembly is enhanced by the protein UNC-34/VASP, which can bind the proline-rich domain of WAVE to increase cell motility (Havrylenko, Noguera, et al., 2015). In addition, the *C. elegans* WASP homologue WSP-1, acting probably downstream of CDC-42, works redundantly with UNC-34 and WAVE to activate ARP2/3 and promote F-actin nucleation (Rohatgi, Ma, et al., 1999; Sawa et al., 2003; Withee, Galligan, et al., 2004; Fig. 2A). Although initially identified and characterized in *Dictyostelium discoideum* and mammals (Blanchoin et al., 2014), work in *C. elegans* has been instrumental in establishing the role of the WAVE/SCAR, UNC-34/VASP, and ARP2/3 pathway in an *in vivo* morphogenetic process.

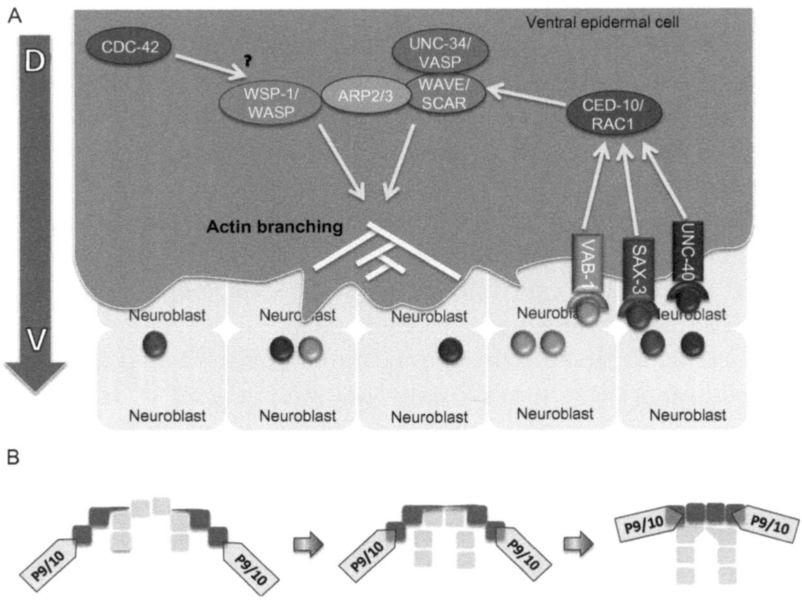

Figure 2 Neuroblasts guide ventral epidermis migration during ventral enclosure. (A) Signaling from neuroblasts promotes actin dynamics in the ventral epidermis. CED-10/Rac1, which acts probably downstream of VAB-1/EphR, SAX-3/Robo, and UNC-40/DCC, activates the WAVE/SCAR complex, and in turn the ARP2/3 complex. UNC-34/VASP binds WAVE/SCAR to promote actin polymerization. WSP-1, probably activated by CDC-42, contributes to a redundant pathway with WAVE/SCAR and UNC-34/VASP, which has not yet been fully investigated. Expression of guidance cues is discussed in the text. Big blue arrow: direction of migration; D, dorsal side; V, ventral side. Small ovals: guidance cues secreted by neuroblasts. (B) Schematic ventral view of the ventral midline, depicting how epidermal ventral cells (P9/10) migrate over a bridge made by neuroblasts. Some neuroblasts (green squares; only key neuroblasts are represented) extend protrusions and translocate over their neighbors (blue squares) to establish a bridge (Ikegami et al., 2012). (A color version of this figure is located at the end of this chapter.)

The outcome of ventral epidermal enclosure depends on an earlier event, the closure of the ventral cleft which is a depression on the ventral surface created by cell ingression during gastrulation. Neuroblasts migrate to fill in the cleft about 1 h before epidermal cell migration (Chisholm & Hardin, 2005). Semiautomated nuclear tracking has shown that neuroblast movements are not stochastic but follow a pattern that correlates with that of ventral epidermal cells before and during ventral enclosure (Giurumescu, Kang, et al., 2012). Thus, neuroblasts play a key role, in particular, to guide pocket cell movements. Precise imaging has revealed that some neuroblasts under the ventral pocket extend protrusions and translocate themselves toward the ventral midline (Fig. 2B). Thereby, they queue up in a row spanning the ventral pocket and form a "bridge" over which the epidermal pocket cells can "walk" before joining together from the two edges of the pocket (Ikegami, Simokat, et al., 2012).

The characterization of partially penetrant ventral enclosure mutants identified the Ephrin, Semaphorin, Robo/Slit, and Netrin guidance proteins, which also mediate neuronal migration and axon pathfinding (Bernadskaya et al., 2012; Chin-Sang, George, et al., 1999; Chin-Sang, Moseley, et al., 2002; George, Simokat, et al., 1998; Ghenea, Boudreau, et al., 2005; Ikegami et al., 2012; Miller & Chin-Sang, 2012; Nakao, Hudson, et al., 2007; Roy, Zheng, et al., 2000; Wang, Roy, et al., 1999). Their study established that ventral enclosure depends on the completion of ventral cleft closure and on signaling between neuroblasts and ventral epidermal cells. Indeed, mutants in these genes often cause a larger, deeper, and more persistent ventral cleft, which subsequently affects the final steps of epidermal cell migration during ventral closure. It suggests that signaling provided by these pathways guides the movement of neuroblasts (Chin-Sang et al., 1999, 2002; George et al., 1998; Ghenea et al., 2005; Ikegami et al., 2012; Nakao et al., 2007; Roy et al., 2000; Wang et al., 1999). In addition, genetic analysis strongly suggests that neuroblasts represent a signaling center for ventral epidermal cell migration. Many of these guidance proteins, such as VAB-1/EphR, EFN-2/Ephrin, SAX-3/Robo, PLX-2/Plexin, MAB-20/Semaphorin-2a, are expressed both in neuroblasts and in the epidermis (George et al., 1998; Ghenea et al., 2005; Ikegami et al., 2012; Nakao et al., 2007; Roy et al., 2000; Wang et al., 1999). In particular, SAX-3 is expressed in the leading cells (Ghenea et al., 2005), and UNC-40 is expected to act cell autonomously (Chan, Zheng, et al., 1996). Finally, rescue experiments show that expression of VAB-1/EphR in P9/10 or in a subset of neuroblasts can rescue mutant lethality, showing a close communication between both cell types (Ikegami et al., 2012). Other proteins, like EFN-1/Ephrin, work nonautonomously in the developing neurons to regulate epidermal morphogenesis (Chin-Sang et al., 1999).

There is a high level of redundancy between these guidance pathways, such that embryos lacking two of them often exhibit more severe enclosure defects. Specifically, the unique *C. elegans* Ephrin receptor VAB-1 can form a coreceptor with SAX-3/Robo to guide cell migration (Ghenea et al., 2005). Interestingly, EFN-4/Ephrin does not interact with VAB-1/EphR but acts instead in a redundant pathway involving the *C. elegans* Semaphorin-2a MAB-20 (Chin-Sang et al., 2002; Wang et al.,

1999). The VAB-1/EphR and SAX-3/Robo pathways are also redundant with the Netrin pathway (Bernadskaya et al., 2012). Given the complexity of the guidance cue system, how these pathways affect precisely neuronal and epidermal cell movements requires further investigation.

How the guidance cues are translated to changes in the orientation and motility of epidermal cells is starting to be revealed. VAB-1/EphR, SAX-3/Robo, and UNC-40/DCC Netrin receptor mutants affect the correct accumulation of CED-10/Rac1 and WVE-1/WAVE at the basolateral membrane of the epidermis (Bernadskaya et al., 2012; Fig. 2A). They display reduced level of actin polarization along the dorsoventral axis of migrating epidermal cells. Although not as pronounced even in double mutants, it is similar to what is observed in WAVE/SCAR mutants, consistent with the fact that the ventral enclosure phenotype of *vab-1/ephR, sax-3/robo*, and *unc-40/DCC* mutants is less severe than those of *wave/scar* and *arp2/3* mutants. It is unclear yet whether there is another redundant pathway to be identified, or whether removing all three guidance pathways (which would certainly be embryonic lethal) would abrogate cell migration.

Once the contralateral ventral cells have met at the ventral midline, they assemble stable junctions through the cadherin–catenin complex, which become capable of resisting mechanical tension. Depletion of zygotic HMR-1/E-cadherin or maternal and zygotic HMP-1/α-catenin leads to unenclosed embryos, which extrude their pharynx and intestine anteriorly (Costa, Raich, et al., 1998). During the sealing process, HMP-1/α-catenin quickly localizes to the point of contact between the filopodia-like extensions of contralateral leading cells, while concomitantly disappearing from the cytoplasm (Raich et al., 1999). This process requires HMR-1/E-cadherin function, as in *hmr-1* mutants, the accumulation of HMP-1 at the cell contact region and the drop in the concentration of HMP-1 cytoplasmic pool are not observed. In these mutants, the leading cells still extend filopodia and meet at the ventral midline like in wild type. However, after failing to make contact at the ventral midline, the leading cells cease their protrusive activity and retract dorsally (Raich et al., 1999). Thus, stable contacts require both cell–cell adhesion via E-cadherin and junction "reinforcement" by connection with the cytoskeleton through α- and β-catenin. In contrast, the sealing of ventral pocket cells does not have a stringent requirement for the cadherin–catenin complex. In *hmr-1/E-cadherin* and *hmp-1/α-catenin* mutant, ventral pocket cells still close albeit incompletely (Raich et al., 1999). These results suggest that the ventral pocket cells may use a cadherin-independent adhesion system or redundant adhesion signals. The building of new junctions and the feedback loop with the leading cell cytoskeleton remain poorly understood.

Many of the guidance cue mutant embryos (*vab-1, sax-3, mab-20, efn-1, efn-4)* undergo apparently normal ventral enclosure, yet they rupture during elongation (Chin-Sang et al., 1999, 2002; George et al., 1998; Ghenea et al., 2005; Nakao et al., 2007). Although this phenomenon is also observed when the ventral epidermal cells are inactivated by laser during enclosure (Williams-Masson et al., 1997), it is not clear if the rupture phenotype results solely from improper epidermal sealing, or if the guidance cues have some impact on the integrity of epidermis at a later stage. Since these mutants affect the correct actin distribution (Bernadskaya et al., 2012), one possibility could be that junction reinforcement through actin-dependent processes becomes defective as tension increases during elongation.

Although superficially similar, the closure of *C. elegans* ventral epidermis differs from dorsal closure of *Drosophila* in several aspects. In *Drosophila*, the dorsal closure is aided mechanically by the amnioserosa (Gorfinkiel et al., 2011; Harden, 2002). It is not known if in *C. elegans*, the neuroblasts or other embryonic tissues exert repeated pulsatile contractions like the amnioserosa cells (Solon, Kaya-Copur, et al., 2009). Although myosin II is enriched at the ventral border of pocket cells in *C. elegans*, it is unclear to what extent the cable akin to that observed at the *Drosophila* dorsal leading edge plays a role in ventral enclosure. In *C. elegans*, mutations in the upstream regulators of myosin II, like *rho-1/rhoA*, *let-502/Rock*, and *mel-11/Mypt*, induce a small proportion of ventral enclosure defective or early ruptured embryos (Fotopoulos, Wernike, et al., 2013). However, the defects are not penetrant enough to conclude that myosin is essential. Precise force measurements will be necessary to elucidate the contribution of neuroblasts, the epidermis, and the supracontractile ring around the pocket during *C. elegans* ventral enclosure.

2. ELONGATION FROM THE LIMA-BEAN TO THE TWOFOLD STAGE

Once ventral enclosure has finished, the embryo elongates from a lima-bean (equivalent to onefold stage) shape to the characteristic worm shape, resulting in a fourfold increase in length and a twofold decrease in diameter approximately since the volume remains constant (Priess & Hirsh, 1986). The whole process takes place with neither cell division nor change in the relative positions of epidermal cells (Sulston et al.,

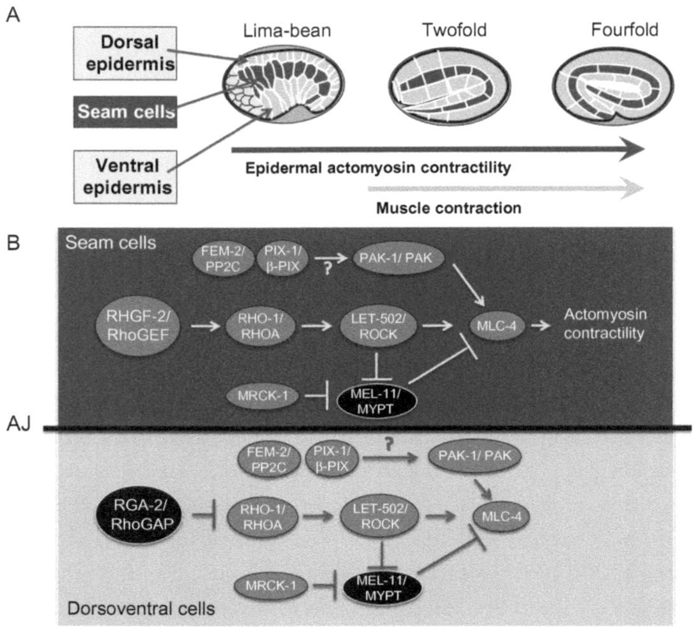

Figure 3 Overview of *C. elegans* embryonic elongation. (A) Schematic drawings of embryos along elongation. Note the change in shape of the seam cells (lateral epidermis; violet) where the junctions shorten in the dorsoventral (DV, circumferential) direction and lengthen in the anteroposterior (AP) direction. Embryos are staged according to their length (lima-bean or onefold to fourfold). Anterior to the left, dorsal up. (B) Molecular pathways mediating activation of nonmuscle myosin II contractility through phosphorylation of the myosin regulatory light chain MLC-4 during elongation. (A color version of this figure is located at the end of this chapter.)

1983). Instead, epidermal cells modify their shapes, as can be most easily observed among seam cells (Fig. 3A). About midway through the process, muscles start to contract; this is essential, since muscle-defective embryos are paralyzed and arrest their elongation at the twofold stage. This phenotype is called Pat, for paralyzed at twofold (Waterston, 1989; Williams & Waterston, 1994). Elongation can be divided in an early and a late phase with respect to the beginning of muscle contractions, each stage being mediated by distinct molecular pathways. In the following paragraphs, we discuss how actomyosin contractility drives elongation as well as how junctions and the cytoskeleton get remodeled during this process.

Genetic and pharmacological studies have shown that actomyosin contractility is essential for early elongation. Actin filaments form circumferential bundles in the dorsal and ventral epidermis but a rather disorganized pattern in seam cells characterized by thinner, shorter, and less-oriented

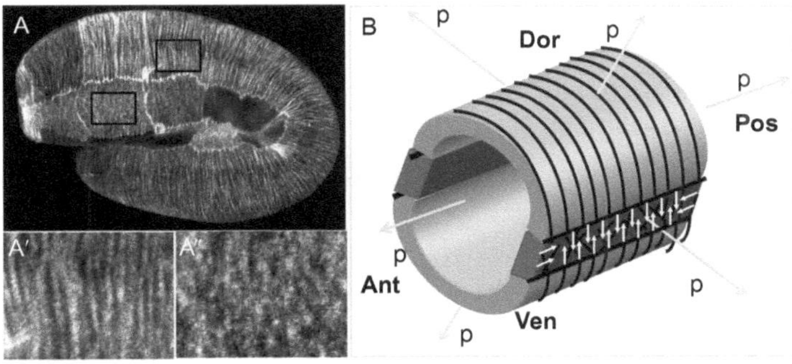

Figure 4 Actin organization and possible role during elongation. (A) Actin filaments in nearly twofold stage embryo are organized in parallel bundles in DV cells (yellow rectangle—dorsal cell hyp7, A′) and in thinner and less circumferentially oriented filaments in lateral seam cells (red rectangle, A″). (B) Hypothetical model for *C. elegans* elongation where the seam cells (violet) pull to increase the hydrostatic pressure and the DV cells (pink) orient the elongation in the direction perpendicular to actin bundle alignment (green lines), p, hydrostatic pressure; Ant, anterior; Pos, posterior; Dor, dorsal; Ven, ventral; arrows, force exerted. *Panel (A–A″) adapted with permission from Gally et al. (2009).* (A color version of this figure is located at the end of this chapter.)

filaments (Fig. 4A; Gally et al., 2009; Priess & Hirsh, 1986). The mechanism that controls the polarized organization of actin filaments in the circumferential direction is not known. Inhibition of actin polymerization with cytochalasin-D either blocks elongation if it is applied prior to the 1.5-fold stage or causes embryos to retract to their preelongation state later on (Priess & Hirsh, 1986). Actomyosin forces are thought to squeeze the embryo radially to decrease their diameter, inducing most likely anterior–posterior (AP) lengthening by an increase in hydrostatic pressure. Indeed, embryos burst if the epidermis is punctured with a laser (Priess & Hirsh, 1986). Nonmuscle myosin II motors are regulated through phosphorylation and dephosphorylation of the regulatory light chain MLC-4 by the LET-502/Rho-binding kinase and MEL-11/Myosin binding subunit of myosin phosphatase, respectively (Gally et al., 2009; Wissmann, Ingles, et al., 1997, 1999; Fig. 3B). LET-502 is the effector of the Rho GTPase RHO-1, which can be activated and inactivated by the *C. elegans* RhoGEF (Rho-Guanine nucleotide Exchange Factor) RHGF-2, and RhoGAP (Rho-GTPase-Activating Protein) RGA-2, respectively (Chan, Rocheleau, et al., 2015; Diogon, Wissler, et al., 2007). Mutations affecting positive regulators of contractility, such as RHGF-2/RhoGEF, LET-502/ROCK, MLC-4, and NMY-1/Myosin II heavy chain, lead to hypoelongation, whereby embryos arrest earlier than or at the twofold stage (Chan et al., 2015;

Gally et al., 2009; Piekny, Johnson, et al., 2003; Shelton, Carter, et al., 1999; Wissmann et al., 1997). By contrast, hypercontractility mutants like *mel-11* or *rga-2* cause embryos to burst during elongation due to increased tension exerted on adherens junctions (Diogon et al., 2007; Wissmann et al., 1999). Activation of myosin II is achieved mainly through the LET-502/Rho kinase, but two additional kinases can contribute to maintain myosin II active. Specifically, the p21-activated kinase PAK-1 and the CDC-42-activated kinase MRCK-1 act in parallel to LET-502, since their loss enhances the *let-502* mutant severity (Gally et al., 2009). Genetic analysis suggests that PAK-1 acts downstream of the phosphatase FEM-2/PP2C and the CDC-42/RAC-specific GEF (Guanine nucleotide Exchange Factor) PIX-1/β-PIX (Martin, Harel, et al., 2014; Vanneste, Pruyne, et al., 2013), whereas MRCK-1 acts upstream of MEL-11 (Gally et al., 2009). How FEM-2 regulates PAK-1 in this scheme is actually unclear (Vanneste et al., 2013). In addition, PIX-1 may more specifically promote head constriction (Martin et al., 2014).

Genetic and molecular data suggest that seam cells concentrate actomyosin forces, while the dorsoventral (DV) cells may remain passive. Indeed, several lines of evidence support this hypothesis. First, the activity of the nonmuscle myosin regulatory light chain MLC-4 is mainly required in seam cells (Gally et al., 2009). Second, MLC-4, MLC-5/myosin essential light chain, and NMY-1 are enriched in seam cells (Gally et al., 2009; Piekny et al., 2003; Shelton et al., 1999). Third, rescue experiments have shown that the positive regulator of contractility RHGF-2/RhoGEF acts specifically in seam cells, whereas the negative regulator RGA-2/RhoGAP is required only in DV cells (Chan et al., 2015; Diogon et al., 2007). Thus, all the players of actomyosin regulation pathway act in a way to keep contractility high in seam cells and low in DV cells (Fig. 3B).

During *Drosophila* gastrulation and germband extension, myosin II shows a pulsatile behavior which is essential for cell shape changes (Martin, Kaschube, et al., 2009; Rauzi, Lenne, et al., 2010). In particular during germband extension, myosin II is planar polarized with higher accumulation in vertical junctions (along DV direction) (Bertet, Sulak, et al., 2004; Blankenship, Backovic, et al., 2006; Rauzi et al., 2010). A planar polarity-dependent junctional shortening and actomyosin pulses are also observed during the convergence–extension of *Xenopus* mesoderm (Shindo & Wallingford, 2014). By contrast, myosin II distribution is not planar polarized (Gally et al., 2009; Piekny et al., 2003) and flows have so far not been reported during *C. elegans* elongation, although myosin II pulses and flows have been described in *C. elegans* zygotes and at gastrulation (Mayer, Depken, et al., 2010; Munro, Nance, et al., 2004; Roh-Johnson, Shemer, et al., 2012). Similarly, the *C. elegans* cadherin–catenin complex is apparently not planar polarized in contrast to what is observed in *Drosophila* germband extension (Blankenship et al., 2006; Costa et al., 1998). Mechanistically, it is not clear how myosin II and adherens junctions are remodeled during early elongation; *C. elegans* may rely on different processes compared to other species.

The preferential extension of a material in one direction can come from higher forces in that direction (anisotropy of forces), or from an anisotropy of its response to isotropic forces. The extension of *Drosophila* germband is induced by anisotropic forces generated by planar-polarized actomyosin network (Bertet et al., 2004), whereas the oriented growth of plants is due to the anisotropic response of the cell wall to the isotropic force generated by turgor pressure (Baskin, 2005). As we argue below, because the myosin network does not seem to be planar polarized up to the twofold stage (Gally et al., 2009; Piekny et al., 2003; Shelton et al., 1999), the growth of plants provides a more relevant comparison with the elongation of *C. elegans* embryos than *Drosophila*. It has been proposed that the alignment of microfibrils in plants around the circumferential direction restricts radial expansion and orients the growth in the longitudinal direction (Baskin, 2005). The *C. elegans* actin filament organization in DV cells is reminiscent of the microfibril arrangement. Moreover, since myosin II activity is downregulated in DV cells, it supports the hypothesis that the DV cells create a passive anisotropic response to the hydrostatic pressure that is similar to the response of microfibrils to turgor pressure (Fig. 4B). While in plants the turgor pressure increase is due to osmosis across semipermeable cell membranes (Pritchard, 2001), in *C. elegans* embryo the increased hydrostatic pressure is likely to originate from actomyosin contractility in seam cells (Fig. 4B). The phenotypes of several mutants affecting junctional proteins thought to anchor actin bundles (Priess & Hirsh, 1986) support the role of actin bundles during elongation. In particular, zygotic *hmp-1/α-catenin* mutants, in which actin bundles detach from the junctional belt, display bulges and cannot elongate. Likewise, loss of ZOO-1/ZO-1 and VAB-9/Claudin homologues affects actin bundle organization in DV cells, leading to deformities and an incomplete elongation (Lockwood, Zaidel-Bar, et al., 2008; Simske, Koppen, et al., 2003). Although it has been proposed that the circumferentially oriented microtubules in DV cells could also provide a passive force to help elongation (Ciarletta, Ben Amar, et al., 2009; Priess & Hirsh, 1986), our more recent data favor the notion that they act by enabling the transport of junctional proteins necessary for elongation (Quintin, Wang, et al., 2016). Perturbation of actin bundle organization in DV cells and computer modeling could help to shed light on their role during elongation.

3. ELONGATION BEYOND THE TWOFOLD STAGE

When the *C. elegans* embryo reaches the 1.7-fold stage, muscle cells located under the dorsal and ventral epidermal cells start to contract. During this period, muscles play a central role, whereas actomyosin contractility in the epidermis is thought to be less important. Indeed, time-shift experiments with a thermosensitive mutant of the LET-502/Rho kinase have suggested that it is only required until the twofold stage (Diogon et al., 2007). Moreover, our laboratory recently unraveled a possible mechanism for how muscles could promote elongation (Zhang, Landmann, et al., 2011), which features a prominent role for mechanical stimulation of the epidermis by muscles through hemidesmosome-like junctions. Below, we discuss how the tension exerted by muscles can help remodel epidermal cell–cell and cell–extracellular matrix junctions.

Muscles attach to the exoskeleton cuticle through transepithelial attachment structures, known as fibrous organelles (FOs), which span the thickness of the dorsal and ventral epidermis in areas of contact; in effect, they act as tendons transmitting the forces exerted by muscle contraction to the cuticle exoskeleton (Francis & Waterston, 1985, 1991). For this reason, muscles are tightly connected to dorsal and ventral epidermal cells. FOs correspond to two hemidesmosome-like junctions (known as CeHDs), one apical and the other basal, connected by intermediate filaments (Zhang & Labouesse, 2010; Fig. 5A). The only evolutionary conserved protein between vertebrate hemidesmosomes and CeHDs is VAB-10A, the homologue of vertebrate Plectin and BPAG1e. VAB-10A is thought to anchor intermediate filaments to membrane plaques on both the apical and basal sides of the epidermis (Bosher, Hahn, et al., 2003), and to interact with nematode-specific transmembrane extracellular matrix receptors, LET-805/Myotactin basally, MUP-4 and MUA-3 apically (Fig. 5A; Bercher, Wahl, et al., 2001; Hahn & Labouesse, 2001; Hong, Elbl, et al., 2001). Animals with defective hemidesmosomes have strong epidermal integrity defects and muscle anchoring defects; furthermore, in their absence, embryos arrest between the 1.7-fold and 2-fold stages of elongation (Bercher et al., 2001; Bosher et al., 2003; Ding, Goncharov, et al., 2003; Ding, King, et al., 2008;

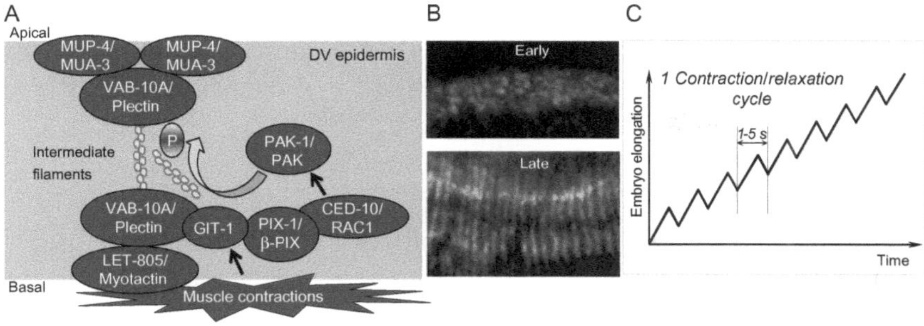

Figure 5 Muscle activity through CeHDs may promote elongation via a ratchet system. (A) CeHD remodeling under muscle contractile tension. Muscle contractions lead to GIT-1 recruitment, which subsequently activates PIX-1, CED-10/Rac1, and PAK-1. In turn, PAK-1 phosphorylates intermediate filaments leading to CeHD maturation (see panel B). (B) Images of a small portion of CeHD (marked with an antibody against the CeHD protein VAB-10A) when muscles start to contract (top) and at the end of elongation (bottom). (C) Elongation after the twofold stage as a ratchet-like process. (A color version of this figure is located at the end of this chapter.)

Hetherington, Gally, et al., 2011; Hong et al., 2001; Woo, Goncharov, et al., 2004; Zahreddine, Zhang, et al., 2010).

Due to the tight connection between muscles and the epidermis and their overall anterior–posterior orientation, muscle contractions create a longitudinal tension that is transmitted to the epidermis; thereby, the dorsal and ventral cells become locally compressed and stretched longitudinally (Zhang et al., 2011). These lateral cyclic stretches could promote elongation beyond the twofold stage. First, our laboratory showed that the tension originating from muscle contractions is felt at CeHDs and induces a mechanotransduction pathway in the epidermis (Fig. 5A). It ultimately activates the CeHD-associated p21-activated kinase homologue PAK-1, through a signaling cascade involving the Git1 homologue GIT-1 (an adaptor protein), its partner the β-Pix homologue PIX-1, and the Rac GTPase CED-10, resulting in the phosphorylation of intermediate filaments and their further recruitment to CeHDs (Zhang et al., 2011). In the absence of muscle contractions, the GIT-1 protein diffuses away from CeHDs; to account for this observation, we suggested that a CeHD component (which remains to be identified) might undergo a conformational change upon tension, triggering its interaction with GIT-1 to unleash the signaling. Prior to muscle contractions, CeHDs form puncta, which progressively evolve into short parallel circumferentially oriented stripes above the muscles, which are likely to provide stronger attachments as the force exerted by muscles increases (Fig. 5B). Since in the absence of muscle contractions, or when

GIT-1, PIX-1, or PAK-1 are knocked down together with a weak mutation in VAB-10A, CeHDs fall apart and do not mature, we proposed that the mechanotransduction pathway promotes CeHD maturation (Zhang et al., 2011).

Imaging the effect of muscle contractions on the epidermis revealed that muscles contract every few seconds (1–5 s), and do not all contract at the same time. Hence, the mechanical stimuli powered by muscles are oscillatory. Although cell shape changes during *C. elegans*, elongation do not appear to involve pulsatile flows of myosin II like in many *Drosophila* cell shape changes, it is nevertheless likely to involve a ratchet process (Fig. 5C).

Taking into consideration that *git-1*, *pix-1*, or *pak-1* mutants do not have an elongation phenotype as severe as that of muscle-defective *pat* mutants, it is conceivable that the muscle tensional input impinges on other processes in addition to CeHD maturation. In particular, since *pat* mutants fail to elongate beyond the twofold stage, tension might promote the shortening of the circumferential actin bundles present in dorsal and ventral cells (shortening is necessary since the embryonic diameter diminishes), and the lengthening of seam-dorsal and seam-ventral adherens junctions. In recent genetic screens, we have identified several genes whose downregulation in a *git-1* mutant results in a twofold arrest (C. Gally & M. Labouesse, unpublished). We anticipate that their characterization will further elucidate how muscles stimulate elongation.

How mechanical tension originating from muscles could stimulate adherens junction remodeling is not known. Lessons from other experimental systems are pointing to several potential directions. In both *Drosophila* embryonic epithelial cells and mammalian cells, endocytosis and the subsequent recycling of E-cadherin (E-cad) contribute to junction remodeling (Lecuit & Yap, 2015; Levayer & Lecuit, 2012). In particular, the nonmuscle myosin II (Myo-II) regulates E-cadherin trafficking by promoting E-cad endocytosis and by facilitating the formation of recycling endosomes containing E-cad and their targeting to adherens junctions (Lecuit & Yap, 2015; Levayer & Lecuit, 2012). Such pathways could be important during the first phase of elongation when junctions oriented along the circumference get shorter (Fig. 3A). Other studies suggest that the neosynthesis of junctional proteins and their lateral diffusion after reaching junctions by exocytosis is required to promote junction elongation (Goldenberg & Harris, 2013). Thus, one possibility for junction lengthening during *C. elegans* elongation could be that muscle contractions somehow trigger the expression of specific genes that promote E-cad synthesis, posttranslational modifications or transport. In particular, sumoylation and phosphorylation of HMR-1/E-cadherin cytoplasmic tail mediate its binding to HMP-2/β-catenin, affecting HMR-1 accumulation at the apical junctions and the formation of a functional adherens junction (Callaci, Morrison, et al., 2015; Choi, Loveless, et al., 2015; Tsur, Bening Abu-Shach, et al., 2015). The dependence of HMR-1 maintenance on clathrin-mediated transport only seen after twofold stage may also be somehow promoted by muscle contractions (Gillard, Shafaq-Zadah, et al., 2015). Finally, recent studies in cell culture have established that adherens junctions are mechanosensitive, and stiffen in response to applied forces by recruiting vinculin and more actin (Chen, Hong, et al., 2015; le Duc, Shi, et al., 2010; Yonemura, Wada, et al., 2010). Moreover, the E-cadherin turnover rate appears to increase with increasing mechanical stress applied to a junction (de Beco, Perney, et al., 2015), suggesting that the trafficking machinery for junctional proteins may be mechanosensitive.

In *C. elegans*, the actin filaments anchored to adherens junctions might somehow transmit the tensional input originating from muscle contraction during elongation (Fig. 6A). Indeed, adherens junctions are likely to be under tension since mutants such as *mel-11* and *rga-2* that hyperactivate non-muscle myosin II cause the embryo to rupture (Diogon et al., 2007; Wissmann et al., 1999). Although vinculin is not expressed in the *C. elegans* epidermis, other proteins might serve a similar function as vinculin (without necessarily direct binding to α-catenin). In particular, several

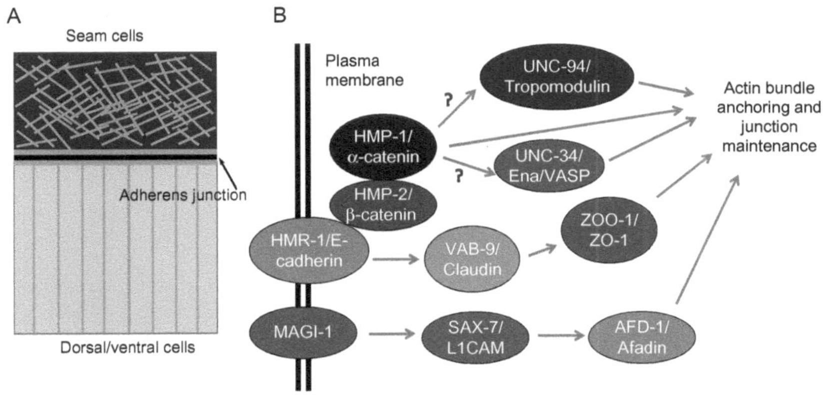

Figure 6 Role of cell junctions during elongation. (A) Junctional proteins are thought to anchor actin bundles in DV cells, green lines represent actin filaments and bundles. (B) Different junction-associated proteins may contribute to actin bundle anchoring and junction maintenance during elongation. (A color version of this figure is located at the end of this chapter.)

actin-binding proteins, such as AFD-1/Afadin (through binding with SAX-7/L1CAM and MAGI-1), UNC-94/Tropomodulin, and UNC-34/Ena, localize to *C. elegans* adherens junctions where they reinforce the attachment between cells and the connection with the actin cytoskeleton (Fig. 6B; Cox-Paulson, Walck-Shannon, et al., 2012; Lynch, Grana, et al., 2012; Sheffield, Loveless, et al., 2007). Their loss does not always induce a strong phenotype but enhances the defects of weak mutants of the cadherin–catenin complex (Cox-Paulson et al., 2012; Lynch et al., 2012; Sheffield et al., 2007). Force measurement on junctions to assess their distribution and evolution during elongation, and further characterization of E-cadherin transport pathways should help define how muscles might stimulate junction lengthening and embryo elongation.

In summary, research on *C. elegans* embryo morphogenesis has not only shown that many pathways found in vertebrates are conserved, but it has also opened perspectives on new mechanisms regulating morphogenetic processes. The next challenge will be to measure the properties of mechanical forces (magnitude, direction, point of application) at the cellular scale, and to understand how different proteins as well as feedback mechanisms regulate those properties. Biophysical methods and computer modeling will also be instrumental to clarify how the cooperation of signaling and mechanical forces shapes the organism.

ACKNOWLEDGMENTS

We thank Martha Soto, François Robin, Flora Llense, Christelle Gally, and Sophie Quintin for comments on the manuscript and for discussions. Our work is supported by a grant from the European Research Council to ML, and by institutional funds from the CNRS and University Pierre et Marie Curie (UPMC) to the IBPS and UMR7622. In addition, M.L. thanks the CNRS and the UPMC for a relocation package.

REFERENCES

Ankeny, R. A. (2001). The natural history of Caenorhabditis elegans research. *Nature Reviews Genetics*, *2*(6), 474–479.

Antoshechkin, I., & Sternberg, P. W. (2007). The versatile worm: Genetic and genomic resources for Caenorhabditis elegans research. *Nature Reviews Genetics*, *8*(7), 518–532.

Baskin, T. I. (2005). Anisotropic expansion of the plant cell wall. *Annual Review of Cell and Developmental Biology*, *21*, 203–222.

Bercher, M., Wahl, J., et al. (2001). mua-3, a gene required for mechanical tissue integrity in Caenorhabditis elegans, encodes a novel transmembrane protein of epithelial attachment complexes. *The Journal of Cell Biology*, *154*(2), 415–426.

Bernadskaya, Y. Y., Wallace, A., et al. (2012). UNC-40/DCC, SAX-3/Robo, and VAB-1/Eph polarize F-actin during embryonic morphogenesis by regulating the WAVE/SCAR actin nucleation complex. *PLoS Genetics*, *8*(8), e1002863.

Bertet, C., Sulak, L., et al. (2004). Myosin-dependent junction remodelling controls planar cell intercalation and axis elongation. *Nature*, *429*(6992), 667–671.
Blanchoin, L., Boujemaa-Paterski, R., et al. (2014). Actin dynamics, architecture, and mechanics in cell motility. *Physiological Reviews*, *94*(1), 235–263.
Blankenship, J. T., Backovic, S. T., et al. (2006). Multicellular rosette formation links planar cell polarity to tissue morphogenesis. *Developmental Cell*, *11*(4), 459–470.
Bone, C. R., Tapley, E. C., et al. (2014). The Caenorhabditis elegans SUN protein UNC-84 interacts with lamin to transfer forces from the cytoplasm to the nucleoskeleton during nuclear migration. *Molecular Biology of the Cell*, *25*(18), 2853–2865.
Bosher, J. M., Hahn, B. S., et al. (2003). The *Caenorhabditis elegans vab-10* spectraplakin isoforms protect the epidermis against internal and external forces. *The Journal of Cell Biology*, *161*(4), 757–768.
Callaci, S., Morrison, K., et al. (2015). Phosphoregulation of the C. elegans cadherin-catenin complex. *The Biochemical Journal*, *472*(3), 339–352.
Chan, B. G., Rocheleau, S. K., et al. (2015). The Rho guanine exchange factor RHGF-2 acts through the Rho-binding kinase LET-502 to mediate embryonic elongation in C. elegans. *The Biochemical Journal*, *405*(2), 250–259.
Chan, S. S., Zheng, H., et al. (1996). UNC-40, a C. elegans homolog of DCC (deleted in colorectal cancer), is required in motile cells responding to UNC-6 netrin cues. *Cell*, *87*(2), 187–195.
Chen, C. S., Hong, S., et al. (2015). alpha-Catenin-mediated cadherin clustering couples cadherin and actin dynamics. *The Journal of Cell Biology*, *210*(4), 647–661.
Chin-Sang, I. D., George, S. E., et al. (1999). The ephrin VAB-2/EFN-1 functions in neuronal signaling to regulate epidermal morphogenesis in C. elegans. *Cell*, *99*(7), 781–790.
Chin-Sang, I. D., Moseley, S. L., et al. (2002). The divergent C. elegans ephrin EFN-4 functions in embryonic morphogenesis in a pathway independent of the VAB-1 Eph receptor. *Development*, *129*(23), 5499–5510.
Chisholm, A. D., & Hardin, J. (2005). Epidermal morphogenesis. *WormBook*, 1–22. http://dx.doi.org/10.1895/wormbook.1.7.1.. http://www.wormbook.org.
Choi, H. J., Loveless, T., et al. (2015). A conserved phosphorylation switch controls the interaction between cadherin and beta-catenin in vitro and in vivo. *Developmental Cell*, *33*(1), 82–93.
Ciarletta, P., Ben Amar, M., et al. (2009). Continuum model of epithelial morphogenesis during Caenorhabditis elegans embryonic elongation. *Philosophical Transactions Series A, Mathematical, Physical, and Engineering Sciences*, *367*(1902), 3379–3400.
Costa, M., Raich, W., et al. (1998). A putative catenin-cadherin system mediates morphogenesis of the Caenorhabditis elegans embryo. *The Journal of Cell Biology*, *141*(1), 297–308.
Cox-Paulson, E. A., Walck-Shannon, E., et al. (2012). Tropomodulin protects alpha-catenin-dependent junctional-actin networks under stress during epithelial morphogenesis. *Current Biology*, *22*(16), 1500–1505.
de Beco, S., Perney, J. B., et al. (2015). Mechanosensitive Adaptation of E-Cadherin Turnover across adherens Junctions. *PLoS One*, *10*(6), e0128281.
Ding, M., Goncharov, A., et al. (2003). C. elegans ankyrin repeat protein VAB-19 is a component of epidermal attachment structures and is essential for epidermal morphogenesis. *Development*, *130*(23), 5791–5801.
Ding, M., King, R. S., et al. (2008). The cell signaling adaptor protein EPS-8 is essential for C. elegans epidermal elongation and interacts with the ankyrin repeat protein VAB-19. *PLoS One*, *3*(10), e3346.
Diogon, M., Wissler, F., et al. (2007). The RhoGAP RGA-2 and LET-502/ROCK achieve a balance of actomyosin-dependent forces in C. elegans epidermis to control morphogenesis. *Development*, *134*(13), 2469–2479.

Fotopoulos, N., Wernike, D., et al. (2013). Caenorhabditis elegans anillin (ani-1) regulates neuroblast cytokinesis and epidermal morphogenesis during embryonic development. *Developmental Biology*, *383*(1), 61–74.

Francis, G. R., & Waterston, R. H. (1985). Muscle organization in Caenorhabditis elegans: Localization of proteins implicated in thin filament attachment and I-band organization. *The Journal of Cell Biology*, *101*(4), 1532–1549.

Francis, R., & Waterston, R. H. (1991). Muscle cell attachment in Caenorhabditis elegans. *The Journal of Cell Biology*, *114*(3), 465–479.

Fridolfsson, H. N., Ly, N., et al. (2010). UNC-83 coordinates kinesin-1 and dynein activities at the nuclear envelope during nuclear migration. *Developmental Biology*, *338*(2), 237–250.

Gally, C., Wissler, F., et al. (2009). Myosin II regulation during C. elegans embryonic elongation: LET-502/ROCK, MRCK-1 and PAK-1, three kinases with different roles. *Development*, *136*(18), 3109–3119.

George, S. E., Simokat, K., et al. (1998). The VAB-1 Eph receptor tyrosine kinase functions in neural and epithelial morphogenesis in C. elegans. *Cell*, *92*(5), 633–643.

Ghenea, S., Boudreau, J. R., et al. (2005). The VAB-1 Eph receptor tyrosine kinase and SAX-3/Robo neuronal receptors function together during C. elegans embryonic morphogenesis. *Development*, *132*(16), 3679–3690.

Gillard, G., Shafaq-Zadah, M., et al. (2015). Control of E-cadherin apical localisation and morphogenesis by a SOAP-1/AP-1/clathrin pathway in C. elegans epidermal cells. *Development*, *142*(9), 1684–1694.

Giurumescu, C. A., Kang, S., et al. (2012). Quantitative semi-automated analysis of morphogenesis with single-cell resolution in complex embryos. *Development*, *139*(22), 4271–4279.

Goldenberg, G., & Harris, T. J. (2013). Adherens junction distribution mechanisms during cell-cell contact elongation in Drosophila. *PLoS One*, *8*(11), e79613.

Gorfinkiel, N., Schamberg, S., et al. (2011). Integrative approaches to morphogenesis: Lessons from dorsal closure. *Genesis*, *49*(7), 522–533.

Hahn, B. S., & Labouesse, M. (2001). Tissue integrity: Hemidesmosomes and resistance to stress. *Current Biology*, *11*(21), R858–R861.

Harden, N. (2002). Signaling pathways directing the movement and fusion of epithelial sheets: Lessons from dorsal closure in Drosophila. *Differentiation*, *70*(4–5), 181–203.

Havrylenko, S., Noguera, P., et al. (2015). WAVE binds Ena/VASP for enhanced Arp2/3 complex-based actin assembly. *Molecular Biology of the Cell*, *26*(1), 55–65.

Heid, P. J., Raich, W. B., et al. (2001). The zinc finger protein DIE-1 is required for late events during epithelial cell rearrangement in C. elegans. *Developmental Biology*, *236*(1), 165–180.

Hetherington, S., Gally, C., et al. (2011). PAT-12, a potential anti-nematode target, is a new spectraplakin partner essential for Caenorhabditis elegans hemidesmosome integrity and embryonic morphogenesis. *Developmental Biology*, *350*(2), 267–278.

Hong, L., Elbl, T., et al. (2001). MUP-4 is a novel transmembrane protein with functions in epithelial cell adhesion in *Caenorhabditis elegans*. *The Journal of Cell Biology*, *154*(2), 403–414.

Ikegami, R., Simokat, K., et al. (2012). Semaphorin and Eph receptor signaling guide a series of cell movements for ventral enclosure in C. elegans. *Current Biology*, *22*(1), 1–11.

le Duc, Q., Shi, Q., et al. (2010). Vinculin potentiates E-cadherin mechanosensing and is recruited to actin-anchored sites within adherens junctions in a myosin II-dependent manner. *The Journal of Cell Biology*, *189*(7), 1107–1115.

Lecuit, T., & Yap, A. S. (2015). E-cadherin junctions as active mechanical integrators in tissue dynamics. *Nature Cell Biology*, *17*(5), 533–539.

Levayer, R., & Lecuit, T. (2012). Biomechanical regulation of contractility: Spatial control and dynamics. *Trends in Cell Biology*, *22*(2), 61–81.

Lockwood, C., Zaidel-Bar, R., et al. (2008). The C. elegans zonula occludens ortholog cooperates with the cadherin complex to recruit actin during morphogenesis. *Current Biology*, *18*(17), 1333–1337.
Lynch, A. M., Grana, T., et al. (2012). A genome-wide functional screen shows MAGI-1 is an L1CAM-dependent stabilizer of apical junctions in C. elegans. *Current Biology*, *22*(20), 1891–1899.
Martin, E., Harel, S., et al. (2014). pix-1 controls early elongation in parallel with mel-11 and let-502 in Caenorhabditis elegans. *PLoS One*, *9*(4), e94684.
Martin, A. C., Kaschube, M., et al. (2009). Pulsed contractions of an actin-myosin network drive apical constriction. *Nature*, *457*(7228), 495–499.
Mayer, M., Depken, M., et al. (2010). Anisotropies in cortical tension reveal the physical basis of polarizing cortical flows. *Nature*, *467*(7315), 617–621.
Meyerzon, M., Fridolfsson, H. N., et al. (2009). UNC-83 is a nuclear-specific cargo adaptor for kinesin-1-mediated nuclear migration. *Development*, *136*(16), 2725–2733.
Miller, M. A., & Chin-Sang, I. D. (2012). Eph receptor signaling in C. elegans. *WormBook*, 1–17.
Munro, E., Nance, J., et al. (2004). Cortical flows powered by asymmetrical contraction transport PAR proteins to establish and maintain anterior-posterior polarity in the early C. elegans embryo. *Developmental Cell*, 7(3), 413–424.
Nakao, F., Hudson, M. L., et al. (2007). The PLEXIN PLX-2 and the ephrin EFN-4 have distinct roles in MAB-20/Semaphorin 2A signaling in Caenorhabditis elegans morphogenesis. *Genetics*, *176*(3), 1591–1607.
Patel, F. B., Bernadskaya, Y. Y., et al. (2008). The WAVE/SCAR complex promotes polarized cell movements and actin enrichment in epithelia during C. elegans embryogenesis. *Developmental Biology*, *324*(2), 297–309.
Piekny, A. J., Johnson, J. L., et al. (2003). The *Caenorhabditis elegans* nonmuscle myosin genes *nmy-1* and *nmy-2* function as redundant components of the *let-502*/Rho-binding kinase and *mel-11*/myosin phosphatase pathway during embryonic morphogenesis. *Development*, *130*(23), 5695–5704.
Priess, J. R., & Hirsh, D. I. (1986). *Caenorhabditis elegans* morphogenesis: The role of the cytoskeleton in elongation of the embryo. *Developmental Biology*, *117*(1), 156–173.
Pritchard, J. (2001). *Turgor pressure. eLS*. http://dxdoi.org/10.1038/npg.els.0001687.
Quintin, S., Wang, S., et al. (2016). Non-centrosomal epidermal microtubules act in parallel to LET-502/ROCK to promote C. elegans elongation. *Development*, *143*(1), 160–173.
Raich, W. B., Agbunag, C., et al. (1999). Rapid epithelial-sheet sealing in the Caenorhabditis elegans embryo requires cadherin-dependent filopodial priming. *Current Biology*, *9*(20), 1139–1146.
Rauzi, M., Lenne, P. F., et al. (2010). Planar polarized actomyosin contractile flows control epithelial junction remodelling. *Nature*, *468*(7327), 1110–1114.
Rohatgi, R., Ma, L., et al. (1999). The interaction between N-WASP and the Arp2/3 complex links Cdc42-dependent signals to actin assembly. *Cell*, *97*(2), 221–231.
Roh-Johnson, M., Shemer, G., et al. (2012). Triggering a cell shape change by exploiting preexisting actomyosin contractions. *Science*, *335*(6073), 1232–1235.
Roy, P. J., Zheng, H., et al. (2000). mab-20 encodes Semaphorin-2a and is required to prevent ectopic cell contacts during epidermal morphogenesis in Caenorhabditis elegans. *Development*, *127*(4), 755–767.
Sawa, M., Suetsugu, S., et al. (2003). Essential role of the C. elegans Arp2/3 complex in cell migration during ventral enclosure. *Journal of Cell Science*, *116*(Pt. 8), 1505–1518.
Sheffield, M., Loveless, T., et al. (2007). C. elegans enabled exhibits novel interactions with N-WASP, Abl, and cell-cell junctions. *Current Biology*, *17*(20), 1791–1796.
Shelton, C. A., Carter, J. C., et al. (1999). The nonmuscle myosin regulatory light chain gene mlc-4 is required for cytokinesis, anterior-posterior polarity, and body morphology during Caenorhabditis elegans embryogenesis. *The Journal of Cell Biology*, *146*(2), 439–451.

Shindo, A., & Wallingford, J. B. (2014). PCP and septins compartmentalize cortical actomyosin to direct collective cell movement. *Science*, *343*(6171), 649–652.

Simske, J. S., Koppen, M., et al. (2003). The cell junction protein VAB-9 regulates adhesion and epidermal morphology in C. elegans. *Nature Cell Biology*, *5*(7), 619–625.

Solon, J., Kaya-Copur, A., et al. (2009). Pulsed forces timed by a ratchet-like mechanism drive directed tissue movement during dorsal closure. *Cell*, *137*(7), 1331–1342.

Soto, M. C., Qadota, H., et al. (2002). The GEX-2 and GEX-3 proteins are required for tissue morphogenesis and cell migrations in C. elegans. *Genes & Development*, *16*(5), 620–632.

Sulston, J. E., Schierenberg, E., et al. (1983). The embryonic cell lineage of the nematode Caenorhabditis elegans. *Developmental Biology*, *100*(1), 64–119.

Tsur, A., Bening Abu-Shach, U., et al. (2015). ULP-2 SUMO protease regulates E-cadherin recruitment to adherens junctions. *Developmental Cell*, *35*(1), 63–77.

Vanneste, C. A., Pruyne, D., et al. (2013). The role of the formin gene fhod-1 in C. elegans embryonic morphogenesis. *Worm*, *2*(3), e25040.

Walck-Shannon, E., Reiner, D., et al. (2015). Polarized Rac-dependent protrusions drive epithelial intercalation in the embryonic epidermis of C. elegans. *Development*, *142*(20), 3549–3560.

Wang, X., Roy, P. J., et al. (1999). Multiple ephrins control cell organization in C. elegans using kinase-dependent and -independent functions of the VAB-1 Eph receptor. *Molecular Cell*, *4*(6), 903–913.

Waterston, R. H. (1989). The minor myosin heavy chain, mhcA, of *Caenorhabditis elegans* is necessary for the initiation of thick filament assembly. *The EMBO Journal*, *8*(11), 3429–3436.

Williams, B. D., & Waterston, R. H. (1994). Genes critical for muscle development and function in *Caenorhabditis elegans* identified through lethal mutations. *The Journal of Cell Biology*, *124*(4), 475–490.

Williams-Masson, E. M., Malik, A. N., et al. (1997). An actin-mediated two-step mechanism is required for ventral enclosure of the C. elegans hypodermis. *Development*, *124*(15), 2889–2901.

Wissmann, A., Ingles, J., et al. (1997). *Caenorhabditis elegans* LET-502 is related to Rho-binding kinases and human myotonic dystrophy kinase and interacts genetically with a homolog of the regulatory subunit of smooth muscle myosin phosphatase to affect cell shape. *Genes & Development*, *11*(4), 409–422.

Wissmann, A., Ingles, J., et al. (1999). The *Caenorhabditis elegans mel-11* myosin phosphatase regulatory subunit affects tissue contraction in the somatic gonad and the embryonic epidermis and genetically interacts with the Rac signaling pathway. *Developmental Biology*, *209*(1), 111–127.

Withee, J., Galligan, B., et al. (2004). Caenorhabditis elegans WASP and Ena/VASP proteins play compensatory roles in morphogenesis and neuronal cell migration. *Genetics*, *167*(3), 1165–1176.

Woo, W. M., Goncharov, A., et al. (2004). Intermediate filaments are required for C. elegans epidermal elongation. *Developmental Biology*, *267*(1), 216–229.

Yonemura, S., Wada, Y., et al. (2010). alpha-Catenin as a tension transducer that induces adherens junction development. *Nature Cell Biology*, *12*(6), 533–542.

Zahreddine, H., Zhang, H., et al. (2010). CRT-1/calreticulin and the E3 ligase EEL-1/HUWE1 control hemidesmosome maturation in *C. elegans* development. *Current Biology*, *20*(4), 322–327.

Zhang, H., & Labouesse, M. (2010). The making of hemidesmosome structures in vivo. *Developmental Dynamics*, *239*(5), 1465–1476.

Zhang, H., Landmann, F., et al. (2011). A tension-induced mechanotransduction pathway promotes epithelial morphogenesis. *Nature*, *471*(7336), 99–103.

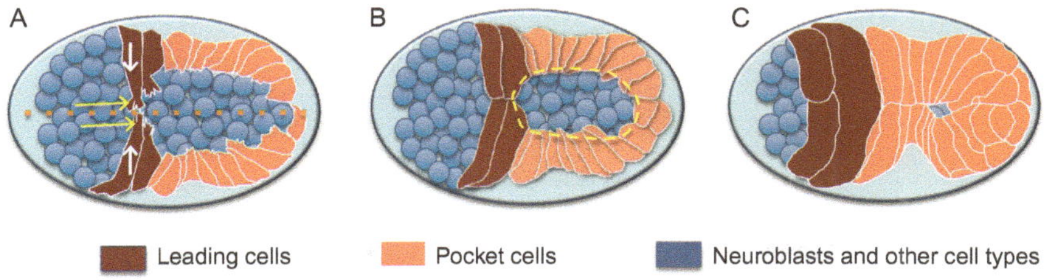

Thanh T.K. Vuong-Brender *et al.*, Figure 1 The two steps of epidermal ventral enclosure. (A) Schematic ventral view of an embryo (anterior to the left). First, the leading cells (dark red) extend long protrusions (yellow arrows), migrate (white arrows show the direction of migration) toward the ventral midline (dotted line), meet and establish adherens junctions. (B) The pocket cells (pink) exert coordinated contraction to close the open pouch (dashed line). (C) End of ventral closure; the head is not totally enclosed (some head epidermal cells anterior to the leading cells are not illustrated). Neuroblasts and other cell types are depicted in blue.

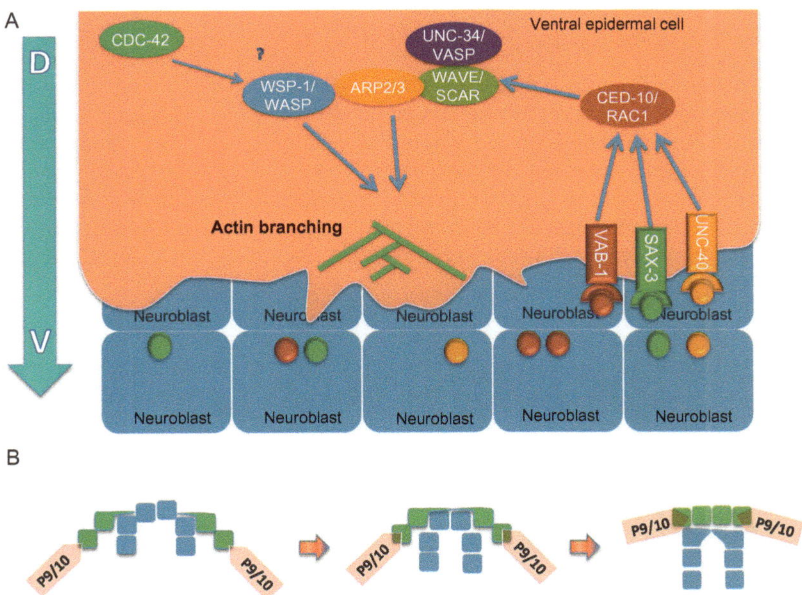

Thanh T.K. Vuong-Brender *et al.*, Figure 2 Neuroblasts guide ventral epidermis migration during ventral enclosure. (A) Signaling from neuroblasts promotes actin dynamics in the ventral epidermis. CED-10/Rac1, which acts probably downstream of VAB-1/EphR, SAX-3/Robo, and UNC-40/DCC, activates the WAVE/SCAR complex, and in turn the ARP2/3 complex. UNC-34/VASP binds WAVE/SCAR to promote actin polymerization. WSP-1, probably activated by CDC-42, contributes to a redundant pathway with WAVE/SCAR and UNC-34/VASP, which has not yet been fully investigated. Expression of guidance cues is discussed in the text. Big blue arrow: direction of migration; D, dorsal side; V, ventral side. Small ovals: guidance cues secreted by neuroblasts. (B) Schematic ventral view of the ventral midline, depicting how epidermal ventral cells (P9/10) migrate over a bridge made by neuroblasts. Some neuroblasts (green squares; only key neuroblasts are represented) extend protrusions and translocate over their neighbors (blue squares) to establish a bridge (Ikegami et al., 2012).

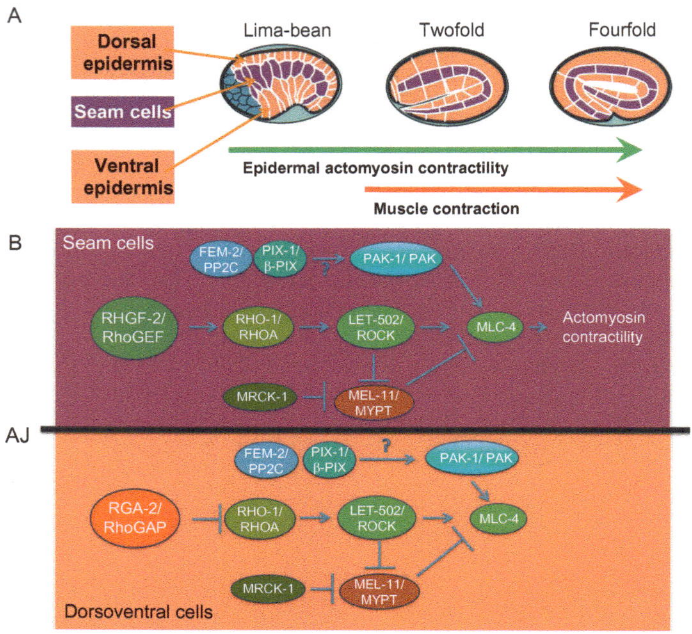

Thanh T.K. Vuong-Brender *et al.*, Figure 3 Overview of *C. elegans* embryonic elongation. (A) Schematic drawings of embryos along elongation. Note the change in shape of the seam cells (lateral epidermis; violet) where the junctions shorten in the dorsoventral (DV, circumferential) direction and lengthen in the anteroposterior (AP) direction. Embryos are staged according to their length (lima-bean or onefold to fourfold). Anterior to the left, dorsal up. (B) Molecular pathways mediating activation of nonmuscle myosin II contractility through phosphorylation of the myosin regulatory light chain MLC-4 during elongation.

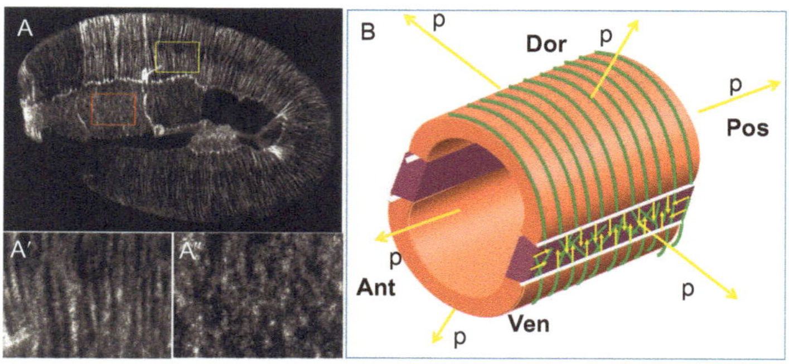

Thanh T.K. Vuong-Brender *et al.*, Figure 4 Actin organization and possible role during elongation. (A) Actin filaments in nearly twofold stage embryo are organized in parallel bundles in DV cells (yellow rectangle—dorsal cell hyp7, A′) and in thinner and less circumferentially oriented filaments in lateral seam cells (red rectangle, A″). (B) Hypothetical model for *C. elegans* elongation where the seam cells (violet) pull to increase the hydrostatic pressure and the DV cells (pink) orient the elongation in the direction perpendicular to actin bundle alignment (green lines), p, hydrostatic pressure; Ant, anterior; Pos, posterior; Dor, dorsal; Ven, ventral; arrows, force exerted. *Panel (A–A″) adapted with permission from Gally et al. (2009).*

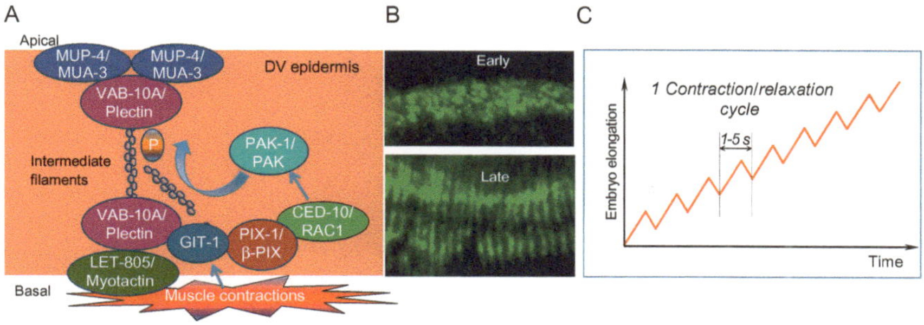

Thanh T.K. Vuong-Brender *et al.*, Figure 5 Muscle activity through CeHDs may promote elongation via a ratchet system. (A) CeHD remodeling under muscle contractile tension. Muscle contractions lead to GIT-1 recruitment, which subsequently activates PIX-1, CED-10/Rac1, and PAK-1. In turn, PAK-1 phosphorylates intermediate filaments leading to CeHD maturation (see panel B). (B) Images of a small portion of CeHD (marked with an antibody against the CeHD protein VAB-10A) when muscles start to contract (top) and at the end of elongation (bottom). (C) Elongation after the twofold stage as a ratchet-like process.

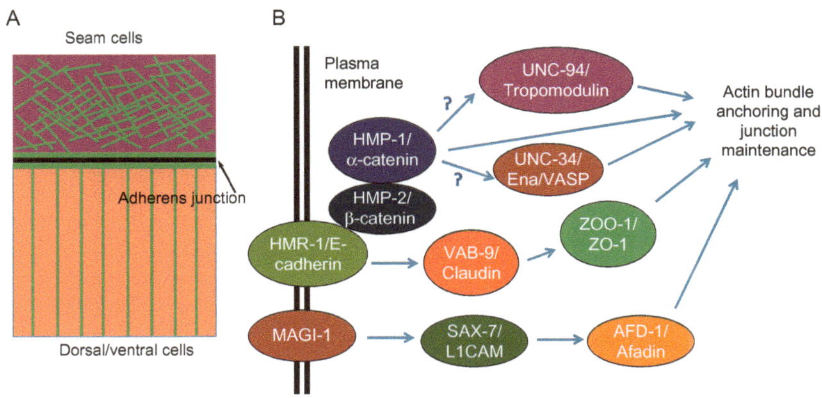

Thanh T.K. Vuong-Brender *et al.*, Figure 6 Role of cell junctions during elongation. (A) Junctional proteins are thought to anchor actin bundles in DV cells, green lines represent actin filaments and bundles. (B) Different junction-associated proteins may contribute to actin bundle anchoring and junction maintenance during elongation.

CHAPTER THIRTY-SIX

The Last 50 Years: Mismeasurement and Mismanagement Are Impeding Scientific Research

Peter A. Lawrence[1]
Department of Zoology, University of Cambridge, Cambridge, United Kingdom
[1]Corresponding author: e-mail address: pal38@cam.ac.uk

Abstract

In the last 50 years, there have been many changes to the substance, conduct, and style of research. Many of these changes have proved disastrous to the life of scientists and to science itself. As a consequence, the near-romantic spirit of adventure and exploration that inspired young scientists of my own and earlier generations has become tarnished. Now, many of us feel beleaguered by bureaucrats and by politicians: they affect our lives profoundly, apparently without an understanding of the way discoveries are made or of the nature of science itself. The core purposes of universities, teaching and research, are being eroded by excessive administration. The number and locations of our publications are counted up like beans and the outcomes are used to rank us, one against another; a process of evaluation that has recast the purposes of publication. Applying for grants takes far too much time from a young scientist's life.

DILBERT © 2015 Scott Adams. Used by permission of UNIVERSAL UCLICK. All rights reserved. (A color version of this figure is located at the end of this chapter.)

Current Topics in Developmental Biology, Volume 116
ISSN 0070-2153
http://dx.doi.org/10.1016/bs.ctdb.2015.12.013

O good old man, how well in thee appears
The constant service of the antique world,
When service sweat for duty, not for meed.
Thou art not for the fashion of these times,
Where none will sweat but for promotion,
And having that do choke their service up
Even with the having

1599 Shakespeare, As You Like It II, ii

A man is nothing now unless he has within him an appreciation of the new era, an era in which it would seem that neither honesty nor truth is very desirable, but in which success is the only touchstone of merit

Trollope (1857)

If he had been quite a different person, he could now have made his position one of great influence, it being the time for sitting on coordinating committees, Anglo-American committees, future policy committees. But he had no concern for finding a place in anything but scientific research itself

Hodges (1983)—referring to Alan Turing

Let me tell you how you become a man of power…of influence: you trade your ideals for self-interest

Morton-Smith (2015)

1966 was the initial issue of CTDB. That year I was a starting postdoc in Charlottesville, Virginia and my first experimental papers on insect development were published in the same year. So, professionally, CTDB and I are almost exact contemporaries. In our 50 years, we have both seen much evolution in the substance, conduct, and style of research. Some of these many changes have proved disastrous to the life of scientists and to science itself. It is not to say there has not been progress, of course there has, but this has been accompanied by an insidious corruption of the practice of research. As a consequence, the near-romantic spirit of adventure and exploration that inspired young scientists of my own and earlier generations has become tarnished and almost extinguished. Now, many of us feel beleaguered by bureaucrats and by politicians: they affect our lives profoundly, apparently without an understanding of the way discoveries are made or of the nature of science itself. The core purposes of universities, teaching and research, are being eroded by excessive administration. The number and locations of our publications are counted up like beans and the outcomes are used to rank us, one against another; a process of evaluation that has recast the purposes of publication. In addition, the granting system is so dysfunctional it could not have been designed—it has evolved from benign beginnings into a clumsy and purblind monster that tramples on innovation and creativity.

Let us have a closer look into seven of these claims:

1. the near-romantic spirit of adventure and exploration that inspired young scientists of my own and earlier generations has become tarnished This is a subjective statement; it describes how we felt then and now feel about scientific research, how we were and are motivated. I have long been impressed with the story of "The Worst Journey in the World" (Cherry-Garrard, 1922) when, in the darkness and cold of an Antarctic winter (1911), a few intrepid men set off on foot to collect one egg of the emperor penguin. The main justification was that such an egg had never before been seen by science. They risked and nearly lost their lives. Yes, their motivation was romantic and proved so, for afterward when they took the egg to the Natural History Museum in London they did not get the heroes' welcome they had expected (Cherry-Garrard, 1922). Later, on the same expedition, Scott and his companions continued to manhaul heavy rocks right up to their death. Common sense dictated they should have abandoned the rocks; but they saw their expedition as scientific and, therefore, these geological specimens symbolized a grander purpose than the aim of being first to reach the South Pole (Scott, 1913). Science was highly rated in the first half of the twentieth century.

Even when I started to do research in 1962, there was an atmosphere of excited curiosity among graduate students; yes there were career thoughts, but few. Such thoughts were secondary to the quest to discover, something, anything. Now, it is relatively rare for students to discuss their projects in the coffee room; much more time is spent worrying about the next career step, about publication and the politics of science. This change has even affected my life, I now get more invitations to talk about scientific politics and careers than about my research.

One consequence of this current pragmatism is that we have lost many gifted people who might otherwise have gone into pure science. Knowing how tough, insecure and financially unrewarded the scientific life is, most bright students prefer other paths such as finance, law, and medicine. The result, particularly in the prosperous regions of the west, is that we have been recruiting young scientists from abroad, including the third world. A large fraction of graduate students in science in the United Kingdom and in the United States did their first degrees outside these countries. I have not been able to recruit a British person to my (little) group for more than a decade, mainly because "the homegrown postgraduate population is pitifully thin" (Dyson, 2015). Nevertheless, when people of different backgrounds and education work together, the clash of perspectives may well enhance the quality of their research.

Box 1

Some data:

University of Cambridge, currently 65% of all graduate students in science are non-UK nationals (information from U of Cambridge).

In the United Kingdom, Chinese students account for 23% (26,860) of the total full-time postgraduate numbers in England, while UK students account for 26% (30,320).

Even in the United States, about half all graduate students in engineering are from other countries (Anderson, 2013). In Australia, about one-third of all graduate students come from elsewhere (University of Western Australia, 2013).

However, there are bright lights in the darkness and one of the most inspiring is the discovery by Emmanuelle Charpentier and Jennifer Doudna and colleagues. Their CRISPR cas9 genetic editing technique is a big advance and many applications to science and human health are expected. Their technique derives from basic research on bacterial immunity by Phillipe Horvath, Sylvain Moineau, Charpentier and others. All of these people researched in a subject that was out of fashion and dismissed as "old microbiology". Yet they persisted and the genetic editing technique is "a great example of a project that came about really through serendipity... with a very little bit of grant support...to not follow the fashion but to follow your curiosity and...discover things you would not have expected." (Breakthrough Prize, 2015). It is my opinion that this discovery was made in spite of the present systems and not because of them.

2. now many of us feel beleaguered by bureaucrats and politicians

Another subjective statement: but there is no doubt that most scientists resent being assessed and ruled by those who do not seem to understand how we work or what we try to achieve. Our new overlords seem to think that innovation is like digging tunnels, it is just a matter of more effort and then we will get further. There seems to be little understanding of the summer lightning of creativity or the inconvenient truth that trying new things carries a big risk of failure. There are many examples that illustrate how our purposes have become despoiled by bureaucrats but one stands out. It is the "impact statement" that we must write when we apply for money to do research. In the United Kingdom, we are forced to "describe societal and economic deliverables and milestones instead of focusing on just scientific deliverables" (BBSRC, 2015). This demand is not only badly written but also impossible to fulfill. Research is investigating the unknown; we cannot predict what we will find. Even less can we predict how what we might find could yield "societal and economic deliverables."

Box 2

the U.S. National Institutes of Health has elevated "significance" to an explicit criterion in funding decisions. It requires that grant reviewers write a paragraph on "impact," which it defines as the likelihood that the proposed work will have a sustained and powerful influence. Especially in fundamental research, which historically underlies the greatest innovation, the people doing the work often cannot themselves anticipate the ways in which it may bring human benefit. Thus, under the guise of an objective assessment of impact, such requirements invite exaggerated claims of the importance of the predictable outcomes—which are unlikely to be the most important ones. This is both misleading and dangerous.

Kirschner (2013)

There are many instances that make my point, three clear cases will do: *First*, when Milstein and Kohler discovered how to make monoclonal antibodies they consulted people (the National Research Development Council) especially employed to assess future "economic deliverables" and to patent discoveries that could be lucrative. After considering monoclonal antibodies, the Council concluded "It is certainly difficult for us to identify any immediate practical applications which could be pursued as a commercial venture" and decided to take no action (WhatisBiotechnology.org, 2013). Remember at the time of their evaluation the discovery had already been made, so that at least was no longer an unknown. Yet they were unable to see any value in it! For the year 2016, the sales attributable to monoclonal antibodies will be worth approximately 58 billion (sic) US dollars (BCCResearchLLC, 2012).

Second, Townes, who invented the laser, was told by the former and current chairmen of his department—both Nobel laureates—that "It isn't going to work. You know it's not going to work, we know it's not going to work." (Townes, 1999). Nevertheless, it did work and even Townes himself could not see much use for it...then!

Third, Turing and colleagues found it difficult to get adequate funds to make what would be the first digital computer (Hodges, 1983).

As we write our impact statements, we know we will be rewarded for writing hogwash and putting aside objectivity. But, even worse, the requirement for impact statements is an outcome of a more widespread corruption: increasingly, we scientists, both inside and outside of our scientific papers, are talking up our work. We are using "the alchemy of spin to transform leaden pieces of information into fool's gold..." (Lawrence, 2001) and as we do, we forsake our scientific principles. Hyping our research has become institutionalized: we do it, our universities do it, our funding bodies do it, and press releases and glossy brochures do it, let's give it up.

3. The core purposes of universities, teaching and research, are being eroded by excessive administration There has been a disproportionate increase in management posts within universities and research institutes including, for example, in human resources, "research support" (Colquhoun, 2007), and health and safety. There has been a phenomenal growth in jobs in various kinds of outreach, for example, befriending donors, relating with the media, and manipulating the presentation of research in order to maximize government support for the university (HEFCE, 2014). Jobs in "metascience" have increased hugely over the last decades. By metascience I mean administration within grant-giving bodies (much of the work consists of considering and rejecting applications) and editorship of journals (most of the work consists of considering and rejecting submitted articles). Within higher education in the United States, "non-classroom costs have ballooned, administrative payrolls being a prime example. The number of employees hired by colleges and universities to manage or administer people, programs and regulations increased 50% faster than the number of instructors between 2001 and 2011," according to the U.S. Department of Education (Belkin & Thurm, 2012). Many gifted young scientists, particularly women, have found these posts to be a wiser option than research. These posts are usually secure and carry pensions and therefore take a large amount of money from the total scientific and teaching effort. They can be contrasted with the ephemeral and insecure support given to many teachers, researchers, and their younger dependents (students, postdocs).

Administration, although its avowed purpose should be to facilitate the work of the researchers and teachers, actually eats up a lot of our time. Reports and evaluations are frequently sought and I am supposed to check, annotate and update all my publications on several independent web-based systems. The process of buying items with our grant money can become enmeshed in complex procedures of procurement, purchasing, and accountability. A main force driving this increase in bureaucracy is the institution's wish to create a paper trail, one of whose purposes is to pass any risks of litigation onto us (Vrountas & Ayer, 2013).

And in a recent and cruel twist, we scientists are increasingly being seen by management as cash cows whose raison d'être is financial (Jump, 2015)—if we do not attract enough grant money, we are losing our jobs. Some who have been fulfilling the primary duties of our posts—teaching and research—are being unfairly humiliated and even dismissed (Garwood, 2012). The recent case of Stefan Grimm who committed suicide after being hounded by the head administrators of Imperial College, London for not winning enough grant money is a terrible example (Colquhoun, 2014; Parr, 2014).

Box 3

In Australia: "*Research as much as teaching has been eroded by the bureaucratic university.... The proliferation of research management bureaucracies in universities over the period is a parallel symptom of a system that values procedure rather than productivity.*" (Murphy, 2013)

In America: "*American academics on average teach twice as much as they research. They teach for eleven hours a week. In colleges they research two hours a week and in research universities five hours a week. The rest of the time (twenty hours plus) is made up of committee meetings, e-mail correspondence and various professional obligations. Time disappears in this vortex—time that academics once would have spent interacting with students, and students with them. Bureaucratic time now continuously eats away at the vocation of the university—until one day it will disappear, unless a new model of the university is conceived.*" (Murphy, 2013; and see Arum, 2011 for a detailed analysis).

In the UK: *"a study by Brighton University, which looked at the remuneration between 1998 and 2009 of 193 vice-chancellors leading 95 UK institutions, has uncovered a real-term pay increase of 59%. On average, vice-chancellors received pay awards that were four times those of lecturers and the differential has widened".* Neil Gorman, former vice-chancellor of Nottingham Trent University, received £623,000 in 2013 (Henry, 2015); a typical University Professor receives about one-tenth of that salary and a graduate student about one fortieth. In Oxford, "*The university's central administrative staff is now almost three times what it was 15 years ago. There was no similar increase in full-time academic staff*" (McDonagh, 2015).

4. The number and locations of our publications are counted up...and then used to rank us, one against another Over the last 30 years, we have been building an "audit society" (Power, 1997) in which almost everything, including the scientific value of individuals and of their published articles, is measured or rather, mismeasured. This matter has been discussed by many (see for example, Gruber, 2014; Lawrence, 2007, 2009; Metze, 2010; Simons, 2008; Wang, Song, & Barabasi, 2013) but we scientists have also been seduced by the meretricious appeal of bibliometrics. There are a number of consequences:

Box 4

A true story of impact: Henry Disney (Department of Zoology, University of Cambridge, *rhld2@hermes.cam.ac.uk*) was on leave from West Cameroon in 1969 in the City of Bath, UK: "*I was struck by one respect in which Bath was less hygienic than the large market town of Kumba. In the latter, people openly urinated against the boundary wall of the market and deposited faeces in the drainage channels. Likewise, dogs deposited excrement everywhere. However, within a few hours the dung beetles rolled up the dung into balls and buried them along with their eggs. By contrast one could not walk 10 metres along the streets of the salubrious City of Bath without passing a deposit of dog faeces. I therefore recorded which flies were visiting these deposits, which fly larvae were developing in them and which of these (such as the lesser housefly* Fannia canicularis*) were also commonly found in the kitchens of our houses. I suggested that the prevalence of dog dung in English cities was directly related to the frequency of human enteric infections in England (Disney, 1972). Zakaria Erzinclioglu read my article and carried out a more extensive study in a north London suburb, confirming and extending my findings (Erzinclioglu, 1981).*

The curiosity of neighbours observing him collecting samples gave rise to his work coming to the attention of one of the London evening newspapers. In the subsequent correspondence it was pointed out that a parasitic infection of the eye, then frequent among children, was also commonly derived from dog dung. The result was the introduction of the requirement for dog owners to collect up their dog's excrement and for local councils to provide bins for its deposition. The end result has been a reduction in the infections derived from dog dung. This undoubted impact of my initial note is not reflected in the citations or impact factors used to measure the value of a scientific publication. Indeed, according to these measures my note was of no significance whatsoever!"

First, so-called top journals are now places of pilgrimages, worth sacrifices, and long painful journeys to reach. The reasons are both pragmatic and mythical. It is only partly true but widely believed that just one paper in *Nature*, right or wrong, can get a young scientist an entrée into a grant or post.

Second, papers in *Nature* or *Science* tend to be so dense and specialized that few can follow them. But this does not matter too much—it is more important to be published than to communicate: "papers have become competitive tokens for insertion into grant-dispensing gambling machines rather than bricks in the edifice of science" (Lawrence & Locke, 1997).

Third, the increasing use of citations as a measure of success has reinforced the hegemony of the top journals. Because citations to these journals give the writer's paper an elitist glow, such citations tend to be preferred over more appropriate citations to a lesser journal and, therefore, papers in the top journals get more citations than they deserve.

5. the granting system is so dysfunctional it could not have been designed For a young researcher it is discouraging to navigate the maze of grants, all with different eligibility requirements, clashing deadlines, and byzantine web-based forms. Typical postdoc grants last 2–3 years and up to 50% of that entire time and effort can be taken up as the young scientists try to negotiate their futures (Germain, 2015; Lawrence, 2009). Note that the odds of winning a research project grant are around 1 in 5 (NIH, 2014) and grant applications take weeks or months to complete. Grant applications are part exercises in fiction, in which a long list of experiments is contrived to look feasible and fashionable. They are fictional because one cannot predict how experiments will turn out and, anyway, it is imprudent to put one's best ideas into these applications as they are likely to be reviewed by competitors. For some grants, one is expected to write a specific timetable for the development of the project up to several years ahead, as if we scientists were soothsayers. Even worse, reports can be demanded and then judged as to how well these plans have been met. This process discourages original research by inhibiting changes in direction and making it less likely that new opportunities will be taken up. Science management should try to remember Max Perutz's famous observation "Creativity in science, as in art, cannot be organised. It arises spontaneously from individual talent. Well-run laboratories can foster it, but hierarchical organisations, inflexible bureaucratic rules, and mountains of futile paperwork can kill it. Discoveries cannot be planned, they pop up, like Puck, in unexpected corners" (Perutz, 2002).

6. mismeasurement has damaged the practice of publication itself Let us look at how scientific articles have changed over the last 50 years. In the mid-1960s, some of my papers were quite easily published in *Nature* as preliminary notes, cut and tailored to reach a "general reader" (who may have actually existed then; the general reader is now a mythical beast). Later,

I published the same pieces of work more completely and at greater length in specialized journals. Now *Nature* letters detail substantial pieces of work but they have to be prepared like dressed crabs and squashed into a tiny space. Figures are reduced to the size of postage stamps and the text is often a packed jumble of acronyms. Indeed, there cannot be many who understand more than a small proportion of the papers in scientific journals, yet that is not so important to the authors—because a paper's main function is not communication but to be a badge of honor. This change of purpose has had fundamental consequences, one is that scientists make less time for reading and comprehending because, in order to survive, they must put so much effort into writing manuscripts and manoeuvring them into journals.

In my opinion, there are other factors, apart from scientific quality, that help papers into publication in the highest ranking journals. These factors have affected the way papers are written, changes that have not benefited readers. For example, journals and editors are in thrall to impact factors. During the time I edited Development, which I did for most of the 50 years under review, the impact factor grew from a somewhat irrelevant measure into a dominating force. The impact factor relates to the average number of citations given to papers in a journal (Garfield, 2006). Editors know there will be more citations if they select papers in areas where many scientists are currently working. This means that manuscripts in unfamiliar fields tend to be rejected, while articles in currently fashionable fields are preferred. Of course, papers that prove later on to have had real quality or novelty attract citations over time, but they can be difficult to identify when submitted and, anyway, for the impact factor only the first 2 years of citations count. Therefore, fashionable papers are a safer bet. Editors tend to like research with medical relevance, such as findings related to a common human genetic disease, as the medical literature is vast, yielding many citations. The widespread conviction that new, exploratory or unusual topics are unlikely to be appreciated by editors has bred a play-safe attitude in the choice of projects. Thus, over the years, the sense of adventure so vital to research has been dampened.

Then, authors also have to get their manuscripts past reviewers and for this it is wise to fill in small gaps in our knowledge (identifying new members of an established genetic or metabolic pathway is a good strategy). Such findings fit in well with current dogma and are likely to be welcomed by reviewers. Another safe strategy is to produce papers that are data heavy and hypothesis light. This is not as difficult as modern methods can yield data relatively easily and then, a mass of results can give an article momentum, making it more difficult for a reviewer to find reasons to reject. But, unfortunately, too much data and too little storyline do not help the reader, who increasingly, cannot see any connection between the hyped claims in the abstract (tailored for the editor) and the mind-numbing piles of numbers and statistics in the result sections (designed to get past the reviewers). Another point for authors to remember: it is not prudent to try to overturn current opinion as this is likely to irritate at least one reviewer. Editors, tending to err toward safety, give too much weight to a negative reviewer and, in practice, a single dissenting voice often vetoes a paper (Indusieumgresium, 2009). Consequently, a fear of antagonizing reviewers has led to an increasing tendency to be diplomatic, to write manuscripts where conclusions are obfuscated and opinion is hidden—this does not help readers either.

Review articles have also been affected because they have two mutually conflicting purposes: first, they should aim to help readers understand a field and second, for the journals, they have become a means to gain citations and increase impact factors. Critical or "biased" reviews are actually more useful for readers as they compare cases and present evidence and arguments. However, editors prefer reviews that are anodyne lists of references because these can be cited as representing a whole field. Consequently, most reviews simply repeat current dogma and fail to examine the evidence behind it.

7. There has been "*an insidious corruption of the practice of research*"
Have these political changes impacted on research itself? I think the answer is yes. It is instructive to look at how the change in the primary purposes of publication (from communication and record to producing tokens that will yield salaries and grants) has affected the way we structure our research. All of us have had to focus our research to produce enough papers to compete and survive. Thus, projects are published as soon as possible and many therefore resemble lab reports rather than fully rounded and completed stories. There are many reasons why projects may not be pursued to a point of clarification, a clarification that would benefit everyone, particularly the reader. Often the person responsible only has support for 2 or 3 years and has to leave; therefore, passing on the project to another person can cause authorship disputes. Also clearing up inconsistencies in research can take too much time; it may be more productive to publish what we have and move on. Consequently, it may be more effective in terms of the numbers of papers to start a new project for each person (for each potential first author). Thus, I think this emphasis on article numbers has helped make papers poorer in quality. And, even more significant, there is an effect on the choice of projects:

few can afford to do really adventurous research these days, it is just too risky. We stick with what we know, it is easier and it is safer.

Signs of Change?

We have to change things and, encouragingly, there are signs of reformation. There are attempts to improve the processes of publication (Adie et al., 2012). Governments have begun to worry that the huge investment in metrics has been a mistake (HEFCE, 2015). Changes are more feasible in private organizations, which can and should be more flexible than government institutions. For example, the Wellcome Trust in the United Kingdom has reduced the average time spent on preparing grant applications. There is a two or three step process of selection, starting with only a short written application. Applications for large grants have been reduced in length, only a brief "vision" of research aims is required. The Wellcome Trust has also begun to break away from too much reliance on metrics and brought back interviews. The HHMI in the United States has asked its fundees to select only their five best papers for evaluation, thus rewarding people for completing research projects rather than aiming to maximize the number of their publications. These are encouraging trends and I think they should be extended to other organizations and taken further (so we can get back to doing research!).

These last two agencies together with the Max Planck Society have started a new journal, *eLife*, that is open access and hopes to break the oligarchy of the three main biomedical journals. It has also begun to change a particular imbalance in the system: reviewers have gained too much power over authors. Power sharing was not the original purpose of peer review but, little by little, reviewers have taken control of authors' research. Thus, if reviewers ask for an experiment, that request is usually passed on unvetted by the editor, even if this experiment has just occurred to the reviewer after little thought and is more of the sort of "wouldn't it be nice if they did this extra experiment"... And the authors feel compelled to undertake the experiment because of their overweening desire to publish in that journal. Consequently, I believe that too much research is now dictated by anonymous reviewers who cannot know the full picture of the lab situation, other projects competing for time, personnel running out of grant support and all the other real life inputs into the formulation of an effective scientific strategy. This situation is being improved by *eLife* because they ask their reviewers to judge the paper as submitted and, while reviewers are encouraged to ask for checks and controls concerning the fabric of the work, they are not allowed to suggest more experiments that would simply extend the research.

Education that is founded on rational argument and scientific research must be the only feasible way to save our planet and our species. We therefore need to keep science at the forefront of all decision making and to continue to overcome the "hegemony of the supernatural" (Mellman & Warren, 2000). It is therefore crucial that the primary purposes of science to understand and to innovate can once again become the overt aims of researchers. To achieve this, our dependence on phony measurements and bureaucracy must be abandoned and our processes must again reward originality and risk taking. Universities and institutes must reduce investment in administration and increase investment in teaching and research. Then young people, many of whom still come into science with a sense of excitement, will no longer become disenchanted. We will need them for our survival.

ACKNOWLEDGMENT

Thanks to José Casal and Jeremy Garwood for advice and the Wellcome Trust, 096645 and 107060, for support.

REFERENCES

Adie, E., Adler, E. M., Ahmad, S., Albertine, K. H., Alberts, B., Amigó, J. M., et al. (2012). *San Francisco Declaration on Research Assessment. Putting science into the assessment of research.* http://www.ascb.org/dora/.

Anderson, S. (2013). *The importance of international students to America.* http://www.nfap.com/pdf/New NFAP Policy Brief The Importance of International Students to America, July 2013.pdf.

Arum, R. a. R. J. (2011). *Academically adrift.* Chicago: University of Chicago Press.

BBSRC. (2015). *About pathways to impact.* http://www.bbsrc.ac.uk/funding/apply/impact/about-pathways-to-impact/.

BCCResearchLLC. (2012). Global sales of therapeutic monoclonal antibodies to reach $57.7 billion in 2016. http://www.bccresearch.com/pressroom/bio/global-sales-therapeutic-monoclonal-antibodies-reach-$57.7-billion-2016.

Belkin, D., & Thurm, S. (2012). Deans list hiring spree fattens college bureaucracy—And tuition. *Wall Street Journal.* December 28.

Breakthrough Prize. (2015). In *2015 Breakthrough prize symposium.* https://www.youtube.com/watch?v=VMlBzXdCtIc.

Cherry-Garrard, A. (1922). *The worst journey in the world.* London: Carroll & Graf.

Colquhoun, D. (2007). More administrator = more efficient administration. *DC's Improbable Science.* http://www.dcscience.net/2007/09/19/more-administrators-more-efficient-administration/.

Colquhoun, D. (2014). Publish and perish at Imperial College London: The death of Stefan Grimm. *DC's Improbable Science.* http://www.dcscience.net/2014/12/01/publish-and-perish-at-imperial-college-london-the-death-of-stefan-grimm/.

Disney, R. H. L. (1972). Some flies associated with dog dung in an English city. *Entomologist's Monthly Magazine, 108,* 93–94.

Dyson, J. (2015). No Theresa May, we need those foreign graduates. *The Guardian*. Sunday 4th January.
Erzinclioglu, Y. Z. (1981). On the Diptera associated with dog dung in London. *London Naturalist, 60*, 45–46.
Garfield, E. (2006). The history and meaning of the journal impact factor. *JAMA, 295*, 90–93.
Garwood, J. (2012). UK academia is dying from over-management: The Queen Mary Scandal. *Lab Times Online*. http://www.labtimes.org/editorial/e_328.lasso.
Germain, R. N. (2015). Healing the NIH-funded biomedical research enterprise. *Cell, 161*, 1485–1491.
Gruber, T. (2014). Academic sell-out: How an obsession with metrics and rankings is damaging academia. *Journal of Marketing for Higher Education, 24*, 165–177.
HEFCE. (2014). *Research excellence framework*. http://www.ref.ac.uk/.
HEFCE. (2015). *The metric tide: Report of the independent review of the role of metrics in research assessment and management*. http://www.hefce.ac.uk/pubs/rereports/Year/2015/metrictide/Title,104463,en.html.
Henry, J. (2015). *Eye-watering' salary rises for university chiefs cannot be justified, says report*. http://www.theguardian.com/education/2015/mar/15/rises-for-university-chiefs-cannot-be-justified.
Hodges, A. (1983). *Alan Turing: The enigma*. London: Burnett Books in Association with Hutchinson.
Indusieumgresium. (2009). *Scientific peer review, ca. 1945*. https://www.youtube.com/watch?v=-VRBWLpYCPY.
Jump, P. (2015). *Grant income targets set at one in six universities, THE poll suggests*. https://www.timeshighereducation.co.uk/news/grant-income-targets-set-one-six-universities-poll-suggests.
Kirschner, M. (2013). A perverted view of "impact". *Science, 340*, 1265.
Lawrence, P. A. (2001). Science or alchemy? *Nature Reviews. Genetics, 2*, 139–142.
Lawrence, P. A. (2007). The mismeasurement of science. *Current Biology, 17*, R583–R585.
Lawrence, P. A. (2009). Real lives and white lies in the funding of scientific research: The granting system turns young scientists into bureaucrats and then betrays them. *PLoS Biology*, 7. e1000197.
Lawrence, P. A., & Locke, M. (1997). A man for our season. *Nature, 386*, 757–758.
McDonagh, M. (2015). *How come our cash-strapped universities can afford so many administrators?* http://www.spectator.co.uk/features/9547622/how-come-our-cash-strapped-universities-can-afford-so-many-administrators/.
Mellman, I., & Warren, G. (2000). The road taken: Past and future foundations of membrane traffic. *Cell, 100*, 99–112.
Metze, K. (2010). Bureaucrats, researchers, editors, and the impact factor—A vicious circle that is detrimental to science. *Clinics (São Paulo, Brazil), 65*, 937–940.
Morton-Smith, T. (2015). *Oppenheimer*. London: Oberon Books.
Murphy, P. (2013). The rise and fall of our bureaucratic universities. *Quadrant Online*. http://quadrant.org.au/magazine/2013/05/the-rise-and-fall-of-our-bureaucratic-universities/.
NIH. (2014). *Success rates*. http://report.nih.gov/success_rates/.
Parr, C. (2014). Imperial College professor Stefan Grimm 'was given grant income target'. In "*Times higher education*". December 3, 2014.
Perutz, M. F. (2002). *I wish I'd made you angrier earlier*. Oxford: Oxford University Press.
Power, M. (1997). *The audit society: Rituals of verification*. Oxford: Oxford University Press.
Scott, R. F. (1913). Scott's last expedition, Vol. I: Being the journals of captain R. F. Scott, R. N., C. V. O. arranged by L. Huxley. London: Smith, Elder & Co.
Simons, K. (2008). The misused impact factor. *Science, 322*, 165.

Townes, C. (1999). *How the laser happened: Adventures of a scientist*. Oxford: Oxford University Press.

Trollope, A. (1857). *Barchester towers*. London: Longman, Brown, Green, Longmans, & Roberts.

University of Western Australia. (2013). *UWA in brief*. http://www.uwa.edu.au/university/governance/uwa-in-brief-2013 - new_div_2394948.

Vrountas, C. T., & Ayer, A. (2013). Policies, timing, documentation, and consistency: Four tools for colleges and universities to manage employee relations and mitigate risks of employment discrimination claims. *URMIA Journal, 2013*, 9–22.

Wang, D., Song, C., & Barabasi, A. L. (2013). Quantifying long-term scientific impact. *Science*, *342*, 127–132.

WhatisBiotechnology.org. (2013). *The story of César Milstein*. http://www.whatisbiotechnology.org/exhibitions/milstein/patents.

DO YOU WANT THE BORING AND AWFUL PROJECT THAT IS LIKELY TO SUCCEED. . .
Dilbert.com DilbertCartoonist@gmail.com
OR THE FUN PROJECT THAT IS CERTAIN TO FAIL AND TAKE YOUR CAREER WITH IT?
11-11-15 © 2015 Scott Adams, Inc. /Dist. by Universal Uclick
YOU CAME HERE TO GIVE BOTH OF THEM TO ME.
HA HA! YOU KNOW ME.

CHAPTER THIRTY-SEVEN

Mechanochemical Interplay Drives Polarization in Cellular and Developmental Systems

Qiyan Mao, Thomas Lecuit[1]
Aix Marseille Université, CNRS, IBDM UMR7288, Marseille, France
[1]Corresponding author: e-mail address: thomas.lecuit@univ-amu.fr

Contents

Abstract

Polarity plays important roles in biological processes, such as motility, differentiation, growth, and pattern formation. One major goal in the field of biological polarity is to understand logics used by signaling networks underlying diverse polarization processes: symmetry breaking, amplification, inhibition, and coordination. In this essay, we explore various polarization processes on cell, tissue, and whole-organism scales, aiming to elucidate how diverse mechanical signals, in addition to chemical cues, can be integrated to a common framework of polarization.

The mechanical part of the state describes the positions, masses, velocities and elastic properties of the cells, and the forces between them… The chemical part of the state is given as the chemical composition of each separate cell… One cannot at the present hope to make any progress with the understanding of such systems except in very simplified cases. The interdependence of chemical and mechanical data adds enormously to the difficulty.

Alan Turing (Turing, 1952)

Current Topics in Developmental Biology, Volume 116
ISSN 0070-2153
http://dx.doi.org/10.1016/bs.ctdb.2015.11.039

1. INTRODUCTION

Polarity is an almost universal property of cells. It is characterized by an asymmetric spatial distribution of substances. Polarity exists on different scales, from biomolecules to the entire organism (Bryant & Mostov, 2008; Li & Bowerman, 2010). Polarity plays important roles in most biological functions, such as motility, differentiation, growth, and pattern formation (St Johnston & Ahringer, 2010). Loss of intra- or intercellular polarity is often prelude to cell death, tissue disintegration, and developmental defects (Wodarz & Näthke, 2007). Thus, it is of little wonder that polarization mechanisms exist ubiquitously in all eukaryotic life forms, and share core features and components between diverse cell and tissue types (Abley et al., 2013; Asnacios & Hamant, 2012; Goehring & Grill, 2013; McCaffrey & Macara, 2009; Slaughter, Smith, & Li, 2009; Williams & Harwood, 2003).

Biological polarity is arguably best understood on the level of single cells, where it is manifested as coordinated distribution of membrane lipids and proteins (Wang, 2009), as well as alignment of the cytoskeleton and organelles (Mullins, 2010). In multicellular systems, intracellular polarity of neighbors often aligns according to a common axis (e.g., apicobasal or planar) (Abley et al., 2013). Another form of multicellular polarity involves vectored distribution of cell identity or positional information within the tissue (Wolpert, 1998).

One major goal is to decipher universal logics and principles of the signaling networks underlying polarization processes. The fundamentals of intracellular polarization processes consist of the following three steps: (a) initial symmetry breaking from uniformity via a polarity signal or stochastic fluctuations; (b) amplification of the initial asymmetry to generate robust polarity; (c) in a multicellular context, alignment of polarity between neighboring cells (Abley et al., 2013; Asnacios & Hamant, 2012; Goehring & Grill, 2013). Mathematical models based on feedback and cross-inhibition circuits can largely capture the emergence of bistability and the essential behaviors of the polarization system (Jilkine & Edelstein-Keshet, 2011; Onsum & Rao, 2009; van der Gucht & Sykes, 2009).

From the biological perspective, it is crucial to identify the physical nature of signals and their interaction involved in the polarization process. Generally, two forms of signals present in the milieu of living tissues: chemical signals and mechanical stimuli (Asnacios & Hamant, 2012; Goehring &

Grill, 2013). The former consists of lipids, ligands, their receptors, polarity protein complexes, small GTPases, and the cytoskeleton (McCaffrey & Macara, 2009). The latter results from mechanical forces present outside and/or inside cells, such as tension, protrusive forces, adhesion, and flow-based advection (Asnacios & Hamant, 2012; Goehring & Grill, 2013). A cell is capable of integrating various polarity signals to partition polarity components and reorient the cytoskeleton. Further, cells can coordinate various forms of polarity information via cell–cell relay or long-range signal propagation to achieve tissue-level polarity (Asnacios & Hamant, 2012; Goehring & Grill, 2013). Importantly, while chemical signals act at a distance relatively slowly via diffusion or active transport, mechanical stimuli propagate almost instantly at a long distance.

In this essay, we aim to highlight the emerging role of mechanical signals in intra- and intercellular polarization processes. Since we will cover a relatively large number of research topics, it is impossible to cite all the relevant literature. We have contrived to refer to pioneering articles, recent advances, and otherwise very useful reviews. We will first outline a conceptual framework for the intracellular chemical polarization process. We will then classify various mechanical signals by their source of origin and explore how they can be incorporated into the intracellular chemical polarization framework. We will further explore how intracellular polarity is influenced by mechanical signals in the context of multicellular tissues. Finally, we will discuss how intercellular polarity can be achieved through an analogous polarization network involving both chemical and mechanical signals.

2. A COMMON LOGIC FOR CHEMICAL POLARIZATION NETWORKS

There are at least three major forms of intracellular polarity: front–rear polarity in migrating cells (or neurons), apicobasal polarity, and planar polarity in epithelial cells. Despite diverse contexts, cells use strikingly similar logics to generate polarity: a self-organizing partition system based on feedback and cross-inhibition circuits to amplify any symmetry-breaking events (Fig. 1C) (McCaffrey & Macara, 2009; Nelson, 2009; Wang, 2009).

A single migrating cell can either spontaneously polarize without any external cues or polarize according to an external chemoattractant gradient (Fig. 1A) (Graziano & Weiner, 2014; Jin, 2013; Swaney, Huang, &

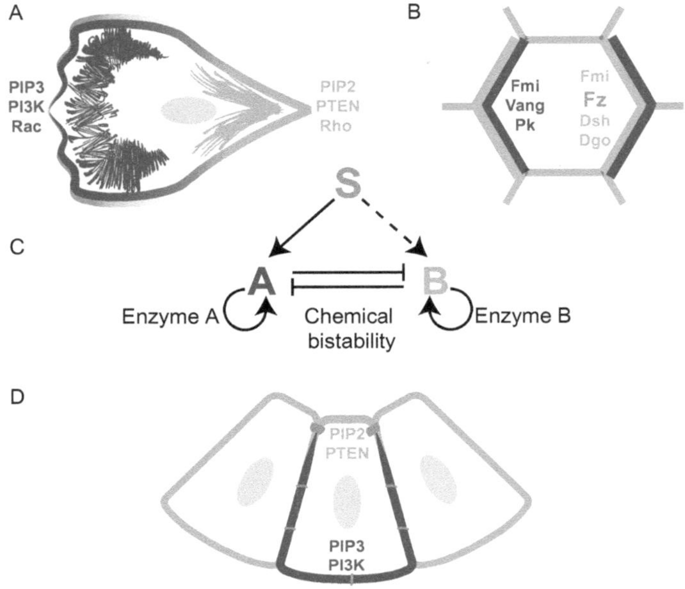

Figure 1 A conceptual chemical polarization network. (A) A migrating cell aligns its front–rear polarity (red-to-blue outline) with the direction of the chemoattractant gradient (green outline). Inside the cell: Red sketches indicate branched, protrusive actin filaments at the front, while blue sketches show bundled, contractile actin filaments at the rear. (B) An epithelial cell displaying planar cell polarity, with Core PCP components highlighted in red (proximal; Fmi-Vang-Pk) and blue (distal; Fmi-Dgo-Dsh). Fz in the distal PCP complex (green) is thought to be the symmetry-breaking signal transported along microtubules. (C) Schematics showing polarization of a two-component chemical system. S: symmetry-breaking cue that can activate either A alone, or first A then B. A and B: two mutually antagonistic, self-amplifying polarity states/components. (D) An epithelial cell displaying apicobasal polarity, with apical (blue) and basolateral (red) surfaces highlighted. Green bars shows cadherin complexes, with two thick ones showing the position of adherens junctions, and several thin ones showing cadherin complexes at the basolateral surfaces. (A color version of this figure is located at the end of this chapter.)

Devreotes, 2010; Wang, 2009). A chemotaxing cell aligns its polarity with the direction of the chemoattractant gradient. This is achieved by breaking the symmetry where membrane-bound receptors sense the highest concentration of the chemoattractant (Iglesias & Devreotes, 2012; Jin, 2013). The initial asymmetry is thought to lie within the subtle level differences in membrane-bound phosphoinositides PI(4,5)P2 (PIP2) and PI(3,4,5)P3 (PIP3)

(Jin, 2013; McCaffrey & Macara, 2009; Shewan, Eastburn, & Mostov, 2011). In a simplified scheme, both PIP2 and PIP3 can self-amplify through recruitment and activation of enzymes that favor their own production at the cost of one another (PTEN and PI3K, respectively) (Arai et al., 2010; Knoch, Tarantola, Bodenschatz, & Rappel, 2014). This partitions the membrane into two stable domains dominated by PIP2 or PIP3 (Graziano & Weiner, 2014; Swaney et al., 2010; Wang, 2009; Williams & Harwood, 2003). Further, phosphoinositides can differentially recruit small GTPases such as Cdc42, Rac, and Rho, as well as their GEFs and GAPs (Graziano & Weiner, 2014; Shewan et al., 2011; Wang, 2009). The antagonistic relationship between those small GTPases through GEF- and GAP-based cross talk further translates into different forms of cytoskeleton occupying distinct spatial domains (Tsujita & Itoh, 2015), thus stabilizing intracellular polarity between a front end dominated by PIP3/PI3K/Rac/protrusive actin brunches, and a rear end dominated by PIP2/PTEN/Rho/contractile actomyosin bundles (Graziano & Weiner, 2014; Shewan et al., 2011; Wang, 2009).

A similar PIP2/PIP3 partition-based mechanism is at play during the establishment of apicobasal polarity in the epithelium (Fig. 1D) (Bryant & Mostov, 2008; Rodriguez-Boulan & Macara, 2014; Veglio, Gamba, Nicodemi, Bussolino, & Serini, 2009). One good example of the *de novo* generation of apical surface is lumen formation (Nelson, 2009; Rodriguez-Boulan & Macara, 2014; Veglio et al., 2009). Prior to the establishment of an apical surface, cells are in contact with each other through cell–cell adhesion or with the basal lamina through cell–matrix adhesion (Veglio et al., 2009). Bound cadherin or integrin molecules recruit and activate PI3K in a positive feedback loop, such that at initial stages, the entire cell surface is occupied by PI3K/PIP3 (Gassama-Diagne et al., 2006; Orlando & Guo, 2009) (Veglio et al., 2009). A nascent apical surface emerges through spontaneous detachment of cell–cell contact, which breaks the PI3K/PIP3 positive feedback, and enables the membrane presence of PIP2 (Veglio et al., 2009). PIP2 then acts as a master regulator of apical identity, subsequently recruiting the Cdc42 GTPase and the apical PAR complex (PAR-3/PAR-6/aPKC) (Martin-Belmonte et al., 2007). Since the apical PAR complex recruits the PIP2-producing PTEN, this forms a positive feedback loop reinforcing the PIP2 domain. Meanwhile, the basolateral surfaces maintain the PI3K/PIP3 positive feedback (Pinal et al., 2006; Shewan et al., 2011). Thus, the plasma membrane is divided into an apical domain dominated by PIP2/PTEN/Cdc42/apical PAR complexes and a basolateral domain occupied by PIP3/PI3K/cell adhesion. Once established, the apicobasal axis is further stabilized by rearrangement of the cortical cytoskeleton and condensation of adhesion complexes into the zonula adherens or zonula occludens (Bryant & Mostov, 2008; Rodriguez-Boulan & Macara, 2014).

Epithelial cells often display another form of polarity orthogonal to the apicobasal axis termed planar cell polarity (PCP). PCP is characterized by the asymmetric distribution of polarity protein complexes and polarized alignment of cytoskeletons along cell–cell junctions (Fig. 1B). PCP is often coordinated over long distance and aligned with the main axis of the developing tissue (Abley et al., 2013; Devenport, 2014). The Core PCP and Fat-PCP pathways are two parallel and coordinated signaling pathways quasi-ubiquitously required for planar cell polarity (Devenport, 2014). The Core PCP pathway causes intracellular partitioning of two polarity complexes, the proximal (Fmi, Vang, and Pk) and distal (Fmi, Fz, Dsh, and Dgo) complex (Devenport, 2014). This is achieved through *cis*-inhibition between the proximal and the distal complexes within one cell and *trans*-activation between the two between neighbor cells (Gray, Roszko, & Solnica-Krezel, 2011; Vladar, Antic, & Axelrod, 2009). The Fat-PCP pathway is thought to convert the tissue-scale ascending gradient of the Golgi kinase Four-jointed and the descending gradient of atypical cadherin Dachsous along the proximodistal axis into intracellular junctional asymmetry in Dachsous-Fat cadherin heterodimers, thus linking global tissue axis to individual cell polarity (Matis & Axelrod, 2013). Both PCP pathways can crosstalk extensively. Components of both PCP signaling pathways are enriched at intercellular junctions along with classic polarity players, such as small GTPases, PAR-3, and the actomyosin cytoskeleton, which are shown to function downstream to PCP signaling (Takeichi, 2014). Additional evidences provide links between Wnt-PCP signaling, PAR complexes (McCaffrey & Macara, 2009), membrane phosphoinositide signaling (Bardet et al., 2013; Slusarski, Corces, & Moon, 1997), and the perijunctional cytoskeleton (Harris & Tepass, 2008; Röper, 2012). Thus, PCP signaling crosstalks with the classic polarity components to establish planar polarity in junctional cytoskeleton and membrane lipid composition.

Numerous theoretical models have been developed to recapitulate the behaviors of such chemical polarization systems (Iglesias & Devreotes, 2012; Jilkine & Edelstein-Keshet, 2011; Meinhardt, 2012; Mori, Jilkine, & Edelstein-Keshet, 2008; Onsum & Rao, 2009). In essence, all of them require the following key components: (a) an initial symmetry-breaking signal, (b) a fast or local activator that self-amplifies, (c) a slow or global inhibitor that restricts or counteracts the propagation of the activator, and in the case of multicellular polarity, (d) a local or global alignment system that coordinates individual cell polarity across the tissue (Fig. 1D). This leads to one important question: can signals other than of chemical nature fulfill the roles of, or interact with any of the key chemical components in the polarization network? The answer is yes.

3. MECHANICAL SIGNALS INSIDE AND OUTSIDE CELLS

During polarization, the redistribution of biochemical polarity components is often coupled with modifications of the cytoskeleton, changes in membrane tension, and remodeling of the cell–cell/cell–ECM adhesion (Levayer & Lecuit, 2012). However, rather than passive readouts of polarity information, cytoskeleton, membrane tension, and the adhesion complexes can actively participate in the symmetry-breaking and polarization process by generating or transmitting mechanical forces (Asnacios & Hamant, 2012; Chanet & Martin, 2014; Goehring & Grill, 2013). In general, mechanical forces that a cell may experience can either come from the extracellular environment and transmitted via cell–cell contact or cell–matrix contact, or from the contractile cortical cytoskeleton (animal cell) or turgor pressure (plant cell) inside the cell.

Cells are generally mechanosensitive, which means that they can sense externally applied mechanical forces from the environment and respond to it by modifying their signaling activities and rearranging their cytoskeleton (Chanet & Martin, 2014; Eyckmans, Boudou, Yu, & Chen, 2011; Mammoto, Mammoto, & Ingber, 2013). Force sensing in cells takes place in three main locations: the cell–cell/cell–matrix adhesion complexes, tension-activated ion channels on the plasma membrane, and the cytoskeleton (Chanet & Martin, 2014; Eyckmans et al., 2011; Mammoto et al., 2013). Environmental forces can trigger mechanotransduction signaling, which leads to changes in distributions of polarity proteins and lipids on the membrane, as well as in the topology, direction, and density of the cytoskeleton, thereby interacting with the chemical polarization network. Since adhesion, membrane tension, and the cytoskeleton are integral components of the mechanotransduction pathways, this allows feedbacks and to amplify, inhibit, or coordinate any initial asymmetry due to mechanical constraints or local mechanical instability (Chanet & Martin, 2014; Eyckmans et al., 2011; Mammoto et al., 2013).

Cells are not merely passive observers of mechanical forces. The molecular motors and the cytoskeleton inside a cell can actively generate forces, most notably protrusive forces by actin or microtubule polymerization (Lansky et al., 2015), and tensile forces by the myosin motor pulling on actin (Heisenberg & Bellaïche, 2013; Lecuit, Lenne, & Munro, 2011). Actin or microtubule polymerization can modulate membrane tension, while the actomyosin contractility can influence cortical tension (Munjal & Lecuit, 2014). While elevated cortical tension equips cells with mechanical instability facilitating mechanical symmetry breaking, increase of membrane tension allows actomyosin cortex displacement away from the plasma membrane and inhibits new actin polymerization elsewhere, thus generating mono-directional actomyosin flows (Houk et al., 2012; Lieber, Yehudai-Resheff, Barnhart, Theriot, & Keren, 2013; Mullins, 2010). Moreover, chemical polarity components, such as PAR complexes and small GTPases, can redistribute intracellularly via bulk actomyosin flows (advection) or molecular motors (motor transport), thus feeding back to the chemical polarization network (Goehring & Grill, 2013; Howard, Grill, & Bois, 2011).

In the following sections, we will illustrate how mechanical forces of both external and internal sources can influence processes such as symmetry-breaking, local amplification, long-range inhibition during cell polarization, as well as the alignment of intracellular polarity between neighbors. In addition, we will further discuss how the role of mechanical forces in the intracellular polarization system can be extrapolated to the intercellular polarization system during embryonic patterning.

4. EXTERNAL MECHANICAL SIGNALS CAN BIAS POLARITY

The immediate environment of a cell consists of the extra cellular matrix (ECM) and neighboring cells, both of which form mechanosensitive connections with the cell, such as focal adhesions and cell–cell adhesions. Cells can sense mechanical properties of the extracellular environment, such as stiffness, compression, and tension. When such mechanical properties are anisotropic, they tend to bias the direction of symmetry breaking and result in long-range alignment of the polarity axis across multiple cell diameters.

4.1 External Mechanical Signals can Bias Polarity of Migrating Cells

The simplest example of external mechanical influence on a single cell comes from so-called "durotaxis," or the ability of cultured fibroblasts to sense and to prefer migrating towards stiffer substratum via regulated persistence of leading edge lamellipodia (Lo, Wang, Dembo, & Wang, 2000). It has been proposed that this might be due to local mechanical sensing by focal adhesions and local actomyosin–paxillin complexes, which can then break the symmetry of actin polymerization and bias its direction towards stiffer substratum (Charras & Sahai, 2014). In the context of collective cell migration, external mechanical cues are mainly transmitted through cell–cell contacts. In order to maintain a mono-directional polarized state with high motility persistence, a cell must keep its contact with neighboring cells via cadherin complexes (Fig. 2A) (Weber et al., 2011), as isolated cells cannot form a single polarized axis. Strikingly, mono-directional polarity can be restored in isolated cells when they are pulled with cadherin-coated beads (Weber et al., 2011). This can be explained either by an instructive, amplifying role of local cadherin complexes on keratin polymerization at the bead–cell interface (Weber et al., 2011), or by a permissive, inhibitory role of increased membrane tension on cytoskeleton polymerization as a consequence of pulling forces.

The latter has been demonstrated in chemotaxing neutrophils (Houk et al., 2012) and migrating fish keratocytes (Lieber et al., 2013). In neutrophils, elevated plasma membrane tension, resulting from actin polymerization at the leading edge, can rapidly inhibit Rac and SCAR/WAVE-dependent actin polymerization elsewhere over the entire cell diameter (Houk et al., 2012). It has been further shown in keratocytes that membrane tension results from a mechanical force balance between the plasma membrane and cytoskeletal forces (Lieber et al., 2013). Together, these studies highlight the role of membrane tension as the long-range cytoskeleton polymerization inhibitor crucial for coordinating polarity at different parts of the cell (Fig. 2B).

4.2 External Mechanical Signals can Reorient Cell Polarity in a Tissue

Mechanical forces across non-migrating cells, such as those in the animal epithelium and plant tissue, can also influence the direction of polarity in individual cells. Since the global pattern of mechanical forces is constrained by tissue geometry, such mechanical inputs often result in global alignment of cell polarity with respect to overall tissue shapes. The output of cell polarity, such as oriented cell division and cell intercalation, can in turn influence tissue shapes, thus feeding back on the mechanical pattern (Fernandez-Gonzalez, Simoes Sde, Röper, Eaton, & Zallen, 2009). Mechanical cross talk between

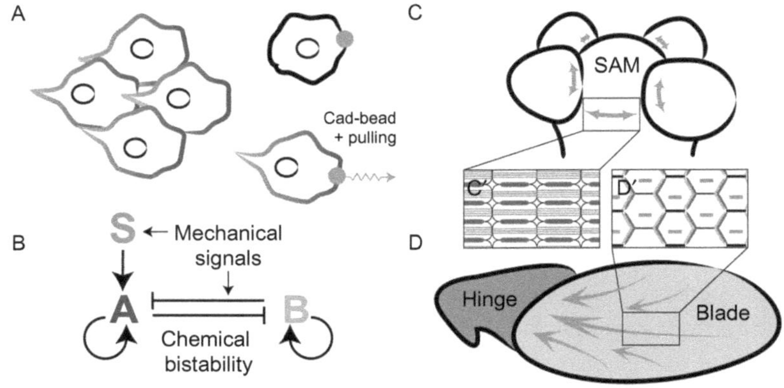

Figure 2 External mechanical signals can bias polarity. (A) Pulling force via cadherin-coated beads rescues migration polarity in isolated *Xenopus* gastrula mesendoderm cells. Red-to-blue color gradient indicates front–rear polarity in collectively migrating cells and single cells with pulling forces applied via cadherin-coated beads, while black indicates no polarity in cells in contact with cadherin-coated beads but without pulling forces. Green dots show magnetic beads coated with C-Cadherin. Green arrow implies pulling force. (B) Schematics showing mechanical signals may either play an instructive role in biasing the direction of symmetry breaking of a two-component system, or a permissive role in enhancing long-range inhibition between two polarity states. S: symmetry-breaking cue. A and B: two mutually antagonistic, self-amplifying polarity states/components. (C) Hypothetical stress pattern in the *Arabidopsis* shoot apical meristem (green arrows). Inset (C′) shows alignment of microtubules (green) and membrane enrichment of Auxin transporter PIN1 (red) along with the direction of mechanical tension. (D) Pattern of cell flow (green arrows) in the *Drosophila* pupal wing blade due to hinge contraction. Gradients of Ds (orange) and Fj (blue) are superimposed. Inset (D′) shows alignment PCP direction or microtubules (green) with proximal (red: Fmi-Vang-Pk) and distal (blue: Fmi-Fz-Dgo-Dsh) components oriented along the proximodistal axis of the wing blade. *Panel (A): Adapted from Weber, Bjerke, and Desimone (2011). Panel (C): Adapted from Heisler et al. (2010). Panel (D): Adapted from Aigouy et al. (2010) and Harumoto et al. (2010).* (A color version of this figure is located at the end of this chapter.)

cell polarity and tissue shape is an area of active investigation in morphogenesis (Legoff & Lecuit, 2015).

One excellent example of stress-induced polarity alignment comes from the shoot apical meristem (SAM), a population of stem cells undergoing complex morphogenesis giving rise to organ primordia with regular spacing (phyllotaxis) (Fig. 2C) (Murray, Jones, Godin, & Traas, 2012). Initiation of organ primordia results from polarized auxin fluxes, the direction of which is determined by the polarized localization of PIN auxin transporters. The polarity of PINs is influenced by the direction of stress between neighboring cells (Asnacios & Hamant, 2012). What mediates directional stress sensing in plant cells? One line of evidence suggests that the plasma membrane directly translates cell wall strains and recruits PIN in a tension-dependent manner (Fig. 2C′) (Nakayama et al., 2012). In parallel, it has been proposed that cortical microtubules respond to anisotropic cell wall strains and regulate oriented cell division and new cellulose fiber deposition (Hamant et al., 2008). The directions of auxin transport and microtubules are likely regulated by a common upstream mechanical signal from the cell walls (Heisler et al., 2010).

Interestingly, microtubules may also play an instrumental role in the long-range alignment of planar cell polarity in animal tissues (Asnacios & Hamant, 2012; Chanet & Martin, 2014; Goehring & Grill, 2013; Siegrist & Doe, 2007). In the wing epithelium of the fruit fly *Drosophila melanogaster*, microtubules orient along the proximal–distal axis, with a small bias of more plus-ends pointing towards the distal end (Fig. 2D). It has been proposed that Frizzled, a Core PCP pathway component, is transported in vesicles along microtubules towards the plus end, resulting in its distal junctional enrichment (Figs. 1B and 2D′) (Shimada, Yonemura, Ohkura, Strutt, & Uemura, 2006), thus breaking the symmetry to establish Core PCP. If microtubules are instructive for the direction of planar cell polarity, what aligns them along the proximal–distal axis? One line of evidences point towards the Fat-PCP pathway. Harumoto and colleagues proposed that the Dachsous gradients bias the direction of microtubule polymerization, such that there is a slight enrichment of microtubules pointing towards the distal end, where Dachsous level is the lowest (Fig. 2D) (Harumoto et al., 2010). However, since microtubules are mechanosensitive, can they also respond to a mechanical signal in the fly wing, similar to what happens in the plant SAM? It has been demonstrated that global anisotropic tension does exist along the proximodistal axis of the pupal wing due to the contraction of the wing hinge at the proximal side (Fig. 2D) (Aigouy et al., 2010; Etournay et al., 2015; Ray et al., 2015). This tension anisotropy is required for the reorientation of Core PCP polarity from an early pattern in the imaginal disc into the final proximodistal pattern in the pupal wing via cell flow (Aigouy et al., 2010; Etournay et al., 2015; Sagner et al., 2012). Intriguingly, microtubule reorientation mimics that of the Core PCP polarity (Fig. 2D′) (Harumoto et al., 2010). Thus, it is plausible that microtubules can sense tension anisotropy in the wing blade to reorient Core PCP polarity. The two potential upstream signals to microtubule orientation, Fat-PCP gradient and mechanical tension during morphogenesis, can also interact (Bosveld et al., 2012).

Mechanical tension was shown to participate in another example of planar cell polarization in the early *Drosophila* embryo. Polarized cell intercalation driven by junction remodeling is dependent upon planar polarized Myosin II enrichment at cell junctions which is itself dependent upon polarized Rho1 activity (Kasza, Farrell, & Zallen, 2014; Munjal, Philippe, Munro, & Lecuit, 2015). It was shown that polarized mechanical tension at cell junctions can reinforce and probably amplify Myosin II planar polarity in neighboring cells (Fernandez-Gonzalez et al., 2009).

5. INTERNAL MECHANICAL SIGNALS CAN INDUCE POLARITY

The interior of the cell is full of microscopic forces in action: the protrusive forces resulting from polymerization of actin and microtubule cytoskeletons, and the tensile forces from myosin motors contracting against cross-linked actin filaments. Distinct forms of internal forces can influence cell polarity in various fashions. On the one hand, internal forces can engage the plasma membrane and modulate membrane tension in a similar way compared with external forces, thus providing a mechanism for coordination of polarity on the cell scale (Fig. 2B) (Houk et al., 2012; Lieber et al., 2013). On the other hand, internal forces are the root cause of mechanical instability in the cytoskeleton, thus providing a built-in mechanism for symmetry breaking. Moreover, internal force-driven actomyosin flows can sustain and amplify cell polarity by transporting polarity cues via advection (Fig. 3). In this section, we will mainly focus on how actomyosin flows control multiple aspects of polarization, including symmetry breaking, local amplification, and long-range inhibition within single cells, or even the alignment of polarity between neighboring cells (Fig. 3D). Finally, we will discuss how other forms of internal flows can be exploited to polarize larger scale tissues.

5.1 Intracellular Actomyosin Flows Induce Intracellular Symmetry Breaking

Myosin motors bind actin filaments to generate contractile forces, which can result in deformation of the actin filaments. Such deformations, when coupled with actin filament remodeling and crosslinker turnover, allow actomyosin networks to behave like a flowing viscoelastic fluid on the

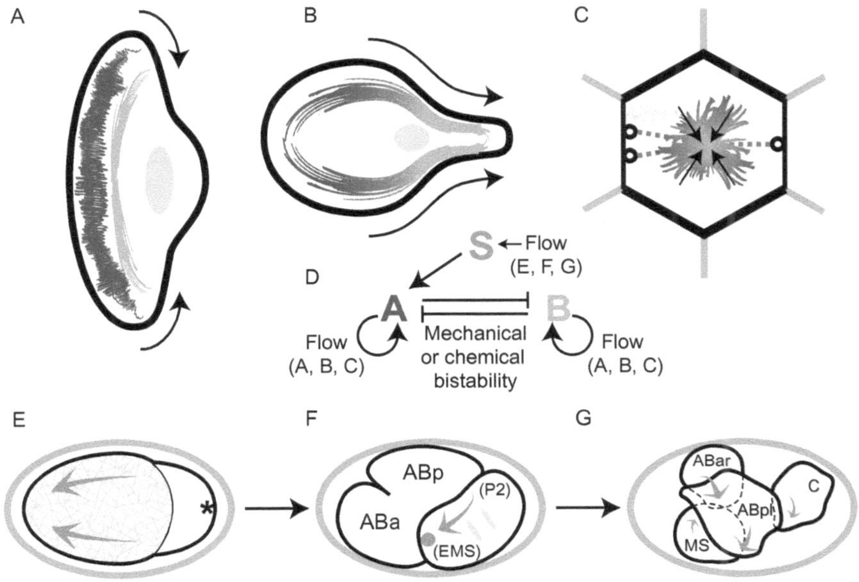

Figure 3 Internal mechanical signals can induce polarity. (A) A polarized, crawling fish keratocyte with actomyosin retrograde flows, characterized by two populations of actin filaments at the front (red: branched, highly protrusive) and the rear (blue: bundled, highly contractile). Black arrow indicates the flow direction. (B) A polarized cell with stable-bleb morphology and with actomyosin retrograde flows. Red-to-blue gradient indicates force gradient of myosin contractility from front to rear. Black arrow indicates the flow direction. (C) An epithelial cell from the *Drosophila* germband ectoderm with oscillatory actomyosin pulse/centripetal flows at the apical cortex. Red-to-blue gradient indicates enrichment of myosin from periphery to center. Black arrow indicates the flow direction. Black circles show site of cadherin endocytosis. Gray dashed lines suggest mechanical coupling between the actomyosin pulse and cadherin endocytosis. (D) Schematics showing how actomyosin flows can amplify two mechanically incompatible regimes of actomyosin networks via feedbacks in the case of (A)–(C). In the case of (E)–(G), flows can induce symmetry breaking. S: symmetry-breaking cue. A and B: two mutually antagonistic, self-amplifying polarity states/components. (E) A one-cell stage *C. elegans* embryo with transient actomyosin flows breaking the AP symmetry. Green arrows indicate the direction of actomyosin flows. Star shows site of sperm entry. (F) A 2 to 4-cell stage *C. elegans* embryo with chiral actomyosin flows breaking the DV symmetry by displacing the ancestral midbody remnant to the ventral side. Green arrows indicate the direction of actomyosin flows. Green dot shows midbody remnant. (G) An 8-cell stage *C. elegans* embryo with chiral actomyosin flows breaking the LR symmetry. Green arrows indicate direction of cell body rotation. *Panel (A): Adapted from Yam et al. (2007). Panel (B): Adapted from Ruprecht et al. (2015). Panel (C): Adapted from Levayer and Lecuit (2013) and Munjal et al. (2015). Panel (E): Adapted from Munro, Nance, and Priess (2004). Panel (F): Adapted from Singh and Pohl (2014). Panel (G): Adapted from Pohl and Bao (2010).* (A color version of this figure is located at the end of this chapter.)

timescale of tens of seconds to minutes (Levayer & Lecuit, 2012). Based on their spatiotemporal characteristics, actomyosin flows can be roughly classified into three categories, namely, sustained flows, oscillatory flows, and transient flows (Fig. 3). Below we will use several examples to illustrate how each flow pattern can contribute to intracellular polarization.

5.1.1 Sustained Flows

Migrating cells, regardless of cell shape or mode of locomotion, often display a continuous displacement of the actomyosin networks from the migrating front towards the rear, termed actomyosin retrograde flows (Étienne et al., 2015). Retrograde flows of actomyosin networks is sustained by polymerization of new actin filaments at the migrating front, and depolymerization of existing actin filaments at the rear (Étienne et al., 2015). The polarized treadmilling of actomyosin cytoskeleton, when coupled with the frictional forces between the cell body and its migrating substratum can generate cell motility.

Actomyosin retrograde flows are best characterized in fish keratocytes (Fig. 3A), where it can be observed in both unpolarized and polarized cells (Theriot & Mitchison, 1991). The unpolarized state of keratocytes exhibits centripetal actomyosin flows from the periphery to the cell body, which depends on Rho-Rok-Myosin driven actomyosin contractility (Yam et al., 2007). The presence of cortical actomyosin contractility indicates an unstable equilibrium, which enables the keratocytes to break their symmetry spontaneously or in response to small mechanical perturbation (Verkhovsky, Svitkina, & Borisy, 1999). In contrast to the neutrophil mode of symmetry breaking driven by initiation of lamellipodia at the leading edge, keratocytes break symmetry by a steady displacement of actomyosin contractile array away from the trailing edge, such that the overall direction of retrograde actomyosin flows changes from centripetal to unidirectional (Yam et al., 2007). Following symmetry breaking, actomyosin retrograde flows form an ascending front-to-rear contractility gradient, promote new actin polymerization at the leading edge by removing actin filaments from the plasma membrane, while favoring tension-dependent myosin loading at the rear where it can catalyze actin network disassembly (Wilson et al., 2010). Thus, actomyosin retrograde flow itself functions as a long-range coordinator of the front and rear polarity in migrating keratocytes. Interestingly, a similar phenomenon is observed in epithelial cells with relaxed actomyosin cytoskeletons, when they can spontaneously switch to a migratory mode following the emergence of actomyosin retrograde flows (Lomakin et al., 2015). Like in keratocytes, this spontaneous front–rear polarity is maintained by continuous polymerization of branched actin networks at the front and myosin-dependent formation of contractile bundles at the rear, due to polarized enrichment of myosin by actomyosin retrograde flows. Since the two distinct actin networks are incompatible structurally, competition for actin between the two induces antagonistic coordination between the front and the rear (Lomakin et al., 2015).

Actomyosin retrograde flows also exist in cells with high cortical contractility, which migrate without lamellipodia and focal adhesions between cells and the matrix (Fig. 3B) (Bergert et al., 2015; Liu et al., 2015; Ruprecht et al., 2015). Dubbed as "stable-blebs," this new form of cell motility shares certain similarities with the blebbing-type motility (Charras & Paluch, 2008), but is different from the later due to the presence of retrograde actomyosin flows and fast persistent motion. High cortical tension triggered by either spatial confinement or chemical stimulation is the prerequisite to stable-bleb formation. Symmetry breaking can either be stochastic (Liu et al., 2015), or in response to a spatial gradient of cortical contractility-inducing chemicals (Ruprecht et al., 2015). Similar to keratocytes, the stable retrograde actomyosin flows are maintained by myosin contractility-mediated rearward displacement of actomyosin cortex, actin polymerization at the leading edge, and actin depolymerization at the trailing edge (Ruprecht et al., 2015). Since myosin contractility induces densification of actomyosin cortex, thus enhancing contractility, this positive feedback likely presents a local amplification of "rearness" crucial for maintaining polarity in stable-bleb cells (Ruprecht et al., 2015).

Maiuri and colleagues compared different migrating cell types, and further generalized the role of actomyosin retrograde flows in maintaining cell polarity. They proposed that a rear-associated polarity cue, when advected by actomyosin retrograde flows in a front-to-rear fashion, can positively reinforce cell polarity and thus the maintenance of the actomyosin retrograde flow itself (Maiuri et al., 2015). This positive feedback between polarity and actomyosin flows can be used to explain the interesting phenomenon of faster cells being more persistent in their migration direction (Maiuri et al., 2015).

5.1.2 Oscillatory Flows

Besides the sustained mono-directional retrograde flows observed in migrating cells, actomyosin flows can also exist in an oscillatory form in epithelial cells, alternating flow directions and driving deformation at multiple locations of the intercellular junctions along the apical cortex (Fig. 3C) (Levayer & Lecuit, 2012). Such multidirectional actomyosin flows are extensively characterized in the intercalating ectodermal cells of *Drosophila* embryos, where cell neighborhood exchange via spatially biased junctional shrinkage and elongation results in large-scale tissue elongation along the embryonic anteroposterior (AP) axis (Bertet, Sulak, & Lecuit, 2004; Blankenship, Backovic, Sanny, Weitz, & Zallen, 2006; Collinet, Rauzi, Lenne, & Lecuit, 2015; David, Tishkina, & Harris, 2010; Fernandez-Gonzalez & Zallen, 2011; Irvine & Wieschaus, 1994; Kasza & Zallen, 2011; Kasza et al., 2014; Levayer & Lecuit, 2013; Rauzi, Lenne, & Lecuit, 2010; Rauzi, Verant, Lecuit, & Lenne, 2008; Zallen & Blankenship, 2008; Zallen & Wieschaus, 2004). Planar polarized junctional remodeling involves spatially biased actomyosin flows preferentially towards junctions perpendicular to the elongation axis (Rauzi et al., 2010), which is thought to emerge from the fluctuations in the coupling strength between the actomyosin networks and junctional E-Cadherin clusters (Levayer & Lecuit, 2013). These fluctuations are a consequence of the step-wise removal of E-Cadherin clusters by endocytosis. Since actomyosin flows facilitate junctional E-Cadherin endocytosis, this could form a positive mechanical feedback between junctional E-Cadherin fluctuations and the bidirectional actomyosin flows, which then aligns and amplifies the polarity of both processes (Fig. 3C) (Levayer & Lecuit, 2013).

One prerequisite for the oscillatory bidirectional actomyosin flows is the transient coalescence of the contractile actomyosin networks, referred to as actomyosin pulses (Fig. 3C) (Levayer & Lecuit, 2012; Martin, Kaschube, & Wieschaus, 2009; Solon, Kaya-Çopur, Colombelli, & Brunner, 2009). It is demonstrated that the pulsatile behavior of actomyosin networks requires myosin phosphocycles and the co-advection of positive and negative regulators of myosin, such as Rho1, Rho-kinase, and Myosin phosphatase (Munjal et al., 2015; Vasquez, Tworoger, & Martin, 2014). Thus, oscillatory bidirectional actomyosin flows also harbor a self-organized amplification/inhibition mechanism.

5.1.3 Transient Flows

A third type of actomyosin flows is a nonrecursive, one-time event resulting from perturbation of a highly contractile actomyosin cortex (Fig. 3E–G). Naturally, this generates an intracellular asymmetry. However, since a transient flow cannot drive amplification of polarity cues through time, cells have to couple the transient flows with additional amplification circuits in order to maintain the polarized state. Beyond promoting cell motility, actomyosin flows can also couple with the chemical polarization network to rapidly transport polarity components along the cell diameter via advection (Goehring & Grill, 2013; Hird & White, 1993; Mayer, Depken, Bois, Jülicher, & Grill, 2010; Munro et al., 2004). One well-characterized case is the symmetry breaking of the one-cell *C. elegans* embryo along the AP axis, in which the membrane-bound anterior PAR complex (PAR-3/PAR-6/PKC-3) is advected by the cortical actomyosin flows towards the anterior end (Fig. 3E) (Munro & Bowerman, 2009). The actomyosin flows initiate after mechanical weakening of the cortex by sperm entry at the posterior end of the embryo (Munro & Bowerman, 2009; Munro et al., 2004) leading to anisotropies in cortical tension (Mayer et al., 2010). Once depleted with the anterior PAR complex, plasma membrane at the posterior end of the embryo can then be occupied by the posterior PAR complex (PAR-1/PAR-2/LGL-1), which forms mutually antagonistic chemical feedbacks with the anterior PARs (Goehring et al., 2011). A recently identified CDC-42/CHIN-1 cross-inhibitory circuit interacts with both PAR complexes and acts redundantly to stabilize AP boundary (Sailer, Anneken, Li, Lee, & Munro, 2015). Moreover, since PAR-2 inhibits myosin accumulation at the posterior, this forms an additional layer of reinforcement of stable AP polarity (Munro et al., 2004).

5.2 Intracellular Actomyosin Flows and Tissue Polarity

In multicellular organisms, polarity exists beyond the cell scale into the tissue scale. The distribution of cell fates in tissues and embryos are rarely stochastic, but rather follows a spatial pattern that aligns with a few axes, such as AP, dorsoventral (DV) and left–right (LR). As described in the previous section, intracellular polarity along the AP axis in the early *C. elegans* embryo is driven by the interplay between mechanical forces generated by cortical actomyosin flows, and chemical components that antagonize and stabilize the spatial AP domains. Asymmetric cell division along the AP axis subsequently establishes the AP intercellular polarity. Can the same logic be extended to the emergence of DV and LR polarities? The *C. elegans* early embryo once again provides elegant answers to this question.

Though characterized by a stereotypical cell fate lineage, early *C. elegans* embryonic development is in fact highly regulative, where major fate determination events depend on intercellular communications between relatively few numbers of early blastomeres (Gönczy & Rose, 2005). In particular, the determination of both DV and LR axes involves rotation of subsequent cell division planes relative to the AP axis, such that contacts between daughter cells become asymmetric along either orthogonal axis (Naganathan, Fürthauer, Nishikawa, Jülicher, & Grill, 2014; Pohl & Bao, 2010; Singh & Pohl, 2014). Recent studies revealed that cortical actomyosin flows, from the cell poles towards the cytokinetic furrow, underlie the rotation of cell division planes away from default orientations (Fig. 3F–G) (Naganathan et al., 2014; Pohl & Bao, 2010; Singh & Pohl, 2014). In both cases, asymmetric cortical contractility between a pair of sister cells induces chiral cortical actomyosin flows with respect to the AP axis, which asymmetrically repositions either the previous division plane (in LR polarization) (Fig. 3G) (Naganathan et al., 2014; Pohl & Bao, 2010) or a physical polarity cue that reorients the mitotic spindle for the new division (the cytokinetic midbody remnant in DV polarization) (Fig. 3F) (Singh & Pohl, 2014). This establishes an asymmetric cellular neighborhood, and enables differential cell-contact signaling to partition cell fates along DV or LR axis. Thus, in an analogous fashion to AP polarization, intracellular mechanical forces break the symmetry along DV or LR axes. It will be of future interest to investigate the mechanochemical origin of such chiral forces involved in cell–cell rearrangement.

5.3 Internal Hydrodynamic Flows can Promote Large-Scale Polarization

So far, we have discussed roles of cortical flows generated by intracellular actomyosin contractility on symmetry-breaking events on the scale of single or few cells, using the *C. elegans* embryo as one example. However, this local symmetry-breaking mechanism based on direct intercellular interaction is often not applicable to other large-scale developmental systems, such as the vertebrate embryo during LR polarization. Consisting of hundreds or thousands of cells when the LR axis is formed, the vertebrate embryo must employ a different strategy of polarization, which involves the long-range diffusion of activator/inhibitor couples of the Nodal signaling pathway (Müller et al., 2012; Nakamura et al., 2006; Schier, 2003). Nevertheless, the initial symmetry-breaking event in vertebrate LR polarization also involves advection of a chemical polarity signal via bulk flows generated by motile cilia in the left–right organizer (LRO) (Basu & Brueckner, 2008). A closer look at the LRO reveals two types of ciliary functions: force generation via beating and force sensing via mechanotransduction (Babu & Roy, 2013). Recently, biased intraciliary calcium oscillations along the LR axis have been observed in the zebrafish LRO as a consequence of mechanotransduction of forces generated by leftward flows (Yuan, Zhao, Brueckner, & Sun, 2015). This initial asymmetry in calcium wave is translated into differential Nodal levels along the LR axis, which is then amplified via downstream local-excitation, global-inhibition mechanism in the nodal signaling cascade (Yuan et al., 2015). Thus, despite the drastic differences in cell number and the nature of the flows, both *C. elegans* and vertebrate embryos break LR symmetry with mechanical forces, which result in an initial internal bias in the spatial distribution of activator/inhibitor couples capable of engaging in self-organizing local-excitation, global-inhibition circuits.

6. CONCLUDING REMARKS

In this essay, we have attempted to present a broad picture of how polarity arises in various cellular and developmental systems through biomechanical interactions. By partitioning polarity signals according to their physical nature, we have summarized how chemical and mechanical signals from both external and internal sources can interact, limit, or reinforce each other to generate robust polarity at a range of scales. Thanks to recent advances in live imaging and quantitative analyses, the complex nature of the mechanochemical interplay, heralded by Alan Turing more than 60 years ago, has emerged as a new frontier in the field of cell and developmental biology.

Many unknowns have thus come forth, a few of which we attempt to list here. First, the symmetry breaking/asymmetry amplification paradigm we have outlined tends to generate stable polarity. However, in the context of living organisms, cells, and tissues need to respond to instantaneous changes in their environments. How do cells and tissues overcome their previous polarity states to adapt to new polarizing signals (Iglesias & Devreotes, 2012)? Second, as cells reside in living tissues, polarity signals from its external and internal environments may sometimes conflict with each other. How do cells resolve and integrate conflicting polarity signals from different sources? Third, though we mainly focused on the interplay between mechanical forces and chemical polarity components at the posttranslational level, it is well established that mechanical forces can influence gene transcription and translation (Chanet & Martin, 2014; Eyckmans et al., 2011).

Are there "mechanogenetic" mechanisms involved in cell and tissue polarity? Last but not least, can we apply our understanding of the mechanochemical polarization framework into the design of synthetic cell and tissues to facilitate self-organized morphogenesis (Chau, Walter, Gerardin, Tang, & Lim, 2012)?

Biomechanical interactions provide a robust framework to organize and remodel subcellular compartments, whole cells, and tissues. Such complex organization emerges through feedbacks and symmetry-breaking events at different scales. We are just seeing the beginning of how biomechanics is at the heart of cellular and tissue organization.

ACKNOWLEDGMENTS

We thank members of our group for very useful discussions and feedback on the chapter. We acknowledge the ERC (Biomecamorph, AdvGrant), CNRS, Aix-Marseille Université (AMU), and the Labex INFORM (ANR-11-LABX-0054) for funding.

REFERENCES

Abley, K., De Reuille, P. B., Strutt, D., Bangham, A., Prusinkiewicz, P., Marée, A. F. M., et al. (2013). An intracellular partitioning-based framework for tissue cell polarity in plants and animals. *Development*, *140*, 2061–2074.

Aigouy, B., Farhadifar, R., Staple, D. B., Sagner, A., Roper, J.-C., Julicher, F., et al. (2010). Cell flow reorients the axis of planar polarity in the wing epithelium of Drosophila. *Cell*, *142*, 773–786.

Arai, Y., Shibata, T., Matsuoka, S., Sato, M. J., Yanagida, T., & Ueda, M. (2010). Self-organization of the phosphatidylinositol lipids signaling system for random cell migration. *Proceedings of the National Academy of Sciences*, *107*, 12399–12404.

Asnacios, A., & Hamant, O. (2012). The mechanics behind cell polarity. *Trends in Cell Biology*, *22*, 584–591.

Babu, D., & Roy, S. (2013). Left-right asymmetry: Cilia stir up new surprises in the node. *Open Biology*, *3*, 130052.

Bardet, P.-L., Guirao, B., Paoletti, C., Serman, F., Léopold, V., Bosveld, F., et al. (2013). PTEN controls junction lengthening and stability during cell rearrangement in epithelial tissue. *Developmental Cell*, *25*, 534–546.

Basu, B., & Brueckner, M. (2008). Cilia multifunctional organelles at the center of vertebrate left-right asymmetry. *Current Topics in Developmental Biology*, *85*, 151–174.

Bergert, M., Erzberger, A., Desai, R. A., Aspalter, I. M., Oates, A. C., Charras, G., et al. (2015). Force transmission during adhesion-independent migration. *Nature Cell Biology*, *17*, 524–529.

Bertet, C., Sulak, L., & Lecuit, T. (2004). Myosin-dependent junction remodelling controls planar cell intercalation and axis elongation. *Nature*, *429*, 667–671.

Blankenship, J. T., Backovic, S. T., Sanny, J. S. P., Weitz, O., & Zallen, J. A. (2006). Multicellular rosette formation links planar cell polarity to tissue morphogenesis. *Developmental Cell*, *11*, 459–470.

Bosveld, F., Bonnet, I., Guirao, B., Tlili, S., Wang, Z., Petitalot, A., et al. (2012). Mechanical control of morphogenesis by Fat/Dachsous/Four-jointed planar cell polarity pathway. *Science*, *336*, 724–727.

Bryant, D. M., & Mostov, K. E. (2008). From cells to organs: Building polarized tissue. *Nature Reviews. Molecular Cell Biology*, *9*, 887–901.
Chanet, S., & Martin, A. C. (2014). Mechanical force sensing in tissues. In *Mechanotransduction* (pp. 317–352). Amsterdam: Elsevier.
Charras, G., & Paluch, E. (2008). Blebs lead the way: How to migrate without lamellipodia. *Nature Reviews. Molecular Cell Biology*, *9*, 730–736.
Charras, G., & Sahai, E. (2014). Physical influences of the extracellular environment on cell migration. *Nature Reviews. Molecular Cell Biology*, *15*, 813–824.
Chau, A. H., Walter, J. M., Gerardin, J., Tang, C., & Lim, W. A. (2012). Designing synthetic regulatory networks capable of self-organizing cell polarization. *Cell*, *151*, 320–332.
Collinet, C., Rauzi, M., Lenne, P.-F., & Lecuit, T. (2015). Local and tissue-scale forces drive oriented junction growth during tissue extension. *Nature Cell Biology*, *17*(10), 1247–1258.
David, D. J. V., Tishkina, A., & Harris, T. J. C. (2010). The PAR complex regulates pulsed actomyosin contractions during amnioserosa apical constriction in Drosophila. *Development*, *137*, 1645–1655.
Devenport, D. (2014). The cell biology of planar cell polarity. *The Journal of Cell Biology*, *207*, 171–179.
Étienne, J., Fouchard, J., Mitrossilis, D., Bufi, N., Durand-Smet, P., & Asnacios, A. (2015). Cells as liquid motors: Mechanosensitivity emerges from collective dynamics of actomyosin cortex. *Proceedings of the National Academy of Sciences*, *112*, 2740–2745.
Etournay, R., Popović, M., Merkel, M., Nandi, A., Blasse, C., Aigouy, B., et al. (2015). Interplay of cell dynamics and epithelial tension during morphogenesis of the Drosophila pupal wing. *Elife*, *4*, e07090.
Eyckmans, J., Boudou, T., Yu, X., & Chen, C. S. (2011). A hitchhiker's guide to mechanobiology. *Developmental Cell*, *21*, 35–47.
Fernandez-Gonzalez, R., Simoes Sde, M., Röper, J. C., Eaton, S., & Zallen, J. A. (2009). Myosin II dynamics are regulated by tension in intercalating cells. *Developmental Cell*, *17*, 736–743.
Fernandez-Gonzalez, R., & Zallen, J. A. (2011). Oscillatory behaviors and hierarchical assembly of contractile structures in intercalating cells. *Physical Biology*, *8*, 045005.
Gassama-Diagne, A., Yu, W., Beest ter, M., Martin-Belmonte, F., Kierbel, A., Engel, J., et al. (2006). Phosphatidylinositol-3,4,5-trisphosphate regulates the formation of the basolateral plasma membrane in epithelial cells. *Nature Cell Biology*, *8*, 963–970.
Goehring, N. W., & Grill, S. W. (2013). Cell polarity: Mechanochemical patterning. *Trends in Cell Biology*, *23*, 72–80.
Goehring, N. W., Trong, P. K., Bois, J. S., Chowdhury, D., Nicola, E. M., Hyman, A. A., et al. (2011). Polarization of PAR proteins by advective triggering of a pattern-forming system. *Science*, *334*, 1137–1141.
Gönczy, P., & Rose, L. S. (2005). Asymmetric cell division and axis formation in the embryo. *WormBook*, *15*, 1–20.
Gray, R. S., Roszko, I., & Solnica-Krezel, L. (2011). Planar cell polarity: Coordinating morphogenetic cell behaviors with embryonic polarity. *Developmental Cell*, *21*, 120–133.
Graziano, B. R., & Weiner, O. D. (2014). Self-organization of protrusions and polarity during eukaryotic chemotaxis. *Current Opinion in Cell Biology*, *30*, 60–67.
Hamant, O., Heisler, M. G., Jönsson, H., Krupinski, P., Uyttewaal, M., Bokov, P., et al. (2008). Developmental patterning by mechanical signals in Arabidopsis. *Science*, *322*, 1650–1655.
Harris, K. P., & Tepass, U. (2008). Cdc42 and Par proteins stabilize dynamic adherens junctions in the Drosophila neuroectoderm through regulation of apical endocytosis. *The Journal of Cell Biology*, *183*, 1129–1143.

Harumoto, T., Ito, M., Shimada, Y., Kobayashi, T. J., Ueda, H. R., Lu, B., et al. (2010). Atypical cadherins Dachsous and Fat control dynamics of noncentrosomal microtubules in planar cell polarity. *Developmental Cell*, *19*, 389–401.

Heisenberg, C.-P., & Bellaïche, Y. (2013). Forces in tissue morphogenesis and patterning. *Cell*, *153*, 948–962.

Heisler, M. G., Hamant, O., Krupinski, P., Uyttewaal, M., Ohno, C., Jönsson, H., et al. (2010). Alignment between PIN1 polarity and microtubule orientation in the shoot apical meristem reveals a tight coupling between morphogenesis and auxin transport. *PLoS Biology*, *8*, e1000516.

Hird, S. N., & White, J. G. (1993). Cortical and cytoplasmic flow polarity in early embryonic cells of Caenorhabditis elegans. *The Journal of Cell Biology*, *121*, 1343–1355.

Houk, A. R., Jilkine, A., Mejean, C. O., Boltyanskiy, R., Dufresne, E. R., Angenent, S. B., et al. (2012). Membrane tension maintains cell polarity by confining signals to the leading edge during neutrophil migration. *Cell*, *148*, 175–188.

Howard, J., Grill, S. W., & Bois, J. S. (2011). Turing's next steps: The mechanochemical basis of morphogenesis. *Nature Reviews. Molecular Cell Biology*, *12*, 392–398.

Iglesias, P. A., & Devreotes, P. N. (2012). Biased excitable networks: How cells direct motion in response to gradients. *Current Opinion in Cell Biology*, *24*, 245–253.

Irvine, K. D., & Wieschaus, E. (1994). Cell intercalation during Drosophila germband extension and its regulation by pair-rule segmentation genes. *Development*, *120*, 827–841.

Jilkine, A., & Edelstein-Keshet, L. (2011). A comparison of mathematical models for polarization of single eukaryotic cells in response to guided cues. *PLoS Computational Biology*, 7, e1001121.

Jin, T. (2013). Gradient sensing during chemotaxis. *Current Opinion in Cell Biology*, *25*, 532–537.

Kasza, K. E., Farrell, D. L., & Zallen, J. A. (2014). Spatiotemporal control of epithelial remodeling by regulated myosin phosphorylation. *PNAS*, *111*, 11732–11737. 201400520.

Kasza, K. E., & Zallen, J. A. (2011). Dynamics and regulation of contractile actin-myosin networks in morphogenesis. *Current Opinion in Cell Biology*, *23*, 30–38.

Knoch, F., Tarantola, M., Bodenschatz, E., & Rappel, W.-J. (2014). Modeling self-organized spatio-temporal patterns of PIP3 and PTEN during spontaneous cell polarization. *Physical Biology*, *11*, 046002.

Lansky, Z., Braun, M., Lüdecke, A., Schlierf, M., Wolde ten, P. R., Janson, M. E., et al. (2015). Diffusible crosslinkers generate directed forces in microtubule networks. *Cell*, *160*, 1159–1168.

Lecuit, T., Lenne, P.-F., & Munro, E. (2011). Force generation, transmission, and integration during cell and tissue morphogenesis. *Annual Review of Cell and Developmental Biology*, *27*, 157–184.

Legoff, L., & Lecuit, T. (2015). Mechanical forces and growth in animal tissues. *Cold Spring Harbor Perspectives in Biology*, 7, a019232.

Levayer, R., & Lecuit, T. (2012). Biomechanical regulation of contractility: Spatial control and dynamics. *Trends in Cell Biology*, *22*, 61–81.

Levayer, R., & Lecuit, T. (2013). Oscillation and polarity of E-Cadherin asymmetries control actomyosin flow patterns during morphogenesis. *Developmental Cell*, *26*, 162–175.

Li, R., & Bowerman, B. (2010). Symmetry breaking in biology. *Cold Spring Harbor Perspectives in Biology*, *2*, a003475.

Lieber, A. D., Yehudai-Resheff, S., Barnhart, E. L., Theriot, J. A., & Keren, K. (2013). Membrane tension in rapidly moving cells is determined by cytoskeletal forces. *Current Biology*, *23*, 1409–1417.

Liu, Y.-J., Le Berre, M., Lautenschlaeger, F., Maiuri, P., Callan-Jones, A., Heuzé, M., et al. (2015). Confinement and low adhesion induce fast amoeboid migration of slow mesenchymal cells. *Cell*, *160*, 659–672.

Lo, C. M., Wang, H. B., Dembo, M., & Wang, Y. L. (2000). Cell movement is guided by the rigidity of the substrate. *Biophysical Journal, 79*, 144–152.
Lomakin, A. J., Lee, K.-C., Han, S. J., Bui, D. A., Davidson, M., Mogilner, A., et al. (2015). Competition for actin between two distinct F-actin networks defines a bistable switch for cell polarization. *Nature Cell Biology, 17*(11), 1435–1445.
Maiuri, P., Rupprecht, J.-F., Wieser, S., Ruprecht, V., Bénichou, O., Carpi, N., et al. (2015). Actin flows mediate a universal coupling between cell speed and cell persistence. *Cell, 161*, 374–386.
Mammoto, T., Mammoto, A., & Ingber, D. E. (2013). Mechanobiology and developmental control. *Annual Review of Cell and Developmental Biology, 29*, 27–61.
Martin, A. C., Kaschube, M., & Wieschaus, E. F. (2009). Pulsed contractions of an actin-myosin network drive apical constriction. *Nature, 457*, 495–499.
Martin-Belmonte, F., Gassama, A., Datta, A., Yu, W., Rescher, U., Gerke, V., et al. (2007). PTEN-mediated apical segregation of phosphoinositides controls epithelial morphogenesis through Cdc42. *Cell, 128*, 383–397.
Matis, M., & Axelrod, J. D. (2013). Regulation of PCP by the Fat signaling pathway. *Genes & Development, 27*, 2207–2220.
Mayer, M., Depken, M., Bois, J. S., Jülicher, F., & Grill, S. W. (2010). Anisotropies in cortical tension reveal the physical basis of polarizing cortical flows. *Nature, 467*, 617–621.
McCaffrey, L. M., & Macara, I. G. (2009). Widely conserved signaling pathways in the establishment of cell polarity. *Cold Spring Harbor Perspectives in Biology, 1*, a001370.
Meinhardt, H. (2012). Turing's theory of morphogenesis of 1952 and the subsequent discovery of the crucial role of local self-enhancement and long-range inhibition. *Interface Focus, 2*, 407–416.
Mori, Y., Jilkine, A., & Edelstein-Keshet, L. (2008). Wave-pinning and cell polarity from a bistable reaction–diffusion system. *Biophysical Journal, 94*, 3684–3697.
Müller, P., Rogers, K. W., Jordan, B. M., Lee, J. S., Robson, D., Ramanathan, S., et al. (2012). Differential diffusivity of nodal and lefty underlies a reaction–diffusion patterning system. *Science, 336*, 721–724.
Mullins, R. D. (2010). Cytoskeletal mechanisms for breaking cellular symmetry. *Cold Spring Harbor Perspectives in Biology, 2*, a003392.
Munjal, A., & Lecuit, T. (2014). Actomyosin networks and tissue morphogenesis. *Development, 141*, 1789–1793.
Munjal, A., Philippe, J.-M., Munro, E., & Lecuit, T. (2015). A self-organized biomechanical network drives shape changes during tissue morphogenesis. *Nature, 524*, 351–355.
Munro, E., & Bowerman, B. (2009). Cellular symmetry breaking during Caenorhabditis elegans development. *Cold Spring Harbor Perspectives in Biology, 1*, a003400.
Munro, E., Nance, J., & Priess, J. R. (2004). Cortical flows powered by asymmetrical contraction transport PAR proteins to establish and maintain anterior-posterior polarity in the early C. elegans embryo. *Developmental Cell, 7*, 413–424.
Murray, J. A. H., Jones, A., Godin, C., & Traas, J. (2012). Systems analysis of shoot apical meristem growth and development: Integrating hormonal and mechanical signaling. *Plant Cell, 24*, 3907–3919.
Naganathan, S. R., Fürthauer, S., Nishikawa, M., Jülicher, F., & Grill, S. W. (2014). Active torque generation by the actomyosin cell cortex drives left-right symmetry breaking. *Elife, 3*, e04165.
Nakamura, T., Mine, N., Nakaguchi, E., Mochizuki, A., Yamamoto, M., Yashiro, K., et al. (2006). Generation of robust left-right asymmetry in the mouse embryo requires a self-enhancement and lateral-inhibition system. *Developmental Cell, 11*, 495–504.
Nakayama, N., Smith, R. S., Mandel, T., Robinson, S., Kimura, S., Boudaoud, A., et al. (2012). Mechanical regulation of auxin-mediated growth. *Current Biology, 22*, 1468–1476.

Nelson, W. J. (2009). Remodeling epithelial cell organization: Transitions between front–rear and apical–basal polarity. *Cold Spring Harbor Perspectives in Biology*, *1*, a000513.
Onsum, M. D., & Rao, C. V. (2009). Calling heads from tails: The role of mathematical modeling in understanding cell polarization. *Current Opinion in Cell Biology*, *21*, 74–81.
Orlando, K., & Guo, W. (2009). Membrane organization and dynamics in cell polarity. *Cold Spring Harbor Perspectives in Biology*, *1*, a001321.
Pinal, N., Goberdhan, D. C. I., Collinson, L., Fujita, Y., Cox, I. M., Wilson, C., et al. (2006). Regulated and polarized PtdIns(3,4,5)P3 accumulation is essential for apical membrane morphogenesis in photoreceptor epithelial cells. *Current Biology*, *16*, 140–149.
Pohl, C., & Bao, Z. (2010). Chiral forces organize left-right patterning in C. elegans by uncoupling midline and anteroposterior axis. *Developmental Cell*, *19*, 402–412.
Rauzi, M., Lenne, P.-F., & Lecuit, T. (2010). Planar polarized actomyosin contractile flows control epithelial junction remodelling. *Nature*, *468*, 1110–1114.
Rauzi, M., Verant, P., Lecuit, T., & Lenne, P.-F. (2008). Nature and anisotropy of cortical forces orienting Drosophila tissue morphogenesis. *Nature Cell Biology*, *10*, 1401.
Ray, R. P., Matamoro-Vidal, A., Ribeiro, P. S., Tapon, N., Houle, D., Salazar-Ciudad, I., et al. (2015). Patterned anchorage to the apical extracellular matrix defines tissue shape in the developing appendages of Drosophila. *Developmental Cell*, *34*, 310–322.
Rodriguez-Boulan, E., & Macara, I. G. (2014). Organization and execution of the epithelial polarity programme. *Nature Reviews. Molecular Cell Biology*, *15*, 225–242.
Röper, K. (2012). Anisotropy of crumbs and aPKC drives myosin cable assembly during tube formation. *Developmental Cell*, *23*, 939–953.
Ruprecht, V., Wieser, S., Callan-Jones, A., Smutny, M., Morita, H., Sako, K., et al. (2015). Cortical contractility triggers a stochastic switch to fast amoeboid cell motility. *Cell*, *160*, 673–685.
Sagner, A. A., Merkel, M. M., Aigouy, B. B., Gaebel, J. J., Brankatschk, M. M., Jülicher, F. F., et al. (2012). Establishment of global patterns of planar polarity during growth of the Drosophila wing epithelium. *Current Biology*, *22*, 1296–1301.
Sailer, A., Anneken, A., Li, Y., Lee, S., & Munro, E. (2015). Dynamic opposition of clustered proteins stabilizes cortical polarity in the C. elegans zygote. *Developmental Cell*, *35*, 131–142.
Schier, A. F. (2003). Nodal signaling in vertebrate development. *Annual Review of Cell and Developmental Biology*, *19*, 589–621.
Shewan, A., Eastburn, D. J., & Mostov, K. (2011). Phosphoinositides in cell architecture. *Cold Spring Harbor Perspectives in Biology*, *3*, a004796.
Shimada, Y., Yonemura, S., Ohkura, H., Strutt, D., & Uemura, T. (2006). Polarized transport of Frizzled along the planar microtubule arrays in Drosophila wing epithelium. *Developmental Cell*, *10*, 209–222.
Siegrist, S. E., & Doe, C. Q. (2007). Microtubule-induced cortical cell polarity. *Genes & Development*, *21*, 483–496.
Singh, D., & Pohl, C. (2014). Coupling of rotational cortical flow, asymmetric midbody positioning, and spindle rotation mediates dorsoventral axis formation in C. elegans. *Developmental Cell*, *28*, 253–267.
Slaughter, B. D., Smith, S. E., & Li, R. (2009). Symmetry breaking in the life cycle of the budding yeast. *Cold Spring Harbor Perspectives in Biology*, *1*, a003384.
Slusarski, D. C., Corces, V. G., & Moon, R. T. (1997). Interaction of Wnt and a frizzled homologue triggers G-protein-linked phosphatidylinositol signalling. *Nature*, *390*, 410–413.
Solon, J., Kaya-Çopur, A., Colombelli, J., & Brunner, D. (2009). Pulsed forces timed by a ratchet-like mechanism drive directed tissue movement during dorsal closure. *Cell*, *137*, 1331–1342.

St Johnston, D., & Ahringer, J. (2010). Cell polarity in eggs and epithelia: Parallels and diversity. *Cell*, *141*, 757–774.
Swaney, K. F., Huang, C.-H., & Devreotes, P. N. (2010). Eukaryotic chemotaxis: A network of signaling pathways controls motility, directional sensing, and polarity. *Annual Review of Biophysics*, *39*, 265–289.
Takeichi, M. (2014). Dynamic contacts: Rearranging adherens junctions to drive epithelial remodelling. *Nature Reviews. Molecular Cell Biology*, *15*, 397–410.
Theriot, J. A., & Mitchison, T. J. (1991). Actin microfilament dynamics in locomoting cells. *Nature*, *352*, 126–131.
Tsujita, K., & Itoh, T. (2015). Phosphoinositides in the regulation of actin cortex and cell migration. *Biochimica et Biophysica Acta (BBA)—Molecular and Cell Biology of Lipids*, *1851*, 824–831.
Turing, A. M. (1952). The chemical basis of morphogenesis. *Philosophical Transactions of the Royal Society of London. Series B, Biological Sciences*, *237*, 37–72.
van der Gucht, J., & Sykes, C. (2009). Physical model of cellular symmetry breaking. *Cold Spring Harbor Perspectives in Biology*, *1*, a001909.
Vasquez, C. G., Tworoger, M., & Martin, A. C. (2014). Dynamic myosin phosphorylation regulates contractile pulses and tissue integrity during epithelial morphogenesis. *The Journal of Cell Biology*, *206*, 435–450.
Veglio, A., Gamba, A., Nicodemi, M., Bussolino, F., & Serini, G. (2009). Symmetry breaking mechanism for epithelial cell polarization. *Physical Review. E, Statistical, Nonlinear, and Soft Matter Physics*, *80*, 031919.
Verkhovsky, A. B., Svitkina, T. M., & Borisy, G. G. (1999). Self-polarization and directional motility of cytoplasm. *Current Biology*, *9*, S1.
Vladar, E. K., Antic, D., & Axelrod, J. D. (2009). Planar cell polarity signaling: The developing cell's compass. *Cold Spring Harbor Perspectives in Biology*, *1*, a002964.
Wang, F. (2009). The signaling mechanisms underlying cell polarity and chemotaxis. *Cold Spring Harbor Perspectives in Biology*, *1*, a002980.
Weber, G. F., Bjerke, M. A., & Desimone, D. W. (2011). A mechanoresponsive cadherin-keratin complex directs polarized protrusive behavior and collective cell migration. *Developmental Cell*, *22*, 104–115.
Williams, H. P., & Harwood, A. J. (2003). Cell polarity and dictyostelium development. *Current Opinion in Microbiology*, *6*, 621–627.
Wilson, C. A., Tsuchida, M. A., Allen, G. M., Barnhart, E. L., Applegate, K. T., Yam, P. T., et al. (2010). Myosin II contributes to cell-scale actin network treadmilling through network disassembly. *Nature*, *465*, 373–377.
Wodarz, A., & Näthke, I. (2007). Cell polarity in development and cancer. *Nature Cell Biology*, *9*, 1016–1024.
Wolpert, L. (1998). Pattern formation in epithelial development: The vertebrate limb and feather bud spacing. *Philosophical Transactions: Biological Sciences*, *353*, 871–875.
Yam, P. T., Wilson, C. A., Ji, L., Hebert, B., Barnhart, E. L., Dye, N. A., et al. (2007). Actin-myosin network reorganization breaks symmetry at the cell rear to spontaneously initiate polarized cell motility. *The Journal of Cell Biology*, *178*, 1207–1221.
Yuan, S., Zhao, L., Brueckner, M., & Sun, Z. (2015). Intraciliary calcium oscillations initiate vertebrate left-right asymmetry. *Current Biology*, *25*, 556–567.
Zallen, J. A., & Blankenship, J. T. (2008). Multicellular dynamics during epithelial elongation. *Seminars in Cell & Developmental Biology*, *19*, 263–270.
Zallen, J. A., & Wieschaus, E. (2004). Patterned gene expression directs bipolar planar polarity in Drosophila. *Developmental Cell*, *6*, 13.

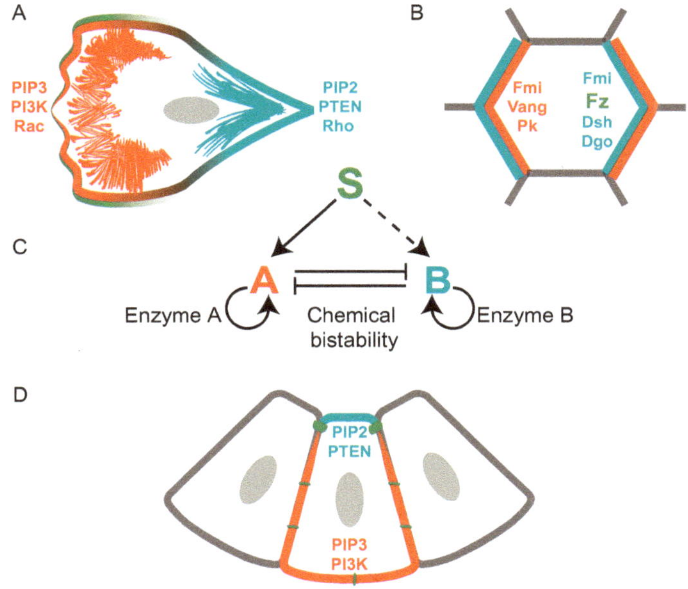

Qiyan Mao and Thomas Lecuit, Figure 1 A conceptual chemical polarization network. (A) A migrating cell aligns its front–rear polarity (red-to-blue outline) with the direction of the chemoattractant gradient (green outline). Inside the cell: Red sketches indicate branched, protrusive actin filaments at the front, while blue sketches show bundled, contractile actin filaments at the rear. (B) An epithelial cell displaying planar cell polarity, with Core PCP components highlighted in red (proximal; Fmi-Vang-Pk) and blue (distal; Fmi-Dgo-Dsh). Fz in the distal PCP complex (green) is thought to be the symmetry-breaking signal transported along microtubules. (C) Schematics showing polarization of a two-component chemical system. S: symmetry-breaking cue that can activate either A alone, or first A then B. A and B: two mutually antagonistic, self-amplifying polarity states/components. (D) An epithelial cell displaying apicobasal polarity, with apical (blue) and basolateral (red) surfaces highlighted. Green bars shows cadherin complexes, with two thick ones showing the position of adherens junctions, and several thin ones showing cadherin complexes at the basolateral surfaces.

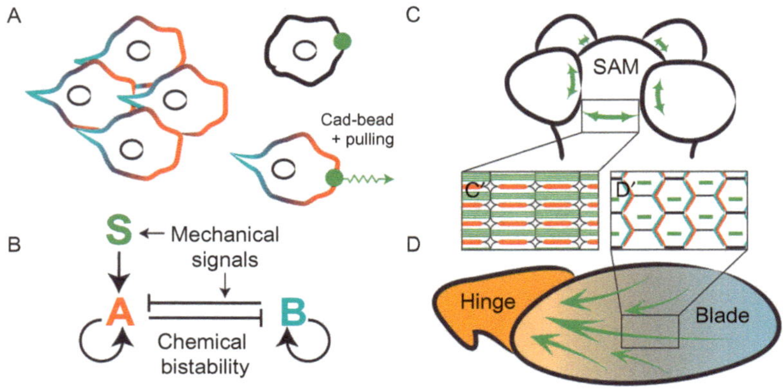

Qiyan Mao and Thomas Lecuit, Figure 2 External mechanical signals can bias polarity. (A) Pulling force via cadherin-coated beads rescues migration polarity in isolated *Xenopus* gastrula mesendoderm cells. Red-to-blue color gradient indicates front–rear polarity in collectively migrating cells and single cells with pulling forces applied via cadherin-coated beads, while black indicates no polarity in cells in contact with cadherin-coated beads but without pulling forces. Green dots show magnetic beads coated with C-Cadherin. Green arrow implies pulling force. (B) Schematics showing mechanical signals may either play an instructive role in biasing the direction of symmetry breaking of a two-component system, or a permissive role in enhancing long-range inhibition between two polarity states. S: symmetry-breaking cue. A and B: two mutually antagonistic, self-amplifying polarity states/components. (C) Hypothetical stress pattern in the *Arabidopsis* shoot apical meristem (green arrows). Inset (C′) shows alignment of microtubules (green) and membrane enrichment of Auxin transporter PIN1 (red) along with the direction of mechanical tension. (D) Pattern of cell flow (green arrows) in the *Drosophila* pupal wing blade due to hinge contraction. Gradients of Ds (orange) and Fj (blue) are superimposed. Inset (D′) shows alignment PCP direction or microtubules (green) with proximal (red: Fmi-Vang-Pk) and distal (blue: Fmi-Fz-Dgo-Dsh) components oriented along the proximodistal axis of the wing blade. *Panel (A): Adapted from Weber, Bjerke, and Desimone (2011). Panel (C): Adapted from Heisler et al. (2010). Panel (D): Adapted from Aigouy et al. (2010) and Harumoto et al. (2010).*

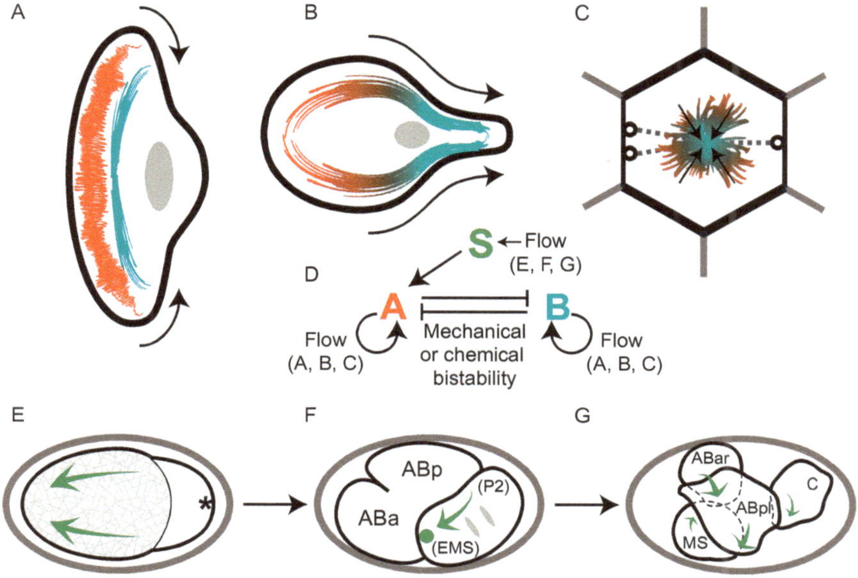

Qiyan Mao and Thomas Lecuit, Figure 3 Internal mechanical signals can induce polarity. (A) A polarized, crawling fish keratocyte with actomyosin retrograde flows, characterized by two populations of actin filaments at the front (red: branched, highly protrusive) and the rear (blue: bundled, highly contractile). Black arrow indicates the flow direction. (B) A polarized cell with stable-bleb morphology and with actomyosin retrograde flows. Red-to-blue gradient indicates force gradient of myosin contractility from front to rear. Black arrow indicates the flow direction. (C) An epithelial cell from the *Drosophila* germband ectoderm with oscillatory actomyosin pulse/centripetal flows at the apical cortex. Red-to-blue gradient indicates enrichment of myosin from periphery to center. Black arrow indicates the flow direction. Black circles show site of cadherin endocytosis. Gray dashed lines suggest mechanical coupling between the actomyosin pulse and cadherin endocytosis. (D) Schematics showing how actomyosin flows can amplify two mechanically incompatible regimes of actomyosin networks via feedbacks in the case of (A)–(C). In the case of (E)–(G), flows can induce symmetry breaking. S: symmetry-breaking cue. A and B: two mutually antagonistic, self-amplifying polarity states/components. (E) A one-cell stage *C. elegans* embryo with transient actomyosin flows breaking the AP symmetry. Green arrows indicate the direction of actomyosin flows. Star shows site of sperm entry. (F) A 2 to 4-cell stage *C. elegans* embryo with chiral actomyosin flows breaking the DV symmetry by displacing the ancestral midbody remnant to the ventral side. Green arrows indicate the direction of actomyosin flows. Green dot shows midbody remnant. (G) An 8-cell stage *C. elegans* embryo with chiral actomyosin flows breaking the LR symmetry. Green arrows indicate direction of cell body rotation. *Panel (A): Adapted from Yam et al. (2007). Panel (B): Adapted from Ruprecht et al. (2015). Panel (C): Adapted from Levayer and Lecuit (2013) and Munjal et al. (2015). Panel (E): Adapted from Munro, Nance, and Priess (2004). Panel (F): Adapted from Singh and Pohl (2014). Panel (G): Adapted from Pohl and Bao (2010).*

CHAPTER THIRTY-EIGHT

The Pluripotency of Neural Crest Cells and Their Role in Brain Development

Nicole M. Le Douarin[*,†,‡,§,1], **Elisabeth Dupin**[†,‡,§]
[*]Collège de France, 3 rue d'Ulm, Paris, France
[†]INSERM U968, Paris, France
[‡]Sorbonne Universités, UPMC Univ Paris 06, UMR S 968, Institut de la Vision, Paris, France
[§]CNRS, UMR 7210, Paris, France
[1]Corresponding author: e-mail address: nicole.le-douarin@college-de-france.fr

Contents

Abstract

The neural crest (NC) is, in the Chordate phylum, an innovation of vertebrates, which exhibits several original characteristics: its component cells are pluripotent and give rise to both ectodermal and mesodermal cell types. Moreover, during the early stages of neurogenesis, the NC cells exert a paracrine stimulating effect on the development of the preotic brain.

1. INTRODUCTION

The neural crest (NC) is an embryonic structure of vertebrates, arising from a territory lying between the neural plate and the adjacent superficial ectoderm, the neural plate border. The NC becomes individualized as the neural tube closes through the fusion of the neural folds, and then resides on

Current Topics in Developmental Biology, Volume 116
ISSN 0070-2153
http://dx.doi.org/10.1016/bs.ctdb.2015.10.008

the dorsal surface of the neural tube. The NC cells (NCC) can be distinguished from both the neural epithelium of the neural tube on one side, and the superficial ectoderm on the other, by their expression of a characteristic suite of transcription factors, including *Snail2/Slug*, *Sox10, Pax7, FoxD3,* and *Sox9*, designated as NC specifier genes (Meulemans & Bronner-Fraser, 2004; Simões-Costa & Bronner, 2015).

The NC can be considered as one of the most specific structure of the vertebrate embryo at several points of view. Its most original character resides in the migratory properties of its component cells: soon after it is formed, the cells of the NC lose their epithelial arrangement, undergo an epithelial–mesenchymal transition and start migrating away from their source, along definite pathways to "invade" the whole developing body. This dispersion results, not only from the progression of the NCC but also from their proliferation, stimulated by growth factors originating from the tissues through which they move and to which they home. This is the case, for example, for glia-derived neurotrophic factor acting via Ret tyrosine kinase receptor, which is the main crucial signaling pathway for the proliferation and migration of the NCC that invade the developing gut and yield the intrinsic neural plexuses of the intestine (Gershon, 1999; Heanue & Pachnis, 2007; Schuchardt, D'Agati, Larsson-Blomberg, Costantini, & Pachnis, 1994); the endothelin-receptor B signaling pathway is also required for enteric nerve cell as well as melanocyte development (Baynash et al., 1994; Hosoda et al., 1994; Le Douarin & Dupin, 2003; Saldana-Caboverde & Kos, 2010).

A second character of the NC is that, as a population, it is *pluripotent*, and yields an impressive number of cell types (see Dupin & Le Douarin, 2014; Le Douarin, 1982; Le Douarin & Dupin, 2003; Le Douarin & Kalcheim, 1999; and references therein). The NC gives rise to the neurons and glial cells of the peripheral nervous system (PNS), to which the neural placodes of the head also contribute to a much lesser extent. The axons of peripheral nerves extend all over the body, notably insuring blood vessel vasomotricity and thus expanding NC-derived cells over all tissues and organs. The NC is also at the origin of endocrine cells: the catecholaminergic cells of the adrenal glands, the calcitonine-producing cells of the thyroid gland in mammals (also designated parafollicular cells) and of the ultimobranchial bodies in birds, the secretory cells of the carotid bodies. Moreover, the NC produces all the melanocytes except those of the pigmented retina.

In addition to these "*neural-melanocytic*" cell types, the NC provides the cephalic region of vertebrates with mesenchymal and skeletogenic type cells that, in all the other parts of the body, are derived from the mesodermal germ layer. In all vertebrates except birds, the NC is at the origin of the dentine and the pulp of the teeth. When they start to migrate, during the migration process, and when they have reached their destination, a number of NCC have been shown to be multipotent and endowed of self-renewal capacity, thus exhibiting stem cell properties (Baroffio, Dupin, & Le Douarin, 1988, 1991; Bronner-Fraser & Fraser, 1988; Kléber et al., 2005; Le Douarin, Calloni, & Dupin, 2008; Morrison, White, Zock, & Anderson, 1999; Sieber-Blum & Cohen, 1980; Stemple & Anderson, 1992; Trentin, Glavieux-Pardanaud, Le Douarin, & Dupin, 2004; for references, Dupin & Coelho-Aguiar, 2013; Dupin & Sommer, 2012).

The capacity to produce, in the head, cell types that, in the other parts of the body, arise from different germ layers (ectoderm and mesoderm) has led to designate the NC as a "*fourth germ layer*," and vertebrates as *quadroblastic*, and not *triblastic*, animals. This notion was already proposed by the discoverer of the NC, Wilhem His in 1868 to designate this particular structure, after Karl von Baer had formulated the germ layer theory (His, 1868), and further discussed by several authors, including Hörstadius (1950), Hall (2000), and Le Douarin and Dupin (2012, 2014).

Another role of the NC, more recently identified in the avian embryo, is the paracrine effect, decisive for the growth and patterning of the fore- and midbrain, which it exerts at an early stage of development.

In this review, two aspects of this remarkable embryonic structure will be further documented and discussed: first, the pluripotency of the NC as a fourth germ layer and at the level of its component cells, some of which endowed with stem cell properties. The second feature of the NC is the crucial role it plays in brain development, the latter being responsible for the extraordinary evolutionary success of the vertebrate phylum.

2. THE NEURAL CREST AS A FOURTH GERM LAYER: EVOLUTIONARY IMPLICATION

The qualification of the NC as a fourth germ layer recently gained further support from the work of Buitrago-Delgado and collaborators, (Buitrago-Delgado, Nordin, Rao, Geary, & LaBonne, 2015) who showed, in *Xenopus* embryos, that, through the process of gastrulation, the NCC have retained pluripotency factors that they directly inherited from the highly pluripotent stem cell of the blastula stage. Among those factors, *Snail1* and *Sox5* are essential for maintaining the broad developmental capacities present in the NCC, and not shared by the other cells of the ectodermal germ layer.

Prior to the work of Buitrago-Delgado and colleagues, it was considered that the NCC were induced, in the course of development, to regain differentiation capacities that had been lost in the ectoderm during the gastrulation process. They could then be considered as an endogenous population of *induced pluripotent stem cells*. According to Buitrago-Delgado et al., the future NCC, unlike their cellular neighbors, kept the pluripotency they had at the blastula stage and which allows them to yield, not only *bona fide* ectodermal derivatives but also cell types that normally arise from the mesodermal germ layer. Moreover, they can be induced to generate endodermal cells if subjected to concentrations of activin that are able to induce mesodermal and endodermal derivatives from the pluripotent *Xenopus* blastula pole cells. The NC can therefore be considered as "an example of cellular neoteny" (Anderson, 1989). Blastula cells are *Snail1* and *Sox5*-positive, and when confronted with high concentrations of activin, a powerful ventralizing factor, they give rise to endoderm but lose this capacity if depleted of *Snail1* and/or *Sox5*.

These results strongly suggest that genes such as *Pax7*, *Snail1,* and *Sox5*, considered as NC factors or NC specifiers, should rather be viewed as pluripotency maintenance factors.

These data can shed light on the evolutionary origin of the NC and on the evolutionary route through which the vertebrate phylum arose among the primitive chordates. The case of the tunicates can be mentioned in this context. On the ground of their genomic sequence they are now considered to be closer to vertebrates than the cephalochordates (Delsuc, Brinkmann, Chourrout, & Philippe, 2006; Pani et al., 2012). The ascidian embryos possess a cell lineage (*a9.49*), from which are derived cells that migrate from the neural plate border and differentiate into pigment cells. These cells express genes such as *Snail, Id, FoxD*, and *Ap2*, and thus have been proposed to belong to a rudimentary NC-like embryonic structure (Abitua, Wagner, Navarrete, & Levine, 2012). These findings are in line with the comparisons of the genomic sequence data that led to consider urochordates more closely related to vertebrates than cephalocordates, as previously believed.

3. THE PLURIPOTENCY AND STEM CELL PROPERTIES OF THE NEURAL CREST CELLS

The longstanding issue of whether the NC comprises multipotent cells or already committed diversified precursors has been tackled at the single cell level through two main approaches: the first one consisted in *in vitro* clonal assays, inspired by the studies of colony-forming units in the hemopoïetic system (for references, Metcalf, 2007). The other one was achieved by labeling single premigratory NCC in the chick embryo *in vivo* (Bronner-Fraser & Fraser, 1988; Krispin, Nitzan, Kassem, & Kalcheim, 2010). The first evidence of NCC multipotency was the finding of common progenitors for unpigmented and pigmented cells in quail NC single cell cultures (Cohen & Königsberg, 1975; Sieber-Blum & Cohen, 1980). Improvements in NCC culture conditions and discovery of phenotypic markers for the diverse NC derivatives led since to a more comprehensive identification of the *in vitro* developmental capacities of individual cells isolated from trunk, cranial, and cardiac NC (Baroffio et al., 1988, 1991; Ito, Morita, & Sieber-Blum, 1993; Sieber-Blum, 1989, 1991; Stemple & Anderson, 1992). These first investigations were extended over four decades (Dupin & Sommer, 2012 for a review). As far as the trunk NC is concerned, these studies showed the presence of multipotent cells in avian and rodent embryos, which were able to yield glial cells, autonomic neurons and myofibroblasts/smooth muscle cells, as well as, at least in the case of avian NCC, melanocytes. Other progenitors gave rise to only glia and melanocytes or glia and myofibroblasts. These multipotent and bipotent trunk NCC were endowed of self-renewal, a *bona fide* property of stem cells, as demonstrated by *in vitro* serial subcloning of NCC progeny (Bittencourt, da Costa, Calloni, Alvarez-Silva, & Trentin, 2013; Stemple & Anderson, 1992; Trentin et al., 2004). Treatment with endothelin-3 (EDN3) and fibroblast growth factor-2 (FGF2) promoted stem cell renewal of avian bipotent progenitors of glia-melanocytes and glia-myofibroblasts, respectively (Bittencourt et al., 2013; Trentin et al., 2004), whereas combination of Wnt and BMP factors was required for the *in vitro* maintenance of mammalian multipotent NC stem cells (Kléber et al., 2005). A number of growth factors and cytokines tested in single NCC cultures, such as TGFβ, BMP4, neuregulin-1, retinoic acid, EDN3 and delta-1, appeared to favor the development of specific NC sublineages in an instructive or permissive manner (Dupin & Le Douarin, 1995; Lahav et al., 1998; Morrison et al., 2000; Shah & Anderson, 1997; Shah, Groves, & Anderson, 1996; Shah, Marchionni, Isaacs, Stroobant, & Anderson, 1994; Trentin et al., 2004).

The large array of NC-derived cell types that arise from the cephalic NC (CNC) comprises not only neural-melanocytic and endocrine cells as in the trunk NC but also the so-called *ectomesenchymal* phenotypes. These CNC-derived mesenchymal phenotypes in the vertebrate head include cartilage and bone cells, adipocytes, tenocytes, and connective tissue cells as well as smooth muscle cells and pericytes, which all originate from the mesoderm in the trunk of the embryo. These features of the CNC have raised the possibility that the ectomesenchyme could originate from a separate NC lineage. To address this question, our laboratory performed a series of *in vitro* cloning studies to investigate the lineage repertoire of individual mesencephalic NCC of the quail embryo. Using a battery of phenotypic markers for six main CNC-derived lineages, i.e., glial cells, neurons, melanocytes, myofibroblasts/smooth muscle cells, chondrocytes, and osteocytes, these studies provided evidence that the vast majority (93%) of the clonogenic CNCC were multipotent progenitors able to give rise to both neural-melanocytic and mesenchymal cell types in diverse combinations (Calloni, Glavieux-Pardanaud, Le Douarin, & Dupin, 2007; Calloni, Le Douarin, & Dupin, 2009) (Fig. 1). Multipotent neural-melanocytic progenitors with skeletogenic ability accounted for more than 90% of the CNCC, including chondrogenic, osteogenic, and osteochondrogenic cells. Altogether, the analysis identified 30 categories of clones, comprising from one to six distinct cell types. Data are consistent with a hierarchical model of the diversification of CNCC wherein restricted progenitors appear lying downstream of a highly multipotent cell, which yielded glia, neurons, melanocytes, myofibroblasts, chondrocytes, and osteocytes (GNMFCO) (Fig. 1). The frequency of this GNMFCO multipotent progenitor was twofold increased after treatment with the morphogen Sonic hedgehog (Shh) (reaching 13% of total clones). In the presence of this factor, it is remarkable that the majority of single CNCC behave as highly multipotent cells with a progeny that comprised five and six different phenotypes. Moreover, Shh also enhanced chondrogenesis in the colonies while not affecting the high proportion of osteogenic progenitors (Calloni et al., 2007, 2009; Dupin et al., 2010) (Fig. 1). In both Shh-treated and untreated cultures, the proportion of CNCC with exclusively neural-melanocytic or mesenchymal potentials was thus very low.

By showing the high prevalence of single CNCC that give rise to both neural and mesenchymal/skeletogenic lineages, these data therefore unequivocally argue against an independent cellular origin of the NC-derived mesenchyme and for a high level of pluripotentiality of the NC progenitor cells.

The *in vivo* experiments designed to address individual NCC fate have also uncovered that the NC contains multipotent cells. Bronner-Fraser and

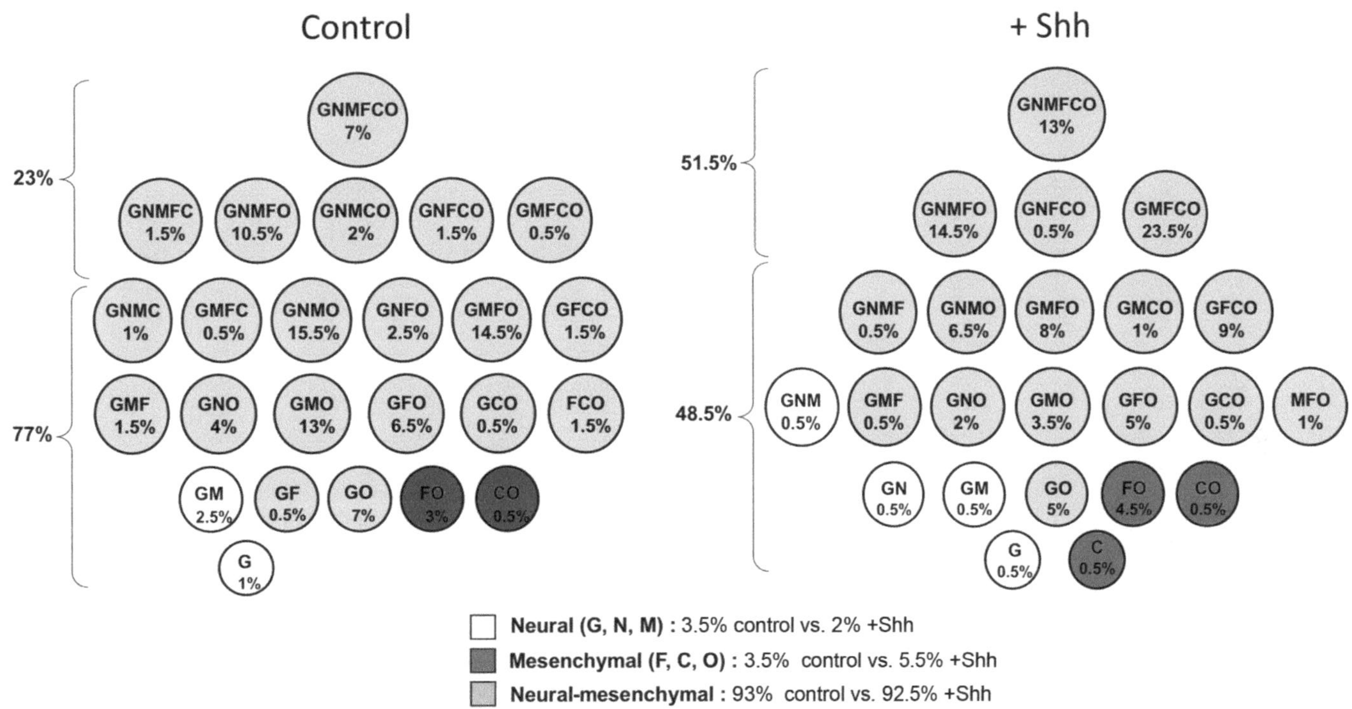

Figure 1 Schematic lineage tree of avian CNCC progenitors and influence of Shh. Quail NCC were isolated from the NCC migratory outgrowth of explanted mesencephalic neural primordium and then cultured as single cells in medium conditions permissive for a diversified cell differentiation. Colonies were phenotypically analyzed after growth for 10 days either in the presence of Shh (109 clones; 52% of initially plated cells) or in control medium (146 clones; 65% of the plated cells). Identified progenitors are classified according to the number of distinct differentiated cell types in their progeny (G, glial cells; N, neurons; M, melanocytes; F, myofibroblasts; C, chondrocytes; O, osteocytes) and their respective frequencies are indicated (% of total clones). In both medium conditions, the vast majority (93%) of CNCC is multipotent and endowed with both neural-melanocytic and mesenchymal (including skeletogenic) potentials (in gray); the frequencies of "only neural" (in yellow) and "only mesenchymal" (in blue), progenitors, were very low. Addition of Shh led to a twofold increase of the proportion of the most multipotent progenitors (with six and five distinct cell types) (accounting for 51.5% vs. 23% of the clones in Shh-treated and untreated cultures, respectively). Shh treatment also promoted the *in vitro* development of progenitors able to produce chondrocytes, without changing the overall frequency of osteogenic multipotent cells. *Adapted from Dupin, Calloni, and Le Douarin (2010).* (A color version of this figure is located at the end of this chapter.)

Fraser were the first to identify multiphenotypic clonal progeny derived from premigratory trunk NCC after fluorescent dye labeling of individual NCC still in the dorsal neural tube of the chick embryo (Bronner-Fraser & Fraser, 1988, 1989). Similar results were recently obtained in the mammalian NC: tracing the fate of single premigratory and migratory NCC *in vivo* using genetic multicolor cell labeling in *R26R-Confetti* mice, revealed that the vast majority of mouse trunk NCC are multipotent (Baggiolini et al., 2015). These results are in contrast with those by Kalcheim and collaborators, who labeled cells *in vivo* in the dorsal midline of the neuroepithelium and found that these premigratory NCC appeared to be fate-restricted at time of their emigration (Krispin et al., 2010; Nitzan et al., 2013). The question as to whether differences in the stage and precise location of the labeled NCC, and also between the labeling methods, can account for these discrepancies, remains unclear. Although the presence of fate-restricted precursors in the early NCC population is not excluded, most *in vivo* and *in vitro* NC lineage analysis as reviewed above, argue for the existence of a diversity of multipotent stem cells in the early NC. The idea that all NCC would be determined at time of migration is also difficult to reconcile with data from transplantation experiments, which have shown that even NCC already settled in PNS ganglia are capable of adopting a new phenotype when grafted back in the early chicken embryo (Ayer-Le Lièvre & Le Douarin, 1982; Dupin, 1984; Schweizer, Ayer-Le Lièvre, & Le Douarin, 1983). In addition, recent investigations on the origin of pigment cells in the chick and mouse embryos have uncovered that a significant fraction of melanocytes arises from Schwann cell precursors (Adameyko et al., 2009; Ernfors, 2010). These precursors lining peripheral nerves and terminals actually behave as bipotent glial-melanocytic NC progenitors, which either differentiate into mature Schwann cells or, if they detach from the nerve fibers, start to activate the molecular program of melanogenesis, migrate to the skin and later contribute to body pigmentation.

Therefore, the presence of multipotent NC stem cells is maintained at late stages of development in various NCC postmigratory sites, such as PNS ganglia and nerves. Remarkably, even in the adult organism, many tissues, including the skin, heart, bone marrow, and many craniofacial and periocular structures were found to be the source of NC-derived multipotent progenitors (reviewed by Achilleos & Trainor, 2012; Dupin & Sommer, 2012; Kaltschmidt, Kaltschmidt, & Widera, 2012; Le Douarin & Dupin, 2014).

4. THE CEPHALIC NEURAL CREST PROMOTES THE DEVELOPMENT OF THE PREOTIC BRAIN

4.1 Ablation of *Hox*-Negative Cephalic Neural Crest Impairs Both Head Skeleton and Anterior Brain Development

Experiments based on the construction of chimeras between embryos of two species of birds, the quail and the chick, have shown that a large part of the skull and the entire facial skeleton originate from the CNC (Couly, Coltey, & Le Douarin, 1993; Le Lièvre, 1974; Le Lièvre & Le Douarin, 1975). The precise level of the CNC that participates in the head skeleton has been determined: it extends from the mid-diencephalon down to the level of rhombomere-2 (r2). The first branchial arch (BA1), which is at the origin of the lower jaw, receives NCC from r1 and r2 with a participation of the NCC exiting r3 (which also provides NCC to BA2) (Fig. 2). The posterior limit of this territory (located between r2 and r3) corresponds to the rostral level of expression of the patterning genes of the *Hox* clusters: meaning that the skeletogenic territory of the head, including the corresponding neural anlage, corresponds to the *Hox*-negative region of the vertebrate embryo (Duboule & Dollé, 1989; Krumlauf, 1993).

Further investigations, carried out in our laboratory during more than two decades, have revealed that this region of the CNC has another role in the development of the vertebrate head since it is essential for the growth and patterning of fore- and midbrain. The initial goal of these investigations was to test the regeneration capacities of the CNC at the premigratory stage and namely, to see if the *Hox*-positive NC could regenerate the rostral, *Hox*-negative one. The answer to this question was negative: the posterior *Hox*-positive region of the CNC did not regenerate the *Hox*-negative adjacent territory and the morphology of the embryos subjected to this operation was largely unexpected: apart from the absence of the facial and cranial skeleton that are normally derived from the NC, the brain of the embryos was severely impacted (Creuzet, Couly, Vincent, & Le Douarin, 2002). The telencephalon did not develop and neither the thalamus nor the optic tectum was formed; the brain remained open, down to the isthmus, due to the lack of growth of the alar plates of the embryonic neural tube at the level of the fore- and midbrain. This was unexpected because the surgery did not involve the neural epithelium of the encephalic vesicles but was strictly restricted to the NC, as shown in Fig. 3.

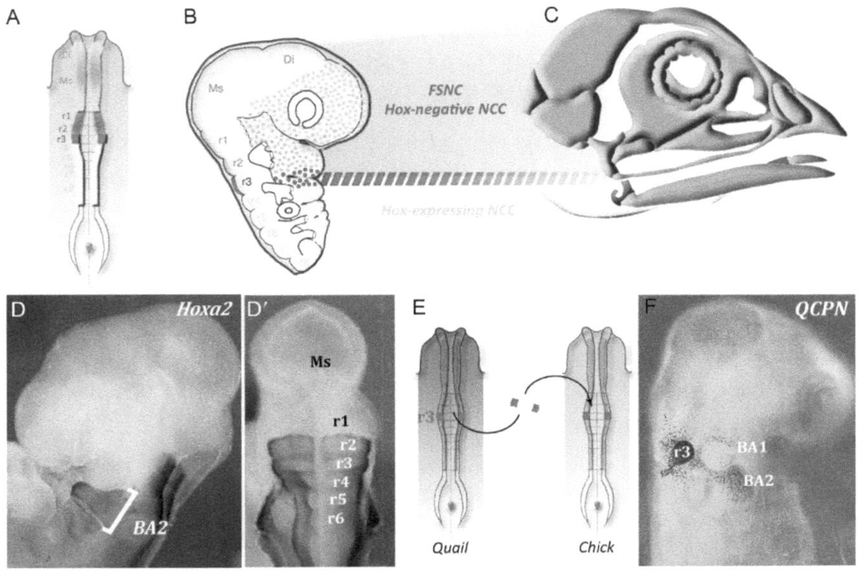

Figure 2 CNC contribution to head skeleton development. Color-coded schematics of premigratory (A) and migratory (B) CNC at the mesencephalic and rhombencephalic axial levels, and their head skeleton derivatives (C). The more rostral *Hox*-negative CNC (down to r2; in green and blue) gives rise to the skull vault and the face, including the nasal capsule, jaws, and entoglossum, whereas more posterior NCC that express *Hox* genes (in yellow) migrate to BA2 and more posterior BAs to produce the hypobranchial skeleton. Between these two domains, r3 NCC (in red) colonize both BA1 and BA2. (D) Whole-mount E3 chick embryos showing that the rostral head and BA1 do not express *Hoxa2*, whereas this gene is abundantly expressed in more posterior BAs (D′). (E) Schematic representation of r3 NC graft from quail to chick at 5ss. (F) One-day postgrafting, quail r3 CNCC (identified using the quail-specific marker QCPN; in brown) has colonized BA1 and BA2 of the host embryo. *Adapted from Le Douarin, Couly, and Creuzet (2012) with permission.* (A color version of this figure is located at the end of this chapter.)

This result uncovered that the development of the head skeleton and that of the brain are strongly associated.

Interestingly, if excision of the rostral part of the *Hox*-negative NC, which we have designated as facial skeletogenic NC (FSNC), is followed by the graft of one-third of its length originating from a stage-matched quail embryo, the development of the head proceeds normally, and the skeletal and mesenchymal structures of the head are made up of quail cells. The result however, is very different when the removed fragment of the FSNC is replaced, partly or completely, by fragments of the NC from the postotic

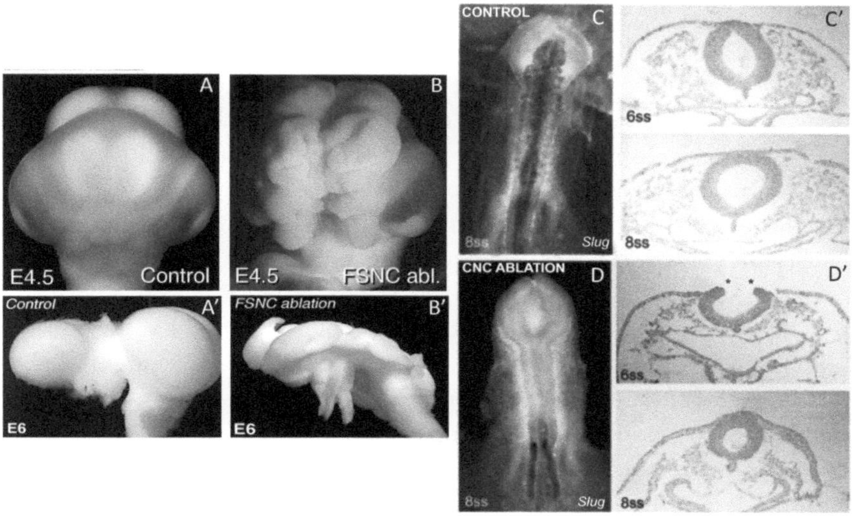

Figure 3 CNC ablation alters head skeleton and brain development. (A and B) Frontal views of E4.5 chick embryos, showing facial defects caused by anterior CNC ablation (B), compared with a control head (A). Whole-mount preparations of the brain of control (A′) and operated (B′) E6 embryos provide evidence that deprivation of NCC led to a wide-open brain with altered partitions of the anterior brain vesicles. The surgical procedure leading to these defects eliminates the migratory NCC that express *Slug* (compare C and D); the NC ablation consists in excision of the premigratory NC without involving the neuroepithelium, as shown on transverse sections of control and operated embryos (C′ and D′, respectively): after resection of the neural folds at 6ss (C′ and D′, upper panels), the neural tube properly closed at 8ss, in both control and operated embryos (C′ and D′, lower panels). (A color version of this figure is located at the end of this chapter.)

Hox-positive level: the morphology of the head was, in these cases, similar to that resulting from the excision of the FSNC alone and was characterized by the absence of skeletal structures and anencephaly (Creuzet, Schuler, Couly, & Le Douarin, 2004).

This series of experiments was completed by the forced expression of *Hoxa2* that normally reaches the more rostral level (Krumlauf, 1993) (see Fig. 2). The morphology of these embryos, examined at embryonic day 3.5 (E3.5), was similar to that resulting from FSNC excision.

The expression of several genes in the embryos subjected to the removal of the skeletogenic CNC was strongly perturbed (Creuzet, Martinez, & Le Douarin, 2006): Shh, whose gene expression is normally restricted to the notochord and floor plate was largely expanded laterally and dorsally, attesting for the ventralization of the basal plates of the neural primordium; in absence of alar plate development, no transcripts of the genes that are normally expressed dorsally in fore- and midbrain, such as *Wnt1* and *Wnt3a*, were present. Another spectacular change in gene expression in these embryos was a strong reduction of the transcripts of *Fgf8* in the so-called "secondary brain organizers": the ANR (anterior neural ridge), and to a lesser extent, the isthmus (also designated MHB for midbrain–hindbrain boundary), and their total absence in BA1 ectoderm. This "ectodermal ridge" of the BA is known to be critical for BA development. In contrast, the production of BMPs in the rostral region of the embryo was neatly increased (Creuzet et al., 2006).

BMPs are known to antagonize the synthesis of FGF8 (Ohkubo, Chiang, & Rubinstein, 2002). The question was then raised as to whether the phenotype of embryos deprived of their FSNC could be rescued by the administration of exogenous FGF8. Heparin acrylic beads impregnated with FGF8 were placed laterally to the neural tube on the rostral region of embryos deprived of their FSNC. The rescue of the embryos was observed provided that r3 was not included in the extirpated NC territory (Creuzet et al., 2004). As mentioned above, the NCC exiting r3 express *Hoxa2* and participate in both BA1 and BA2. The cells colonizing the former lose their *Hox* gene expression after reaching the *Hox*-negative environment of BA1. In contrast those colonizing BA2 maintain *Hoxa2* when they have migrated to the *Hox*-positive environment that prevails at this level. Examination the NCC exiting r3 after the excision of the FSNC in presence and absence of FGF8-beads showed a vigorous proliferation in the case where exogenous FGF8 was provided: the r3-NCC that migrated rostrally and invaded the cephalic area could replace the NCC normally issued from the more rostral levels of the NC. The cells originating from r3 thus demonstrated a strong regenerative capacity that exists only in *Hox* gene-free NCC evolving in a similarly *Hox* gene-free environment.

The conclusion that can be drawn from this first part of this analysis is the fact that the construction of the vertebrate head obeys rules that are different from those governing the rest of the body: *Hox* genes, in this area of the embryo exert a strong inhibitory effect on the mesenchymal derivatives of the NC, which do neither proliferate nor differentiate in the presence of *Hox* gene products.

4.2 How Does the Cephalic Neural Crest Regulate the Production of FGF8 by Secondary Brain Organizers and BA1 Ectoderm?

The CNC is known to be the site of BMP antagonists such as Gremlin and Noggin (Tzahor et al., 2003). We put forward the hypothesis that removal of the NC at early stages might be responsible for an increase in BMP4 and BMP7 production that, in turn, decreases FGF8 synthesis in the secondary brain organizers and BA1 ectoderm (Ohkubo et al., 2002). One way to test this hypothesis was to suppress the production of the anti-BMP factors by RNA interference on the one hand, and to increase it by transducing mRNA of these factors to the premigratory CNC, on the second hand. Both operations were conducted through *in ovo* electroporation associated with GFP reporter transduction into the FSNC. In order to strictly restrict changes in gene expression to the neural folds, without any contamination of the brain anlage, the electroporation was done in the cranial neural folds of quail embryos that were then resected and grafted in stage-matched chicken, of which the same part of the neural folds had been removed. The results of these experiments were striking: increasing the production of the BMP antagonists by forced expression of *gremlin* and *noggin* resulted in a highly significant enlargement of the brain (Creuzet, 2009). In contrast, treatment by RNA interference, inhibiting the production of Gremlin and Noggin proteins, produced the reverse effect and led to a strong reduction in forebrain size, with disorganized interhemispheric structures, absence of olfactory bulbs and nerves and atrophy of the choroid plexus. Knocking-down *gremlin* and *noggin* in the FSNC also severely decreased *Fgf8* expression in the ANR and isthmus and, later on, caused significant microcephaly of the telencephalon, thalamus, and optic tectum (Fig. 4) (Creuzet, 2009).

These results are in line with the view that the CNC plays a key role in the development of the brain by regulating the amount of FGF8 produced by the secondary brain organizers, via the secretion of BMP antagonists: removal of the CNC results in the increase of BMP production in the cephalic region, which decreases the amount of FGF8 that, at these stages, is critical for the development of the telencephalon and the dorsal part of the di- and mesencephalic vesicles.

5. CONCLUDING REMARKS

The contribution of the NC to the vertebrate body has been studied in the first half of the twentieth century, mainly on the amphibian embryo. This body of work was reviewed by Sven Hörstadius in 1950 (Hörstadius, 1950). Up to the 1970s, only a few studies were devoted to this structure in embryos of higher vertebrates. Cell marking techniques that could be applied to the avian embryo, available to experimentation in the egg during the complete period of its development, provided insights into the contribution of this structure, essentially known before to yield the

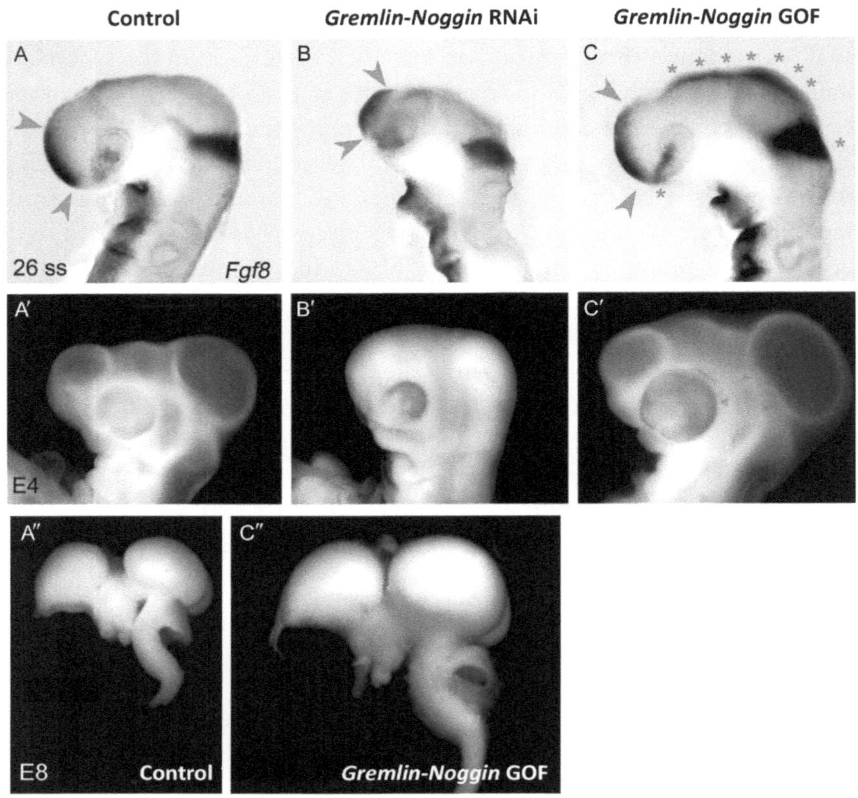

Figure 4 BMP inhibitors in the CNC control *Fgf8* expression and forebrain development. (A–C) Expression of *Fgf8* in the ANR (red arrowheads) at E2.5 (26ss) in control embryos (A); in embryos submitted to *gremlin* and *noggin* knockdown by RNA interference (RNAi), *Fgf8* expression domain is reduced (B); conversely, it is extended after gain-of-function (GOF) for *gremlin* and *noggin* (C) in the NC. Lateral views of embryonic heads at E4 show hypoplasic anterior head after *gremlin* and *noggin* knockdown (B′), whereas the head is enlarged after forced expression of these genes (C′), as compared with the untreated embryo (A′). The size of the brain at E8 is dramatically increased following GOF for *gremlin* and *noggin* in the NC (C″, compared to the control brain in A″). *Adapted from Le Douarin et al. (2012) with permission.* (A color version of this figure is located at the end of this chapter.)

progenitors of the PNS and the melanocytes, and to contribute to the amphibian jaws and skull bones (see Le Douarin, 1982 for references).

The use of tritiated-thymidine to label NCC in the chick embryo was promoted by James Weston (Weston, 1963), but was rapidly abandoned for several reasons, that concern the technique itself, such as its lability and the fact that the radioactive label can be transmitted through leakage to undesired neighboring territories. The method, devised by one of us

(Le Douarin, 1969, 1973), and based on the construction of chimeras between avian embryos of two closely related species, the chick and the Japanese quail, turned out to be particularly suitable to study the fate of the cells endowed with extensive migratory properties such as those of the NC. Its use revealed the paramount contribution of the NCC to the construction of the vertebrate head, the pathways followed by the NCC and the various derivatives arising from the different levels of the neural axis. Quail–chick chimeras of NC territories also showed the large diversity of cell types derived from this highly pluripotent structure. Moreover, Le Douarin and colleagues devised grafting experiments showing the plasticity of the NCC and the influence of the environmental signals on their fate. The results obtained in two decades of work in our and others' laboratories, were summarized in the 1982 and 1999 monographs (Le Douarin, 1982; Le Douarin & Kalcheim, 1999).

The embryological data thus collected, together with other considerations led two authors, Gans and Northcutt (1983), to propose that the NC played a major role in the evolution of the vertebrate phylum by promoting the development of a "new head" including a brain, which was an important asset to the extraordinary radiation of this group of animals. This seminal article triggered the interest of a number of investigators, and, in contrast to the early time of NC investigations, the NC is now a subject of active research (see for example recent monographs devoted to the NC, e.g., Hall, 2009; Nelms & Labosky, 2010; Trainor, 2014).

In the present article, we have chosen three characters of the NC that seemed of interest because they emphasize its originality. The first concerns the evolutionary aspect related to the origin of this vertebrate innovation, on which the work by Carole LaBonne and coworkers shed a new light (Buitrago-Delgado et al., 2015). This work presents a novel and argumented concept according to which the NC is a "neotenic" structure, which has conserved over the gastrulation process the original pluripotency that characterizes the cells of the blastula, thus making the NC a reservoir of "pluripotent embryonic stem cells" in which the potentialities of both ectoderm and mesoderm are still present. This work opens a novel avenue of research on the molecular mechanisms through which the NC progenitors were "arrested" in their evolution while acquiring migratory properties, whereas their neighbors segregated into divergent lineages and developed into *bona fide* ectodermal and mesodermal cells.

The second point discussed in this article concerns the potentialities of the NCC in terms of differentiation. It is clear that NCC possess a considerable array of developmental options, as shown by the large body of clonal investigations carried out by several laboratories during the last three decades. Cells able to yield NC-derived cell types in clonal culture have been found widely distributed in the adult body. Their proliferation and differentiation needs to be under strict control. The rupture of this control is likely to be responsible for the onset of various types of highly metastatic tumors that reflect the diverse potentialities of the NCC and their extensive capacity to migrate, like melanoma neuroblastoma, or Ewing sarcoma (Powell, O'Brien, Ford, & Artinger, 2014).

The last point developed in this chapter discusses an unexpected role of the NC shown to exert a critical influence on the development of the preotic brain: removal of the cephalic NC responsible for the formation of the skull results in the absence of the telencephalon, thalamus, and optic tectum. The development of these structures from the alar plates of the rostral neural plate, turned out to depend upon the production of BMP antagonists such as Gremlin and Noggin by the cephalic NC. The increase of BMP signaling resulting from the NC excision decreases expression of *Fgf8* by the secondary brain organizers, particularly in the ANR. A new important role of the NC as a paracrine regulator of early brain development is thus demonstrated.

ACKNOWLEDGMENTS

Research in authors' laboratory was supported by Centre National de la Recherche Scientifique (CNRS), Institut National de la Santé et de la Recherche Médicale (INSERM), Institut National du Cancer (INCa), and Fondation Bettencourt-Schueller. The authors thank Michèle Scaglia for help in the preparation of chapter and bibliography.

REFERENCES

Abitua, P. B., Wagner, E., Navarrete, I. A., & Levine, M. (2012). Identification of a rudimentary neural crest in a non-vertebrate chordate. *Nature*, *492*(427), 104–107.

Achilleos, A., & Trainor, P. A. (2012). Neural crest stem cells: Discovery, properties and potential for therapy. *Cell Research*, *22*(2), 288–304.

Adameyko, I., Lallemend, F., Aquino, J. B., Pereira, J. A., Topilko, P., Müller, T., et al. (2009). Schwann cell precursors from nerve innervation are a cellular origin of melanocytes in skin. *Cell*, *139*(2), 366–379.

Anderson, D. J. (1989). Cellular 'neoteny': A possible developmental basis for chromaffin cell plasticity. *Trends in Genetics*, *5*(6), 174–178.

Ayer-Le Lièvre, C. S., & Le Douarin, N. M. (1982). The early development of cranial sensory ganglia and the potentialities of their component cells studied in quail-chick chimeras. *Developmental Biology*, *94*(2), 291–310.

Baggiolini, A., Varum, S., Mateos, J. M., Bettosini, D., John, N., Bonalli, M., et al. (2015). Premigratory and migratory neural crest cells are multipotent *in vivo*. *Cell Stem Cell*, *16*(3), 314–322.

Baroffio, A., Dupin, E., & Le Douarin, N. M. (1988). Clone-forming ability and differentiation potential of migratory crest cells. *Proceedings of the National Academy of Sciences of the United States of America*, *85*(14), 5325–5329.

Baroffio, A., Dupin, E., & Le Douarin, N. M. (1991). Common precursors for neural and mesectodermal derivatives in the cephalic neural crest. *Development*, *112*(1), 301–305.

Baynash, A. G., Hosoda, K., Giaid, A., Richardson, J. A., Emoto, N., Hammer, R. E., et al. (1994). Interaction of endothelin-3 with endothelin-B receptor is essential for development of epidermal melanocytes and enteric neurons. *Cell*, *79*(7), 1277–1285.

Bittencourt, D. A., da Costa, M. C., Calloni, G. W., Alvarez-Silva, M., & Trentin, A. G. (2013). Fibroblast growth factor 2 promotes the self-renewal of bipotent glial smooth muscle neural crest progenitors. *Stem Cells and Development*, *384*(1), 13–25.

Bronner-Fraser, M., & Fraser, S. E. (1988). Cell lineage analysis reveals multipotency of some avian neural crest cells. *Nature*, *335*(186), 161–164.

Bronner-Fraser, M., & Fraser, S. (1989). Developmental potential of avian trunk neural crest cells *in situ*. *Neuron*, *3*(6), 755–766.

Buitrago-Delgado, E., Nordin, K., Rao, A., Geary, L., & LaBonne, C. (2015). Shared regulatory programs suggest retention of blastula-stage potential in neural crest cells. *Science*, *348*(6241), 1332–1335.

Calloni, G. W., Glavieux-Pardanaud, C., Le Douarin, N. M., & Dupin, E. (2007). Sonic Hedgehog promotes the development of multipotent neural crest progenitors endowed with both mesenchymal and neural potentials. *Proceedings of the National Academy of Sciences of the United States of America*, *104*(50), 19879–19884.

Calloni, G. W., Le Douarin, N. M., & Dupin, E. (2009). High frequency of cephalic neural crest cells shows coexistence of neurogenic, melanogenic, and osteogenic differentiation capacities. *Proceedings of the National Academy of Sciences of the United States of America*, *106*(22), 8947–8952.

Cohen, A. M., & Königsberg, I. R. (1975). A clonal approach to the problem of neural crest determination. *Developmental Biology*, *46*(2), 263–280.

Couly, G. F., Coltey, P. M., & Le Douarin, N. M. (1993). The triple origin of skull in higher vertebrates: A study in quail-chick chimeras. *Development*, *117*(2), 409–429.

Creuzet, S. E. (2009). Regulation of pre-otic brain development by the cephalic neural crest. *Proceedings of the National Academy of Sciences of the United States of America*, *106*(37), 15774–15779.

Creuzet, S., Couly, G., Vincent, C., & Le Douarin, N. M. (2002). Negative effect of Hox gene expression on the development of the neural crest-derived facial skeleton. *Development*, *129*(18), 4301–4313.

Creuzet, S. E., Martinez, S., & Le Douarin, N. M. (2006). The cephalic neural crest exerts a critical effect on forebrain and midbrain development. *Proceedings of the National Academy of Sciences of the United States of America*, *103*(38), 14033–14038.

Creuzet, S., Schuler, B., Couly, G., & Le Douarin, N. M. (2004). Reciprocal relationships between Fgf8 and neural crest cells in facial and forebrain development. *Proceedings of the National Academy of Sciences of the United States of America*, *101*(14), 4843–4847.

Delsuc, F., Brinkmann, H., Chourrout, D., & Philippe, H. (2006). Tunicates and not cephalochordates are the closest living relatives of vertebrates. *Nature*, *439*(7079), 965–968.

Duboule, D., & Dollé, P. (1989). The structural and functional organization of the murine HOX gene family resembles that of *Drosophila* homeotic genes. *EMBO Journal*, *8*(5), 1497–1505.

Dupin, E. (1984). Cell division in the ciliary ganglion of quail embryos in situ and after back-transplantation into the neural crest migration pathways of chick embryos. *Developmental Biology*, *105*(2), 288–299.

Dupin, E., Calloni, G. W., & Le Douarin, N. M. (2010). The cephalic neural crest of amniote vertebrates is composed of a large majority of precursors endowed with neural, melanocytic, chondrogenic and osteogenic potentialities. *Cell Cycle*, *9*(2), 238–249.

Dupin, E., & Coelho-Aguiar, J. M. (2013). Isolation and differentiation properties of neural crest stem cells. *Cytometry Part A*, *83*(1), 38–47.

Dupin, E., & Le Douarin, N. M. (1995). Retinoic acid promotes the differentiation of adrenergic cells and melanocytes in quail neural crest cultures. *Developmental Biology*, *168*(2), 529–548.

Dupin, E., & Le Douarin, N. M. (2014). The neural crest, a multifaceted structure of the vertebrates. *Birth Defects Research Part C, Embryo Today: Reviews*, *102*(3), 187–209.

Dupin, E., & Sommer, L. (2012). Neural crest progenitors and stem cells: From early development to adulthood. *Developmental Biology*, *366*(1), 83–95.

Ernfors, P. (2010). Cellular origin and developmental mechanisms during the formation of skin melanocytes. *Experimental Cell Research*, *316*(8), 1397–1407.

Gans, C., & Northcutt, G. (1983). Neural crest and the origin of vertebrates. A new head. *Science*, *200*(6), 268–273.

Gershon, M. D. (1999). Endothelin and the development of the enteric nervous system. *Clinical and Experimental Pharmacology & Physiology*, *26*(12), 985–988.

Hall, B. K. (2000). The neural crest as a fourth germ layer and vertebrates as quadroblastic not triploblastic. *Evolution & Development*, *2*(1), 3–5.

Hall, B. K. (2009). *The neural crest and neural crest cells in vertebrate development and évolution*. New York: Springer.

Heanue, T. A., & Pachnis, V. (2007). Enteric nervous system development and Hirschsprung's disease: Advances in genetic and stem cell studies. *Nature Reviews. Neuroscience*, *8*(6), 466–479.

His, W. (1868). *Untersuchungen über die erste Anlage des Wirbeltierleibes. Die erste Entwicklung des Hühnchens im Ei*. Leipzig: F. C. W. Vogel.

Hörstadius, S. (1950). *The neural crest: Its properties and derivatives in the light of experimental research*. Oxford: Oxford University Press.

Hosoda, K., Hammer, R. E., Richardson, J. A., Baynash, A. G., Cheung, J. C., Giaid, A., et al. (1994). Targeted and natural (piebald-lethal) mutations of endothelin-B receptor gene produce megacolon associated with spotted coat color in mice. *Cell*, *79*(7), 1267–1276.

Ito, K., Morita, T., & Sieber-Blum, M. (1993). *In vitro* clonal analysis of mouse neural crest development. *Developmental Biology*, *157*(2), 517–525.

Kaltschmidt, B., Kaltschmidt, C., & Widera, D. (2012). Adult craniofacial stem cells: Sources and relation to the neural crest. *Stem Cell Review*, *8*(3), 658–671.

Kléber, M., Lee, H. Y., Wurdak, H., Buchstaller, J., Riccomagno, M. M., Ittner, L. M., et al. (2005). Neural crest stem cell maintenance by combinatorial Wnt and BMP signaling. *Journal of Cell Biology*, *169*(2), 309–320.

Krispin, S., Nitzan, E., Kassem, Y., & Kalcheim, C. (2010). Evidence for a dynamic spatiotemporal fate map and early fate restrictions of premigratory avianneural crest. *Development*, *137*(4), 585–595.

Krumlauf, R. (1993). *Hox* genes and pattern formation in the branchial region of the vertebrate head. *Trends in Genetics*, *9*(4), 106–112.

Lahav, R., Dupin, E., Lecoin, L., Glavieux, C., Champeval, D., Ziller, C., et al. (1998). Endothelin 3 selectively promotes survival and proliferation of neural crest-derived glial and melanocytic precursors. *Proceedings of the National Academy of Sciences of the United States of America*, *95*(24), 14214–14219.

Le Douarin, N. (1969). Details of the interphase nucleus in Japanese quail (*Coturnix coturnix japonica*). *Bulletin Biologique de la France et de la Belgique*, *103*(3), 435–452.

Le Douarin, N. (1973). A biological cell labeling technique and its use in experimental embryology. *Developmental Biology, 30*(1), 217–222.
Le Douarin, N. (1982). *The neural crest.* Cambridge: Cambridge University Press.
Le Douarin, N. M., Calloni, G. W., & Dupin, E. (2008). The stem cells of the neural crest. *Cell Cycle,* 7(8), 1013–1019.
Le Douarin, N. M., Couly, G., & Creuzet, S. E. (2012). The neural crest is a powerful regulator of pre-otic brain development. *Developmental Biology, 366*(1), 74–82.
Le Douarin, N. M., & Dupin, E. (2003). Multipotentiality of the neural crest. *Current Opinion in Genetics and Development, 13*(5), 529–536.
Le Douarin, N. M., & Dupin, E. (2012). The neural crest in vertebrate evolution. *Current Opinion in Genetics & Development, 22*(4), 381–389.
Le Douarin, N. M., & Dupin, E. (2014). The neural crest, a fourth germ layer of the vertebrate embryo: Significance in chordate evolution. In P. Trainor (Ed.), *Evolution, development and disease. Neural crest cells* (1st ed., pp. 3–26). London: Elsevier/Academic Press. Chapter 1.
Le Douarin, N. M., & Kalcheim, C. (1999). *The neural crest* (2nd ed.). Cambridge: Cambridge University Press.
Le Lièvre, C. (1974). Role of mesectodermal cells arising from the cephalic neural crest in the formation of the branchial arches and visceral skeleton. *Journal of Embryology and Experimental Morphology, 31*(2), 453–477.
Le Lièvre, C. S., & Le Douarin, N. M. (1975). Mesenchymal derivatives of the neural crest: Analysis of chimeric quail and chick embryos. *Journal of Embryology and Experimental Morphology, 34*(1), 125–154.
Metcalf, D. (2007). Concise review: Hematopoietic stem cells and tissue stem cells: Current concepts and unanswered questions. *Stem Cells, 25*(10), 2390–2395.
Meulemans, D., & Bronner-Fraser, M. (2004). Gene-regulatory interactions in neural crest evolution and development. *Developmental Cell,* 7(3), 291–299.
Morrison, S. J., Perez, S. E., Qiao, Z., Verdi, J. M., Hicks, C., Weinmaster, G., et al. (2000). Transient notch activation initiates an irreversible switch from neurogenesis to gliogenesis by neural crest stem cells. *Cell, 101*(5), 499–510.
Morrison, S. J., White, P. M., Zock, C., & Anderson, D. J. (1999). Prospective identification, isolation by flow cytometry, and *in vivo* self-renewal of multipotent mammalian neural crest stem cells. *Cell, 96*(5), 737–739.
Nelms, B. L., & Labosky, P. A. (2010). *Transcriptional control of neural crest development.* Vanderbilt University San Rafael (CA): Morgan & Claypool Life Sciences.
Nitzan, E., Krispin, S., Pfaltzgraff, E. R., Klar, A., Labosky, P. A., & Kalcheim, C. (2013). A dynamic code of dorsal neural tube genes regulates the segregation between neurogenic and melanogenic neural crest cells. *Development, 140*(11), 2269–2279.
Ohkubo, Y., Chiang, C., & Rubinstein, J. L. (2002). Coordinate regulation and synergistic actions of BMP4, SHH and FGF8 in the rostral prosencephalon regulate morphogenesis of the telencephalic and optic vesicles. *Neuroscience, 111*(1), 1–17.
Pani, A. M., Mullarkey, E. E., Aronowicz, J., Assimacopoulos, S., Grove, E. A., & Lowe, C. J. (2012). Ancient deuterostome origins of vertebrate brain signalling centres. *Nature, 483*(7389), 289–294.
Powell, D. R., O'Brien, J. H., Ford, H. L., & Artinger, K. B. (2014). Neural crest cells and cancer: Insights into tumor progression. In P. Trainor (Ed.), *Evolution, development and disease. Neural crest cells* (1st ed., pp. 335–357). London: Elsevier/Academic Press. Chapter 16.
Saldana-Caboverde, A., & Kos, L. (2010). Roles of endothelin signaling in melanocyte development and melanoma. *Pigment Cell & Melanoma Research, 23*(2), 160–170.
Schuchardt, A., D'Agati, V., Larsson-Blomberg, L., Costantini, F., & Pachnis, V. (1994). Defects in the kidney and enteric nervous system of mice lacking the tyrosine kinase receptor Ret. *Nature, 367*(6461), 380–383.

Schweizer, G., Ayer-Le Lièvre, C., & Le Douarin, N. M. (1983). Restrictions of developmental capacities in the dorsal root ganglia during the course of development. *Cell Differentiation*, *13*(3), 191–200.
Shah, N. M., & Anderson, D. J. (1997). Integration of multiple instructive cues by neural crest stem cells reveals cell-intrinsic biases in relative growth factor responsiveness. *Proceedings of the National Academy of Sciences of the United States of America*, *94*(21), 11369–11374.
Shah, N. M., Groves, A. K., & Anderson, D. J. (1996). Alternative neural crest cell fates are instructively promoted by TGFbeta superfamily members. *Cell*, *85*(3), 331–343.
Shah, N. M., Marchionni, M. A., Isaacs, I., Stroobant, P., & Anderson, D. J. (1994). Glial growth factor restricts mammalian neural crest stem cells to a glial fate. *Cell*, *77*(3), 349–360.
Sieber-Blum, M. (1989). Commitment of neural crest cells to the sensory neuron lineage. *Science*, *243*(4898), 1608–1611.
Sieber-Blum, M. (1991). Role of the neurotrophic factors BDNF and NGF in the commitment of pluripotent neural crest cells. *Neuron*, *6*(6), 949–955.
Sieber-Blum, M., & Cohen, A. M. (1980). Clonal analysis of quail neural crest cells: They are pluripotent and differentiate in vitro in the absence of noncrest cells. *Developmental Biology*, *80*(1), 96–106.
Simões-Costa, M., & Bronner, M. E. (2015). Establishing neural crest identity: A gene regulatory recipe. *Development*, *142*(2), 242–257.
Stemple, D. L., & Anderson, D. J. (1992). Isolation of a stem cell for neurons and glia from the mammalian neural crest. *Cell*, *71*(6), 973–985.
Trainor, P. A. (Ed.), (2014). *Neural crest cells evolution, development and disease.* (1st ed.). San Diego: Academic Press.
Trentin, A., Glavieux-Pardanaud, C., Le Douarin, N. M., & Dupin, E. (2004). Self-renewal capacity is a widespread property of various types of neural crest precursor cells. *Proceedings of the National Academy of Sciences of the United States of America*, *101*(13), 4495–4500.
Tzahor, E., Kempf, H., Mootoosamy, R. C., Poon, A. C., Abzhanov, A., Tabin, C. J., et al. (2003). Antagonists of Wnt and BMP signaling promote the formation of vertebrate head muscle. *Genes & Development*, *17*(24), 3087–3099.
Weston, J. A. (1963). A radioautographic analysis of the migration and localization of trunk neural crest cells in the chick. *Developmental Biology*, *6*, 279–310.

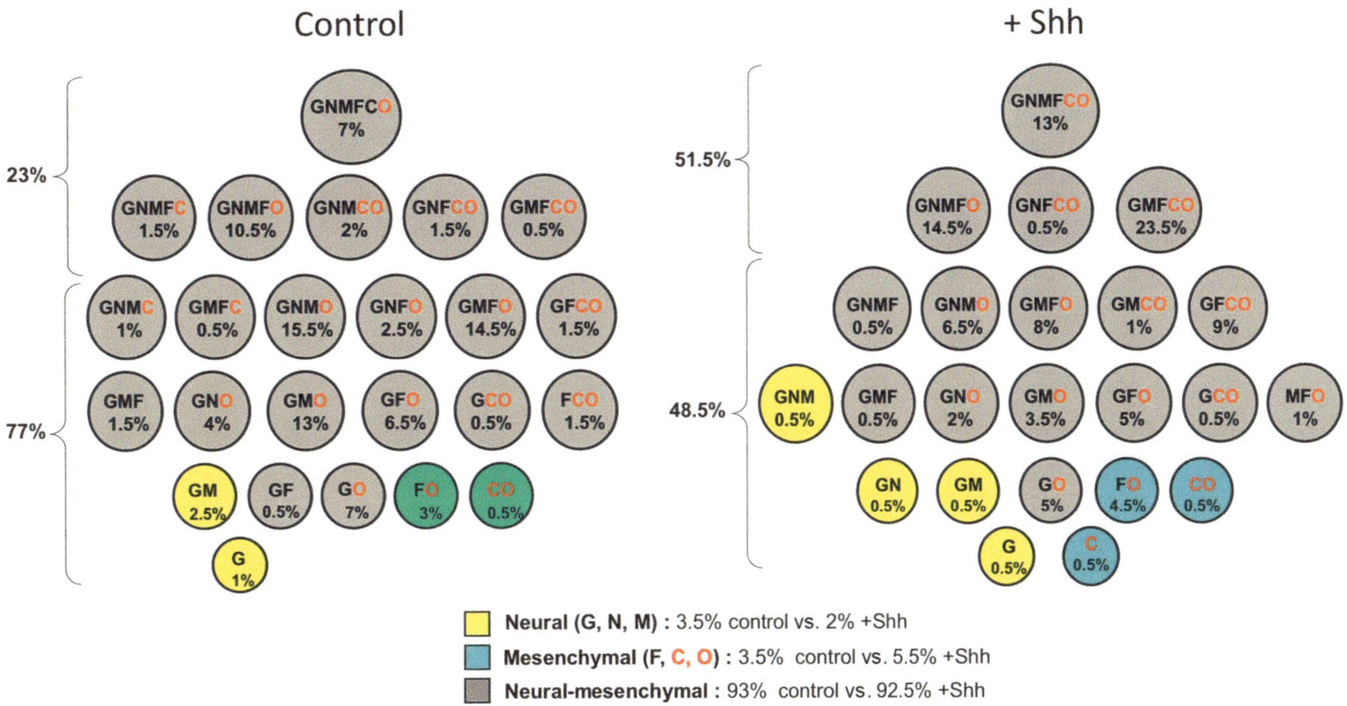

Nicole M. Le Douarin and Elisabeth Dupin, Figure 1 Schematic lineage tree of avian CNCC progenitors and influence of Shh. Quail NCC were isolated from the NCC migratory outgrowth of explanted mesencephalic neural primordium and then cultured as single cells in medium conditions permissive for a diversified cell differentiation. Colonies were phenotypically analyzed after growth for 10 days either in the presence of Shh (109 clones; 52% of initially plated cells) or in control medium (146 clones; 65% of the plated cells). Identified progenitors are classified according to the number of distinct differentiated cell types in their progeny (G, glial cells; N, neurons; M, melanocytes; F, myofibroblasts; C, chondrocytes; O, osteocytes) and their respective frequencies are indicated (% of total clones). In both medium conditions, the vast majority (93%) of CNCC is multipotent and endowed with both neural-melanocytic and mesenchymal (including skeletogenic) potentials (in gray); the frequencies of "only neural" (in yellow) and "only mesenchymal" (in blue), progenitors, were very low. Addition of Shh led to a twofold increase of the proportion of the most multipotent progenitors (with six and five distinct cell types) (accounting for 51.5% vs. 23% of the clones in Shh-treated and untreated cultures, respectively). Shh treatment also promoted the *in vitro* development of progenitors able to produce chondrocytes, without changing the overall frequency of osteogenic multipotent cells. *Adapted from Dupin, Calloni, and Le Douarin (2010).*

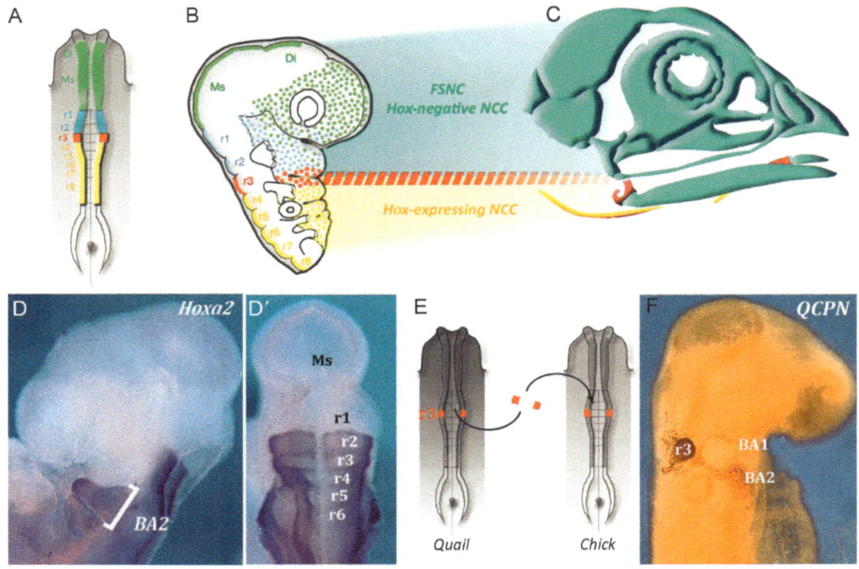

Nicole M. Le Douarin and Elisabeth Dupin, Figure 2 CNC contribution to head skeleton development. Color-coded schematics of premigratory (A) and migratory (B) CNC at the mesencephalic and rhombencephalic axial levels, and their head skeleton derivatives (C). The more rostral *Hox*-negative CNC (down to r2; in green and blue) gives rise to the skull vault and the face, including the nasal capsule, jaws, and entoglossum, whereas more posterior NCC that express *Hox* genes (in yellow) migrate to BA2 and more posterior BAs to produce the hypobranchial skeleton. Between these two domains, r3 NCC (in red) colonize both BA1 and BA2. (D) Whole-mount E3 chick embryos showing that the rostral head and BA1 do not express *Hoxa2*, whereas this gene is abundantly expressed in more posterior BAs (D′). (E) Schematic representation of r3 NC graft from quail to chick at 5ss. (F) One-day postgrafting, quail r3 CNCC (identified using the quail-specific marker QCPN; in brown) has colonized BA1 and BA2 of the host embryo. *Adapted from Le Douarin, Couly, and Creuzet (2012) with permission.*

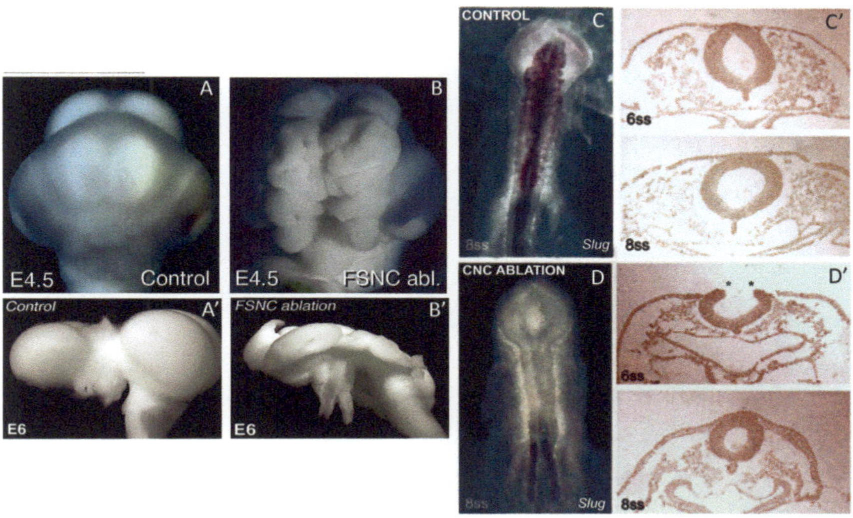

Nicole M. Le Douarin and Elisabeth Dupin, Figure 3 CNC ablation alters head skeleton and brain development. (A and B) Frontal views of E4.5 chick embryos, showing facial defects caused by anterior CNC ablation (B), compared with a control head (A). Whole-mount preparations of the brain of control (A′) and operated (B′) E6 embryos provide evidence that deprivation of NCC led to a wide-open brain with altered partitions of the anterior brain vesicles. The surgical procedure leading to these defects eliminates the migratory NCC that express *Slug* (compare C and D); the NC ablation consists in excision of the premigratory NC without involving the neuroepithelium, as shown on transverse sections of control and operated embryos (C′ and D′, respectively): after resection of the neural folds at 6ss (C′ and D′, upper panels), the neural tube properly closed at 8ss, in both control and operated embryos (C′ and D′, lower panels).

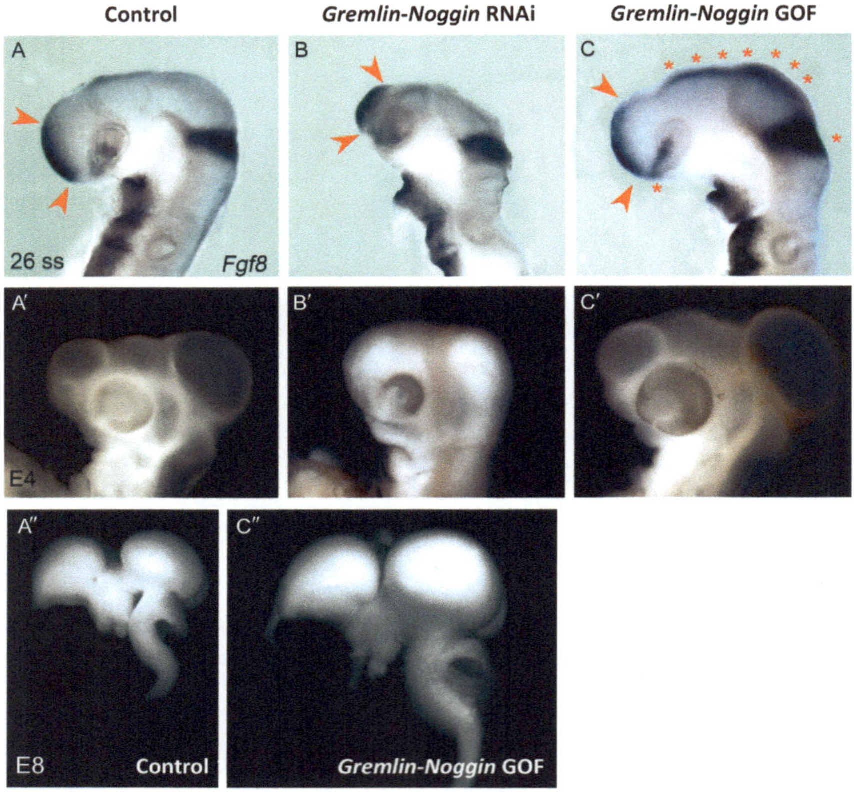

Nicole M. Le Douarin and Elisabeth Dupin, Figure 4 BMP inhibitors in the CNC control *Fgf8* expression and forebrain development. (A–C) Expression of *Fgf8* in the ANR (red arrowheads) at E2.5 (26ss) in control embryos (A); in embryos submitted to *gremlin* and *noggin* knockdown by RNA interference (RNAi), *Fgf8* expression domain is reduced (B); conversely, it is extended after gain-of-function (GOF) for *gremlin* and *noggin* (C) in the NC. Lateral views of embryonic heads at E4 show hypoplasic anterior head after *gremlin* and *noggin* knockdown (B′), whereas the head is enlarged after forced expression of these genes (C′), as compared with the untreated embryo (A′). The size of the brain at E8 is dramatically increased following GOF for *gremlin* and *noggin* in the NC (C″, compared to the control brain in A″). *Adapted from Le Douarin et al. (2012) with permission.*

CHAPTER THIRTY-NINE

Germ Plasm Biogenesis—An Oskar-Centric Perspective

Ruth Lehmann[1]
Howard Hughes Medical Institute (HHMI), Department of Cell Biology, Kimmel Center for Biology and Medicine of the Skirball Institute, New York University School of Medicine, New York, USA
[1]Corresponding author: e-mail address: ruth.lehmann@med.nyu.edu

Contents

Abstract

Germ granules are the hallmark of all germ cells. These membrane-less, electron-dense structures were first observed over 100 years ago. Today, their role in regulating and processing transcripts critical for the establishment, maintenance, and protection of germ cells is well established, and pathways outlining the biochemical mechanisms and physical properties associated with their biogenesis are emerging.

In all sexually reproducing organisms, germ cells, the cells that form the next generation, are marked by the inheritance of specialized RNA particles, collectively referred to as germ granules. In some organisms, like *Drosophila*, *C. elegans*, *Xenopus*, or zebrafish, germ granule components are synthesized during oogenesis and assemble into granules that are passed onto the egg and embryo. As the embryo starts to develop, those cells that inherit the

Current Topics in Developmental Biology, Volume 116
ISSN 0070-2153
http://dx.doi.org/10.1016/bs.ctdb.2015.11.024

maternally derived germ granules will develop as germ cells. In other organisms, like mice or humans, which lack such maternally inherited germ cell determinants, germ cells are specified by cell–cell signaling and inductive events. Despite these differences, some of the same, conserved germ granule components are expressed in all germ cells. This has led to the hypothesis that the organization and function of germ granules is intimately related to germ cell fate.

In contrast to evolutionarily conserved, transcription-based mechanisms that control cell and tissue specification of somatic cell types, no "transcriptional master regulators of germ cell fate" have been identified. Instead, the conserved components of germ cells include: *structural, scaffolding proteins*, such as the Tudor class of proteins; *RNA-binding and processing proteins*, such as the ATP-dependent helicase Vasa and the Piwi class of Argonaute proteins that bind and processes small RNAs; and *translational regulators*, like the conserved Zn finger proteins Nanos and Dazl (Voronina, Seydoux, Sassone-Corsi, & Nagamori, 2011). Furthermore, these components are assembled into larger ribonucleoprotein (RNP) particles with strikingly similar electron-dense appearance suggesting that not just the molecules *per se,* but their intracellular organization into large protein machines plays a critical role in all aspects of germ cell biology. Germ granules are present throughout the germ line cycle, where they are found in different cellular locations and have varied molecular composition (Voronina et al., 2011). In *Drosophila*, at least four different types of germ granules have been described and associated with distinct functions in germ line development (Fig. 1): The *Balbiani body* (likely analogous to the *mitochondrial cloud* found in early frog oocytes) is found in the early, growing oocyte, where it is closely associated with mitochondria, Golgi, centrioles, and ER-rich fusome material (Cox & Spradling, 2003). It has been functionally linked to oocyte specification and initial microtubule organization. *Nuage* is associated with nuclear pores on the cytoplasmic side of the nurse cell nucleus and is the hub for processing small, piwi-interacting (pi) RNAs in defense of transposable elements (Pek, Patil, & Kai, 2012). *Sponge bodies* are found in the nurse cells and the oocyte. They are large, ER-rich structures that lack ribosomes but contain many components that are also found in P bodies (Wilsch-Brauninger, Schwarz, & Nusslein-Volhard, 1997). Consistent with a role in RNA storage, sponge body architecture is highly dynamic depending on environmental conditions (Snee & Macdonald, 2009). *Polar granules* make up the germ plasm at the posterior pole of the mature egg and early embryo. They contain mRNA transcripts as well as piRNAs required to establish, maintain, and protect the germ line of the next generation. Consistent with their role in primordial germ cell identity and function, polar granules are associated with ribosomes and mitochondria (Illmensee, Mahowald, & Loomis, 1976). Despite their morphological resemblance as membrane-less, RNA-rich granules and the identification of granule specific and shared components, it remains unclear how the structure of different granule types relates to their function in germ cell biology. Recent results from genetic and molecular analysis as well as structural and biophysical studies are beginning to shed new light on these.

In this review, I will focus on the germ plasm and germ granule biology of one species, *Drosophila melanogaster*. Germ cells of *Drosophila* have been studied for over a century and pioneered the study of germ plasm. Early morphological studies have been complemented by functional and molecular analysis and can now be connected to the striking structural properties of germ granules.

This essay summarizes some of the original findings that led to the identification of germ plasm components (see also excellent reviews on the topic written by Tony Mahowald one of the early leaders of germ plasm analysis (Mahowald, 2001), and more recently by Gao and Arkov (2013)) and will highlight new observations that have shaped our cellular, molecular, and structural understanding of the process of germ granule biogenesis under the control of the germ plasm determinant Oskar.

1. GENETIC ANALYSIS OF GENES REQUIRED FOR GERM PLASM ASSEMBLY

The first mutants in germ plasm assembly were identified because of their abdominal segmentation defects in screens for maternal effect genes required for embryonic pattern formation (Boswell & Mahowald, 1985; Lehmann & Nusslein-Volhard, 1986; Schupbach & Wieschaus, 1989). Subsequently, it was found that such mutant embryos also lack germ plasm and germ cells. The genetics agreed well with transplantation experiments that showed that germ plasm, when transplanted to an ectopic site, is not only able to induce germ cell formation but can also generate a second abdomen at the new location (Illmensee & Mahowald, 1974; Sander, 1975). Thus, information about germ cells and posterior embryonic patterning are contained in the germ plasm. Maternal effect genes with a similar role in embryonic abdomen formation were collectively referred to as "posterior group genes." Most posterior group genes affect germ plasm and germ cell formation as well as abdominal development. Only two posterior group genes, *nanos* and *pumilio,* specifically affect abdominal development by repressing the translation of maternal *hunchback* in the future abdominal region, thereby allowing the abdomen to form (Barker, Wang, Moore, Dickinson, & Lehmann, 1992; Irish, Lehmann, & Akam, 1989; Lehmann & Nusslein-Volhard, 1991; Struhl, Johnston, & Lawrence, 1992). While *nos* and *pum* are not directly involved in germ cell formation, they do play an important role in primordial germ cell specification and development (Asaoka-Taguchi, Yamada, Nakamura, Hanyu, & Kobayashi, 1999; Kobayashi, Yamada, Asaoka, & Kitamura, 1996).

Although phenotypically the posterior group genes are very similar, molecular and genetic analysis revealed a key role for *oskar (osk)* in germ plasm organization. First, the pattern of *osk* RNA distribution during oogenesis foreshadows events that lead to germ plasm biogenesis (Ephrussi, Dickinson, & Lehmann, 1991; Kim-Ha, Smith, & Macdonald, 1991). Second, mutations in Oskar protein affect the enrichment of other posterior group RNAs and/or proteins at the posterior pole. Finally, mislocalization of *osk* RNA to the anterior pole and expression of Oskar protein at this ectopic location are sufficient to instruct germ plasm assembly, leading to the formation of ectopic germ cells and a second abdomen (Ephrussi & Lehmann, 1992). The latter finding informed the design of genetic epistasis experiments that distinguished between those genes regulating *osk* (upstream genes) and those genes that depend on *osk* for their posterior localization and function (downstream genes; Fig. 3). The "upstream" group includes genes required for the establishment of oocyte polarity, genes involved in the processing and localization of *osk* RNA, and genes that control the translation and stability of Oskar protein. The "downstream" group includes genes that act together with Oskar in germ plasm assembly and also "effector genes" whose products are not involved in germ plasm assembly but are localized to the germ plasm and have diverse functions in primordial germ cell formation, germ cell specification, and migration as well as abdominal patterning (Figs. 2 and 3).

In the following, I will review how oocyte specification and establishment of oocyte polarity lead to the spatial restriction of *osk* RNA, how Oskar protein synthesis is regulated, and how Oskar together with other posterior group genes and effector RNAs assemble into a functional germ plasm that instructs the next generation.

2. OOCYTE SPECIFICATION AND DYNEIN-MEDIATED RNA TRANSPORT

Biogenesis of germ plasm is intimately linked to successive polarizing events leading from the asymmetric division of a germ line stem cell (GSC) to a mature egg that harbors a prepattern of the embryonic axes (Fig. 1A). In *Drosophila*, GSCs reside in a somatic niche, which supports GSC maintenance (Spradling et al., 1997). GSCs divide asymmetrically to self-renew and produce a differentiating daughter. This daughter undergoes four rounds of mitosis to produce a 16-cell cyst. During each division, the sisters do not complete cytokinesis and remain connected by cytoplasmic bridges, the ring canals (Fig. 1B). Through the ring canals weaves the fusome, an endoplasmic reticulum-rich structure associated with cytoskeletal components and microtubules. As a consequence of this division strategy and an apparent asymmetry of fusome material, a bias is created that orients the microtubule network toward one of the cells with four ring canals (Fig. 1C; Roth & Lynch, 2009). This cell remains in meiosis and becomes the oocyte. The other 15 cells develop into polyploid nurse cells that feed RNA, protein, and even mitochondria into the transcriptionally silent oocyte (Fig. 1C′). As part of the Balbiani body, a Microtubule Organizing Center (MTOC) forms in the oocyte. The MTOC nucleates microtubules that direct dynein motor-mediated transport from the nurse cells to the oocyte (Fig. 1D). Mutants in the dynein cargo adapters, Egalitarian and Bicaudal-D, fail to specify an oocyte and all 16 cells take on equal, nurse cell fate (Mach & Lehmann, 1997; Ran, Bopp, & Suter, 1994). Large-scale RNA *in situ* analysis shows that 17% of all germ line expressed genes are specifically enriched in the oocyte during stage 1–7 of oogenesis (Jambor et al., 2015). Among these are many RNAs involved in germ plasm assembly and function including *osk* (Ephrussi et al., 1991; Kim-Ha et al., 1991). The transport of *osk* RNA into the oocyte is mediated by a 67 nt stem loop in the *osk* 3′UTR, the oocyte entry signal (OES), which shares functional similarity with the K10 oocyte localization signal (Figs. 1D and 2B and D; Jambor, Mueller, Bullock, & Ephrussi, 2014; Serano & Cohen, 1995). The OES and the K10 localization signals mediate dynein-dependent, microtubule minus-end-directed transport dependent on a short dsRNA stem. It is likely that these and RNAs with localization signals of similar structure are transported into the oocyte in direct association with Egalitarian (Dienstbier, Boehl, Li, & Bullock, 2009).

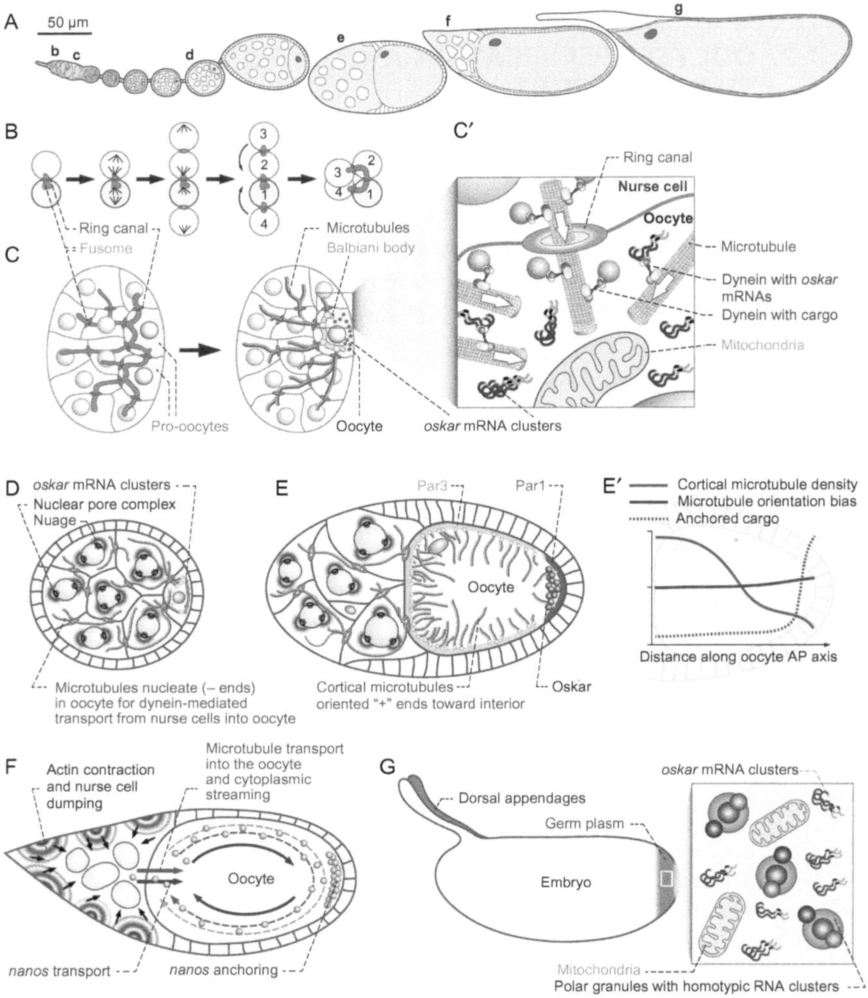

Figure 1 Establishment of anterior–posterior polarity during *Drosophila* oogenesis. (A) Ovariole with developing egg chambers. Each ovary contains 12–16 ovarioles. At the tip (left, b and c regions) is the germarium, where 2–3 germ line stem cells (GSCs) generate a cytoblast that matures into an egg chamber consisting of the oocyte and 15 nurse cells surrounded by somatic follicle cells (stage 1 egg chamber). Daily stem cell division produces a chain of maturing egg chambers (right, g). Small letters on top refer to stages depicted in other figure panels. (B) Division pattern of the cystoblast, the GSC daughter in the germarium. During four synchronous divisions, cells remain connected via ring canals with fusome material traversing the cyst. The first two cells born are connected by four ring canals, the next two by three, the next four by two, and the last eight by one ring canal each. Note that the earlier born cells inherit proportionally more fusome material. (C) 16-cell cyst in germarium, all cells are connected by ring canals

(Continued)

3. OOCYTE POLARIZATION AND KINESIN-MEDIATED TRANSPORT OF *osk* RNA WITHIN THE OOCYTE

Beginning at stage 7, the MTOC is disassembled and replaced by a noncentromeric microtubule network that nucleates from the egg cortex (Fig. 1E). This reorganization of the microtubule network is initiated by a signal sent by the posterior follicle cells (Gonzalez-Reyes, Elliott, & St Johnston, 1995; Roth, Neuman-Silberberg, Barcelo, & Schupbach, 1995). While the nature of this signal is still unknown, it is thought to trigger the targeted degradation of Par-6 and atypical Protein Kinase C complex at

Figure 1—Cont'd and fusomes. The two cells with four ring canals "pro-oocytes" are positioned posteriorly within the cluster and enter into meiosis (left), only one of the two cells remains in meiosis and arrests in meiotic prophase (right). All other cells develop into polyploid nurse cells, which continue to be connected to the oocyte by ring canals. All centrosomes migrate into the oocyte and a microtubule network nucleate from the oocyte to the nurse cells. Dynein-mediated transport brings cargo and RNAs from the nurse cell into the oocyte. Densely packed mitochondria and RNA form the Balbiani body (yellow in C and C′). (C′) Oocyte cytoplasm indicating microtubule polarity and RNA transport particles. (D) Stage 5 egg chamber. Nurse cells and oocyte are surrounded by somatic follicle cells. Continued dynein-mediated RNA transport from the nurse cells into oocyte. Microtubules are oriented to nucleate (− pole) from oocyte and extend into nurse cells. Nuage: RNA-rich granules form around nurse cell nucleus often in close proximity of nuclear pore. (E) Stage 10A egg chamber. Mutually exclusive interactions between Par-1 (blue) at the posterior pole and the Par-3/Bazooka complex (yellow) at the lateral and anterior cortex establish polarity within the oocyte. Microtubules nucleate from lateral and anterior cortex extending into the oocyte and are inhibited at the posterior pole by Par-1. Kinesin motors transport *osk* RNA to the posterior pole, where it is translated. Dynein-mediated transport continues to bring *osk* RNA from the nurse cells to the oocyte. Somatic follicle cells surround the egg chamber. (E′) Summary of microtubule organization and *osk* RNA transport dynamics during stage 10A. Microtubule network is denser at the anterior than posterior of oocyte (green). A slight bias of kinesin-mediated microtubule transport (red line) leads to enrichment of *osk* RNA at posterior pole over time (red stippled). (F) Stage 10B/11, cytoplasmic streaming. The actin meshwork is disassembled in the oocyte and microtubules align in bundles (green). This leads to a streaming process where RNAs and proteins are mixed within the oocyte and additional RNAs/proteins are drawn into the oocyte (green arrow). Effector RNAs like *nanos* (red) become tethered at posterior pole. (G) Mature egg. Germ plasm is assembled and polar granules of similar size to mitochondria are found in a precise region at the posterior pole. Effector RNAs are localized to the granules, Osk, Vasa, Tud, and Aub proteins are essential components of the granules but *osk* RNA is not associated with polar granules (see Fig. 3E). *Panel (C′): After Roth and Lynch (2009). Panel (E′): After Parton et al. (2011).* (A color version of this figure is located at the end of this chapter.)

the posterior pole by Slmb, the SCF ubiquitin ligase (Morais-de-Sa, Mukherjee, Lowe, & St Johnston, 2014). This allows recruitment of Par-1 kinase to the posterior pole excludes Par-3/Bazooka, thereby subdividing the oocyte cortex into an anterior-lateral and a posterior region (Huynh, Petronczki, Knoblich, & St Johnston, 2001; Shulman, Benton, & St Johnston, 2000). As a consequence, microtubules nucleate from the anterior and lateral cortex, with a higher microtubule concentration at the anterior, extending plus ends in all directions. While these microtubules mediate both kinesin (+ pole directed) and dynein-mediated (− pole directed) transport, *osk* travels only by kinesin-dependent transport (Brendza, Serbus, Duffy, & Saxton, 2000; Parton et al., 2011; Zimyanin et al., 2008). Direct live observation of *osk* RNA transport particles and mathematical modeling suggest that microtubule +ends are oriented with a slight bias toward posterior and that, over time, this bias leads to the accumulation of *osk* RNA at the posterior pole (Fig. 1E and F; Khuc Trong, Doerflinger, Dunkel, St Johnston, & Goldstein, 2015; Zimyanin et al., 2008).

An isotropic actin meshwork maintains microtubule organization until stage 10B when this meshwork disassembles and microtubules align in parallel bundles close to the cortex of the oocyte and initiate minus-end-directed, kinesin-dependent ooplasmic streaming (Dahlgaard, Raposo, Niccoli, & St Johnston, 2007; Serbus, Cha, Theurkauf, & Saxton, 2005). Ooplasmic streaming along the microtubule bundles mixes nurse cell and oocyte contents that are not yet anchored in the oocyte (Fig. 1F). Active nurse cell-to-oocyte transport ceases and nurse cells "dump" their contents into the oocyte as the actin cytoskeleton within the nurse cells constricts followed by nurse cell death (Fig. 1G). While *osk* RNA localization at the posterior begins well before ooplasmic streaming and nurse cell dumping, green fluorescent protein (GFP) tagging of *osk* RNA with the MS2/MCP system allowed observation of the localization process live (multiple MS2 RNA stem loops inserted into *osk* RNA were bound by MS2 coat protein fused to a fluorescent protein like green fluorescent protein (GFP)). This revealed a 45% increase in *osk* RNA accumulation in the oocyte after stage 10 (Sinsimer, Jain, Chatterjee, & Gavis, 2011; Snee, Harrison, Yan, & Macdonald, 2007). This suggests that *osk* RNA is continuously produced and transported into the oocyte even after the microtubule network dependent, directed transport into and within the oocyte ceases. Late accumulation and sustained maintenance of *osk* RNA are mediated by the RNA-binding protein Rump and its associated factor Lost, the actin cytoskeleton, and require Oskar protein translation (see below) (Fig. 1G; Babu, Cai, Bahri, Yang, & Chia, 2004; Sinsimer et al., 2011; Suyama, Jenny, Curado, Pellis-van Berkel, & Ephrussi, 2009).

4. *CIS*-ACTING SEQUENCES AND RNA TRANSPORT PARTICLES

Initially all data suggested that the 3′UTR of *osk* was both necessary and sufficient for posterior localization (Kim-Ha, Webster, Smith, & Macdonald, 1993). However, these experiments were carried out in *osk* mutants that lacked protein but produced stable RNA. With the identification of *osk* RNA null mutations, it became clear that additional elements outside the *osk* 3′UTR are needed for posterior localization. These additional elements were found in a "Spliced *Oskar* Localization Element" (SOLE) that forms when sequences proximal and distal to the first intron fuse during splicing (Fig. 2A and B; Hachet & Ephrussi, 2004; Simon, Masiewicz, Ephrussi, & Carlomagno, 2015). The SOLE element (Fig. 2C) associates with the exon junction complex (EJC) components Mago Nashi, Y14 and eIF4AIII at the 5′splice site in the nucleus of the nurse cells, where *osk* is transcribed (Ghosh, Marchand, Gaspar, & Ephrussi, 2012; Hachet & Ephrussi, 2004). Barentz, an EJC component, and the dsRNA-binding protein Staufen join the complex, possibly via the widened grove formed by the A′ helix in SOLE in the nurse cell cytoplasm to form large transport particles (Macchi et al., 2003; Micklem, Adams, Grunert, & St Johnston, 2000; Simon et al., 2015; St Johnston, Beuchle, & Nusslein-Volhard, 1991; van Eeden, Palacios, Petronczki, Weston, & St Johnston, 2001; Fig. 2A and B). In these transport particles, *osk RNA* is thought to self-associate via palindromic sequences, located in the dimerization domain (DD), contained within the distal loop of the OES stem-loop element (Fig. 2D; Jambor, Brunel, & Ephrussi, 2011). RNA–RNA hybrid formation between palindromic loops apparently enables "hitchhiking" of *osk* RNA lacking the SOLE element on RNA molecules with intact SOLE elements. This model provides an explanation why constructs containing only the *osk* 3′UTR localize properly when tested in the background of *protein* null mutations, which produce stable, spliced RNA (Kim-Ha et al., 1993; Munro, Kwon, Schnapp, & St Johnston, 2006). Additional elements besides the DD "kissing loop" are found in other regions of the 3′UTR and play roles in *osk* RNA localization and anchoring (Kim-Ha et al., 1993).

Curiously, *osk* RNA travels in a dynein-dependent manner into the oocyte after which it depends on kinesin motors to get to the posterior pole. Mutants in EJC components, as well as Tropomyosin and Staufen alter *osk* RNA transport bias in the oocyte toward the anterior but do not affect dynein-mediated transport into the oocyte (Erdelyi, Michon, Guichet, Glotzer, & Ephrussi, 1995; Micklem et al., 2000; Zimyanin et al., 2008). Thus, splicing and EJC/Tropomyosin/Staufen association with *osk* RNA define kinesin motor preference of the transport particle. This type of transport seems necessary only within the oocyte and could indeed be a hindrance for efficient dynein-mediated transport from the nurse cells into the oocyte. How such a switch in motor preference is achieved remains to be determined. Since kinesin-directed transport is only observed within the oocyte, it is possible that the "DD-kissing-loop," which is located at the distal tip of the stem that mediates dynein-directed oocyte entry, could "passively" pull RNA associated with EJC/SOLE transport complex into the oocyte. Here, abundant kinesin motors could switch the balance. Whether such regulation occurs and, if so, how other parts of the *osk 3′UTR* known to be required for posterior localization interact with EJC loaded RNA molecules remain to be seen.

5. TRANSLATIONAL CONTROL OF *OSK* RNA

During transport, *osk* RNA is packaged into transport particles that contain EJC components, the RNA-binding proteins Staufen and the translational repressor Bruno. Staufen contains five double-stranded RNA-binding domains, two of which have been implicated in microtubule-dependent transport of *osk* RNA and in translational activation of *osk* RNA upon localization at the posterior pole (Micklem et al., 2000). Bruno is a RRM (RNA recognition motif) protein that binds to multiple regions in the 3′UTR of *osk,* termed Bruno response elements (BREs; Kim-Ha, Kerr, & Macdonald, 1995; Snee, Benz, Jen, & Macdonald, 2008). Bruno represses translation of *osk* during transport and prior to its localization to the posterior pole in a process that involves the 5′cap-binding protein Cup (Nakamura, Sato, & Hanyu-Nakamura, 2004; Wilhelm, Hilton, Amos, & Henzel, 2003). Two mechanisms have been proposed to achieve *osk* translational repression. One mechanism is based on the interaction of Bruno with Cup. In this scenario, initiation of translation is inhibited by Cup competing with the 5′UTR cap-binding protein and translational mediator eIF4E

(Kinkelin, Veith, Grunwald, & Bono, 2012; Wilhelm et al., 2003). The other model proposes that Bruno, independently of Cup, packages oligomerized *osk* RNA molecules into large silencing particles (Chekulaeva, Hentze, & Ephrussi, 2006). Both mechanisms may coexist, the first acting at the level of ribosome engagement, and the second consistent with the large transport particles observed *in vivo*.

Precise control of *osk* translational activation is critical for germ plasm assembly. During transport, *osk* should not be translated prematurely as this can lead to patterning defects (Smith, Wilson, & Macdonald, 1992). Once *osk* RNA becomes localized, it is equally important that this translational repression is released, so sufficient quantities of Oskar protein can be synthesized for germ plasm biogenesis. Indeed, overexpression of the *osk* 3′UTR alone is sufficient to activate translation, suggesting that the concentration of repressors, such as Bruno, is limiting and may be titrated as concentration of *osk* RNA at the pole increases (Kanke & Macdonald, 2015; Smith et al., 1992). Interestingly, proteins involved in *osk* transport and translational repression such as Staufen and Bruno have also been implicated in translational activation (Kim et al., 2015; Micklem et al., 2000). One of the three regions with BRE-binding sites (C region) is required for *osk* translation. A role of Bruno in translational activation may depend on the Bruno phosphorylation state (Reveal et al., 2010; Yoshida, Muller, Wodarz, & Ephrussi, 2004). This scenario is reminiscent of the mechanism that releases translational repression of actin RNA by phosphorylation of zip code-binding protein 1 (ZBP1) at the leading edge of mammalian tissue culture cells (Huttelmaier et al., 2005). Sequences with binding specificity similar to ZBP1, a class of conserved KH (K homology) domain proteins that also includes the insulin growth factor II mRNA-binding protein (IMP), as well as its *Xenopus laevis* homologue, Vg1 RNA and endoplasmic reticulum-associated protein (VERA)/Vg1 RNA-binding protein, have been identified in the *osk* 3′UTR and are required for *osk* transport and translation. While the *Drosophila* IMP protein binds to these sequences, it is apparently not required for their regulation (Munro et al., 2006). In general, *trans*-acting sequences mediating *osk* translation are found in the distal, 3′ end of the 3′UTR, close to the polyadenylation signal. Consistent, proteins involved in polyadenylation such as the cytoplasmic poly-A-binding protein Orb (Castagnetti & Ephrussi, 2003) and Poly-A-binding protein PABP, which binds to A-rich sequences (ARS) are required for *oskar* translation (Fig. 2B; Vazquez-Pianzola, Urlaub, & Suter, 2011).

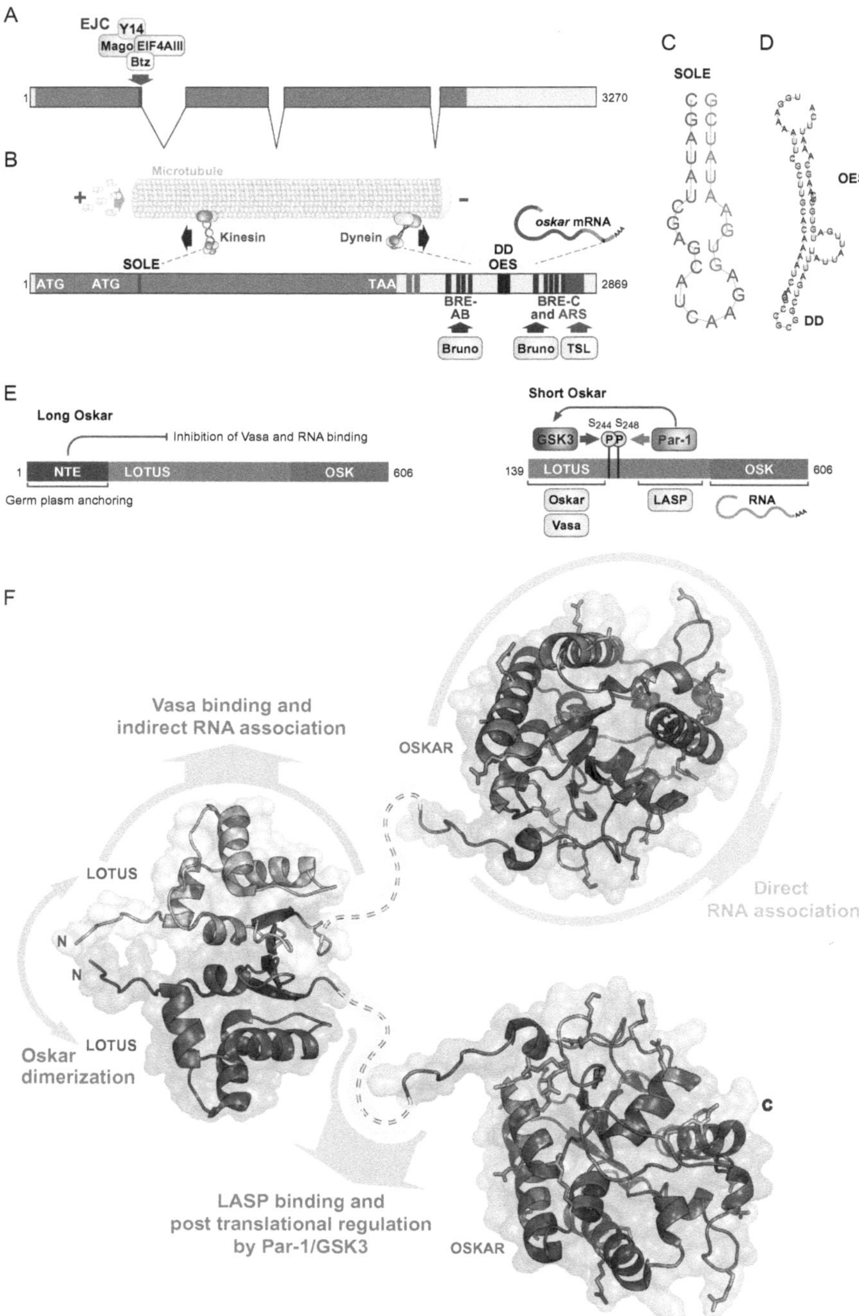

Figure 2 See legend on next page.

6. A NONCODING RNA FUNCTION FOR *OSK*

In addition to its role in Oskar protein synthesis, *oskar* RNA also has a role as a noncoding RNA (Jenny et al., 2006). This noncoding RNA function maps to the C-terminal end of the *oskar* 3′UTR and largely overlaps with Bruno- and ARS-binding sequences (Fig. 2B; Kanke et al., 2015; Ryu & Macdonald, 2015). *osk* RNA null mutants show a strikingly different, much earlier oogenesis phenotype than Oskar protein null mutants that seems unrelated to germ plasm organization. In *osk* RNA null mutants, oogenesis is arrested early and prior to the stage when Oskar protein is first synthesized, and compaction of the oocyte nucleus is defective. Consistent with *osk* RNA sequestering Bruno protein, the phenotype of the noncoding RNA is partially rescued by lowering Bruno levels (Kanke et al., 2015). These results argue against an instructive role of *osk* noncoding RNA in oogenesis. Rather, they favor the idea that binding sites in the *osk* 3′UTR sequester Bruno and possibly other *osk* RNA-binding proteins and thereby

Figure 2—Cont'd Oskar RNA and protein history. (A) Structure of nascent *osk* RNA. Exon junction complex components bind to sequences flanking the first intron, which is critical for kinesin-mediated *osk* RNA transport. (B) *Osk* RNA transport particles are transported along microtubules by dynein (directed toward − end of microtubule) and kinesin (directed toward + end of microtubule) motors. The oocyte entry signal (OES) (D) regulates the dynein-mediated transport of *osk* RNA into the early oocyte. (C) The Spliced Oskar Localization Element (SOLE) forms as a consequence of first intron splicing, where 18 nt from the first and 10 nt from the second exon merge into a stem loop structure. The DD palindromic sequence element (D) is located in the loop of the OES stem loop. Via RNA–RNA "kissing interaction" this element promotes RNA transport particle formation and hitchhiking of RNA lacking the SOLE element on those with the element. A, B, C regions are rich of Bruno regulatory elements (BRE), which repress RNA translation (blue), and IMP-binding elements, which promote translation and localization (red). Binding elements close to the most distal part of the 3′UTR are important for translational activation, via adenine-rich sequences (ARS) that mediate polyadenylation. (E) Structure of Oskar isoforms. Left: Long Oskar is a 606-amino acid protein with an N-terminal extension (NTE) required for actin filament nucleation and enhanced endocytosis at the posterior pole. Long Oskar suppresses short Oskar functions. Right: Short Oskar starts at Methionine 139. The Lotus domain dimerizes and interacts with Vasa (see also F), the C-terminal OSK domain resembles a SGNH hydrolase domain; however, the enzymatic amino acids are missing. Instead, the OSK domain is thought to interact with effector RNAs. (F) Structure diagram of Lotus and OSK domain. The Lotus domain forms a homodimer of two winged helix structures and interacts with Vasa, while the OSK domain forms a globular structure with basic and hydrophilic amino acids exposed to the surface, which are thought to interact with RNA. (A color version of this figure is located at the end of this chapter.)

prevent inappropriate binding of these factors to off-target sites in RNAs with roles in early oogenesis.

7. OSKAR PROTEIN STRUCTURE AND FUNCTION

Once *osk* RNA reaches the posterior pole, it is translated into two protein isoforms: Long Oskar, which is translated at the first Methionine codon and encodes a protein of 606 amino acids (Fig. 2E); and short Oskar which is initiated at the second Methionine codon and encodes a protein of 467 amino acids (Fig. 2F; Markussen, Michon, Breitwieser, & Ephrussi, 1995; Rongo, Gavis, & Lehmann, 1995). Despite one being an N-terminal extension (NTE) of the other, the two proteins have distinct localization patterns and strikingly different functions. Long Oskar localizes to the oocyte and embryo posterior cortex and is associated with endosomes, while short Oskar is an integral part of the polar granules (Tanaka & Nakamura, 2008; Vanzo, Oprins, Xanthakis, Ephrussi, & Rabouille, 2007). Long Oskar has been implicated in the formation of an extended actin meshwork at the posterior pole and implicated in increased endocytic activity (Tanaka, Kato, Matsuda, Hanyu-Nakamura, & Nakamura, 2011). While interaction partners specific to the NTE of Long Oskar have not been identified, loss of long Oskar causes germ plasm to be less tightly associated with the posterior pole (Rongo et al., 1997; Vanzo & Ephrussi, 2002). This and the fact that long Oskar can act in *trans* to increase the localization efficiency of short Oskar suggest that enhanced actin nucleation and endocytosis somehow contribute to the tethering of germ plasm. In support Lasp, an actin-binding protein, which interacts with a region of Oskar shared by long and short isoforms, affects *oskar* RNA localization (Suyama et al., 2009). An Oskar-dependent actin network may also attract additional Par-1 to the posterior pole, which, as part of a positive feedback loop, stabilizes the microtubule network that transports more *oskar* RNA to the posterior (Doerflinger, Benton, Torres, Zwart, & St Johnston, 2006; Zimyanin, Lowe, & St Johnston, 2007). Despite these roles of long Oskar in stabilizing *osk* RNA and protein localization, its function is dispensable for embryonic patterning and germ cell formation, although fewer germ cells form in long Oskar mutants.

Short Oskar, in contrast, is absolutely required for germ plasm assembly, germ cell formation, and normal posterior patterning (Markussen et al., 1995; Rongo et al., 1995; Vanzo & Ephrussi, 2002). Indeed, short Oskar expressed ectopically at the anterior pole is capable of inducing pole cell formation and abdominal patterning at the new location independent of posteriorly produced Oskar (Tanaka & Nakamura, 2008). A number of interaction partners for short Oskar have been identified. These include Vasa, an ATP-dependent helicase, Valois, a methyltransferase-associated WD repeat protein, and Lasp (Anne, 2010; Breitwieser, Markussen, Horstmann, & Ephrussi, 1996; Jeske et al., 2015; Suyama et al., 2009). Recently, two studies provided insight into the function of Oskar by determining the structure of two domains in Oskar (Fig. 2F): a domain at the N terminus of short Oskar (139–240), which was initially identified by bioinformatics and termed LOTUS domain due to its presence in Limkain, Oskar, and Tudor containing proteins 5 and 7; and the C-terminal "OSK" domain, which resembles a SGNH hydrolase or lipase domain found in mammalian platelet-activating factor acetylhydrolase and GDSL-like lipases (Jeske et al., 2015; Yang et al., 2015). The studies found that the LOTUS domain forms a winged-helix-turn-helix structure with three-to-four alpha helices (depending on the size of the fragment used for crystallization) as wings emerging from two antiparallel beta strands. In contrast to other winged helix structures, the Osk N-terminal LOTUS domain homodimerizes masking the helices that in other winged helix proteins bind DNA. The Oskar LOTUS domain binds to the RNA helicase Vasa, an important component of germ plasm. The C-terminal "OSK" domain forms a globular structure composed of central beta sheets surrounded by alpha helices with extended loops. While the structure of this domain resembles lipase-fold structures, three of the four residues required for enzymatic activity are missing, making it highly unlikely that the OSK domain functions as a hydrolase (Jeske et al., 2015; Yang et al., 2015). Instead, the structure reveals a number of basic, positively charged residues at its surface suggesting chemical property more typical for RNA-binding proteins. Indeed, *in vitro* experiments suggest that this domain can bind to RNA and demonstrate specific interaction with the *osk* and *nos* 3′UTRs that can be disrupted by mutations in positively charged amino acids in the OSK domain (Yang et al., 2015). Accordingly, these key residues are in proximity to genetically identified point mutations in Osk that strongly affect posterior pattern and germ plasm assembly. Furthermore, *in vivo* pull-down experiments from embryo extracts after UV crosslinking, which stabilizes RNA–protein interactions, revealed significant interactions between Osk and *nos*, *pgc* and, to a lesser extent, *gcl* RNA *in vivo* (Jeske et al., 2015). All three RNAs are known to be localized to the germ plasm.

8. ASSEMBLY OF GERM PLASM

Germ plasm assembly initiated by Oskar requires the function of Vasa and Tudor. While homologs of these two proteins are widely conserved as universal germ cell markers, *oskar* evolved within the order of insects (Ewen-Campen, Srouji, Schwager, & Extavour, 2012). Like *osk*, females mutant for *vasa* or *tudor* produce embryos that lack germ cells and fail to form an abdomen (Boswell & Mahowald, 1985; Dehghani & Lasko, 2015). Vasa, a DEAD-box RNA helicase, has been proposed to be involved in translational activation of RNA localized to the posterior pole (Lasko & Ashburner, 1988). However, while Vasa binds the translation factor eIF5E, direct evidence for a role of Vasa in binding to or regulating the translation of germ plasm localized RNAs (Johnstone & Lasko, 2004). In addition to its role in germ plasm assembly, Vasa also plays an important role in piRNA biogenesis, likely as part of the piRNA processing machinery that is assembled in the nuage of the nurse cells (for reviews on piRNA processing see: Malone & Hannon, 2009; Pek et al., 2012; Siomi, Sato, Pezic, & Aravin, 2011). Tudor is the founding member of the Tudor domain family of proteins (Golumbeski, Bardsley, Tax, & Boswell, 1991). Tudor encodes a large protein of 2515 amino acids that contains 11 copies of the Tudor domain motif (Arkov, Wang, Ramos, & Lehmann, 2006). Structural analysis of a single Tudor domain revealed that the core Tud barrel-like structure consisting of four β strands that forms an aromatic cage, is extended N-and C-terminal helices, folded into an oligonucleotide-binding fold, commonly called the extended Tudor (eTud) domain. The ~120 amino acid eTud domain is predicted to be the common structural motif found in germ line Tudor proteins. Tudor domain 11 specifically recognizes symmetrically dimethylated Arginines (sDMAs) within the Argonaute family protein Aubergine (Liu et al., 2010). Point mutations in Tudor that abolish sDMA peptide binding or mutations of the corresponding arginines in Aubergine abolish Aubergine localization to the germ plasm (Liu et al., 2010; Webster et al., 2015). With its many domains, it is likely that Tudor acts as a scaffold for germ plasm assembly by interacting with sDMA-modified target proteins (Fig. 3A; Ren et al., 2014; Thomson & Lasko, 2004). sDMA modification of target proteins could be achieved specifically at the posterior pole through the localization of the sDMA methyltransferase Capsuleen and its associated protein Valois to the germ plasm, possibly via direct interaction of Valois with Oskar (Anne, Ollo, Ephrussi, & Mechler, 2007; Gonsalvez, Rajendra, Tian, & Matera, 2006). Another potential interactor is pyruvate kinase (Pyk), which contains sDMAs modification, localizes to polar granules and interacts with Tudor (Thomson, Liu, Arkov, Lehmann, & Lasko, 2008). Interestingly, not only Pyk but several enzymes in the glycolytic pathway localize specifically to the germ plasm and affect germ cell formation (Gao et al., 2015). It has therefore been suggested that Tudor may

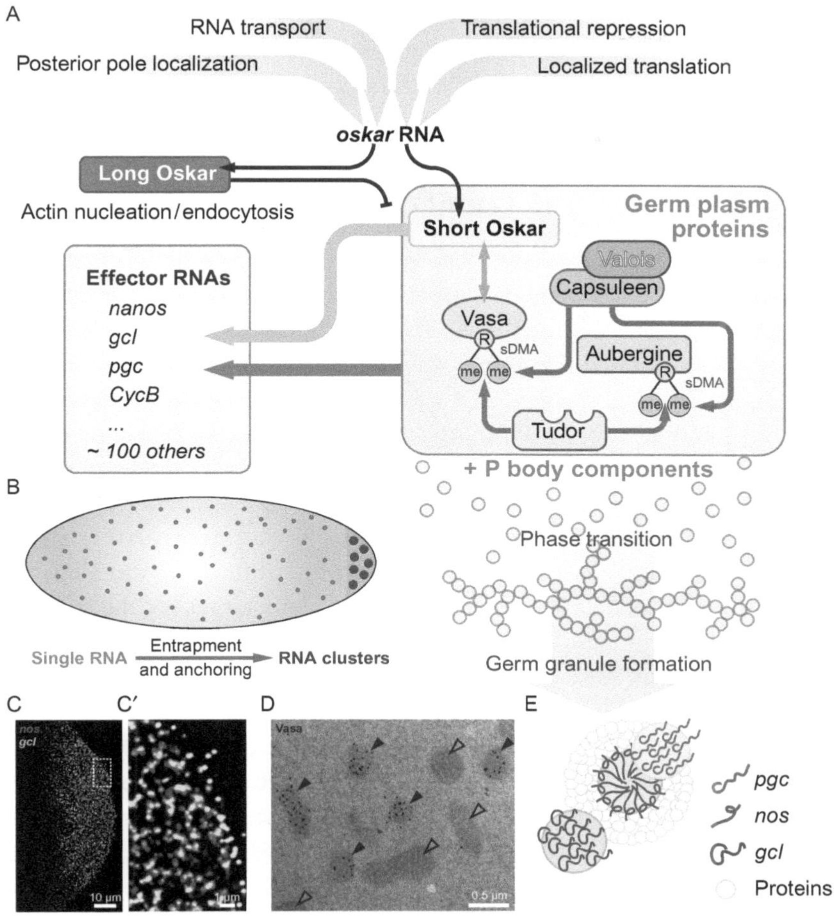

Figure 3 Germ plasm assembly. (A) Multiple controls described in Figs. 1 and 2 feed into the localization and translation of *osk*. Osk protein recruits other components of the germ plasm such as Vasa, Tudor, and Aubergine. Symmetric dimethylation of Arginine (sDMA) in Vasa and Aubergine by the methyltransferase Capsuleen allows binding to Tudor domain proteins (Kirino, Vourekas, Kim, et al., 2010; Kirino, Vourekas, Sayed, et al., 2010; Liu et al., 2010; Webster et al., 2015). This together with liquid to gel phase transitions could provide a possible basis for the dense assembly of polar granules. (B) Model for phase transition that occurs at posterior pole when concentration of germ plasm protein and RNAs become high (Brangwynne, 2013). (C) SIM image of *nos* and *gcl* RNA hybridization using Stellaris© RNA *in situ* hybridization probes. (C′) Higher magnification to reveal particle overlap. (D) Electron micrograph of germ plasm with polar granules marked by Vasa (filled arrow) and mitochondria (open arrow). (E) Organization of homotypic RNA clusters within polar granules according to Trcek et al. (2015). (A color version of this figure is located at the end of this chapter.)

recruit glycolytic enzymes to the germ plasm to provide a local source of ATP for helicases like Vasa. Taken together, sDMA modification-mediated binding of proteins to the Tudor scaffold could lead to a robust network assembly of proteins in the germ plasm (Fig. 3A and B).

9. EFFECTOR RNA LOCALIZATION

More than 100 different, maternally synthesized mRNAs have been reported by *in situ* hybridization to be localized to the germ plasm (Jambor et al., 2015; Lecuyer et al., 2007). A function in germ cell biology has been demonstrated for only a few of these. The best-studied effector RNA is *nanos*. Nanos encodes a zinc-finger protein that together with the RNA-binding protein Pumilio acts as a translational repressor (Curtis et al., 1997). Targets of Nos/Pum-mediated translational repression include the maternally provided *hunchback* (*hb*) RNAs (Sonoda & Wharton, 1999). Elimination of *hb* RNA from the posterior domain is necessary for normal abdomen formation (Irish et al., 1989; Struhl, 1989). Nos and Pum also affect early germ cell development, and one of the target for this function is *CyclinB* RNA, which is localized to the germ plasm and translationally repressed by Nos and Pum (Hayashi, Hayashi, & Kobayashi, 2004; Kadyrova, Habara, Lee, & Wharton, 2007). PGCs do not divide until this translational block is relieved at the end of embryogenesis. Other targets include the cell death-promoting gene *hid* and other targets that mediate defects in germ cell migration, PGC survival, and PGC chromatin remodeling remain to be identified (Sato et al., 2007). Other germ granule localized RNAs whose functions are known include *germ cell-less*, which encodes a BTB protein that is required for the formation of PGCs (Cinalli & Lehmann, 2013; Jongens, Hay, Jan, & Jan, 1992), and *polar granule component*, which encodes a small protein that inhibits transcription in early germ cells by preventing the recruitment P-TEFb kinase complex to chromatin (Hanyu-Nakamura, Sonobe-Nojima, Tanigawa, Lasko, & Nakamura, 2008; Martinho, Kunwar, Casanova, & Lehmann, 2004; Nakamura, Amikura, Mukai, Kobayashi, & Lasko, 1996).

The localization efficiency for effector RNAs is low and estimated at 2.5–4% depending on the RNA studied (Bergsten & Gavis, 1999; Little, Sinsimer, Lee, Wieschaus, & Gavis, 2015; Trcek et al., 2015). A more detailed understanding of how posteriorly localized RNAs reach the germ plasm has come from *in vivo* studies using GFP and the MS2/MCP RNA labeling system. These studies revealed that *nanos* RNAs reach the germ plasm during the late stage of oogenesis starting during cytoplasmic streaming and continuing throughout nurse cell dumping (Forrest & Gavis, 2003; Ganguly, Williams, Palacios, & Goldstein, 2012). Once at the posterior pole, persistent microtubule-dependent trafficking keeps RNAs associated with the germ plasm (Sinsimer, Lee, Thiberge, & Gavis, 2013). These observations have led to a model by which RNAs destined for posterior localization have an affinity for the posterior germ plasm and are trapped by this specialized cytoplasm (Fig. 3).

The localization patterns of other RNAs, such as *CycB*, *gcl*, and *pgc* seem to follow similar patterns (Little et al., 2015; Trcek et al., 2015). In all cases studied, the 3′UTR was sufficient to recapitulate the posterior RNA localization pattern (Rangan et al., 2009). Single-molecule analysis of *nos*, *cyclinB*, *pgc*, and *gcl* RNA has revealed that RNAs first arrive in the germ plasm as single molecules and then organize into larger RNA particles associated with polar granules (Figs. 1A and 3B; Little et al., 2015; Trcek et al., 2015). Structural illumination microscopy, a super resolution method, further showed that RNAs form homotypic clusters within polar granules and that these RNA clusters have unique positions within the granule, with *nos* and *cycB* RNA more toward the center of the granule and *pgc* and *gcl* tend to be more peripheral (Fig. 3C, C′, and E; Trcek et al., 2015). While large *osk* RNA clusters are found in the germ plasm, these do not associate with polar granules (Fig. 1G; Little et al., 2015; Trcek et al., 2015; Vanzo et al., 2007). It is presently unclear how this organization is achieved as protein components of the germ granule, such as Vasa, Oskar, Tudor, and Aubergine, show uniform distribution throughout the granule (Fig. 3D; Little et al., 2015; Trcek et al., 2015). One intriguing possibility is that RNAs self-associate once they are localized. Indeed, the concentration of effector RNAs in the germ plasm is about 8- to 12-fold higher than their concentration in the rest of the embryo (Little et al., 2015; Trcek et al., 2015). This increase in concentration depends on the germ plasm proteins. RNA-binding proteins have been shown to undergo phase transitions from a soluble to viscous state dependent on their concentration and association with RNA (Brangwynne, 2013; Lin, Protter, Rosen, & Parker, 2015; Zhang et al., 2015). Thus, RNAs may be trapped by germ plasm aggregates, which create specific, homotypic RNA clusters due to RNA–protein as well as RNA–RNA recognition (Fig. 3B). Such an organization could provide an efficient mechanism to organize a large number of localized RNAs in an RNA-specific manner.

Like *osk*, other germ plasm RNAs are also translationally silent during transport (Gavis & Lehmann, 1994). 3′UTR sequences from different effector RNAs fused to GFP reporter transcripts showed that the respective 3′UTRs are sufficient to recapitulate the posterior RNA localization pattern, to mediate translational repression of unlocalized RNA, and to translationally derepress RNA upon localization (Rangan et al., 2011). For *nos* RNA, translational repression is achieved through *cis*-acting sequences in the *nos* 3′UTR and binding of the hnRNP protein Glorund during oogenesis and the SAM domain protein Smaug together with Argonaute 1 during embryogenesis (Andrews, Snowflack, Clark, & Gavis, 2011; Crucs, Chatterjee, & Gavis, 2000; Gavis & Lehmann, 1994; Kalifa, Huang, Rosen, Chatterjee, & Gavis, 2006; Pinder & Smibert, 2013; Smibert, Wilson, Kerr, & Macdonald, 1996). The onset of translation is strikingly different depending on the effector RNA: *nos* RNA is translated as soon as it becomes localized to the posterior pole, *gcl* is translated as nuclei reach the embryo cortex, *pgc* is translated when germ cells form, and *cycB* is not translated until the end of embryogenesis (Clark, Wyckoff, & Gavis, 2000; Dalby & Glover, 1993; Hanyu-Nakamura et al., 2008; Jongens, Ackerman, Swedlow, Jan, & Jan, 1994). Curiously, the position of a specific RNA cluster with respect to the germ granule seems not to correlate with the timing of translational derepression. Indeed, *nos* RNA, which is located in the center of the granule is translated first, while *pgc* and *gcl*, located more peripherally are translated later (Trcek et al., 2015). This is in contrast with the organization of RNA in sponge bodies, another large RNA–protein granule found at the anterior of oocytes. Here translational activity is correlated with position, as translationally active *gurken* RNA is located more peripherally, and quiescent *bicoid* RNA is positioned in the center of the sponge body (Delanoue, Herpers, Soetaert, Davis, & Rabouille, 2007). This type of regulation has also been proposed for stress granules and P bodies. While polar granules share RNA decay and translational regulators with sponge bodies, stress granules, and P bodies, the access of specific RNAs to RNA regulators cannot be controlled by position within the granule alone. Thus, it remains unclear how RNA clusters are organized with respect to each other, how position is controlled within the germ plasm and what dictates translational timing.

10. SUMMARY AND OUTLOOK

- Regulation of *osk* exemplifies how the nuclear history of an RNA and cell polarity controls RNA localization and subcellular translation. Osk protein has multiple roles in the assembly and anchoring of germ plasm components at the egg cortex. In contrast to Vasa, Tudor, and Nanos, Osk is not a conserved component of germ plasm. Interestingly in other species, seemingly unrelated proteins, like Buckyball in fish and the Pgl proteins in *C. elegans,* seem to play similar roles in the initial steps of germ granule assembly (Bontems et al., 2009; Hanazawa, Yonetani, & Sugimoto, 2011). With the emerging concept that phase transitions and proteins aggregates play an important role in RNA granule assembly and biology, these functional similarities may be grounded in common biophysical properties.

- Electron-dense granules of the germ line share many components that are also common to other cytoplasmic, membrane-less granules like P bodies and stress granules. However, certain proteins are specific identifiers for each granule class. While Osk is critical for the assembly of germ plasm, it is not needed or present in the nuage, which requires Vasa and a different set of Tudor proteins as well as the Argonaute proteins Ago3 and Aubergine. Similarly, sponge bodies rely on the Exuperantia protein (Wilsch-Brauninger et al., 1997). While all germ line granules are involved in the regulation of RNA, the processes that are controlled are strikingly different. Nuage is required for the processing of small pi (Piwi interacting) RNAs involved in the recognition and defense against transposable elements, while sponge bodies have a major role in the localization and regulation of RNAs encoding signaling factors. Germ plasm regulates the translation of RNAs destined for the germ line of the next generation. How protein composition is determined and RNAs are sorted between these compartments and whether there is direct exchange between specific components like Vasa, Tudor, or Aubergine, is presently unclear. The Tudor class of proteins may play a critical role, as scaffold, for different germ line RNPs whereby individual Tudor domains recognize and bind specific proteins depending on context and modification.
- Enrichment of germ plasm proteins and effector RNAs creates a steep transition between aggregated germ plasm components at the posterior pole and a low concentration of these components throughout the rest of the embryo. Such a transition may involve a phase shift from a fluid to a more gel-like state, as suggested for other RNP particles. How such a gel-like state could facilitate homotypic RNA cluster formation and how within these clusters micro domains may regulate translational timing remains to be determined.

ACKNOWLEDGMENTS

My special thanks go to Dr. Alexey Soshnev for generating the figures. I thank members of my lab for discussion and particularly Thomas Hurd, Lacy Barton, Juhee Pae, and Tatjana Trcek for comments on the chapter and Tatjana Trcek and Fengxia Liang for images in Fig. 3. Paul Wasserman's encouragement and patience were critical in completing this review. Due to space limitation, I apologize for neglecting to cite all pertinent literature. This work was supported by NIH R01/R37HD41900 and R.L. is a HHMI investigator.

REFERENCES

Andrews, S., Snowflack, D. R., Clark, I. E., & Gavis, E. R. (2011). Multiple mechanisms collaborate to repress nanos translation in the Drosophila ovary and embryo. *RNA, 17*, 967–977.

Anne, J. (2010). Targeting and anchoring Tudor in the pole plasm of the Drosophila oocyte. *PLoS One, 5*. e14362.

Anne, J., Ollo, R., Ephrussi, A., & Mechler, B. M. (2007). Arginine methyltransferase Capsuleen is essential for methylation of spliceosomal Sm proteins and germ cell formation in Drosophila. *Development, 134*, 137–146.

Arkov, A. L., Wang, J. Y., Ramos, A., & Lehmann, R. (2006). The role of Tudor domains in germline development and polar granule architecture. *Development, 133*, 4053–4062.

Asaoka-Taguchi, M., Yamada, M., Nakamura, A., Hanyu, K., & Kobayashi, S. (1999). Maternal Pumilio acts together with Nanos in germline development in Drosophila embryos. *Nature Cell Biology, 1*, 431–437.

Babu, K., Cai, Y., Bahri, S., Yang, X., & Chia, W. (2004). Roles of Bifocal, Homer, and F-actin in anchoring Oskar to the posterior cortex of Drosophila oocytes. *Genes & Development, 18*, 138–143.

Barker, D. D., Wang, C., Moore, J., Dickinson, L. K., & Lehmann, R. (1992). Pumilio is essential for function but not for distribution of the Drosophila abdominal determinant Nanos. *Genes & Development, 6*, 2312–2326.

Bergsten, S. E., & Gavis, E. R. (1999). Role for mRNA localization in translational activation but not spatial restriction of nanos RNA. *Development, 126*, 659–669.

Bontems, F., Stein, A., Marlow, F., Lyautey, J., Gupta, T., Mullins, M. C., et al. (2009). Bucky ball organizes germ plasm assembly in zebrafish. *Current Biology, 19*, 414–422.

Boswell, R. E., & Mahowald, A. P. (1985). tudor, a gene required for assembly of the germ plasm in *Drosophila melanogaster*. *Cell, 43*, 97–104.

Brangwynne, C. P. (2013). Phase transitions and size scaling of membrane-less organelles. *The Journal of Cell Biology, 203*, 875–881.

Breitwieser, W., Markussen, F. H., Horstmann, H., & Ephrussi, A. (1996). Oskar protein interaction with Vasa represents an essential step in polar granule assembly. *Genes & Development, 10*, 2179–2188.

Brendza, R. P., Serbus, L. R., Duffy, J. B., & Saxton, W. M. (2000). A function for kinesin I in the posterior transport of oskar mRNA and Staufen protein. *Science, 289*, 2120–2122.

Castagnetti, S., & Ephrussi, A. (2003). Orb and a long poly(A) tail are required for efficient oskar translation at the posterior pole of the Drosophila oocyte. *Development, 130*, 835–843.

Chekulaeva, M., Hentze, M. W., & Ephrussi, A. (2006). Bruno acts as a dual repressor of oskar translation, promoting mRNA oligomerization and formation of silencing particles. *Cell, 124*, 521–533.

Cinalli, R. M., & Lehmann, R. (2013). A spindle-independent cleavage pathway controls germ cell formation in Drosophila. *Nature Cell Biology, 15*, 839–845.

Clark, I. E., Wyckoff, D., & Gavis, E. R. (2000). Synthesis of the posterior determinant Nanos is spatially restricted by a novel cotranslational regulatory mechanism. *Current Biology*, *10*, 1311–1314.
Cox, R. T., & Spradling, A. C. (2003). A Balbiani body and the fusome mediate mitochondrial inheritance during Drosophila oogenesis. *Development*, *130*, 1579–1590.
Crucs, S., Chatterjee, S., & Gavis, E. R. (2000). Overlapping but distinct RNA elements control repression and activation of nanos translation. *Molecular Cell*, *5*, 457–467.
Curtis, D., Treiber, D. K., Tao, F., Zamore, P. D., Williamson, J. R., & Lehmann, R. (1997). A CCHC metal-binding domain in Nanos is essential for translational regulation. *The EMBO Journal*, *16*, 834–843.
Dahlgaard, K., Raposo, A. A., Niccoli, T., & St Johnston, D. (2007). Capu and Spire assemble a cytoplasmic actin mesh that maintains microtubule organization in the Drosophila oocyte. *Developmental Cell*, *13*, 539–553.
Dalby, B., & Glover, D. M. (1993). Discrete sequence elements control posterior pole accumulation and translational repression of maternal cyclin B RNA in Drosophila. *The EMBO Journal*, *12*, 1219–1227.
Dehghani, M., & Lasko, P. (2015). In vivo mapping of the functional regions of the DEAD-box helicase Vasa. *Biology Open*, *4*, 450–462.
Delanoue, R., Herpers, B., Soetaert, J., Davis, I., & Rabouille, C. (2007). Drosophila Squid/hnRNP helps Dynein switch from a gurken mRNA transport motor to an ultrastructural static anchor in sponge bodies. *Developmental Cell*, *13*, 523–538.
Dienstbier, M., Boehl, F., Li, X., & Bullock, S. L. (2009). Egalitarian is a selective RNA-binding protein linking mRNA localization signals to the dynein motor. *Genes & Development*, *23*, 1546–1558.
Doerflinger, H., Benton, R., Torres, I. L., Zwart, M. F., & St Johnston, D. (2006). Drosophila anterior-posterior polarity requires actin-dependent PAR-1 recruitment to the oocyte posterior. *Current Biology*, *16*, 1090–1095.
Ephrussi, A., Dickinson, L. K., & Lehmann, R. (1991). Oskar organizes the germ plasm and directs localization of the posterior determinant nanos. *Cell*, *66*, 37–50.
Ephrussi, A., & Lehmann, R. (1992). Induction of germ cell formation by oskar. *Nature*, *358*, 387–392.
Erdelyi, M., Michon, A. M., Guichet, A., Glotzer, J. B., & Ephrussi, A. (1995). Requirement for Drosophila cytoplasmic tropomyosin in oskar mRNA localization. *Nature*, *377*, 524–527.
Ewen-Campen, B., Srouji, J. R., Schwager, E. E., & Extavour, C. G. (2012). Oskar predates the evolution of germ plasm in insects. *Current Biology*, *22*, 2278–2283.
Forrest, K. M., & Gavis, E. R. (2003). Live imaging of endogenous RNA reveals a diffusion and entrapment mechanism for nanos mRNA localization in Drosophila. *Current Biology*, *13*, 1159–1168.
Ganguly, S., Williams, L. S., Palacios, I. M., & Goldstein, R. E. (2012). Cytoplasmic streaming in Drosophila oocytes varies with kinesin activity and correlates with the microtubule cytoskeleton architecture. *Proceedings of the National Academy of Sciences of the United States of America*, *109*, 15109–15114.
Gao, M., & Arkov, A. L. (2013). Next generation organelles: Structure and role of germ granules in the germline. *Molecular Reproduction and Development*, *80*, 610–623.
Gao, M., Thomson, T. C., Creed, T. M., Tu, S., Loganathan, S. N., Jackson, C. A., et al. (2015). Glycolytic enzymes localize to ribonucleoprotein granules in Drosophila germ cells, bind Tudor and protect from transposable elements. *EMBO Reports*, *16*, 379–386.
Gavis, E. R., & Lehmann, R. (1994). Translational regulation of nanos by RNA localization. *Nature*, *369*, 315–318.
Ghosh, S., Marchand, V., Gaspar, I., & Ephrussi, A. (2012). Control of RNP motility and localization by a splicing-dependent structure in oskar mRNA. *Nature Structural & Molecular Biology*, *19*, 441–449.

Golumbeski, G. S., Bardsley, A., Tax, F., & Boswell, R. E. (1991). tudor, a posterior-group gene of *Drosophila melanogaster*, encodes a novel protein and an mRNA localized during mid-oogenesis. *Genes & Development*, *5*, 2060–2070.

Gonsalvez, G. B., Rajendra, T. K., Tian, L., & Matera, A. G. (2006). The Sm-protein methyltransferase, dart5, is essential for germ-cell specification and maintenance. *Current Biology*, *16*, 1077–1089.

Gonzalez-Reyes, A., Elliott, H., & St Johnston, D. (1995). Polarization of both major body axes in Drosophila by gurken-torpedo signalling. *Nature*, *375*, 654–658.

Hachet, O., & Ephrussi, A. (2004). Splicing of oskar RNA in the nucleus is coupled to its cytoplasmic localization. *Nature*, *428*, 959–963.

Hanazawa, M., Yonetani, M., & Sugimoto, A. (2011). PGL proteins self associate and bind RNPs to mediate germ granule assembly in *C. elegans*. *The Journal of Cell Biology*, *192*, 929–937.

Hanyu-Nakamura, K., Sonobe-Nojima, H., Tanigawa, A., Lasko, P., & Nakamura, A. (2008). Drosophila Pgc protein inhibits P-TEFb recruitment to chromatin in primordial germ cells. *Nature*, *451*, 730–733.

Hayashi, Y., Hayashi, M., & Kobayashi, S. (2004). Nanos suppresses somatic cell fate in Drosophila germ line. *Proceedings of the National Academy of Sciences of the United States of America*, *101*, 10338–10342.

Huttelmaier, S., Zenklusen, D., Lederer, M., Dictenberg, J., Lorenz, M., Meng, X., et al. (2005). Spatial regulation of beta-actin translation by Src-dependent phosphorylation of ZBP1. *Nature*, *438*, 512–515.

Huynh, J. R., Petronczki, M., Knoblich, J. A., & St Johnston, D. (2001). Bazooka and PAR-6 are required with PAR-1 for the maintenance of oocyte fate in Drosophila. *Current Biology*, *11*, 901–906.

Illmensee, K., & Mahowald, A. P. (1974). Transplantation of posterior polar plasm in Drosophila. Induction of germ cells at the anterior pole of the egg. *Proceedings of the National Academy of Sciences of the United States of America*, *71*, 1016–1020.

Illmensee, K., Mahowald, A. P., & Loomis, M. R. (1976). The ontogeny of germ plasm during oogenesis in Drosophila. *Developmental Biology*, *49*, 40–65.

Irish, V., Lehmann, R., & Akam, M. (1989). The Drosophila posterior-group gene nanos functions by repressing hunchback activity. *Nature*, *338*, 646–648.

Jambor, H., Brunel, C., & Ephrussi, A. (2011). Dimerization of oskar 3′ UTRs promotes hitchhiking for RNA localization in the Drosophila oocyte. *RNA*, *17*, 2049–2057.

Jambor, H., Mueller, S., Bullock, S. L., & Ephrussi, A. (2014). A stem-loop structure directs oskar mRNA to microtubule minus ends. *RNA*, *20*, 429–439.

Jambor, H., Surendranath, V., Kalinka, A. T., Mejstrik, P., Saalfeld, S., & Tomancak, P. (2015). Systematic imaging reveals features and changing localization of mRNAs in Drosophila development. *eLife*, *4*, e05003.

Jenny, A., Hachet, O., Zavorszky, P., Cyrklaff, A., Weston, M. D., Johnston, D. S., et al. (2006). A translation-independent role of oskar RNA in early Drosophila oogenesis. *Development*, *133*, 2827–2833.

Jeske, M., Bordi, M., Glatt, S., Muller, S., Rybin, V., Muller, C. W., et al. (2015). The crystal structure of the Drosophila germline inducer Oskar identifies two domains with distinct vasa helicase- and RNA-binding activities. *Cell Reports*, *12*, 587–598.

Johnstone, O., & Lasko, P. (2004). Interaction with eIF5B is essential for Vasa function during development. *Development*, *131*, 4167–4178.

Jongens, T. A., Ackerman, L. D., Swedlow, J. R., Jan, L. Y., & Jan, Y. N. (1994). Germ cell-less encodes a cell type-specific nuclear pore-associated protein and functions early in the germ-cell specification pathway of Drosophila. *Genes & Development*, *8*, 2123–2136.

Jongens, T. A., Hay, B., Jan, L. Y., & Jan, Y. N. (1992). The germ cell-less gene product: A posteriorly localized component necessary for germ cell development in Drosophila. *Cell, 70*, 569–584.
Kadyrova, L. Y., Habara, Y., Lee, T. H., & Wharton, R. P. (2007). Translational control of maternal Cyclin B mRNA by Nanos in the Drosophila germline. *Development, 134*, 1519–1527.
Kalifa, Y., Huang, T., Rosen, L. N., Chatterjee, S., & Gavis, E. R. (2006). Glorund, a Drosophila hnRNP F/H homolog, is an ovarian repressor of nanos translation. *Developmental Cell, 10*, 291–301.
Kanke, M., Jambor, H., Reich, J., Marches, B., Gstir, R., Ryu, Y. H., et al. (2015). oskar RNA plays multiple noncoding roles to support oogenesis and maintain integrity of the germline/soma distinction. *RNA, 21*, 1096–1109.
Kanke, M., & Macdonald, P. M. (2015). Translational activation of oskar mRNA: Reevaluation of the role and importance of a 5′ regulatory element [corrected]. *PLoS One, 10*. e0125849.
Khuc Trong, P., Doerflinger, H., Dunkel, J., St Johnston, D., & Goldstein, R. E. (2015). Cortical microtubule nucleation can organise the cytoskeleton of Drosophila oocytes to define the anteroposterior axis. *eLife, 4*, e06088.
Kim, G., Pai, C. I., Sato, K., Person, M. D., Nakamura, A., & Macdonald, P. M. (2015). Region-specific activation of oskar mRNA translation by inhibition of Bruno-mediated repression. *PLoS Genetics, 11*. e1004992.
Kim-Ha, J., Kerr, K., & Macdonald, P. M. (1995). Translational regulation of oskar mRNA by bruno, an ovarian RNA-binding protein, is essential. *Cell, 81*, 403–412.
Kim-Ha, J., Smith, J. L., & Macdonald, P. M. (1991). oskar mRNA is localized to the posterior pole of the Drosophila oocyte. *Cell, 66*, 23–35.
Kim-Ha, J., Webster, P. J., Smith, J. L., & Macdonald, P. M. (1993). Multiple RNA regulatory elements mediate distinct steps in localization of oskar mRNA. *Development, 119*, 169–178.
Kinkelin, K., Veith, K., Grunwald, M., & Bono, F. (2012). Crystal structure of a minimal eIF4E-Cup complex reveals a general mechanism of eIF4E regulation in translational repression. *RNA, 18*, 1624–1634.
Kirino, Y., Vourekas, A., Kim, N., de Lima Alves, F., Rappsilber, J., Klein, P. S., et al. (2010). Arginine methylation of vasa protein is conserved across phyla. *The Journal of Biological Chemistry, 285*, 8148–8154.
Kirino, Y., Vourekas, A., Sayed, N., de Lima Alves, F., Thomson, T., Lasko, P., et al. (2010). Arginine methylation of Aubergine mediates Tudor binding and germ plasm localization. *RNA, 16*, 70–78.
Kobayashi, S., Yamada, M., Asaoka, M., & Kitamura, T. (1996). Essential role of the posterior morphogen nanos for germline development in Drosophila. *Nature, 380*, 708–711.
Lasko, P. F., & Ashburner, M. (1988). The product of the Drosophila gene vasa is very similar to eukaryotic initiation factor-4A. *Nature, 335*, 611–617.
Lecuyer, E., Yoshida, H., Parthasarathy, N., Alm, C., Babak, T., Cerovina, T., et al. (2007). Global analysis of mRNA localization reveals a prominent role in organizing cellular architecture and function. *Cell, 131*, 174–187.
Lehmann, R., & Nusslein-Volhard, C. (1986). Abdominal segmentation, pole cell formation, and embryonic polarity require the localized activity of oskar, a maternal gene in Drosophila. *Cell, 47*, 141–152.
Lehmann, R., & Nusslein-Volhard, C. (1991). The maternal gene nanos has a central role in posterior pattern formation of the Drosophila embryo. *Development, 112*, 679–691.
Lin, Y., Protter, D. S., Rosen, M. K., & Parker, R. (2015). Formation and maturation of phase-separated liquid droplets by RNA-binding proteins. *Molecular Cell, 60*, 208–219.

Little, S. C., Sinsimer, K. S., Lee, J. J., Wieschaus, E. F., & Gavis, E. R. (2015). Independent and coordinate trafficking of single Drosophila germ plasm mRNAs. *Nature Cell Biology*, *17*, 558–568.

Liu, H., Wang, J. Y., Huang, Y., Li, Z., Gong, W., Lehmann, R., et al. (2010). Structural basis for methylarginine-dependent recognition of Aubergine by Tudor. *Genes & Development*, *24*, 1876–1881.

Macchi, P., Kroening, S., Palacios, I. M., Baldassa, S., Grunewald, B., Ambrosino, C., et al. (2003). Barentsz, a new component of the Staufen-containing ribonucleoprotein particles in mammalian cells, interacts with Staufen in an RNA-dependent manner. *The Journal of Neuroscience*, *23*, 5778–5788.

Mach, J. M., & Lehmann, R. (1997). An Egalitarian-BicaudalD complex is essential for oocyte specification and axis determination in Drosophila. *Genes & Development*, *11*, 423–435.

Mahowald, A. P. (2001). Assembly of the Drosophila germ plasm. *International Review of Cytology*, *203*, 187–213.

Malone, C. D., & Hannon, G. J. (2009). Small RNAs as guardians of the genome. *Cell*, *136*, 656–668.

Markussen, F. H., Michon, A. M., Breitwieser, W., & Ephrussi, A. (1995). Translational control of oskar generates short OSK, the isoform that induces pole plasma assembly. *Development*, *121*, 3723–3732.

Martinho, R. G., Kunwar, P. S., Casanova, J., & Lehmann, R. (2004). A noncoding RNA is required for the repression of RNApolII-dependent transcription in primordial germ cells. *Current Biology*, *14*, 159–165.

Micklem, D. R., Adams, J., Grunert, S., & St Johnston, D. (2000). Distinct roles of two conserved Staufen domains in oskar mRNA localization and translation. *The EMBO Journal*, *19*, 1366–1377.

Morais-de-Sa, E., Mukherjee, A., Lowe, N., & St Johnston, D. (2014). Slmb antagonises the aPKC/Par-6 complex to control oocyte and epithelial polarity. *Development*, *141*, 2984–2992.

Munro, T. P., Kwon, S., Schnapp, B. J., & St Johnston, D. (2006). A repeated IMP-binding motif controls oskar mRNA translation and anchoring independently of *Drosophila melanogaster* IMP. *The Journal of Cell Biology*, *172*, 577–588.

Nakamura, A., Amikura, R., Mukai, M., Kobayashi, S., & Lasko, P. F. (1996). Requirement for a noncoding RNA in Drosophila polar granules for germ cell establishment. *Science*, *274*, 2075–2079.

Nakamura, A., Sato, K., & Hanyu-Nakamura, K. (2004). Drosophila cup is an eIF4E binding protein that associates with Bruno and regulates oskar mRNA translation in oogenesis. *Developmental Cell*, *6*, 69–78.

Parton, R. M., Hamilton, R. S., Ball, G., Yang, L., Cullen, C. F., Lu, W., et al. (2011). A PAR-1-dependent orientation gradient of dynamic microtubules directs posterior cargo transport in the Drosophila oocyte. *The Journal of Cell Biology*, *194*, 121–135.

Pek, J. W., Patil, V. S., & Kai, T. (2012). piRNA pathway and the potential processing site, the nuage, in the Drosophila germline. *Development, Growth & Differentiation*, *54*, 66–77.

Pinder, B. D., & Smibert, C. A. (2013). microRNA-independent recruitment of Argonaute 1 to nanos mRNA through the Smaug RNA-binding protein. *EMBO Reports*, *14*, 80–86.

Ran, B., Bopp, R., & Suter, B. (1994). Null alleles reveal novel requirements for Bic-D during Drosophila oogenesis and zygotic development. *Development*, *120*, 1233–1242.

Rangan, P., DeGennaro, M., Jaime-Bustamante, K., Coux, R. X., Martinho, R. G., & Lehmann, R. (2009). Temporal and spatial control of germ-plasm RNAs. *Current Biology*, *19*, 72–77.

Rangan, P., Malone, C. D., Navarro, C., Newbold, S. P., Hayes, P. S., Sachidanandam, R., et al. (2011). piRNA production requires heterochromatin formation in Drosophila. *Current Biology*, *21*, 1373–1379.

Ren, R., Liu, H., Wang, W., Wang, M., Yang, N., Dong, Y. H., et al. (2014). Structure and domain organization of Drosophila Tudor. *Cell Research*, *24*, 1146–1149.

Reveal, B., Yan, N., Snee, M. J., Pai, C. I., Gim, Y., & Macdonald, P. M. (2010). BREs mediate both repression and activation of oskar mRNA translation and act in trans. *Developmental Cell*, *18*, 496–502.

Rongo, C., Broihier, H. T., Moore, L., Van Doren, M., Forbes, A., & Lehmann, R. (1997). Germ plasm assembly and germ cell migration in Drosophila. *Cold Spring Harbor Symposia on Quantitative Biology*, *62*, 1–11.

Rongo, C., Gavis, E. R., & Lehmann, R. (1995). Localization of oskar RNA regulates oskar translation and requires Oskar protein. *Development*, *121*, 2737–2746.

Roth, S., & Lynch, J. A. (2009). Symmetry breaking during Drosophila oogenesis. *Cold Spring Harbor Perspectives in Biology*, *1*, a001891.

Roth, S., Neuman-Silberberg, F. S., Barcelo, G., & Schupbach, T. (1995). cornichon and the EGF receptor signaling process are necessary for both anterior-posterior and dorsal-ventral pattern formation in Drosophila. *Cell*, *81*, 967–978.

Ryu, Y. H., & Macdonald, P. M. (2015). RNA sequences required for the noncoding function of oskar RNA also mediate regulation of Oskar protein expression by Bicoid Stability Factor. *Developmental Biology*, *407*, 211–223.

Sander, K. (1975). Pattern specification in the insect embryo. *Ciba Foundation Symposium*, 241–263.

Sato, K., Hayashi, Y., Ninomiya, Y., Shigenobu, S., Arita, K., Mukai, M., et al. (2007). Maternal Nanos represses hid/skl-dependent apoptosis to maintain the germ line in Drosophila embryos. *Proceedings of the National Academy of Sciences of the United States of America*, *104*, 7455–7460.

Schupbach, T., & Wieschaus, E. (1989). Female sterile mutations on the second chromosome of *Drosophila melanogaster*. I. Maternal effect mutations. *Genetics*, *121*, 101–117.

Serano, T. L., & Cohen, R. S. (1995). A small predicted stem-loop structure mediates oocyte localization of Drosophila K10 mRNA. *Development*, *121*, 3809–3818.

Serbus, L. R., Cha, B. J., Theurkauf, W. E., & Saxton, W. M. (2005). Dynein and the actin cytoskeleton control kinesin-driven cytoplasmic streaming in Drosophila oocytes. *Development*, *132*, 3743–3752.

Shulman, J. M., Benton, R., & St Johnston, D. (2000). The Drosophila homolog of *C. elegans* PAR-1 organizes the oocyte cytoskeleton and directs oskar mRNA localization to the posterior pole. *Cell*, *101*, 377–388.

Simon, B., Masiewicz, P., Ephrussi, A., & Carlomagno, T. (2015). The structure of the SOLE element of oskar mRNA. *RNA*, *21*, 1444–1453.

Sinsimer, K. S., Jain, R. A., Chatterjee, S., & Gavis, E. R. (2011). A late phase of germ plasm accumulation during Drosophila oogenesis requires lost and rumpelstiltskin. *Development*, *138*, 3431–3440.

Sinsimer, K. S., Lee, J. J., Thiberge, S. Y., & Gavis, E. R. (2013). Germ plasm anchoring is a dynamic state that requires persistent trafficking. *Cell Reports*, *5*, 1169–1177.

Siomi, M. C., Sato, K., Pezic, D., & Aravin, A. A. (2011). PIWI-interacting small RNAs: The vanguard of genome defence. *Nature Reviews. Molecular Cell Biology*, *12*, 246–258.

Smibert, C. A., Wilson, J. E., Kerr, K., & Macdonald, P. M. (1996). smaug protein represses translation of unlocalized nanos mRNA in the Drosophila embryo. *Genes & Development*, *10*, 2600–2609.

Smith, J. L., Wilson, J. E., & Macdonald, P. M. (1992). Overexpression of oskar directs ectopic activation of nanos and presumptive pole cell formation in Drosophila embryos. *Cell*, *70*, 849–859.

Snee, M., Benz, D., Jen, J., & Macdonald, P. M. (2008). Two distinct domains of Bruno bind specifically to the oskar mRNA. *RNA Biology*, *5*, 1–9.

Snee, M. J., Harrison, D., Yan, N., & Macdonald, P. M. (2007). A late phase of Oskar accumulation is crucial for posterior patterning of the Drosophila embryo, and is blocked by ectopic expression of Bruno. *Differentiation; Research in Biological Diversity*, *75*, 246–255.

Snee, M. J., & Macdonald, P. M. (2009). Dynamic organization and plasticity of sponge bodies. *Developmental Dynamics*, *238*, 918–930.

Sonoda, J., & Wharton, R. P. (1999). Recruitment of Nanos to hunchback mRNA by Pumilio. *Genes & Development*, *13*, 2704–2712.

Spradling, A. C., de Cuevas, M., Drummond-Barbosa, D., Keyes, L., Lilly, M., Pepling, M., et al. (1997). The Drosophila germarium: Stem cells, germ line cysts, and oocytes. *Cold Spring Harbor Symposia on Quantitative Biology*, *62*, 25–34.

St Johnston, D., Beuchle, D., & Nusslein-Volhard, C. (1991). Staufen, a gene required to localize maternal RNAs in the Drosophila egg. *Cell*, *66*, 51–63.

Struhl, G. (1989). Differing strategies for organizing anterior and posterior body pattern in Drosophila embryos. *Nature*, *338*, 741–744.

Struhl, G., Johnston, P., & Lawrence, P. A. (1992). Control of Drosophila body pattern by the hunchback morphogen gradient. *Cell*, *69*, 237–249.

Suyama, R., Jenny, A., Curado, S., Pellis-van Berkel, W., & Ephrussi, A. (2009). The actin-binding protein Lasp promotes Oskar accumulation at the posterior pole of the Drosophila embryo. *Development*, *136*, 95–105.

Tanaka, T., Kato, Y., Matsuda, K., Hanyu-Nakamura, K., & Nakamura, A. (2011). Drosophila Mon2 couples Oskar-induced endocytosis with actin remodeling for cortical anchorage of the germ plasm. *Development*, *138*, 2523–2532.

Tanaka, T., & Nakamura, A. (2008). The endocytic pathway acts downstream of Oskar in Drosophila germ plasm assembly. *Development*, *135*, 1107–1117.

Thomson, T., & Lasko, P. (2004). Drosophila tudor is essential for polar granule assembly and pole cell specification, but not for posterior patterning. *Genesis*, *40*, 164–170.

Thomson, T., Liu, N., Arkov, A., Lehmann, R., & Lasko, P. (2008). Isolation of new polar granule components in Drosophila reveals P body and ER associated proteins. *Mechanisms of Development*, *125*, 865–873.

Trcek, T., Grosch, M., York, A., Shroff, H., Lionnet, T., & Lehmann, R. (2015). Drosophila germ granules are structured and contain homotypic mRNA clusters. *Nature Communications*, *6*, 7962.

van Eeden, F. J., Palacios, I. M., Petronczki, M., Weston, M. J., & St Johnston, D. (2001). Barentsz is essential for the posterior localization of oskar mRNA and colocalizes with it to the posterior pole. *The Journal of Cell Biology*, *154*, 511–523.

Vanzo, N. F., & Ephrussi, A. (2002). Oskar anchoring restricts pole plasm formation to the posterior of the Drosophila oocyte. *Development*, *129*, 3705–3714.

Vanzo, N., Oprins, A., Xanthakis, D., Ephrussi, A., & Rabouille, C. (2007). Stimulation of endocytosis and actin dynamics by Oskar polarizes the Drosophila oocyte. *Developmental Cell*, *12*, 543–555.

Vazquez-Pianzola, P., Urlaub, H., & Suter, B. (2011). Pabp binds to the osk 3′UTR and specifically contributes to osk mRNA stability and oocyte accumulation. *Developmental Biology*, *357*, 404–418.

Voronina, E., Seydoux, G., Sassone-Corsi, P., & Nagamori, I. (2011). RNA granules in germ cells. *Cold Spring Harbor Perspectives in Biology*, *3*, a002774.

Webster, A., Li, S., Hur, J. K., Wachsmuth, M., Bois, J. S., Perkins, E. M., et al. (2015). Aub and Ago3 are recruited to Nuage through Two mechanisms to form a ping-pong complex assembled by Krimper. *Molecular Cell*, *59*, 564–575.

Wilhelm, J. E., Hilton, M., Amos, Q., & Henzel, W. J. (2003). Cup is an eIF4E binding protein required for both the translational repression of oskar and the recruitment of Barentsz. *The Journal of Cell Biology*, *163*, 1197–1204.
Wilsch-Brauninger, M., Schwarz, H., & Nusslein-Volhard, C. (1997). A sponge-like structure involved in the association and transport of maternal products during Drosophila oogenesis. *The Journal of Cell Biology*, *139*, 817–829.
Yang, N., Yu, Z., Hu, M., Wang, M., Lehmann, R., & Xu, R. M. (2015). Structure of Drosophila Oskar reveals a novel RNA binding protein. *Proceedings of the National Academy of Sciences of the United States of America*, *112*, 11541–11546.
Yoshida, S., Muller, H. A., Wodarz, A., & Ephrussi, A. (2004). PKA-R1 spatially restricts Oskar expression for Drosophila embryonic patterning. *Development*, *131*, 1401–1410.
Zhang, H., Elbaum-Garfinkle, S., Langdon, E. M., Taylor, N., Occhipinti, P., Bridges, A. A., et al. (2015). RNA controls PolyQ protein phase transitions. *Molecular Cell*, *60*, 220–230.
Zimyanin, V. L., Belaya, K., Pecreaux, J., Gilchrist, M. J., Clark, A., Davis, I., et al. (2008). In vivo imaging of oskar mRNA transport reveals the mechanism of posterior localization. *Cell*, *134*, 843–853.
Zimyanin, V., Lowe, N., & St Johnston, D. (2007). An oskar-dependent positive feedback loop maintains the polarity of the Drosophila oocyte. *Current Biology*, *17*, 353–359.

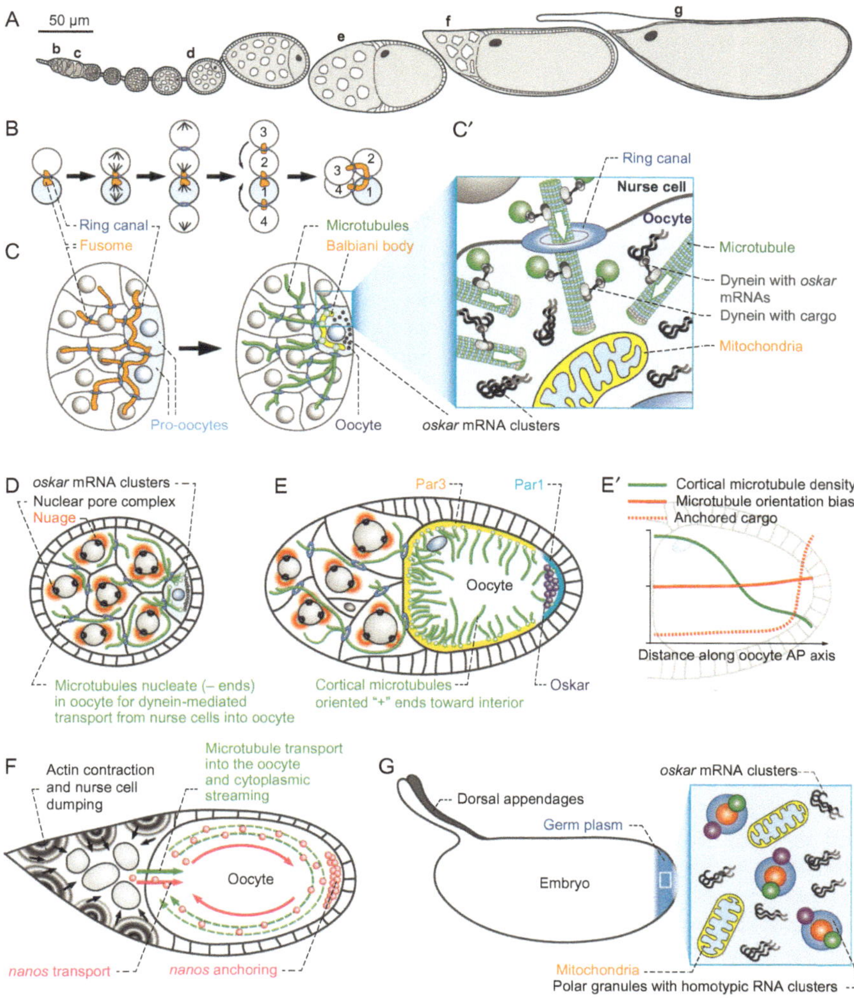

Ruth Lehmann, Figure 1 Establishment of anterior–posterior polarity during *Drosophila* oogenesis. (A) Ovariole with developing egg chambers. Each ovary contains 12–16 ovarioles. At the tip (left, b and c regions) is the germarium, where 2–3 germ line stem cells (GSCs) generate a cytoblast that matures into an egg chamber consisting of the oocyte and 15 nurse cells surrounded by somatic follicle cells (stage 1 egg chamber). Daily stem cell division produces a chain of maturing egg chambers (right, g). Small letters on top refer to stages depicted in other figure panels. (B) Division pattern of the cystoblast, the GSC daughter in the germarium. During four synchronous divisions, cells remain connected via ring canals with fusome material traversing the cyst. The first two cells born are connected by four ring canals, the next two by three, the next four by two, and the last eight by one ring canal each. Note that the earlier born cells inherit proportionally more fusome material. (C) 16-cell cyst in germarium, all cells are

(Continued)

Ruth Lehmann, Figure 1—Cont'd connected by ring canals and fusomes. The two cells with four ring canals "pro-oocytes" are positioned posteriorly within the cluster and enter into meiosis (left), only one of the two cells remains in meiosis and arrests in meiotic prophase (right). All other cells develop into polyploid nurse cells, which continue to be connected to the oocyte by ring canals. All centrosomes migrate into the oocyte and a microtubule network nucleate from the oocyte to the nurse cells. Dynein-mediated transport brings cargo and RNAs from the nurse cell into the oocyte. Densely packed mitochondria and RNA form the Balbiani body (yellow in C and C′). (C′) Oocyte cytoplasm indicating microtubule polarity and RNA transport particles. (D) Stage 5 egg chamber. Nurse cells and oocyte are surrounded by somatic follicle cells. Continued dynein-mediated RNA transport from the nurse cells into oocyte. Microtubules are oriented to nucleate (− pole) from oocyte and extend into nurse cells. Nuage: RNA-rich granules form around nurse cell nucleus often in close proximity of nuclear pore. (E) Stage 10A egg chamber. Mutually exclusive interactions between Par-1 (blue) at the posterior pole and the Par-3/Bazooka complex (yellow) at the lateral and anterior cortex establish polarity within the oocyte. Microtubules nucleate from lateral and anterior cortex extending into the oocyte and are inhibited at the posterior pole by Par-1. Kinesin motors transport *osk* RNA to the posterior pole, where it is translated. Dynein-mediated transport continues to bring *osk* RNA from the nurse cells to the oocyte. Somatic follicle cells surround the egg chamber. (E′) Summary of microtubule organization and *osk* RNA transport dynamics during stage 10A. Microtubule network is denser at the anterior than posterior of oocyte (green). A slight bias of kinesin-mediated microtubule transport (red line) leads to enrichment of *osk* RNA at posterior pole over time (red stippled). (F) Stage 10B/11, cytoplasmic streaming. The actin meshwork is disassembled in the oocyte and microtubules align in bundles (green). This leads to a streaming process where RNAs and proteins are mixed within the oocyte and additional RNAs/proteins are drawn into the oocyte (green arrow). Effector RNAs like *nanos* (red) become tethered at posterior pole. (G) Mature egg. Germ plasm is assembled and polar granules of similar size to mitochondria are found in a precise region at the posterior pole. Effector RNAs are localized to the granules, Osk, Vasa, Tud, and Aub proteins are essential components of the granules but *osk* RNA is not associated with polar granules (see Fig. 3E). *Panel (C′): After Roth and Lynch (2009). Panel (E′): After Parton et al. (2011).*

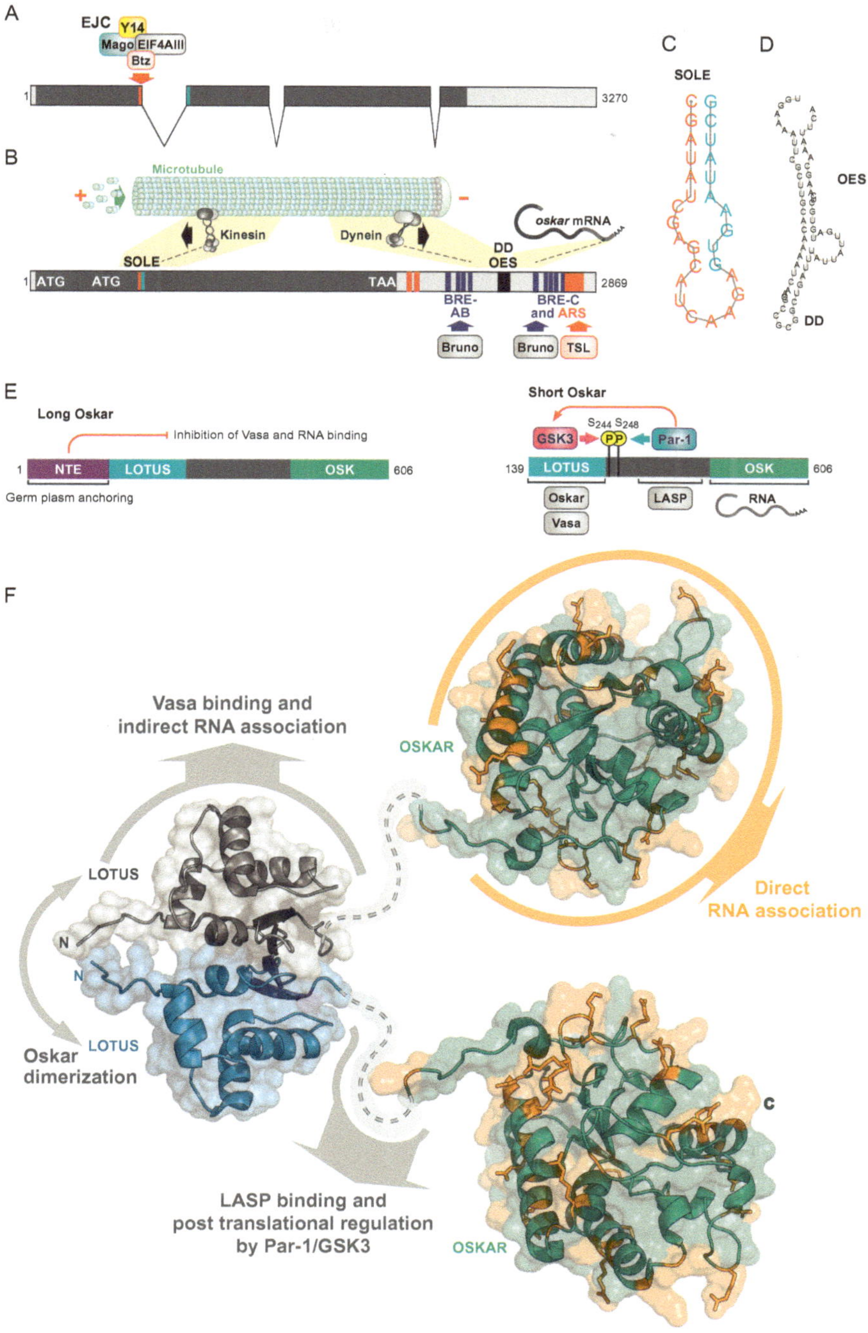

Ruth Lehmann, Figure 2 See legend on next page.

Ruth Lehmann, Figure 2—Cont'd Oskar RNA and protein history. (A) Structure of nascent *osk* RNA. Exon junction complex components bind to sequences flanking the first intron, which is critical for kinesin-mediated *osk* RNA transport. (B) *Osk* RNA transport particles are transported along microtubules by dynein (directed toward − end of microtubule) and kinesin (directed toward + end of microtubule) motors. The oocyte entry signal (OES) (D) regulates the dynein-mediated transport of *osk* RNA into the early oocyte. (C) The Spliced Oskar Localization Element (SOLE) forms as a consequence of first intron splicing, where 18 nt from the first and 10 nt from the second exon merge into a stem loop structure. The DD palindromic sequence element (D) is located in the loop of the OES stem loop. Via RNA–RNA "kissing interaction" this element promotes RNA transport particle formation and hitchhiking of RNA lacking the SOLE element on those with the element. A, B, C regions are rich of Bruno regulatory elements (BRE), which repress RNA translation (blue), and IMP-binding elements, which promote translation and localization (red). Binding elements close to the most distal part of the 3′UTR are important for translational activation, via adenine-rich sequences (ARS) that mediate polyadenylation. (E) Structure of Oskar isoforms. Left: Long Oskar is a 606-amino acid protein with an N-terminal extension (NTE) required for actin filament nucleation and enhanced endocytosis at the posterior pole. Long Oskar suppresses short Oskar functions. Right: Short Oskar starts at Methionine 139. The Lotus domain dimerizes and interacts with Vasa (see also F), the C-terminal OSK domain resembles a SGNH hydrolase domain; however, the enzymatic amino acids are missing. Instead, the OSK domain is thought to interact with effector RNAs. (F) Structure diagram of Lotus and OSK domain. The Lotus domain forms a homodimer of two winged helix structures and interacts with Vasa, while the OSK domain forms a globular structure with basic and hydrophilic amino acids exposed to the surface, which are thought to interact with RNA.

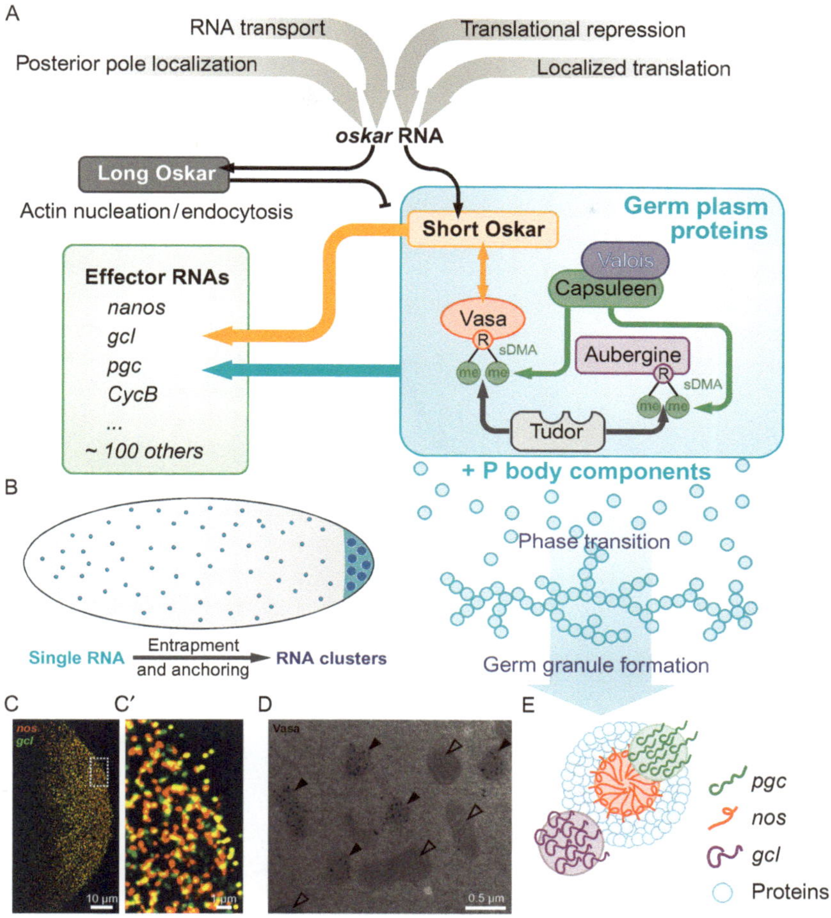

Ruth Lehmann, Figure 3 Germ plasm assembly. (A) Multiple controls described in Figs. 1 and 2 feed into the localization and translation of *osk*. Osk protein recruits other components of the germ plasm such as Vasa, Tudor, and Aubergine. Symmetric dimethylation of Arginine (sDMA) in Vasa and Aubergine by the methyltransferase Capsuleen allows binding to Tudor domain proteins (Kirino, Vourekas, Kim, et al., 2010; Kirino, Vourekas, Sayed, et al., 2010; Liu et al., 2010; Webster et al., 2015). This together with liquid to gel phase transitions could provide a possible basis for the dense assembly of polar granules. (B) Model for phase transition that occurs at posterior pole when concentration of germ plasm protein and RNAs become high (Brangwynne, 2013). (C) SIM image of *nos* and *gcl* RNA hybridization using Stellaris© RNA *in situ* hybridization probes. (C′) Higher magnification to reveal particle overlap. (D) Electron micrograph of germ plasm with polar granules marked by Vasa (filled arrow) and mitochondria (open arrow). (E) Organization of homotypic RNA clusters within polar granules according to Trcek et al. (2015).

CHAPTER FORTY

Emerging Modeling Concepts and Solutions in Stem Cell Research

Dmitri Papatsenko[*,†], **Ihor R. Lemischka**[*,†,‡,1]
[*]Department of Regenerative and Developmental Biology, Icahn School of Medicine at Mount Sinai, New York, USA
[†]Black Family Stem Cell Institute, Mount Sinai School of Medicine, New York, USA
[‡]Department of Pharmacology and System Therapeutics, Mount Sinai School of Medicine, Systems Biology Center New York, New York, USA
[1]Corresponding author: e-mail address: ihor.lemischka@mssm.edu

Contents

Abstract

Modern stem cell research, as well as other fields of contemporary biology involves quantitative sciences in many ways. Identifying candidates for key differentiation or reprogramming factors, tracing global transcriptome changes, or finding drugs is now broadly involves bioinformatics and biostatistics. However, the next key step, understanding the underlying reasons and establishing causal links leading to differentiation or reprogramming requires qualitative and quantitative biological models describing complex biological systems. Currently, quantitative modeling is a challenging science, capable to deliver rather modest results or predictions. What model types are the most popular and what features of stem cell behavior they are capturing? What new insights do we expect from the computational modeling of stem cells in the foreseeable future? Current review attempts to approach these essential questions by considering published quantitative models and solutions emerging in the area of stem cell research.

Current Topics in Developmental Biology, Volume 116
ISSN 0070-2153
http://dx.doi.org/10.1016/bs.ctdb.2015.11.040

1. PREDICTIVE STEM CELL BIOLOGY AND QUANTITATIVE MODELS

For centuries biology was rather a descriptive science, which nevertheless produced large bodies of descriptive data, served to found prominent theories (like the Darwin's theory). In the modern times, tremendous amount of genome-wide data at the molecular level have been collected and awaits its turn. These theories are expected to explain, for instance, conversion of stem cell phenotypes during differentiation or reprogramming given a set of predefined (and yet untested experimentally) conditions, such as removal of key pluripotency factors or changing thresholds of cytokines in the stem cell culturing media. Despite the difficulty of such tasks, the *predictive* stem cell biology, utilizing conceptual biological models is slowly gaining its strength.

Quantitative modeling of stem cell behavior involves processing input information by a system, represented either by a gene network or by an empirically derived set of rules mimicking the network/cell responses. Predefined conditions (such as knockdowns of selected pluripotency factors) correspond to inputs and the resulting, *predicted* by the model cell fate changes represent the outputs. Limited numbers of quantitative models have been constructed in the field of stem cells, yet fewer have been tested *in vitro* and *in vivo* with various degree of success. Known models, mentioned in the current and related reviews (Herberg & Roeder, 2015) explain alternative cell fate commitment in differentiation of embryonic stem cells (ESCs) (Herberg et al., 2014; Xu et al., 2014) or hematopoietic stem cells (HSCs) (Krinner et al., 2013; Tian & Smith-Miles, 2014), principles of stem cell self-renewal (Herberg et al., 2015), reprogramming (Buganim et al., 2012), or even spatial patterning in the context of developing embryos (Bessonnard et al., 2014; Junker et al., 2014; Krupinski et al., 2011).

2. NETWORK-BASED AND CELL-BASED MODELS

Quantitative models in systems biology may be focused on different aspects of cell function at different levels, including interaction networks, activity flows, process description, or entity description (Le Novere, 2015). In stem cell research, models are focused on gene regulatory networks (interaction networks), involving transcriptional regulation (process description) or cellular properties (entity relationships) involving cell signaling, proliferation, and differentiation.

One popular approach is to focus on simplified pluripotency gene regulatory networks (PGRN), described by systems of ordinary differentiation equations (ODEs). Solutions of such ODEs describe equilibrium states (also called dynamic attractors) achieved by the network components and may be interpreted as certain cell fates (Fig. 1A). In this case, concentrations of proteins within a single-cell serve as variables and cells are not interacting with each other, assuming that the network converges to the equilibrium in a cell-autonomous fashion. Such models were introduced for differentiation

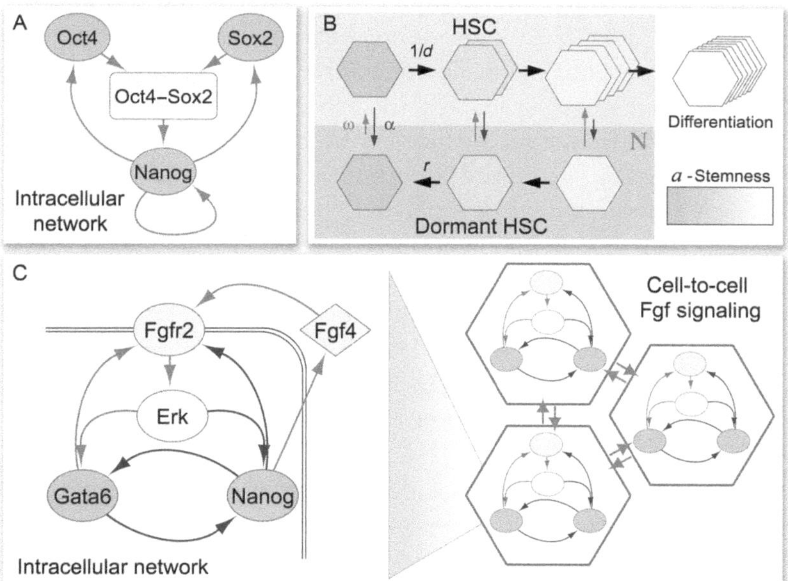

Figure 1 Network-focused, cell-focused, and multiscale models. (A) An early model for the core pluripotency circuit explaining formation of high and low populations of Nanog cells. (B) A model for proliferation of hematopoietic cells. Red and blue arrows show probabilities of transition from and to dormant HSC subpopulation. Green color demonstrates the degree of stemness; N, the number of HSCs is the model output. Letters are model parameters describing probabilities of cell conversion from one state to another. (C) A multiscale model for specification of Epi and PrE. At the lower level, the model takes into account intracellular network (left), at the higher level, model considers cell-to-cell signaling via Fgf4. The red arrows show activation and the blue arrows show repression. (A color version of this figure is located at the end of this chapter.)

of erythroid and myeloid lineages in the case of HSC, and differentiation of trophectoderm, epiblast (Epi), and primitive endoderm (PrE) in the case of ESC (Chickarmane & Peterson, 2008; Chickarmane et al., 2009; Glauche et al., 2010).

Another modeling approach focuses on properties of a cell as a whole and minimizes network perturbations within single cells to single values or even to a Boolean ON/OFF terms (Fig. 1B). However, in the context of the cell-focused models, stem cells may proliferate, differentiate, and communicate with other cells (or cell compartments, such as stem cell niches). Cell-focused models may also take into account lineage history and include details of epigenetic status, also descried via simple terms (ON/OFF) at the level of a whole cell. Thus, in a cell-focused model proposed by Roeder and coworkers for HSC (Roeder & Loeffler, 2002), the cells may proliferate, differentiate, and change their properties (parameters) stochastically, with a certain probability. The model assumed that dormant HSCs do not proliferate, but can eventually change their state and start differentiation and proliferation, give rise to other cells, including proliferating HSCs and dormant HSCs. Model output included total number of HSCs (variables), which may be stable, exhausted over time or oscillate with certain periodicity. Further development of the cell-focused HSC model incorporated details of human granulopoiesis and signaling factors (granulocyte colony-stimulating factor, G-CSF), affecting differentiation, but not details on organization of intracellular networks (Krinner et al., 2013). The model output represented the number of leukocyte cells, which was explored *in silico* by simulating cell damaging in HSC compartment by chemotherapy (model input). Scenarios predicted *in silico* were evaluated based on the corresponding clinical data.

Whole-cell properties may include gene activity status and gene activity potential, which includes epigenetic control of gene activation via histone modification or promoter DNA methylation. In a model proposed for differentiation and reprogramming of ESC by Artyomov and coworkers (Artyomov et al., 2010), properties of whole cells (resulting in certain cell fates) were considered in the context of cell lineages in a form of hierarchical trees. In that model, the cell fates depended on gene activity and epigenetic properties described via simple ON/OFF terms at the cellular level. The model assumed that stochastic switching of gene activities may lead to changes of epigenetic status and changes of cell phenotype; the epigenetic status may be inherited from ancestor cell, then recalculated and passed to the progeny. Given specific rules the model predicts distribution of phenotypes, which may be then compared to experimental data.

3. MULTISCALE MODELS INVOLVING GENE NETWORKS AND CELL SIGNALING

The two modeling strategies described above demonstrate two different biological scales or levels of modeling, one, focused on gene network properties and another one, focused on the properties of entire cells. At present, multiscale models are being developed, which take into account different biological scales (or levels) and combine cell properties with the properties of gene networks (Chapman, 2009; Degenring et al., 2004). Description of behavior of cell ensembles, such as colonies of ESCs or differentiating cell layers in vertebrate blastocyst may include both time and space domains into the multiscale models (Herberg & Roeder, 2015).

A recent multiscale model describing ESC heterogeneity in culture incorporates both, the intracellular core pluripotency network (interactions between Oct4, Sox2, and Nanog) and interactions between ES cells mediated by FGF/Erk signaling. In its turn, FGF/Erk is incorporated into the intracellular network (Fig. 1C). Supporting their model by live imaging of ESC colonies the authors came to a conclusion that each colony contains self-renewing (SR) cluster of cells, located primarily at the center of colony (Herberg et al., 2015). A model with a similar focus has been developed for early patterning of mouse blastocyst (Bessonnard et al., 2014). This model, developed by Bessonard and coworkers also incorporates FGF/Erk signaling as a linkage between the network level and cellular level and predicts the output ratios of embryonic tissues, which is then compared with the ratios observed *in vivo*. In both models the negative auto regulation via FGF/Erk signaling appears to produce bistable (Herberg et al., 2015) or even multistable (Bessonnard et al., 2014) dynamic systems, in which the dynamic point attractors (equilibrium system states) may correspond to either inner cell mass (ICM), Epi, or PrE cell fates (Bessonnard et al., 2014).

Another established biological model related to stem cells describes early differentiation of morula cells into ICM and trophectoderm (Strumpf et al., 2005). Initial quantitative models dedicated to this topic considered interaction of core pluripotency factors with major trophectoderm determinant, Cdx2 in the context of intracellular network only (Chickarmane & Peterson, 2008). The model assumed the presence of mutual repression between Oct4 and Cdx2 in the context of PGRN and the corresponding bistability, in which one stable system's state (cell fate) corresponds to ICM and another one to trophectoderm. PrE–Epi bistable switch in the same model was attributed to mutual repression between Nanog and Gata6. More recently, a multiscale model for blastocyst differentiation has been developed by Krupinski and coworkers (Krupinski et al., 2011); the model integrated properties of PGRN (limited largely by the mutual repression between Oct4 and Cdx2) as well as the cell properties represented by asymmetric cell division, possible maternal polarization signals and differential cell adhesion leading to cell segregation. Alternative scenarios of ICM and trophectoderm differentiation have been tested using this model, such as maternal predetermination versus self-organization (latter appeared to be more likely).

4. DETERMINISTIC VERSUS STOCHASTIC SCENARIOS, HETEROGENEITY OF STEM CELLS

Sorting ESCs based on gene expression revealed their heterogeneity with respect to major pluripotency factors such as Nanog, Rex1, and many others (Kalmar et al., 2009; Papatsenko et al., 2015; Toyooka et al., 2008). The idea of stem cell heterogeneity appears to be essential in the context of potential "priming" of stem cells toward a given lineage with the consequent differentiation or, alternatively, returning to pluripotency due to consequent fluctuations of gene expression. Expression of the same gene may vary from one cell to another due to intrinsic noise or due to extrinsic factors, caused by the cell's surroundings (Hilfinger & Paulsson, 2011; Mettetal et al., 2006; Singer et al., 2014). In quantitative models, presence of noise or variation in the gene expression is usually combined with causal relationships, such as regulatory network links, so the stochastic model component is added to the deterministic one. The two components may be considered as the model dimension (Degenring et al., 2004), where determinism and fixed-point (or trajectories) stable state attractors dominate on one end of the scale (no noise, determinism) and clouds of probable states (high noise masking determinism) dominate the other. Balance between the two model components produces fuzzy attractors or predicts substantial variations, which are nevertheless placed around the point of equilibrium (Fig. 2A–C).

Detected variations in expression of Nanog were introduced into deterministic network-focused model combining core PGRN components Oct4, Sox2, and Nanog (Glauche et al., 2010). Outcome of that model represented distributions of probable cell states (clouds), which were then compared with distributions observed in the cell sorting assays (Kalmar et al., 2009). Positive autoregulation of Nanog in that model produced

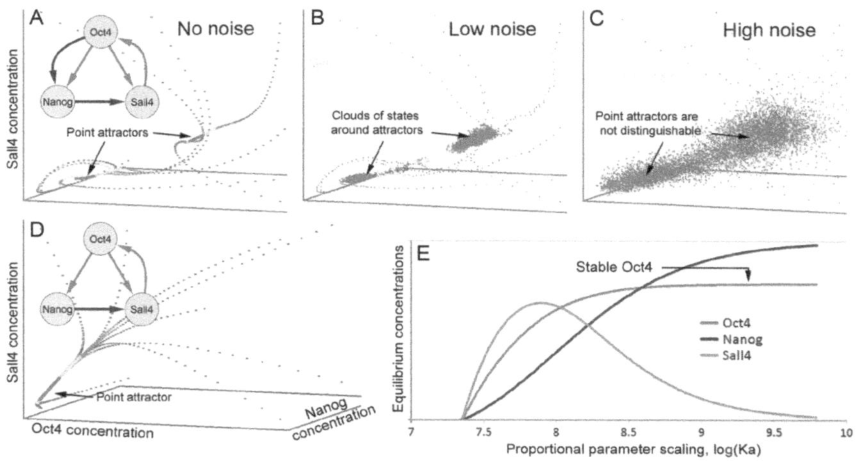

Figure 2 Deterministic, stochastic models and robust models. (A)–(C) An example of a bistable model, which explains formation of two populations of Nanog in ESC cultures. In this model, Oct4 is assumed to be a dual regulator of Nanog. The red arrows in the network motif correspond to activation and the blue arrows correspond to a repression. (A) A deterministic scenario (see the network motif), in which noise is absent; the two point attractors correspond to two stable states (e.g., high and low Nanog). (B) The same deterministic model with adding low noise in the gene expression (a stochastic component). The attractors are visible as clouds of states. (C) The same deterministic system with high level of noise, the stochastic component prevails and the attractors disappear, solution represents a single cloud of probable states. (D) A similar model explaining stability of a single state, such as observed in the case of cells grown in 2i-LIF media. The network motif represents incoherent feedforward loop with a feedback. (E) Bifurcation diagram of the model shown on (D); proportional scaling of all regulatory parameters (*X* axis, binding constants) has little effect on equilibrium concentration of Oct4. The model appears to be robustness to proportional scaling of all parameters. (A color version of this figure is located at the end of this chapter.)

two such "cloud-states," two distribution peaks, which were interpreted as Nanog-high and Nanog-low cell states, previously observed in the cell sorting assays.

Experimental measurements of gene expression in single HSCs for multiple genes using quantitative RT-PCR have identified variation in the expression of many genes, which was assumed to be essential for priming these HSCs to differentiation. Teles and coworkers (Teles et al., 2013) have recently explored such data in order to identify genes, which may be essential for transition from SR HSCs to primed, erythroid-committed progenitors. The model included stochastic components described via different rates of transcription bursts for different genes (Mettetal et al., 2006) and deterministic components, described via causal relationships between HSC genes detected in the single-cell measurements. The authors have found that expression bursts of Gata2 define commitment-permissive states, where commitment may occur through stochastically turning on several gene combinations, including Gata2 and Mpo. Earlier model assumed that HSCs differentiate to hematopoietic multipotent progenitor cell, where mutual repression between Gata1/2 and Pu1 specifies erythroid and lymphoid lineages correspondingly (Chickarmane et al., 2009). Alternative biological models describing commitment to myeloid and lymphoid lineages have been recently described elsewhere (Nimmo et al., 2015).

Special interest represents models dedicated to reprogramming of somatic cells to induced pluripotent cells (Takahashi & Yamanaka, 2006). It has been suggested that forced expression of reprogramming factors (Oct4, Sox2, Klf4, c-Myc, OSKM) in somatic cells leads to stochastic reactivation of pluripotency genes and pathways (Hanna et al., 2009; Takahashi & Yamanaka, 2006; Yamanaka, 2009). During such early, stochastic phase of reprogramming different components of PGRN may be turned ON. Later on, this leads to reactivation of the entire core PGRN and the consequent deterministic phase, where the cells acquire pluripotency. Most frequent markers of the early stochastic phase were Sall4 and Lin28 by Sox2, while expression of Oct4 and Nanog was detected at later reprogramming stages (Buganim et al., 2012).

The examples above demonstrate stochastic and deterministic components in various models describing stem cell behavior and how stochastic processes may overcome or break the limits imposed by deterministic relationships between the genes in regulatory networks.

5. MODULARITY, ROBUSTNESS, AND FEEDBACK CONTROL IN PLURIPOTENCY GENE NETWORKS

In the case of gene regulatory network or any other system with established causal relationships, the key minimal element sufficient to make a quantitative model may be a network motif, a small network combining two to three (rarely more) key genes and regulatory links, known from biology or high-throughput assays. Most of the current models invoke known network motifs and model solutions previously known from systems biology and network motif theory (Milo et al., 2002; Shen-Orr et al., 2002). Feedforward and feedback regulatory loops, toggle switches, and their combinations have been noticed in the context of nearly all deterministic model components for ESCs and HSCs (Herberg & Roeder, 2015). However, differentiation as well as reprogramming involves many steps and the future quantitative models will need to incorporate several models in a row, like the models proposed for erythroid (Pu1–Gata1) or PrE–Epi (Gata6–Nanog) toggle switches. It has been suggested that integrated modular design models may come into the play, where collections of simple step-specific models may be synchronized and integrated together (Le Novere, 2015). In this case, the modularity of PGRN may or may not reflect biological properties, but it may be a necessary methodological step on the road to integrated models.

Evaluation of quantitative models may be carried out using experimental data, however, amount of open parameters (typically corresponding to unknown rates of gene regulation) is sometimes difficult to control, so the resulting models may fit any kind of data, which is known as overfitting problem. One way to overcome this problem is to identify network motifs, which would produce similar solutions, sustaining substantial changes of parameters (rates of gene regulation). One such model has been recently proposed for ESCs (Papatsenko et al., 2015). The model explains stabilization of presumably high-Nanog state via regulatory loop incorporating Oct4, Nanog, and Sall4. Architecture of the proposed network motif conforms to incoherent feedforward loop with a feedback (iFFL-FB, Fig. 2D) (Goentoro et al., 2009; Papatsenko, 2009) and this particular architecture sustains proportional scaling of all regulatory parameters. Indeed, the iFFL-FB combines positive and negative regulatory links, which compensate each other influences such that the equilibrium concentration of the major pluripotency factor Oct4 remains the same within broad range of the parameter values (orders of magnitude, Fig. 2E).

One critical element in the described model for Oct4 (as well as in many other models) is feedback control of the core pluripotency machinery from factors, which otherwise (in differentiation) appear rather as downstream targets in the context of PGRN. Importance of such feedback control and its role in the formation of stable pluripotency states has been explored by MacArthur and coworkers (MacArthur et al., 2012). The authors have established that downregulation of Nanog in a cell line carrying rescue Dox-inducible *Nanog* gene leads to the loss of pluripotency with substantial delay. Upon removal of Nanog, there is a time window, in which the major core PGRN factors Oct4, Sox2, and Klf4 still maintain their expression. In that study the authors attributed the observed robustness of PGRN to multiple feedback control loops, acting in parallel fashion. Apparently, such architectural features ensure stability of cell states, including pluripotency states and the parallel control serves as a major source of robustness (sustainability to parameter changes), at least in the context of ESC PGRN.

Feedback control via multiple parallel circuits may be a feature reaching far beyond stem cell field and supplying robustness to many (if not all) biological systems residing in steady states. In the case of stem cells, the feedback control may be achieved at the level of intracellular networks, such as Oct4–Nanog–Sall4, or at the level of cells, communicating via Fgf, Wnt, Bmp, and other signaling pathways in a paracrine fashion, as described above in the example of Fgf/Erk/Nanog loop.

6. KEY PROBLEMS AND FUTURE DIRECTIONS

Very often, construction of quantitative models is limited by biological knowledge available for a given system. In a certain sense, qualitative coarse-grain biological models come first and then updated by quantitative models with better level of detail. Biological models fuzzy in the part of the core factors Oct4, Nanog, Sox2, Esrrb, Tbx3, and their connectivity with other genes in the context of PGRN. Hierarchy of PGRN has been suggested, but remains inconsistent among various sources (Buganim et al., 2012; Papatsenko et al., 2015). Two recent studies established minimal PGRNs, which would explain gene expression in the core factor knockdown studies via Boolean network models (Dunn et al., 2014; Xu et al., 2014). It turned out that in each case the minimal network included around 12–15 genes and 20–30 regulatory links; however, the consistency between the two reconstructed networks on the part of overlapping nodes was low (8 shared nodes, ~50%) and on the part of regulatory connectivity was almost nonexistent (3–5 common regulatory links, ~20%). Nevertheless, the both studies explained available experimental data with a similar degree of fitness. This example illustrates well the current state of biological knowledge on the part of PGRN architecture and the role of the core factors (empirically shown to be reprogramming factors); identifying exact architecture of PGRN for both ESCs and HSCs remains a dilemma. This is not mentioning problems related to modeling such complex networks (dozens of nodes—variables) were the same network architecture may produce a broad spectrum of completely different dynamic solutions (different sets of dynamic attractors) (Herberg & Roeder, 2015; Papatsenko & Lemischka, 2015).

Since the exact architectures of PGRNs, based on available data remain obscure, new approaches to generate and analyze both genome-wide and single-cell data may be necessary in order to establish deterministic relationships among the genes. In many knockdown studies, measurements are carried out after 2–3 days or more (Nishiyama et al., 2013), thus masking direct gene responses and resulting in many indirect targets. An alternative approach would be to measure early responses (Ivanova et al., 2006; Nishiyama et al., 2009) and find targets responding within 12–24 h after removal of a regulator (Martello et al., 2012). Networks and deterministic models produced using this approach may be substantially different, but they will better reveal direct links in PGRNs.

Studies of variations in gene expression and introduction of stochastic models require very precise measurements of mRNA levels in single cells using RT-PCR or deep sequencing. However, such measurements introduce technical noise, which may be comparable with the biological noise itself. In a recent study by Grun and coworkers (Grun et al., 2014), a precise assessment of technical versus biological noise has been done for ESCs. Using DNA bar-coding based on unique molecular identifiers (UMIs) (Kivioja et al., 2011), the authors were able to estimate the absolute number of transcripts in different cells as the number of different UMIs for a given transcript. This study has shown that biological noise caused by variation of gene expression exceeds technical noise produced by various amplification and sequencing techniques.

Finally, new imaging methods may have great impact for building quantitative models. In a recent study of ESC heterogeneity, Herberg and coworkers (Herberg et al., 2015) introduced an imaging approach, which allowed the authors to monitor gene expression not only in single cells, but explore spatial gene expression patterns in the context of entire ESC colonies. Such spatial imaging data at a single-cell resolution has been already being utilized in model organisms, such as fly, mouse embryos, or other developmental systems (Junker et al., 2014; Surkova et al., 2013).

REFERENCES

Artyomov, M. N., et al. (2010). A model for genetic and epigenetic regulatory networks identifies rare pathways for transcription factor induced pluripotency. *PLoS Computational Biology*, *6*, e1000785.

Bessonnard, S., et al. (2014). Gata6, Nanog and Erk signaling control cell fate in the inner cell mass through a tristable regulatory network. *Development*, *141*, 3637–3648.

Buganim, Y., et al. (2012). Single-cell expression analyses during cellular reprogramming reveal an early stochastic and a late hierarchic phase. *Cell*, *150*, 1209–1222.

Chapman, S. (2009). Multiscale mathematical modelling in medicine and biology. In *18th World IMACS*.

Chickarmane, V., & Peterson, C. (2008). A computational model for understanding stem cell, trophectoderm and endoderm lineage determination. *PLoS One*, *3*, e3478.

Chickarmane, V., et al. (2009). Computational modeling of the hematopoietic erythroid-myeloid switch reveals insights into cooperativity, priming, and irreversibility. *PLoS Computational Biology*, *5*, e1000268. Epub 1002009 Jan 1000223.

Degenring, D., et al. (2004). Discrete event, multi-level simulation of metabolite channeling. *Biosystems*, *75*, 29–41.

Dunn, S. J., et al. (2014). Defining an essential transcription factor program for naive pluripotency. *Science*, *344*, 1156–1160.

Glauche, I., et al. (2010). Nanog variability and pluripotency regulation of embryonic stem cells—Insights from a mathematical model analysis. *PLoS One*, *5*, e11238.

Goentoro, L., et al. (2009). The incoherent feedforward loop can provide fold-change detection in gene regulation. *Molecular Cell*, *36*, 894–899.

Grun, D., et al. (2014). Validation of noise models for single-cell transcriptomics. *Nature Methods*, *11*, 637–640.

Hanna, J., et al. (2009). Direct cell reprogramming is a stochastic process amenable to acceleration. *Nature*, *462*, 595–601. Epub 2009 Nov 2008.

Herberg, M., & Roeder, I. (2015). Computational modelling of embryonic stem-cell fate control. *Development*, *142*, 2250–2260.

Herberg, M., et al. (2014). A model-based analysis of culture-dependent phenotypes of mESCs. *PLoS One*, *9*, e92496.

Herberg, M., et al. (2015). Image-based quantification and mathematical modeling of spatial heterogeneity in ESC colonies. *Cytometry. Part A*, *87*, 481–490.

Hilfinger, A., & Paulsson, J. (2011). Separating intrinsic from extrinsic fluctuations in dynamic biological systems. *Proceedings of the National Academy of Sciences of the United States of America*, *108*, 12167–12172.

Ivanova, N., et al. (2006). Dissecting self-renewal in stem cells with RNA interference. *Nature*, *442*, 533–538.

Junker, J. P., et al. (2014). A predictive model of bifunctional transcription factor signaling during embryonic tissue patterning. *Developmental Cell*, *31*, 448–460.

Kalmar, T., et al. (2009). Regulated fluctuations in nanog expression mediate cell fate decisions in embryonic stem cells. *PLoS Biology*, 7, e1000149. Epub 1002009 Jul 1000147.

Kivioja, T., et al. (2011). Counting absolute numbers of molecules using unique molecular identifiers. *Nature Methods*, *9*, 72–74.

Krinner, A., et al. (2013). Merging concepts—Coupling an agent-based model of hematopoietic stem cells with an ODE model of granulopoiesis. *BMC Systems Biology*, 7, 117.

Krupinski, P., et al. (2011). Simulating the mammalian blastocyst—Molecular and mechanical interactions pattern the embryo. *PLoS Computational Biology*, 7, e1001128.

Le Novere, N. (2015). Quantitative and logic modelling of molecular and gene networks. *Nature Reviews. Genetics*, *16*, 146–158.

MacArthur, B. D., et al. (2012). Nanog-dependent feedback loops regulate murine embryonic stem cell heterogeneity. *Nature Cell Biology*, *14*, 1139–1147.

Martello, G., et al. (2012). Esrrb is a pivotal target of the Gsk3/Tcf3 axis regulating embryonic stem cell self-renewal. *Cell Stem Cell*, *11*, 491–504.

Mettetal, J. T., et al. (2006). Predicting stochastic gene expression dynamics in single cells. *Proceedings of the National Academy of Sciences of the United States of America*, *103*, 7304–7309. Epub 2006 Apr 7328.

Milo, R., et al. (2002). Network motifs: Simple building blocks of complex networks. *Science*, *298*, 824–827.

Nimmo, R. A., et al. (2015). Primed and ready: Understanding lineage commitment through single cell analysis. *Trends in Cell Biology*, *25*, 459–467.
Nishiyama, A., et al. (2009). Uncovering early response of gene regulatory networks in ESCs by systematic induction of transcription factors. *Cell Stem Cell*, *5*, 420–433.
Nishiyama, A., et al. (2013). Systematic repression of transcription factors reveals limited patterns of gene expression changes in ES cells. *Scientific Reports*, *3*, 1390.
Papatsenko, D. (2009). Stripe formation in the early fly embryo: Principles, models, and networks. *Bioessays*, *31*, 1172–1180.
Papatsenko, D., & Lemischka, I. R. (2015). NetExplore: A web server for modeling small network motifs. *Bioinformatics*, *31*, 1860–1862.
Papatsenko, D., et al. (2015). Single-cell analyses of ESCs reveal alternative pluripotent cell states and molecular mechanisms that control self-renewal. *Stem Cell Reports*, *5*, 207–220.
Roeder, I., & Loeffler, M. (2002). A novel dynamic model of hematopoietic stem cell organization based on the concept of within-tissue plasticity. *Experimental Hematology*, *30*, 853–861.
Shen-Orr, S. S., et al. (2002). Network motifs in the transcriptional regulation network of Escherichia coli. *Nature Genetics*, *31*, 64–68.
Singer, Z. S., et al. (2014). Dynamic heterogeneity and DNA methylation in embryonic stem cells. *Molecular Cell*, *55*, 319–331.
Strumpf, D., et al. (2005). Cdx2 is required for correct cell fate specification and differentiation of trophectoderm in the mouse blastocyst. *Development*, *132*, 2093–2102. Epub 2005 Mar 2023.
Surkova, S., et al. (2013). Quantitative imaging of gene expression in Drosophila embryos. *Cold Spring Harbor Protocols*, *2013*, 488–497.
Takahashi, K., & Yamanaka, S. (2006). Induction of pluripotent stem cells from mouse embryonic and adult fibroblast cultures by defined factors. *Cell*, *126*, 663–676.
Teles, J., et al. (2013). Transcriptional regulation of lineage commitment—A stochastic model of cell fate decisions. *PLoS Computational Biology*, *9*, e1003197.
Tian, T., & Smith-Miles, K. (2014). Mathematical modeling of GATA-switching for regulating the differentiation of hematopoietic stem cell. *BMC Systems Biology*, *8*(Suppl. 1), S8.
Toyooka, Y., et al. (2008). Identification and characterization of subpopulations in undifferentiated ES cell culture. *Development*, *135*, 909–918.
Xu, H., et al. (2014). Construction and validation of a regulatory network for pluripotency and self-renewal of mouse embryonic stem cells. *PLoS Computational Biology*, *10*, e1003777.
Yamanaka, S. (2009). Elite and stochastic models for induced pluripotent stem cell generation. *Nature*, *460*, 49–52. http://dx.doi.org/10.1038/nature08180.

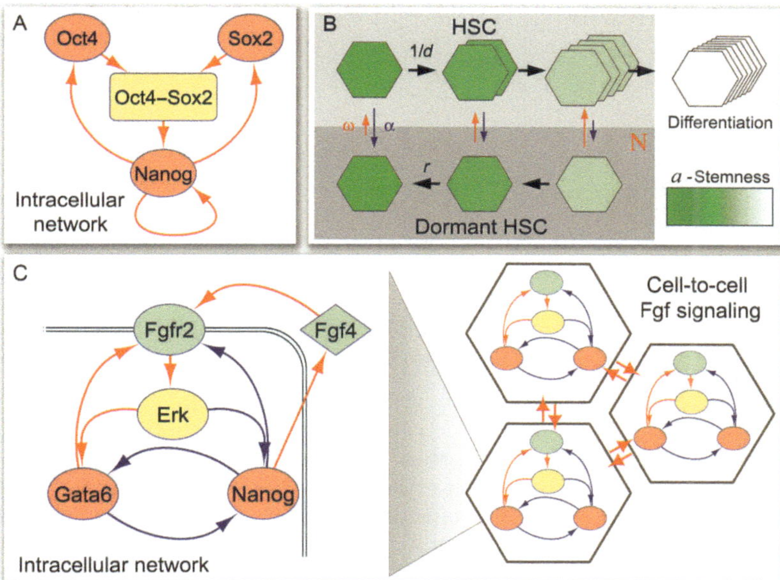

Dmitri Papatsenko and Ihor R. Lemischka, Figure 1 Network-focused, cell-focused, and multiscale models. (A) An early model for the core pluripotency circuit explaining formation of high and low populations of Nanog cells. (B) A model for proliferation of hematopoietic cells. Red and blue arrows show probabilities of transition from and to dormant HSC subpopulation. Green color demonstrates the degree of stemness; N, the number of HSCs is the model output. Letters are model parameters describing probabilities of cell conversion from one state to another. (C) A multiscale model for specification of Epi and PrE. At the lower level, the model takes into account intracellular network (left), at the higher level, model considers cell-to-cell signaling via Fgf4. The red arrows show activation and the blue arrows show repression.

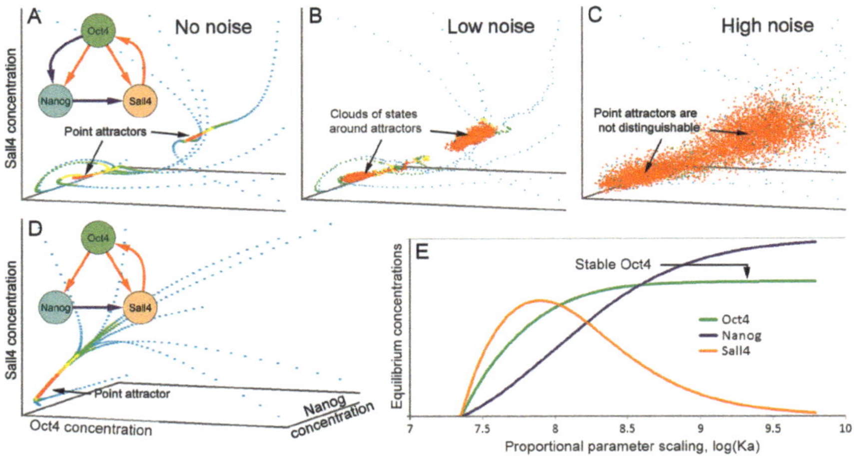

Dmitri Papatsenko and Ihor R. Lemischka, Figure 2 Deterministic, stochastic models and robust models. (A)–(C) An example of a bistable model, which explains formation of two populations of Nanog in ESC cultures. In this model, Oct4 is assumed to be a dual regulator of Nanog. The red arrows in the network motif correspond to activation and the blue arrows correspond to a repression. (A) A deterministic scenario (see the network motif), in which noise is absent; the two point attractors correspond to two stable states (e.g., high and low Nanog). (B) The same deterministic model with adding low noise in the gene expression (a stochastic component). The attractors are visible as clouds of states. (C) The same deterministic system with high level of noise, the stochastic component prevails and the attractors disappear, solution represents a single cloud of probable states. (D) A similar model explaining stability of a single state, such as observed in the case of cells grown in 2i-LIF media. The network motif represents incoherent feedforward loop with a feedback. (E) Bifurcation diagram of the model shown on (D); proportional scaling of all regulatory parameters (*X* axis, binding constants) has little effect on equilibrium concentration of Oct4. The model appears to be robustness to proportional scaling of all parameters.

INDEX

Note: Page numbers followed by "*f*" indicate figures, and "*t*" indicate tables.

D

E

I

Q

R

S

T

U

V

W

X

Z

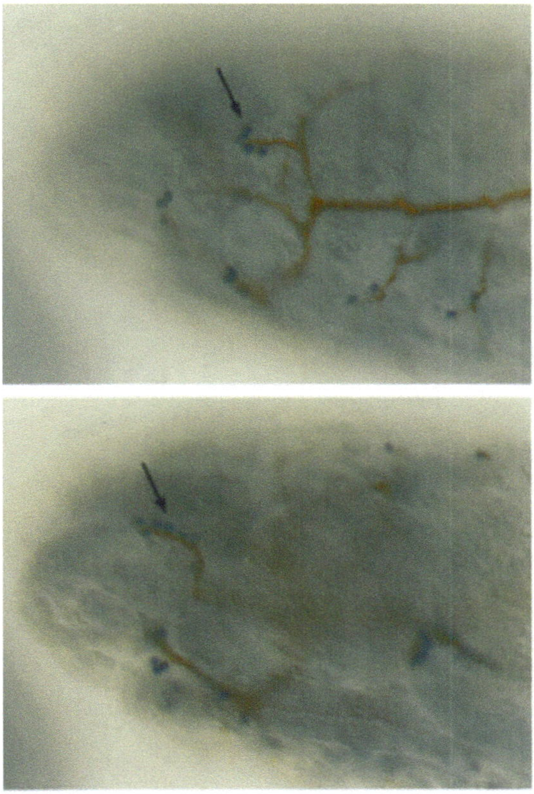

Markus Affolter, Figure 1 Head region of a *Drosophila* embryo showing the tracheal lumen (brown) and the nuclei of tracheal cells expressing dSRF (blue). *Taken from Affolter et al. (1994).*

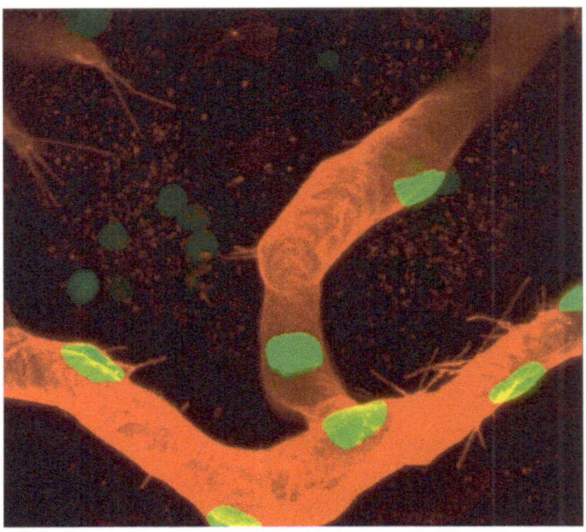

Markus Affolter, Figure 3 Blood vessels in the head of a zebrafish embryo expressing membrane-tethered RFP and nuclear GFP. *Picture taken by Anna Lenard.*

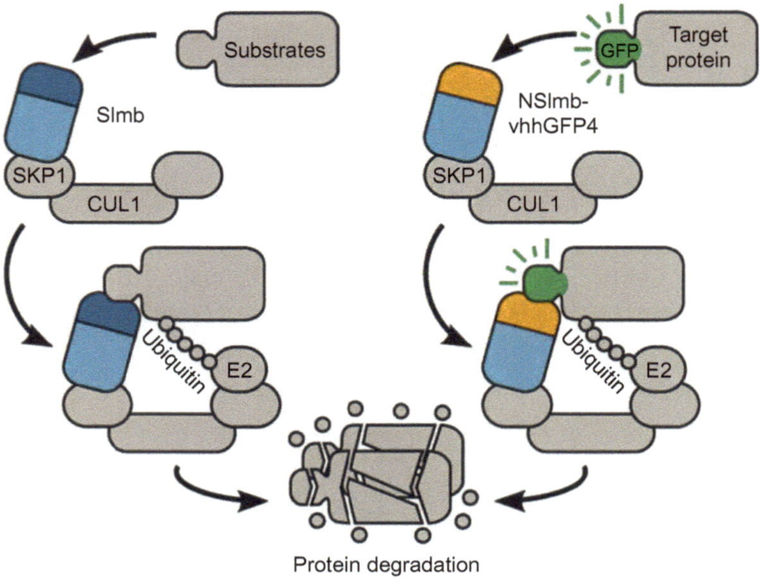

Markus Affolter, Figure 4 Protein degradation using an anti-GFP nanobody (vhhGFP4) to target GFP fusion proteins. *Taken from Caussinus et al. (2012).*

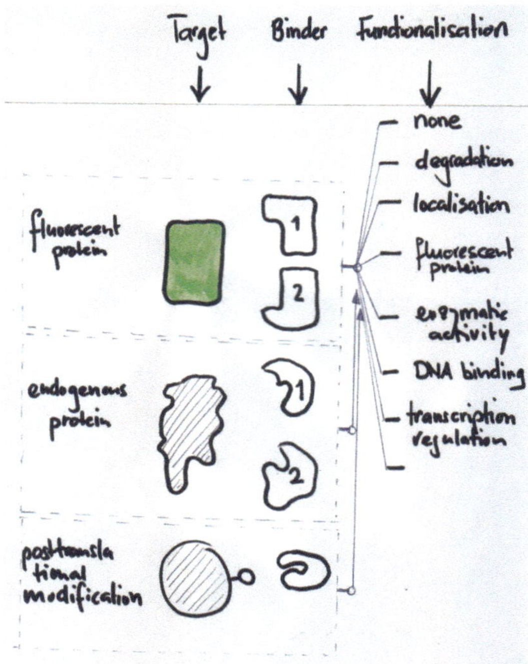

Markus Affolter, Figure 5 Protein binders (nanobodies, DARPINs, and others) can be isolated against fluorescent proteins, endogenous proteins, or against protein post-translational modifications. Such protein binders can be functionalized by fusing them with different protein domains. For more information, see Helma et al. (2015), Kaiser et al. (1844), and Pluckthun (2015).

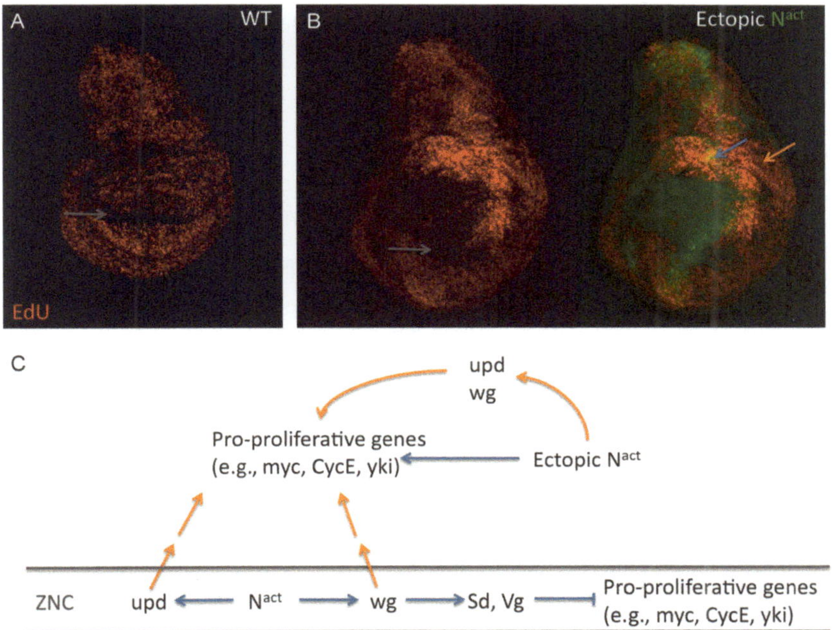

Diana M. Ho and Spyros Artavanis-Tsakonas, Figure 1 (A and B) Proliferation patterns in late third instar *Drosophila* wing imaginal discs stained with EdU. In wild-type discs (A), the ZNC (gray arrow) can be clearly distinguished by lack of EdU incorporation. When activated Notch (N^{act}) is expressed along the anterior–posterior boundary (labeled with GFP), the ZNC (gray arrow) expands and hyperproliferation occurs in the distal wing pouch and hinge regions. Both cell-autonomous (blue arrow) and non-cell-autonomous (orange arrow) components can be seen. (C) Simplified schematic of Notch gene interactions driving proliferation in the wing disc. Blue arrows represent cell-autonomous events, whereas orange arrows represent non-cell-autonomous events. See text for details.

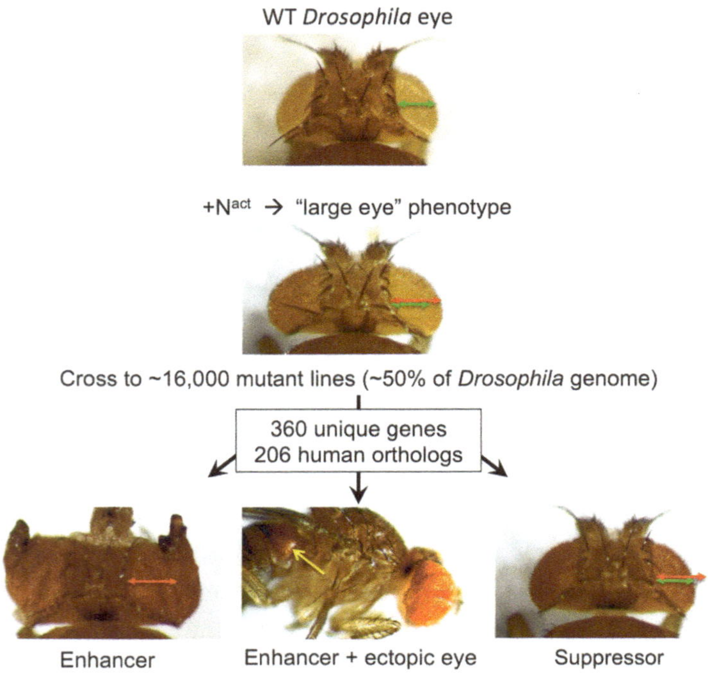

Diana M. Ho and Spyros Artavanis-Tsakonas, Figure 2 Schematic of and representative phenotypes from our genetic screen for modifiers of Notch-induced proliferation in the *Drosophila* eye. Flies ectopically expressing activated Notch (N^{act}) in the developing eye, which results in a "large eye" phenotype, were crossed to ~16,000 mutant lines representing ~50% of the *Drosophila* genome. Three hundred and sixty unique modifier genes, which caused either enhancement or suppression of the N^{act} phenotype, sometimes in conjunction with ectopic eyes, were identified, of which 206 had clear human orthologs.

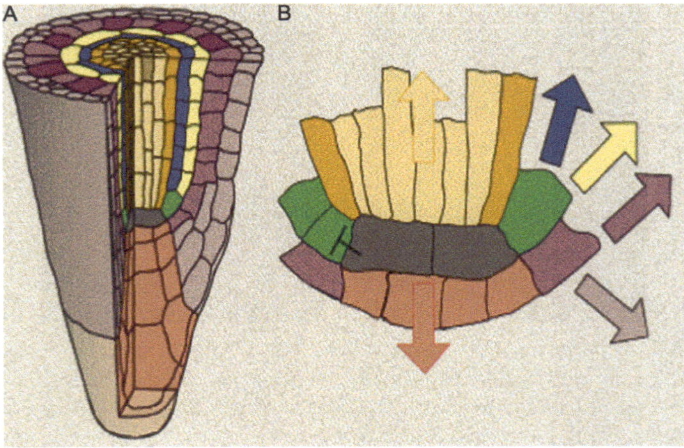

Philip N. Benfey, Figure 1 Stem cell niche in the root tip. (A and B) The four stem cell populations are adjacent to the quiescent center (QC) in gray. The cells immediately above the QC in beige give rise to the vascular tissue, the green cells give rise to the ground tissue (endodermis and cortex), the purple cells give rise to the epidermis and lateral root cap, and the pink cells give rise to the columella root cap. *From Benfey and Scheres (2000).*

Philip N. Benfey, Figure 2 Division of the cortex/endodermal initial. The stem cell that gives rise to the cortex and endodermis divides first in the transverse (anticlinal) orientation to regenerate itself and a daughter, which divides in the longitudinal (periclinal) orientation to give the first cells of the endodermis and cortex lineages.

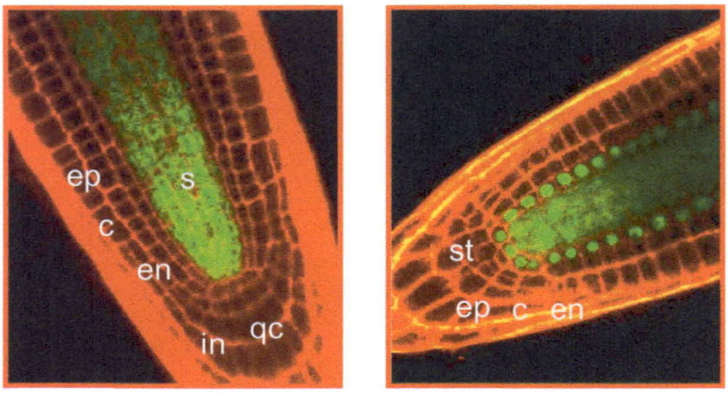

Philip N. Benfey, Figure 3 Expression of SHORT-ROOT (SHR) mRNA and protein. Left: the SHR promoter driving GFP. Right: the SHR promoter driving the SHR cDNA fused to GFP. SHR protein moves from the vascular tissue to the adjacent QC and endodermis. *From Nakajima, Sena, Nawy, and Benfey (2001).*

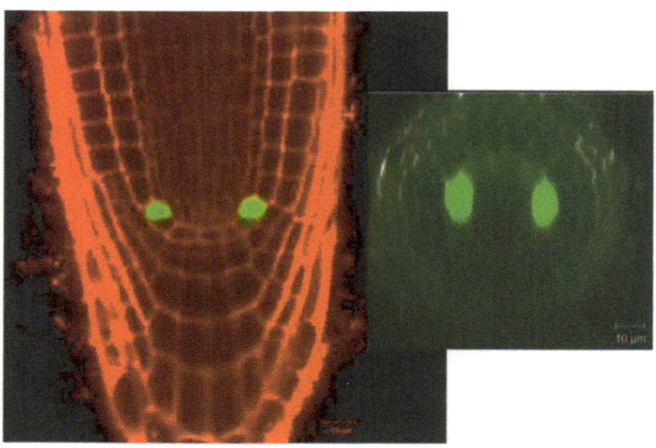

Philip N. Benfey, Figure 4 Expression of CYCD6-1. The CYCD6-1 promoter driving GFP exhibits expression in the cortex/endodermal initials prior to their periclinal division. *From Sozzani et al. (2010).*

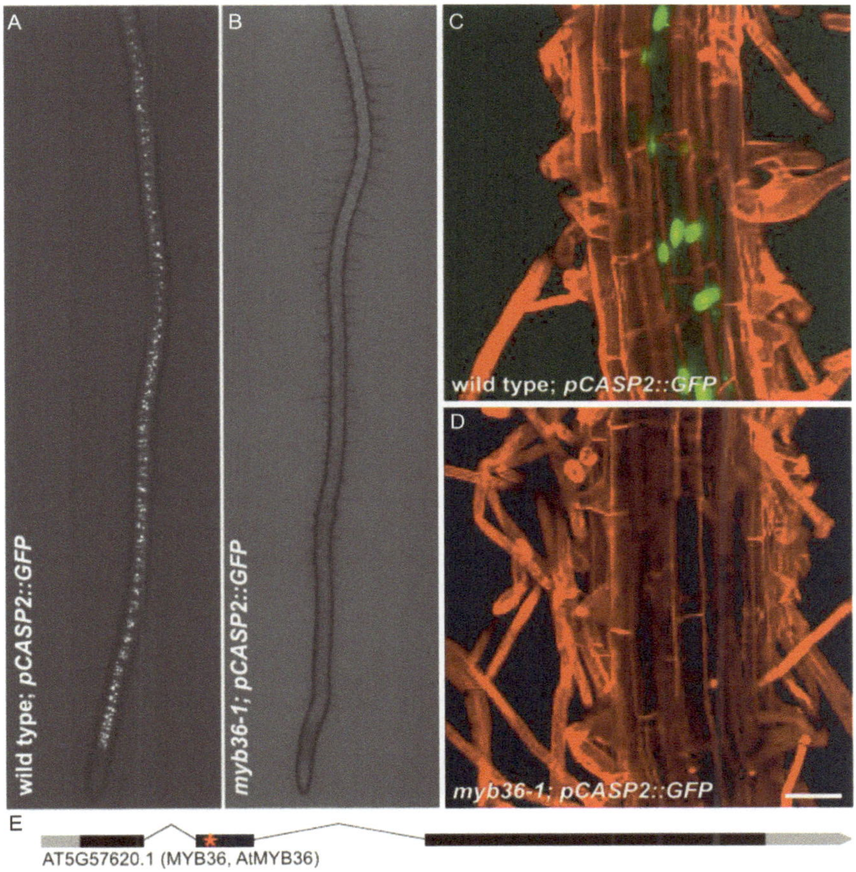

Philip N. Benfey, Figure 5 Characterization of a mutation in the MYB36 gene. (A and C) Expression of the CASP 2 gene is specific to the endodermis. (B and D) Expression is greatly reduced in the *myb36* mutant. (E) Structure of the MYB36 locus. *From Liberman, Sparks, Moreno-Risueno, Petricka, and Benfey (2015).*

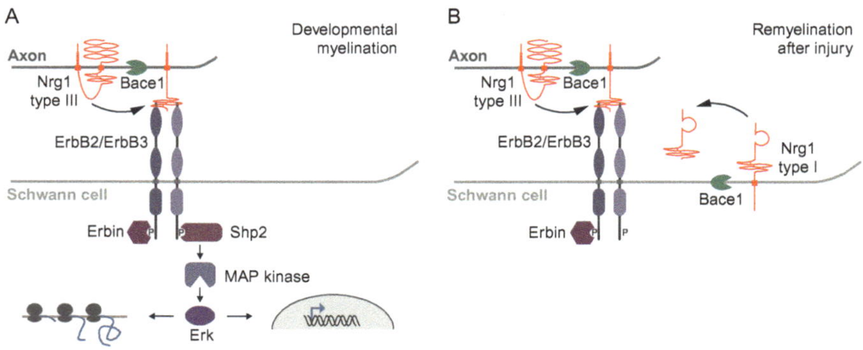

Carmen Birchmeier and David L.H. Bennett, Figure 1 Nrg1 signaling during developmental myelination and remyelination. Developmental myelination (A) depends on type III Nrg1 that is produced by peripheral sensory and motoneurons and presented in the axonal membrane. Nrg1 needs to be processed by Bace1 for full signaling activity. ErbB2/ErbB3 receptors on Schwann cells recognize this signal and transmit it. Erbin, an intracellular ErbB2 interacting protein, is important for full activation of ErbB receptors. Many signaling components act downstream of Nrg1/ErbB, and we display here Shp2 and Erk1/2, because mutations of Nrg1/ErbB, Shp2, and Erk1/2 all produce very similar or even identical changes in Schwann cell development and myelination. Activation of MAPK, i.e., expression of an activated variant, can substitute for Nrg1/ErbB3/Shp2 during Schwann cell development and myelination. MAPK represents thus an important readout of Nrg1/ErbB signaling. Transcriptional and translational responses to the Nrg1 signal are observed in developmental myelination. Remyelination after injury (B) reuses the Nrg1 signaling network, and remyelination and nerve repair are severely delayed in the absence of Nrg1. Two Nrg1 isoforms participate in remyelination, axonal type III and Schwann cell-derived type I. Full ErbB activation and efficient regeneration depend again on Bace1 and Erbin.

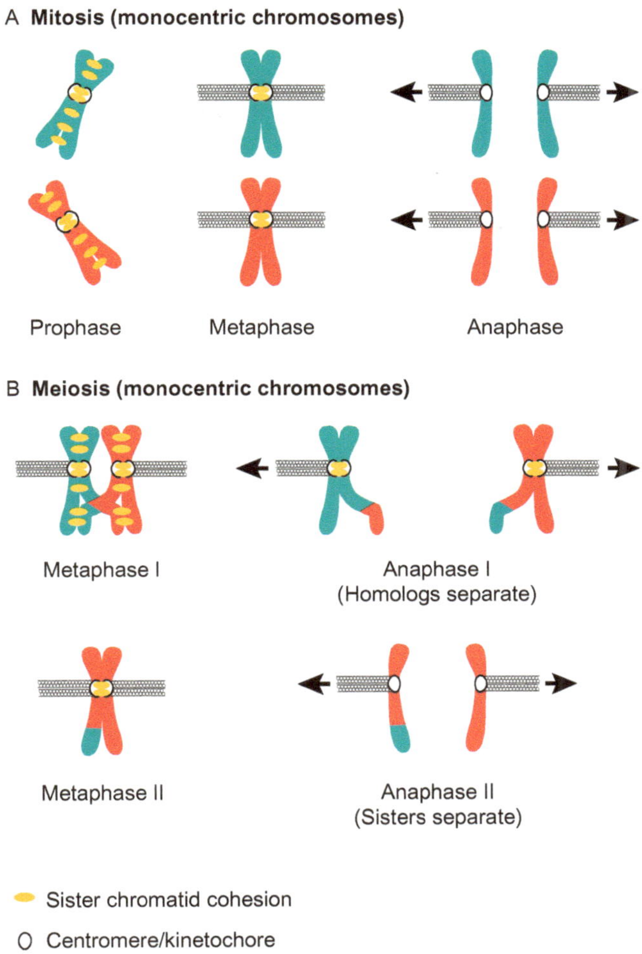

Aaron F. Severson *et al.*, Figure 1 Distinct patterns of chromosome segregation occur during meiosis and mitosis. The progression of a single pair of homologous chromosomes (blue and red) through mitosis (A) and meiosis (B) is shown. Before either division, DNA replication duplicates both homologs. Subsequently, each homolog is composed of two sister chromatids tethered by sister chromatid cohesion (yellow ovals). (A) In many organisms, cohesion along chromosome arms is released during mitotic prophase. At metaphase, sisters remain linked by centromeric cohesion. Bipolar attachment of sister kinetochores to microtubules from opposite spindle poles aligns each homolog on the metaphase plate. In anaphase, centromeric cohesion is released and sister chromatids segregate toward opposite poles. (B) During prophase of meiosis I, homologous chromosomes become linked by reciprocal exchange of DNA during crossover recombination. Both sister kinetochores of one homolog capture microtubules from the same spindle pole; the sister kinetochores of the other homolog attach to microtubules from the opposite spindle pole (for simplicity, microtubule contact with both sister chromatid kinetochores is not depicted). In anaphase of meiosis I, release of cohesion between chromosome arms allows homologs to separate; sisters remain tethered by centromeric cohesion. In anaphase II, centromeric cohesion is released, allowing sisters to separate.

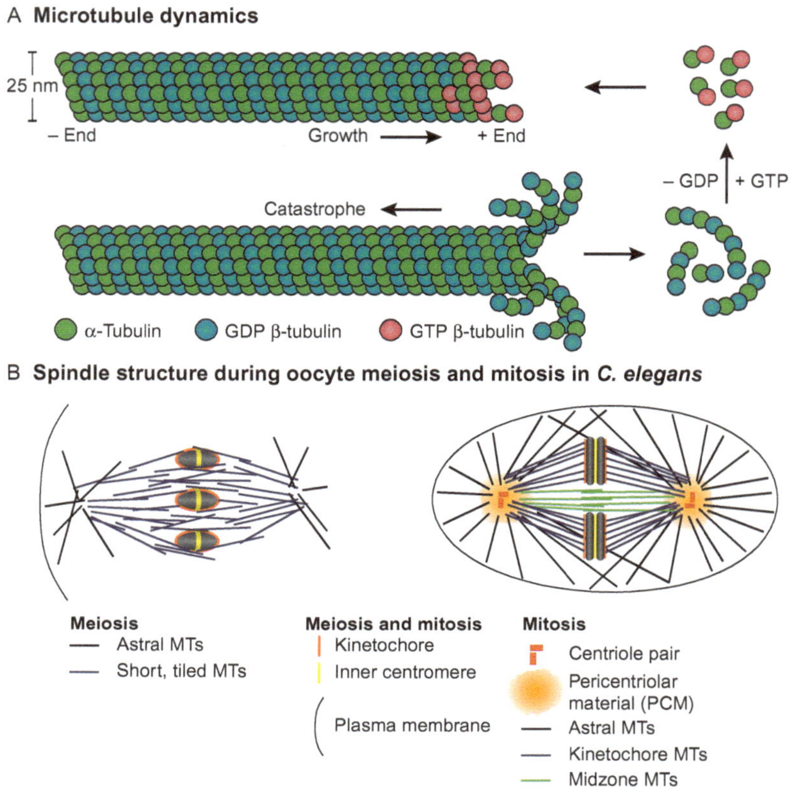

Aaron F. Severson *et al.*, Figure 2 Microtubule dynamics and spindle structures during oocyte meiosis and mitosis. (A) Microtubules are hollow tubes 25 nm in diameter formed by lateral association of 13 protofilaments, linear polymers composed α- and β-tubulin heterodimers. Due to the ordered assembly of α/β-tubulin heterodimers, microtubules have intrinsic polarity. One end (the minus end) has α-tubulin as the terminal subunit. This end is relatively stable. The other end (the plus end) has β-tubulin as the terminal subunit. This highly dynamic end switches between rapid growth and shrinkage depending on whether the exposed β-tubulin subunits are bound to GTP (growth) or GDP (shrinkage). During growth, GTP-tubulin adds to the plus end, forming a GTP cap. GTP is hydrolyzed to GDP, and if GDP-β-tubulin is exposed on the plus end of a microtubule, catastrophic disassembly can occur. (B) Cartoons of spindle structure during meiosis and mitosis in *C. elegans*. During mitosis, long, radial microtubule arrays are nucleated in the pericentriolar material surrounding the centrioles. Some microtubules (blue) from each spindle pole are captured by kinetochores (red). Ultimately, these kinetochore/microtubule attachments align each chromosome on the metaphase plate. Other overlapping, antiparallel microtubules (green) become bundled together in a region called the midzone, a process that stabilizes mitotic spindle structure. In contrast, centrioles are degraded prior to assembly of the oocyte meiotic spindle. Short, often tiled, microtubules form near chromosomes and ultimately coalesce into a bipolar structure.

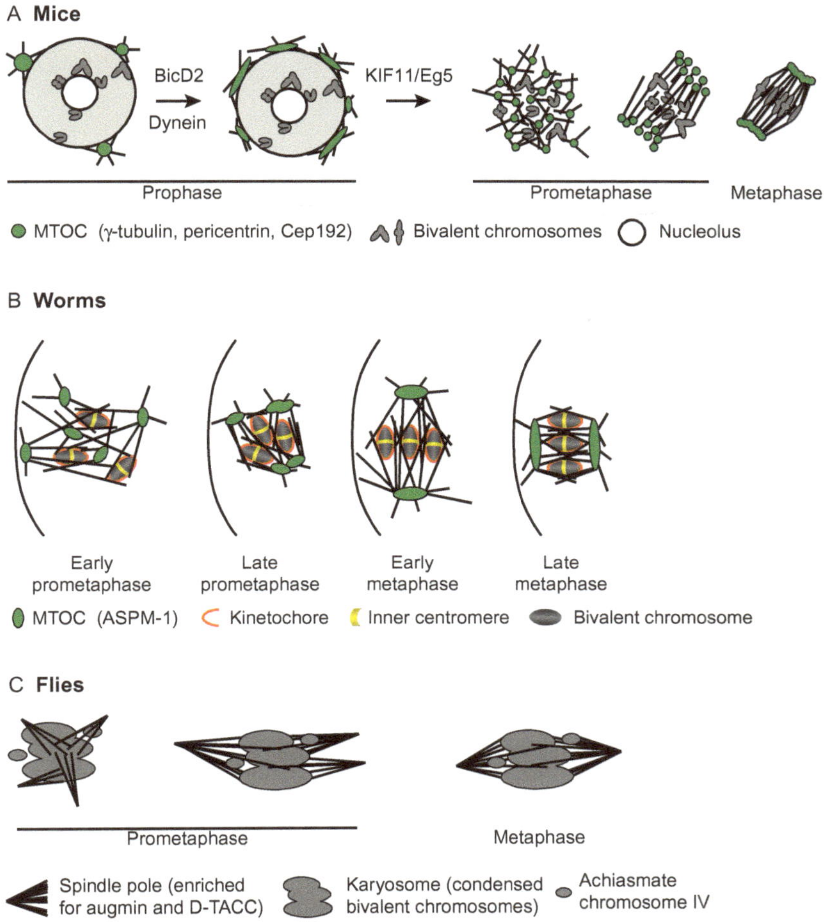

Aaron F. Severson *et al.*, Figure 3 Assembly of meiotic spindles in mouse, worm, and fly oocytes. Live imaging of microtubules and chromosomes during oocyte meiosis in these three model organisms has revealed similarities and differences in spindle assembly. When spindle assembly begins, homolog pairs have undergone crossover recombination to form bivalents, with the exception of *Drosophila* chromosome 4, which does not recombine. (A) In mouse oocytes, small microtubule-organizing centers (MTOCs) form near the nuclear envelope prior to NEB. Subsequently, MTOCs are stretched into thin ribbons, often becoming fragmented. This stretching process requires dynein and BicD2, which anchors dynein at the nuclear envelope. After NEB, the Eg5 homolog KIF11 promotes further MTOC fragmentation, yielding an average of approximately 26 MTOCs per oocyte. These MTOCs then coalesce as bivalents congress, forming a bipolar spindle by metaphase. (B) At NEB in *C. elegans* oocytes, most bivalents are located near the nuclear envelope. Following NEB, a cloud of short microtubules assembles around the bivalents, which subsequently cluster together. Concurrently, several foci of the microtubule-scaffolding protein ASPM-1 appear, presumably present at small MTOCs. The ASPM-1 foci coalesce into two poles as bivalents congress onto the metaphase plate. Initially, the bipolar metaphase spindle is oriented parallel to the oocyte cortex, but then rotates to be perpendicular to the cortex. During rotation, the MTOCs widen as the spindle shortens dramatically. (C) In fly oocytes, the recombined bivalents form a compact structure called a karyosome. After NEB, microtubules accumulate around the karyosome and several dynamic, pole-like structures form. These microtubule foci coalesce to form the poles of the bipolar meiotic spindle. Augmin, γ-tubulin, and D-TACC are enriched at the poles.

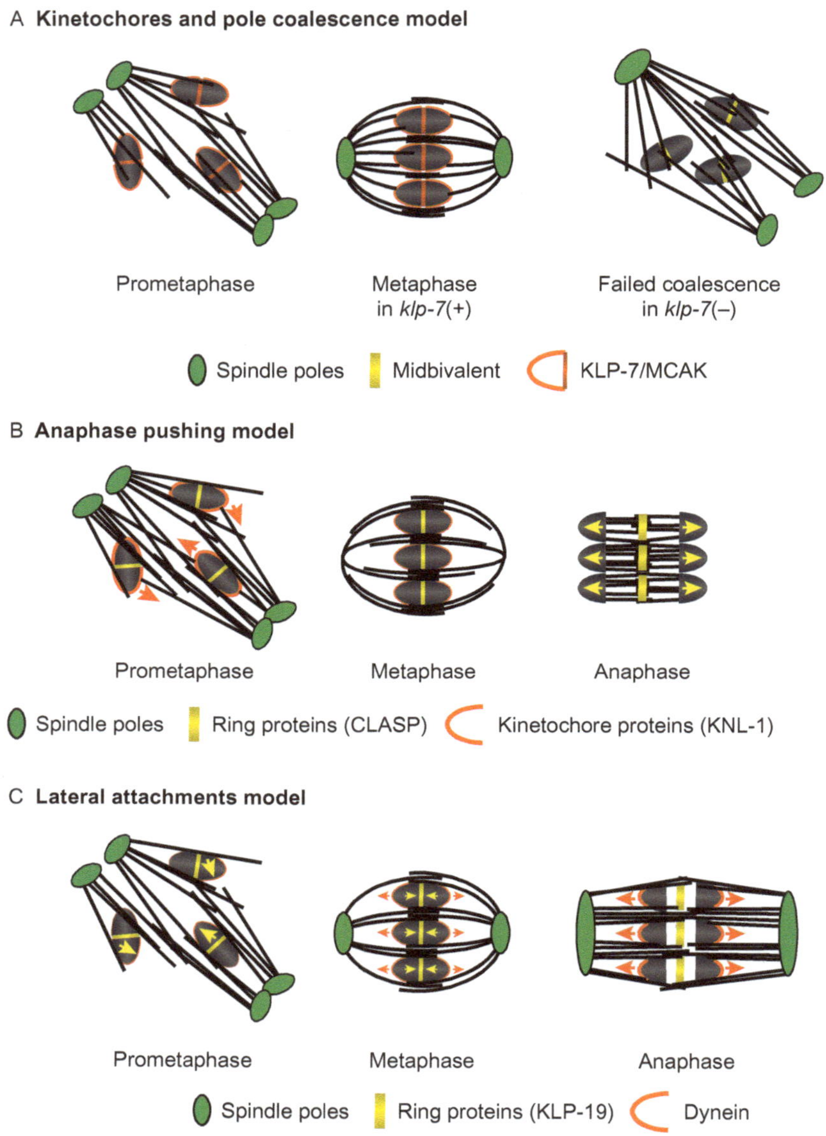

Aaron F. Severson *et al.*, Figure 4 Models for bipolar spindle assembly and chromosome congression and segregation in *C. elegans*. Three models are shown that address pole assembly (A) and chromosome congression and segregation (B and C) in *C. elegans* oocytes. (A) Proper microtubule–kinetochore attachments permit the coalescence of early pole foci to form a bipolar oocyte spindle. In this model, bivalent alignment and spindle bipolarity require that the two homologs of each bivalent attach specifically to microtubules from opposite spindle poles. Incorrect attachments (see text) are

(Continued)

Aaron F. Severson *et al.*, Figure 4—Cont'd eliminated by the microtubule depolymerase KLP-7/MCAK. Failure to disrupt these incorrect attachments in mutants with reduced KLP-7 function results in an imbalance in spindle tension that interferes with the coalescence of early pole foci and thus results in the assembly of multipolar spindles. (B) Microtubule–kinetochore interactions orient and align bivalents on the metaphase spindle, but at the metaphase-to-anaphase transition, the polar microtubule arrays disassemble, and a new population of microtubules forms between homologs and pushes them apart independently of kinetochores. The assembly of these microtubules is mediated by CLASP and other factors present in ring-shaped structures between each homolog pair at metaphase. When homologs separate, the rings are left behind at the metaphase plate. (C) Lateral interactions between microtubules and microtubule motor proteins mediate both chromosome congression to the metaphase plate and anaphase segregation to the poles. The chromokinesin KLP-19, localized to midbivalent rings, interacts with spindle microtubules to produce a polar ejection force that aligns bivalents at the metaphase plate. Subsequently, during anaphase, polar microtubules interact with dynein, a minus-end directed microtubule motor that accumulates at increasing levels on the poleward/lateral regions of bivalents as meiosis progresses. Dynein-directed motility mediates the poleward movement of chromosomes during anaphase.

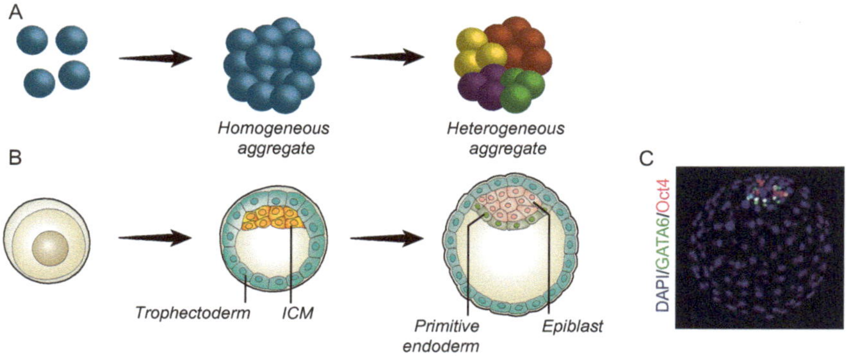

Alessia Deglincerti *et al.*, Figure 1 (A) Self-organization is the process by which a population of seemingly homogeneous cells (homogenous aggregate) gives rise to cells with heterogeneous properties (heterogeneous aggregate). (B and C) The developing embryo self-organizes to give rise to cells of different lineages: after a first distinction between trophectoderm (blue) and inner cell mass (ICM) (orange), cells in the ICM further differentiate into hypoblast (the precursor of primitive endoderm, green) and epiblast (which will give rise to the embryo proper, pink).

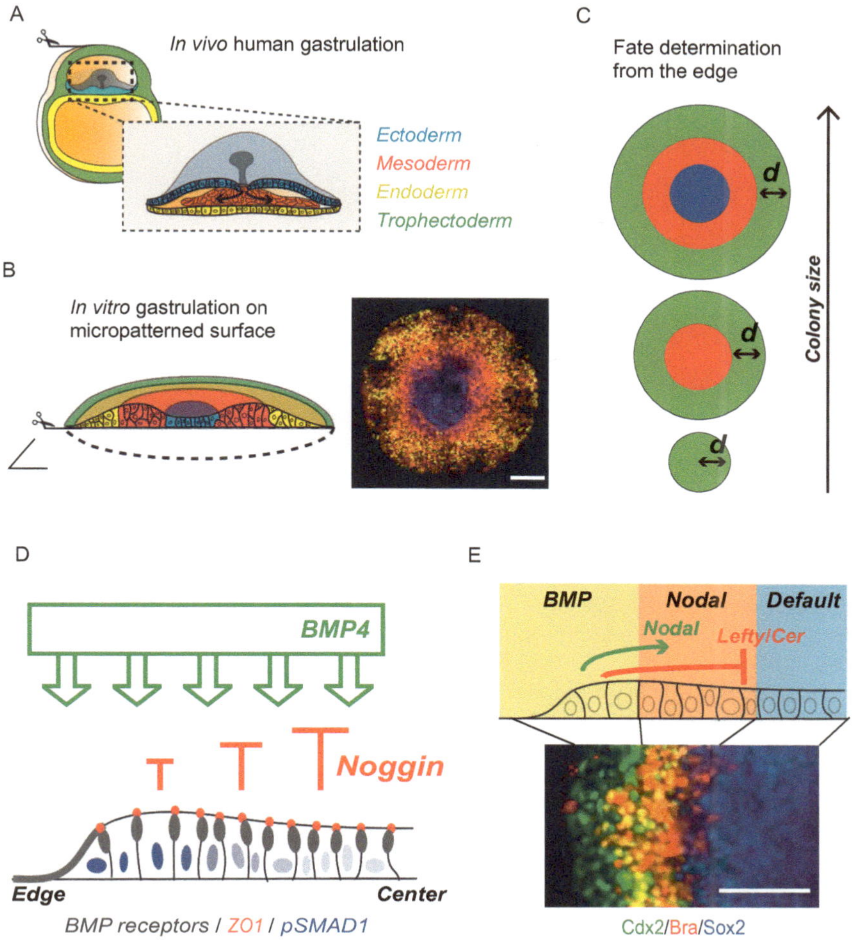

Alessia Deglincerti *et al.*, Figure 2 Self-organization of embryonic germ layers. (A) Schematic of a gastrulating human embryo (~ day postfertilization 15) showing the epiblast (light blue), primitive endoderm (yellow), and trophectoderm (green). Inset: Higher magnification representation of the epiblast. Gastrulation starts by formation of the primitive streak (gray). Cells ingress through the streak to form definitive endoderm (yellow) and mesoderm (red). Cells that do not go through the streak become ectoderm (dark blue). (B) *In vitro* gastrulation on micropatterned surfaces. Human ESCs are plated on 1 mm circular substrates. Forty-eight hours after BMP4 presentation, cell fates are radially organized (from center to edge) into ectoderm (blue), mesoderm (red), endoderm (yellow), and trophectoderm (green). A schematic representation of fate organization is shown on the left, while a picture of differentiating cells expressing markers of

(Continued)

Alessia Deglincerti *et al.*, Figure 2—Cont'd the three germ layers is shown on the right. Scale bar 200 μm. (C) Schematic of micropatterned colonies of different sizes: from top to bottom, 1000, 500, and 200 μm diameter. Green: trophectoderm, red: mesendoderm, blue: ectoderm. As the colony size decreases, only the outer fate is induced via an edge-sensing mechanism. (D) Interplay between BMP4 and Noggin in micropatterned colonies. BMP4 is applied homogeneously. BMP receptors (gray) are apically localized in cells at the edge and laterally localized in cells in the center of the colony, below adherens junction (labeled by ZO1, red). Noggin is expressed in response to BMP4 and contributes to inhibition of BMP4 signal at the center of the colony (red arrows). As a result, BMP4 signaling as detected by pSMAD1 intensity, is higher at the edge than at the center (blue nuclei). (E) The domain of maximal BMP4 signaling will give rise to trophectoderm. These cells produce Nodal and its inhibitors that diffuse inward toward the center of the colony. The cells that integrate both Nodal and BMP4 will generate mesendoderm, while the cells that do not respond to either BMP4 or Nodal will give rise to ectoderm. Scale bar 100 μm.

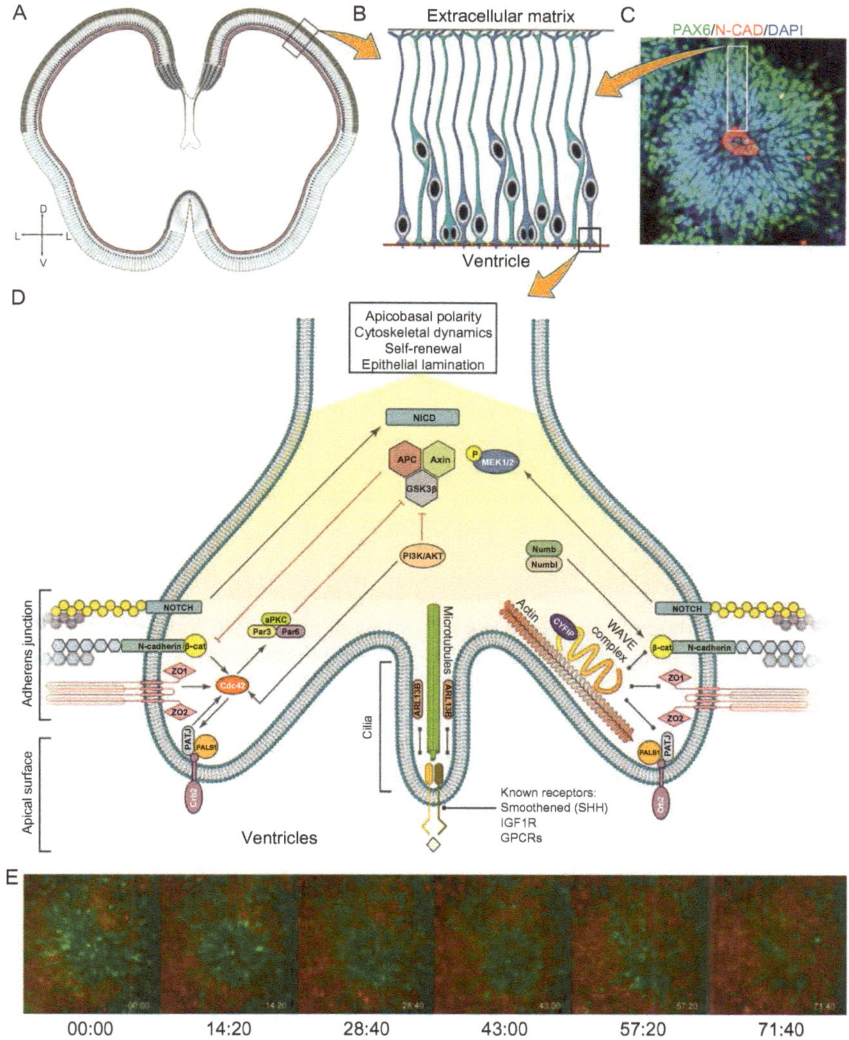

Alessia Deglincerti *et al.*, Figure 3 Self-organization of neural rosettes. (A) Schematic of the cortical neuroepithelium in a postconception week 5 human fetus. The neuroepithelium is comprised of radial glial cells. (B) Higher magnification view of the boxed area in (A) representing a schematic of radial glial cells lining the ventricle. The apical surface is the closest to the ventricle while the basal surface is away from it and in contact with the extracellular matrix. (C) Rosette structure at day 22 after neural induction in human ESCs. The neuroepithelium is identified by expression of PAX6 (green), while the central lumen structure (corresponding to the ventricles *in vivo*) is lined by N-cadherin (red), a component of adherens junction. This organization resembles the *in vivo* situation. (D) Schematic of radial glia end feet and high-magnification view of the boxed

(Continued)

Alessia Deglincerti *et al.*, Figure 3—Cont'd region in (B). The various pathways involved in establishment and maintenance of apico-basal polarity are displayed, as reviewed in Table 1. (E) Human ESC-derived neuroepithelium dynamically self-organizes into proliferative and nonproliferative zones. The pictures are still frames from Movie 2 (http://dx.doi.org/10.1016/bs.ctdb.2015.11.010) and depict a human ESC line engineered with FUCCI markers. Green cells in this montage represent actively proliferating radial glia, while the red cells are mostly postmitotic neurons. Initially, green cells are spread across the entire rosette. Over the course of 48 h, radial glia come to lie next to the rosette lumen forming a proliferative zone resembling the "ventricular zone" of the embryonic cortex. Conversely, the postmitotic neurons come to lie on the outside of the rosette; this basal part resembles the "cortical plate" of the embryonic cortex.

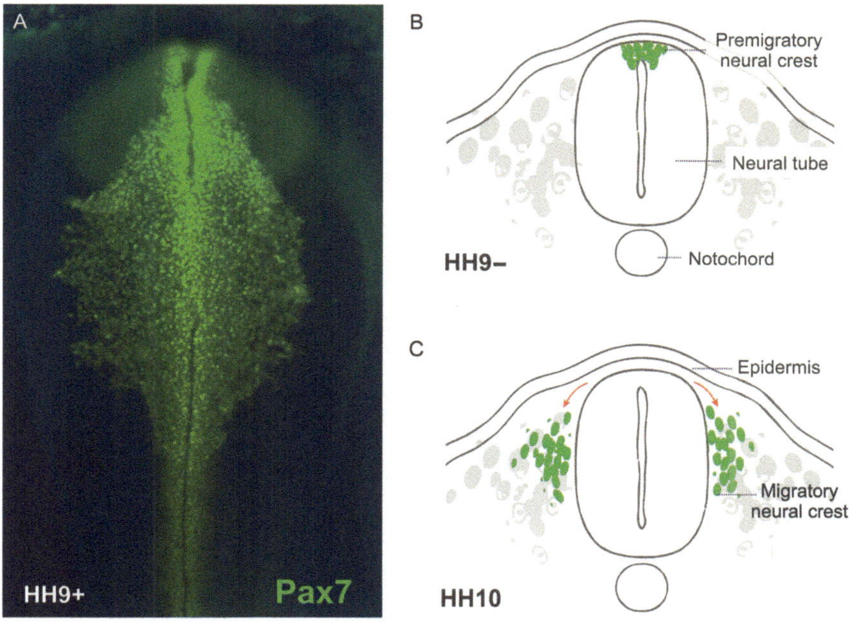

Marianne E. Bronner and Marcos Simões-Costa, Figure 1 Neural crest cells delaminate from the neural tube and undergo extensive migration. (A) Immunohistochemistry with an antibody for neural crest marker Pax7 highlights neural crest cells delaminating from the neural tube of a stage HH9 chicken embryo. (B) Diagram representing a cross section of a neurulating chicken embryo. At the stage represented, the neural cells occupy the dorsal portion of the neural tube. (C) Neural crest cells eventually undergo epithelial-to-mesenchymal transition to delaminate from the neural tube and undergo extensive migration.

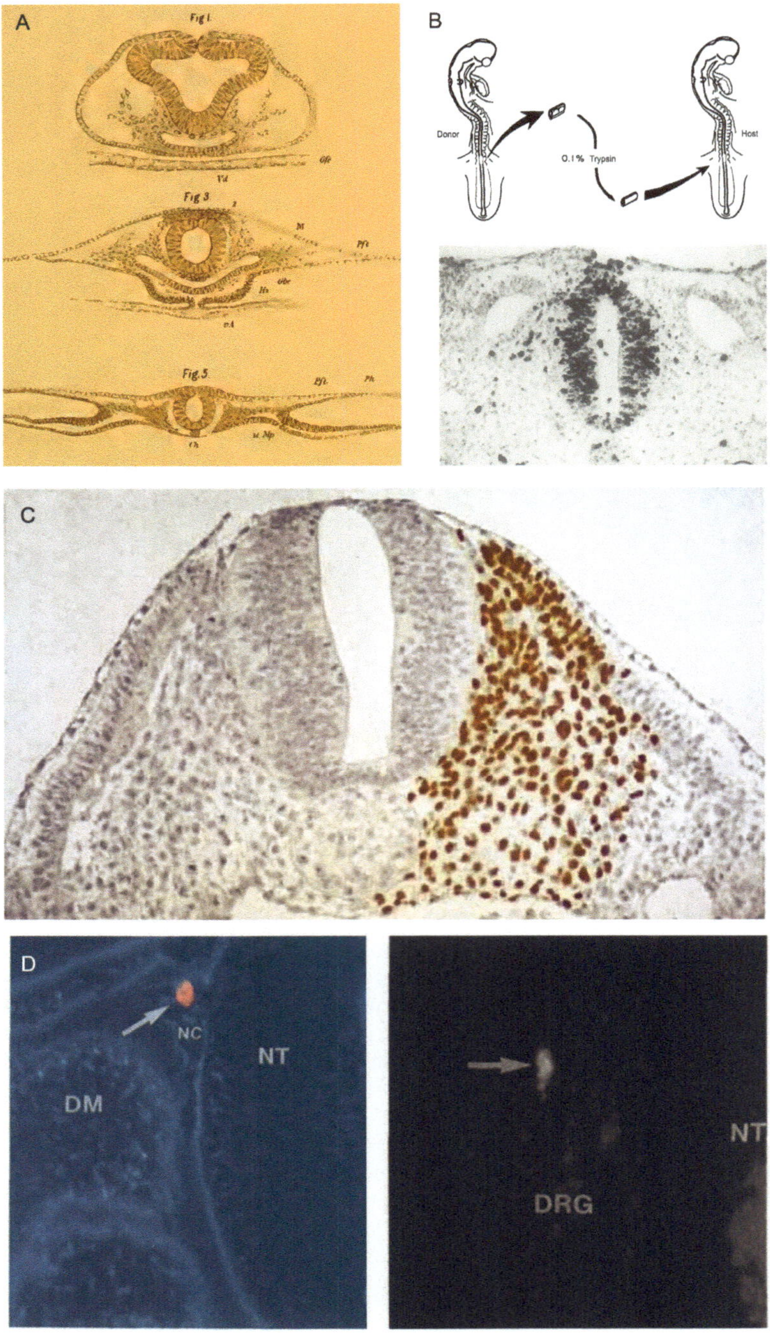

Marianne E. Bronner and Marcos Simões-Costa, Figure 2 See legend on opposite page.

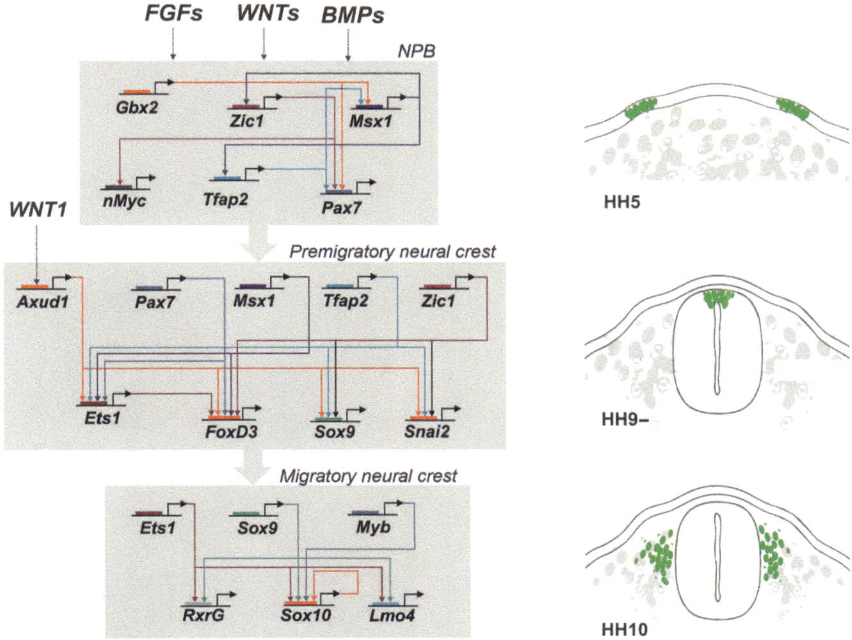

Marianne E. Bronner and Marcos Simões-Costa, Figure 3 A gene regulatory network (GRN) controls neural crest formation. A complex genetic program comprised of numerous transcription factors underlies neural crest development. These factors are arranged in distinct hierarchical modules corresponding to the different stages of neural crest formation. First, in the neural plate border, the interplay of signaling systems such as Wnts, BMPs, and FGFs activates the neural plate border specifier genes, which include *Gbx2*, *Msx1*, *Zic1*, *Tfap2*, and *Pax7*. These genes drive the transition of the GRN to the neural crest specification module. Signaling systems such as Wnts also feed into the system by activating transcription factor Axud1, which interacts with PAX7 and MSX1 to activate expression of neural crest specifier genes (*FoxD3*, *Ets1*, *Sox9*, etc.) in the dorsal neural folds. The neural crest specifier genes endow the neural crest with its defining features and activate the process of epithelial-to-mesenchymal transition. In the migratory neural crest, there is activation of additional factors that will affect migration and drive differentiation in multiple cell types.

Marianne E. Bronner and Marcos Simões-Costa, Figure 2 Classical experiments in neural crest research utilizing avian embryos. (A) Reproduction of the original drawings of His, including the first depiction of neural crest cells (His, 1868). The neural crest is identified with a "Z" for "Zwischenstrang" or cord between the neural and nonneural ectoderm. (B) Transplant experiments by Weston (1963), where neural tubes from radioactively labeled embryos were transplanted to unlabeled hosts. This allowed for the tracking of neural crest cells after delamination and identification of distinct migratory pathways. (C) The chick–quail chimeras of Nicole Le Douarin. Grafting of quail tissue in chicken embryos permitted extensive fate mapping of progenitor cells (Le Douarin, 2004). This approach was used to map migratory pathways and derivatives of the neural crest at distinct axial levels of the avian embryo (Le Douarin, 1982). (D) Clonal analysis of neural crest cells in the developing embryo. Labeling of single neural crest cells with vital dyes showed that many individual progenitor cells are multipotent and differentiate into diverse cell types *in vivo* (Fraser & Bronner-Fraser, 1991). *Panel (B) Reprinted from Weston (1963). Copyright (1963), with permission from Elsevier. Panel (C) Reprinted from Le Douarin (2004). Copyright (2004), with permission from Elsevier. Panel (D) Reprinted from Fraser and Bronner-Fraserer (1991). Copyright (1991), with permission from Company of Biologists.*

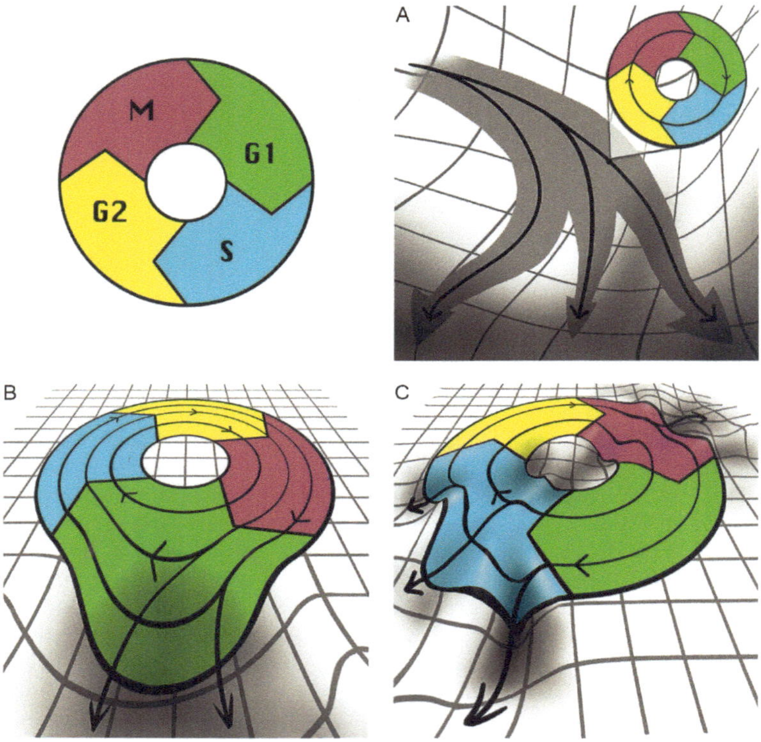

Corey Bunce and Blanche Capel, Figure 1 The integration of cell cycle and differentiation pathways depicted as progenitor cell cycles overlaid on epigenetic landscapes. The slope of the landscape represents a cell's propensity to differentiate. (A) The common view, wherein cell movement along a developmental trajectory is independent of its cell cycle dynamics. (B) In phases with "priming" or "competence" windows, changes in path length alter the probability of differentiation. (C) The complexity of nuclear remodeling during S and M phase makes a rugged landscape, where cells are highly susceptible to differentiation cues, suggesting that considerable regulatory programs must be employed to maintain a progenitor cell trajectory. *Figure drawn by Amanda Bunce.*

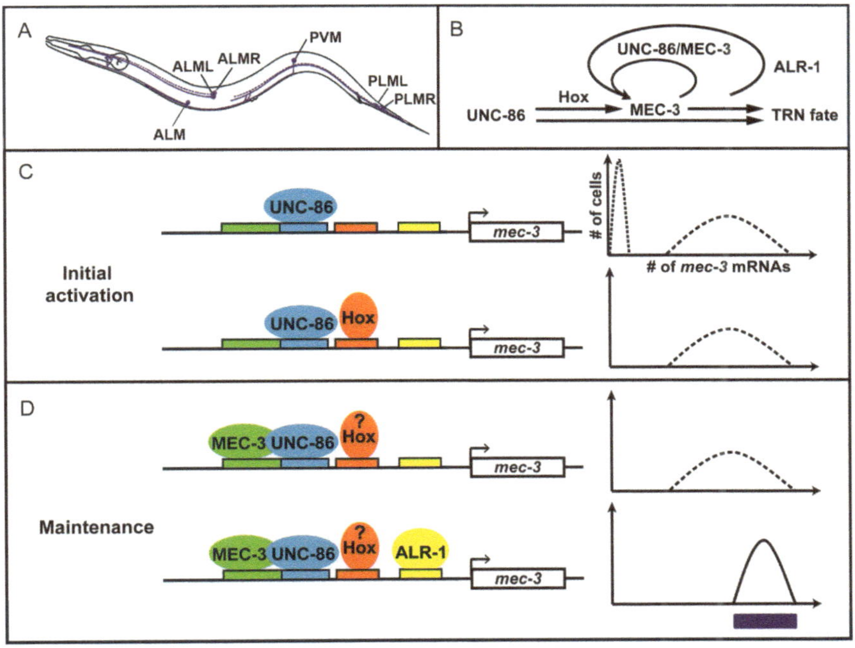

Chaogu Zheng and Martin Chalfie, Figure 1 Guarantors ensure the activation and maintained expression of the terminal selector *mec-3* in the TRNs. (A) The six TRNs in *C. elegans*. (B) Proteins involved in TRN specification. UNC-86 initially activates *mec-3*, facilitated by Hox proteins; and the expressed MEC-3 forms a heteromer with UNC-86 that autoregulates *mec-3* and activates the TRN fate. UNC-86/MEC-3 also activates *alr-1*, which potentiates *mec-3* autoregulation. (C) and (D) Modes of *mec-3* transcription in the absence and presence of guarantors. The schematic plots represent quantitative measurements of *mec-3* mRNA molecules in populations of cells using smFISH. (C) During the initiation phase, UNC-86 alone cannot fully activate *mec-3* in all TRNs. In Hox mutants, a binary loss of *mec-3* expression leads to the failure of TRN fate acquisition in some of the cells. (D) *mec-3* autoregulation requires not only UNC-86/MEC-3 but also ALR-1 to reduce the variability and optimize *mec-3* expression. Whether Hox proteins are also needed for the *mec-3* maintenance is unclear. The binding sites of the guarantors are found adjacent to the essential UNC-86/MEC-3 binding site in the proximal promoter of *mec*-3. *Panel (A): Adapted from Bounoutas and Chalfie (2007) with kind permission from Springer Science+Business Media and from Chalfie and Sulston (1981), Copyright (1981), with permission from Elsevier.*

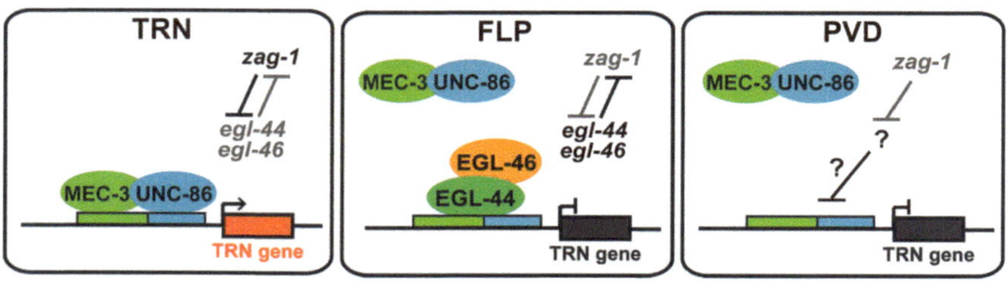

Chaogu Zheng and Martin Chalfie, Figure 2 ZAG-1 protects the activity of UNC-86/MEC-3 heteromer. In the TRNs, UNC-86/MEC-3 complex activates the expression of TRN terminal differentiation genes by binding to similar *cis*-regulatory motifs in the promoters. In FLP neurons, EGL-44/EGL-46 complex displaces UNC-86/MEC-3 from its binding sites and thus prevent the expression of TRN fate. Presumably, unknown factor(s) block the activation of TRN genes in PVD neurons. ZAG-1 silences those inhibitors in the TRN neurons but is repressed by them in at least FLPs. This mutual inhibition also enables the switch of cell fate between neurons that express similar terminal selectors.

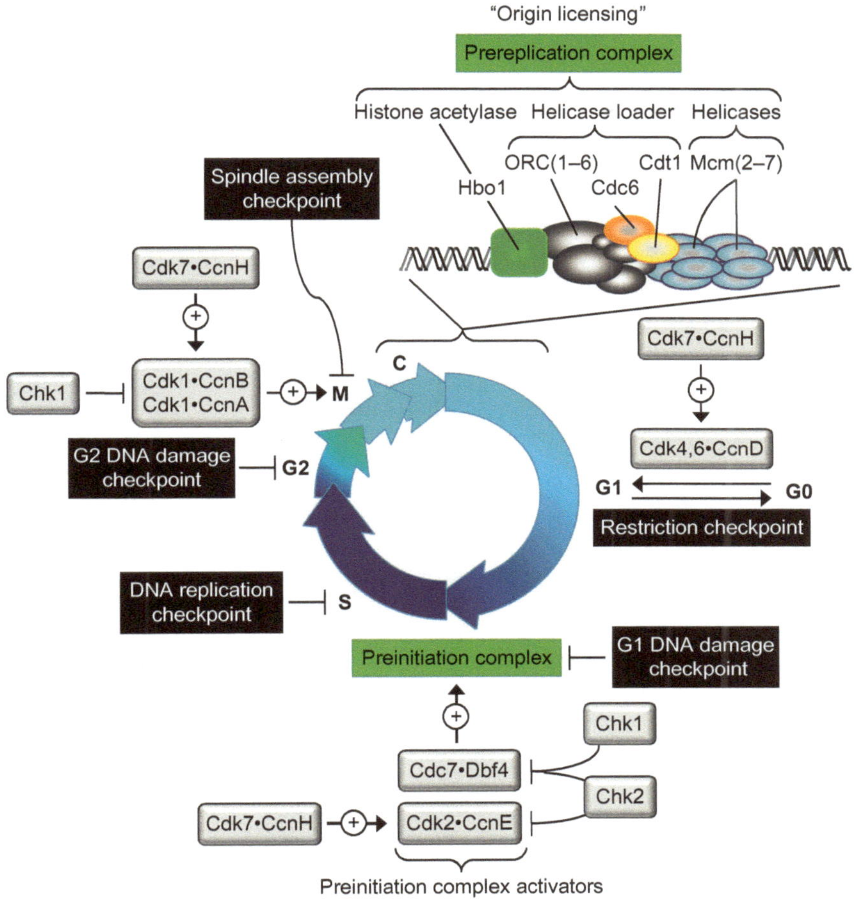

Melvin L. DePamphilis, Figure 1 Regulation of DNA replication during mitotic cell cycles. Nuclear DNA replication in the metazoa begins during the anaphase to G1 phase transition of the mitotic cell cycle with the assembly of prereplication complexes onto the DNA in chromatin (Baldinger & Gossen, 2009; Noguchi, Vassilev, Ghosh, Yates, & DePamphilis, 2006). In mammals, this event, termed "origin licensing," begins by assembly of the six subunit "origin recognition complex" [ORC(1–6)] through independent nuclear localization of Orc1, ORC(2–5), and Orc6 (Ghosh, Vassilev, Zhang, Zhao, & DePamphilis, 2011). The BAH domain in human Orc1 facilitates the ability of Orc1 to activate replication origins *in vivo* and triggers assembly of ORC(1–6) (Kara, Hossain, Prasanth, & Stillman, 2015; Noguchi et al., 2006). Mutations in the BAH domain of Orc1 link histone H4 methylation to origin licensing and to Meier–Gorlin syndrome, a genetic defect leading to primordial dwarfism (Kuo et al., 2012). Cdc6 stabilizes the binding of ORC(1–6) to DNA and allows recruitment of Cdt1 to form a "helicase loader." The six subunit "minichromosome maintenance complex" (Mcm2–7) is the eukaryotic replicative DNA helicase. Each replication fork requires one MCM helicase; thus two or

(Continued)

Melvin L. DePamphilis, Figure 1—Cont'd more are loaded at each origin of bidirectional replication. *In vivo*, the Hbo1/Kat7 histone acetylase activity is also essential origin licensing, because assembly of prereplication complexes is facilitated by histone acetylation (Miotto & Struhl, 2010; Unnikrishnan, Gafken, & Tsukiyama, 2010). However, prereplication complexes cannot unwind DNA; they must first be converted into preinitiation complexes by addition of 10 other protein (Bell & Kaguni, 2013; Tanaka & Araki, 2013). Only then can the MCM helicase unwind the double helix and allow DNA polymerases to synthesize the nascent DNA strand. DNA replication is regulated directly by eight protein kinases (Depamphilis, de Renty, Ullah, & Lee, 2012), five of which are cyclin-dependent kinases (Cdk) that require a cyclin (Ccn) partner. Cdk7·CcnH is essential to activate Cdk1, Cdk2, Cdk4, and Cdk6. Cdk4·CcnD and Cdk6·CcnD drive cells from a quiescent state (G0) into a proliferative state. Cdk1·CcnB1 and Cdk1·CcnA2 initiate and maintain mitosis. Cdk2·CcnE and the Dbf4-dependent kinase Cdc7 are the engines that drive the assembly and activation of preinitiation complexes. Cdk2·CcnA2 prevents DNA rereplication during S phase. Cdk1·CcnA2 prevents premature initiation during G2 phase. The remaining two protein kinases are components of the cell's DNA damage response. Five checkpoints ensure the accuracy of mitotic cell division (DePamphilis & Bell, 2010). In the absence of sufficient nutrients, growth factors or optimal temperature, cells in G1 phase enter a quiescent state (G0). This transition is regulated by the "restriction checkpoint," which consists of two signal detection pathways (Foster, Yellen, Xu, & Saqcena, 2010). One senses growth factors, and the other senses nutrients. The G1 to G0 transition is regulated by action of Cdk4·CcnD and Cdk6·CcnD. When cells are arrested in G0, their DNA replication machines are dismantled, and they eventually enter a senescent state from which they cannot reenter the mitotic cell cycle (Campisi, 2013). Conversely, once a cell passes the restriction checkpoint, it is committed to division; it must divide, or terminally differentiate into a polyploid cell, or die. Nuclear DNA damage is assessed before beginning S phase by the G1 DNA damage checkpoint, during S phase by the DNA replication checkpoint and before beginning M phase by the G2 DNA damage checkpoint, but in reality there are only two DNA damage response systems that basically detect excess single-stranded DNA complexed with replication protein-A that results from different problems. DNA damage and stalled replication forks are detected by the ATR kinase which then activates the Chk1 kinase to phosphorylate and inhibit Cdc7 and Cdc25 (a phosphatase required for activation of Cdk1 and Cdk2). Double-stranded DNA breaks are detected by the ATM kinase, which then activates Cdk2 kinase to phosphorylate Cdc7, Cdc25, and tumor suppressor Tp53. Many other proteins are also targeted. Finally, the spindle assembly checkpoint prevents the onset of anaphase until the kinetochores are attached correctly to the mitotic spindle (Lara-Gonzalez, Westhorpe, & Taylor, 2012).

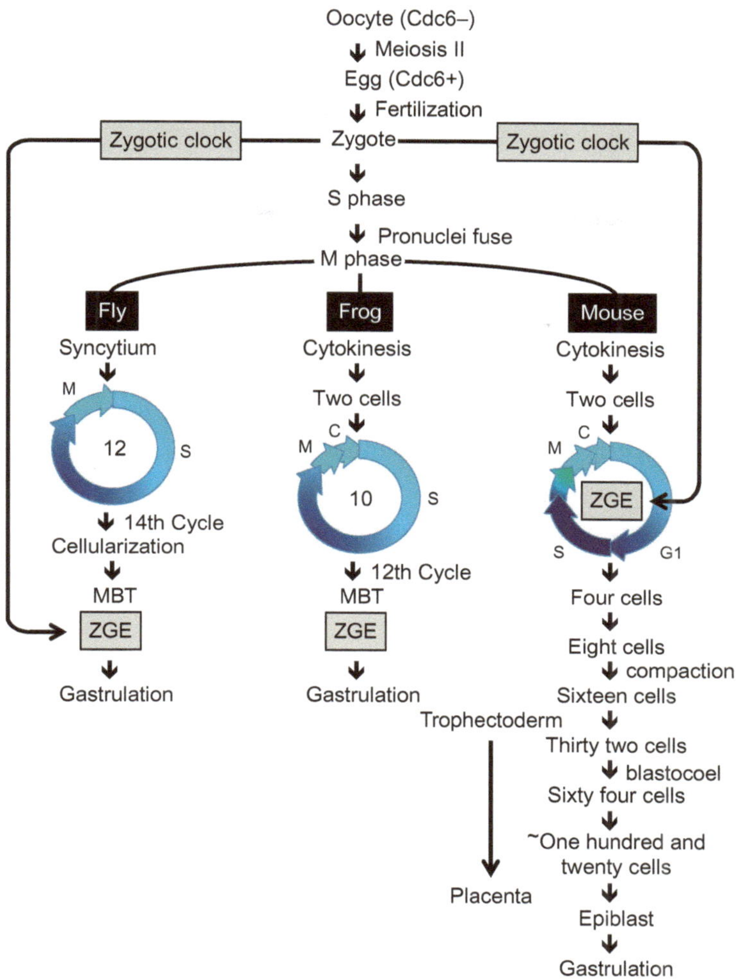

Melvin L. DePamphilis, Figure 2 Developmentally programmed changes in the mitotic cell cycle. The first mitotic cell division is universal, as exemplified in *Drosophila melanogaster* (fly), *Xenopus laevis* (frog), and *Mus musculus* (mouse). Oocyte maturation into eggs includes expression of the Cdc6 protein that is essential for initiation of nuclear DNA replication. The second mitotic cell division distinguishes flies from frogs and mammals from nonmammals. Flies and other insects continue to rapidly undergo cycles of DNA replication (S phase) followed by mitosis (M phase) without undergoing cell cleavage. Frogs, fish, and echinoderms continue to rapidly undergo cycles of DNA replication followed by mitosis followed by cell cleavage (cell division in the absence of cell growth). Mice and other mammals continue to undergo slow cycles of DNA replication followed by mitosis followed by cell cleavage. However, mammalian cell cleavage appears to include G phases. The most significant change is that zygotic gene expression (ZGE), the onset of transcription and the degradation of maternally inherited RNA, occurs in two-cell embryos of mice and the 4- to 16-cell embryos of rabbits, cows, pigs, and humans. Nevertheless, activation of the bulk of ZGE in both flies and mice is determined by the time elapsed after fertilization (a "zygotic clock"), rather than by DNA replication or mitosis. The differences in cell division cycles between mammals and nonmammals might well be related to the fact that mammals need to produce both a placenta and an embryo concurrently (O'Farrell, Stumpff, & Su, 2004).

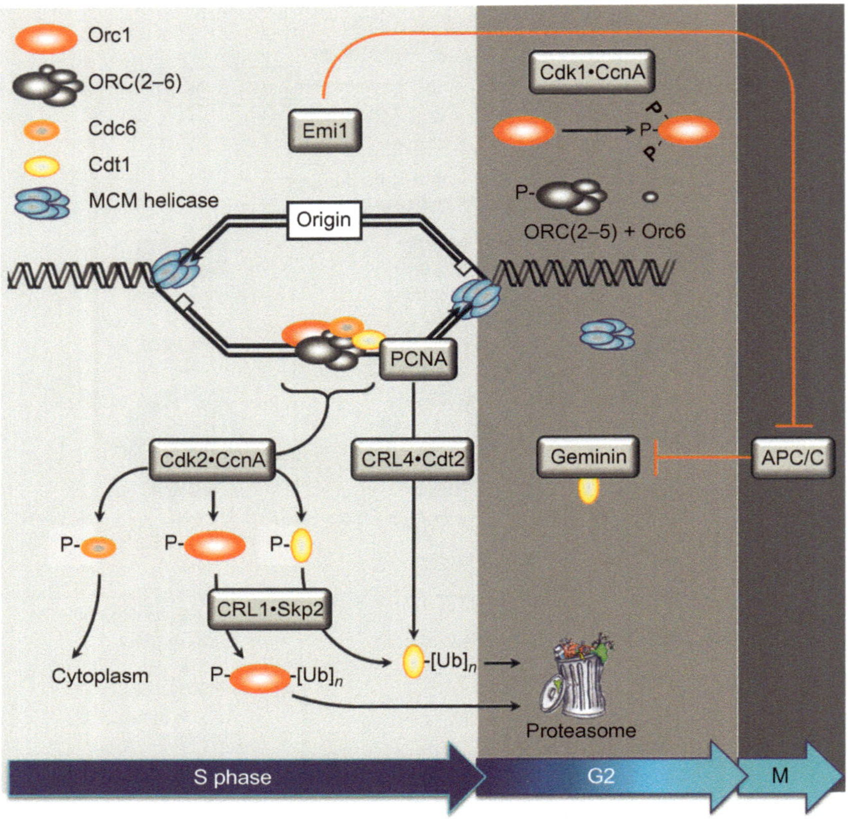

Melvin L. DePamphilis, Figure 3 Multiple concerted pathways prevent DNA rereplication. Once metazoan cells have entered S phase, seven concerted pathways can inactivate the helicase loader, thereby preventing rereplication of DNA prior to mitosis. Upon S phase entry, Cdk2·CcnA phosphorylates Orc1, Cdc6, and Cdt1, thereby inhibiting their ability to form helicase loaders. In addition, Cdc6-P is localized to the cytoplasm where it cannot facilitate origin licensing. Orc1-P is ubiquitinated and targeted for destruction, with subsequent dissociation of ORC(2–5) and Orc6 from chromosomes. Cdt1-P is ubiquitinated by the CRL1·Skp2 ubiquitin ligase. Cdt1 that is not phosphorylated is ubiquitinated by the CRL4·Cdt2 ubiquitin ligase bound to PCNA and DNA. PCNA is the eukaryotic sliding clamp protein that facilitates DNA synthesis by DNA polymerases δ and ε. The ubiquitinated proteins are then degraded by the 26S proteasome. Geminin is a protein that binds specifically to Cdt1 and inhibits its activity. These activities are available from S through early M phase. As cells exit mitosis, geminin and cyclin A are ubiquitinated by the anaphase promoting complex [originally termed cyclosome (APC/C)], an activity that is inhibited during S to early M phase by the APC/C-specific inhibitor Emi1. Reassembly of the helicase loader occurs by reassembly of Orc1, ORC(2–5), Orc6, and Cdc6 onto chromatin (Ghosh et al., 2011; Kara et al., 2015; Sonneville et al., 2012), an event that appears to be triggered by the Orc1 subunit (Noguchi et al., 2006). Geminin binding to Cdt1 and CDK-dependent phosphorylation of Cdc6 prevents targeting of Cdt1 and Cdc6 for degradation by the APC/C, thereby allowing them to participate in origin licensing (Ballabeni et al., 2004; Mailand & Diffley, 2005).

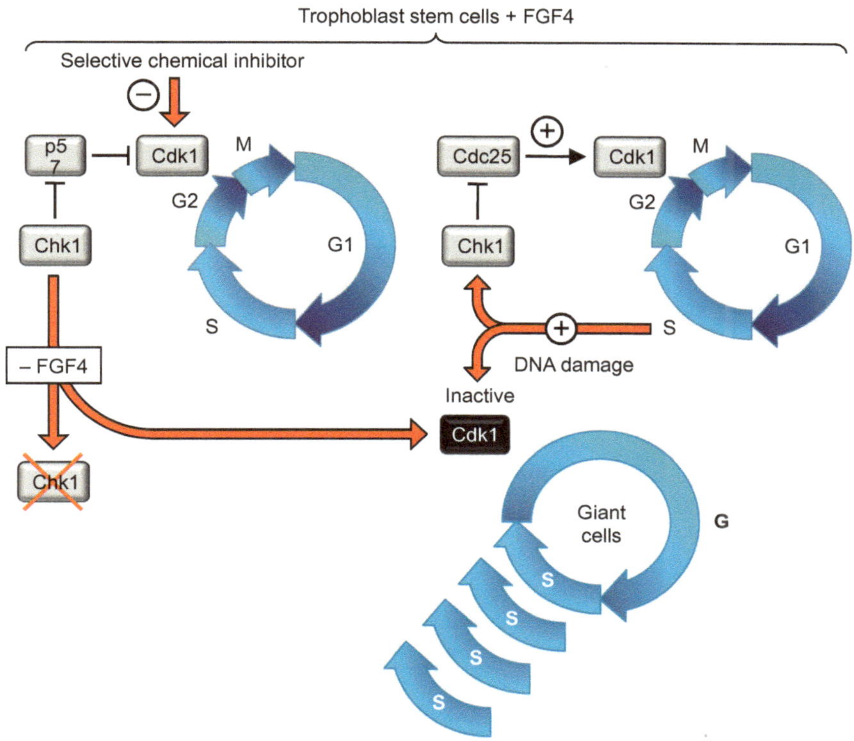

Melvin L. DePamphilis, Figure 4 Endoreplication is triggered in trophoblast cells by selective inhibition of Cdk1 activity. Developmentally programmed endoreplication in the metazoa occurs when Cdk1 activity is suppressed so that cells cannot enter mitosis (Edgar, Zielke, & Gutierrez, 2014; Fox & Duronio, 2013; Zielke, Edgar, & DePamphilis, 2013). This results in multiple S phases without an intervening mitosis or cytokinesis to produce giant cells with a single enlarged nucleus containing as many as several hundred copies of each chromosome. In the absence of FGF4, trophoblast cells in mouse blastocysts are triggered to differentiate into giant cells. In the presence of FGF4, the same result can be elicited either by selective inhibition of Cdk1 with RO3306, or by suppressing or inactivating geminin (de Renty, Kaneko, & DePamphilis, 2014; Ullah, Kohn, Yagi, Vassilev, & DePamphilis, 2008). The effect of FGF4 deprivation is to reduce the level of geminin, which results in the loss of checkpoint kinase-1 (Chek1/Chk1). Since Chk1 phosphorylation of the CDK-specific inhibitor Cdkn1c/p57 targets it for ubiquitin-dependent degradation, loss of Chk1 results in p57 inhibition of Cdk1 activity, which leads to endoreplication (Ullah, de Renty, & Depamphilis, 2011). Ironically, DNA damage in trophoblast cells also results in endoreplication, because DNA damage activates Chk1, which then phosphorylates and inactivates Cdc25, a phosphatase that is essential to maintain Cdk1 in its active form.

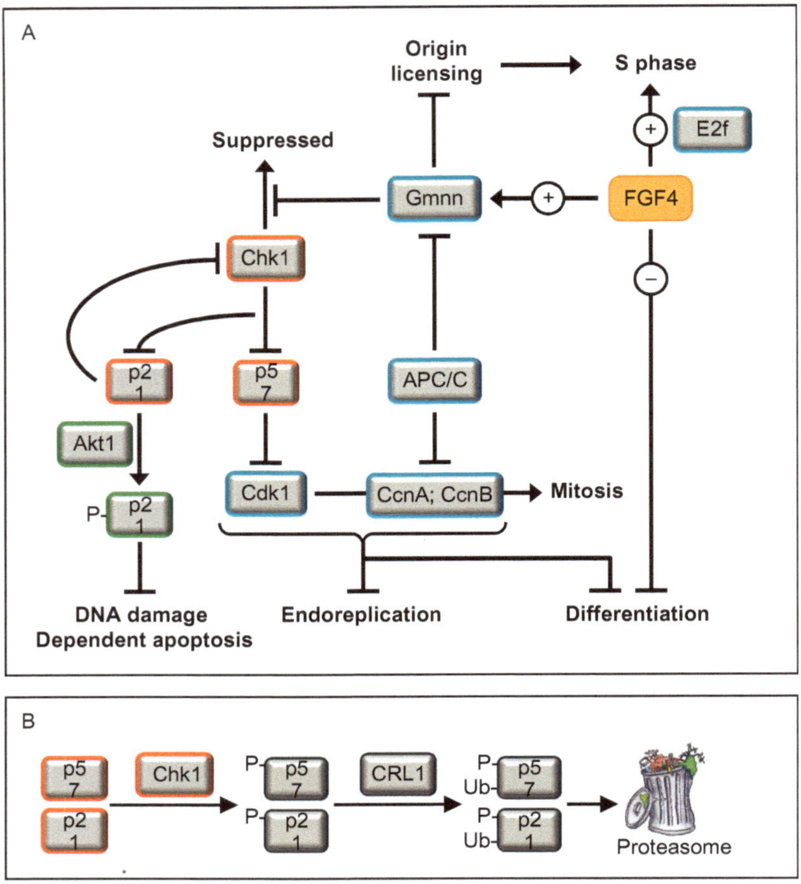

Melvin L. DePamphilis, Figure 5 The FGF4 signal transduction pathway governs trophoblast proliferation and differentiation. FGF4 (and probably other mitogenic proteins as well) is essential for trophoblast proliferation. This mitogenic activity is likely mediated by E2F-dependent gene expression (Chen et al., 2010), and possibly directed at regulating the activity of the anaphase-promoting complex (APC) (Yang et al., 2012). FGF4 deprivation results in downregulation of geminin activity to a level that maintains endocycles (de Renty, Kaneko, & DePamphilis, 2014), but that does not prevent downregulation of Chk1 protein. The loss of Chk1 kinase activity results in expression of two CDK-specific inhibitors, p57 and p21 (Ullah et al., 2011). The p57 protein prevents the onset of mitosis by selectively inhibiting Cdk1 activity, thereby triggering the first round of endoreplication (Hattori, Davies, Anson-Cartwright, & Cross, 2000; Ullah et al., 2008). This event activates the G1-phase APC·Cdh1 ubiquitin ligase, which targets geminin, cyclin B, and cyclin A for degradation, thereby allowing licensing of replication origins and the onset of S-phase without passing through mitosis (Ullah, Lee, Lilly, & DePamphilis, 2009). Inhibition of Cdk1 triggers both endoreplication and TSC differentiation. In the absence of p57, FGF4 deprivation produces multinucleated TGCs, revealing the existence of alternative mechanisms to trigger TSC differentiation (Ullah et al., 2008). Endocycles also require p57, which is expressed during G-phase and then suppressed during S-phase to allow sequential assembly and activation of pre-replication complexes (Ullah et al., 2008). Geminin maintains endocycles by preventing DNA rereplication. The p21/Cdkn1a protein localizes to the cytoplasm in TGCs where it prevents DNA damage induced apoptosis (de Renty, DePamphilis, & Ullah, 2014). It might also maintain suppression of Chk1 by reducing *Chk1* RNA levels (Gottifredi, Karni-Schmidt, Shieh, & Prives, 2001), as observed during FGF4 deprivation (Fig. 3).

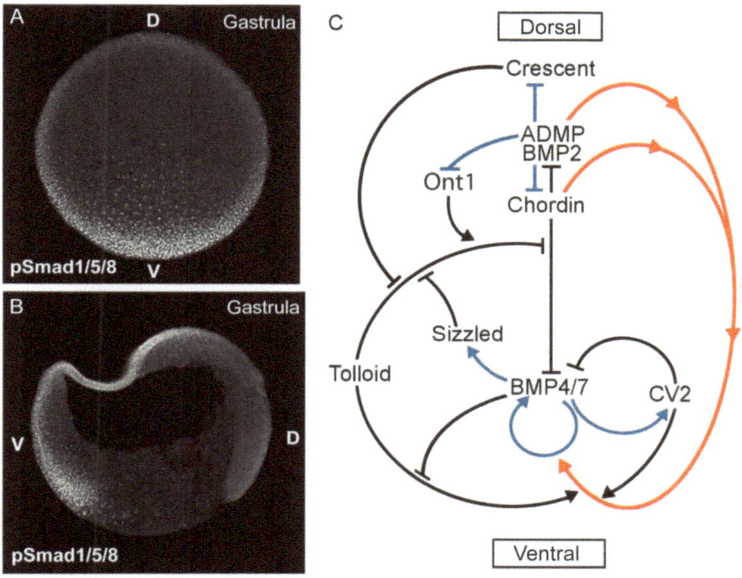

Edward M. De Robertis and Yuki Moriyama, Figure 1 A gradient of nuclear phospho-Smad1 is established by a biochemical network of secreted proteins that interact with each other. (A) Horizontal optical section at gastrula (stage 11) showing a ventral to dorsal gradient of BMP activity using pSmad1/5/8 antibody as the readout. D, dorsal; V, ventral. (B) Gastrula (stage 11) embryo sectioned sagittally, showing higher BMP signaling in ventral nuclei. (C) The D–V morphogenetic signaling pathway results from a series of direct protein–protein interactions between Chordin and other partners (black arrows), transcriptional regulation (blue arrows), and protein flux (red arrows). The entire embryo participates in forming the BMP gradient, which results from the dueling activities of the dorsal and ventral signaling centers. *Images from Plouhinec, Zakin, Moriyama, and De Robertis (2013).*

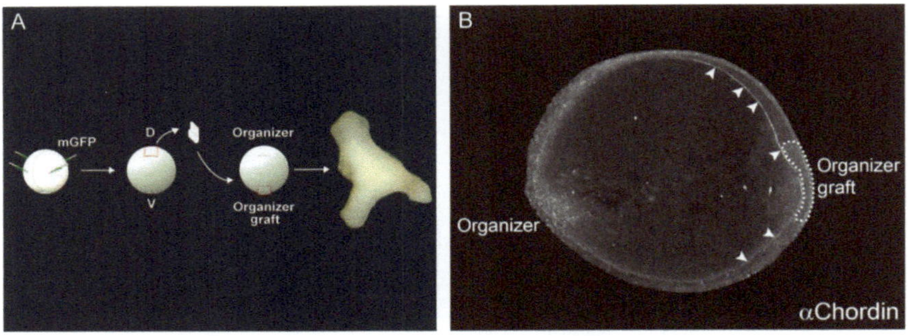

Edward M. De Robertis and Yuki Moriyama, Figure 3 A long-range gradient of Chordin protein emanating from transplanted organizer tissue. (A) Diagram of the experimental procedure; membrane-targeted *mGFP* was injected in the donor embryo (four injections at the four-cell stage) to trace the lineage of grafted Spemann organizer tissue. (B) Embryo transplanted at stage 10 and fixed at stage 12 stained with anti-Chordin antibody shown in optical transverse section. A second gradient of Chordin was observed with Chordin protein staining in Brachet's cleft extending a considerable distance from the graft (arrowheads). The GFP fluorescence of the transplanted organizer cells is outlined by the dotted line. Chordin protein diffuses long-distance from the graft through Brachet's cleft ECM. *Data from Plouhinec et al. (2013).*

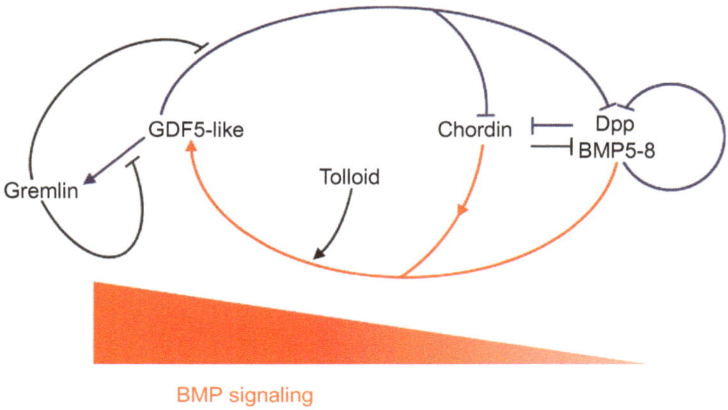

Edward M. De Robertis and Yuki Moriyama, Figure 4 The directive axis of the cnidarian *Nematostella vectensis* is established by a Chordin/BMP gradient. As in *Xenopus*, BMP activity along the body axis results from a series of direct protein–protein interactions between Chordin and other protein partners (black arrows), transcriptional regulation (blue arrows), and protein flux (red arrows) in *Nematostella*. *Diagram based on Genikhovich et al. (2015).*

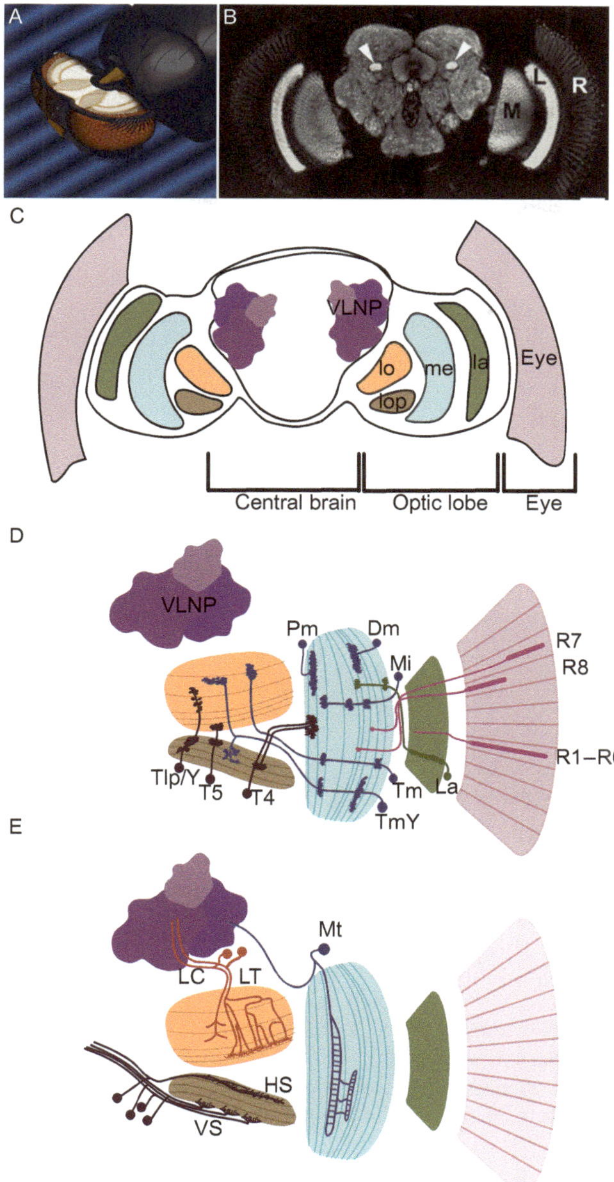

Nathalie Nériec and Claude Desplan, Figure 1 The visual system of *Drosophila*. (A) The fly brain in the head. (B) Neuropils of the optic lobes and central brain. (C) Cross section of the fly brain indicating the different parts of the visual system: eye, lamina (la), medulla (me), lobula (lo), lobula plate (lop), and ventrolateral neuropils (VLNPs). (D) Interneurons of the optic lobe: outer and inner photoreceptors (pink), lamina neurons (La, green), medulla interneurons (dark blue), distal (Dm) and proximal medulla (Pm); medulla intrinsic (Mi); unicolumnar transmedullary (Tm) or multicolumnar TmY; lobula plate interneurons (T4, T5, Tlp, and Y neurons, dark brown). (E) Projections neurons: medulla tangential (Mt, dark blue); lobula columnar (LC, red) and tangential/tree-like (LT, red); lobula plate tangential cells: HS and VS (dark brown). *Panel (A) adapted from Spalthoff, Gerdes, and Kurtz (2012). Panel (B) from Krzeptowski et al. (2014). Panels (D and E) adapted from Erclik, Hartenstein, McInnes, and Lipshitz (2009).*

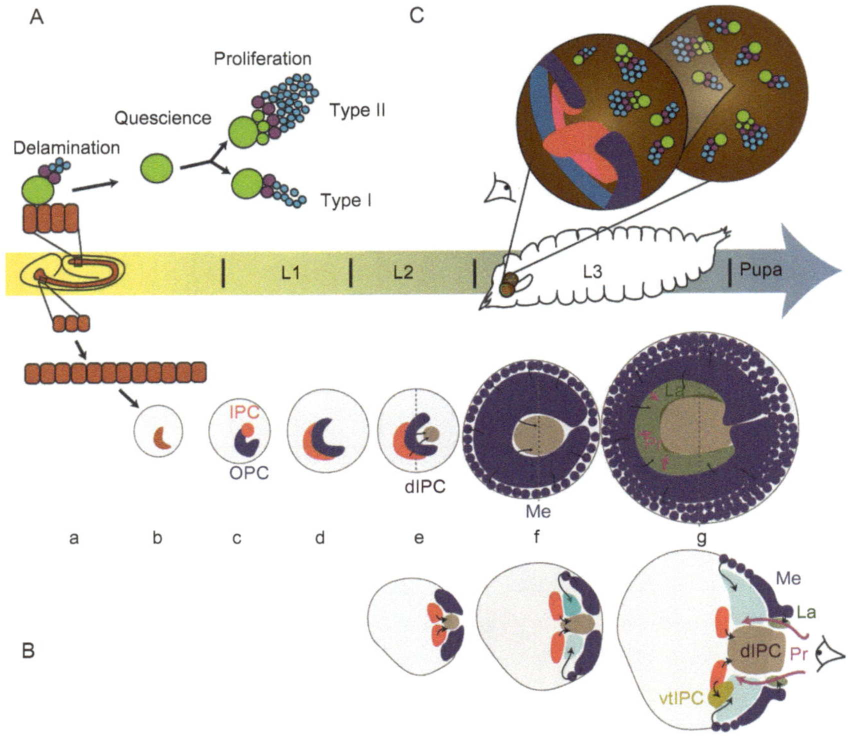

Nathalie Nériec and Claude Desplan, Figure 2 The development of the optic lobe of *Drosophila*. (A) Central brain: neuroblasts (green) delaminate from the neuroepithelium (red) during the embryonic stages and generate GMCs (purple) and larval neurons (blue) before they become quiescent. At L1, Type I and Type II neuroblasts are reactivated and exhibit different modes of proliferation to generate adult neurons. (B) Optic lobe: (a) Cells in the procephalic regions generate the optic placode neuroepithelium (red), which invaginates (b). (c) The optic placode splits into inner proliferation center (IPC, pink) and outer proliferation center (OPC, blue) and adopt a crescent shape (d). (e) Cells migrate out of the IPC to form the dIPC (light brown). (f) The medial edge of the OPC becomes neuroblasts that will produce the medulla (round blue cells). (g) The inner OPC generates lamina progenitors (La, green) that will produce lamina neurons upon induction by photoreceptors (Pr, pink arrows). The ventral tip of the IPC generates lobula complex neurons (vtIPC, dark yellowish green). Lower diagrams are cross sections at the dashed line in upper diagrams. (C) 3D representation of the brain at L3. Central brain neuroblasts and their progeny, OPC (blue) and IPC (pink). The eye symbol shows the view angle of panel (B). *Panels (A and C) adapted from Homem and Knoblich (2012). Panel (B) adapted from Nassif, Noveen, and Hartenstein (2003).*

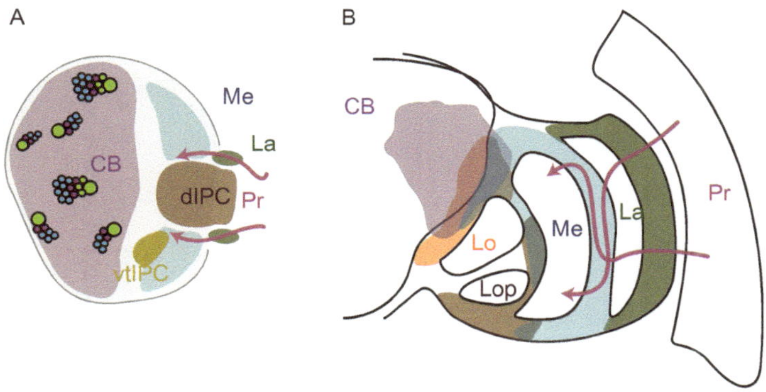

Nathalie Nériec and Claude Desplan, Figure 3 Linking development to adult fate. Fate of the different neuronal populations found at the L3 larval stage (A) in the adult visual system (B). Pr, photoreceptors; La, lamina; Me, medulla; Lo, lobula; Lop, lobula plate; dIPC, distal IPC; vtIPC, ventral tip of the IPC; CB, central brain.

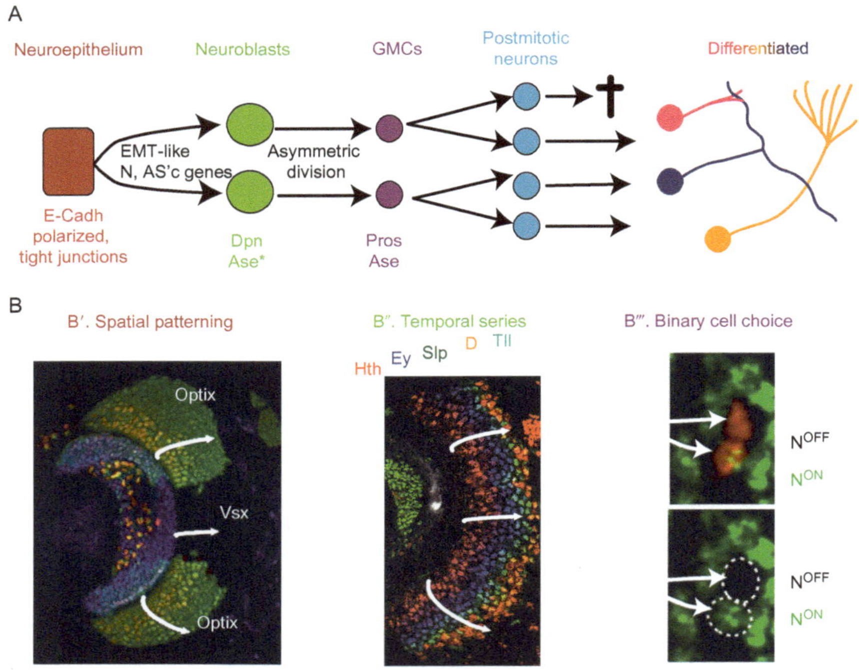

Nathalie Nériec and Claude Desplan, Figure 4 The making of a functional neuronal network. (A) Common mechanisms for neurogenesis: polarized neuroepithelial cells (red) undergo EMT-like transition to become neuroblasts (green) which generate GMCs (purple). GMCs produce postmitotic neurons (blue) that form adult neuronal circuits. * Dpn is found in all neuroblasts while Ase is found in most neuroblasts (see text). (B) Common mechanisms to generate neuronal diversity: (B′) Spatial patterning of the neuroepithelium illustrated by the expression of Optix and Vsx (Gold & Brand, 2014). (B″) Temporal patterning of neuroblasts. In the OPC, neuroblasts sequentially express a series of transcription factors, Hth (red), Ey (blue), Slp (green), D (orange), and Tll (cyan) as they age (arrows) (Li, Chen, & Desplan, 2013; Li, Erclik, et al., 2013). (B‴) Notch-dependent binary cell fate choices during the division of GMCs. Only one of the two daughter cells (red clone, dashed line) receives Notch activity and expresses Apterous (green), while the other does not (Li, Chen, et al., 2013; Li, Erclik, et al., 2013).

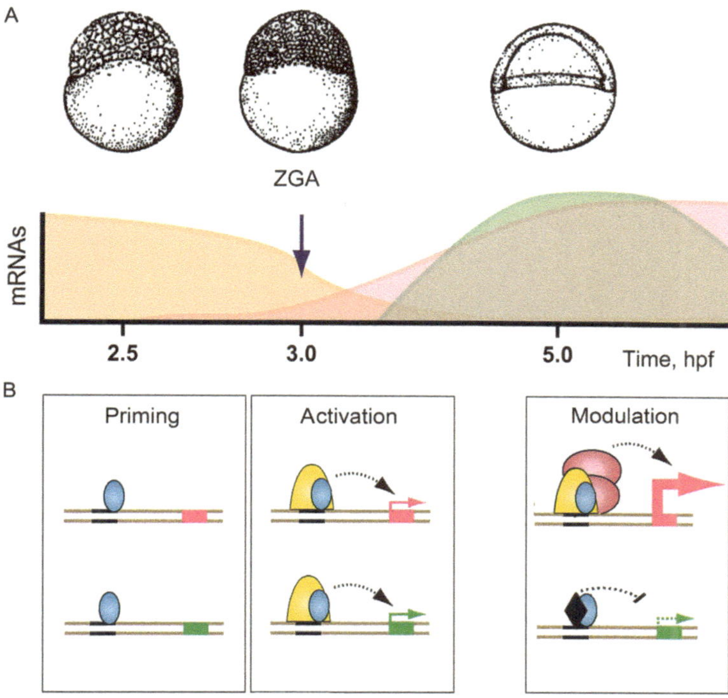

Daria Onichtchouk and Wolfgang Driever, Figure 1 Zygotic genome activation. (A) Schematic drawing of ZGA in zebrafish: starting from 1000 cell stage (middle embryo at the top), early zygotic genes are switched on, and some maternal mRNAs degrade. Hpf, hours postfertilization. (B) Hypothetical scheme of action of zygotic gene activators. Already before ZGA, zygotic genome activators (small oval) bind to defined regions in the genome and prime the neighboring genes for expression (left box, two example genes are shown). Ubiquitous cofactors (RNA polymerase or chromatin modifier enzyme; half oval) are attracted to zygotic genome activator binding site and switch on gene expression (middle box). Later, the binding sites for zygotic genome activators are occupied by additional tissue-specific factors, which either repress (gene at bottom) or enhance (gene at top) activation of transcription (right box).

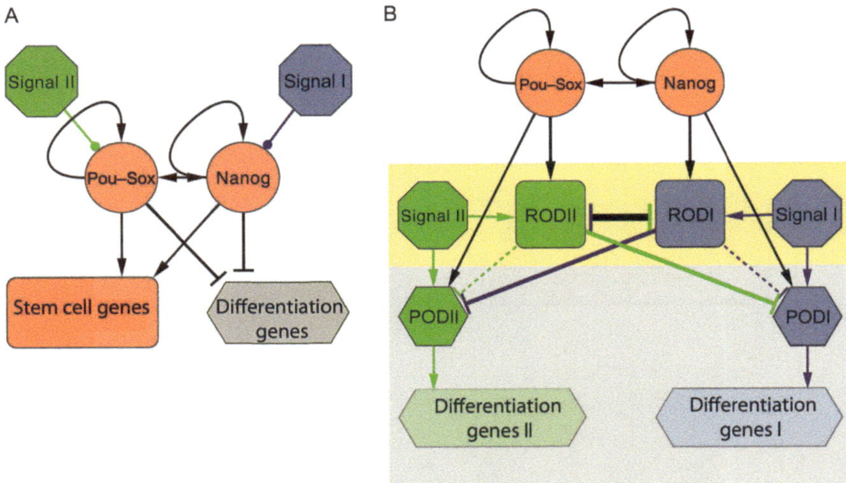

Daria Onichtchouk and Wolfgang Driever, Figure 2 Models of pluripotency. (A) Bistable switch model in ES cell (Chickarmane et al., 2006). Signals I and II can be activatory (arrowheads) or repressory (indicated by rounded arrowheads). The model describes two stable states. In the pluripotent state, in response to environmental signals, Pou5f1, SoxB1, and Nanog genes are switched on, they sustain their own expression by means of positive feedback loops, and the "pluripotency circuit" (Pou–Sox–Nanog) is active. In this state, stem cell genes are on and differentiation genes are switched off. When the positive feedback is interrupted by differentiation signals, the differentiation genes are on. (B) Simplified hypothetical model for transient pluripotency and links to developmental control in blastula to gastrula stage embryo cells: ROD, inter-(and/or intra-)territorial repressor of differentiation; POD, promoter of differentiation; signals I and II, different tissue specific patterning and differentiation signals. In the late blastula pluripotent state, signals I and II together with Pou–Sox–Nanog control RODII and RODI expression such that their activity is balanced. In this situation, PODs and differentiation genes are off (gray-shaded area). In committed differentiating cell in a specific germ layer or tissue, one RODI is off, POD starts to be expressed, induces differentiation genes, and presumably shuts down the pluripotency circuit.

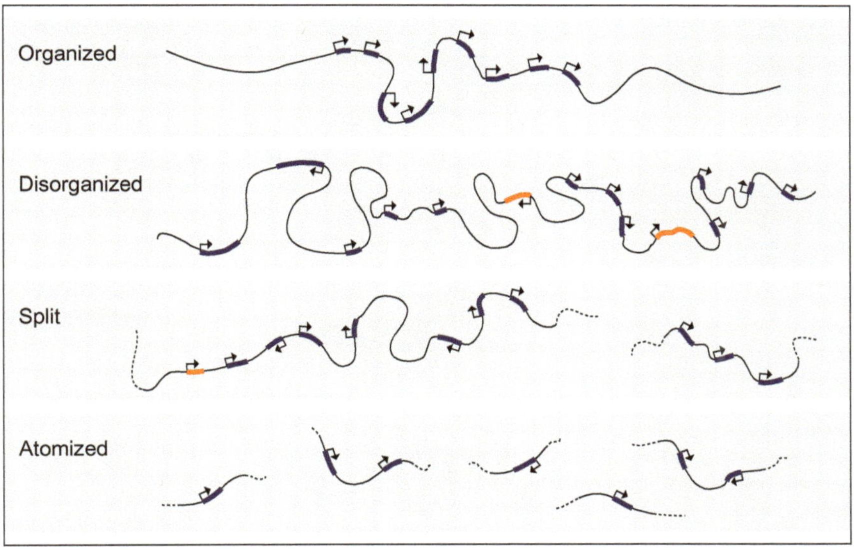

Fabrice Darbellay and Denis Duboule, Figure 1 The genomic organization of *Hox* genes varies in different bilaterian animals. *Hox* genes can be found either in tightly organized gene clusters, ca. 100 kb in length, containing few repetitive elements and no foreign genes (organized) or in much larger linkage group, ca. 1 mb in size, including repeats and sometimes foreign genes (disorganized). Such genomic clusters can also be split in various subclusters (split) or even in single unit as in atomized *Hox* (non)clusters (atomized). *Adapted from Duboule (2007).*

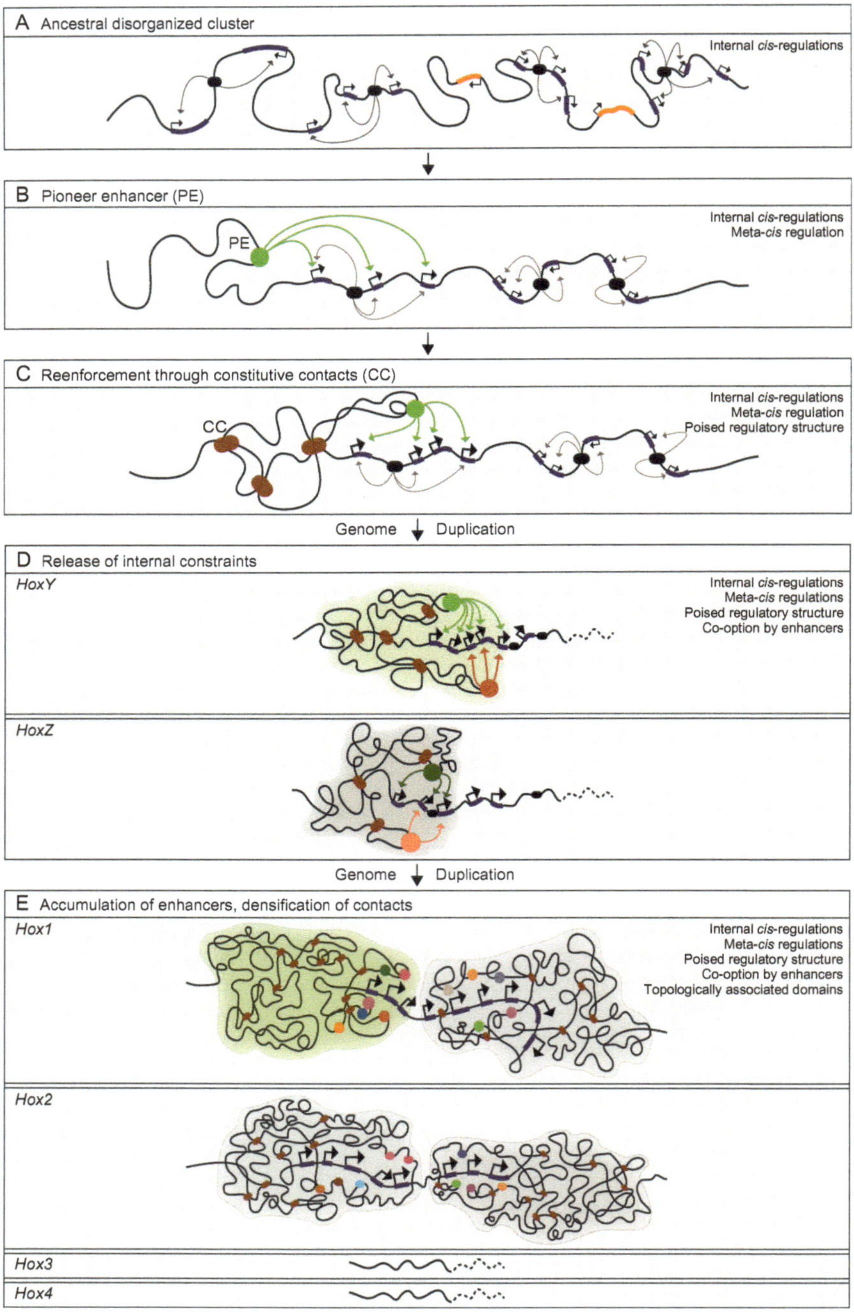

Fabrice Darbellay and Denis Duboule, Figure 3 See legend on opposite page.

Fabrice Darbellay and Denis Duboule, Figure 3 TAD-dependent emergence of multiple global regulations at *Hox* loci and concomitant consolidation of the *Hox* gene clusters in jaw vertebrates. (A) Initially, an ancestral disorganized *Hox* genes (blue) cluster implements its collinear regulation along the rostral to caudal axis through series of internal regulatory sequences (*cis*-regulation; black). The cluster may also contain unrelated genes (orange). In (B), some genes of the cluster become coordinately regulated by the evolution of a strong enhancer located outside the cluster itself, such as not to interfere with the internal regulatory organization. As a consequence, the transcriptional response (the adaptive value) becomes slowly optimized via the progressive genomic rearrangement of the target genes (start of consolidation), in order to optimize their response to this new regulation. This initial meta-*cis* regulation through a "pioneer enhancer" (PE) triggers the formation of a local 3D architecture. (C) Should this coordinated transcriptional control be of some adaptive value, the 3D structure may start to be stabilized or strengthened by the evolution of constitutive interactions (brown ovals), which will help and improve the interactions between the pioneer enhancer and its targets. This global (meta-*cis*) regulation may however interfere with some of the constraints imposed by the internal collinear mechanism. (D) This antagonism between *cis* and meta-*cis* regulations is partly released after genome duplications, which allows some functional compensation to occur between the duplicated clusters and thus restores more flexibility in evolving global regulations. As a consequence of genome duplication, this 3D structure will thus facilitate the evolution of yet other global enhancers on each duplicate (*HoxY* and *HoxZ*), due to the preexistence of a poised transcriptional read out of high adaptive value. This process of "regulatory priming" (Gonzalez et al., 2007) will continue as a virtuous circle and progressively lead to the maximal compaction of the target *Hox* gene cluster, with all genes in the same transcriptional orientation, very few repeats and no longer any foreign genes interspersed. (E) At the same time, large and constitutive regulatory domains will be observed on both sides, reflecting the successful cooption of many global enhancers that accompanied the emergence of most jaw vertebrate innovations (see the text). These domains correspond to topologically associating domains (TADs) (Dixon et al., 2012; Nora et al., 2012). In (B)–(D), only one of the flanking regions of the cluster is shown for sake of simplicity, even though the same mechanism may apply to the other. In (E), only two clusters are shown instead of the four derived from both rounds of genome duplications.

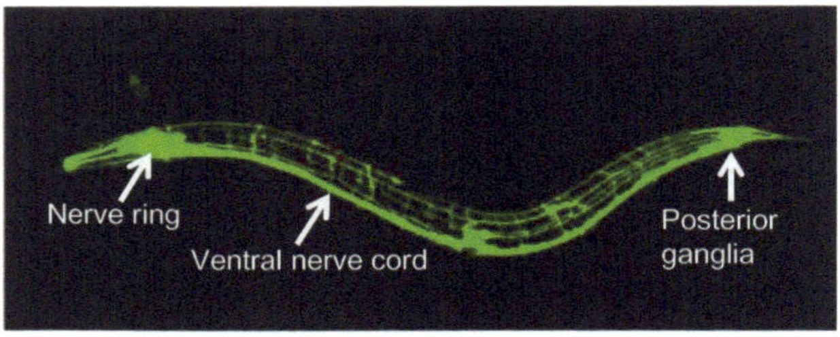

Scott W. Emmons and Dominick P. Purpura, Figure 1 The nerve bundles of the *C. elegans* nervous system, visualized by a reporter gene that expresses the fluorescent protein GFP in all of the neurons. *Image from Hang Ung, Jean-Louis Bessereau laboratory, France.*

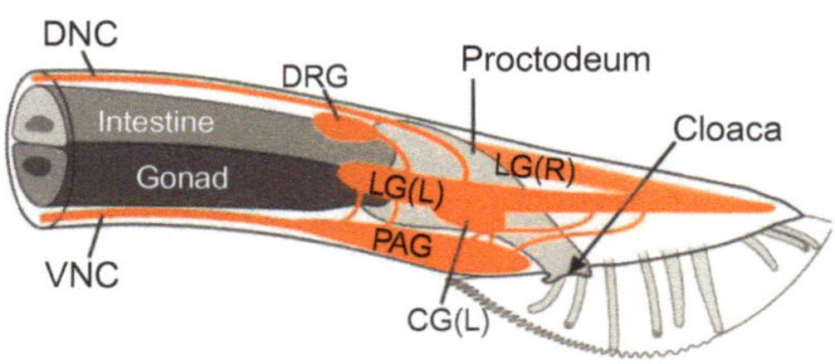

Scott W. Emmons and Dominick P. Purpura, Figure 2 Ganglia in the *C. elegans* male tail. CG(L), left cloacal ganglion; DNC, dorsal nerve cord; DRG, dorsorectal ganglion; LG(L), left lumbar ganglion; LG(R), right lumbar ganglion; PAG, preanal ganglion; VNC, ventral nerve cord; right cloacal ganglion not shown. Cell bodies are distributed in each of the six ganglia. Synapses can be found throughout, but are situated primarily within the PAG. Neurons outside this ganglion send processes through commissures into it. *Image from WormAtlas, http://www.wormatlas.org.*

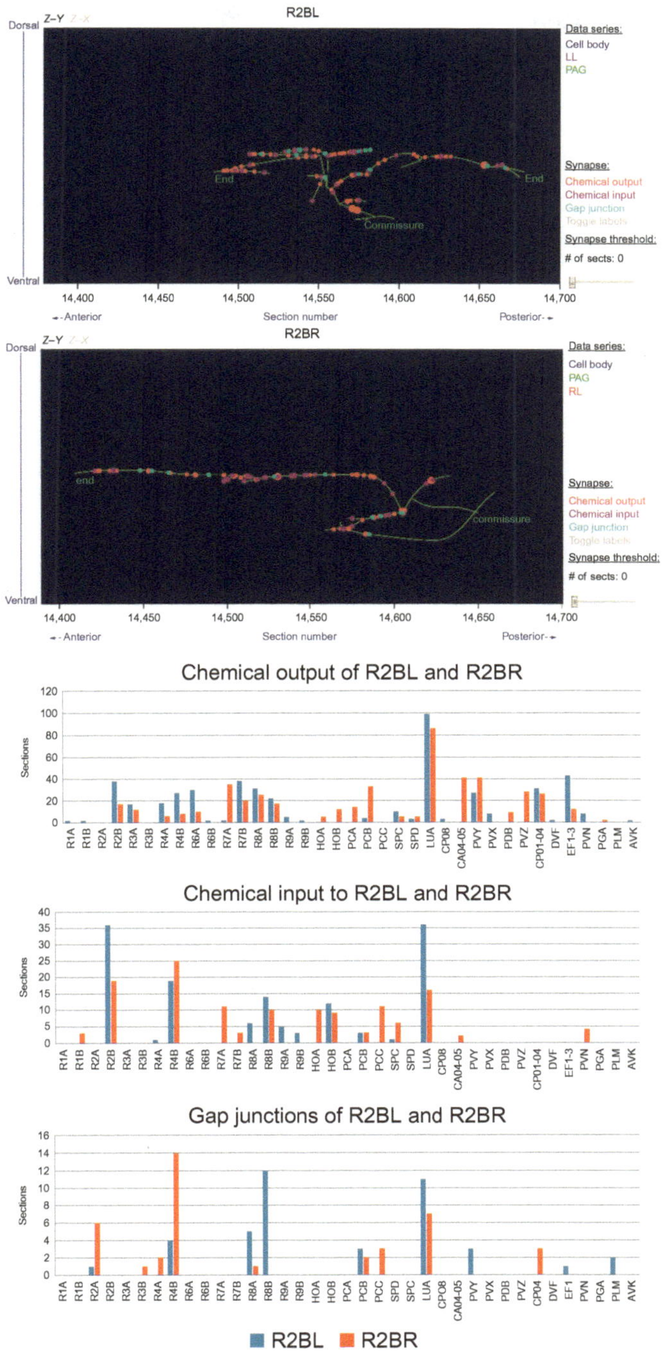

Scott W. Emmons and Dominick P. Purpura, Figure 3 See legend on next page.

Scott W. Emmons and Dominick P. Purpura, Figure 3 Homologous neurons have dissimilar branching structures in the preanal ganglion but similar sets of synaptic partners. Upper panels: the processes and synapses in the PAG of a pair of homologous neurons, the ray sensory neurons R2BL and R2BR. The cell bodies of these neurons are, respectively, in the left and right lumbar ganglia (Fig. 2). Pink circles show sites of chemical synaptic output, magenta circles are chemical input, and blue circles are gap junctions. These maps are from Wormwiring.org. Lower panels: histograms comparing the synaptic relationships made by R2BL and R2BR. The neurons listed along the bottom are the 68 neurons (out of 170) with synaptic interaction of some kind with one or both of these two neurons. Data for left/right homologs of R2BL and R2BR partners are combined; e.g., R1A represents the interaction with R1AL plus R1AR. The vertical axis gives the total morphological strength of interaction in number of EM serial sections. Note the differences in scales.

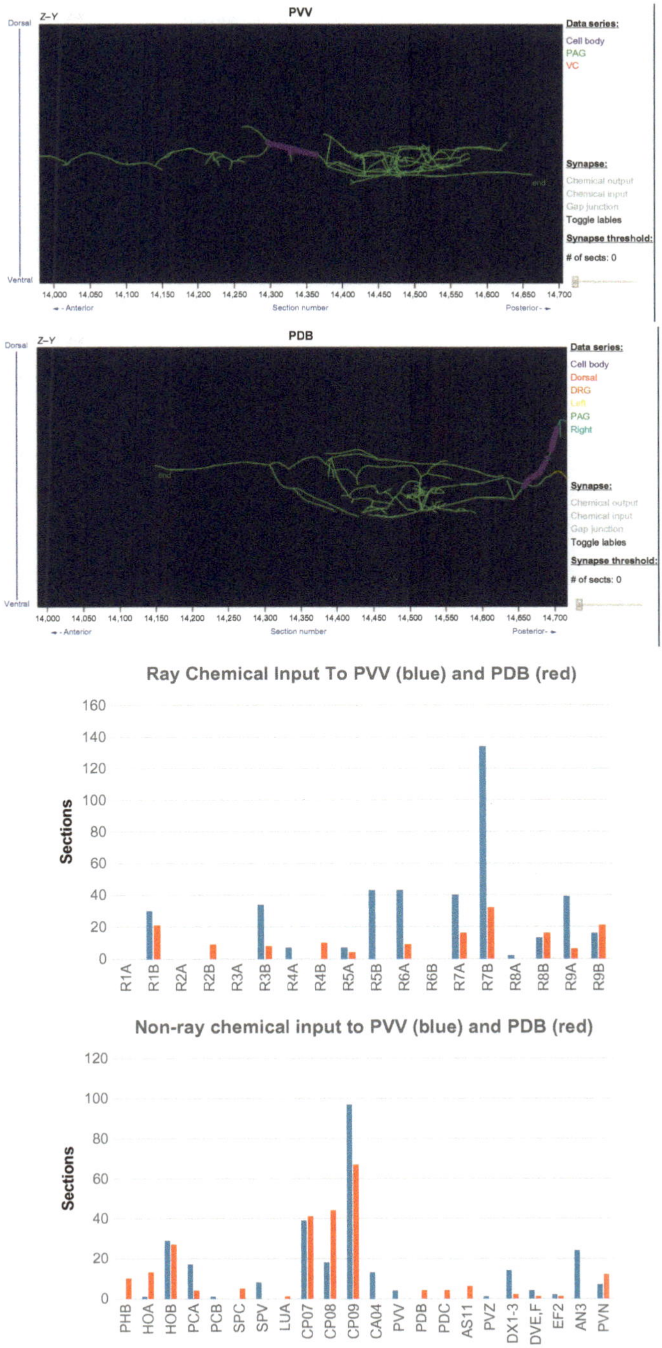

Scott W. Emmons and Dominick P. Purpura, Figure 4 See legend on next page.

Scott W. Emmons and Dominick P. Purpura, Figure 4 PVV and PDB are distinct neurons with similar patterns of synaptic input. Top panels show the overlapping branching structures of PVV and PDB in the preanal ganglion. PVV extends further anteriorly into the ventral cord, while PDB sends a process posteriorly around the end of the tail to the dorsal side. Both have output onto bodywall muscles. The blue boxes are the cell bodies. The horizontal axes show the serial section number of the transverse serial sections (A–P axis); the vertical axis shows position along the D–V axis; the maps are flattened along the L–R axis. The top histogram shows connectivity data for all the ray neurons; R1A(L/R), R2A(L/R), R3A(L/R), and R6B(L/R) make no connection to either PVV or PDB. The bottom histogram shows data for all the remaining neurons that make a chemical connection to either neuron. Data for left/right homologs and other equivalent neurons are combined.

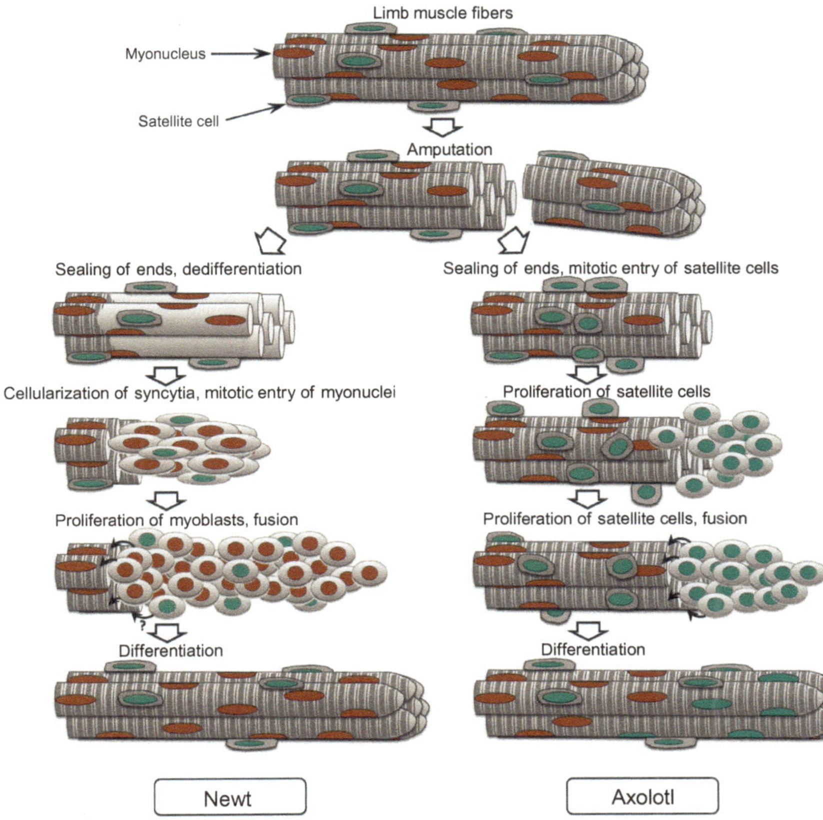

Manfred Frasch, Figure 1 Schematic comparison of limb muscle regeneration via muscle fiber fragmentation versus satellite cell activation. Examples of the distal ends of seven myofibers before, during, and after muscle regeneration are shown from top to bottom. In newt limbs (left), muscle fibers dedifferentiate following amputation and fragment into mononucleate cells, which enter the cell cycle, proliferate, and fuse with the damaged muscle fibers and, potentially, with one another. The contribution of satellite cells to muscle regeneration in this context is not known. In axolotl limbs (and likewise, *Xenopus* tadpole tails; right), amputation activates satellite cells, which proliferate and fuse to regenerate muscles.

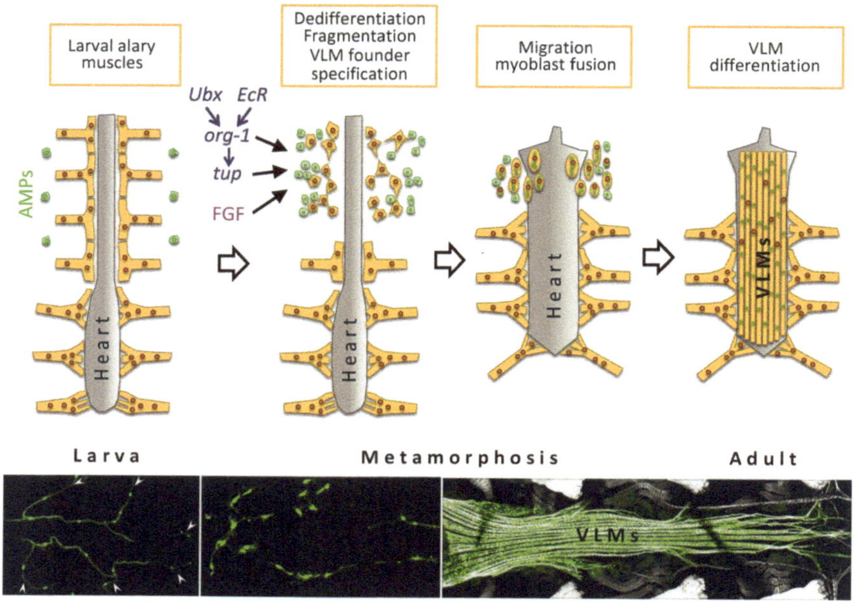

Manfred Frasch, Figure 2 Transdifferentiation of *Drosophila* larval alary muscles into ventral longitudinal muscles of the adult heart. Top: Schematic drawings of the fragmentation of specific larval alary muscles into mononucleate cells, which become specified as founder cells of the ventral longitudinal muscles (VLMs) of the adult heart. Upon myoblast fusion with fusion-competent myoblasts (derived from adult muscle progenitors [AMPs], green), they undergo morphogenesis and differentiate into VLMs. The known genetic inputs are depicted as described in the text. Bottom: Anterior three pairs of larval alary muscles (arrowheads) are shown on the left, fragmented mononucleate cells from these muscles in the center, and fully differentiated VLMs on the right. (All three panels show an *org-1* reporter in green and the right hand panel in addition shows F-actin in white.)

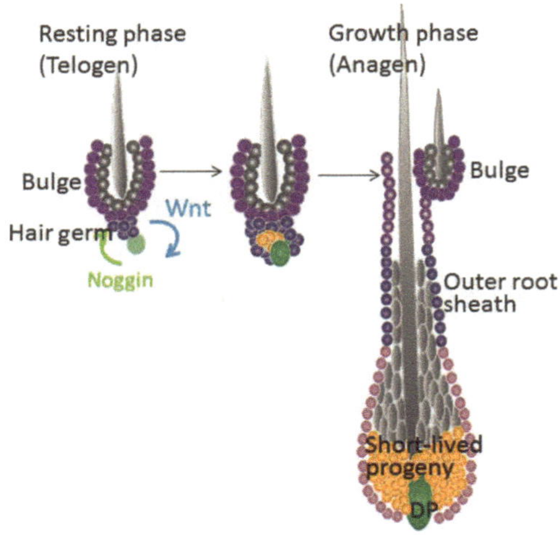

Elaine Fuchs, Figure 1 Schematic of the resting and growth phases of the hair follicle. The stem cells reside in the outer layer of the bulge niche, which anchors the hair from the previous hair cycle. The bulge cells at the base are referred to as the hair germ, which during telogen, is abutted next to the dermal papilla (DP), the mesenchymal stimulus of the hair follicle. Following a buildup of BMP inhibitors (Noggin) in the DP and WNTs in the hair germ, the stem cells at the base are activated and generate the short-lived progeny (orange), which ultimately will differentiate into the hair and its channel. *This figure is patterned after Figure S5 in Hsu, Li, and Fuchs (2014b). Elsevier Press.*

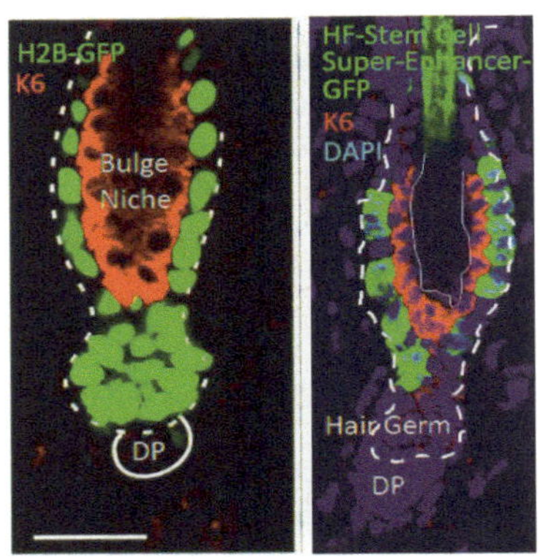

Elaine Fuchs, Figure 2 Immunofluorescence images showing a resting phase hair follicle, replete with bulge stem cells and keratin 6+ inner bulge nonstem cells. At left, bulge and hair germ cells are label-retaining and marked by H2B-GFP used in a pulse-chase labeling experiment. At right, bulge is labeled with GFP driven by a HF stem cell-specific super-enhancer epicenter element. *Left panel is Figure 5 Frame C of Hsu et al. (2011). Elsevier Press. Right panel is Figure 2e of Adam et al. (2015). Nature Press.*

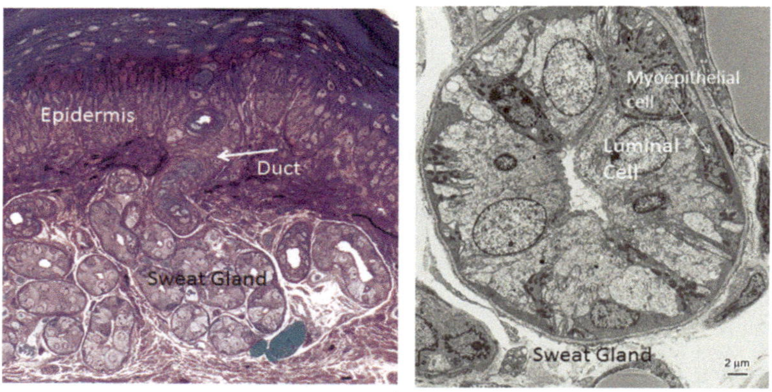

Elaine Fuchs, Figure 3 Semi-thin section and ultrastructure of a sweat gland from the mouse paw pad. The gland is a bilayered epithelium consisting of myoepithelial and luminal epithelial cells. The duct opens out onto the surface of the epidermis. *This figure is courtesy of H.A. Pasolli.*

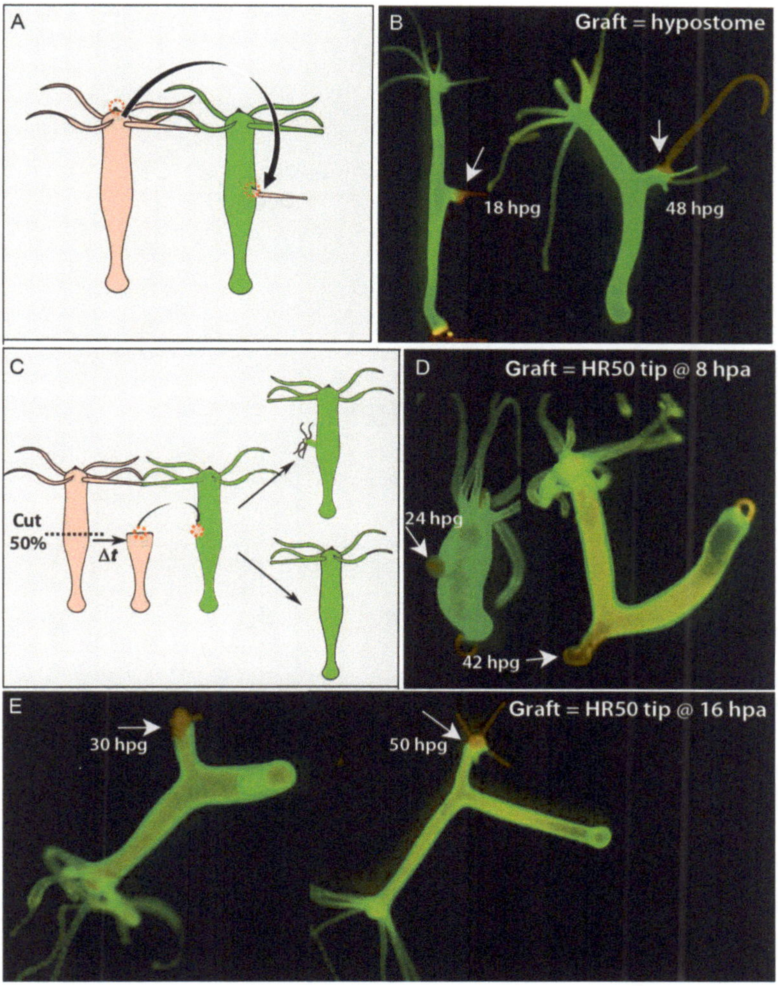

Matthias C. Vogg *et al.*, Figure 1 Homeostatic- and regeneration-induced head organizers in *Hydra* as initially characterized by Browne (1909). Reproduction of Ethel Browne's original experiments demonstrating the ability of the *Hydra* head or the *Hydra* head-regenerating tip to induce a complete ectopic axis (Browne, 1909). Here, we used as hosts two *AEP* actin::eGFP transgenic strains that constitutively expresses GFP either in the endodermal (Wittlieb, Khalturin, Lohmann, Anton-Erxleben, & Bosch, 2006) or in the ectodermal (Anton-Erxleben, Thomas, Wittlieb, Fraune, & Bosch, 2009) epithelial cells, named Endo-GFP and Ecto-GFP, respectively. (A) A fragment of head bearing a tentacle was dissected from a nonfluorescent *Hydra AEP* and immediately transplanted to the lateral side of the body column of the Endo-GFP host. As in Browne's experiment, the tentacle was used as a marker, which does not have organizing activity itself. (B) An ectopic axis that contains endodermal fluorescent cells from the host is rapidly induced, observed here at 18 and 48 hours post-grafting (hpg, arrows). (C) A nonfluorescent *Hydra AEP* was bisected at midgastric position (50%), the lower part was allowed to regenerate for 8 (D) or 16 (E) hours, and the tip was transplanted to the lateral side of the body column of a Ecto-GFP host. (D) Ectopic axis induced by a 8 h postamputation (hpa) graft visualized at 24 and 42 hpg (arrows). (E) Ectopic axis induced by a 16 hpa graft visualized at 30 and 50 hpa (arrows). In (D and E), the progressive growth and patterning of the ectopic axis is obvious. However, the development of the structure induced by the 16 hpa graft is more advanced than the one induced by the 8 hpa graft, illustrating the fact that the inductive capability of the tip increases within the first 12 h of head regeneration (MacWilliams, 1983b).

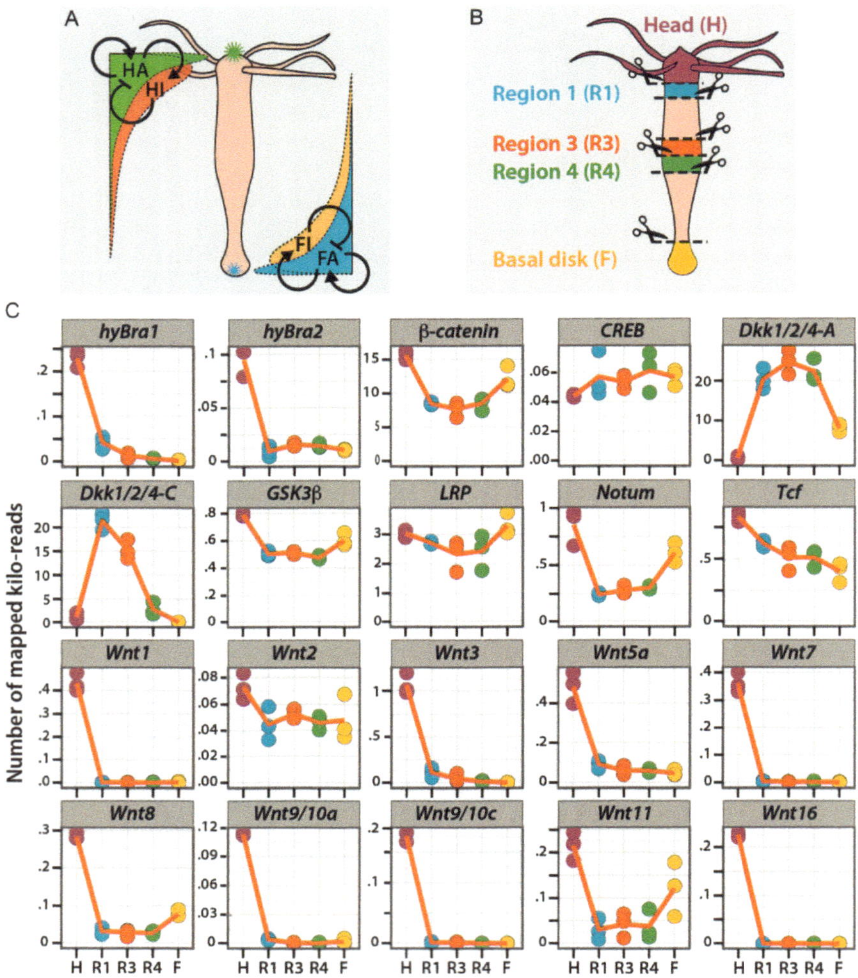

Matthias C. Vogg *et al.*, Figure 2 Molecular components of organizer maintenance in homeostatic *Hydra*. (A) Schematic representation of the two pairs of graded biological activities, head activation (HA, green) and head inhibition (HI, red), foot activation (HA, blue) and foot inhibition (FI, golden) as detected by lateral transplantation experiments (see Shimizu, 2012). Note their inverted graded distribution, maximal at the apical pole for HA/HI, at the basal pole for FA/FI. Green and blue stars represent head and foot organizers, respectively. (B) Scheme depicting the procedure used to perform quantitative RNA-seq transcriptomics at five distinct levels along the *Hydra* body axis (for details, see in Wenger, Buzgariu, & Galliot, 2015). (C) RNA-seq profiles of components, putative regulators and target genes of the Wnt/β-catenin pathway, involved in the maintenance and regeneration of the head organizer (Chera et al., 2009; Chera, Ghila, Wenger, & Galliot, 2011; Chera, Kaloulis, & Galliot, 2007; Guder et al., 2006; Hobmayer et al., 2000; Lengfeld et al., 2009; Nakamura, Tsiairis, Ozbek, & Holstein, 2011; Petersen et al., 2015; Technau et al., 2000). The color code for each position corresponds to colors selected in panel (B). Note that wnt signaling in *Hydra* specifies head identity whereas it specifies tail identity in planarians (see Fig. 3).

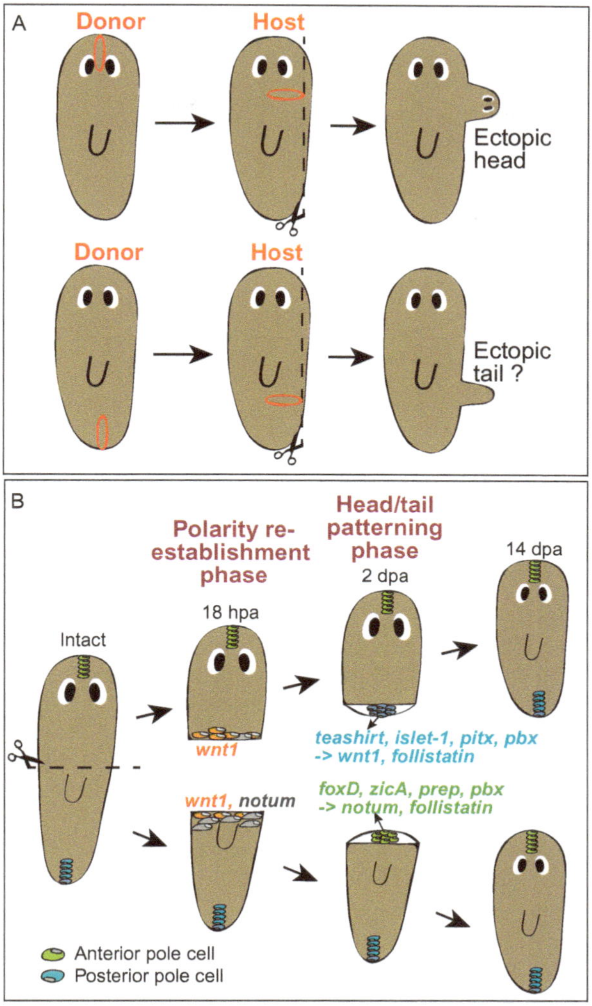

Matthias C. Vogg *et al.*, Figure 3 Anterior and posterior poles might represent putative head and tail organizers in homeostatic and regenerating planarians. (A) Transplantation experiment according to Saito, Koinuma, Watanabe, and Agata (2003). Tissue of the region between the eyes was transplanted into a host planarian that was subsequently cut laterally. Head outgrowth was observed in transplanted animals, suggesting that the transplanted tissue has head organizer activity. A similar scenario is expected to be true for the transplantation of midline tail tissue: the outgrowth of a tail on the host would indicate organizer activity of the tail transplant. (B) Amputation at any level along the animal induces *wnt1* expression at anterior- and posterior-facing wound sites at 18 hours postamputation (hpa) (Petersen & Reddien, 2009). At anterior-facing wound sites, the activity of Wnt1 is antagonized by the secreted Wnt inhibitor Notum, which triggers

(Continued)

Matthias C. Vogg *et al.*, Figure 3—Cont'd anterior polarity reestablishment (Petersen & Reddien, 2011). In contrary, at posterior wounds, active Wnt signaling is necessary for posterior polarity reestablishment. Note that at 18 hpa *notum* and *wnt1* expression is induced in fully differentiated cells, whereas at 48 hpa on the posterior side, the transcription factor genes *islet-1* (Hayashi et al., 2011), *pitx* (Currie & Pearson, 2013; Marz, Seebeck, & Bartscherer, 2013), *pbx* (Blassberg, Felix, Tejada-Romero, & Aboobaker, 2013; Chen, Wang, & Reddien, 2013), and *teashirt* (Owen, Wagner, Chen, Petersen, & Reddien, 2015; Reuter et al., 2015) (blue) are upregulated in neoblasts that subsequently differentiate in pole cells that express *wnt1* (Petersen & Reddien, 2009) and *follistatin* (Gavino, Wenemoser, Wang, & Reddien, 2013). Similarly on the anterior side, the transcription factor genes *foxD, zicA, prep,* and *pbx* (Blassberg et al., 2013; Chen et al., 2013; Felix & Aboobaker, 2010; Scimone, Lapan, & Reddien, 2014; Vasquez-Doorman & Petersen, 2014; Vogg et al., 2014) are upregulated in neoblasts that provide the pole cells expressing *notum* (Petersen & Reddien, 2011) and *follistatin* (Gavino et al., 2013; Roberts-Galbraith & Newmark, 2013).

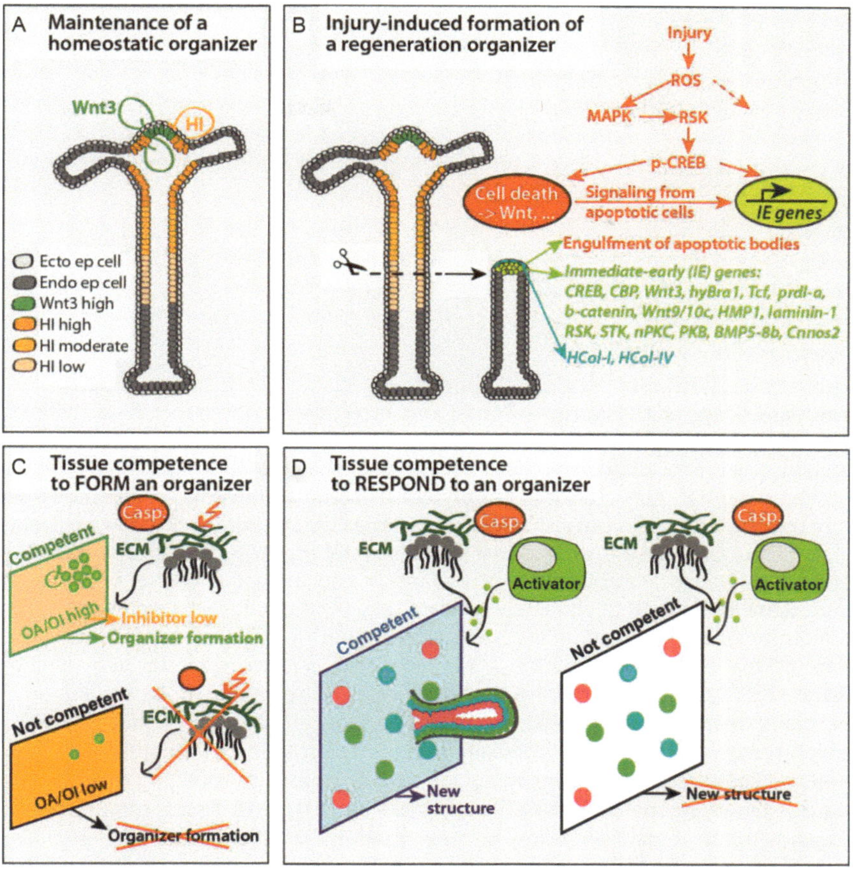

Matthias C. Vogg *et al.*, Figure 4 Organizer maintenance, organizer *de novo* formation upon injury, and competence to respond to organizer activity in an adult organism. (A) Organizer activity is maintained in homeostatic *Hydra* thanks to the head activator Wnt3 that maintains active β-catenin signaling in the head through a positive feedback loop (Nakamura et al., 2011). Note that head inhibitor (HI), which is expressed under the control of Wnt3 and antagonizes Wnt3 signaling, is not characterized yet. (B) *De novo* formation of a head organizer in head-regenerating tips after midgastric bisection relies on the asymmetric activation of the MAPK/CREB pathway (Galliot, Welschof, Schuckert, Hoffmeister, & Schaller, 1995; Kaloulis, Chera, Hassel, Gauchat, & Galliot, 2004), which together with injury-induced ROS signaling, triggers local cell death in the interstitial lineage. Apoptotic cells transiently release signaling molecules, including Wnt3, which activates the β-catenin pathway in surrounding cells, inducing a wave of mitotic division (Chera et al., 2009). At the same time, the engulfment of the apoptotic bodies by the endodermal epithelial cells (green) is followed by the upregulation of immediate-early (IE) genes, including *wnt3* that contains CREB response elements (CREs) in its regulatory sequence (Galliot, 2013; Nakamura et al., 2011). (C) Tissue competence to form an

(Continued)

Matthias C. Vogg *et al.*, Figure 4—Cont'd organizer relies on the respective activities of Activator (OA) and Inhibitor (OI) components, where OA are extracellular signal(s) and OI inhibitors that restrict OA extracellular activity. Only tissues with a high OA/OI ratio can form an organizer. Activators (i.e., morphogens) can be secreted as soluble molecules from the producing cells but also released from the ECM by proteolytic cleavage (Fleury et al., 2006). Once in the extracellular space, several mechanisms can promote their long-range graded activity such as the diffusion and the accumulation of proteases or protease inhibitors that ultimately regulate spatially the stability of proteins involved in morphogenesis (Inomata et al., 2013). Note that in *Hydra*, injury-induced modifications of the extracellular matrix (ECM, red flash) are necessary for regeneration (Sarras, 2012), possibly to increase the OA/OI ratio in the vicinity of the wound, up to the point where tissues become competent to form an organizer. Similarly in adult zebrafish regenerating its fins, signaling from the apoptotic cells and/or from the apoptotic bodies is necessary for the upregulation of IE genes in surrounding cells and subsequent tissue regeneration (Vriz, Reiter, & Galliot, 2014). (D) Tissue competence to respond to an organizer relies on the strength of the intracellular signaling pathway(s) activated by the activator. Tissues with a high signaling activity (blue background) are able to induce ectopic structures in response to OA (left), while tissues with a low signaling activity (white background) are unable to respond (right). Casp., caspase activation.

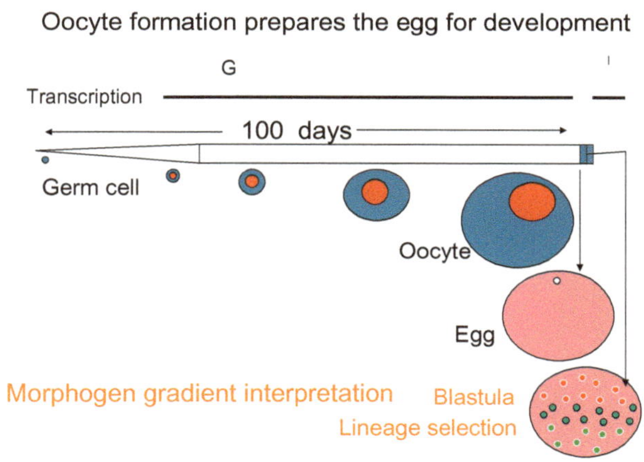

J.B. Gurdon, Figure 1 A diagram to illustrate the relationship of an oocyte to the embryo in *Xenopus*.

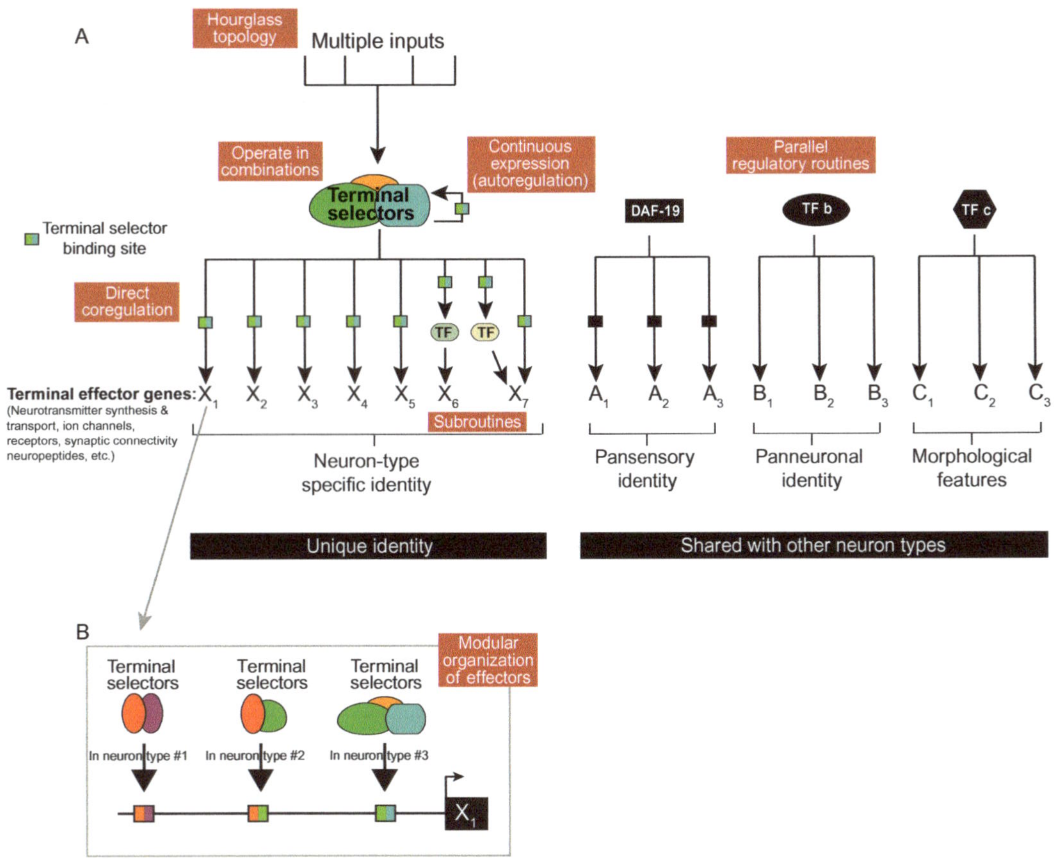

Oliver Hobert, Figure 1 See legend on opposite page.

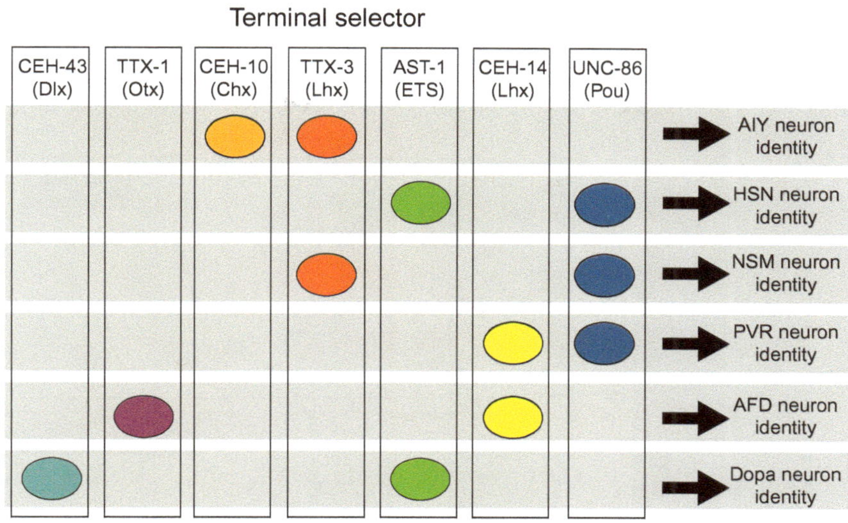

Oliver Hobert, Figure 2 Combinatorial activity of terminal selectors. Selected examples are shown, some not showing the complete set of factors (e.g., dopaminergic neurons are controlled by the combined action of AST-1, CEH-43 plus a Pbx-type homeodomain factor). Note that all except one factor (AST-1) are homeodomain transcription factors.

Oliver Hobert, Figure 1 Key features of gene expression programs in mature neuron types. (A) Terminal selector regulons. Red squares highlight key concepts. Sustained expression of terminal selectors is often, but not always ensured by autoregulation. Small rectangles illustrate binding sites for transcription factors and indicate direct regulation of effector genes by terminal selectors. Terminal selectors shown on the left control neuron-type specific identity features that assign a unique identity to a neuron. See Table 2 for an exemplary list of effector genes. Parallel regulatory routines such as those that regulate pansensory features (via DAF-19) are controlled by factors that could also considered to be terminal selectors; but they do not assign unique identities. Morphology regulators control generic aspect of a neuron's morphology, such as placement of axons/dendrites into specific fascicles or axo/dendritic polarity. TF, transcription factor. (B) Modular organization of terminal effector genes. X_1 is a representative of any effector gene expressed in more than one neuron type (e.g., the vesicular transporter of glutamate, which is expressed in all glutamatergic neurons, but not in GABA or cholinergic neurons). This schematic also indicates the key principle of combinatorial "reusage" of the same terminal selector in different neuron types, as schematized in more detail in Fig. 2.

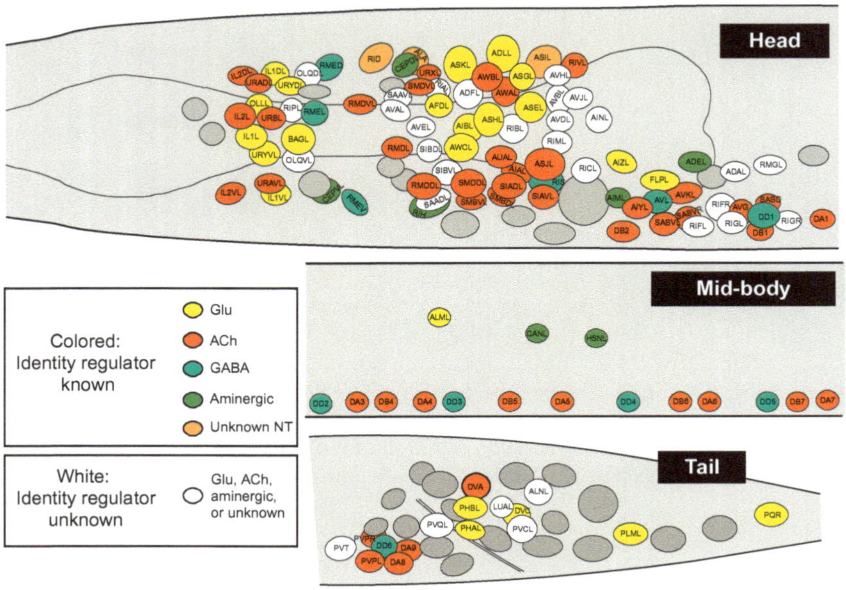

Oliver Hobert, Figure 3 An overview of identity regulator in an L1 stage animal. Colored neurons are those in which terminal selector-type transcription factors have been identified that affect multiple identity features, including (in those cases known) the neurotransmitter identity of the neuron (color-coded). The only currently ambiguous cases included in this figure are the command interneurons, in which one factor (*unc-3*) affects neurotransmitter identity, but not other identity features, while another factor (*unc-42*) controls other identity features, but not neurotransmitter identity (Pereira et al., 2015). Pharyngeal neurons and postembryonically generated neurons are not shown. References: Altun, Chen, Wang, and Hall (2009), Cassata et al. (2000), Doitsidou et al. (2013), Etchberger et al. (2007), Gonzalez-Barrios et al. (2015), Gordon and Hobert (2015), Guillermin, Castelletto, and Hallem (2011), Jin et al. (1994), Hutter (2003), Kim, Kim, and Sengupta (2010), Kratsios et al. (2015, 2011), Lanjuin, VanHoven, Bargmann, Thompson, and Sengupta (2003), Nokes et al. (2009), Pereira et al. (2015), Satterlee et al. (2001), Sengupta et al. (1994), Serrano-Saiz et al. (2013), Sze, Zhang, Li, and Ruvkun (2002), Van Buskirk and Sternberg (2010), Zhang et al. (2014), and Zhang et al. (2002).

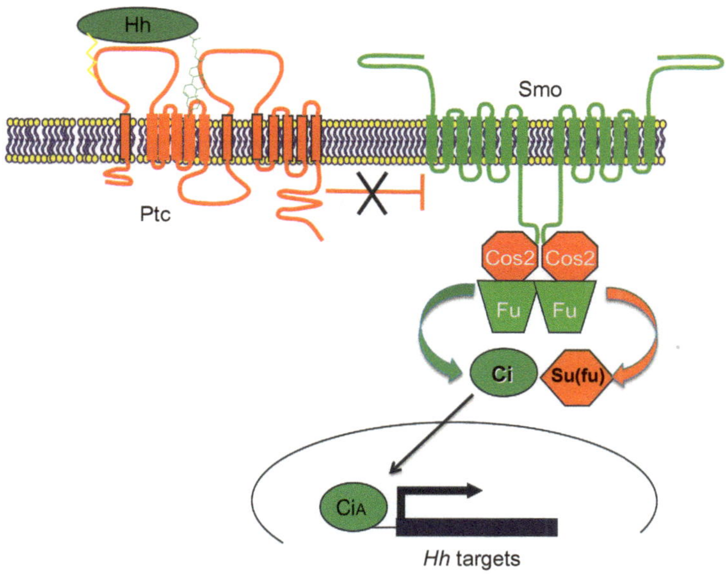

Philip W. Ingham, Figure 1 Simplified schema of the *Drosophila* Hh signal transduction pathway showing the core components. Negative regulators and repressive interactions are shown in red, positive regulators and activating interactions are shown in green.

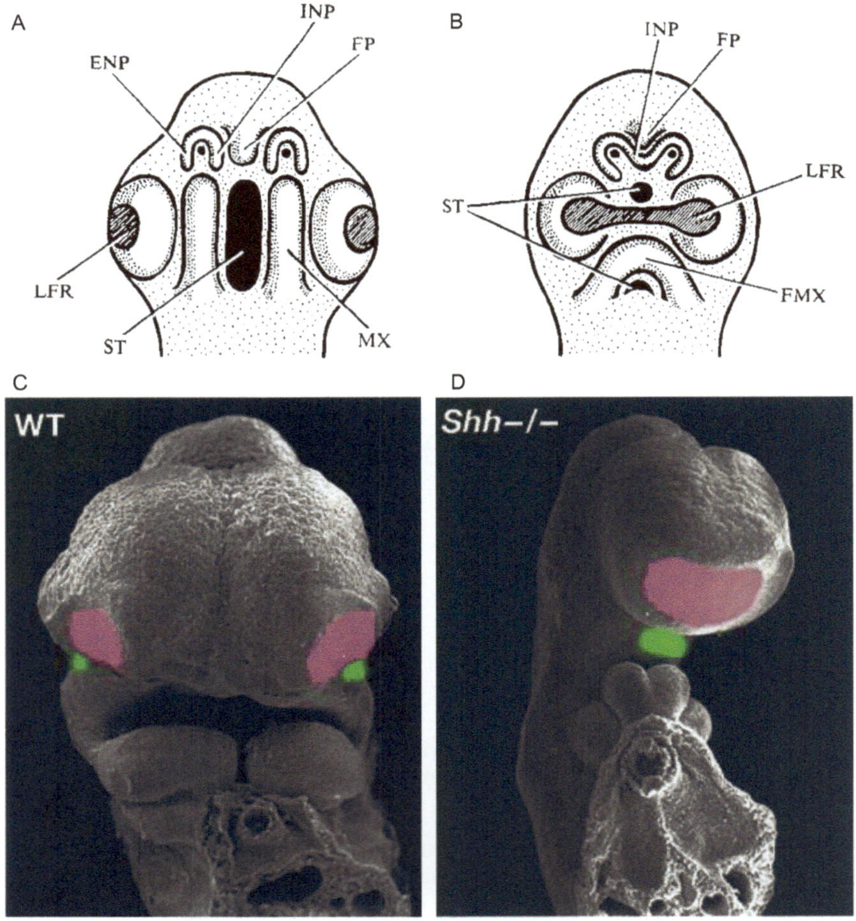

Philip W. Ingham, Figure 4 (A and B) Schematic representations of the facial structures in a wild-type chick embryo (A) and a *talpid3* mutant chick embryo (B). ENP, external nasal process; FMX, fused maxillary processes; FP, frontal process; INP, internal nasal process; LFR, lens-forming region; MX, maxillary process; ST, stomodaeum. (C and D) Scanning electron micrographs of a wild-type (C) and Shh (D) mutant mouse embryo. Note the similarities between the *talpid3* and *Shh* mutant phenotypes. *Panels (A) and (B): Reprinted from Ede and Kelly (1964), Copyright (1964), with permission from The Company of Biologists. Panels (C) and (D): Reprinted from Aota, K., et al. Mouse Shh is required for prechordal plate maintenance during brain and craniofacial morphogenesis.* Developmental Biology, 327, *106–120, Copyright (2009), with permission from Elsevier.*

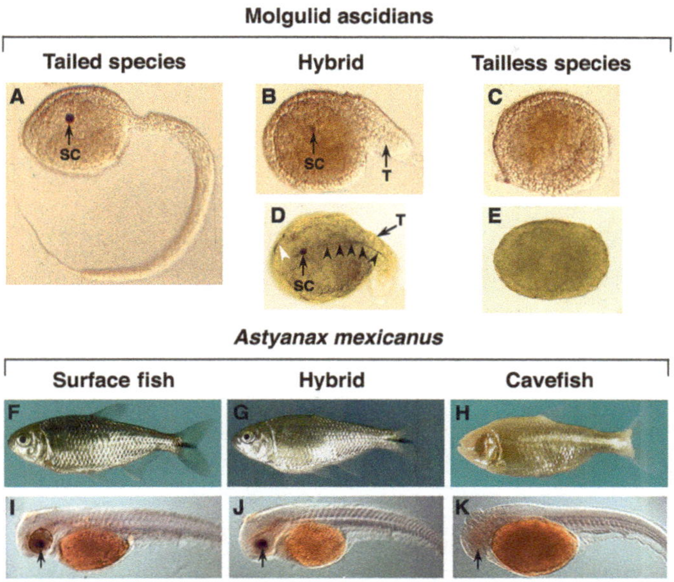

William R. Jeffery, Figure 1 The Molgulid ascidian and *Astyanax mexicanus* models. *Top*: molgulid ascidians. (A) Tailed urodele larva of *Molgula oculata* (the tailed species). (B) Short-tailed urodele larva of a cross between *Molgula occulta* (the tailless species) eggs and *M. oculata* (the tailed species) sperm. (C) Tailless anural larva of the *M. occulta* (the tailless species). (D and E) Alpha-acetylated tubulin antibody staining showing the restoration of anterior and tail nerves, which is absent in the tailless species larva (E), in a tailless × tailed species hybrid (D). SC, brain sensory cell. White arrowhead: part of a restored anterior nerve. Black arrowheads: restored tail nerve. *Bottom*: *A. mexicanus*. (F) Adult surface fish with eyes and pigment. (G) Adult surface fish × cavefish hybrid with eyes and pigment. (H) Adult cavefish lacking eyes and pigment. (I) Larval surface fish showing *alphaA-crystallin* gene expression in the lens. (J) Larval surface fish × cavefish hybrid showing *alphaA-crystallin* gene expression in the lens. (K) Larval cavefish lacking *alphaA-crystallin* gene expression in the lens. Arrows: lens. *Panels (A) and (B) from Swalla and Jeffery (1996), (C) and (D) from Y. Yamamoto and W.R. Jeffery (unpublished), and (I)–(K) from Ma, Parkhurst, and Jeffery (2014).*

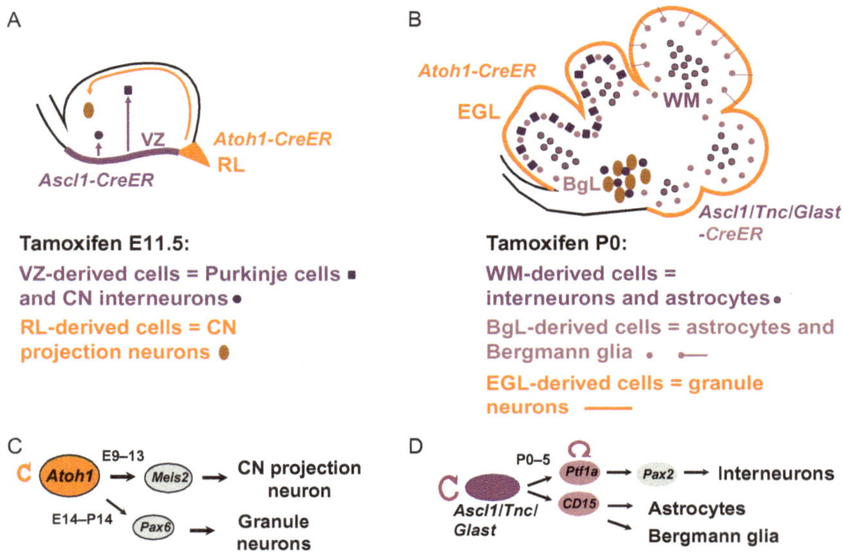

Alexandra L. Joyner, Figure 1 Genetic inducible fate-mapping studies define distinct progenitor pools in the cerebellum separated spatially and/or in time. (A and B) Schematic drawings of midsagittal sections of cerebella at E11.5 and P0 are shown. The cells that can be labeled using a reporter allele such as *R26*^{Yfp} and an *Atoh1-CreER* transgene or *Ascl1*^{CreER} knock-in allele (*Ascl1-CreER*) are indicated by orange and purple, respectively, and the cells they give rise to when tamoxifen is given at E11.5 (A) or (P1) are indicated below. In (B), Purkinje cells are only shown in the anterior lobules (folds) and glial fibers of Bergmann glial progenitors only in one posterior lobule for simplicity. At E12.5–17.5, the rhombic lip gives rise to the external granule cell layer (orange) shown in (B) made up of *Atoh1*-expressing progenitors that can divide until P16. Three knock-in alleles have been used to mark progenitors in the WM and BgL (*Ascl1*^{CreER}, *Tnc*^{CreER}, *Glast*^{CreER}). Round arrows indicate proliferating progenitors. VZ, ventricular zone; RL, rhombic lip; CN, cerebellar nuclei; EGL, external granule cell layer; BgL, Bergmann glial layer; WM, white matter. (C and D) The *Atoh1* and *Ascl1/Tnc/Glast* lineages are show. *Data are based on Fleming et al. (2013), Machold and Fishell (2005), Parmigiani et al. (2015), and Sudarov et al. (2011).*

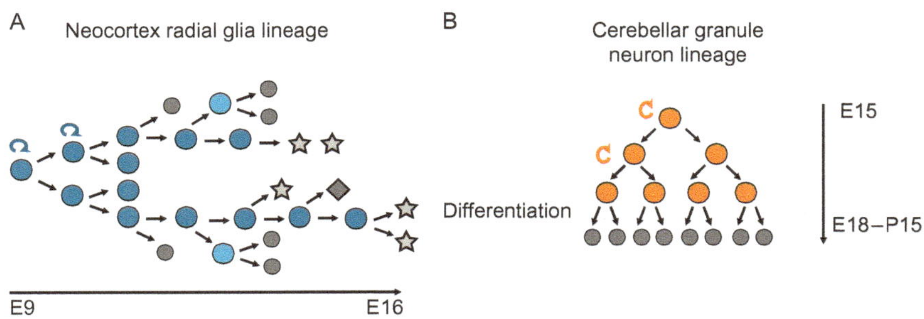

Alexandra L. Joyner, Figure 2 Clonal analyses of glutamatergic progenitors in the forebrain (neocortex) and cerebellum reveal different modes of generating neurons and glia. (A) Radial glial progenitors (blue circles) in the ventricular zone of the neocortex initially divide symmetrically to expand the progenitors and then divide mainly asymmetrically to produce a progenitor and a differentiated projection neuron (gray circle), astrocyte (star), or oligodendrocyte (triangle), or a progenitor with more limited proliferation capacity (light blue) that gives rise often to two neurons. Some symmetric divisions produce two neurons or astrocytes (differentiation divisions). Round arrows indicate symmetrical divisions producing two progenitors. Gliogenesis (production of astrocytes and oligodendrocytes) occurs after neurogenesis. (B) In the cerebellum, granule cell progenitors in the external granule cell layer (EGL in Fig. 1B) divide only symmetrically to expand the number of progenitors and then to produce postmitotic granule neurons. *Data are based on Espinosa and Luo (2008), Gao et al. (2014), and Legue et al. (2015).*

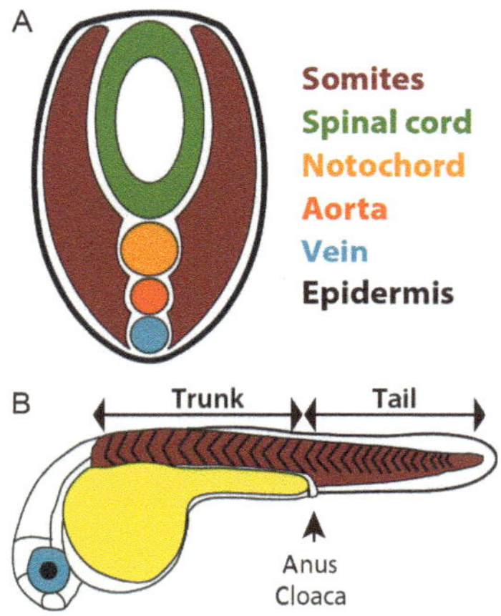

David Kimelman, Figure 1 Morphological features of the vertebrate posterior body. (A) A generic cross section through the postanal region of a vertebrate embryo. In the preanal region, the pronephric tubules and gut tube would also be present. (B) The posterior of the embryo is divided into preanal trunk and postanal tail. Shown is a zebrafish embryo but the same division applies to all vertebrates.

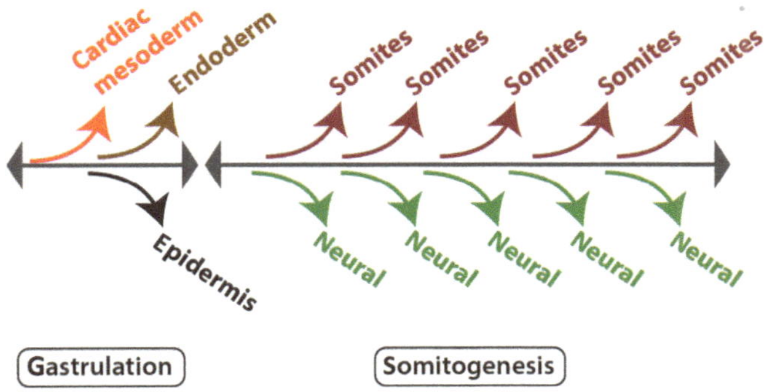

David Kimelman, Figure 2 Maintenance of the neuromesodermal progenitors. Lineage labeling studies in the mouse show that while many germ layer decisions are made early, a neuromesodermal progenitor population continues to produce both neural and mesodermal cells during the somitogenesis stages (Tzouanacou, Wegener, Wymeersch, Wilson, & Nicolas, 2009). The times at which the gastrula stage fate decisions are made are approximate.

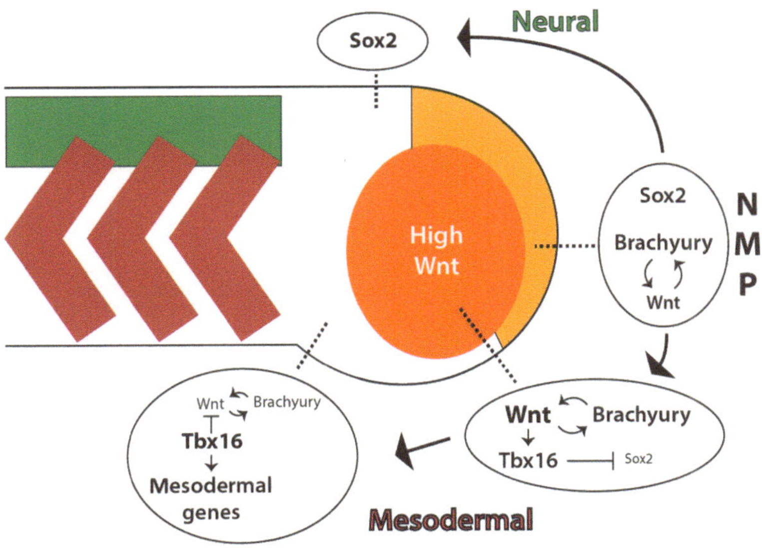

David Kimelman, Figure 3 Model for regulation of mesoderm formation in zebrafish. Sox2-expressing NMP cells are in a zone of moderate Wnt signaling (light orange) that maintains Brachyury. As cells leave this region, if they are not exposed to continue Wnt signaling, they express Sox2 but not Brachyury and become neural. Instead, if they are exposed to high levels of Wnt (dark orange), they express Tbx16, which represses Sox2. Tbx16 also represses Wnts so as cells continue to move anteriorly (left side), they shut off Brachyury and turn on mesodermal differentiation genes, thus locking in the mesodermal choice.

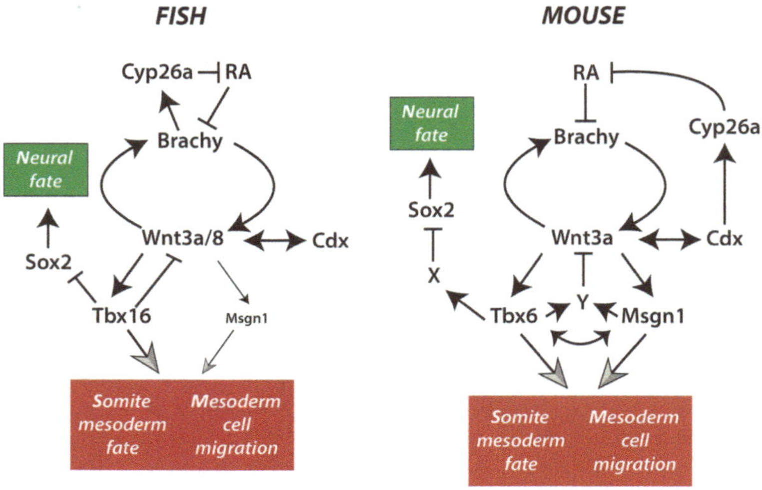

David Kimelman, Figure 4 Comparison of the regulatory logic of fish and mouse posterior cell fate choices. A schematic showing the regulatory relationships among signaling and transcription factors discussed in this review. Tbx6 is shown here to cause repressive effects indirectly, but as with Tbx16, it may turn out to be able to directly repress target genes. Fgf is omitted for simplicity.

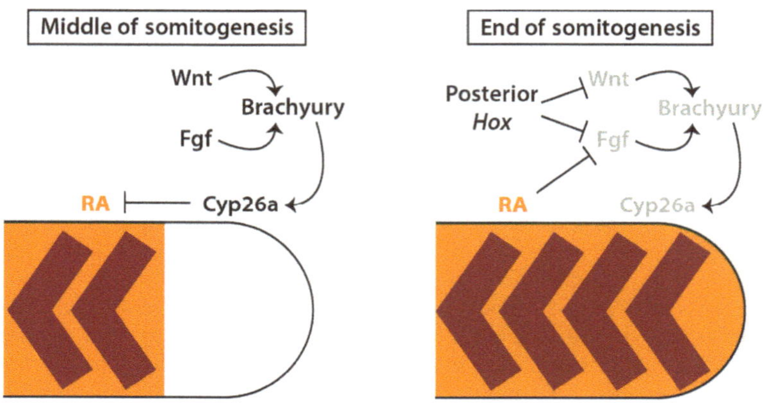

David Kimelman, Figure 5 Model for termination of axis extension in chick. During somitogenesis, the progenitor domain excludes RA. Near the end of somitogenesis, the most posterior *Hox* genes, possibly together with other molecular changes, causes RA to expand posteriorly, thus inhibiting the progenitor cell fate.

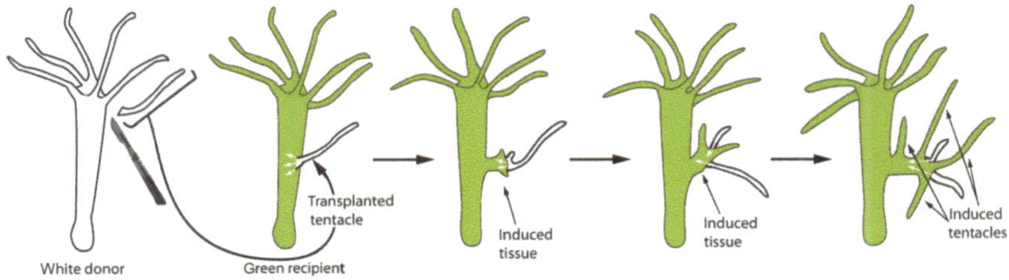

Thomas B. Kornberg, Figure 1 Grafting experiments reveal developmental induction. A tentacle transplanted from a white hydra to the stalk of a green recipient induces the host to produce ectopic tentacles.

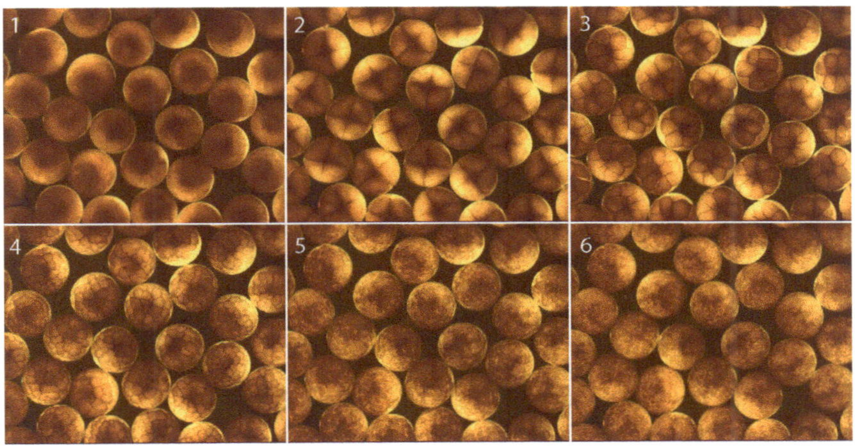

Thomas B. Kornberg, Figure 2 *Xenopus* embryos undergoing cleavage stage divisions. Fertilization of this field of embryos was initiated simultaneously, and the synchrony and regularity of the early stages of development are apparent in these six frames of a movie taken by J. Smith.

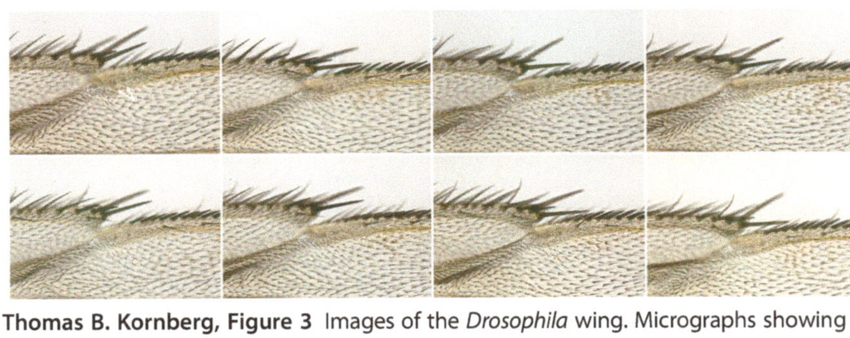

Thomas B. Kornberg, Figure 3 Images of the *Drosophila* wing. Micrographs showing a portion of the wing margin of wings dissected from eight *Drosophila* adults. The orientation, relative size, and color of all the cuticular elements are reproducibly patterned in these specimens. Arrows point to two sensillae, sense organs of the peripheral nervous system.

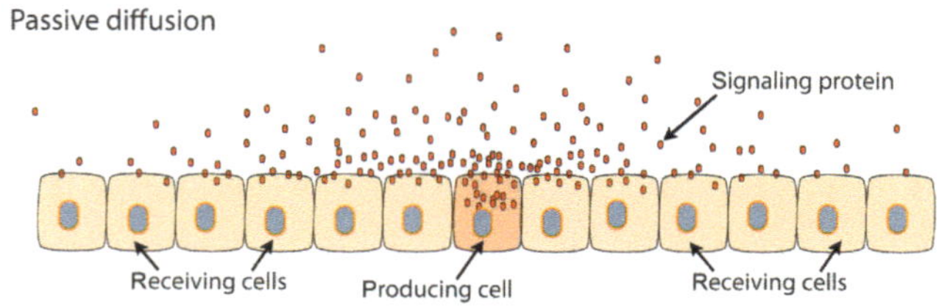

Thomas B. Kornberg, Figure 4 Passive diffusion of secreted signaling protein. This drawing of a hypothetical epithelial sheet depicts a single source cell that secretes signaling protein into extracellular space, generating a distribution of signaling protein that declines with increasing distance from the source.

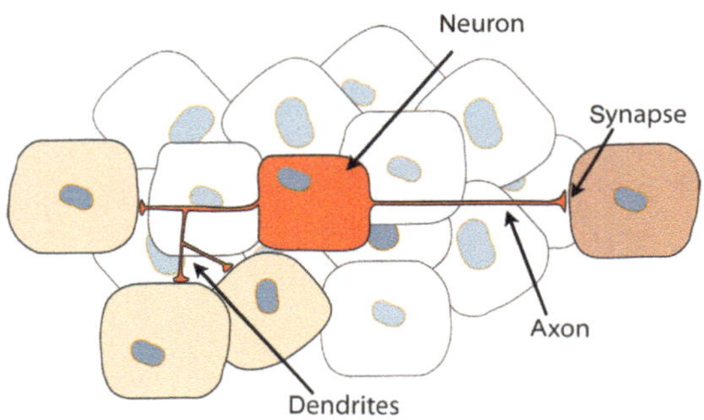

Thomas B. Kornberg, Figure 5 Cytoneme-mediated dispersion of signaling protein. This drawing of a hypothetical epithelial sheet depicts a single source cell that released signaling protein only at sites of direct contact between cytonemes that extend from cells that receive the signaling protein. Signaling protein moves across the epithelial sheet in a cytoneme-associated form and is extracellular only at the point of transfer where cytonemes make direct contact.

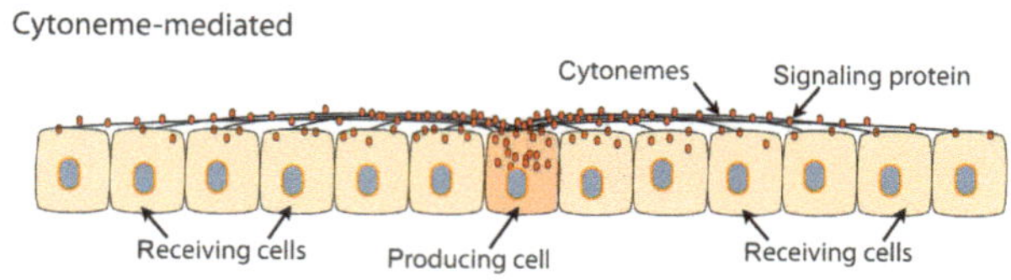

Thomas B. Kornberg, Figure 6 Synaptic signaling by a neuron. This drawing of a hypothetical neuron with an axon and dendritic extensions, making multiple synaptic contacts with target cells.

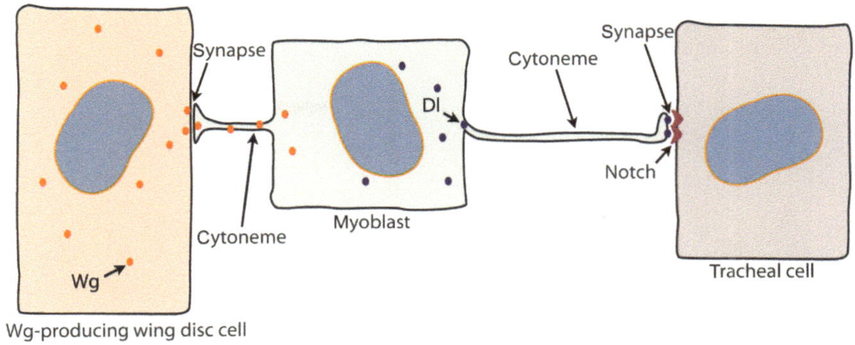

Thomas B. Kornberg, Figure 7 Synaptic signaling by columnar, tracheal, and myoblast cells of the *Drosophila* wing imaginal disc. The three cell types depicted are a wing disc columnar epithelial cell (light red) that expresses Wg, a myoblast (light blue) that picks up Wg from the disc cell and expresses Dl, and a tracheal cell (light purple) that expresses Notch and picks up Dl from the myoblast. Exchange of Wg and signaling by Dl takes place at the synapses at the tips of the two types of myoblast cytonemes.

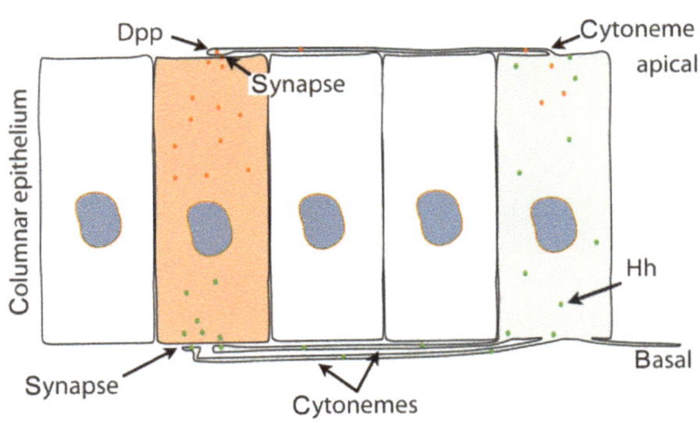

Thomas B. Kornberg, Figure 8 Synaptic signaling by columnar cells of the *Drosophila* wing imaginal disc. This drawing of a portion of the wing disc columnar epithelium depicts a Hh-expressing cell (light green) that extends Hh-carrying cytonemes from its basal surface and apical cytonemes that pick up and transport Dpp from a Dpp-expressing cell (light red). Exchange of Dpp and Hh takes place at the synapses at the tips of the cytonemes.

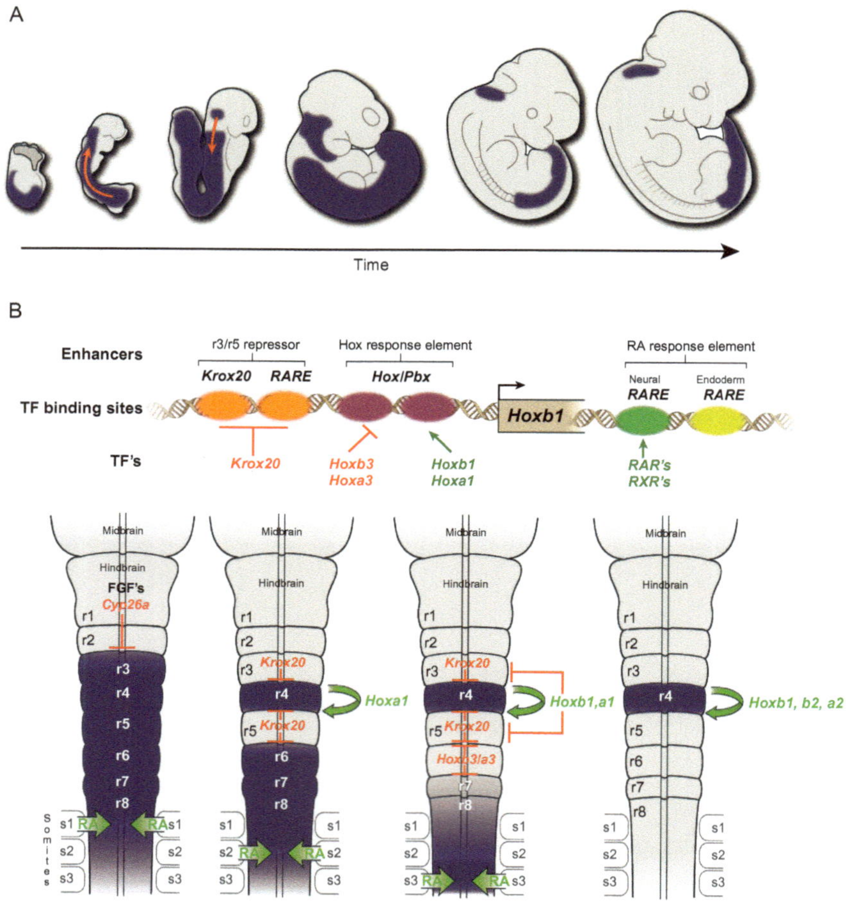

Robb Krumlauf, Figure 1 Temporal dynamics in the segmentally restricted expression and regulation of *Hoxb1* in r4. (A) Sketches of *in situ* expression profiles of *Hoxb1* in mouse embryos showing changes over time. (B) At the top is a diagram of the *Hoxb1* gene with enhancer regions, binding sites, and regulatory inputs from transcription factors. At the bottom is a drawing of how the initial broad domain of *Hoxb1* becomes progressively restricted to r4 through the input of repressive and activation inputs.

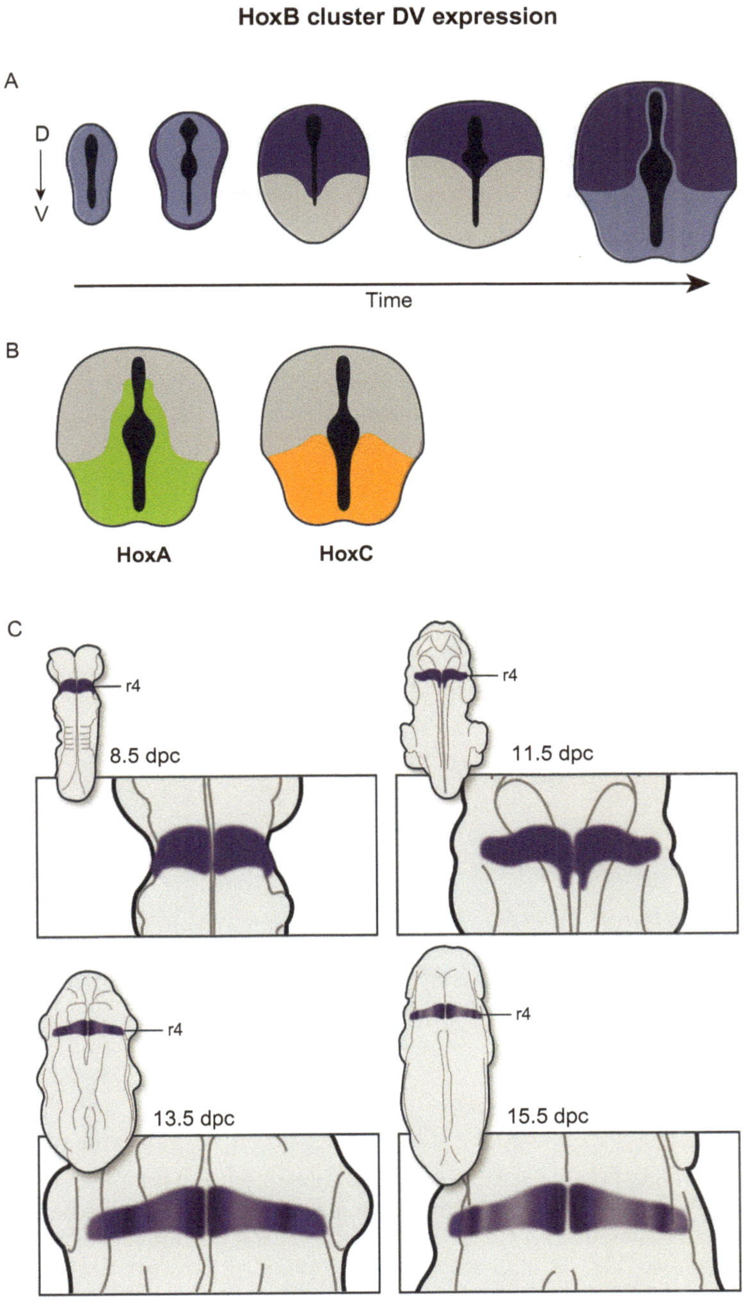

Robb Krumlauf, Figure 2 Dorsoventral changes in *Hox* expression in the neural tube. (A) Cross-sectional diagrams of the developing neural tube showing how *HoxB* genes become enriched in dorsal regions. (B) Comparison of the different dorsoventral distributions of genes from separate Hox clusters. (C) Sketch showing that the r4 expression of *Hoxb1* becomes restricted to small longitudinal stripes of neurons in r4.

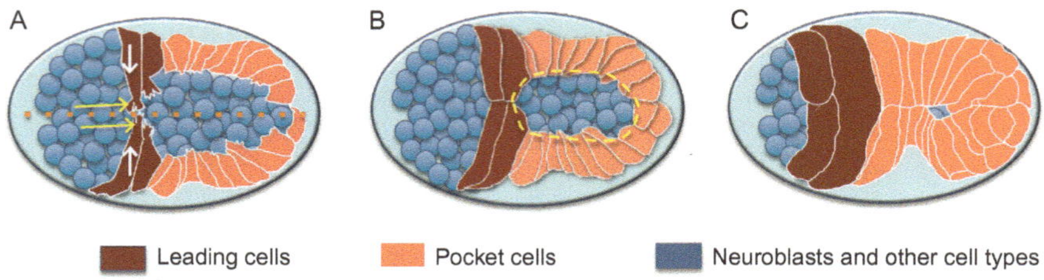

Thanh T.K. Vuong-Brender *et al.*, Figure 1 The two steps of epidermal ventral enclosure. (A) Schematic ventral view of an embryo (anterior to the left). First, the leading cells (dark red) extend long protrusions (yellow arrows), migrate (white arrows show the direction of migration) toward the ventral midline (dotted line), meet and establish adherens junctions. (B) The pocket cells (pink) exert coordinated contraction to close the open pouch (dashed line). (C) End of ventral closure; the head is not totally enclosed (some head epidermal cells anterior to the leading cells are not illustrated). Neuroblasts and other cell types are depicted in blue.

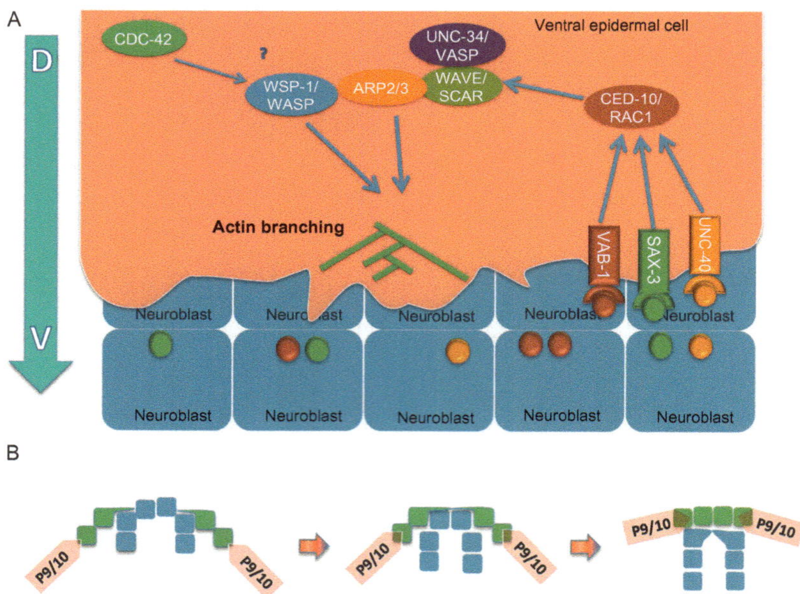

Thanh T.K. Vuong-Brender *et al.*, Figure 2 Neuroblasts guide ventral epidermis migration during ventral enclosure. (A) Signaling from neuroblasts promotes actin dynamics in the ventral epidermis. CED-10/Rac1, which acts probably downstream of VAB-1/EphR, SAX-3/Robo, and UNC-40/DCC, activates the WAVE/SCAR complex, and in turn the ARP2/3 complex. UNC-34/VASP binds WAVE/SCAR to promote actin polymerization. WSP-1, probably activated by CDC-42, contributes to a redundant pathway with WAVE/SCAR and UNC-34/VASP, which has not yet been fully investigated. Expression of guidance cues is discussed in the text. Big blue arrow: direction of migration; D, dorsal side; V, ventral side. Small ovals: guidance cues secreted by neuroblasts. (B) Schematic ventral view of the ventral midline, depicting how epidermal ventral cells (P9/10) migrate over a bridge made by neuroblasts. Some neuroblasts (green squares; only key neuroblasts are represented) extend protrusions and translocate over their neighbors (blue squares) to establish a bridge (Ikegami et al., 2012).

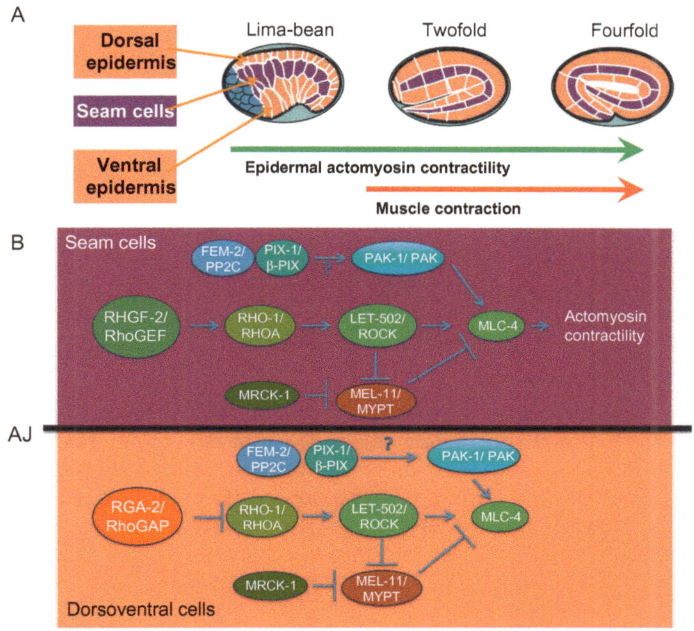

Thanh T.K. Vuong-Brender *et al.*, Figure 3 Overview of *C. elegans* embryonic elongation. (A) Schematic drawings of embryos along elongation. Note the change in shape of the seam cells (lateral epidermis; violet) where the junctions shorten in the dorsoventral (DV, circumferential) direction and lengthen in the anteroposterior (AP) direction. Embryos are staged according to their length (lima-bean or onefold to fourfold). Anterior to the left, dorsal up. (B) Molecular pathways mediating activation of nonmuscle myosin II contractility through phosphorylation of the myosin regulatory light chain MLC-4 during elongation.

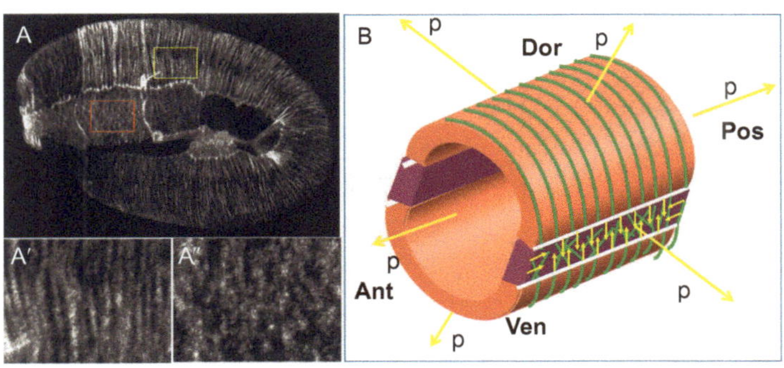

Thanh T.K. Vuong-Brender *et al.*, Figure 4 Actin organization and possible role during elongation. (A) Actin filaments in nearly twofold stage embryo are organized in parallel bundles in DV cells (yellow rectangle—dorsal cell hyp7, A′) and in thinner and less circumferentially oriented filaments in lateral seam cells (red rectangle, A″). (B) Hypothetical model for *C. elegans* elongation where the seam cells (violet) pull to increase the hydrostatic pressure and the DV cells (pink) orient the elongation in the direction perpendicular to actin bundle alignment (green lines), p, hydrostatic pressure; Ant, anterior; Pos, posterior; Dor, dorsal; Ven, ventral; arrows, force exerted. *Panel (A–A″) adapted with permission from Gally et al. (2009).*

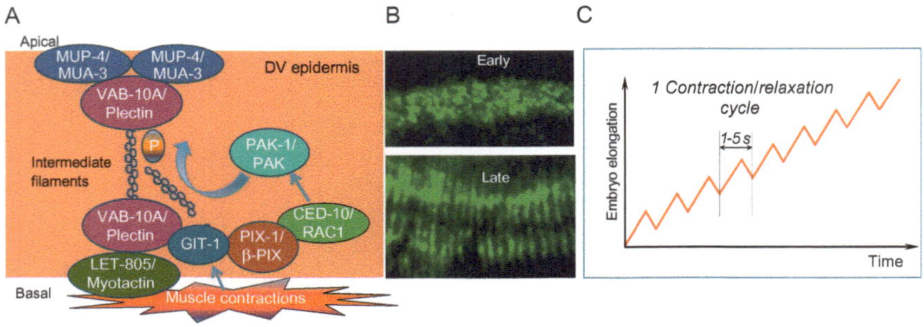

Thanh T.K. Vuong-Brender *et al.*, Figure 5 Muscle activity through CeHDs may promote elongation via a ratchet system. (A) CeHD remodeling under muscle contractile tension. Muscle contractions lead to GIT-1 recruitment, which subsequently activates PIX-1, CED-10/Rac1, and PAK-1. In turn, PAK-1 phosphorylates intermediate filaments leading to CeHD maturation (see panel B). (B) Images of a small portion of CeHD (marked with an antibody against the CeHD protein VAB-10A) when muscles start to contract (top) and at the end of elongation (bottom). (C) Elongation after the twofold stage as a ratchet-like process.

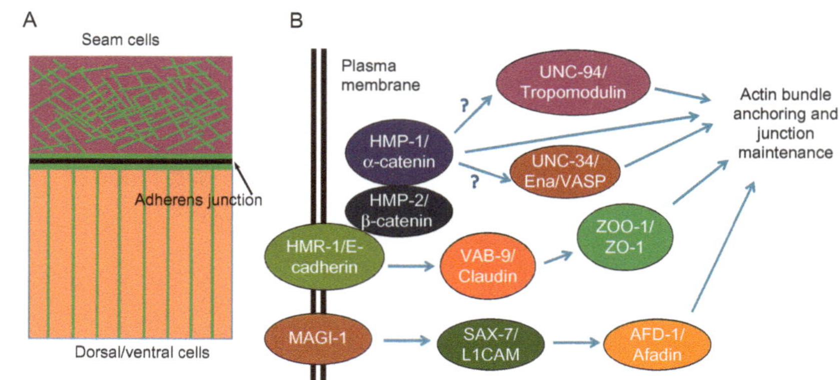

Thanh T.K. Vuong-Brender *et al.*, Figure 6 Role of cell junctions during elongation. (A) Junctional proteins are thought to anchor actin bundles in DV cells, green lines represent actin filaments and bundles. (B) Different junction-associated proteins may contribute to actin bundle anchoring and junction maintenance during elongation.

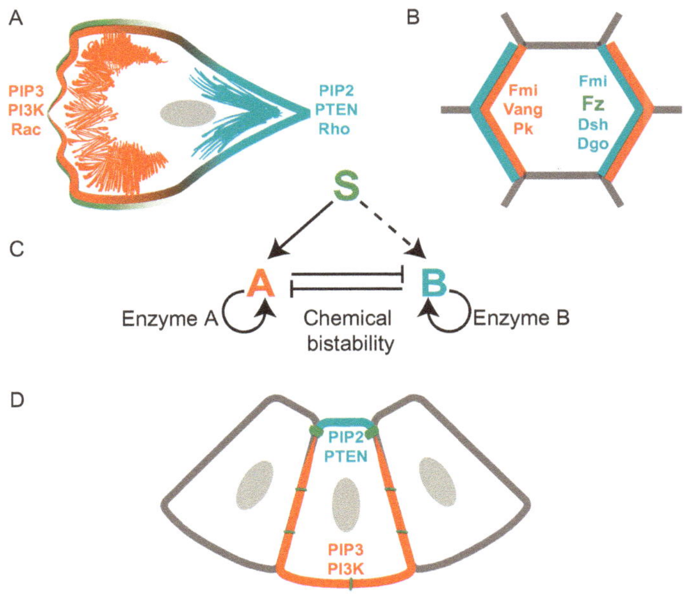

Qiyan Mao and Thomas Lecuit, Figure 1 A conceptual chemical polarization network. (A) A migrating cell aligns its front–rear polarity (red)-to-blue outline) with the direction of the chemoattractant gradient (green outline). Inside the cell: Red sketches indicate branched, protrusive actin filaments at the front, while blue sketches show bundled, contractile actin filaments at the rear. (B) An epithelial cell displaying planar cell polarity, with Core PCP components highlighted in red (proximal; Fmi-Vang-Pk) and blue (distal; Fmi-Dgo-Dsh). Fz in the distal PCP complex (green) is thought to be the symmetry-breaking signal transported along microtubules. (C) Schematics showing polarization of a two-component chemical system. S: symmetry-breaking cue that can activate either A alone, or first A then B. A and B: two mutually antagonistic, self-amplifying polarity states/components. (D) An epithelial cell displaying apicobasal polarity, with apical (blue) and basolateral (red) surfaces highlighted. Green bars shows cadherin complexes, with two thick ones showing the position of adherens junctions, and several thin ones showing cadherin complexes at the basolateral surfaces.

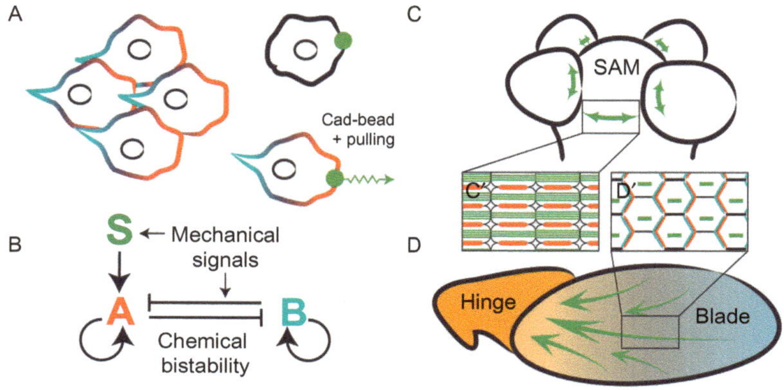

Qiyan Mao and Thomas Lecuit, Figure 2 External mechanical signals can bias polarity. (A) Pulling force via cadherin-coated beads rescues migration polarity in isolated *Xenopus* gastrula mesendoderm cells. Red-to-blue color gradient indicates front–rear polarity in collectively migrating cells and single cells with pulling forces applied via cadherin-coated beads, while black indicates no polarity in cells in contact with cadherin-coated beads but without pulling forces. Green dots show magnetic beads coated with C-Cadherin. Green arrow implies pulling force. (B) Schematics showing mechanical signals may either play an instructive role in biasing the direction of symmetry breaking of a two-component system, or a permissive role in enhancing long-range inhibition between two polarity states. S: symmetry-breaking cue. A and B: two mutually antagonistic, self-amplifying polarity states/components. (C) Hypothetical stress pattern in the *Arabidopsis* shoot apical meristem (green arrows). Inset (C′) shows alignment of microtubules (green) and membrane enrichment of Auxin transporter PIN1 (red) along with the direction of mechanical tension. (D) Pattern of cell flow (green arrows) in the *Drosophila* pupal wing blade due to hinge contraction. Gradients of Ds (orange) and Fj (blue) are superimposed. Inset (D′) shows alignment PCP direction or microtubules (green) with proximal (red: Fmi-Vang-Pk) and distal (blue: Fmi-Fz-Dgo-Dsh) components oriented along the proximodistal axis of the wing blade. *Panel (A): Adapted from Weber, Bjerke, and Desimone (2011). Panel (C): Adapted from Heisler et al. (2010). Panel (D): Adapted from Aigouy et al. (2010) and Harumoto et al. (2010).*

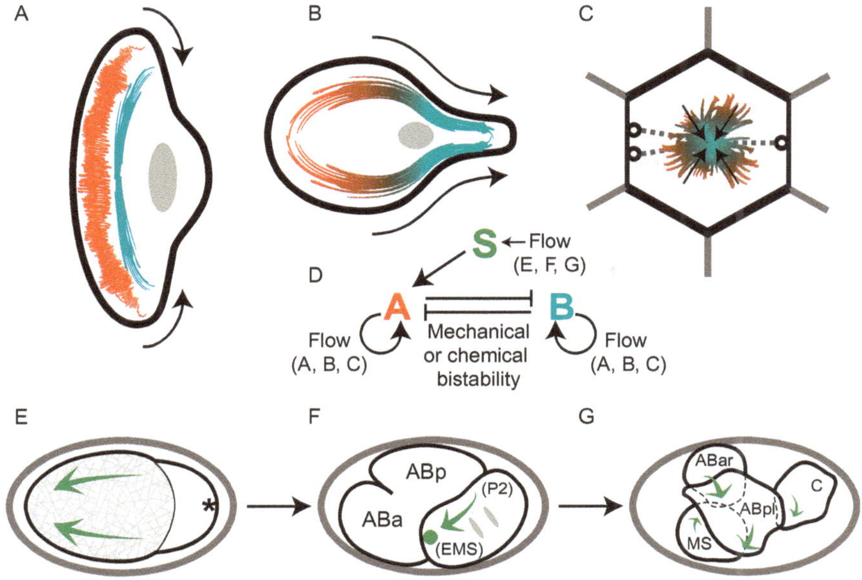

Qiyan Mao and Thomas Lecuit, Figure 3 Internal mechanical signals can induce polarity. (A) A polarized, crawling fish keratocyte with actomyosin retrograde flows, characterized by two populations of actin filaments at the front (red: branched, highly protrusive) and the rear (blue: bundled, highly contractile). Black arrow indicates the flow direction. (B) A polarized cell with stable-bleb morphology and with actomyosin retrograde flows. Red-to-blue gradient indicates force gradient of myosin contractility from front to rear. Black arrow indicates the flow direction. (C) An epithelial cell from the *Drosophila* germband ectoderm with oscillatory actomyosin pulse/centripetal flows at the apical cortex. Red-to-blue gradient indicates enrichment of myosin from periphery to center. Black arrow indicates the flow direction. Black circles show site of cadherin endocytosis. Gray dashed lines suggest mechanical coupling between the actomyosin pulse and cadherin endocytosis. (D) Schematics showing how actomyosin flows can amplify two mechanically incompatible regimes of actomyosin networks via feedbacks in the case of (A)–(C). In the case of (E)–(G), flows can induce symmetry breaking. S: symmetry-breaking cue. A and B: two mutually antagonistic, self-amplifying polarity states/components. (E) A one-cell stage *C. elegans* embryo with transient actomyosin flows breaking the AP symmetry. Green arrows indicate the direction of actomyosin flows. Star shows site of sperm entry. (F) A 2 to 4-cell stage *C. elegans* embryo with chiral actomyosin flows breaking the DV symmetry by displacing the ancestral midbody remnant to the ventral side. Green arrows indicate the direction of actomyosin flows. Green dot shows midbody remnant. (G) An 8-cell stage *C. elegans* embryo with chiral actomyosin flows breaking the LR symmetry. Green arrows indicate direction of cell body rotation. *Panel (A): Adapted from Yam et al. (2007). Panel (B): Adapted from Ruprecht et al. (2015). Panel (C): Adapted from Levayer and Lecuit (2013) and Munjal et al. (2015). Panel (E): Adapted from Munro, Nance, and Priess (2004). Panel (F): Adapted from Singh and Pohl (2014). Panel (G): Adapted from Pohl and Bao (2010).*

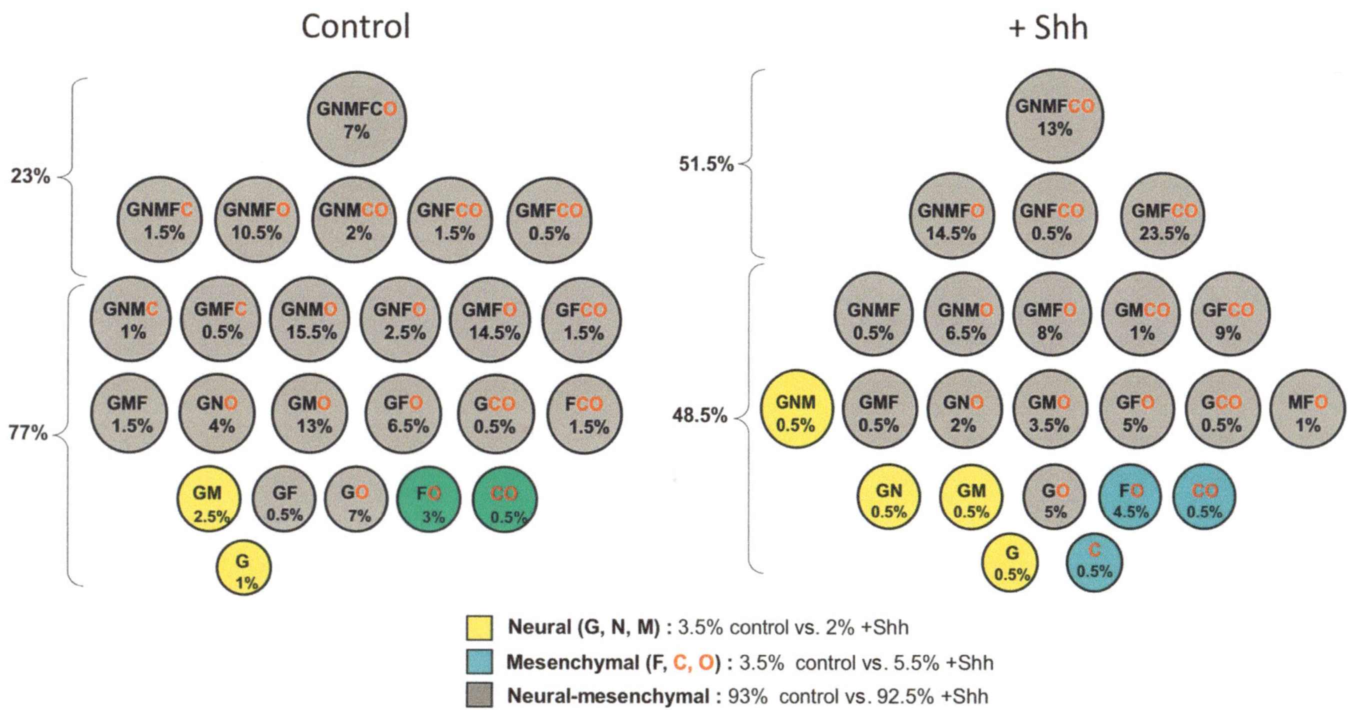

Nicole M. Le Douarin and Elisabeth Dupin, Figure 1 Schematic lineage tree of avian CNCC progenitors and influence of Shh. Quail NCC were isolated from the NCC migratory outgrowth of explanted mesencephalic neural primordium and then cultured as single cells in medium conditions permissive for a diversified cell differentiation. Colonies were phenotypically analyzed after growth for 10 days either in the presence of Shh (109 clones; 52% of initially plated cells) or in control medium (146 clones; 65% of the plated cells). Identified progenitors are classified according to the number of distinct differentiated cell types in their progeny (G, glial cells; N, neurons; M, melanocytes; F, myofibroblasts; C, chondrocytes; O, osteocytes) and their respective frequencies are indicated (% of total clones). In both medium conditions, the vast majority (93%) of CNCC is multipotent and endowed with both neural-melanocytic and mesenchymal (including skeletogenic) potentials (in gray); the frequencies of "only neural" (in yellow/white) and "only mesenchymal" (in blue/dark gray), progenitors, were very low. Addition of Shh led to a twofold increase of the proportion of the most multipotent progenitors (with six and five distinct cell types) (accounting for 51.5% vs. 23% of the clones in Shh-treated and untreated cultures, respectively). Shh treatment also promoted the *in vitro* development of progenitors able to produce chondrocytes, without changing the overall frequency of osteogenic multipotent cells. *Adapted from Dupin, Calloni, and Le Douarin (2010).*

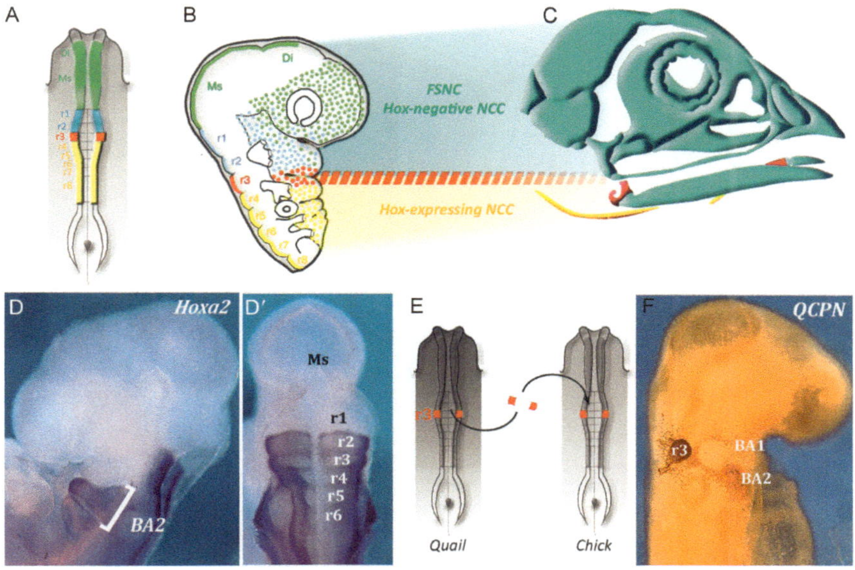

Nicole M. Le Douarin and Elisabeth Dupin, Figure 2 CNC contribution to head skeleton development. Color-coded schematics of premigratory (A) and migratory (B) CNC at the mesencephalic and rhombencephalic axial levels, and their head skeleton derivatives (C). The more rostral *Hox*-negative CNC (down to r2; in green and blue) gives rise to the skull vault and the face, including the nasal capsule, jaws, and entoglossum, whereas more posterior NCC that express *Hox* genes (in yellow) migrate to BA2 and more posterior BAs to produce the hypobranchial skeleton. Between these two domains, r3 NCC (in red) colonize both BA1 and BA2. (D) Whole-mount E3 chick embryos showing that the rostral head and BA1 do not express *Hoxa2*, whereas this gene is abundantly expressed in more posterior BAs (D′). (E) Schematic representation of r3 NC graft from quail to chick at 5ss. (F) One-day postgrafting, quail r3 CNCC (identified using the quail-specific marker QCPN; in brown) has colonized BA1 and BA2 of the host embryo. *Adapted from Le Douarin, Couly, and Creuzet (2012) with permission.*

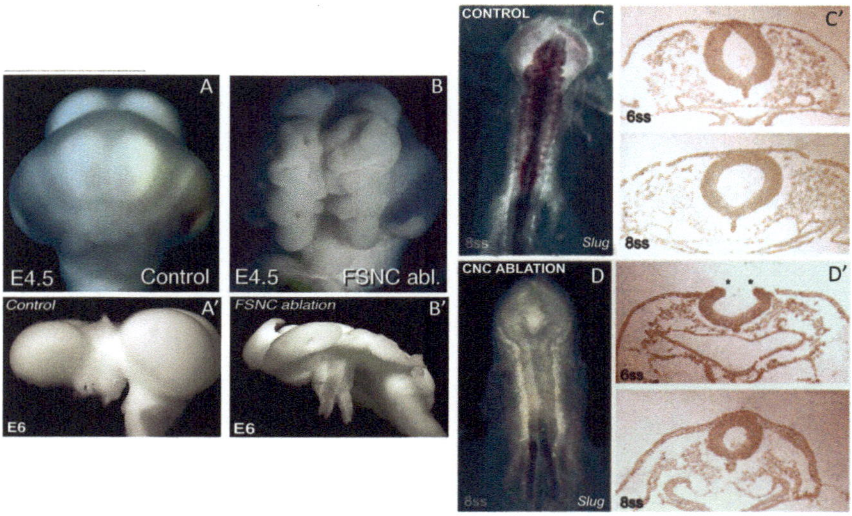

Nicole M. Le Douarin and Elisabeth Dupin, Figure 3 CNC ablation alters head skeleton and brain development. (A and B) Frontal views of E4.5 chick embryos, showing facial defects caused by anterior CNC ablation (B), compared with a control head (A). Whole-mount preparations of the brain of control (A′) and operated (B′) E6 embryos provide evidence that deprivation of NCC led to a wide-open brain with altered partitions of the anterior brain vesicles. The surgical procedure leading to these defects eliminates the migratory NCC that express *Slug* (compare C and D); the NC ablation consists in excision of the premigratory NC without involving the neuroepithelium, as shown on transverse sections of control and operated embryos (C′ and D′, respectively): after resection of the neural folds at 6ss (C′ and D′, upper panels), the neural tube properly closed at 8ss, in both control and operated embryos (C′ and D′, lower panels).

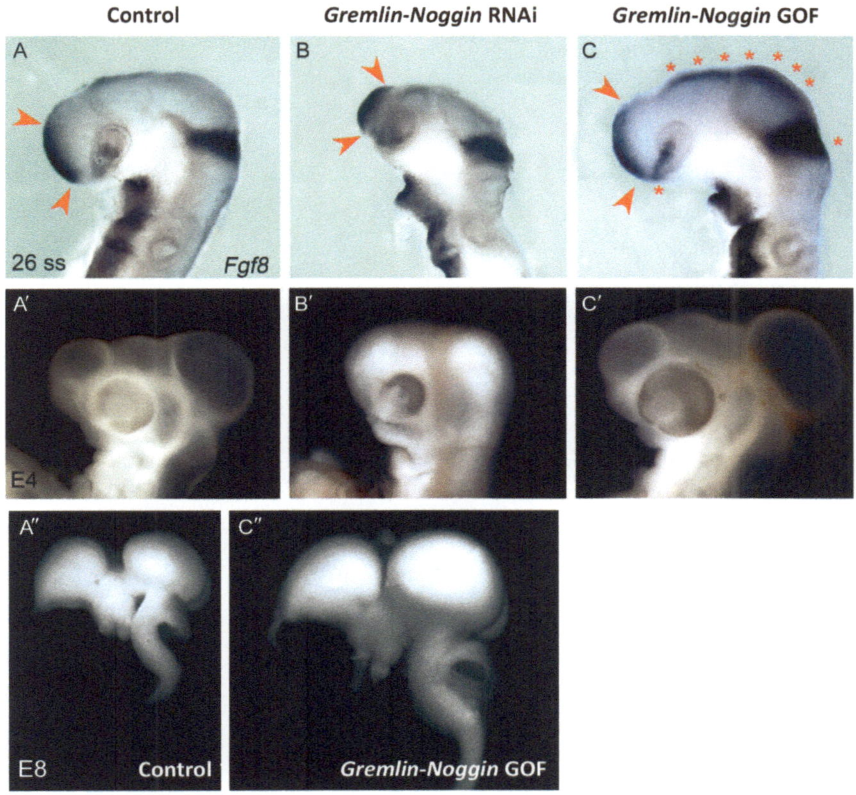

Nicole M. Le Douarin and Elisabeth Dupin, Figure 4 BMP inhibitors in the CNC control *Fgf8* expression and forebrain development. (A–C) Expression of *Fgf8* in the ANR (red arrowheads) at E2.5 (26ss) in control embryos (A); in embryos submitted to *gremlin* and *noggin* knockdown by RNA interference (RNAi), *Fgf8* expression domain is reduced (B); conversely, it is extended after gain-of-function (GOF) for *gremlin* and *noggin* (C) in the NC. Lateral views of embryonic heads at E4 show hypoplasic anterior head after *gremlin* and *noggin* knockdown (B′), whereas the head is enlarged after forced expression of these genes (C′), as compared with the untreated embryo (A′). The size of the brain at E8 is dramatically increased following GOF for *gremlin* and *noggin* in the NC (C″, compared to the control brain in A″). *Adapted from Le Douarin et al. (2012) with permission.*

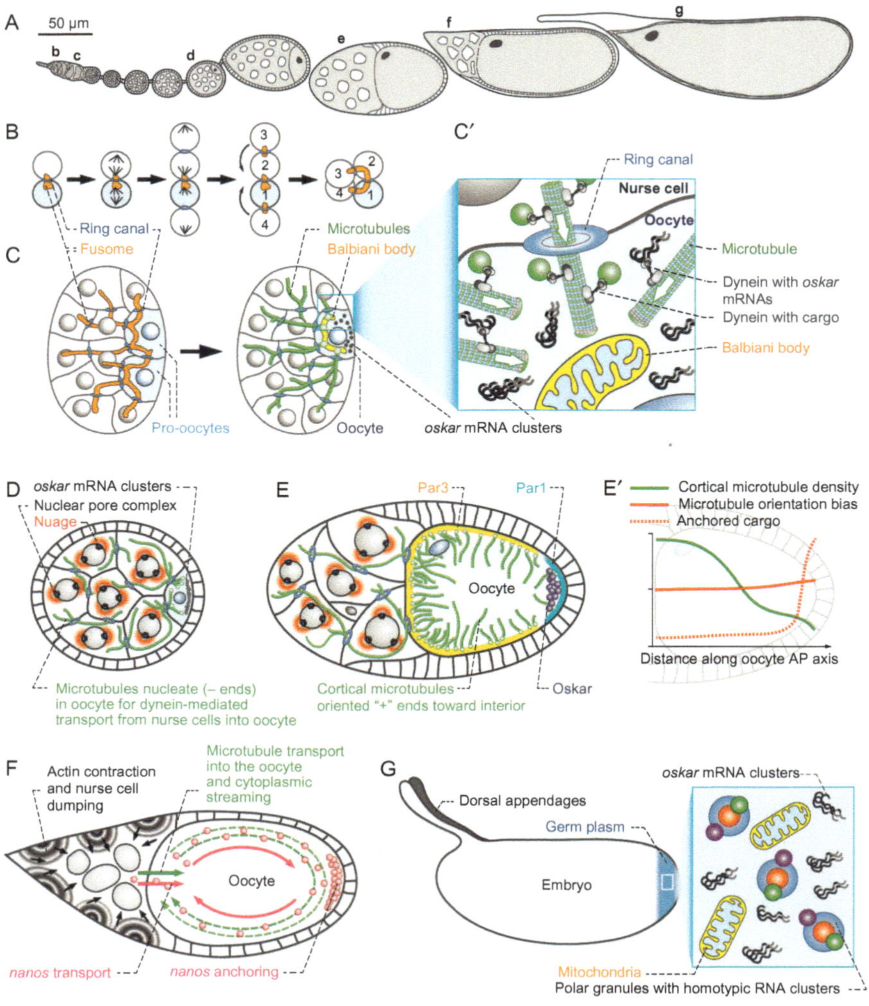

Ruth Lehmann, Figure 1 Establishment of anterior–posterior polarity during *Drosophila* oogenesis. (A) Ovariole with developing egg chambers. Each ovary contains 12–16 ovarioles. At the tip (left, b and c regions) is the germarium, where 2–3 GSCs generate a cytoblast that matures into a egg chamber consisting of the oocyte and 15 nurse cells surrounded by somatic follicle cells (stage 1 egg chamber). Daily stem cell division produces a chain of maturing egg chambers that mature to the fully developed egg (right, g). Small letters on top refer to stages depicted in other figure panels. (B) Division pattern of the cystoblast, the GSC daughter in the germarium. After division cells remain connected via ring canals with fusome material traversing the cyst. The first two cells born are connected by four ring canals, the next two by three, the next four by two, and the last eight by one ring canal each. Note that the earlier born cells inherit proportionally more fusome material. (C) 16-cell cyst in germarium, all cells are

(Continued)

Alessia Deglincerti *et al.*, Figure 2—Cont'd connected by ring canals and fusomes. The two cells with four ring canals "pro-oocytes" are positioned posteriorly within the cluster and enter into meiosis (left), only one of the two cells remains in meiosis and arrests in meiotic prophase (right). All other cells develop into polyploid nurse cells, which continue to be connected to the oocyte by ring canals. All centrosomes migrate into the oocyte and a microtubule network nucleate from the oocyte to the nurse cells. Dynein-mediated transport brings cargo and RNAs from the nurse cell into the oocyte. Densely packed mitochondria and RNA form the Balbiani body (yellow in C and C′). (C′) Oocyte cytoplasm indicating microtubule polarity and RNA transport particles. (D) Stage 5 egg chamber. Nurse cells and oocyte are surrounded by somatic follicle cells. Continued dynein-mediated RNA transport from the nurse cells into oocyte. Microtubules are oriented to nucleate (− pole) from oocyte and extend into nurse cells. Nuage: RNA-rich granules form around nurse cell nucleus often in close proximity of nuclear pore. (E) Stage 10A egg chamber. Mutually exclusive interactions between Par-1 (blue) at the posterior pole and the Par-3/Bazooka (yellow) complex at the lateral and anterior cortex establish polarity within the oocyte. Microtubules nucleate from lateral and anterior cortex extending into the oocyte and are inhibited at the posterior pole by Par-1. Kinesin motors transport *osk* RNA to the posterior pole, where it is translated. Dynein-mediated transport continues to bring *osk* RNA from the nurse cells to the oocyte. Somatic follicle cells surround the egg chamber. (E′) Summary of microtubule organization and *osk* RNA transport dynamics during stage 10A. Microtubule network is denser at the anterior than posterior of oocyte (green). A slight bias of kinesin-mediated microtubule transport (red line) leads to enrichment of *osk* RNA at posterior pole over time (red stippled). (F) Stage 10B/11, cytoplasmic streaming. The actin meshwork is disassembled in the oocyte and microtubules align in bundles (green). This leads to a streaming process where RNAs and proteins are mixed within the oocyte and additional RNAs/proteins are drawn into the oocyte (green arrow). Effector RNAs like *nanos* (red) become tethered at posterior pole. (G) Mature egg. Germ plasm is assembled and polar granules of similar size to mitochondria are found in a precise region at the posterior pole. Effector RNAs are localized to the granules, Osk, Vasa, Tud, and Aub proteins are essential part of the granules but *osk* RNA is not associated with polar granules (see Fig. 3E). *Panel (C′): After Roth and Lynch (2009). Panel (E′): After Parton et al. (2011).*

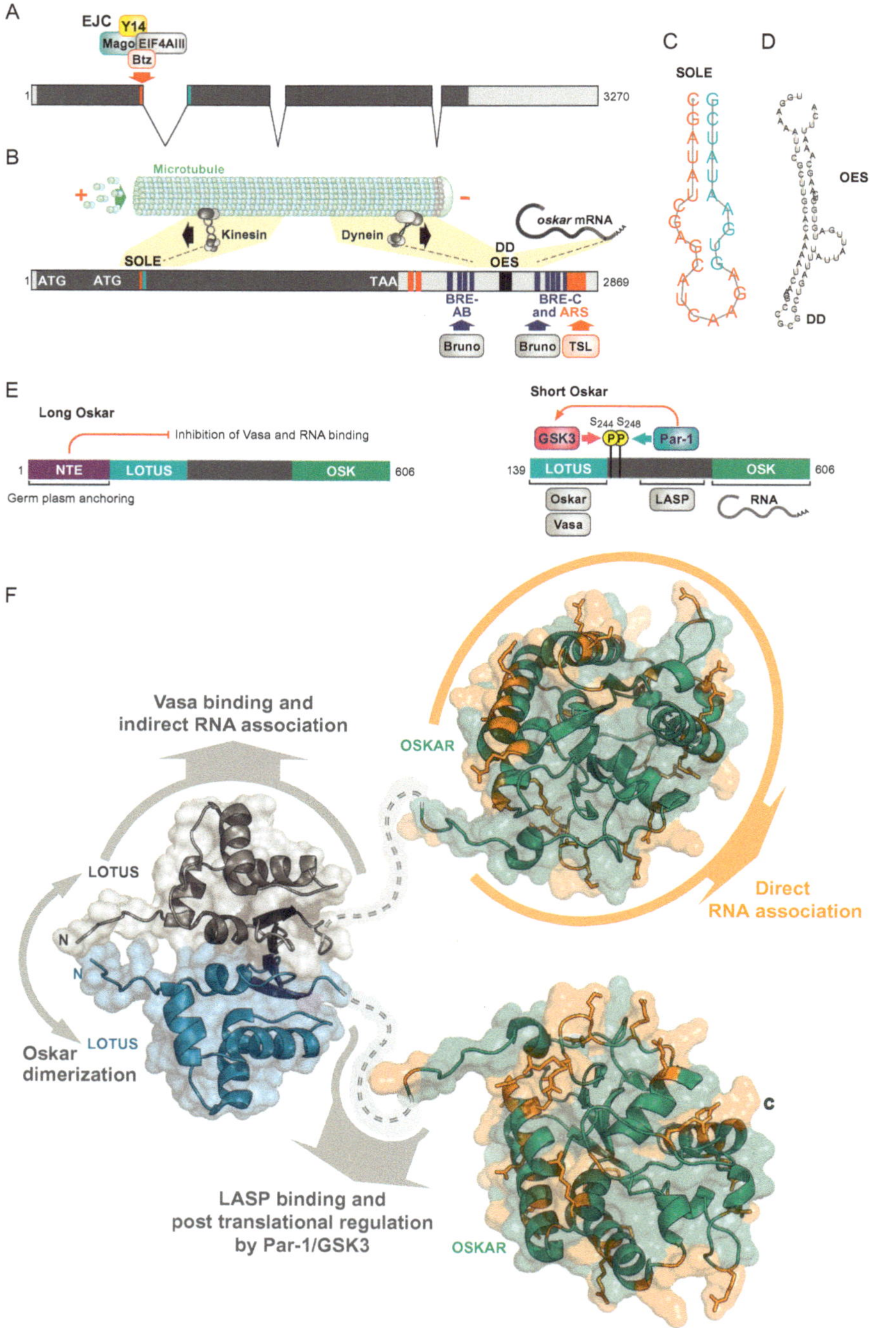

Ruth Lehmann, Figure 2 See legend on opposite page.

Ruth Lehmann, Figure 2—Cont'd Oskar RNA and protein history. (A) Structure of nascent *osk* RNA. Exon junction complex components bind to sequences flanking the first intron, which is critical for kinesin-mediated *osk* RNA transport. (B) *Osk* RNA transport particles are transported along microtubule by dynein (directed toward − end of microtubule) and kinesin (directed toward + end of microtubule) motors. The oocyte entry signal (D) regulates the dynein-mediated transport of *osk* RNA into the early oocyte. (C) The Spliced Oskar Localization Element (SOLE) forms as a consequence of first intron splicing, where 18 nt from the first and 10 nt from the second exon merge into a stem loop structure. The DD palindromic sequence element (D) is located in the loop of the OES stem loop. Via RNA–RNA "kissing interaction" this element promotes RNA transport particle formation and hitchhiking of RNA lacking the SOLE element on those with the element. AB, C region are regions rich of Bruno regulatory elements (BRE), which repress RNA translation, and IMP-binding elements, which promote translation and localization. Binding elements close to the most distal part of the 3′UTR are important for translational activation, via adenine-rich sequences (ARS) that mediate polyadenylation. (E) Structure of Oskar isoforms. Left: Long Oskar is a 606-amino acid protein, the N-terminal extension (NTE) is required for actin filament nucleation and enhanced endocytosis at the posterior pole. Somehow with NTE suppresses short Oskar functions. Right: Short Oskar start at the second Methionine 139. The Lotus domain dimerizes and interacts with Vasa (see also F), the C-terminal OSK domain resembles a SGNH hydrolase domain; however, the enzymatic amino acids are missing. Instead the OSK domain is thought to interact with effector RNAs. (F) Structure diagram of Lotus and OSK domain. The Lotus domain forms a homodimer of two winged helix structures and interacts with Vasa, while the OSK domain forms a globular structure with basic and hydrophilic amino acids exposed to the surface, which are thought to interact with RNA.

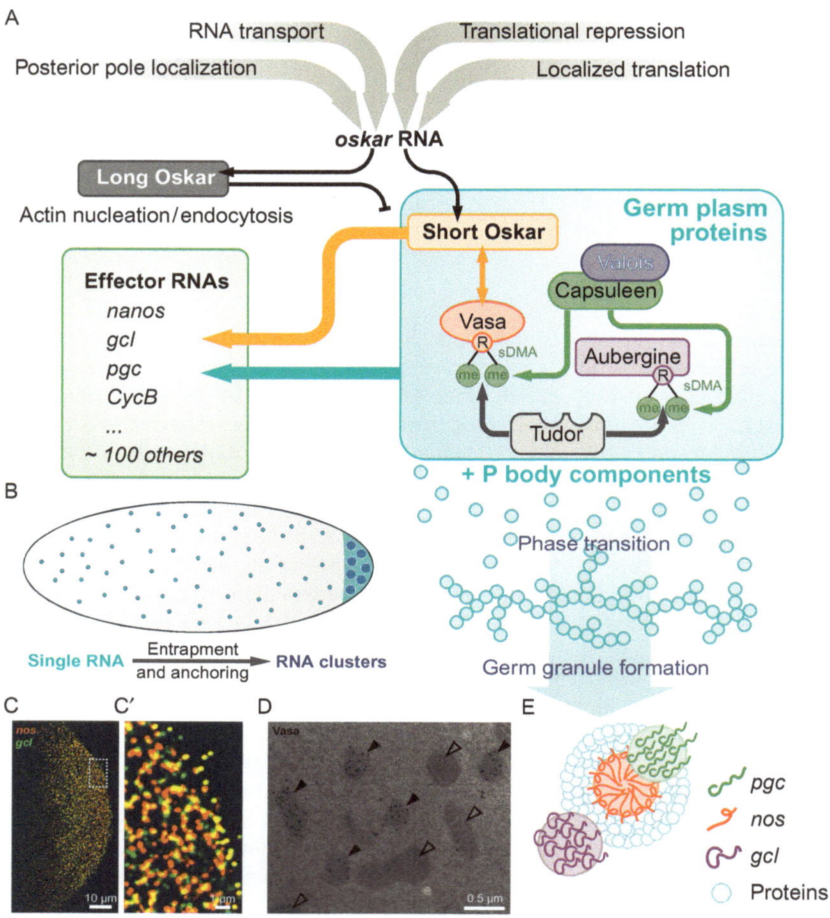

Ruth Lehmann, Figure 3 Germ plasm assembly. (A) Multiple controls described in Figs. 1 and 2 feed into the localization and translation of *osk*. Osk protein recruits other components of the germ plasm such as Vasa, Tudor, and Aubergine. Symmetric dimethylation of Arginine (sDMA) in Vasa and Aubergine by the methyltransferase Capsuleen allows binding to Tudor domain proteins (Kirino, Vourekas, Kim, et al., 2010; Kirino, Vourekas, Sayed, et al., 2010; Liu et al., 2010; Webster et al., 2015). This together with liquid to gel phase transitions could provide a possible basis for the dense assembly of polar granules. (B) Model for phase transition that occurs at posterior pole when concentration of germ plasm protein and RNAs become high (Brangwynne, 2013). (C) SIM image of *nos* and *gcl* RNA hybridization using Stellaris© RNA *in situ* hybridization probes. (C′) Higher magnification to reveal particle overlap. (D) Germ plasm with polar granules marked by Vasa (filled arrow) and mitochondria (open arrow). (E) Organization of homotypic RNA clusters within polar granules according to Trcek et al. (2015).

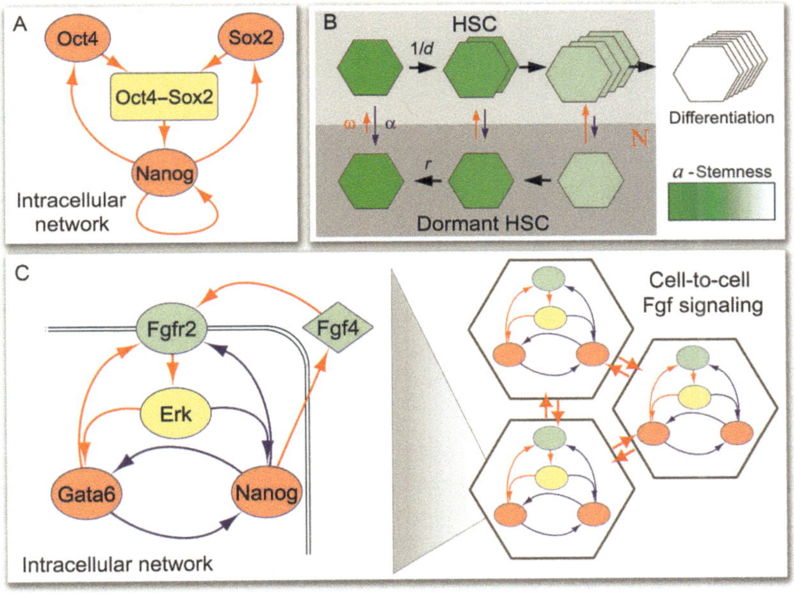

Dmitri Papatsenko and Ihor R. Lemischka, Figure 1 Network-focused, cell-focused, and multiscale models. (A) An early model for the core pluripotency circuit explaining formation of high and low populations of Nanog cells. (B) A model for proliferation of hematopoietic cells. Red and blue arrows show probabilities of transition from and to dormant HSC subpopulation. Green color demonstrates the degree of stemness; N, the number of HSCs is the model output. Letters are model parameters describing probabilities of cell conversion from one state to another. (C) A multiscale model for specification of Epi and PrE. At the lower level, the model takes into account intracellular network (left), at the higher level, model considers cell-to-cell signaling via Fgf4. The red arrows show activation and the blue arrows show repression.

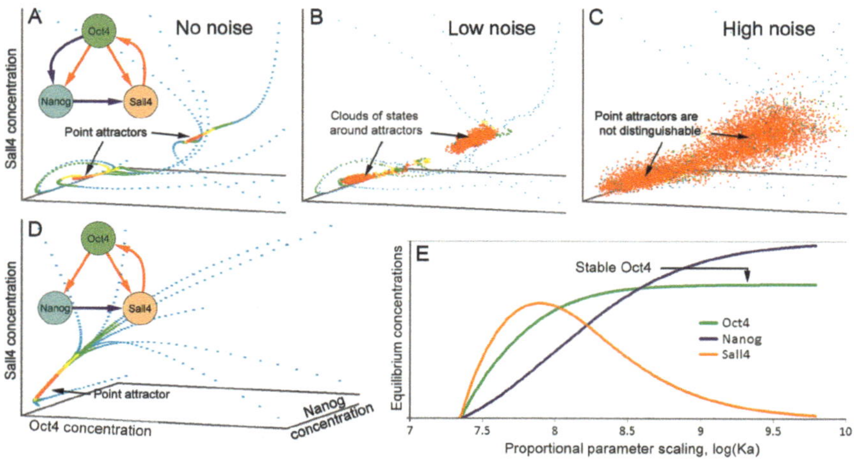

Dmitri Papatsenko and Ihor R. Lemischka, Figure 2 Deterministic, stochastic models and robust models. (A)–(C) An example of a bistable model, which explains formation of two populations of Nanog in ESC cultures. In this model, Oct4 is assumed to be a dual regulator of Nanog. The red arrows in the network motif correspond to activation and the blue arrows correspond to a repression. (A) A deterministic scenario (see the network motif), in which noise is absent; the two point attractors correspond to two stable states (e.g., high and low Nanog). (B) The same deterministic model with adding low noise in the gene expression (a stochastic component). The attractors are visible as clouds of states. (C) The same deterministic system with high level of noise, the stochastic component prevails and the attractors disappear, solution represents a single cloud of probable states. (D) A similar model explaining stability of a single state, such as observed in the case of cells grown in 2i-LIF media. The network motif represents incoherent feedforward loop with a feedback. (E) Bifurcation diagram of the model shown on (D); proportional scaling of all regulatory parameters (*X* axis, binding constants) has little effect on equilibrium concentration of Oct4. The model appears to be robustness to proportional scaling of all parameters.

FSC
www.fsc.org
MIX
Papier | Fördert
gute Waldnutzung
FSC® C083411

Kolibri 360 GmbH
Ferdinand-Jühlke-Straße 7
99095 Erfurt, Deutschland
produktsicherheit@kolibri360.de